Unilateral Variational Analysis in Banach Spaces

Part II: Special Classes of Functions and Sets

Unilateral Variational Analysis in Banach Spaces

Part II: Special Classes of Functions and Sets

Lionel Thibault
University of Montpellier, France

World Scientific

NEW JERSEY • LONDON • SINGAPORE • BEIJING • SHANGHAI • HONG KONG • TAIPEI • CHENNAI • TOKYO

Published by

World Scientific Publishing Co. Pte. Ltd.
5 Toh Tuck Link, Singapore 596224
USA office: 27 Warren Street, Suite 401-402, Hackensack, NJ 07601
UK office: 57 Shelton Street, Covent Garden, London WC2H 9HE

British Library Cataloguing-in-Publication Data
A catalogue record for this book is available from the British Library.

UNILATERAL VARIATIONAL ANALYSIS IN BANACH SPACES
(In 2 Parts)
Part I: General Theory
Part II: Special Classes of Functions and Sets

ISBN 978-981-125-816-9 (Set_hardcover)
ISBN 978-981-125-817-6 (Set_ebook for institutions)
ISBN 978-981-125-818-3 (Set_ebook for individuals)

ISBN 978-981-125-494-9 (Part I_hardcover)

ISBN 978-981-125-495-6 (Part II_hardcover)

For any available supplementary material, please visit
https://www.worldscientific.com/worldscibooks/10.1142/12797#t=suppl

Printed in Singapore

To Janine and Sylvain

To the memory of my parents Arntz and Germaine

To Janie and Sylvain

To the memory of my parents Anna and Germaine

Contents

Part I

Part II

List of Figures

Preface

The study in depth of *unilateral properties* in mathematical analysis apparently began at the end of the 19th century with the 1899 Baire's thesis (Thèse de doctorat ès sciences mathématiques). Given a real-valued function f of n real variables, under the term of *maximum of* f *at a point* P_0 and with the notation $M[f, P_0]$, R. Baire introduced in his thesis [**61**, p. 4] what is nowadays called the upper limit (or limit superior) of f at P_0 usually denoted as $\limsup_{P \to P_0} f(P)$ (or $\overline{\lim}_{P \to P_0} f(P)$). Considering in [**61**] a decreasing sequence $(\rho_n)_n$ of positive reals tending to 0 and the nondecreasing sequence $(M_n)_n$ of suprema of f on $B[P_0, \rho_n]$, Baire defined $M[f, P_0]$ as the limit of M_n as $n \to \infty$. The property $M[f, P_0] = f(P_0)$, or equivalently the inequality $M[f, P_0] \leq f(P_0)$, was observed by Baire [**61**] to be equivalent to the fact that for each $\varepsilon > 0$ there exists some $\rho > 0$ such that $f(P) < f(P_0) + \varepsilon$ for all $P \in B[P_0, \rho]$. Baire then wrote: "La fonction possède donc au point P_0 l'une des deux propriétés dont l'ensemble constitue la continuité"(English translation: "The function thus has one of the two properties which together constitute the continuity"). That *unilateral* property was then called the *upper semicontinuity* of f at P_0 by Baire in [**61**, p. 6]. The *minimum* $m[f, P_0]$ *of* f *at* P_0 and its *lower semicontinuity* at P_0 were defined similarly in [**61**], and various properties of $M[f, \cdot]$ and $m[f, \cdot]$ and of the semicontinuity notions were established. Through those concepts Baire carried out in his dissertation thesis the thorough analysis of his celebrated classification of diverse classes of discontinuous functions. By analogy, for a multimapping $M : X \rightrightarrows Y$ from a metric space X with closed values into a compact metric space Y, C. Kuratowski said in his 1932 paper [**640**, p. 148] that M is *upper* (resp. *lower*) *semicontinuous* at x_0 whenever for each sequence $(x_n)_n$ converging to x_0 one has $\operatorname{Ls} M(x_n) \subset M(x_0)$ (resp. $M(x_0) \subset \operatorname{Li} M(x_n)$). For a sequence of sets $(S_n)_n$ in a metric space, the set $\operatorname{Ls} S_n$ (resp. $\operatorname{Li} S_n$) is taken in [**640**] as the set of $\lim y_n$ with $y_n \in S_n$ for infinitely many n (resp. for large n). In a Euclidean space, the former set was considered much earlier in 1887 with the same above form under the name of *"limit"* by P. Painlevé [**779**, p. 123]. In the same year 1887, G. Peano [**782**, p. 302] instead called for a family of figures $M(x)$, in finite dimensions, *"limit"* of $M(x)$ as $x \to x_0$ the set of y such that $\lim\limits_{x \to x_0} d(y, M(x)) = 0$, and this is nowadays known to coincide with the usual limit inferior $\operatorname*{Lim\,inf}\limits_{x \to x_0} M(x)$; the notation $\lim(M, x_0)$ was used by Peano in his other monograph [**784**, p. 302] published in 1908. In this same 1908 monograph [**784**, p. 237] Peano also employed, with the notation $\operatorname{Lm}(M, x_0)$, the set of y such that 0 is a limit point of the mapping $x \mapsto d(y, M(x))$ as $x \to x_0$, which corresponds to the limit superior $\operatorname*{Lim\,sup}\limits_{x \to x_0} M(x)$. We refer to Section 1.10 in Chapter 1 for comments. In what follows and throughout the text, instead of the notations $\operatorname{Ls} M(x_n)$ and $\operatorname{Li} M(x_n)$, we will utilize the today more usual ones $\operatorname*{Lim\,sup}_{n\to\infty} M(x_n)$ and $\operatorname*{Lim\,inf}_{n\to\infty} M(x_n)$. Those concepts are clearly *unilateral semilimit* notions for a multimapping M and they generate *unilateral semicontinuities* for the multimapping by means of the inclusions $\operatorname*{Lim\,sup}_{x\to x_0} M(x) \subset M(x_0)$ and $M(x_0) \subset \operatorname*{Lim\,inf}_{x\to x_0} M(x)$.

Previously to [**640**], instead of the classic tangent line to a regular curve or tangent plane to a regular surface, G. Bouligand employed the Lim sup to put into light another viewpoint for the concept of tangent. For a set S in a finite-dimensional Euclidean space X and a point $x_0 \in S$, Bouligand considered in his 1928 paper [**157**, p. 29] the set of elements belonging (with the preceding notation) to $x_0+\operatorname*{Lim\,sup}_{n\to\infty} \frac{1}{t_n}(S-x_0)$ for some $t_n \downarrow 0$; the notation Δ was used there for that set. Bouligand provided an analysis of sets S without point x_0 where the corresponding set Δ coincides with the whole space X. Several other studies involving this set Δ were carried out by Bouligand in a series of works [**158, 159, 160, 161, 162**] and by others influenced in the 1930s by his works as, for example, the authors of [**217, 367, 368, 714, 826, 867**]; see comments in Section 2.9 in Chapter 2. This concept of tangent was previously considered in 1908 by G. Peano [**784**, p. 331] in the form $\operatorname{Lm}\big(x_0 + \frac{1}{t}(S - x_0), 0\big)$ for $t \downarrow 0$ (see Penao's notation Lm recalled above), and it was denoted there $\operatorname{Tang}(S, x_0)$. In his earlier 1887 manuscript [**782**, p. 305] Peano used another notion as tangent, namely the set of elements y such that $\lim_{t\downarrow 0} d\big(y - x_0, \frac{1}{t}(S - x_0)\big) = 0$, which corresponds, with the nowadays notation, to elements belonging to $x_0 + \operatorname*{Lim\,inf}_{n\to\infty} \frac{1}{t_n}(S - x_0)$ for every $t_n \downarrow 0$. In the 1908 manuscript [**784**, p. 335] Peano also stated in terms of $\operatorname{Tang}(S, x_0)$ necessary optimality conditions when x_0 is a maximizer of a function under the constraint set S. Both above concepts appeared as *unilateral* viewpoints of tangents in geometry and mathematical analysis.

Unilateral derivates of real-valued functions were already present in U. Dini's 1878 book [**339**]. Given a real-valued function f on an interval in $\mathbb{R}$ the *right-hand* (resp. *left-hand*) *derivate* of f at x is defined at the page 66 of Dini's book [**339**] as the limit in $\mathbb{R} \cup \{-\infty, +\infty\}$ of $\frac{f(x+\delta)-f(x)}{\delta}$ as δ tends to 0 with $\delta > 0$ (resp. with $\delta < 0$); Dini used there the terminology "*derivata della funzione* $f(x)$ *a destra* (resp. *a sinistra*) *di* x". The *right-hand upper* (resp. *lower*) *derivate* of f at x is defined in A. Denoy's paper [**327**, p. 144-145] and in S. Saks' book [**876**, p. 108] as the upper (resp. lower) limit of the above difference quotient as δ tends to 0 with $\delta > 0$; the left-hand upper (resp. lower) derivate is defined there similarly. Denjoy in [**327**, p. 146] and Saks in [**876**, p. 108] called these four quantities the *unilateral extreme derivates*, and Saks also said *unilateral Dini derivates*. For a locally Lipschitz function f on $X = \mathbb{R}^N$, if x_0 is a minimizer of f on $S \subset X$ and $\mathcal{T}$ denotes the above tangent set in the sense of Bouligand (or equivalently the Peano tangent set $\operatorname{Tang}(S, x_0)$), the following "*necessary optimality condition*" holds:

$$\underline{d}_D^+ f(x_0; x - x_0) \geq 0 \quad \text{for all } x \in \mathcal{T},$$

where by $\underline{d}_D^+ f(x_0; h)$ we denote the above right-hand lower Dini derivate at 0 of the one-variable function $t \mapsto f(x + th)$. It is worth noticing that the *distributional derivative* does not allow to state an optimality condition in such a situation. In many cases in mathematics and diverse sciences, the set S arises as a closed subset of a Banach space X in the form

$$S = \{x \in X : g_i(x) \leq 0 \; i \in I, h_j(x) = 0 \; j \in J\},$$

where I and J are finite sets. When the functions g_i are all null and the functions f, h_j are continuously differentiable, under suitable conditions the classical Lagrangian optimality condition is well-known, that is, there exist reals $(\lambda_j)_{j \in J}$ such that

$$Df(x_0) + \sum_{j \in J} \lambda_j Dh_j(x_0) = 0.$$

The great progress in this direction with nondifferentiable functions was achieved with the concept of subdifferential of a convex function $\varphi : X \to \mathbb{R}$ independently introduced in 1963 by J. J. Moreau [**738**] and R. T. Rockafellar [**844**] as given by the *unilateral* description

$$\partial\varphi(x) = \{v \in \mathbb{R}^N : \langle v, y - x \rangle \leq \varphi(y) - \varphi(x), \; \forall y \in \mathbb{R}^N\}.$$

In addition to its rich calculus similar to that of differential calculus, this subdifferential is known, and this will be seen later, to allow, for example with $S = \{x : g_i(x) \leq 0, \; i \in I\}$ in the above minimization problem, to obtain under certain conditions and the convexity of the functions f, g_j that there exist non-negative reals $(\lambda_j)_{j \in J}$ such that

$$\partial f(x_0) + \sum_{j \in J} \lambda_j \partial f_j(x_0) \ni 0.$$

Such an optimality condition, for the minimization problem with nondifferentiable nonconvex locally Lipschitz functions f, g_i, h_j, was proved with the subdifferential $\partial\varphi(x)$ (also called generalized gradient) introduced by F. H. Clarke in his 1973 thesis [**232**] for a nondifferentiable nonconvex function φ. As it will be presented in the treatise, this subdifferential also enjoys a full calculus and coincides with that of Moreau-Rockafellar when φ is convex. Two other fundamental subdifferentials, yielding optimality conditions as above and enjoying rich calculus, were introduced by B. S. Mordukhovich [**720**] and A. D. Ioffe [**515, 517**] respectively. All those subdifferentials offer efficient *unilateral first order variation* tools for the analysis of nondifferentiable functions, and they possess rich calculus. According to the necessity and advantages pointed out in the 1960s by Moreau and Rockafellar to work, for the analysis of unilateral properties, with nondifferentiable functions taking values in the extended line $\mathbb{R} \cup \{-\infty, +\infty\}$, the subdifferential theories of convex and nonconvex functions were carried out with such extended real-valued functions.

One of the main aims of the book is to develop the theory of the unilateral concepts: semilimits of variable sets, semicontinuities of multimappings, tangent cones, normal cones, subdifferentials. A second main aim is to present various basic classes of sets and functions which can be seen as *subregular* in a unilateral viewpoint via the aforementioned subdifferentials, and to study the smallness of sets of points which are singular relative to subdifferentials. A third main aim is to analyze in depth the very large class of prox-regular sets, that is, sets whose metric projection is continuous near a reference point inside. There exist diverse books/lectures devoted (or partially devoted) to the analysis of nondifferentiable convex functions as [**35, 43, 45, 55, 72, 78, 134, 146, 155, 214, 384, 494, 529, 643, 649, 744, 801, 852, 1000**]. Chapter 3 furnishes a comprehensive presentation of the general theory of extended real-valued convex functions on locally convex spaces, and many results related to these functions either appear here for the first time in a book or are established with new simple proofs. After the theory of semilimits and semicontinuities for multimappings in Chapter 1, tangent cones and Clarke subdifferential are completely studied in Chapter 2 in the setting of normed spaces. The Mordukhovich limiting subdifferential in Asplund spaces and the Ioffe approximate subdifferential in Banach spaces are developed in details in Chapter 4 and Chapter 5, respectively. A systematic exposition of mean value inequalities for nondifferentiable functions is given in Chapter 6. The theory of metric regularity in Banach spaces for mappings and multimappings is the topic of Chapter 7. Subsmooth sets and subsmooth functions are investigated in Chapter 8, and the subdifferential determination property of such functions and others is studied in Chapter 9. Three important other classes of functions and their very useful properties are developed in details: semiconvex functions in Chapter 10, and primal lower regular and prox-regular functions in Chapter 11. The smallness of sets of subdifferentially singular points of certain such functions is the subject of Chapters 12 and 13. The study of differential properties of distance functions and metric projection in Chapter 14 is made in a very large context. Many results of that chapter are employed in the systematic exposition of prox-regular sets presented in Chapters 15, 16, 17 and 18. Each chapter finishes with a section called "*Comments*" which contains discussions and references regarding papers related to the results presented in the chapter in question. When a theorem is due to authors of a same paper (that is, co-authors) this is indicated as Theorem [**Author and Author'**], while we use the form Theorem [**Author; Author'**] for the case of authors of distinct papers; the same convention is adopted for certain examples, lemmas and propositions. For convenience of the reader, we recalled in Appendix diverse basic results used in the text and related to the theories: Topology, Functional Analysis, Convex Sets in Finite Dimensions, Measure Theory, Differential Geometry. The bibliography contains most of the contributions which are pertinent to the topics in the book.

Certain chapters in the book were the subject of lectures given in the 2010s at the "Université de Montpellier", and they benefited from comments of some students. I am also indebted to my colleague Florent Nacry for many valuable suggestions.

Montpellier
September, 2021

Lionel Thibault

CHAPTER 8

Subsmooth functions and sets

To facilitate the reading we recall our notation that $\mathbb{R}_+$ is the set of non-negative real numbers $[0, +\infty[$ and $\mathbb{N}$ is the set of positive integers $1, 2, \cdots, n, \cdots$. For an extended real-valued function $f : X \to \mathbb{R} \cup \{-\infty, +\infty\}$ by $\operatorname{dom} f$ and $\operatorname{epi} f$ we denote its effective domain and its epigraph respectively, that is,

$$\operatorname{dom} f := \{x \in X : f(x) < +\infty\} \quad \text{and} \quad \operatorname{epi} f := \{(x, r) \in X \times \mathbb{R} : f(x) \le r\}.$$

The indicator function of a subset S of X is denoted by Ψ_S, so

$$\Psi_S(x) = 0 \text{ if } x \in S \quad \text{and} \quad \Psi_S(x) = +\infty \text{ if } x \in X \setminus S.$$

When X is a metric space, d_S or $d(\cdot, S)$ stands for the usual distance function from S; for $x \in X$ and $r > 0$ the closed (resp. open) ball centered at x with radius r is still denoted by $B[x, r]$ (resp. $B(x, r)$). If X is a normed space, its closed unit ball (resp. open unit ball) is denoted by $\mathbb{B}_X$ (resp. $\mathbb{U}_X$), that is,

$$\mathbb{B}_X := B[0, 1] \quad \text{and} \quad \mathbb{U}_{\mathbb{X}} := B(0, 1);$$

whenever there is no ambiguity, we remove the subscript X and simply write $\mathbb{B}$ (resp. $\mathbb{U}$).

In Part I the strict Fréchet differentiability of functions and mappings have been utilized very often. It will be still central in many places of this Part II. We will see below in this section that such a property enjoys the remarkable characterization that a continuous real-valued function $f : X \to \mathbb{R}$ on a normed space X is strictly differentiable at $\overline{x} \in X$ provided that for each $\varepsilon > 0$ there is $\delta > 0$ such that for all $u, v \in B(\overline{x}, \delta)$ with $u \neq v$ and all $z \in]u, v[:= \{tu + (1-t)v : t \in]0, 1[\}$

$$-\varepsilon \le \frac{f(u) - f(z)}{\|u - z\|} + \frac{f(v) - f(z)}{\|v - z\|} \le \varepsilon.$$

The aim of this chapter is to study the large class of nonsmooth extended real-valued functions $f : X \to \mathbb{R} \cup \{+\infty\}$ which satisfy (for $\overline{x} \in \operatorname{dom} f$) the unilateral left-side inequality above.

8.1. Definition and first properties of subsmooth functions

We saw in Lemma 4.48(b) that a continuous convex function $f : U \to \mathbb{R}$ is strictly Fréchet differentiable at $\overline{x} \in U$ if and only if

$$\lim_{t \downarrow 0} \frac{f(\overline{x} + th) + f(\overline{x} - th) - 2f(\overline{x})}{t} = 0$$

uniformly with respect to $h \in \mathbb{B}_X$ (or equivalently with respect to $h \in \mathbb{S}_X$). In the case of a general mapping, a characterization in the same line holds true as shown in the following theorem.

THEOREM 8.1 (L. Veselý and L. Zajíček: characterization of strict Fréchet differentiability). Let U be a nonempty open set of a normed space X and $G : U \to Y$ be a mapping from U into a Banach space Y which is continuous at $\overline{x} \in U$. The following assertions are equivalent:
(a) the mapping G is strictly Fréchet differentiable at the point $\overline{x}$;
(b) for every $\varepsilon > 0$ there is $\delta > 0$ with $B(\overline{x}, \delta) \subset U$ such that for all $h \in \mathbb{B}_X$, $x \in U$, $r, s > 0$ with $x + rh \in B(\overline{x}, \delta)$, $x - sh \in B(\overline{x}, \delta)$ one has

$$\left\| \frac{G(x+rh) - G(x)}{r} - \frac{G(x) - G(x-sh)}{s} \right\| \le \varepsilon;$$

(c) for every $\varepsilon > 0$ there is $\delta > 0$ with $B(\overline{x}, \delta) \subset U$ such that for all $h \in \mathbb{S}_X$, $x \in U$, $r, s > 0$ with $x + rh \in B(\overline{x}, \delta)$, $x - sh \in B(\overline{x}, \delta)$ one has

$$\left\| \frac{G(x+rh) - G(x)}{r} - \frac{G(x) - G(x-sh)}{s} \right\| \le \varepsilon;$$

(d) for every $\varepsilon > 0$ there is $\delta > 0$ with $B(\overline{x}, \delta) \subset U$ such that for all $u, v \in B(\overline{x}, \delta)$ with $u \neq v$ and all $z \in]u, v[:= \{tu + (1-t)v : t \in]0, 1[\}$ one has

$$\left\| \frac{G(u) - G(z)}{\|u - z\|} - \frac{G(z) - G(v)}{\|z - v\|} \right\| \le \varepsilon.$$

We establish first the following lemma.

LEMMA 8.2. Let U be a nonempty open set of a normed space X and $G : U \to Y$ be a mapping from U into a Banach space Y. Assume that the property (c) in the above theorem is satisfied. Then

$$\lim_{t \downarrow 0} t^{-1}\big(G(\overline{x} + th) - G(\overline{x})\big)$$

exists uniformly with respect to $h \in \mathbb{S}_X$.

PROOF. Without loss of generality, we may suppose $U = X$. Fix any $\varepsilon > 0$ and take $\delta > 0$ given by the property in (c) of the theorem. Fix any $h \in \mathbb{S}_X$ and consider any $0 < \tau < t < \delta$. Then, putting $Q_\tau(h) := \tau^{-1}\big(G(\overline{x} + \tau h) - G(\overline{x})\big)$ we have

$$\left\| \frac{G(\overline{x} + th) - G(\overline{x} + \tau h)}{t - \tau} - \frac{G(\overline{x} + \tau h) - G(\overline{x})}{\tau} \right\| \le \varepsilon,$$

or equivalently

$$\|G(\overline{x} + th) - G(\overline{x} + \tau h) - (t - \tau)Q_\tau(h)\| \le \varepsilon(t - \tau).$$

Observing that $G(\overline{x}+th) - G(\overline{x}+\tau h) - (t-\tau)Q_\tau(h) = G(\overline{x}+th) - G(\overline{x}) - tQ_\tau(h)$, we derive that

$$\|G(\overline{x} + th) - G(\overline{x}) - tQ_\tau(h)\| \le \varepsilon(t - \tau) \le \varepsilon t, \text{ hence } \|Q_t(h) - Q_\tau(h)\| \le \varepsilon.$$

The latter clearly implies the assertion of the lemma by completeness of Y. □

PROOF OF THEOREM 8.1. Suppose again (without loss of generality) that $U = X$. The assertions (b), (c) and (d) are easily seen to be pairwise equivalent, and (a) clearly implies (c).

Suppose that (c) is satisfied. By Lemma 8.2 above, for each $h \in X$ put

$$\Lambda(h) := \lim_{t \downarrow 0} t^{-1}\big(G(\overline{x} + th) - G(\overline{x})\big).$$

Clearly, the equality $\Lambda(rh) = r\Lambda(h)$ holds for all reals $r \geq 0$. Fix any $h_1, h_2 \in X$ with $h_1 \neq h_2$ and fix also any real $\varepsilon > 0$. For $\varepsilon' := 2\varepsilon/\|h_1 - h_2\| > 0$ there exists by (c) some $\delta > 0$ such that for any $t \in]0, \delta[$

$$\left\| \frac{G(\overline{x} + 2th_1) - G(\overline{x} + t(h_1 + h_2))}{t\|h_1 - h_2\|} - \frac{G(\overline{x} + t(h_1 + h_2)) - G(\overline{x} + 2th_2)}{t\|h_1 - h_2\|} \right\| \leq \varepsilon',$$

or equivalently

$$\left\| \frac{G(\overline{x} + 2th_1) - G(\overline{x} + t(h_1 + h_2))}{t} - \frac{G(\overline{x} + t(h_1 + h_2)) - G(\overline{x} + 2th_2)}{t} \right\| \leq 2\varepsilon.$$

The latter amounts to saying, with notation in the proof of the above lemma, that $\|Q_{2t}(h_1) + Q_{2t}(h_2) - Q_t(h_1 + h_2)\| \leq \varepsilon$. By the above lemma choose some $\delta' \in]0, \delta[$ such that for all $t \in]0, \delta'[$

$$\|Q_t(h_1 + h_2) - \Lambda(h_1 + h_2)\| \leq \varepsilon, \ \|Q_{2t}(h_i) - \Lambda(h_i)\| \leq \varepsilon/2, \ i = 1, 2.$$

Therefore, for all $t \in]0, \delta'[$ we obtain $\|\Lambda(h_1) + \Lambda(h_2) - \Lambda(h_1 + h_2)\| \leq 3\varepsilon$, which yields that $\Lambda(h_1 + h_2) = \Lambda(h_1) + \Lambda(h_2)$. The latter equality combined with the positive homogeneity of Λ easily entails that Λ is linear.

On the other hand, the uniform convergence on $\mathbb{S}_X$ of the family $(Q_t)_{t>0}$ to Λ (as $t \downarrow 0$) is equivalent to the existence of a function $\eta : \mathbb{R} \to [0, +\infty]$ with $t^{-1}\eta(t) \to 0$ (as $t \downarrow 0$) such that $\|G(\overline{x} + h) - G(\overline{x}) - \Lambda(h)\| \leq \eta(\|h\|)$ for all $h \in X$. This combined with the continuity of G at $\overline{x}$ implies the continuity of Λ, and hence G is Fréchet differentiable at $\overline{x}$.

Finally, let us show the strict Fréchet differentiability. Fix any $\varepsilon > 0$ and choose some $\delta > 0$ satisfying (c) and such that (by the Fréchet differentiability)

$$\|G(\overline{x} + h) - G(\overline{x}) - \Lambda(h)\| \leq \varepsilon\|h\| \quad \text{for all } h \in B(0, \delta).$$

Fix any $x, y \in B(\overline{x}, \delta/4)$ with $x \neq y$ and put $u := (x - y)/\|x - y\|$, $r := \|x - y\|$, $s := \delta/4$. It ensues that $\|(y - su) - \overline{x}\| < \delta$, hence

$$\|G(y - su) - G(\overline{x}) - \Lambda(y - su - \overline{x})\| \leq \varepsilon\|y - su - \overline{x}\|,$$

$$\|G(y) - G(\overline{x}) - \Lambda(y - \overline{x})\| \leq \varepsilon\|y - \overline{x}\|.$$

Both inequalities yield

$$\|G(y) - G(y - su) - \Lambda(su)\| \leq \varepsilon(\|y - su - \overline{x}\| + \|y - \overline{x}\|) \leq \varepsilon(2\|y - \overline{x}\| + s) \leq 3\varepsilon s,$$

which gives

$$\|s^{-1}\big(G(y) - G(y - su)\big) - \Lambda(u)\| \leq 3\varepsilon.$$

On the other hand, by the choice of δ from (c) we also have

$$\left\| \frac{G(x) - G(y)}{\|x - y\|} - \frac{G(y) - G(y - su)}{s} \right\| \leq \varepsilon.$$

It then follows that

$$\left\| \frac{G(x) - G(y)}{\|x - y\|} - \Lambda(u) \right\| \leq 4\varepsilon, \text{ or equivalently } \|G(x) - G(y) - \Lambda(x - y)\| \leq 4\varepsilon\|x - y\|,$$

which translates the strict Fréchet differentiability of G at $\overline{x}$ and finishes the proof of the theorem. □

In the case of a real-valued function f the inequality (c) in the above theorem characterizing the strict Fréchet differentiability at $\overline{x}$ can be rewritten as

$$-\varepsilon \le \frac{f(x+rh)-f(x)}{r} + \frac{f(x-sh)-f(x)}{s} \le \varepsilon.$$

As we will see along this chapter, functions satisfying the left-side inequality alone enjoy remarkable and useful properties. This yields to the following definition.

DEFINITION 8.3. Let U be a nonempty open set of a normed space X. An extended real-valued function $f : U \to \mathbb{R} \cup \{+\infty\}$ is *subsmooth at a point* $\overline{x} \in U$ provided that for every real $\varepsilon > 0$ there is a real $\delta > 0$ (depending on ε and $\overline{x}$) with $B(\overline{x}, \delta) \subset U$ such that, for all $h \in \mathbb{S}_X$, $x \in U$, $r, s > 0$ with $x+rh \in B(\overline{x}, \delta) \cap \operatorname{dom} f$, $x - sh \in B(\overline{x}, \delta) \cap \operatorname{dom} f$ one has

$$-\varepsilon \le \frac{f(x+rh)-f(x)}{r} + \frac{f(x-sh)-f(x)}{s}.$$

When f is subsmooth at any $\overline{x} \in U_0$ for an open set $U_0 \subset U$, one says that it is *subsmooth on* U_0. The function f is *subsmooth near a point* if it subsmooth on an open neighborhood $U_0 \subset U$ of this point.

If $f \equiv +\infty$ on an open set $U_0 \subset U$ (resp. near a point $\overline{x} \in U$), then f is obviously subsmooth on U_0 (resp. near $\overline{x}$). The following features are also clear from the definition.

PROPOSITION 8.4. Let X be a normed space.
(a) Any function on an open set of X which is strictly differentiable at a point is subsmooth at that point.
(b) Given an open convex set U of X, any extended real-valued convex function $f : U \to \mathbb{R} \cup \{+\infty\}$ is subsmooth on U.

It is also worth pointing out that the above definition of subsmoothness of f at $\overline{x}$ is equivalent to requiring that, for every real $\varepsilon > 0$ there exists a real $\delta > 0$ with $B(\overline{x}, \delta) \subset U$ such that, for all $u, v \in B(\overline{x}, \delta) \cap \operatorname{dom} f$ with $u \neq v$ and all $z \in]u, v[$ one has

$$-\varepsilon \le \frac{f(u)-f(z)}{\|u-z\|} + \frac{f(v)-f(z)}{\|v-z\|}. \tag{8.1}$$

This obviously entails the following property.

PROPOSITION 8.5. Let U be a nonempty open set of a normed space X and $f : U \to \mathbb{R} \cup \{+\infty\}$ be a proper function. If f is subsmooth at a point $\overline{x} \in U$, then there exists some $\delta > 0$ such that $B(\overline{x}, \delta) \cap \operatorname{dom} f$ is a convex set.

The characterization (8.1) of subsmoothness also leads to introduce the corresponding uniform and one-sided notions.

DEFINITION 8.6. Let U be a nonempty open subset of a normed space X and $f : U \to \mathbb{R} \cup \{+\infty\}$ be an extended real-valued function. The function f is said to be *uniformly subsmooth on a nonempty open set* $U_0 \subset U$ if for every real $\varepsilon > 0$ there exists a real $\delta > 0$ such that, for all $u \neq v$ in $U_0 \cap \operatorname{dom} f$ with $\|u - v\| < \delta$ and all $z \in]u, v[\cap U$ one has

$$-\varepsilon \le \frac{f(u)-f(z)}{\|u-z\|} + \frac{f(v)-f(z)}{\|v-z\|}.$$

The function f is *uniformly subsmooth near a point* in U if it is uniformly subsmooth on an open neighborhood of this point.

Similarly, one says that f is *one-sided subsmooth* at $\overline{x} \in \operatorname{dom} f$ if for every real $\varepsilon > 0$ there exists a real $\delta > 0$ with $B(\overline{x}, \delta) \subset U$ such that, for every $v \in B(\overline{x}, \delta) \cap \operatorname{dom} f$ with $v \neq \overline{x}$ and for every $z \in]\overline{x}, v[$ one has

$$-\varepsilon \leq \frac{f(\overline{x}) - f(z)}{\|\overline{x} - z\|} + \frac{f(v) - f(z)}{\|v - z\|}.$$

The uniform equi-subsmoothness is defined in a similar way.

DEFINITION 8.7. Given a family $(f_i)_{i \in I}$ of functions from the open set U into $\mathbb{R} \cup \{+\infty\}$ and a family $(U_i)_{i \in I}$ of open subsets of U, one says that this family of functions is *uniformly equi-subsmooth relative to* $(U_i)_{i \in I}$ when for every real $\varepsilon > 0$ there exists a real $\delta > 0$ such that for each $i \in I$ one has

$$-\varepsilon \leq \frac{f_i(u) - f_i(z)}{\|u - z\|} + \frac{f_i(v) - f_i(z)}{\|v - z\|}$$

for all $u \neq v$ in $U_i \cap \operatorname{dom} f_i$ with $\|u - v\| < \delta$ and all $z \in]u, v[\cap U$. If all the sets U_i coincide with a same open set $U_0 \subset U$, one simply says that the family of functions $(f_i)_{i \in I}$ is *uniformly equi-subsmooth on* U_0. □

Of course, the uniform subsmoothness implies subsmoothness, which in turn implies one-sided subsmoothness. It is also clear that the three properties of subsmoothness, one-sided subsmoothness and uniform subsmoothness are stable under sum of finitely many functions.

EXAMPLE 8.8. Consider the function $f := -|\cdot|$ on $\mathbb{R}$. For any $v \neq 0$ in $\mathbb{R}$ and any z strictly between 0 and v, we have

$$\frac{f(0) - f(z)}{|z|} + \frac{f(v) - f(z)}{|v - z|} = 1 + \frac{-|v| + |z|}{|v - z|} \geq 0,$$

so f is one-sided subsmooth at the point 0 (and hence one-sided subsmooth at any point in $\mathbb{R}$).

However, observing that

$$\frac{f(1/n) - f(0)}{1/n} + \frac{f(-1/n) - f(0)}{1/n} = -2,$$

we see that f is not subsmooth at 0. □

The following strict differentiability result follows directly from Definition 8.3 and Theorem 8.1.

PROPOSITION 8.9. Let U be a nonempty open set of a normed space X and $f : U \to \mathbb{R}$ be a real-valued function which is continuous at $\overline{x} \in U$. Then f is strictly Fréchet differentiable at $\overline{x}$ if and only if both functions f and $-f$ are subsmooth at the point $\overline{x}$.

The next proposition extends to subsmooth functions a property already seen for convex functions in Proposition 2.178(a).

PROPOSITION 8.10. Let U be a nonempty open set of a normed space X and $f : U \to \mathbb{R}$ be a real-valued function which is subsmooth at $\overline{x}$. Then f is strictly Fréchet differentiable at $\overline{x}$ if and only if it is Fréchet differentiable at $\overline{x}$.

PROOF. Only the implication $\Leftarrow$ needs to be justified. Assume that f is Fréchet differentiable at $\overline{x}$. Without loss of generality we may suppose that $U = X$ along with $f(\overline{x}) = 0$ and $Df(\overline{x}) = 0$. Choose a real $\delta > 0$ such that the inequality in Definition 8.3 is satisfied for $\varepsilon' := \varepsilon/2$ in place of ε and such that for all $x \in B(\overline{x}, \delta)$ one has

$$|f(x)| = |f(x) - f(\overline{x}) - Df(\overline{x})(x - \overline{x})| \le (\varepsilon/8)\|x - \overline{x}\|.$$

Take any $x, y \in B(\overline{x}, \delta/2)$ with $x \neq y$ and set $h := (x-y)/\|x-y\|$. Putting $r := \delta/2$ and noting that $y = x - \|y - x\|\, h$, by Definition 8.3 we have

(8.2)

$$-\varepsilon/2 \le \frac{f(x+rh) - f(x)}{r} + \frac{f(y) - f(x)}{\|y - x\|}, \; -\varepsilon/2 \le \frac{f(x) - f(y)}{\|x - y\|} + \frac{f(y - rh) - f(y)}{r}.$$

Further, the above inequality given by the Fréchet differentiability of f at $\overline{x}$ entails that

$$\max\{|f(x)|, |f(y)|, |f(x + rh)|, |f(y - rh)|\} \le (\varepsilon/8)2r = \varepsilon r/4,$$

which in turn ensures that

$$|r^{-1}(f(x+rh) - f(x))| \le \varepsilon/2 \quad \text{and} \quad |r^{-1}(f(y - rh) - f(y))| \le \varepsilon/2.$$

The latter inequalities combined with the inequalities in (8.2) yield

$$\frac{f(x) - f(y)}{\|x - y\|} \le \varepsilon \quad \text{and} \quad -\varepsilon \le \frac{f(x) - f(y)}{\|x - y\|},$$

which translates the strict Fréchet differentiability of f at $\overline{x}$ (with $Df(\overline{x}) = 0$). $\square$

Anyone of Propositions 8.5 and 8.9 tells us in particular that any $\mathcal{C}^1$ function $f : U \to \mathbb{R}$ on an open set U is subsmooth on U. A similar result provides a first example of families of uniformly equi-subsmooth functions. Given a nonempty open set U of a normed space X, a family of mappings $(G_i)_{i \in I}$ from U into a normed space Y is said to be *uniformly equi-continuous relative to a family* $(U_i)_{i\in I}$ *of open subsets* of U when for any $\varepsilon > 0$ there exists $\delta > 0$ such that for any $i \in I$ one has

(8.3) $$\|G_i(x') - G_i(x)\| \le \varepsilon \quad \text{for all } x, x' \in U_i \text{ with } \|x' - x\| < \delta.$$

PROPOSITION 8.11. Let U be an open set of a normed space X and $(f_i)_{i\in I}$ be a family of functions from U into $\mathbb{R}$. Let $(U_i)_{i\in I}$ be a family of open convex subsets of U such that for each $i \in I$ the function f_i is differentiable on U_i and such that the family of derivatives $(Df_i)_{i\in I}$ is uniformly equi-continuous relative to $(U_i)_{i\in I}$. Then the family of functions $(f_i)_{i\in I}$ is uniformly equi-subsmooth relative to $(U_i)_{i\in I}$.

PROOF. Take any real $\varepsilon > 0$ and choose a real $\delta > 0$ such that $\|Df_i(x') - Df_i(x)\| < \varepsilon$ for any $i \in I$ and any $x, x' \in U_i$ with $\|x' - x\| < \delta$. Fix any $i \in I$ and take any $u, v \in U_i$ with $\|u - v\| < \delta$. Consider $z \in]u, v[$ and note that with $\nu := (v - u)/\|v - u\|$, $u_t := z + t(u - z)$ and $v_t := z + t(v - z)$

$$\begin{aligned}
&\frac{f_i(u) - f_i(z)}{\|u - z\|} + \frac{f_i(v) - f_i(z)}{\|v - z\|} \\
&= \int_0^1 \left\langle Df_i(u_t), \frac{u - z}{\|u - z\|} \right\rangle dt + \int_0^1 \left\langle Df_i(v_t), \frac{v - z}{\|v - z\|} \right\rangle dt \\
&= \int_0^1 \langle Df_i(v_t) - Df_i(u_t), \nu \rangle \, dt \ge -\varepsilon,
\end{aligned}$$

where the latter inequality is due to the fact that for every $t \in [0,1]$ one has $u_t, v_t \in U_i$ with $\|v_t - u_t\| = t\|v-u\| < \delta$. This justifies the desired uniform equi-subsmoothness property. □

Subsmooth functions can be characterized via a Jensen-like inequality.

PROPOSITION 8.12. Let $f : U \to \mathbb{R} \cup \{+\infty\}$ be a function on a nonempty open set U of a normed space X. The function f is subsmooth at $\overline{x} \in U$ (resp. one-sided subsmooth at $\overline{x} \in \operatorname{dom} f$) if and only if for every real $\varepsilon > 0$ there exists a real $\delta > 0$ with $B(\overline{x}, \delta) \subset U$ such that, for all $x, y \in B(\overline{x}, \delta)$ and all $t \in]0,1[$ the inequality

$$f(tx + (1-t)y) \le tf(x) + (1-t)f(y) + \varepsilon t(1-t)\|x-y\| \tag{8.4}$$

holds (resp. the inequality holds with $y = \overline{x}$ and all $x \in B(\overline{x}, \delta)$).

Similarly, f is uniformly subsmooth on an open set $U_0 \subset U$ if and only if for every real $\varepsilon > 0$ there exists a real $\delta > 0$ such that for any $x, y \in U_0$ with $\|x-y\| < \delta$ and any $t \in]0,1[$ with $tx + (1-t)y \in U$ the above inequality is satisfied.

PROOF. We only justify the equivalence for the subsmoothness property, the case of either one-sided or uniform smoothness is similar. Let $u, v \in \operatorname{dom} f$ with $u \neq v$ and $z \in]u, v[\cap U$ with $z = tu + (1-t)v$ and $t \in]0,1[$. Since $u - z = (1-t)(u-v)$ and $v - z = t(v-u)$, we note that

$$\frac{f(u) - f(z)}{\|u-z\|} + \frac{f(v) - f(z)}{\|v-z\|} = \frac{tf(u) + (1-t)f(v) - f(z)}{t(1-t)\|u-v\|}.$$

From this and (8.1) the implication $\Leftarrow$ follows. The reverse implication being obtained in an analogous way, the equivalence is justified. □

REMARK 8.13. Functions satisfying the inequality (8.4) are also called *approximately convex at* $\overline{x}$ in the literature, so the above proposition says that the approximate convexity coincides with the subsmoothness property. □

REMARK 8.14. Let be given an open set U of a normed space X, a family $(f_i)_{i\in I}$ of functions from U into $\mathbb{R} \cup \{+\infty\}$ and a family of open subsets $(U_i)_{i\in I}$ of U. The above arguments also show that this family of functions is *uniformly equi-subsmooth relative to* $(U_i)_{i\in I}$ if and only if for every real $\varepsilon > 0$ there exists a real $\delta > 0$ such that for each $i \in I$ one has

$$f_i(tx + (1-t)y) \le tf_i(x) + (1-t)f_i(y) + \varepsilon t(1-t)\|x-y\|$$

for any $x, y \in U_i$ with $\|x-y\| < \delta$ and any $t \in]0,1[$ with $tx + (1-t)y \in U$. In particular the family $(f_i)_{i\in I}$ is uniformly equi-subsmooth relative to $(U_i)_{i\in I}$ whenever any set U_i is convex and any function f_i is convex on U_i. □

Through Proposition 8.12 we can characterize uniformly subsmooth functions via modulus functions.

DEFINITION 8.15. A function $\omega : [0, +\infty[\to [0, +\infty]$ continuous at 0 with $\omega(0) = 0$ will be called a *modulus function*.

Given a function $f : U \to \mathbb{R} \cup \{+\infty\}$ which is uniformly subsmooth on an open set $U_0 \subset U$, define $\omega_f : [0, +\infty[\to [0, +\infty]$ by $\omega_f(0) = 0$ and for $r > 0$ by

$$\omega_f(r) = \sup_{(t,x,y) \in G_r(f)} \frac{\big(f(tx + (1-t)y) - tf(x) - (1-t)f(y)\big)^+}{t(1-t)\|x-y\|},$$

where

$$G_r(f) := \{(t, x, y) : x, y \in U_0 \cap \operatorname{dom} f, 0 < \|x - y\| \le r, t \in]0, 1[, tx + (1 - t)y \in U\}$$

and the above supremum is 0 (as usual) when $G_r(f) = \emptyset$. In the case when there is some $r_0 > 0$ such that $G_{r_0}(f) = \emptyset$, or equivalently $G_r(f) = \emptyset$ for any $r \in]0, r_9]$, we see that $\omega_f(r) = 0$ for every $r \in [0, r_0]$. Under the uniform subsmoothness of f on U_0 we also see that ω_f is a (nondecreasing) modulus function and

$$f(tx + (1 - t)y) \le tf(x) + (1 - t)f(y) + t(1 - t)\|x - y\|\omega_f(\|x - y\|)$$

for all $x, y \in U_0$ and $t \in]0, 1[$ with $tx + (1 - t)y \in U$. This justifies the implication $\Longrightarrow$ of Proposition 8.16 below. The converse implication being obvious we have the following equivalence:

PROPOSITION 8.16. Let $f : U \to \mathbb{R} \cup \{+\infty\}$ be a function on a nonempty open set U of a normed space X. The function f is uniformly subsmooth on an open set $U_0 \subset U$ if and only if it is semiconvex on U_0 in the sense that there exists a modulus function $\omega : [0, +\infty[\to [0, +\infty]$ such that

$$f(tx + (1 - t)y) \le tf(x) + (1 - t)f(y) + t(1 - t)\|x - y\|\omega(\|x - y\|)$$

for all $x, y \in U_0$ and $t \in]0, 1[$ with $tx + (1 - t)y \in U$.

Semiconvex functions with respect to a given modulus function will be studied in detail in Chapter 10.

REMARK 8.17. Considering as above $G_r(f_i)$ with U_i in place of U_0 and defining $\overline{\omega}(\cdot) := \sup_{i \in I} \omega_{f_i}(\cdot)$ we see that $\overline{\omega}(\cdot)$ is a modulus function whenever the family of functions $(f_i)_{i \in I}$ from an open set U of the normed space X into $\mathbb{R} \cup \{+\infty\}$ is uniformly equi-subsmooth relative to a family of open subsets $(U_i)_{i \in I}$ of U. Then, as above we see that a family of functions $(f_i)_{i \in I}$ from U into $\mathbb{R} \cup \{+\infty\}$ is uniformly equi-subsmooth relative a family of open subsets $(U_i)_{i \in I}$ of U if and only if there is a modulus function $\omega : [0, +\infty[\to [0, +\infty]$ such that for each $i \in I$

$$f_i(tx + (1 - t)y) \le tf_i(x) + (1 - t)f_i(y) + t(1 - t)\|x - y\|\omega(\|x - y\|)$$

for all $x, y \in U_i$ and $t \in]0, 1[$ with $tx + (1 - t)y \in U$. □

The next proposition proves that any function which is subsmooth at a point $\overline{x}$ and bounded from above near $\overline{x}$ is Lipschitz near $\overline{x}$.

PROPOSITION 8.18. Let $f : U \to \mathbb{R} \cup \{+\infty\}$ be a function on an open set U of a normed space X which is subsmooth at a point $\overline{x} \in U$ and bounded from above near $\overline{x}$. Then f is Lipschitz continuous near $\overline{x}$.

PROOF. Take $\varepsilon = 1$ and by Proposition 8.12 take a real $\delta_0 > 0$ with $B(\overline{x}, \delta_0) \subset U$ such that f is bounded above on $B(\overline{x}, \delta_0)$ and such that (8.4) is satisfied for all $x, y \in B(\overline{x}, \delta_0)$. We note that f is finite on $B(\overline{x}, \delta_0)$. Taking any $x \in B(\overline{x}, \delta_0)$ and setting $u := 2\overline{x} - x$, we see that $\overline{x} = (1/2)x + (1/2)u$ with $u \in B(\overline{x}, \delta_0)$, thus

$$f(\overline{x}) \le \frac{1}{2}f(x) + \frac{1}{2}f(u) + \frac{1}{4}\|u - x\|.$$

Since f is bounded above on the ball $B(\overline{x}, \delta_0)$, it follows that f is also bounded from below on this ball. We can then choose an upper bound $\mu > 0$ of $|f|$ on the ball $B(\overline{x}, \delta_0)$.

Now put $\delta := \delta_0/2$ and fix any $x, y \in B(\overline{x}, \delta)$ with $x \neq y$. Putting $t := \frac{\|y-x\|}{\delta+\|y-x\|}$ and $z := y + \delta\frac{y-x}{\|y-x\|}$, and noting that $z \in B(\overline{x}, \delta_0)$, it ensues that

$$f(y) = f(tz + (1-t)x) \le tf(z) + (1-t)f(x) + t(1-t)\|z-x\|,$$

hence (since $t(1-t) \le t \le \|y-x\|/\delta$ and $\|z-x\| \le 3\delta$)

$$f(y) - f(x) \le t\big(f(z) - f(x)\big) + 3\|y-x\| \le (3 + \frac{2\mu}{\delta})\|y-x\|,$$

which translates the Lipschitz property of f on $B(\overline{x}, \delta)$. □

COROLLARY 8.19. Let U be a nonempty open set of a Banach space X and $f : U \to \mathbb{R} \cup \{+\infty\}$ be a lower semicontinuous function. If f is subsmooth at a point $\overline{x} \in \operatorname{int}(\operatorname{dom} f)$, then f is Lipschitz continuous near $\overline{x}$.

PROOF. Without loss of generality (putting $g(x) := f(x + \overline{x}) - f(\overline{x})$) we may suppose that $\overline{x} = 0$ and $f(\overline{x}) = 0$. Let $\delta_0 > 0$ be such that $B(0, \delta_0) \subset \operatorname{dom} f$ and such that the condition (8.4) holds with $\varepsilon := 1$. Let $\delta := \delta_0/2$ and for each integer n, as in the proof of Theorem 2.161, put $V_n := \{x \in W : f(x) \le n\}$ with $W := B(0, \delta_0)$. Noting that $W = \bigcup_{n\in\mathbb{N}} V_n$ (and keeping in mind that $B(0, \delta_0)$ is open in the complete space X), Baire theorem tells us that $\operatorname{int} V_k \neq \emptyset$ for some $k \in \mathbb{N}$. Choose $a \in W$ and $r \in]0, \delta_0[$ such that $B(a, 2r) \subset V_k$. We have $-a \in W$ and for each $x \in B(0, r)$ there is some $y_x \in B(a, 2r)$ such that $x = (1/2)(-a) + (1/2)y_x$, hence

$$f(x) \le \frac{1}{2}f(-a) + \frac{1}{2}f(y_x) + \frac{1}{4}\|y_x + a\| \le \frac{1}{2}\big(f(-a) + k + 2\delta_0\big).$$

The function f is then bounded from above near the point $\overline{x} = 0$, so it is Lipschitz continuous near this point according to Proposition 8.18. □

Consider now the case of subsmooth functions over intervals of the real line.

PROPOSITION 8.20. Let I be an open interval of $\mathbb{R}$ and $f : I \to \mathbb{R} \cup \{+\infty\}$ be a proper lower semicontinuous subsmooth function. Then for any reals $r < s$ with $[r, s] \subset \operatorname{dom} f$, the function f is locally Lipschitz on $]r, s[$ and the restriction of f to $[r, s]$ is continuous, so f is continuous on the right at r and on the left at s.

PROOF. Let $[r, s] \subset I$ with $r < s$. We already know by Corollary 8.19 above that f is (locally Lipschitz) continuous on $]r, s[$. Let us prove, for example, that f is continuous on the right at r. Taking $\varepsilon = 1$, choose $\delta > 0$ with $I \cap B(r, 2\delta) \subset I$ such that

$$f(\theta x + (1-\theta)y) \le \theta f(x) + (1-\theta)f(y) + \theta(1-\theta)|x-y|$$

for all $x, y \in I \cap B(r, 2\delta)$, and $\theta \in]0, 1[$. Putting $\sigma := \min\{s, r + \delta\}$ we obtain for all $t \in]r, \sigma[$

$$f(t) \le \frac{t-r}{\sigma-r}f(\sigma) + \frac{\sigma-t}{\sigma-r}f(r) + \frac{(t-r)(\sigma-t)}{(\sigma-r)^2}|\sigma-r|,$$

and hence $\limsup_{t\downarrow r} f(t) \le f(r)$. This and the lower semicontinuity of f guarantee that f is continuous on the right at r as desired. □

8.2. Directional derivatives and subdifferentials of subsmooth functions

This section analyzes directional derivatives and subdifferentials of subsmooth functions.

8.2.1. General properties of derivatives and subdifferentials. Let us begin with some properties of the differential quotient. Let U be a nonempty open set of a normed space X and $f : U \to \mathbb{R}\cup\{+\infty\}$ be a function which is subsmooth at a point $\overline{x} \in U$. Fix any real $\varepsilon > 0$. Let $\delta > 0$ with $B(\overline{x}, \delta) \subset U$ for which condition (8.4) is fulfilled. Fix any $x \in B(\overline{x}, \delta) \cap \operatorname{dom} f$ and any $h \in X$. Let $0 < s < t$ with $t\|h\| < \delta - \|x - \overline{x}\|$ and let any $r > 0$ with $r\|h\| < \delta - \|x - \overline{x}\|$. Observing that

$$x = (r+s)^{-1}s(x - rh) + (r+s)^{-1}r(x + sh),$$

we have

$$f(x) \le \frac{s}{r+s} f(x - rh) + \frac{r}{r+s} f(x + sh) + \frac{\varepsilon r s}{r+s}\|h\|,$$

which is equivalent to the following first *slope ε-inequality*:

$$-r^{-1}[f(x - rh) - f(x)] \le s^{-1}[f(x + sh) - f(x)] + \varepsilon\|h\| \tag{8.5}$$

for $r, s > 0$ with $\|h\| \max\{r, s\} < \delta - \|x - \overline{x}\|$.

Similarly, from the equality

$$x + sh = \frac{s}{t}(x + th) + \left(1 - \frac{s}{t}\right) x$$

we obtain

$$f(x + sh) \le \frac{s}{t} f(x + th) + \left(1 - \frac{s}{t}\right) f(x) + \varepsilon s \left(1 - \frac{s}{t}\right) \|h\|,$$

which in turn is equivalent the following second *slope ε-inequality*:

$$s^{-1}[f(x + sh) - f(x)] \le t^{-1}[f(x + th) - f(x)] + \varepsilon \left(1 - \frac{s}{t}\right) \|h\| \tag{8.6}$$

for reals $0 < s < t$ with $t\|h\| < \delta - \|x - \overline{x}\|$.

PROPOSITION 8.21. Let U be a nonempty open set of a normed space X and $f : U \to \mathbb{R} \cup \{+\infty\}$ be a function which is subsmooth at a point $\overline{x} \in U$. The following hold.

(a) For each real $\varepsilon > 0$ there is $\delta > 0$ with $B(\overline{x}, 2\delta) \subset U$ such that for any $x \in B(\overline{x}, \delta) \cap \operatorname{dom} f$, any $h \in X$ and any $t > 0$ with $t\|h\| < \delta$

$$\underline{d}_H^+ f(x; h) \le \limsup_{s \downarrow 0} s^{-1}[f(x + sh) - f(x)] \le t^{-1}[f(x + th) - f(x)] + \varepsilon\|h\|.$$

(b) If $\overline{x} \in U$, the directional derivative

$$f'(\overline{x}; h) := \lim_{t \downarrow 0} t^{-1}[f(\overline{x} + th) - f(\overline{x})]$$

exists in $\mathbb{R}\cup\{-\infty, +\infty\}$ for any direction $h \in X$, and the function $f'(\overline{x}; \cdot)$ is convex and positively homogeneous.

PROOF. Let any $\varepsilon > 0$ and let $\delta_0 > 0$ with $B(\overline{x}, \delta_0) \subset U$ such that the condition (8.4) is fulfilled. Set $\delta := \delta_0/2$ and fix any $x \in B(\overline{x}, \delta) \cap \operatorname{dom} f$ and any $h \in X$. Take $0 < s < t$ with $t\|h\| < \delta$ and write according to (8.6) that

$$s^{-1}[f(x + sh) - f(x)] \le t^{-1}[f(x + th) - f(x)] + \varepsilon \left(1 - \frac{s}{t}\right) \|h\|.$$

Fixing t we deduce as $s \downarrow 0$ that

$$(8.7)\quad \underline{d}_H^+ f(x;h) \le \limsup_{s\downarrow 0} s^{-1}[f(x+sh)-f(x)] \le t^{-1}[f(x+th)-f(x)] + \varepsilon\|h\|,$$

which justifies (a). On the other hand, keeping $h \in X$ and $t > 0$ with $t\|h\| < \delta$ in the second inequality in (8.7) and choosing $x = \overline{x}$, we obtain by passing to the limit inferior as $t \downarrow 0$ that

$$\limsup_{s\downarrow 0} s^{-1}[f(\overline{x}+sh)-f(\overline{x})] \le \liminf_{t\downarrow 0} t^{-1}[f(\overline{x}+th)-f(\overline{x})] + \varepsilon\|h\|.$$

This being true for all $\varepsilon > 0$, the desired limit giving $f'(\overline{x};h)$ exists in $\mathbb{R} \cup \{-\infty,+\infty\}$.

The positive homogeneity being obvious, it remains to show the convexity of $f'(\overline{x};\cdot)$. Fix any (h,α) and (h',β) in $X \times \mathbb{R}$ and satisfying $f'(\overline{x};h) < \alpha$ and $f'(\overline{x};h') < \beta$. Take any $\varepsilon > 0$ and choose $\delta > 0$ such that the condition (8.4) holds and such that for all $0 < t < \delta$

$$(2t)^{-1}[f(\overline{x}+2th)-f(\overline{x})] < \alpha \quad \text{and} \quad (2t)^{-1}[f(\overline{x}+2th')-f(\overline{x})] < \beta.$$

Take any $t > 0$ with $t\max\{\|h\|,\|h'\|\} < \delta/2$. It ensues that

$$f(\overline{x}+th+th') \le \frac{1}{2}f(\overline{x}+2th) + \frac{1}{2}f(\overline{x}+2th') + \frac{\varepsilon t}{2}\|h-h'\|,$$

or otherwise written

$$\begin{aligned} t^{-1}[f(\overline{x}+th+th')-f(\overline{x})] \le {} & (2t)^{-1}[f(\overline{x}+2th)-f(\overline{x})] \\ & + (2t)^{-1}[f(\overline{x}+2th')-f(\overline{x})] + (\varepsilon/2)\|h-h'\|, \end{aligned}$$

which entails

$$t^{-1}[f(\overline{x}+th+th')-f(\overline{x})] < \alpha+\beta+(\varepsilon/2)\|h-h'\|.$$

Consequently, $f'(\overline{x};h+h') \le \alpha+\beta$, so $f'(\overline{x};\cdot)$ is convex. □

Before stating the result concerning the subdifferential, we need a lemma.

LEMMA 8.22. Let U be a nonempty open set of a normed space X and $f : U \to \mathbb{R}\cup\{+\infty\}$ be a function.
(a) If f is subsmooth at a point $\overline{x} \in U$, then for any real $\varepsilon > 0$ there exists a real $\delta > 0$ with $B(\overline{x},\delta) \subset U$ such that, for each $x \in B(\overline{x},\delta) \cap \operatorname{dom} f$ and for each $(u,r) \in \operatorname{epi} f$ with $\|u-x\| < \delta$, one has

$$(8.8)\qquad \big(u-x, r-f(x)+\varepsilon\|u-x\|\big) \in T^C\big(\operatorname{epi} f;(x,f(x))\big).$$

(b) If f is uniformly subsmooth on an open subset $U_0 \subset U$, then for any real $\varepsilon > 0$ there exists a real $\delta > 0$ such that the inclusion (8.8) holds for any $x,u \in U_0 \cap \operatorname{dom} f$ with $\|u-x\| < \delta$ and any real $r \ge f(u)$

PROOF. Fix any real $\varepsilon > 0$. Under the assumption in (a) (resp. in (b)) choose a real $\delta_0 > 0$ satisfying $B(\overline{x},\delta_0) \subset U$ as well as the condition (8.4) in Proposition 8.12 and put $\delta := \delta_0/2$ (resp. choose a real $\delta > 0$ satisfying the condition for uniform subsmoothness in Proposition 8.12 similar to (8.4)). Take any $x \in B(\overline{x},\delta)$ with $f(x)$ finite and take any $(u,r) \in \operatorname{epi} f$ with $\|u-x\| < \delta$ (resp. take any $x,u \in U_0 \cap \operatorname{dom} f$ with $\|x-u\| < \delta$ and any real $r \ge f(u)$). If $u = x$, the result is obvious since $T^C(\operatorname{epi} f;(x,f(x))$ is an epigraph set containing $(0,0)$. Suppose that $\|u-x\| > 0$. Take any sequence $(x_n,r_n)_n$ in $\operatorname{epi} f$ converging to $(x,f(x))$ and any sequence $(t_n)_n$ in $]0,+\infty[$ tending to 0. Fix an integer N such that $t_n < 1$ and

$\|x_n - x\| < \delta$ for all $n \geq N$ (resp. $t_n < 1$, $\|x_n - u\| < \delta$ and $x_n + t_n(u - x_n) \in U$ for all $n \geq N$). Putting $z_n := x_n + t_n(u - x_n)$, the condition (8.4) in Proposition 8.12 (resp. the condition for uniform subsmoothness in Proposition 8.12) tells us that, for all $n \geq N$

$$f(z_n) - \varepsilon t_n \|u - x_n\| \leq t_n r + (1 - t_n) r_n,$$

and from this we get that

$$(x_n, r_n) + t_n\big(u - x_n, r - r_n + \varepsilon\|u - x_n\|\big) \in \operatorname{epi} f.$$

This implies the desired inclusion

$$\big(u - x, r - f(x) + \varepsilon\|u - x\|\big) \in T^C\big(\operatorname{epi} f; (x, f(x))\big)$$

and finishes the proof. □

REMARK 8.23. (a) The above proof also shows for $S \subset X$ that if $f : X \to \mathbb{R} \cup \{+\infty\}$ satisfies with $\varepsilon \geq 0$ the inequality

$$f(tx + (1 - t)y) \leq t f(x) + (1 - t) f(y) + \varepsilon t \|x - y\|$$

for all $x, y \in S$ and $t \in]0, 1[$, then

$$\big(u - x, r - f(x) + \varepsilon\|u - x\|\big) \in T^C\big((S \times \mathbb{R}) \cap \operatorname{epi} f; (x, f(x))\big),$$

for every $x \in S \cap \operatorname{dom} f$ and every $(u, r) \in \operatorname{epi} f$ with $u \in S$.
(b) In particular, if S is a nonempty open set $U \subset X$, then

$$\big(u - x, r - f(x) + \varepsilon\|u - x\|\big) \in T^C\big(\operatorname{epi} f; (x, f(x))\big),$$

for every $x \in U \cap \operatorname{dom} f$ and every $(u, r) \in \operatorname{epi} f$ with $u \in U$. □

REMARK 8.24. Let $(f_i)_{i \in I}$ be a family of functions from an open set U of a normed space X into $\mathbb{R} \cup \{+\infty\}$ and let $(U_i)_{i \in I}$ be a family of open subsets of U. Assume that this family of functions is uniformly equi-subsmooth relative to $(U_i)_{i \in I}$. Using Remark 8.14 in place of Proposition 8.12 in the proof of Lemma 8.22 it is not difficult to see that for any $\varepsilon > 0$ there exists $\delta > 0$ such that for each $i \in I$ the inclusion

$$\big(u - x, r - f_i(x) + \varepsilon\|u - x\|\big) \in T^C\big(\operatorname{epi} f_i; (x, f_i(x))\big)$$

holds for any $x, u \in U_i \cap \operatorname{dom} f_i$ with $\|u - x\| < \delta$ and any real $r \geq f_i(u)$. □

We can now establish the result showing in particular the coincidence of Fréchet and Clarke subdifferentials of f at any point where the subsmoothness property is satisfied.

THEOREM 8.25 (H.V. Ngai, D.T. Luc and M. Théra: coincidence of subdifferentials of subsmooth functions). Let U be a nonempty open set of a normed space X and $f : U \to \mathbb{R} \cup \{+\infty\}$ be a function which is subsmooth at a point $\overline{x} \in U$. The following hold:
(a) For each real $\varepsilon > 0$ there exists a real $\delta > 0$ with $B(\overline{x}, 2\delta) \subset U$ such that for every $(x, x^*) \in \operatorname{gph} \partial_C f$ with $\|x - \overline{x}\| < \delta$ one has

$$\langle x^*, h \rangle \leq f(x + h) - f(x) + \varepsilon\|h\| \quad \text{for all } h \in B(0, \delta). \tag{8.9}$$

(b) If $\overline{x} \in \operatorname{dom} f$, the following subdifferential regularity

$$\partial_C f(\overline{x}) = \partial_F f(\overline{x}) = \{x^* \in X^* : \langle x^*, h \rangle \leq f'(\overline{x}; h), \quad \forall h \in X\}$$

holds true at the point $\overline{x}$.

PROOF. Take any real $\varepsilon > 0$. Choose a real $\delta_0 > 0$ given by Lemma 8.22(a) above and put $\delta := \delta_0/2$. Consider any $x \in B(\overline{x}, \delta) \cap \mathrm{Dom}\, \partial_C f$ and any $x^* \in \partial_C f(x)$, which is equivalent to $\langle x^*, h\rangle - r \leq 0$ for all $(h, r) \in T^C(\mathrm{epi}\, f; (x, f(x)))$. For any $h \in B(0, \delta)$ with $x + h \in \mathrm{dom}\, f$, Lemma 8.22 yields

$$\big(h, f(x+h) - f(x) + \varepsilon\|h\|\big) \in T^C\big(\mathrm{epi}\, f; (x, f(x))\big),$$

thus (keeping in mind that $f(x+h) = +\infty$ when $x + h \notin \mathrm{dom}\, f$) we obtain

$$\langle x^*, h\rangle \leq f(x+h) - f(x) + \varepsilon\|h\| \quad \text{for all } h \in B(0, \delta),$$

which translates the desired first property (a) of the theorem.

Now assume that $\overline{x} \in \mathrm{dom}\, f$. The latter property also tells us in particular with $x = \overline{x}$ and $x^* \in \partial_C f(\overline{x})$ that, for every real $\varepsilon > 0$ there is a real $\delta > 0$ such that $\langle x^*, h\rangle \leq f(\overline{x}+h) - f(\overline{x}) + \varepsilon\|h\|$ for all $h \in B(0, \delta)$, so $x^* \in \partial_F f(\overline{x})$. We derive that $\partial_C f(\overline{x}) = \partial_F f(\overline{x})$ since the inclusion $\partial_F f(\overline{x}) \subset \partial_C f(\overline{x})$ always holds.

On the other hand, setting $\Delta := \{x^* \in X^* : \langle x^*, h\rangle \leq f'(\overline{x}; h),\ \forall h \in X\}$ it is obvious that $\partial_F f(\overline{x}) \subset \Delta$ (keep in mind that $f'(\overline{x}; \cdot)$ exists by Proposition 8.21(b)). Conversely, let $x^* \in \Delta$. Fix any $\varepsilon > 0$. Taking $\delta > 0$ given by Proposition 8.21(a) we obtain that

$$\langle x^*, h\rangle \leq f(\overline{x}+h) - f(\overline{x}) + \varepsilon\|h\|$$

for all $h \in X$ with $\|h\| < \delta$, which means that $x^* \in \partial_F f(\overline{x})$. This justifies the inclusion $\Delta \subset \partial_F f(\overline{x})$, so (b) is established and the proof is complete. $\square$

REMARK 8.26. Let $f : U \to \mathbb{R} \cup \{+\infty\}$ be a function on an open set U of a normed space X satisfying for a real $\varepsilon \geq 0$ the inequality

$$f(tx + (1-t)y) \leq tf(x) + (1-t)f(y) + \varepsilon t\|x - y\|$$

for all $x, y \in U$ and $t \in]0, 1[$. Under this assumption, the proof of Theorem 8.25 combined with Remark 8.23(b) also shows that for any $x^* \in \partial_C f(x)$ with $x \in U$ one has

$$\langle x^*, y - x\rangle \leq f(y) - f(x) + \varepsilon\|x - y\|$$

for all $y \in U$. $\square$

The following first corollary is a direct consequence of Theorem 8.25(b).

COROLLARY 8.27. *Let U be a nonempty open set of a normed space X and $f : U \to \mathbb{R}$ be a function which is subsmooth at a point $\overline{x} \in U$. If f is Gâteaux differentiable at $\overline{x}$, then $\partial_C f(\overline{x}) = \{Df(\overline{x})\}$.*

Clearly, Theorem 8.25(b) also ensures the following tangential regularity.

COROLLARY 8.28. *Let U be a nonempty open set of a normed space X and $f : U \to \mathbb{R}$ be a locally Lipschitz function. If f is subsmooth at a point $\overline{x} \in U$, then it is tangentially regular at $\overline{x}$.*

REMARK 8.29. Unlike locally Lipschitz subsmooth functions, a locally Lipschitz function which is one-sided subsmooth at $\overline{x}$ may fail to be tangentially regular at $\overline{x}$. The same function $f := -|\cdot|$ on $\mathbb{R}$ in Example 8.8 is one-sided subsmooth at $\overline{x} := 0$ according to this example, but it is not tangentially regular at $\overline{x} = 0$. $\square$

Similarly to convex functions we have the following closedness property for the C-subdifferential of subsmooth functions.

PROPOSITION 8.30. Let $f : U \to \mathbb{R} \cup \{+\infty\}$ be a function on an open set U of a normed space X which is subsmooth at $\overline{x} \in U$ and lower semicontinuous at $\overline{x}$. Then for any net $(x_j)_{j\in J}$ in U converging to $\overline{x}$ and any norm-bounded net $(x_j^*)_{j\in J}$ converging weakly* to x^* in X^* with $x_j^* \in \partial_C f(x_j)$ for all $j \in J$ one has $x^* \in \partial_C f(\overline{x})$.

PROOF. Take any real $\varepsilon > 0$ and by Theorem 8.25(a) choose a real $\delta > 0$ with $B(\overline{x}, 2\delta) \subset U$ such that for every $(u, u^*) \in \operatorname{gph} \partial_C f$ with $\|u - \overline{x}\| < \delta$ one has

$$\langle u^*, h\rangle \le f(u+h) - f(u) + \varepsilon\|h\| \quad \text{for all } h \in B(0,\delta).$$

Consider any $y \in B(\overline{x}, \delta)$. There exists some $j_0 \in J$ such that for every $j \in J$ with $j \succeq j_0$ one has both $\|x_j - \overline{x}\| < \delta$ and $\|y - x_j\| < \delta$. Therefore, for each $j \in J$ with $j \succeq j_0$ we may take $u = x_j$ and $h = y - x_j$ to obtain

$$\langle x_j^*, y - x_j\rangle \le f(y) - f(x_j) + \varepsilon\|y - x_j\|,$$

which entails by the lower semicontinuity of f at $\overline{x}$ and by the boundedness of $(x_j^*)_{j\in J}$ that

$$\langle x^*, y - \overline{x}\rangle \le f(y) - f(\overline{x}) + \varepsilon\|y - \overline{x}\|.$$

It results that $x^* \in \partial_F f(\overline{x}) \subset \partial_C f(\overline{x})$. □

We can then deduce the lower semicontinuity of $d(0, \partial_C f(\cdot))$ at subsmoothness points.

PROPOSITION 8.31. Let $f : U \to \mathbb{R} \cup \{+\infty\}$ be a function on an open set U of a normed space X which is subsmooth at $\overline{x} \in U$ and lower semicontinuous at $\overline{x}$. Then the function $x \mapsto d(0, \partial_C f(x))$ is lower semicontinuous at $\overline{x}$.

PROOF. We proceed as in the proof of Proposition 3.35. We may suppose that $\lambda := \liminf_{x\to\overline{x}} d(0, \partial_C f(x)) < +\infty$. Choose a sequence $(x_n)_n$ in U converging to $\overline{x}$ such that

$$\lim_{n\to\infty} d(0, \partial f(x_n)) = \lambda.$$

There exists $N \in \mathbb{N}$ such that for each $n \ge N$ we have $d(0, \partial f(x_n)) < +\infty$, so we can choose $x_n^* \in \partial_C f(x_n)$ satisfying $\|x_n^*\| \le d(0, \partial_C f(x_n)) + (1/n)$. The sequence $(x_n^*)_{n\ge N}$ is bounded in X^*, hence it admits a subnet $(x_{s(j)}^*)_{j\in J}$ weakly* converging in X^* to some x^*. Proposition 8.30 yields that $x^* \in \partial_C f(\overline{x})$, hence by weak* semicontinuity of the dual norm

$$d(0, \partial_C f(\overline{x})) \le \|x^*\| \le \liminf_{j\in J} d(0, \partial_C f(x_{s(j)})) = \lambda = \liminf_{x\to\overline{x}} d(0, \partial_C f(x)),$$

which confirms the desired lower semicontinuity property. □

Under the uniform subsmoothness of f, we obtain much more than the property in Theorem 8.25(a).

PROPOSITION 8.32. Let U be a nonempty open set of a normed space X and $f : U \to \mathbb{R} \cup \{+\infty\}$ be a function which is uniformly subsmooth on an open set $U_0 \subset U$. Then for each real $\varepsilon > 0$ there exists a real $\delta > 0$ such that for every $y \in U_0$ and every $(x, x^*) \in \operatorname{gph} \partial_C f$ with $x \in U_0$ and $\|x - y\| < \delta$ one has

$$\langle x^*, y - x\rangle \le f(y) - f(x) + \varepsilon\|y - x\|. \tag{8.10}$$

PROOF. Let any real $\varepsilon > 0$. Choose a real $\delta > 0$ given by Lemma 8.22(b). Take any $y \in U_0$ and any $(x, x^*) \in \operatorname{gph} \partial_C f$ with $x \in U_0$ and $\|x - y\| < \delta$. The inclusion $x^* \in \partial_C f(x)$ means $(x^*, -1) \in N^C(\operatorname{epi} f; (x, f(x)))$. If $f(y) < +\infty$, Lemma 8.22 tells us that

$$\big(y - x, f(y) - f(x) + \varepsilon\|y - x\|\big) \in T^C\big(\operatorname{epi} f; (x, f(x))\big),$$

hence we obtain

$$\langle x^*, y - x\rangle \le f(y) - f(x) + \varepsilon\|y - x\|.$$

Trivially, the latter inequality still holds if $f(y) = +\infty$, so the proof is complete. □

REMARK 8.33. Let U be a nonempty open set of a normed space X and $(f_i)_{i \in I}$ be a family of functions from U into $\mathbb{R} \cup \{+\infty\}$ which is uniformly equi-subsmooth relative to a family of open subsets $(U_i)_{i \in I}$ of U. Using Remark 8.24 instead of Lemma 8.22(b) we obtain that, for every real $\varepsilon > 0$ there exists $\delta > 0$ such that for each $i \in I$ one has

$$\langle x^*, y - x\rangle \le f_i(y) - f_i(x) + \varepsilon\|y - x\|$$

for any $y \in U_i$ and any $(x, x^*) \in \operatorname{gph} \partial_C f_i$ with $x \in U_i$ and $\|x - y\| < \delta$. □

Theorem 8.1 established a characterization of strict Fréchet differentiability of a mapping G at a point $\overline{x}$ through, for $h \in \mathbb{B}_X$, the difference of ratios

$$\frac{G(x + rh) - G(x)}{r} - \frac{G(x) - G(x - sh)}{s}$$

involving x near $\overline{x}$ and both r and s. When G is a continuous function f which is subsmooth at $\overline{x}$, the next lemma provides a similar characterization of the Fréchet differentiability (or equivalently, strict Fréchet differentiability by Proposition 8.10) of f at $\overline{x}$ through the ratio $t^{-1}[f(\overline{x} + th) + f(\overline{x} - th) - 2f(\overline{x})]$ involving merely the reference point $\overline{x}$, as seen for convex functions in Lemma 4.48. The lemma will be used in Theorem 8.35 below.

LEMMA 8.34. Let U be a nonempty open set of a normed space X and $f : U \to \mathbb{R}$ be a real-valued function which is subsmooth at $\overline{x} \in U$ and continuous at $\overline{x}$. The following assertions hold.
(a) The function f is Fréchet differentiable at $\overline{x}$ if and only if

$$\lim_{t \downarrow 0} \frac{f(\overline{x} + th) + f(\overline{x} - th) - 2f(\overline{x})}{t} = 0$$

uniformly with respect to $h \in \mathbb{B}_X$ (or equivalently, with respect to $h \in \mathbb{S}_X$).
(b) The function f is Gâteaux differentiable at $\overline{x}$ if and only if for each $\overline{h} \in X$ with $\|\overline{h}\| = 1$

$$\lim_{t \downarrow 0} \frac{f(\overline{x} + t\overline{h}) + f(\overline{x} - t\overline{h}) - 2f(\overline{x})}{t} = 0.$$

PROOF. (a) Suppose first that f is Fréchet differentiable at $\overline{x}$ and fix $r > 0$ such that $B(\overline{x}, r) \subset U$. There exists a function $\eta :]0, r[\times X \to \mathbb{R}$ with $\sup_{h \in \mathbb{B}_X} |\eta(t, h)| \to 0$ as $t \downarrow 0$ such that $f(\overline{x} + th) - f(\overline{x}) = tDf(\overline{x})h + t\eta(t, h)$ for all $t \in]0, r[$ and $h \in \mathbb{B}_X$. Consequently,

$$t^{-1}|f(\overline{x} + th) + f(\overline{x} - th) - 2f(\overline{x})| = |\eta(t, h) + \eta(t, -h)| \le 2 \sup_{v \in \mathbb{B}_X} |\eta(t, v)|,$$

which justifies the implication $\Rightarrow$ of the lemma.

Now, let us suppose that $\lim_{t\downarrow 0} t^{-1}\rho(t) = 0$, where

$$\rho(t) := \sup_{h\in\mathbb{B}_X} |f(\overline{x}+th) + f(\overline{x}-th) - 2f(\overline{x})|.$$

Since f is subsmooth at $\overline{x}$ and continuous at this point, it is Lipschitz on some ball $B(\overline{x},r) \subset U$ (see Proposition 8.18). Fix some $\overline{x}^* \in \partial_C f(\overline{x})$. Take any real $\varepsilon > 0$ and, by Theorem 8.25(a) choose a positive real $\delta < r$ such that

$$\text{(8.11)} \qquad \langle \overline{x}^*, x-\overline{x}\rangle \le f(x) - f(\overline{x}) + \varepsilon\|x-\overline{x}\|, \quad \text{for all } x \in B(\overline{x},\delta).$$

Choose a positive real $\delta_0 < \delta$ such that $t^{-1}\rho(t) \le \varepsilon$ for all $t \in]0,\delta_0[$. Considering any $t \in]0,\delta_0[$, we derive from (8.11) that, for all $h \in \mathbb{B}_X$

$$\rho(t) \ge f(\overline{x}+th) + f(\overline{x}-th) - 2f(\overline{x}) \ge f(\overline{x}+th) - f(\overline{x}) - t\langle \overline{x}^*, h\rangle - \varepsilon t,$$

and hence by (8.11) again

$$-\varepsilon \le \frac{f(\overline{x}+th) - f(\overline{x}) - t\langle \overline{x}^*, h\rangle}{t} \le t^{-1}\rho(t) + \varepsilon \le 2\varepsilon.$$

This tells us that f is Fréchet differentiable at $\overline{x}$ (with $\overline{x}^*$ as Fréchet derivative at $\overline{x}$), so the proof of (a) is finished.

(b) For any fixed $\overline{h} \in X$ with $\|\overline{h}\| = 1$, a slight modification of the above arguments with the use of $K := \{-\overline{h}, \overline{h}\}$ in place of $\mathbb{B}_X$ and $\overline{x} + t\overline{h}$ in place of x establishes the assertion (b). □

Given a multimapping $M : U \rightrightarrows Y$ between two sets U and Y, recall that a mapping $\zeta : S \to Y$ is a *selection* of M on a set $S \subset \operatorname{Dom} M$ whenever $\zeta(x) \in M(x)$ for all $x \in S$. When $S = \operatorname{Dom} M$, one just says that ζ is a selection of M. The next theorem extends to subsmooth functions the characterizations of differentiability of continuous convex functions by means of a continuous selection of the subdifferential in Theorem 3.108 and Corollaries 3.109 and 3.110.

THEOREM 8.35 (differentiability of subsmooth function via continuous selection of subdifferential). Let U be a nonempty open set of a normed space X and $f : U \to \mathbb{R}$ be a function which is subsmooth at $\overline{x} \in U$ and continuous at $\overline{x}$. The following are equivalent:

(a) the function f is Fréchet (resp. Gâteaux) differentiable at $\overline{x}$;

(b) any selection $\zeta(\cdot)$ of $\partial_C f$ is norm-norm (resp. norm-weak*) continuous at $\overline{x}$;

(c) there exists an open neighborhood $U_0 \subset U$ of $\overline{x}$ and a selection $\zeta : U_0 \to X^*$ of $\partial_C f$ on U_0 which is norm-norm (resp. norm-weak*) continuous at $\overline{x}$.

PROOF. We prove the theorem only for the Fréchet differentiability; the other case is similar. We note first that the implication (b)⇒(c) is obvious.

Let us show (c)⇒(a). Let U_0 and ζ be given by (c), and take any real $\varepsilon > 0$. By continuity of ζ at $\overline{x}$ and by Theorem 8.25(a) there exists a real $\delta > 0$ with $B(\overline{x}, 2\delta) \subset U_0$ such that $\|\zeta(x) - \zeta(\overline{x})\| < \varepsilon$ for any $x \in B(\overline{x},\delta)$ and such that

$$\langle x^*, h\rangle \le f(x+h) - f(x) + \varepsilon\|h\|$$

for any $x \in B(\overline{x},\delta)$, any $x^* \in \partial_C f(x)$ and any $h \in B(0,\delta)$. Fixing any $x \in B(\overline{x},\delta)$, we deduce that

$$\text{(8.12)} \; \langle \zeta(\overline{x}), x-\overline{x}\rangle \le f(x)-f(\overline{x})+\varepsilon\|x-\overline{x}\|, \; \langle \zeta(x), \overline{x}-x\rangle \le f(\overline{x})-f(x)+\varepsilon\|x-\overline{x}\|.$$

The latter inequality along with the fact that $\|\zeta(x) - \zeta(\overline{x})\| < \varepsilon$ yields

$$f(x)-f(\overline{x}) \le \langle \zeta(\overline{x}), x-\overline{x}\rangle + \langle \zeta(x)-\zeta(\overline{x}), x-\overline{x}\rangle + \varepsilon\|x-\overline{x}\| \le \langle \zeta(\overline{x}), x-\overline{x}\rangle + 2\varepsilon\|x-\overline{x}\|,$$

which combined with the first inequality in (8.12) gives

$$|\, f(x) - f(\overline{x}) - \langle \zeta(\overline{x}), x - \overline{x}\rangle \,| \le 2\varepsilon \|x - \overline{x}\|.$$

This translates the Fréchet differentiability of f at $\overline{x}$, that is, (a) holds.

Finally, let us prove (a)⇒(b). By Proposition 8.18 the function f is Lipschitz near $\overline{x}$, so without loss of generality we may suppose that f is Lipschitz on U with constant $\gamma > 0$. Let $\zeta : U \to X^*$ be any selection of $\partial_C f$ on U. Fix any real $\varepsilon > 0$. By Theorem 8.25 choose a real $\delta > 0$ with $B(\overline{x}, 2\delta) \subset U$ such that for any $x, y \in B(\overline{x}, \delta)$ and $x^* \in \partial_C f(x)$

$$\langle x^*, y - x\rangle \le f(y) - f(x) + \varepsilon \|y - x\|. \tag{8.13}$$

Since f is strictly Fréchet differentiable at $\overline{x}$ by Proposition 8.10 and Proposition 8.18, we have $Df(\overline{x}) = \zeta(\overline{x})$, so $t^{-1}\eta(t) \to 0$ as $t \downarrow 0$, where

$$\eta(t) := \sup_{h \in \mathbb{B}_X} |f(\overline{x} + th) - f(\overline{x}) - t\langle \zeta(\overline{x}), h\rangle| \quad \text{for all } t \in]0, \delta[.$$

Now choose some positive real $r < \delta$ such that $r^{-1}\eta(r) < \varepsilon$. Taking any $x \in B(\overline{x}, \delta)$ and any $h \in \mathbb{B}_X$ we derive from (8.13) and from the definition of $\eta(\cdot)$ that

$$\begin{aligned}\langle \zeta(x), \overline{x} + rh - x\rangle &\le f(\overline{x} + rh) - f(x) + \varepsilon \|\overline{x} + rh - x\| \\ &\le f(\overline{x}) + r\langle \zeta(\overline{x}), h\rangle + \eta(r) - f(x) + \varepsilon\|x - \overline{x}\| + \varepsilon r,\end{aligned}$$

which gives

$$r\langle \zeta(x) - \zeta(\overline{x}), h\rangle \le \langle \zeta(x), x - \overline{x}\rangle + f(\overline{x}) - f(x) + \eta(r) + \varepsilon\|x - \overline{x}\| + \varepsilon r,$$

and hence

$$\langle \zeta(x) - \zeta(\overline{x}), h\rangle \le r^{-1}[\gamma\|x - \overline{x}\| + |f(x) - f(\overline{x})|] + r^{-1}\eta(r) + \varepsilon r^{-1}\|x - \overline{x}\| + \varepsilon.$$

Choosing a positive real $\delta_0 < \delta$ such that $r^{-1}[\,\gamma\|x - \overline{x}\| + |f(x) - f(\overline{x})|\,] < \varepsilon$ and $r^{-1}\|x - \overline{x}\| < 1$ for all $x \in B(\overline{x}, \delta_0)$, it ensues that $\langle \zeta(x) - \zeta(\overline{x}), h\rangle \le 4\varepsilon$ for all $x \in B(\overline{x}, \delta_0)$ and all $h \in \mathbb{B}_X$. It results that $\|\zeta(x) - \zeta(\overline{x})\| \le 4\varepsilon$ for all $x \in B(\overline{x}, \delta_0)$, which confirms the norm-norm continuity of $\zeta(\cdot)$ at $\overline{x}$. The proof of the theorem is then complete. □

8.2.2. Submonotonicity of subdifferentials. Let any real $\varepsilon > 0$. Taking with $\varepsilon' := \varepsilon/2$ a real $\delta > 0$ given by Theorem 8.25(a), we see that, for all $(x_i, x_i^*) \in \operatorname{gph} \partial_C f$, $i = 1, 2$, with $x_i \in B(\overline{x}, \delta)$

$$\langle x_1^*, x_2 - x_1\rangle \le f(x_2) - f(x_1) + \varepsilon'\|x_2 - x_1\|,\ \langle x_2^*, x_1 - x_2\rangle \le f(x_1) - f(x_2) + \varepsilon'\|x_1 - x_2\|,$$

and hence $\langle x_1^* - x_2^*, x_1 - x_2\rangle \ge -\varepsilon\|x_1 - x_2\|$. This property of the multimapping $\partial_C f$ for such a subsmooth function f is clearly weaker than the usual monotonicity property. We formalize it as a definition.

Definition 8.36. Let U be a nonempty open set of a normed space X and $M : X \rightrightarrows X^*$ be a multimapping from U into the topological dual X^* of X. One says that M is *submonotone* at a point $\overline{x} \in U$ provided that for any real $\varepsilon > 0$ there exists a real $\delta > 0$ with $B(\overline{x}, \delta) \subset U$ such that for all $x, y \in B(\overline{x}, \delta) \cap \operatorname{Dom} M$ and $x^* \in M(x)$ and $y^* \in M(y)$, one has

$$\langle y^* - x^*, y - x\rangle \ge -\varepsilon\|y - x\|. \tag{8.14}$$

When M is submonotone at any point of a nonempty open set $U_0 \subset U$, one says that M is *submonotone on* U_0.

When the above inequality holds true with $x = \overline{x} \in U \cap \operatorname{Dom} M$ and all $y \in B(\overline{x}, \delta) \cap \operatorname{Dom} M$, $y^* \in M(y)$, $x^* \in M(\overline{x})$, one says that M is *one-sided submonotone* at $\overline{x}$. The multimapping M is one-sided submonotone on an open set $U_0 \subset U$ if it is one-sided submonotone at any point in $U_0 \cap \operatorname{Dom} M$.

We say that M is *uniformly submonotone on an open set* $U_0 \subset U$ when for any real $\varepsilon > 0$ there exists a real $\delta > 0$ such that the inequality (8.14) is fulfilled for all $x, y \in U_0 \cap \operatorname{Dom} M$ with $\|y - x\| < \delta$ and all $x^* \in M(x)$ and $y^* \in M(y)$. The multimapping M is *uniformly submonotone near a point* in U if it is uniformly submonotone on an open neighborhood of this point.

REMARK 8.37. Given an open set U of a normed space X, a family of multimappings $(M_i)_{i \in I}$ from U into X^* is called *uniformly equi-submonotone relative to a family of open subsets* $(U_i)_{i \in I}$ *of* U provided that for every $\varepsilon > 0$ there is $\delta > 0$ such that for each $i \in I$ the inequality (8.14) is satisfied for all $x, y \in U_i \cap \operatorname{Dom} M_i$ with $\|y - x\| < \delta$ and all $x^* \in M_i(x)$ and $y^* \in M_i(y)$. When all the sets U_i coincide with a same open set U_0, one simply says that the family of multimappings is *uniformly equi-submonotone on* U_0. □

With notation of the above definition, let $M_0 : U \rightrightarrows X^*$ be another multimapping whose graph is included and sequentially $\|\cdot\| \times w(X^*, X)$ dense in $\operatorname{gph} M$. It is clear that M is submonotone at $\overline{x} \in U$ if and only if M_0 is submonotone at $\overline{x}$. It is also worth pointing out that the sum of two multimappings from $U \subset X$ into X^* is clearly submonotone (resp. one-sided submonotone) at a point whenever both are submonotone (resp. one-sided submonotone) at that point. Further, submonotonicity obviously implies one-sided submonotonicity.

We will focus our analysis on the submonotonicity (resp. one-sided submonotonicity) of subdifferentials. The next theorem shows that the subsmoothness at $\overline{x}$ of lower semicontinuous functions on a Banach space is characterized by the submonotonicity at $\overline{x}$ of their subdifferentials. Before establishing the theorem let us give an example pointing out the difference between submonotonicity and one-sided submonotonicity even for subdifferential.

EXAMPLE 8.38 (J.E. Spingarn example). Consider the real-valued locally Lipschitz function $f : \mathbb{R}^2 \to \mathbb{R}$ defined by

$$f(s,t) = \begin{cases} |t| & \text{if } s \le 0 \\ |t| - s^2 & \text{if } s \ge 0 \text{ and } |t| \ge s^2 \\ (s^4 - t^2)/2s^2 & \text{if } s > 0 \text{ and } |t| \le s^2. \end{cases}$$

Put $x_n := (1/n, 1/n^2)$ and $y_n := (1/n, -1/n^2)$. By the gradient representation of Clarke subdifferential (see Theorem 2.83) one easily sees (through the third line in the definition of f) that

$$x_n^* := (2/n, -1) \in \partial_C f(x_n) \quad \text{and} \quad y_n^* := (2/n, 1) \in \partial_C f(y_n).$$

It follows that

$$\frac{\langle x_n^* - y_n^*, x_n - y_n \rangle}{\|x_n - y_n\|} = -2 \quad \text{for all } n \in \mathbb{N},$$

so $\partial_C f$ is not submonotone at $\overline{x} := (0,0)$.

However, noting via Theorem 2.83 again

$$\partial_C f(\overline{x}) = [(0,-1), (0,1)]$$

(the line segment between $(0,-1)$ and $(0,1)$) one can verify that $\partial_C f$ is one-sided submonotone at $\overline{x}$. $\square$

THEOREM 8.39 (subdifferential and tangential characterizations of subsmooth functions). Let U be a nonempty open set of a Banach space X and $f: U \to \mathbb{R} \cup \{+\infty\}$ be a proper function which is lower semicontinuous near $\overline{x} \in U$. Let ∂ be a subdifferential on X with ∂f included in the Clarke one and satisfying the properties **Prop.1-$\cdots$-Prop.4** in Section 6.2. The following assertions are equivalent:
(a) the function f is subsmooth at $\overline{x}$;
(b) for any real $\varepsilon > 0$ there is $\delta > 0$ with $B(\overline{x},\delta) \subset U$ such that for all $x \in B(\overline{x},\delta) \cap \operatorname{dom} f$ and $y \in B(\overline{x},\delta)$, one has

$$\underline{d}_H^+ f(x; y-x) \le f(y) - f(x) + \varepsilon\|y-x\|;$$

(c) for any real $\varepsilon > 0$ there is $\delta > 0$ with $B(\overline{x},\delta) \subset U$ such that for any $x, y \in B(\overline{x},\delta) \cap \operatorname{dom} f$ with the sum $\underline{d}_H^+ f(x; y-x) + \underline{d}_H^+ f(y; x-y)$ well defined, one has

$$\underline{d}_H^+ f(x; y-x) + \underline{d}_H^+ f(y; x-y) \le \varepsilon\|x-y\|;$$

(d) for any real $\varepsilon > 0$ there is $\delta > 0$ with $B(\overline{x},\delta) \subset U$ such that for any $y \in B(\overline{x},\delta)$, $x \in B(\overline{x},\delta) \cap \operatorname{Dom} \partial f$, and $x^* \in \partial f(x)$, one has

$$\langle x^*, y-x \rangle \le f(y) - f(x) + \varepsilon\|y-x\|;$$

(e) the multimapping ∂f is submonotone at $\overline{x}$, that is, for any real $\varepsilon > 0$ there is $\delta > 0$ such that for all $x, y \in B(\overline{x},\delta) \cap \operatorname{Dom} \partial f$, $x^* \in \partial f(x)$, and $y^* \in \partial f(y)$, one has

$$\langle x^* - y^*, x-y \rangle \ge -\varepsilon\|x-y\|.$$

PROOF. We may assume that $\overline{x} \in \operatorname{cl}(\operatorname{dom} f)$.

Assume that f is subsmooth at $\overline{x}$. Take any real $\varepsilon > 0$. Since ∂f is contained in the Clarke subdifferential by assumption, Theorem 8.25(a) furnishes $\delta > 0$ with $B(\overline{x},\delta) \subset U$ such that for any $y \in B(\overline{x},\delta)$, $x \in B(\overline{x},\delta) \cap \operatorname{Dom} \partial f$ and $x^* \in \partial f(x)$ we have

$$\langle x^*, y-x \rangle \le f(y) - f(x) + \varepsilon\|x-y\|.$$

Further, for $\delta > 0$ sufficiently small, by Proposition 8.21(a) we can write, for any $x \in B(\overline{x},\delta) \cap \operatorname{dom} f$ and $y \in B(\overline{x},\delta)$

$$\underline{d}_H^+ f(x; y-x) \le f(y) - f(x) + \varepsilon\|x-y\|.$$

The implications (a) $\Rightarrow$ (d) and (a) $\Rightarrow$ (b) are then justified. We also note that the implications (b) $\Rightarrow$ (c) and (d) $\Rightarrow$ (e) are evident.

It remains to show (e) $\Rightarrow$ (a) and (c) $\Rightarrow$ (a). Suppose that (e) (resp. (c)) holds. Fix any real $\varepsilon > 0$, take $\delta > 0$ given by (e) (resp. (c)) and set $V := B(\overline{x},\delta)$. Consider any $x, y \in V \cap \operatorname{dom} f$ with $x \neq y$ and $s, t \in]0,1[$ with $s + t = 1$. Put $z := sx + ty$ and take any reals $\rho < r < f(z)$. Repeating the arguments in the last part of the proof of Theorem 6.68, applying twice the Zagrodny mean value theorem produces $s_n \to s$ with $s_n > 0$, $c \in [x, z[\cap \operatorname{dom} f$, $c_n \to_f c$ with $c_n \in V \cap \operatorname{Dom} \partial f$, and $c_n^* \in \partial f(c_n)$ (resp. $c \in [x, z[\cap \operatorname{dom} f$ and $c_n \to_f c$ with $c_n \in V \cap \operatorname{dom} f$), and $d_n \in V \cap \operatorname{Dom} \partial f$, $d_n^* \in \partial f(d_n)$ (resp. $d_n \in V \cap \operatorname{dom} f$), and all those elements satisfy

$$\left\langle c_n^*, \frac{d_n - c_n}{\|d_n - c_n\|} \right\rangle > \frac{\rho - f(x)}{t\|x-y\|} = \frac{s_n\rho - s_n f(x)}{s_n t\|x-y\|}$$

$$\left(\text{resp. } \underline{d}_H^+ f(c_n; \frac{d_n - c_n}{\|d_n - c_n\|}) > \frac{\rho - f(x)}{t\|x-y\|} = \frac{s_n\rho - s_n f(x)}{s_n t\|x-y\|}\right),$$

as well as

$$\left\langle d_n^*, \frac{c_n - d_n}{\|d_n - c_n\|} \right\rangle > \frac{t\rho - tf(y)}{s_n t\|x-y\|} \quad \left(\text{resp. } \underline{d}_H^+ f\left(d_n; \frac{c_n - d_n}{\|d_n - c_n\|}\right) > \frac{t\rho - tf(y)}{s_n t\|x-y\|}\right).$$

Taking the assumption in (e) (resp. in (c)) into account, it follows that

$$\varepsilon \geq \left\langle c_n^* - d_n^*, \frac{d_n - c_n}{\|d_n - c_n\|} \right\rangle > \frac{(s_n + t)\rho - s_n f(x) - tf(y)}{s_n t\|x-y\|},$$

$$\left(\text{resp. } \varepsilon \geq \frac{1}{\|d_n - c_n\|}\left(\underline{d}_H^+ f(c_n; d_n - c_n) + \underline{d}_H^+ f(d_n; c_n - d_n)\right) \right.$$
$$\left. > \frac{(s_n + t)\rho - s_n f(x) - tf(y)}{s_n t\|x-y\|}\right),$$

which implies

$$\varepsilon s_n t\|x-y\| > (s_n + t)\rho - s_n f(x) - tf(y).$$

Since $s_n \to s$ as $n \to \infty$ and ρ is arbitrarily less than $f(z)$, it results that

$$f(z) \leq sf(x) + tf(y) + \varepsilon st\|x-y\|.$$

This completes the proof. □

REMARK 8.40. It is also worth noticing that in Asplund spaces one can directly verify that the property (d) holds with C-subdifferential if and only if it holds true with F-subdifferential. The implication $\Rightarrow$ is evident. For the converse, assume that X is an Asplund space and that f is lower semicontinuous and that the property (d) in Theorem 8.39 holds with $\partial_F f$. Take any real $\varepsilon > 0$ and let $\delta > 0$ be given by this property (d). It is trivial that the property (d) still holds for $\partial_L f$. Further, for $x \in B(\overline{x}, \delta) \cap \operatorname{Dom} \partial_L f$ and $z^* \in \partial_L^\infty f(x)$ there are a sequence $(x_n)_n$ in $B(\overline{x}, \delta)$ with $(x_n f(x_n)) \to (x, f(x))$, a sequence of reals $(t_n)_n$ tending to 0 with $t_n > 0$ and a sequence $(x_n^*)_n$ with $x_n^* \in \partial_F f(x_n)$ such that $t_n x_n^* \to z^*$ weakly* as $n \to \infty$. Then for each n we have for all $y \in B(\overline{x}, \delta) \cap \operatorname{dom} f$

$$\langle t_n x_n^*, y - x_n \rangle \leq t_n f(y) - t_n f(x_n) + t_n \varepsilon \|y - x_n\|,$$

hence $\langle z^*, y - x \rangle \leq 0$. From this we easily see that the property (d) holds for $\partial_L f + \partial_L^\infty f$. Since $\partial_C f(x) = \overline{\text{co}}^*\big(\partial_L f(x) + \partial_L^\infty f(x)\big)$ for all $x \in B(\overline{x}, \delta)$, it ensues that the property (d) holds for $\partial_C f$. In fact, the above arguments work with any subdifferential ∂ for which the representation formula

$$\partial_C f(x) = \overline{\text{co}}^* \left({}^{seq}\operatorname*{Lim\,sup}_{u \to_f x} \partial f(u) + {}^{seq}\operatorname*{Lim\,sup}_{t \downarrow 0, u \to_f x} t\partial f(u) \right)$$

is satisfied for lower semicontinuous functions f on the Banach space X. □

In the case when f is locally Lipschitz, the completeness of the space X is not needed and much more simpler arguments as in Proposition 2.182 yield the following.

PROPOSITION 8.41. Let U be a nonempty open subset of a normed space X and $f : U \to \mathbb{R}$ be a locally Lipschitz function. For $\overline{x} \in U$ the following are equivalent:

(a) the function f is subsmooth at $\overline{x}$;
(b) for each real $\varepsilon > 0$ there exists a real $\delta > 0$ with $B(\overline{x}, \delta) \subset U$ such that

$$\langle x^*, y - x \rangle \leq f(y) - f(x) + \varepsilon \|y - x\|,$$

for all $x, y \in B(\overline{x}, \delta)$ and all $x^* \in \partial_C f(x)$;
(c) the multimapping $\partial_C f$ is submonotone at $\overline{x}$.

PROOF. The implications (a)⇒(b) and (b)⇒(c) follow directly as above. Now suppose (c), that is, $\partial_C f$ is submonotone at $\overline{x}$. Fix any real $\varepsilon > 0$ and fix $\delta > 0$ with $V := B(\overline{x}, \delta) \subset U$ such that for any $x_i \in V$ and $x_i^* \in \partial_C f(x_i)$ $i = 1, 2$, one has

$$\langle x_1^* - x_2^*, x_1 - x_2 \rangle \geq -\varepsilon \|x_1 - x_2\|.$$

Fix any $x, y \in V$ with $x \neq y$ and any $s, t \in]0, 1[$ with $s+t = 1$. Putting $z := sx+ty$, by the mean value equality (see Theorem 2.180) there are $c \in]x, z[$, $d \in]z, y[$, $c^* \in \partial_C f(c)$ and $d^* \in \partial_C f(d)$ such that

$$f(z) - f(x) = \langle c^*, z - x \rangle \quad \text{and} \quad f(z) - f(y) = \langle d^*, z - y \rangle.$$

Noting that $z - x = t(y - x)$ and $z - y = s(x - y)$, multiplying the first equality by s and the second by t and adding together yield

$$\begin{aligned} f(z) - sf(x) - tf(y) &= st\langle c^* - d^*, y - x \rangle \\ &= st\frac{\|y - x\|}{\|d - c\|}\langle c^* - d^*, d - c \rangle \leq \varepsilon st\|y - x\|. \end{aligned}$$

This and Proposition 8.12 tell us that the function f is subsmooth at $\overline{x}$. □

The same above arguments justify the following similar equivalences.

PROPOSITION 8.42. Let f and $(f_i)_{i\in I}$ be real-valued locally Lipschitz functions on a nonempty open convex subset U of a normed space X. Given a nonempty open subset $U_0 \subset U$ and a family $(U_i)_{i\in I}$ of open subsets of U the following are equivalent:
(a) the function f is uniformly subsmooth on U_0 (resp. the family $(f_i)_{i\in I}$ is uniformly equi-subsmooth relative to $(U_i)_{i\in I}$);
(b) for each real $\varepsilon > 0$ there exists a real $\delta > 0$ such that (resp. such that for each $i \in I$)

$$\langle x^*, y - x \rangle \leq f(y) - f(x) + \varepsilon \|y - x\| \text{ (resp. } \langle x^*, y - x \rangle \leq f_i(y) - f_i(x) + \varepsilon \|y - x\| \text{)}$$

for any $x, y \in U_0$ (resp. any $x, y \in U_i$) with $\|x - y\| < \delta$ and any $x^* \in \partial_C f(x)$ (resp. any $x^* \in \partial_C f_i(x)$);
(c) the multimapping $\partial_C f$ is uniformly submonotone on U_0 (resp. the family of multimappings $(\partial_C f_i)_{i\in I}$ from U into X^* is uniformly equi-submonotone relative to $(U_i)_{i\in I}$).

PROOF. The implication (a)⇒(b) is a direct consequence of Proposition 8.32 while (b)⇒(c) is obvious. For the remaining implication (c)⇒(a) it suffices to proceed like in the proof of Proposition 8.41. □

We provide now two additional subdifferential characterizations of uniform subsmoothness of locally Lipschitz functions via modulus functions.

PROPOSITION 8.43. Let U be a nonempty open convex subset of a normed space X and $f : U \to \mathbb{R}$ be a locally Lipschitz function. Given a nonempty open subset $U_0 \subset U$ the following are equivalent:
(a) the function f is uniformly subsmooth on U_0;
(b) there are a real $\theta > 0$ and a modulus function $\omega : [0, +\infty[\to [0, +\infty[$ of class $\mathcal{C}^1$ on $]0, +\infty[$ with $t\omega'(t) \to 0$ as $t \downarrow 0$ such that

$$\langle x^*, y - x\rangle \le f(y) - f(x) + \|y - x\|\omega(\|y - x\|)$$

for all $x, y \in U_0$ with $\|x - y\| \le \theta$ and all $x^* \in \partial_C f(x)$;
(c) there are a real $\theta > 0$ and a modulus function $\omega : [0, +\infty[\to [0, +\infty[$ of class $\mathcal{C}^1$ on $]0, +\infty[$ with $t\omega'(t) \to 0$ as $t \downarrow 0$ such that

$$\langle x^* - y^*, x - y\rangle \ge -\|x - y\|\omega(\|x - y\|)$$

for all $x, y \in U_0$ with $\|x - y\| \le \theta$, all $x^* \in \partial_C f(x)$ and all $y^* \in \partial_C f(y)$.

PROOF. The implication (b)⇒(c) is evident, and (c) implies (a) according to the implication (c)⇒(a) in Proposition 8.42. It remains to show (a)⇒(b). Assume that (a) is satisfied. For any $x, y \in U_0$ and $x^* \in X^*$ put

$$g(x, y, x^*) := \begin{cases} \big(f(y) - f(x) - \langle x^*, y - x\rangle\big)/\|y - x\| & \text{if } x \ne y \\ 0 & \text{if } x = y, \end{cases}$$

and for every real $t > 0$ put

$$\zeta(t) := \inf\{g(x, y, x^*) : x, y \in U_0, \|x - y\| \le t, x^* \in \partial_C f(x)\}.$$

Put also $\zeta(0) = 0$. By the implication (a)⇒(b) in Proposition 8.42, for each $\varepsilon > 0$ there is $\delta > 0$ such that

$$\zeta(t) \ge -\varepsilon \quad \text{for all } t \in [0, \delta[.$$

Then the function $\xi : [0, \infty[\to [0, +\infty]$ defined by $\xi(t) := \max\{-\zeta(t), 0\}$ is continuous at 0 with $\xi(0) = 0$. By Lemma 4.24 there is a real $\theta > 0$ and a continuously derivable function $\varphi : [0, \theta] \to [0, +\infty[$ with $\varphi(0) = \varphi'_+(0) = 0$ such that

$$\varphi(t) \ge t\xi(t) \quad \text{for all } t \in [0, \theta].$$

Let us extend φ to $[0, +\infty[$ by putting

$$\varphi(t) := \varphi(\theta) + \varphi'_-(\theta)(t - \theta) \quad \text{for all } t \in]\theta, +\infty[.$$

Then $\varphi : [0, +\infty[\to [0, +\infty[$ is of class $\mathcal{C}^1$. Take any $x, y \in U_0$ with $\|x - y\| \le \theta$ and any $x^* \in \partial_C f(x)$. Suppose $x \ne y$ and put $t = \|x - y\|$. Since $t \in]0, \theta]$ we have

$$g(x, y, x^*) \ge \zeta(t) \ge -\xi(t) \ge -\frac{\varphi(t)}{t} = -\frac{\varphi(\|x - y\|)}{\|x - y\|},$$

so $f(y) - f(x) - \langle x^*, y - x\rangle \ge -\varphi(\|x - y\|)$. This latter inequality is still trivially true when $x = y$. Consequently, to get (b) it suffices to define $\omega : [0, +\infty[\to [0, +\infty[$ by $\omega(0) = 0$ and $\omega(t) = \varphi(t)/t$ for every real $t > 0$. □

The next proposition shows in particular the equivalence in finite dimensions between the subsmoothness near a point and the lower $\mathcal{C}^1$ property near that point.

PROPOSITION 8.44. Let U be a nonempty open subset of a normed space X and $f : U \to \mathbb{R}$ be a locally Lipschitz function. For $\overline{x} \in U$ the following are equivalent:
(a) the function f is uniformly subsmooth near $\overline{x}$;

(b) there are an open convex neighborhood $V \subset U$ of $\overline{x}$ and a modulus function $\omega : [0,+\infty[\to [0,+\infty[$ of class $\mathcal{C}^1$ on $]0,+\infty[$ with $t\,\omega'(t) \to 0$ as $t \downarrow 0$ such that

$$\langle x^*, y-x\rangle \le f(y) - f(x) + \|y-x\|\omega(\|y-x\|)$$

for all $x, y \in V$ and all $x^* \in \partial_C f(x)$;
(c) there are an open convex neighborhood $V \subset U$ of $\overline{x}$ and a modulus function $\omega : [0,+\infty[\to [0,+\infty[$ of class $\mathcal{C}^1$ on $]0,+\infty[$ with $t\,\omega'(t) \to 0$ as $t \downarrow 0$ such that

$$\langle x^* - y^*, x-y\rangle \ge -\|x-y\|\omega(\|x-y\|)$$

for all $x, y \in V$, all $x^* \in \partial_C f(x)$ and all $y^* \in \partial_C f(y)$.

If X is finite-dimensional, one can add anyone of (d) and (e) below to the list of equivalences:
(d) the function f is subsmooth near $\overline{x}$;
(e) the function f is lower $\mathcal{C}^1$ near $\overline{x}$, that is, there exist a compact metric space T, an open neighborhood $V \subset U$ of $\overline{x}$ and a continuous function $\Phi : V \times T \to \mathbb{R}$ such that $D_1\Phi(\cdot,\cdot)$ exists and is continuous on $V \times T$, and such that

$$f(x) = \max_{t\in T} \Phi(x,t) \quad \text{for all } x \in V.$$

PROOF. The equivalences (a)$\Leftrightarrow$(b)$\Leftrightarrow$(c) follow easily from Proposition 8.43 while the implication (a)$\Rightarrow$(d) is evident. Assume now that X is finite-dimensional and f is subsmooth near $\overline{x}$. There exists a real $r > 0$ and an open set $V \subset U$ containing $B[\overline{x}, r]$ such that f is subsmooth at each point in V. Let any real $\varepsilon > 0$. For each $u \in B[\overline{x}, r]$ choose a real $\delta_u > 0$ with $B(u, 2\delta_u) \subset V$ such that for all $x, y \in B(u, 2\delta_u)$ and all $x^* \in \partial_C f(x)$

$$\langle x^*, y-x\rangle \le f(y) - f(x) + \varepsilon\|y-x\|. \tag{8.15}$$

By compactness of $B[\overline{x}, r]$ there are $u_1, \cdots, u_m$ in $B[\overline{x}, r]$ such that the balls $B(u_i, \delta_{u_i})$ cover $B[\overline{x}, r]$. Denote $\delta := \min\{\delta_{u_1}, \cdots, \delta_{u_m}\} > 0$ and take any $x, y \in B(\overline{x}, r)$ with $\|x-y\| < \delta$ and any $x^* \in \partial_C f(x)$. Choose $k \in \{1, \cdots, m\}$ such that $x \in B(u_k, \delta_{u_k})$. Then both x, y belong to $B(u_k, 2\delta_{u_k})$, so by (8.15) we have

$$\langle x^*, y-x\rangle \le f(y) - f(x) + \varepsilon\|y-x\|.$$

This justifies the uniform subsmoothness of f on $B(\overline{x}, r)$, so (d)$\Rightarrow$(a) holds true.

Assume again that X is finite-dimensional and fix a Euclidean norm $\|\cdot\|$ on X associated to an inner product $\langle\cdot,\cdot\rangle$. Let us first show (b)$\Rightarrow$(e). Let V and ω be given by (b), so the function $\xi : [0,+\infty[\to [0,+\infty[$, given by $\xi(t) := t\,\omega(t)$ for all $t \in [0,+\infty[$, is continuously derivable on $[0,+\infty[$ with $\xi(0) = \xi'_+(0) = 0$. Choose a real $r > 0$ such that $B[\overline{x}, r] \subset V$ and put

$$T := \{(y, y^*) \in X \times X : y \in B[\overline{x}, r], y^* \in \partial_C f(y)\}.$$

From the local boundedness of $\partial_C f$ we easily see that T is a (nonempty) compact subset of $X \times X$. Further, the function $\Phi : V \times T \to \mathbb{R}$ defined by

$$\Phi(x, (y, y^*)) := f(y) + \langle y^*, x-y\rangle + \xi(\|x-y\|)$$

is continuous on $V \times T$ and $D_1\varphi(\cdot,\cdot)$ exists and is continuous on $V \times T$ according to the above properties of the function ξ. Since $f(x) = \max\limits_{(y,y^*)\in T} \Phi(x, (y, y^*))$ for all $x \in V$, we have shown (b)$\Rightarrow$(e). Let us finally prove (e)$\Rightarrow$(a). Let T, V, Φ be as given by (e). Choose a real $r > 0$ such that $B[\overline{x}, r] \subset V$. Fix any real $\varepsilon > 0$. The mapping $D_1\Phi(\cdot,\cdot)$ being uniformly continuous on the compact set $B[\overline{x}, r] \times T$, there is a real $\delta > 0$ such that for all $(x,t),(y,\tau)$ in $B[\overline{x},r]\times T$ with $\|x-y\|+d(t,\tau)<\delta$ one has $\|D_1\Phi(x,t)-D_1\Phi(y,\tau)\|\le\varepsilon$. Fix any $x,y\in B(\overline{x},r)$ with $\|x-y\|<\delta$ and any $t\in T(x):=\{\tau\in T:\Phi(x,\tau)=f(x)\}$. We note that

$$
\begin{aligned}
&\langle D_1\Phi(x,t),y-x\rangle\\
&=\Phi(y,t)-\Phi(x,t)-\int_0^1\langle D\Phi_1(x+s(y-x),t)-D\phi_1(x,t),y-x\rangle\,ds\\
&\le f(y)-f(x)+\|y-x\|\int_0^1\|D\phi_1(x+s(y-x),t)-D_1\Phi(x,t)\|\,ds\\
&\le f(y)-f(x)+\varepsilon\|y-x\|.
\end{aligned}
$$

From this and Theorem 2.198 we deduce that $\langle x^*,y-x\rangle\le f(y)-f(x)+\varepsilon\|y-x\|$ for all $x^*\in\partial_C f(x)$, which translates the uniform subsmoothness of f on $B(\overline{x},r)$. The implication (e)$\Rightarrow$(a) then holds, and the proof is finished. $\square$

Concerning the distance function, given a set S and $\overline{x}\in S$ the next proposition shows that a relative particular Jensen type inequality of the distance function d_S entails the submonotonicity of $\partial_C d_S$ at $\overline{x}$ relative to S.

PROPOSITION 8.45. Let S be a subset of a normed space X and $\overline{x}\in S$. Consider the assertions:
(a) For every real $\varepsilon>0$ there exists $\delta>0$ such that

$$d_S\big(tx+(1-t)y\big)\le\varepsilon t(1-t)\|x-y\|\quad\text{for all } x,y\in S\cap B(\overline{x},\delta),t\in\,]0,1[.$$

(b) For every real $\varepsilon>0$ there exists $\delta>0$ such that

$$\langle x^*,y-x\rangle\le\varepsilon\|y-x\|\quad\text{for all } x,y\in S\cap B(\overline{x},\delta),x^*\in\partial_C d_S(x).$$

(c) The multimapping $\partial_C d_S$ is submonotone at $\overline{x}$ relative to S, that is, for every $\varepsilon>0$ there exists $\delta>0$ such that

$$\langle x^*-y^*,x-y\rangle\ge-\varepsilon\|x-y\|\text{ for all } x,y\in S\cap B(\overline{x},\delta),x^*\in\partial_C d_S(x),y^*\in\partial_C d_S(y).$$

The implications (a) $\Rightarrow$ (b) $\Leftrightarrow$ (c) hold.

PROOF. The equivalence (b) $\Leftrightarrow$ (c) is trivial since $0\in\partial_C d_S(y)$ for any $y\in S$. Suppose that (a) holds and take any $\varepsilon>0$. Let $\delta>0$ be given by (a). Fix any $x,y\in S\cap B(\overline{x},\delta)$. Proposition 2.95(a) along with the Lipschitz property of d_S tells us that

$$d_S^\circ(x;y-x)=\limsup_{S\ni x'\to x,t\downarrow 0}t^{-1}d_S(x'+t(y-x')).$$

On the other hand, for any $t\in\,]0,1[$ and any $x'\in B(\overline{x},r)$ with $r:=\delta-\|x-\overline{x}\|>0$, we have $t^{-1}d_S(x'+t(y-x'))\le\varepsilon(1-t)\|y-x'\|$. It results that $d_S^\circ(x;y-x)\le\varepsilon\|y-x\|$, which is equivalent to $\langle x^*,y-x\rangle\le\varepsilon\|y-x\|$ for all $x^*\in\partial_C d_S(x)$. $\square$

We use Proposition 8.41 and Proposition 8.42 to establish the assertion (c) in the next proposition.

PROPOSITION 8.46. Let X and Y be two normed spaces and U be a nonempty open set in X.
(a) For any real $\lambda>0$ and for two functions $f_1,f_2:U\to\mathbb{R}\cup\{+\infty\}$ which are subsmooth at $\overline{x}\in U$ (resp. uniformly subsmooth on an open set $V\subset U$), the functions λf_1 and f_1+f_2 are subsmooth at $\overline{x}$ (resp. uniformly subsmooth on V).
(b) If $A:X\to Y$ is a continuous linear mapping and $g:Y\to\mathbb{R}\cup\{+\infty\}$ is

subsmooth at $A\overline{x}$ (resp. uniformly subsmooth on an open set $W \supset A(V)$, where V is an open set of X), then $g \circ A$ is subsmooth at $\overline{x}$ (resp. uniformly subsmooth on the open set V).
(c) If $G : U \to Y$ is of class $\mathcal{C}^1$ near $\overline{x}$ and if $g : Y \to \mathbb{R}$ is Lipschitz near $G(\overline{x})$ and subsmooth at $G(\overline{x})$, then $g \circ G$ is subsmooth at $\overline{x}$.
(d) If $G : U \to Y$ is Lipschitz and differentiable on an open convex set $V \subset U$ with DG uniformly continuous on V and if $g : Y \to \mathbb{R}$ is Lipschitz and uniformly subsmooth on an open convex set $W \supset G(V)$, then the function $g \circ G$ is uniformly subsmooth on V.

PROOF. The assertions (a) and (b) follow from Proposition 8.12. Concerning (c) put $\overline{y} := G(\overline{x})$ and choose a real $\delta > 0$ such that g is Lipschitz on $B(\overline{y}, \delta)$ with Lipschitz constant $\gamma > 0$ and G is Lipschitz on $B(\overline{x}, \delta)$ with the same Lipschitz constant γ. Fix any $\varepsilon > 0$ and put $\varepsilon' := \varepsilon/(2\gamma)$. By Proposition 8.41 shrinking δ if necessary, we have $\langle y^*, y' - y\rangle \le g(y') - g(y) + \varepsilon'\|y' - y\|$ for all $y, y' \in B(\overline{y}, \delta)$ and $y^* \in \partial_C g(y)$. Choose a positive real $\delta_0 < \delta$ such that $G(B(\overline{x}, \delta_0)) \subset B(\overline{y}, \delta)$ and $\|DG(u') - DG(u)\| \le \varepsilon'$ for all $u, u' \in B(\overline{x}, \delta_0)$. Take any $x, x' \in B(\overline{x}, \delta_0)$ and any $x^* \in \partial_C(g \circ G)(x)$. There exists $y^* \in \partial_C g(G(x))$ with $x^* = y^* \circ DG(x)$ (see Theorem 2.135). It follows that

$$\begin{aligned}\langle x^*, x' - x\rangle &= \langle y^*, DG(x)(x' - x)\rangle \\ &= \langle y^*, G(x') - G(x)\rangle - \left\langle y^*, \int_0^1 \big(DG(x + t(x' - x)) - DG(x)\big)(x' - x)dt\right\rangle \\ &\le g(G(x')) - g(G(x)) + \varepsilon'\|G(x') - G(x)\| + \varepsilon'\|y^*\|\,\|x' - x\|,\end{aligned}$$

which gives $\langle x^*, x' - x\rangle \le (g \circ G)(x') - (g \circ G)(x) + \varepsilon\|x' - x\|$. This tells us by Proposition 8.41 again that $g \circ G$ is subsmooth at $\overline{x}$.

The proof of the assertion (d) is similar. □

8.2.3. Subdifferential characterizations of one-sided subsmooth functions. This subsection provides various characterizations similar to those of Proposition 8.41 for one-sided subsmoothness property of functions. The approach requires first two lemmas. Recall that a multimapping M between two metric spaces T and Y is bounded near a point $\overline{t} \in T$ if $M(V)$ is bounded for some neighborhood V of $\overline{t}$.

LEMMA 8.47. Let U be a nonempty open subset of a normed space X and $M : U \rightrightarrows X^*$ be a multimapping which is bounded near a point $\overline{x} \in \operatorname{Dom} M$ and $\|\cdot\| - w^*$ outer semicontinuous at $\overline{x}$.
(a) If M is one-sided submonotone at $\overline{x}$, then for any $u \in \mathbb{S}_X$, for any net $(x_j)_{j\in J}$ in $U \setminus \{\overline{x}\}$ converging to $\overline{x}$ with $\|x_j - \overline{x}\|^{-1}(x_j - \overline{x}) \to u$ and for any net $(x_j^*)_j$ converging weakly* to x^* in X^* with $x_j^* \in M(x_j)$ for all $j \in J$, one has

$$\langle x^*, u\rangle = \sigma\big(M(\overline{x}), u\big),$$

where $\sigma\big(M(\overline{x}), \cdot\big)$ is the support function of $M(\overline{x})$.
(b) The latter implication is an equivalence whenever X is finite-dimensional.

PROOF. (a) Assume that M is one-sided submonotone at $\overline{x}$ and let u, $(x_j)_j$ and $(x_j^*)_j$ as above. We note that there is some $j_0 \in J$ such that $(x_j^*)_{j\succeq j_0}$ is bounded in X^*. Then, according to the one-sided submonotonicity property and the local boundedness of M it ensues that, for any $y^* \in M(\overline{x})$

$$\langle x^* - y^*, u\rangle = \lim_{j\in J} \left\langle x_j^* - y^*, \frac{x_j - \overline{x}}{\|x_j - \overline{x}\|}\right\rangle \geq 0.$$

Since $x^* \in M(\overline{x})$ by outer semicontinuity of M at $\overline{x}$, it follows that

$$\langle x^*, u\rangle = \sup_{y^*\in M(\overline{x})} \langle y^*, u\rangle = \sigma(M(\overline{x}), u)$$

as desired.
(b) Now assume that X is finite-dimensional and that M is not one-sided submonotone at $\overline{x}$. There exists a real $\varepsilon > 0$, a sequence $(x_n)_n$ in $U \setminus \{\overline{x}\}$ converging to $\overline{x}$, sequences $(x_n^*)_n$ and $(y_n^*)_n$ with $x_n^* \in M(x_n)$ and $y_n^* \in M(\overline{x})$ such that

$$\left\langle x_n^* - y_n^*, \frac{x_n - \overline{x}}{\|x_n - \overline{x}\|}\right\rangle \leq -\varepsilon \quad \text{for all } n \in \mathbb{N}.$$

Since X is finite-dimensional and M is bounded near $\overline{x}$, we may and do suppose that $\|x_n - \overline{x}\|^{-1}(x_n - \overline{x}) \to u$ with $\|u\| = 1$ and that $x_n^* \to x^*$ and $y_n^* \to y^*$. By outer semicontinuity of M at $\overline{x}$, we have both x^* and y^* in $M(\overline{x})$. It results that

$$\langle x^*, u\rangle \leq \langle y^*, u\rangle - \varepsilon \leq \sigma\big(M(\overline{x}), u\big) - \varepsilon,$$

which contradicts the property in (a). The converse implication in (a) is then justified. □

The second lemma shows the tangential regularity of locally Lipschitz functions with one-sided submonotone Clarke subdifferentials. Its proof uses the above lemma.

LEMMA 8.48. *Let $f : U \to \mathbb{R}$ be a locally Lipschitz function on an open set U of a normed space X. If $\partial_C f$ is one-sided submonotone at $\overline{x} \in U$, then f is tangentially regular at $\overline{x}$.*

PROOF. Fix any $u \in \mathbb{S}_X$ (if $X = \{0\}$ there is nothing to prove). Since f is locally Lipschitz, there exists a sequence $(t_n)_n$ tending to 0 with $t_n > 0$ such that $\underline{d}_H^+ f(\overline{x}; u) = \lim_{n\to\infty} t_n^{-1}[f(\overline{x} + t_n u) - f(\overline{x})]$. By the mean value theorem, for each $n \in \mathbb{N}$, there exists some $\theta_n \in]0, 1]$ and $x_n^* \in \partial_C f(\overline{x} + t_n\theta_n u)$ such that $t_n^{-1}[f(\overline{x} + t_n u) - f(\overline{x})] = \langle x_n^*, u\rangle$ (see Theorem 2.180). Take a subnet $(x_{s(j)}^*)_{j\in J}$ converging weakly* to some x^* (keep in mind that $\partial_C f$ is bounded near $\overline{x}$ since f is locally Lipschitz). Then, noting that $z_n := \overline{x} + t_n\theta_n u \to \overline{x}$ with $\|z_n - \overline{x}\|^{-1}(z_n - \overline{x}) \to u$ as $n \to \infty$, Lemma 8.47(a) ensures that

$$\underline{d}_H^+ f(\overline{x}; u) = \lim_{j\in J}\langle x_{s(j)}^*, u\rangle = \langle x^*, u\rangle = \sigma\big(\partial_C f(\overline{x}), u\big) = f^\circ(\overline{x}; u).$$

This being true for all $u \in \mathbb{S}_X$, it ensues that f is tangentially regular at $\overline{x}$. □

Let us provide an example showing that the reverse implication in the above lemma is false.

EXAMPLE 8.49 (J.E. Spingarn example). Consider an even function $f : \mathbb{R} \to \mathbb{R}$ with $f(0) = 0$ which satisfies the following properties:
(i) $f(1/n) = 1/n - 1/n^2$ for every integer $n \geq 2$;
(ii) for each integer $n \geq 2$ the usual derivative f' exists on $]1/(n+1), 1/n[$ with f' continuous on $]1/(n+1), 1/n[$ and f' decreasing on $]1/(n+1), 1/n[$, $f'_+(1/(n+1)) = 1$

and $f'_-(1/n) = 0$;
(iii) $f(x) = 1/4$ for all $x \geq 1/2$.

The function f is Lipschitz, and noting that $|x| - x^2 \leq f(x) \leq |x|$ for all x it ensues that

$$f'(0;h) = |h| \quad \text{for all } h \in \mathbb{R}.$$

Further, the gradient representation theorem (see Theorem 2.83) allows us to see that $\partial_C f(0) = [-1,1]$, which gives $f^\circ(0;h) = |h|$ for all $h \in \mathbb{R}$. The function f is then tangentially regular at $\overline{x} := 0$, even tangentially regular at any $x \in \mathbb{R}$.

On the other hand, taking $x_n := 1/n$, $x_n^* := 0 \in \partial_C f(x_n)$ and $\overline{x}^* := 1 \in \partial_C f(0)$ we see that

$$\frac{(x_n^* - \overline{x}^*)(x_n - \overline{x})}{|x_n - \overline{x}|} = -1,$$

so $\partial_C f$ is not one-sided submonotone at $\overline{x}$. □

With Lemma 8.48 in particular at hands, we can now establish subdifferential characterizations of one-sided subsmoothness on an open set for tangentially regular locally Lipschitz functions.

PROPOSITION 8.50. Let U be a nonempty open subset of a normed space X and $f : U \to \mathbb{R}$ be a locally Lipschitz function. Consider the following assertions:
(a) The function f is one-sided subsmooth on U and tangentially regular on U.
(b) For each $\overline{x} \in U$ and each real $\varepsilon > 0$ there exists a real $\delta > 0$ with $B(\overline{x}, \delta) \subset U$ such that

$$\langle x^*, y - x \rangle \leq f(y) - f(x) + \varepsilon \|y - x\|,$$

for all $x, y \in B(\overline{x}, \delta)$ with either $x = \overline{x}$ or $y = \overline{x}$ and for all $x^* \in \partial_C f(x)$.
(c) The multimapping $\partial_C f$ is one-sided submonotone on U.

Then (a) $\Rightarrow$ (b) $\Rightarrow$ (c), and both implications are equivalences whenever X is finite-dimensional.

PROOF. (a)$\Rightarrow$(b). Fix any $\overline{x} \in U$ and any real $\varepsilon > 0$, and take $\delta > 0$ such that the property for one-sided subsmoothness in Proposition 8.12 is satisfied. Let any $x \in B(\overline{x}, \delta)$ and any $t \in]0,1[$. Since $\overline{x} + t(x - \overline{x}) = tx + (1-t)\overline{x}$, we have

$$f(\overline{x} + t(x - \overline{x})) - f(\overline{x}) \leq t[\,f(x) - f(\overline{x}) + \varepsilon(1-t)\|x - \overline{x}\|\,].$$

Dividing by t and taking the limit inferior as $t \downarrow 0$ give

$$\underline{d}_H^+ f(\overline{x}; x - \overline{x}) \leq f(x) - f(\overline{x}) + \varepsilon \|x - \overline{x}\|.$$

Similarly, with $s := 1 - t$ and $s \downarrow 0$ one obtains the same inequality with x and $\overline{x}$ mutually changed. Further, the tangential regularity of f on U tells us that $\underline{d}_H^+ f(x;\cdot) = f^\circ(x;\cdot)$ and $\underline{d}_H^+ f(\overline{x};\cdot) = f^\circ(\overline{x};\cdot)$. This and both inequalities concerning $\underline{d}_H^+ f(\cdot;\cdot)$ yield the property in (b).
(b)$\Rightarrow$(c). Fix any $\overline{x} \in U$ and any real $\varepsilon > 0$. Let $\delta > 0$ satisfying the property in (b) with $\varepsilon/2$ in place of ε. Applying the related inequality once with $y = \overline{x}$ and with x and again with $x = \overline{x}$ and with $y = x$, and adding the resulting inequalities we obtain

$$\langle x^* - \overline{x}^*, x - \overline{x} \rangle \geq -\varepsilon \|x - \overline{x}\|$$

for all $x^* \in \partial_C f(x)$ and all $\overline{x}^* \in \partial_C f(\overline{x})$. This translates the one-sided submonotonicity of $\partial_C f$ on U.

Now assume that X is finite-dimensional and that (c) holds. The tangential regularity of f follows from Lemma 8.48. Let us show the one-sided subsmoothness of f on U. Fix any $\overline{x} \in U$ and any real $\varepsilon > 0$. By (c) choose $\delta > 0$ with $B(\overline{x}, \delta) \subset U$ such that

$$\langle x^* - u^*, x - \overline{x}\rangle \geq -(\varepsilon/2)\|x - \overline{x}\|, \tag{8.16}$$

for all $x \in B(\overline{x}, \delta)$, all $x^* \in \partial_C f(x)$ and all $u^* \in \partial_C f(\overline{x})$. By the $\|\cdot\| - \|\cdot\|$-upper semicontinuity of $\partial_C f$ (keep in mind that X is finite-dimensional) we may also suppose that

$$\partial_C f(z) \subset \partial_C f(\overline{x}) + B(0, \varepsilon/2) \quad \text{for all } z \in B(\overline{x}, \delta). \tag{8.17}$$

Now fix any $x \in B(\overline{x}, \delta)$ with $x \neq \overline{x}$ and any $t \in]0, 1[$, and set $x_t := tx + (1-t)\overline{x}$. By the mean value theorem (see Theorem 2.180) there are $z_1 \in [x, x_t[$ and $z_1^* \in \partial_C f(z_1)$ along with $z_2 \in [\overline{x}, x_t[$ and $z_2^* \in \partial_C f(z_2)$ such that

$$\langle z_1^*, x_t - x\rangle = f(x_t) - f(x) \quad \text{and} \quad \langle z_2^*, x_t - \overline{x}\rangle = f(x_t) - f(\overline{x}).$$

Multiplying the first equality by t and the second by $(1-t)$, and adding the resulting equalities we get (noting that $x_t - x = (1-t)(\overline{x} - x)$ and $x_t - \overline{x} = t(x - \overline{x})$)

$$tf(x) + (1-t)f(\overline{x}) - f(x_t) = t(1-t)\langle z_1^* - z_2^*, x - \overline{x}\rangle. \tag{8.18}$$

By (8.17) choose some $\overline{x}^* \in \partial_C f(\overline{x})$ such that

$$\|\overline{x}^* - z_2^*\| \leq \varepsilon/2, \tag{8.19}$$

and note by (8.16) that

$$\left\langle z_1^* - \overline{x}^*, \frac{z_1 - \overline{x}}{\|z_1 - \overline{x}\|}\right\rangle \geq -\varepsilon/2.$$

From the latter inequality and from (8.19) we deduce through the equality

$$\frac{z_1 - \overline{x}}{\|z_1 - \overline{x}\|} = \frac{x - \overline{x}}{\|x - \overline{x}\|}$$

that we have

$$\left\langle z_1^* - z_2^*, \frac{x - \overline{x}}{\|x - \overline{x}\|}\right\rangle \geq -\varepsilon.$$

Combining this with (8.18) it results that

$$tf(x) + (1-t)f(\overline{x}) - f(x_t) \geq -\varepsilon t(1-t)\|x - \overline{x}\|,$$

which translates (by Proposition 8.12) the one-sided subsmoothness of f on U. □

REMARK 8.51. The proof of the above implication (c)⇒(a) shows that the locally Lipschitz function f is one-sided subsmooth at $\overline{x} \in U$ whenever $\partial_C f$ is one-sided submonotone at $\overline{x}$ and X is finite-dimensional. □

In addition to the one-sided subsmoothness property, another notion of interest is that of semismoothness for locally Lipschitz functions.

DEFINITION 8.52. Let U be a nonempty open subset of a normed space X and $f : U \to \mathbb{R}$ be a locally Lipschitz function. One says that f is *semismooth* (or *Mifflin semismooth*) at a point $\overline{x} \in U$ if for any $u \in \mathbb{S}_X$, any sequence $(x_n)_n$ in $U \setminus \{\overline{x}\}$ with

$$\lim_{n\to\infty} x_n = \overline{x} \quad \text{and} \quad \lim_{n\to\infty} \left\| \frac{x_n - \overline{x}}{\|x_n - \overline{x}\|} - u \right\| = 0,$$

one has

$$\langle x_n^*, u\rangle \to f'(\overline{x}; u) \quad \text{as } n \to \infty,$$

for any sequence $(x_n^*)_n$ with $x_n^* \in \partial_C f(x_n)$ for all $n \in \mathbb{N}$.

When f is semismooth at any point in an open set U_0 of U, one says that f is semismooth on U_0.

REMARK 8.53. Although the semismoothness of a locally Lipschitz function f requires the existence of $f'(\overline{x};\cdot)$, such a function f may fail to be tangentially regular at $\overline{x}$. The same Lipschitz function $f := -|\cdot|$ in Example 8.8 and Remark 8.29 is semismooth on $\mathbb{R}$ but not tangentially regular at $\overline{x} = 0$. Therefore, it is both semismooth and one-sided subsmooth on $\mathbb{R}$ but not tangentially regular at $\overline{x} = 0$.

Further, since $\partial_C f$ is obviously submonotone at any point in $\mathbb{R} \setminus \{0\}$, from Proposition 8.50 we derive that $\partial_C f$ is not one-sided submonotone at the origin. This can also be easily checked, taking $x_n := 1/n$, $x_n^* := -1 \in \partial_C f(x_n)$ and $\overline{x}^* := 1 \in \partial_C f(0)$ and noting that $\frac{(x_n^*-\overline{x}^*)(x_n-0)}{|x_n-0|} = -2$ for all $n \in \mathbb{N}$. □

For a locally Lipschitz function, the semismoothness property is satisfied at a point whenever the Clarke subdifferential of the function is one-sided submonotone at that point, as implied by the assertion (a) in the following proposition.

PROPOSITION 8.54. Let U be a nonempty open set of a normed space X and $f : U \to \mathbb{R}$ be a locally Lipschitz function. The following hold:
(a) The function f is semismooth at $\overline{x} \in U$ and tangentially regular at $\overline{x}$ whenever $\partial_C f$ is one-sided submonotone at $\overline{x}$.
(b) If f is one-sided subsmooth on U and tangentially regular on U, then f is semismooth on U.

PROOF. (a) Assume that $\partial_C f$ is one-sided submonotone at $\overline{x}$, so f is tangentially regular at $\overline{x}$ by Lemma 8.48. Take any $u \in \mathbb{S}_X$, any sequence $(x_n)_n$ in $U \setminus \{\overline{x}\}$ converging to $\overline{x}$ with $\|x_n - \overline{x}\|^{-1}(x_n - \overline{x}) \to u$ as $n \to \infty$, and any sequence $(x_n^*)_n$ with $x_n^* \in \partial_C f(x_n)$ for all $n \in \mathbb{N}$. Since $\partial_C f$ is one-sided submonotone at $\overline{x}$, we have

$$\liminf_{n\to\infty} \inf_{z^* \in \partial_C f(x_n)} \left\langle z^*, \frac{x_n - \overline{x}}{\|x_n - \overline{x}\|} \right\rangle \geq \limsup_{n\to\infty} \sup_{y^* \in \partial_C f(\overline{x})} \left\langle y^*, \frac{x_n - \overline{x}}{\|x_n - \overline{x}\|} \right\rangle .$$

This combined with the boundedness of $(x_n^*)_n$ and the continuity of $f^\circ(\overline{x};\cdot)$ yields

$$\liminf_{n\to\infty} \langle x_n^*, u\rangle \geq f^\circ(\overline{x};u).$$

Further, the upper semicontinuity of $f^\circ(\cdot;u)$ at $\overline{x}$ assures us that

$$\limsup_{n\to\infty} \langle x_n^*, u\rangle \leq \limsup_{n\to\infty} f^\circ(x_n;u) \leq f^\circ(\overline{x};u).$$

We deduce that

$$\langle x_n^*, u\rangle \longrightarrow f^\circ(\overline{x};u) = f'(\overline{x};u),$$

so f is semismooth at $\overline{x}$.
(b) The assertion (b) follows from (a) and from the implication (a)⇒(c) in Proposition 8.50. □

In the context of finite-dimensional normed spaces, the implication in the assertion (a) in the above proposition is an equivalence.

PROPOSITION 8.55. Let U be a nonempty open set of a finite-dimensional normed space X. Let $f : U \to \mathbb{R}$ be a locally Lipschitz function and let $\overline{x} \in U$. Then f is semismooth at $\overline{x}$ and tangentially regular at $\overline{x}$ if and only if $\partial_C f$ is one-sided submonotone at $\overline{x}$.

PROOF. According to Proposition 8.54(a), we only need to prove the implication $\Rightarrow$. So, assume that f is semismooth at $\overline{x}$ and tangentially regular at $\overline{x}$. It suffices to show that the sequential property in (a) in Lemma 8.47 is satisfied for the multimapping $\partial_C f$. Take any $u \in \mathbb{S}_X$, any sequence $(x_n)_n$ in $U\setminus\{\overline{x}\}$ converging to $\overline{x}$ with $\|x_n - \overline{x}\|^{-1}(x_n - \overline{x}) \to u$ as $n \to \infty$, and any sequence $(x_n^*)_n$ converging to x^* with $x_n^* \in \partial_C f(x_n)$ for all $n \in \mathbb{N}$. By outer semicontinuity of $\partial_C f$ at $\overline{x}$ we have $x^* \in \partial_C f(\overline{x})$. Further, by the semismoothness of f at $\overline{x}$

$$\langle x^*, u\rangle = \lim_{n\to\infty} \langle x_n^*, u\rangle = f'(\overline{x}; u),$$

so $\langle x^*, u\rangle = f^\circ(\overline{x}; u) = \sigma\big(\partial_C f(\overline{x}), u\big)$, which is the desired property. $\square$

The next corollary is a direct consequence of the above proposition and of Proposition 8.50.

COROLLARY 8.56. let U be a nonempty open set of a finite-dimensional normed space X and $f : U \to \mathbb{R}$ be a locally Lipschitz function. The following assertions are equivalent:
(a) the function f is one-sided subsmooth on U and tangentially regular on U;
(b) the subdifferential multimapping $\partial_C f$ is one-sided submonotone on U;
(c) the function f is semismooth on U and tangentially regular on U.

8.3. Subsmooth sets

Given a nonempty closed set S of a normed space X, the subsmoothness property in (8.1) for its indicator function Ψ_S at $\overline{x} \in S$ is evidently equivalent to the convexity of $S \cap B(\overline{x}, \delta)$ for some $\delta > 0$ (see also Proposition 8.5). Now suppose that X is a Banach space. By Theorem 8.39 the subsmoothness of the indicator function of the closed set S at $\overline{x} \in S$ amounts to saying that $\partial_C \Psi_C(\cdot) = N^C(S; \cdot)$ is submonotone at $\overline{x}$. This means that, for each $\varepsilon > 0$ there exists a real $\delta > 0$ such that for all $x_i \in S \cap B(\overline{x}, \delta)$ and all $x_i^* \in N^C(S; x_i)$, $i = 1, 2$, one has

$$\langle x_1^* - x_2^*, x_1 - x_2\rangle \geq -\varepsilon\|x_1 - x_2\|,$$

which by the positive homogeneity of the C-normal cone gives for every real $t > 0$

$$\langle tx_1^* - tx_2^*, x_1 - x_2\rangle \geq -\varepsilon\|x_1 - x_2\|, \text{ or equivalently } \langle x_1^* - x_2^*, x_1 - x_2\rangle \geq -\frac{\varepsilon}{t}\|x_1 - x_2\|,$$

so $\langle x_1^* - x_2^*, x_1 - x_2 \geq 0$ by taking the limit as $t \to +\infty$. Using Theorem 6.68 we see again that the set $S \cap B(\overline{x}, \delta)$ is convex.

Clearly, from a geometric point of view, the convexity property of $S \cap B(\overline{x}, \delta)$ is not enough relevant for (locally) nonconvex sets; the property is not fulfilled even for $\mathcal{C}^2$-submanifolds.

8.3.1. Definition of subsmooth sets and general properties. Taking into account the above analysis, by virtue of the positive homogeneity of $N^C(S; \cdot)$ we define subsmooth sets through the submonotonicity of the truncation of this multimapping $N^C(S; \cdot)$ with the closed unit ball.

DEFINITION 8.57. A subset S of a normed space $(X, \|\cdot\|)$ is called *subsmooth at a point* $\overline{x} \in S$ if the multimapping $N^C(S; \cdot) \cap \mathbb{B}_{X^*}$ is submonotone at $\overline{x}$, or equivalently provided that for each $\varepsilon > 0$ there exists $\delta > 0$ such that

$$\langle x^* - y^*, x - y\rangle \geq -\varepsilon\|x - y\| \tag{8.20}$$

for all $x, y \in S \cap B(\overline{x}, \delta)$, all $x^* \in N^C(S; x) \cap \mathbb{B}_{X^*}$ and all $y^* \in N^C(S; y) \cap \mathbb{B}_{X^*}$. The set S is called *subsmooth* when it is subsmooth at every point in S.

When for each $\varepsilon > 0$ there is $\delta > 0$ such that (8.20) is satisfied for all $x, y \in S$ with $\|x - y\| < \delta$ and all $x^* \in N^C(S; x) \cap \mathbb{B}_{X^*}$ and all $y^* \in N^C(S; x) \cap \mathbb{B}_{X^*}$, one says that the set S is *uniformly subsmooth.* The set S is *uniformly subsmooth near* $\overline{x} \in S$ if there exists an open neighborhood U of $\overline{x} \in S$ such that the set $S \cap U$ is uniformly subsmooth.

Clearly, by the inclusion $0 \in N^C(S; y) \cap \mathbb{B}_{X^*}$ for $y \in S$ the set S is subsmooth at $\overline{x}$ (resp. the set S is uniformly subsmooth) if and only if for every $\varepsilon > 0$ there is $\delta > 0$ such that

$$\langle x^*, y - x\rangle \le \varepsilon \|y - x\| \tag{8.21}$$

for all $x, y \in S \cap B(\overline{x}; \delta)$ and all $x^* \in N^C(S; x) \cap \mathbb{B}_{X^*}$ (resp. for all $x, y \in S$ with $\|x - y\| < \delta$ and all $x^* \in N^C(S; x) \cap \mathbb{B}_{X^*}$). Similarly, the set S is uniformly subsmooth near $\overline{x}$ if there is some open neighborhood U of $\overline{x}$ such that for each $\varepsilon > 0$ there exists $\delta > 0$ for which (8.21) holds for all $x, y \in S \cap U$ with $\|x - y\| < \delta$ and all $x^* \in N^C(S; x) \cap \mathbb{B}_{X^*}$.

REMARK 8.58. (a) Clearly, the three above concepts of uniform subsmoothness, uniform subsmoothness near $\overline{x}$, and subsmoothness at $\overline{x}$ are invariant with respect to equivalent norms.
(b) Any open set is uniformly subsmooth, since the C-normal cone is zero at any of its points.
(c) Any set is uniformly subsmooth near anyone of its interior points.
(d) If a set is uniformly subsmooth, then it is uniformly subsmooth near each one of its points.
(e) The uniform subsmoothness of a set near a point entails its subsmoothness at that point. □

Now observe that the monotonicity of the normal cone of a convex set directly gives the following result.

PROPOSITION 8.59. Any convex set of a normed space is uniformly subsmooth.

Since $N^C(S_1 \times \cdots \times S_m; (x_1, \cdots, x_m)) = N^C(S_1; x_1) \times \cdots \times N^C(S_m; x_m)$ the following result is also obvious.

PROPOSITION 8.60. For each $k = 1, \cdots, m$ let S_k be a set in a normed space X_k which is subsmooth at a point $\overline{x}_k \in S_k$. Then the set $S_1 \times \cdots \times S_m$ is subsmooth at $(\overline{x}_1, \cdots, \overline{x}_m)$.

The uniform equi-subsmoothness for families of sets need also to be defined.

DEFINITION 8.61. A family $(S_i)_{i \in I}$ of sets of a normed space X is said to be *uniformly equi-subsmooth* if for any $\varepsilon > 0$ there is some $\delta > 0$ such that for any $i \in I$, any $x, y \in S_i$, any $x^* \in N^C(S_i; x) \cap \mathbb{B}_{X^*}$ and any $y^* \in N^C(S_i; y) \cap \mathbb{B}_{X^*}$ one has

$$\langle x^* - y^*, x - y\rangle \ge -\varepsilon \|x - y\|.$$

Clearly, this is equivalent to require that for any $\varepsilon > 0$ there is $\delta > 0$ such that for each $i \in I$ the inequality

$$\langle x^*, y - x\rangle \le \varepsilon \|y - x\|$$

holds for all $x, y \in S_i$ and all $x^* \in N^C(S_i; x) \cap \mathbb{B}_{X^*}$.

The following proposition is obvious.

PROPOSITION 8.62. Let $(S_i)_{i\in I}$ be a family of sets of a normed space X. If all the sets S_i are convex, then $(S_i)_{i\in I}$ is a family of sets uniformly equi-subsmooth.

Using (8.21) we show that $N^C(S;\cdot)$ is sequentially norm-to-weak* closed at $\overline{x}$.

PROPOSITION 8.63. Let S be a subset of a Banach space $(X, \|\cdot\|)$ which is subsmooth at $\overline{x} \in S$. Then the multimapping $N^C(S;\cdot)$ is sequentially norm-to-weak* closed at the point $\overline{x}$.

PROOF. Let any sequences $(x_n)_n$ in S converging to $\overline{x}$ and $(x_n^*)_n$ in X^* weak-star converging to $x^* \in X^*$ with $x_n^* \in N^C(S;x_n)$ for all $n \in \mathbb{N}$. Since X is a Banach space, there is a real $\beta > 0$ such that $\|x_n^*\| \le \beta$ for all $n \in \mathbb{N}$. Take any real $\varepsilon > 0$. There is a real $\delta > 0$ such that (8.21) is satisfied with $\beta^{-1}\varepsilon$ in place of ε. Let $n_0 \in \mathbb{N}$ be such that for every integer $n \ge n_0$ one has $x_n \in B(\overline{x}, \delta)$. Then for each integer $n \ge n_0$ we see that for every $y \in S \cap B(\overline{x}, \delta)$

$$\langle \beta^{-1}x_n^*, y - x_n\rangle \le \beta^{-1}\varepsilon\|y - x_n\|,$$

hence $\langle x_n^*, y-x_n\rangle \le \varepsilon\|y-x_n\|$. Taking the limit as $n \to \infty$ ensures that $\langle x^*, y-\overline{x}\rangle \le \varepsilon\|y-\overline{x}\|$ for every $y \in S\cap B(\overline{x}, \delta)$. This entails that $x^* \in N^F(S;\overline{x})$, thus in particular $x^* \in N^C(S;\overline{x})$, which justifies the desired closedness property of $N^C(S;\cdot)$ at $\overline{x}$. □

Requiring $y := \overline{x}$ in (8.20) yields with a radial counterpart of the above concept. The precise definition is as follows:

DEFINITION 8.64. A subset S of a normed space $(X, \|\cdot\|)$ is called *one-sided subsmooth at* a point $\overline{x} \in S$ if for each $\varepsilon > 0$ there exists $\delta > 0$ such that

$$\langle x^* - \overline{x}^*, x - \overline{x}\rangle \ge -\varepsilon\|x - \overline{x}\| \tag{8.22}$$

for all $x \in S \cap B(\overline{x}, \delta)$, all $x^* \in N^C(S;x) \cap \mathbb{B}_{X^*}$ and all $\overline{x}^* \in N^C(S;\overline{x}) \cap \mathbb{B}_{X^*}$. The set S is called *one-sided subsmooth* when it is one-sided subsmooth at every point in S.

REMARK 8.65. (a) It is clear that the definition of one-sided subsmooth sets is unchanged if any equivalent norm on X is used in place of $\|\cdot\|$.
(b) If S is subsmooth at $\overline{x}$, then it is one-sided subsmooth at $\overline{x}$.
(c) The natural definition of uniform one-sided subsmoothness of the set S obviously yields to the above notion of uniform subsmoothness for S. □

Since 0 always belong to the C-normal cone of a set at any point in the set, we note that the above definition of one-sided subsmoothness obviously amounts to requiring for any $\varepsilon > 0$ the existence of some $\delta > 0$ such that for all $x \in S \cap B(\overline{x}, \delta)$, all $x^* \in N^C(S;x) \cap \mathbb{B}_{X^*}$, and all $\overline{x}^* \in N^C(S;\overline{x}) \cap \mathbb{B}_{X^*}$ both inequalities

$$\langle x^*, \overline{x} - x\rangle \le \varepsilon\|\overline{x} - x\| \quad \text{and} \quad \langle \overline{x}^*, x - \overline{x}\rangle \le \varepsilon\|x - \overline{x}\|$$

are satisfied.

If only the second one of the two latter inequalities is required, we obtain another concept that we call hemi-subsmoothness.

DEFINITION 8.66. A subset S of a normed space $(X, \|\cdot\|)$ is called *hemi-subsmooth at* a point $\overline{x} \in S$ if for each $\varepsilon > 0$ there exists $\delta > 0$ such that

$$\langle \overline{x}^*, x - \overline{x}\rangle \le \varepsilon\|x - \overline{x}\|, \tag{8.23}$$

for all $x \in S \cap B(\overline{x}, \delta)$ and all $\overline{x}^* \in N^C(S;\overline{x}) \cap \mathbb{B}_{X^*}$.

Obviously a set S is subsmooth (resp. one-sided subsmooth, hemi-subsmooth) at $\overline{x} \in S$ if and only if there is some neighborhood W of $\overline{x}$ such that $S \cap W$ enjoys the same property at $\overline{x}$. The similar equivalence also holds for the uniform subsmoothness of S near $\overline{x} \in S$.

Taking the positive homogeneity of $N^C(S;x)$ into account, we directly obtain:

PROPOSITION 8.67. Let S be a nonempty subset of a normed space X and let $\overline{x} \in S$. The following hold:
(a) The set S is subsmooth at $\overline{x}$ if and only if for any reals $r > 0$ and $\varepsilon > 0$ there exists a real $\delta > 0$ such that

$$\langle x^* - y^*, x - y \rangle \geq -\varepsilon \|x - y\|, \tag{8.24}$$

for all $x, y \in S \cap B(\overline{x}, \delta)$, all $x^* \in N^C(S;x) \cap r\mathbb{B}_{X^*}$ and all $y^* \in N^C(S;y) \cap r\mathbb{B}_{X^*}$.
(b) The set S is one-sided subsmooth at $\overline{x}$ if and only if for any reals $r > 0$ and $\varepsilon > 0$ there exists a real $\delta > 0$ such that

$$\langle x^* - \overline{x}^*, x - \overline{x} \rangle \geq -\varepsilon \|x - \overline{x}\|, \tag{8.25}$$

for all $x \in S \cap B(\overline{x}, \delta)$, all $x^* \in N^C(S;x) \cap r\mathbb{B}_{X^*}$ and all $\overline{x}^* \in N^C(S;\overline{x}) \cap r\mathbb{B}_{X^*}$.
(c) A similar equivalence holds for uniform subsmoothness of S (resp. uniform subsmoothness of S near $\overline{x}$, hemi-subsmoothness of S at $\overline{x}$).

We already observed above that convex sets are obviously subsmooth (even uniformly subsmooth). In fact, a remarkable class of examples of subsmooth sets is given by inverse images of convex sets with $\mathcal{C}^1$ mappings with surjective derivatives between Banach spaces. The proof will use the next lemma.

LEMMA 8.68. Let $A : X \to Y$ be a continuous linear mapping between two Banach spaces X, Y and let $G : X \to Y$ be a mapping which is of class $\mathcal{C}^1$ near a point $\overline{x} \in X$.
(a) If there is a real $s > 0$ satisfying $s\mathbb{U}_Y \subset A(\mathbb{B}_X)$, then for any $x^* \in X^*$ and $y^* \in Y^*$ with $x^* = y^* \circ A$ one has

$$s\|y^*\| \leq \|x^*\|.$$

(b) If there is a real $s > 0$ satisfying $s\mathbb{B}_Y \subset DG(\overline{x})(\mathbb{B}_X)$, then for any real $\eta > 0$ there is an open neighborhood U of $\overline{x}$ such that for every $x \in U$ the inclusion $s'\mathbb{B}_Y \subset DG(x)(\mathbb{B}_X)$ holds with $s' := (1+\eta)^{-1}s$, and hence in particular for all $x \in U$, $x^* \in X^*$ and $y^* \in Y^*$ with $x^* = y^* \circ DG(x)$ one has

$$s\|y^*\| \leq (1+\eta)\|x^*\|.$$

PROOF. (a) Fix any $x^* \in X^*$ and $y^* \in Y^*$ such that $x^* = y^* \circ A$. Take any $v \in \mathbb{U}_Y$ and choose $u \in \mathbb{B}_X$ such that $sv = A(u)$. We notice that

$$\langle y^*, sv \rangle = \langle y^*, A(u) \rangle = \langle x^*, u \rangle \leq \|x^*\|.$$

This being true for all $v \in \mathbb{U}_Y$, we obtain that $s\|y^*\| \leq \|x^*\|$.
(b) Put $A := DG(\overline{x})$ and $s' := (1+\eta)^{-1}s$ and choose an open convex neighborhood U of $\overline{x}$ over which G is $\mathcal{C}^1$ and such that $\|DG(x) - DG(\overline{x})\| < s - s'$ for all $x \in U$. Fix any $x \in U$ and put $\Lambda := DG(x)$. By Remark 7.25 we deduce that $s'\mathbb{U}_X \subset \Lambda(\mathbb{B}_X)$. Therefore, taking any $x^* \in X^*$ and $y^* \in Y^*$ such that $x^* = y^* \circ DG(x)$, the assertion (a) tells us that $s'\|y^*\| \leq \|x^*\|$, which justifies the assertion (b). □

REMARK 8.69. Notice that (a) in the above lemma still holds if X, Y are normed spaces. □

PROPOSITION 8.70. (a) If a subset M of a Banach space X is a $\mathcal{C}^1$-submanifold at $m_0 \in M$, then it is subsmooth at m_0.
(b) If $G : X \to Y$ is a mapping between Banach spaces which is of class $\mathcal{C}^1$ near a point $\overline{x} \in X$ with $DG(\overline{x})$ surjective and if C is a convex set of Y containing $G(\overline{x})$, then the set $G^{-1}(C)$ is subsmooth at $\overline{x}$.

PROOF. (b) We begin by proving (b). By the Banach-Schauder open mapping theorem (see Theorem C.3) there is a real $s > 0$ such that $s\mathbb{B}_Y \subset DG(\overline{x})(\mathbb{B}_Y)$. Then by the above lemma there are an open neighborhood U of $\overline{x}$ and a real $\gamma > 0$ such that for each $x \in U$ the continuous linear mapping $DG(x)$ is open and $\|y^*\| \le \gamma\|x^*\|$ for all $x^* \in X^*$ and $y^* \in Y^*$ satisfying $x^* = y^* \circ DG(x)$. Now fix any $\varepsilon > 0$ and choose an open convex neighborhood $U_0 \subset U$ of $\overline{x}$ such that $\|DG(x') - DG(x)\| \le \varepsilon/\gamma$ for all $x, x' \in U_0$. Consider any $x, u \in U_0 \cap G^{-1}(C)$ and $x^* \in N^C\big(G^{-1}(C); x\big) \cap \mathbb{B}_{X^*}$. We know by Theorem 3.174 that there is $y^* \in N^C(C; G(x))$ such that $x^* = y^* \circ DG(x)$, so $\|y^*\| \le \gamma$ by the choice of γ. Since $\langle y^*, G(u) - G(x)\rangle \le 0$, we deduce that

$$\begin{aligned}\langle x^*, u - x\rangle &= \langle y^*, DG(x)(u - x)\rangle \\ &= \langle y^*, G(u) - G(x)\rangle - \left\langle y^*, \int_0^1 (DG(x + t(u - x)) - DG(x))(u - x)dt\right\rangle \\ &\le \|y^*\|(\varepsilon/\gamma)\|u - x\|,\end{aligned}$$

hence $\langle x^*, u - x\rangle \le \varepsilon\|u - x\|$. This establishes the subsmoothness of $G^{-1}(C)$ at $\overline{x}$.
(a) By definition of submanifold (see E.8 in Appendix E) there is a closed vector subspace E of X such that there exist open neighborhoods U of m_0 in X and V of zero in X along with a $\mathcal{C}^1$ diffeomorphism $\varphi : U \to V$ with $\varphi(m_0) = 0$ such that $\varphi(M \cap U) = E \cap V$. Shrinking the open neighborhood V of zero if necessary, we may suppose that it is convex. Then, by (b) proved above the set M is subsmooth at m_0, which finishes the proof. □

The properties of uniform subsmoothness and hemi-smoothness for sets can be characterized via modulus functions. Recall that $\omega : [0, +\infty[\to [0, +\infty]$ is a modulus function (see Definition 8.15) if $\omega(0) = 0$ and $\omega(t) \to 0$ as $t \downarrow 0$.

PROPOSITION 8.71. Let S be a nonempty set of a normed space X.
(a) The set S is uniformly subsmooth (resp. uniformly subsmooth near a point $\overline{x} \in S$) if and only if there exists a modulus function $\omega : [0, +\infty[\to [0, +\infty]$ (resp. there exist an open neighborhood U of $\overline{x}$ and a modulus function $\omega : [0, +\infty[\to [0, +\infty[$ of class $\mathcal{C}^1$ on $]0, +\infty[$ with $t\,\omega'(t) \to 0$ as $t \downarrow 0$) such that

$$\langle x^* - y^*, x - y\rangle \ge -\|x - y\|\omega(\|x - y\|)$$

for all $x, y \in S$ (resp. $x, y \in S \cap U$), all $x^* \in N^C(S; x) \cap \mathbb{B}_{X^*}$ and all $y^* \in N^C(S; y) \cap \mathbb{B}_{X^*}$.
(b) The set S is hemi-subsmooth at a point $\overline{x} \in S$ if and only if there exists a modulus function $\omega : [0, +\infty[\to [0, +\infty]$ such that

$$\langle \overline{x}^*, x - \overline{x}\rangle \le \|x - \overline{x}\|\omega(\|x - \overline{x}\|)$$

for all $x \in S$ and all $\overline{x}^* \in N^C(S; \overline{x}) \cap \mathbb{B}_{X^*}$.
(c) If X is finite-dimensional, the set S is uniformly subsmooth near $\overline{x}$ if and only if it is subsmooth near $\overline{x}$.

PROOF. (a) The implication $\Leftarrow$ is obvious. Let us first prove the reverse implication in the case when S is uniformly subsmooth. Define $\omega : [0, +\infty[\to [0, +\infty]$ by $\omega(0) := 0$ and for $t > 0$

$$\omega(t) := \sup\left\{ \frac{\langle x^* - y^*, y - x\rangle^+}{\|x - y\|} : 0 < \|x - y\| \le t, (x, x^*), (y, y^*) \in \operatorname{gph} N^C(S; \cdot) \cap \mathbb{B} \right\},$$

where (as usual) $r^+ := \max\{0, r\}$ for $r \in \mathbb{R}$ and where we use the convention that the supremum is 0 whenever the set over which it is taken is empty, that is, S is a singleton. Clearly, the definition of uniform subsmoothness of S guarantees that $\omega(t) \to 0$ as $t \downarrow 0$ and by the very definition of ω the inequality in the proposition holds true for all $x, y \in S$, $x^* \in N^C(S; x) \cap \mathbb{B}$ and $y^* \in N^C(S; y) \cap \mathbb{B}$. The case of uniform subsmoothness near $\overline{x}$ follows from what precedes and from Lemma 4.24.
(b) Similarly, for the implication $\Rightarrow$ it suffices to define $\omega : [0, +\infty[\to [0, +\infty]$ by $\omega(0) := 0$ and for $t > 0$

$$\omega(t) := \sup\left\{ \frac{\langle x^*, x - \overline{x}\rangle^+}{\|x - \overline{x}\|} : x \in S, 0 < \|x - \overline{x}\| \le t, x^* \in N^C(S; \overline{x}) \cap \mathbb{B} \right\},$$

and to argue like in (a).
(c) For the implication (c)$\Rightarrow$(a) when X is finite-dimensional, it suffices to proceed like in the proof of the implication (d)$\Rightarrow$(a) in Proposition 8.44. □

REMARK 8.72. We will see in Chapter 15 that the inequality in (a) of the above proposition with the particular modulus $\sigma|\cdot|^2$ characterizes the fundamental class of closed subsets of Hilbert spaces, known as prox-regular sets. □

The hemi-subsmoothness of a set entails its tangential regularity.

PROPOSITION 8.73. Let S be a subset of a normed space X and let $\overline{x} \in S$.
(a) The subsmoothness of S at $\overline{x}$ entails its one-sided subsmoothness at $\overline{x}$, which in turn entails the hemi-subsmoothness at $\overline{x}$.
(b) If the set S is hemi-subsmooth at $\overline{x}$, then it enjoys the normal regularity

$$N^C(S; \overline{x}) = N^F(S; \overline{x}),$$

and hence it is tangentially regular at $\overline{x}$.
If in addition X is a Banach space, then one has

$$N^C(S; \overline{x}) = N^A(S; \overline{x}) = N^F(S; \overline{x}).$$

PROOF. The assertion (a) is evident. To justify (b), assume that S is hemi-subsmooth at $\overline{x}$. Take any $\overline{x}^* \in N^C(S; \overline{x})$ and let $r > \|\overline{x}^*\|$. Fix any real $\varepsilon > 0$ and choose $\delta > 0$ satisfying the property related to hemi-subsmoothness in Proposition 8.67. Fixing any $x \in S \cap B(\overline{x}, \delta)$ we have

$$\langle \overline{x}^*, x - \overline{x}\rangle \le \varepsilon \|x - \overline{x}\|.$$

This tells us that $\overline{x}^*$ is a Fréchet normal of S at $\overline{x}$, that is, the inclusion $N^C(S; \overline{x}) \subset N^F(S; \overline{x})$ holds. In fact, the latter inclusion is an equality since the reverse inclusion always holds. This equality is also known (see Corollary 4.10) to entail the tangential regularity of S at $\overline{x}$.

Finally, when X is a Banach space, the additional equality with $N^A(S; \overline{x})$ follows from what precedes since the inclusions $N^F(S; \cdot) \subset N^A(S; \cdot) \subset N^C(S; \cdot)$ are always true in Banach spaces. □

Converses of above assertions will be discussed later.

Local subsmoothness of sets can be characterized with C-subdifferentials of distance functions. The following lemma will be useful for that.

LEMMA 8.74. Let S be a subset of a Banach space X and let $\overline{x} \in S$. If $\partial_C d_S(\cdot)$ in place of $N^C(S;\cdot)$ satisfies (8.22) for all $x \in S \cap B(\overline{x},\delta)$, $x^* \in \partial_C d_S(x)$, $\overline{x}^* \in \partial_C d_S(\overline{x})$, then

$$\partial_C d_S(\overline{x}) = \partial_F d_S(\overline{x}) \quad \text{and} \quad N^C(S;\overline{x}) = N^F(S;\overline{x}).$$

PROOF. Fix any $\overline{x}^* \in \partial_C d_S(\overline{x})$. Then for any real $\varepsilon > 0$ there exists $\delta > 0$ such that for every $x \in S \cap B(\overline{x},\delta)$ we have $\langle -\overline{x}^*, x-\overline{x}\rangle \geq -\varepsilon\|x-\overline{x}\|$, since $0 \in \partial_C d_S(x)$. This implies that $\overline{x}^* \in N^F(S;\overline{x})$. Moreover, the inclusion $\overline{x}^* \in \partial_C d_S(\overline{x})$ ensures that $\|\overline{x}^*\| \leq 1$. This combined with the equality $\partial_F d_S(\overline{x}) = N^F(S;\overline{x}) \cap \mathbb{B}_{X^*}$ (see Proposition 4.30) entails that $\partial_C d_S(\overline{x}) \subset \partial_F d_S(\overline{x})$. The reverse inclusion being always true, it ensues that the first equality $\partial_C d_S(\overline{x}) = \partial_F d_S(\overline{x})$ of the proposition is established.

Concerning the second equality, observe first that the first equality ensures in particular that $\partial_F d_S(\overline{x})$ is $w(X^*,X)$-closed, and hence for every real $r > 0$ the convex set

$$N^F(S;\overline{x}) \cap r\mathbb{B}_{X^*} = r\partial_F d_S(\overline{x})$$

is $w(X^*,X)$-closed. The space X being a Banach space, the Krein-Šmulian theorem (see Theorem C.4 in Appendix C) guarantees that the convex set $N^F(S;\overline{x})$ is $w(X^*,X)$-closed. Further, the first equality again combined with the equalities (see Proposition 2.95 and Proposition 4.30)

$$N^C(S;\overline{x}) = \mathrm{cl}_{w^*}\left(\mathbb{R}_+\partial_C d_S(\overline{x})\right) \quad \text{and} \quad N^F(S;\overline{x}) = \mathbb{R}_+\partial_F d_S(\overline{x})$$

gives the equality

$$N^C(S;\overline{x}) = \mathrm{cl}_{w^*}\left(N^F(S;\overline{x})\right).$$

This and the above $w(X^*,X)$-closedness of $N^F(S;\overline{x})$ justifies the desired second equality $N^C(S;\overline{x}) = N^F(S;\overline{x})$. □

PROPOSITION 8.75. Let S be a nonempty subset of a Banach space X and U be a nonempty open set of X with $U \cap S \neq \emptyset$.
(A) The following are equivalent:
(a) the set S is subsmooth at each point of $S \cap U$;
(b) for any $\overline{x} \in S \cap U$ and any $\varepsilon > 0$ there is $\delta > 0$ such that (8.20) holds on $S \cap B(\overline{x},\delta)$ with $\partial_C d_S$ in place of $N^C(S;\cdot) \cap \mathbb{B}_{X^*}$.
(B) The set S is uniformly subsmooth if and only if for each $\varepsilon > 0$ there exists a real $\delta > 0$ such that $\langle x^* - y^*, x-y\rangle \geq -\varepsilon\|x-y\|$ for all $x, y \in S$ with $\|x-y\| < \delta$, $x^* \in \partial_C d_S(x)$ and $y^* \in \partial_C d_S(y)$.

PROOF. We prove only (A), since (B) follows in the same way. The implication (a)⇒(b) is evident. To prove the converse, suppose that (b) holds. By Lemma 8.74 we have, for all $x \in S \cap U$

$$\partial_C d_S(x) = \partial_F d_S(x) \quad \text{and} \quad N^C(S;x) = N^F(S;x).$$

Both equalities combined with the equality $\partial_F d_S(x) = N^F(S;x) \cap \mathbb{B}_{X^*}$ (for $x \in S$) yield $N^C(S;x) \cap \mathbb{B}_{X^*} = \partial_C d_S(x)$ for all $x \in S \cap U$. This and the assumption (b) entail that (a) holds true. □

Alternative characterizations of local subsmoothness of sets in Asplund spaces can be established via Fréchet normals or via subdifferentials of distance functions. A lemma is needed first, and it has its own interest.

LEMMA 8.76. Let S be a set of an Asplund space which is closed near $\overline{x} \in S$. The following assertions are equivalent:
(a) for any $\varepsilon > 0$ there is $\delta > 0$ such that (8.20) holds on $S \cap B(\overline{x}, \delta)$ with $\partial_F d_S$ in place of $N^C(S; \cdot) \cap \mathbb{B}_{X^*}$;
(b) for any $\varepsilon > 0$ there is $\delta > 0$ such that (8.20) holds on $S \cap B(\overline{x}, \delta)$ with $\partial_L d_S$ in place of $N^C(S; \cdot) \cap \mathbb{B}_{X^*}$;
(c) for any $\varepsilon > 0$ there is $\delta > 0$ such that (8.20) holds on $S \cap B(\overline{x}, \delta)$ with $\partial_C d_S$ in place of $N^C(S; \cdot) \cap \mathbb{B}_{X^*}$.

PROOF. The implication (c)$\Rightarrow$(a) follows directly from the inclusion $\partial_F d_S(\cdot) \subset \partial_C d_S(\cdot)$. Let us show the implication (a)$\Rightarrow$(b). Fix any real $\varepsilon > 0$ and by (a) choose $\delta > 0$ such that for all $u, v \in S \cap B(\overline{x}, \delta)$, $u^* \in \partial_F d_S(u)$ and $v^* \in \partial_F d_S(v)$

$$\langle u^* - v^*, u - v \rangle \geq -\varepsilon \|u - v\|.$$

Fix any $x, y \in S \cap B(\overline{x}, \delta)$, $x^* \in \partial_L d_S(x)$, $y^* \in \partial_L d_S(y)$. By Theorem 4.85(a) we know that there are $x_n \to_S x$, $y_n \to_S y$, $(x_n^*)_n$ and $(y_n^*)_n$ converging weakly* to x^* and y^* respectively, with $x_n^* \in \partial_F d_S(x_n)$ and $y_n^* \in \partial_F d_S(y_n)$. For n large enough we have $x_n, y_n \in S \cap B(\overline{x}, \delta)$, and hence by the above inequality

$$\langle x_n^* - y_n^*, x_n - y_n \rangle \geq -\varepsilon \|x_n - y_n\|.$$

Passing to the limit as $n \to \infty$, it ensues that

$$\langle x^* - y^*, x - y \rangle \geq -\varepsilon \|x - y\|,$$

which corresponds to (b).

To finish the proof, it remains to prove that (b)$\Rightarrow$(c). Note that for each real $\varepsilon > 0$ and for any fixed $x, y \in S \cap B(\overline{x}, \delta)$, the set of $(x^*, y^*) \in X^* \times X^*$ satisfying the inequality $\langle x^* - y^*, x - y \rangle \geq -\varepsilon \|x - y\|$ is convex and weakly* closed in $X^* \times X^*$. The result in (c) then follows from the equality $\partial_C d_S(x) = \overline{\mathrm{co}}^* \big(\partial_L d_S(x)\big)$ (see Corollary 4.122). $\square$

We are now able to characterize local subsmoothness of sets via the Fréchet normal cone in Asplund space.

PROPOSITION 8.77. Let S be a nonempty closed set of an Asplund space X and U be a nonempty open set of X with $U \cap S \neq \emptyset$. The following are equivalent:
(a) the set S is subsmooth at each point of $S \cap U$;
(b) the multimapping $N^L(S \cdot) \cap \mathbb{B}_{X^*}$ is submonotone at each point of $S \cap U$;
(c) The multimapping $N^F(S \cdot) \cap \mathbb{B}_{X^*}$ is submonotone at each point of $S \cap U$.

PROOF. First, we note that the implications (a)$\Rightarrow$(b) and (b)$\Rightarrow$(c) follow directly from the inclusions

$$N^F(S; x) \cap \mathbb{B}_{X^*} \subset N^L(S; x) \cap \mathbb{B}_{X^*} \subset N^C(S; x) \cap \mathbb{B}_{X^*}, \text{ for all } x \in S.$$

On the other hand, (c) implies (a) by Proposition 8.75(A) according to the equality $\partial_F d_S(x) = N^F(S; \cdot) \cap \mathbb{B}_{X^*}$ for any $x \in S$ and to the implication (a)$\Rightarrow$(c) in Lemma 8.76, as easily seen. $\square$

8.3.2. Subsmoothness of sets versus Shapiro property. By Proposition 8.73 we know that a set S is tangentially regular at a point in S whenever it is one-sided subsmooth at this point. Let us now compare the notions of subsmoothness and one-sided subsmoothness of sets with other concepts.

DEFINITION 8.78. Let S be a nonempty subset of a normed space X and let $\overline{x} \in S$.
(a) The set S is said to satisfy the *Shapiro k-order contact property* ($k \in \mathbb{N}$) at $\overline{x}$, if for every real $\varepsilon > 0$ there exists a real $\delta > 0$ such that for all $x_1, x_2 \in S \cap B(\overline{x}, \delta)$ one has

$$\operatorname{dist}(x_2 - x_1, T^B(S; x_1)) \leq \varepsilon \|x_1 - x_2\|^k.$$

(b) The set S is called *nearly radial* at $\overline{x}$ if for every real $\varepsilon > 0$ there exists a real $\delta > 0$ such that for all $x \in S \cap B(\overline{x}, \delta)$ one has

$$\operatorname{dist}(\overline{x} - x, T^B(S; x)) \leq \varepsilon \|x - \overline{x}\|,$$

(that is, the inequality in (a) holds for $k = 1$ with $x_2 = \overline{x}$).

Given a nonempty convex cone Q in a normed space $(X, \|\cdot\|)$, we have seen in Lemma 2.249 that for any $u \in X$

$$d(u, Q) = \max_{x^* \in Q^\circ \cap \mathbb{B}_{X^*}} \langle x^*, u \rangle, \tag{8.26}$$

where $Q^\circ := \{x^* \in X^* : \langle x^*, x \rangle \leq 0, \ \forall x \in Q\}$.

REMARK 8.79. One can also show that the effective domain of the support function $\sigma(\cdot, Q)$ is $\operatorname{dom} \sigma(\cdot, Q) = Q^\circ$ and apply the equality

$$d(x, S) = \sup_{x^* \in \mathbb{B}_{X^*}} \big(\langle x^*, x \rangle - \sigma(x^*, S)\big)$$

established in Proposition 2.180(a). In this way, the above equality (8.26) can also be obtained from Proposition 2.180. □

Before proving the next theorem we also need a lemma related to the distance to the Bouligand-Peano tangent cone.

LEMMA 8.80. Let S be a subset of a finite-dimensional normed space X and let $\overline{x} \in S$. Then one has

$$\lim_{S \ni x \to \overline{x}} \operatorname{dist}\left(\frac{x - \overline{x}}{\|x - \overline{x}\|}, T^B(S; \overline{x})\right) = 0.$$

PROOF. Suppose that the equality in the lemma is not true. Then there exists a real $\varepsilon > 0$ and a sequence $(x_n)_n$ in $S \setminus \{\overline{x}\}$ converging to $\overline{x}$ such that for all $n \in \mathbb{N}$

$$\operatorname{dist}\left(\frac{x_n - \overline{x}}{\|x_n - \overline{x}\|}, T^B(S; \overline{x})\right) \geq \varepsilon.$$

Since X is finite-dimensional, for some subsequence $(x'_n)_n$ of $(x_n)_n$ there is some $h \in X$ such that $h_n := \|x'_n - \overline{x}\|^{-1}(x'_n - \overline{x}) \to h$ as $n \to \infty$, so (as easily seen) $h \in T^B(S; \overline{x})$. Passing to the limit as $n \to \infty$ in the inequality $\operatorname{dist}(h_n, T^B(S; \overline{x})) \geq \varepsilon$, we obtain

$$\operatorname{dist}(h, T^B(S; \overline{x})) \geq \varepsilon,$$

which contradicts the inclusion $h \in T^B(S; \overline{x})$. □

THEOREM 8.81 (local subsmoothness versus Shapiro property). Let S be a nonempty set of a normed space X and U be a nonempty open set of X with $U \cap S \neq \emptyset$. The following hold:
(a) The set S is subsmooth at each point in $S \cap U$ if and only if it is tangentially regular at each point in $S \cap U$ and satisfies the Shapiro first order contact property at each point in $S \cap U$.
(b) If X is an Asplund space and S is closed, then S is subsmooth at each point in $S \cap U$ if and only if it satisfies the Shapiro first order contact property at each point in $S \cap U$.
(c) If S is one-sided subsmooth at each point in $S \cap U$, then it is tangentially regular at each point in $S \cap U$ and nearly radial at each point in $S \cap U$. The converse also holds whenever X is finite-dimensional.

PROOF. (a) To show the "necessity"part, let us suppose that S is subsmooth at each point in $S \cap U$. By Proposition 8.73 we know that S is tangentially regular at each point in $S \cap U$. Fix any $\overline{x} \in S \cap U$ and let us prove that the Shapiro first order contact property is satisfied at $\overline{x}$. Consider any real $\varepsilon > 0$. By definition of subsmoothness of sets there exists $\delta > 0$ such that

$$\langle x_1^* - x_2^*, x_1 - x_2 \rangle \geq -\varepsilon \|x_1 - x_2\|,$$

for all $x_i \in S \cap B(\overline{x}, \delta)$ and all $x_i^* \in N^C(S; x_i) \cap \mathbb{B}_{X^*}$ for $i = 1, 2$. Fixing $x_1, x_2 \in S \cap B(\overline{x}, \delta)$ and taking $x_2^* = 0$, we obtain

$$\sup_{x_1^* \in N^C(S;x_1) \cap \mathbb{B}_{X^*}} \langle x_1^*, x_2 - x_1 \rangle \leq \varepsilon \|x_1 - x_2\|.$$

Further, from the tangential regularity we have $T^B(S; x_1) = T^C(S; x_1)$. Then, keeping in mind that $N^C(S; x_1)$ is the polar cone of the closed convex cone $T^C(S, x_1)$, the latter inequality combined with Lemma 2.249 (recalled in (8.26)) yields

$$\operatorname{dist}\left(x_2 - x_1, T^B(S; x_1)\right) \leq \varepsilon \|x_1 - x_2\|,$$

which translates the Shapiro first order contact property for S at $\overline{x}$.

Conversely, to show the "sufficiently"part, let us assume that S is tangentially regular at each point in $S \cap U$ and satisfies the Shapiro first order contact property at each point in $S \cap U$. Fix any $\overline{x} \in S \cap U$ and let us show that S is subsmooth at $\overline{x}$. Consider any real $\varepsilon > 0$. By definition of Shapiro first order property there exists a real $\delta > 0$ with $B(\overline{x}, \delta) \subset U$ such that for all $x_i \in S \cap B(\overline{x}, \delta)$, $i = 1, 2$,

$$\max\{\operatorname{dist}\left(x_2 - x_1, T^B(S; x_1)\right), \operatorname{dist}\left(x_1 - x_2, T^B(S; x_2)\right)\} \leq \frac{\varepsilon}{2} \|x_1 - x_2\|. \tag{8.27}$$

Since $T^B(S; x_i) = T^C(S; x_i)$ for $i = 1, 2$, by virtue of Lemma 2.249 we deduce that

$$\max \left\{ \sup_{x_1^* \in N^C(S;x_1) \cap \mathbb{B}} \langle x_1^*, x_2 - x_1 \rangle, \sup_{x_2^* \in N^C(S;x_2) \cap \mathbb{B}} \langle x_2^*, x_1 - x_2 \rangle \right\} \leq \frac{\varepsilon}{2} \|x_1 - x_2\|.$$

This entails that, for all $x_i^* \in N^C(S; x_i) \cap \mathbb{B}_{X^*}$ with $x_i \in S \cap B(\overline{x}, \delta)$

$$\langle x_1^* - x_2^*, x_1 - x_2 \rangle \geq -\varepsilon \|x_1 - x_2\|,$$

hence S is subsmooth at $\overline{x}$.
(b) Assume that X is an Asplund space. According to (a) we only have to show the "sufficiency"part. So, let us suppose that S satisfies the Shapiro first order contact property at each point in $S \cap U$ and let us show that S is subsmooth at each point in $S \cap U$. By virtue of the implication (c)$\Rightarrow$(a) in Proposition 8.77,

it is enough to show that the truncated Fréchet normal cone $N^F(S;\cdot)\cap\mathbb{B}_{X^*}$ is submonotone at each point in $S\cap U$. Fix any $\overline{x}\in S\cap U$ and any real $\varepsilon>0$. As in the proof of "sufficiency"part of (a), there is a real $\delta>0$ with $B(\overline{x},\delta)\subset U$ such that (8.27) holds with $\varepsilon/3$ in place of $\varepsilon/2$. Consider any $x_i\in S\cap B(\overline{x},\delta)$ and any $x_i^*\in N^F(S;x_i)\cap\mathbb{B}_{X^*}$, $i=1,2$. From the latter inequality in (8.27) with $\varepsilon/3$ (since its first member is strictly less than $(\varepsilon/2)\|x_1-x_2\|$ for $x_1\neq x_2$) we may choose $v_i\in T^B(S;x_i)$ and $e_i\in X$ (for $i=1,2$) such that

$$x_2-x_1=v_1+e_1 \quad\text{and}\quad x_1-x_2=v_2+e_2,$$

with $\|e_i\|\le(\varepsilon/2)\|x_1-x_2\|$. Note that $\langle x_i^*,v_i\rangle\le 0$ since the Fréchet normal cone $N^F(S;x)$ is always included in the (negative) polar cone of $T^B(S;x)$ for $x\in S$. Using this and the inequality $\|x_i^*\|\le 1$ (for $i=1,2$), it follows that

$$\langle x_1^*,x_1-x_2\rangle\ge-\frac{\varepsilon}{2}\|x_1-x_2\| \quad\text{and}\quad \langle x_2^*,x_2-x_1\rangle\ge-\frac{\varepsilon}{2}\|x_1-x_2\|.$$

Adding these inequalities give

$$\langle x_1^*-x_2^*,x_1-x_2\rangle\ge-\varepsilon\|x_1-x_2\|,$$

which is the desired submonotonicity of $N^F(S;\cdot)\cap\mathbb{B}_{X^*}$.

(c) Assume that S is one-sided subsmooth at each point in $S\cap U$. By Proposition 8.73 the set S is tangentially regular at each point in $S\cap U$. To show that S is nearly radial at any point $\overline{x}\in S\cap U$, it suffices to proceed in the same way as in the above proof of the "necessity"part of (a), setting $x_2=\overline{x}$ (and $x_1=x$) to conclude that the property in Definition 8.78 holds true.

Suppose now that X is finite-dimensional and that S is tangentially regular at each point in $S\cap U$ and nearly radial at each point in $S\cap U$. Let us fix any $\overline{x}\in S\cap U$ and show that S is one-sided subsmooth at $\overline{x}$. Take any real $\varepsilon>0$. By Definition 8.78(b) and by tangential regularity of S at points in $S\cap U$ there is a real $\delta'>0$ with $B(\overline{x},\delta')\subset U$ such that for all $x\in S\cap B(\overline{x},\delta')$

$$\operatorname{dist}\left(\overline{x}-x,T^C(S;x)\right)\le\frac{\varepsilon}{2}\|x-\overline{x}\|.$$

Since $N^C(S;x)$ is the polar of the convex cone $T^C(S;x)$, Lemma 2.249 gives for all $x\in S\cap B(\overline{x},\delta')$

$$\sup_{x^*\in N^C(S;x)\cap\mathbb{B}}\langle x^*,\overline{x}-x\rangle\le\frac{\varepsilon}{2}\|x-\overline{x}\|. \tag{8.28}$$

On the other hand, by Lemma 8.80 there exists $\delta\in]0,\delta'[$ such that for all $x\in S\cap B(\overline{x},\delta)$ with $x\neq\overline{x}$

$$\operatorname{dist}\left(\frac{x-\overline{x}}{\|x-\overline{x}\|},T^B(S;\overline{x})\right)\le\frac{\varepsilon}{2}.$$

Since $T^B(S;\overline{x})=T^C(S:\overline{x})$, using again Lemma 2.249 we obtain

$$\sup_{u^*\in N^C(S;\overline{x})\cap\mathbb{B}}\langle u^*,x-\overline{x}\rangle\le\frac{\varepsilon}{2}\|x-\overline{x}\| \quad\text{for all } x\in S\cap B(\overline{x},\delta).$$

From this and (8.28) we see that

$$\langle x^*-u^*,x-\overline{x}\rangle\ge-\varepsilon\|x-\overline{x}\|$$

for all $x\in S\cap B(\overline{x},\delta)$, all $x^*\in N^C(S;x)$ and all $u^*\in N^C(S;\overline{x})$. This means that S is one-sided subsmooth at $\overline{x}$, and the proof is finished. □

The proofs of (a) and (c) in the above theorem also work for the subsmoothness (resp. one-sided subsmoothness) property at a fixed point $\overline{x}$ provided that S is tangentially regular near $\overline{x}$. We state this in the proposition:

PROPOSITION 8.82. Let S be a set of a normed space X which is tangentially regular near a point $\overline{x} \in S$. The following hold:
(a) The set S is subsmooth at $\overline{x}$ if and only if it satisfies the Shapiro first order contact property at $\overline{x}$.
(b) If S is one-sided subsmooth at $\overline{x}$, then it is nearly radial at $\overline{x}$. The converse also holds whenever X is finite-dimensional.

8.4. Epi-Lipschitz subsmooth sets

In this section we will show that epi-Lipschitz subsmooth sets can be seen as epigraphs of Lipschitz subsmooth functions.

Let S be a subset of the normed space X which is closed near $\overline{x} \in \operatorname{bdry} S$ and has the interior tangent property at this point in a direction $h \in X$ with $h \neq 0$. By Theorem 2.20 there exists a topologically complemented closed vector hyperplane E of $\mathbb{R}h$ (so, $X = E \oplus \mathbb{R}h$), an open neighborhood W of $\overline{x}$ in X, and a function $f : E \to \mathbb{R}$ Lipschitz continuous near $\pi_E\overline{x}$ (with E endowed with the induced norm and $x = \pi_E x + (\pi_h x)h$ with $\pi_E x \in E$ and $\pi_h x \in \mathbb{R}$) such that

$$W \cap S = W \cap \{u + rh : u \in E, r \in \mathbb{R}, f(u) \leq r\}.$$

We recall that such a function f is a *locally Lipschitz representative of S around $\overline{x}$*. Consider the linear isomorphism $A : E \times \mathbb{R} \to X$ defined by $A(u, s) := u \oplus sh$, so that $\Lambda := A^{-1} : X \to E \times \mathbb{R}$ is given $\Lambda(x) = (\pi_E x, \pi_h x)$ and $W \cap S$ can be rewritten as

$$W \cap S = W \cap A(\operatorname{epi} f).$$

We endow $E \times \mathbb{R}$ with the product norm $(u, s) \mapsto (\|u\|^2 + s^2)^{1/2}$ and we define $F : E \to X$ by $F(u) := u \oplus f(u)h$, so that $F(u) = A(u, f(u))$. If $\gamma > 0$ denotes a Lipschitz constant of f in an open neighborhood V of $\overline{u} := \pi_E\overline{x}$ with $F(V) \subset W$, then it is easily seen that

$$\|F(u_1) - F(u_2)\| \leq \alpha\|u_1 - u_2\| \quad \text{for all } u_1, u_2 \in V, \tag{8.29}$$

where

$$\alpha := \|A\|(\gamma^2 + 1)^{1/2}. \tag{8.30}$$

From the equality $W \cap S = W \cap \Lambda^{-1}(\operatorname{epi} f)$ we also note by the isomorphism property of Λ that, for all $u \in V \cap F^{-1}(S)$

$$T^C(S; F(u)) = \Lambda^{-1}\big(T^C(\operatorname{epi} f; (u, f(u)))\big)$$

and

$$N^C(S; F(u)) = \Lambda^*\big(N^C(\operatorname{epi} f; (u, f(u)))\big). \tag{8.31}$$

LEMMA 8.83. Let X be a normed space and S be a subset which is closed near a point $\overline{x} \in \operatorname{bdry} S$ and epi-Lipschitz at $\overline{x}$ in a direction $h \neq 0$. Assume that $N^C(S; \cdot) \cap \mathbb{B}_{X^*}$ is submonotone (resp. one-sided submonotone) at $\overline{x}$. Then, for every locally Lipschitz representative $f : E \to \mathbb{R}$ of S around $\overline{x}$ (where E is a topological vector complement of $\mathbb{R}h$ in X), the Clarke subdifferential $\partial_C f$ is submonotone (resp. one-sided submonotone) at $\pi_E\overline{x}$.

PROOF. Keep the above notation and let V be an open neighborhood of $\overline{u} := \pi_E\overline{x}$ over which f is Lipschitz with constant $\gamma > 0$ and such that $F(V) \subset W$.

Let us first establish the result for submonotonicity. From the submonotonicity of $N^C(S;\cdot)\cap\mathbb{B}_{X^*}$ at $\overline{x}$ it is easily seen that $N^C(S;\cdot)\cap r\mathbb{B}_{X^*}$ is submonotone at $\overline{x}$ for any real $r > 0$. Take $r := \|\Lambda^*\|(\gamma^2+1)^{1/2}$, where as above Λ^* denotes the adjoint of $\Lambda := A^{-1}$. In order to prove that $\partial_C f$ is submonotone at $\overline{u} := \pi_E\overline{x}$, take any real $\varepsilon > 0$ and set $\varepsilon_1 := \alpha^{-1}\varepsilon$, where α is given by (8.30). There exists a real $\delta > 0$ with $B(\overline{x},\delta) \subset W$ such that for all $x_i \in B(\overline{x},\delta)$ and all $x_i^* \in N^C(S;x_i)\cap r\mathbb{B}_{X^*}$, $i = 1,2$,

$$\langle x_1^* - x_2^*, x_1 - x_2\rangle \geq -\varepsilon_1\|x_1 - x_2\|. \tag{8.32}$$

Choose a positive real $\delta_1 \leq \alpha^{-1}\delta$ such that $B(\overline{u},\delta_1) \subset V$. Then by (8.29) we have $F(B(\overline{u},\delta_1)) \subset B(\overline{x},\delta)$. Take any $u_1, u_2 \in B(\overline{u},\delta_1)$ and any $u_i^* \in \partial_C f(u_i)$ for $i = 1,2$. Then $(u_i^*,-1) \in N^C\big(\operatorname{epi} f;(u_i,f(u_i))\big)$. Using the equality $W\cap S = W\cap\Lambda^{-1}(\operatorname{epi} f)$ and the points

$$x_i := F(u_i) = A(u_i,f(u_i)) \in B(\overline{x},\delta),\ \ i = 1,2,$$

we obtain $x_i^* := \Lambda^*((u_i^*,-1)) \in N^C(S;x_i)$, $i = 1,2$. Since $\|u_i^*\| \leq \gamma$ for $i = 1,2$ (according to the γ-Lipschitz property of f on V), it ensues that

$$\|x_i^*\| \leq \|\Lambda^*\|(\|u_i^*\|^2+1)^{1/2} \leq r, \quad \text{so } x_i^* \in N^C(S;x_i)\cap r\mathbb{B}_{X^*}.$$

This combined with (8.32) yields

$$\langle\Lambda^*(u_1^*,-1) - \Lambda^*(u_2^*,-1), F(u_1) - F(u_2)\rangle \geq -\varepsilon\alpha^{-1}\|F(u_1) - F(u_2)\|.$$

Since

$$\begin{aligned}\langle\Lambda^*(u_1^*,-1) - \Lambda^*(u_2^*,-1), F(u_1) - F(u_2)\rangle &= \langle(u_1^* - u_2^*,0),(u_1,f(u_1)) - (u_2,f(u_2))\rangle\\ &= \langle u_1^* - u_2^*, u_1 - u_2\rangle,\end{aligned}$$

the above inequality and (8.29) yield

$$\langle u_1^* - u_2^*, u_1 - u_2\rangle \geq -\varepsilon\|u_1 - u_2\|.$$

This justifies the submonotonicity property of $\partial_C f$ at $\overline{u}$.

The case of one-sided submonotonicity property is obtained by replacing u_2 by $\overline{u}$ and $u_2^* \in \partial_C f(u_2)$ by $\overline{u}^* \in \partial_C f(\overline{u})$ in the above arguments. □

A similar lemma holds for uniform submonotonicity near a point.

LEMMA 8.84. Let X be a normed space and S be a subset which is closed near a point $\overline{x} \in \operatorname{bdry} S$ and epi-Lipschitz at $\overline{x}$ in a direction $h \neq 0$. Assume that $N^C(S;\cdot)\cap\mathbb{B}_{X^*}$ is uniformly submonotone near $\overline{x}$. Then, for every locally Lipschitz representative $f : E \to \mathbb{R}$ of S around $\overline{x}$ (where E is a topological vector complement of $\mathbb{R}h$ in X), the Clarke subdifferential $\partial_C f$ is uniformly submonotone near $\pi_E\overline{x}$.

PROOF. Again keep the above notation with $W\cap S = W\cap A(\operatorname{epi} f)$ and let V be an open neighborhood of $\overline{u} := \pi_E\overline{x}$ over which f is Lipschitz with constant $\gamma > 0$ and such that $F(V) \subset W$. Let $W_0 \subset W$ be an open convex neighborhood of $\overline{x}$ over which the multimapping $N^C(S;\cdot)\cap\mathbb{B}_{X^*}$ is uniformly submonotone and let $V_0 \subset V$ be an open convex neighborhood of $\overline{u}$ such that $F(V_0) \subset W_0$. Recall that $\alpha := \|A\|(\gamma^2+1)^{1/2}$ (see (8.30)), and as in the previous lemma set $r := \|\Lambda^*\|(\gamma^2+1)^{1/2}$. Note that the multimapping $N^C(S;\cdot)\cap r\mathbb{B}_{X^*}$ is uniformly submonotone on W_0. Let us show that $\partial_C f(\cdot)$ is uniformly submonotone on V_0. Let any real $\varepsilon > 0$. There exists a real $\delta > 0$ such that

$$(8.33) \qquad \langle x_1^* - x_2^*, x_1 - x_2\rangle \geq -\alpha^{-1}\varepsilon\|x_1 - x_2\|$$

for all $x_1, x_2 \in W_0 \cap S$ with $\|x_1 - x_2\| < \delta$ and all $x_i^* \in N^C(S;\cdot) \cap r\mathbb{B}_{X^*}$. Set $\delta_0 := \alpha^{-1}\delta$ and take any $u_1, u_2 \in V_0$ with $\|u_1 - u_2\| < \delta_0$ and any $u_i^* \in \partial_C f(u_i)$, $i = 1, 2$, so $(u_i^*, -1) \in N^C(\operatorname{epi} f; (u_i, f(u_i)))$. Putting $x_i := F(u_i) = A(u_i, f(u_i))$, we see that $x_i \in W_0 \cap \Lambda^{-1}(\operatorname{epi} f)$. Observing that $W_0 \cap S = W_0 \cap \Lambda^{-1}(\operatorname{epi} f)$ (since $W_0 \subset W$), it ensues that $x_i \in W_0 \cap S$ and $x_i^* := \Lambda^*(u^*, -1) \in N^C(S, x_i)$. Further, the inequality $\|u_i^*\| \leq \gamma$ gives

$$\|x_i^*\| \leq \|\Lambda^*\|(\|u_i^*\|^2 + 1)^{1/2} \leq r.$$

The mapping F being α-Lipschitz on V we also observe that

$$\|x_1 - x_2\| = \|F(u_1) - F(u_2)\| \leq \alpha\|u_1 - u_2\| < \delta.$$

We deduce from (8.33) that

$$\langle \Lambda^*(u_1^*, -1) - \Lambda^*(u_2^*, -1), F(u_1) - F(u_2)\rangle \geq -\varepsilon\alpha^{-1}\|F(u_1) - F(u_2)\|.$$

Writing

$$\begin{aligned}\langle \Lambda^*(u_1^*, -1) - \Lambda^*(u_2^*, -1), F(u_1) - F(u_2)\rangle &= \langle (u_1^* - u_2^*, 0), (u_1, f(u_1)) - (u_2, f(u_2))\rangle \\ &= \langle u_1^* - u_2^*, u_1 - u_2\rangle,\end{aligned}$$

and using (8.29) it results that

$$\langle u_1^* - u_2^*, u_1 - u_2\rangle \geq -\varepsilon\|u_1 - u_2\|,$$

which confirms the uniform submonotonicity property of $\partial_C f$ on V_0. □

THEOREM 8.85 (subsmooth functional representation of epi-Lipschitz subsmooth set). Let S be a subset of a normed space X which is closed near $\overline{x} \in \operatorname{bdry} S$ and epi-Lipschitz at $\overline{x}$ in a direction $h \neq 0$. The following are equivalent:
(a) the set S is subsmooth at $\overline{x}$ (resp. uniformly subsmooth near $\overline{x}$);
(b) every locally Lipschitz representative $f : E \to \mathbb{R}$ of S is subsmooth at $\pi_E\overline{x}$ (resp. uniformly subsmooth near $\pi_E\overline{x}$), where E is a topological vector complement of $\mathbb{R}h$ in X;
(c) some locally Lipschitz representative $f : E \to \mathbb{R}$ of S is subsmooth at $\pi_E\overline{x}$ (resp. uniformly subsmooth near $\pi_E\overline{x}$).

PROOF. The implication (b)⇒(c) is evident and the implication (a)⇒(b) follows from Lemma 8.83 and from Proposition 8.41 (resp. Lemma 8.84 and Proposition 8.42). It remains to prove (c)⇒(a). Suppose that f is subsmooth at $\overline{u} := \pi_E\overline{x}$ (resp. uniformly subsmooth near $\overline{u}$). Let V be an open neighborhood of $\overline{u}$ and W an open neighborhood of $\overline{x}$ as in what precedes Lemma 8.83 and such that f is Lipschitz with constant $\gamma > 0$ on V and $F(V) \subset W$ (and in the case of uniform subsmoothness near $\overline{u}$, choose open convex neighborhoods $V_0 \subset V$ and $W_0 \subset W$ of $\overline{u}$ and $\overline{x}$ respectively with $F(V_0) \subset W_0$ such that f is uniformly subsmooth on V_0, and choose also an open convex neighborhood $W_0' \subset W_0$ of $\overline{x}$ such that $\pi_E(W_0') \subset V_0$). Take any real $\varepsilon > 0$ and set

$$\varepsilon_1 := \frac{\varepsilon}{2\|A^*\| \cdot \|\Lambda\|}.$$

According to Theorem 8.25 or Proposition 8.41 (resp. Proposition 8.42), there exists a real $\delta_1 > 0$ with $B(\overline{x}, \delta_1) \subset V$ such that (resp. a real $\delta_1 > 0$ such that)

$$f(u_2) \geq f(u_1) + \langle u_1^*, u_2 - u_1 \rangle - \varepsilon_1 \|u_2 - u_1\|, \tag{8.34}$$

for all $u_1, u_2 \in B(\overline{u}, \delta_1)$ (resp. $u_1, u_2 \in V_0$ with $\|u_1 - u_2\| < \delta_1$) and all $u_1^* \in \partial_C f(u_1)$. Choose a positive real $\delta \leq \delta_1/\|\Lambda\|$ so that $B(\overline{x}, \delta) \subset W$ (resp. a positive real $\delta \leq \delta_1/\|\Lambda\|$). Take any $x_i \in S \cap B(\overline{x}, \delta)$ (resp. $x_i \in S \cap W_0'$ with $\|x_1 - x_2\| < \delta$) and any $x_i^* \in N^C(S; x_i) \cap \mathbb{B}_{X^*}$, $i = 1, 2$. We claim that

$$\langle x_1^*, x_2 - x_1 \rangle \leq \frac{\varepsilon}{2} \|x_1 - x_2\|.$$

To this end, put $u_i := \pi_E x_i$ and $t_i := \pi_h x_i$, so that $x_i = A(u_i, t_i)$ and $t_i \geq f(u_i)$, $i = 1, 2$. If $t_1 > f(u_1)$, then $x_1 \in \operatorname{int} S$, and hence $N^C(S; x_1) = \{0\}$, so the inequality of the claim holds trivially. We may then suppose that $t_1 = f(u_1)$ and $x_1^* \neq 0$. Note that $x_1^* \in \Lambda^*\big(N^C(\operatorname{epi} f; (x_1, f(x_1)))\big)$ according to the isomorphic property of Λ. There exists by definition of C-subdifferential a real $\lambda > 0$ and $u_1^* \in \partial_C f(u_1)$ such that

$$x_1^* = \lambda \Lambda^*(u_1^*, -1). \tag{8.35}$$

Then $\lambda(u_1^*, -1) = A^* x_1^*$, and hence $\lambda \leq \|A^*\|$ since $\|x_1^*\| \leq 1$. Further,

$$\|u_1 - u_2\| \leq \|\Lambda\| \cdot \|x_1 - x_2\|,$$

so by the choice of δ we see that $u_1, u_2 \in B(\overline{u}, \delta_1)$ (resp. $\|u_1 - u_2\| < \delta_1$ and $u_1, u_2 \in V_0$). Therefore, (8.34) along with the inequality $t_2 \geq f(u_2)$ gives

$$\langle (u_1^*, -1), (u_2, t_2) - (u_1, f(u_1)) \rangle \leq \varepsilon_1 \|u_1 - u_2\| \leq \varepsilon_1 \|\Lambda\| \cdot \|x_1 - x_2\|.$$

It ensues that

$$\langle \lambda(u_1^*, -1), \Lambda(x_2) - \Lambda(x_1) \rangle \leq \varepsilon_1 \|A^*\| \cdot \|\Lambda\| \cdot \|x_1 - x_2\| = \frac{\varepsilon}{2} \|x_1 - x_2\|,$$

which by (8.35) yields $\langle x_1^*, x_2 - x_1 \rangle \leq (\varepsilon/2)\|x_1 - x_2\|$ as stated in the claim. By symmetry we also have $\langle x_2^*, x_1 - x_2 \rangle \leq (\varepsilon/2)\|x_2 - x_1\|$. Adding the two latter inequalities we obtain

$$\langle x_1^* - x_2^*, x_1 - x_2 \rangle \geq -\varepsilon \|x_1 - x_2\|,$$

and this translates the subsmoothness of the set S at $\overline{x}$ (resp. the uniform subsmoothness of $S \cap W_0'$). □

As a direct consequence of the above theorem and Proposition 8.41 we have the following corollary.

COROLLARY 8.86. *Let X be a normed space and $f : X \to \mathbb{R}$ be a function which is Lipschitz near $\overline{x} \in X$. The following are equivalent:*
(a) the epigraphical set $\operatorname{epi} f$ is subsmooth at $(\overline{x}, f(\overline{x}))$ (resp. uniformly subsmooth near $(\overline{x}, f(\overline{x}))$);
(b) the function f is subsmooth at $\overline{x}$ (resp. uniformly subsmooth near $\overline{x}$);
(c) the multimapping $\partial_C f$ is submonotone at $\overline{x}$ (resp. uniformly submonotone near $\overline{x}$).

Through Corollary 8.86 we provide an epi-Lipschitz set in $\mathbb{R}^2$ which is tangentially regular at a point and fails to be subsmooth at that point.

EXAMPLE 8.87. Example 8.49 furnished a Lipschitz function $f : \mathbb{R} \to \mathbb{R}$ which is tangentially regular at $\overline{x}$ (even at any $x \in \mathbb{R}$) such that the multimapping $\partial_C f$ is not one-sided submonotone at $\overline{x} = 0$, hence in particular f is not subsmooth at $\overline{x}$ by Proposition 8.41. Then the epi-Lipschitz set $\operatorname{epi} f$ is tangentially regular at any of its points but not subsmooth at $(\overline{x}, f(\overline{x}))$ according to Corollary 8.86. □

PROPOSITION 8.88. Let S be a subset of a normed space X which is closed near $\overline{x} \in \operatorname{bdry} S$ and epi-Lipschitz at $\overline{x}$. Then S is uniformly subsmooth near $\overline{x}$ if and only if there exist a nonzero vector $h \in X$, a topological complement vector subspace E with $X = E \oplus \mathbb{R}h$, an open neighborhood W of $\overline{x}$ and a modulus function $\omega : [0, +\infty[\to [0, +\infty[$ of class $\mathcal{C}^1$ on $]0, +\infty[$ with $t\,\omega'(t) \to 0$ as $t \downarrow 0$ such that

$$\langle x^*, y - x \rangle \le \|\pi_E y - \pi_E x\| \, \omega(\|\pi_E y - \pi_E x\|)$$

for all $x, y \in S \cap W$ and all $x^* \in N^C(S; x)$.

If in addition X is finite-dimensional, then the latter property is also equivalent to the subsmoothness of the set S near $\overline{x}$.

PROOF. First let us suppose that the property in the proposition holds. We may suppose that W is convex. Fix any $\varepsilon > 0$. Since ω is continuous at 0 with $\omega(0) = 0$, there is $\eta > 0$ such that $\omega(t) \le \varepsilon/(1 + \|\pi_E\|)$ for all $t \in [0, \eta]$. Choose $\delta > 0$ such that $\|\pi_E z\| \le \eta$ whenever $\|z\| \le \delta$. Take any $x, y \in S \cap W$ with $\|y - x\| < \delta$ and any $x^* \in N^C(S; x) \cap \mathbb{B}_{X^*}$. We have

$$\begin{aligned}\langle x^*, y - x \rangle &\le \|\pi_E y - \pi_E x\| \, \omega(\|\pi_E y - \pi_E x\|) = \|\pi_E(y - x)\| \, \omega(\|\pi_E(y - x)\|) \\ &\le \frac{\varepsilon}{1 + \|\pi_E\|} \|\pi_E(y - x)\| \le \varepsilon \|y - x\|,\end{aligned}$$

which tells us that the set S is uniformly subsmooth near the point $\overline{x}$ and justifies the implication $\Leftarrow$.

To prove the converse implication $\Rightarrow$, let us assume that S is epi-Lipschitz and uniformly subsmooth near $\overline{x}$. Choose a nonzero vector $h \in X$, a complement vector subspace E with $X = E \oplus \mathbb{R}h$, a Lipschitz function $f : E \to \mathbb{R}$ and an open neighborhood W of $\overline{x}$ such that

$$W \cap S = W \cap A(\operatorname{epi} f),$$

where $A : E \times \mathbb{R} \to X$ is given by $A(u, r) = u \oplus rh$ as stated in the beginning of this section. By Theorem 8.85 the function f is uniformly subsmooth near $\pi_E \overline{x}$, so by Proposition 8.44 there exist an open neighborhood V of $\pi_E \overline{x}$ and a modulus function $\omega : [0, +\infty[\to [0, +\infty[$ of class $\mathcal{C}^1$ on $]0, +\infty[$ with $t\,\omega'(t) \to 0$ as $t \downarrow 0$ such that for all $u, v \in V$, $u^* \in \partial f(u)$ and $r \ge f(v)$

$$r \ge f(u) + \langle u^*, v - u \rangle - \|v - u\| \, \omega(\|v - u\|),$$

or equivalently

$$\langle (u^*, -1), (v, r) - (u, f(u)) \rangle \le \|v - u\| \, \omega(\|v - u\|). \tag{8.36}$$

Let $W_0 \subset W$ be an open neighborhood of $\overline{x}$ such that $\pi_E(W_0) \subset V$. Let us show the inequality in the proposition with $\omega_0(\cdot) := \|A\| \, \omega(\cdot)$. Consider any $x, y \in S \cap W_0$ and any $x^* \in N^C(S; x) \cap \mathbb{B}_{X^*}$. Set $u := \pi_E x$ and $v := \pi_E y$. If $\pi_h x > f(u)$, then $x \in \operatorname{int} S$, which entails that $x^* = 0$, so the inequality in the proposition is satisfied since $\omega_0(\cdot) \ge 0$. Suppose that $\pi_h x = f(u)$. We may also suppose that $x^* \neq 0$, for otherwise the desired inequality is trivial. Since A is an isomorphism and f is Lipschitz, by definition of C-subdifferential there is some real $t > 0$ and $u^* \in \partial_C f(u)$ such that $A^*(x^*) = t(u^*, -1)$, or equivalently $t(u^*, -1) = x^* \circ A$. This gives $t\left(\|u^*\|^2 + 1\right)^{1/2} \le \|A\|$, hence $t \le \|A\|$. Further, by (8.36) we have

$$t^{-1}\langle A^*(x^*), (v, \pi_h y) - (u, f(u))\rangle \le \|\pi_E y - \pi_E x\| \, \omega(\|\pi_E y - \pi_E x\|),$$

which combined with the above inequality $t \le \|A\|$ yields

$$\langle x^*, A(\pi_E y, \pi_h y) - A(\pi_E x, \pi_h x)\rangle \le \|A\| \, \|\pi_E y - \pi_E x\| \, \omega(\|\pi_E y - \pi_E x\|),$$

and this means with $\omega_0(\cdot) := \|A\| \, \omega(\cdot)$ as above

$$\langle x^*, y - x\rangle \le \|\pi_E y - \pi_E x\| \, \omega_0(\|\pi_E y - \pi_E x\|).$$

This justifies the desired implication.

Finally, the situation when X is finite-dimensional follows from what precedes and from Proposition 8.71(c). □

The next result concerns the functional representation of epi-Lipschitz one-sided subsmooth sets. Comparing with Theorem 8.85, two differences need to be emphasized:

- the statements are local (and not pointwise);
- the equivalence is proved only in finite dimensions.

PROPOSITION 8.89. *Let S be a subset of a finite-dimensional normed space X which is closed near $\overline{x} \in \operatorname{bdry} S$ and epi-Lipschitz at $\overline{x}$ in a direction $h \neq 0$. The following are equivalent:*
(a) the set S is one-sided subsmooth near $\overline{x}$;
(b) every locally Lipschitz representative $f : E \to \mathbb{R}$ of S is one-sided subsmooth near $\pi_E\overline{x}$ and tangentially regular near $\pi_E\overline{x}$, where E is a vector complement subspace of $\mathbb{R}h$ in the space X;
(c) some locally Lipschitz representative $f : E \to \mathbb{R}$ of S is one-sided subsmooth near $\pi_E\overline{x}$ and tangentially regular near $\pi_E\overline{x}$.

PROOF. The implication (b)⇒(c) is obvious. To prove (a)⇒(b), suppose that S is one-sided subsmooth near $\overline{x}$ and let f be any locally Lipschitz representative of S around $\overline{x}$. Lemma 8.83 tells us that $\partial_C f$ is one-sided submonotone in a neighborhood of $\pi_E\overline{x}$. From (c)⇒(a) in Proposition 8.50 we obtain (b). Finally, the implication (c)⇒(a) follows in the same way as in Theorem 8.85. It suffices to use (b) in Proposition 8.50 in place of (b) in Proposition 8.41 to obtain the desired one-sided subsmoothness near $\overline{x}$ of $N^C(S; \cdot) \cap \mathbb{B}$. □

The next corollary directly follows from the above proposition and from Corollary 8.56.

COROLLARY 8.90. *Let X be a finite-dimensional normed space and $f : X \to \mathbb{R}$ be a function which is Lipschitz near $\overline{x} \in X$. The following are equivalent:*
(a) the epigraphical set $\operatorname{epi} f$ is one-sided subsmooth near $(\overline{x}, f(\overline{x}))$;
(b) the function f is one-sided subsmooth near $\overline{x}$ and tangentially regular near $\overline{x}$;
(c) the multimapping $\partial_C f$ is one-sided submonotone near $\overline{x}$.

8.5. Metrically subsmooth sets

Given a subset S of a normed space X we know that $\partial d_S(x)$ is included in $N^C(S;x)$ for all $x \in S$. Then, a weaker notion of subsmoothness of sets is obtained with the use of the Clarke subdifferential of the distance function to S in (8.57) instead of the truncation of the Clarke normal cone with the closed unit ball.

DEFINITION 8.91. A set S of a normed space $(X, \|\cdot\|)$ is called *metrically subsmooth* at $\overline{x} \in S$ when for every $\varepsilon > 0$ there exists some $\delta > 0$ such that

$$\langle x^* - y^*, x - y\rangle \geq -\varepsilon\|x - y\|$$

for all $x, y \in S \cap B(\overline{x}, \delta)$, all $x^* \in \partial_C d_S(x)$ and all $y^* \in \partial_C d_S(y)$. When the property holds at any $\overline{x} \in S$ we say that S is *metrically subsmooth*. The set S is called *uniformly metrically subsmooth* if for each $\varepsilon > 0$ there is a real $\delta > 0$ such that the above inequality is satisfied for all $x, y \in S$ with $\|x - y\| < \delta$, all $x^* \in \partial_C d_S(x)$ and all $y^* \in \partial_C d_S(y)$. We also say that S is *uniformly metrically subsmooth near* $\overline{x} \in S$ whenever there is some neighborhood U of $\overline{x}$ such that the set $S \cap U$ is uniformly metrically subsmooth.

Since $0 \in \partial_C d_S(y)$ for any $y \in S$, it is easily seen that S is metrically subsmooth at $\overline{x} \in S$ if and only if for any $\varepsilon > 0$ there exists some $\delta > 0$ such that

$$\langle x^*, y - x\rangle \leq \varepsilon\|y - x\| \tag{8.37}$$

for all $x, y \in B(\overline{x}, \delta) \cap S$ and all $x^* \in \partial_C d_S(x)$.

We also notice, according to the inclusion $\partial_C d_S(x) \subset N^C(S;x) \cap \mathbb{B}_{X^*}$ for any $x \in S$, that any set which is subsmooth (at $\overline{x} \in S$) is metrically subsmooth (at $\overline{x}$). Further, from Proposition 8.75(A) (resp. Proposition 8.75(B)) we immediately obtain the equivalence in (b) (resp. in (c)) below.

PROPOSITION 8.92. Let $(X, \|\cdot\|)$ be a normed space and U be an open subset of X with $S \cap U \neq \emptyset$.
(a) If the set S is subsmooth (resp. subsmooth at $\overline{x} \in S$), then it is metrically subsmooth (resp. metrically subsmooth at $\overline{x}$).
(b) If X is a Banach space, then the set S is subsmooth at every point in $S \cap U$ if and only if it is metrically subsmooth at every point in $S \cap U$.
(c) If X is a Banach space, then the set S in X is uniformly subsmooth (resp. uniformly subsmooth near a point $\overline{x} \in S$) if and only if it is uniformly metrically subsmooth (resp. uniformly metrically subsmooth near $\overline{x}$).

We will see below in Example 8.101 that the converse of the assertion (a) in the above proposition does not hold even in finite dimensions.

First we show that the concept of metric subsmoothness does not depend on the norm.

PROPOSITION 8.93. The concept of metrical subsmoothness does not depend on the norm on X in the sense that, for any norm $\|\cdot\|_1$ on X equivalent to $\|\cdot\|$, the set S is metrically subsmooth at $\overline{x} \in S$ with respect to the norm $\|\cdot\|$ if and only if it is metrically subsmooth at $\overline{x}$ with respect to $\|\cdot\|_1$.

PROOF. Fix two constants $\alpha, \beta > 0$ such that $\alpha\|x\|_1 \leq \|x\| \leq \beta\|x\|_1$ for all $x \in X$. Denote by $\mathrm{dist}_1(S, \cdot)$ the distance function to S with respect to the norm

$\|\cdot\|_1$. Then

$$\alpha\,\mathrm{dist}_1(S,x) \le \mathrm{dist}\,(S,x) \le \beta\,\mathrm{dist}_1(S,x) \quad \text{for all } x \in X$$

and these inequalities through Proposition 2.95(a) entail for each $x \in S$

(8.38)

$$\alpha\,(\mathrm{dist}_1(\cdot,S))^\circ(x;h) \le (\mathrm{dist}\,(\cdot,S))^\circ(x;h) \le \beta(\mathrm{dist}_1(\cdot,S))^\circ(x;h) \quad \text{for all } h \in X.$$

Suppose that S is metrically subsmooth at $\overline{x}$ with respect to the norm $\|\cdot\|$. Fix any $\varepsilon > 0$. By definition of metrical subsmoothness there exists $\delta > 0$ such that for all $u, v \in S \cap B(\overline{x},\delta)$ and all $u^* \in \partial_C \mathrm{dist}\,(\cdot,S)(u)$ one has

$$\langle u^*, v-u\rangle \le \frac{\alpha\varepsilon}{\beta}\|u-v\|. \tag{8.39}$$

Fix any $x, y \in S \cap B(\overline{x},\delta)$ and $x^* \in \partial_C \mathrm{dist}_1(\cdot,S)(x)$. According to the first inequality of (8.38) we have $\alpha x^* \in \partial_C \mathrm{dist}\,(\cdot,S)(x)$, and hence by (8.39) we obtain $\langle \alpha x^*, y-x\rangle \le \frac{\alpha\varepsilon}{\beta}\|y-x\|$, which entails $\langle x^*, y-x\rangle \le \varepsilon\|y-x\|_1$. This ensures that S is metrically subsmooth at $\overline{x}$ with respect to the norm $\|\cdot\|_1$.

The reverse implication holds in the same way. □

Considering normed spaces $(X_k, \|\cdot\|_k)$ and sets $S_k \subset X_k$, with $k = 1,\cdots,m$, and endowing $X_1 \times \cdots \times X_m$ with the sum norm, we know (and it is easily seen) that $d_S(x_1,\cdots,x_m) = d_{S_1}(x_1) + \cdots + d_{S_m}(x_m)$, where $S := S_1 \times \cdots \times S_m$. From this and Proposition 8.93 above we deduce the following:

PROPOSITION 8.94. For each $k = 1,\cdots,m$ let S_k be a set in a normed space X_k which is metrically subsmooth at a point $\overline{x}_k \in S_k$. Then the set $S_1 \times \cdots \times S_m$ is metrically subsmooth at $(\overline{x}_1,\cdots,\overline{x}_m)$.

We saw in Proposition 8.73 that a subsmooth set at $\overline{x}$ is tangentially regular at $\overline{x}$. The same property is shown, in the following proposition, to hold under the metric subsmoothness. In fact it is even true for metrically hemi-subsmooth set. When we require (8.22) (resp. (8.23)) to be satisfied with $\partial_C \mathrm{dist}\,(\cdot,S)(\cdot)$ in place of $N^C(S;\cdot)\cap\mathbb{B}$, we say that the set S is *metrically one-sided subsmooth* (resp. *metrically hemi-subsmooth*) at $\overline{x} \in S$. Like the subsmoothness property, the metric subsmoothness clearly entails the metric one-sided subsmoothness, which in turn implies the metric hemi-subsmoothness.

PROPOSITION 8.95. Every set S in a normed space X which is metrically hemi-subsmooth (metrically one-sided subsmooth, or metrically subsmooth) at $\overline{x} \in S$ is tangentially regular at the point $\overline{x}$.

PROOF. Fix any $x^* \in \partial_C d_S(\overline{x})$ and any $h \in T^B(S;\overline{x})$. By definition of metric hemi-subsmoothness, for each $\varepsilon > 0$ there exists some $\delta > 0$ such that

$$\langle x^*, x-\overline{x}\rangle \le \varepsilon\|x-\overline{x}\| \quad \text{for all } x \in S \cap B(\overline{x},\delta). \tag{8.40}$$

Choose, by definition of $T^B(S;\overline{x})$, a sequence $(t_n)_n$ tending to 0 with $t_n > 0$ and a sequence $(h_n)_n$ converging to h such that $\overline{x} + t_n h_n \in S$ for all n. For n large enough we have $\overline{x} + t_n h_n \in B(\overline{x},\delta)$, and hence according to (8.40) we obtain $\langle x^*, h_n\rangle \le \varepsilon\|h_n\|$, which entails $\langle x^*, h\rangle \le \varepsilon\|h\|$. This being true for every $\varepsilon > 0$, we get $\langle x^*, h\rangle \le 0$. Thus $\partial_C d_S(\overline{x}) \subset (T^B(S;\overline{x}))^\circ$, where the second member of the inclusion denotes the negative polar cone of $T^B(S;\overline{x})$. The equality $N^C(S;\overline{x}) = \mathrm{cl}_{w^*}\big(\mathbb{R}_+\partial d_S(\overline{x})\big)$ and the $w(X^*,X)$-closedness of $(T^B(C;\overline{x}))^\circ$ then yield $N^C(S;\overline{x}) \subset (T^B(S;\overline{x}))^\circ$. Using the equality $N^C(S;\overline{x}) = \big(T^C(S;\overline{x})\big)^\circ$ and the $w(X,X^*)$-closedness of $T^C(S;\overline{x})$, we can write $\big(T^B(C;\overline{x})\big)^{oo} \subset T^C(S;\overline{x})$, and hence $T^B(S;\overline{x}) \subset T^C(S;\overline{x})$. This means that the set S is tangentially regular at $\overline{x}$ since the reverse inclusion of the latter inclusion always holds. □

Example 8.102 below will provide an example of a closed set tangentially regular at a point $\overline{x}$ which fails to be metrically subsmooth at $\overline{x}$. In contrast, we will see in Proposition 8.105 that in finite dimensions a closed set is tangentially regular at a point $\overline{x}$ if and only if it is hemi-subsmooth at $\overline{x}$ (or equivalently metrically hemi-subsmooth at $\overline{x}$ by Proposition 8.98).

The next proposition examines when one of the two points x or y in the definition of metric subsmoothness is allowed to be outside S (but near $\overline{x}$).

PROPOSITION 8.96. Let S be a set of a normed space X which is closed near $\overline{x} \in S$. Then the following assertions (a) and (b) are equivalent:
(a) the set S is metrically subsmooth at $\overline{x}$;
(b) for any $\varepsilon > 0$ there exists some $\delta > 0$ such that for all $y \in B(\overline{x},\delta)$, all $x \in S \cap B(\overline{x},\delta)$, and all $x^* \in \partial_C d_S(x)$

$$\langle x^*, y - x\rangle \le d_S(y) + \varepsilon\|y - x\|.$$

If X is Asplund, each one of the above assertions is equivalent to:
(c) for any $\varepsilon > 0$ there exists some $\delta > 0$ such that for all $x \in B(\overline{x},\delta)$, all $x^* \in \partial_C d_S(x)$, and all $y \in S \cap B(\overline{x},\delta)$ one has

$$\langle x^*, y - x\rangle \le d_S(x) + \varepsilon\|y - x\|.$$

PROOF. The implication (b) ⇒ (a) is obvious. Assume that (a) holds and fix any $\varepsilon > 0$. For $\varepsilon' := \varepsilon/2$ choose by definition of metric subsmoothness some $\delta > 0$ such that for all $u, u' \in S \cap B(\overline{x}, 2\delta)$ and all $u^* \in \partial_C d_S(u)$ one has $\langle u^*, u' - u\rangle \le \varepsilon'\|u' - u\|$. Fix any $y \in B(\overline{x},\delta)$, any $x \in S \cap B(\overline{x},\delta)$, and any $x^* \in \partial_C d_S(x)$. Take any $z \in S \cap B(\overline{x}, 2\delta)$. Then

$$\begin{aligned}\langle x^*, y - x\rangle &= \langle x^*, y - z\rangle + \langle x^*, z - x\rangle \\ &\le \|y - z\| + \varepsilon'\|z - x\| \\ &\le (1 + \varepsilon')\|y - z\| + \varepsilon'\|y - x\|.\end{aligned}$$

Taking the infimum over all $z \in S \cap B(\overline{x}, 2\delta)$ we obtain

$$\langle x^*, y - x\rangle \le (1 + \varepsilon')\mathrm{dist}\,(y, S \cap B(\overline{x}, 2\delta)) + \varepsilon'\|y - x\|$$

and observing that $\mathrm{dist}\,(y, S \cap B(\overline{x}, 2\delta)) = \mathrm{dist}\,(y, S)$ by the inclusion $y \in B(\overline{x},\delta)$ (see Lemma 2.219) we may write

$$\begin{aligned}\langle x^*, y - x\rangle &\le (1 + \varepsilon')d_S(y) + \varepsilon'\|y - x\| \\ &\le d_S(y) + 2\varepsilon'\|y - x\|,\end{aligned}$$

the second inequality being due to the fact that $d_S(y) \le \|y - x\|$ since $x \in S$. The last inequality translates the property (b) since $2\varepsilon' = \varepsilon$, so we have proved (a)⇒(b).

The implication (c) ⇒ (a) is obvious. Assume that X is Asplund and that (a) holds. Without loss of generality we may suppose that S is closed. Fix any $\varepsilon > 0$ and take some $\varepsilon' > 0$ such that $2\varepsilon' + \varepsilon'(2 + \varepsilon') < \varepsilon$. By definition of metric subsmoothness, choose some $\delta > 0$ such that for all $u, u' \in S \cap B(\overline{x}, 3\delta)$ and $u^* \in \partial_C d_S(u)$ one has

$$\langle u^*, u' - u \rangle \le \varepsilon' \|u' - u\|. \tag{8.41}$$

Fix any $x \in B(\overline{x}, \delta) \setminus S$ and $y \in S \cap B(\overline{x}, \delta)$. Suppose that $\partial_F d_S(x) \neq \emptyset$ and take any $x^* \in \partial_F d_S(x)$. Applying Lemma 4.88 with any positive $\varepsilon'' < \min\{\delta, \varepsilon', \varepsilon' \mathrm{dist}\,(x, S)\}$ in place of ε, we obtain some $v \in S$ and $v^* \in \partial_F d_S(v)$ such that

$$\|v - x\| < \varepsilon'' + d_S(x) < (1 + \varepsilon') d_S(x) \quad \text{and} \quad \|v^* - x^*\| < \varepsilon'. \tag{8.42}$$

Observe that by the first inequality of (8.42) we have

$$\|v - \overline{x}\| \le \|v - x\| + \|x - \overline{x}\| < \varepsilon'' + d_S(x) + \|x - \overline{x}\|,$$

and hence by the inclusions $\overline{x} \in S$ and $x \in B(\overline{x}, \delta)$ we obtain

$$\|v - \overline{x}\| < \varepsilon'' + 2\|x - \overline{x}\| < 3\delta. \tag{8.43}$$

Keeping in mind that $y \in S \cap B(\overline{x}, \delta)$, we see that

$$\begin{aligned} \langle x^*, y - x \rangle &\le \langle v^*, y - x \rangle + \varepsilon' \|y - x\| \\ &= \langle v^*, y - v \rangle + \langle v^*, v - x \rangle + \varepsilon' \|y - x\| \\ &\le \varepsilon' \|y - v\| + \|v - x\| + \varepsilon' \|y - x\|, \end{aligned}$$

the first inequality being due to the last inequality of (8.42) and the second one being due to (8.41) and (8.43) and to the fact that $\|v^*\| \le 1$. Taking the second inequality of the first part of (8.42) into account it ensues that

$$\begin{aligned} \langle x^*, y - x \rangle &\le 2\varepsilon' \|y - x\| + (1 + \varepsilon') \|v - x\| \\ &\le 2\varepsilon' \|y - x\| + (1 + \varepsilon') d_S(x) + \varepsilon'(1 + \varepsilon') d_S(x), \end{aligned}$$

which gives (since $y \in S$)

$$\begin{aligned} \langle x^*, y - x \rangle &\le 2\varepsilon' \|y - x\| + d_S(x) + \varepsilon'(2 + \varepsilon') \|y - x\| \\ &= (2\varepsilon' + \varepsilon'(2 + \varepsilon')) \|y - x\| + d_S(x). \end{aligned}$$

The choice of ε' yields

$$\langle x^*, y - x \rangle \le d_S(x) + \varepsilon \|y - x\| \tag{8.44}$$

and this is satisfied for any $x \in B(\overline{x}, \delta)$ since the case when $x \in S \cap B(\overline{x}, \delta)$ follows from (8.41). The definition of any limiting subgradient at x (as the weak-star limit of some sequence of Fréchet subgradients at points x_n converging strongly to x) assures us that (8.44) continues to hold for any $x \in B(\overline{x}, \delta)$ and any $x^* \in \partial_L d_S(x)$, and of course for any $x^* \in \mathrm{co}\, \partial_L d_S(x)$. Take now any $y \in S \cap B(\overline{x}, \delta)$, any $x \in B(\overline{x}, \delta)$ and any $x^* \in \partial_C d_S(x)$. Since $\partial_C d_S(x) = \overline{\mathrm{co}}^*(\partial_L d_S(x))$ (see Corollary 4.122) and since the function $\langle \cdot, y - x \rangle$ is weak* continuous, it follows that the inequality is still true for such x^*, x, and y, that is, the property (c) is obtained. This finishes the proof of the proposition. □

The same arguments with $N^C(S; x) \cap \mathbb{B}_{X^*}$ in place of $\partial_C d_S(x)$ in the proof of the implication (a) ⇒ (b) in the above proposition yields the following equivalence for subsmooth sets.

PROPOSITION 8.97. For any set S of a normed space X which is closed near $\overline{x} \in S$, the assertions (a) and (b) below are equivalent:
(a) the set S is subsmooth at $\overline{x}$;
(b) for any $\varepsilon > 0$ there exists some $\delta > 0$ such that for all $y \in B(\overline{x}, \delta)$, all $x \in S \cap B(\overline{x}, \delta)$, and all $x^* \in N^C(S; x) \cap \mathbb{B}_{X^*}$

$$\langle x^*, y - x \rangle \leq d_S(y) + \varepsilon \|y - x\|.$$

A property similar to (b) of Proposition 8.96 also holds for metric hemi-subsmooth sets. Such sets have been seen in Proposition 8.95 to be tangentially regular. In fact the corresponding characterization below of hemi-subsmooth sets will allow us to prove more, in the sense that we even have the stronger property of tangential regularity of the distance function d_S. We will also be able to show that hemi-subsmoothness and metric hemi-subsmoothness are the same property.

PROPOSITION 8.98. Let S be a subset of a normed space X which is closed near $\overline{x} \in S$. Consider the following assertions.
(a) The set S is metrically hemi-subsmooth at $\overline{x}$.
(b) The set S is hemi-subsmooth at $\overline{x}$.
(c) For any $\varepsilon > 0$ there exists some $\delta > 0$ such that for all $x \in B(\overline{x}, \delta)$ and all $u^* \in \partial_C d_S(\overline{x})$ one has

$$\langle u^*, x - \overline{x} \rangle \leq d_S(x) + \varepsilon \|x - \overline{x}\|.$$

(d) The distance function d_S is Clarke-Fréchet regular at $\overline{x}$, in the sense that

$$\partial_C d_S(\overline{x}) = \partial_F d_S(\overline{x}).$$

(e) The distance function d_S is tangentially regular at $\overline{x}$, that is,

$$d_S^\circ(\overline{x}; \cdot) = d'_S(\overline{x}; \cdot).$$

(f) $N^C(S; \overline{x}) = N^F(S; \overline{x})$ and $N^C(S; \overline{x}) \cap \mathbb{B}_{X^*} = \partial_C d_S(\overline{x})$.

Then (b) $\Rightarrow$ (a) $\Leftrightarrow$ (c) $\Rightarrow$ (d) $\Rightarrow$ (e). If X is a Banach space, the implications (a) $\Leftrightarrow$ (b) and (d) $\Rightarrow$ (f) also hold.

PROOF. The implications (b) $\Rightarrow$ (a) and (c) $\Rightarrow$ (a) are obvious and (a) $\Rightarrow$ (c) is obtained like for (a) $\Rightarrow$ (b) in Proposition 8.96. The property (c) entails that any $x^* \in \partial_C d_S(\overline{x})$ is a Fréchet subgradient of d_S at $\overline{x}$, and hence the equality $\partial_C d_S(\overline{x}) = \partial_F d_S(\overline{x})$. This means that (c) $\Rightarrow$ (d) holds.

Suppose now that (d) is satisfied. It is not difficult to see that for any function f and $x, h \in X$ one has $\sup\limits_{x^* \in \partial_F f(x)} \langle x^*, h \rangle \leq f^B(x; h)$, and hence by (d)

$$(d_S)^\circ(\overline{x}; h) = \sup_{x^* \in \partial_C d_S(\overline{x})} \langle x^*, h \rangle \leq (d_S)^B(\overline{x}; h).$$

The reverse inequality being always true, we obtain the directional regularity of d_S at $\overline{x}$, that is, (d) $\Rightarrow$ (e) is shown.

Let us prove (f) under (d) and the completeness of X. Since we know from Proposition 4.30 that $\partial_F d_S(\overline{x}) = N^F(S; \overline{x}) \cap \mathbb{B}_{X^*}$, the equality in (d) assures us that

$$\partial_C d_S(\overline{x}) = N^F(S; \overline{x}) \cap \mathbb{B}_{X^*}. \tag{8.45}$$

Thanks to the weak-star closedness of the Clarke subdifferential of a function, the latter equality yields that $N^F(S; \overline{x}) \cap \mathbb{B}_{X^*}$ is weak-star closed and hence $N^F(S; \overline{x})$ is weak-star closed as well, according to the Krein-Šmulian theorem (see Theorem C.4 in Appendix C) since $N^F(S;\overline{x})$ is a convex cone of X^*. This weak-star closedness property along with (8.45) give

$$N^C(S;\overline{x}) = \mathrm{cl}_{w^*}[\mathbb{R}_+\partial_C d_S(\overline{x})] = \mathrm{cl}_{w^*}[N^F(S;\overline{x})] = N^F(S;\overline{x}), \tag{8.46}$$

that is, $N^C(S;\overline{x}) = N^F(S;\overline{x})$. Combining the latter equality with

$$N^F(S;\overline{x}) \cap \mathbb{B}_{X^*} = \partial_F d_S(\overline{x}) = \partial_C d_S(\overline{x}),$$

we see that $N^C(S;\overline{x}) \cap \mathbb{B}_{X^*} = \partial_C d_S(\overline{x})$. So the implication (d)⇒(f) holds.

It remains, when X is a Banach space, to establish (a) ⇒ (b). Under (a) and the completeness of X we know by what precedes that (f) holds, and hence $\partial_C d_S(\overline{x}) = N^C(S;\overline{x}) \cap \mathbb{B}_{X^*}$. Consequently, the metric hemi-subsmoothness of S at $\overline{x}$ implies its hemi-subsmoothness at $\overline{x}$ as well. The proof is then complete. □

The next theorem provides in addition to Proposition 8.96 some other characterizations of metric subsmoothness in the context of Asplund space but its interest essentially rests on the important characterizations furnished by (e) and (f) when the space X is Hilbert or finite-dimensional.

THEOREM 8.99 (metric subsmoothness in Asplund/Hilbert spaces). *Assume that X is an Asplund space and S is a subset of X which is closed near $\overline{x} \in S$. Then the following assertions are equivalent:*
(a) the set S is metrically subsmooth at $\overline{x}$;
(b) the multimapping $\partial_F d_S$ is submonotone at $\overline{x}$ relative to the set S;
(b') for every $\varepsilon > 0$ there is $\delta > 0$ such that $\langle x^*, y - x\rangle \le \varepsilon\|y - x\|$ for all $x, y \in S \cap B(\overline{x}, \delta)$ and $x^* \in \partial_F d_S(x)$;
(c) the multimapping $\partial_L d_S$ is submonotone at $\overline{x}$ relative to the set S;
(c') for every $\varepsilon > 0$ there is $\delta > 0$ such that $\langle x^*, y - x\rangle \le \varepsilon\|y - x\|$ for all $x, y \in S \cap B(\overline{x}, \delta)$ and $x^* \in \partial_L d_S(x)$;
(d) the multimapping $N^F(S;\cdot) \cap \mathbb{B}_{X^*}$ is submonotone at $\overline{x}$;
(d') for every $\varepsilon > 0$ there is $\delta > 0$ such that $\langle x^*, y - x\rangle \le \varepsilon\|y - x\|$ for all $x, y \in S \cap B(\overline{x}, \delta)$ and $x^* \in N^F(S;x) \cap \mathbb{B}_{X^*}$;

If in addition X is a Hilbert space, the following assertions may be added to the list of equivalences:
(e) for any $\varepsilon > 0$ there exists some $\delta > 0$ such that, for all $y \in X$, $x, u \in S \cap B(\overline{x}, \delta)$ with $u \in \mathrm{Proj}_S(y)$ one has

$$\langle y - u, x - u\rangle \le \varepsilon\|y - u\|\,\|x - u\|;$$

(f) the multimapping $N^P(S;\cdot) \cap \mathbb{B}$ is submonotone at $\overline{x}$;
(g) the multimapping $\partial_P d_S$ is submonotone at $\overline{x}$ relative to the set S;
(g') for every $\varepsilon > 0$ there is $\delta > 0$ such that $\langle x^*, y - x\rangle \le \varepsilon\|y - x\|$ for all $x, y \in S \cap B(\overline{x}, \delta)$ and $x^* \in \partial_P d_S(x)$.

If X is finite-dimensional, then anyone of all the above properties is equivalent to:
(h) The multimapping $N^L(S;\cdot) \cap \mathbb{B}$ is submonotone at $\overline{x}$.

PROOF. The equivalence between (a), (b), and (c) is a reformulation of Lemma 8.76, while the equivalences (b)⇔(b'), (c)⇔(c'), (d)⇔(d'), and (g)⇔(g') are evident. The equivalence between (b) and (d) follows from the equality $\partial_F d_S(x) = N^F(S;x) \cap \mathbb{B}$ for all $x \in S$.

Assume now that X is a Hilbert space. Without loss of generality we may suppose that S is closed. We know (see (b) and (d) in Proposition 4.136) that any proximal normal vector of the form $\|y-u\|^{-1}(y-u)$, for $u \in \operatorname{Proj}_S(y)$ with $d_S(y) > 0$, is a unit Fréchet normal vector of S at u and this justifies the implication (d) $\Rightarrow$ (e). Taking into account the definition of the proximal normal cone it is easily seen that (e) entails that the multimapping $N^P(S;\cdot) \cap \mathbb{B}$ is submonotone at $\overline{x}$, which is exactly (e) $\Rightarrow$ (f). By Proposition 4.153 we know that $\partial_P d_S(x) = N^P(S;x) \cap \mathbb{B}$ for all $x \in S$. Therefore, the assertion (g) is just a reformulation of (f). Let us show that (g) implies (c). For any $x \in S$ near $\overline{x}$ we know (see Proposition 4.158) that

$$\partial_L d_S(x) = {}^{\mathrm{seq}} \operatorname*{Lim\,sup}_{S \ni v \to x} \partial_P d_S(v). \tag{8.47}$$

Fix now any $\varepsilon > 0$. By the assumption (g) there exists some $\delta > 0$ such that $\langle w^*, y-w \rangle \le \varepsilon \|y-w\|$ for all $y, w \in S \cap B(\overline{x}, \delta)$ and $w^* \in \partial_P d_S(w)$. Fix any $y, x \in S \cap B(\overline{x}, \delta)$ and $x^* \in \partial_L d_S(x)$. According to (8.47) there are a sequence $(x_n)_n$ in S converging to x and a sequence $(x_n^*)_n$ converging weakly to x^* with $x_n^* \in \partial_P d_S(x_n)$. For n large enough we have $\langle x_n^*, y - x_n \rangle \le \varepsilon \|y - x_n\|$, and hence $\langle x^*, y-x \rangle \le \varepsilon \|y-x\|$. So the equivalence of any assertion among (a) to (d) with anyone of (e) to (g) is established under the Hilbert assumption of X.

The implication (h)$\Rightarrow$(d) being obvious, suppose that (d) holds and X is finite-dimensional. Without loss of generality, we may suppose that the norm of X is a Euclidean norm and we may identify X^* with X through the Euclidean inner product. Fix any $\varepsilon > 0$ and take $\delta > 0$ such that $\langle w, x-y \rangle \le \frac{\varepsilon}{2} \|x-y\|$ for all $x, y \in S \cap B(\overline{x}, \delta)$ and $w \in N^F(S;y) \cap \mathbb{B}$. Fix any $x, u \in S \cap B(\overline{x}, \delta)$ and any $v \in N^L(S;u)$ with $\|v\| = 1$. By definition of $N^L(S;u)$, there exist some sequence $(u_n)_n$ in $S \cap B(\overline{x}, \delta)$ converging to u and some sequence $(v_n)_n$ converging to v with $v_n \in N^F(S;u_n)$. Then for any integer n sufficiently large we have

$$\left\langle \frac{1}{\|v_n\|} v_n, x - u_n \right\rangle \le (\varepsilon/2) \|x - u_n\|,$$

which yields

$$\langle v, x-u \rangle \le (\varepsilon/2)\|x-u\| \quad \text{for all } x, u \in S \cap B(\overline{x}, \delta) \text{ and } v \in N^L(S;u) \cap \mathbb{S}.$$

This property is easily seen to entail the submonotonicity of the multimapping $N^L(S;\cdot) \cap \mathbb{B}$ at $\overline{x}$. So the equivalence between (h) and (d) is established provided that X is finite-dimensional. The proof is then complete. □

REMARK 8.100. Roughly speaking, the property (e) in Theorem 8.99 above means that the angle between a unit proximal normal vector $x^* \in N^P(S;x)$ and $(y-x)/\|y-x\|$ is not much less than $\pi/2$ for $x, y \in S$ sufficiently close to $\overline{x}$. □

We use Theorem 8.99(e) in the next two examples. The first one is an example in $\mathbb{R}^2$ of a metrically subsmooth set at a point $\overline{x}$ which fails to be subsmooth at $\overline{x}$.

EXAMPLE 8.101 (A.S. Lewis, D.R. Luke and J. Malick example). Let $f : [-1,1] \to \mathbb{R}$ be the continuous even (f(-t)=f(t)) function defined by $f(1/2^n) = 1/4^n$ and f is affine on $[1/2^{n+1}, 1/2^n]$ for all $n \in \mathbb{N}$. Considering S as the graph of f, $S := \operatorname{gph} f$, we see by Theorem 8.99(e) that S is metrically subsmooth at $\overline{x} = (0,0)$ since the angle between a unit proximal normal vector $x^* \in N^P(S;x)$ and $(y-x)/\|y-x\|$ is not much less than $\pi/2$ for $x, y \in S$ sufficiently close to $\overline{x}$, as illustrated in Figure 8.1.

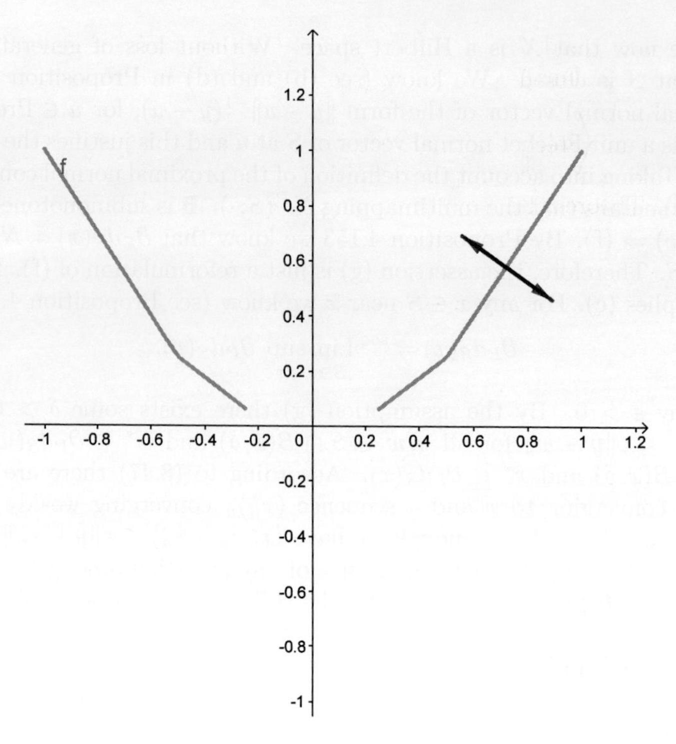

FIGURE 8.1. A metrically subsmooth set at $\overline{x}$ but not subsmooth at $\overline{x}$.

On the other hand, setting $x_n := (1/2^n, 1, /4^n)$ we see that $N^C(S; x_n) = \mathbb{R}^2$ since $T^C(S; x_n) = \{(0,0)\}$ (as illustrated in Figure 8.1). So, taking $u_n := (\overline{x} - x_n)/\|\overline{x} - x_n\|$ we have

$$u_n \in N^C(S; x_n), \quad \left\langle u_n, \frac{\overline{x} - x_n}{\|\overline{x} - x_n\|} \right\rangle = 1, \quad S \ni x_n \to \overline{x}.$$

Consequently, the set S is not subsmooth at $\overline{x}$. □

The second example provides a set in $\mathbb{R}^2$ which is tangentially regular at a point $\overline{x}$ but not metrically subsmooth at $\overline{x}$.

EXAMPLE 8.102 (A.S. Lewis, D.R. Luke and J. Malick example). Let $f : \mathbb{R} \to \mathbb{R} \cup \{+\infty\}$ be the lower semicontinuous even function defined by $f(0) = 0$, $f(1) = 1/4$, $f(t) = +\infty$ if $t > 1$ and

$$f(t) = \frac{1}{2^n}\left(t - \frac{1}{2^n}\right) \quad \text{if } t \in \left[\frac{1}{2^n}, \frac{1}{2^{n-1}}\right[, \ n \in \mathbb{N}.$$

Define S as the epigraph of f, $S := \operatorname{epi} f$, and $\overline{x} = (0,0)$. We claim that S is tangentially regular at $\overline{x}$ but it fails to be metrically subsmooth at $\overline{x}$.

First, it is clear that

$$T^B(S; \overline{x}) = \mathbb{R}^2_+ \quad \text{and} \quad \operatorname*{Lim\,inf}_{S' \ni x \to \overline{x}} T^B(S; x) = \mathbb{R}^2_+,$$

where $S' := S \setminus \{\overline{x}\}$, so $T^B(S; \overline{x}) = \operatorname*{Lim\,inf}_{S \ni x \to \overline{x}} T^B(S; x) = \mathbb{R}^2_+$. Since $\operatorname*{Lim\,inf}_{S \ni x \to \overline{x}} T^B(S; x) = T^C(S; \overline{x})$ (see Theorem 2.235), it ensues that S is tangentially regular at $\overline{x}$.

On the other hand, for $x_n := (1/2^n, 0)$ we have $u_n := (-1, 0) \in N^P(S; x_n)$, as seen in Figure 8.2. Further,

$$\left\langle u_n, \frac{x_{n+1} - x_n}{\|x_{n+1} - x_n\|} \right\rangle = 1 \quad \text{with} \quad S \ni x_n \to \overline{x}.$$

It follows by Theorem 8.99(e) that S is not metrically subsmooth at $\overline{x}$. □

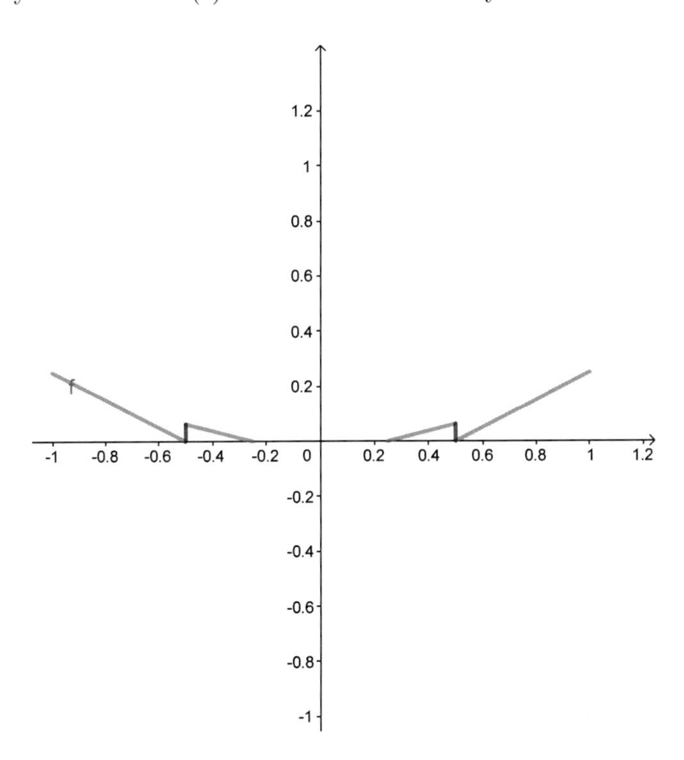

FIGURE 8.2. A tangentially regular set at $\overline{x}$ but not metrically subsmooth at $\overline{x}$.

The next proposition compares the metric subsmoothness of a set with the Jensen type inequality of the distance function involved in (a) of Proposition 8.45.

PROPOSITION 8.103. Let S be a subset of a normed space X and $\overline{x} \in S$. Consider the assertions:
(a) For any $\varepsilon > 0$ there exists $\delta > 0$ such that for all $x \in B(\overline{x}, \delta)$, $x^* \in \partial_C d_S(x)$ and $y \in S \cap B(\overline{x}, \delta)$

$$d_S(x) + \langle x^*, y - x \rangle \leq \varepsilon \|y - x\|.$$

(b) For every $\varepsilon > 0$ there exists $\delta > 0$ such that

$$d_S(tx + (1-t)y) \leq \varepsilon t(1-t)\|x - y\|$$

for all $t \in [0, 1]$ and all $x, y \in S \cap B(\overline{x}, \delta)$.
(c) The set S is metrically subsmooth at $\overline{x}$.

The implications (a) ⇒ (b) ⇒ (c) hold. If X is an Asplund space, these implications are equivalences.

PROOF. First, we note that the implication (b) $\Rightarrow$ (c) follows directly from Proposition 8.45. Now, assume that (a) holds and fix any $\varepsilon > 0$. By (a) there is $\delta > 0$ such that for all $x' \in B(\overline{x}, \delta)$, $x^* \in \partial_C d_S(x')$ and $y \in S \cap B(\overline{x}, \delta)$

$$d_S(x') + \langle x^*, y - x' \rangle \le (\varepsilon/2)\|y - x'\|.$$

Let any $x, y \in S \cap B(\overline{x}, \delta)$ with $x \neq y$ and any $t \in]0, 1[$, and set $u := tx + (1-t)y$. By the mean value theorem with C-subdifferential (see Theorem 2.180) there are $z \in [x, u]$, $z^* \in \partial_C d_S(z)$ such that with $\alpha := \|y - z\|^{-1}\|u - x\|$

$$d_S(u) = \langle z^*, u - x \rangle = \alpha \langle z^*, y - z \rangle,$$

which entails by the above inequality that

$$d_S(u) \le -\alpha d_S(z) + \frac{\varepsilon}{2}\alpha\|y - z\| = -\alpha d_S(z) + \frac{\varepsilon}{2}\|u - x\| = -\alpha d_S(z) + \frac{\varepsilon}{2}(1-t)\|y - x\|.$$

It results that

$$d_S(u) \le \frac{\varepsilon}{2}(1-t)\|y - x\|. \tag{8.48}$$

Similarly, there are $\zeta \in [u, y]$ and $\zeta^* \in \partial_C d_S(\zeta)$ such that with $\beta := \|x - \zeta\|^{-1}\|u - y\|$

$$d_S(u) = \langle \zeta^*, u - y \rangle \le -\beta d_S(\zeta) + \frac{\varepsilon}{2}\beta\|y - \zeta\| = -\beta d_S(\zeta) + \frac{\varepsilon}{2}t\|y - x\|,$$

which implies that

$$d_S(u) \le (\varepsilon/2)t\|y - x\|. \tag{8.49}$$

Multiplying (8.48) by t and (8.49) by $(1-t)$, and adding the new inequalities yield

$$d_S(u) \le \varepsilon t(1-t)\|y - x\|,$$

which justifies the implication (a) $\Rightarrow$ (b).

Now, assume that X is an Asplund space and that (c) holds. Let any $\varepsilon > 0$. There is $\delta > 0$ such that for any $z, z' \in S \cap B(\overline{x}, \delta)$ and $z^* \in \partial_C d_S(z)$ one has $\langle z^*, z' - z \rangle \le (\varepsilon/4)\|z' - z\|$. Fix any $y \in S \cap B(\overline{x}, \delta)$ and any $(x, x^*) \in \operatorname{gph} \partial_F d_S(x)$ with $x \in B(\overline{x}, \delta) \setminus S$. By Lemma 4.88 there are $u_n \in S \cap B(\overline{x}, \delta)$, $u_n^* \in \partial_F d_S(u_n)$ such that

$$\|x - u_n\| \to d_S(x) \quad \text{and} \quad \|x^* - u_n^*\| \to 0 \quad \text{as } n \to \infty.$$

Further, one has $\langle x^*, x - u_n \rangle \to d_S(x)$ according to Proposition 4.32. Write with $A_n := \langle x^*, x - u_n \rangle + \langle x^* - u_n^*, u_n - y \rangle$ that

$$\langle x^*, x - y \rangle = \langle x^*, x - u_n \rangle + \langle x^*, u_n - y \rangle = A_n + \langle u_n^*, u_n - y \rangle,$$

and note by what precedes that

$$\langle u_n^*, u_n - y \rangle \ge -(\varepsilon/4)\|u_n - y\| \ge -(\varepsilon/4)\|x - y\| - (\varepsilon/4)\|u_n - x\|.$$

Since $\|u_n - x\| \to d_S(x)$ and $d_S(x) > 0$, we have $(\varepsilon/4)\|u_n - x\| \le (\varepsilon/4)d_S(x) + (\varepsilon/4)d_S(x)$ for n large enough, say $n \ge N$. Noting also that $A_n \to d_S(x)$, we may choose an integer $k \ge N$ such that $A_k \ge d_S(x) - (\varepsilon/4)d_S(x)$. Taking $n = k$ in the preceding inequalities, it follows that

$$\begin{aligned}\langle x^*, x - y \rangle &\ge d_S(x) - \frac{\varepsilon}{4}d_S(x) - \frac{\varepsilon}{4}\|x - y\| - \frac{\varepsilon}{2}d_S(x) \\ &\ge d_S(x) - \varepsilon\|x - y\|.\end{aligned}$$

This ensures that $\langle x^*, x - y \rangle \ge d_S(x) - \varepsilon\|y - x\|$ for all $(x, x^*) \in \operatorname{gph} \partial_F d_S(x)$ with $x \in B(\overline{x}, \delta)$, since the case $x \in S \cap B(\overline{x}, \delta)$ is obvious by the chose of δ above. Remembering the definition of L-subdifferential, we deduce that the inequality holds true for all $(x, x^*) \in \operatorname{gph} \partial_L d_S(x)$ with $x \in B(\overline{x}, \delta)$. Finally, the equality $\partial_C d_S(v) = \overline{\operatorname{co}}^* \partial_L d_S(v)$ for every $v \in X$ (see Corollary 4.122), assures us that $\langle x^*, x-y\rangle \geq d_S(x) - \varepsilon\|y-x\|$ for all $y \in S \cap B(\overline{x}, \delta)$, $x \in B(\overline{x}, \delta)$ and $x^* \in \partial_C d_S(x)$, and this finishes the proof of the proposition. □

Putting together Proposition 8.92, Proposition 8.103, Lemma 8.76, Proposition 8.77 and Theorem 8.99 translates and summarizes the situation of local subsmoothness of sets in Asplund spaces as follows.

PROPOSITION 8.104. Assume that X is an Asplund space and U is an open set of X. Let S be a subset of X which is closed near each point of $S \cap U$. Then the following assertions are equivalent:
(a) the set S is metrically subsmooth at each point of $S \cap U$;
(b) the set S is subsmooth at each point of $S \cap U$;
(c) for every $x \in S \cap U$ and every $\varepsilon > 0$ there exists a real $\delta > 0$ with $B(x, \delta) \subset U$ such that $d_S(tx_1 + (1-t)x_2) \leq \varepsilon t(1-t)\|x_1 - x_2\|$ for all $t \in [0,1]$ and $x_1, x_2 \in S \cap B(x, \delta)$;
(d) the multimapping $\partial_F d_S$ is submonotone at each point of $S \cap U$ relative to the set S;
(e) the multimapping $\partial_L d_S$ is submonotone at each point of $S \cap U$ relative to the set S;
(f) the multimapping $N_F(S; \cdot) \cap \mathbb{B}_{X^*}$ is submonotone at each point of $S \cap U$;
(g) the multimapping $N_L(S; \cdot) \cap \mathbb{B}_{X^*}$ is submonotone at each point of $S \cap U$.

Let us establish in addition to Proposition 8.98 two other characterizations of hemi-subsmooth sets. These characterizations will also allow us to show that in the finite-dimensional setting the hemi-subsmoothness of S at $\overline{x}$ is equivalent to its tangential regularity at $\overline{x}$.

PROPOSITION 8.105. Let S be a set of a normed space X and let $\overline{x} \in S$. The following are equivalent:
(a) the set S is hemi-subsmooth at $\overline{x}$;
(b) for every $\overline{x}^* \in N^C(S; \overline{x})$, every $\beta > 0$, and every $\varepsilon > 0$ there exists $\delta > 0$ such that (8.23) holds for all $x \in S \cap B(\overline{x}, \delta)$ and all $x^* \in N^C(S; \overline{x})$ with $\|x^* - \overline{x}^*\| \leq \beta$;
(c) for every $\overline{x}^* \in N^C(S; \overline{x})$ there is $\beta > 0$ such that for any $\varepsilon > 0$ there exists $\delta > 0$ such that (8.23) holds for all $x \in S \cap B(\overline{x}, \delta)$ and $x^* \in N^C(S; \overline{x})$ with $\|x^* - \overline{x}^*\| \leq \beta$;
(d) for every $\overline{x}^* \in N^C(S; \overline{x})$ and for any $\varepsilon > 0$ there exists $\delta > 0$ such that (8.23) holds for all $x \in S \cap B(\overline{x}, \delta)$ and $x^* \in N^C(S; \overline{x})$ with $\|x^* - \overline{x}^*\| \leq \varepsilon$.

If the space X is finite-dimensional, then one may also add to the list of equivalences the property (e) below:
(e) the set S is tangentially regular at $\overline{x}$.

PROOF. Suppose that (a) holds. Fix $\overline{x}^* \in N^C(S; \overline{x})$, $\beta > 0$ and $\varepsilon > 0$. By (a) choose some $\delta > 0$ such that

$$\langle x^*, x - \overline{x}\rangle \leq \frac{\varepsilon}{\beta + \|\overline{x}^*\|}\|x - \overline{x}\|$$

for all $x \in S \cap B(\overline{x}, \delta)$ and $x^* \in N^C(S; \overline{x}) \cap \mathbb{B}_{X^*}$. Then for any $x \in S \cap B(\overline{x}, \delta)$ and any $x^* \in N^C(S; \overline{x})$ with $\|x^* - \overline{x}^*\| \leq \beta$, we have $\|\rho^{-1}x^*\| \leq 1$, for $\rho := \beta + \|\overline{x}^*\|$, and hence $\langle \rho^{-1}x^*, x - \overline{x}\rangle \leq \rho^{-1}\varepsilon\|x - \overline{x}\|$, that is, $\langle x^*, x - \overline{x}\rangle \leq \varepsilon\|x - \overline{x}\|$. This means that (b) holds.

The fact that (b) implies both (c) and (d) is obvious. To see that (c) (resp. (d)) entails (a), assume (c) (resp. (d)) and take some $\beta > 0$ corresponding to the choice $\overline{x}^* = 0$ in (c) (resp. and take the choice $\overline{x}^* = 0$ in (d)). Fix any $\varepsilon > 0$ and by (c) (resp. (d)) choose $\delta > 0$ such that $\langle x^*, x - \overline{x} \rangle \leq \varepsilon\beta \|x - \overline{x}\|$ for any $x \in S \cap B(\overline{x}, \delta)$ and $x^* \in N^C(S; \overline{x})$ with $\|x^*\| \leq \beta$ (resp. $\langle x^*, x - \overline{x} \rangle \leq \varepsilon^2 \|x - \overline{x}\|$ for any $x \in S \cap B(\overline{x}, \delta)$ and $x^* \in N^C(S; \overline{x})$ with $\|x^*\| \leq \varepsilon$). Anyone of the two latter properties is easily seen to give $\langle x^*, x - \overline{x} \rangle \leq \varepsilon \|x - \overline{x}\|$ for all $x \in S \cap B(\overline{x}, \delta)$ and $x^* \in N^C(S; \overline{x}) \cap \mathbb{B}$, that is, (a) is fulfilled.

Now assume that X is finite-dimensional. We note first that (a) $\Rightarrow$ (e) is true by Proposition 8.73(b). Conversely, suppose that (a) does not hold. Then, there exist a real $\varepsilon > 0$, a sequence $(x_n)_n$ in S converging to $\overline{x}$ and a sequence $(x_n^*)_n$ in $N^C(S; \overline{x}) \cap \mathbb{B}_{X^*}$ such that $\langle x_n^*, x_n - \overline{x} \rangle > \varepsilon \|x_n - \overline{x}\|$ for all $n \in \mathbb{N}$. Extracting subsequences, we may suppose that $x_n^* \to x^*$ and $\|x_n - \overline{x}\|^{-1}(x_n - \overline{x}) \to h$. Clearly, $x^* \in N^C(S; \overline{x})$ and $\langle x^*, h \rangle \geq \varepsilon$, hence $h \notin T^C(S; \overline{x})$. Further, the convergence $\|x_n - \overline{x}\|^{-1}(x_n - \overline{x}) \to h$ also ensures that $h \in T^B(S; \overline{x})$, so S is not tangentially regular at $\overline{x}$, which finishes the proof. □

The same arguments for the above equivalences (a) $\Leftrightarrow$ (b) $\Leftrightarrow$ (c) $\Leftrightarrow$ (d) also yield:

PROPOSITION 8.106. Let S be a set of a normed space X and let $\overline{x} \in S$. The following are equivalent:
(a) the set S is subsmooth at $\overline{x}$;
(b) for every $\overline{x}^* \in N^C(S; \overline{x})$, every $\beta > 0$, and every $\varepsilon > 0$ there exists $\delta > 0$ such that for all $x, y \in S \cap B(\overline{x}, \delta)$ and all $x^* \in N^C(S; x)$ with $\|x^* - \overline{x}^*\| \leq \beta$ one has

$$\langle x^*, y - x \rangle \leq \varepsilon \|y - x\|; \tag{8.50}$$

(c) for every $\overline{x}^* \in N^C(S; \overline{x})$ there is $\beta > 0$ such that for any $\varepsilon > 0$ there exists $\delta > 0$ such that (8.50) holds for all $x, y \in S \cap B(\overline{x}, \delta)$ and $x^* \in N^C(S; x)$ with $\|x^* - \overline{x}^*\| \leq \beta$;
(d) for every $\overline{x}^* \in N^C(S; \overline{x})$ and for any $\varepsilon > 0$ there exists $\delta > 0$ such that (8.50) holds for all $x, y \in S \cap B(\overline{x}, \delta)$ and $x^* \in N^C(S; x)$ with $\|x^* - \overline{x}^*\| \leq \varepsilon$.

8.6. Subsmoothness of a set and α-far property of the C-subdifferential of its distance function

For a closed se t S of a normed space X and for any $x \in X \setminus S$ we know (see Proposition 4.31 that $\|x^*\| = 1$ for any $x^* \in \partial_F d_S(x)$, and also for $x^* \in \partial_L d_S(x)$ whenever X is finite-dimensional. So, given $\overline{x} \in \operatorname{bdry} S$ the following question arises: Is there a real $\delta > 0$ such that zero is kept far away from the C-subdifferential on $B(\overline{x}, \delta) \setminus S$, that is, $0 \notin \partial_C d_S(x)$ for all $x \in B(\overline{x}, \delta) \setminus S$? For some sets the answer is negative. Consider in $\mathbb{R}$ the closed set $S := \{0\} \cup \{1/n : n \in \mathbb{N}\}$ and $0 \in \operatorname{bdry} S$. Putting $\mu_n := \frac{1}{2}\left(\frac{1}{n} + \frac{1}{n+1}\right)$, we see that

$$d_S(x) = \frac{1}{n} - x \text{ if } x \in \left[\mu_n, \frac{1}{n}\right[\quad \text{and} \quad d_S(x) = x - \frac{1}{n+1} \text{ if } x \in \left]\frac{1}{n+1}, \mu_n\right].$$

The gradient representation theorem of C-subdifferential (see Theorem 2.83) enables us to say that $\partial_C d_S(\mu_n) = [-1, 1] \ni 0$. Consequently, there is no neighborhood U of $\overline{x} := 0$ such that the set S enjoys the above desirable positively far property of zero from the C-subdifferential of d_S on $U \setminus S$.

Suppose now that X is a finite-dimensional Euclidean space with inner product $\langle \cdot, \cdot \rangle$ and associated norm $\| \cdot \|$ and that S is a closed subset which is metrically subsmooth at $\overline{x} \in \operatorname{bdry} S$. Let any real $\varepsilon \in]0,1[$. By metric subsmoothness choose a real $\delta > 0$ such that

$$\langle z_1^* - z_2^*, z_1 - z_2 \rangle \geq -\varepsilon \|z_1 - z_2\|$$

for all $z_i \in S \cap B(\overline{x}, 2\delta)$ and $z_i^* \in \partial_C d_S(z_i)$, i $= 1, 2$. Fix any $x \in B(\overline{x}, \delta) \setminus S$. By Proposition 4.125(b) we know that

$$\partial_L d_S(x) = \frac{1}{d_S(x)}(x - \operatorname{Proj}_S x).$$

Take $x_1^*, x_2^* \in \partial_L d_S(x)$, so $x_i^* = (x - u_i)/d_S(x)$ with $u_i \in \operatorname{Proj}_S x$ for $i = 1, 2$. Then $x_i^* \in N^P(S; u_i) \cap \mathbb{B}_X = \partial_P d_S(u_i)$ (see Proposition 4.153), which yields

$$\langle x_1^* - x_2^*, u_1 - u_2 \rangle \geq -\varepsilon \|u_1 - u_2\|$$

since $\|u_i - \overline{x}\| \leq 2\|x - \overline{x}\| < 2\delta$. Put $t := d_S(x)$ and note that

$$\begin{aligned} \langle x_1^* - x_2^*, u_1 - u_2 \rangle &= \langle x_1^* - x_2^*, x - tx_1^* - x + tx_2^* \rangle \\ &= -t\|x_1^* - x_2^*\|^2 = -2t + 2t\langle x_1^*, x_2^* \rangle, \end{aligned}$$

where the last equality is due to the fact that $\|x_i^*\| = 1$. Note also that $\|u_1 - u_2\| \leq \|u_1 - x\| + \|u_2 - x\| = 2t$. We derive that

$$-2t + 2t\langle x_1^*, x_2^* \rangle \geq -2t\varepsilon, \text{ or equivalently } \langle x_1^*, x_2^* \rangle \geq 1 - \varepsilon.$$

Since $\partial_C d_S(x) = \operatorname{co}(\partial_L d_S(x))$, we easily see that the latter inequality still holds for $x_i^* \in \partial_C d_S(x)$, that is, for every $x \in B(\overline{x}, \delta) \setminus S$

$$\langle x_1^*, x_2^* \rangle \geq 1 - \varepsilon \quad \text{for all } x_1^*, x_2^* \in \partial_C d_S(x), \tag{8.51}$$

which entails in particular

$$\|x^*\| \geq \sqrt{1 - \varepsilon} \quad \text{for all } x^* \in \partial_C d_S(x). \tag{8.52}$$

Our aim in this section is to extend in some sense both properties (8.51) and (8.52) to metrically subsmooth sets of Hilbert spaces.

DEFINITION 8.107. Let S be a nonempty closed set of a normed space $(X, \|\cdot\|)$ with $S \neq X$ and let $\alpha > 0$. We say that the origin is kept α-far away from the C-subdifferential of d_S on a set $Q \subset X \setminus S$ if

$$\alpha \leq \inf_{x \in Q} \operatorname{dist}(0, \partial_C d_S(x)).$$

If the above inequality holds true for some real $\alpha > 0$, we say that the origin is kept positively far away from the C-subdifferential of d_S on Q. When Q is a singleton set $\{u\}$, we just say that the origin is kept α-far away from $\partial_C d_S(u)$.

Recall that for any closed set S of a Hilbert space H and $x \in H \setminus S$ one has by Proposition 2.206 that

(8.53)

$$\partial_C d_S(x) = \bigcap_{\eta > 0} \overline{\operatorname{co}} \left(\frac{1}{d_S(x)}(x - \operatorname{Proj}_{S,\eta} x) \right) = \frac{x}{d_S(x)} - \frac{1}{d_S(x)} \bigcap_{\eta > 0} \overline{\operatorname{co}} \left(\operatorname{Proj}_{S,\eta} x \right),$$

where $\operatorname{Proj}_{S,\eta} x := \{u \in S : \|x - u\| \leq d_S(x) + \eta\}$.

Taking into account the latter formula, let us start with the following lemma.

LEMMA 8.108. Let S be a nonempty closed set in a Hilbert space H and let $x \in H \setminus S$.
(a) Assume that there exist two reals $\alpha > 0$, $\eta_x > 0$ and a function $\theta_x :]0, \eta_x[\to \mathbb{R}$ with $\lim_{\eta \downarrow 0} \theta_x(\eta) = 0$ such that for each $\eta \in]0, \eta_x[$

$$\langle x_1^*, x_2^* \rangle \geq \alpha^2 + \theta_x(\eta)$$

for all x_1^*, x_2^* in $(x - \operatorname{Proj}_{S,\eta} x) / d_S(x)$. Then the origin is kept α-far away from $\partial_C d_S(x)$, that is, $\operatorname{dist}(0, \partial_C d_S(x)) \geq \alpha$.
(b) As a partial converse, if $\operatorname{dist}(0, \partial_C d_S(x)) \geq \alpha$, then one has

$$\langle x_1^*, x_2^* \rangle \geq 2\alpha^2 - 1 \quad \text{for all } x_1^*, x_2^* \in \partial_C d_S(x).$$

PROOF. (a) For each $\eta \in]0, \eta_x[$ it is clear that for each $x_1^* \in (x - \operatorname{Proj}_{S,\eta} x) / d_S(x)$ the inequality $\langle x_1^*, x_2^* \rangle \geq \alpha^2 + \theta_x(\eta)$ holds for all x_2^* in $\overline{\mathrm{co}}\left(\frac{1}{d_S(x)}(x - \operatorname{Proj}_{S,\eta} x)\right)$ since any set $\{u^* \in H : \langle x_1^*, u^* \rangle \geq t\}$ is closed and convex, hence by (8.53) the inequality holds for all $x_2^* \in \partial_C d_S(x)$. Fixing $x_2^* \in \partial_C d_S(x)$, the same argument gives that the above inequality holds for all x_1^*, x_2^* in $\partial_C d_S(x)$ according to (8.53) again. In particular, for each $x^* \in \partial_C d_S(x)$ we have $\|x^*\|^2 \geq \alpha^2 + \theta_x(\eta)$ for all $0 < \eta < \eta_x$, and hence $\|x^*\| \geq \alpha$.
(b) Under the assumption in (b), for any x_1^*, x_2^* in $\partial_C d_S(x)$ the inclusion $(x_1^* + x_2^*)/2 \in \partial_C d_S(x)$ due to the convexity of $\partial_C d_S(x)$ ensures that

$$\alpha^2 \leq \left\| \frac{x_1^* + x_2^*}{2} \right\|^2 = \frac{\|x_1^*\|^2}{4} + \frac{\|x_2^*\|^2}{4} + \frac{1}{2}\langle x_1^*, x_2^* \rangle \leq \frac{1}{2} + \frac{1}{2}\langle x_1^*, x_2^* \rangle,$$

which justifies the inequality in (b). □

REMARK 8.109. The inequality in (b) above is *sharp*. Indeed, for the closed set $S = \operatorname{epi}(-|\cdot|)$ in $\mathbb{R}^2$ we obtained in Example 2.96(b) that $d_S(r, s) = ||r| + s|$ for $(r, s) \notin S$. Consequently, for each $(r, s) \in \mathbb{R}^2 \setminus S$

$$\partial_C d_S(r, s) = \begin{cases} \{\frac{1}{\sqrt{2}}(1, -1)\} & \text{if } r < 0 \\ \{\frac{1}{\sqrt{2}}(1 - 2t, -1) : t \in [0, 1]\} & \text{if } r = 0 \\ \{\frac{1}{\sqrt{2}}(-1, -1)\} & \text{if } r > 0, \end{cases}$$

so $\inf_{(r,s) \notin S} \operatorname{dist}((0, 0), \partial_C d_S(r, s)) = 1/\sqrt{2}$. Setting $\alpha := 1/\sqrt{2}$ the origin is exactly α-far away from the C-subdifferential of d_S on $X \setminus S$. We note for $(0, s) \notin S$ that both $x_1^* := \frac{1}{\sqrt{2}}(-1, -1)$ and $x_2^* := \frac{1}{\sqrt{2}}(1, -1)$ are in $\partial_C d_S(0, s)$ and satisfy the equality $\langle x_1^*, x_2^* \rangle = 0 = 2\alpha^2 - 1$. This confirms the sharpness of the inequality in (b) in Lemma 8.108. □

Recall that, given $r > 0$ and a subset S of a normed space X, the open r-tube around S is the set

$$\operatorname{Tube}_r(S) := \{x \in X : 0 < d_S(x) < r\}.$$

THEOREM 8.110 (A. Jourani and E. Vilches). Let S be a nonempty closed set in a Hilbert space H with $S \neq H$. The following hold.
(a) If S is metrically subsmooth at a point $\overline{x} \in \operatorname{bdry} S$, then for each $\varepsilon \in]0, 1[$ there exists a real $\delta > 0$ such that $\sqrt{1 - \varepsilon} \leq \operatorname{dist}(0, \partial_C d_S(x))$ for all $x \in B(\overline{x}, \delta) \setminus S$.
(b) If S is uniformly subsmooth, then for each $\varepsilon \in]0, 1[$ there exists a real $r > 0$

such that the origin is kept $\sqrt{1-\varepsilon}$-far away from the C-subdifferential of d_S on the open r-tube $\mathrm{Tube}_r(S)$, that is,

$$\sqrt{1-\varepsilon} \le \inf_{x\in \mathrm{Tube}_r(S)} \mathrm{dist}(0, \partial_C d_S(x)).$$

PROOF. (a) Fix $\varepsilon \in]0,1[$ and choose $0 < \delta_0 < 2$ such that $\langle z_1^* - z_2^*, z_1 - z_2\rangle \ge -\varepsilon\|z_1 - z_2\|$ for all $z_i \in S \cap B(\overline{x}, \delta_0)$ and $z_i^* \in \partial_C d_S(z_i)$, for $i = 1,2$. Put $\delta := \delta_0/4$ and fix any $x \in B(\overline{x}, \delta) \setminus S$. Put also $\overline{\eta} := \delta$ and fix any positive real $\eta < \overline{\eta}$. For $i = 1,2$ take $x_i^* = (x - z_i)/d_S(x)$ with $z_i \in \mathrm{Proj}\,_{S,\eta} x$, so $\|x - z_i\| \le \|x - u\| + \eta$ for all $u \in S$. The Ekeland variational principle (see Theorem 2.221) furnishes some $u_i \in S$ such that
(i) $\|z_i - u_i\| \le \sqrt{\eta}$, $\|u_i - x\| + \sqrt{\eta}\|z_i - u_i\| \le \|z_i - x\|$
(ii) $\|u_i - x\| \le \|u - x\| + \sqrt{\eta}\|u - u_i\|$ for all $u \in S$.
By Lemma 2.213 (of penalization) the point u_i is a global minimizer on H of the function $u \mapsto (1+\sqrt{\eta})d_S(u) + \|u - x\| + \sqrt{\eta}\|u - u_i\|$. Noticing that $u_i \ne x$ (because $u_i \in S$ and $x \notin S$) it ensues that

$$0 \in (1+\sqrt{\eta})\partial_C d_S(u_i) + \frac{u_i - x}{\|u_i - x\|} + \sqrt{\eta}\mathbb{B}_H,$$

hence there is $b_i \in \mathbb{B}_H$ such that

$$u_i^* := \frac{x - u_i}{\|x - u_i\|} + \sqrt{\eta} b_i \in (1+\sqrt{\eta})\partial_C d_S(u_i).$$

It results that

$$x_i^* - u_i^* = \frac{x - u_i}{d_S(x)} - \frac{x - u_i}{\|x - u_i\|} + \frac{u_i - z_i}{d_S(x)} - \sqrt{\eta} b_i,$$

which entails that

$$\begin{aligned}\|x_i^* - u_i^*\| &\le \left(\frac{1}{d_S(x)} - \frac{1}{\|x - u_i\|}\right)\|x - u_i\| + \frac{\|u_i - z_i\|}{d_S(x)} + \sqrt{\eta}\\ &= \frac{\|x - u_i\|}{d_S(x)} - 1 + \frac{\|u_i - z_i\|}{d_S(x)} + \sqrt{\eta}.\end{aligned}$$

This combined with the inequalities $\|u_i - x\| \le \|z_i - x\| \le d_S(x) + \eta$ and $\|z_i - u_i\| \le \sqrt{\eta}$ gives

$$\|x_i^* - u_i^*\| \le \frac{\eta}{d_S(x)} + \frac{\sqrt{\eta}}{d_S(x)} + \sqrt{\eta} \le \sqrt{\eta}\left(\frac{2}{d_S(x)} + 1\right).$$

Observing that

$$\|u_i - \overline{x}\| \le \|u_i - x\| + \|x - \overline{x}\| < d_S(x) + \eta + \delta < \eta + 2\delta < 4\delta,$$

we also have

$$\langle u_1^* - u_2^*, u_1 - u_2\rangle \ge -\varepsilon(1+\sqrt{\eta})\|u_1 - u_2\|. \tag{8.54}$$

Further, from the definition of u_i^* we see that

$$\langle u_1^* - u_2^*, u_1 - u_2\rangle \le \left\langle \frac{x - u_1}{\|x - u_1\|} - \frac{x - u_2}{\|x - u_2\|}, u_1 - u_2\right\rangle + 2\sqrt{\eta}\|u_1 - u_2\|,$$

so writing

$$\begin{aligned}
&\left\langle \frac{x-u_1}{\|x-u_1\|} - \frac{x-u_2}{\|x-u_2\|}, u_1 - u_2 \right\rangle \\
&= \left\langle \frac{x-u_1}{\|x-u_1\|} - \frac{x-u_2}{\|x-u_2\|}, (u_1 - x) - (u_2 - x) \right\rangle \\
&= -\|x-u_1\| - \|x-u_2\| + \left\langle \frac{x-u_1}{\|x-u_1\|}, x-u_2 \right\rangle + \left\langle \frac{x-u_2}{\|x-u_2\|}, x-u_1 \right\rangle \\
&= -\|x-u_1\| - \|x-u_2\| + (\|x-u_2\| + \|x-u_1\|) \left\langle \frac{x-u_1}{\|x-u_1\|}, \frac{x-u_2}{\|x-u_2\|} \right\rangle,
\end{aligned}$$

we deduce according to the equality $(x-u_i)/\|x-u_i\| = u_i^* - \sqrt{\eta} b_i$ that

$$\begin{aligned}
&\langle u_1^* - u_2^*, u_1 - u_2 \rangle \\
&\le (\|x-u_1\| + \|x-u_2\|)[-1 + \langle u_1^* - \sqrt{\eta} b_1, u_2^* - \sqrt{\eta} b_2 \rangle] + 2\sqrt{\eta}\|u_1 - u_2\| \\
&\le (-1 + 5\sqrt{\eta} + \langle u_1^*, u_2^* \rangle)(\|x-u_1\| + \|x-u_2\|).
\end{aligned}$$

Combining with (8.54) both the latter inequality and the inequality

$$-\varepsilon(1+\sqrt{\eta})\|u_1 - u_2\| \ge -\varepsilon(1+\sqrt{\eta})(\|x-u_1\| + \|x-u_2\|),$$

and dividing by $\|x-u_1\| + \|x-u_2\|$ it follows that

$$-1 + 5\sqrt{\eta} + \langle u_1^*, u_2^* \rangle \ge -\varepsilon(1+\sqrt{\eta}),$$

or equivalently

$$\langle u_1^*, u_2^* \rangle \ge 1 - \varepsilon(1+\sqrt{\eta}) - 5\sqrt{\eta}. \tag{8.55}$$

On the other hand, we have

$$\begin{aligned}
\langle u_1^*, u_2^* \rangle &= \langle u_1^* - x_1^*, u_2^* - x_2^* \rangle + \langle u_1^* - x_1^*, x_2^* \rangle + \langle x_1^*, u_2^* - x_2^* \rangle + \langle x_1^*, x_2^* \rangle \\
&\le \eta \left(\frac{2}{d_S(x)} + 1 \right)^2 + 2\sqrt{\eta} \left(\frac{2}{d_S(x)} + 1 \right) \left(\frac{\eta}{d_S(x)} + 1 \right) + \langle x_1^*, x_2^* \rangle \\
&\le 3\sqrt{\eta} \left(\frac{2}{d_S(x)} + 1 \right)^2 + \langle x_1^*, x_2^* \rangle.
\end{aligned}$$

From the latter inequality and (8.55) it results that

$$\begin{aligned}
\langle x_1^*, x_2^* \rangle &\ge 1 - \varepsilon(1+\sqrt{\eta}) - 5\sqrt{\eta} - 3\sqrt{\eta} \left(\frac{2}{d_S(x)} + 1 \right)^2 \\
&\ge 1 - \varepsilon - 9\sqrt{\eta} \left(\frac{2}{d_S(x)} + 1 \right)^2,
\end{aligned}$$

thus Lemma 8.108(a) guarantees that the origin is kept $\sqrt{1-\varepsilon}$-far away from the C-subdifferential of d_S on $B(\overline{x}, \delta) \setminus S$.

(b) Fix any $\varepsilon \in]0,1[$ and choose $0 < \delta_0 < 2$ such that $\langle z_1^* - z_2^*, z_1 - z_2 \rangle \ge -\varepsilon\|z_1 - z_2\|$ for all $z_i \in S$ with $\|z_1 - z_2\| < \delta_0$ and all $z_i^* \in \partial_C d_S(z_i)$, for $i = 1, 2$. Put $\delta := \delta_0/4$ and fix any $x \in \mathrm{Tube}_\delta(S)$. There is some $\overline{x} \in \mathrm{bdry}\, S$ such that $x \in B(\overline{x}, \delta) \setminus S$. Then the proof in (a) shows that $\sqrt{1-\varepsilon} \le \mathrm{dist}(0, \partial_C d_S(x))$, which justifies (b). □

8.7. Preservation of subsmoothness under operations

Consider the function $g : \mathbb{R} \to \mathbb{R}$ defined by $g(0) = 0$ and

$$g(x) := |x|^3 \left(1 - \cos \frac{1}{x}\right) \quad \text{for all } x \in \mathbb{R} \setminus \{0\}, \tag{8.56}$$

and consider also the linear mapping $A : \mathbb{R}^2 \to \mathbb{R}^2$ with $A(x,r) := (x,0)$ for all $(x,r) \in \mathbb{R}^2$. The function g being (easily seen to be) of class $\mathcal{C}^1$ on $\mathbb{R}$, it is locally Lipschitz and subsmooth on $\mathbb{R}$, so by Corollary 8.86 the set $S := \operatorname{epi} g$ is subsmooth (and hence metrically subsmooth). We observe that

$$A^{-1}(S) = \left(\{0\} \cup \left\{\pm \frac{1}{2k\pi} : k \in \mathbb{N}\right\}\right) \times \mathbb{R} =: Q,$$

and the latter set is not subsmooth at $(0,0)$, since it is not even tangentially regular at $(0,0)$ due to the fact that

$$T^C(Q;(0,0)) = \{0\} \times \mathbb{R} \quad \text{and} \quad T^B(Q;(0,0)) = \mathbb{R} \times \mathbb{R}.$$

This says that the subsmoothness (resp. metric subsmoothness) property is not preserved by inverse image with a continuous linear mapping.

On the other hand, considering the closed convex set $S' := \mathbb{R} \times \{0\}$ we also see that $S \cap S' = \big(\{0\} \cup \{\pm \frac{1}{2k\pi} : k \in \mathbb{N}\}\big) \times \{0\}$, and the latter set is not even tangentially regular at $(0,0)$. This is a counterexample for the preservation of subsmoothness (resp. metric subsmoothness) under intersection.

Accordingly, such desired preservation properties require additional conditions.

DEFINITION 8.111. Let $G : X \to Y$ be a mapping between two normed spaces and let S be a subset of Y. Suppose that G is of class $\mathcal{C}^1$ near $\overline{x} \in G^{-1}(S)$. We say that the inverse image set-representative $G^{-1}(S)$ has the (local) *truncated C-normal cone inverse image property* near $\overline{x}$ with a real constant $\gamma > 0$ provided there is a neighborhood U of $\overline{x}$ such that

$$N^C(G^{-1}(S);x) \cap \mathbb{B}_{X^*} \subset DG(x)^* \left(N^C(S;G(x)) \cap (\gamma \mathbb{B}_{Y^*})\right) \quad \text{for all } x \in U \cap G^{-1}(S), \tag{8.57}$$

where $DG(x)^*$ denotes the adjoint of the derivative mapping $DG(x)$ of G at x. If in place of the C-normal cones the above inclusion holds true with a normal cone $\mathcal{N}(\cdot,\cdot)$ in both members, one says that the truncated normal cone inverse image property at $\overline{x}$ holds with respect to the normal cone $\mathcal{N}(\cdot,\cdot)$. We will also say, for short, that the $\mathcal{N}$-normal cone inverse image property is satisfied at $\overline{x}$. When G is of class $\mathcal{C}^1$ on an open set containing $G^{-1}(S)$ and the inclusion (8.57) with C-normal (resp. L-normal, etc) cones holds with the same real $\gamma > 0$ for all $x \in G^{-1}(S)$, one says that the set-representative has the *global truncated C-normal (resp. $\mathcal{N}$-normal) cone inverse image property* with constant γ.

Similarly, subdifferential of distance function can be employed in place of normal cone. This corresponds to the following:

DEFINITION 8.112. Let G and S be as in Definition 8.111. We say that the inverse image set-representative $G^{-1}(S)$ has the (local) *linear inclusion property of C-subdifferential of distance from inverse image* near $\overline{x}$ (relative to $G^{-1}(S)$) with a real constant $\gamma > 0$, when there exists a neighborhood U of $\overline{x}$ such that

$$\partial_C d\left(\cdot, G^{-1}(S)\right)(x) \subset \gamma\, DG(x)^* \left(\partial_C d(\cdot,S)(G(x))\right) \tag{8.58}$$

for all $x \in U \cap G^{-1}(S)$. One defines analogously the linear inclusion property of L-subdifferential of distance function from inverse image. If G is of class $\mathcal{C}^1$ on an open set containing $G^{-1}(S)$ and if the inclusion (8.58) with C-subdifferential (resp. L-subdifferential, etc) holds with the same real $\gamma > 0$ for all $x \in G^{-1}(S)$, one says that the *global linear inclusion property is satisfied with constant* γ *for* C*-subdifferential (resp. L-subdifferential) of distance function to* $G^{-1}(S)$.

Note that anyone of properties (8.57) and (8.58) holds if and only if it holds for all $x \in U \cap \operatorname{bdry}\big(G^{-1}(S)\big)$.

The other important concept of metric subregularity related to the distance function from the set $G^{-1}(S)$ does not require the differentiability of the mapping G. We recall (see (7.23)) that the mapping G is *metrically subregularly transversal* at $\overline{x}$ to the set S, or the *metric subregularity qualification condition* is satisfied at $\overline{x}$ for the inverse image of the set S by G, if there exists some real constant $\gamma > 0$ and some neighborhood U of $\overline{x}$ such that

$$d(x, G^{-1}(S)) \le \gamma\, d(G(x), S) \quad \text{for all } x \in U. \tag{8.59}$$

We already know by Proposition 2.138 that condition (8.59) entails that property (8.58) holds true with C-subdifferential of distance function. We use this in the following theorem.

THEOREM 8.113 (subsmoothness of inverse images). Let $G : X \to Y$ be a mapping between normed spaces X and Y and let S be a subset of the space Y. Assume that G is of class $\mathcal{C}^1$ near $\overline{x} \in G^{-1}(S)$. The following hold.
(a) If the set S is subsmooth at $G(\overline{x})$ and if the truncated C-normal cone inverse image property is satisfied for $G^{-1}(S)$ near $\overline{x}$, then $G^{-1}(S)$ is subsmooth at $\overline{x}$.
(b) If the set S is metrically subsmooth at $G(\overline{x})$ and if the linear inclusion property (8.58) for C-subdifferential of distance function to $G^{-1}(S)$ holds (which is the case in particular whenever the mapping G is metrically subregularly transversal at $\overline{x}$ to the set S of the space Y), then the set $G^{-1}(S)$ is metrically subsmooth at $\overline{x}$.

PROOF. (a) Fix any $\varepsilon > 0$. Take some neighborhood U' of $\overline{x}$ over which (8.57) holds for some constant real number $\gamma > 0$ and over which the mapping G is Lipschitz continuous with Lipschitz constant $\beta > 0$. By definition of subsmooth set, choose a neighborhood V of $G(\overline{x})$ such that for all $y, y' \in S \cap V$ and $y^* \in N^C(S; y) \cap \mathbb{B}_{Y^*}$ we have

$$\langle y^*, y' - y\rangle \le (2\beta\gamma)^{-1}\varepsilon\|y' - y\|. \tag{8.60}$$

Take a convex neighborhood $U \subset U'$ of $\overline{x}$ such that $G(U) \subset V$ and such that, by the continuity of the derivative mapping $DG(\cdot)$, we have $\|DG(x') - DG(x)\| \le (2\gamma)^{-1}\varepsilon$ for all $x, x' \in U$. Fix any $x, x' \in U \cap G^{-1}(S)$ and $x^* \in N^C(G^{-1}(S); x) \cap \mathbb{B}_{X^*}$. By (8.57) there exists some $y^* \in N^C(S; G(x)) \cap (\gamma\mathbb{B}_{Y^*})$ such that $x^* = y^* \circ DG(x)$. Write

$$G(x') - G(x) = DG(x)(x' - x) + \int_0^1 \big(DG(x + t(x' - x)) - DG(x)\big)(x' - x)\, dt.$$

Then we have

$$\begin{aligned}
&\langle x^*, x' - x\rangle \\
&= \langle y^*, DG(x)(x' - x)\rangle \\
&= \langle y^*, G(x') - G(x)\rangle - \textstyle\int_0^1 \langle y^*, \big(DG(x + t(x' - x)) - DG(x)\big)(x' - x)\rangle\, dt \\
&\le \langle y^*, G(x') - G(x)\rangle + (2\gamma)^{-1}\varepsilon\|y^*\|\|x' - x\|.
\end{aligned}$$

Taking (8.60) and the inequality $\|y^*\| \le \gamma$ into account, we obtain

$$\langle x^*, x' - x\rangle \le (2\beta)^{-1}\varepsilon\|G(x') - G(x)\| + (\varepsilon/2)\|x' - x\|,$$

and hence according to the Lipschitz continuous behavior of G with Lipschitz constant β over U

$$\langle x^*, x' - x\rangle \le \varepsilon\|x' - x\|.$$

This means that the set $G^{-1}(S)$ is subsmooth at $\overline{x}$.

(b) Let $\varepsilon > 0$. As above, fix some open neighborhood U' of $\overline{x}$ for which (8.58) holds with some constant $\gamma > 0$ and over which G is Lipschitz continuous with some Lipschitz constant $\beta > 0$. According to the definition of metric subsmoothness of S take some neighborhood V of $G(\overline{x})$ such that for all $y, y' \in S \cap V$ and $y^* \in \partial_C d_S(G(x))$ we have

$$\langle y^*, y' - y\rangle \le (2\beta\gamma)^{-1}\varepsilon\|y' - y\|. \tag{8.61}$$

For $Q := G^{-1}(S)$, by (8.58) we have for all $x \in U' \cap Q$

$$\partial_C d_Q(x) \subset \gamma\, DG(x)^*(\partial_C d_S(G(x))). \tag{8.62}$$

Take a convex neighborhood U of $\overline{x}$ with $U \subset U'$ and such that

$$\|DG(x') - DG(x)\| \le (2\gamma)^{-1}\varepsilon \quad \text{for all } x, x' \in U.$$

Of course, the inclusion (8.62) holds for all $x \in Q \cap U$. Fix any $x, x' \in Q \cap U$ and $x^* \in \partial_C d_Q(x)$. By (8.62) there exists some $y^* \in \partial_C d_S(G(x))$ such that $x^* = \gamma(y^* \circ DG(x))$. As in (a) above, writing

$$G(x') - G(x) = DG(x)(x' - x) + \int_0^1 \left(DG(x + t(x' - x)) - DG(x)\right)(x' - x)\,dt$$

gives

$$\begin{aligned}
&\langle x^*, x' - x\rangle \\
&= \gamma\langle y^*, G(x') - G(x)\rangle - \gamma\textstyle\int_0^1 \langle y^*, \left(DG(x + t(x' - x)) - DG(x)\right)(x' - x)\rangle\,dt \\
&\le \gamma\langle y^*, G(x') - G(x)\rangle + (2\gamma)^{-1}\gamma\varepsilon\|y^*\|\|x' - x\|.
\end{aligned}$$

Invoking (8.61) and the inequality $\|y^*\| \le 1$, we see that

$$\langle x^*, x' - x\rangle \le (2\beta)^{-1}\varepsilon\|G(x') - G(x)\| + (\varepsilon/2)\|x' - x\|,$$

and hence

$$\langle x^*, x' - x\rangle \le \varepsilon\|x' - x\|.$$

We then obtain that the set $Q = G^{-1}(S)$ is metrically subsmooth at $\overline{x}$. Finally, the case of metric subregularity of G follows from Proposition 2.138 (as said just before the statement of the theorem). □

Let $S := \{x \in X : g_1(x) \le 0, \cdots, g_m(x) \le 0\}$ and $\overline{x} \in S$, where X is a normed space and $g_1, \cdots, g_m : X \to \mathbb{R}$ are functions which are of class $\mathcal{C}^1$ on an open neighborhood U of $\overline{x}$. Let $K := \{1, \cdots, m\}$ and $K(x) := \{k \in K : g_k(x) = \max_{j \in K} g_j(x)\}$ for each x. Assume that there is a real $\sigma > 0$ such that for each $x \in U$ there is $\overline{v} \in X$ (depending on x) for which $\langle Dg_k(x), \overline{v}\rangle \le -\sigma$ for every $k \in K(x)$. Defining $G : X \to \mathbb{R}^m$ by $G(x) := (g_1(x), \cdots, g_m(x))$ and putting $S' := \mathbb{R}^m_+$ we see that $S = G^{-1}(S')$ and G is of class $\mathcal{C}^1$ on U. We claim that the truncated C-normal cone inverse image property (8.57) is satisfied. Indeed, let any $x \in U \cap G^{-1}(S')$. We may suppose that $x \in \operatorname{bdry}(G^{-1}(S'))$. Take any $x^* \in N^C(G^{-1}(S')) \cap \mathbb{B}_{X^*}$. We know (see Proposition 2.196) that there is $\lambda \in \mathbb{R}^m_+$

such that $x^* = \sum_{k=1}^m \lambda_k Dg_k(x) = DG(x)(\lambda)$ and such that $\lambda_k = 0$ for $k \notin K(x)$, that is, $\lambda \in N^C(\mathbb{R}_+^m; G(x))$. Endowing $\mathbb{R}^m$ with its natural Euclidean norm and considering the vector $\overline{v}$ given by the assumption we can write

$$\langle x^*, -\overline{v} \rangle = - \sum_{k \in K(x)} \lambda_k \langle Dg_k(x), \overline{v} \rangle \geq \sigma \sum_{k \in K(x)} \lambda_k \geq \sigma \|\lambda\|,$$

which ensures that $\|x^*\| \geq \sigma \|\lambda\|$, and hence $\|\lambda\| \leq 1/\sigma$. It ensues with $\gamma := 1/\sigma > 0$ that for all $x \in U \cap G^{-1}(S'))$ one has

$$N^C(G^{-1}(S'); x) \cap \mathbb{B}_{X^*} \subset DG(x)^* (N^C(S'; G(x)) \cap (\gamma \mathbb{B}_{R^m})),$$

which translates the desired truncated inverse image property. Theorem 8.113(a) then tells us that the set $S = G^{-1}(S')$ is subsmooth at $\overline{x}$. In fact, the next proposition says more and remove the $\mathcal{C}^1$ property of g_i. The uniform subsmoothness of sublevel sets (even the uniform equi-subsmoothness of families of such sets) will be studied in Proposition 8.135.

Proposition 8.114. Let $K := \{1, \cdots, m\}$ and $S = \{x \in X : g_k(x) \leq 0,\ \forall k \in K\}$ be a subset of a normed space X, where the functions $g_1, \ldots, g_m : X \to \mathbb{R}$ are locally Lipschitz on a neighborhood U of a point $\overline{x} \in S$. Assume that the functions $g_1, \cdots, g_m$ are subsmooth at $\overline{x}$ and assume also that the following generalized Slater condition holds: there exists a real $\sigma > 0$ such that for each $x \in U \cap \operatorname{bdry} S$ there exists a vector $\overline{v} \in \mathbb{B}_X$ (depending on x) for which for every $k \in K(x) := \{k \in K : g_k(x) = \max_{j \in K} g_j(x)\}$

$$g_k^\circ(x; \overline{v}) \leq -\sigma$$

(or equivalently $\langle x^*, \overline{v} \rangle \leq -\sigma$ for all $x^* \in \partial_C g_k(x)$). Then the set S is subsmooth at $\overline{x}$.

Proof. Define $g : X \to \mathbb{R}$ by $g(x) := \max_{k \in K} g_k(x)$ for all $x \in X$ and note that $S = \{x \in X : g(x) \leq 0\}$. By Proposition 2.190 we have

$$\partial_C g(x) \subset \operatorname{co} \left(\bigcup_{k \in K(x)} \partial_C g_k(x) \right) \quad \text{for all } x \in U \cap \operatorname{bdry} S. \tag{8.63}$$

This inclusion and the assumption on $\overline{v}$ give us

$$0 \notin \partial_C g(x) \quad \text{for all } x \in U \cap \operatorname{bdry} S.$$

By Proposition 2.196 one has

$$N^C(S; x) \subset \mathbb{R}_+ \partial_C g(x) \quad \text{for all } x \in U \cap \operatorname{bdry} S. \tag{8.64}$$

Take any $\varepsilon > 0$ and put $\varepsilon' := \varepsilon\sigma$. The subsmoothness assumption allows us by Proposition 8.41 to choose $\delta > 0$ with $B(\overline{x}, \delta) \subset U$ such that for any $x, y \in S \cap B(\overline{x}, \delta)$, for any $k \in K$, for any $x^* \in \partial_C g_k(x)$, and for any $y^* \in \partial_C g_k(y)$

$$\langle x^* - y^*, x - y \rangle \geq -\varepsilon' \|x - y\|.$$

Fix any $x \in S \cap B(\overline{x}, \delta)$ with $x \in \operatorname{bdry} S$ and any $u^* \in N^C(S; x) \cap \mathbb{B}_{X^*}$. By (8.64) choose a real $\alpha \geq 0$ and $x^* \in \partial_C g(x)$ such that $u^* = \alpha x^*$. From (8.63) there are $x_k^* \in \partial g_k(x)$ and $\lambda_k \geq 0$ with $\lambda_k = 0$ if $k \notin K(x)$ and with $\sum_{k \in K} \lambda_k = 1$, such that $x^* = \sum_{k \in K} \lambda_k x_k^*$. Fix any $y \in S \cap B(\overline{x}, \delta)$. By the mean value equality with C-subdifferential choose for each $k \in K(x)$ some $z_k := x + t_k(y - x)$ with $t_k \in]0, 1[$

and some $z_k^* \in \partial_C g_k(z_k)$ such that $g_k(y) - g_k(x) = \langle z_k^*, y - x\rangle$. Then for each $k \in K(x)$ writing

$$\begin{aligned} 0 \geq g_k(y) - g_k(x) &= \langle z_k^*, y - x\rangle \\ &= \langle z_k^* - x_k^*, y - x\rangle + \langle x_k^*, y - x\rangle \\ &= \frac{1}{t_k}\langle z_k^* - x_k^*, z_k - x\rangle + \langle x_k^*, y - x\rangle, \end{aligned}$$

we obtain that

$$0 \geq -\frac{1}{t_k}\varepsilon'\|z_k - x\| + \langle x_k^*, y - x\rangle = -\varepsilon'\|y - x\| + \langle x_k^*, y - x\rangle,$$

which means that $\langle x_k^*, y - x\rangle \leq \varepsilon\sigma\|y - x\|$. Recalling that $\lambda_k = 0$ if $k \notin K(x)$ it ensues that $\langle x^*, y - x\rangle \leq \varepsilon\sigma\|y - x\|$. On the other hand, using again the equality $\lambda_k = 0$ if $k \notin K(x)$ as well as the assumption on $\overline{v} \in \mathbb{B}_X$, we also have $\langle x^*, \overline{v}\rangle \geq -\sigma$. Therefore, the above equality $u^* = \alpha x^*$ gives

$$1 \geq \|u^*\| \geq \langle u^*, -\overline{v}\rangle = -\alpha\langle x^*, \overline{v}\rangle \geq \alpha\sigma,$$

so $\alpha \leq 1/\sigma$. It follows that

$$\langle u^*, y - x\rangle = \alpha\langle x^*, y - x\rangle \leq \varepsilon\|y - x\|,$$

and this inequality still holds when $x \in B(\overline{x}, \delta) \cap \operatorname{int} S$. This confirms that the set S is subsmooth at $\overline{x}$. □

We already observed in Proposition 8.92(c) that the uniform subsmoothness and the uniform metric subsmoothness of a set coincide in Banach spaces. The next theorem provides a result of preservation of such a property for inverse images.

THEOREM 8.115 (uniform subsmoothness of inverse images). Let $G : X \to Y$ be a mapping between normed spaces X and Y and let S be a uniformly subsmooth subset of Y. Assume that G is Lipschitz on $G^{-1}(S)$ and differentiable on an open enlargement of $G^{-1}(S)$ with DG uniformly continuous therein. If the global truncated C-normal cone inverse image property (resp. the global linear inclusion property for C-subdifferential of distance function to $G^{-1}(S)$) is satisfied with a same constant $\gamma > 0$, then the set $G^{-1}(S)$ is uniformly subsmooth (resp. uniformly metrically subsmooth).

PROOF. We make the proof under the global truncated C-normal cone inverse image property. Let $\beta > 0$ be a Lipschitz constant of G over $G^{-1}(S)$ and let $r > 0$ be such that on $U_r(G^{-1}(S)) := \{x \in X : d(x, G^{-1}(S)) < r\}$ the mapping G is differentiable with DG uniformly continuous therein. Fix any $\varepsilon > 0$. By definition of uniformly subsmooth set, choose $\eta > 0$ such that for any $y, y' \in S$ with $\|y' - y\| \leq \eta$ and any $y^* \in N^C(S; y) \cap \mathbb{B}_{Y^*}$ we have

$$\langle y^*, y' - y\rangle \leq (2\beta\gamma)^{-1}\varepsilon\|y' - y\|. \tag{8.65}$$

Choose $\delta \in]0, r[$ such that $\|G(x') - G(x)\| \leq \eta$ for all $x, x' \in G^{-1}(S)$ with $\|x' - x\| \leq \delta$ and such that $\|DG(x') - DG(x)\| \leq (2\gamma)^{-1}\varepsilon$ for all $x, x' \in U_r(G^{-1}(S))$ with $\|x' - x\| \leq \delta$. Then fix any $x, x' \in G^{-1}(S)$ with $\|x' - x\| \leq \delta$ and any $x^* \in N^C(G^{-1}(S); x) \cap \mathbb{B}_{X^*}$. By (8.57) there exists some $y^* \in N(S; G(x)) \cap (\gamma\mathbb{B}_{Y^*})$

such that $x^* = y^* \circ DG(x)$. We continue like in the proof of Theorem 8.113(a). Write

$$G(x') - G(x) = DG(x)(x' - x) + \int_0^1 (DG(x + t(x' - x)) - DG(x))\,(x' - x)\,dt,$$

and note for every $t \in [0, 1]$ that $x + t(x' - x) \in U_r(G^{-1}(S))$ since

$$d_{G^{-1}(S)}(x + t(x' - x)) \le \|x + t(x' - x) - x\| = t\|x' - x\| \le \delta < r.$$

Then we have

$$\begin{aligned} &\langle x^*, x' - x\rangle \\ &= \langle y^*, DG(x)(x' - x)\rangle \\ &= \langle y^*, G(x') - G(x)\rangle - \textstyle\int_0^1 \langle y^*, (DG(x + t(x' - x)) - DG(x))\,(x' - x)\rangle\,dt \\ &\le \langle y^*, G(x') - G(x)\rangle + (2\gamma)^{-1}\varepsilon\|y^*\|\|x' - x\|. \end{aligned}$$

Using (8.65) and the inequality $\|y^*\| \le \gamma$, it ensues that

$$\langle x^*, x' - x\rangle \le (2\beta)^{-1}\varepsilon\|G(x') - G(x)\| + (\varepsilon/2)\|x' - x\| \le \varepsilon\|x' - x\|,$$

which confirms that the set $G^{-1}(S)$ is uniformly subsmooth. □

We turn now to the case when $DG(\overline{x})$ is surjective. Under such a surjectivity assumption, we obtain the assertion (a) below of preservation of metric subsmoothness of inverse images. The assertion (c) is concerned with the uniform equi-subsmoothness.

PROPOSITION 8.116. *Let $G : X \to Y$ be a mapping between Banach spaces X and Y and let S be a subset of Y. Let also $(S_i)_{i\in I}$ be a family of subsets of Y and $(G_i)_{i\in I}$ be a family of mappings from X into Y such that $G_i^{-1}(S_i) \ne \emptyset$ for all $i \in I$.*
(a) Assume that G is of class $\mathcal{C}^1$ near $\overline{x} \in G^{-1}(S)$ with $DG(\overline{x})$ surjective and that the set S is subsmooth (resp. metrically subsmooth) at $G(\overline{x})$. Then the set $G^{-1}(S)$ is subsmooth (resp. metrically subsmooth) at $\overline{x}$.
(b) Assume that S is uniformly subsmooth, G is Lipschitz on $G^{-1}(S)$ and differentiable on an open enlargement $U_r(G^{-1}(S))$ with DG uniformly continuous therein. Assume also that there is a real $\rho > 0$ such that

$$\rho\mathbb{B}_Y \subset DG(x)(\mathbb{B}_X) \quad \text{for all } x \in G^{-1}(S).$$

Then the set $G^{-1}(S)$ is uniformly subsmooth.
(c) Assume that the family of sets $(S_i)_{i\in I}$ is uniformly equi-subsmooth, that for each $i \in I$ the mapping G_i is Lipschitz on $G_i^{-1}(S_i)$ with a common Lipschitz constant $\gamma > 0$, and that there is $r \in]0, +\infty]$ such that G_i is differentiable on the open enlargement $U_r(G_i^{-1}(S_i))$ with $(DG_i)_{i\in I}$ uniformly equi-continuous relative to the family of open sets $(U_r(G_i^{-1}(S_i)))_{i\in I}$. Assume also that there is a real $\rho > 0$ such that for each $i \in I$

$$\rho\mathbb{B}_Y \subset DG_i(x)(\mathbb{B}_X) \quad \text{for all } x \in G_i^{-1}(S_i).$$

Then the family of sets $(G_i^{-1}(S_i))_{i\in I}$ is uniformly equi-subsmooth.

PROOF. (a) By Corollary 7.75 the mapping G is metrically regularly transversal at $\overline{x}$ to the subset S of the space Y. Thus, the metric subsmoothness of S at $G(\overline{x})$ entails the metric subsmoothness of $G^{-1}(S)$ at $\overline{x}$ according to Theorem 8.113(b).

Now suppose that S is subsmooth at $G(\overline{x})$. Choose by Lemma 8.68(b) a real $\gamma > 0$ and an open neighborhood U of $\overline{x}$ where G is $\mathcal{C}^1$ and such that for each $x \in U$ the continuous linear mapping $DG(x)$ is surjective and $\|y^*\| \le \gamma\|x^*\|$ for all $x^* \in X^*$ and $y^* \in Y^*$ satisfying the equality $x^* = y^* \circ DG(x)$. For each $x \in U \cap G^{-1}(S)$, it ensues by the surjectivity of $DG(x)$ and by Theorem 3.174(a) that

$$N^C(G^{-1}(S); x) \subset DG(x)^*\big(N^C(S; G(x))\big). \tag{8.66}$$

Fix any $x \in U \cap G^{-1}(S)$ and any $x^* \in N^C(G^{-1}(S); x) \cap \mathbb{B}_{X^*}$. Then there is an element $y^* \in N^C(S; G(x))$ such that $x^* = DG(x)^*(y^*) = y^* \circ DG(x)$. It results by the choice of U that $\|y^*\| \le \gamma$, which gives

$$N^C(G^{-1}(S); x) \cap \mathbb{B}_{X^*} \subset DG(x)^*\big(N^C(S; G(x)) \cap (\gamma \mathbb{B}_{X^*})\big).$$

The subsmoothness of $G^{-1}(S)$ at $\overline{x}$ then follows from Theorem 8.113(a).
(c) Let us prove (c). Fix any $\varepsilon > 0$ and put $\varepsilon' := \varepsilon\rho/(1+\gamma))$. Choose a real $\delta > 0$ such that for each $i \in I$ we have $\langle y^*, y' - y\rangle \le \varepsilon' \|y' - y\|$ for all $y, y' \in S_i$ with $\|y' - y\| < \delta$ and all $y^* \in N^C(S_i; y) \cap \mathbb{B}_{Y^*}$. By the uniform equi-continuity of the family $(DG_i)_{i \in I}$ relative to $(U_r(G_i^{-1}(S_i)))_{i \in I}$ there is a positive real $\delta' < \min\{r, \delta/\gamma\}$ such that for any $i \in I$ and any $x, x' \in U_r(G_i^{-1}(S_i))$ with $\|x' - x\| < \delta'$ one has $\|DG_i(x') - DG_i(x)\| < \varepsilon'$. Fix any $i \in I$, any $x, x' \in G_i^{-1}(S_i)$ with $\|x' - x\| < \delta'$ and any $x^* \in N^C(G_i^{-1}(S_i); x) \cap \mathbb{B}_{X^*}$. As in (a) the surjectivity of $DG_i(x)$ entails by Theorem 3.174(a)

$$N^C(G_i^{-1}(S_i); x) \subset DG_i(x)^*\big(N^C(S_i; G(x))\big),$$

hence there exists some $y^* \in N^C(S_i; G_i(x))$ (depending on i, x) such that $x^* = y^* \circ DG_i(x)$. For any $b \in \mathbb{B}_Y$ taking by assumption some $u \in \mathbb{B}_X$ such that $\rho b = DG_i(x)(u)$, we obtain

$$\rho\langle y^*, b\rangle = \langle y^*, DG_i(x)(u)\rangle = \langle x^*, u\rangle \le 1,$$

thus $\|y^*\| \le 1/\rho$. Now, putting $z_t := x + t(x' - x)$ we have for every $t \in [0,1]$ that $\|z_t - x\| = t\|x' - x\| < \delta'$ and $z_t \in U_r(G_i^{-1}(S_i))$ since

$$\operatorname{dist}\big(z_t, G_i^{-1}(S_i)\big) \le \|z_t - x\| = t\|x' - x\| < \delta' < r.$$

Noticing that $\|G_i(x') - G_i(x)\| \le \gamma\|x' - x\|\gamma\delta' < \delta$ with $G_i(x'), G_i(x) \in S_i$, it ensues that

$$\begin{aligned}
\langle x^*, x' - x\rangle &= \langle y^*, DG_i(x)(x' - x)\rangle \\
&= \langle y^*, G_i(x') - G_i(x)\rangle - \int_0^1 \langle y^*, \big(DG_i(z_t) - DG_i(x)\big)(x' - x)\rangle\, dt \\
&\le (\varepsilon'/\rho)\|G_i(x') - G_i(x)\| + (\varepsilon'/\rho)\|x' - x\|,
\end{aligned}$$

which implies that

$$\langle x^*, x' - x\rangle \le \frac{\varepsilon'\gamma}{\rho}\|x' - x\| + \frac{\varepsilon'}{\rho}\|x' - x\| \le \varepsilon\|x' - x\|$$

according to the γ-Lipschitz assumption of G_i on $G_i^{-1}(S_i)$. The uniform equi-subsmoothness of the family of sets $(G_i^{-1}(S_i))_{i \in I}$ is established.
(b) The assertion (b) is clearly a particular case of (c). □

The next two corollaries apply Theorem 8.113 to graphs of certain basic multimappings.

COROLLARY 8.117. Let $G : X \to Y$ be a mapping between Banach spaces which is of class $\mathcal{C}^1$ near $\overline{x} \in X$. Let S be a subset of Y which is closed near a point $\overline{y} \in S$ and subsmooth (resp. metrically subsmooth) at $\overline{y}$. Then the graph of the multimapping $x \mapsto G(x) - S$ is subsmooth (resp. metrically subsmooth) at $(\overline{x}, G(\overline{x}) - \overline{y})$.

PROOF. Denoting by Γ the graph of the multimapping in the corollary, we see that $\Gamma = g^{-1}(S)$, where $g : X \times Y \to Y$ denotes the mapping defined by $g(x, y) := G(x) - y$. It is clear that g of class $\mathcal{C}^1$ near $(\overline{x}, \overline{y})$ with $Dg(\overline{x}, \overline{y})(u, v) = DG(\overline{x})(u) - v$, so $Dg(\overline{x}, \overline{y})$ is surjective. Assuming that the set S is subsmooth (resp. metrically subsmooth) at $\overline{y}$, Proposition 8.116(a) tells us that $\Gamma = g^{-1}(S)$ is subsmooth (resp. metrically subsmooth) at $(\overline{x}, g(\overline{x}, \overline{y}))$, that is, at $(\overline{x}, G(\overline{x}) - \overline{y})$. $\square$

COROLLARY 8.118. Let $g : X \to Y$ be a mapping between Banach spaces which is of class $\mathcal{C}^1$ near a point $\overline{x} \in X$ with $Dg(\overline{x})$ surjective. Let Z be another Banach space and $M : Y \rightrightarrows Z$ be a multimapping from Y into Z whose graph is closed near $(\overline{y}, \overline{z})$ in $\operatorname{gph} M$ and subsmooth (resp. metrically subsmooth) at $(\overline{y}, \overline{z})$, where $\overline{y} := g(\overline{x})$. Then the graph of the multimapping $M \circ g$ is subsmooth (resp. metrically subsmooth) at $(\overline{x}, \overline{z})$.

PROOF. Since $z \in M(g(x)) \Leftrightarrow (g(x), z) \in \operatorname{gph} M$, we see that $\operatorname{gph}(M \circ g) = G^{-1}(\operatorname{gph} M)$, where $G : X \times Z \to Y \times Z$ is defined by $G(x, z) := (g(x), z)$. The mapping G is obviously of class $\mathcal{C}^1$ near $(\overline{x}, \overline{z})$ with $DG(\overline{x}, \overline{z})(u, w) = (Dg(\overline{x})(u), w)$. We then see that $DG(\overline{x}, \overline{z})$ is surjective, hence Proposition 8.116(a) guarantees the desired subsmoothness (resp. metric subsmoothness) property of $\operatorname{gph}(M \circ g)$. $\square$

The next proposition provides another example extending the one in Proposition 8.70(b). We prove first two lemmas which have their own great interest.

LEMMA 8.119. Let C and D be closed convex sets of Banach spaces X and Y respectively and let $A : X \to Y$ be a continuous linear mapping. Let $\overline{x} \in C$ and $\overline{y} \in D$ such that

$$0 \in \operatorname{core}\big(A(C - \overline{x}) - (D - \overline{y})\big).$$

Then there exist reals $s > 0, \delta > 0$ and open neighborhoods U of $\overline{x}$ and V of $\overline{y}$ such that, for all $u \in U \cap C$, $v \in V \cap D$ and for every continuous linear mapping $\Lambda : X \to Y$ with $\|\Lambda - A\| < \delta$, one has

$$s\mathbb{B}_Y \subset \Lambda((C - u) \cap \mathbb{B}_X) - (D - v).$$

PROOF. For any continuous linear mapping $\Lambda : X \to Y$ and any $u \in C$ and $v \in D$, define $M_{u,v}^{\Lambda} : X \rightrightarrows Y$ by $M_{u,v}^{\Lambda}(x) := \Lambda(x) - (D - v)$ if $x \in C - u$ and $M_{u,v}^{\Lambda}(x) = \emptyset$ otherwise. Observe that $\operatorname{gph} M_{u,v}^{\Lambda}$ is closed convex and $0 \in M_{u,v}^{\Lambda}(0)$. Setting $M := M_{\overline{x},\overline{y}}^{A}$, we note in addition that $0 \in \operatorname{core} M(X)$ by the core-assumption of the statement. The Robinson-Ursescu theorem (see Theorem 7.26) says that there is a real $r > 0$ such that $r\mathbb{B}_Y \subset M(\mathbb{B}_X)$ (keep in mind that $0 \in M(0)$), which means

$$r\mathbb{B}_Y \subset A((C - \overline{x}) \cap \mathbb{B}_X) - (D - \overline{y}).$$

Take any reals $\eta > 0$ and $0 < r' < r$. Fix any $v \in D$ with $\|v - \overline{y}\| < (r - r')/3$, any $u \in C$ with $\|u - \overline{x}\| < \eta$ and $\|A\|\,\|u - \overline{x}\| < (r - r')/3$, and any continuous linear mapping $\Lambda : X \to Y$ with $(1 + \eta)\|\Lambda - A\| < (r - r')/3$. For every $b \in \mathbb{B}_Y$ there are $d \in D$ and $c \in C$ with $\|c - \overline{x}\| \le 1$ such that $rb = A(c - \overline{x}) - (d - \overline{y})$, and hence

$$rb = \Lambda(c - u) + (A - \Lambda)(c - u) + A(u - \overline{x}) - (d - v) + (\overline{y} - v).$$

This and the inequality $\|(A-\Lambda)(c-u)\| \leq (1+\eta)\|A-\Lambda\| < (r-r')/3$ ensure that

$$r\mathbb{B}_Y \subset \Lambda\big((C-u)\cap(1+\eta)\mathbb{B}_X\big) - (D-v) + (r-r')\mathbb{B}_Y,$$

or equivalently

$$r'\mathbb{B}_Y + (r-r')\mathbb{B}_Y \subset M^{\Lambda}_{u,v}((1+\eta)\mathbb{B}_X) + (r-r')\mathbb{B}_Y.$$

Taking support functions (see (1.22)) yields with $\alpha := 1+\eta$

$$r'\mathbb{B}_Y \subset \mathrm{cl}_Y\, M^{\Lambda}_{u,v}(\alpha\mathbb{B}_X) = \alpha\,\mathrm{cl}_Y\left(\frac{1}{\alpha}M^{\Lambda}_{u,v}(\alpha\mathbb{B}_X)\right) \subset \alpha\,\mathrm{cl}_Y\, M^{\Lambda}_{u,v}(\mathbb{B}_X),$$

where the latter inclusion is due to the fact that $\frac{1}{\alpha}M^{\Lambda}_{u,v}(\alpha\mathbb{B}_X) \subset M^{\Lambda}_{u,v}(\mathbb{B}_X)$ since the graph of $M^{\Lambda}_{u,v}$ is convex with $0 \in M^{\Lambda}_{u,v}(0)$. By Lemma 7.24 we obtain $r'\mathbb{U}_Y \subset (1+\eta)M^{\Lambda}_{u,v}(\mathbb{B}_X)$. In conclusion, for any positive real $s < r'/(1+\eta)$ we get $s\mathbb{B}_Y \subset M^{\Lambda}_{u,v}(\mathbb{B}_X)$, that is,

$$s\mathbb{B}_Y \subset \Lambda((C-u)\cap\mathbb{B}_X) - (D-v),$$

which finishes the proof of the lemma. □

REMARK 8.120. In fact, the above proof reveals the following. Given a closed convex set C of a Banach space X, a closed convex set D of a normed space Y, $\overline{x} \in C$, $\overline{y} \in D$ and a real $r > 0$ such that

$$r\mathbb{B}_Y \subset A((C-\overline{x})\cap\mathbb{B}_X) - (D-\overline{y}),$$

where $A : X \to Y$ is a continuous linear mapping, then for any real $0 < s < r$ there exits a real $\delta > 0$ such that

$$s\mathbb{B}_Y \subset \Lambda((C-u))\cap\mathbb{B}_X) - (D-v)$$

for all $u \in C \cap B(\overline{x},\delta)$, $v \in D\cap B(\overline{y},\delta)$ and $\Lambda \in B(A,\delta)$. □

The next lemma is an extension of the assertions (a) in Lemma 8.68 for a continuous linear mapping $\Lambda : X \to Y$ between two Banach spaces X, Y. Here, we significantly weaken the assumption $s\mathbb{U}_Y \subset \Lambda(\mathbb{B}_X)$ in $s\mathbb{U}_Y \subset \Lambda((C-u))\cap\mathbb{B}_X) - (D-v)$, where $C \subset X$ and $D \subset Y$ are two closed convex sets containing u and v respectively. We recall that $\mathbb{U}_Y$ denotes the *open unit ball* of Y.

LEMMA 8.121. Let C and D be closed convex sets of normed spaces X and Y respectively and let $u \in C$ and $v \in D$. Let $\Lambda : X \to Y$ be a continuous linear mapping for which there is a real $s > 0$ such that

$$s\mathbb{U}_Y \subset \Lambda((C-u))\cap\mathbb{B}_X) - (D-v).$$

Then given $u^* \in N(C;u)$, $v^* \in N(D;v)$ and $x^* = v^*\circ\Lambda + u^*$, one has

$$\|v^*\| \leq s^{-1}\|x^*\| \quad \text{and} \quad \|u^*\| \leq (1+s^{-1}\|\Lambda^*\|)\|x^*\|.$$

PROOF. Take any $b \in \mathbb{U}_Y$ and choose $d \in D$ and $c \in C$ with $\|c-u\| \leq 1$ such that $sb = -\Lambda(c-u) + (d-v)$. We then have

$$\begin{aligned} s\langle v^*, b\rangle &= \langle v^*\circ\Lambda, -c+u\rangle + \langle v^*, d-v\rangle \\ &= \langle x^*, u-c\rangle + \langle u^*, c-u\rangle + \langle v^*, d-v\rangle \\ &\leq \langle x^*, u-c\rangle \leq \|x^*\|. \end{aligned}$$

This being true for any $b \in \mathbb{U}_Y$, it follows that $s\|v^*\| \leq \|x^*\|$. Further, using this in the equality $u^* = x^* - \Lambda^*(v^*)$ gives $\|u^*\| \leq (1+s^{-1}\|\Lambda^*\|)\|x^*\|$. □

We can now prove now the promised extension of Proposition 8.70(b).

PROPOSITION 8.122. Let C and D be closed convex sets of Banach spaces X and Y respectively and let $g : X \to Y$ be mapping which is continuously differentiable near a point $\overline{x} \in C \cap g^{-1}(D)$. Assume that, for $\overline{y} := g(\overline{x})$ the Robinson qualification condition

$$0 \in \operatorname{core}\big(Dg(\overline{x})(C - \overline{x}) - (D - \overline{y})\big)$$

is satisfied. Then the set $C \cap g^{-1}(D)$ is subsmooth at $\overline{x}$.

PROOF. Put $S := C \cap g^{-1}(D)$ and note that S is closed near $\overline{x}$. By Theorem 7.95, Lemma 8.119 and Lemma 8.121 there are a real $\gamma > 0$ and an open neighborhood U of $\overline{x}$ over which g is of class $\mathcal{C}^1$ and such for every $x \in U \cap S$ one has

$$N^C(S;x) \cap \mathbb{B}_{X^*} \subset Dg(x)^*\big(N(D;g(x)) \cap \gamma\mathbb{B}_{Y^*}\big) + N(C;x) \cap \gamma\mathbb{B}_{X^*}.$$

Define $G : X \to X \times Y$ by $G(x) = (x, g(x))$ for all $x \in X$, so $S = G^{-1}(Q)$, where $Q := C \times D$, and G is of class $\mathcal{C}^1$ near $\overline{x}$ with $DG(x)(u) = (u, Dg(x)u)$ for all $u \in X$ and x in some open neighborhood U of $\overline{x}$ where g is $\mathcal{C}^1$. Fix any $x \in U \cap S$ and note that $DG(x)^* : X^* \times Y^* \to X^*$ is given by $DG(x)^*(u^*, v^*) = u^* + Dg(x)^*(v^*)$ for all $(u^*, v^*) \in X^* \times Y^*$. Therefore, with $y := g(x)$, $\Lambda := Dg(x)$ and $L := DG(x)$ we have

$$\begin{aligned} L^*\big(N(Q;G(x)) \cap (\gamma(\mathbb{B}_{X^*} \times \mathbb{B}_{Y^*}))\big) &= L^*\big((N(C;x) \cap \gamma\mathbb{B}_{X^*}) \times (N(D;y) \cap \gamma\mathbb{B}_{Y^*})\big) \\ &= N(C;x) \cap \gamma\mathbb{B}_{X^*} + \Lambda^*\big(N(D;y) \cap \gamma\mathbb{B}_{Y^*}\big), \end{aligned}$$

which by what precedes yields

$$N^C(S;x) \cap \mathbb{B}_{X^*} \subset DG(x)^*\big(N(Q : G(x)) \cap (\gamma(\mathbb{B}_{X^*} \times \mathbb{B}_{Y^*}))\big).$$

Theorem 8.113(a) allows us to conclude that the set S is subsmooth at $\overline{x}$. □

Subsmoothness of more usual structured optimization constraint sets follows under the Mangasarian-Fromovitz qualification condition.

COROLLARY 8.123. Let $g_1, \cdots, g_m$ be functions from a Banach space X into $\mathbb{R}$ and $G : X \to Y$ be a mapping from X into a Banach space Y, and let $S = \{x \in X : g_1(x) \le 0, \cdots, g_m(x) \le 0, G(x) = 0\}$. Assume that $g_1, \cdots, g_m$ and G are of class $\mathcal{C}^1$ near a point $\overline{x} \in S$ and assume the following Mangasarian-Fromovitz qualification condition: the derivative $DG(\overline{x})$ is surjective and there exists a vector $\overline{v} \in X$ such that $DG(\overline{x})\overline{v} = 0$ and $\langle Dg_k(\overline{x}), \overline{v}\rangle < 0$ for all $k \in K(\overline{x}) := \{k \in K : g_k(\overline{x}) = 0\}$, where $K := \{1, \cdots, m\}$. Then the set S is subsmooth at $\overline{x}$.

PROOF. By the Banach-Schauder open mapping theorem (see Theorem C.3 in Appendix C) choose a real $\rho > 0$ such that $\rho\mathbb{B}_Y \subset DG(\overline{x})(\mathbb{B}_X)$. Write $K(\overline{x}) := \{k_1, \cdots, k_p\}$ with distinct k_i and define $g : X \to \mathbb{R}^p \times Y$ by

$$g(x) = (g_{k_1}(x), \cdots, g_{k_p}(x), G(x)) \quad \text{for all } x \in X.$$

Put $D := (-\mathbb{R}_+)^p \times \{0_Y\}$ and $S_0 := g^{-1}(D)$, so $U \cap S = U \cap S_0$ for some neighborhood U of $\overline{x}$. Choose a real $\sigma > 0$ such that

$$\eta_i := -\sigma\langle Dg_{k_i}(\overline{x}), \overline{v}\rangle - \|Dg_{k_i}(\overline{x})\| - \rho > 0 \quad \text{for any } i = 1, \cdots, p.$$

Consider any $(\zeta, y) \in \mathbb{B}_{\mathbb{R}^p} \times \mathbb{B}_Y$. By the choice of ρ there is some $h \in \mathbb{B}_X$ such that $\rho y = DG(\overline{x})h$, hence $\rho y = DG(\overline{x})(h + \sigma\overline{v})$ since $DG(\overline{x})\overline{v} = 0_Y$ by assumption. For each $i \in \{1, \cdots, p\}$ putting $\xi_i := -\sigma\langle Dg_{k_i}(\overline{x}), \overline{v}\rangle - Dg_{k_i}(\overline{x})h + \rho\zeta_i$ we notice that $\xi_i \geq \eta_i > 0$ and $\rho\zeta_i = \langle Dg_{k_i}(\overline{x}), h + \sigma\overline{v}\rangle + \xi_i$, hence

$$\rho(\zeta, y) = Dg(\overline{x})(h + \sigma\overline{v}) + (\xi, 0_Y).$$

Consequently, $\rho(\mathbb{B}_{\mathbb{R}^p} \times \mathbb{B}_Y) \subset Dg(\overline{x})(X) + \mathbb{R}^p_+ \times \{0_Y\}$, which means that the Robinson qualification condition in Proposition 8.122 is fulfilled with $\overline{x}$ and $g(\overline{x}) = (0_{\mathbb{R}^p}, 0_Y)$, hence the set S_0 is subsmooth at $\overline{x}$. It results that the set S is subsmooth at $\overline{x}$ as well. □

A similar result for uniform equi-subsmoothness also holds true by adapting the above arguments of Proposition 8.122 and the arguments in Proposition 8.116(c).

PROPOSITION 8.124. Let $(C_i)_{i\in I}$ and $(D_i)_{i\in I}$ be two families of closed convex sets of Banach spaces X and Y respectively and let $(g_i)_{i\in I}$ be a family of mappings from X into Y such that every g_i is γ-Lipschitz on $Q_i := C_i \cap g_i^{-1}(D_i)$ with a common Lipschitz constant $\gamma > 0$. Assume that there is $r \in]0, +\infty]$ with G_i differentiable on the r-open enlargement $U_r(Q_i)$ for every $i \in I$ and such that the family $(Dg_i)_{i\in I}$ is uniformly equi-continuous relative to the family of open sets $(U_r(Q_i))_{i\in I}$. Assume also that there is a real $\rho > 0$ such that for every $i \in I$ the Robinson qualification condition

$$\rho\mathbb{B}_Y \subset Dg_i(x)(C - x) - \big(D - g_i(x)\big) \quad \text{for all } x \in Q_i,$$

is satisfied. Then the family of sets $(Q_i)_{i\in I}$ is uniformly equi-subsmooth.

The uniform equi-subsmoothness of families of sublevel sets will be studied in Proposition 8.135.

Regarding the intersection of finitely many sets, we need to translate the conditions in (8.57) and (8.59).

DEFINITION 8.125. Let $S_1, \cdots, S_m$ be a finite system of sets of X and $\overline{x} \in \bigcap_{i=1}^{m} S_i$. We say that this system of sets satisfies the *truncated C-normal cone intersection property* near $\overline{x}$ if there are a positive real constant γ and a neighborhood U of $\overline{x}$ such that

$$(8.67)\quad N^C\Big(\bigcap_{i=1}^{m} S_i; x\Big) \cap \mathbb{B}_{X^*} \subset N^C(S_1; x) \cap (\gamma\mathbb{B}^*_X) + \cdots + N^C(S_m; x) \cap (\gamma\mathbb{B}^*_X)$$

for all $x \in U \cap S_1 \cap \cdots \cap S_m$. Obviously, the above property holds if and only if it holds for all $x \in U \cap S_1 \cap \cdots \cap S_m$ which lies in bdry $(\bigcap_{i=1}^{m} S_i)$.

The *metric subregular intersection property*, or the *metric subregular transversality*, of the system of sets $(S_i)_{i=1}^m$ at $\overline{x}$ has been stated in a similar way (see Definition 7.42) requiring that there exist a real $\gamma > 0$ and a neighborhood U of $\overline{x}$ such that

$$(8.68)\qquad d\Big(u, \bigcap_{i=1}^{m} S_i\Big) \leq \gamma[d(u, S_1) + \cdots + d(u, S_m)] \quad \text{for all } u \in U.$$

PROPOSITION 8.126. Let $S_1, \cdots, S_m$ be a finite system of sets of a normed space X and let $\overline{x} \in \bigcap_{i=1}^{m} S_i$. The following hold.
(a) If the sets $S_1, \cdots, S_m$ are subsmooth at $\overline{x}$ and if the truncated C-normal cone intersection property is satisfied for these sets near $\overline{x}$, then the intersection $\bigcap_{i=1}^{m} S_i$ is subsmooth at $\overline{x}$.
(b) If the sets $S_1, \cdots, S_m$ are metrically subsmooth at $\overline{x}$ and satisfy the metrically subregular transversality property at $\overline{x}$, then $\bigcap_{i=1}^{m} S_i$ is metrically subsmooth at $\overline{x}$.

PROOF. Consider the normed space $Y := X \times \cdots \times X$ endowed with the sum norm (that is, $\|(x_1, \cdots, x_m)\| = \|x_1\| + \cdots + \|x_m\|$) and consider the subset $S := S_1 \times \cdots \times S_m$ of Y. Defining the continuous linear mapping $A : X \to Y$ by $A(x) := (x, \cdots, x)$ for all $x \in X$, we see that $S_1 \cap \cdots \cap S_m = A^{-1}(S)$.
(a) We know (see Theorem 2.2) that

$$N^C(S; A(x)) = N^C(S_1; x) \times \cdots \times N^C(S_m; x). \tag{8.69}$$

Note that $\mathbb{B}_{Y^*} = \mathbb{B}_{X^*} \times \cdots \times \mathbb{B}_{X^*}$ since the dual norm in Y^* is the box norm related to the dual norm in X^* (the norm of Y being the sum norm). Fix a neighborhood U of $\overline{x}$ and a positive constant γ such that (8.67) holds. Observing that $A^*(x_1^*, \cdots, x_m^*) = x_1^* + \cdots + x_m^*$ for any $(x_1^*, \cdots, x_m^*) \in Y^*$, and using (8.69) we see that

$$N^C(A^{-1}(S); x) \cap \mathbb{B}_{X^*} \subset A^*(N^C(S; A(x)) \cap \gamma\mathbb{B}_{Y^*}) \quad \text{for all } x \in U \cap A^{-1}(S),$$

that is, the inverse image representative $A^{-1}(S)$ has the truncated normal cone inverse image property near $\overline{x}$. The set S being easily seen to inherit the subsmoothness property at $\overline{x}$ from the ones of S_i, $i = 1, \cdots, m$, it follows from (a) in Theorem 8.113 that the set $A^{-1}(S)$ is subsmooth at $\overline{x}$.
(b) Obviously, the definition of the sum norm yields that

$$d(y, S) = d(y_1, S_1) + \cdots + d(y_m, S_m) \quad \text{for all } y = (y_1, \cdots, y_m) \in Y.$$

The metrically subregular transversality property of the sets $S_1, \cdots, S_m$ with the constant $\gamma > 0$ and the neighborhood U of $\overline{x}$ may then be translated as

$$d(x, A^{-1}(S)) \le \gamma\, d(A(x), S) \quad \text{for all } x \in U.$$

The property (b) of the corollary is then a consequence of (b) in Theorem 8.113, and this completes the proof. □

Given subsets C and D of normed spaces X and Y respectively, Proposition 8.122 established the subsmoothness of $C \cap g^{-1}(D)$ under the convexity of both sets C and D and the $\mathcal{C}^1$ property of g near $\overline{x}$. The next proposition provides a similar result under the subsmoothness of C and D. Extending Definitions 8.111 and 8.125, and taking a subdifferential ∂, we will say that the representative set $C \cap g^{-1}(D)$ has the *truncated ∂-normal cone property near* $\overline{x} \in C \cap g^{-1}(D)$ if there exist a real $\gamma > 0$ and a neighborhood U of $\overline{x}$ such that for all $x \in U \cap C \cap g^{-1}(D)$

$$N^\partial(C \cap g^{-1}(D); x) \subset N^\partial(C; x) \cap (\gamma\mathbb{B}_{X^*}) + Dg(x)^* \big(N^\partial(D; g(x)) \cap (\gamma\mathbb{B}_{Y^*})\big). \tag{8.70}$$

This is in the spirit of the *metric subregular qualification condition* requiring (as defined in (7.25)) a real $\gamma > 0$ and a neighborhood U of $\overline{x}$ such that

$$d(x, C \cap g^{-1}(D)) \le \gamma[d_C(x) + d_D(g(x)] \quad \text{for all } x \in U. \tag{8.71}$$

In fact, endowing $X \times Y$ with the sum norm and defining $G : X \to X \times Y$ by $G(x) = (x, g(x))$, we see that (8.71) is equivalent to the metric subregular transversality at $\overline{x}$ of G to the set $C \times Y$. So, we obtain the assertions (b) (resp. (a)) below by applying (b) (resp. proceeding as for (a) in) Theorem 8.130.

PROPOSITION 8.127. Let $g : X \to Y$ be a mapping between normed spaces X and Y and let C and D be subsets of X and Y respectively. Assume that g is of class $\mathcal{C}^1$ near $\overline{x} \in C \cap g^{-1}(D)$. The following hold.
(a) If the sets C and D are subsmooth at $\overline{x}$ and $G(\overline{x})$ respectively and if the truncated C-normal cone inverse image property is satisfied for $C \cap g^{-1}(D)$ near $\overline{x}$,

then $C \cap g^{-1}(D)$ is subsmooth at $\overline{x}$.
(b) If the sets C and D are metrically subsmooth at $\overline{x}$ and $G(\overline{x})$ respectively and if the metric subregular qualification condition is satisfied at $\overline{x}$ for $C \cap g^{-1}(D)$, then the set $C \cap g^{-1}(D)$ is metrically subsmooth at $\overline{x}$.

8.8. Metric subregularity under metric subsmoothness

Our next aim is to provide conditions for the metric subregularity in presence of subsmoothness. Let us prove first the following lemma. It is in the line of Lemma 4.88, but it differs from it in two aspects. On the one hand, the first assertion is concerned with any Banach space and not just the Asplund one and the proof works, for example, with the Clarke or Ioffe approximate subdifferential of the distance function instead of the Fréchet normal cone. On the other hand, in the case of an Asplund space no element is required to be in the Fréchet subdifferential at the point outside the set.

LEMMA 8.128. Let S be a closed set of a Banach space X and $x \in X \setminus S$. Let also ∂ be a subdifferential on X such that $0 \in \partial f(\overline{x}) + \partial g(\overline{x})$ whenever $\overline{x}$ is a minimizer of $f + g$ and $f : X \to \mathbb{R}$ is locally Lipschitz and $g : X \to \mathbb{R}$ is convex continuous. Then for any $\varepsilon > 0$ there exist some $u \in S$ and $u^* \in \partial d_S(u)$ such that

$$\|u - x\| \le (1 + \varepsilon(1 + \varepsilon))d_S(x) \quad \text{and} \quad \langle u^*, x - u\rangle \ge \frac{1-\varepsilon}{1+\varepsilon}\|x - u\|.$$

If X is an Asplund space (resp. a Hilbert space), then for any $\varepsilon > 0$ there is $u \in S$ and $u^* \in N^F(S; u)$ (resp. $u^* \in N^P(S; u)$) with $\|u^*\| = 1$ such that the above inequalities hold.

PROOF. Fix any positive $\varepsilon' < \min\{\varepsilon, \varepsilon d_S(x), \varepsilon\sqrt{d_S(x)}\}$. Choose some $x' \in S$ satisfying $\|x' - x\| \le d_S(x) + (\varepsilon')^2$, that is,

$$\|x' - x\| \le \|y - x\| + (\varepsilon')^2 \quad \forall y \in S.$$

According to the Ekeland variational principle applied to the function $y \mapsto \|y - x\|$ over the complete metric space S, there exists some $u \in S$ such that $\|u - x'\| \le \varepsilon'$ and

$$\|u - x\| \le \|y - x\| + \varepsilon'\|y - u\| \quad \forall y \in S.$$

Since the function $y \mapsto \|y - x\| + \varepsilon'\|y - u\|$ is Lipschitz continuous on X with $(1 + \varepsilon')$ as Lipschitz constant, the latter inequality yields (see Lemma 2.213)

$$\|u - x\| \le \|y - x\| + \varepsilon'\|y - u\| + (1 + \varepsilon')d_S(y) \quad \forall y \in X,$$

that is, the point u is a minimizer on the whole space X of the function in y given by the second member of the inequality. The three functions involved in that second member being Lipschitz continuous, according to the assumption on the subdifferential ∂ and to the Moreau-Rockafellar subdifferential sum rule for convex functions we have

$$0 \in \partial\| \cdot - x\|(u) + \varepsilon'\mathbb{B} + (1 + \varepsilon')\partial d_S(u),$$

that is, $0 = v^* + \varepsilon' b^* + (1 + \varepsilon')u^*$ for some $v^* \in \partial\| \cdot - x\|(u)$, $b^* \in \mathbb{B}$, and $u^* \in \partial d_S(u)$. Observing (by (2.23)) that $\langle v^*, u - x\rangle = \|u - x\|$ (and $\|v^*\| = 1$), we see that

$$(1 + \varepsilon')\langle u^*, x - u\rangle + \varepsilon'\langle b^*, x - u\rangle = \|u - x\|,$$

and hence

$$(1 + \varepsilon')\langle u^*, x - u\rangle \ge (1 - \varepsilon')\|x - u\|,$$

which furnishes the second inequality of the statement of the lemma.

Regarding the first one, it suffices to write

$$\|u - x\| \le \|u - x'\| + \|x' - x\| \le \varepsilon' + d_S(x) + (\varepsilon')^2$$
$$\le \varepsilon d_S(x) + d_S(x) + \varepsilon^2 d_S(x).$$

The proof of the first assertion is complete.

Assume now that X is an Asplund space (resp. a Hilbert space). Clearly, we may suppose $\varepsilon \in]0,1[$. Fix any $\eta \in]0,\varepsilon[$. Let $u \in S$ and $u^* \in \partial_L d_S(u)$ satisfying the inequalities obtained above with η in place of ε. By Theorem 85(a) (resp. Proposition 4.158) take sequences $(u_n)_n$ in S converging to u and $(u_n^*)_n$ converging weakly* to u^* with $u_n^* \in \partial_F d_S(u_n)$ (resp. $u_n^* \in \partial_P d_S(u_n)$). Since $d_S(x) > 0$, there is some $k \in \mathbb{N}$ with $d_S(u_k) > 0$ such that u_k and u_k^* satisfies the same inequalities with ε in place of η. Then, $u_k^* \neq 0$ and putting $v := u_k$ and $v^* := u_k^*/\|u_k^*\|$ and recalling that $\partial_F d_S(v) \subset N^F(S;v)$ by Proposition 4.30 (resp. $\partial_P d_S(v) \subset N^P(S;v)$ by Proposition 4.153), we have $v^* \in N^F(S;v)$ (resp. $v^* \in N^P(S;v)$) with $\|v^*\| = 1$ and

$$\langle v^*, x - v\rangle \ge \frac{1}{\|u_k^*\|}\frac{1-\varepsilon}{1+\varepsilon}\|x - v\| \ge \frac{1-\varepsilon}{1+\varepsilon}\|x - v\|,$$

where the latter inequality is due to the fact $\|u_k^*\| \le 1$. We conclude that the elements v and v^* satisfy the desired properties in the Asplund (resp. Hilbert) space setting. $\square$

REMARK 8.129. In general, even for a closed convex set S one cannot find $u \in S$ satisfying the first inequality in the lemma and $u^* \in N(S;u)$ with $\|u^*\| = 1$ satisfying $\langle u^*, x - u\rangle = \|x - u\|$. Indeed, take a nonreflexive Banach space X, so by James theorem there exists $a^* \in \mathbb{S}_{X^*}$ not attaining its norm on the unit ball. Then for $S := \{y \in X : \langle a^*, y\rangle \le 1\}$, for any $u \in S$ and any $u^* \in N(S;u)$ with $\|u^*\| = 1$, we have $u^* = a^*$ (by definition of S), so (since a^* does not attain its norm on $\mathbb{B}_X$) we derive that

$$\langle u^*, x - u\rangle = \langle a^*, x - u\rangle < \|x - u\|,$$

which confirms the remark. $\square$

THEOREM 8.130 (metric subregular transversality under metric subsmoothness). Let $G : X \to Y$ be a mapping from a Banach space X into a Banach space Y and let S be a subset of Y. Assume that G is of class $\mathcal{C}^1$ near $\overline{x} \in G^{-1}(S)$ and that S is closed near $G(\overline{x})$. Then the following hold.

(a) If the metric subregularity condition (8.59) is satisfied with some real $\gamma > 0$ over some neighborhood U of $\overline{x}$, then the linear inclusion property of C-subdifferential of distance from inverse image (8.58) is fulfilled with the same constant γ over $U' \cap S$ for some neighborhood U' of $\overline{x}$.

If in addition to the metric subregularity condition (8.59) both spaces X, Y are Asplund, then the linear inclusion property of L-subdifferential of distance from inverse image is satisfied near $\overline{x}$ (relative to $G^{-1}(S)$.

(b) If S is metrically subsmooth at $G(\overline{x})$ and if (8.58) holds with a real constant $\gamma > 0$, then for any positive real number $\varepsilon < 1$ satisfying $1-\varepsilon > \varepsilon(1+\varepsilon)(1+\varepsilon(1+\varepsilon))$ there exists some neighborhood U' of $\overline{x}$ such that for all $x \in U'$

$$d(x, G^{-1}(S)) \le \frac{\gamma(1+\varepsilon)}{1-\varepsilon-\varepsilon(1+\varepsilon)(1+\varepsilon(1+\varepsilon))} d(G(x), S).$$

(c) If S is subsmooth at $G(\overline{x})$ with X, Y Asplund spaces and if for a real $\gamma > 0$ there exists a neighborhood U of $\overline{x}$ such that for all $x \in U \cap G^{-1}(S)$

$$N^F\left(G^{-1}(S); x\right) \cap \mathbb{B}_{X^*} \subset \gamma DG(x)^*\left(N^C(S; G(x)) \cap \mathbb{B}_{Y^*}\right), \tag{8.72}$$

then the same conclusion in (b) holds.

PROOF. Concerning the assertion (a), the linear inclusion property for C-subdifferential follows from Proposition 2.138 while for L-subdifferential it follows from Theorem 4.85.

Let us prove (b) and (c). Without loss of generality, we may suppose that S is closed and G is continuous on X. Let γ, U, and ε be as in the statement of (b) (resp. (c)). We may suppose that G is Lipschitz on U with some Lipschitz constant $\beta > 0$ and that $\|DG(x_1) - DG(x_2)\| \le (2\gamma)^{-1}\varepsilon$ for all $x_1, x_2 \in U$. The set S being metrically subsmooth (resp. subsmooth) at $G(\overline{x})$, by (b) of Proposition 8.96 (resp. by (b) of Proposition 8.97) there exists some $\delta > 0$ such that $B(\overline{x}, \delta) \subset U$ and

$$\langle v^*, y - v\rangle \le d_S(y) + (2\beta\gamma)^{-1}\varepsilon\|y - v\| \tag{8.73}$$

for all $y \in B(G(\overline{x}), \delta)$, $v \in S \cap B(G(\overline{x}), \delta)$, and $v^* \in \partial_C d_S(v)$ (resp. $v^* \in N^C(S; v) \cap \mathbb{B}_{Y^*}$). Fix any positive $\eta < \delta$ and such that $\eta\beta(2 + \varepsilon(1 + \varepsilon)) < \delta$ and $G(B(\overline{x}, \eta)) \subset B(G(\overline{x}), \delta)$. Fix any $x \in B(\overline{x}, \eta) \setminus G^{-1}(S)$. By the above lemma there exist $u \in G^{-1}(S)$ and $u^* \in \partial_C d(\cdot, G^{-1}(S))(u)$ (resp. $u^* \in \partial_F d(\cdot, G^{-1}(S))(u)$) such that

$$\|u - x\| \le (1 + \varepsilon(1 + \varepsilon))d(x, G^{-1}(S)) \quad \text{and} \quad \langle u^*, x - u\rangle \ge \frac{1 - \varepsilon}{1 + \varepsilon}\|x - u\|. \tag{8.74}$$

Note that

$$\|G(x) - G(\overline{x})\| \le \beta\|x - \overline{x}\| < \eta\beta < \delta. \tag{8.75}$$

We also note that

$$\begin{aligned}\|u - \overline{x}\| &\le \|x - \overline{x}\| + (1 + \varepsilon(1 + \varepsilon))d(x, G^{-1}(D)) \\ &\le (2 + \varepsilon(1 + \varepsilon))\|x - \overline{x}\| \le \eta(2 + \varepsilon(1 + \varepsilon)),\end{aligned}$$

so $u \in B(\overline{x}, \delta)$. This ensures that

$$\|G(u) - G(\overline{x})\| \le \beta\|u - \overline{x}\| < \eta\beta(2 + \varepsilon(1 + \varepsilon)) < \delta. \tag{8.76}$$

Choose by the property (8.58) (resp. (8.72)) some $v^* \in \partial_C d_S(G(u))$ (resp. $v^* \in N^C(S; G(u)) \cap \mathbb{B}_{Y^*}$) such that we have $u^* = \gamma v^* \circ DG(u)$. By (8.75) and (8.76) and by the inclusion $G(u) \in S$ we may invoke (8.73) to write

$$\langle v^*, G(x) - G(u)\rangle \le d_S(G(x)) + (2\beta\gamma)^{-1}\varepsilon\|G(x) - G(u)\|,$$

and hence

$$\langle v^*, G(x) - G(u)\rangle \le d_S(G(x)) + (2\gamma)^{-1}\varepsilon\|x - u\|. \tag{8.77}$$

Then according to the second inequality of (8.74) and to the evident inequality $\operatorname{dist}(x, G^{-1}(S)) \le \|x - u\|$ (because $u \in G^{-1}(S)$) we have

$$\begin{aligned}&\frac{1 - \varepsilon}{1 + \varepsilon}\operatorname{dist}(x, G^{-1}(S)) \\ &\le \langle u^*, x - u\rangle = \langle \gamma v^* \circ DG(u), x - u\rangle \\ &= \gamma\langle v^*, G(x) - G(u)\rangle - \gamma\int_0^1 \langle v^* \circ [DG(u + t(x - u)) - DG(u)], x - u\rangle\, dt,\end{aligned}$$

which ensures by (8.77) and by the inequality $\|v^*\| \le 1$ that

$$\frac{1-\varepsilon}{1+\varepsilon}\mathrm{dist}\,(x, G^{-1}(S)) \le \gamma\,\mathrm{dist}\,(G(x), S) + \varepsilon\|x - u\|$$
$$\le \gamma\,\mathrm{dist}\,(G(x), S) + \varepsilon(1+\varepsilon(1+\varepsilon))\mathrm{dist}\,(x, G^{-1}(S)),$$

the second inequality being due to the first inequality of (8.74). It ensues that

$$[\frac{1-\varepsilon}{1+\varepsilon} - \varepsilon(1+\varepsilon(1+\varepsilon))]\mathrm{dist}\,(x, G^{-1}(S)) \le \gamma\,\mathrm{dist}\,(G(x), S).$$

The latter inequality continuing to hold for $x \in G^{-1}(S)$, we conclude that for all $x \in B(\overline{x}, \eta)$

$$\mathrm{dist}\,(x, G^{-1}(S)) \le \frac{\gamma(1+\varepsilon)}{1-\varepsilon-\varepsilon(1+\varepsilon)(1+\varepsilon(1+\varepsilon))}\mathrm{dist}\,(G(x), S).$$

□

Consider (as related to (8.59)) the rate (or modulus) $\mathrm{subreg}_{\cdot,S}[G](\overline{x})$ of metric subregularity at $\overline{x}$ of the mapping G with constraint $S \subset Y$, or metric subregular transversality of G at $\overline{x}$ to the set S, that is,

$$\mathrm{subreg}_{\cdot,S}[G](\overline{x}) = \inf_{U \in \mathcal{N}(\overline{x})} \mathrm{subreg}_{\cdot,S}[G]_U,$$

where $\mathcal{N}(\overline{x})$ denotes the collection of all neighborhoods of $\overline{x}$ and

$$\mathrm{subreg}_{\cdot,S}[G]_U := \sup_{x \in U \setminus G^{-1}(S)} \mathrm{dist}\,(x, G^{-1}(S))/\mathrm{dist}\,(G(x), S).$$

The metric subregularity property of the mapping G with constraint $S \subset Y$ at $\overline{x}$, or metric subregular transversality of G at $\overline{x}$ to the set S, then means that $\mathrm{subreg}_{\cdot,S}[G](\overline{x}) < +\infty$.

The assertion (b) of Theorem 8.130 also leads us to consider in a similar way the following constant related to the property (8.58)

$$\mathrm{subdist}_{\cdot,S}[G](\overline{x}) := \inf_{U \in \mathcal{N}(\overline{x})} \mathrm{subdist}_{\cdot,S}[G]_U,$$

where $\mathrm{subdist}_{\cdot,S}[G]_U$ is the infimum of real numbers $\gamma > 0$ such that (8.58) is satisfied for all $x \in U \cap G^{-1}(S)$.

Theorem 8.130 then admits the following direct corollary.

COROLLARY 8.131. Let X, Y be Banach spaces, S be a subset of Y and $G : X \to Y$ be a mapping which is of class $\mathcal{C}^1$ near $\overline{x} \in G^{-1}(S)$ and such that S is closed near $G(\overline{x})$.
(I) Assume that S is metrically subsmooth at $G(\overline{x})$. Then the following hold.
(a) The mapping G is metrically subregularly transversal at $\overline{x}$ to the set S (that is, metrically subregular at $\overline{x}$ with respect to the subset S of the image space Y) if and only if the property (8.58) is fulfilled.
(b) One has the equality between the above constants at $\overline{x}$, that is,

$$\mathrm{subreg}_{\cdot,S}[G](\overline{x}) = \mathrm{subdist}_{\cdot,S}[G](\overline{x}).$$

(II) Assume now that both spaces X, Y are Asplund and S is subsmooth at $G(\overline{x})$. Then the mapping G is metrically subregularly transversal at $\overline{x}$ to the set S of the image space Y if and only if (8.72) holds for some real constant $\gamma > 0$; in fact, $\mathrm{subreg}_{\cdot,S}[G](\overline{x})$ coincides with the infimum of all reals $\gamma > 0$ satisfying condition (8.72) in Theorem 8.130.

For the intersection of a family of sets $S_1, \cdots, S_m$ of a Banach space X which are all closed near a point $\overline{x}$ of that intersection, we already defined (see Definition 7.42) the *rate (or modulus)* $\operatorname{subreg}_\cap [S_1, \cdots, S_m](\overline{x})$ *of metric subregularity at* $\overline{x}$ *for the system of sets*, or *metric subregular transversality of the system of sets at* $\overline{x}$. Note that this constant clearly coincides with the infimum over all neighborhoods U of $\overline{x}$ of

$$\sup_{x \in U \setminus \cap_{i=1}^m \overline{S_i}} \frac{\operatorname{dist}(x, S_1 \cap \cdots \cap S_m)}{\operatorname{dist}(x, S_1) + \cdots + \operatorname{dist}(x, S_m)};$$

its finiteness translates the existence of $\gamma \in]0, +\infty[$ such that (8.68) holds over a neighborhood U of $\overline{x}$. Consider also the other constant

$$\operatorname{subdist}_\cap [S_1, \cdots, S_m](\overline{x}) := \inf_{U \in \mathcal{N}(u_0)} \operatorname{subdist}_\cap [S_1, \cdots, S_m]_U,$$

where $\operatorname{subdist}_\cap [S_1, \cdots, S_m]_U$ is the infimum of real numbers $\gamma > 0$ such that the inclusion

$$\partial_C \operatorname{dist}\left(\cdot, \bigcap_{i=1}^m S_i\right)(x) \subset \gamma[\partial_C \operatorname{dist}(\cdot, S_1)(x) + \cdots + \partial_C \operatorname{dist}(\cdot, S_m)(x)] \tag{8.78}$$

is satisfied for all $x \in U \cap S_1 \cap \cdots \cap S_m$.

Proceeding like in the proof of Proposition 8.126 one obtains the following result.

THEOREM 8.132 (metric subregular transversality of sets under metric subsmoothness). Let $S_1, \cdots, S_m$ be finitely many sets of a Banach space X which are closed near $\overline{x} \in \bigcap_{i=1}^m S_i$.
(I) Assume that each set S_i is metrically subsmooth at $\overline{x}$. Then the following hold.
(a) The sets $S_1, \cdots, S_m$ are metrically subregularly tranversal at $\overline{x}$ if and only if the property (8.78) is fulfilled with some real $\gamma > 0$ and some neighborhood U of the point $\overline{x}$.
(b) One has the equality between the foregoing constants at $\overline{x}$, that is,

$$\operatorname{subreg}_\cap[S_1, \cdots, S_m](\overline{x}) = \operatorname{subdist}_\cap[S_1, \cdots, S_m](\overline{x}).$$

(II) Assume that X is an Asplund space and each set S_i is subsmooth at $\overline{x}$. Then sets $S_1, \cdots, S_m$ are metrically subregularly transversal at $\overline{x}$ if and only if there are a real $\gamma > 0$ and a neighborhood U of $\overline{x}$ such that

$$N^F(S; x) \cap \mathbb{B}_{X^*} \subset N^C(S_1; x) \cap \gamma\mathbb{B}_{X^*} + \cdots + N^C(S_m; x) \cap \gamma\mathbb{B}_{X^*}$$

for all $x \in U \cap \operatorname{bdry} S$.

Similar conditions with the coderivative can be used to study the metric subregularity of multimappings with subsmooth graphs. Let us say that a multimapping $M : X \to Y$ between two normed spaces satisfies the *truncated* C*-coderivative condition* at a point $\overline{x} \in \operatorname{Dom} M$ for $\overline{y} \in M(\overline{x})$ provided there exist a real $\gamma > 0$ and neighborhoods U and V of $\overline{x}$ and $\overline{y}$ respectively such that

$$N^C\left(M^{-1}(y); x\right) \cap \mathbb{B}_{X^*} \subset \gamma\, D_C^* M(x, y)(\mathbb{B}_{Y^*}) \tag{8.79}$$

for all $x \in U \cap \operatorname{Dom} M$ and $y \in M(x) \cap V$.

Before giving the result with this C-coderivative condition, let us establish the metric subregularity under a similar condition with the C-subdifferential of the distance from the graph of M. For this, recall by Proposition 7.20(b) that a multimapping $M : X \rightrightarrows Y$ (between normed spaces) is metrically subregular at a point $\overline{x}$ for $\overline{y} \in M(\overline{x})$ if and only if it is graphically metrically subregular at $\overline{x}$ for $\overline{y}$, that is, there exist (see Definition 7.19) a real $\gamma > 0$ and a neighborhood U of $\overline{x}$ such that

$$d(x, M^{-1}(\overline{y})) \le \gamma\, d((x,\overline{y}), \operatorname{gph} M) \quad \text{for all } x \in U. \tag{8.80}$$

THEOREM 8.133 (metric subregularity of multimappings under metric subsmoothness: I). Let $M : X \rightrightarrows Y$ be a multimapping between Banach spaces whose graph is metrically subsmooth at $(\overline{x},\overline{y}) \in \operatorname{gph} M$ and closed near $(\overline{x},\overline{y})$. Then M is metrically subregular at $\overline{x}$ for $\overline{y}$ if and only if there exist a real $\gamma > 0$ and a neighborhood U of $\overline{x}$ in X such for any $x \in U \cap M^{-1}(\overline{y})$

$$\partial_C d(\cdot, M^{-1}(\overline{y}))(x) \subset \gamma\, \Pi_{X^*}\big(\partial_C d(\cdot, \operatorname{gph} M)(x,\overline{y})\big),$$

where we recall that the mapping $\Pi_{X^*} : X^* \times Y^* \to X^*$ is the projector defined by $\Pi_{X^*}(x^*, y^*) := x^*$.

PROOF. Suppose first that M is metrically subregular at $\overline{x}$ for $\overline{y}$. By the characterization (8.80) there are a real $\gamma > 0$ and an open neighborhood U of $\overline{x}$ such that

$$d(x, M^{-1}(\overline{y})) \le \gamma d((x,\overline{y}), \operatorname{gph} M) \quad \text{for all } x \in U.$$

Putting $g(x) := \gamma\, d((x,\overline{y}), \operatorname{gph} M)$, Lemma 2.137 tells us that $\partial_C d_{M^{-1}(\overline{y})}(x) \subset \partial_C g(x)$ for all $x \in U \cap M^{-1}(\overline{y})$. On the other hand, for the continuous linear mapping $A : X \to X \times Y$ defined by $A(u) := (u, 0)$ for all $u \in X$ we have $\partial_C g(x) \subset \gamma\, A^*\big(\partial_C d_{\operatorname{gph} M}(x,\overline{y})\big)$. Since A^* coincides with Π_{X^*}, we deduce that for all $x \in U \cap M^{-1}(\overline{y}))$

$$\partial_C d(\cdot, M^{-1}(\overline{y}))(x) \subset \gamma\, \Pi_{X^*}\big(\partial_C d(\cdot, \operatorname{gph} M)(x,\overline{y})\big).$$

Conversely, suppose that the latter property holds and set $S_1 := \operatorname{gph} M$ and $S_2 := X \times \{\overline{y}\}$, so $S := S_1 \cap S_2 = M^{-1}(\overline{y}) \times \{\overline{y}\}$ and $d_S(x,y) = d_{M^{-1}(\overline{y})}(x) + \|y - \overline{y}\|$, where $X \times Y$ is equipped with the sum norm. For every $(x,y) \in (U \times Y) \cap S$ we see that $y = \overline{y}$ and $x \in M^{-1}(\overline{y})$, and (see Proposition 2.108)

$$\partial_C d_S(x,y) \subset \partial_C d_{M^{-1}(\overline{y})}(x) \times \mathbb{B}_{Y^*}.$$

Fix any $(x,y) \in (U \times Y) \cap S$ and any $(x^*, y^*) \in \partial_C d_S(x,y)$. It follows that $y^* \in \mathbb{B}_{Y^*}$ and $x^* \in \partial_C d_{M^{-1}(\overline{y})}(x)$. Thus, by assumption there exists $v^* \in Y^*$ such that $(x^*, v^*) \in \gamma \partial_C d_{S_1}(x,\overline{y})$. Since

$$(x^*, y^*) = (x^*, v^*) + (0, y^* - v^*) \in \gamma \partial_C d_{S_1}(x,\overline{y}) + (1+\gamma)\partial_C d_{S_2}(x,\overline{y}),$$

we derive that $\partial_C d_S(x,y) \subset (1+\gamma)[\partial_C d_{S_1}(x,y) + \partial_{S_2}(x,y)]$ for all $(x,y) \in (U \times Y) \cap S$. Taking any real $\gamma'' > \gamma'$ it results by Theorem 8.132(b) that there exist neighborhoods U' and V' of $\overline{x}$ and $\overline{y}$ such that for all $x \in U'$ and $y \in V'$

$$d((x,y), S_1 \cap S_2) \le \gamma''[d((x,y), \operatorname{gph} M) + d((x,y), X \times \{\overline{y}\})].$$

This gives with $y = \overline{y}$ that for all $x \in U'$

$$d(x, M^{-1}(\overline{y})) \le \gamma'' d((x,\overline{y}), \operatorname{gph} M) \le \gamma'' d(\overline{y}, M(x)),$$

which translates the metric subregularity of M at $\overline{x}$ for $\overline{y}$. □

THEOREM 8.134 (metric subregularity of multimappings under subsmoothness: II). Let $M : X \rightrightarrows Y$ be a multimapping between Asplund spaces whose graph is subsmooth at $(\overline{x},\overline{y}) \in \operatorname{gph} M$ and closed near $(\overline{x},\overline{y})$. Then M is metrically subregular at $\overline{x}$ for $\overline{y}$ if and only if there exist a real $\gamma > 0$ and a neighborhood U of $\overline{x}$ such that

$$N^F\left(M^{-1}(\overline{y}); x\right) \cap \mathbb{B}_{X^*} \subset \gamma\, D_C^* M(x, \overline{y})^*(\mathbb{B}_{Y^*})$$

for all $x \in U \cap \operatorname{bdry}\left(M^{-1}(\overline{y})\right)$.

PROOF. Endow $X \times Y$ with the sum norm and note that $\mathbb{B}_{X^* \times Y^*} = \mathbb{B}_{X^*} \times \mathbb{B}_{Y^*}$. Then for $x \in M^{-1}(\overline{y})$ the inclusion

$$\Pi_{X^*}\left(\partial_C d(\cdot, \operatorname{gph} M)(x, \overline{y})\right) \subset \Pi_{X^*}\left(N^C(\operatorname{gph} M; (x, \overline{y})) \cap (\mathbb{B}_{X^*} \times \mathbb{B}_{Y^*})\right)$$

holds. According to this inclusion and the equality

$$N^F(M^{-1}(\overline{y}); x) \cap \mathbb{B}_{X^*} = \partial_F d(\cdot, M^{-1}(\overline{x}))(\overline{y}),$$

Proposition 8.133 clearly shows that the metric subregularity of M at $\overline{x}$ for $\overline{y}$ implies the condition of the proposition.

Let us show the converse implication. As in the proof of Theorem 8.133 set $S_1 := \operatorname{gph} M$ and $S_2 := X \times \{\overline{y}\}$, so $S := S_1 \cap S_2 = M^{-1}(\overline{y}) \times \{\overline{y}\}$. For every $(x, y) \in (U \times Y) \cap S$ we see that $y = \overline{y}$ and $x \in M^{-1}(\overline{y})$, hence

$$N^F(S; (x, y)) \cap \mathbb{B}_{X^* \times Y^*} = \left(N^F(M^{-1}(\overline{y}); x)\right) \cap (\mathbb{B}_{X^*} \times \mathbb{B}_{Y^*}).$$

Fix any $(x, y) \in (U \times Y) \cap S$ and any $(x^*, y^*) \in N^F(S; (x, y)) \cap \mathbb{B}_{X^* \times Y^*}$. The second inclusion can be rewritten by the latter equality as $y^* \in \mathbb{B}_{Y^*}$ and $x^* \in N^F\left(M^{-1}(\overline{y}); x\right) \cap \mathbb{B}_{X^*}$. Thus, by assumption there exists $b^* \in \mathbb{B}_{Y^*}$ such that $(x^*, -\gamma b^*) \in N^C(\operatorname{gph} M; (x, \overline{y}))$. Writing $(x^*, y^*) = (x^*, -\gamma b^*) + (0, y^* + \gamma b^*)$, it ensues with $\gamma' := 1 + \gamma$ that

$$(x^*, y^*) \in N^C(S_1; (x, y)) \cap \gamma' \mathbb{B}_{X^* \times Y^*} + N^C(S_2; (x, y)) \cap \gamma' \mathbb{B}_{X^* \times Y^*}.$$

Taking any real $\gamma'' > \gamma'$ and applying Corollary 8.131(c) we obtain as in the end of the proof of Theorem 8.133 above that M is metrically subregular at $\overline{x}$ for $\overline{y}$ with constant γ''. □

8.9. Equi-subsmoothness of sets and subdifferential of their distance functions

This section is concerned with additional conditions for the property of uniform equi-subsmoothness of sets and with closedness properties of normal cones and subdifferentials of distance functions under uniform equi-subsmoothness conditions.

In addition to examples in Propositions 8.62, 8.116 and 8.124, we establish first a result for the uniform equi-subsmoothness for families of sublevel sets. This result is a uniformization of Proposition 8.114.

PROPOSITION 8.135. Let I be a nonempty set and $m \in \mathbb{N}$. For each $i \in I$ let $g_{1,i}, \cdots, g_{m,i}$ be m locally Lipschitz functions from a normed space X into $\mathbb{R}$ and let

$$S_i := \{x \in X : g_{1,i}(x) \le 0, \cdots, g_{m,i}(x) \le 0\}$$

that we assume to be a nonempty set. Assume that there exists $r \in]0, +\infty]$ such that the families of functions $(g_{1,i})_{i \in I}, \cdots, (g_{m,i})_{i \in I}$ are uniformly equi-subsmooth relative to the family of open sets $(U_r(S_i))_{i \in I}$ (which holds in particular when for each $k \in \{1, \cdots, m\}$ every function $g_{k,i}$ is differentiable on $U_r(S_i)$ with $(Dg_{k,i})_{i \in I}$ uniformly equi-continuous relative to $(U_r(S_i))_{i \in I}$). Assume also that the following generalized Slater condition holds: there exists a real $\sigma > 0$ such that for each $i \in I$ and each $x \in \operatorname{bdry} S_i$ there exists a vector $\overline{v} \in \mathbb{B}_X$ (depending on i, x) for which

$$\langle x^*, \overline{v} \rangle \geq \sigma$$

for every $k \in K(x) := \{k \in K : g_{k,i}(x) = \max_{j \in K} g_{j,i}(x)\}$ and every $x^* \in \partial_C g_{k,i}(x)$, where $K := \{1, \cdots, m\}$.

Then the family of sets $(S)_{i \in I}$ is uniformly equi-subsmooth.

PROOF. (I) For each $i \in I$ define the function $g_i : X \to \mathbb{R}$ by $g_i(x) := \max_{k \in K} g_{k,i}(x)$ for all $x \in X$ and observe that $S_i = \{x \in X : g_i(x) \leq 0\}$. By Proposition 2.190 we have for each $i \in I$

$$\partial_C g_i(x) \subset \operatorname{co}\left(\bigcup_{k \in K(x)} \partial_C g_{k,i}(x)\right) \quad \text{for all } x \in \operatorname{bdry} S_i. \tag{8.81}$$

This inclusion and the assumption on $\overline{v}$ give us for each $i \in I$

$$0 \notin \partial_C g_i(x) \quad \text{for all } x \in \operatorname{bdry} S_i.$$

By Proposition 2.196 one deduces that for each $i \in I$

$$N^C(S_i; x) \subset \mathbb{R}_+ \partial_C g_i(x) \quad \text{for all } x \in \operatorname{bdry} S_i. \tag{8.82}$$

Take any $\varepsilon > 0$ and set $\varepsilon' := \varepsilon\sigma$. The uniform equi-subsmoothness assumption for the families of functions allows us by Proposition 8.41 to choose $\delta \in]0, r[$ such that for any $i \in I$, for any $x, y \in U_r(S_i)$, for any $k \in K$, for any $x^* \in \partial_C g_{k,i}(x)$, and for any $y^* \in \partial_C g_{k,i}(y)$

$$\langle x^* - y^*, x - y \rangle \geq -\varepsilon' \|x - y\|.$$

Fix any $i \in I$, any $x \in \operatorname{bdry} S_i$ and any $u^* \in N^C(S_i; x) \cap \mathbb{B}_{X^*}$. By (8.82) choose a real $\alpha \geq 0$ and $x^* \in \partial_C g_i(x)$ (both depending on i) such that $u^* = \alpha x^*$. From (8.81) there are $x_k^* \in \partial g_{k,i}(x)$ and $\lambda_k \geq 0$ with $\lambda_k = 0$ if $k \notin K(x)$ and with $\sum_{k \in K} \lambda_k = 1$, such that $x^* = \sum_{k \in K} \lambda_k x_k^*$.

(II) Fix any $y \in S_i$ with $0 < \|y - x\| < \delta$. Fix for a moment $k \in K(x)$ and define the locally Lipschitz function $\varphi_k : \mathbb{R} \to \mathbb{R}$ by $\varphi_k(t) := g_{k,i}(x + t(y - x))$, and note that it is Lipschitz on $[0, 1]$. Denote by N a Lebesgue negligible subset of $[0, 1]$ such that at each $t \in [0, 1] \setminus N$ the function φ_k is derivable at t, so $\varphi_k'(t) \in \partial_C(g_{k,i} \circ G)(t)$ with $G(t) = x + t(y - x) =: z(t)$. Then for each $t \in [0, 1] \setminus N$ there exists some $z_k^*(t) \in \partial_C g_{k,i}(z(t))$ such that $\varphi_k'(t) = \langle z_k^*(t), y - x \rangle$. Note also for each $t \in [0, 1]$ that $z(t) \in U_r(S_i)$ since $d_{S_i}(z(t)) \leq \|z(t) - x\| = t\|y - x\| < r$. It results that

$$\begin{aligned}
0 \geq g_{k,i}(y) - g_{k,i}(x) &= \int_0^1 \langle z_k^*(t), y - x \rangle \, dt \\
&= \int_0^1 \langle z_k^*(t) - x_k^*, y - x \rangle \, dt + \langle x_k^*, y - x \rangle \\
&= \int_0^1 \frac{1}{t} \langle z_k^*(t) - x_k^*, z(t) - x \rangle \, dt + \langle x_k^*, y - x \rangle,
\end{aligned}$$

which yields

$$0 \geq -\int_0^1 \frac{1}{t} \varepsilon' \|z(t) - x\| \, dt + \langle x_k^*, y - x \rangle = -\varepsilon' \|y - x\| + \langle x_k^*, y - x \rangle. \tag{8.83}$$

Recalling that $\lambda_k = 0$ if $k \notin K(x)$, we deduce that $\langle x^*, y-x\rangle \le \varepsilon\sigma\|y-x\|$. Using again the equality $\lambda_k = 0$ if $k \notin K(x)$ and using the assumption on $\overline{v} \in \mathbb{B}_X$, we also have $\langle x^*, \overline{v}\rangle \ge \sigma$. Therefore, the above equality $u^* = \alpha x^*$ gives

$$1 \ge \|u^*\| \ge \langle u^*, \overline{v}\rangle = \alpha\langle x^*, \overline{v}\rangle \ge \alpha\sigma,$$

so $\alpha \le 1/\sigma$. It follows that

$$\langle u^*, y-x\rangle = \alpha\langle x^*, y-x\rangle \le \varepsilon\|y-x\|,$$

and this inequality still holds when $x \in \operatorname{int} S_i$. This confirms that the family of sets $(S_i)_{i\in I}$ is uniformly equi-subsmooth. □

REMARK 8.136. Let us provide another way to see (8.83) above. Keep the part (I) in the proof of Proposition 8.135 and let us modify the beginning of the part (II) as follows. Fix any $y \in S_i$ with $0 < \|y-x\| < \delta$. For each $k \in K(x)$ choose by the mean value equality in Theorem 2.180 some $z_k := x + t_k(y-x)$ with $t_k \in]0,1[$ and some $z_k^* \in \partial_C g_{k,i}(x)$ (both depending on i) such that $g_{k,i}(y) - g_{k,i}(x) = \langle z_k^*, y-x\rangle$. For each $k \in K(x)$, observing that $z_k \in U_r(S_i)$ since $d_{S_i}(z_k) \le \|z_k - x\| = t_k\|y - x\| < r$, we can write

$$\begin{aligned} 0 \ge g_{k,i}(y) - g_{k,i}(x) &= \langle z_k^*, y-x\rangle \\ &= \langle z_k^* - x_k^*, y-x\rangle + \langle x_k^*, y-x\rangle \\ &= \frac{1}{t_k}\langle z_k^* - x_k^*, z_k - x\rangle + \langle x_k^*, y-x\rangle. \end{aligned}$$

This entails that

$$0 \ge -\frac{1}{t_k}\varepsilon'\|z_k - x\| + \langle x_k^*, y-x\rangle = -\varepsilon'\|y-x\| + \langle x_k^*, y-x\rangle,$$

which is (8.83), so the proof of Proposition 8.83 can be continued as above. □

The next proposition allows us (as already done for the local subsmoothness of a set in Proposition 8.96(c)) to take into account points x outside the sets S_i when working with uniformly equi-subsmooth families of sets.

PROPOSITION 8.137. Let $(S_i)_{i\in I}$ be a family of nonemty closed sets of an Asplund space X. This family is uniformly equi-subsmooth if and only if for any $\varepsilon > 0$ there exists $\delta > 0$ such that for any $i \in I$, any $y \in S_i$, any $x \in X$ with $\|x-y\| < \delta$ and any $x^* \in \partial_C d_{S_i}(x)$, one has

$$\langle x^*, y-x\rangle \le d_{S_i}(x) + \varepsilon\|y-x\|.$$

PROOF. Suppose first that the family is uniformly equi-subsmooth. Fix any real $\varepsilon > 0$ and take some $\varepsilon' > 0$ such that $2\varepsilon' + \varepsilon'(2+\varepsilon') < \varepsilon$. By definition of uniform equi-subsmoothness, choose some $\delta > 0$ such that for any $i \in I$, for any $u, v \in S_i$ with $\|u-v\| < 3\delta$ and $u^* \in N^C(S_i; u) \cap \mathbb{B}_{X^*}$ one has

$$\langle u^*, v-u\rangle \le \varepsilon'\|v-u\|. \tag{8.84}$$

Fix any $i \in I$. Take any pair (x, y) such that $x \in (X \setminus S_i) \cap \operatorname{Dom} \partial_F d_{S_i}$, $y \in S_i$ and $\|x-y\| < \delta$. Let any $x^* \in \partial_F d_{S_i}(x)$. Following the proof of (a)⇒(c) in Proposition 8.96, let us take a positive real $\varepsilon_i < \min\{\delta, \varepsilon', \varepsilon' d_{S_i}(x)\}$ (depending on i). We use Lemma 4.88 with ε_i in place of ε to get some $v \in S_i$ and $v^* \in \partial_F d_{S_i}(v)$ (both depending on i) such that

$$\|v-x\| < \varepsilon_i + d_{S_i}(x) < (1+\varepsilon')d_{S_i}(x) \quad \text{and} \quad \|v^* - x^*\| < \varepsilon'. \tag{8.85}$$

From the first inequality in (8.85) we notice that

$$\|v-y\| \le \|v-x\| + \|x-y\| < \varepsilon_i + d_{S_i}(x) + \|x-y\|,$$

and hence by the inclusion $y \in S_i$

$$\|v-y\| < \varepsilon_i + 2\|x-y\| < 3\delta. \tag{8.86}$$

It results that

$$\begin{aligned}\langle x^*, y-x\rangle &\le \langle v^*, y-x\rangle + \varepsilon'\|y-x\| \\ &= \langle v^*, y-v\rangle + \langle v^*, v-x\rangle + \varepsilon'\|y-x\| \\ &\le \varepsilon'\|y-v\| + \|v-x\| + \varepsilon'\|y-x\|,\end{aligned}$$

the first inequality being due to the last inequality in (8.85) and the second one being due to (8.84) and (8.86) and to the fact that $\|v^*\| \le 1$. From the second inequality in the first part of (8.85) it ensues that

$$\begin{aligned}\langle x^*, y-x\rangle &\le 2\varepsilon'\|y-x\| + (1+\varepsilon')\|v-x\| \\ &\le 2\varepsilon'\|y-x\| + d_{S_i}(x) + \varepsilon'(2+\varepsilon')d_{S_i}(x),\end{aligned}$$

which gives (since $y \in S_i$)

$$\begin{aligned}\langle x^*, y-x\rangle &\le 2\varepsilon'\|y-x\| + d_{S_i}(x) + \varepsilon'(2+\varepsilon')\|y-x\| \\ &= (2\varepsilon' + \varepsilon'(2+\varepsilon'))\|y-x\| + d_{S_i}(x).\end{aligned}$$

The choice of ε' yields

$$\langle x^*, y-x\rangle \le d_S(x) + \varepsilon\|y-x\|. \tag{8.87}$$

Clearly, this inequality is still satisfied for any pair (x,y) such that $x \in \operatorname{Dom}\partial_F d_{S_i}$, $y \in S_i$ and $\|x-y\| < \delta$ since the case when $x \in S_i$ with $\|x-y\| < \delta$ follows from (8.84). Now take any pair $(x,y) \in X \times S_i$ with $\|x-y\| < \delta$. Any element in $\partial_L d_{S_i}(x)$ being the weak* limit of some sequence of Fréchet subgradients at points x_n converging strongly to x, we see that (8.87) continues to hold for any $x^* \in \partial_L d_S(x)$, and hence also for any $x^* \in \overline{\text{co}}^*(\partial_L d_{S_i}(x)) = \partial_C d_{S_i}(x)$ (see Corollary 4.122 for the equality). This justifies the implication $\Rightarrow$.

Conversely, suppose that the property in the proposition holds. For each $i \in I$, this implies in particular that S_i is metrically subsmooth. This ensures for each $x \in S_i$ that $\partial_C d_{S_i}(x) = N^C(S_i; x) \cap \mathbb{B}_{X^*}$ according to the implication (a)$\Rightarrow$(f) in Proposition 8.98. This combined with the property of the proposition yields that the family of sets $(S_i)_{i\in I}$ is uniformly equi-subsmooth. □

We start now with two closedness properties.

PROPOSITION 8.138. *Let E be a metric space and let $(S(q))_{q\in E}$ be a family of nonempty closed sets of an Asplund space X which is uniformly equi-subsmooth and let $\eta \in [0, +\infty[$. Let $Q \subset E$, $q_0 \in \operatorname{cl} Q$ and $x \in S(q_0)$. Then, for any net $(q_j)_{j\in J}$ in Q converging to q_0 with $d_{S(q_j)}(x) \underset{j\in J}{\to} 0$, for any net $(x_j)_{j\in J}$ converging to x in $(X, \|\cdot\|)$, and for any net $(x_j^*)_{j\in J}$ converging weakly* to x^* in X^* with $x_j^* \in \eta\partial_C d_{S(q_j)}(x_j)$, one has $x^* \in \eta\partial_C d_{S(q_0)}(x)$.*

PROOF. We may suppose $\eta > 0$. Take any real $\varepsilon > 0$. By Proposition 8.137 above choose a real $\delta > 0$ such that for all $q \in E$, $v \in S(q)$, $u \in X$ with $\|u-v\| < \delta$ and all $u^* \in \partial_C d_{S(q)}(u)$

$$\langle u^*, v-u\rangle \le d_{S(q)}(u) + \varepsilon\|v-u\|. \tag{8.88}$$

Fix any net $(q_j)_{j\in J}$ in Q converging to q_0 in E with $d_{S(q_j)}(x) \underset{j\in J}{\to} 0$, and any net $(x_j)_{j\in J}$ in X converging strongly to x, where $(J, \preccurlyeq)$ is a directed preordered set. Fix also any net $(x_j^*)_{j\in J}$ converging weakly* in X^* to x^* such that $x_j^* \in \eta\partial_C d_{S(q_j)}(x_j)$. Fix $y \in B(x, \frac{\delta}{2}) \cap S(q_0)$. For each $n \in \mathbb{N}$ and each $j \in J$, choose some $y_{j,n} \in S(q_j)$ such that

$$\|y_{j,n} - y\| \le d_{S(q_j)}(y) + \frac{1}{n}.$$

Endowing $J \times \mathbb{N}$ with the product preorder which is obviously directed, the family $(y_{j,n})_{(j,n)\in J\times\mathbb{N}}$ is a net in X. Since

$$d_{S(q_j)}(y) + \frac{1}{n} \underset{(j,n)\in J\times\mathbb{N}}{\longrightarrow} 0,$$

we have $\|y_{j,n} - y\| \underset{(j,n)\in J\times\mathbb{N}}{\longrightarrow} 0$, that is, $y_{j,n} \underset{(j,n)\in J\times\mathbb{N}}{\longrightarrow} y$ strongly in H, and hence there exists $j_0 \in J$ and $n_0 \in \mathbb{N}$ such that for all $(j,n) \in J \times \mathbb{N}$ with $j \succcurlyeq j_0$ and $n \ge n_0$ we have $y_{j,n} \in B(x, \frac{\delta}{2})$. Put $x_{j,n} := x_j$ for all $(j,n) \in J \times \mathbb{N}$. Obviously $x_{j,n} \underset{(j,n)\in J\times\mathbb{N}}{\longrightarrow} x$ strongly in X $\left(\text{because } x_j \underset{j\in J}{\to} x\right)$. So, we may also suppose that $x_{j,n} \in B(x, \frac{\delta}{2})$ for all $(j,n) \in J \times \mathbb{N}$, with $j \succcurlyeq j_0$ and $n \ge n_0$. Thus, for all $(j,n) \in J \times \mathbb{N}$ with $j \succcurlyeq j_0$ and $n \ge n_0$ we have

$$\|y_{j,n} - x\| < \frac{\delta}{2} \text{ and } \|x_{j,n} - x\| < \frac{\delta}{2}.$$

Set $x_{j,n}^* := x_j^*$ and $q_{j,n} := q_j$ for all $(j,n) \in J\times\mathbb{N}$. The net $(q_{j,n})_{(j,n)\in J\times\mathbb{N}}$ converges to q_0 and the net $(x_{j,n}^*)_{(j,n)\in J\times\mathbb{N}}$ converges weakly* to x^* in X^*. Thanks to the latter inequalities above, for all $(j,n) \in J \times \mathbb{N}$ with $j \succcurlyeq j_0$ and $n \ge n_0$ we have $\|y_{j,n} - x_{j,n}\| < \delta$ with $y_{j,n} \in S(q_j)$, and hence according to (8.88)

$$\begin{aligned}\langle \eta^{-1}x_{j,n}^*, y_{j,n} - x_{j,n}\rangle &\le d_{S(q_j)}(x_{j,n}) + \varepsilon\|y_{j,n} - x_{j,n}\| \\ &\le d_{S(q_j)}(x) + \|x_{j,n} - x\| + \varepsilon\|y_{j,n} - x_{j,n}\|.\end{aligned}$$

Since the net $(\eta^{-1}x_{j,n}^*)_{(j,n)\in J\times\mathbb{N}}$ is bounded (by the real number 1), we may pass to the limit to obtain

$$\langle \eta^{-1}x^*, y - x\rangle \le \varepsilon\|y - x\|$$

for all $y \in B(x, \frac{\delta}{2}) \cap S(q_0)$. This entails that $\eta^{-1}x^* \in N^F(S(q_0); x)$. Further, $\eta^{-1}x_{j,n}^* \in \mathbb{B}$ for all $(j,n) \in J \times \mathbb{N}$ and this ensures $\eta^{-1}x^* \in \mathbb{B}$. Thus, $\eta^{-1}x^* \in N_{S(q_0)}^F(x) \cap \mathbb{B}$, so $\eta^{-1}x^* \in \partial_F d_{S(q_0)}(x) \subset \partial_C d_{S(q_0)}(x)$. The proof is then complete. □

The second proposition provides a partial upper semicontinuity property.

PROPOSITION 8.139. Let E be a metric space and let $(S(q))_{q\in E}$ be a family of nonempty closed sets of a normed space X which is uniformly equi-subsmooth. Let $Q \subset E$, $q_0 \in \operatorname{cl} Q$ and $x \in S(q_0)$. Then for any net $(q_j)_{j\in J}$ in Q converging to q_0 with $d_{S(q_j)}(x) \underset{j\in J}{\to} 0$, for any net $(x_j)_{j\in J}$ converging to x in $(X, \|\cdot\|)$, one has for every $h \in X$

$$\limsup_{j\in J} \sigma\big(h, \partial_C d_{S(q_j)}(x_j)\big) \le \sigma\big(h, \partial_C d_{S(q_0)}(x)\big).$$

PROOF. Fix any $h \in X$. Let $(q_j)_j$ and $(x_j)_j$ be as in the statement. Extracting a subnet if necessary, we may suppose that

$$\limsup_{j \in J} \sigma\big(h, \partial_C d_{S(q_j)}(x_j)\big) = \lim_{j \in J} \sigma\big(h, \partial_C d_{S(q_j)}(x_j)\big).$$

For each j, the weak* compactness of the convex set $\partial_C d_{Q(q_j)}(x_j)$ ensures the existence of some $x_j^* \in \partial d_{S(q_j)}(x_j)$ such that

$$\langle x_j^*, h \rangle = \sigma\big(h, \eta \partial_C d_{S(q_j)}(x_j)\big).$$

Since $\|x_j^*\| \le 1$, a subnet of $(x_j^*)_j$ (that we do not relabel) converges weakly* to some x^* in X^*. It results that

$$\langle x^*, h \rangle = \limsup_{j \in J} \sigma\big(h, \partial_C d_{S(q_j)}(x_j)\big). \tag{8.89}$$

On the other hand, Proposition 8.138 tells us that $x^* \in \partial_C d_{S(q_0)}(x)$. The latter inclusion combined with (8.89) yields

$$\limsup_{j \in J} \sigma\big(h, \partial_C d_{S(q_j)}(x_j)\big) \le \sigma\big(h, \partial_C d_{S(q_0)}(x)\big),$$

which completes the proof. □

The following corollary is a direct consequence of the previous proposition. It is often involved in the study of existence of solutions of differential inclusions involving normal cones. Before stating the corollary, recall that for an extended real $\rho \in]0, +\infty]$ and two subsets S, S' of a normed space X, the pseudo ρ-excess of S over S' is defined by (see (1.48))

$$\widehat{\mathrm{exc}}_\rho(S, S') := \sup_{u \in S \cap \rho \mathbb{B}_X} d(u, S'),$$

where we employ the usual convention that the latter supremum is zero whenever $S \cap \rho \mathbb{B}_X = \emptyset$.

COROLLARY 8.140. Let (Q, d) be a metric space and $(S(q))_{q \in Q}$ be a family of nonempty closed sets of an Asplund space X which are uniformly equi-subsmooth. Let W be a subset of $Q \times Q$ containing the diagonal set. Assume that there are an extended real $\rho \in]0, +\infty]$ and a function $\vartheta : W \to [0, +\infty[$ satisfying $\vartheta(q_0, q) \to 0$ as $q \to q_0$ with $(q_0, q) \in W$ and such that

$$\widehat{\mathrm{exc}}_\rho(S(q), S(q')) \le \vartheta(q, q')$$

for every $(q, q') \in W$. Then for any sequence $(q_n)_n$ in Q converging to q with $(q, q_n) \in W$, any sequence $(x_n)_n$ in X converging to $x \in S(q) \cap \rho \mathbb{B}_X$, and any $h \in X$, we have

$$\limsup_{n \to \infty} \sigma\big(h, \partial_C d_{S(q_n)}(x_n)\big) \le \sigma\big(h, \partial_C d_{S(q)}(x)\big).$$

8.10. Further results

We begin this section with the following Ngai-Luc-Théra problem. Let $f : X \to \mathbb{R} \cup \{+\infty\}$ be a lower semicontinuous function which is subsmooth at $\overline{x}$ (or equivalently which satisfies (8.4)) and let any real $\varepsilon > 0$. Does there exist a real $\delta > 0$ such that, for each $y \in B(\overline{x}, \delta)$ there is a lower semicontinuous *convex* function $g_y : B(\overline{x}, \delta) \to \mathbb{R} \cup \{+\infty\}$ satisfying the inequality $|f(x) - g_y(x)| \le \varepsilon \|x - y\|$ for all $x \in B(\overline{x}, \delta) \cap \operatorname{dom} f$?

8.10.1. ε-Localization of subsmooth functions by convex functions. Let $f : X \to \mathbb{R} \cup \{+\infty\}$ be a proper lower semicontinuous function on a Banach space X such that, for some real $\varepsilon \geq 0$, one has for all $x, x' \in X$ and $t \in]0,1[$

$$f(tx + (1-t)x') \leq tf(x) + (1-t)f(x') + \varepsilon\|x - x'\|.$$

For each $y \in X$ define $f_{\varepsilon,y}(x) := f(x) + \varepsilon\|x - y\|$ for all $x \in X$. For simplicity, denote by $f^{**}_{\varepsilon,y}$ the restriction to X of the Legendre-Fenchel biconjugate $f^{**}_{\varepsilon,y}$ of $f_{\varepsilon,y}$. By Remark 8.26 note that for any $x_0 \in \operatorname{Dom}\partial_C f \neq \emptyset$ (see Proposition 6.50) and any $x_0^* \in \partial_C f(x_0)$ we have for all $u \in X$

$$\begin{aligned} &\langle x_0^*, u - x_0\rangle \leq f(u) - f(x_0) + \varepsilon\|u - x_0\| \\ (8.90) \qquad &\leq f(u) - f(x_0) + \varepsilon\|u - y\| + \varepsilon\|y - x_0\| = f_{\varepsilon,y}(u) - f(x_0) + \varepsilon\|y - x_0\|. \end{aligned}$$

Our first aim is to estimate the value $|f(x) - f^{**}_{\varepsilon,y}(x)|$ where it is well-defined.

For any proper function $g : X \to \mathbb{R} \cup \{+\infty\}$ we know (or we easily see) that the restriction to X of the Legendre-Fenchel biconjugate g^{**}, that we still denote by g^{**} for simplicity (as above), satisfies the inequality $g^{**} \leq g$. Further, when g is minorized by a continuous affine function, g^{**} is the greatest proper lower semicontinuous convex function minorizing g. This and (8.90) ensure that the function $f^{**}_{\varepsilon,y}$ is the greatest proper lower semicontinuous convex function minoring $f_{\varepsilon,y}$, so in particular

$$(8.91) \qquad f^{**}_{\varepsilon,y}(x) \leq f_{\varepsilon,y}(x) = f(x) + \varepsilon\|x - y\| \quad \text{for all } x \in X.$$

Take any $x \in \operatorname{Dom}\partial_C f$. Choosing $x^* \in \partial_C f(x)$ we have by (8.90)

$$\langle x^*, u - x\rangle \leq f(u) - f(x) + \varepsilon\|u - y\| + \varepsilon\|y - x\| \quad \text{for all } u \in X,$$

which is equivalent to

$$\langle x^*, u\rangle - f(u) - \varepsilon\|u - y\| \leq \langle x^*, x\rangle - f(x) + \varepsilon\|x - y\| \quad \text{for all } u \in X,$$

otherwise stated,

$$(8.92) \qquad \langle x^*, x\rangle - f^*_{\varepsilon,y}(x^*) \geq f(x) - \varepsilon\|x - y\|,$$

and hence

$$(8.93) \qquad f^{**}_{\varepsilon,y}(x) \geq f(x) - \varepsilon\|x - y\|.$$

Now let us fix any $x \in X \setminus \operatorname{Dom}\partial_C f$ and show that the latter inequality still holds. Since $0 \notin \partial_C f(x)$, the point x cannot be a local minimum of f with $f(x) < +\infty$. Therefore, either $V \cap \operatorname{dom} f = \emptyset$ for some open neighborhood V of x or there exists a sequence $(x_n)_n$ converging to x with $f(x_n) < f(x)$ for all $n \in \mathbb{N}$.

Case I. $V \cap \operatorname{dom} f = \emptyset$ for some open neighborhood V of x.
Consider the function $g : X \to \mathbb{R} \cup \{+\infty\}$ defined by $g(u) = f^{**}_{\varepsilon,y}(u)$ if $u \in \operatorname{cl}_X(\operatorname{dom} f)$ and $g(u) = +\infty$ otherwise. Since $f^{**}_{\varepsilon,y} \leq f_{\varepsilon,y}$ and $\operatorname{dom} f_{\varepsilon,y} = \operatorname{dom} f$, it follows that $f^{**}_{\varepsilon,y}(\cdot) \leq g(\cdot) \leq f_{\varepsilon,y}(\cdot)$. The function g being in addition proper lower semicontinuous convex, we deduce that $f^{**}_{\varepsilon,y} = g$ since (as seen above) $f^{**}_{\varepsilon,y}$ is the greatest proper lower semicontinuous convex function minorizing $f_{\varepsilon,y}$. It ensues that $f^{**}_{\varepsilon,y}(x) = g(x) = +\infty$ since $x \notin \operatorname{cl}_X(\operatorname{dom} f)$, and hence (8.93) holds true.

Case II. There exists $x_n \to x$ with $f(x_n) < f(x)$ for all n.
Fix any $n \in \mathbb{N}$ and choose a real r_n with $f(x_n) < r_n < f(x)$. Applying the Zagrodny mean value theorem (see Theorem 6.52(ii)) on $[x_n, x]$ with r_n in place of r, we get

$c_n \neq x$ with $c_n \in [x_n, x] + (1/n)\mathbb{B}$ and $c_n^* \in \partial_C f(c_n)$ such that $\langle c_n^*, x - c_n \rangle > 0$. Since by (8.92)

$$\langle c_n^*, c_n \rangle - f_{\varepsilon,y}^*(c_n^*) \geq f(c_n) - \varepsilon \|c_n - y\|,$$

it follows that

$$\langle c_n^*, x \rangle - f_{\varepsilon,y}^*(c_n^*) \geq f(c_n) - \varepsilon \|c_n - y\|.$$

This gives $f_{\varepsilon,y}^{**}(x) \geq f(c_n) - \varepsilon \|c_n - y\|$. Noting that $c_n \to x$ and using the lower semicontinuity of f, we derive that

$$f_{\varepsilon,y}^{**}(x) \geq f(x) - \varepsilon \|x - y\|,$$

which confirms the inequality (8.93) in this second case.

Both inequalities (8.91) and (8.93) then hold true for all $x \in X$. We have then proved the following result.

PROPOSITION 8.141. Let X be a Banach space, let a real $\varepsilon \geq 0$ and let $f : X \to \mathbb{R} \cup \{+\infty\}$ be a proper lower semicontinuous function such that

$$f(tx + (1-t)x') \leq tf(x) + (1-t)f(x') + \varepsilon t(1-t)\|x - x'\|$$

for all $x, x' \in X$ and $t \in]0,1[$. Then for every $y \in X$ one has

$$f(x) - \varepsilon \|x - y\| \leq f_{\varepsilon,y}^{**}(x) \leq f(x) + \varepsilon \|x - y\| \quad \text{for all } x \in X.$$

One can derive the following corollary.

COROLLARY 8.142 (H.V. Ngai, D.T. Luc and M. Théra). Let X be a Banach space, let $f : X \to \mathbb{R} \cup \{+\infty\}$ be a lower semicontinuous function and let $\overline{x} \in \operatorname{cl}(\operatorname{dom} f)$. Then f is subsmooth at $\overline{x}$ if and only if for each real $\varepsilon > 0$ there exists a real $\delta > 0$ such that for each $y \in B(\overline{x}, \delta)$ there is a proper lower semicontinuous convex function $g_y : X \to \mathbb{R} \cup \{+\infty\}$ satisfying

$$f(x) - \varepsilon \|x - y\| \leq g_y(x) \leq f(x) + \varepsilon \|x - y\| \quad \text{for all } x \in B(\overline{x}, \delta).$$

PROOF. Assume first that f is subsmooth at $\overline{x}$ and take $\varepsilon > 0$. By Proposition 8.12 choose a real $\delta > 0$ such that

$$f(tx + (1-t)x') \leq tf(x) + (1-t)f(x') + \varepsilon t(1-t)\|x - x'\|$$

for all $x, x' \in B[\overline{x}, \delta]$ and $t \in]0,1[$. Applying Proposition 8.141 to the function $F := f + \Psi_{B[\overline{x},\delta]}$, we see that the function $g_y := F_{\varepsilon,y}^{**}$ fulfills the desired inequalities in the corollary.

Let us prove the converse. Fix any real $\varepsilon > 0$ and let $\delta > 0$ be such that the inequalities in the corollary hold with $\varepsilon/2$ in place of ε. Take any $x, x' \in B(\overline{x}, \delta)$ and $t \in]0,1[$, and put $z_t := tx + (1-t)x' \in B(\overline{x}, \delta)$. Then for the convex function $\varphi := g_{z_t}$ we have $\varphi(z_t) = f(z_t)$ and

$$f(x) + \frac{\varepsilon}{2}(1-t)\|x - x'\| \geq \varphi(x) \text{ and } f(x') + \frac{\varepsilon}{2}t\|x - x'\| \geq \varphi(x').$$

It ensues by the convexity of φ that

$$tf(x) + (1-t)f(x') + \varepsilon t(1-t)\|x - x'\| \geq \varphi(z_t) = f(z_t),$$

which justifies (by Proposition 8.12) the subsmoothness of f and finishes the proof. □

8.10.2. Metric regularity of subsmooth-like multimappings. In addition to the metric subregularity results in Theorems 8.133 and 8.134 for multimappings with metrically subsmooth graphs, the analysis in this subsection will involve the class of multimappings with normally $\omega(\cdot)$-regular graphs.

DEFINITION 8.143. Let X be a Banach space, S be a subset of X, $\omega : \mathbb{R}_+ \to \mathbb{R}_+$ be a function with $\omega(0) = 0$, V be a nonempty open subset of X. One says that S is *normally* $\omega(\cdot)$*-regular relative to* V whenever for all $x, x' \in S \cap V$, for all $x^* \in N^C(S; x)$,

$$\langle x^\star, x' - x\rangle \le \|x^*\|\, \omega(\|x' - x\|).$$

Particular sets of the above class are quantified as follows:

DEFINITION 8.144. Let X be a Banach space, S be a subset of X, $\overline{x} \in S$, $\sigma, \delta \in]0, +\infty[$. One says that S is (σ,δ)*-subsmooth at* $\overline{x}$ if for all $x, x' \in S \cap B(\overline{x}, \delta)$, for all $x^* \in N^C(S; x)$,

$$\langle x^*, x' - x\rangle \le \sigma\, \|x^\star\|\, \|x' - x\|\,.$$

A closed set $S \subset X$ which is (σ,δ)-subsmooth at $\overline{x} \in S$ with $\sigma, \delta \in]0, +\infty[$, is obviously normally $\omega(\cdot)$-regular relative to $B(\overline{x}, \delta)$ for $\omega : \mathbb{R}_+ \to \mathbb{R}_+$ defined by

$$\omega(t) = \sigma t \quad \text{for all } t \in \mathbb{R}_+.$$

LEMMA 8.145. Let $M : X \rightrightarrows Y$ be a multimapping between Banach spaces X, Y with closed graph and let $(a, b) \in X \times Y$. Assume that there are $\delta, \gamma \in]0, +\infty[$ such that

$$\gamma\, d(b, M(a)) < \delta < d(a, M^{-1}(b)).$$

Then for any real $\varepsilon > 0$ there exist $(u, v) \in \operatorname{gph} M$ such that

$$\|u - a\| < \delta, \quad 0 < \|v - b\| < \delta/\gamma,$$

and

$$\|v - b\| \le \|y - b\| + \frac{1}{\gamma}(\|x - u\| + \varepsilon\|y - v\|)\ \forall (x, y) \in \operatorname{gph} M.$$

PROOF. By the left-hand inequality in the assumption we can choose $b' \in M(a)$ such that

$$\|b' - b\| < \delta/\gamma. \tag{8.94}$$

Consider the lower semicontinuous function $f : X \times Y \to \mathbb{R} \cup \{+\infty\}$ defined by $f(x, y) := \|y - b\| + \Psi_{\operatorname{gph} M}(x, y)$ for all $(x, y) \in X \times Y$ and note that

$$f(a, b') < \frac{\delta}{\gamma} \le \inf_{X \times Y} f + \frac{\delta}{\gamma}.$$

Endowing $X \times Y$ with the norm $\|(x, y)\|_\varepsilon := \|x\| + \varepsilon\|y\|$, the Ekeland variational principle with the function f furnishes a pair $(u, v) \in \operatorname{gph} M$ such that

$$\|v - b\| \le \|y - b\| + \frac{1}{\gamma}(\|x - u\| + \varepsilon\|y - v\|)\ \forall (x, y) \in \operatorname{gph} M$$

$$f(u, v) \le f(a, b'), \quad \|(u, v) - (a, b')\|_\varepsilon < \delta.$$

The above inequality on the left gives $\|v - b\| < \|b' - b\| < \delta/\gamma$ according to (8.94), while the other inequality implies that $\|u - a\| < \delta$. Further, the latter inequality combined with the right-hand inequality in the assumption ensures that $\|u - a\| < d(a, M^{-1}(b))$, hence $v \neq b$ since $u \in M^{-1}(v)$. This finishes the proof. $\square$

THEOREM 8.146 (metric regularity under subsmooth-like property). Let X, Y be Banach spaces, $M : X \rightrightarrows Y$ be a multimapping with closed graph, Q be a nonempty subset of $\operatorname{gph} M$, $\omega : [0, +\infty[\to [0, +\infty[$ be an upper semicontinuous nondecreasing (modulus) function with $\omega(0) = 0$. Assume that:
(i) there exist $\alpha, \beta, \rho \in]0, +\infty[$ with

$$\beta > \frac{3\alpha}{\rho} + \left(1 + \frac{1}{\rho}\right) \omega \left(\sqrt{4\alpha^2 + \left(\beta - \frac{\alpha}{\rho}\right)^2} \right)$$

such that for all $(\overline{x}, \overline{y}) \in Q$,

$$\overline{y} + \beta \mathbb{U}_Y \subset M(\overline{x} + \alpha \mathbb{B}_X);$$

(ii) the set $\operatorname{gph} M$ is normally $\omega(\cdot)$-regular relative to $\bigcup_{(a,b)\in Q} B\left((a, b), \sqrt{\alpha^2 + \beta^2}\right)$.

Then, there exists a real $\gamma \in [0, \rho[$ such that for every $(\overline{x}, \overline{y}) \in Q$, there exists a real $\delta > 0$ satisfying for all $x \in B(\overline{x}, \delta)$ and for all $y \in B(\overline{y}, \delta)$

$$d(x, M^{-1}(y)) \le \gamma\, d(y, M(x)).$$

In particular, one has $\sup_{(x,y)\in Q} \operatorname{reg}[M](x \mid y) \le \rho$.

PROOF. Set $V = \bigcup_{(a,b)\in Q} B\left((a, b), \sqrt{\alpha^2 + \beta^2}\right)$. By contradiction, assume that for each $\gamma \in [0, \rho[$, there is $(\overline{x}, \overline{y}) \in Q$ such that for each real $\delta > 0$, there are $x \in B(\overline{x}, \delta)$ and $y \in B(\overline{y}, \delta)$ satisfying

$$d(x, M^{-1}(y)) > \gamma d(y, M(x)).$$

Let $(\varepsilon_n)_{n\in\mathbb{N}}$ be a nonincreasing sequence of $]0, \rho[$ with $\varepsilon_n \downarrow 0$. For each $n \in \mathbb{N}$, set

$$\theta_n := \frac{1}{\rho - \varepsilon_n} \left(\alpha + \frac{1}{n}\right) + \frac{1}{n}.$$

Obviously, $(\theta_n)_{n\in\mathbb{N}}$ is a nonincreasing sequence which satisfies $\theta_n \to \alpha/\rho$. Since $\beta > \alpha/\rho$, there is $n_0 \in \mathbb{N}$ such that for any integer $n \ge n_0$ we have $\theta_n < \beta$. Choose $N \in \mathbb{N}$ with $N \ge n_0$ such that for every integer $n \ge N$,

$$\left(\alpha + \frac{2}{n}\right)^2 + \theta_n^2 < \alpha^2 + \beta^2. \tag{8.95}$$

Fix any integer $n \ge N$. There are $(\overline{x}_n, \overline{y}_n) \in Q$, $x_n \in X$, $y_n \in Y$ with

(8.96)

$$\|\overline{x}_n - x_n\| < \min\left\{\beta, \frac{1}{\rho - \varepsilon_n}, \frac{1}{n}\right\} \quad \text{and} \quad \|\overline{y}_n - y_n\| < \min\left\{\beta, \frac{1}{\rho - \varepsilon_n}, \frac{1}{n}\right\}$$

and

$$d(x_n, M^{-1}(y_n)) > (\rho - \varepsilon_n) d(y_n, M(x_n)).$$

According to Lemma 8.145 there is $(u_n, v_n) \in \operatorname{gph} M$ such that

$$\|u_n - x_n\| < d(x_n, M^{-1}(y_n)), \tag{8.97}$$

$$0 < \|v_n - y_n\| < \frac{d(x_n, M^{-1}(y_n))}{\rho - \varepsilon_n} \tag{8.98}$$

and

$$(0,0) \in \{0\} \times \partial \|\cdot\|(v_n - y_n) + \frac{1}{\rho - \varepsilon_n}(\mathbb{B}_{X^*} \times \varepsilon_n \mathbb{B}_{Y^*}) + N^C(\operatorname{gph} M; (u_n, v_n)).$$

Since $v_n - y_n \neq 0$, there is some $z_n^* \in Y^*$ such that

$$\|z_n^*\| = 1 \quad \text{and} \quad \langle z_n^*, v_n - y_n \rangle = \|v_n - y_n\| \tag{8.99}$$

and there is some $(x_n^*, y_n^*) \in N^C(\operatorname{gph} M; (u_n, v_n))$ satisfying

$$(x_n^*, y_n^*) \in (0, -z_n^*) + \frac{1}{\rho - \varepsilon_n}(\mathbb{B}_{X^*} \times \varepsilon_n \mathbb{B}_{Y^*}).$$

It follows that

$$\|x_n^*\| \leq \frac{1}{\rho - \varepsilon_n} \quad \text{and} \quad y_n^* \in -z_n^* + \frac{\varepsilon_n}{\rho - \varepsilon_n}\mathbb{B}_{Y^*}. \tag{8.100}$$

Since $(\overline{x}_n, \overline{y}_n) \in Q$, we can apply our assumption to obtain

$$B(\overline{y}_n, \beta) \subset M(B[\overline{x}_n, \alpha]). \tag{8.101}$$

Using (8.96), we know that $y_n \in B(\overline{y}_n, \beta)$, so there is some $z_n \in B[\overline{x}_n, \alpha]$ such that $y_n \in M(z_n)$. In particular, we have

$$z_n \in M^{-1}(y_n) \cap B[\overline{x}_n, \alpha].$$

Keeping in mind (8.96), the latter inclusion entails

$$d(x_n, M^{-1}(y_n)) \leq \|x_n - z_n\| \leq \|x_n - \overline{x}_n\| + \|\overline{x}_n - z_n\| \leq \frac{1}{n} + \alpha. \tag{8.102}$$

From (8.97), (8.96) and the latter inequality, it ensues that

$$\|u_n - \overline{x}_n\| \leq \|u_n - x_n\| + \|x_n - \overline{x}_n\| < d(x_n, M^{-1}(y_n)) + \frac{1}{n} \leq \frac{2}{n} + \alpha. \tag{8.103}$$

Using (8.97), (8.98) and (8.102), we obtain

$$\begin{aligned} \|v_n - \overline{y}_n\| \leq \|v_n - y_n\| + \|y_n - \overline{y}_n\| &< \frac{d(x_n, M^{-1}(y_n))}{\rho - \varepsilon_n} + \frac{1}{n} \\ &\leq \frac{\frac{1}{n} + \alpha}{\rho - \varepsilon_n} + \frac{1}{n} = \theta_n. \end{aligned} \tag{8.104}$$

Set

$$\zeta_n := (\beta - \theta_n)\frac{v_n - y_n}{\|v_n - y_n\|}.$$

Thanks to the definition of ζ_n, the choice of n and the inequality (8.104), we can check that $v_n - \zeta_n \in B(\overline{y}_n, \beta)$. According to this inclusion and to the inclusion (8.101), we get $v_n - \zeta_n \in M(B[\overline{x}_n, \alpha])$, so there exists $w_n \in B[\overline{x}_n, \alpha]$ such that $v_n - \zeta_n \in M(w_n)$. By (8.103), let us observe that

$$\|w_n - u_n\| \leq \|w_n - \overline{x}_n\| + \|\overline{x}_n - u_n\| < 2\alpha + \frac{2}{n}.$$

Endow $X \times Y$ with the 2-norm. It is clear that

$$\|(w_n, v_n - \zeta_n) - (\overline{x}_n, \overline{y}_n)\|^2 = \|w_n - \overline{x}_n\|^2 + \|v_n - \zeta_n - \overline{y}_n\|^2 < \alpha^2 + \beta^2.$$

According to (8.95), (8.103) and (8.104), we get

$$\begin{aligned}\|(u_n, v_n) - (\overline{x}_n, \overline{y}_n)\|^2 &= \|u_n - \overline{x}_n\|^2 + \|v_n - \overline{y}_n\|^2 \\ &\le \left(\alpha + \frac{2}{n}\right)^2 + \theta_n^2 < \alpha^2 + \beta^2.\end{aligned}$$

In particular, we have the inclusions

$$(w_n, v_n - \zeta_n) \in V \quad \text{and} \quad (u_n, v_n) \in V.$$

Further, from the normal $\omega(\cdot)$-regularity of $\operatorname{gph} M$ relative to V, we have

$$\langle (x_n^*, y_n^*), (w_n, v_n - \zeta_n) - (u_n, v_n) \rangle \le \|(x_n^*, y_n^*)\| \, \omega(\|(w_n, v_n - \zeta_n) - (u_n, v_n)\|),$$

and from (8.100) it can be easily checked that

$$\|(x_n^*, y_n^*)\| \le \|x_n^*\| + \|y_n^*\| \le 1 + \frac{1 + \varepsilon_n}{\rho - \varepsilon_n}.$$

Hence, it follows

$$\langle x_n^*, w_n - u_n \rangle \le \langle y_n^*, \zeta_n \rangle + \left(1 + \frac{1 + \varepsilon_n}{\rho - \varepsilon_n}\right) \omega(\|(w_n - u_n, -\zeta_n)\|). \tag{8.105}$$

From the inequality in (8.100) we also have

$$\langle x_n^*, w_n - u_n \rangle \ge - \|x_n^*\| \, \|w_n - u_n\| \ge - \frac{1}{\rho - \varepsilon_n} \left(2\alpha + \frac{2}{n}\right).$$

Note by the inclusion in (8.100), by (8.99) and by the definition of ζ_n that there is $b_n^* \in \mathbb{B}_{Y^*}$ such that

$$\langle y_n^*, \zeta_n \rangle = \left\langle -z_n^\star + \frac{\varepsilon_n}{\rho - \varepsilon_n} b_n^*, \zeta_n \right\rangle \le \theta_n - \beta + \frac{\varepsilon_n}{\rho - \varepsilon_n} (\beta - \theta_n).$$

Note also that (thanks to the nondecreasing property of ω)

$$\omega(\|(w_n - u_n, -\zeta_n)\|) \le \omega\left(\left[\left(2\alpha + \frac{2}{n}\right)^2 + (\beta - \theta_n)^2\right]^{\frac{1}{2}}\right).$$

Coming back to (8.105), we deduce with $r_n := \left[(2\alpha + \frac{2}{n})^2 + (\beta - \theta_n)^2\right]^{\frac{1}{2}}$

$$-\frac{1}{\rho - \varepsilon_n}\left(2\alpha + \frac{2}{n}\right) \le -\beta + \theta_n + \frac{\varepsilon_n}{\rho - \varepsilon_n}(\beta - \theta_n) + \left(1 + \frac{1 + \varepsilon_n}{\rho - \varepsilon_n}\right) \omega(r_n).$$

Passing to the lim sup as $n \to \infty$ and using the upper semicontinuity of ω, we get

$$-\frac{2\alpha}{\rho} \le -\beta + \frac{\alpha}{\rho} + \left(1 + \frac{1}{\rho}\right) \omega\left(\left[4\alpha^2 + \left(\beta - \frac{\alpha}{\rho}\right)^2\right]^{\frac{1}{2}}\right),$$

and this cannot hold true. □

Corollary 8.147. Let $M : X \rightrightarrows Y$ be a multimapping between Banach spaces whose graph is closed and (σ, δ)-subsmooth at $(\overline{x}, \overline{y}) \in \operatorname{gph} M$ for some reals $\sigma \in]0, 1[$ and $\delta > 0$. Assume that there are positive reals α, β satisfying $2\alpha\sigma < \beta\sqrt{1 - \sigma^2}$ such that

$$\overline{y} + \beta \mathbb{U}_Y \subset M(\overline{x} + \alpha \mathbb{B}_X).$$

Then the multimapping M is metrically regular at $\overline{x}$ for $\overline{y}$.

EXERCISE 8.148. Derive the corollary from the above theorem, noticing that the inequality in the assumption is equivalent to $\beta > \sigma\sqrt{4\alpha^2+\beta^2}$. □

8.11. Comments

Submonotone multimappings have been introduced by J. E. Spingarn in his 1981 paper [**898**] (submitted November 1979) under the name of "strictly submonotone" multimappings; "submonotone" multimappings in [**898**] corresponds to multimappings called one-sided submonotone in the book. Spingarn showed in [**898**] that the Clarke subdifferential of a locally Lipschitz function $f : \mathbb{R}^n \to \mathbb{R}$ is submonotone on $\mathbb{R}^n$ (in the sense of the book) if and only if it is lower-$\mathcal{C}^1$, that is, for each point $\overline{x} \in \mathbb{R}^n$ there exist a compact set T, an open neighborhood V of $\overline{x}$ and a continuous function $\varphi : V \times T \to \mathbb{R}$ such that $D_1\varphi(\cdot,\cdot)$ exists and is continuous on $V \times T$, and such that $f(x) = \max_{t\in T}\varphi(x,t)$; this characterization will be proved in Theorem 10.35. This result also means that, for the locally Lipschitz function $f : \mathbb{R}^n \to \mathbb{R}$, the Clarke subdifferential $\partial_C f$ is submonotone if and only if f is subsmooth on $\mathbb{R}^n$ in the sense of the book. Independently, H. V. Ngai, D. T. Luc and M. Théra introduced in their 2000 paper [**767**] the class of "approximate convex functions" on a normed space as functions satisfying the condition (8.4). Such functions coincide with subsmooth ones by Proposition 8.12. Various important and significant results, in particular the subdifferential determination property, are proved by these authors in [**767**]. We will see in Chapter 10 that approximate convex functions are in the line of semiconvex functions. We used the term of "subsmooth functions" mainly because of Theorem 8.1, but also due to the fact that "approximately convex functions" were previously defined as another concept by D. H. Hyers and S. M. Ulam in their 1952 paper [**507**]. In [**504**] D. H. Hyers investigated the following question of S. M. Ulam: Given a function f which satisfies the "linear" functional equation $f(x+y) = f(x) + f(y)$ only approximately in a certain sense, does there exist a linear function g which approximates f? Of course, "linear" has to be understood there as additive. Hyers showed in his 1941 paper [**504**] the following: Given a real $\varepsilon > 0$ and an ε-linear (in fact ε-additive) mapping $f : X \to X'$ between Banach spaces X and X', in the sense $\|f(x+y) - f(x) - f(y)\| \le \varepsilon$, then there exists a linear (in fact additive) mapping $g : X \to X'$ which approximates f with amount ε, that is, $\|f(x) - g(x)\| \le \varepsilon$ for all $x \in X$; further the mapping g is unique and it is continuous on X whenever f is continuous at some point. Similar questions have been also investigated, in the 1945 and 1947 papers [**505, 506**] of D. H. Hyers and S. M. Ulam, and in the 1946 paper of D. G. Bourgin [**175**], for approximate ε-isometries, that is, mappings T between two metric spaces with $|d(T(x),T(y)) - d(x,y)| < \varepsilon$. All those papers naturally led D. H. Hyers and S. M. Ulam to study the similar problem when additivity or isometry is replaced by convexity. In their paper [**507**] published in 1952, they declared a function $f : C \to \mathbb{R}$ (defined on a convex set C) to be approximately convex with amount $\varepsilon > 0$ (or ε-convex) whenever $f(tx + (1-t)y) \le tf(x) + (1-t)f(y) + \varepsilon$ for all $x, y \in C$ and $t \in [0,1]$. Hyers and Ulam proved in [**507**] that for any function $f : U \to \mathbb{R}$ from an open convex set U of $\mathbb{R}^n$ which is approximately convex (in their sense) with amount ε there exist a convex function $g : U \to \mathbb{R}$ such that

$$|f(x) - g(x)| \le \kappa_n \varepsilon \quad \text{for all } x \in U, \tag{8.106}$$

where κ_n is a universal constant depending only on n (an exact value is even given). Other important results on approximately convex functions and sets can be found in the paper by S. J. Dilworth, R. Howard and J. W. Roberts [**336**]; therein a closed set S is defined as approximately convex if its distance function d_S is approximately convex within amount $\varepsilon = 1$ (the case of any $\varepsilon > 0$ is reduced to $\varepsilon = 1$ with the new equivalent norm $\varepsilon^{-1}\|\cdot\|$). Investigations on lower and upper bounds for possible constants κ_n in (8.106) have been realized by several authors, see, for example, [**186, 208, 223, 336, 337, 338, 449, 644**] and references therein; in particular the impossibility of extension of (8.106) in infinite dimensions is shown in [**208, 644, 186**]. It is also worth pointing out that, taking into account both definitions of Hyers and Ulam and of Ngai, Luc and Théra, the concept of (e,δ)-convex functions have been considered by Z. Páles [**781**] as functions f such that

$$f(tx + (1-t)y) \le tf(x) + (1-t)f(y) + \varepsilon t(1-t)\|x-y\| + \delta.$$

A large class of diverse notions of approximate convexity is presented by J. Makó and Z. Páles [**683**]. For the use of approximate convex functions on the unit ball in the theory of geometry of Banach spaces we refer to the paper [**187**] by F. Cabello Sánchez, J. M. F. Castillo and P. L. Papini and to references therein.

Subsmooth sets in Banach spaces were introduced and largely studied by D. Aussel, A. Daniilidis and L. Thibault in the 2005 paper [**49**]. The concept was motivated by the necessity of a first-order viewpoint of the studies realized by R. A. Poliquin and R. T. Rokafellar [**809**] for the subdifferential characterization of prox-regular functions and by R. A. Poliquin, R. T. Rockafellar and L. Thibault [**813**] concerning the (second-order) hypomonotonicity property of the truncated (Clarke) normal cone $N^C(S;\cdot)\cap\mathbb{B}$ of a prox-regular set. Prox-regularity of sets and hypomonotonicity of their truncated normal cones will be developed in Chapter 15. The development of the present chapter on subsmooth functions and sets mainly followed L. Thibault's 2019 survey [**925**].

Theorem 8.1 is due to L. Veselý and L. Zajíček [**956**] and its proof given in the manuscript follows the main arguments in [**956**, Proposition 3.7]. Its importance led us to present it as a theorem.

Corollary 8.19 was observed in [**767**, Proposition 3.2].

Proposition 8.21 and Theorem 8.25 were first established by H. V. Ngai, D. T. Luc and M. Théra in [**767**], when f is in addition lower semicontinuous, with different approaches based on Theorem 3.4 in [**767**] proving that for such a function f and for each $\varepsilon > 0$ there is some real $\delta > 0$ such that for each $y \in B(\overline{x},\delta)$ there exists a proper lower semicontinuous function $\varphi_y : X \to \mathbb{R}\cup\{+\infty\}$ such that

$$|f(x) - \varphi_y(x)| \le \varepsilon\|x-y| \quad \text{for all } x \in B(\overline{x},\delta)\cap \operatorname{dom} f.$$

The proofs in the manuscript for Proposition 8.21 and Theorem 8.25 follow the approach utilized by L. Thibault and D. Zagrodny [**927**, Proposition 3.4]. Proposition 8.10 was established in Proposition 3.4 of L. Zajíček's 2008 paper [**995**].

The implication (c)⇒(a) in Theorem 8.35 seems to have been first obtained by L. Zajíček in [**995**, Lemma 5.4]; the other implications were given in [**925**]. When the continuous function f is convex (instead of subsmooth) the equivalences in the theorem were established for the first time in the 1968 paper [**30**] by E. Asplund and R. T. Rockafellar. Such a study for the differentiability of a norm (as well as other particular functions) was previously thoroughly realized by V. S. Šmulian in a series of celebrated 1939-1941 papers [**893, 894, 895, 896, 897**].

Example 8.38 and the development therein are due to J. E. Spingarn [**898**, p. 84]. The subdifferential characterizations (d)-(e) of subsmooth functions in Theorem 8.39 were independently proved by by A. Daniilidis, F. Jules and M. Lassonde [**311**] and by H. V. Ngai and J.-P. Penot [**769**]: they were previously established for locally Lipschitz functions by A. Daniilidis and P. Georgiev [**309**]. The other characterizations (b)-(d) in Theorem 8.39 were noticed in Thibault's survey [**925**, Theorem 4.15]. Proposition 8.43 was also given in [**925**, Proposition 4.19]. The implication (a)⇒(b) in Proposition 8.43 corrects the statement and proof of the implication (i)⇒(ii) in [**49**, Lemma 4.4] where a gap occurred. The equivalence (iii)⇔(iv) in [**49**, Theorem 4.5] must then be replaced by the equivalence (b)⇔(c) in Proposition 8.44. Accordingly, Corollary 4.18 in [**49**] has also to be replaced by Proposition 8.88 in this survey. The equivalence (d)⇔(e) in Proposition 8.44 is due to J. E. Spingarn [**898**, Theorem 3.9]: the proof here uses ideas in [**49**].

One-sided submonotone multimappings are called semi-submonotone in [**49**]: The name "one-sided submonotone"seems to be more appropriate in the sense that it translates the property of the definition. Lemmas 8.47 and 8.48 and Example 8.49 are taken from J. E. Spingarn [**898**, p. 83]. Proposition 8.50 and its proof are taken from the paper [**49**] by D. Aussel, A. Daniliidis and L. Thibault. The concept of semismooth functions in Definition 8.52 was introduced by R. Mifflin [**709**]. Proposition 8.55 is due to Spingarn [**898**, proposition 2.4]: the proof given here follows the approach in [**49**].

Except Proposition 8.71 (appeared in [**925**, Proposition 8.65]), all the results in Section 8.3 devoted to subsmooth sets are taken from D. Aussel, A. Daniilidis and L. Thibault [**49**]. The idea of the use of Krein-Šmulian theorem for the equality $N^C(S;\overline{x}) = N^F(S;\overline{x})$ in Lemma 8.74 is taken from the paper [**1004**] by X. Y. Zheng and K. F. Ng. The subsmooth version of Theorem 8.85 as well as its proof are taken from the paper [**49**] of D. Aussel, A. Daniilidis and L. Thibault.

Metrically subsmooth sets were defined by A. Daniilidis and L. Thibault [**314**]. Property (e) in Theorem 8.99 was considered by A. S. Lewis, D. R. Luke and J. Malick in Definition 4.4 of their 2009 paper [**659**] under the name of *super regularity*. Example 8.101 was provided by Lewis, Luke and Malick at the end of Section 4 in their paper [**659**, Example 4.7] and Example 8.102 appeared in the same paper [**659**, Example 4.7]. Property (c) in Proposition 8.105 with the L-normal cone $N^L(S;\overline{x})$ in place of $N^C(S;\overline{x})$ was used, for an Asplund space X, by A. Jourani under the name of *weak-regularity* of S at $\overline{x}$ in his 2006 paper [**577**] (see Corollary 4.2 therein; diverse other results related to the weak regularity of sets can be found there). The implication (a)⇒(b) in Proposition 8.103 and its proof are due to H. V. Ngai and J.-P. Penot who proved that result with the implication (e)⇒(a) in [**768**, Theorem 4.10]. The implication (c)⇒(a) in Proposition 8.103 appeared in the implication (c)⇒(e) of [**768**, Corollary 11]. When for any $\varepsilon > 0$ the distance function d_S satisfies for $\overline{x} \in S$ the property (8.4) in Proposition 8.12 on some ball $B(\overline{x}, \delta)$ (resp. the relative similar property (b) in Proposition 8.103 on $S \cap B(\overline{x}, \delta)$ for some $\delta > 0$), Ngai and Penot [**768**] defined the set S as approximately convex (resp. intrinsically approximately convex) at $\overline{x}$. The name of approximately convex set was employed much earlier in 1967 in the study of Chebyshev sets by L. Vlasov [**960**] to refer to a set S for which $\mathrm{Proj}_S x$ is a nonempty convex set for every $x \in X$; note that it was also used with another meaning in [**336**].

Lemma 8.108 and Theorem 8.110 as well as their proofs are due to A. Jourani and E. Vilches [**591**].

The truncated normal cone inverse image property in (8.57) as well as the linear inclusion property of subdifferential of distance from inverse image in (8.58) near a point was introduced by A. Daniilidis and L. Thibault [**314**]. These concepts were utilized in [**314**] to prove the results in Theorem 8.113 and Proposition 8.126. Functions of type (8.56) were considered in [**5**, p. 471] to illustrate the non-preservation of prox-regularity of sets (see Chapter 15) under usual operations. Lemma 8.128 is in the spirit of Lemma 3.7 of Aussel, Daniilidis and Thibault [**49**] and of Theorem 3.1 of X. Y. Zheng and K. F. Ng [**1004**]; its proof combines the main ideas of these papers. Theorem 8.130 and Theorem 8.132 are taken from [**314**]. Under the Asplund property of the Banach space X, it was also proved by Zheng and Ng in [**1004**, Theorem 4.2] that a point $\overline{x}$ is metrically subregular for a system of closed sets $S_1, \cdots, S_m$ of X with $\overline{x} \in S := \bigcap_{i=1}^m S_i$ if and only if there are reals $\gamma > 0$ and $\delta > 0$ such that for all $x \in S \cap B(\overline{x}, \delta)$

$$N^F(S; x) \cap \mathbb{B}_{X^*} \subset N^L(S_1; x) \cap \gamma\mathbb{B}_{X^*} + \cdots + N^L(S_m; x) \cap \gamma\mathbb{B}_{X^*}; \tag{8.107}$$

the equality between the infimum of such real constants $\gamma > 0$ and the number $\mathrm{subreg}_{\cap}[S_1, \cdots, S_m](\overline{x})$ is also shown in the same theorem in [**1004**]. The condition (8.107) is slightly different from the one in the assertion (II) in Theorem 8.132. Theorem 8.133 is another result providing a necessary and sufficient condition for the metric subregularity of multimappings with metric subsmooth graphs. In the same framework of Asplund property of X and Y, the sufficiency of the condition in Theorem 8.134 for the metric subregularity of a multimapping M was established by Zheng and Ng in [**1005**, Theorem 4.8] under a weakened partial subsmoothness property of $\mathrm{gph}\, M$. In [**1005**] one can also find other sufficient (resp. necessary) conditions for the subregularity of a multimapping with various types of subsmoothness of its graph.

The proof of Proposition 8.135 utilizes ideas in the proof of [**5**, Theorem 4.1] by S. Adly, F. Nacry and L. Thibault. Proposition 8.137 as well as Propositions 8.138 and 8.139 are taken from [**925**]; similar results were previously obtained by T. Haddad, J. Noel and L. Thibault [**457**].

Corollary 8.142 is due to H. V. Ngai, D. T. Luc and M. Théra [**767**] as well as a very large part of the development preceding Proposition 8.141. Lemma 8.145 follows X. Y. Zheng and K. F. Ng [**1006**, Lemma 3.1]. Theorem 8.146 is taken from S. Adly, F. Nacry and L. Thibault [**7**] and its proof is for a large part adaptations of ideas from the paper [**1006**] by X. Y. Zheng and K. F. Ng. Corollary 8.147 was previously established by X. Y. Zheng and Q. H. He [**1003**]. In those papers [**1006**] and [**1003**] one can find diverse other results concerning metric inequalities for multimappings whose graphs enjoy certain subsmooth-like properties.

CHAPTER 9

Subdifferential determination

Let $f, g : U \to \mathbb{R} \cup \{+\infty\}$ be two extended real-valued functions defined a nonempty open convex subset U of a Banach space X and let ∂ be a subdifferential. The subdifferential determination is concerned with the following question: Are the functions f and g equal up to an additive constant whenever

$$\partial f(x) = \partial g(x) \quad \text{for all } x \in U?$$

Considering the pathological Lipschitz function $f :]0,1[\to \mathbb{R}$ in (5.20) in the form $f(x) = \int_0^x \mathbf{1}_S(t)\,dt$ for all $x \in]0,1[$ (with S as given therein), keeping in mind that $\partial_C f(x) = [0,1]$ for all $x \in]0,1[$, and defining $g :]0,1/2[\to \mathbb{R}$ with $g(x) := f(x+(1/2))$ for all $x \in J :=]0,1/2[$, we see that $\partial_C g(x) = \partial_C f(x)$ for all $x \in J$ but $f-g$ is not a constant on J. Thus, certain conditions on one of the functions will be required when the C-subdifferential is involved. What about the F-subdifferential?

9.1. Denjoy function

Let $f, g : U \to \mathbb{R}$ be two locally Lipschitz functions on an open convex subset U of $\mathbb{R}^m$. Assume that $\partial_F f(x) \subset \partial_F g(x)$ for all $x \in U$. Then by the Rademacher theorem the locally Lipschitz function $\varphi := f - g$ is Fréchet differentiable outside a Lebesgue negligible subset $N \subset U$ with $D_F\varphi(x) = 0$ for all $x \in U \setminus N$. This is known to entail (and it is not difficult to see via the Fubini theorem) that φ is constant, thus $f = g + c$ on U for some real constant c. Below we will see that the situation is different with continuous functions in the sense that there exist continuous functions $f, g :]0,1[\to \mathbb{R}$ which fail to be equal up to an additive constant whereas $\partial_F f(x) = \partial_F g(x)$ for all $x \in]0,1[$.

Let P be a *compact perfect* set in $\mathbb{R}$ (that is, a compact set in $\mathbb{R}$ with no isolated points) such that $\mathrm{int}_{\mathbb{R}} P = \emptyset$ (for example, a set of Cantor type). Let $a = \min P$ and $b = \max P$. Let $(I_n)_{n\in\mathbb{N}}$ be an *enumeration* of the sequence of disjoint nonempty open intervals whose union is $]a,b[\setminus P$, the enumeration $(I_n)_{n\in\mathbb{N}}$ being done in such a way that $\mathrm{meas}\,(I_{n+1}) \le \mathrm{meas}\,(I_n)$ for every $n \in \mathbb{N}$. Let a_n, b_n be such that $I_n =]a_n, b_n[$ for every $n \in \mathbb{N}$. Let $\sum r_n$ be an *absolutely convergent* real series. For any nonempty open interval $]\alpha,\beta[$ in $\mathbb{R}$ we will write $\sum\limits_{I_n \subset]\alpha,\beta[} r_n$ to denote the sum of the series over all integers n such that $I_n \subset]\alpha,\beta[$.

LEMMA 9.1 (A. Denjoy). Under the above data, for each $n \in \mathbb{N}$ let $f_n : [a_n, b_n] \to \mathbb{R}$ be a continuous function such that $f_n(a_n) = 0$ and $f_n(b_n) = r_n$, and $\sup_{x\in[a_n,b_n]} |f_n(x)| \to 0$ as $n \to \infty$. Let $f : [a, b] \to \mathbb{R}$ be defined by

$$f(x) = \begin{cases} \sum_{I_n\subset]a,x[} r_n & \text{if } x \in P \\ \sum_{I_n\subset]a,x[} r_n + f_p(x) & \text{if } x \in]a_p, b_p[. \end{cases}$$

Then the function f is continuous on $[a, b]$.

PROOF. Notice that we have $f(x) = \sum_{I_n\subset]a,a_p[} r_n + f_p(x)$ for every $x \in]a_p, b_p[$ along with by definition of $f_p(a_p)$ and $f_p(b_p)$

$$f(a_p) = \sum_{I_n\subset]a,a_p[} r_n = \sum_{I_n\subset]a,a_p[} r_n + f_p(a_p),\ f(b_p) = \sum_{I_n\subset]a,b_p[} r_n = \sum_{I_n\subset]a,a_p[} r_n + f_p(b_p).$$

This gives for any $x \in [a_p, b_p]$

$$f(x) = \sum_{I_n\subset]a,a_p[} r_n + f_p(x),$$

from which we see that

$$|f(x') - f(x)| = |f_p(x') - f_p(x)| \quad \text{for any } x, x' \in [a_p, b_p].$$

Let $\varepsilon > 0$. Choose $N_1 \in \mathbb{N}$ such that $\sup_{x\in[a_n,b_n]} |f_n(x)| < \varepsilon/2$ for all $n \geq N_1$. Then for any integer $n \geq N_1$ and any $x, x' \in [a_n, b_n]$ we have

$$|f(x') - f(x)| = |f_n(x') - f_n(x)| \leq |f_n(x')| + |f_n(x)| < \varepsilon.$$

For each integer $n \leq N_1$ choose a real $\delta_n > 0$ such that

$$\Big(x, x' \in [a_n, b_n] \text{ and } |x' - x| < \delta_n\Big) \Longrightarrow |f_n(x') - f_n(x)| < \varepsilon,$$

and denote $\delta' := \min\{\delta_1, \cdots, \delta_{N_1}\}$. Then for any integer $n \in \mathbb{N}$

$$\Big(x, x' \in [a_n, b_n] \text{ and } |x' - x| < \delta'\Big) \Longrightarrow |f(x') - f(x)| < \varepsilon. \tag{9.1}$$

Choose an integer N_2 such that $\sum_{n\geq N_2} |r_n| < \varepsilon$, and put

$$\delta'' := \min\{b_k - a_k : 1 \leq k \leq N_2\}.$$

Given any $a < u < u' < b$ with $u' - u < \delta''$ we have

$$\Big(]a_n, b_n[\subset]u, u'[\Big) \Longrightarrow n \geq N_2,$$

so

$$\sum_{I_n\subset]u,u'[} |r_n[\leq \sum_{n\geq N_2} |r_n| < \varepsilon. \tag{9.2}$$

Denote $\delta := \min\{\delta', \delta''\}$ and consider any $x, x' \in [a, b]$ with $0 < x' - x < \delta$. If both $x, x' \in [a_n, b_n]$ for some $n \in \mathbb{N}$, by (9.1) we have $|f(x') - f(x)| < \varepsilon$. So, suppose that x, x' are not both contained in a same $[a_n, b_n]$. Since in this case

$$\sum_{I_n\subset]a,x'[} r_n = \sum_{I_n\subset]a,x[} r_n + \sum_{I_n\subset]x,x'[} r_n + \omega r_p,$$

where $\omega = 0$ if $x \in P$ and $\omega = 1$ if $x \in]a_p, b_p[$, we see that

$$f(x') - f(x) = \sum_{I_n \subset]x,x'[} r_n + \omega(f_p(b_p) - f_p(x)) + \omega'(f_q(x') - f_q(a_q)), \tag{9.3}$$

where

$$\omega = \begin{cases} 0 & \text{if } x \in P \\ 1 & \text{if } x \in]a_p, b_p[\end{cases} \quad \text{and} \quad \omega' = \begin{cases} 0 & \text{if } x' \in P \\ 1 & \text{if } x' \in]a_q, b_q[\end{cases},$$

and where we recall that $f_p(b_p) = r_p$ and $f_p(a_p) = 0$. If $x \in]a_p, b_p[$ (resp. $x' \in]a_q, b_q[$), note that $b_p - x < x' - x$ (resp. $x' - a_q < x' - x$) since x, x' are not both included in $]a_p, b_p[$ (resp. in $]a_q, b_q[$). Therefore, by (9.1), (9.2) and (9.3) we have $|f(x') - f(x)| < 3\varepsilon$. It results that in any case for $x, x' \in [a, b]$

$$\|x' - x\| < \delta \implies |f(x') - f(x)| < 3\varepsilon,$$

which translates the uniform continuity of f on $[a, b]$. □

Now assume that $[a, b] = [0, 1]$ and that $P \subset [0, 1]$ is the classical *Cantor ternary set* with meas $(P) = 0$. Recall that this set can be constructed in the following way. Remove the open middle third interval $J_{1,1} :=]\frac{1}{3}, \frac{2}{3}[$ from the segment $[0, 1]$, and denote P_1 the closed set formed by the union of the 2 remaining segments. At the second step remove the open third middle open interval from each remaining segment, that is, remove the open intervals

$$J_{2,1} := \left]\frac{1}{3^2}, \frac{2}{3^2}\right[, \ J_{2,2} = \left]\frac{3}{3^2}, \frac{3 \times 2}{3^2}\right[,$$

and denote $P_2 := P_1 \setminus (J_{2,1} \cup J_{2,2})$. Inductively, at the k-th step remove the 2^{k-1} open middle third intervals $J_{k,1}, \cdots, J_{k,2^{k-1}}$ from the 2^{k-1} segments forming P_{k-1}, and denote P_k the union of the 2^k remaining segments from P_{k-1}. Then, the Cantor ternary set is given by $P = \bigcap_{k \in \mathbb{N}} P_k$.

As in what precedes Lemma 9.1, for the disjoint open intervals forming $]0, 1[\setminus P$, keep the enumeration $(I_n)_{n \in \mathbb{N}}$ with meas $(I_{n+1}) \leq$ meas (I_n) and continue to denote by a_n and b_n the left and right extremities of I_n, that is, $I_n =]a_n, b_n[$. Put $\ell_n :=$ meas $(I_n) = b_n - a_n$. Let $(h_n)_{n \in \mathbb{N}}$ be an increasing sequence of positive reals tending to $+\infty$ and such that the series $\sum h_n \ell_n$ is convergent, for example $h_n = \dfrac{1}{R_n^\alpha}$ with $R_n := \sum_{k \geq n} \ell_k$ and a constant $\alpha \in]0, 1[$ (independent of n).

Consider an increasing continuous function $\varphi : [0, 1] \to \mathbb{R}$ with $\varphi(0) = 0$, $\varphi(1) = 1$, differentiable on $]0, 1[$ and with $D^+\varphi(0) = D^-\varphi(1) = +\infty$, where we recall that $D^+\varphi(0)$ (resp. $D^-\varphi(1)$) denotes the right-hand (resp. left-hand) derivate of φ at 0 (resp. at 1); for example

$$\varphi(x) = \frac{2}{\pi} \arcsin \sqrt{x} \quad \text{for all } x \in [0, 1].$$

Since the function $x \mapsto \varphi(x)/x$ is continuous on $]0, 1]$ with $\varphi(x)/x \to +\infty$ as $x \downarrow 0$, its minimum μ on $]0, 1]$ is attained, so (keeping in mind $\varphi(1)/1 = 1$)

$$0 < \mu := \min_{x \in]0,1]} \frac{\varphi(x)}{x} \leq 1.$$

For each $n \in \mathbb{N}$ define $f_n : [a_n, b_n] \to \mathbb{R}$ by

$$f_n(x) = h_n \ell_n \varphi\left(\frac{x - a_n}{\ell_n}\right) \quad \text{for all } x \in [a_n, b_n],$$

so f_n is a continuous increasing function on $[a_n, b_n]$ with $f_n(a_n) = 0$ and $f_n(b_n) = h_n \ell_n$. Further, f_n is differentiable on $]a_n, b_n[$, and $D^+ f_n(a_n) = D^- f_n(b_n) = +\infty$.

The function f involved in Lemma 9.1 becomes $f : [0,1] \to \mathbb{R}$ with

$$f(x) = \begin{cases} \sum\limits_{I_n \subset]0,x[} h_n \ell_n & \text{if } x \in P \\ \sum\limits_{I_n \subset]0,x[} h_n \ell_n + f_p(x) & \text{if } x \in]a_p, b_p[, \end{cases}$$

and by Lemma 9.1 it is continuous on $[0,1]$ since $\sup_{[a_n,b_n]} |f_n| \to 0$ as $n \to \infty$.

LEMMA 9.2 (A. Denjoy). With the data and notations above, for the continuous function f one has that $Df(x)$ exists and is finite for any $x \in]0,1[\setminus P$ and $Df(x) = +\infty$ for any $x \in P$.

PROOF. By definition of f for any $x \in]a_p, b_p[$ we have

$$f(x) = \sum_{I_n \subset]0,a_p[} h_n \ell_n + f_p(x),$$

so f is differentiable on $]a_p, b_p[$ with $Df(u) = Df_p(u)$ for every $u \in]a_p, b_p[$. Further, we also see that

$$D^+ f(a_p) = D^+ f_p(a_p) = +\infty \quad \text{and} \quad D^- f(b_p) = D^- f_p(b_p) = +\infty.$$

Fix any $u \in P \setminus \{a_n : n \in \mathbb{N}\}$ with $u < 1$. Consider any $x \in [0,1]$ with $x > u$ and note that $P \cap]u, x[\neq \emptyset$. Take any real $A > 0$. Let $N \in \mathbb{N}$ be such that

$$h_N > \mu^{-1} A \quad \text{and} \quad a_N > u \tag{9.4}$$

(which is possible since P is the Cantor ternary set). Put

$$\delta := \min\{a_k - u : k \in \{1, \cdots, N\},\, a_k > u\} > 0.$$

Since $u \in P$ and $u + \delta \in P$, we have

$$\Big( ]a_n, b_n[\cap [u, u+\delta] \neq \emptyset \Big) \Rightarrow \Big( ]a_n, b_n[\subset]u, u+\delta[\text{ and } n > N \Big). \tag{9.5}$$

Let any $x \in]u, u+\delta[$.

If $x \in P$, then by (9.4) and (9.5)

$$f(x) - f(u) = \sum_{I_n \subset]u,x[} h_n \ell_n > \mu^{-1} A \sum_{I_n \subset]u,x[} \ell_n = \mu^{-1} A(x-u) > A(x-u), \tag{9.6}$$

where the equality $\sum\limits_{I_n \subset]u,x[} \ell_n = x - u$ is due to the fact that

$$\operatorname{meas}(]u,x[) = \operatorname{meas}(]u,x[\setminus P) = \sum_{I_n \subset]u,x[} \operatorname{meas}(I_n) \quad \text{since } u, x \in P.$$

Suppose $x \notin P$. Choose $p \in \mathbb{N}$ such that $x \in]a_p, b_p[$. Then (9.5) assures us that we have $]a_p, b_p[\subset]u, u+\delta[$ with $a_p > u$ (since $u \notin \{a_n : n \in \mathbb{N}\}$) along with $]a_n, b_n[\subset]u, u+\delta[$ and $n > N$ whenever $]a_n, b_n[\subset]u, a_p[$. It follows by (9.4) that

$$f(x) - f(u) = \sum_{I_n \subset]u,a_p[} h_n \ell_n + f_p(x) > \mu^{-1} A(a_p - u) + f_p(x) > A(a_p - u) + f_p(x).$$

Since $\left(\dfrac{x-a_p}{\ell_p}\right)^{-1}\dfrac{f_p(x)}{h_p\ell_p} \geq \mu$, we have $f_p(x) \geq \mu h_p(x-a_p) > A(x-a_p)$, hence

$$f(x) - f(u) > A(a_p - u) + A(x - a_p) = A(x - u).$$

This combined with (9.6) gives

$$\frac{f(x) - f(u)}{x - u} > A \quad \text{for every } x \in]u, u + \delta[,$$

then $D^+ f(u) = +\infty$.

Similarly, one shows that $D^- f(u) = +\infty$ for any $u \in P \setminus \{b_k : k \in \mathbb{N}\}$ with $u \neq 0$. The proof of the lemma is complete. □

Now let us recall a well-known description of the *Cantor-Lebesgue singular function*. Consider the compact sets P_k and the open intervals $J_{k,i}$ involved above in the definition of the Cantor ternary set P (before Lemma 9.2). For each $k \in \mathbb{N}$ define $\zeta_k : [0,1] \to [0,1]$ as the non-decreasing continuous function given by $\zeta_k(0) = 0$, $\zeta_k(1) = 1$, ζ_k is affine with coefficient $1/2^k$ on each of the 2^k segments forming P_k and constant on each of the $2^k - 1$ open intervals forming $]0,1[\setminus P_k$. This sequence of functions $(\zeta_k)_{k\in\mathbb{N}}$ is known (as easily seen) to satisfy

$$\sup_{x\in[0,1]} |\zeta_{k+1}(x) - \zeta_k(x)| \leq \frac{1}{2} \sup_{x\in[0,1]} |\zeta_k(x) - \zeta_{k-1}(x)| \leq \cdots \leq \frac{1}{2^k},$$

so it converges uniformly on $[0,1]$ to a continuous function $\zeta : [0,1] \to [0,1]$, and ζ is non-decreasing on $[0,1]$. Further, the function ζ is constant on each open interval $J_{k,i}$ for each $k \in \mathbb{N}$ and each $1 \leq i \leq 2^{k-1}$, hence ζ is differentiable on $]0,1[\setminus P$ with $D\zeta(x) = 0$ for every $x \in]0,1[\setminus P$.

Considering the function f in Lemma 9.2 we see that the function $g := f + \zeta$ is differentiable on $]0,1[\setminus P$ with $Df(x) = Dg(x)$ for each $x \in]0,1[\setminus P$, and since ζ is non-decreasing on $[0,1]$ the equality $Df(x) = Dg(x) = +\infty$ holds for every $x \in P$. Denoting by f_0 and g_0 the restrictions on $]0,1[$ of the opposites of the functions f and g respectively, it is clear that $Df_0(x) = Dg_0(x)) = -\infty$ for every $x \in P\setminus\{0,1\}$, hence $\partial f_0(x) = \partial g_0(x) = \emptyset$ for every $x \in P \setminus \{0,1\}$. We then obtain the following result.

PROPOSITION 9.3. For the above continuous functions $f_0, g_0 :]0,1[\to \mathbb{R}$, one has

$$\partial_F f_0(x) = \partial_F g_0(x) \quad \text{for all } x \in]0,1[,$$

while the function $f_0 - g_0$ fails to be constant on $]0,1[$.

9.2. Subdifferentially and directionally stable functions

Given an extended real-valued function $f : U \to \mathbb{R} \cup \{+\infty\}$ defined on a nonempty open set U of a normed space X, we recall that its (standard) *directional derivative* $f'(x;h)$ at a point $x \in \operatorname{dom} f$ in the direction $h \in X$ is defined by (see Definition 2.62)

$$f'(x;h) := \lim_{t\downarrow 0} t^{-1}\big(f(x+th) - f(x)\big)$$

when the limits exists in $\mathbb{R} \cup \{-\infty, +\infty\}$.

Throughout the rest of the chapter, any subdifferential that we will consider is assumed to satisfy the properties **Prop.1-$\cdots$-Prop.4** in Section 6.2.

9.2.1. Subdifferentially and directionally stable functions, properties and examples.

DEFINITION 9.4. Let U be a nonempty open convex set of a Banach space X and $f : U \to \mathbb{R} \cup \{+\infty\}$ be a function which is lower semicontinuous on U. Let also ∂ be a subdifferential. The function f is said to be *∂-subdifferentially and directionally stable* on U provided that for each $u \in \operatorname{Dom} \partial f$ and each $v \in \operatorname{dom} f$ with $v \neq u$ the following properties (i)-(iii) hold:
(i) the restriction to $[0, 1]$ of the function $t \mapsto f(u+t(v-u))$ is finite and continuous;
(ii) for any $t \in [0, 1[$ the directional derivative $f'(u + t(v - u); v - u)$ exists and is less than $+\infty$;
(iii) for each fixed $y \in [u, v[$ and for each number $\varepsilon > 0$, there exists some $r_0 \in]0, 1[$ such that for any $w = y + r(v - y)$ with $r \in]0, r_0]$ and for every sequence $(x_n, x_n^*)_n$ in $\operatorname{gph} \partial f$ with $x_n \to x_0 \in [y, w[$

$$\liminf_{n\to\infty} \langle x_n^*, w - x_n\rangle \leq f'(y; w - x_0) + \varepsilon\|w - x_0\|. \tag{9.7}$$

When there is no risk of ambiguity, we will simply say that f is *subdifferentially and directionally stable* on U, or *sds* on U for short.

REMARK 9.5. (a) We may replace lim inf in (9.7) by lim sup.
Indeed, choose a subsequence (with an infinite subset $K \subset \mathbb{N}$) such that

$$\liminf_{n\to\infty} \langle x_n^*, w - x_n\rangle = \lim_{k\in K} \langle x_k^*, w - x_k\rangle.$$

By (9.7) we have

$$\lim_{k\in K} \langle x_k^*, w - x_k\rangle \leq f'(y; w - x_0) + \varepsilon\|w - x_0\|,$$

which finishes the verification.
(b) If f is ∂-subdifferentially and directionally stable on U, then it is also clearly ∂'-subdifferentially and directionally stable on U for any subdifferential ∂' satisfying $\partial' f(\cdot) \subset \partial f(\cdot)$. □

We first observe the preservation of this property under the sum of finitely many functions.

PROPOSITION 9.6. Let U be a nonempty open convex set of a Banach space X and let $f_1, f_2 : U \to \mathbb{R} \cup \{+\infty\}$ be two lower semicontinuous functions which are ∂-subdifferentially and directionally stable on U with $\operatorname{dom} f_1 \cap \operatorname{dom} f_2 \neq \emptyset$. Assume that for every $x \in U$

$$\partial(f_1 + f_2)(x) \subset \partial f_1(x) + \partial f_2(x).$$

Then the function $f_1 + f_2$ is ∂-subdifferentially and directionally stable on U.

PROOF. Denoting $f := f_1 + f_2$, we obviously see that $\operatorname{dom} f = \operatorname{dom} f_1 \cap \operatorname{dom} f_2$, and hence $\operatorname{dom} f \neq \emptyset$. Take any $u \in \operatorname{Dom} \partial f$ and $v \in \operatorname{dom} f$ with $v \neq u$. By the subdifferential inclusion assumption, we have $u \in \operatorname{Dom} \partial f_i$ and $v \in \operatorname{dom} f_i$ for $i = 1, 2$. It clearly results that conditions (i) and (ii) in Definition 9.4 hold for the function f since they hold by assumptions for the functions f_1 and f_2. Fix now any $y \in [u, v[$ and any real $\varepsilon > 0$. Choose some $r_0 \in]0, 1[$ satisfying the condition (iii) in Definition 9.4 for f_1 and f_2 with $\varepsilon/2$ in place of ε and with the limit superior according to Remark 9.5(a). Take any $r \in]0, r_0]$ and set $w := y + r(v - y)$. Consider also any sequence $(x_n, x_n^*)_n$ in $\operatorname{gph} \partial f$ with $x_n \to x_0 \in [y, w[$. By the inclusion

$\partial f(\cdot) \subset \partial f_1(\cdot) + \partial f_2(\cdot)$ there are $x_{i,n}^* \in \partial f_i(x_n)$ such that $x_n^* = x_{1,n}^* + x_{2,n}^*$ for every $n \in \mathbb{N}$. By the choice of r_0 we have for $i = 1, 2$

$$\limsup_{n\to\infty} \langle x_{i,n}^*, w - x_n \rangle \leq f_i'(y; w - x_0) + \frac{\varepsilon}{2}\|w - x_0\|.$$

Further, the definition of directional derivative and the condition (ii) in Definition 9.4 ensure that

$$f'(y; w - x_0) = f_1'(y; w - x_0) + f_2'(y; w - x_0).$$

Using this after adding the above inequalities corresponding to $i = 1, 2$, we obtain after rearrangement

$$\liminf_{n\to\infty} \langle x_n^*, w - x_n \rangle \leq f'(y; w - x_0) + \varepsilon\|w - x_0\|,$$

which is the condition (iii) for f in Definition 9.4. □

REMARK 9.7. Of course, similar preservations also hold under suitable assumptions for precomposition or postcomposition. For example, let $A : X \to Y$ be a continuous linear mapping between two Banach spaces and $f : V \to \mathbb{R} \cup \{+\infty\}$ be a lower semicontinuous function on an open convex set $V \subset Y$ with $V \supset A(U)$, where U is an open convex set of X with $A(U) \cap \operatorname{dom} f \neq \emptyset$. If

$$\partial(f \circ A)(x) \subset A^*\big(\partial f(Ax)\big) \quad \text{for all } x \in U,$$

then the function $f \circ A$ is ∂-subdifferentially and directionally stable on U. □

We proceed now to illustrating the class of subdifferentially and directionally stable functions by providing diverse examples of such functions. We begin by proving that lower semicontinuous convex functions are subdifferentially and directionally stable. The result can also be seen as a particular case of subsmooth functions in Proposition 9.10 below. We present it independently since the proof is simpler and puts in light certain basic arguments.

PROPOSITION 9.8. *Let U be a nonempty open convex set of a Banach space X. Then any proper lower semicontinuous convex function $f : U \to \mathbb{R} \cup \{+\infty\}$ is subdifferentially and directionally stable on U.*

PROOF. Take any $u, v \in \operatorname{dom} f$. We know that the convexity of f entails that (i) in Definition 9.4 is satisfied. Further, the same convexity property guarantees that, for $t \in [0, 1[$, the directional derivative $f'(u + t(v - u); v - u)$ exists and, for $s \in]0, 1 - t[$,

$$f'(u + t(v - u); v - u) \leq s^{-1}[f(u + (s + t)(v - u)) - f(u + t(v - u))] < +\infty.$$

It remains to prove (iii) in Definition 9.4. Fix any u, v and y as in the statement of Definition 9.4 and fix also any $u^* \in \partial f(u)$. The function $\varphi : \mathbb{R} \to \mathbb{R} \cup \{+\infty\}$ defined by

$$\varphi(s) = \begin{cases} +\infty & \text{if } s \in]1, +\infty[\\ f(u) + s\langle u^*, v - u \rangle & \text{if } s \in]-\infty, 0[\\ f(u + s(v - u)) & \text{if } s \in [0, 1] \end{cases}$$

is convex, hence it is locally Lipschitz on $]-\infty, 1[$. The right-hand derivative φ'_+ of φ is then finite on $]-\infty, 1[$ and it is also upper semicontinuous on $]-\infty, 1[$ since

$$\varphi'_+(s) = \inf_{\lambda>0} \lambda^{-1}\big(\varphi(s + \lambda) - \varphi(s)\big),$$

and hence the restriction of the function $s \mapsto f'(u+s(v-u); v-u)$ is finite and upper semicontinuous on $[0,1[$. Fix any real number $\varepsilon > 0$ and choose s_0 satisfying $y = u + s_0(v-u)$. Since $f'(y; v-u)$ is finite, there is some $\lambda \in]0, 1-s_0[$ such that for all $s \in]s_0, s_0+\lambda]$

$$\|v-u\|^{-1} f'(u+s(v-u); v-u) \le \|v-u\|^{-1} f'(y; v-u) + \varepsilon. \tag{9.8}$$

Set $r_0 := \lambda \in]0,1]$ and, for any fixed $r \in]0, r_0]$, consider the point $w := y + r(v-u)$ and a sequence $(x_n, x_n^*)_n$ as in the statement of Definition 9.4. By convexity of f, we have

$$\langle x_n^*, w - x_n \rangle \le f(w) - f(x_n),$$

and this yields by the lower semicontinuity and the convexity of f that

$$\liminf_{n\to\infty} \langle x_n^*, w - x_n \rangle \le f(w) - f(x_0) \le f'(w; w - x_0). \tag{9.9}$$

On the other hand, (9.8), with $s := s_0 + r \in]s_0, s_0 + \lambda]$, gives (by the equality $w = u + s(v-u)$) that

$$\|w - x_0\|^{-1} f'(w; w - x_0) \le \|w - x_0\|^{-1} f'(y; w - x_0) + \varepsilon,$$

or equivalently,

$$f'(w; w - x_0) \le f'(y; w - x_0) + \varepsilon \|w - x_0\|.$$

The latter inequality combined with (9.9) entails the desired inequality

$$\liminf_{n\to\infty} \langle x_n^*, w - x_n \rangle \le f'(y; w - x_0) + \varepsilon \|w - x_0\|,$$

which finishes the proof of the proposition. □

As a second example the next proposition shows that C-subregular (or equivalently, tangentially regular) locally Lipschitz functions are subdifferentially and directionally stable.

PROPOSITION 9.9. Let U be a nonempty open convex set of a Banach space X and $f : U \to \mathbb{R}\{+\infty\}$ be a locally Lipschitz function. Assume that ∂f in included in the Clarke subdifferential of f. Then f is ∂-subdifferentially and directionally stable over U provided it is C-subregular on U, or equivalently tangentially regular on U; the converse implication also holds whenever $\operatorname{Dom} \partial f = U$ (which holds for ∂_C and ∂_A in any Banach space, and for ∂_L in Asplund space).

PROOF. Suppose first that f is C-subregular (that is, tangentially regular) on U. Obviously, (i) and (ii) in Definition 9.4 are fulfilled. Let u, v and $y := u + s_0(v-u)$ with $0 \le s_0 < 1$ as in the statement of Definition 9.4 and let $\varepsilon > 0$. Since the function $f^\circ(\cdot; v-u)$ is finite and upper semicontinuous on U, we may choose some $\lambda \in]0, 1-s_0[$ such that for all $s \in]s_0, s_0+\lambda]$

$$\|v-u\|^{-1} f'(u+s(v-u); v-u) \le \|v-u\|^{-1} f^\circ(y; v-u) + \varepsilon. \tag{9.10}$$

Putting $r_0 := (1-s_0)^{-1}\lambda$ in $]0,1[$, and for any $r \in]0, r_0]$ considering $w := y + r(v-u)$ and $(x_n, x_n^*)_n$ as in the statement of Definition 9.4 with $x_n \to x_0 \in [y, w[$, we obtain by the upper semicontinuity of $f^\circ(\cdot;\cdot)$ on $U \times X$

$$\liminf_{n\to\infty} \langle x_n^*, w - x_n \rangle \le \liminf_{n\to\infty} f^\circ(x_n; w - x_n) \le f^\circ(x_0; w - x_0). \tag{9.11}$$

Choosing $\delta \in [0, r[$ such that $x_0 = y + \delta(v-y)$, we also have for $s := s_0 + \delta(1-s_0)$

$$x_0 = u + s(v-u) \quad \text{along with} \quad s_0 < s \le s_0 + \lambda,$$

and hence taking (9.10) into account it results that

$$\|v-u\|^{-1}f^{\circ}(x_0;v-u) \le \|v-u\|^{-1}f^{\circ}(y;v-u)+\varepsilon,$$

which is equivalent to

$$f^{\circ}(x_0;w-x_0) \le f^{\circ}(y;w-x_0)+\varepsilon\|w-x_0\|$$

according to the equality $\|v-u\|^{-1}(v-u) = \|w-x_0\|^{-1}(w-x_0)$. The property (iii) in Definition 9.4 follows from the latter inequality, from (9.11), and from the C-subregularity of f on U.

To prove the converse, assume that $\operatorname{Dom}\partial f = U$ and that f is subdifferentially and directionally stable on U. Fix $v \in U$ and a nonzero vector $h \in X$, and take any real $\varepsilon > 0$. Note first by (ii) in Definition 9.4 that $f'(u;h)$ exists. Choose by definition of $f^{\circ}(u;h)$ a sequence $(z_n)_n$ in U with $z_n \to u$ and a sequence $(t_n)_n$ in $]0,1[$ with $t_n \downarrow 0$ such that

$$f^{\circ}(u;h) = \lim_{n\to\infty} t_n^{-1}[f(z_n+t_nh)-f(z_n)].$$

For each $n \in \mathbb{N}$ the mean value theorem gives $x_{n,k} \underset{k}{\to} u_n \in [z_n, z_n+t_nh[$ and $x^*_{n,k} \in \partial f(x_{n,k})$ such that

$$t_n^{-1}[f(z_n+t_nh)-f(z_n)] \le \lim_{k\to\infty} \langle x^*_{n,k}, h\rangle$$

(see Theorem 6.52(i)). Then it is not difficult to find a sequence of integers $(k_n)_n$ satisfying $x_{n,k_n} \to u$ as $n \to \infty$ along with

$$\lim_{k\to\infty} \langle x^*_{n,k}, h\rangle \le \langle x^*_{n,k_n}, h\rangle + t_n.$$

Putting $x_n := x_{n,k_n}$ and $x^*_n := x^*_{n,k_n}$, applying (iii) in Definition 9.4 with $v = u+\rho h \in U$ and $\rho \in]0,1[$ and with $y = u \in \operatorname{Dom}\partial f$ (here we utilize the assumption $\operatorname{Dom}\partial f = U$) we obtain some $r \in]0,1[$ such that for $w = u + r(v-u)$, we have (according to the boundedness of $(x^*_n)_n$)

$$f^{\circ}(u;h) \le \liminf_{n\to\infty}\langle x^*_n, h\rangle = (r\rho)^{-1}\liminf_{n\to\infty}\langle x^*_n, w-x_n\rangle \le f'(u;h)+\varepsilon\|h\|.$$

This being true for all $\varepsilon > 0$, it ensues that $f^{\circ}(u;h) \le f'(u;h)$, which means that f is C-subregular at u. The proof is then finished. □

Subsmooth functions with convex effective domains are also subdifferentially and directionally stable.

PROPOSITION 9.10. Let U be a nonempty open convex set in a Banach space X and $f : U \to \mathbb{R}\cup\{+\infty\}$ be a lower semicontinuous subsmooth function on U with $\operatorname{dom} f$ convex and nonempty. Assume that ∂f is included in the Clarke subdifferential of f. Then f is ∂-subdifferentially and directionally stable on U.

PROOF. Let $u \in \operatorname{Dom}\partial f$ and $v \in \operatorname{dom} f$ with $v \neq u$. According to the convexity of $\operatorname{dom} f$ and to Proposition 8.20, the restriction to $[0,1]$ of the function $t \mapsto f(u+t(v-u))$ is finite and continuous, which is the property (i) in Definition 9.4. Further, both assertions (a) and (b) in Proposition 8.21 guarantee that, for each $t \in [0,1[$ the directional derivative $f'(u+t(v-u);v-u)$ exists and is less than $+\infty$, which corresponds to the property (ii) in Definition 9.4.

Now fix any real $\varepsilon > 0$ and any $y \in [u, v[$. Choose $\delta > 0$ with $B(y, \delta) \subset U$ such that by Proposition 8.12

$$f(sx' + (1-s)y') \leq sf(x') + (1-s)f(y') + \frac{\varepsilon}{4}s(1-s)\|x' - y'\|$$

for all $x', y' \in B(y, \delta)$ and $s \in]0, 1[$. For any $t > 0$ with $t\|v - u\| < \delta/2$ we obtain by Proposition 8.21(a) that

$$(9.12)\quad f'(y + t(v-u); v-u) \leq t^{-1}[f(y + 2t(v-u)) - f(y + t(v-u))] + \frac{\varepsilon}{4}\|v - u\|.$$

We also observe on the one hand that $f'(y; v-u) < +\infty$ since condition (ii) in Definition 9.4 is already justified above. On the other hand, if $y \neq u$ we have $f'(y; v-u) > -\infty$ according to the local Lipschitz property on $]0, 1[$ of $t \mapsto f(u + t(v-u))$ (see Proposition 8.20), and if $y = u$ we still have $f'(y; v-u) > -\infty$ taking into account the fact that $\emptyset \neq \partial f(u) \subset \partial_C f(u)$ as well as Theorem 8.25(b). Consequently, $f'(y; v-u)$ is finite. This and the positive homogeneity of $f'(y; \cdot)$ easily gives the equality

$$\lim_{t \downarrow 0} t^{-1}[f(y + 2t(v-u)) - f(y + t(v-u))] = f'(y; v-u).$$

From the latter equality and from the inequality (9.12) there exists some $r_0 \in]0, 1[$ with $r_0\|v - u\| < \delta/4$ such that for all $t \in]0, r_0]$

$$(9.13)\qquad f'(y + t(v-u); v-u) \leq f'(y; v-u) + \frac{\varepsilon}{2}\|v - u\|.$$

Now fix any $r \in]0, r_0]$ and set $w := y + r(v-y)$. Consider any sequence (x_n, x_n^*) in $\operatorname{gph} \partial f$ with $(x_n)_n$ converging to some $x_0 \in [y, w[$. Theorem 8.25(a) ensures that for n large enough

$$\langle x_n^*, w - x_n \rangle \leq f(w) - f(x_n) + \frac{\varepsilon}{4}\|w - x_n\|,$$

and hence using the lower semicontinuity of f we obtain

$$\liminf_{n \to \infty} \langle x_n^*, w - x_n \rangle \leq f(w) - f(x_0) + \frac{\varepsilon}{4}\|w - x_0\|.$$

Taking (8.5) into account, the latter inequality yields

$$(9.14)\qquad \liminf_{n \to \infty} \langle x_n^*, w - x_n \rangle \leq f'(w; w - x_0) + \frac{\varepsilon}{2}\|w - x_0\|.$$

On the other hand, writing

$$w = y + r(v - y) = y + s(v - u)$$

for some $s \in]0, r[$, the inequality (9.13) with $t = s$ gives

$$f'(w; v-u) \leq f'(y; v-u) + \frac{\varepsilon}{2}\|v - u\|,$$

and this inequality is equivalent to

$$f'(w; w - x_0) \leq f'(y; w - x_0) + \frac{\varepsilon}{2}\|w - x_0\|.$$

The latter combined with (9.14) entails that

$$\liminf_{n \to \infty} \langle x_n^*, w - x_n \rangle \leq f'(y; w - x_0) + \varepsilon\|w - x_0\|,$$

which translates the condition (iii) in Definition 9.4. The proof is then finished. □

Example 9.6 tells us that C-subregular (or tangentially regular) locally Lipschitz functions may fail to be subsmooth. The following corollary is then of great interest.

COROLLARY 9.11. Let U be a nonempty open convex set of a Banach space X and $f_1 : U \to \mathbb{R} \cup \{+\infty\}$ be a proper lower semicontinuous subsmooth function on U with $\operatorname{dom} f_1$ convex. Let $f_2 : U \to \mathbb{R}$ be a C-subregular (or tangentially regular) locally Lipschitz function. Then $f_1 + f_2$ is subdifferentially and directionally stable on U.

PROOF. We know that $\partial_C(f_1 + f_2)(x) \subset \partial_C f_1(x) + \partial_C f_2(x)$ for all $x \in U$ (see Theorem 2.98). Then the corollary directly follows from Propositions 9.6, 9.9 and 9.10. $\square$

9.2.2. Subdifferential determination of subdifferentially and directionally stable functions. We prove now the following theorem from which we will deduce that subdifferentially and directionally stable extended real-valued functions are subdifferentially determined.

THEOREM 9.12 (L. Thibault and D. Zagrodny: estimates under enlarged subdifferential inclusions of sds functions). Let U be a nonempty open convex set of a Banach space X and ∂_1 and ∂_2 be two subdifferentials. Let γ be a non-negative real number. Let $g : U \to \mathbb{R} \cup \{+\infty\}$ be a lower semicontinuous function which is ∂_2-subdifferentially and directionally stable on U and let $f : U \to \mathbb{R} \cup \{+\infty\}$ be a proper lower semicontinuous function. The following hold:
(a) If

$$\partial_1 f(x) \subset \partial_2 g(x) + \gamma \mathbb{B}_{X^*} \quad \text{for all } x \in U, \tag{9.15}$$

then $\operatorname{dom} f = \operatorname{dom} g$ and for all $u \in \operatorname{dom} g$ and $v \in U$ the following inequalities hold

$$g(v) - g(u) - \gamma\|v - u\| \le f(v) - f(u) \le g(v) - g(u) + \gamma\|v - u\|.$$

(b) Further, the converse also holds provided $\partial_1 f(x) \subset \partial_C f(x)$ and $\partial_C g(x) \subset \partial_2 g(x)$ for all $x \in U$.

PROOF. (a) Note first that $\operatorname{Dom} \partial_1 f \neq \emptyset$ (see Proposition 6.50). Fix any $u \in \operatorname{Dom} \partial_1 f$ and any $v \in U$. Thanks to the assumption (9.15) we have $u \in \operatorname{Dom} \partial_2 g$, and hence in particular $g(u)$ is finite.

Step I. Our first step is to prove that

$$f(v) - f(u) \le g(v) - g(u) + \gamma\|v - u\|. \tag{9.16}$$

The inequality is obvious when $v = u$ or when $g(v) = +\infty$. So, suppose $v \in \operatorname{dom} g$ along with $v \neq u$, and fix any real $\varepsilon > 0$.

Consider any $z \in [u, v[\cap \operatorname{dom} f$, so $z \neq v$. Choose $\rho(z) \in]0, 1[$ (as r_0 depending on z) satisfying the property (iii) in Definition 9.4 with $y = z$ and with $\varepsilon/2$ in place of ε and such that, by definition of $g'(z; v-z)$ and by the property (ii) in Definition 9.4, for all $t \in]0, \rho(z)]$

$$g'(z; v - z) \le t^{-1}[g(z + t(v - z)) - g(z)] + \frac{\varepsilon}{2}\|v - z\|.$$

For each $r \in]0, \rho(z)]$ setting $w_{z,r} := z + r(v - z)$ we obtain

$$g'(z; w_{z,r} - z) \le g(w_{z,r}) - g(z) + \frac{\varepsilon}{2}\|w_{z,r} - z\|. \tag{9.17}$$

We proceed now to show that the following inequality holds

$$f(w_{z,r}) - f(z) \le g(w_{z,r}) - g(z) + (\gamma + \varepsilon)\|w_{z,r} - z\|. \tag{9.18}$$

Fix any real $\lambda \le f(w_{z,r})$. The sequential Zagrodny mean value theorem (see Theorem 6.52) tells us that there exist $c \in [z, w_{z,r}[$, $U \ni c_n \to c$ and $c_n^* \in \partial_1 f(c_n)$ such that

$$\|w_{z,r} - z\|^{-1}(\lambda - f(z)) \le \|w_{z,r} - c\|^{-1} \lim_{n\to\infty} \langle c_n^*, w_{z,r} - c_n\rangle. \tag{9.19}$$

By assumption (9.15), we have $c_n^* = a_n^* + \gamma b_n^*$ with $a_n^* \in \partial_2 g(c_n)$ and $\|b_n^*\| \le 1$, and hence

$$\liminf_{n\to\infty} \langle c_n^*, w_{z,r} - c_n\rangle \le \liminf_{n\to\infty} \langle a_n^*, w_{z,r} - c_n\rangle + \gamma\|w_{z,r} - c\|. \tag{9.20}$$

Further, the choice of r above and (9.7) in Definition 9.4 give

$$\liminf_{n\to\infty} \langle a_n^*, w_{z,r} - c_n\rangle \le g'(z; w_{z,r} - c) + \frac{\varepsilon}{2}\|w_{z,r} - c\|.$$

Since $\|w_{z,r} - c\|^{-1}(w_{z,r} - c) = \|w_{z,r} - z\|^{-1}(w_{z,r} - z)$, using (9.17), (9.19) and (9.20) it ensues that

$$\|w_{z,r} - z\|^{-1}(\lambda - f(z)) \le \|w_{z,r} - z\|^{-1}\left(g(w_{z,r}) - g(z) + \frac{\varepsilon}{2}\|w_{z,r} - z\|\right) + \frac{\varepsilon}{2} + \gamma,$$

which implies

$$\lambda - f(z) \le g(w_{z,r}) - g(z) + (\gamma + \varepsilon)\|w_{z,r} - z\|.$$

This last inequality confirms that (9.18) is true since the inequality holds for every real $\lambda \le f(w_{z,r})$.

Now, we put

$$\sigma := \sup\{t \in]0,1] : f(u + t(v-u)) - f(u) \le g(u + t(v-u)) - g(u) + (\gamma+\varepsilon)\|t(v-u)\|\}$$

and $y := u + \sigma(v - u)$. According to (9.18) with u in place of z, the set defining σ is nonempty. Consequently, the continuity of the restriction of g to the segment line $[u, v]$ (see (i) in Definition 9.4) and the lower semicontinuity of f ensure that the supremum is attained. Further, by Definition 9.4(i) again we have $g(y) < +\infty$, thus $f(y) < +\infty$. We claim that $y = v$. Otherwise $y \in [u, v[$ and $\sigma \in]0, 1[$. Put

$$w'_y := w_{y,\rho(y)} = y + \rho(y)(v - y).$$

Then (9.18) with y in place of z gives

$$f(w'_y) - f(y) \le g(w'_y) - g(y) + (\gamma + \varepsilon)\|w'_y - y\|. \tag{9.21}$$

Since the supremum defining σ is attained, we also have

$$f(y) - f(u) \le g(y) - g(u) + (\gamma + \varepsilon)\|y - u\|.$$

Adding the latter inequality and (9.21) and noting that $\|w'_y - y\| + \|y - u\| = \|w'_y - u\|$ (because $y \in [u, w'_y]$), it results that

$$f(w'_y) - f(u) \le g(w'_y) - g(u) + (\gamma + \varepsilon)\|w'_y - u\|,$$

that is,

$$f(u + \theta(v-u)) - f(u) \le g(u + \theta(v-u)) - g(u) + (\gamma + \varepsilon)\|\theta(v - u)\|,$$

where $\theta := \sigma + \rho(y)(1 - \sigma)$. This contradicts the definition of σ, and hence $y = v$. The inequality (9.16) is then established.

Step II. We prove that $\operatorname{dom} f = \operatorname{dom} g$.
Fix any $u \in \operatorname{dom} f$ and choose a sequence $(u_n)_n$ in $\operatorname{Dom} \partial_1 f$ with $u_n \to u$ and $f(u_n) \to f(u)$ (see Proposition 6.50). By (9.16) we have for every $v \in \operatorname{dom} g$

$$f(v) - f(u_n) \le g(v) - g(u_n) + \gamma\|v - u_n\|,$$

and hence $\operatorname{dom} g \subset \operatorname{dom} f$. The latter inequality also ensures by the lower semicontinuity of g that

$$f(v) - f(u) \le g(v) - g(u) + \gamma\|v - u\|, \tag{9.22}$$

which obviously entail $u \in \operatorname{dom} g$. It ensues that $\operatorname{dom} f = \operatorname{dom} g$ and that by (9.22)

$$f(v) - f(u) \le g(v) - g(u) + \gamma\|v - u\| \text{ for all } u, v \in \operatorname{dom} g = \operatorname{dom} f. \tag{9.23}$$

Step III. Considering u and v in $\operatorname{dom} g$ and interchanging u and v in (9.23) we obtain

$$g(v) - g(u) - \gamma\|v - u\| \le f(v) - f(u),$$

and this inequality still holds for any $v \in U$ and any $u \in \operatorname{dom} g = \operatorname{dom} f$. We have then shown that for all $u \in \operatorname{dom} g$ and $v \in U$

$$g(v) - g(u) - \gamma\|v - u\| \le f(v) - f(u) \le g(v) - g(u) + \gamma\|v - u\|,$$

which is the inequality in (a) and finishes the proof of (a).

(b) Now let us prove the converse under the assumption $\partial_1 f(\cdot) \subset \partial_C f(\cdot)$ and $\partial_2 g(\cdot) \supset \partial_C g(\cdot)$ on U. Fix any $x \in \operatorname{Dom} \partial_1 f$ and any $h \in X$. Observe that $f^\uparrow(x; 0) = 0$ since $f^\uparrow(x'; 0) \in \{-\infty, 0\}$ for all $x' \in \operatorname{dom} f$ (see Theorem 2.72). Observe also that the inequality in (a) entails that $y \to_f x$ if and only if $y \to_g x$, where we recall that $y \to_f x$ means $y \to x$ along with $f(y) \to f(x)$. Choose a real $r > 0$ such that

$$(x + r\mathbb{B}_X) +]0, r[(h + r\mathbb{B}_X) \subset U,$$

and fix any $\varepsilon \in]0, r[$. Then for any $y \in (x + \varepsilon\mathbb{B}_X) \cap \operatorname{dom} f$, $v \in h + \varepsilon\mathbb{B}_X$ and $t \in]0, \varepsilon[$, we derive from the inequality in (a) that

$$t^{-1}[f(y + tv) - f(y)] \le t^{-1}[g(y + tv) - g(y)] + \gamma\|v\|,$$

which ensures that for any $\eta \in]0, \varepsilon[$

$$\limsup_{t \downarrow 0, y \to_f x} \inf_{v \in h + \eta\mathbb{B}} t^{-1}[f(y+tv) - f(y)] \le \limsup_{t \downarrow 0, y \to_g x} \inf_{v \in h + \eta\mathbb{B}} t^{-1}[g(y+tv) - g(y)] + \gamma\|h\| + \gamma\varepsilon.$$

It results from this inequality and from Theorem 2.72 that

$$f^\uparrow(x; h) \le g^\uparrow(x; h) + \gamma\|h\| + \gamma\varepsilon.$$

This being true for all $\varepsilon > 0$, it follows that

$$f^\uparrow(x; h) \le g^\uparrow(x; h) + \gamma\|h\|.$$

On the one hand, it follows that $g^\uparrow(x; 0) = 0$, and on the other hand, by the Moreau-Rockafellar theorem (see Theorem 2.105) applied at the origin to the sum $g^\uparrow(x; \cdot) + \gamma\|\cdot\|$, we easily deduce that

$$\partial_C f(x) \subset \partial_C g(x) + \gamma\mathbb{B}_{X^*}.$$

This finishes the proof according to the assumptions

$$\partial_1 f(\cdot) \subset \partial_C f(\cdot) \quad \text{and} \quad \partial_2 g(\cdot) \supset \partial_C g(\cdot)$$

concerning the subdifferentials ∂_1 and ∂_2. □

REMARK 9.13. Of course, the equality $\partial_2 g(\cdot) = \partial_C g(\cdot)$ (and hence the required inclusion in (b) of the theorem) is fulfilled whenever ∂_2 is the Clarke subdifferential. It also holds for any subdifferential provided that g is convex. Another important example where that equality still holds is given by the inclusions $\partial_F g(\cdot) \subset \partial_2 g(\cdot) \subset \partial_C g(\cdot)$ and the subsmoothness of g according to Theorem 8.25. When g is locally Lipschitz, the condition

$$\{x^* \in X^* : \langle x^*, h\rangle \le g^B(x;h),\ \forall h \in X\} \subset \partial_2 g(x) \subset \partial_C g(x)$$

also obviously entails the equality $\partial_C g(x) = \partial_2 g(x)$ whenever g is C-subregular (or equivalently, tangentially regular) on U. □

A first corollary follows with $\gamma = 0$.

COROLLARY 9.14 (subdifferential determination under sds property). Let U be a nonempty open convex set of a Banach space X and ∂_1 and ∂_2 be two subdifferentials. Let $g : U \to \mathbb{R} \cup \{+\infty\}$ be a lower semicontinuous function which is ∂_2-subdifferentially and directionally stable on U and let $f : U \to \mathbb{R} \cup \{+\infty\}$ be a proper lower semicontinuous function. Assume that

$$\partial_1 f(x) \subset \partial_2 g(x) \quad \text{for all } x \in U.$$

Then there exists a constant $C \in \mathbb{R}$ such that

$$f(x) = g(x) + C \quad \text{for all } x \in U.$$

Taking Proposition 9.8 into account, we directly obtain the following second corollary of Theorem 9.12.

COROLLARY 9.15. Let U be a nonempty open convex set of a Banach space X, let $f : U \to \mathbb{R} \cup \{+\infty\}$ be a proper lower semicontinuous function and let $g : U \to \mathbb{R} \cup \{+\infty\}$ be a proper lower semicontinuous convex function. Let ∂ be a subdifferential such that $\partial f(\cdot) \subset \partial_C f(\cdot)$ on U. Then, given a real $\gamma \ge 0$, the inclusion

$$\partial f(x) \subset \partial g(x) + \gamma \mathbb{B}_{X^*} \quad \text{for all } x \in U$$

holds if and only if $\operatorname{dom} f = \operatorname{dom} g$ and for all $u \in \operatorname{dom} g$ and $v \in U$

$$g(v) - g(u) - \gamma\|v - u\| \le f(v) - f(u) \le g(v) - g(u) + \gamma\|v - u\|$$

It is worth noticing that taking $\gamma = 0$ in Corollary 9.15 furnishes another proof of the Moreau-Rockafellar Theorem 3.204. In fact, this allows to state the result for convex functions over open convex sets.

COROLLARY 9.16. Let U be a nonempty open convex set of a Banach space X and let $f, g : U \to \mathbb{R} \cup \{+\infty\}$ be two proper lower semicontinuous convex functions such that

$$\partial f(x) \subset \partial g(x) \quad \text{for all } x \in U.$$

Then there exists a constant $C \in \mathbb{R}$ such that

$$f(x) = g(x) + C \quad \text{for all } x \in U.$$

9.3. Essentially directionally smooth functions and their subdifferential determination

The class of subdifferentially and directionally stable functions does not require any condition by means of null sets in its definition while two locally Lipschitz continuous functions (of several real variables) whose gradients coincide almost everywhere are easily seen to be equal up to an additive constant. This section defines and studies an additional class involving some features from measure theory.

9.3.1. Essentially directionally smooth functions, properties and examples.

DEFINITION 9.17. Let U be a nonempty open convex set of a Banach space X, let ∂ be a subdifferential, and let $\mu > 0$ be a positive real. A proper lower semicontinuous function $f : U \to \mathbb{R} \cup \{+\infty\}$ is said to be *essentially ∂, μ-directionally smooth* on U, provided that for each $u \in \operatorname{Dom} \partial f$ and each $v \in \operatorname{dom} f$ with $v \neq u$ the following conditions are satisfied:
(i) the function $f_{u,v}(t) := f(u + t(v - u))$ is finite and continuous on the closed interval $[0, 1]$;
(ii) there are reals $0 = t_0 < \cdots < t_p = 1$ such that the function $f_{u,v}(\cdot)$ is absolutely continuous on every closed interval contained in the set $[0, 1] \setminus \{t_0, t_1, \cdots, t_p\}$;
(iii) there is a subset $T \subset [0, 1]$ of full Lebesgue measure in $[0, 1]$ (that is, of Lebesgue measure equal to 1) such that for every $t \in T$ and every sequence $(x_n, x_n^*)_n$ in $\operatorname{gph} \partial f$ with $x_n \to x(t) := u + t(v - u)$ there exists some $w \in]x(t), v]$ satisfying the inequality

$$\limsup_{n\to\infty} \langle x_n^*, w - x_n \rangle \le \|w - x(t)\| \big(\|w - v\|^{-1} f'_{u,v}(t; 1) + \mu \big).$$

The case where the condition (iii) holds for all reals $\mu > 0$ is of interest and arises in diverse classes of functions, as we will see below.

DEFINITION 9.18. When f in the above definition is ∂, μ-directionally smooth on U for every real $\mu > 0$, one says that it is *∂-directionally smooth* on the open convex set U. When convenient, we will say for short that f is *eds* on U.

The first example is the class of lower semicontinuous convex functions.

PROPOSITION 9.19. Let U be a nonempty open convex set of a Banach space X and $f : U \to \mathbb{R} \cup \{+\infty\}$ be a proper lower semicontinuous convex function. The following hold:
(a) Given $u \in \operatorname{Dom} \partial f$ and $v \in \operatorname{dom} f$ with $v \neq u$, then for any $t \in]0, 1[$ there exists some $\tau_0 \in]0, 1 - t[$ such that for any $\tau \in]0, \tau_0]$, for any $x_0 \in [u + t(v - u), u + (t + \tau)(v - u)[$, and for any sequence $(x_n, x_n^*)_n$ with $x_n \in U$, $x_n^* \in \partial f(x_n)$, and $x_n \to x_0$ as $n \to \infty$, one has

$$\begin{aligned} &\limsup_{n\to\infty} \langle x_n^*, u + (t + \tau)(v - u) - x_n \rangle \\ &\le \frac{\|u + (t + \tau)(v - u) - x_0\|}{\|v - u\|} f'_{u,v}(t; 1) + \mu \|u + (t + \tau)(v - u) - x_0\|, \end{aligned}$$

where $f_{u,v}(r) := f(u + r(v - u))$ for all $r \in [0, 1]$.
(b) The function f is essentially ∂-directionally smooth on U, that is, essentially ∂, μ-directionally smooth on U for every real $\mu > 0$.

PROOF. (a) Take any real $\mu > 0$. Let $u \in \operatorname{Dom} \partial f$ and $v \in \operatorname{dom} f$ with $v \neq u$. Choose $u^* \in \partial f(u)$ and as in the proof of Proposition 9.8 consider the function $\varphi : \mathbb{R} \to \mathbb{R}$ defined by

$$\varphi(s) = \begin{cases} +\infty & \text{if } s \in]1, +\infty[\\ f(u) + s\langle u^*, v-u\rangle & \text{if } s \in]-\infty, 0[\\ f(u+s(v-u)) & \text{if } s \in [0,1]. \end{cases}$$

This function φ is convex on $\mathbb{R}$ and finite on $]-\infty, 1[$, hence it is locally Lipschitz on $]-\infty, 1[$. The right-hand derivative $\varphi'(\cdot; 1)$ of φ is then finite on $]-\infty, 1[$ and it is also upper semicontinuous on $]-\infty, 1[$ since

$$\varphi'(s;1) = \inf_{\lambda>0} \lambda^{-1}\big(\varphi(s+\lambda) - \varphi(s)\big).$$

Therefore, the restriction of the function $s \mapsto f'(u+s(v-u); v-u)$ is finite and upper semicontinuous on $[0,1[$. Fix any real $t \in [0,1[$. By the latter upper semicontinuity property, we may choose some $\tau_0 \in]0, 1-t[$ such that for every $\tau \in]0, \tau_0]$

$$f'(u+(t+\tau)(v-u); v-u) \le f'(u+t(v-u); v-u) + \mu\|v-u\|. \tag{9.24}$$

Now fix any $\tau \in]0, \tau_0]$ and set $x(s) := u + s(v-u)$ for all $s \in [0,1]$. Take any $x_0 \in [x(t), x(t+\tau)[$ and any sequence $(x_n, x_n^*)_{n\in\mathbb{N}}$ in $\operatorname{gph} \partial f$ with $x_n \to x_0$ as $n \to \infty$. According to the convexity of f we have

$$\langle x_n^*, x(t+\tau) - x_n\rangle \le f(x(t+\tau)) - f(x_n) \quad \text{for all } n \in \mathbb{N},$$

which implies by the lower semicontinuity and convexity of f that

$$\limsup_{n\to\infty} \langle x_n^*, x(t+\tau) - x_n\rangle \le f(x(t+\tau)) - f(x_0) \le f'(x(t+\tau); x(t+\tau) - x_0).$$

This and the positive homogeneity of $f'(x(t+\tau); \cdot)$ yield

$$\limsup_{n\to\infty} \langle x_n^*, x(t+\tau) - x_n\rangle \le \frac{\|x(t+\tau) - x_0\|}{\|v-u\|} f'(x(t+\tau); v-u),$$

which combined with (9.24) ensures that

$$\begin{aligned} &\limsup_{n\to\infty} \langle x_n^*, x(t+\tau) - x_n\rangle \\ &\le \frac{\|x(t+\tau) - x_0\|}{\|v-u\|}\Big(f'(x(t); v-u) + \mu\|v-u\|\Big) \\ &= \frac{\|x(t+\tau) - x_0\|}{\|v-u\|} f'(x(t); v-u) + \mu\|x(t+\tau) - x_0\|, \end{aligned}$$

so (a) is established.

(b) The condition (iii) in Definition 9.17 is ensured by the above assertion (a). Further, for $u \in \operatorname{Dom} \partial f$ and $v \in \operatorname{dom} f$ with $v \neq u$, by convexity the function $f_{u,v}$ (as given in the statement) is finite and continuous on the closed interval $[0,1]$ as well as locally Lipschitz on the open interval $]0,1[$. Then the conditions (i) and (ii) in Definition 9.17 are clearly satisfied. This finishes the proof. □

The second example is relative to C-subregular locally Lipschitz continuous functions. In fact, it corresponds to the larger class of segmentwise essentially subregular functions.

DEFINITION 9.20. Let $f : U \to \mathbb{R}$ be a locally Lipschitz continuous function on an open set U of a Banach space X. One says that f is *segmentwise essentially subregular* on U if, for any $u, v \in U$ with $u \neq v$ and $x(t) := u + t(v - u)$ one has

$$f^\circ(x(t); v - u) = f'(x(t); v - u)$$

for Lebesgue almost all $t \in]0, 1[$.

EXAMPLE 9.21. (a) Obviously, any C-subregular (that is, tangentially regular) locally Lipschitz continuous function on the open convex set U is segmentwise essentially subregular on U.
(b) It is known that any locally absolutely continuous mapping $z : I \to X$ from an interval $I \subset \mathbb{R}$ into a reflexive Banach space X is differentiable almost everywhere; so for any locally Lipschitz function φ on an open set of X including $z(I)$, both derivatives $\frac{d(\varphi \circ z)}{dt}(t)$ and $z't) := \frac{dz}{dt}(t)$ exist for Lebesgue almost every $t \in I$. A locally Lipschitz continuous function $f : U \to \mathbb{R}$ from an open set U of a reflexive Banach space X is said to be *sound* if for any absolutely continuous mapping $z : [0, 1] \to X$ with $z([0, 1]) \subset U$ one has for almost every $t \in]0, 1[$

$$\frac{d}{dt}(f \circ z)(t) = \langle \zeta, z'(t) \rangle \ \forall \zeta \in \partial_C f(z(t));$$

as easily seen, according to the equality $\big(\partial_C f(x)\big)(h) = [-f^\circ(x; -h), f^\circ(x; h)]$, it is equivalent to requiring that for almost every $t \in]0, 1[$

$$f^\circ(z(t); -z'(t)) = -f^\circ(z(t); z'(t)). \tag{9.25}$$

Then any such locally Lipschitz sound function is segmentwise essentially subregular in the sense of (a).
(c) Modifying slightly (9.25), a locally Lipschitz continuous function $f : U \to \mathbb{R}$ on an open set U of a Banach space X is called *arcwise essentially smooth* on U provided that, for any locally Lipschitz continuous *essentially smooth* mapping $z :]0, 1[\to X$ with $z(]0, 1[) \subset U$, the set

$$\{t \in]0, 1[: f^\circ(z(t); -z'(t)) \neq -f^\circ(z(t); z'(t))\}$$

has Lebesgue measure null. The locally Lipschitz continuous mapping $z(\cdot)$ is *essentially smooth* if for each $e \in \mathbb{R}$ (or equivalently $e \in \{-1, 1\}$) the set

$$\left\{ r \in]0, 1[: \lim_{\rho \to r; t \downarrow 0} t^{-1}[z(\rho + te) - z(\rho)] \text{ does not exists in } X \right\}$$

is Lebesgue null in $\mathbb{R}$.

Locally Lipschitz arcwise essentially smooth functions are segmentwise essentially subregular.
(d) The segmentwise essential subregularity property is preserved by positively linear combination as well as by maximum of finitely many functions. Indeed, let $f, g, : U \to \mathbb{R}$ be two locally Lipschitz continuous functions which are segmentwise essentially subregular on the open convex set U of the normed space X. Given any real constant $\lambda > 0$, the property is clear for the function λf. Now fix any $u, v \in U$ with $v \neq u$ and put $x(t) := u + t(v - u)$. There exists a subset $T \subset]0, 1[$ with Lebesgue measure equal to 1 such that for every $t \in T$

$$f^\circ(x(t); v - u) = f'(x(t); v - u) \quad \text{and} \quad g^\circ(x(t); v - u) = g'(x(t); v - u).$$

Fix any $t \in T$.

Concerning the sum, writing

$$\begin{aligned}(f+g)^\circ(x(t);v-u) &\le f^\circ(x(t),v-u)+g^\circ(x(t);v-u)\\ &= f'(x(t);v-u)+g'(x(t);v-u)\\ &= (f+g)'(x(t);v-u)\\ &\le (f+g)^\circ(x(t);v-u),\end{aligned}$$

we see that $(f+g)^\circ(x(t);v-u) = (f+g)'(x(t);v-u)$. Therefore, $f+g$ is segmentwise essentially subregular.

Now put $\varphi = \max\{f,g\}$. If $f(x(t)) > g(x(t))$, then $f(u) > g(u)$ for all u in a neighborhood $U_t \subset U$ of $x(t)$. This entails that $\varphi(u) = f(u)$ for all $u \in U_t$, so $\varphi^\circ(x(t);\cdot) = f^\circ(x(t);\cdot)$ and $f'(x(t);v-u) = \varphi'(x(t):v-u)$, hence

$$\varphi^\circ(x(t);v-u) = \varphi'(x(t);v-u).$$

The case $g(x(t)) > f(x(t))$ is similar. Finally, suppose $f(x(t)) = g(x(t))$. Accordingly, we know by Proposition 2.190 that

$$\varphi^\circ(x(t);\cdot) \le \max\{f^\circ(x(t);\cdot), g^\circ(x(t);\cdot)\}$$

along with

$$\varphi^B(x(t);\cdot) \ge \max\{f^B(x(t);\cdot), g^B(x(t);\cdot)\}.$$

The equality $f^\circ(x(t);v-u) = f'(x(t);v-u)$ entails that

$$f^\circ(x(t);v-u) = f^B(x(t);v-u).$$

Using this and the analogous counterpart for g, we obtain by what precedes that $\varphi^\circ(x(t);v-u) \le \varphi^B(x(t);v-u)$, which in turn implies that $\varphi'(x(t);v-u)$ exists and coincides with $\varphi^\circ(x(t);v-u)$. Consequently, for any $t \in T$ we obtain that

$$\varphi^\circ(x(t);v-u) = \varphi'(x(t);v-u),$$

which confirms that the function φ is segmentwise essentially subregular. □

PROPOSITION 9.22. Let U be a nonempty open convex set of a Banach space X and $f : U \to \mathbb{R}$ be a locally Lipschitz function which is segmentwise essentially subregular on U. Let ∂ be a subdifferential with ∂f included in $\partial_C f$. The following hold:
(a) Given $u \in \operatorname{Dom} \partial f$ and $v \in \operatorname{dom} f$ with $v \neq u$, then there exists some set $T \subset]0,1[$ of Lebesgue measure equal to 1 such that for any $t \in T$, for any $w \in]x(t),v]$ with $x(t) := u+t(v-u)$, and for any sequence $(x_n, x_n^*)_n$ with $x_n \in U$, $x_n^* \in \partial f(x_n)$, and $x_n \to x(t)$ as $n \to \infty$, one has

$$\limsup_{n\to\infty} \langle x_n^*, w - x(t)\rangle \le \frac{\|w-x(t)\|}{\|v-u\|} f'_{u,v}(t;1),$$

where $f_{u,v}(r) := f(u+r(v-u))$ for all $r \in [0,1]$.
(b) The function f is essentially ∂-directionally smooth on U, that is, essentially ∂,μ-directionally smooth on U for every real $\mu > 0$.

PROOF. (a) Fix $u \in \operatorname{Dom} \partial f$ and $v \in U$ with $v \neq u$. Put $x(t) := u+t(v-u)$ for all $t \in [0,1]$. The segmentwise essential subregularity property of f furnishes some set $T \subset]0,1[$ of full Lebesgue measure such that

$$f^\circ(x(t);v-u) = f'(x(t);v-u) \quad \text{for all } t \in T. \tag{9.26}$$

Take any $t \in T$ and any $w \in]x(t), v]$, and take also any sequence $(x_n, x_n^*)_n$ in $\mathrm{gph}\, \partial f$ with $x_n \to x(t)$ as $n \to \infty$. By the upper semicontinuity property of $f^\circ(\cdot;\cdot)$ on $U \times X$ (see Proposition 2.74(a)) and by the inclusion $\partial f(\cdot) \subset \partial_C f(\cdot)$ we have

$$\begin{aligned}\limsup_{n\to\infty} \langle x_n^*, w - x_n\rangle &\le \limsup_{n\to\infty} f^\circ(x_n; w - x_n) \\ &\le f^\circ(x(t); w - x(t)) = \frac{\|w - x(t)\|}{\|v - u\|} f^\circ(x(t); v - u).\end{aligned}$$

Combining this with (9.26) we obtain the assertion (a).
(b) The assertion (a) obviously entails the condition (iii) in Definition 9.17 while the other conditions (i) and (ii) in that definition directly follow from the local Lipschitz property of f on U. □

From the previous proposition we can show the preservation of essential directional smoothness by addition with a segmentwise essentially subregular locally Lipschitz function.

COROLLARY 9.23. Let U be a nonempty open convex set of a Banach space X, let $f_1 : U \to \mathbb{R} \cup \{+\infty\}$ be a proper lower semicontinuous function which is essentially ∂, μ-directionally smooth on U for some $\mu > 0$, and let $f_2 : U \to \mathbb{R}$ be a locally Lipschitz function which is segmentwise essentially subregular on U. Assume that $\partial f_2 \subset \partial_C f_2$ and $\partial(f_1 + f_2)(x) \subset \partial f_1(x) + \partial f_2(x)$ for all $x \in U$. Then the function $f_1 + f_2$ is essentially ∂, μ-directionally smooth on U.

PROOF. Clearly, conditions (i) and (ii) in Definition 9.17 hold true. Consider any $u \in \mathrm{Dom}\, \partial(f_1 + f_2)$ and any $v \in \mathrm{dom}\,(f_1 + f_2)$. By the subdifferential sum rule assumption one has $u \in \mathrm{Dom}\, \partial f_1 \cap \mathrm{Dom}\, \partial f_2$, and obviously $v \in \mathrm{dom}\, f_1$. Take a set $T \subset]0, 1[$ of full Legesgue measure satisfying condition (iii) in Definition 9.17 for f_1 and such that (a) in Proposition 9.22 holds for the function f_2. Fix any $t \in T$ and any sequence $(x_n, x_n^*)_n$ in $\mathrm{gph}\, \partial(f_1 + f_2)$ with $x_n \to x(t) := u + t(v - u)$ as $n \to \infty$. There exist $x_{i,n}^* \in \partial f_i(x_n)$, with $i = 1, 2$, such that $x_n^* = x_{1,n}^* + x_{2,n}^*$. By (iii) in Definition 9.17 we have some $w \in]x(t), v]$ such that

$$\limsup_{n\to\infty} \langle x_{1,n}^*, w - x_n\rangle \le \frac{\|w - x(t)\|}{\|v - u\|} f'_{u,v}(t; 1) + \mu \|w - x(t)\|.$$

Proposition 9.22(a) also gives that

$$\limsup_{n\to\infty} \langle x_{2,n}^*, w - x_n\rangle \le \frac{\|w - x(t)\|}{\|v - u\|} g'_{u,v}(t; 1).$$

It ensues that

$$\begin{aligned}\limsup_{n\to\infty} \langle x_n^*, w - x_n\rangle &\le \limsup_{n\to\infty} \langle x_{1,n}^*, w - x_n\rangle + \limsup_{n\to\infty} \langle x_{2,n}^*, w - x_n\rangle \\ &\le \frac{\|w - x(t)\|}{\|v - u\|} f'_{u,v}(t; 1) + \mu \|w - x(t)\| + \frac{\|w - x(t)\|}{\|v - u\|} g'_{u,v}(t; 1) \\ &= \frac{\|w - x(t)\|}{\|v - u\|} (f + g)'_{u,v}(t; 1) + \mu \|w - x(t)\|,\end{aligned}$$

which translates condition (iii) in Definition 9.17 for the function $f + g$. We conclude that $f + g$ is essentially ∂, μ-directionally smooth as desired. □

We noticed in Example 9.21(a) above that C-subregular locally Lipschitz functions are segmentwise essentially subregular. The next proposition shows that the opposite of any C-subregular locally Lipschitz function is also segmentwise essentially subregular.

PROPOSITION 9.24. Let U be a nonempty open convex set of a Banach space X and $f : U \to \mathbb{R}$ be a locally Lipschitz continuous function which is C-subregular on U. Then the opposite function $-f$ is segmentwise essentially subregular on U.

PROOF. Let any $u, v \in U$ with $v \neq u$ and let $x_{u,v}(t) := u + t(v - u)$ for all $t \in [0, 1]$. By local Lipschitz property of $f_{u,v} := f \circ x_{u,v}$ there exists a set $T \subset]0, 1[$ of full Lebesgue measure (that is, of Lebesgue measure 1) such that the function $f_{u,v}$ is derivable at every point in T. Fix any $t \in T$. By derivability of $f_{u,v}$ at t we have on one hand

$$\begin{aligned}
-\frac{d}{dt} f_{u,v}(t) &= -\lim_{\tau \downarrow 0} \frac{f(u + (t - \tau)(v - u)) - f(u + t(v - u))}{-\tau} \\
&= \lim_{\tau \downarrow 0} \frac{f(x_{u,v}(t) + \tau(u - v)) - f(x_{u,v}(t))}{\tau} \\
&= f'(x_{u,v}(t); u - v)
\end{aligned}$$

and on the other hand

$$\begin{aligned}
-\frac{d}{dt} f_{u,v}(t) &= -\lim_{\tau \downarrow 0} \frac{f(u + (t + \tau)(v - u)) - f(u + t(v - u))}{\tau} \\
&= \lim_{\tau \downarrow 0} \frac{(-f)(x_{u,v}(t) + \tau(v - u)) - (-f)(x_{u,v}(t))}{\tau} \\
&= (-f)'(x_{u,v}(t); v - u),
\end{aligned}$$

hence $f'(x_{u,v}(t); u-v) = (-f)'(x_{u,v}(t); v-u)$. Then, recalling by the Lipschitz property of f that $(-f)^\circ(y; h) = f^\circ(y; -h)$, and keeping in mind that f is subregular, we can write

$$\begin{aligned}
(-f)^\circ(x_{u,v}(t); v - u) = f^\circ(x_{u,v}(t); u - v) &= f'(x_{u,v}(t); u - v) \\
&= (-f)'(x_{u,v}(t); v - u).
\end{aligned}$$

This confirms the segmentwise essential s of the function $-f$. □

The following corollary establishes the essential directional smoothness property of DC functions.

COROLLARY 9.25. Let U be an open convex set of a Banach space X and let the subdifferential ∂ be either ∂_C or ∂_A (or ∂_L if X is an Asplund space). Let $f : U \to \mathbb{R} \cup \{+\infty\}$ be a DC function with $f = f_1 - f_2$, where $f_1 : U \to \mathbb{R} \cup \{+\infty\}$ is a proper lower semicontinuous convex function and $f_2 : U \to \mathbb{R}$ is a finite-valued continuous convex function. Then the function f is essentially ∂-directionally smooth on U.

PROOF. The function f_2 being convex continuous on U, it is locally Lipschitz and C-subregular on the open convex set U. Proposition 9.24 tells us that $-f_2$ is segmentwise essentially subregular on U. Since f_1 is essentially ∂-directionally smooth on U according to Proposition 9.19, it results from Corollary 9.23 that the function $f = f_1 + (-f_2)$ is essentially ∂-directionally smooth on U. □

9.3.2. Subdifferential determination of essentially directionally smooth functions.

THEOREM 9.26 (L. Thibault and D. Zagrodny: estimates under enlarged subdifferential inclusions of eds functions). Let U be a nonempty open convex set of a Banach space X and ∂ be a subdifferential. Let $\gamma \geq 0$ be a non-negative real number. Let $f, g : U \to \mathbb{R} \cup \{+\infty\}$ be two proper lower semicontinuous functions which are essentially ∂, μ-directionally smooth on U for some real $\mu > 0$. Assume that

$$\partial f(x) \subset \partial g(x) + \gamma \mathbb{B}_{X^*} \quad \text{for all } x \in U. \tag{9.27}$$

Then $\operatorname{dom} f \subset \operatorname{dom} g$ and for all $u, v \in \operatorname{dom} f$ the following inequalities hold

$$g(v) - g(u) - (\gamma + \mu)\|v - u\| \leq f(v) - f(u) \leq g(v) - g(u) + (\gamma + \mu)\|v - u\|.$$

Further, for any $u \in \operatorname{Dom} \partial f$ and $v \in \operatorname{dom} g$ the set $[u, v] \cap \operatorname{dom} f$ is a closed interval in X.

PROOF. Notice first that $\operatorname{Dom} \partial f$ is nonempty by Proposition 6.50 and that we have $\operatorname{Dom} \partial f \subset \operatorname{Dom} \partial g$ by (9.27). Notice also that, without loss of generality, we may and do suppose that $\operatorname{dom} f$ is not a singleton. Fix any $u \in \operatorname{Dom} \partial f$ and any $v \in \operatorname{dom} f \cap \operatorname{dom} g$ (such elements v exist since $\operatorname{Dom} \partial f \subset \operatorname{Dom} \partial g$).

Step 1. In a first step we show the inequality

$$f(v) - f(u) \leq g(v) - g(u) + (\gamma + \mu)\|v - u\|. \tag{9.28}$$

Indeed, suppose on the contrary that

$$f(v) - f(u) > g(v) - g(u) + (\gamma + \mu)\|v - u\|. \tag{9.29}$$

Let us set for every $t \in [0, 1]$

$$x(t) := u + t(v - u), \;\; f_{u,v}(t) := f(x(t)), \;\; g_{u,v}(t) := g(x(t)).$$

Choose reals $0 = t_0 < \cdots < t_p = 1$ and $r_0 < \frac{1}{2} \min_{1 \leq i \leq p}(t_i - t_{i-1})$ such that the property (ii) in Definition 9.17 holds for both $f_{u,v}$ and $g_{u,v}$. By this and by the property (iii) in Definition 9.17 for g, take also a subset $T \subset]0, 1[$ of full Lebesgue measure (that is, of Lebesgue measure equal to 1) such that for each $\theta \in T$

(α) both derivatives $f'_{u,v}(\theta; 1)$ and $g'_{u,v}(\theta; 1)$ exist,

(β) for every sequence $(z_n, z_n^*)_n$ in $\operatorname{gph} \partial g$ with $z_n \to u + \theta(v - u)$ there is a $\zeta \in]u + \theta(v - u), v]$ for which

$$\limsup_{n \to \infty} \langle z_n^*, \zeta - z_n \rangle \leq \left\|\zeta - \big(u + \theta(v - u)\big)\right\| \Big(\|v - u\|^{-1} g'_{u,v}(\theta; 1) + \mu\Big).$$

By (9.29) and by the continuity on $[0, 1]$ of the functions $f_{u,v}$ and $g_{u,v}$ (see (i) in Definition 9.17), there exists a positive real $r < r_0$ such that

$$f(v) - f(u) - g(v) - g(u) - (\gamma + \mu)\|v - u\| \tag{9.30}$$

$$> \sum_{i=1}^{p} \Big\{ \big|f_{u,v}(t_{i-1} + r) - f_{u,v}(t_{i-1})\big| + \big|f_{u,v}(t_i) - f_{u,v}(t_i - r)\big|$$

$$+ \big|g_{u,v}(t_{i-1} + r) - g_{u,v}(t_{i-1})\big| + \big|g_{u,v}(t_i) - g_{u,v}(t_i - r)\big| \Big\} \tag{9.31}$$

$$+ 2pr(\gamma + \mu)\|v - u\|.$$

Fix any $t \in T$ and choose a sequence $(\tau_n)_n$ tending to 0 with $0 < \tau_n < 1 - t$. We claim that

$$(9.32) \quad \lim_{n\to\infty} \frac{1}{\tau_n\|v-u\|}\Big[f_{u,v}(t+\tau_n) - f_{u,v}(t) + g_{u,v}(t) - g_{u,v}(t+\tau_n)\Big] \le \gamma + \mu$$

(note that the limit exists since $f'_{u,v}(t;1)$ and $g'_{u,v}(t;1)$ exist). Suppose that (9.32) is not true, that is, there is a real $\delta > 0$ such that for n sufficiently large, say $n \ge n_0$,

$$(9.33) \quad \frac{1}{\tau_n\|v-u\|}\Big[f_{u,v}(t+\tau_n) - f_{u,v}(t) + g_{u,v}(t) - g_{u,v}(t+\tau_n)\Big] \ge \gamma + \mu + 3\delta.$$

Since the lower semicontinuous function f is finite at $x(t)$ and $x(t+\tau_n)$ by (i) in Definition 9.17, the Zagrodny mean value inequality theorem (see Theorem 6.52) furnishes for each fixed integer $n \ge n_0$

$$x_{n,0} \in [x(t), x(t+\tau_n)[, \quad x_{n,k} \underset{k\to\infty}{\longrightarrow} x_{n,0}, \quad \text{and } y^*_{n,k} \in \partial f(x_{n,k})$$

such that (see (ii) in Theorem 6.52)

$$\begin{aligned}&\langle y^*_{n,k}, x(t+\tau_n) - x_{n,k}\rangle \\ &\ge \frac{\|x(t+\tau_n) - x_{n,0}\|}{\|x(t+\tau_n) - x(t)\|}\Big[f_{u,v}(t+\tau_n) - f_{u,v}(t)\Big] - \delta\|x(t+\tau_n) - x_{n,0}\|\end{aligned}$$

and such that (see (i) in Theorem 6.52)

$$\langle y^*_{n,k}, x(t+\tau_n) - x(t)\rangle \ge f_{u,v}(t+\tau_n) - f_{u,v}(t) - \delta\|x(t+\tau_n) - x(t)\|.$$

For each integer $n \ge n_0$ and each integer $k \ge 1$, by the assumption (9.27) we can choose

$$(9.34) \qquad x^*_{n,k} \in \partial g(x_{n,k}) \quad \text{such that } \|x^*_{n,k} - y^*_{n,k}\| \le \gamma.$$

Fiw any $w \in]x(t), v]$ and choose an integer $n_1 \ge n_0$ such that $x(t+\tau_n) \in]x(t), w[$ for all $n \ge n_1$. Fix now any integer $n \ge n_1$. It ensues that

$$\begin{aligned}\langle y^*_{n,k}, w - x_{n,k}\rangle \ge\ & \frac{\|w - x(t+\tau_n)\| + \|x(t+\tau_n) - x_{n,0}\|}{\|x(t+\tau_n) - x(t)\|}\Big[f_{u,v}(t+\tau_n) - f_{u,v}(t)\Big] \\ & - \delta\Big(\|w - x(t+\tau_n)\| + \|x(t+\tau_n) - x_{n,0}\|\Big).\end{aligned}$$

Noting that $\|w - x(t+\tau_n)\| + \|x(t+\tau_n) - x_{n,0}\| = \|w - x_{n,0}\|$ (since $x(t+\tau_n)$ lies in $[x_{n,0}, w[$) we obtain

$$(9.35) \quad \langle y^*_{n,k}, w - x_{n,k}\rangle \ge \frac{\|w - x_{n,0}\|}{\|x(t+\tau_n) - x(t)\|}\Big[f_{u,v}(t+\tau_n) - f_{u,v}(t)\Big] - \delta\|w - x_{n,0}\|.$$

Note also that the equality $\tau_n\|v-u\| = \|x(t+\tau_n) - x(t)\|$ and (9.33) give

$$(9.36) \quad f_{u,v}(t+\tau_n) - f_{u,v}(t) \ge g_{u,v}(t+\tau_n) - g_{u,v}(t) + (\gamma+\mu+3\delta)\|x(t+\tau_n) - x(t)\|.$$

According to (9.34) the inequalities (9.35) and (9.36) imply that for every integer k we have

$$(9.37) \quad \begin{aligned}\langle x^*_{n,k}, w - x_{n,k}\rangle + \gamma\|w - x_{n,k}\| \ge\ & \frac{\|w - x_{n,0}\|}{\|x(t+\tau_n) - x(t)\|}\Big[g_{u,v}(t+\tau_n) - g_{u,v}(t)\Big] \\ & + (\gamma+\mu+3\delta)\|w - x_{n,0}\| - \delta\|w - x_{n,0}\|.\end{aligned}$$

From the inclusions $x_{n,0} \in [x(t), x(t+\tau_n)[$ and $x(t+\tau_n) \in [x(t), w[$ we note that $x_{n,0} \neq w$. Since $x_{n,k} \to x_{n,0}$ as $k \to \infty$, we then have for k sufficiently large

$$(9.38) \qquad \gamma\|w - x_{n,0}\| - \gamma\|w - x_{n,k}\| > -\delta\|w - x_{n,0}\|.$$

Choose some integer $k_n \geq 1$ such that (9.38) holds with $k = k_n$ and such that $\|x_{n,k_n} - x_{n,0}\| < \tau_n$. Put $x_n := x_{n,k_n}$ and $x_n^* := x_{n,k_n}^*$, so by (9.37) and by (9.38) with $k = k_n$

$$\begin{aligned}\langle x_n^*, w - x_n\rangle &\geq \frac{\|w - x_{n,0}\|}{\|x(t+\tau_n) - x(t)\|}\Big[g_{u,v}(t+\tau_n) - g_{u,v}(t)\Big] + (\mu+\delta)\|w - x_{n,0}\| \\ &= \frac{\|w - x_{n,0}\|}{\|v-u\|}\frac{1}{\tau_n}\Big[g_{u,v}(t+\tau_n) - g_{u,v}(t)\Big] + (\mu+\delta)\|w - x_{n,0}\|.\end{aligned}$$

Therefore, we see that for every $w \in]x(t), v]$

$$\limsup_{n\to\infty}\langle x_n^*, w - x_n\rangle \geq \|w - x(t)\|\Big(\|v-u\|^{-1}g'_{u,v}(t;1) + (\mu+\delta)\Big),$$

which is in contradiction with the item (β) in the choice of the set T. This confirms the claim that the inequality (9.32) holds true.

It follows, according to the property (ii) in Definition 9.17, that for almost every $t \in T \bigcap (\bigcup_{i=1}^p [t_{i-1}+r, t_i - r])$ we have

$$f'_{u,v}(t;1) \leq g'_{u,v}(t;1) + (\gamma+\mu)\|v-u\|.$$

Using this inequality and the absolute continuity property of $g_{u,v}$ and $f_{u,v}$ on each closed interval $[t_{i-1}+r, t_i - r]$, and using the equality

$$\begin{aligned}&g_{u,v}(1) - g_{u,v}(0) \\ &= \sum_{i=1}^p\Big(g_{u,v}(t_i - r) - g_{u,v}(t_{i-1}+r)\Big) + \sum_{i=1}^p\Big\{\Big(g_{u,v}(t_{i-1}+r) - g_{u,v}(t_{i-1})\Big) \\ &\quad + \Big(g_{u,v}(t_i) - g_{u,v}(t_i - r)\Big)\Big\},\end{aligned}$$

we deduce that

$$\begin{aligned}&g_{u,v}(1) - g_{u,v}(0) \\ &\geq \sum_{i=1}^p\Big(f_{u,v}(t_i - r) - f_{u,v}(t_{i-1}+r)\Big) + 2pr(\gamma+\mu)\|v-u\| - (\gamma+\mu)\|v-u\| \\ &\quad - \sum_{i=1}^p\Big\{\big|g_{u,v}(t_{i-1}+r) - g_{u,v}(t_{i-1})\big| + \big|g_{u,v}(t_i) - g_{u,v}(t_i - r)\big| \\ &\quad + \big|f_{u,v}(t_{i-1}+r) - f_{u,v}(t_{i-1})\big| + \big|f_{u,v}(t_i) - f_{u,v}(t_i - r)\big|\Big\} \\ &\quad + \sum_{i=1}^p\Big\{f_{u,v}(t_i) - f_{u,v}(t_i - r) + f_{u,v}(t_{i-1}+r) - f_{u,v}(t_{i-1})\Big\}.\end{aligned}$$

From this we see that

$$\begin{aligned}&g_{u,v}(1) - g_{u,v}(0) \\ &\geq 2pr(\gamma+\mu)\|v-u\| - (\gamma+\mu)\|v-u\| + f_{u,v}(1) - f_{u,v}(0) \\ &\quad - \sum_{i=1}^p\Big\{\big|g_{u,v}(t_{i-1}+r) - g_{u,v}(t_{i-1})\big| + \big|g_{u,v}(t_i) - g_{u,v}(t_i - r)\big| \\ &\quad + \big|f_{u,v}(t_{i-1}+r) - f_{u,v}(t_{i-1})\big| + \big|f_{u,v}(t_i) - f_{u,v}(t_i - r)\big|\Big\}.\end{aligned}$$

The latter inequality combined with (9.31) gives

$$g_{u,v}(1) - g_{u,v}(0)$$
$$> -f_{u,v}(1) + f_{u,v}(0) + g_{u,v}(1) - g_{u,v}(0) + 4pr(\gamma + \mu)\|v - u\| + f_{u,v}(1) - f_{u,v}(0),$$

which is clearly a contradiction. The inequality (9.28) is then demonstrated.

Step 2. Now fix any $u, v \in \operatorname{dom} f$. By the graphical density of $\operatorname{Dom} \partial f$ in $\operatorname{dom} f$ (see Proposition 6.50) choose a sequence $(u_n)_n$ in $\operatorname{Dom} \partial f$ such that $u_n \to u$ and $f(u_n) \to f(u)$. By the inequality (9.28) we have for every n

$$f(v) - f(u_n) \le g(v) - g(u_n) + (\gamma + \mu)\|v - u_n\|,$$

hence taking the lower semicontinuity of g into account we get

$$f(v) - f(u) \le g(v) - g(u) + (\gamma + \mu)\|v - u\|. \tag{9.39}$$

This inequality tells us in particular that $g(u) < +\infty$, so

$$\operatorname{dom} f \subset \operatorname{dom} g.$$

Combining (9.39) with the new inequality obtained after interchanging $u, v \in \operatorname{dom} f$ therein, we obtain that for all $u, v \in \operatorname{dom} f$

$$g(v) - g(u) - (\gamma + \mu)\|v - u\| \le f(v) - f(u) \le g(v) - g(u) + (\gamma + \mu)\|v - u\|.$$

Step 3. It remains to show that $[u, v] \cap \operatorname{dom} f$ is a closed interval in X for $u \in \operatorname{Dom} \partial f$ and $v \in \operatorname{dom} g$.

Fix any $u \in \operatorname{Dom} \partial f$ and $v \in \operatorname{dom} g$. We may and do suppose that $v \neq u$, otherwise the property is trivial. Define

$$s := \sup\{t \in [0, 1] : u + t(v - u) \in \operatorname{dom} f\}.$$

By (i) in Definition 9.17 of essentially ∂, μ-directional smoothness of f, we have $[u, u + s(v - u)[\subset \operatorname{dom} f$. Choose a sequence $(v_n)_n$ in $[u, u + s(v - u)[$ converging to $u + s(v - u)$ as $n \to \infty$. The inequality (9.39) gives for every integer n

$$f(v_n) - f(u) \le g(v_n) - g(u) + (\gamma + \mu)\|v_n - u\|. \tag{9.40}$$

On the other hand, observing that $u \in \operatorname{Dom} \partial g$ according to the assumption (9.27) we see that the restriction of g to $[u, v]$ is finite and continuous by (i) in Definition 9.17. Consequently,

$$\lim_{n \to \infty} g(v_n) = g(u + s(v - u)),$$

so passing to the limit in (9.40) and using the lower semicontinuity of f, we obtain

$$f(u + s(v - u)) - f(u) \le g(u + s(v - u)) - g(u) + (\gamma + \mu)\|v - u\|. \tag{9.41}$$

We have already noticed that g is finite on $[u, v]$, so $g(u + s(v - u)) < +\infty$. This and (9.41) entail that $f(u + s(v - u)) < +\infty$, or equivalently $u + s(v - u) \in \operatorname{dom} f$. It results that

$$[u, v] \cap \operatorname{dom} f = [u, u + s(v - u)],$$

which confirms that $[u, v] \cap \operatorname{dom} f$ is a closed interval in X. The theorem is completely proved. □

COROLLARY 9.27 (subdifferential determination of eds functions). *Let U be a nonempty open convex set of a Banach space X and ∂ be a subdifferential. Let $f, g : U \to \mathbb{R} \cup \{+\infty\}$ be two proper lower semicontinuous functions which are essentially ∂-directionally smooth on U. Assume that*

$$\partial f(x) = \partial g(x) \quad \text{for all } x \in U.$$

Then there exists a real constant C such that

$$f(x) = g(x) + C \quad \text{for all } x \in U.$$

PROOF. By assumption both functions f, g are essentially ∂, μ-directionally smooth for every real $\mu > 0$. Since $\gamma = 0$, we may interchange f, g to obtain from Theorem 9.26 that $\operatorname{dom} f = \operatorname{dom} g$. Choose $u_0 \in \operatorname{dom} f$. Theorem 9.26 again (with $\gamma = 0$) says that for any $\mu > 0$ one has for all $v \in \operatorname{dom} f = \operatorname{dom} g$

$$f(v) - f(u_0) - \mu\|v - u_0\| \leq g(v) - g(u_0) \leq f(v) - f(u_0) + \mu\|v - u_0\|.$$

This being true for every real $\mu > 0$ we obtain for all $v \in \operatorname{dom} f = \operatorname{dom} g$

$$f(v) - f(u_0) = g(v) - g(u_0),$$

and the latter equality is still valid for all $v \in U$ since $\operatorname{dom} f = \operatorname{dom} g$. This justifies the statement of the corollary. □

The second corollary directly follows from Theorem 9.26 and Corollary 9.25. It concerns the case of DC functions.

COROLLARY 9.28. Let U be a nonempty open convex set of a Banach space X and let the subdifferential ∂ be either ∂_C or ∂_A (or ∂_L if the space X is Asplund). Let $\gamma \geq 0$ be a non-negative real number. Let $f : U \to \mathbb{R} \cup \{+\infty\}$ be a proper lower semicontinuous function and $g : U \to \mathbb{R} \cup \{+\infty\}$ be a DC function with $g = g_1 - g_2$, where $g_1 : U \to \mathbb{R} \cup \{+\infty\}$ is a proper lower semicontinuous convex function and $g_2 : U \to \mathbb{R}$ is a finite-valued continuous convex function. Assume that

$$\partial f(x) \subset \partial g(x) + \gamma \mathbb{B}_{X^*} \quad \text{for all } x \in U.$$

Then $\operatorname{dom} f \subset \operatorname{dom} g = \operatorname{dom} g_1$ and for all $u, v \in \operatorname{dom} f$ the following inequalities hold

$$g(v) - g(u) - \gamma\|v - u\| \leq f(v) - f(u) \leq g(v) - g(u) + \gamma\|v - u\|.$$

9.4. Comments

The subdifferential determination of extended real-valued functions (that is, taking values in $\mathbb{R} \cup \{+\infty\}$) started with J. J. Moreau who proved in his 1965 paper [**741**] the following theorem: If two proper lower semicontinuous convex functions $f, g : H \to \mathbb{R} \cup \{+\infty\}$ on a Hilbert space H are such that $\partial f(x) = \partial g(x)$ for all $x \in H$, then they are equal up to an additive constant, that is, there exists a real constant C such that

$$f(\cdot) = g(\cdot) + C.$$

The theorem was extended to Banach spaces by R. T. Rockafellar in his papers [**846, 851**]. For more details and comments on subdifferential determination of convex functions see Sections 3.17 and 3.24 in Chapter 3. The terminology "*subdifferential determination*" for that result was utilized by Moreau [**744**]. Concerning nonconvex extended real-valued functions, the first subdifferential determination theorem in that context was provided in 1991 by R. A. Poliquin [**805**, Theorem 4.1] for functions from $\mathbb{R}^n$ into $\mathbb{R} \cup \{+\infty\}$ which are *primal lower nice*; see Chapter 11 for the definition of such functions under the name of *primal lower regular functions*, and see Theorem 11.42 for Poliquin's statement. Note that previous results but for merely finite-valued locally Lipschitz continuous functions were established under various subregularity assumptions: by R. T. Rockafellar [**859**] under Clarke subregularity in finite dimensions, by R. Correa and A. Jofre [**290**] in Banach spaces under either Mifflin semismoothness or Clarke subregularity.

Apart the aforementioned finite-dimensional theorem by R. A. Poliquin, the first infinite-dimensional results of subdifferential determination for nonconvex extended real-valued functions on Banach spaces were provided by L. Thibault and D. Zagrodny in their 1995 paper [**926**] (submitted in 1993). In that paper [**926**] it was shown: Given a real $\varepsilon > 0$ and two proper lower semicontinuous functions $f, g : U \to \mathbb{R} \cup \{+\infty\}$ on an open convex set U of a Banach space X such that g is convex and

$$\partial_C f(x) \subset \partial_C g(x) + \varepsilon \mathbb{B}_{X^*} \quad \text{for all } x \in U,$$

then $\operatorname{dom} f = \operatorname{dom} g$ and

$$g(v) - g(u) - \varepsilon\|v - u\| \le f(v) - f(u) \le g(v) - g(u) + \varepsilon\|v - u\|$$

for all $v \in U$ and $u \in \operatorname{dom} f$. By means of that result, the Poliquin's aforementioned subdifferential determination theorem for primal lower nice functions on $\mathbb{R}^n$ was extended in the same paper [**926**] to the setting of Hilbert spaces; see Section 11.6 for more comments. The classes of *subdifferentially and directionally stable* functions and *essentially* ∂, μ*-directionally smooth* functions, were introduced by Thibault and Zagrodny in [**927**] and [**928**] respectively.

As it can be seen in Definition 9.17 and as suggested by the name, essentially directionally smooth functions involve, in their definition and behavior, certain features from measure theory. Let us present a rapid history of definitions of classes of nonsmooth functions employing measure theory and for which the subdifferential determination property was proved. M. Valadier [**949, 950**] declared that a locally Lipschitz continuous function $f : U \to \mathbb{R}$ on an open set U of $\mathbb{R}^n$ is *sound* ("saine" in Valadier's French terminology) provided that for any absolutely continuous mapping $x : [0,1] \to \mathbb{R}^n$ with $x([0,1]) \subset U$, one has for almost every $t \in]0,1[$ that the vector $x'(t)$ is geometrically orthogonal to the affine hull of $\partial_C f(x(t))$, which can be expressed as

$$\langle \zeta_1, x'(t)\rangle = \langle \zeta_2, x'(t)\rangle \;\; \forall \zeta_1, \zeta_2 \in \partial_C f(x(t)).$$

Since $f \circ x$ is differentiable almost everywhere on $[0,1]$, it is equivalent to requiring that for almost every $t \in]0,1[$

$$\frac{d}{dt}(f \circ x)(t) = \langle \zeta, x'(t)\rangle \;\; \forall \zeta \in \partial_C f(x(t)). \tag{9.42}$$

As noticed by Valadier, one of great interests of such functions is that they fulfill the *dynamic descent property* in the following sense: Given a locally Lipschitz *sound* function $f : \mathbb{R}^n \to \mathbb{R}$ and the dynamic system on $[0,T]$

$$\begin{cases} x'(t) \in -\partial_C f(x(t)) \\ x(0) = x_0 \in \mathbb{R}^n, \end{cases}$$

for any absolutely continuous solution $x(\cdot)$, the function $f \circ x$ is nonincreasing on $[0,T]$. Indeed, it suffices to observe by (9.42) that for almost every $t \in]0,T[$

$$\frac{d}{dt}(f \circ x)(t) = \langle -x'(t), x'(t)\rangle = -\|x'(t)\|^2,$$

hence $f \circ x$ is nonincreasing. Many other interesting properties of sound functions (as for example, stability for usual operations "opposite, sum, product, quotient, maximum") are established in [**949**]. The formulation taken in the definition given in Example 9.21(b) is just that in (9.42) with any reflexive Banach space in place of $\mathbb{R}^n$. The stability results of Valadier [**949**] (with the same arguments therein) are still valid with any reflexive Banach space, and the dynamic descent property still holds for any Hilbert space in place of $\mathbb{R}^n$. Recalling (see Proposition 2.76) that a locally Lipschitz function is strictly Gâteaux differentiable at a point if and only if its C-subdifferential at that point is reduced to a singleton, Proposition 3 in Valadier's paper [**949**] (related to sound functions of one real variable) deserves to be stated as follows:

PROPOSITION 9.29 (M. Valadier [**949**]). Let $f : I \to \mathbb{R}$ be a locally Lipschitz continuous function on an open interval $I \subset \mathbb{R}$. The following are equivalent:
(a) the function f is sound;
(b) for almost every $t \in I$ the C-subdifferential $\partial_C f(t)$ is a singleton;
(c) the function f is strictly differentiable almost everywhere in I.

Using the concept of Haar null sets (see Chapter 13), J. M. Borwein and W. B. Moors defined two other general types of specific locally Lipschitz functions: the class of essentially smooth Lipschitz functions on general Banach spaces in [**137**] and the class of arcwise essentially smooth Lipschitz functions on $\mathbb{R}^n$ in [**135**]. The diverse steps were the following. Given an open set U of a separable Banach space X, J. M. Borwein said in his 1991 paper [**124**, p. 68] that a locally Lipschitz continuous function $f : U \to \mathbb{R}$ is *essentially strictly differentiable* provided that the set of points $u \in U$ where f is not strictly Gâteaux differentiable is Haar null in X. J. M. Borwein and W. B. Moors [**135**, p. 323] employed the terminology "*essentially smooth*" (or *smooth almost everywhere*) for such functions. A set N in a general Banach space Y is Haar null if there exist in X a universally (Radon) measurable set $B \supset N$ and a Radon probability measure P (not necessarily unique) on Y such that $P(y + B) = 0$ for all $y \in Y$. That definition of essentially smooth functions corresponds to the characterization (c) in Proposition 9.29 above for sound functions of one real variable. J. M. Borwein [**124**, Proposition 4.4] established the C-subdifferential determination property for essentially smooth locally Lipschitz functions on open sets of separable Banach spaces; another proof was given by J. M. Borwein and W. B. Moors [**135**, Proposition 4.2]. The arguments in [**124**] and [**135**] are based on certain features for upper semicontinuous compact-valued multimappings (called *usco* in those papers, for short). One can find in [**124, 135**] many basic properties of *usco* multimappings and many examples of essentially smooth Lipschitz functions along with conditions for essential smoothness of distance functions to subsets. Borwein and Moors implicitly noticed in [**135**, p. 329] that the locally Lipschitz function f on the open set U of the separable Banach space X is essentially smooth if and only if for every unit vector $h \in X$ the set

$$\{u \in U : f^{\circ}(u; -h) \neq -f^{\circ}(u; h)\} \tag{9.43}$$

is Haar null in X. They even used that characterization in the proof of Theorem 4.3 (p. 327) in the same paper [**135**]. So, Borwein and Moors extended to general Banach spaces the definition of essentially smooth locally Lipschitz functions as those for which the set in (9.43) is Haar null for each unit vector h. Via *usco* multimappings Borwein and Moors [**137**, Theorem 3.5] extended the subdifferential determination property for essentially smooth locally Lipschitz functions on general Banach spaces.

Regarding the other concept of *arcwise essentially smooth functions*, Borwein and Moors in [**137**, p. 326] (see also [**136**]) defined a locally Lipschitz function $f : U \to \mathbb{R}$ on an open set U in $\mathbb{R}^n$ to be *arcwise essentially smooth* if for any locally Lipschitz mapping $x :]0,1[\to U$ (with $x(t) = (x_1(t), \cdots, x_n(t))$, whose each component x_i is essentially smooth, the set

$$\{t \in]0,1[: f^\circ(x(t); -x'(t)) \neq -f^\circ(x(t); x'(t))\}$$

is Lebesgue null in $\mathbb{R}$. The definition of arcwise essentially smooth functions appear in the line of that of Valadier's sound functions. In addition to the fact that locally Lipschitz continuous functions which are essentially smooth are shown to be subdifferentially determined in [**135, 137**], several other basic results on those functions can be found in those papers. The article [**108**] by J. Bolte and E. Pauwels also contains many further results principally related to sound functions (called there *path differentiable functions*) with applications to the study of behavior of diverse optimization algorithms. Other types of essential smoothness for locally Lipschitz continuous functions along with their subdifferential determination can be found in J. F. Edmond and L. Thibault's paper [**375**]. It was also shown by L. Thibault and N. Zlateva [**930**] that *tangentially regular directionally Lipschitz* functions are subdifferentially determined.

The concept of essential directional smoothness was introduced by L. Thibault and D. Zagrodny [**928**], mainly in order to generalize to specific lower semicontinuous functions the aforementioned subdifferential determination results established in [**124, 135, 137**] for Lipschitz finite-valued essentially smooth functions.

The specific function f in Lemma 9.1 and its continuity appeared in the 1915 paper [**327**, p. 115-118] by A. Denjoy while the function f in Lemma 9.2 and the function $f + \zeta$ along with the properties of their derivates were given by A. Denjoy in pages 167-170 of his same paper [**327**]. Denjoy's paper [**327**] also contains diverse fundamental results on non-differentiable real-valued functions of one real variable; those results are used in a series of Denjoy's papers [**326, 328, 329, 330, 331**]. Concerning subdifferentially and directionally stable functions, all the results in Section 9.2 as well as their proofs are taken from the paper [**927**] by L. Thibault and D. Zagrodny. Regarding essentially directionally smooth functions, Section 9.3 reproduced the results and proofs in Thibault and Zagrodny's paper [**928**].

In [**589**] one can find subdifferential determination/integration results under the nonvacuity property

$$\partial f(q) \cap \partial g(q) \neq \emptyset \quad \text{for all } q \in Q,$$

where Q is a dense set in the open set U. More generally, let reals $\mu > 0$ and $\gamma \geq 0$, and let be given a ∂, μ-essentially smooth function $f : U \to \mathbb{R} \cup \{+\infty\}$ on an open convex set U of a Banach space X and a continuous convex function $g : U \to \mathbb{R}$. If for some dense set Q in U

$$\big(\partial f(q) + \gamma \mathbb{B}_{X^*}\big) \cap \partial g(q) \neq \emptyset \quad \text{for all } q \in Q,$$

then [**589**, Theorem 4.1] shows that f is finite on U and for all $u, v \in U$

$$f(v) - f(u) - (\gamma + \mu)\|v - u\| \leq g(v) - g(u) \leq f(v) - f(u) + (\gamma + \mu)\|v - u\|.$$

We also refer to Section 11.4 for the subdifferential determination of primal lower regular functions and to Section 11.6 for other comments.

CHAPTER 10

Semiconvex functions

10.1. Semiconvex functions

In Unilateral Variational Analyis the class of extended real-valued semiconvex functions plays an important role and as we will see, such functions enjoy many properties similar to those of convex functions.

10.1.1. Semiconvexity, moduli of semiconvexity.

Modulus functions $\omega(\cdot)$ have been presented in Definition 8.15. They arise in particular in the definition of semiconvex functions. In Proposition 8.16 the semiconvexity property has been already used. Here we formalize it in a general definition.

DEFINITION 10.1. Let U be a nonempty convex set of a normed space $(X, \|\cdot\|)$ and let $\omega : [0, +\infty[\to [0, +\infty]$ be a nondecreasing function with $\omega(0) = 0$ which is continuous at 0, or equivalently $\lim_{r \downarrow 0} \omega(r) = 0$. A function $f : U \to \mathbb{R} \cup \{-\infty, +\infty\}$ is said to be *$\omega(\cdot)$-semiconvex on U* whenever for all $t \in]0, 1[$, $x, y \in U$, and all reals $\alpha > f(x)$, $\beta > f(y)$ one has

$$f(tx + (1-t)y) \le t\alpha + (1-t)\beta + t(1-t)\|x-y\|\omega(\|x-y\|).$$

The function $\omega(\cdot)$ is called *a modulus of semiconvexity* of f on U. When, for some real constant $\sigma \ge 0$, the modulus $\omega(\cdot)$ is of the form $\omega(r) = \frac{1}{2}\sigma r$ for every real $r \ge 0$, one says $\omega(\cdot)$ is *a linear modulus of semiconvexity*, and f is *linearly semiconvex on U with coefficient σ*.

When there exist reals $\sigma \ge 0$ and $\gamma > 0$ such that $\omega(r) = \frac{1}{1+\gamma}\sigma r^{\gamma}$ for every real $r \ge 0$, one says that f is *semiconvex of power type on U with power $1+\gamma$ and coefficient σ*; so semiconvex functions of power type 2 (and coefficient σ) coincide with linearly semiconvex functions (with coefficient σ).

If the opposite function $-f$ is $\omega(\cdot)$-semiconvex, the function f is called *$\omega(\cdot)$-semiconcave*. Of course, whenever f is $\omega(\cdot)$-semiconvex (resp. $\omega(\cdot)$-semiconcave) relative to a norm $\|\cdot\|$, then it is $\beta^{-1}\omega(\beta^{-1}\cdot)$-semiconvex (resp. $\beta^{-1}\omega(\beta^{-1}\cdot)$-semiconcave) relative to any norm $|\!|\!|\cdot|\!|\!| \ge \beta\|\cdot\|$ with $\beta > 0$.

Functions which are $\omega(\cdot)$-semiconvex (resp. $\omega(\cdot)$-semiconcave) on U for some $\omega(\cdot)$ (which is nondecreasing with $\lim_{r \downarrow 0} \omega(r) = 0$) will be called semiconvex (resp. semiconcave) functions on U. So, the semiconvexity of a function does not depend on the (equivalent) norm of the space.

Clearly, f is $\omega(\cdot)$-semiconvex (resp. $\omega(\cdot)$-semiconcave) on the open convex set U if and only if its restriction to any open interval $I_{u,h} := U \cap (u + \mathbb{R}h)$ has the same property for any $u \in U$ and any nonzero $h \in X$, or in other words, $t \mapsto f(u + th)$ is $\omega(\cdot)$-semiconvex (resp. $\omega(\cdot)$-semiconcave) on $I_{u,h}$.

Below whenever the $\omega(\cdot)$-semiconvexity or $\omega(\cdot)$-semiconcavity is concerned, the modulus function $\omega(\cdot) : [0,+\infty[\to [0,+\infty]$ will be assumed (unless otherwise stated) to be a *nondecreasing function which is continuous at* 0 *with* $\omega(0)=0$.

Obviously, when the values of f on U are in $\mathbb{R}\cup\{+\infty\}$ (resp. in $\mathbb{R}\cup\{-\infty\}$), then the $\omega(\cdot)$-semiconvexity (resp. $\omega(\cdot)$-semiconcavity) of f on U amounts to

$$f(tx+(1-t)y) \le tf(x)+(1-t)f(y)+t(1-t)\|x-y\|\omega(\|x-y\|)$$
$$\big(\text{resp. } f(tx+(1-t)y) \ge tf(x)+(1-t)f(y)-t(1-t)\|x-y\|\omega(\|x-y\|)\big)$$

for all $x,y\in U$ and $t\in]0,1[$.

It is worth emphasizing that the above definition obviously entails that $\{x\in U : f(x)<+\infty\}$ is a convex subset of X whenever f is $\omega(\cdot)$-semiconvex on U and $\omega(\cdot)$ is finite on $[0,\operatorname{diam} U[$ (resp. on $[0,\operatorname{diam} U]$) if $\operatorname{diam} U$ is not attained (resp. if $\operatorname{diam} U$ is attained). If $f : U\to\mathbb{R}\cup\{-\infty,+\infty\}$ is semiconvex (resp. semiconcave) on a neighborhood of each point of an open set U, one says that f is *locally semiconvex (resp. locally semiconcave) on* the open set U.

It is also worth pointing out that the indicator function Ψ_S of a set S in X is $\omega(\cdot)$-semiconvex on X with $\omega(\cdot)$ finite on $[0,\operatorname{diam} S]$ if and only if the set S is convex (as easily seen from the definition).

REMARK 10.2. For a modulus function $\omega(\cdot)$ as in Definition 10.1, defining δ as the supremum of all reals $t>0$ such that $\omega(t)$ is finite, it is easily seen that $\omega(\cdot)$ is finite on $[0,\delta[$ and takes on merely the value $+\infty$ on $]\delta,+\infty[$. □

If $\omega : [0,+\infty[\to[0,+\infty]$ is a nondecreasing modulus function with the property $\lim_{r\downarrow 0}\omega(r)=0$, then, for $\theta(0):=0$ and $\theta(r):=\limsup_{t\to r}\omega(t)$ if $r>0$, the function $\theta(\cdot)$ is nondecreasing and upper semicontinuous on $[0,+\infty[$ with $\omega(\cdot)\le\theta(\cdot)$; so whenever a function is semiconvex with respect to a modulus, it is also semiconvex with respect to an upper semicontinuous modulus. The following lemma provides a more precise result.

LEMMA 10.3. Let $\omega : [0,+\infty[\to[0,+\infty]$ be a nondecreasing function continuous at 0 with $\omega(0)=0$. The following hold:
(a) If $\omega(\cdot)$ takes on merely finite values, then there exists a nondecreasing finite continuous function $\theta : [0,+\infty[\to[0,+\infty[$ such that

$$\omega(r)\le\theta(r)\le\omega(2r)\quad\text{for all } r\ge 0;$$

the function θ can even be chosen to be locally absolutely continuous on $]0,+\infty[$.
(b) If $\omega(\cdot)$ takes on the value $+\infty$ somewhere, then, for each real $\delta>0$ with $\omega(\delta)<+\infty$, there exists a nondecreasing upper semicontinuous function $\theta : [0,+\infty[\to[0,+\infty]$ which is finite and continuous on $[0,\delta[$ with $\theta(0)=0$ and such that $\omega(r)\le\theta(r)$ for all $r\ge 0$, and $\theta(r)\le\omega(2r)$ for all $r\ne\delta$ in $[0,\delta[\cup]\delta_0,+\infty[$, where $\delta_0:=\sup\{t>0 : \omega(t)<+\infty\}$.

PROOF. (a) The function ω being finite and monotone, $I(t):=\int_t^{2t}\omega(s)\,ds$ is well defined, for every $t\ge 0$, and I is locally absolutely continuous on $[0,+\infty[$. Setting $\theta(0):=0$ and $\theta(t):=I(t)/t$ for $t>0$, we see that $\omega(t)\le\theta(t)\le\omega(2t)$, hence in particular θ is continuous at 0. The function θ is also locally absolutely continuous on $]0,+\infty[$ and for almost every $t\in]0,+\infty[$ the derivate $\theta'(t)$ satisfies

$$\begin{aligned}\theta'(t) &= t^{-1}[2\omega(2t)-\omega(t)]-t^{-2}I(t)=t^{-1}[2\omega(2t)-\omega(t)]-t^{-1}\theta(t)\\ &\ge t^{-1}[2\omega(2t)-\omega(t)]-t^{-1}\omega(2t)=t^{-1}[\omega(2t)-\omega(t)]\ge 0.\end{aligned}$$

This guarantees that θ is nondecreasing on $[0,+\infty[$.
(b) From the assumption $\lim_{r\downarrow 0}\omega(r)=0$ the set of $\delta>0$ with $\omega(\delta)<+\infty$ is nonempty. Fix such a real $\delta>0$. Putting $\omega_0(r)=\omega(r)$ if $r\in[0,\delta]$ and $\omega_0(r)=\omega(\delta)$ if $r>\delta$, the function $\omega_0(\cdot)$ fulfills the properties in (a), so there exists a nondecreasing continuous function $\theta_0:[0,+\infty[\to[0,+\infty[$ with $\theta_0(0)=0$ and such that $\omega_0(r)\le\theta_0(r)\le\omega_0(2r)$ for all $r\ge 0$. Setting, $\theta_1(r)=\theta_0(r)$ if $r\in[0,\delta[$ and $\theta_1(r)=+\infty$ if $r\ge\delta$, the function θ_1 fulfills the properties in the assertion (b). □

As seen in the following definition, in the presence of semiconvexity the least modulus of semiconvexity exists.

DEFINITION 10.4. Given a nonempty open convex set U of a normed space $(X,\|\cdot\|)$ and an extended real-valued proper function $f:U\to\mathbb{R}\cup\{+\infty\}$, one defines the function $\omega_f:[0,+\infty[\to[0,+\infty]$ by $\omega_f(0):=0$ and, for $r>0$,

$$\omega_f(r):=\sup\Big\{\Big(\frac{f(tx+(1-t)y)-tf(x)-(1-t)f(y)}{t(1-t)\|x-y\|}\Big)^+ :\\ x,y\in U\cap\operatorname{dom}f,\ 0<\|x-y\|\le r,\ t\in]0,1[\Big\},$$

where by convention the above supremum is zero whenever the set over which it is taken is empty. The function ω_f is clearly nondecreasing and f is semiconvex on U if and only if ω_f is continuous at 0, that is, $\lim_{t\downarrow 0}\omega_f(t)=0$; in this case ω_f is the least modulus of semiconvexity of f on U. The function ω_f is then called the *least (or minimal) modulus of semiconvexity of the function* f on U.

Another useful expression of ω_f is the following:

PROPOSITION 10.5. Let U be a nonempty open convex set of a normed space X and $f:U\to\mathbb{R}\cup\{+\infty\}$ be a proper function. Then for every real $r>0$ one has

$$\omega_f(r):=\sup\Big\{\Big(\frac{f(x)-f(x-\sigma h)}{\sigma}-\frac{f(x+\tau h)-f(x)}{\tau}\Big)^+ :\\ \|h\|=1, x\in U,\sigma,\tau>0, x-\sigma h, x+\tau h\in\operatorname{dom}f,\sigma+\tau\le r\Big\},$$

again with the convention that the above supremum is zero whenever the set over which it is taken is empty.

PROOF. For $x\in U$, $h\in\mathbb{S}_X$, $u:=x+\tau h\in\operatorname{dom}f$, $v:=x-\sigma h\in\operatorname{dom}f$ and $t:=\sigma/(\sigma+\tau)$ with $\sigma,\tau>0$, on the one hand we have $1-t=\tau/(\sigma+\tau)$, $x=tu+(1-t)v$, $\|u-v\|=\sigma+\tau$, and on the other hand it is easily checked that

$$\frac{f(x)-f(x-\sigma h)}{\sigma}-\frac{f(x+\tau h)-f(x)}{\tau}=\frac{f(tu+(1-t)v)-tf(u)-(1-t)f(v)}{t(1-t)\|u-v\|}.$$

From this we easily see by symmetry that the two sets over which the suprema are taken in the proposition and Definition 10.4 coincide. □

It is clear by Proposition 8.12 that any semiconvex function on an open convex set U of a normed space is subsmooth at any point of U where f is finite. In fact, the least modulus of semiconvexity was the main tool to prove the following in Proposition 8.16: Given a nonempty open convex set U of a normed space X, a proper function $f:U\to\mathbb{R}\cup\{+\infty\}$ is uniformly subsmooth on U if and only if it is $\omega(\cdot)$-semiconvex on U for some modulus $\omega(\cdot)$.

The behavior of semiconvex functions taking both the values $-\infty$ and $+\infty$ is quite similar to that of convex functions taking both those values. Before stating some properties translating the behavior of such functions, recall (see Definition 2.55) that the core of a subset S of the vector space X is defined by

$$\operatorname{core} S := \{x \in S : \forall y \in X, \exists r > 0, \forall t \in [-r, r],\ x + ty \in S\}.$$

PROPOSITION 10.6. Let U be a convex set of a normed space X and $f : U \to \mathbb{R} \cup \{-\infty, +\infty\}$ be an $\omega(\cdot)$-semiconvex function. Assume that $\omega(\cdot)$ takes on finite values on $[0, \operatorname{diam} U[$ if $\operatorname{diam} U$ is not attained (resp. on $[0, \operatorname{diam} U]$ if $\operatorname{diam} U$ is attained). The following properties hold:
(a) The set $\{u \in U : f(u) < +\infty\}$ is convex.
(b) If $u, v \in U$ are such that $f(u) = -\infty$ and $f(v) < +\infty$, then $f(tu+(1-t)v) = -\infty$ for all $t \in]0, 1]$.
(c) If f is lower semicontinuous on U and takes on the value $-\infty$ at some point $u \in U$, then f is finite at no point in U.
(d) If $f(u) = -\infty$ for some $u \in U$, then

$$f(x) = -\infty \quad \text{for all } x \in \operatorname{core}\{u' \in U : f(u') < +\infty\}.$$

PROOF. The assertion (a) has been already observed and it follows directly from the definition. Concerning (b), let $u \in U$ with $f(u) = -\infty$ and let $v \in U$ with $f(v) < +\infty$. Fix $t \in]0, 1[$ and some real $\beta > f(v)$. For all $\alpha \in \mathbb{R}$ we have $f(u) < \alpha$, hence

$$f(tu + (1-t)v) \le t\alpha + (1-t)\beta + t(1-t)\|u - v\|\omega(\|u - v\|),$$

which implies the desired equality $f(tu + (1-t)v) = -\infty$ of (b).

Now suppose that f is lower semicontinuous on U and $f(u) = -\infty$ for some $u \in U$. If there exists some $v \in U$ where f is finite, then we would have $f(tu + (1-t)v) = -\infty$ for all $t \in]0, 1[$ by (b), hence $f(v) \le \liminf_{t \downarrow 0} f(tu + (1-t)v) = -\infty$, which would contradict the finiteness of $f(v)$. This justifies the assertion (c).

To prove (d), fix $u \in U$ with $f(u) = -\infty$ and take any $x \in \operatorname{core}\{u' \in U : f(u') < +\infty\}$. We may suppose that $x \neq u$ since otherwise we obviously have $f(x) = -\infty$. By the definition of the core of a set, there is a real $r > 0$ such that $]x - r(u - x), x + r(u - x)[$ is included in $\{u' \in U : f(u') < +\infty\}$. This yields that there exists some v with $f(v) < +\infty$ such that $x \in]u, v[$, so (b) guarantees that $f(x) = -\infty$. □

Taking into account the above proposition, we will focus the study on semiconvex (resp. semiconcave) functions which do not take on values $-\infty$ (resp. $+\infty$).

10.1.2. Semiconvexity of diverse types of functions. As we will see, the least modulus of continuity of the derivative of a differentiable function can be used to study the semiconvexity of such a function. Let us recall that modulus.

DEFINITION 10.7. Given a mapping $F : U \to V$ between two metric spaces (U, d_U) and (V, d_V), we recall that the *(least) modulus of continuity of F relative to a nonempty subset S of U* is the function $\omega^{\mathfrak{c}}_{F,S}(\cdot)$ defined on $[0, +\infty[$ by

$$\omega^{\mathfrak{c}}_{F,S}(r) := \sup\{d_V(F(x), F(x')) : x, x' \in S, d_U(x, x') \le r\} \text{ for all } r \in [0, +\infty[.$$

When $S = U$, we will just write $\omega^{\mathfrak{c}}_F(\cdot)$ and say the *(least) modulus of continuity of F on U*.

Obviously, $\omega_F^c(\cdot)$ is nondecreasing on $[0,+\infty[$. Further, for all $x, x' \in U$

$$d_V(F(x), F(x')) \le \omega_F^c\big(d_U(x,x')\big) \quad \text{with } \omega_F^c(0) = 0,$$

so F is uniformly continuous on U if and only if $\omega_F^c(\cdot)$ is continuous at 0.

The following classical results concerning the modulus of continuity will be helpful.

PROPOSITION 10.8. Let U be a nonempty convex set of a normed space $(X, \|\cdot\|)$ and $F : U \to V$ be a mapping from U into a metric space (V, d_V). Let ω_F^c be the modulus of continuity of F on U. The following assertions are equivalent:
(a) the mapping F is uniformly continuous on U;
(b) the modulus $\omega_F^c(\cdot)$ is continuous at 0;
(c) the modulus $\omega_F^c(\cdot)$ is finite-valued, subadditive and uniformly continuous on $[0,+\infty[$.

PROOF. We already observed above the equivalence between the assertions (a) and (b), and the assertion (c) clearly implies (b). It remains to show that (a) entails (c). Suppose that (a) is fulfilled. Fix any reals $r, s \ge 0$ and take arbitrary $x, x' \in U$ with $\|x - x'\| \le r + s$. Suppose without loss of generality that $r \le s$ and consider the element $x'' \in [x, x']$ with $\|x - x''\| = \min\{r, \|x - x'\|\}$. Then $\|x'' - x'\| = \|x - x'\| - \min\{r, \|x - x'\|\}$, so $\|x'' - x'\| \le s$. Note also that $x'' \in U$ according to the convexity of U. Consequently, we have

$$d_V(F(x), F(x')) \le d_V(F(x), F(x'')) + d_V(F(x''), F(x')) \le \omega_F^c(r) + \omega_F^c(s),$$

which yields $\omega_F^c(r+s) \le \omega_F^c(r) + \omega_F^c(s)$, that is, ω_F^c is subadditive on $[0,+\infty[$.

On the other hand, since $\lim_{t\downarrow 0} \omega_F^c(t) = 0$ (according to the implication (a) $\Rightarrow$ (b)), we can take $r > 0$ such that $\omega_F^c(\cdot)$ is finite on $[0, r]$. From this and the subadditivity of $\omega_F^c(\cdot)$ we see that $\omega_F^c(\cdot)$ is also finite on $[r, 2r]$ (write any $t \in [r, 2r]$ as $t = r + (t - r)$ and note that r and $t - r$ are in $[0, r]$). By induction we obtain that $\omega_F^c(\cdot)$ is finite on the whole interval $[0,+\infty[$. Further, for $r, s \ge 0$ with $r \le s$, the subadditivity of $\omega_F^c(\cdot)$ ensures that

$$|\omega_F^c(s) - \omega_F^c(r)| = \omega_F^c(s) - \omega_F^c(r) \le \omega_F^c(s - r) = \omega_F^c(|s - r|),$$

so for arbitrary $r, s \ge 0$ we have $|\omega_F^c(s) - \omega_F^c(r)| \le \omega_F^c(|s-r|)$, and the latter inequality guarantees that $\omega_F^c(\cdot)$ is uniformly continuous on $[0,+\infty[$ since $\lim_{t\downarrow 0} \omega_F^c(t) = 0$. All the properties in (c) are then justified and the proof is complete. □

As shown below, $\mathcal{C}^1$ functions whose derivatives are uniformly continuous are both semiconvex and semiconcave.

PROPOSITION 10.9. Let U be a nonempty open convex set of a normed space X and $f : U \to \mathbb{R}$ be a $\mathcal{C}^1$ function on U.
(a) If Df is uniformly continuous on U, then both functions f and $-f$ are $\omega(\cdot)$-semiconvex on U, where $\omega(\cdot)$ is the (least) modulus on U of continuity of the derivative mapping $Df : U \to X^*$ (hence $\omega(\cdot)$ is finite and uniformly continuous on $[0,+\infty[$).
(b) If Df is Lipschitz continuous on U with K as a Lipschitz constant, then f is linearly semiconvex and linearly semiconcave with coefficient $2K$ on U.

PROOF. (a) Let $\omega(\cdot)$ be the least modulus of continuity of the mapping Df on U, that is,

$$\omega(r) := \sup\{\|Df(u) - Df(v)\| : u, v \in U,\ \|u - v\| \le r\} \quad \text{for all } r \in [0,+\infty[.$$

Note that $\omega(\cdot)$ is nondecreasing, and (according to the assumption of uniform continuity of Df on U) continuous at 0 with $\omega(0) = 0$. Fix $x, y \in U$ and $s, t \in]0, 1[$ with $s + t = 1$, and put $z := sx + ty$. There exist $c_1 \in [x, z]$ and $c_2 \in [z, y]$ such that

$$\begin{aligned} f(z) - sf(x) - tf(y) &= s\big(f(z) - f(x)\big) + t\big(f(z) - f(y)\big) \\ &= s\langle Df(c_1), z - x\rangle + t\langle Df(c_2), z - y\rangle \\ &= st\langle Df(c_1) - Df(c_2), y - x\rangle \leq st\|y - x\|\omega(\|y - x\|), \end{aligned}$$

where the inequality is due to the nondecreasing property of $\omega(\cdot)$ and to the fact that $\|c_1 - c_2\| \leq \|y - x\|$. The function f is then semiconvex on U with modulus $\omega(\cdot)$. Changing f in $-f$, we obtain that f is $\omega(\cdot)$-semiconcave too. Further, the uniform continuity of $\omega(\cdot)$ on $[0, +\infty[$ is a consequence of Proposition 10.8.
(b) The assertion (b) follows directly from (a). □

Pointwise supremum of semiconvex functions with the same $\omega(\cdot)$ is semiconvex.

PROPOSITION 10.10. Let U be a nonempty convex set of the normed space X and $(f_i)_{i\in I}$ be a family of extended real-valued semiconvex (resp. semiconcave) functions from U into $\mathbb{R}\cup\{-\infty, +\infty\}$ with the same modulus $\omega(\cdot)$ of semiconvexity (resp. semiconcavity). Then the pointwise supremum (resp. infimum) function f, defined for all $x \in U$ by

$$f(x) := \sup_{i\in I} f_i(x) \quad (\text{resp. } f(x) := \inf_{i\in I} f_i(x)),$$

is $\omega(\cdot)$-semiconvex (resp. $\omega(\cdot)$-semiconcave on U).

PROOF. We prove the result in the context of semiconvexity. Fix $s, t \in]0, 1[$ with $s + t = 1$, $x, y \in U$, and reals $\alpha > f(x)$ and $\beta > f(y)$. For $z := sx + ty$ we have, for each $i \in I$, that $f_i(x) < \alpha$ and $f_i(y) < \beta$, yielding

$$f_i(z) \leq s\alpha + t\beta + st\|x - y\|\omega(\|x - y\|).$$

Taking the supremum over $i \in I$ gives

$$f(z) \leq s\alpha + t\beta + st\|x - y\|\omega(\|x - y\|).$$

This translates the $\omega(\cdot)$-semiconvexity of f on U. □

A first corollary is related to the infimal convolution.

COROLLARY 10.11. Let $f, g : X \to \mathbb{R} \cup \{-\infty, +\infty\}$ be two extended real-valued functions on a normed space X such that $f(y) + g(x - y)$ is well defined for all $x, y \in X$ (as, in particular, it is the case when either $g(X) \subset \mathbb{R}$ or $f(X), g(X) \subset \mathbb{R} \cup \{-\infty\}$). Assume that g is $\omega(\cdot)$-semiconcave. Then the infimal convolution function $f \Box g$ is $\omega(\cdot)$-semiconcave.

PROOF. Since $(f \Box g)(x) = \inf_{y\in X} \big(f(y) + g(x - y)\big)$, it is enough to observe that each function $g(\cdot - y)$ is $\omega(\cdot)$-semiconcave and to apply Proposition 10.10. □

The second corollary concerns families of continuously differentiable functions.

COROLLARY 10.12. Let $(f_i)_{i\in I}$ be a family of functions from a nonempty open convex set U of a normed space X into $\mathbb{R}$ and let

$$f(x) = \sup_{i\in I} f_i(x) \quad \left(\text{resp. } f(x) = \inf_{i\in I} f_i(x)\right) \quad \text{for all } x \in U.$$

(a) If the functions f_i are differentiable on U and the family $(Df_i)_{i\in I}$ is equi-uniformly continuous on U (in the sense that the derivatives Df_i have a common modulus of uniform continuity $\omega(\cdot)$), then the function f is semiconvex (resp. semiconcave) on the set U (with $\omega(\cdot)$ as a modulus of semiconvexity (resp. semiconcavity)).
(b) If the functions f_i are differentiable on U and the derivatives Df_i are equi-Lipschitz on U with K as a common Lipschitz constant, then f is linearly semiconvex and linearly semiconcave on U with a common coefficient $2K$.

PROOF. Let $\omega(\cdot)$ be a common modulus of uniform continuity (of the mappings $(Df_i)_{i\in I}$) which is nondecreasing in the case of (a) (resp. let $\omega(t) = Kt$ for all $t \geq 0$ in the case of (b)). The corollary then follows from the above proposition. □

Another corollary is related to the eigenvalues of symmetric real matrices. For a square symmetric real matrix A of order n, denoting by $(\lambda_1(A), \cdots, \lambda_n(A))$ the vector of its eigenvalues in ascending order, where eigenvalues are repeated according to their multiplicity. The sum of the smallest k eigenvalues of A with repetition according to multiplicity is $\sum_{j=1}^{k} \lambda_j(A)$. Similarly, the real $\sum_{j=n-k+1}^{n} \lambda_j(A)$ is the sum of the largest k eigenvalues of A with repetition according to multiplicity.

COROLLARY 10.13. Let U be an open convex set of a normed space X and let $A(x)$ be an $n \times n$ symmetric real matrix whose entries are differentiable functions on U. Assume that the derivatives of the entries of $A(x)$ are uniformly continuous (resp. Lipschitz continuous) on U. Then for each integer $k = 1, \cdots, n$ the sum function of the smallest k eigenvalues of $A(x)$, with repetition according to multiplicity, is semiconcave (resp. linearly semiconcave) on U.

Similarly, the sum of the largest k eigenvalues, with repetition according to multiplicity, is semiconvex (resp. linearly semiconvex) on U.

Further, in the case of uniform continuity of the derivatives of the entries of $A(x)$, the above sum functions admit a modulus function of semiconcavity or semiconvexity respectively on U which is finite and uniformly continuous on $[0, +\infty[$.

PROOF. If we denote by $\sigma_k(x)$ the sum of the smallest k eigenvalues of $A(x)$ with repetition according to multiplicity, we know that

$$\sigma_k(x) = \min\left\{ \sum_{i=1}^{k} \langle A(x)v_i, v_i\rangle : \|v_1\| = \cdots = \|v_k\| = 1,\ \langle v_i, v_j\rangle = 0 \text{ for } i \neq j \right\},$$

where $\langle\cdot,\cdot\rangle$ denotes here the Euclidean inner product of $\mathbb{R}^n$. So, the above corollary easily yields that the function σ_k is semiconcave (resp. linearly semiconcave) on U.

In fact, we know by Proposition 10.8 that the (least) modulus of continuity $\omega(\cdot)$ of $DA(\cdot)$ on U is finite and uniformly continuous on $[0, +\infty[$. From this and the above expression on σ_k we easily see through the above corollary again that $n\omega(\cdot)$ is a modulus of semiconcavity of σ_k.

Finally, concerning the semiconvexity in x of the sum $\tau_k(x)$ of the largest k eigenvalues, it is enough to observe that $\tau_k(x) + \sigma_{n-k}(x) = \mathrm{tr}\big(A(x)\big)$ for $1 \leq k \leq n-1$, where $\mathrm{tr}\big(A(x)\big)$ denotes the trace of $A(x)$. □

10.1.3. Sup-representation of linearly semiconvex functions. When the space is Hilbert, the semiconvexity with linear modulus (or equivalently the semiconvexity of power type with power 2) corresponds to the convexity up to a square, as established in the next theorem. It is also related to the pointwise supremum of some families of smooth functions. Before proving the theorem we need a lemma of the type of Proposition 3.77(a).

LEMMA 10.14. Let U be a nonempty open convex set U of a Banach space X and $\varphi : U \to \mathbb{R} \cup \{+\infty\}$ be a proper lower semicontinuous convex function. Then φ is the pointwise supremum on U of a collection of continuous affine functions; in fact

$$\varphi(x) = \sup_{(u,u^*)\in \operatorname{gph} \partial\varphi} \big(\langle u^*, x-u\rangle + \varphi(u)\big) \quad \text{for all } x \in U.$$

PROOF. Fix a pair (x_0, x_0^*) in $\operatorname{gph} \partial\varphi$ that we know to be nonempty (see Proposition 6.50).

Define the lower semicontinuous convex function $\phi : U \to \mathbb{R}\cup\{+\infty\}$ by $\phi(x) := \sup_{(u,u^*)\in \operatorname{gph} \partial\varphi} \big(\langle u^*, x-u\rangle + \varphi(u)\big)$ and note that $\phi(x) \le \varphi(x)$ for every $x \in U$ and $\phi(y) = \varphi(y)$ for every $y \in \operatorname{Dom} \partial\varphi$, so in particular $\phi(x_0) = \varphi(x_0)$ and ϕ is proper. Then, for any $y \in \operatorname{Dom} \partial\varphi$ and any $y^* \in \partial\varphi(y)$ we have by definition of ϕ that, for all $x \in U$

$$\langle y^*, x-y\rangle + \phi(y) = \langle y^*, x-y\rangle + \varphi(y) \le \phi(x),$$

so $y^* \in \partial\phi(y)$. It ensues that $\partial\varphi(y) \subset \partial\phi(y)$ for every $y \in \operatorname{Dom} \partial\varphi$, hence $\partial\varphi(x) \subset \partial\phi(x)$ for all $x \in U$. Theorem 3.204 guarantees that there is a real constant c such that $\phi(\cdot) = \varphi(\cdot) + c$, thus $\phi = \varphi$ since $\phi(x_0) = \varphi(x_0) \in \mathbb{R}$. □

THEOREM 10.15 ($\mathcal{C}^\infty$ sup-representation of linearly semiconvex functions on Hilbert space). Let U be a convex set of a Hilbert space H endowed with the norm $\|\cdot\|$ associated with its inner product and let f be a function from U into $\mathbb{R}\cup\{+\infty\}$ (resp. $\mathbb{R}\cup\{-\infty\}$).
(A) The function f is linearly semiconvex (resp. linearly semiconcave) on U with coefficient $\sigma \ge 0$ if and only if the function $f + \frac{\sigma}{2}\|\cdot\|^2$ is convex (resp. $f - \frac{\sigma}{2}\|\cdot\|^2$ is concave) on U.
(B) Suppose that U is in addition open and f is lower (resp. upper) semicontinuous on U and finite at some point of U. Then the following assertions are equivalent:
(a) the function f is linearly semiconvex (resp. linearly semiconcave) on U;
(b) the function f is the sum on U of a lower semicontinuous convex (resp. upper semicontinuous concave) function g_1 and a $\mathcal{C}^{1,1}$ function g_2 with Dg_2 Lipschitz on the whole set U;
(c) there exists a family $(f_t)_{t\in T}$ of $\mathcal{C}^\infty$-functions from U into $\mathbb{R}$ such that the derivatives $(Df_t)_{t\in T}$ are equi-Lipschitz on U and, for all $x \in U$,

$$f(x) = \sup_{t\in T} f_t(x) \quad \left(\text{resp. } f(x) = \inf_{t\in T} f_t(x)\right);$$

(d) there exists a family $(f_t)_{t\in T}$ of $\mathcal{C}^{1,1}$-functions from U into $\mathbb{R}$ such that the derivatives $(Df_t)_{t\in T}$ are equi-Lipschitz on U and, for all $x \in U$,

$$f(x) = \sup_{t\in T} f_t(x) \quad \left(\text{resp. } f(x) = \inf_{t\in T} f_t(x)\right).$$

PROOF. We consider only the semiconvex situation. Denote by $\langle\cdot,\cdot\rangle$ the inner product in H.
(A) For all $t \in]0,1[$, all $x, y \in U$, the following inequalities are equivalent:

$$f(tx+(1-t)y) + \frac{\sigma}{2}\|tx+(1-t)y\|^2 \le tf(x) + \frac{\sigma t}{2}\|x\|^2 + (1-t)f(y) + \frac{\sigma(1-t)}{2}\|y\|^2$$

$$f(tx+(1-t)y) \le tf(x)+\frac{\sigma t}{2}\|x\|^2+(1-t)f(y)+\frac{\sigma(1-t)}{2}\|y\|^2-\frac{\sigma}{2}\|tx+(1-t)y\|^2$$

$$f(tx+(1-t)y) \le tf(x)+(1-t)f(y)+\frac{\sigma t}{2}\|x\|^2+\frac{\sigma(1-t)}{2}\|y\|^2 \\ -\frac{\sigma t^2}{2}\|x\|^2-\sigma t(1-t)\langle x,y\rangle-\frac{\sigma(1-t)^2}{2}\|y\|^2$$

$$f(tx+(1-t)y) \le tf(x)+(1-t)f(y)+\frac{\sigma t(1-t)}{2}\|x\|^2+\frac{\sigma(1-t)t}{2}\|y\|^2-\sigma t(1-t)\langle x,y\rangle$$

and hence the convexity of $f+\frac{\sigma}{2}\|\cdot\|^2$ on U is equivalent to the inequality

$$f(tx+(1-t)y) \le tf(x)+(1-t)f(y)+\frac{1}{2}\sigma t(1-t)\|x-y\|^2,$$

for all $t \in]0,1[$ and $x, y \in U$.
(B) If f is linearly semiconvex, there is by (A) a real $\sigma \ge 0$ such that the lower semicontinuous function $g_1 := f+(\sigma/2)\|\cdot\|^2$ is convex on U. Considering the function $g_2 := -(\sigma/2)\|\cdot\|^2$ we see that (a) $\Rightarrow$ (b). On the other hand, suppose that $f = g_1+g_2$ with g_1, g_2 as in the statement of (b). According to Proposition 10.9(b) the Lipschitz property of Dg_2 assures that g_2 is linearly semiconvex on U, and from this and the convexity of g_1 it follows that $f = g_1+g_2$ is linearly semiconvex on U. So, the equivalence between (a) and (b) holds true.

Let us show the implication (a) $\Rightarrow$ (c). Suppose that f is linearly semiconvex. By (A) there exists a constant $\sigma \ge 0$ such that $g := f+(\sigma/2)\|\cdot\|^2$ is convex on U. As a proper lower semicontinuous convex function, the function g is, by the previous Lemma 10.14, the pointwise supremum of a family $(g_t)_{t\in T}$ of continuous affine functions; say, $g_t := \langle a_t, \cdot\rangle+\beta_t$ with $a_t \in H$ and $\beta_t \in \mathbb{R}$. Putting $f_t := g_t-(\sigma/2)\|\cdot\|^2$, the functions f_t are $\mathcal{C}^\infty$ and their derivatives Df_t are equi-Lipschitz on U (with σ as a Lipschitz constant) since $\nabla f_t(x) = a_t - \sigma x$. Since $f(x) = \sup_{t\in T} f_t(x)$ for all $x \in U$, the implication (a) $\Rightarrow$ (c) is established.

The implication (c) $\Rightarrow$ (d) being obvious, it remains to prove (d) $\Rightarrow$ (a). Suppose (d) and denote by K a common Lipschitz constant of the derivatives $(Df_t)_{t\in T}$. By Proposition 10.9(b) the functions f_t are $\omega(\cdot)$-semiconvex with $\omega(r) = Kr$, so Proposition 10.10 guarantees that f is $\omega(\cdot)$-semiconvex, that is, linearly semiconvex on U. The desired implication (d) $\Rightarrow$ (a) is established and the proof is finished. □

The characterization (b) in (B) does not hold for semiconvex functions of power type with power less than 2.

EXAMPLE 10.16. Fix any real $1<\gamma<2$ and choose any real $\alpha>1$ such that $\alpha(\gamma-1)<1$. Put $r_n := \sum_{k=n}^{\infty}(1/k^\alpha)$ and define the function g on $\mathbb{R}$ by $g(t)=0$ if either $t \le 0$ or $t \ge r_1$, and

$$g(t) = -(t-r_n)^\beta \quad \text{if } r_n \le t < r_{n-1},$$

where $\beta := \gamma-1$, and observe that g is integrable on $\mathbb{R}$. Consider the function f given by

$$f(x) = \int_0^x g(s)\,ds.$$

We prove that f is semiconvex on $\mathbb{R}$ with power γ but f cannot be written as the sum of a convex function and a $\mathcal{C}^1$ function.

Claim 1. $u^\beta + v^\beta \geq (u+v)^\beta$ for all $u, v \geq 0$.
Indeed, since $\beta - 1 = \gamma - 2 < 0$ we have for $0 < u \leq v$

$$\left(\frac{u}{u+v}\right)^\beta + \left(\frac{v}{u+v}\right)^\beta = \frac{u}{u+v}\left(\frac{u}{u+v}\right)^{\beta-1} + \frac{v}{u+v}\left(\frac{v}{u+v}\right)^{\beta-1} \geq \frac{u}{u+v} + \frac{v}{u+v} = 1,$$

which yields $u^\beta + v^\beta \geq (u+v)^\beta$.

Claim 2. $g(u) - g(v) \leq (v-u)^\beta$ for all $u \leq v$ in $\mathbb{R}$.
Suppose that $0 < u < v < r_1$. There are $q \geq p > 1$ with $r_q \leq u < r_{q-1}$ and $r_p \leq v < r_{p-1}$, so $g(u) = -(u - r_q)^\beta$ and $g(v) = -(v - r_p)^\beta$. If $u - r_q \leq v - r_p$, then by Claim 1

$$g(u) - g(v) = (v - r_p)^\beta - (u - r_q)^\beta \leq (v - u + (r_q - r_p))^\beta \leq (v-u)^\beta,$$

where the latter inequality holds because $r_q - r_p \leq 0$ and $\beta > 0$. Since the inequality between the second member and the fourth is obviously true when $v - r_p < u - r_q$, we have $g(u) - g(v) \leq (v-u)^\beta$ for all $0 < u < v < r_1$.

We also observe that the inequality in the claim is still obviously true if either $v \geq r_1$ or $v \leq 0$. It remains to look at the case $u \leq 0 < v < r_1$. In such a case choosing p such that $r_p \leq v < r_{p-1}$, we have $(v - r_p)^\beta \leq (v-u)^\beta$ since $u < r_p$. This means $g(u) - g(v) \leq (v-u)^\beta$.

Claim 3. f is semiconvex of power type on $\mathbb{R}$ with power γ.
Fix $x < y$ in $\mathbb{R}$ and write for $t \in]0,1[$ and $\Delta := f(tx + (1-t)y) - tf(x) - (1-t)f(y)$

$$\begin{aligned}\Delta &= t[f\big(x + (1-t)(y-x)\big) - f(x)] + (1-t)[f\big(y - t(y-x)\big) - f(y)] \\ &= t(1-t)(y-x)\int_0^1 [g\big(x + s(1-t)(y-x)\big) - g\big(y - st(y-x)\big)]\, ds.\end{aligned}$$

Since $x + s(1-t)(y-x) \leq y - st(y-x)$ for all $t \in [0,1]$, we deduce from Claim 2

$$\Delta \leq t(1-t)(y-x)\int_0^1 (1-s)^\beta (y-x)^\beta\, ds = \frac{t(1-t)}{\gamma}(y-x)^\gamma,$$

which translates that f is semiconvex of power type on $\mathbb{R}$ with power γ.

Claim 4. f cannot be written as the sum of a convex function and a $\mathcal{C}^1$ function.
Suppose that there are on $\mathbb{R}$ a convex function Φ and a $\mathcal{C}^1$ function Λ such that $f = \Phi + \Lambda$. Since the function f is derivable at each point of $U := \mathbb{R} \setminus (\{0\} \cup \{r_n : n \in \mathbb{N}\})$, so is Φ. Denote by φ the derivative of Φ on U and by λ the derivative of Λ on $\mathbb{R}$, hence $g = \varphi + \lambda$ on U. The function φ is nondecreasing on U and by continuity of λ at each r_n we have

$$\sum_{n=1}^{\infty}\big(g(r_n^+) - g(r_n^-)\big) = \sum_{n=1}^{\infty}\big(\varphi(r_n^+) - \varphi(r_n^-)\big).$$

On one hand by the nondecreasing property of φ we have $\varphi(r_{n+1}^+) \leq \varphi(r_n^-)$, and this guarantees that

$$\sum_{n=1}^{N}\big(\varphi(r_n^+) - \varphi(r_n^-)\big) \leq \sum_{n=1}^{N}\big(\varphi(r_n^+) - \varphi(r_{n+1}^+)\big) = \varphi(r_1^+) - \varphi(r_{N+1}^+) \leq \varphi(r_1^+) - \varphi(0^+),$$

hence $\sum\limits_{n=1}^{\infty}\big(\varphi(r_n^+) - \varphi(r_n^-)\big)$ is finite.

On the other hand, by definition of g

$$\sum_{n=1}^{\infty}\big(g(r_n^+)-g(r_n^-)\big)=\sum_{n=1}^{\infty}-g(r_n^-)=\sum_{n=1}^{\infty}1/n^{\alpha\beta}=+\infty,$$

which is in contradiction with what precedes. $\square$

From (A) of Theorem 10.15 we immediately obtain:

PROPOSITION 10.17. Let U be an open convex set of a Hilbert space H endowed with the norm $\|\cdot\|$ associated with its inner product and let f be a lower semicontinuous (resp. upper semicontinuous) function from U into $\mathbb{R}\cup\{+\infty\}$ (resp. $\mathbb{R}\cup\{-\infty\}$). The function f is linearly semiconvex (resp. linearly semiconcave) on U with coefficient $\sigma\geq 0$ if and only if for all $x,y\in U$

$$f\left(\frac{1}{2}(x+y)\right)\leq\frac{1}{2}f(x)+\frac{1}{2}f(y)+\frac{1}{8}\sigma\|x-y\|^2$$

$$\left(\text{resp. } f\left(\frac{1}{2}(x+y)\right)\geq\frac{1}{2}f(x)+\frac{1}{2}f(y)-\frac{1}{8}\sigma\|x-y\|^2\right).$$

In the finite-dimensional framework, as will be seen in the next theorem, the topological compactness of the set T in Theorem 10.15 can be required as well as the continuity with respect to both x and the parameter t. Before stating the proposition, let us define first the concept of lower $\mathcal{C}^k$ functions.

DEFINITION 10.18. Let U be an open set of a finite-dimensional normed space X and $f:U\to\mathbb{R}$ be a continuous function. For $k\in\mathbb{N}\cup\{+\infty\}$, one says that f is *lower $\mathcal{C}^k$ near a point $\overline{x}\in U$* provided there exist a compact set T, an open neighborhood $V\subset U$ of $\overline{x}$, and a function $g:T\times V\to\mathbb{R}$ such that:
(i) the functions $g(t,\cdot)$ are k-times differentiable on V;
(ii) the function $g(\cdot,\cdot)$ as well as all the derivatives in x of order not greater than k are continuous jointly in (t,x) on $T\times V$;
(iii) for all $x\in V$,

$$f(x)=\max_{t\in T}g(t,x).$$

If f is lower $\mathcal{C}^k$ near each point of U, one says that it is lower $\mathcal{C}^k$ on U.

Lower $\mathcal{C}^1$ functions have been utilized in Proposition 8.44(e) where it has been shown that given a locally Lipschitz function f on an open set U of a finite-dimensional normed space and a point $\overline{x}\in U$ one has the equivalences:

$$\begin{aligned} f \text{ lower } \mathcal{C}^1 \text{ near } \overline{x} &\Leftrightarrow f \text{ subsmooth near } \overline{x}\\ &\Leftrightarrow f \text{ uniformly subsmooth near } \overline{x}\\ &\Leftrightarrow f\ \omega(\cdot)-\text{semiconvex near } \overline{x} \text{ for some } \omega(\cdot).\end{aligned}$$

Here we characterize, in the Euclidean space setting, the linear semiconvexity of a function near a point as the lower $\mathcal{C}^\infty$ property of the function near that point.

THEOREM 10.19 (lower $\mathcal{C}^\infty$ characterization of linear semiconvexity of continuous function in finite dimensions). Let U be an open set of a finite-dimensional Euclidean vector space $(X,\|\cdot\|)$ and $f:U\to\mathbb{R}$ be a continuous function. For $\overline{x}\in U$, the following properties are equivalent:
(a) the function f is linearly semiconvex near $\overline{x}$;

(b) there exist a real constant $c \geq 0$, a compact topological space T and continuous mappings $b : T \to X$ and $d : T \to \mathbb{R}$, and a neighborhood $V \subset U$ of $\overline{x}$ such that

$$f(x) = \max_{t \in T} \big(-c\|x\|^2 + \langle b(t), x\rangle + d(t)\big) \quad \text{for all } x \in V;$$

(c) the function f is lower $\mathcal{C}^\infty$ near $\overline{x}$;
(d) the function f is lower $\mathcal{C}^2$ near $\overline{x}$.

PROOF. Suppose that f is linearly semiconvex near $\overline{x}$ and denote by $\langle \cdot, \cdot \rangle$ the inner product from which the Euclidean norm $\|\cdot\|$ is defined. By Theorem 10.15(A) there exist an open ball $W := B(\overline{x}, 2\delta)$ with $W \subset U$ and a constant $c \geq 0$ such that the continuous function $\varphi := f + c\|\cdot\|^2$ is convex on W, so (shrinking W if necessary) we can suppose that φ is Lipschitz with some Lipschitz constant γ on W. Consequently, putting $W_0 := B[\overline{x}, \delta]$, the set $T := \{(v, v^*) \in W_0 \times X : v^* \in \partial\varphi(v)\}$ is compact as a closed subset of $W_0 \times \gamma\mathbb{B}_X$ since $\partial\varphi(W_0) \subset \gamma\mathbb{B}_X$. Further, for all $x \in V := B(\overline{x}, \delta)$, we have $\varphi(x) = \max_{(v,v^*)\in T} \big(\langle v^*, x - v\rangle + \varphi(v)\big)$, hence

$$f(x) = \max_{(v,v^*)\in T} \big(-c\|x\|^2 + \langle b(v, v^*), x\rangle + d(v, v*)\big) \quad \text{for all } x \in V,$$

where $b(v, v*) := v^*$ and $d(v, v^*) := f(v) + c\|v\|^2 - \langle v^*, v\rangle$. This justifies the implication (a) $\Rightarrow$ (b).

On the other hand, the implications (b) $\Rightarrow$ (c) and (c) $\Rightarrow$ (d) are evident. Now suppose (d) and consider V, T and g as given by Definition 10.18. Choose an open bounded convex neighborhood W of $\overline{x}$ such that $W_0 := \operatorname{cl} W \subset V$, so W_0 is compact. Put $\gamma := \max_{(t,x)\in T\times W_0} \|D_x^2 g(t, x)\|$. Then, for $g_t := g(t, \cdot)$, all the derivatives $(Dg_t)_{t\in T}$ are equi-Lipschitz on W (with γ as a common Lipschitz constant), hence by Proposition 10.9(b) all the functions g_t are $\omega(\cdot)$-semiconvex on W with $\omega(r) = \gamma r$. From this and Proposition 10.10 we conclude that f is $\omega(\cdot)$-semiconvex on W, hence linearly semiconvex on W, which justifies that the assertion (c) entails (a). □

10.1.4. Composite stability for semiconvexity and distance function. The connexion with lower $\mathcal{C}^1$ functions will be investigated later after the description of subdifferentials of $\omega(\cdot)$-semiconvex functions by means of the modulus function $\omega(\cdot)$. Let us examine composite stability properties.

PROPOSITION 10.20. Let X be a normed space. The following hold.
(a) Any convex combination of $\omega(\cdot)$-semiconvex (resp. $\omega(\cdot)$-semiconcave) functions from a convex set U of X into $\mathbb{R} \cup \{+\infty\}$ (resp. into $\mathbb{R} \cup \{-\infty\}$) is $\omega(\cdot)$-semiconvex (resp. $\omega(\cdot)$-semiconcave) on U.
(b) Let A be a continuous linear mapping between X and a normed space Y and let $f : V \to \mathbb{R} \cup \{+\infty\}$ (resp. $f : V \to \mathbb{R} \cup \{-\infty\}$) be an $\omega(\cdot)$-semiconvex (resp. $\omega(\cdot)$-semiconcave) function on a convex set V of Y. For any fixed $b \in Y$ and for $\omega_A(r) = \|A\|\omega(\|A\|r)$ for every real $r \geq 0$, the function $x \mapsto f(b + Ax)$ is $\omega_A(\cdot)$-semiconvex (resp. $\omega_A(\cdot)$-semiconcave) on the convex set $U := A^{-1}(-b + V)$.
(c) If in addition in (b) the modulus $\omega(\cdot)$ is concave, then the function $x \mapsto f(b+Ax)$ is $K\omega(\cdot)$-semiconvex for $K := \|A\| \max\{1, \|A\|\}$.

PROOF. The assertion (a) is evident. Concerning (b), it suffices to prove the result relative to semiconvex functions. Fix $b \in Y$ and put $g(x) := f(b + Ax)$ for all $x \in U$. Note that the function $\omega_A(\cdot)$ in the statement is nondecreasing and continuous at 0 with $\omega_A(0) = 0$. Take now $x, x' \in U$, $s, t \in]0, 1[$ with $s + t = 1$, and write

$$\begin{aligned} g(sx + tx') &= f(sb + sAx + tb + tAx') \\ &\le sf(b + Ax) + tf(b + Ax') + st\|A(x - x')\|\omega(\|A(x - x')\|) \\ &\le sg(x) + tg(x') + st\|A\|\,\|x - x'\|\omega(\|A\|\|x - x'\|), \end{aligned}$$

where the last inequality is due to the nondecreasing property of $\omega(\cdot)$. So,

$$g(sx + tx') \le sg(x) + tg(x') + st\|x - x'\|\omega_A(\|x - x'\|),$$

which justifies the $\omega_A(\cdot)$-semiconvexity of the function g on U.

To prove (c), assume the (modulus) function $\omega(\cdot)$ is concave. By (b) the above function g is semiconvex with respect to the modulus $r \mapsto \|A\|\omega(\|A\|r)$. On the other hand, putting $C := \max\{1, \|A\|\}$, by the nonincreasing property of $r \mapsto r^{-1}\big(\omega(r) - \omega(0)\big)$ (due to the concavity of $\omega(\cdot)$), for all $r > 0$, we have $(Cr)^{-1}\omega(Cr) \le r^{-1}\omega(r)$, or equivalently $\omega(Cr) \le C\omega(r)$. Thus for all $r > 0$, we obtain through the nondecreasing property of $\omega(\cdot)$

$$\omega(\|A\|r) \le \omega(Cr) \le C\omega(r),$$

so the function g is $C\|A\|\omega(\cdot)$-semiconvex. □

In addition to (b) in Proposition 10.20 above, similar stability results also hold for other superpositions as below.

PROPOSITION 10.21. Let U and V be nonempty convex sets of two normed spaces X and Y respectively, let $f : V \to \mathbb{R} \cup \{+\infty\}$ (resp. $f : V \to \mathbb{R} \cup \{-\infty\}$) be a function which is semiconvex (resp. semiconcave) on V with modulus $\omega^f(\cdot)$. Let I be an interval with $+\infty$ as right endpoint and containing the effective range $f\big(f^{-1}(\mathbb{R})\big)$ of f. Let also $g : I \to \mathbb{R}$ be a nondecreasing function which is K_g-Lipschitz continuous on I.
(a) If g is convex (resp. concave), then $g \circ f$ is semiconvex (resp. semiconcave) on V with modulus $K_g\omega^f(\cdot)$, where $(g \circ f)(y) = +\infty$ by convention if $f(y) = +\infty$ (resp. $(g \circ f)(y) = -\infty$ if $f(y) = -\infty$).
(b) If g is semiconvex (resp. semiconcave) with modulus $\omega^g(\cdot)$ and if f is finite and K_f-Lipschitz continuous on V, then $g \circ f$ is semiconvex (resp. semiconcave) on V, with $K_f\omega^g(K_f\,\cdot) + K_g\omega^f(\cdot)$ as a modulus of semiconvexity (resp. semiconcavity).
(c) If f is finite and K_f-Lipschitz continuous on V, and if U is open and $F : U \to Y$ is a differentiable mapping with $F(U) \subset V$ and such that DF is uniformly continuous on U along with $\|DF(x)\| \le K_F$ for all $x \in U$, then $f \circ F$ is semiconvex (resp. semiconcave) on U with a modulus of semiconvexity (resp. semiconcavity) $\omega(\cdot)$ given by $\omega(r) = K_F\omega^f(K_F r) + K_f\mu_F(2r)$, where $\mu_F(\cdot)$ denotes the (least) modulus of continuity of the mapping DF on U.

PROOF. (a) Fix $y, y' \in V$ where f is finite, $s, t \in]0, 1[$ with $s + t = 1$, and put $z = sy + ty'$. Note that $f(z)$ is finite by the semiconvexity (resp. semiconcavity) of the function f.

Suppose first that f is semiconvex and g is convex. The nondecreasing property and the Lipschitz continuity of g guarantee that

$$\begin{aligned} g(f(z)) &\le g\big(sf(y) + tf(y') + st\|y - y'\|\omega^f(\|y - y'\|)\big) \\ &\le g\big(sf(y) + tf(y')\big) + K_g st\|y - y'\|\omega^f(\|y - y'\|). \end{aligned} \tag{10.1}$$

The convexity of g then gives

$$g(f(z)) \le sg(f(y)) + tg(f(y')) + K_g st\|y-y'\|\omega^f(\|y-y'\|),$$

which translates the desired semiconvexity of $g \circ f$, since $\operatorname{dom} g \circ f = \operatorname{dom} f$.

Suppose now that f is semiconcave and g is concave. Then

$$f(z) + st\|y-y'\|\omega^f(\|y-y'\|) \ge sf(y) + tf(y'),$$

which by the Lipschitz and nondecreasing properties of g gives

$$\begin{aligned} K_g st\|y-y'\|\omega^f(\|y-y'\|) + g(f(z)) &\ge g\big(f(z) + st\|y-y'\|\omega^f(\|y-y'\|)\big) \\ &\ge g\big(sf(y) + tf(y')\big). \end{aligned} \tag{10.2}$$

By concavity of g, we obtain

$$K_g st\|y-y'\|\omega^f(\|y-y'\|) + g(f(z)) \ge sg(f(y)) + tg(f(y')),$$

so $g \circ f$ is semiconcave on V with $K_g\omega^f(\cdot)$ as modulus of semiconcavity.

(b) Under the additional Lipschitz property of f, we deduce from (10.1) and from the semiconvexity of g

$$\begin{aligned} g(f(z)) &\le sg(f(y)) + tg(f(y')) + st\|f(y)-f(y')\|\omega^g(\|f(y)-f(y')\|) \\ &\quad + K_g st\|y-y'\|\omega^f(\|y-y'\|) \\ &\le s(g(f(y)) + tg(f(y')) + st\|y-y'\|\big(K_f\omega^g(K_f\|y-y'\|) + K_g\omega^f(\|y-y'\|)\big), \end{aligned}$$

where the latter inequality uses the nondecreasing property of ω^g.

If instead g is semiconcave, (10.2) ensures

$$\begin{aligned} &K_g st\|y-y'\|\omega^f(\|y-y'\|) + g(f(z)) \\ &\ge sg(f(y)) + tg(f(y')) - st\|f(y)-f(y')\|\omega^g(\|f(y)-f(y')\|) \\ &\ge sg(f(y)) + tg(f(y')) - K_f st\|y-y'\|\omega^g(K_f\|y-y'\|), \end{aligned}$$

hence $g \circ f$ is semiconcave on V with modulus of semiconcavity $\omega(\cdot)$ given by $\omega(r) = K_g\omega^f(r) + K_f\omega^g(K_f r)$ for all $r \in [0, +\infty[$.

(c) Consider now $x, x' \in U$, $s, t \in]0,1[$ with $s+t=1$, and put $z := sx+tx'$. Denote by $\mu_F(\cdot)$ the (least) modulus of continuity of DF on U and observe that

$$\begin{aligned} &\|F(z) - sF(x) - tF(x')\| = \|s\big(F(z)-F(x)\big) + t\big(F(z)-F(x')\big)\| \\ &= \left\| st\int_0^1 \big(DF(x+t\theta(x'-x)) - DF(x'+s\theta(x-x'))\big)(x'-x)\,d\theta \right\| \\ &\le st\|x-x'\|\mu_F(2\|x-x'\|). \end{aligned}$$

Using this and the Lipschitz and semiconvexity properties of f, we see that

$$\begin{aligned} &f(F(z)) - sf(F(x)) - tf(F(x')) \\ &\le f(sF(x)+tF(x')) - sf(F(x)) - tf(F(x')) + K_f\|F(z) - sF(x) - tF(x')\| \\ &\le st\|F(x)-F(x')\|\omega^f(\|F(x)-F(x')\|) + stK_f\|x-x'\|\mu_F(2\|x-x'\|), \end{aligned}$$

which entails

$$\begin{aligned} &f(F(z)) - sf(F(x)) - tf(F(x')) \\ &\le st\|x-x'\|\left(K_F\omega^f(K_F\|x-x'\|) + K_f\mu_F(2\|x-x'\|)\right). \end{aligned}$$

This says that $f \circ F$ is $\omega(\cdot)$-semiconvex for the appropriate $\omega(\cdot)$.

If f is semiconcave (instead of being semiconvex), the same reasoning shows that $f \circ F$ is semiconcave with the same above modulus $\omega(\cdot)$. □

Distance functions from subsets of Hilbert spaces have particular semiconcavity properties.

PROPOSITION 10.22. Let S be a nonempty subset of a Hilbert space H and let $\|\cdot\|$ be the norm associated with the inner product of H. The following hold.
(a) The square distance function d_S^2 is linearly semiconcave with coefficient 2.
(b) For any convex set U with $U \cap (S + B(0,\delta)) = \emptyset$ for some real $\delta > 0$, the distance function d_S is linearly semiconcave with coefficient $1/\delta$ on U. So d_S is locally linearly semiconcave on $H \setminus S$.
(c) If S is the union of a collection of closed balls with a common radius $r > 0$, then on each convex set U included in $\mathrm{cl}\,(H \setminus S)$, the distance function d_S is linearly semiconcave with coefficient $1/r$.
(d) For any $x \in S$ with $N^P(S;x) \neq \{0\}$ the distance function d_S is not semiconcave near the point x.

PROOF. (a) It is enough to observe that $d_S^2(x) = \|x\|^2 + \inf_{y\in S}\left(\|y\|^2 - 2\langle x, y\rangle\right)$ and that the function in x equal to the infimum above is concave, and to apply the characterization in Theorem 10.15(A) of linearly semiconcave functions in the Hilbert setting.
(b) Fix a convex set U such that $U \cap (S + B(0,\delta)) = \emptyset$. Then, for $f(\cdot) := d_S^2(\cdot)$, we have $\{f(x) : x \in U\} \subset [\delta^2, +\infty[$. Further, the function $g : [\delta^2, +\infty[\to \mathbb{R}$ with $g(r) = \sqrt{r}$ is nondecreasing, concave, and $\frac{1}{2\delta}$-Lipschitz continuous on $[\delta^2, +\infty[$ since for all $r_1, r_2 \in [\delta^2, +\infty[$

$$|g(r_1) - g(r_2)| = |\sqrt{r_1} - \sqrt{r_2}| = \frac{1}{\sqrt{r_1} + \sqrt{r_2}}|r_1 - r_2| \le \frac{1}{2\delta}|r_1 - r_2|.$$

Since the function f is linearly semiconcave on U with coefficient 2 according to the assertion (a) above, we obtain through the assertion (a) of Proposition 10.21 that $d_S = g \circ f$ is linearly semiconcave on U with coefficient $1/\delta$.
(c) Let $(a_i)_{i\in I}$ be a family of H such that $S = \bigcup_{i\in I} B[a_i, r]$ and let a convex set U included in $\mathrm{cl}\,(H \setminus S)$. Put $S_i := B[a_i, r]$ for each $i \in I$. Note also that, for each $i \in I$, we have $d_{\{a_i\}}^2(x) \ge r^2$ for all $x \in U$, hence by (b) the function $d_{\{a_i\}}(\cdot) = \|\cdot - a_i\|$ is linearly semiconcave on U with coefficient $1/r$. Since $d_{S_i}(x) = \|x - a_i\| - r$ for all $x \in U$, each function d_{S_i} is also linearly semiconcave on U with coefficient $1/r$. Observing that $d_S(x) = \inf_{i\in I} d_{S_i}(x)$ for all $x \in U$, Proposition 10.10 tells us that the function d_S is linearly semiconcave on U with the same coefficient $1/r$.
(d) Let $x \in S$ with $N^P(S;x) \neq \{0\}$. From the definition of the proximal normal cone there exists some $y \notin S$ such that $x = P_S(y)$, hence for $h := (y - x)/\|y - x\|$ we have $d_S(x + rh) = r$ for all $r \in [0, d_S(y)]$. We deduce, for all $r \in]0, d_S(y)]$,

$$d_S(x + rh) + d_S(x - rh) - 2d_S(x) \ge r = r\|h\|,$$

hence

$$\liminf_{r \downarrow 0} \frac{d_S(x + rh) + d_S(x - rh) - 2d_S(x)}{r\|h\|} \ge 1.$$

This ensures that d_S is not semiconcave near x. □

In addition to the above proposition, it is worth pointing out that we will see in Chapter 15 that a closed set S of a Hilbert space H (more generally, of some uniformly convex space X, see Chapter 18) is convex whenever the distance function d_S or the square distance function d_S^2 is $\omega(\cdot)$-semiconvex on the whole space. So, on such spaces the semiconvexity of d_S or d_S^2 on the whole space characterizes the convexity of the closed set S. We will also see that the semiconvexity of d_S^2 on open convex sets of an open enlargement $U_r(S) := \{x \in X : d_S(x) < r\}$ characterizes another important concept for the closed set S which will be developed in Chapter 15: *the prox-regularity*.

10.1.5. Lipschitz continuity of semiconvex functions. Any semiconvex function on an open convex set U being subsmooth, Proposition 8.18 ensures that it is Lipschitz near a point in U where it is finite if and only if it is bounded above near this point. We provide in the next proposition a boundedness condition for a global Lipschitz property of semiconvex functions on convex sets.

PROPOSITION 10.23. Let $f : X \to \mathbb{R} \cup \{+\infty\}$ be a function on a normed space $(X, \|\cdot\|)$.
(a) Let U be a convex set of X. Assume, for some real $r > 0$, that f is $\omega(\cdot)$-semiconvex on $U + r\mathbb{B}$ and bounded on $U + r\mathbb{B}$, and that $\{\|x - y\|\omega(\|x - y\|) : x, y \in U + r\mathbb{B}\}$ is bounded. Then f is Lipschitz on U.
(b) If f is $\omega(\cdot)$-semiconvex near a point $\overline{x} \in X$ and bounded from above near $\overline{x}$, then f is Lipschitz continuous near $\overline{x}$.

PROOF. The assertion (b) follows as said above from Proposition 8.18. Let us show (a). Let μ be a real upper bound of $|f|$ on $U + r\mathbb{B}$ such that $\|x - y\|\omega(\|x - y\|) \le \mu$ for all $x, y \in U + r\mathbb{B}$. Fix $x, y \in U$ with $x \neq y$. Put $z := y + r\frac{y-x}{\|y-x\|}$ and observe that $z \in U + r\mathbb{B}$. The $\omega(\cdot)$-semiconvexity of f guarantees that, for $t := \frac{\|y-x\|}{r+\|y-x\|}$

$$f(y) = f(tz + (1-t)x) \le tf(z) + (1-t)f(x) + t(1-t)\|z - x\|\omega(\|z - x\|),$$

hence (since $t(1-t) \le t \le \|y - x\|/r$)

$$f(y) - f(x) \le t\big(f(z) - f(x)\big) + \frac{\mu}{r}\|y - x\| \le \mu\left(\frac{2}{r} + \frac{1}{r}\right)\|y - x\| = \frac{3\mu}{r}\|y - x\|,$$

which translates the Lipschitz property of f on U. □

COROLLARY 10.24. Assume that the normed space X is finite-dimensional. Let $f : U \to \mathbb{R}$ be a function which is finite on the open convex set U of X and which is $\omega(\cdot)$-semiconvex (resp. $\omega(\cdot)$-semiconcave) on U. Then f is locally Lipschitz continuous on U.

PROOF. We establish the result in the case of semiconvexity of f. Without loss of generality, we may suppose that $X = \mathbb{R}^n$. Fix $\overline{x} \in U$ and choose a closed cube U_0 (with nonempty interior) centered at $\overline{x}$ and included in U and such that $\omega(\cdot)$ is finite on $[0, \operatorname{diam} U_0]$ according to the continuity of $\omega(\cdot)$ at 0 and the equality $\omega(0) = 0$. Denoting by δ the diameter of U_0, that is, $\delta := \operatorname{diam} U_0$, we see that for all $x, y \in U_0$ and $s, t \in [0, 1]$ with $s + t = 1$

$$f(sx + ty) \le sf(x) + tf(y) + st\|x - y\|\omega(\|x - y\|) \le sf(x) + tf(y) + \frac{1}{4}\delta\omega(\delta).$$

Let $x_1, \cdots, x_{2^n}$ be the vertices of U_0 and $\beta := \max\limits_{1 \le i \le 2^n} f(x_i)$. If x_i and x_j are two consecutive vertices of U_0 (that is, x_i and x_j are in a same one-dimensional face of

U_0), then $f(sx_i + tx_j) \le \beta + \frac{1}{4}\delta\omega(\delta)$, guaranteeing that f is bounded from above on the 1-dimensional faces of U_0. Repeating the procedure by taking any convex combination of two points lying on two different 1-dimensional faces, we obtain that f is bounded from above by $\beta + \frac{1}{4}\delta\omega(\delta) + \frac{1}{4}\delta\omega(\delta)$, say $\beta + \frac{1}{2}\delta\omega(\delta)$, on the 2-dimensional faces of U_0. The iteration up to the n-dimensional faces of U_0 yields that f is bounded from above on the whole cube U_0. By (b) in Proposition 10.23 above we conclude that f is Lipschitz continuous near $\overline{x}$. □

Now consider the particular case of one real variable semiconvex functions.

PROPOSITION 10.25. *Let I be an interval of $\mathbb{R}$ and $f : I \to \mathbb{R} \cup \{+\infty\}$ be an $\omega(\cdot)$-semiconvex function. Then for any $r, s \in I \cap \operatorname{dom} f$ with $r < s$ the function f is locally Lipschitz continuous on $]r, s[$, upper semicontinuous on the right at r, and upper semicontinuous on the left at s.*

PROOF. The local Lipschitz property on $]r, s[$ follows directly from the corollary above. By continuity of ω at 0 with $\omega(0) = 0$, we can fix $\tau > 0$ such that $\omega(\cdot)$ is bounded from above on $[0, \tau]$. Choosing $r' \in]r, s[$ such that $s - r' < \tau$, we have for any $t \in]r', s[$

$$f(t) \le \frac{t - r'}{s - r'} f(s) + \frac{s - t}{s - r'} f(r') + \frac{(t - r')(s - t)}{(s - r')^2}(s - r')\omega(s - r'),$$

hence $\limsup_{t \uparrow s} f(t) \le f(s)$ since $\omega(\cdot)$ is bounded from above on $[0, \tau]$, justifying the upper semicontinuity on the left of f at s.

Now choose $s' \in]r, s[$ such that $s' - r < \tau$. As above we have, for all $t \in]r, s'[$,

$$f(t) \le \frac{t - r}{s' - r} f(s) + \frac{s' - t}{s' - r} f(r) + \frac{(t - r)(s' - t)}{(s' - r)^2}(s' - r)\omega(s' - r),$$

hence $\limsup_{t \downarrow r} f(t) \le f(r)$, that is, f is upper semicontinuous on the right at r. □

The following proposition shows that semiconvexity is reduced to convexity whenever $\omega(r)/r \to 0$ as $r \downarrow 0$. So, the concept of semiconvexity of power type with power $1 + \gamma$ is useful only when $0 < \gamma \le 1$.

PROPOSITION 10.26. *Let $f : X \to \mathbb{R} \cup \{+\infty\}$ be an $\omega(\cdot)$-semiconvex function on a normed space X. If $\omega(r)/r \to 0$ as $r \downarrow 0$, then f is convex.*

PROOF. Fix $x, y \in \operatorname{dom} f$ with $x \ne y$. According to Proposition 10.20 we may suppose $X = \mathbb{R}$ and $x = 0$, $y = 1$. Fix any $t \in]0, 1[$. Consider any integer $n > 2$ and note that for each integer $1 < p < n$

$$f\left(\frac{p}{n}\right) \le \frac{1}{2}\left(f\left(\frac{p-1}{n}\right) + f\left(\frac{p+1}{n}\right)\right) + \frac{1}{2n}\omega\left(\frac{2}{n}\right),$$

or equivalently

$$n\left(f\left(\frac{p}{n}\right) - f\left(\frac{p-1}{n}\right)\right) \le n\left(f\left(\frac{p+1}{n}\right) - f\left(\frac{p}{n}\right)\right) + \frac{1}{2}\omega\left(\frac{2}{n}\right).$$

By induction, for any other integer $0 \le q < n - p$ we have

$$n\left(f\left(\frac{p}{n}\right) - f\left(\frac{p-1}{n}\right)\right) \le n\left(f\left(\frac{p+q}{n}\right) - f\left(\frac{p+q-1}{n}\right)\right) + \frac{1}{2}q\omega\left(\frac{2}{n}\right). \tag{10.3}$$

The function f being locally Lipschitz continuous on $]0,1[$ (see Corollary 10.24), consider two reals $r < s$ in $]0,1[$ which are Lebesgue points of f. Take two sequences of integers $(p_n)_n$ and $(q_n)_n$ with $p_n + q_n < n$ and $p_n > 1$, and such that

$$r \in \left]\frac{p_n}{n} - \frac{1}{n}, \frac{p_n}{n}\right] \text{ and } s \in \left]\frac{p_n + q_n}{n} - \frac{1}{n}, \frac{p_n + q_n}{n}\right].$$

Since $(q_n/n)_n$ is bounded in $\mathbb{R}$, the equality $q_n\omega(2/n) = (2q_n/n)(n/2)\omega(2/n)$ guarantees that $q_n\omega(2/n) \to 0$. Combining this with (10.3) and with the fact that r and s are Lebesgue points of f, we obtain $f'(r) \le f'(s)$, so f is convex on $]0,1[$. Since, by the previous proposition, f is upper semicontinuous on the right at 0 and on the left at 1, the semiconvexity of f is easily seen to be preserved on the whole interval $[0,1]$. This finishes the proof. □

10.2. Subdifferentials and derivatives of semiconvex functions

In this section we proceed to the study of properties of subdifferentials of semiconvex functions.

10.2.1. Directional derivatives and subdifferentials. Let us start with some facts concerning the differential quotient. The case of a function which is semiconvex on a convex set U of X with values in $\mathbb{R} \cup \{+\infty\}$ can be reduced, for directional derivatives and subdifferentials, to the case of a function which is semiconvex on the whole space X by putting the value $+\infty$ outside U. So, consider a function $f : X \to \mathbb{R} \cup \{+\infty\}$ which is $\omega(\cdot)$-semiconvex on X and consider $x \in \operatorname{dom} f$ and $h \in X$. Let $0 < s < t$ and let any $r > 0$. Since

$$x = s(r+s)^{-1}(x - rh) + r(r+s)^{-1}(x + sh),$$

we have

$$f(x) \le \frac{s}{r+s} f(x - rh) + \frac{r}{r+s} f(x + sh) + \frac{rs}{r+s}\|h\|\omega((r+s)\|h\|),$$

or equivalently the following first *slope inequality* holds:

(10.4)
$$-r^{-1}[f(x - rh) - f(x)] \le s^{-1}[f(x + sh) - f(x)] + \|h\|\omega((r+s)\|h\|) \quad \text{for } r, s > 0.$$

In the same way the equality

$$x + sh = \frac{s}{t}(x + th) + \left(1 - \frac{s}{t}\right)x$$

gives

$$f(x + sh) \le \frac{s}{t} f(x + th) + \left(1 - \frac{s}{t}\right) f(x) + s\left(1 - \frac{s}{t}\right)\|h\|\omega(t\|h\|),$$

or equivalently the following second *slope inequality* holds:

(10.5)
$$s^{-1}[f(x + sh) - f(x)] \le t^{-1}[f(x + th) - f(x)] + \left(1 - \frac{s}{t}\right)\|h\|\omega(t\|h\|) \text{ for } 0 < s < t.$$

PROPOSITION 10.27. Let X be a normed space and $f : X \to \mathbb{R} \cup \{+\infty\}$ (resp. $f : X \to \mathbb{R} \cup \{-\infty\}$) be a function which is semiconvex (resp. semiconcave) near a point x at which f is finite. Then the directional derivative

$$f'(x; h) := \lim_{t \downarrow 0} t^{-1}[f(x + th) - f(x)]$$

exists in $\mathbb{R}\cup\{-\infty,+\infty\}$ for any direction $h \in X$, and the function $f'(x;\cdot)$ is convex (resp. concave) and positively homogeneous.

If f is $\omega(\cdot)$-semiconvex on X, then for any $h \in X$ and for any real $t > 0$ and any real $\varepsilon > 0$

$$-t^{-1}[f(x-th)-f(x)]-\|h\|\omega((t+\varepsilon)\|h\|)\leq f'(x;h)\leq t^{-1}[f(x+th)-f(x)]+\|h\|\omega(t\|h\|).$$

PROOF. We prove the result in the case f is semiconvex near x. Without loss of generality, we may suppose f is $\omega(\cdot)$-semiconvex on the whole space X. Take any $h \in X$ and $0 < s < t$ and write according to (10.5)

$$s^{-1}[f(x+sh)-f(x)] \leq t^{-1}[f(x+th)-f(x)] + \left(1-\frac{s}{t}\right)\|h\|\omega(t\|h\|).$$

Fixing t and making $s \downarrow 0$, the latter inequality gives

$$\limsup_{s\downarrow 0} s^{-1}[f(x+sh)-f(x)] \leq t^{-1}[f(x+th)-f(x)] + \|h\|\omega(t\|h\|)$$

and this entails

$$\limsup_{s\downarrow 0} s^{-1}[f(x+sh)-f(x)] \leq \liminf_{t\downarrow 0} t^{-1}[f(x+th)-f(x)].$$

So, the desired limit exists in $\mathbb{R}\cup\{-\infty,+\infty\}$ and fixing $t > 0$ we see from the second inequality above that

$$f'(x;h) \leq t^{-1}[f(x+th)-f(x)] + \|h\|\omega(t\|h\|).$$

Further, taking any reals $t > 0$ and $\varepsilon > 0$ and using (10.4) and the nondecreasing property of $\omega(\cdot)$ we have for every $s \in]0,\varepsilon[$

$$-t^{-1}[f(x-th)-f(x)] \leq s^{-1}[f(x+sh)-f(x)] + \|h\|\omega((t+\varepsilon)\|h\|),$$

which gives as $s \downarrow 0$

$$-t^{-1}[f(x-th)-f(x)] - \|h\|\omega((t+\varepsilon)\|h\|) \leq f'(x;h).$$

The positive homogeneity being obvious, it remains to show the convexity of $f'(x;\cdot)$. Fix any (h,α) and (h',β) in $X\times\mathbb{R}$ and satisfying $f'(x;h) < \alpha$ and $f'(x;h') < \beta$. Choose some $\delta > 0$ such that for all $0 < t < \delta$

$$(2t)^{-1}[f(x+2th)-f(x)] < \alpha \quad \text{and} \quad (2t)^{-1}[f(x+2th')-f(x)] < \beta.$$

For any $0 < t < \delta$, we note that

$$f(x+th+th') \leq \frac{1}{2}f(x+2th) + \frac{1}{2}f(x+2th') + \frac{t}{2}\|h-h'\|\omega(2t\|h-h'\|),$$

or equivalently

$$\begin{aligned} t^{-1}[f(x+th+th')-f(x)] \leq (2t)^{-1}[f(x+2th)-f(x)]+(2t)^{-1}[f(x+2th')-f(x)] \\ +(1/2)\|h-h'\|\omega(2t\|h-h'\|), \end{aligned}$$

which implies

$$t^{-1}[f(x+th+th')-f(x)] < \alpha+\beta+(1/2)\|h-h'\|\omega(2t\|h-h'\|).$$

We deduce $f'(x;h+h') \leq \alpha+\beta$, hence $f'(x;\cdot)$ is convex. $\square$

The following lemma is useful for the study of subdifferentials of semiconvex functions.

LEMMA 10.28. Let $f : X \to \mathbb{R} \cup \{-\infty, +\infty\}$ be an $\omega(\cdot)$-semiconvex function on the normed space $(X, \|\cdot\|)$ and let $x \in X$ with $|f(x)| < +\infty$. Consider any $(u, r) \in \operatorname{epi} f$ with $\omega(\|u - x\|) < +\infty$. Then one has

$$\big(u - x, r - f(x) + \|u - x\|\omega(\|u - x\|)\big) \in T^B\big(\operatorname{epi} f; (x, f(x))\big),$$

and, for each real $s > 1$ with $\omega(s\|u - x\|) < +\infty$,

$$\big(u - x, r - f(x) + \|u - x\|\omega(s\|u - x\|)\big) \in T^C\big(\operatorname{epi} f; (x, f(x))\big);$$

in particular, whenever $\omega(\cdot)$ is upper semicontinuous at $\|u - x\|$ one has

$$\big(u - x, r - f(x) + \|u - x\|\omega(\|u - x\|)\big) \in T^C\big(\operatorname{epi} f; (x, f(x))\big).$$

PROOF. Let $x \in X$ such that $|f(x)| < +\infty$ and let $(u, r) \in \operatorname{epi} f$ such that $\omega(\|u - x\|) < +\infty$.

For each $n \in \mathbb{N}$ putting $t_n := 1/n$, we have from the $\omega(\cdot)$-semiconvexity of f

$$\begin{aligned} f(x + t_n(u - x)) &\le (1 - t_n)f(x) + t_n r + t_n(1 - t_n)\|u - x\|\omega(\|u - x\|) \\ &\le (1 - t_n)f(x) + t_n r + t_n\|u - x\|\omega(\|u - x\|), \end{aligned}$$

thus

$$\big(x, f(x)\big) + t_n\big(u - x, r - f(x) + \|u - x\|\omega(\|u - x\|)\big) \in \operatorname{epi} f,$$

which guarantees the inclusion

$$\big(u - x, r - f(x) + \|u - x\|\omega(\|u - x\|)\big) \in T^B(\operatorname{epi} f; (x, f(x))).$$

Now suppose that $s > 1$ is such that $\omega(s\|u - x\|) < +\infty$. If $u = x$, the result is obvious since $T^C(\operatorname{epi} f; (x, f(x))$ is an epigraph set containing $(0, 0)$. So suppose that $\|u - x\| > 0$. Consider $(x_n, r_n)_n$ in $\operatorname{epi} f$ converging to $(x, f(x))$ and $(t_n)_n$ in $]0, +\infty[$ tending to 0. Fix an integer N such that $t_n < 1$ and $\|u - x_n\| < s\|u - x\|$ for all $n \ge N$. Putting $z_n := x_n + t_n(u - x_n)$, by the non-decreasing property of $\omega(\cdot)$ and by the $\omega(\cdot)$-semiconvexity of f we have, for all $n \ge N$,

$$\begin{aligned} f(z_n) - t_n\|u - x_n\|\omega(s\|u - x\|) &\le f(z_n) - t_n\|u - x_n\|\omega(\|u - x_n\|) \\ &\le t_n r + (1 - t_n)r_n, \end{aligned}$$

hence

$$\big(x_n, r_n\big) + t_n\big(u - x_n, r - r_n + \|u - x_n\|\omega(s\|u - x\|)\big) \in \operatorname{epi} f.$$

This justifies the inclusion

$$\big(u - x, r - f(x) + \|u - x\|\omega(s\|u - x\|)\big) \in T^C\big(\operatorname{epi} f; (x, f(x))\big).$$

Finally, suppose that $\omega(\cdot)$ is upper semicontinuous at $\|u - x\|$. Choose a sequence $(s_n)_n$ in $]1, +\infty[$ tending to 1. According to the upper semicontinuity of $\omega(\cdot)$ at $\|u - x\|$ there is some $N \in \mathbb{N}$ such that $\omega(s_n\|u - x\|) < +\infty$ for all $n \ge N$, so the second result proved above tells us that

$$\big(u - x, r - f(x) + \|u - x\|\omega(s_n\|u - x\|)\big) \in T^C\big(\operatorname{epi} f; (x, f(x))\big) \quad \text{for all } n \ge N.$$

Further, the inequality $\omega(\|u - x\|) \le \omega(s_n\|u - x\|)$ (due to the nondecreasing property of $\omega(\cdot)$) combined with the upper semicontinuity of $\omega(\cdot)$ at $\|u - x\|$ entails that $\omega(s_n\|u - x\|) \to \omega(\|u - x\|)$ as $n \to \infty$. The closedness property of the Clarke tangent cone then gives

$$\big(u - x, r - f(x) + \|u - x\|\omega(\|u - x\|)\big) \in T^C(\operatorname{epi} f; (x, f(x)),$$

which finishes the proof of the lemma. □

Through the above lemma we can provide a description of subdifferentials of semiconvex functions in the line of the inequality characterizing subdifferentials of convex functions (resp. subsmooth functions). This description shows in particular that the Clarke and Fréchet subdifferentials of such functions coincide.

THEOREM 10.29 (description of subdifferential of semiconvex function). Let $f : X \to \mathbb{R} \cup \{+\infty\}$ be an $\omega(\cdot)$-semiconvex function on a normed space $(X, \|\cdot\|)$ and $x \in \operatorname{dom} f$. Then for any real $s > 0$

$$\begin{aligned}\partial_C f(x) = \partial_F f(x) &= \{x^* \in X^* : \langle x^*, h\rangle \le f'(x; h)\ \forall h \in X\}\\ &= \{x^* \in X^* : \langle x^*, h\rangle \le f(x+h) - f(x) + \|h\|\omega(\|h\|)\ \forall h \in X\}\\ &= \{x^* \in X^* : \langle x^*, h\rangle \le f(x+h) - f(x) + \|h\|\omega(\|h\|)\ \forall h \in s\mathbb{B}_X\}.\end{aligned}$$

If f is linearly semiconvex, then the five sets above coincide with the variational proximal subdifferential $\partial_{VP} f(x)$.

PROOF. Consider any $s \in]0, +\infty]$. Fix any x^* in the set Δ_s given by

$$\Delta_s := \{u^* \in X^* : \langle u^*, h\rangle \le f(x+h) - f(x) + \|h\|\omega(\|h\|)\ \forall h \in s\mathbb{B}_X\},$$

with the convention $s\mathbb{B}_X = X$ for $s = +\infty$. Take any $\varepsilon > 0$ and choose by the equality $\lim_{r\downarrow 0} \omega(r) = 0$ some positive real $\delta < s$ such that $\omega(r) < \varepsilon$ for all $r \in [0, \delta[$. Then for all $u \in B(x, \delta)$ we have with $h = u - x$

$$\langle x^*, u - x\rangle \le f(u) - f(x) + \varepsilon\|u - x\|,$$

which means $x^* \in \partial_F f(x)$, so $\Delta_\infty \subset \Delta_s \subset \partial_F f(x) \subset \partial_C f(x)$.

Further, for any $x^* \in \partial_F f(x)$, or equivalently $(x^*, -1) \in N^F\big(\operatorname{epi} f; (x, f(x))\big)$, we have, for all $h \in X$ with $x + h \in \operatorname{dom} f$ and $\omega(\|h\|) < +\infty$,

$$\langle x^*, h\rangle - \big(f(x+h) - f(x) + \|h\|\omega(\|h\|)\big) \le 0$$

since $\big(h, f(x+h) - f(x) + \|h\|\omega(\|h\|)\big) \in T^B(\operatorname{epi} f; (x, f(x)))$ according to Lemma 10.28. The latter inequality being still true if either $f(x+h)$ or $\omega(\|h\|)$ is $+\infty$, it follows that

$$\langle x^*, h\rangle \le f(x+h) - f(x) + \|h\|\omega(\|h\|) \quad \text{for all } h \in X,$$

which means $x^* \in \Delta_\infty$. Consequently, we have $\Delta_\infty = \Delta_s = \partial_F f(x)$.

Let us prove the inclusion $\partial_C f(x) \subset \partial_F f(x)$. Fix $x^* \in \partial_C f(x)$, which is equivalent to $\langle x^*, h\rangle - r \le 0$ for all $(h, r) \in T^C\big(\operatorname{epi} f; (x, f(x))\big)$. For any $h \in X$ with $x + h \in \operatorname{dom} f$ and $\omega(2\|h\|) < +\infty$, Lemma 10.28 again tells us that

$$\big(h, f(x+h) - f(x) + \|h\|\omega(2\|h\|)\big) \in T^C\big(\operatorname{epi} f; (x, f(x))\big),$$

thus

$$\langle x^*, h\rangle \le f(x+h) - f(x) + \|h\|\omega(2\|h\|),$$

which obviously ensures $x^* \in \partial_F f(x)$. So, we have proved the equalities

$$\Delta_\infty = \Delta_s = \partial_F f(x) = \partial_C f(x).$$

Moreover, for $\Lambda := \{x^* \in X^* : \langle x^*, \cdot\rangle \le f'(x; \cdot)\}$, the inequality in Proposition 10.27 with $t = 1$ guarantees that $\Lambda \subset \Delta_\infty$, and the reverse inclusion $\Delta_\infty \subset \Lambda$ is easily seen from the definition of Δ_∞. This means that $\Lambda = \Delta_\infty$, and the equality between the five sets of the proposition is established.

Now assume that f is linearly semiconvex, that is, there exists some real constant $\sigma \geq 0$ such that $\omega(r) = (\sigma r)/2$ for all $r \in [0, +\infty[$. Then the equality between the first and the fourth set of the proposition becomes

$$\partial_F f(x) = \{x^* \in X^* : \langle x^*, h\rangle \leq f(x+h) - f(x) + \frac{\sigma}{2}\|h\|^2 \ \forall h \in X\},$$

ensuring $\partial_F f(x) \subset \partial_{VP} f(x)$ hence $\partial_F f(x) = \partial_{VP} f(x)$. This completes the proof. $\square$

If the semiconvex function f is Hadamard (or Gâteaux) differentiable, the equality $\partial_C f(x) = \{x^* \in X^* : \langle x^*, h\rangle \leq f'(x;h), \ \forall h \in X\}$ in Theorem 10.29 furnishes the following corollary. The result in the corollary can also be seen as a consequence of Corollary 8.27.

COROLLARY 10.30. Let U be a nonempty open convex set of a normed space X and let $f : U \to \mathbb{R}$ be a function which is $\omega(\cdot)$-semiconvex on U. If f is Gâteaux differentiable at a point $\overline{x} \in U$, then $\partial_C f(\overline{x}) = \{Df(\overline{x})\}$.

The next corollary provides certain specific features for Fréchet subdifferentials of continuous semiconvex functions.

COROLLARY 10.31. Let U be a nonempty open convex set of a normed space X and let $f : U \to \mathbb{R}$ be a function which is continuous on U and $\omega(\cdot)$-semiconvex on U. The following hold.
(a) For all $x \in U$, one has $\partial_F f(x) \neq \emptyset$.
(b) The function f is Fréchet differentiable at a point $x \in U$ if and only if $\partial_F(-f)(x)$ is nonempty.
(c) If $\partial_F f(x)$ is a singleton set for a point $x \in U$, then f is strictly Hadamard differentiable at x.
(d) The graph of $\partial_F f$ is $\|\cdot\| \times w^*$ closed relative to $U \times X^*$.
(e) For each $x \in U$ one has the inclusion

$$\left\{\lim_{n\to\infty} D_G f(x_n) : x_n \underset{\mathrm{Dom}\, D_G f}{\to} x, ((x_n - x)/\|x_n - x\|)_n \text{ converges in } X\right\} \subset \mathrm{bdry}\, \partial_F f(x),$$

where $\lim_{n\to\infty} D_G f(x_n)$ is the strong limit in X^*.

In particular, for every $x \in U$ the inclusion

$$\left\{\lim_{n\to\infty} D_G f(x_n) : x_n \underset{\mathrm{Dom}\, D_G f}{\to} x\right\} \subset \mathrm{bdry}\, \partial_F f(x)$$

holds whenever X is finite-dimensional.

PROOF. We note that f is locally Lipschitz continuous on U according to Proposition 10.23.
(a) Since $\partial_C f(x) \neq \emptyset$ by the local Lipschitz property of f on U, the non-emptiness of $\partial_F f(x)$ follows from Theorem 10.29 above.
(b) Under the assumption in (b), both sets $\partial_F(-f)(x)$ and $\partial_F f(x)$ are nonempty by (a), hence f is Fréchet differentiable at x (see Proposition 4.13).
(c) Under the assumption in (c), the set $\partial_C f(x)$ is a singleton, then the locally Lipschitz function f is strictly Hadamard differentiable at x (see Proposition 2.76).
(d) Let $(x_i, x_i^*)_{i\in I}$ be a net in $\mathrm{gph}\, \partial_F f$ which $\|\cdot\| \times w^*$ converges to $(x, x^*) \in U \times X^*$. Choose a real $\gamma \geq 0$ such that f is γ-Lipschitz on $B(x, \delta) \subset U$ for some $\delta > 0$, and take also δ such that $\omega(\cdot)$ is bounded on $[0, 2\delta]$. Putting $\omega_1(r) = \omega(r)$ if $r \in [0, 2\delta]$

and $\omega_1(r) = \omega(2\delta)$ if $r > 2\delta$, the function $\omega_1(\cdot)$ is a nondecreasing finite modulus function on $[0, +\infty[$ with $\lim_{r\downarrow 0} \omega_1(r) = 0$ and f is obviously $\omega_1(\cdot)$-semiconvex on $B(x, \delta)$. By Lemma 10.3 let $\omega_0 : [0, +\infty[\to [0, +\infty[$ be a nondecreasing finite continuous modulus function with $\omega_1(\cdot) \le \omega_0(\cdot) \le \omega_1(2\,\cdot)$, so the function f is $\omega_0(\cdot)$-semiconvex on $B(x, \delta)$. Choose some $i_0 \in I$ such that $x_i \in B(x, \delta)$ for all $i \succeq i_0$. For each $u \in B(x, \delta)$, by Theorem 10.29 we have

$$\langle x_i^*, u - x_i\rangle \le f(u) - f(x_i) + \|u - x_i\|\omega(\|u - x_i\|) \quad \text{for all } i \succeq i_0,$$

hence using the boundedness of $(x_i^*)_{i\succeq i_0}$ (due to the Lipschitz property of f) and the continuity of $\omega_0(\cdot)$, we obtain by taking the limit

$$\langle x^*, u - x\rangle \le f(u) - f(x) + \|u - x\|\omega_0(\|u - x\|).$$

This clearly entails $x^* \in \partial_F f(x)$, showing (d).
(e) Take x^* in the first member of (e) and choose $(x_n)_n$ converging to x with $\|D_G f(x_n) - x^*\| \to 0$ and $\big((x - x_n)/\|x - x_n\|\big)$ converging to some $v \in X$ with $\|v\| = 1$. Since $D_G(x_n) \in \partial_C f(x_n) = \partial_F f(x_n)$ for large n, we have $x^* \in \partial_F f(x)$ by the above assertion (d). Take $v^* \in X^*$ such that $\|v^*\| = 1$ and $\langle v^*, v\rangle = 1$. By Theorem 10.29 again, for each real $r > 0$, we have, for large n,

$$\begin{aligned} &f(x_n) - f(x) - \langle x^* - rv^*, x_n - x\rangle \\ &= f(x_n) - f(x) + \langle D_G f(x_n), x - x_n\rangle + \langle x^* - D_G f(x_n), x - x_n\rangle - r\langle v^*, x - x_n\rangle \\ &\le \|x - x_n\|\omega(\|x - x_n\|) + \|x^* - D_G f(x_n)\|\,\|x - x_n\| - r\langle v^*, x - x_n\rangle. \end{aligned}$$

Consequently,

$$\liminf_{n\to\infty} \frac{1}{\|x_n - x\|}[f(x_n) - f(x) - \langle x^* - rv^*, x_n - x\rangle] \le -r,$$

which entails $x^* - rv^* \notin \partial_F f(x)$ for all $r > 0$. We then conclude $x^* \in \operatorname{bdry} \partial_F f(x)$. □

We already noticed before the statement of Proposition 10.6 that (by Proposition 8.12) semiconvex functions on an open convex set U of a normed space are uniformly subsmooth on U, hence in particular subsmooth on U. From Proposition 8.30 we directly derive the following proposition (which can be also easily seen from the characterization in Theorem 10.29 that $x^* \in \partial_C f(x)$ if and only if $\langle x^*, y - x\rangle \le f(y) - f(x) + \|y - x\|\omega(\|y - x\|)$ for all $y \in X$).

Proposition 10.32. *Let U be an open convex set in a normed space X and $f : U \to \mathbb{R} \cup \{+\infty\}$ be a proper lower semicontinuous $\omega(\cdot)$-semiconvex function. Let ∂ be either the C-subdifferential, or the A-subsdifferential or the the F-subdifferential. Then, the graph $\operatorname{gph} \partial f$ is sequentially $\|\cdot\| \times w^*$ closed; in fact, for any net $(x_j, x_j^*)_{j\in J}$ in $\operatorname{gph} \partial f$ $\|\cdot\| \times w^*$ converging to $(\overline{x}, x^*)$ with $(x_j^*)_{j\in J}$ norm-bounded one has $x^* \in \partial f(\overline{x})$.*

Similarly, the next proposition is a direct consequence of Proposition 8.31.

Proposition 10.33. *Let U be an open convex set in a normed space X and $f : U \to \mathbb{R} \cup \{+\infty\}$ be a proper lower semicontinuous $\omega(\cdot)$-semiconvex function. Let ∂ be either the C-subdifferential, or the A-subsdifferential or the the F-subdifferential. Then the function $x \mapsto d(0, \partial f(x))$ is lower semicontinuous on U.*

Next, we show how the characterization of subgradients of semiconvex functions in Theorem 10.29 allows us to obtain estimates of differences of values of f and the infimum of f.

PROPOSITION 10.34. Let X be a normed space and $f : X \to \mathbb{R} \cup \{+\infty\}$ be a proper function with $A_f := \operatorname{Argmin} f \neq \emptyset$. Assume that f is $\omega(\cdot)$-semiconvex and that the modulus of semiconvexity $\omega(\cdot)$ is upper semicontinuous. Then, for all $x \in \operatorname{Dom} \partial_C f$, one has

$$f(x) - \inf_X f \leq d(x, A_f)\,\big(d(0, \partial_C f(x)) + \omega(d(x, A_f))\big).$$

In the particular case when f is convex, one then has, for all $x \in \operatorname{Dom} \partial f$,

$$f(x) - \inf_X f \leq d(x, A_f)\, d(0, \partial f(x)).$$

PROOF. Fix an arbitrary $x \in \operatorname{Dom} \partial_C f$ and fix also any $x^* \in \partial_C f(x)$. Let a sequence $(y_n)_{n\in\mathbb{N}}$ in A_f with $\|y_n - x\| \to d(x, A_f)$. Since $f(y_n) = \inf_X f$, the characterization of subgradients in Theorem 10.29 gives

$$f(x) - \inf_X f \leq \langle x^*, x - y_n\rangle + \|y_n - x\|\omega(\|y_n - x\|) \leq \|x^*\|\,\|y_n - x\| + \|y_n - x\|\omega(\|y_n - x\|),$$

so, using the upper semicontinuity of $\omega(\cdot)$, it results that

$$f(x) - \inf_X f \leq \|x^*\|\, d(x, A_f) + d(x, A_f)\omega(d(x, A_f)).$$

The latter inequality being true for every $x^* \in \partial_C f(x)$, we conclude that

$$f(x) - \inf_X f \leq d(x, A_f)\,\big(d(0, \partial_C f(x)) + \omega(d(x, A_f))\big).$$

The case when f is convex follows directly with $\omega(\cdot) = 0$. □

Theorem 10.19 shows that, in a finite-dimensional Euclidean space, the class of locally linearly semiconvex functions coincide with that of locally lower $\mathcal{C}^2$ functions. With the description of subgradients in Theorem 10.29 at hand we can now study in the Euclidean setting the link between general semiconvex functions and lower $\mathcal{C}^1$ functions.

THEOREM 10.35 (lower $\mathcal{C}^1$ characterization of semiconvexity of finite-valued function in finite dimensions). Let U be a nonempty open set of a finite-dimensional Euclidean vector space $(X, \|\cdot\|)$ and $f : U \to \mathbb{R}$ be a real-valued function. For $\overline{x} \in U$, the following properties are equivalent:
(a) the function f is semiconvex near $\overline{x} \in U$;
(b) the function f is subsmooth at each point of a neighborhood of $\overline{x}$;
(c) the function f is Lipschitz near $\overline{x}$ and there exist an open convex neighborhood $U_0 \subset U$ of $\overline{x}$ and an even $\mathcal{C}^1$ function $\theta : \mathbb{R} \to [0, +\infty[$ with $\theta(0) = \theta'(0) = 0$ and such that, for all $x, y \in U_0$ and $y^* \in \partial_C f(y)$,

$$f(x) \geq f(y) + \langle y^*, x - y\rangle - \theta(\|x - y\|);$$

(d) the function f is lower $\mathcal{C}^1$ near $\overline{x}$.

PROOF. Suppose that f is subsmooth at each point of an open neighborhood $U_0 \subset U$ of $\overline{x}$. Choose an open bounded convex neighborhood V of $\overline{x}$ with $W := \operatorname{cl} V \subset U_0$. We show that f is uniformly subsmooth on V. Fix an arbitrary real

$\varepsilon > 0$. For each $x \in W$, choose (by subsmoothness of f at x) a real $\delta_x > 0$ such that, for all $t \in]0,1[$, $x', x'' \in B(x, 2\delta_x)$,

$$f(tx' + (1-t)x'') \le tf(x') + (1-t)f(x'') + \varepsilon t(1-t)\|x' - x''\|.$$

By compactness of W there exist $x_1, \cdots, x_m \in W$ such that $W \subset \bigcup_{i=1}^m B(x_i, \delta_{x_i})$. Put $\delta := \min_{1\le i\le m} \delta_{x_i} > 0$ and take any $x', x'' \in V$ with $\|x' - x''\| < \delta$. There is some $1 \le k \le m$ such that $x' \in B(x_k, \delta_{x_k})$. Since $\|x'' - x'\| < \delta$, we have $x'' \in B(x_k, 2\delta_{x_k})$, hence for every $t \in]0,1[$

$$f(tx' + (1-t)x'') \le tf(x') + (1-t)f(x'') + \varepsilon t(1-t)\|x' - x''\|.$$

The function f is then uniformly subsmooth on V, thus Proposition 8.16 confirms that f is semiconvex on V. This establishes (b) $\Rightarrow$ (a), hence the equivalence (a) $\Leftrightarrow$ (b) holds since the implication (a) $\Rightarrow$ (b) is evident.

Suppose (d), that is, f is lower $\mathcal{C}^1$ near $\overline{x}$, and choose V, T and g as given by Definition 10.18. As in the proof of Theorem 10.19 choose an open bounded convex neighborhood W of $\overline{x}$ such that $W_0 := \operatorname{cl} W \subset V$, so W_0 is compact. For each real $r \ge 0$, putting

$$\omega(r) := \max\{\|Dg_t(x') - Dg_t(x'')\| : t \in T,\ x', x'' \in W_0,\ \|x' - x''\| \le r\},$$

where $g_t := g(t, \cdot)$, it is easy to see that $\omega(\cdot)$ is a nondecreasing (finite-valued) modulus function with $\lim_{r\downarrow 0} \omega(r) = 0$. All the derivatives $(Dg_t)_{t\in T}$ are then uniformly continuous on W with $\omega(\cdot)$ as a common modulus of continuity. By Proposition 10.9(b) all the functions g_t are $\omega(\cdot)$-semiconvex on W. From this and Proposition 10.10 we conclude that f is $\omega(\cdot)$-semiconvex on W, which justifies that the property (d) entails (a).

Now let us suppose (a) and prove (c). Let $W \subset U$ be an open convex neighborhood of $\overline{x}$ such that f is $\omega(\cdot)$-semiconvex on W for some nondecreasing modulus $\omega : [0, +\infty[\to [0, +\infty]$ with $\lim_{r\downarrow 0} \omega(r) = 0$. By Corollary 10.24 (shrinking W if necessary) we may suppose that f is Lipschitz on W and by the description of subdifferential of semiconvex function in Theorem 10.29, for all $x, y \in W$ and $y^* \in \partial_C f(x)$,

$$f(x) \ge f(y) + \langle y^*, x - y\rangle - \|y - x\|\omega(\|x - y\|).$$

Choose some real $\delta > 0$ such that $B(\overline{x}, 2\delta) \subset W$ and $\omega(\cdot)$ is finite-valued on $[0, 2\delta]$. Putting $\sigma(r) := r\omega(r)$ for $0 \le r \le 2\delta$ and $\sigma(r) := 2\delta\,\omega(2\delta)$ for $r \ge 2\delta$, the function σ is a finite-valued non-negative nondecreasing function on $[0, +\infty[$ with $\sigma(0) = \sigma'_+(0) = 0$. By Lemma 4.24 there is some even $\mathcal{C}^1$ function $\theta : \mathbb{R} \to [0, +\infty[$ such that $\theta(0) = \theta'(0) = 0$ and $\theta(r) \ge \sigma(r)$ for all $r \ge 0$, hence $\theta(r) \ge r\omega(r)$ for all $r \in [0, 2\delta]$. It results that, for all $x, y \in U_0 := B(\overline{x}, \delta)$ and $y^* \in \partial_C f(y)$,

$$f(x) \ge f(y) + \langle y^*, x - y\rangle - \theta(\|x - y\|),$$

which translates the property (c).

It remains to show that (c) implies (d). Let θ and $U_0 \subset U$ as given by (c) and such that f is Lipschitz on U_0 with $\gamma \ge 0$ as a Lipschitz constant therein. Choose some real $\delta > 0$ such that $V := B[\overline{x}, \delta] \subset U_0$ and put $T := \{(y, y^*) \in V \times X : y^* \in \partial_C f(y)\}$. We know that the graph of the multimapping $\partial_C f$ is closed relative to $U_0 \times X$ with $\partial_C f(U_0) \subset \gamma\mathbb{B}_X$ (see Proposition 2.74(a)), so T is compact in $X \times X$. Further, it easy to see that the function $(x, y) \mapsto \theta(\|y - x\|)$ is $\mathcal{C}^1$ on $X \times X$ according to the $\mathcal{C}^1$ property of θ with $\theta(0) = \theta'(0) = 0$. Consequently, considering the function $g : T \times U_0 \to \mathbb{R}$ with $g(t, x) := f(y) + \langle y^*, x - y\rangle - \theta(\|y - x\|)$ for all $t := (y, y^*) \in T$ and $x \in U_0$, the function $g(\cdot, \cdot)$ as well as its derivative in the second variable are continuous on $T \times U_0$ and

$$f(x) = \max_{t \in T} g(t, x) \quad \text{for all } x \in V.$$

This means that f is lower $\mathcal{C}^1$ near $\overline{x}$ and the proof is finished. □

10.2.2. Properties under linear semiconvexity and linear semiconcavity. In Proposition 10.9 we saw that any differentiable function on U whose derivative is uniformly continuous on U is simultaneously semiconvex and semiconcave. A much more important result related to the converse is provided in the next theorem.

THEOREM 10.36 ($\mathcal{C}^{1,0}$ property of both linear semiconvex and semiconcave continuous functions in normed space). *Let U be a nonempty open set of a normed space X and $f : U \to \mathbb{R}$ be a continuous function which is semiconvex and semiconcave near $\overline{x} \in U$ with the same modulus $\omega(\cdot)$. Then f is differentiable near $\overline{x}$ and its derivative is uniformly continuous near $\overline{x}$. More precisely, there exists $\delta > 0$ such that f is differentiable on $B(\overline{x}, \delta)$, its derivative Df is uniformly continuous on $B(\overline{x}, \delta)$, and the modulus of continuity of Df is at each $r \geq 0$ bounded above by $4\omega(2r)$.*

PROOF. Suppose that f and $-f$ are $\omega(\cdot)$-semiconvex around $\overline{x}$. Then there exists $\delta > 0$ and $\sigma > 0$ such that f is Lipschitz continuous on $B(\overline{x}, \delta) \subset U$ and simultaneously f and $-f$ are $\omega(\cdot)$-semiconvex on $B(\overline{x}, \delta)$. The non vacuity of $\partial_C f(x)$ and $\partial_C(-f)(x)$ along with Theorem 10.29 imply that $\partial_F f(x)$ and $\partial_F(-f)(x)$ are nonempty for every $x \in B(\overline{x}, \delta)$. The function f is then Fréchet differentiable on $B(\overline{x}, \delta)$ and by Theorem 10.29 again

$$|f(x+h) - f(x) - \langle Df(x), h\rangle| \leq \|h\|\omega(\|h\|) \tag{10.6}$$

for all $x \in B(\overline{x}, \delta)$ and $h \in B(0, \delta - \|x - \overline{x}\|)$. Fix $t \in]0, \frac{\delta}{2}[$ and $h \in X$ with $\|h\| = 1$, and let $x, y \in B(\overline{x}, \delta/4)$, $x \neq y$. We deduce from (10.6)

$$\begin{aligned}
\langle Df(x), h\rangle &\geq t^{-1}[f(x) - f(x - th)] - \omega(t) \\
&= t^{-1}[f(x) - f(y)] + t^{-1}[f(y) - f(x - th)] - \omega(t) \\
&\geq t^{-1}[\langle Df(y), x - y\rangle - \|x - y\|\omega(\|x - y\|)] - \omega(t) \\
&\quad + t^{-1}[-\langle Df(y), x - th - y\rangle - \|x - th - y\|\omega(\|x - th - y\|)] \\
&\geq -t^{-1}\|x - y\|\omega(\|x - y\|) + \langle Df(y), h\rangle \\
&\quad - t^{-1}\|x - th - y\|\omega(\|x - th - y\|) - \omega(t).
\end{aligned}$$

For $t = \|x - y\|$, we obtain by the nondecreasing property of $\omega(\cdot)$

$$\begin{aligned}
\langle Df(x), h\rangle &\geq \langle Df(y), h\rangle - 2\omega(\|x - y\|) - 2\omega(2\|x - y\|)) \\
&\geq \langle Df(y), h\rangle - 4\omega(2\|x - y\|).
\end{aligned}$$

Since h is arbitrary with $\|h\| = 1$, we get

$$\|Df(x) - Df(y)\| \leq 4\omega(2\|x - y\|) \quad \text{for all } x, y \in B(\overline{x}, \delta/4),$$

justifying the uniform continuity of Df on $B(\overline{x}, \delta/4)$ along with the fact that the value at r of the modulus of uniform continuity of Df is bounded from above by the real $4\omega(2r)$. □

In the Hilbert setting we have a much more precise result for linearly semiconvex and semiconcave functions.

THEOREM 10.37 ($\mathcal{C}^{1,1}$ characterization of both linear semiconvexity and semiconcavity in Hilbert space). Let $f : U \to \mathbb{R}$ be a real-valued continuous function on a nonempty open convex set U of a Hilbert space H.
(A) Given a real $\sigma \geq 0$, the following assertions are equivalent:
(a) the function f is linearly semiconvex and linearly semiconcave over U with σ as a same coefficient both for the linear semiconvexity and linear semiconcavity;
(b) the function f is differentiable on U and its derivative Df is σ-Lipschitz continuous on the open convex set U, that is, for all $x, y \in U$

$$\|Df(x) - Df(y)\| \leq \sigma\|x - y\|;$$

(c) the function f is differentiable on U and for all $x, y \in U$

$$|\langle Df(x) - Df(y), x - y\rangle| \leq \sigma\|x - y\|^2.$$

(B) The function f is linearly semiconvex and linearly semiconcave near $\overline{x} \in U$ if and only if f is $\mathcal{C}^{1,1}$ near $\overline{x}$.

PROOF. It suffices to prove (A) since (B) follows from the equivalence (a)$\Leftrightarrow$(b) and from Proposition 10.9.
(a)$\Rightarrow$(b) Suppose that f is linearly semiconvex and linearly semiconcave on U with the coefficient $\sigma \geq 0$, that is, both functions $f + \frac{\sigma}{2}\|\cdot\|^2$ and $-f + \frac{\sigma}{2}\|\cdot\|^2$ are convex on U according to Theorem 10.15(A).

Step 1. Fix any $\overline{x} \in U$ and take $\delta > 0$ such that $B(\overline{x}, 2\delta) \subset U$. Theorem 10.36 tells us that f is $\mathcal{C}^1$ on $B(\overline{x}, 2\delta)$. Our objective in this Step 1 is to show that Df is σ-Lipschitz continuous on $B(\overline{x}, \delta)$. So, fix any $x, x' \in B(\overline{x}, \delta)$ with $x \neq x'$. Then for any $h, k \in \delta\mathbb{B}$ we have

$$\begin{aligned}\langle Df(x) + \sigma x, h\rangle &\leq f(x + h) + \frac{\sigma}{2}\|x + h\|^2 - f(x) - \frac{\sigma}{2}\|x\|^2 \\ \langle -Df(x) + \sigma x, k\rangle &\leq -f(x + k) + \frac{\sigma}{2}\|x + k\|^2 + f(x) - \frac{\sigma}{2}\|x\|^2,\end{aligned}$$

and adding the inequalities we obtain

$$\begin{aligned}\langle Df(x), h - k\rangle &\leq f(x + h) - f(x + k) + \frac{\sigma}{2}\big(\|x + h\|^2 + \|x + k\|^2 - 2\|x\|^2\big) \\ &\quad - \sigma\langle x, h + k\rangle \\ &= f(x + h) - f(x + k) + \frac{\sigma}{2}(\|h\|^2 + \|k\|^2) \\ &= f(x + h) - f(x + k) + \frac{\sigma}{2}\big(\|h - k\|^2 + 2\langle h, k\rangle\big).\end{aligned}$$

Similarly, for any $h', k' \in \delta\mathbb{B}$ we have

$$\langle Df(x'), h' - k'\rangle \leq f(x' + h') - f(x' + k') + \frac{\sigma}{2}\big(\|h' - k'\|^2 + 2\langle h', k'\rangle\big).$$

Adding together the latter inequalities yield

$$\begin{aligned}&\langle Df(x), h - k\rangle + \langle Df(x'), h' - k'\rangle \\ &\leq \{f(x + h) - f(x + k) + f(x' + h') - f(x' + k')\} \\ &\quad + \sigma\{\langle h, k\rangle + \langle h', k'\rangle\} + \frac{\sigma}{2}\{\|h - k\|^2 + \|h' - k'\|^2\}.\end{aligned} \tag{10.7}$$

Fix any $b \in H$ with $\|b\| = 1$ and put $q = \|x - x'\|b$. We proceed to choose h, k, h', k' with $h - k = q$ in such a way that on the one hand $x + h = x' + k'$ and $x + k = x' + h'$ in order to make null the first sum in brackets in (10.7), and on the other hand $h + h' = 0$ in order to make null also the second sum in brackets in (10.7). So, we choose h, k, h', k' such that

$$h - k = q,\ x + h = x' + k',\ x + k = x' + h',\ h + h' = 0,$$

otherwise stated, we choose

$$h = \frac{1}{2}(x' - x) + \frac{q}{2},\ k = \frac{1}{2}(x' - x) - \frac{q}{2},\ h' = -\frac{1}{2}(x' - x) - \frac{q}{2},\ k' = -\frac{1}{2}(x' - x) + \frac{q}{2}.$$

Concerning the inner products $\langle h, k\rangle$ and $\langle h', k'\rangle$ we observe indeed

$$\langle h, k\rangle = \langle h, h - q\rangle = \left\|h - \frac{q}{2}\right\|^2 - \frac{1}{4}\|q\|^2 = \frac{1}{4}\|x - x'\|^2 - \frac{1}{4}\|q\|^2 = 0,$$

$$\text{and} \quad \langle h', k'\rangle = \langle -h, -k\rangle = 0.$$

We then deduce from (10.7)

$$\langle Df(x) - Df(x'), q\rangle \le \sigma\|q\|^2, \quad \text{that is, } \langle Df(x) - Df(x'), b\rangle \le \sigma\|x - x'\|.$$

Since b is arbitrary with $\|b\| = 1$, we conclude that $\|Df(x) - Df(x')\| \le \sigma\|x - x'\|$, that is, Df is σ-Lipschitz continuous on $B(\overline{x}, \delta)$.

Step 2. Take now any $x, y \in U$. By compactness of $[x, y]$ we have $\rho := \inf_{u \in [x,y]} d(u, X \setminus U) > 0$. Fix a positive real $\delta < \rho/2$, so $[x, y] + B(0, 2\delta) \subset U$. Let $x_0, \cdots, x_n \in [x, y]$ with $x_0 = x$, $x_n = y$ and $\|x_{i-1} - x_i\| < \delta$ for $i = 1, \cdots, n$. By Step 1 for each $i = 1, \cdots, n$ we have $\|Df(x_{i-1}) - Df(x_i)\| \le \sigma\|x_{i-1} - x_i\|$, hence

$$\|Df(x) - Df(y)\| \le \sum_{i=1}^{n}\|Df(x_{i-1}) - Df(x_i)\| \le \sigma\sum_{i=1}^{n}\|x_{i-1} - x_i\| = \sigma\|x - y\|.$$

This justifies the σ-Lipschitz property of Df over U. The implication (a)⇒(b) is then established.

(b)⇒(c) This implication is trivial, since the σ-Lipschitz property of $Df(\cdot)$ on U trivially gives for all $x, y \in U$

$$|\langle Df(x) - Df(y), x - y\rangle| \le \|Df(x) - Df(y)\|\,\|x - y\| \le \sigma\|x - y\|^2.$$

(c)⇒(a) For any $x, y \in U$, the inequality in (c) is equivalent to

$$-\sigma\|x - y\|^2 \le \langle Df(x) - Df(y), x - y\rangle \text{ and } \langle Df(x) - Df(y), x - y\rangle \le \sigma\|x - y\|^2,$$

which in turn, with $f_+ := f + (\sigma/2)\|\cdot\|^2$ and $f_- := -f + (\sigma/2)\|\cdot\|^2$, is equivalent to

$$\langle Df_+(x) - Df_+(y), x - y\rangle \ge 0 \quad \text{and} \quad \langle -Df_-(x) - Df_-(y), x - y\rangle \ge 0.$$

This tells us that, under (c), both functions f_+ and f_- are conve on U. Therefore, (c) implies (a), so the proof is finished. □

10.2.3. Subdifferential and tangential characterizations of semiconvex functions. Like for convex functions, tangential and subdifferential characterizations of semiconvex (resp. semiconcave) functions can be established.

THEOREM 10.38 (**tangential and subdifferential characterizations of semiconvexity**). Let U be a nonempty open convex set of a Banach space X and $\omega : [0, +\infty[\to [0, +\infty]$ be a nondecreasing upper semicontinuous function which is continuous at 0 with $\omega(0) = 0$. Let also $f : U \to \mathbb{R} \cup \{+\infty\}$ be a proper lower semicontinuous function and ∂ be a subdifferential on X with ∂f included in the Clarke one and satisfying the properties **Prop.1-Prop.2** in Section 6.2. Consider the following assertions:
(a) The function f is $\omega(\cdot)$-semiconvex on U;
(b) for any $x \in U \cap \operatorname{dom} f$, one has

$$\underline{d}_H^+ f(x; y - x) \le f(y) - f(x) + \|y - x\|\omega(\|y - x\|);$$

(c) for any $x, y \in U \cap \operatorname{dom} f$ with the sum $f^B(x; y - x) + f^B(y; x - y)$ well defined, one has

$$\underline{d}_H^+ f(x; y - x) + \underline{d}_H^+ f(y; x - y) \le 2\|x - y\|\omega(\|x - y\|);$$

(d) for any $y \in U$, $x \in U \cap \operatorname{Dom} \partial f$, and $x^* \in \partial f(x)$, one has

$$\langle x^*, y - x\rangle \le f(y) - f(x) + \|y - x\|\omega(\|y - x\|);$$

(e) for all $x, y \in U \cap \operatorname{Dom} \partial f$, $x^* \in \partial f(x)$, and $y^* \in \partial f(y)$, one has

$$\langle x^* - y^*, x - y\rangle \ge -2\|x - y\|\omega(\|x - y\|).$$

Then (a) implies (b) and (d), (b) implies (c), and (d) implies (e), that is,

$$\begin{array}{ccccc} \text{(d)} & \Leftarrow & \text{(a)} & \Rightarrow & \text{(b)} \\ \Downarrow & & & & \Downarrow \\ \text{(e)} & & & & \text{(c)} \end{array}$$

Further each of assertions (b), (c), (d) or (e) entails that f is semiconvex on U with modulus $2\omega(\cdot)$. So, f is $K\omega(\cdot)$-semiconvex on U for some constant $K \ge 0$ if and only if anyone of properties (b), (c), (d) or (e) holds with a modulus $K'\omega(\cdot)$ for a constant $K' \ge 0$.

PROOF. It will be convenient to use the notation $f^B(x; \cdot)$ instead of $\underline{d}_H^+ f(x; \cdot)$.

Assume that f is $\omega(\cdot)$-semiconvex on U. Since ∂f is included in the Clarke subdifferential, by the description of the Clarke subdifferential of semiconvex functions in Theorem 10.29, for $y \in U$, $x \in U \cap \operatorname{Dom} \partial f$ and $x^* \in \partial f(x)$ we have

$$\langle x^*, y - x\rangle \le f(y) - f(x) + \|x - y\|\omega(\|x - y\|).$$

On the other hand, for any $x \in U \cap \operatorname{dom} f$ and $y \in U$ we know that the semiconvexity of f gives (see the second slope inequality (10.5))

$$t^{-1}[f(x + t(y - x)) - f(x)] \le f(y) - f(x) + (1 - t)\|x - y\|\omega(\|x - y\|) \text{ for all } t \in]0, 1[.$$

We deduce

$$f^B(x; y - x) \le f(y) - f(x) + \|x - y\|\omega(\|x - y\|).$$

We have then proved the desired implications (a) $\Rightarrow$ (d) and (a) $\Rightarrow$ (b). Further, the implications (b) $\Rightarrow$ (c) and (d) $\Rightarrow$ (e) are evident.

It remains to show (e) $\Rightarrow$ (a) and (c) $\Rightarrow$ (a). Suppose that (e) (resp. (c)) holds. Fix $x, y \in U \cap \operatorname{dom} f$ with $x \neq y$ and $s, t \in]0, 1[$. Put $z := sx + ty$ and take any reals $\rho < r < f(z)$. Repeating the arguments in the last part of the proof of

Theorem 6.68, applying twice the Zagrodny mean value theorem produces $s_n \to s$ with $s_n > 0$, $c \in [x, z[\cap \operatorname{dom} f$, $c_n \to_f c$, and $c_n^* \in \partial f(c_n)$ (resp. $c \in [x, z[\cap \operatorname{dom} f$ and $c_n \to_f c$ with $c_n \in U \cap \operatorname{dom} f$), and $d_n \in U \cap \operatorname{Dom} \partial f$, $d_n^* \in \partial f(d_n)$ (resp. $d_n \in U \cap \operatorname{dom} f$) with $\operatorname{dist}(d_n, [c_n, y]) \to 0$, and all those elements satisfy

$$\left\langle c_n^*, \frac{d_n - c_n}{\|d_n - c_n\|} \right\rangle > \frac{\rho - f(x)}{t\|x - y\|} = \frac{s_n\rho - s_n f(x)}{s_n t\|x - y\|}$$

$$\left(\text{resp. } f^B(c_n; \frac{d_n - c_n}{\|d_n - c_n\|}) > \frac{\rho - f(x)}{t\|x - y\|} = \frac{s_n\rho - s_n f(x)}{s_n t\|x - y\|} \right),$$

as well as

$$\left\langle d_n^*, \frac{c_n - d_n}{\|d_n - c_n\|} \right\rangle > \frac{t\rho - tf(y)}{s_n t\|x - y\|} \quad \left(\text{resp.} f^B(d_n; \frac{c_n - d_n}{\|d_n - c_n\|}) > \frac{t\rho - tf(y)}{s_n t\|x - y\|} \right).$$

We deduce according to the assumption in (e) (resp. in (c)) that

$$2\omega(\|d_n - c_n\|) \geq \left\langle c_n^* - d_n^*, \frac{d_n - c_n}{\|d_n - c_n\|} \right\rangle > \frac{(s_n + t)\rho - s_n f(x) - tf(y)}{s_n t\|x - y\|},$$

$$\left(\text{resp. } 2\omega(\|d_n - c_n\|) \geq \frac{1}{\|d_n - c_n\|} (f^B(c_n; d_n - c_n) + f^B(d_n; c_n - d_n)) \right.$$
$$\left. > \frac{(s_n + t)\rho - s_n f(x) - tf(y)}{s_n t\|x - y\|} \right),$$

which ensures

$$2s_n t\|x - y\|\omega(\|d_n - c_n\|) > (s_n + t)\rho - s_n f(x) - tf(y).$$

Further, the properties $\operatorname{dist}(d_n, [c_n, y]) \to 0$ and $c_n \to c \in [x, y]$ allow us to extract from $(d_n)_n$ a subsequence (that we do not relabel) converging to some d and $d \in [x, y]$. The upper semicontinuity and the nondecreasing property of $\omega(\cdot)$ ensure

$$\limsup_{n \to \infty} \omega(\|d_n - c_n\|) \leq \omega(\|d - c\|) \leq \omega(\|y - x\|).$$

Since $s_n \to s$ as $n \to \infty$ and ρ is arbitrarily less than $f(z)$, we conclude that

$$f(z) \leq sf(x) + tf(y) + 2st\|x - y\|\omega(\|x - y\|).$$

This finishes the proof. □

In the case of locally Lipschitz functions one also has the following characterization through a selection of subgradients.

Proposition 10.39. Let $\omega : [0, +\infty[\to [0, +\infty]$ be a nondecreasing concave function which is continuous at 0 with $\omega(0) = 0$. Let U be a convex set of a normed space X and $f : U \to \mathbb{R}$ be a real-valued function on U. If for each $x \in U$ there exists some $\zeta(x) \in X^*$ such that

$$\langle \zeta(x), y - x \rangle \leq f(y) - f(x) + \|y - x\|\omega(\|y - x\|) \quad \text{for all } y \in U,$$

then f is $\omega(\cdot)$-semiconvex on U and for any $x^* \in \partial_C f(x)$

$$\langle x^*, y - x \rangle \leq f(y) - f(x) + \|y - x\|\omega(\|y - x\|) \quad \text{for all } y \in U.$$

So, if f is locally Lipschitz continuous on the open convex set U, then the following three assertions are pairwise equivalent:

(a) f is $\omega(\cdot)$-semiconvex on U;
(b) for any $x \in U$ and $x^* \in \partial_C f(x)$

$$\langle x^*, y - x \rangle \le f(y) - f(x) + \|y - x\| \omega(\|y - x\|) \quad \text{for all } y \in U;$$

(c) for each $x \in U$ there exists some $\zeta(x) \in X^*$ such that

$$\langle \zeta(x), y - x \rangle \le f(y) - f(x) + \|y - x\| \omega(\|y - x\|) \quad \text{for all } y \in U.$$

PROOF. Assume the existence of $\zeta(\cdot)$ satisfying the inequality of the proposition. Fix any $x, y \in U$ and any $t \in]0, 1[$. The inequality of the proposition yield on the one hand

$$\begin{aligned} f(x + t(y - x)) \le f(y) &+ \langle \zeta(x + t(y - x)), x + t(y - x) - y \rangle \\ &+ \|x + t(y - x) - y\| \omega(\|x + t(y - x) - y\|), \end{aligned}$$

that is, for $\zeta_t := \zeta(x + t(y - x))$

$$f((1 - t)x + ty) \le f(y) + (1 - t)\langle \zeta_t, x - y \rangle + (1 - t)\|y - x\| \omega((1 - t)\|y - x\|),$$

and on the other hand

$$f((1 - t)x + ty) \le f(x) + t\langle \zeta_t, y - x \rangle + t\|y - x\| \omega(t\|y - x\|).$$

Adding the last two inequalities after multiplication by t and $(1 - t)$ respectively gives

$$f((1-t)x+ty) \le (1-t)f(x)+tf(y)+t(1-t)\|y-x\| \big(\omega(t\|y-x\|)+\omega((1-t)\|y-x\|)\big).$$

Using the concavity of the function $\omega(\cdot)$ we obtain

$$f((1 - t)x + ty) \le (1 - t)f(x) + tf(y) + t(1 - t)\|y - x\| \omega(\|y - x\|),$$

which means that f is $\omega(\cdot)$-semiconvex on U. Further, the inequality concerning any $x^* \in \partial_C f(x)$ following from Theorem 10.29, and hence the first part of the proposition is then established.

Assume now that f is locally Lipschitz continuous. If it is $\omega(\cdot)$-semiconvex, then for each $x \in U$ and each $x^* \in \partial_C f(x)$ the $\omega(\cdot)$-semiconvexity entails by Theorem 10.29 that

$$\langle x^*, y - x \rangle \le f(y) - f(x) + \|y - x\| \omega(\|y - x\|) \quad \text{for all } y \in U,$$

that is, the implication (a) $\Rightarrow$ (b) is true. The implication (b) $\Rightarrow$ (c) is obvious since for each $x \in U$ the local Lipschitz continuity of f ensures that $\partial_C f(x) \ne \emptyset$ for all $x \in U$. The last implication (c) $\Rightarrow$ (a) follows from the first part above, the proof of the proposition is finished. □

10.3. Max-representation and extension of Lipschitz semiconvex functions

One of the main aims of this section is to provide representation of proper lower semicontinuous semiconvex function as pointwise maximum of smooth functions. The study has already begun with pointwise supremum representation in the Hilbert setting in the previous section through Theorem 10.15 and with max-representation in finite dimensions through Theorem 10.19. This section establishes diverse max-representations in infinite dimensions.

10.3.1. Max-representation with quadratic/differentiable functions. We first continue the study with the following theorem of max-representation in the Hilbert framework. The theorem is in the line of Theorem 10.15 but with a max-representation instead of a sup-representation.

THEOREM 10.40 (quadratic max-representation of continuous linear semiconvex functions in Hilbert space). *Let $(H, \|\cdot\|)$ be a Hilbert space and $f : U \to \mathbb{R}$ be a real-valued function defined on an open convex subset U of H. The following assertions are equivalent:*
(a) the function f is continuous and linearly semiconvex on U;
(b) there exist a real constant $c \geq 0$, and families $(b_t)_{t\in T}$ in H and $(d_t)_{t\in T}$ in $\mathbb{R}$ such that

$$f(x) = \max_{t\in T} \big(-c\|x\|^2 + \langle b_t, x\rangle + d_t \big) \quad \text{for all } x \in U;$$

(c) there exists a family $(F_t)_{t\in T}$ of differentiable functions from H into $\mathbb{R}$ such that

$$f(x) = \max_{t\in T} F_t(x) \quad \text{for all } x \in U$$

and such that the derivatives $(DF_t)_{t\in T}$ are equi-Lipschitz on H.

Further, if U is bounded and f is Lipschitz on U, then the family $(F_t)_{t\in T}$ can be taken to be equi-Lipschitz on U.

PROOF. Suppose that (a) holds and let $c \geq 0$ be such that, for $\omega(r) := cr$ for all $r \in [0, +\infty[$, the function f is $\omega(\cdot)$-semiconvex on U. Since f is locally Lipschitz on U (see Proposition 10.23), we may choose $\zeta(a) \in \partial_C f(a)$ for each $a \in U$. For

$$F_a(x) := -c\|x - a\|^2 + \langle \zeta(a), x - a\rangle + f(a),$$

the characterization of subdifferentials of semiconvex functions (see Theorem 10.29) gives $F_a(x) \leq f(x)$ for all $a, x \in U$, hence, since $F_x(x) = f(x)$, we see that $f(x) = \max_{a\in U} F_a(x)$ for all $x \in U$. Further, the family of derivatives $(DF_a)_{a\in U}$ is equi-Lipschitz on H since $\nabla F_a(x) = -2cx + \zeta(a)$ for all $x \in H$. This justifies the implications (a) $\Rightarrow$ (b) and (a) $\Rightarrow$ (c). If in addition f is Lipschitz on U with γ as a Lipschitz constant and U is bounded, say $U \subset \rho\mathbb{B}$, then $\|a\| \leq \rho$ and $\|\zeta(a)\| \leq \gamma$ for all $a \in U$, so it is easily seen that the family $(F_a)_{a\in U}$ is equi-Lipschitz on U.

The assertion (b) obviously implies (a). Under the assumption of (c), denoting by $c \geq 0$ a common Lipschitz constant of DF_t on U, we see that all the functions F_t are $\omega(\cdot)$-semiconvex on U, for $\omega(r) = cr$. The implication (c) $\Rightarrow$ (a) then follows directly from Proposition 10.10. □

The next step is the investigation of max-representation of semiconvex functions in the framework of Banach space (instead of Hilbert space). Suppose that U is an open convex set of a Banach space X and that $f : U \to \mathbb{R}$ is a real-valued continuous function which is semiconvex with respect to a modulus function $\omega(\cdot)$ which is nondecreasing with $\lim_{r\downarrow 0} \omega(r) = 0 = \omega(0)$. We know (see Proposition 10.23) that such a function is locally Lipschitz on U. Further, observe that, for any $a \in \operatorname{Dom} \partial_C f$ and any $a^* \in \partial_C f(a)$, we have by Theorem 10.29

$$\langle a^*, x - a\rangle + f(a) - \|x - a\|\omega(\|x - a\|) \leq f(x) \quad \text{for all } x \in U$$

with equality if $x = a$. Since $\partial_C f(x) \neq \emptyset$ for each $x \in U$, we see that

$$(10.8)\quad f(x) = \max_{(a,a^*)\in \operatorname{gph} \partial_C f} \big(\langle a^*, x-a\rangle + f(a) - \|x-a\|\omega(\|x-a\|)\big) \quad \text{for all } x \in U.$$

The function $\|\cdot\|\omega(\cdot)$ is nonsmooth in general. However, in the case $\omega(\cdot)$ is concave we will see below that $\|\cdot-a\|\,\omega(\|\cdot-a\|)$ can be replaced by a differentiable function. This objective requires first the following lemma:

LEMMA 10.41. Let $r_0 \in]0,+\infty]$ and $\omega : [0,r_0[\to \mathbb{R}$ be a concave function with $\omega(0)=0$. Then

$$2\int_0^r \omega(t)\,dt \ge r\omega(r) \quad \text{for all } r \in [0,r_0[.$$

PROOF. Fix an arbitrary $r \in]0,r_0[$. For every $0 < t \le r$ we have

$$\omega(t) \ge \frac{t}{r}\omega(r) + \left(1-\frac{t}{r}\right)\omega(0), \text{ that is } \omega(t) \ge t\big(r^{-1}\omega(r)\big),$$

hence $\int_0^r \omega(t)\,dt \ge (1/2)r\,\omega(r)$. The latter inequality is still obviously true for $r=0$. $\square$

Now, assuming, for the continuous $\omega(\cdot)$-semiconvex function $f : U \to \mathbb{R}$, that $\omega(\cdot)$ is concave on $[0,r_0[$ for some $+\infty \ge r_0 \ge \operatorname{diam} U$ and considering, according to Lemma 10.41 the function φ defined by

$$\varphi(x) := 2\int_0^{\|x\|} \omega(t)\,dt \quad \text{for all } x \in X \text{ with } \|x\| < r_0, \tag{10.9}$$

we see from (10.8) that

$$f(x) = \max_{(a,a^*)\in \operatorname{gph}\partial_C f} \big(\langle a^*, x-a\rangle + f(a) - \varphi(x-a)\big) \quad \text{for all } x \in U. \tag{10.10}$$

If the concave function $\omega(\cdot)$ is finite on $[0,r_0[$, we know that it is continuous on $]0,r_0[$ hence continuous on $[0,r_0[$. So, whenever the norm $\|\cdot\|$ is Fréchet differentiable off zero, it is easily checked that the function φ is Fréchet differentiable on $B(0,r_0)$ with

$$D\varphi(0) = 0 \quad \text{and} \quad D\varphi(x) = 2\omega(\|x\|)(D\|\cdot\|)(x) \quad \text{for } 0 < \|x\| < r_0;$$

for $r_0 = +\infty$ the convention $B(0,r_0) = X$ is used. Under the finiteness of $\omega(\cdot)$ on $[0,r_0[$ and the Fréchet differentiability of the norm off zero, (10.10) provides a representation of f on U as a pointwise maximum of Fréchet differentiable functions on U.

We have then proved in particular the following result:

PROPOSITION 10.42. Let U be a nonempty open convex set of a Banach space $(X,\|\cdot\|)$ and $f : U \to \mathbb{R}$ be a continuous function which is semiconvex with respect to a nondecreasing modulus function $\omega(\cdot)$ with $\lim_{r\downarrow 0}\omega(r) = 0 = \omega(0)$. Assume that the norm is Fréchet differentiable off zero and that $\omega(\cdot)$ is concave and finite on $[0,r_0[$ for some $+\infty \ge r_0 \ge \operatorname{diam} U$. Then the following hold:
(a) The function φ, defined on $B(0,r_0)$ by

$$\varphi(x) := 2\int_0^{\|x\|} \omega(t)\,dt \quad \text{for all } x \in B(0,r_0),$$

is Fréchet differentiable on $B(0,r_0)$ with

$$D\varphi(0) = 0 \quad \text{and} \quad D\varphi(x) = 2\omega(\|x\|)\big(D\|\cdot\|\big)(x) \quad \text{for } 0 < \|x\| < r_0.$$

(b) For every $x \in U$, one has

$$f(x) = \max_{(a,a^*)\in \operatorname{gph}\partial_C f} \big(\langle a^*, x-a\rangle + f(a) - \varphi(x-a)\big),$$

and each function $x \mapsto \langle a^*, x-a\rangle + f(a) - \varphi(x-a)$ is Fréchet differentiable on U.

10.3.2. Max-representation in uniformly convex space. The assertion (b) in Proposition 10.42 and the analysis of arguments yielding to its justification lead first to the study of the existence of a concave modulus $\tilde{\omega}(\cdot)$ such that the $C\omega(\cdot)$-semiconvexity of f, for some real constant $C > 0$, is equivalent to its $\tilde{\omega}(\cdot)$-semiconvexity; such a property obviously holds true whenever $\omega_f(\cdot) \le \tilde{\omega}(\cdot) \le K\omega(\cdot)$ for some real constant $K > 0$, where $\omega_f(\cdot)$ denotes the least modulus of semiconvexity of f on U. On the other hand, more smoothness (than the Fréchet differentiability) for the function φ in Proposition 10.42 will yield a better representation for f.

Before proceeding with the existence of concave modulus functions, let us introduce certain functions which can serve to test the semiconvexity of a function.

DEFINITION 10.43. Let U be a nonempty open convex set of a normed space X and $f : U \to \mathbb{R} \cup \{+\infty\}$ be an extended real-valued proper function. We define the following *control functions of subgradients of* f:

$$\underline{\omega}_f(0) = \underline{\omega}_{f,\partial}(0) = \overline{\omega}_{f,\partial}(0) = 0$$

and, for every real $r > 0$,

$$\underline{\omega}_f(r) := \sup_{x \in \operatorname{dom} f} \inf_{x^* \in X^*} \sup \left\{ \left(\frac{\langle x^*, h\rangle - f(x+h) + f(x)}{\|h\|} \right)^+ : x+h \in U, 0 < \|h\| \le r \right\},$$

$$\underline{\omega}_{f,\partial}(r) := \sup_{x \in \operatorname{Dom} \partial_C f} \inf_{x^* \in \partial_C f(x)} \sup_{\substack{x+h \in U \\ 0<\|h\|\le r}} \left(\frac{\langle x^*, h\rangle - f(x+h) + f(x)}{\|h\|} \right)^+,$$

$$\overline{\omega}_{f,\partial}(r) := \sup_{x \in \operatorname{Dom} \partial_C f} \sup_{x^* \in \partial_C f(x)} \sup_{\substack{x+h \in U \\ 0<\|h\|\le r}} \left(\frac{\langle x^*, h\rangle - f(x+h) + f(x)}{\|h\|} \right)^+,$$

with the convention that any above supremum is zero whenever the set over which it is taken is empty according to the feature that $\rho^+ \ge 0$ for any extended real ρ.

If C is a nonempty set of X and $f : C \to \mathbb{R} \cup \{+\infty\}$ is finite at some point in C, we put

$$\underline{\omega}_{f,C}(\cdot) := \underline{\omega}_{f_C}(\cdot), \ \ \underline{\omega}_{f,C,\partial}(\cdot) := \underline{\omega}_{f_C,\partial}(\cdot) \ \text{ and } \ \overline{\omega}_{f,C,\partial}(\cdot) := \overline{\omega}_{f_C,\partial}(\cdot),$$

where f_C is the function defined on X by $f_C(x) = f(x)$ if $x \in C$ and $f_C(x) = +\infty$ if $x \in X \setminus C$.

PROPOSITION 10.44. Let U be a nonempty open convex set of a normed space X and $f : U \to \mathbb{R} \cup \{+\infty\}$ be a proper function. The following hold:
(a) One has $\omega_f(\cdot) \le 2\underline{\omega}_f(\cdot)$, where $\omega_f(\cdot)$ is the least modulus of semiconvexity of f.
(b) If the function f is semiconvex on U, then $\overline{\omega}_{f,\partial}(\cdot) \le \omega_f(\cdot)$.
(c) If X is a Banach space and f is lower semicontinuous, then f is semiconvex on U if and only if the equality $\lim_{r \downarrow 0} \overline{\omega}_{f,\partial}(r) = 0$ holds.
(d) If f is finite and locally Lipschitz on U, then $\underline{\omega}_f(\cdot) \le \underline{\omega}_{f,\partial}(\cdot) \le \overline{\omega}_{f,\partial}(\cdot)$.

PROOF. (a) Fix any real $r > 0$ such that the set over which the right supremum in the definition of $\underline{\omega}_f(r)$ is nonempty. Consider any $x \in \operatorname{dom} f$, $h \in X$ with

$\|h\| = 1$, and $\sigma, \tau > 0$ with $x - \sigma h \in U$, $x + \tau h \in U$ and $\sigma + \tau \le r$. For each $x^* \in X^*$ and for $q(x, x^*) := \sup \left\{ \left(\frac{\langle x^*, v\rangle - f(x+v) + f(x)}{\|v\|} \right)^+ : 0 < \|v\| \le r \right\}$, we have

$$\langle x^*, h\rangle - \frac{f(x+\tau h) - f(x)}{\tau} \le q(x, x^*), \quad \langle x^*, -h\rangle - \frac{f(x - \sigma h) - f(x)}{\sigma} \le q(x, x^*),$$

hence adding the latter inequalities gives

$$\frac{f(x) - f(x - \sigma h)}{\sigma} - \frac{f(x + \tau h) - f(x)}{\tau} \le 2q(x, x^*).$$

Since $q(x, x^*) \ge 0$, it results that

$$\left(\frac{f(x) - f(x - \sigma h)}{\sigma} - \frac{f(x + \tau h) - f(x)}{\tau} \right)^+ \le 2q(x, x^*).$$

Taking the supremum over (h, σ, τ) with $\|h\| = 1$, $\sigma, \tau > 0$ with $x - \sigma h \in U$, $x + \tau h \in U$ and $\sigma + \tau \le r$, then the infimum over $x^* \in X^*$ and then the supremum over $x \in \operatorname{dom} f$, we obtain according to Proposition 10.5, $\omega_f(r) \le 2\underline{\omega}_f(r)$.
(b) If the function f is semiconvex on U, then f is semiconvex with respect to $\omega_f(\cdot)$ as a modulus of semiconvexity on U (see Definition 10.4), so (b) follows from the description of Clarke subdifferential of semiconvex functions in Theorem 10.29.
(c) If f is semiconvex, we know that $\lim_{r \downarrow 0} \omega_f(r) = 0$, so the assertion (b) above tells us that $\lim_{r \downarrow 0} \overline{\omega}_{f,\partial}(r) = 0$. Suppose now that $\lim_{r \downarrow 0} \overline{\omega}_{f,\partial}(r) = 0$. On one hand $\overline{\omega}_{f,\partial}(\cdot)$ is a nondecreasing function which is continuous at 0 with $\overline{\omega}_{f,\partial}(0) = 0$, and on the other hand the definition of $\overline{\omega}_{f,\partial}(\cdot)$ tells us that, for all $x, y \in U$ with $x \in \operatorname{Dom} \partial_C f$ and all $x^* \in \partial_C f(x)$,

$$\langle x^*, y - x\rangle \le f(y) - f(x) + \|x - y\| \, \overline{\omega}_{f,\partial}(\|x - y\|).$$

By Lemma 10.3(b) choose a nondecreasing upper semicontinuous function ω from $[0, +\infty[$ into $[0, +\infty]$ with $\lim_{r \downarrow 0} \omega(r) = 0$ and such that $\overline{\omega}_{f,\partial}(r) \le \omega(r)$ for all $r \ge 0$. Then, for all $x, y \in U$ with $x \in \operatorname{Dom} \partial_C f$ and all $x^* \in \partial_C f(x)$,

$$\langle x^*, y - x\rangle \le f(y) - f(x) + \|x - y\| \, \omega(\|x - y\|).$$

This combined with the theorem on subdifferential characterization of lower semicontinuous semiconvex functions (see Theorem 10.38) guarantees that f is semiconvex with respect to $2\omega(\cdot)$ as a modulus of semiconvexity on U.
(d) Concerning (d) it is enough to recall that $\partial_C f(x) \ne \emptyset$, for all $x \in U$, whenever f is finite and locally Lipschitz on the open set U, so $\operatorname{dom} f = \operatorname{Dom} \partial_C f$. $\square$

In the presence of local Lipschitz property, we have the following equivalences.

PROPOSITION 10.45. Let U be a nonempty open convex set of a normed space X and $f : U \to \mathbb{R}$ be a real-valued locally Lipschitz function. The following assertions are equivalent:
(a) the function f is semiconvex on U;
(b) there exists a mapping $\zeta : U \to X^*$ such that for every real $\varepsilon > 0$, there exists some $\delta > 0$ such that for any $x \in U$

$$\langle \zeta(x), y - x\rangle \le f(y) - f(x) + \varepsilon\|y - x\| \quad \text{for all } y \in U \text{ with } \|y - x\| \le \delta;$$

(c) for every real $\varepsilon > 0$, there exists some $\delta > 0$ such that for each $x \in U$ there is some $x^* \in X^*$ for which

$$\langle x^*, y - x\rangle \le f(y) - f(x) + \varepsilon\|y - x\| \quad \text{for all } y \in U \text{ with } \|y - x\| \le \delta;$$

(d) for every real $\varepsilon > 0$, there exists some $\delta > 0$ such that for each $x \in U$ there is some $x^* \in \partial_C f(x)$ for which

$$\langle x^*, y - x\rangle \le f(y) - f(x) + \varepsilon\|y - x\| \quad \text{for all } y \in U \text{ with } \|y - x\| \le \delta;$$

(e) for every real $\varepsilon > 0$, there exists some $\delta > 0$ such that for every $x \in U$ and every $x^* \in \partial_C f(x)$ one has

$$\langle x^*, y - x\rangle \le f(y) - f(x) + \varepsilon\|y - x\| \quad \text{for all } y \in U \text{ with } \|y - x\| \le \delta.$$

PROOF. Assume that f is locally Lipschitz on U.

If f is semiconvex on U, say with respect to some modulus $\omega(\cdot)$, choosing some $\zeta(x) \in \partial_C f(x)$ for each $x \in U$ according to the non-emptiness of $\partial_C f(x)$ due to the local Lipschitz property of f, the description in Theorem 10.29 guarantees that the property (b) holds true. Under the same semiconvexity of f, considering the least modulus of semiconvexity $\omega_f(\cdot)$ of f, Definition 10.4 says that $\lim_{r\downarrow 0}\omega_f(r) = 0$, so the inequality $\overline{\omega}_{f,\partial}(\cdot) \le \omega_f(\cdot)$ in (b) of the previous proposition implies that $\overline{\omega}_{f,\partial}(r) \to 0$ as $r \downarrow 0$, which is the property (e). We then deduce that (a) $\Rightarrow$ (b) $\Rightarrow$ (c) and (a) $\Rightarrow$ (e) $\Rightarrow$ (d) $\Rightarrow$ (c) since the implications (b) $\Rightarrow$ (c) and (e) $\Rightarrow$ (d) $\Rightarrow$ (c) are obvious.

Finally, the property (c) means that $\underline{\omega}_f(r) \to 0$ as $r \downarrow 0$, hence the inequality $\omega_f(\cdot) \le 2\underline{\omega}_f(\cdot)$ in (a) of the previous proposition ensures that $\omega_f(r) \to 0$ as $r \downarrow 0$. This combined with Definition 10.4 justifies the implication (c) $\Rightarrow$ (a) and finishes the proof of the proposition. $\square$

When f is locally Lipschitz on U, another useful function needs to be associated with the upper Dini directional derivative.

DEFINITION 10.46. Let U be an open convex set of the normed space X and $f : U \to \mathbb{R}$ be a real-valued locally Lipschitz function. For every real $r \ge 0$, we define the *control difference function of the upper Dini directional derivative* of f at $x \in U$ by

$$\widehat{\omega}_f(r) := \sup\left\{\left(\overline{d}_D^+ f(x;h) - \overline{d}_D^+ f(x+sh;h)\right)^+ : \|h\| = 1, 0 \le s \le r, x, x+sh \in U\right\},$$

with the usual convention that the above supremum is zero whenever the set over which it is taken is empty.

PROPOSITION 10.47. Let U be a nonempty open convex set of the normed space X and $f : U \to \mathbb{R}$ be a real-valued locally Lipschitz function. The following hold:
(a) the function $\widehat{\omega}_f$ is nondecreasing and subadditive on $[0, +\infty[$ with $\widehat{\omega}_f(0) = 0$, and hence it is finite on $[0, \infty[$ whenever it is continuous at 0.
(b) The connexion between $\widehat{\omega}_f(\cdot)$ and the function $\omega_f(\cdot)$ in Definition 10.4 is given by the inequalities

$$\omega_f(r) \le \widehat{\omega}_f(r) \le 2\omega_f(r) \quad \text{for all } r \ge 0;$$

so the locally Lipschitz function f is semiconvex on U if and only if $\lim_{r\downarrow}\widehat{\omega}_f(r) = 0$.

PROOF. We may suppose that the space X is not reduced to zero.
(a) The equality $\widehat{\omega}_f(0) = 0$ is evident. Let $r_1, r_2 \ge 0$. Take any $x \in U$, $h \in X$ with $\|h\| = 1$, and $s \in [0, r_1 + r_2]$ such that $x + sh \in U$. Write $s = s_1 + s_2$ with $0 \le s_i \le r_i$

for $i = 1, 2$, and observe that $x + s_i h \in U$. By the inequality $(a+b)^+ \leq a^+ + b^+$ in $\mathbb{R}$, we then have

$$\begin{aligned}
&\left(\overline{d}_D^+ f(x;h) - \overline{d}_D^+ f(x+sh;h)\right)^+ \\
&\leq \left(\overline{d}_D^+ f(x;h) - \overline{d}_D^+ f(x+s_1;h)\right)^+ + \left(\overline{d}_D^+ f(x+s_1 h;h) - \overline{d}_D^+ f((x+s_1 h)+s_2 h;h)\right)^+ \\
&\leq \widehat{\omega}_f(r_1) + \widehat{\omega}_f(r_2),
\end{aligned}$$

hence $\widehat{\omega}_f(r_1 + r_2) \leq \widehat{\omega}_f(r_1) + \widehat{\omega}_f(r_2)$.

If $\widehat{\omega}_f$ is continuous at 0, then it is finite on some nontrivial interval with 0 as left-point and hence finite on the whole interval $[0, +\infty[$ thanks to the subadditivity property.

(b) To prove the first inequality in (b), fix any $x \in U$, $h \in X$ with $\|h\| = 1$, and $\sigma, \tau > 0$ with $x - \sigma h, x + \tau h \in U$ and $\sigma + \tau \leq r$. Theorem 6.2(c) furnishes some $\sigma_0 \in]0, \sigma[$ and $\tau_0 \in]0, \tau[$ such that $\sigma^{-1}[f(x) - f(x - \sigma h)] \leq \overline{d}_D^+ f(x - \sigma_0 h; h)$ and $\tau^{-1}[f(x + \tau h) - f(x)] \geq \overline{d}_D^+ f(x + \tau_0 h; h)$, and this yields

$$\begin{aligned}
&\left(\frac{f(x) - f(x - \sigma h)}{\sigma} - \frac{f(x + \tau h) - f(x)}{\tau}\right)^+ \\
&\leq \left(\overline{d}_D^+ f(x - \sigma_0 h; h) - \overline{d}_D^+ f(x - \sigma_0 h + (\sigma_0 + \tau_0) h; h)\right)^+ \leq \widehat{\omega}_f(r),
\end{aligned}$$

which gives $\omega_f(r) \leq \widehat{\omega}_f(r)$ according to Proposition 10.5.

To prove the second inequality, fix any $x \in U$, $r > 0$, $h \in X$ with $\|h\| = 1$, and $s \in]0, r]$ with $x + sh \in U$. Choose sequences $(\sigma_n)_n$ and $(\tau_n)_n$ in $]0, +\infty[$ tending to 0 and such that

$$\overline{d}_D^+ f(x;h) = \lim_{n\to\infty} \sigma_n{}^{-1}[f(x + \sigma_n h) - f(x)],$$

$$\overline{d}_D^+ f(x + sh; h) = \lim_{n\to\infty} \tau_n^{-1}[f(x + sh + \tau_n h) - f(x + sh)],$$

and note that $x + \sigma_n h, x + \tau_n h, x + sh + \tau_n h \in U$ for large n. The latter equalities combined with Proposition 10.5 give on the one hand

$$\begin{aligned}
&\left(\overline{d}_D^+ f(x;h) - \frac{f(x+sh) - f(x)}{s}\right)^+ \\
&= \lim_{n\to\infty} \left(\frac{f(x + \sigma_n h) - f(x)}{\sigma_n} - \frac{f(x+sh) - f(x + \sigma_n h)}{s - \sigma_n}\right)^+ \leq \omega_f(r)
\end{aligned}$$

and on the other hand

$$\begin{aligned}
&\left(\frac{fx+sh) - f(x)}{s} - \overline{d}_D^+ f(x+sh;h)\right)^+ \\
&= \lim_{n\to\infty} \left(\frac{f(x+sh) - f(x+\tau_n h)}{s - \tau_n} - \frac{f(x+sh+\tau_n h) - f(x+sh)}{\tau_n}\right)^+ \leq \omega_f(r).
\end{aligned}$$

We deduce $\left(\overline{d}_D^+ f(x;h) - \overline{d}_D^+ f(x+sh;h)\right)^+ \leq 2\omega_f(r)$ thanks again to the inequality $(a+b)^+ \leq a^+ + b^+$ in $\mathbb{R}$, so $\widehat{\omega}_f(r) \leq 2\omega_f(r)$ as desired.

Finally, the equivalence property concerning the semiconvexity of f is a consequence of the above inequalities $\omega_f(\cdot) \leq \widehat{\omega}_f(\cdot)) \leq 2\omega_f(\cdot)$ and of Definition 10.4. $\square$

Through the above proposition we can tackle the question of existence of concave modulus function in the translation of semiconvexity of a function. The following lemma is useful for that.

LEMMA 10.48. Let $\omega : [0,+\infty[\to [0,+\infty[$ be a nondecreasing finite-valued (modulus) function with $\lim_{r\downarrow 0}\omega(r) = 0 = \omega(0)$. Assume that $\omega(\cdot)$ is subadditive, that is, $\omega(r+s) \leq \omega(r)+\omega(r)$ for all $r, s \geq 0$. Then the least finite-valued concave function $\omega_0(\cdot)$ on $[0,+\infty[$ majorizing $\omega(\cdot)$ exists. Further, $\omega_0(\cdot)$ is a nondecreasing finite-valued continuous subadditive concave (modulus) function from $[0,+\infty[$ into $[0,+\infty[$ such that

$$\omega_0(\lambda r) \leq (\lambda+1)\omega(r) \quad \text{for all } \lambda \geq 0,\ r \geq 0,$$

hence in particular $\omega(\cdot) \leq \omega_0(\cdot) \leq 2\omega(\cdot)$ on $[0,+\infty[$.

PROOF. Fix any real $r_0 > 0$ and consider the affine function defined by $a(r) := \omega(r_0) + \big(\omega(r_0)/r_0\big)\, r$ for all $r \in [0,+\infty[$. For any integer $k \in \mathbb{N}$ and any $r \in [(k-1)r_0, kr_0]$, since $\omega(\cdot)$ is nondecreasing and subadditive, we have

$$\omega(r) \leq \omega(kr_0) \leq k\omega(r_0) = \omega(r_0) + (k-1)\omega(r_0) \leq \omega(r_0) + \frac{r}{r_0}\omega(r_0) = a(r),$$

where the first and last inequalities are due to the inclusion $r \in [(k-1)r_0, kr_0]$. Consequently, the affine (thus concave) function $a(\cdot)$ majorizes $\omega(\cdot)$, hence the set of finite-valued concave functions on $[0,+\infty[$ majorizing $\omega(\cdot)$ is nonempty and its pointwise infimum $\omega_0(\cdot)$ is finite-valued and concave; and the latter property ensures the continuity of $\omega_0(\cdot)$ on $]0,+\infty[$.

Now take any real $\lambda > 0$. It results from the definitions of $\omega_0(\cdot)$ and $a(\cdot)$ that $\omega_0(\lambda r_0) \leq a(\lambda r_0) = (1+\lambda)\omega(r_0)$. The inequality $\omega_0(\lambda r_0) \leq (\lambda+1)\omega(r_0)$ being true for any $r_0 > 0$ we obtain, for all $r > 0$

$$\omega_0(\lambda r) \leq (\lambda+1)\omega(r), \quad \text{hence in particular } \omega(r) \leq \omega_0(r) \leq 2\omega(r). \tag{10.11}$$

Therefore, $\omega_0(r) \to 0$ as $r \to 0$ with $r > 0$ and since $\omega_0(\cdot)$ is concave with $\omega_0(\cdot) \geq 0$, it ensues that $\omega_0(0) = 0 = \lim_{r\downarrow 0}\omega_0(r)$, so $\omega_0(\cdot)$ is continuous on $[0,+\infty[$ and the inequalities in (10.11) hold true for all $r \geq 0$.

To show that $\omega_0(\cdot)$ is nondecreasing, suppose by contradiction that there are $0 < r_1 < r_2$ with $\omega_0(r_2) < \omega_0(r_1)$. For every $r > r_2$, observing that $r_2 = \frac{r_2-r_1}{r-r_1}r + \frac{r-r_2}{r-r_1}r_1$ the concavity of $\omega_0(\cdot)$ gives

$$\omega_0(r_2) \geq \frac{r_2-r_1}{r-r_1}\omega_0(r) + \frac{r-r_2}{r-r_1}\omega_0(r_1),$$

which is equivalent to $r\big(\omega_0(r_2) - \omega_0(r_1)\big) - r_1\omega_0(r_2) + r_2\omega_0(r_1) \geq (r_2 - r_1)\omega_0(r)$. Since $\omega_0(r_2) - \omega_0(r_1) < 0$, we deduce $\omega_0(r) \to -\infty$ as $r \to +\infty$, and this contradicts the fact that $\omega_0(\cdot) \geq 0$.

Finally, the subadditivity of $\omega_0(\cdot)$ is a consequence of Lemma 10.49 below. □

The lemma involved in the above proof concerning the subadditivity of concave modulus has its own interest. The subadditivity property will be even established for any non-negative concave function on $[0,+\infty[$ null at 0.

LEMMA 10.49. Let $\theta : [0,+\infty[\to [0,+\infty[$ be a non-negative function.
(a) If $t \mapsto t^{-1}\theta(t)$ is nonincreasing on $]0,+\infty[$, then θ is subadditive on $]0,+\infty[$.
(b) If θ is concave with $\theta(0) = 0$, then θ is subadditive on $[0,+\infty[$.

PROOF. (a) To justify (a) it suffices, taking $r, s > 0$, to see by the nonincreasing property of $t \mapsto \theta(t)/t$ that

$$\theta(r+s) = r\frac{\theta(r+s)}{r+s} + s\frac{\theta(r+s)}{r+s} \leq r\frac{\theta(r)}{r} + s\frac{\theta(s)}{s} = \theta(r) + \theta(s).$$

(b) Concerning (b), because of the concavity of θ, the function $t \mapsto t^{-1}\big(\theta(t) - \theta(0)\big) = t^{-1}\theta(t)$ is nonincreasing on $]0, +\infty[$, hence θ is subadditive on $]0, +\infty[$ according to (a). This justifies the subadditivity on $[0, +\infty[$ since the subadditivity inequality is obviously true if, in the above inequality, either r or s is null. $\square$

REMARK 10.50. It is worth noting that the non-negative function θ on $]0, +\infty[$ with $\theta(t) := \sqrt{t} + 1/t$ satisfies the condition of decreasing property of $t \mapsto \theta(t)/t$ but θ is neither concave nor convex. $\square$

PROPOSITION 10.51. Let $\omega : [0, +\infty[\to [0, +\infty[$ be a nondecreasing finite-valued (modulus) function with $\lim_{r \downarrow 0} \omega(r) = 0 = \omega(0)$. Let U be an open convex set of the normed space X and $f : U \to \mathbb{R}$ be a real-valued continuous function which is $\omega(\cdot)$-semiconvex on U. Then there exists a nondecreasing finite-valued continuous concave (modulus) function $\omega_0 : [0, +\infty[\to [0, +\infty[$ such that f is $\omega_0(\cdot)$-semiconvex on U and $\omega_0(\cdot) \leq 4\omega_f(\cdot) \leq 4\omega(\cdot)$, where $\omega_f(\cdot)$ is the least modulus of semiconvexity of the function f.

PROOF. By Proposition 10.23 the $\omega(\cdot)$-semiconvex function f is locally Lipschitz, hence from Proposition 10.47 the function $\widehat{\omega}_f : [0, +\infty[\to [0, +\infty[$ (in Definition 10.46) is a nondecreasing finite-valued subadditive (modulus) function with $\omega_f(\cdot) \leq \widehat{\omega}_f(\cdot) \leq 2\omega_f(\cdot)$. By Lemma 10.48 there exists a nondecreasing finite-valued continuous concave (modulus) function $\omega_0 : [0, +\infty[\to [0, +\infty[$ such that $\widehat{\omega}_f(\cdot) \leq \omega_0(\cdot) \leq 2\widehat{\omega}_f(\cdot)$. Then $\omega_0(\cdot) \leq 4\omega_f(\cdot) \leq 4\omega(\cdot)$ and f is $\omega_0(\cdot)$-semiconvex since $\omega_f(\cdot) \leq \omega_0(\cdot)$. $\square$

Now we can come back to the function φ (given by $\varphi(x) = 2\int_0^{\|x\|} \omega(t)\,dt$) considered in the beginning of the section. Our purpose here is to study conditions ensuring the uniform continuity or Hölder property for the Fréchet derivative of the function φ. Before proving the next lemma related to that property, let us observe the following:

LEMMA 10.52. For any nonzero x, y in a normed space $(X, \|\cdot\|)$, one has

$$\left| \frac{y}{\|y\|} - \frac{x}{\|x\|} \right| \leq \frac{2}{\max\{\|x\|, \|y\|\}} \|y - x\|.$$

PROOF. Writing

$$\left| \frac{y}{\|y\|} - \frac{x}{\|x\|} \right| = \left| \frac{1}{\|y\|}(y - x) + \big(\frac{1}{\|y\|} - \frac{1}{\|x\|}\big)x \right| \leq \frac{1}{\|y\|}\|y - x\| + \frac{|\,\|y\| - \|x\|\,|}{\|y\|\,\|x\|}\|x\|,$$

we see that the first member of the desired inequality is not greater than $(2/\|y\|)\|y - x\|$ and by symmetry not greater than $(2/\|x\|)\|y - x\|$ as well. Thus the inequality of the lemma holds true. $\square$

LEMMA 10.53. Let X be a vector space endowed with a norm $\|\cdot\|$ which is differentiable off zero and such that the derivative of the norm is uniformly continuous on the unit sphere $\mathbb{S}_X$, and let $\omega_1 : [0, +\infty[\to [0, +\infty[$ be the least modulus of continuity relative to $\mathbb{S}_X$ of the derivative of the norm $\|\cdot\|$. Let ω :

$[0, +\infty[\to [0, +\infty[$ be a nondecreasing continuous concave function with $\omega(0) = 0$. For $r \geq 0$, let

$$\lambda(r) := \sup_{t \geq r} \omega(t)\omega_1(2r/t), \ \mu(r) := 3\omega(r) + \lambda(r), \ \eta(r) := 2 \sup_{0 \leq t \leq r} \mu(t),$$

with the convention $\lambda(0) = 0$. Let also the function $\varphi : X \to [0, +\infty[$ defined by

$$\varphi(x) := 2 \int_0^{\|x\|} \omega(t)\, dt \quad \text{for all } x \in X.$$

(A) Under the above assumptions the following hold:
(a1) The function φ is Fréchet differentiable on X and the modulus of continuity $\omega^c_{D\varphi}(\cdot)$ of its derivative $D\varphi$ on X satisfies

$$\omega^c_{D\varphi}(r) := \sup_{x,y \in X, \|y-x\| \leq r} \|D\varphi(y) - D\varphi(x)\| \leq \eta(r) \leq +\infty \quad \text{for all } r \geq 0.$$

(a2) If in addition the function $\omega(\cdot)$ is bounded on $[0, +\infty[$, then the derivative of φ is uniformly continuous on X and $\omega^c_{D\varphi}(\cdot)$ is bounded on $[0, +\infty[$; further, for some real constant $C_1 > 0$,

$$\omega^c_{D\varphi}(r) \leq C_1 \left(\omega(\sqrt{r}) + \omega_1(2\sqrt{r})\right) \quad \text{for all } r \geq 0.$$

(B) Assume in the remaining of the lemma that there is $0 < \beta \leq 1$ such that the least modulus of continuity ω_1 on $\mathbb{S}_X$ of the derivative of the norm satisfies $\omega_1(r) \leq Cr^\beta$ for all $r \in [0, +\infty[$.
(b) If $\omega(\cdot)$ is bounded on $[0, +\infty[$ and $\omega(r) \leq c_2 r$ for all $r \geq 0$, then for some real constant $C_2 \geq 0$

$$\omega^c_{D\varphi}(r) \leq C_2 r^\beta \quad \text{for all } r \geq 0.$$

(c) If $\beta = 1$, then for some real constant $C_3 \geq 0$

$$\omega^c_{D\varphi}(r) \leq C_3 \omega(r) \quad \text{for all } r \geq 0.$$

(d) If $0 < \alpha \leq \beta$ and $\omega(r) \leq c_4 r^\alpha$ for all $r \geq 0$, then for some real constant $C_4 \geq 0$

$$\omega^c_{D\varphi}(r) \leq C_4 r^\alpha \quad \text{for all } r \geq 0.$$

PROOF. The case when the non-negative nondecreasing continuous concave function ω is the null function being trivial, we suppose that $\omega(r) > 0$ for all $r > 0$.
(A) (a1) Denoting by $\Lambda(x)$ the Fréchet derivative of the norm $\|\cdot\|$ at $x \neq 0$, we already observed in Proposition 10.42 that the function φ is Fréchet differentiable on X with

$$D\varphi(0) = 0 \quad \text{and} \quad D\varphi(x) = 2\omega(\|x\|)\Lambda(x) \quad \text{for all } x \in X \setminus \{0\}.$$

Consider $x, y \in X$ with $0 < \|x - y\| \leq r$ and $\|x\| \leq \|y\|$. If $x = 0$, then

$$\begin{aligned} \|D\varphi(y) - D\varphi(x)\| &= 2\omega(\|y\|)\|\Lambda(y)\| \leq 2\omega(\|y\|) \\ &= 2\omega(\|y - x\|) \leq 2\omega(r) \leq \mu(r) \leq \eta(r). \end{aligned}$$

Suppose that $x, y \neq 0$. We have

$$\begin{aligned} \|D\varphi(x) - D\varphi(y)\| &= 2\,\|\,\omega(\|y\|)\Lambda(y) - \omega(\|x\|)\Lambda(x)\,\| \\ &= 2\left\|\,\omega(\|y\|)\big(\Lambda(y) - \Lambda(x)\big) + \big(\omega(\|y\|) - \omega(\|x\|)\big)\Lambda(x)\,\right\| \\ &\leq 2\big(\omega(\|y\|) - \omega(\|x\|)\big)\,\|\Lambda(x)\| + 2\omega(\|y\|)\,\|\Lambda(y) - \Lambda(x)\| \\ &\leq 2\omega(\|y - x\|) + 2\omega(\|y\|)\,\|\Lambda(y) - \Lambda(x)\|, \end{aligned}$$

where the last inequality is due to the subadditivity (see Lemma 10.49) and the nondecreasing property of $\omega(\cdot)$ and to the equality $\|\Lambda(x)\| = 1$. If $\|y\| \le \|y-x\|$, then using the inequality $\|\Lambda(y) - \Lambda(x)\| \le 2$ we obtain

$$\|D\varphi(y) - D\varphi(x)\| \le 2\omega(\|y-x\|) + 4\omega(\|y-x\|) \le 6\omega(r) \le 2\mu(r) \le \eta(r).$$

On the other hand, if $\|y\| > \|y-x\|$, we have (since $\Lambda(x) = \Lambda(x/\|x\|)$)

$$\begin{aligned}\|D\varphi(y) - D\varphi(x)\| &\le 2\omega(\|y-x\|) + 2\omega(\|y\|) \left\|\Lambda(\frac{y}{\|y\|}) - \Lambda(\frac{x}{\|x\|})\right\| \\ &\le 2\omega(\|y-x\|) + 2\omega(\|y\|)\,\omega_1\left(\left\|\frac{y}{\|y\|} - \frac{x}{\|x\|}\right\|\right) \\ &\le 2\omega(\|y-x\|) + 2\omega(\|y\|)\,\omega_1\left(\frac{2}{\|y\|}\|y-x\|\right),\end{aligned}$$

where the last inequality is due to Lemma 10.52 above. Taking the definitions of λ and μ into account, we deduce that

$$\|D\varphi(y) - D\varphi(x)\| \le 2\omega(\|y-x\|) + 2\lambda(\|y-x\|) \le 2\mu(\|y-x\|) \le \eta(r),$$

which completes the proof of the inequality

$$\omega^c_{D\varphi}(r) \le \eta(r) \quad \text{for all } r \ge 0. \tag{10.12}$$

(a2) Observing that $\|\Lambda(y) - \Lambda(x)\| \le 2$ for all $x, y \in \mathbb{S}_X$, we see that $\omega_1(r) \le 2$ for all $r \ge 0$ since ω_1 is the least modulus of continuity of the derivative of the norm $\|\cdot\|$ on the unit sphere $\mathbb{S}_X$. This says that $\omega_1(\cdot)$ is bounded on $[0, +\infty[$. Then assuming that $\omega(\cdot)$ is bounded on $[0, +\infty[$, the functions $\lambda(\cdot)$, $\mu(\cdot)$, $\eta(\cdot)$ are bounded on $[0, +\infty[$ as well.

Fix any $r \in]0, 1[$ and any $t \ge r$. For $K := \sup_{s\ge 0} \omega(s)$, we have

$$\omega(t)\omega_1(2r/t) \le K\,\omega_1(2\sqrt{r}) \quad \text{if } t \ge \sqrt{r}$$

and

$$\omega(t)\omega_1(2r/t) \le \omega(\sqrt{r})\omega_1(2) \quad \text{if } r \le t < \sqrt{r},$$

so for $K_1 := \max\{K, \omega_1(2)\}$ we obtain $\lambda(r) \le K_1\big(\omega(\sqrt{r}) + \omega_1(2\sqrt{r})\big)$. Noting that $r < \sqrt{r}$ (since $0 < r < 1$), we have also $\omega(r) \le \omega(\sqrt{r})$, hence for $K_2 := 3 + K_1$, we get $\mu(r) \le K_2\big(\omega(\sqrt{r}) + \omega_1(2\sqrt{r})\big)$ and

$$\eta(r) = 2 \sup_{s\in[0,r]} \mu(s) \le 2K_2 \sup_{s\in[0,r]} \big(\omega(\sqrt{s}) + \omega_1(2\sqrt{s})\big) \le 2K_2\big(\omega(\sqrt{r}) + \omega_1(2\sqrt{r})\big).$$

Putting $C_1 := \max\{2K_2, (\sup_{t\ge 1} \eta(t))/\omega(1)\}$, it results, for every real $r \ge 0$, that $\eta(r) \le C_1\big(\omega(\sqrt{r}) + \omega_1(2\sqrt{r})\big)$, thus $\omega^c_{D\varphi}(r) \le C_1\big(\omega(\sqrt{r}) + \omega_1(2\sqrt{r})\big)$ according to (10.12). All the properties in the assertion (a) then follow.

(B) (b) Assume that $\omega(\cdot)$ is bounded on $[0, +\infty[$ and $\omega(s) \le c_2 s$ for all $s \ge 0$. Fix any $r \in]0, 1[$ and any $t \ge r$. For $t \in]r, 1]$ we have $t^{1-\beta} \le 1$ since $1 - \beta \ge 0$, hence

$$\omega(t)\omega_1(2r/t) \le c_2 t C 2^\beta r^\beta t^{-\beta} \le c_2 C 2^\beta r^\beta;$$

on the other hand, for $t \ge 1$, putting $K := \sup_{s\ge 0} \omega(s)$ we have

$$\omega(t)\omega_1(2r/t) \le K C 2^\beta r^\beta.$$

It ensues that, for $K_1 := 2^\beta C \max\{c_2, K\}$, we have $\lambda(r) \le K_1 r^\beta$ and $\mu(r) \le (K_1 + 3c_2) r^\beta$, where the inequality for $\mu(r)$ uses the fact that $c_2 r \le c_2 r^\beta$ since $0 < r < 1$ and $0 < \beta \le 1$.

Since $\mu(\cdot)$ is bounded on $[0, +\infty[$, for $K_2 := \max\{K_1 + 3c_2, \sup_{t\geq 0} \mu(t)\}$, we obtain $\mu(r) \leq K_2 r^\beta$ for all $r \geq 0$, thus

$$\omega^c_{D\varphi}(r) \leq \eta(r) \leq 2K_2 r^\beta \quad \text{for all } r \geq 0,$$

which gives (b) with $C_2 := 2K_2$.
(c) Since $\omega(\cdot)$ is concave and $\omega(0) = 0$, the function $t \mapsto \omega(t)/t$ is nonincreasing on $]0, +\infty[$. So, assuming $\beta = 1$ hence $\omega_1(s) \leq Cs$ for all $s > 0$ (by the assumption concerning $\omega_1(\cdot)$), we see that, for $r > 0$ and $t \geq r$,

$$\omega(t)\omega_1(2r/t) \leq \omega(t)C(2r/t) = 2Cr\big(\omega(t)/t\big) \leq 2Cr\big(\omega(r)/r\big) = 2C\omega(r),$$

which yields, for all $r > 0$, according to (10.12)

$$\omega^c_{D\varphi}(r) \leq \eta(r) \leq 2(3 + 2C)\omega(r).$$

It is then enough to take $C_3 = 2(3 + 2C)$.
(d) Finally, assume that $0 < \alpha \leq \beta$ and $\omega(s) \leq c_4 s^\alpha$ for all $s \geq 0$. Fix any $r > 0$. Writing, for all $t \geq r$,

$$\omega(t)\omega_1(2r/t) \leq c_4 t^\alpha C 2^\beta r^\beta t^{-\beta} = 2^\beta c_4 C (r/t)^{\beta-\alpha} r^\alpha \leq 2^\beta c_4 C r^\alpha,$$

ensures that $\lambda(r) \leq 2^\beta c_4 C r^\alpha$, hence $\eta(r) \leq 2c_4(3 + 2^\beta C)r^\alpha$. This combined with (10.12) justifies the desired inequality in (d). □

We can now prove some maximum type representations of semiconvex functions defined on uniformly convex Banach spaces. Recall that a norm $\|\cdot\|$ on a vector space X is *uniformly convex*, or the *normed space* $(X, \|\cdot\|)$ *is uniformly convex*, when for any real $\varepsilon \in]0, 2]$ there exists a real $\eta > 0$ such that for all $x, y \in \mathbb{S}_X$ with $\|x - y\| \geq \eta$ one has

$$\left\|\frac{1}{2}(x + y)\right\| \leq 1 - \eta. \tag{10.13}$$

Defining the *modulus of convexity* $\delta_{\|\cdot\|} : [0, 2] \to [0, 1]$ of the norm by

$$\delta_{\|\cdot\|}(\varepsilon) = \inf\left\{1 - \left\|\frac{1}{2}(x + y)\right\| : x, y \in \mathbb{S}_X, \|x - y\| \geq \varepsilon\right\}, \tag{10.14}$$

we see that $\|\cdot\|$ is uniformly convex if and only if $\delta_{\|\cdot\|}(\varepsilon) > 0$ for every $0 < \varepsilon \leq 2$. We also recall that the *modulus of smoothness of a norm* $\|\cdot\|$ on a vector space X is the function $\varrho_{\|\cdot\|} : [0, +\infty[\to [0, +\infty[$ defined by

$$\varrho_{\|\cdot\|}(r) := \sup\left\{\frac{\|x + ry\| + \|x - ry\|}{2} - 1 : x, y \in \mathbb{S}_X\right\} \quad \text{for all } r \geq 0. \tag{10.15}$$

The norm $\|\cdot\|$ is *uniformly smooth* when $\lim\limits_{r\downarrow 0} \varrho_{\|\cdot\|}(r)/r = 0$. The modulus of smoothness is said to be *of power type* $1 + \beta$ with $\beta > 0$ provided there exists some real constant $C > 0$ such that $\varrho_{\|\cdot\|}(r) \leq Cr^{1+\beta}$ for all $r \geq 0$; of course, in such a case the norm is uniformly smooth. We recall that any uniformly convex Banach space admits an equivalent norm which is uniformly smooth with modulus of smoothness of power type $1 + \beta$ with $0 < \beta \leq 1$; see Subsection 18.1.2 in Chapter 18 where the above concepts are recalled and where diverse features of such spaces and norms can be found. Further, given a Gâteaux differentiable (off zero) norm $\|\cdot\|$ on a vector space X, Proposition 18.42 in the same chapter tells us that, for the modulus of continuity $\omega_1(\cdot)$ relative to the sphere $\mathbb{S}_X$ of the derivative of the norm one has

$$\omega_1(r) \le \frac{2}{r}\varrho_{\|\cdot\|}(r) \quad \text{for all } r \in]0,+\infty[. \tag{10.16}$$

THEOREM 10.54 (J. Duda and L. Zajíček: $\mathcal{C}^{1,0}$ max-representation of continuous semiconvex function in uniformly convex space). *Let X be a uniformly convex Banach space, U be a nonempty open bounded convex subset of X and $f: U \to \mathbb{R}$ be a real-valued function on U. The following assertions are equivalent:*
(a) the function f is continuous on U and semiconvex on U with respect to a nondecreasing finite-valued modulus function from $[0,+\infty[\to [0,+\infty[$ (which is null at 0 and continuous at 0);
(b) there exists a family $(F_t)_{t\in T}$ of differentiable functions from X into $\mathbb{R}$ with derivatives $(DF_t)_{t\in T}$ equi-uniformly continuous on X and such that

$$f(x) = \max_{t\in T} F_t(x) \quad \text{for all } x \in U.$$

Further, if the semiconvex function f is globally Lipschitz on U, the family $(F_t)_{t\in T}$ can be taken to be equi-Lipschitz on X.

PROOF. As recalled above, X admits an equivalent norm $\|\cdot\|$ which is uniformly smooth with modulus of smoothness $\varrho_{\|\cdot\|}(\cdot)$ of power type $1+\beta$ for some $0<\beta\le 1$, that is, there is some constant $C_0>0$ such that $\varrho_{\|\cdot\|}(r) \le C_0 r^{1+\beta}$ for all $r \ge 0$. Thus, according to (10.16) for some real constant $C>0$, the least modulus of continuity $\omega_1(\cdot)$ relative to $\mathbb{S}_X$ of the derivative of the norm $\|\cdot\|$ satisfies the inequality $\omega_1(r) \le Cr^\beta$ for all $r \ge 0$.

Suppose that f is continuous on U and semiconvex on U with respect to some nondecreasing modulus from $[0,+\infty[$ into $[0,+\infty[$ which is continuous and null at zero. By Proposition 10.51 there exists a nondecreasing continuous concave modulus $\omega_0 : [0,+\infty[\to [0,+\infty[$ with $\omega_0(0)=0$ with respect to which f is $\omega_0(\cdot)$-semiconvex on U. Putting $\omega(r) := \min\{\omega_0(r), \omega_0(\operatorname{diam} U)\}$, the function $\omega(\cdot)$ is also a nondecreasing continuous concave modulus with $\omega(0)=0$ and f is clearly $\omega(\cdot)$-semiconvex on U. Since $\omega(\cdot)$ is bounded on $[0,+\infty[$, the assertion (a) of Lemma 10.53 above tells us that the function φ in that lemma (with $\varphi(x) = 2\int_0^{\|x\|}\omega(t)\,dt$) is differentiable on X and the derivative of φ is uniformly continuous on X with a modulus of continuity $\eta(\cdot)$ given by $\eta(r) := C_1\big(\omega(\sqrt{r}) + \omega_1(2\sqrt{r})\big)$ for some real constant $C_1>0$ and all $r\ge 0$, where ω_1 is the least modulus of continuity on $\mathbb{S}_X$ of $\|\cdot\|$. Choosing some $\zeta(a) \in \partial_C f(a)$ for each $a \in T := U$ (according to the local Lipschitz property of f on U) and putting

$$F_a(x) := \langle \zeta(a), x-a\rangle + f(a) - \varphi(x-a) \quad \text{for all } x \in X,$$

each function F_a is differentiable on X and its derivative is uniformly continuous on X with $\eta(\cdot)$ as a modulus of uniform continuity. Further, the characterization of subdifferentials of semiconvex functions (see Theorem 10.29) readily yields $f(x) = \max_{a\in T} F_a(x)$ for all $x\in U$. The implication (a) $\Rightarrow$ (b) is then established.

If in addition f is globally Lipschitz on U with K as a Lipschitz contant, then for any $a\in T$ we see through Proposition 10.42(a) that, for all $x\in U$,

$$\|DF_a(x)\| \le \|\zeta(a)\| + \|D\varphi(x-a)\| \le K + 2\omega(\|x-a\|)\|\Lambda(x-a)\| \le K + 2\omega_0(\operatorname{diam} U),$$

where $\Lambda(0)=0$ and $\Lambda(u) = (D\|\cdot\|)(u)$ for $u\neq 0$. This confirms the equi-Lipschitz property of $(F_a)_{a\in T}$ on the whole space X.

Finally, under the property in (b) we see that all the functions F_t are semiconvex with respect to a common modulus. The implication (b) $\Rightarrow$ (a) then follows directly from Proposition 10.10. $\square$

REMARK 10.55. If the semiconvex function f is globally Lipschitz on U, then as we saw in the proof of Theorem 10.54, all the functions F_a with $a \in U$ are Lipschitz on X with a common Lipschitz constant. Further, the derivatives of all functions F_a are uniformly continuous on X with $\eta(\cdot)$ as a common modulus of continuity, hence all the functions F_a are semiconvex with respect to a same modulus. Consequently, the function $F : X \to \mathbb{R}$ with $F(x) := \sup_{a \in U} F_a(x)$, for all $x \in X$, is globally Lipschitz and semiconvex on the whole space X. $\square$

Under a growth condition on the modulus $\omega(\cdot)$, the next theorem provides a max-representation when the open set is not required to be bounded.

THEOREM 10.56 (J. Duda and L. Zajíček: $\mathcal{C}^{1,\alpha}$ max-representation of continuous semiconvex functions in uniformly convex space). Let $f : U \to \mathbb{R}$ be a real-valued function defined on an open convex subset U of a uniformly convex Banach space X. Let $\|\cdot\|$ be an equivalent norm on X which is uniformly smooth with modulus of smoothness of power type $1+\beta$ with $0 < \beta \le 1$. Let $\omega : [0, +\infty[\to [0, +\infty[$ be a real-valued nondecreasing modulus function with $\lim_{r \downarrow 0} \omega(r) = 0 = \omega(0)$.

(A) If $\beta = 1$, then the assertions (a) and (b) below are equivalent:
(a) the function f is continuous on U and semiconvex on U with respect to a modulus $c\omega(\cdot)$ for some real constant $c > 0$;
(b) there exists a family $(F_t)_{t \in T}$ of differentiable functions from X into $\mathbb{R}$ such that

$$f(x) = \max_{t \in T} F_t(x) \quad \text{for all } x \in U$$

and such that, for some constant $K > 0$, the derivatives $(DF_t)_{t \in T}$ are equi-uniformly continuous on X with $K\omega(\cdot)$ as a modulus of continuity.

(B) Let $\alpha \in]0, \beta]$. Then the assertions (a') and (b') below are equivalent:
(a') the function f is continuous and semiconvex of power type on U with power $1 + \alpha$;
(b') there exists a family $(F_t)_{t \in T}$ of differentiable functions from X into $\mathbb{R}$ such that

$$f(x) = \max_{t \in T} F_t(x) \quad \text{for all } x \in U$$

and such that, for some real $K > 0$, the derivatives $(DF_t)_{t \in T}$ are equi-Hölderian on X with power α and the same coefficient K, otherwise stated

$$\|DF_t(x_1) - DF_t(x_2)\| \le K\|x_1 - x_2\|^{\alpha} \quad \text{for all } x_1, x_2 \in X.$$

Further, if U is bounded and f is globally Lipschitz on U, then the family $(F_t)_{t \in T}$ can be taken to be equi-Lipschitz on X.

PROOF. By assumption, for some real constant $C > 0$, the least modulus of continuity $\omega_1(\cdot)$ of the derivative of the norm $\|\cdot\|$ satisfies the inequality $\omega_1(r) \le Cr^{\beta}$ for all $r \ge 0$. Under the property (a') consider a real $c > 0$ such that f is $\omega_0(\cdot)$-semiconvex on U for $\omega_0(r) := cr^{\alpha}$. On the contrary, under the property (a) choose, by Proposition 10.51, a nondecreasing finite continuous concave modulus function $\omega_0 : [0, +\infty[\to [0, +\infty[$ with $\omega_0(0) = 0$ with respect to which f is $\omega_0(\cdot)$-semiconvex on U and such that $\omega_0(\cdot) \le 4C_0\omega(\cdot)$ for some constant $C_0 > 0$. Under the property (a) or (b), assertions (c) and (d) of Lemma 10.53 above tell us that the function φ in that lemma (with $\varphi(x) = 2\int_0^{\|x\|} \omega_0(t)\,dt$) is differentiable on X and the derivative of φ is uniformly continuous on X with a modulus of continuity $\eta(\cdot) = C_1\omega_0(\cdot)$ for some real constant $C_1 > 0$. The derivative of φ has thus also, as a modulus of continuity, $4C_0C_1\omega(\cdot)$ under (a) and the function $r \mapsto C_1 r^\alpha$ under (a'). Choosing some $\zeta(a) \in \partial_C f(a)$ for each $a \in T := U$ (according to the local Lipschitz property of f on U) and putting

$$F_a(x) := \langle \zeta(a), x - a\rangle + f(a) - \varphi(x - a) \quad \text{for all } x \in X,$$

each function F_a is differentiable on X and its derivative is uniformly continuous on X, with as a modulus of continuity, $4C_0C_1\omega(\cdot)$ under (a) and the function $r \mapsto C_1 r^\alpha$ under (a'). Further, the characterization of subgradients of semiconvex functions (see Theorem 10.29) readily yields $f(x) = \max_{a\in T} F_a(x)$ for all $x \in U$. The implications (a) $\Rightarrow$ (b) and (a') $\Rightarrow$ (b') are then proved.

If in addition U is bounded, in the above development we can take in place of ω_0 the bounded concave function $\tilde{\omega}_0(\cdot)$ with $\tilde{\omega}_0(r) := \min\{\omega_0(r), \omega_0(\operatorname{diam} U)\}$. So, assuming that f is globally Lipschitz on U with L as a Lipschitz contant, then (as in the proof of Theorem 10.54) for any $a \in T$ we see, for all $x \in X$, that

$$\|DF_a(x)\| \le \|\zeta(a)\| + \|D\varphi(x-a)\| \le L + 2\tilde{\omega}_0(\|x-a\|)\|\Lambda(x-a)\| \le L + 2\omega_0(\operatorname{diam} U),$$

where $\Lambda(0) = 0$ and $\Lambda(u) = (D\|\cdot\|)(u)$ for $u \neq 0$. This confirms the equi-Lipschitz property of $(F_a)_{a\in T}$ on the whole space X.

Finally, under the property (b) or (b') we see that all the functions F_a are semiconvex with respect to a common modulus. The implications (b) $\Rightarrow$ (a) and (b') $\Rightarrow$ (a') then follow directly from Proposition 10.10. $\square$

REMARK 10.57. We observe that the case of Hilbert space H in Theorem 10.40(c) follows also from Theorem 10.56. Indeed, the norm of H being uniformly smooth with modulus of smoothness of power type 2, we can apply (B) of Theorem 10.56 with $\alpha = \beta = 1$. $\square$

PROPOSITION 10.58. Let X be a uniformly convex Banach space endowed with a uniformly smooth norm $\|\cdot\|$ with modulus of smoothness of power type $1 + \beta$ with $0 < \beta \le 1$. Let A be a nonempty bounded subset of X and $f : A \to \mathbb{R}$ be a real-valued function which is Lipschitz on A and let $\omega : [0, +\infty[\to [0, +\infty[$ be a nondereasing finite-valued concave function with $\lim_{r\downarrow 0} \omega(r) = 0 = \omega(0)$. Assume that there is some real $K \ge 0$ such that, for each $a \in A$, there exists some $\zeta(a) \in X^*$ with $\|\zeta(a)\| \le K$ satisfying the following $\omega(\cdot)$-subgradient inequality:

$$\langle \zeta(a), x - a\rangle \le f(x) - f(a) + \|x - a\|\omega(\|x - a\|) \quad \text{for all } x \in A.$$

Then the following assertions hold:
(a) The function f can be extended on X to a function F which is globally Lipschitz on X and semiconvex on X.
(b) If $\beta = 1$, the Lipschitz extension function F can be taken to be semiconvex on X with $C\omega(\cdot)$ as a modulus of semiconvexity for some $C > 0$.
(c) If for some constants $c > 0$ and $\alpha \in]0, \beta]$, the modulus $\omega(\cdot)$ is of the form $\omega(r) = cr^\alpha$ for all $r \ge 0$, the Lipschitz extension function F can be taken to be semiconvex of power type α, that is, semiconvex with modulus of the form Cr^α.

PROOF. (a) Putting $\omega_0(r) := \min\{\omega(r), \omega(\operatorname{diam} A)\}$, the function $\omega_0(\cdot)$ is continuous and it is also a nondecreasing concave modulus with $\omega_0(0) = 0$. Since $\omega_0(\cdot)$ is bounded on $[0, +\infty[$, the assertion (a) of Lemma 10.53 tells us that, for

$\varphi(x) = 2\int_0^{\|x\|} \omega_0(t)\,dt$ for all $x \in X$, the function φ is differentiable on X and the derivative of φ is uniformly continuous on X with a modulus of continuity $\eta(\cdot)$. Further, by Lemma 10.41 we have $\varphi(x) \ge \|x\|\omega_0(\|x\|)$ for all $x \in X$. Putting, for $a \in A$,

$$F_a(x) := \langle \zeta(a), x - a\rangle + f(a) - \varphi(x-a) \quad \text{for all } x \in X,$$

each function F_a is differentiable on X and its derivative is uniformly continuous on X with $\eta(\cdot)$ as a common modulus of (uniform) continuity. On the other hand, for each $a \in A$, we have for all $x \in A$

$$\begin{aligned}\langle \zeta(a), x-a\rangle + f(a) - \varphi(x-a) &\le \langle \zeta(a), x-a\rangle + f(a) - \|x-a\|\omega_0(\|x-a\|)\\ &= \langle \zeta(a), x-a\rangle + f(a) - \|x-a\|\omega(\|x-a\|) \le f(x),\end{aligned}$$

hence $f(x) = \max_{a\in A} F_a(x)$ for all $x \in A$.

Now observe, for any $a \in A$, that the boundedness assumption on $\zeta(\cdot)$ yields, for all $x \in X$, according to Proposition 10.42(a)

$$\|DF_a(x)\| \le \|\zeta(a)\| + \|D\varphi(x-a)\| \le K + 2\omega_0(\|x-a\|)\|\Lambda(x-a)\| \le K + 2\omega(\operatorname{diam} A),$$

where $\Lambda(0) = 0$ and $\Lambda(u) = (D\|\cdot\|)(u)$ for $u \neq 0$. This shows the equi-Lipschitz property of $(F_a)_{a\in A}$ on the whole space X, thus the function $F : X \to \mathbb{R}$ with $F(x) = \sup_{a\in A} F_a(x)$ for all $x \in X$ is Lipschitz on X and coincides with f on A. All the functions F_a are also semiconvex on X with respect to a common modulus, which entails that F is semiconvex on X according to Proposition 10.10.

(b)-(c) Assume that $\beta = 1$ (resp. $\omega(r) = r^\alpha$ with $0 < \alpha \le \beta$). The assertion (c) (resp. (d)) of Lemma 10.53 tells us that the derivative of φ is uniformly continuous on X with $K\omega(\cdot)$ as a modulus of continuity for some contant $K > 0$. This combined with the arguments in (a) above justifies the assertions (b) and (c). □

THEOREM 10.59 (J. Duda and L. Zajíček: Lipschitz semiconvex extension). Let X be a uniformly convex Banach space, U be a nonempty open bounded convex set of X and $f : U \to \mathbb{R}$ be a real-valued function which is Lipschitz on U and semiconvex on U with respect to a finite-valued modulus.

(A) The function f can be extended on X to a function F which is globally Lipschitz on X and semiconvex on X.

(B) Let $0 < \beta \le 1$ be such that X has an equivalent uniformly smooth norm with modulus of power type $1+\beta$ and let $\omega : [0,+\infty[\to [0,+\infty[$ be a nondecreasing finite-valued modulus function of f with $\lim_{r\downarrow 0}\omega(r) = 0$.

(a) If $\beta = 1$, the Lipschitz extension function F can be taken to be $K\,\omega(\cdot)$-semiconvex on X for some real constant $K > 0$.

(b) If for some constants $c > 0$ and $\alpha \in\,]0,\beta]$, the modulus of semiconvexity $\omega(\cdot)$ of f on U is of the form $\omega(r) = cr^\alpha$ for all $r \ge 0$, the Lipschitz extension function F can be taken to be semiconvex of power type α, that is, semiconvex with modulus of the form Cr^α.

PROOF. The assertion (A) has been already observed in Remark 10.55. Let us also justify that with other arguments which will also be useful for the other assertions. Choose an equivalent norm on X which is of power type with power $1+\beta$ with $\beta \in\,]0,1]$ and we choose by Proposition 10.51 some nondereasing finite concave function $\omega_0 : [0,+\infty[\to [0,+\infty[$ with $\lim_{r\downarrow 0}\omega_0(r) = 0 = \omega_0(0)$ and with respect to which f is semiconvex on U. Since f is Lipschitz on the open set U we can select, for each $a \in U$, some $\zeta(a) \in \partial_C f(a)$ and clearly $\|\zeta(a)\| \le K$, where

$K \geq 0$ is a Lipschitz constant of f on U. The property in (A) then follows from Proposition 10.58(a) with $A := U$ and from Proposition 10.29.

Concerning the assertions in (B) keep β as given in (B). In the case of (b) put $\omega_0(r) := cr^\alpha$ with the constant $c > 0$ as in (b), and in the case of (a) choose by Proposition 10.51 a finite concave modulus $\omega_0(\cdot)$ such that f is $\omega_0(\cdot)$-semiconvex on U and $\omega_0(\cdot) \leq 4\omega(\cdot)$. Choosing $\zeta(a)$ as above and applying (c) or (b) in Proposition 10.58 with $A := U$, we obtain (c) and (b) of the present theorem. □

10.4. Semiconvex multimappings

The concept of semiconvexity can also be defined for multimappings.

Definition 10.60. Let $M : U \rightrightarrows Y$ be a multimapping from a convex set U of a normed space X into a normed vector space Y. One says that M is $\omega(\cdot)$-*semiconvex* on U provided

$$tM(x) + (1-t)M(x') \subset M(tx + (1-t)x') + \frac{1}{2}t(1-t)\|x - x'\|\omega(\|x - x'\|)\mathbb{B}_Y$$

for all $x, x' \in U$ and $t \in]0,1[$. When U is the whole space X, one just says that the multimapping M is semiconvex.

Evidently, taking $x' = x$ it follows from the inclusion above that the set $M(x)$ is convex in Y for any $x \in U$ whenever the multimapping is semiconvex on U. We also see through the same inclusion that the effective domain $\operatorname{Dom} M$ (relative to U) is convex in X.

The first result below concerns the infimum over images of a semiconvex multimapping. We recall that the infimum of a function over the empty set is $+\infty$.

Proposition 10.61. Let $M : U \rightrightarrows Z$ be a multimapping from a convex set U of a normed space X into a normed space Z and let $f : U \times V \times W \to \mathbb{R}$ and $g : U \times Z \to \mathbb{R}$ be real-valued functions, where V is a convex set of a normed space Y and W is a convex set of Z containing $M(U)$. Assume that M is $\omega(\cdot)$-semiconvex on U. The following hold:

(a) If f is convex on $U \times V \times W$ and the functions $f(x, y, \cdot)$ for $(x, y) \in U \times V$ are equi-Lipschitz on W with K_f as a common Lipschitz constant, then, for the function $\Phi : U \times V \to \mathbb{R} \cup \{-\infty, +\infty\}$ defined by

$$\Phi(x, y) := \inf_{z \in M(x)} f(x, y, z) \quad \text{for all } (x, y) \in U \times V,$$

one has, for all reals $\alpha > \Phi(x, y)$ and $\alpha' > \Phi(x', y')$ with $x, x' \in U$, $y, y' \in V$, and for $s, t > 0$ with $s + t = 1$,

$$\Phi(sx + tx', sy + ty') \leq s\alpha + t\alpha' + \frac{1}{2}K_f st\|x - x'\|\omega(\|x - x'\|).$$

(b) If g is convex on $U \times W$ and the functions $g(x, \cdot)$ for $x \in U$ are equi-Lipschitz on W with K_g as a common Lipschitz constant, then the infimum value function $\varphi : U \to \mathbb{R} \cup \{-\infty, +\infty\}$ defined by

$$\varphi(x) := \inf_{z \in M(x)} g(x, z) \quad \text{for all } x \in U,$$

is semiconvex on U with $K_f\omega(\cdot)$ as a modulus of semiconvexity.

PROOF. (a) Let $s,t \in]0,1[$ with $s+t=1$, $x,x' \in U$, $y,y' \in V$ and let reals $\alpha > \Phi(x,y)$ and $\alpha' > \Phi(x',y')$. Choose by definition of Φ some $z \in M(x)$ and $z' \in M(x')$ such that $\alpha > f(x,y,z)$ and $\alpha' > f(x',y',z')$. The $\omega(\cdot)$-semiconvexity of the multimapping M gives some $v \in M(sx+tx')$ such that $\|sz+tz'-v\| \leq \frac{1}{2}st\|x-x'\|\omega(\|x-x'\|)$. Note that $z,z',v \in W$ since $M(U) \subset W$ by assumption. From the convexity of f on $U \times V \times W$ we have

$$s\alpha + t\alpha' > f(sx+tx', sy+ty', sz+tz'),$$

hence from the K_f-Lipschitz property of $f(sx+tx', sy+ty', \cdot)$ on W we obtain

$$\begin{aligned} s\alpha + t\alpha' &> f(sx+tx', sy+ty', v) - \frac{1}{2}K_f st\|x-x'\|\omega(\|x-x'\|) \\ &\geq \Phi(sx+tx', sy+ty') - \frac{1}{2}K_f st\|x-x'\|\omega(\|x-x'\|), \end{aligned}$$

which is the desired inequality in (a).
(b) Putting $f(x,y,z) := g(x,z)$ for all $(x,y,z) \in U \times V \times W$ we see that (b) follows from (a). □

The next proposition studies the distance function to images of a semiconvex multimapping.

PROPOSITION 10.62. Let $M : U \rightrightarrows Y$ be a multimapping from a convex set U of a normed space X into a normed space Y.
(a) If the multimapping M is $\omega(\cdot)$-semiconvex on U, then for all $x,x' \in U$, $y,y' \in Y$, and for $s,t > 0$ with $s+t=1$, one has

$$d(sy+ty', M(sx+tx')) \leq sd(y, M(x)) + td(y', M(x')) + \frac{1}{2}st\|x-x'\|\omega(\|x-x'\|).$$

(b) Conversely, the latter inequality entails that the multimapping M is semiconvex on U with $\lambda\omega(\cdot)$ as a modulus of semiconvexity for any real $\lambda > 1$ whenever the images of M are closed in Y.
(c) Suppose that $M(x)$ is closed for all $x \in U$ and that Y is a reflexive Banach space. Then the multimapping M is $\omega(\cdot)$-semiconvex on U if and only if the inequality property in (a) holds true.

PROOF. (a) Putting $Z := Y$ and $f(x,y,z) := \|y-z\|$ for all $x \in X$, $y \in Y$, and $z \in Z$, we see that f is convex on $X \times Y \times Z$ and that the functions $f(x,y,\cdot)$ are equi-Lipschitz on $Z := Y$ with 1 as a common Lipschitz constant. For

$$\Phi(x,y) := \inf_{z \in M(x)} f(x,y,z) = \inf_{z \in M(x)} \|y-z\| = d(y, M(x)),$$

the assertion (a) in the above proposition ensures the desired inequality.
(b) Fix any real $\lambda > 1$. Consider $x,x' \in U$ and $s,t > 0$ with $s+t=1$, and take any $y \in M(x)$ and $y' \in M(x')$. If $\omega(\|x-x'\|) = 0$, then the inequality in (a) says that $d(sy+ty', M(sx+tx')) \leq 0$, which means according the closedness of $M(sx+tx')$

$$sy+ty' \in M(sx+tx') = M(sx+tx') + \frac{\lambda}{2}st\|x-x'\|\omega(\|x-x'\|)\mathbb{B}_Y.$$

If $\omega(\|x-x'\|) > 0$, then the inequality in (a) again yields $d(sy+ty', M(sx+tx')) < (\lambda/2)st\|x-x'\|\omega(\|x-x'\|)$ hence

$$sy+ty' \in M(sx+tx') + \frac{\lambda}{2}st\|x-x'\|\omega(\|x-x'\|)\mathbb{B}_Y.$$

So, the latter inclusion holds in any case, thus

$$sM(x) + tM(x') \subset M(sx + tx') + \frac{\lambda}{2} st\|x - x'\|\omega(\|x - x'\|)\mathbb{B}_Y,$$

which translates the desired semiconvexity of M in (b).
(c) Now suppose that Y is a reflexive Banach space, the inequality in (a) is satisfied and the images of M are closed. Taking $x = x' \in U$ and taking $y, y' \in M(x)$, the inequality in (a) gives $d(sy + ty', M(x)) \le 0$ for $s, t > 0$ with $s + t = 1$, hence $sy + ty' \in M(x)$ since $M(x)$ is closed. The sets $M(x)$ are then convex and closed. This combined with the reflexivity of Y guarantees, for $x'' \in X$ and $y'' \in Y$, that $d(y'', M(x''))$ is attained whenever $M(x'')$ is nonempty, hence in particular $d(y'', M(x'')) \le r$ with $r \ge 0$ if and only if $y'' \in M(x'') + r\mathbb{B}_Y$. Consequently, for $x, x' \in U$, $y \in M(x)$, $y' \in M(x')$, $s, t > 0$ with $s+t = 1$, noting that $M(sx+tx') \neq \emptyset$, the inequality

$$d(sy + ty', M(sx + tx')) \le \frac{1}{2} st\|x - x'\|\omega(\|x - x'\|)$$

produced by (a) entails $sy + ty' \in M(sx + tx') + (1/2)st\|x - x'\|\omega(\|x - x'\|)\mathbb{B}_Y$, so M is $\omega(\cdot)$-semiconvex on U. This justifies the implication $\Leftarrow$ in (c) and the reverse implication is due to the assertion (a). □

10.5. Comments

The concept of *paraconvexity* was introduced in 1979 by S. Rolewicz [**869**] by means of the nondecreasing function $\alpha(t) = ct^2$. Some months later, Rolewicz [**870**] extended the concept, under the name of γ-paraconvexity, to the case where the function $\alpha(\cdot)$ is of the form $\alpha(t) = ct^\gamma$ with $\gamma > 1$. In the above papers, a function f on a convex set U of a normed space X is defined by Rolewicz to be *γ-paraconvex* provided there exists some constant $c \ge 0$ such that for all $t \in]0, 1[$ and $x, y \in U$

$$f(tx + (1 - t)y) \le tf(x) + (1 - t)f(y) + c\|x - y\|^\gamma.$$

This form does not involve, as in Definition 10.1 in the book, the expression $t(1-t)$ in the third term in the right-hand side of the inequality above. Nevertheless, in [**869, 870**] it is proved that for $\gamma \in]1, 2]$, the function f is γ-paraconvex if and only if there exists a constant $c \ge 0$ such that for all $t \in]0, 1[$, $x, y \in U$

$$f(tx + (1 - t)y) \le tf(x) + (1 - t)f(y) + c \min\{t, 1 - t\}\|x - y\|^\gamma.$$

Then, given a nondecreasing function $\alpha : [0, +\infty[\to [0, +\infty]$ continuous at 0 with $\alpha(t)/t \to 0$ as $t \downarrow 0$, Rolewicz [**872**] in 2000 defined the function f to be:

(i) *$\alpha(\cdot)$-paraconvex* if there exists some constant $c \ge 0$ such that for all $t \in]0, 1[$ and $x, y \in U$

$$f(tx + (1 - t)y) \le tf(x) + (1 - t)f(y) + c\alpha(\|x - y\|);$$

(ii) *strongly $\alpha(\cdot)$-paraconvex* if there exists a constant $c \ge 0$ such that for all $t \in]0, 1[$, $x, y \in U$

$$f(tx + (1 - t)y) \le tf(x) + (1 - t)f(y) + c \min\{t, 1 - t\}\alpha(\|x - y\|).$$

Since $\frac{1}{2} \min\{t, 1 - t\} \le t(1 - t) \le \min\{t, 1 - t\}$ for all $t \in]0, 1[$, the latter form of strong $\alpha(\cdot)$-paraconvexity (resp. strong γ-paraconvexity) is equivalent to Definition 10.1 of *semiconvexity* in the book with the modulus $\omega(t) = c'\alpha(t)/t$ (resp. $\omega(t) = c't^{\gamma-1}$). for some constant $c' \ge 0$. The study of paraconvex functions by Rolewicz (as said in page 293 of Rolewicz [**872**]) was essentially motivated by properties of the infimum value function $\varphi(x) = \inf_{y \in M(x)} f(y)$. So, in [**869**] and [**870**] paraconvex multimappings $M : X \rightrightarrows Y$ (between normed spaces) were already introduced and investigated. The semiconcavity of optimal value functions in optimal control theory was previously noticed in 1978 by M. M. Hrustalev [**503**], and the role of semiconcavity in Hamilton-Jacobi equations was pointed out by S. N. Kruzhkov [**636**] in 1960 and A. Douglis [**356**] in 1961.

Many subclasses of paraconvex and semiconvex functions were studied for diverse purposes in the literature. In 1981 J. E. Spingarn [**898**] (submitted November 1979) introduced and deeply analyzed multimappings which are submonotone along with locally Lipschitz functions on $\mathbb{R}^n$ whose Clarke subdifferentials are submonotone. Spingarn [**898**] also introduced and studied the class of *lower* $\mathcal{C}^1$*-functions*. A favorable subclass of that of Spingarn was considered in 1982 by R. T. Rockafellar [**859**] under the name *lower* $\mathcal{C}^2$*-functions*. In his 1983 paper [**958**] (submitted in 1981) J.-P. Vial defined and studied *weakly convex* locally Lipschitz functions. Lower $\mathcal{C}^1$, lower $\mathcal{C}^2$ and weakly convex functions on $\mathbb{R}^n$ are strongly connected with *sublinearizable functions* introduced in 1973 by R. Janin [**555, 556**]. The definitions and comments related to those classes can be found in Section 8.11 of comments in Chapter 8. Several fundamental results on those functions on infinite dimensional spaces can be found in L. Zajíček's 2008 paper [**995**] and in the 2009 paper [**363**] by J. Duda and L. Zajíček.

In addition to the above works of Rolewicz related to paraconvexity, the similar class of semiconcave/semiconvex functions was independently investigated by many authors, mainly because of the role of these functions in the study of value functions in optimal control theory, of properties of viscosity solutions of Hamilton-Jacobi-Bellman equations, etc. Semiconcave functions with modulus $\omega(t) = ct^\gamma$ where $\gamma > 0$ were used by I. Capuzzo Dolcettta and I. Ishii [**205**] in 1984, and by P. Cannarsa and H. M. Soner [**202**] in 1987. Semiconcave functions with a general modulus $\omega(\cdot)$, as presented in Definition 10.1 in the manuscript, were involved by G. Alberti, L. Ambrosio and P. Cannarsa [**14**] and by L. Ambrosio, P. Cannarsa and H. M. Soner [**17**]. Definition 10.1 with a general modulus was stated in the same form by Ambrosio, Cannarsa and Soner [**14**, p. 4]. Semiconcave functions on subsets of $\mathbb{R}^n$ are largely studied in the 2004 book of P. Cannarsa and C. Sinestrari [**201**, Chapters II-IV]. We also refer to P. Cannarsa's 1989 paper [**196**], and to the 1999 papers [**11, 12**] by P. Albano and P. Cannarsa related to singularities of semiconcave functions.

The assertion (a) of Proposition 10.9 was proved in Theorem 4 of Rolewicz [**870**] when $\omega(\cdot)$ is a modulus of power type. It was also established in finite dimensions in Proposition 2.1.2 and Theorem 3.3.7 in Cannarsa and Sinestrari [**201**]. Corollary 10.13 is taken from Cannarsa and Sinestrari [**201**, Proposition 2.2.4]. The arguments in the proof of Theorem 10.15 were already present in R. T. Rockafellars's paper [**859**] and J.-P. Vial's paper [**958**]. Example 10.16 corresponds to Example 1 in Rolewicz [**870**]; another example for the same feature also appeared in Proposition 2.1.4 of the book [**201**] by Cannarsa and Sinestrari. Our arguments for this Example 10.16 follow [**870**] and use some ideas of [**201**]. Definition 10.18 of *lower* $\mathcal{C}^k$ *functions* is due to R. T. Rockafellar [**859**] as the extension of the concept of *lower* $\mathcal{C}^1$ *functions* of J. E. Spingarn. A similar statement of Theorem 10.19 was established by Rockafellar [**859**]. Proposition 10.21 appeared in Rolewicz

[**869, 870**]. Proposition 10.22 is taken from Cannarsa and Sinsestrari's book [**201**, Proposition 2.2.2]. The arguments for the Lipschitz property of semiconvex functions in Proposition 10.23 follow those for convex functions. Corollary 10.24 in finite dimensions along its proof correspond to Proposition 2.1 in G. Alberti, L. Ambrosio and P. Cannarsa [**14**] and to Theorem 2.1.7 in the book of Cannarsa and Sinestrari [**201**]. Proposition 10.26 appeared in Theorem 3 of Rolewicz [**870**] for multimappings which are paraconvex with a power greater than 2. The proof of Proposition 10.26 follows the arguments of Lemma 4 in [**870**].

Our description of all subdifferentials of extended real-valued semiconvex functions in Theorem 10.29 follows A. Jourani 1996 [**573**] where these results were first established for paraconvex functions with moduli of power type. Note that the particular case of the description

$$\partial_F f(x) = \{x^* \in X^* : \langle x^*, h\rangle \le f(x+h) - f(x) + \|h\|\omega(\|h\|),\, \forall h \in X\}$$

in Theorem 10.29 of the Fréchet subdifferential was previously shown in 1992 by Alberti, Ambrosio and Cannarsa [**14**, Proposition 2.1]. Assertion (e) in Corollary 10.31 is an adaptation to Hilbert spaces of assertion (b) of Proposition 3.3.4 in Cannarsa and Sinestrari's book [**201**]. Theorem 10.36 was established by Rolewicz [**870**] (see Theorem 4 therein). Our proof follows the one of A. Jourani, L. Thibault and D. Zagrodny [**589**, Proposition 6.1, Theorem 6.1]. The arguments for the Step 1 of the implication (a)$\Rightarrow$(b) in the proof of Theorem 10.37 have been communicated to us by P. Redont [**834**]. The subdifferential characterization of extended real-valued semiconvex functions in Theorem 10.38 was proved by A. Jourani [**573**] for paraconvex functions through an efficient adaptation of the proof of the corresponding characterization of convex functions by R. Correa, A. Jofre and L. Thibault [**292, 293**]. In Theorem 10.38 we also provide the new characterization in terms of the Bouligand/Hadamard directional derivatives.

Theorem 10.56 and Theorem 10.59 are due to J. Duda and L. Zajíček in their 2009 paper [**363**]; the manuscript followed their proofs. The control functions in Definitions 10.43 and 10.46 were introduced in [**363**]. Statements and proofs of Proposition 10.44, Proposition 10.47, Lemma 10.53 and Proposition 10.58 are taken from the same paper [**363**] by Duda and Zajíček. Lemma 10.48 seems to be essentially due to S. B. Stechkin; the proof given here uses strongly the main arguments in N. V. Efimov's paper [**378**]. The statement and proof of Theorem 10.54 correspond to Theorem 5.5 in Duda and Zajíček [**363**]; a similar result can also be found through Theorem 22 in H. V. Ngai and J.-P. Penot paper [**769**].

Proposition 10.61 is at the origin of Rolewicz's studies in [**869, 870**]. Proposition 10.62 is new.

CHAPTER 11

Primal lower regular functions and prox-regular functions

In Theorem 10.38 we have seen that a proper lower semicontinuous function $f : U \to \mathbb{R} \cup \{+\infty\}$ is $K\,\omega(\cdot)$-semiconvex, for some constant $K > 0$, on an open convex set U of a Banach space X if and only if for some constant $K' > 0$ one has

$$\langle x^*, y - x\rangle \leq f(y) - f(x) + K'\|y - x\|\,\omega(\|y - x\|), \tag{11.1}$$

for all $y \in U$, $x \in U \cap \operatorname{Dom} \partial_C f$ and $x^* \in \partial_C f(x)$. The present chapter will relax in some ways the above inequality.

11.1. s-Lower regular functions

The rest term $K'\|y - x\|\,\omega(\|y - x\|)$ in the right-hand side of (11.1) does not depend at all on the subgradient x^*. The goal of this section is to study other fundamental classes of functions involving rests in the type of (11.1) but depending also on subgradients of the functions.

11.1.1. Primal lower and s-lower regular functions. We start with the following first natural simple relaxation of (11.1) with a rest term depending affinely on the norm of the subgradient.

Definition 11.1. Given a normed space $(X, \|\cdot\|)$ and a real $s > 0$, we say that a function $f : X \to \mathbb{R} \cup \{+\infty\}$ is *s-lower regular on an open convex set U* of X with $U \cap \operatorname{dom} f \neq \emptyset$, when it is lower semicontinuous on U and there exists some real coefficient $c \geq 0$ such that for all $x \in U \cap \operatorname{Dom} \partial_C f$ and for all $x^* \in \partial_C f(x)$ we have

$$f(y) \geq f(x) + \langle x^*, y - x\rangle - c\,(1 + \|x^*\|)\,\|x - y\|^{s+1}, \ \forall y \in U. \tag{11.2}$$

The real constant $c \geq 0$ is called a *coefficient of s-lower regularity* of f on U.

When $s = 1$, the function f is called *primal lower regular* (or *plr*, for short) on the open convex set U.

If $\overline{x} \in \operatorname{cl}(\operatorname{dom} f)$ is such that f is s-lower regular (resp. primal lower regular) on some open convex neighborhood U of $\overline{x}$, one says that f is *s-lower regular at $\overline{x}$* (resp. *primal lower regular* or *plr at $\overline{x}$*).

Remark 11.2. If the above inequality (11.2) holds with a real $s \geq 1$ for all y in a neighborhood of a point $\bar{x} \in \operatorname{cl}(\operatorname{dom} f)$, then it also holds with $s = 1$ for all y in some neighborhood of $\overline{x}$ (as easily seen). So, f is primal lower regular at a point $\overline{x} \in \operatorname{cl}(\operatorname{dom} f)$ whenever it is s-lower regular at $\overline{x} \in \operatorname{cl}(\operatorname{dom} f)$ with some real number $s \geq 1$. □

Before considering various examples of s-lower regular functions, let us notice another equivalent way to define s-lower regular functions at a point.

PROPOSITION 11.3. Let $(X, \|\cdot\|)$ be a normed space. A lower semicontinuous function $f : X \to \mathbb{R} \cup \{+\infty\}$ is s-lower regular at $\overline{x} \in \mathrm{cl}\,(\mathrm{dom}\, f)$ in the sense of Definition 11.1 if and only if there are constant real numbers $\delta_0 > 0$, $c_0 > 0$ and $q_0 \geq 0$, such that for all $x \in B(\overline{x}, \delta_0) \cap \mathrm{Dom}\, \partial_C f$, for all $q \geq q_0$ and all $x^* \in \partial_C f(x)$ with $\|x^*\| \leq c_0 q$, one has

$$f(x') \geq f(x) + \langle x^*, x' - x\rangle - \frac{q}{s}\|x' - x\|^s \tag{11.3}$$

for each $x' \in B(\overline{x}, \delta_0)$.

PROOF. Suppose first that f is s-lower regular at $\overline{x}$ in the sense of Definition 11.1, that is, there exist reals $c \geq 0$ and $\delta > 0$ such that (11.1) holds for c and $U := B(\overline{x}, \delta)$. Set $\delta_0 := \delta$, $q_0 := 2sc$ and

$$c_0 := \begin{cases} 1/(2sc) & \text{if } c > 0 \\ 1 & \text{if } c = 0. \end{cases}$$

Consider any $x' \in B(\overline{x}, \delta_0)$, $x \in B(\overline{x}, \delta_0) \cap \mathrm{Dom}\, \partial_C f$ and $x^* \in \partial_C f(x)$ with $\|x^*\| \leq c_0 q$ for some $q \geq q_0$. If $c = 0$, then trivially $c(1 + \|x^*\|) \leq q/s$. So, suppose that $c > 0$. If $\|x^*\| \leq 1$, we have

$$c(1 + \|x^*\|) \leq 2c = q_0/s \leq q/s.$$

If $\|x^*\| > 1$, then $c_0 q \geq \|x^*\| > 1$ and these inequalities entail

$$c(1 + \|x^*\|) \leq 2cc_0 q = q/s.$$

Therefore, in any case we obtain from (11.1) that

$$\langle x^*, x' - x\rangle \leq f(x') - f(x) + \frac{q}{s}\|x' - x\|^s,$$

which means that the inequality of the proposition holds true.

Conversely, suppose that the inequality of the proposition holds with $\delta_0 > 0$, $c_0 > 0$ and $q_0 \geq 0$. Put $\delta := \delta_0$ and $c := \max\left\{\dfrac{1}{sc_0}, \dfrac{q_0}{s}\right\}$. Fix any $x' \in B(\overline{x}, \delta)$, $x \in B(\overline{x}, \delta) \cap \mathrm{dom}\, \partial_C f$ and $x^* \in \partial_C f(x)$. If $\dfrac{\|x^*\|}{c_0} \geq q_0$, the equality $\|x^*\| = c_0 \dfrac{\|x^*\|}{c_0}$ allows us to take $q := \dfrac{\|x^*\|}{c_0} \geq q_0$ to obtain

$$\langle x^*, x' - x\rangle \leq f(x') - f(x) + \frac{\|x^*\|}{sc_0}\|x' - x\|^s,$$

and hence

$$\langle x^*, x' - x\rangle \leq f(x') - f(x) + c(1 + \|x^*\|)\|x' - x\|^s$$

because by definition of c we have

$$\frac{1}{sc_0}\|x^*\| \leq c\|x^*\| \leq c(1 + \|x^*\|).$$

If $\|x^*\|/c_0 < q_0$, that is, $\|x^*\| < c_0 q_0$, then we may take $q = q_0$ to get from the inequality of the proposition that

$$\langle x^*, x' - x\rangle \leq f(x') - f(x) + \frac{q_0}{s}\|x' - x\|^s.$$

Since $q_0/s \leq c$ by the definition of c, the latter inequality implies that

$$\langle x^*, x' - x\rangle \leq f(x') - f(x) + c(1 + \|x^*\|)\|x' - x\|^s.$$

Consequently, f is s-lower regular at $\overline{x}$ in the sense of Definition 11.1 and the proof is complete. □

Proper lower semicontinuous convex or linearly semiconvex functions from a normed space X into $\mathbb{R}\cup\{+\infty\}$ are obviously primal lower regular on X. The next proposition shows that continuous primal lower regular functions coincide with continuous locally linearly semiconvex functions on Hilbert space.

PROPOSITION 11.4. Let $f : U \to \mathbb{R}$ be a real-valued lower semicontinuous function, where U is an open convex set of a Hilbert space H. The following are equivalent:
(a) the function f is locally linearly semiconvex and locally Lipschitz continuous on U;
(b) the function f is locally linearly semiconvex and continuous on U;
(c) the function f is locally bounded from above on U and primal lower regular at any point of U.

PROOF. Clearly, (a) implies (b). Suppose now f is locally linearly semiconvex and continuous on U. Let $\overline{x} \in U$. By Theorem 10.15(A) there exists $\delta > 0$ and $c \geq 0$ such that $f + c\|\cdot\|^2$ is convex on $B(\overline{x}, \delta)$. Since f is continuous, we may suppose without loss of generality that f is bounded from above on $B(\overline{x}, \delta)$. Choose $x \in B(\overline{x}, \delta)$ and $\zeta \in \partial_C f(x)$. Note that

$$\zeta + 2cx \in \partial_C f(x) + \nabla(c\|\cdot\|^2)(x) = \partial_C(f + c\|\cdot\|^2)(x)$$

since $c\|\cdot\|^2$ is $\mathcal{C}^1$ on H, and then $\zeta + 2cx \in \partial_C(f + c\|\cdot\|^2)(x)$. Thus, for all $x' \in B(\overline{x}, \delta)$,

$$\langle \zeta + 2cx, x' - x\rangle \leq f(x') + c\|x'\|^2 - f(x) - c\|x\|^2,$$

or equivalently

$$\langle \zeta, x' - x\rangle \leq f(x') - f(x) + c\|x' - x\|^2,$$

which guarantees that f is primal lower regular at $\overline{x}$. So, (b) $\Rightarrow$ (c) since the local boundedness property of f near $\overline{x}$ has been already observed above.

Now suppose that (c) holds and let $\overline{x} \in U$. Since f is lower semicontinuous, there exist some positive number δ_0 and some $\gamma \in \mathbb{R}$ such that $f(x) \geq \gamma$ for all $x \in B(\overline{x}, \delta_0)$. By assumption, we can find $0 < \delta_1 < \delta_0$ and $\beta \in \mathbb{R}$ such that $f(x) \leq \beta$ for all $x \in B(\overline{x}, \delta_1)$. This means that f is locally bounded on U. So, take $\beta \geq 0$ and $\delta_1 > 0$ such that $|f(x)| \leq \beta$ for all $x \in B(\overline{x}, \delta_1)$. Since f is primal lower regular at $\overline{x}$, there exist $0 < \delta < \delta_1$ and $c \geq 0$ such that for all $x \in B(\overline{x}, \delta)$ and all $\zeta \in \partial_C f(x)$ the inequality

$$\langle \zeta, x' - x\rangle \leq f(x') - f(x) + c(1 + \|\zeta\|)\|x' - x\|^2 \tag{11.4}$$

is valid for all $x' \in B(\overline{x}, \delta)$. Choose some real number $r > 0$ such that $cr < 1$ and $B(\overline{x}, 2r) \subset B(\overline{x}, \delta)$. Considering $x \in B(\overline{x}, r) \cap \operatorname{Dom} \partial_C f$ and $\zeta \in \partial_C f(x)$, for $x' = x + rb$ with $b \in \mathbb{B}$, the latter inequality gives

$$\langle \zeta, rb\rangle \leq 2\beta + c(1 + \|\zeta\|)r^2.$$

Hence $\|\zeta\| \leq r^{-1}(1 - cr)^{-1}(2\beta + cr^2) =: \kappa$. This entails that f is Lipschitz continuous on $B(\overline{x}, r)$ with κ as a Lipschitz constant according to the subdifferential characterization of Lipschitz property in Theorem 6.55. This justifies the local Lipschitz continuity of f on U. It remains to establish the semiconvexity of f near $\overline{x}$.

Using in (11.4) the κ-Lipschitz continuity of f on $B(\overline{x}, r)$, we get for all $\zeta \in \partial_C f(x)$, with $x \in B(\overline{x}, r)$,

$$\langle \zeta, x' - x \rangle \leq f(x') - f(x) + c(1+\kappa)\|x' - x\|^2$$

for all $x' \in B(\overline{x}, r)$. So, setting $c' := c(1+\kappa)$ we obtain

$$\langle (\zeta_1 + 2c'x_1) - (\zeta_2 + 2c'x_2), x_1 - x_2 \rangle \geq 0$$

for all $\zeta_i \in \partial_C f(x_i)$ with $x_i \in B(\overline{x}, r)$. By the subdifferential characterization of convex functions in Theorem 6.68 the latter inequality is equivalent to the convexity of $f + c'\|\cdot\|^2$ on $B(\overline{x}, r)$. Consequently, the implication (c) $\Rightarrow$ (a) is established. $\square$

EXAMPLE 11.5. Given any function which is $\mathcal{C}^1$ near a point $\overline{x}$ but not $\mathcal{C}^{1,1}$ near $\overline{x}$, we see that f is subsmooth at the point $\overline{x}$ but fails to be primal lower regular at this point $\overline{x}$. $\square$

11.1.2. Convexly composite functions. Assume that f is finite and Fréchet differentiable on an open convex set U of a normed space X and that the derivative Df is Hölder continuous on U with exponent $s > 0$ and coefficient $c \geq 0$, that is, f is of class $\mathcal{C}^{1,s}$ on U. Then for all $x, y \in U$ we have

$$\begin{aligned} f(x) - f(y) &= \langle Df(x), x - y \rangle + \int_0^1 \langle Df(y + t(x-y)) - Df(x), x - y \rangle \, dt \\ &\leq \langle Df(x), x - y \rangle + c\|x - y\|^{1+s} \int_0^1 (1-t)^s \, dt, \end{aligned}$$

and the latter inequality is equivalent to

$$f(y) \geq f(x) + \langle Df(x), y - x \rangle - \frac{c}{s+1}\|x - y\|^{s+1}.$$

Since $\partial_C f(x) = \{Df(x)\}$, we see that the function f is s-lower regular on U whenever the derivative of f exists on U and is Hölderian therein with exponent s.

Other examples of s-lower regular functions are some convexly composite functions as established in the next proposition via the main idea in the above arguments.

DEFINITION 11.6. When $G : U \to Y$ is a mapping from an open set U of a Banach space X into a Banach space Y which is $\mathcal{C}^1$ on U (resp. strictly differentiable at $\overline{x} \in U$) and $g : Y \to \mathbb{R} \cup \{+\infty\}$ is a proper lower semicontinuous convex function, one says that the function $g \circ G$ is *convexly composite on* U (resp. *convexly composite at* $\overline{x}$). If G is $\mathcal{C}^{1,s}$ (resp. $\mathcal{C}^2$ etc), one says that $g \circ G$ is *convexly* $\mathcal{C}^{1,s}$*-composite* (resp. $\mathcal{C}^2$*-composite* etc).

The convexly composite function $g \circ G$ is said to be *qualified* at $\overline{x} \in U \cap \operatorname{dom}(g \circ G)$ provided the (so-called) Robinson qualification condition

$$\mathbb{R}_+\big(\operatorname{dom} g - G(\overline{x})\big) - DG(\overline{x})(X) = Y \tag{11.5}$$

holds for $g \circ G$ at $\overline{x}$.

THEOREM 11.7 (subdifferential of convexly composite functions). Let X and Y be two Banach spaces, $G : X \to Y$ be a mapping which is strictly Fréchet differentiable at $\overline{x} \in X$, and $g : Y \to \mathbb{R} \cup \{+\infty\}$ be a lower semicontinuous convex function which is finite at $G(\overline{x})$. Assume that the Robinson qualification condition

(11.5) is satisfied. Then the subdifferentials and horizon subdifferentials at $\overline{x}$ of the composition $g \circ G$ are given by

$$\partial_F(g \circ G)(\overline{x}) = \partial_A(g \circ G)(\overline{x}) = \partial_C(g \circ G)(\overline{x}) = DG(\overline{x})^*\big(\partial g(G(\overline{x}))\big),$$

$$\partial_F^\infty(g \circ G)(\overline{x}) = \partial_A^\infty(g \circ G)(\overline{x}) = \partial_C^\infty(g \circ G)(\overline{x}) = DG(\overline{x})^*\big(\partial^\infty g(G(\overline{x}))\big).$$

If, instead of the strict Fréchet differentiability, G is continuously differentiable near $\overline{x}$, then for some neighborhood U of $\overline{x}$ one has for all $x \in U$

$$\partial_F(g \circ G)(x) = \partial_A(g \circ G)(x) = \partial_C(g \circ G)(x) = DG(x)^*\big(\partial g(G(x))\big)$$

as well as the similar equalities with the respective horizon subdifferentials at x.

Proof. Define $G_0 : X \times \mathbb{R} \to Y \times \mathbb{R}$ by $G_0(u,t) = (G(u),t)$ and note that $\mathrm{epi}\,(g \circ G) = G_0^{-1}(\mathrm{epi}\, g)$. Putting $\overline{r} := g(G(\overline{x}))$, it is not difficult to see that (11.5) is equivalent to

$$\mathbb{R}_+\big(\mathrm{epi}\, g - (G(\overline{x}), \overline{r})\big) - DG_0(\overline{x}, \overline{r})(X \times \mathbb{R}) = Y \times \mathbb{R}.$$

Putting $\Lambda := DG_0(\overline{x}, \overline{r})$ and noting that $\mathrm{epi}\, g$ is closed and convex, we obtain according to Theorem 7.95 that, for $S := \mathrm{epi}\,(g \circ G)$ the equalities

$$N^F\big(S; (\overline{x}, \overline{r})\big) = N^A\big(S; (\overline{x}, \overline{r})\big) = N^C\big(S; (\overline{x}, \overline{r})\big) = \Lambda^*\big(N(\mathrm{epi}\, g; (G(\overline{x}), \overline{r}))\big)$$

are valid. Setting $A := DG(\overline{x})$ it is clear that

$$\Lambda(h,\tau) := (Ah, \tau)\ \forall (h,\tau) \in X \times \mathbb{R} \text{ and } \Lambda^*(y^*, \tau) = (A^* y^*, \tau)\ \forall (y^*, \tau) \in Y^* \times \mathbb{R}.$$

Then for $s \geq 0$ one has $(x^*, -s) \in N^F\big(S; (\overline{x}, \overline{r})\big) = N^C\big(S; (\overline{x}, \overline{r})\big)$ if and only if $(x^*, -s) = (A^* y^*, -s)$ for some $y^* \in Y^*$ with $(y^*, -s) \in N\big(\mathrm{epi}\, g; (G(\overline{x}), \overline{r})\big)$, otherwise stated,

$$\partial_F(g \circ G)(\overline{x}) = \partial_A(g \circ G)(\overline{x}) = \partial_C(g \circ G)(\overline{x}) = A^*\big(\partial g(G(\overline{x}))\big),$$

$$\partial_F^\infty(g \circ G)(\overline{x}) = \partial_A^\infty(g \circ G)(\overline{x}) = \partial_C^\infty(g \circ G)(\overline{x}) = A^*\big(\partial^\infty g(G(\overline{x}))\big).$$

This confirms the first part of the statement of the theorem.

The case when G is continuously differentiable on a neighborhood of $\overline{x}$ follows from what precedes and from Lemma 11.8 below. □

The lemma is in the line of Proposition 7.71.

Lemma 11.8. Let X and Y be two Banach spaces, $G : X \to Y$ be a mapping which is continuously Fréchet differentiable on a neighborhood U of $\overline{x} \in X$, and $g : Y \to \mathbb{R} \cup \{+\infty\}$ be a lower semicontinuous convex function which is finite at $G(\overline{x})$. Assume that the Robinson qualification condition (11.5) is satisfied at $\overline{x}$. Then there exists an open neighborhood $U_0 \subset U$ of $\overline{x}$ such that the Robinson qualification condition

$$\mathbb{R}_+\big(\mathrm{dom}\, g - G(x)\big) - DG(x)(X) = Y$$

holds for any $x \in U_0 \cap G^{-1}(\mathrm{dom}\, g)$.

Proof. As already said above in the proof of Theorem 11.7, the Robinson qualification condition at $\overline{x}$ is equivalent to

$$\mathbb{R}_+\big(\mathrm{epi}\, g - G_0(\overline{x}, \overline{r})\big) - DG_0(\overline{x}, \overline{r})(X \times \mathbb{R}) = Y \times \mathbb{R},$$

where $\overline{r} := g(G(\overline{x}))$ and $G_0 : X \times \mathbb{R} \to Y \times \mathbb{R}$ is defined by $G_0(x,r) := (G(x), r)$ for all $(x,r) \in X \times \mathbb{R}$. By Theorem 7.70(b) there exists an open neighborhood $U_0 \subset U$ of $\overline{x}$ in X, a neighborhood V of zero in Y and reals $\varepsilon > 0$ and $\gamma > 0$ such that

$$d\big((u,r), G_0^{-1}(\operatorname{epi} g + (y,t))\big) \le \gamma d(G_0(u,r) - (y,t), \operatorname{epi} g)\big) \tag{11.6}$$

for all $(u,r) \in U_0\times\,]\overline{r}-\varepsilon, \overline{r}+\varepsilon[$ and $(y,t) \in V\times\,]-\varepsilon,\varepsilon[$. Shrinking U_0 if necessary, we may suppose that $G(U_0) \subset\,]\overline{r}-\varepsilon, \overline{r}+\varepsilon[$. Fix any $x \in U_0 \cap G^{-1}(\operatorname{dom} g)$ and set $\rho := G(x) \in\,]\overline{r}-\varepsilon, \overline{r}+\varepsilon[$. Since G_0 is strictly Fréchet differentiable at (x,ρ), Theorem 7.70(b) again combined with (11.6) entails that

$$\mathbb{R}_+\big(\operatorname{epi} g - G_0(x,\rho)\big) - DG_0(x,\rho)(X \times \mathbb{R}) = Y \times \mathbb{R},$$

which is equivalent to

$$\mathbb{R}_+\big(\operatorname{dom} g - G(x)\big) - DG(x)(X) = Y,$$

which is the conclusion of the lemma. □

Convexly composite functions also arise in the form $f + g \circ G$ with lower semicontinuous convex functions $f : X \to \mathbb{R} \cup \{+\infty\}$, $g : Y \to \mathbb{R}\{+\infty\}$ and a mapping $G : X \to Y$ strictly differentiable at $\overline{x}$, plus a qualification condition as stated in the following corollary.

COROLLARY 11.9. Let X and Y be two Banach spaces at $G : X \to Y$ be a mapping which is strictly Fréchet differentiable and $\overline{x} \in X$. Let $f : X \to \mathbb{R} \cup \{+\infty\}$ and $g : Y \to \mathbb{R} \cup \{+\infty\}$ be two lower semicontinuous convex functions which are finite at $\overline{x}$ and $G(\overline{x})$ respectively. Assume that the Robinson qualification condition

$$\mathbb{R}_+[(\operatorname{dom} g - G(\overline{x})) - DG(\overline{x})(\operatorname{dom} f - \overline{x})] = Y \tag{11.7}$$

is satisfied. Then the subdifferentials and horizon subdifferentials at $\overline{x}$ of the function $f + g \circ G$ are given by

$$\partial_F(f+g\circ G)(\overline{x}) = \partial_A(f+g\circ G)(\overline{x}) = \partial_C(f+g\circ G)(\overline{x}) = \partial f(\overline{x}) + DG(\overline{x})^*\big(\partial g(G(\overline{x}))\big),$$

$$\partial_F^\infty(f+g\circ G)(\overline{x}) = \partial_A^\infty(f+g\circ G)(\overline{x}) = \partial_C^\infty(f+g\circ G)(\overline{x}) = \partial^\infty f(\overline{x}) + DG(\overline{x})^*\big(\partial^\infty g(G(\overline{x}))\big).$$

If, instead of the strict Fréchet differentiability, G is continuously differentiable near $\overline{x}$, then for some neighborhood U of $\overline{x}$ one has for all $x \in U$

$$\partial_F(f+g\circ G)(x) = \partial_A(f+g\circ G)(x) = \partial_C(f+g\circ G)(x) = \partial f(x) + DG(x)^*\big(\partial g(G(x))\big)$$

as well as the similar equalities with the respective horizon subdifferentials at x.

PROOF. Let f, g be as above and $G : X \to Y$ be strictly differentiable at $\overline{x}$ (resp. of class $\mathcal{C}^1$ near $\overline{x}$) and satisfying (11.7). Considering the lower semicontinuous convex function $\varphi : X \times Y \to \mathbb{R} \cup \{+\infty\}$ defined by $\varphi(x,y) = f(x) + g(y)$, we know that $\partial\varphi(x,y) = (\partial f(x)) \times (\partial g(y))$. It is also easily seen that $\partial^\infty\varphi(x,y) = (\partial^\infty f(x)) \times (\partial^\infty g(y))$. On the other hand, the mapping $\Phi : X \to X \times Y$ with $\Phi(x) = (x, G(x))$ is clearly strictly differentiable at $\overline{x}$ (resp. of class $\mathcal{C}^1$ near $\overline{x}$) and $f + g \circ G = \varphi \circ \Phi$. Further, by Lemma 11.10 below we have that

$$\mathbb{R}_+[(\operatorname{dom}\varphi - \Phi(\overline{x})) - D\Phi(\overline{x})(X \times Y)] = X \times Y,$$

so applying Theorem 11.9 and rearranging furnish the desired equalities of the corollary. □

LEMMA 11.10. Let $A : X \to Y$ be a linear mapping between vector spaces and let C and D be subsets of X and Y containing the zeros of X and Y respectively. Then $\mathbb{R}_+[D - A(C)] = Y$ if and only if

$$\mathbb{R}_+ \left[(C \times D) - \widehat{A}(X)\right] = X \times Y,$$

where $\widehat{A} : X \to X \times Y$ is defined by $\widehat{A}(x) = (x, Ax)$ for all $x \in X$.

PROOF. Assume first $\mathbb{R}_+[D - A(C)] = Y$. To show the equality related to $C \times D$ it suffices to prove the inclusion of the right-hand side into the left-hand one. Take any non-zero $(u, v) \in X \times Y$. By assumption there exist some real $\lambda > 0$, some $y \in D$ and some $x \in C$ such that $v - Au = \lambda(y - Ax)$. Denote $x' := x - \lambda^{-1}u$, so $u = \lambda(x - x')$. It ensues that

$$v = Au + \lambda(y - Ax) = \lambda(Ax - Ax') + \lambda(y - Ax) = \lambda(y - Ax'),$$

and hence $(u, v) = \lambda\big((x, y) - (x', Ax')\big) = \lambda\left((x, y) - \widehat{A}(x')\right)$, which justifies the desired inclusion.

Conversely, assume that the equality related to $C \times D$ is satisfied. Take any nonzero $v \in Y$. There exist a real $\lambda > 0$, a pair $(x, y) \in C \times D$ and a point $x' \in X$ such that

$$(0, v) = \lambda\big((x, y) - (x', Ax')\big),$$

so $x' = x$ and $v = \lambda(y - Ax') = \lambda(y - Ax)$. It ensues that $Y \subset \mathbb{R}_+[D - A(C)]$, and this inclusion is an equality. $\square$

In addition to Theorem 11.7, the proof of the s-lower regularity property for qualified convexly $\mathcal{C}^{1,s}$-composite functions requires some lemmas. The first lemma provides another useful way to translate the qualification of $g \circ G$.

LEMMA 11.11. Let X and Y be two Banach spaces, $G : X \to Y$ be a mapping which is differentiable at $\overline{x} \in X$, and $g : Y \to \mathbb{R} \cup \{+\infty\}$ be a lower semicontinuous convex function which is finite at $G(\overline{x})$. Then anyone of conditions (a) and (b) below is equivalent to the Robinson qualification condition

$$\mathbb{R}_+\big(\operatorname{dom} g - G(\overline{x})\big) - DG(\overline{x})(X) = Y$$

for $g \circ G$ at $\overline{x}$:
(a) There is some real $\sigma > 0$ such that

$$\sigma\mathbb{B}_Y \subset \{g(\cdot) \leq 1 + g(G(\overline{x}))\} - G(\overline{x}) - DG(\overline{x})(\mathbb{B}_X).$$

(b) For every real $\rho > 0$ there is some real $\sigma > 0$ such that

$$\sigma\mathbb{B}_Y \subset \big(\{g(\cdot) \leq \rho + g(G(\overline{x}))\} - G(\overline{x})\big) \cap \rho\mathbb{B}_Y - DG(\overline{x})(\rho\mathbb{B}_X).$$

PROOF. The condition (b) of the lemma obviously implies (a) which itself readily entails the Robinson qualification condition for $g \circ G$ at $\overline{x}$. To prove that the Robinson qualification condition implies (b), let us consider for $\overline{y} := G(\overline{x})$ and $A := DG(\overline{x})$ the multimapping M from $X \times Y \times \mathbb{R}$ into Y defined by

$$M(x, y, r) = \{y - \overline{y} - A(x)\} \quad \text{if } g(y) \leq r \quad \text{and} \quad M(x, y, r) = \emptyset \quad \text{otherwise.}$$

We easily see that the graph of M is closed and convex and that $\operatorname{Range} M = \operatorname{dom} g - \overline{y} - A(X)$.

The latter equality yields that the Robinson qualification condition for $g \circ G$ at $\overline{x}$ can be translated as $\mathbb{R}_+(\operatorname{Range} M) = Y$, and this entails that $0 \in \operatorname{core}(\operatorname{Range} M)$.

Since $0 \in M(0, \overline{y}, g(\overline{y}))$, we may deduce from the Robinson-Ursescu theorem (see Theorem 7.26) that, for every neighborhood U of zero in $X \times Y \times \mathbb{R}$ we have

$$0 \in \operatorname{int}\big(M\big((0, \overline{y}, g(\overline{y})) + U\big)\big).$$

For every real $\rho > 0$, taking $U = \rho\mathbb{B}_X \times \rho\mathbb{B}_Y \times [-\rho, \rho]$ we obtain some real $\sigma > 0$ such that

$$\begin{aligned}\sigma\mathbb{B}_Y &\subset M\big(\rho\mathbb{B}_X \times (\overline{y} + \rho\mathbb{B}_Y) \times [-\rho + g(\overline{y}), \rho + g(\overline{y})]\big)\\ &\subset \big(\{g(\cdot) \le \rho + g(\overline{y})\} - \overline{y}\big) \cap \rho\mathbb{B}_Y - A(\rho\mathbb{B}_X),\end{aligned}$$

which ensures the condition (b). □

The second lemma concerns an estimation of the norm of subgradients of $g \circ G$.

LEMMA 11.12. Let X and Y be two Banach spaces, $G : X \to Y$ be a mapping which is continuously differentiable near $\overline{x} \in X$, and $g : Y \to \mathbb{R} \cup \{+\infty\}$ be a lower semicontinuous convex function which is finite at $G(\overline{x})$. Assume that there are reals $\rho > 0$, $\sigma > 0$ such that

$$\sigma\mathbb{B}_Y \subset \big(\{g \le \rho + g(G(\overline{x}))\} - G(\overline{x})\big) - DG(\overline{x})(\rho\mathbb{B}_X). \tag{11.8}$$

Then there exists a neighborhood U of $\overline{x}$ such that, for all $x \in U$ and $x^* = y^* \circ DG(x)$ with $y^* \in \partial g(G(x))$, one has

$$\sigma\|y^*\| \le 2\rho(1 + \|x^*\|).$$

PROOF. By lower semicontinuity of g at $G(\overline{x})$ and continuity of G and DG at $\overline{x}$ choose a neighborhood U of $\overline{x}$ such that G is differentiable on U and for every $x \in U$

$$g(G(\overline{x})) - (\rho/2) \le g(G(x)) \text{ and } \|G(x) - G(\overline{x})\| + \rho\|DG(x) - DG(\overline{x})\| < \sigma/4.$$

Fix $x \in U$ and $x^* = y^* \circ DG(x)$ with $y^* \in \partial g(G(x))$. Consider any $y \in \mathbb{B}_Y$. By assumption there are $b \in \mathbb{B}_X$ and $y' \in Y$ with $g(y') \le \rho + g(G(\overline{x}))$ such that $\sigma y = y' - G(\overline{x}) - A(\rho b)$ where $A := DG(\overline{x})$. Then writing

$$\begin{aligned}\langle y^*, \sigma y\rangle &= \langle y^*, y' - G(\overline{x})\rangle - \langle y^*, A(\rho b)\rangle\\ &= \langle y^*, y' - G(x)\rangle + \langle y^*, G(x) - G(\overline{x})\rangle - \rho\langle y^* \circ DG(x), b\rangle\\ &\quad - \rho\langle y^* \circ \big(DG(\overline{x}) - DG(x)\big), b\rangle,\end{aligned}$$

we see from the inclusion $y^* \in \partial g(G(x))$ that

$$\begin{aligned}\sigma\langle y^*, y\rangle &\le g(y') - g(G(x)) + \|y^*\|\,\|G(x) - G(\overline{x})\| + \rho\|x^*\|\\ &\quad + \rho\|y^*\|\,\|DG(x) - DG(\overline{x})\|\\ &\le g(y') - g(G(\overline{x})) + (\rho/2) + \|y^*\|\,\|G(x) - G(\overline{x})\| + \rho\|x^*\|\\ &\quad + \rho\|y^*\|\,\|DG(x) - DG(\overline{x})\|.\end{aligned}$$

It follows that

$$\begin{aligned}\sigma\langle y^*, y\rangle &\le (3\rho/2) + \big(\|G(x) - G(\overline{x})\| + \rho\|DG(x) - DG(\overline{x})\|\big)\,\|y^*\| + \rho\|x^*\|\\ &\le (3\rho/2) + (\sigma/4)\|y^*\| + \rho\|x^*\|.\end{aligned}$$

Finally, we get $(3\sigma/4)\|y^*\| \le (3\rho/2)(1 + \|x^*\|)$, that is, $\sigma\|y^*\| \le 2\rho(1 + \|x^*\|)$. □

We are now in a position to prove the s-lower regularity property for qualified convexly $\mathcal{C}^{1,s}$-composite functions through arguments similar to those used above for $\mathcal{C}^{1,s}$ functions.

THEOREM 11.13 (s-lower regularity of $\mathcal{C}^{1,s}$-composite functions). Let $f = g \circ G$ be a qualified convexly $\mathcal{C}^{1,s}$-composite function at $\overline{x} \in \operatorname{dom} f$. Then f is s-lower regular at $\overline{x}$; further there exists a neighborhood U of $\overline{x}$, such that for every $r > 0$ the restriction of f to $\operatorname{proj}_X\big((U \times r\mathbb{B}_{X^*}) \cap \operatorname{gph} \partial_C f\big)$ is Lipschitz continuous.

PROOF. By assumptions there exist a real $c > 0$ and a convex open neighborhood U of $\overline{x}$ over which the formula in Theorem 11.7 holds true and over which G is c-Lipschitzian and DG is Hölderian with exponent s and coefficient c. We also know by Lemma 11.11 that there are some reals $\sigma > 0$ and $\rho > 0$ such that

$$\sigma\mathbb{B}_Y \subset \big(\{g \le \rho + g\big(G(\overline{x})\big)\} - G(\overline{x})\big) - DG(\overline{x})(\rho\mathbb{B}_X).$$

Then by Lemma 11.12 we may also suppose that, shrinking U if necessary, one has

$$\sigma\|y^*\| \le 2\rho(1 + \|x^*\|), \tag{11.9}$$

for any $x^* = y^* \circ DG(x)$ with $x \in U$ and $y^* \in \partial_C g(G(x))$.

Fix any $x \in U \cap \operatorname{Dom} \partial_C f$ and $x^* \in \partial_C f(x)$. By Theorem 11.7, there is some $y^* \in \partial_C g(G(x))$ such that $x^* = y^* \circ DG(x)$. For any $u \in U$ we can write

$$\begin{aligned} f(x) - f(u) &\le \langle y^*, G(x) - G(u)\rangle \\ &= \left\langle y^*, DG(x)(x-u) + \int_0^1 [DG(u + t(x-u)) - DG(x)](x-u)\,dt \right\rangle \\ &= \langle x^*, x - u\rangle + \left\langle y^*, \int_0^1 [DG(u + t(x-u)) - DG(x)](x-u)\,dt \right\rangle. \end{aligned}$$

From this and (11.9) we see that

$$f(x) - f(u) \le \langle x^*, x-u\rangle + \sigma^{-1} 2\rho c(1 + \|x^*\|)\|x-u\|^{s+1} \int_0^1 (1-t)^s\,dt,$$

which means that

$$f(u) \ge f(x) + \langle x^*, u - x\rangle - \frac{2c\rho}{\sigma(s+1)}(1 + \|x^*\|)\|x-u\|^{s+1},$$

which translates the s-lower regularity of f on U.

It remains to show the Lipschitz property. Fix any real $r > 0$ and suppose that $x^* = y^* \circ DG(x)$ is in $r\mathbb{B}_{X^*}$. Then for every (u, u^*) in $(U \times r\mathbb{B}_{X^*}) \cap \operatorname{gph} \partial_C f$, we have by the convexity of g and by (11.9)

$$\begin{aligned} f(x) - f(u) = g(G(x)) - g(G(u)) &\le \langle y^*, G(x) - G(u)\rangle \\ &\le \sigma^{-1} 2c\rho(1 + r)\|x - u\|. \end{aligned}$$

This and the symmetry between (x, x^*) and (u, u^*) guarantee the desired Lipschitz property of the proposition and finishes its proof. □

As a particular example of qualified convexly composite functions we have:

PROPOSITION 11.14. Let U be an open convex set of a Banach space X, $g : X \to \mathbb{R} \cup \{+\infty\}$ be a proper lower semicontinuous convex function, and $\varphi : U \to \mathbb{R}$ be a real-valued continuously differentiable (resp. $\mathcal{C}^{1,s}$) function on U. Then the function f with $f(x) = g(x) + \varphi(x)$ for all $x \in U$ is convexly composite (resp. convexly $\mathcal{C}^{1,s}$-composite) on U and qualified at any point of $U \cap \operatorname{dom} g$.

PROOF. Put $Y := X \times \mathbb{R}$ and define $G : U \to Y$ and $\widehat{g} : Y \to \mathbb{R} \cup \{+\infty\}$ by

$$G(x) := (x, \varphi(x)) \quad \text{for all } x \in U \quad \text{and} \quad \widehat{g}(x,r) := g(x) + r \quad \text{for all } (x,r) \in Y.$$

The function $\widehat{g}$ is convex and lower semicontinuous with $\operatorname{dom} \widehat{g} = (\operatorname{dom} g) \times \mathbb{R}$, the mapping G is continuously differentiable (resp. $\mathcal{C}^{1,s}$) on U, and we have $f(x) = \widehat{g} \circ G(x)$ for all $x \in U$ as easily seen. Let us check the Robinson qualification condition for $\widehat{g} \circ G$. Fix any $\overline{x} \in U \cap \operatorname{dom} g$ and note that $DG(\overline{x})h = (h, D\varphi(\overline{x})h)$ for all $h \in X$. Taking any $(x,r) \in Y = X \times \mathbb{R}$, we have

$$(x,r) = \big((\overline{x},\rho) - (\overline{x},\varphi(\overline{x}))\big) - (-x, D\varphi(\overline{x})(-x)) = \big((\overline{x},\rho) - G(\overline{x})\big) - DG(\overline{x})(-x),$$

for $\rho := r - D\varphi(\overline{x})x + \varphi(\overline{x})$. Further $(\overline{x},\rho) \in \operatorname{dom} \widehat{g}$, so the equality

$$X \times \mathbb{R} = \big(\operatorname{dom} \widehat{g} - G(\overline{x})\big) - DG(\overline{x})(X)$$

holds, hence the Robinson qualification condition is satisfied for $\widehat{g} \circ G$ at $\overline{x}$ as desired. □

REMARK 11.15. Instead of s-lower regularity with a real $0 < s \le 1$, we could deal with the concept of lower $\omega(\cdot)$-regularity where $\omega : [0,+\infty[\to [0,+\infty[$ is a nondecreasing upper semicontinuous function with $\omega(0) = 0$. Define a lower semicontinuous function $f : X \to \mathbb{R} \cup \{+\infty\}$ as $\omega(\cdot)$-*lower regular* near $\overline{x} \in \operatorname{cl}(\operatorname{dom} f)$ provided there are an open convex neighborhood U of $\overline{x}$ and reals $c, \gamma > 0$ such that

$$f(y) \ge f(x) + \langle x^*, y - x\rangle - c(1 + \|x^*\|)\|x - y\|\omega(\gamma\|x - y\|)$$

for all $y \in U$, $x \in U \cap \operatorname{Dom} \partial_C f$, and $x^* \in \partial_C f(x)$. Concerning such a concept, it can be proved as above that any convexly composite function $g \circ G$ is lower $\omega(\cdot)$-regular near $\overline{x} \in \operatorname{dom}(g \circ G)$ whenever the mapping G is of class $\mathcal{C}^{1,\omega(\cdot)}$ on an open convex neighborhood U of $\overline{x}$, that is, the derivative DG is uniformly continuous on U and the modulus of continuity of DG relative to U is bounded from above by $r\omega(\cdot)$ for some real constant $r > 0$. Statements similar to the next results below can also be obtained in that framework. We did not develop that viewpoint since the study of differentiability properties seem to be more natural with Moreau s-envelope of s-lower regular functions than the Moreau $\omega(\cdot)$-envelope of $\omega(\cdot)$-lower regular ones. Other remarks concerning $\omega(\cdot)$-lower regular functions can be found in the section of comments. □

11.1.3. Coincidence of subdifferentials of s-lower regular functions.

The next theorem shows in particular that the concept of s-lower regularity is invariant with respect to diverse subdifferentials.

THEOREM 11.16 (coincidence of subdifferentials under s-lower regularity). Let X be a Banach space, U be an open convex set of X, and $f : X \longrightarrow \mathbb{R} \cup \{+\infty\}$ be an extended real-valued function which is lower semicontinuous on U with $U \cap \operatorname{dom} f \neq \emptyset$. Let a real $s > 0$. Then the following assertions hold:
(a) If f is s-lower regular on U, then for all $x \in U$, one has

$$\partial_F f(x) = \partial_L f(x) = \partial_A f(x) = \partial_C f(x).$$

If in addition $s = 1$, then the above sets are also equal to $\partial_{VP} f(x)$ for $x \in U$.
(b) If X is Asplund, then f is s-lower regular on U with coefficient $c \ge 0$ if and only if for all $x \in U \cap \operatorname{Dom} \partial_F f$ and for all $x^* \in \partial_F f(x)$ one has

$$f(y) \ge f(x) + \langle x^*, y - x\rangle - c(1 + \|x^*\|)\|x - y\|^{s+1}, \ \forall y \in U.$$

If X is a Hilbert space, then f is primal lower regular on U with coefficient $c \geq 0$ if and only if the above property holds with $s = 1$ and with $\partial_P f$ in place of $\partial_F f$.
(c) The function f is s-lower regular (resp. primal lower regular when X is Hilbertian) on U with coefficient $c \geq 0$ if and only if for all $x \in U \cap \operatorname{Dom} \partial f$ and for all $x^* \in \partial f(x)$ one has

$$f(y) \geq f(x) + \langle x^*, y - x \rangle - c(1 + \|x^*\|)\|x - y\|^{s+1}, \ \forall y \in U$$

(resp. the same inequality with $s = 1$), where ∂ denotes any subdifferential satisfying $\partial_F f(\cdot) \subset \partial f(\cdot) \subset \partial_C f(\cdot)$ (resp. satisfying $\partial_P f(\cdot) \subset \partial f(\cdot) \subset \partial_C f(\cdot)$).

PROOF. (a) Concerning the assertion (a) we only need to show that $\partial_C f(x)$ is included in $\partial_F f(x)$ (resp. $\partial_{VP} f(x)$) for all $x \in U$. Let $x \in U \cap \operatorname{Dom} \partial_C f$ and $x^* \in \partial_C f(x)$. The conclusion is then trivial with $s = 1$ for $\partial_{VP} f(x)$. Concerning the case for $\partial_F f(x)$, since f is s-lower regular on U, there exists a real $c \geq 0$ such that

$$f(y) + c(1 + \|x^*\|)\|x - y\|^{s+1} \geq f(x) + \langle x^*, y - x \rangle, \forall y \in U.$$

Fix any $\varepsilon > 0$. We can choose a real $\delta > 0$ such that $B(x, \delta) \subset U$ and $c(1 + \|x^*\|)(\delta)^s \leq \varepsilon$. It results that

$$f(y) + \varepsilon\|x - y\| \geq f(x) + \langle x^*, y - x \rangle, \ \forall y \in B(x, \delta),$$

and this translates the inclusion $x^* \in \partial_F f(x)$.
(b) Only the implication ($\Leftarrow$) needs to be proved since the reverse one is obvious. Assume that X is an Asplund space and take a real $c \geq 0$ such that for all $x \in U \cap \operatorname{Dom} \partial_F f$ and for all $x^* \in \partial_F f(x)$

$$f(y) \geq f(x) + \langle x^*, y - x \rangle - c(1 + \|x^*\|)\|x - y\|^{s+1}, \forall y \in U. \tag{11.10}$$

Step I. First, let us prove that $\partial_F f(x)$ is weakly* closed for any $x \in U$. Fix any real $r > 0$. Suppose that $\partial_F f(x) \cap r\mathbb{B}_{X^*}$ is nonempty. Let $(x_j^*)_{j \in J}$ be a net in $\partial_F f(x) \cap r\mathbb{B}_{X^*}$ converging weakly* to $x^* \in X^*$. The weak* lower semicontinuity of the dual norm $\|\cdot\|$ in X^* ensures that

$$\|x^*\| \leq \liminf_{j \in J} \|x_j^*\| \leq r.$$

On the other hand, from (11.10) we see that for all $y \in U$

$$\langle x_j^*, y - x \rangle \leq f(y) - f(x) + c(1 + r)\|x - y\|^{s+1},$$

hence taking the limit we obtain

$$\langle x^*, y - x \rangle \leq f(y) - f(x) + c(1 + r)\|x - y\|^{s+1}.$$

From this inequality, it is easily seen as in the proof of (a) that $x^* \in \partial_F f(x)$, thus $x^* \in \partial_F f(x) \cap r\mathbb{B}_{X^*}$. Consequently, for any $x \in U$, the set $\partial_F f(x) \cap r\mathbb{B}_{X^*}$ is weakly* closed for all reals $r > 0$, so the convex set $\partial_F f(x)$ is weakly* closed in X^* according to the Krein-Šmulian theorem (see Theorem C.4 in Appendix C).

Step II. Fix any $x \in U \cap \operatorname{Dom} \partial_C f$ and define the sets

$$V := \left\{ {}^{w^*}\lim x_n^* : x_n^* \in \partial_F f(x_n), x_n \underset{n \to \infty}{\longrightarrow_f} x \right\} = \partial_L f(x),$$

and

$$V_0 := \left\{ {}^{w^*}\lim \sigma_n x_n^* : x_n^* \in \partial_F f(x_n), x_n \underset{n \to \infty}{\longrightarrow_f} x, \sigma_n \downarrow 0 \right\} = \partial_L^\infty f(x),$$

where ${}^{w^*}\lim x_n^*$ denotes the limit of $(x_n^*)_n$ with respect to the weak* topology of X^*. Fix any $x^* \in V$, so there exist a sequence $(x_n)_n$ such that $(x_n, f(x_n))_n$ converges strongly to $(x, f(x))$, and a sequence $(x_n^*)_n$ in X^* converging weakly* to x^* with $x_n^* \in \partial_F f(x_n)$. The sequence $(x_n^*)_n$ is then bounded, say there exists a real $\gamma > 0$ such that $\|x_n^*\| \le \gamma$ for all n. Further, for n large enough $x_n \in U$, so from (11.10) we have

$$f(y) \ge f(x_n) + \langle x_n^*, y - x_n\rangle - c(1 + \|x_n^*\|)\|x_n - y\|^{s+1}, \forall y \in U,$$

hence

$$f(y) \ge f(x_n) + \langle x_n^*, y - x_n\rangle - c(1 + \gamma)\|x_n - y\|^{s+1}, \forall y \in U.$$

Taking the limit as $n \to \infty$, we get

$$f(y) \ge f(x) + \langle x^*, y - x\rangle - c(1 + \gamma)\|x - y\|^{s+1}, \forall y \in U. \tag{11.11}$$

Consider now an arbitrary element $x_0^* \in V_0$, so there exist a sequence $(x_n)_n$ such that $(x_n, f(x_n))_n$ converges strongly to $(x, f(x))$, a sequence $(\sigma_n)_n$ in $]0, +\infty[$ with $\sigma_n \to 0$ and a sequence $(x_n^*)_n$ in X^* with $(\sigma_n x_n^*)_n$ converging weakly* to x_0^* with $x_n^* \in \partial_F f(x_n)$ for all n. Let a real $\gamma' > 0$ such that $\|\sigma_n x_n^*\| \le \gamma'$ for all n. For n large enough, $x_n \in U$ thus from (11.10)

$$f(y) \ge f(x_n) + \langle x_n^*, y - x_n\rangle - c(1 + \|x_n^*\|)\|x_n - y\|^{s+1}, \forall y \in U,$$

and this implies

$$\sigma_n f(y) \ge \sigma_n f(x_n) + \langle \sigma_n x_n^*, y - x_n\rangle - \sigma_n c(1 + \|x_n^*\|)\|x_n - y\|^{s+1}, \forall y \in U.$$

This ensures that, for n large enough,

$$\sigma_n f(y) \ge \sigma_n f(x_n) + \langle \sigma_n x_n^*, y - x_n\rangle - \sigma_n c\|x_n - y\|^{s+1} - \gamma' c\|x_n - y\|^{s+1}, \forall y \in U,$$

so taking the limit as $n \to \infty$ gives

$$0 \ge \langle x_0^*, y - x\rangle - c\gamma'\|x - y\|^{s+1}, \forall y \in U \cap \operatorname{dom} f.$$

The latter inequality and (11.11) yield

$$f(y) \ge f(x) + \langle x^* + x_0^*, y - x\rangle - c(1 + \gamma + \gamma')\|x - y\|^{s+1}, \forall y \in U \cap \operatorname{dom} f,$$

so for $c' := c(1 + \gamma + \gamma')$ (depending on x^* and x_0^*)

$$f(y) \ge f(x) + \langle x^* + x_0^*, y - x\rangle - c'\|x - y\|^{s+1}, \forall y \in U.$$

From this inequality it is easily seen as in (a) that $x^* + x_0^* \in \partial_F f(x)$, hence $V + V_0 \subset \partial_F f(x)$. Since $\partial_F f(x)$ is convex and weakly* closed according to Step I, it results that

$$\overline{\mathrm{co}}^*[V + V_0] \subset \partial_F f(x) \subset \partial_C f(x).$$

These inclusions and the equality $\partial_C f(x) = \overline{\mathrm{co}}^*[V + V_0]$ (see Theorem 4.120(b)) entail that $\partial_C f(x) = \partial_F f(x)$ for all $x \in U$. Consequently, the inequality (11.10) holds for any $(x, x^*) \in \operatorname{gph} \partial_C f$ with $x \in U$, which means that f is s-lower regular in the sense of Definition 11.1.

The arguments are similar for $\partial_P f$ when X is Hilbert and $s = 1$.

(c) Since $\partial_F f(x) \subset \partial f(x) \subset \partial_C f(x)$ (resp. $\partial_P f(x) \subset \partial f(x) \subset \partial_C f(x)$), for all $x \in U$, the assertion (c) follows from (a) and (b). □

11.1.4. Subdifferential characterization of s-lower regular functions. The next theorem in this section characterizes the s-lower regularity of a function in terms of a property of monotonicity type of its subdifferential. Let us first establish the following lemma. It is proved in a context much more general than the statement that we need here. The general statement will be used later for other results.

LEMMA 11.17. Let $(X, \|\cdot\|)$ be a normed space and $f : X \to \mathbb{R} \cup \{+\infty\}$ be an extended real-valued function with $f(\overline{x}) < +\infty$. Let r be a positive number such that f is bounded from below over $B[\overline{x}, r]$ by some real α. Let reals $s > 0$, $\beta \in \mathbb{R}$, $\gamma \geq 0$ and $\theta \geq 0$. For each real $c \geq 0$, let

$$F_{\beta,c}(x^*, x, y) := f(y) + \beta\langle x^*, x - y\rangle + c(1 + \gamma\|x^*\|)\|x - y\|^{s+1},$$

for all $x, y \in X$ and $x^* \in X^*$. Let any real c_0 with

$$c_0\gamma \geq \frac{|\beta|4^{s+1}}{(2^{s+1} - 1)r^s} \quad \text{such that} \quad c_0 > \frac{4^{s+1}}{(2^{s+1} - 1)r^{s+1}}(f(\overline{x}) + \theta - \alpha).$$

Then, for any real $c \geq c_0$, for any $x^* \in X^*$ and for any $x \in B[\overline{x}, \frac{r}{4}]$, every point $u \in B[\overline{x}, r]$ such that

$$F_{\beta,c}(x^*, x, u) \leq \inf_{y \in B[\overline{x}, r]} F_{\beta,c}(x^*, x, y) + \theta$$

must belong to $B(\overline{x}, \frac{3r}{4})$.

PROOF. Fix $s > 0$, $x \in B[\overline{x}, \frac{r}{4}]$, $x^* \in X^*$, and fix also any real $c \geq c_0$. Take any $y \in B[\overline{x}, r]$ with $\|y - \overline{x}\| \geq \frac{3r}{4}$. Since

$$\|x - y\| \geq \|\overline{x} - y\| - \|x - \overline{x}\| \geq \frac{r}{2},$$

we observe that

$$\|x - y\|^{s+1} - \|x - \overline{x}\|^{s+1} \geq \frac{2^{s+1}r^{s+1}}{4^{s+1}} - \frac{r^{s+1}}{4^{s+1}} = \frac{(2^{s+1} - 1)r^{s+1}}{4^{s+1}}.$$

Then, for $F(y) := F_{\beta,c}(x^*, x, y)$ we have

$$\begin{aligned}
&F(y) - F(\overline{x}) - \theta \\
\geq &f(y) - f(\overline{x}) - \theta + \beta\langle x^*, \overline{x} - y\rangle + c(1 + \gamma\|x^*\|)(\|x - y\|^{s+1} - \|x - \overline{x}\|^{s+1}) \\
\geq &\alpha - f(\overline{x}) - \theta - r|\beta|\|x^*\| + c(1 + \gamma\|x^*\|)\frac{(2^{s+1} - 1)r^{s+1}}{4^{s+1}} \\
= &\left(\alpha - f(\overline{x}) - \theta + c\frac{(2^{s+1} - 1)r^{s+1}}{4^{s+1}}\right) + r\|x^*\|\left(c\gamma\frac{(2^{s+1} - 1)r^s}{4^{s+1}} - |\beta|\right),
\end{aligned}$$

so for $\eta := \alpha - f(\overline{x}) - \theta + c\frac{(2^{s+1}-1)r^{s+1}}{4^{s+1}} > 0$ we obtain $F(y) - \eta \geq F(\overline{x}) + \theta$, which finishes the proof of the lemma. □

THEOREM 11.18 (subdifferential characterization of s-lower regularity). Let a real $s > 0$ and $f : X \to \mathbb{R} \cup \{+\infty\}$ be an extended real-valued function on a Banach space X which is finite at $\overline{x}$ and lower semicontinuous near $\overline{x}$. The following are equivalent:

(a) the function f is s-lower regular at $\overline{x}$;

(b) there exist reals $\varepsilon > 0$ and $c \geq 0$ such that for all $x_i^* \in \partial_C f(x_i)$ with $\|x_i - \overline{x}\| < \varepsilon$, $i = 1, 2$, one has

$$\langle x_1^* - x_2^*, x_1 - x_2 \rangle \geq -c(1 + \|x_1^*\| + \|x_2^*\|)\|x_1 - x_2\|^{s+1}.$$

If in addition X is an Asplund space (resp. X is a Hilbert space and $s = 1$), then the following assertion is also equivalent to the s-lower regularity of the function f at the point $\overline{x}$:
(c) the inequality in (b) is fulfilled with $\partial_L f$ or ∂_F (resp. $\partial_P f$) in place of $\partial_C f$.

PROOF. First, we show that (a) $\Rightarrow$ (b). Suppose that f is s-lower regular on some ball $B(\overline{x}, \varepsilon)$ with some coefficient $c \geq 0$. Then, for $x_i \in X$ with $\|x_i - \overline{x}\| < \varepsilon$ and $x_i^* \in \partial_C f(x_i)$, $i = 1, 2$, we have by Definition 11.1

$$f(x_1) \geq f(x_2) + \langle x_2^*, x_1 - x_2 \rangle + c(1 + \|x_2^*\|)\|x_1 - x_2\|^{s+1}$$
$$f(x_2) \geq f(x_1) + \langle x_1^*, x_2 - x_1 \rangle + c(1 + \|x_1^*\|)\|x_1 - x_2\|^{s+1},$$

and adding these inequalities we obtain

$$\begin{aligned}\langle x_1^* - x_2^*, x_1 - x_2 \rangle &\geq -c(2 + \|x_1^*\| + \|x_2^*\|)\|x_1 - x_2\|^{s+1} \\ &\geq -2c(1 + \|x_1^*\| + \|x_2^*\|)\|x_1 - x_2\|^{s+1}.\end{aligned}$$

So, the implication (a) $\Rightarrow$ (b) holds true with $\partial_C f$, and hence also (a) implies all the other assertions.

Let us prove the reverse implication. Denote by ∂f anyone of the subdifferentials involved in (b) and (c) with the appropriate space. Let $\varepsilon > 0, c \geq 0$ be such that the assertion (b) is fulfilled and f is lower semicontinuous on $B(\overline{x}, \varepsilon)$. Let $0 < \varepsilon' < \min\left\{\varepsilon, \dfrac{1}{c^{1/s}}\right\}$ be such that $\alpha := \inf\limits_{B[\overline{x}, \varepsilon']} f$ is finite (according to the lower semicontinuity property of f). We fix a real

$$c_0 \geq \frac{4^{s+1}}{(2^{s+1} - 1)(\varepsilon')^s} \quad \text{with} \quad c_0 > \frac{4^{s+1}}{(2^{s+1} - 1)(\varepsilon')^{s+1}} \left(f(\overline{x}) + 1 - \alpha\right)$$

and a real $c' > \max\left\{c_0, \frac{2c}{s+1-(s+1)c(\varepsilon')^s}\right\}$. Let $u \in \operatorname{Dom} \partial f \cap B(\overline{x}, \frac{\varepsilon'}{4})$ and $u^* \in \partial f(u)$. We define

$$\varphi(x) := f(x) + \langle u^*, u - x \rangle + c' \left(1 + \|u^*\|\right) \|x - u\|^{s+1} \quad \text{for all } x \in X$$

and

$$\overline{\varphi}(x) := \begin{cases} \varphi(x) & \text{if } x \in B[\overline{x}, \varepsilon'] \\ +\infty & \text{if } x \in X \setminus B[\overline{x}, \varepsilon'], \end{cases} \tag{11.12}$$

so clearly $\overline{\varphi}$ is lower semicontinuous on X (since f is lower semicontinuous on $B[\overline{x}, \varepsilon']$).

Let (ε_n) be a sequence of real numbers which converges to 0 with $0 < \varepsilon_n < \min\left\{\frac{1}{4}, (\frac{\varepsilon'}{4})^2\right\}$. For every $n \in \mathbb{N}$, choose $u_n \in X$ such that

$$\overline{\varphi}(u_n) < \inf_X \overline{\varphi} + \varepsilon_n.$$

Applying the last lemma with $\beta = \theta = 1$ we obtain that $u_n \in B(\overline{x}, \frac{3\varepsilon'}{4})$ for all $n \in \mathbb{N}$. By the Ekeland variational principle, for each $n \in \mathbb{N}$, there exists $x_n \in X$ such that

$$\|x_n - u_n\| \leq \sqrt{\varepsilon_n},\ \overline{\varphi}(x_n) < \inf_X \overline{\varphi} + \varepsilon_n,\ \overline{\varphi}(x_n) = \inf_{x \in X} \left\{\overline{\varphi}(x) + \sqrt{\varepsilon_n}\|x - x_n\|\right\},$$

then
$$\|x_n - \overline{x}\| < \varepsilon' \text{ and } 0 \in \partial(\overline{\varphi} + \sqrt{\varepsilon_n}\|\cdot - x_n\|)(x_n).$$
Since $\varphi = \overline{\varphi}$ on $B[\overline{x}, \varepsilon']$ and $x_n \in B(\overline{x}, \varepsilon')$, we deduce $0 \in \partial(\varphi + \sqrt{\varepsilon_n}\|\cdot - x_n\|)(x_n)$ hence by the fuzzy sum rule for the subdifferential ∂ and the $\sqrt{\varepsilon_n}$-Lipschitz property of the function $\sqrt{\varepsilon_n}\|\cdot\|$, we derive that there are x'_n, x''_n with $\|x'_n - x_n\| < \sqrt{\varepsilon_n}$, $|f(x'_n) - f(x_n)| < \sqrt{\varepsilon_n}$ and $\|x''_n - x_n\| < \sqrt{\varepsilon_n}$ such that
$$0 \in \partial f(x'_n) - u^* + c'(1 + \|u^*\|)\partial_C(\|\cdot - u\|^{s+1})(x''_n) + 2\sqrt{\varepsilon_n}\mathbb{B}_{X^*},$$
which furnishes some $x^*_n \in \partial f(x'_n)$ and $y^*_n \in -u^* + c'(1 + \|u^*\|)\partial_C(\|\cdot - u\|^{s+1})(x''_n)$ with
$$\|x^*_n + y^*_n\| \le 2\sqrt{\varepsilon_n}. \tag{11.13}$$
Set $z^*_n := \dfrac{y^*_n + u^*}{c'(1 + \|u^*\|)} \in \partial_C(\|\cdot - u\|^{s+1})(x''_n)$ and note that (as easily seen through the subdifferential of the convex function $\|\cdot - u\|^{s+1}$)
$$\langle z^*_n, x''_n - u\rangle = (s+1)\|x''_n - u\|^{s+1} \text{ and } \|z^*_n\| \le (s+1)\|x''_n - u\|^s. \tag{11.14}$$
On the other hand,
$$\|x''_n - u\| \le \|x''_n - x_n\| + \|x_n - u_n\| + \|u_n - \overline{x}\| + \|\overline{x} - u\| < 2\sqrt{\varepsilon_n} + \frac{3\varepsilon'}{4} + \|\overline{x} - u\|$$
and $\frac{3\varepsilon'}{4} + \|\overline{x} - u\| < \varepsilon'$, so there exists some integer n_0 such that, for all $n \ge n_0$, $\|x''_n - u\| < \varepsilon'$ and $\|z^*_n\| \le (s+1)(\varepsilon')^s$. Fix any $n \ge n_0$. From the equality $y^*_n = -u^* + c'(1 + \|u^*\|)z^*_n$ we see that
$$\|y^*_n\| \le \|u^*\| + c'(s+1)(\varepsilon')^s(1 + \|u^*\|),$$
and from the inequality
$$\|x^*_n\| \le \|x^*_n + y^*_n\| + \|y^*_n\|$$
and (11.13) we also see that
$$\|x^*_n\| \le 2\sqrt{\varepsilon_n} + \|u^*\| + c'(s+1)(\varepsilon')^s(1 + \|u^*\|). \tag{11.15}$$
Further, the assertion (b), with $x^*_1 = x^*_n$ and $x^*_2 = u^*$, ensures that
$$\langle u^* - x^*_n, u - x'_n\rangle \ge -c\,(1 + \|u^*\| + \|x^*_n\|)\,\|u - x'_n\|^{s+1}.$$
Putting $\mu := c'(s+1)(1 + \|u^*\|)(\varepsilon')^s$ and writing by (11.14) and (11.13)
$$\begin{aligned}&\langle u^* - x^*_n, u - x'_n\rangle\\ &= \langle c'(1 + \|u^*\|)z^*_n - y^*_n - x^*_n, u - x''_n\rangle + \langle c'(1 + \|u^*\|)z^*_n - y^*_n - x^*_n, x''_n - x'_n\rangle\\ &\le -(s+1)c'(1 + \|u^*\|)\|x''_n - u\|^{s+1} + 2\sqrt{\varepsilon_n}\|x''_n - u\| + (\mu + 2\sqrt{\varepsilon_n})\|x''_n - x'_n\|,\end{aligned}$$
it results that
$$\begin{aligned}&-(s+1)c'(1 + \|u^*\|)\|x''_n - u\|^{s+1} + 2\sqrt{\varepsilon_n}\|u - x''_n\| + (\mu + 2\sqrt{\varepsilon_n})\|x''_n - x'_n\|\\ &\ge -c(1 + \|u^*\| + \|x^*_n\|)\|u - x'_n\|^{s+1}.\end{aligned}$$
Noticing that there is some real λ_n between $\|x''_n - u\|$ and $\|x'_n - u\|$ such that
$$\begin{aligned}\big|\,\|x'_n - u\|^{s+1} - \|x''_n - u\|^{s+1}\big| &= |(s+1)\lambda^s_n(\,\|x'_n - u\| - \|x''_n - u\|)|\\ &\le (s+1)\lambda^s_n\|x'_n - x''_n\|,\end{aligned}$$
with $\gamma_n := 2\sqrt{\varepsilon_n}\|u - x''_n\| + \big(\mu + 2\sqrt{\varepsilon_n} + c'(s+1)^2\lambda^s_n(1 + \|u^*\|)\big)\|x'_n - x''_n\|$ we obtain
$$\big((s+1)c'(1 + \|u^*\|) - c(1 + \|u^*\| + \|x^*_n\|)\big)\|u - x'_n\|^{s+1} \le \gamma_n \tag{11.16}$$

along with $\gamma_n \to 0$ as $n \to \infty$ (since $(\lambda_n)_n$ is bounded and $\|x'_n - x''_n\| \le 2\sqrt{\varepsilon_n}$). Further, the inequality (11.15) implies

$$\begin{aligned}&(s+1)c'(1+\|u^*\|) - c(1+\|u^*\|+\|x^*_n\|)\\ &\ge (s+1)c'(1+\|u^*\|) - c(1+\|u^*\|) - c\big(2\sqrt{\varepsilon_n}+\|u^*\|+c'(s+1)(\varepsilon')^s(1+\|u^*\|)\big)\\ &> (s+1)c'(1+\|u^*\|) - c(1+\|u^*\|) - c\big(1+\|u^*\|+c'(s+1)(\varepsilon')^s(1+\|u^*\|)\big)\\ &= (1+\|u^*\|)\big((s+1)c' - 2c - (s+1)cc'(\varepsilon')^s\big),\end{aligned}$$

so by (11.16) we get

$$(1+\|u^*\|)\big((s+1)c' - 2c - (s+1)cc'(\varepsilon')^s\big)\|u - x'_n\|^{s+1} \le \gamma_n. \tag{11.17}$$

By the choice of c' we have

$$c' > \frac{2c}{(s+1)-(s+1)c(\varepsilon')^s}, \quad \text{or equivalently } (s+1)c' - 2c - (s+1)cc'(\varepsilon')^s > 0,$$

then it follows from (11.17) that

$$\lim_{n\to\infty} x'_n = u, \quad \text{hence} \quad \lim_{n\to\infty} u_n = u.$$

Further, we know that $\varphi(u_n) \le \inf_{x\in B[\overline{x},\varepsilon']} \inf \varphi(x) + \varepsilon_n$, or equivalently

$$\begin{aligned}&f(u_n) + \langle u^*, u - u_n\rangle + c'(1+\|u^*\|)\|u_n - u\|^{s+1}\\ &\le \inf_{x\in B[\overline{x},\varepsilon']} \big\{f(x) + \langle u^*, u - x\rangle + c'(1+\|u^*\|)\|x-u\|^{s+1}\big\} + \varepsilon_n.\end{aligned}$$

Since f is lower semicontinuous and $\lim_{n\to\infty} u_n = u$, the latter inequality ensures that

$$f(u) \le \liminf_{n\to\infty} f(u_n) \le \inf_{x\in B[\overline{x},\varepsilon']} \big\{f(x) + \langle u^*, u - x\rangle + c'(1+\|u^*\|)\|x-u\|^{s+1}\big\},$$

and so

$$f(u) \le f(x) + \langle u^*, u - x\rangle + c'(1+\|u^*\|)\|x-u\|^{s+1}, \ \forall x \in B\left(\overline{x}, \frac{\varepsilon'}{4}\right).$$

We then conclude that f is s-lower regular at $\overline{x}$ by definition in the case ∂ is ∂_C and by Theorem 11.16 in the other cases. □

For primal lower regular functions (that is, for $s = 1$) a useful corollary of monotonicity can be derived by means of a specific truncation of $\partial_C f$. Given a function $f : X \to \mathbb{R}\cup\{+\infty\}$ on a Banach space X, a subdifferential ∂ and two extended reals $\varepsilon, \rho > 0$, we denote by $T^f_{\varepsilon,\rho}$ the multimapping whose graph is defined by

$$\operatorname{gph} T^f_{\varepsilon,\rho} := \{(x, x^*) \in \operatorname{gph}\partial f : \|x - \overline{x}\| < \varepsilon, \|x^*\| \le \rho\}. \tag{11.18}$$

When the subdifferential ∂ needs to be emphasized we will write $T^{f,\partial}_{\varepsilon,\rho}$.

COROLLARY 11.19. *Let H be a Hilbert space and $f : H \to \mathbb{R}\cup\{+\infty\}$ be a function which is primal lower regular at $\overline{x} \in \operatorname{dom} f$. Then there exist reals $\varepsilon_0 > 0$ and $c > 0$ such that for any reals $0 < \varepsilon \le \varepsilon_0$ and $\rho \ge 1$ the multimapping $c\rho I + T^f_{\varepsilon,\rho}$ is monotone, where I denotes the identity mapping on H, where $T^f_{\varepsilon,\rho}$ is the truncation of the C-subdifferential $\partial_C f$.*

PROOF. By Theorem 11.18 there are reals $c_0, \varepsilon_0 > 0$ such that for any $(x_i, v_i) \in \operatorname{gph} \partial_C f$ with $\|x_i - \overline{x}\| < \varepsilon_0$, for $i = 1, 2$, one has

$$\langle v_1 - v_2, x_1 - x_2\rangle \geq -c_0(1 + \|v_1\| + \|v_2\|)\|x_1 - x_2\|^2.$$

Putting $c := 3c_0$ and fix any reals $0 < \varepsilon \leq \varepsilon_0$ and $\rho \geq 1$. For any $(x_i, v_i) \in \operatorname{gph} \partial_C f$ with $\|x_i - \overline{x}\| < \varepsilon_0$, for $i = 1, 2$, we deduce that

$$\langle v_1 - v_2, x_1 - x_2 \geq -c\rho\|x_1 - x_2\|^2,$$

or equivalently

$$\langle (c\rho x_1 + v_1) - (c\rho x_2 + v_2, x_1 - x_2\rangle \geq 0.$$

This justifies the monotonicity of $c\rho I + T^f_{\varepsilon,\rho}$. □

Like for convex or semiconvex functions, s-lower regular functions enjoy a closure property for graphs of their subdifferential.

PROPOSITION 11.20. Let U be an open convex set of a Banach space X and $f : X \to \mathbb{R} \cup \{+\infty\}$ be a lower semicontinuous function on U with $U \cap \operatorname{dom} f \neq \emptyset$ and such that f is s-lower regular on U. Let $x \in U \cap \operatorname{Dom} \partial_C f$ and $(x_j)_{j\in J}$ be a net converging strongly to x in X, and let $(x_j^*)_{j\in J}$ be a bounded net of X^* converging weakly star to some x^* in X^* with $x_j^* \in \partial_C f(x_j)$. Then $\lim_{j\in J} f(x_j) = f(x)$ and with $c > 0$ given by (11.2) one has

$$\langle x^*, x' - x\rangle \leq f(x') - f(x) + c(1 + \|x^*\|)\|x' - x\|^s \ \forall x' \in U, \text{ and } x^* \in \partial_F f(x) \subset \partial_C f(x).$$

In particular $\partial_F f(x)$ is weakly* closed in X^*, and if $s = 2$ the set $\partial_{VP} f(x)$ is also weakly* closed in X^*.

PROOF. Take $c \geq 0$ given (11.2), that is, such that

$$f(x') \geq f(x'') + \langle x^*, x' - x''\rangle - c(1 + \|x^*\|)\|x' - x''\|^s$$

whenever $x', x'' \in U$ and $x^* \in \partial_C f(x'')$. Choose a real $\gamma > 0$ such that $\|x_j^*\| \leq \gamma$ for all $j \in J$ (according to the boundedness assumption) and choose also some $j_0 \in J$ such that $x_j \in U$ for all $j \succeq j_0$ (since $x_j \to x$). Fixing $x' \in U$, this yields, for all $j \succeq j_0$,

$$\begin{aligned} f(x') &\geq f(x_j) + \langle x_j^*, x' - x_j\rangle - c(1 + \|x_j^*\|)\|x' - x_j\|^s \\ &\geq f(x_j) + \langle x_j^*, x' - x_j\rangle - c(1 + \gamma)\|x' - x_j\|^s. \end{aligned} \tag{11.19}$$

Taking the limit inferior, it follows that

$$f(x') \geq f(x) + \langle x^*, x' - x\rangle - c(1 + \gamma)\|x' - x\|^s,$$

which is the desired inequality, and from this it is easily seen that $x^* \in \partial_F f(x) \subset \partial_C f(x)$. Further, putting $x' = x$ in (11.19), we obtain

$$f(x) \geq f(x_j) + \langle x_j^*, x - x_j\rangle - c(1 + \gamma)\|x - x_j\|^s$$

for $j \succeq j_0$, which allows us to write

$$f(x) \geq \limsup_{j\in J} f(x_j) \geq \liminf_{j\in J} f(x_j) \geq f(x),$$

that is, $\lim_{j\in J} f(x_j) = f(x)$.

It remains to show that $\partial_F f(x)$ and $\partial_{VP} f(x)$ (if $s = 2$) are weakly* closed. Both sets being convex, the Krein-Šmulian theorem (see Theorem C.4 in Appendix C) guarantees that they are weakly* closed since the arguments above ensure that the weak* limit of every bounded net of each one of these sets remains therein. □

Using the above closedness property and proceeding as in Proposition 8.31 we obtain:

PROPOSITION 11.21. Let U be an open convex set of a Banach space X and $f : X \to \mathbb{R} \cup \{+\infty\}$ be a lower semicontinuous function on U with $U \cap \operatorname{dom} f \neq \emptyset$ and such that f is s-lower regular on U. Then, for ∂ as ∂_F, or ∂_A or ∂_C the function $x \mapsto d(0, \partial f(x))$ is lower semicontinuous on U.

11.2. Moreau s-envelope

Given a normed space X and two extended real-valued functions $f, g : X \to \mathbb{R} \cup \{+\infty\}$, we recall (see Definition 4.26) that the infimal convolution function of f and g is defined by

$$(f \square g)(x) = \inf_{y \in X} \{f(y) + g(x - y)\} \quad \text{for all } x \in X.$$

In particular, when $g = \frac{1}{2\lambda}\|\cdot\|^2$ with $\lambda > 0$, we obtain the important concepts of (standard) Moreau envelope and proximal mapping with index λ as presented in Definition 3.262, that is,

$$e_\lambda f(x) := \inf_{y \in X} \left(f(y) + \frac{1}{2\lambda}\|x - y\|^2 \right),$$

$$\operatorname{Prox}_\lambda f(x) := \operatorname*{Argmin}_{y \in X} \left(f(y) + \frac{1}{2\lambda}\|x - y\|^2 \right) := \left\{ u \in X : e_\lambda f(x) = f(u) + \frac{1}{2\lambda}\|x - u\|^2 \right\}.$$

Recall that if $\operatorname{Prox}_\lambda f(x)$ is a singleton, its unique element is denoted in Definition 3.262 by $P_\lambda f(x)$.

In this section we will study diverse properties of the more general concepts of Moreau s-envelope and s-proximal mapping with index $\lambda > 0$.

DEFINITION 11.22. Let $f : X \longrightarrow \mathbb{R} \cup \{+\infty\}$ be an extended real-valued function and $\lambda > 0$ be a positive real number. For any real $s > 1$, we define the *Moreau s-envelope* $e_{\lambda,[s]} f$ and the *s-proximal multimapping* $\operatorname{Prox}_{\lambda,[s]} f$ *of* f *with index* λ as

$$e_{\lambda,[s]} f(x) := \inf_{y \in X} \left(f(y) + \frac{1}{s\lambda}\|x - y\|^s \right),$$

$$\operatorname{Prox}_{\lambda,[s]} f(x) := \operatorname*{Argmin}_{y \in X} \left(f(y) + \frac{1}{s\lambda}\|x - y\|^s \right).$$

When $\operatorname{Prox}_{\lambda,[s]} f(x))$ is a singleton, its unique element will be denoted by $P_{\lambda,[s]} f(x)$.

Given a subset W of X, the s-Moreau envelope and s-proximal multimapping of f *relative to* W *and with index* λ are defined similarly by

$$e_{\lambda,[s],W} f(x) := \inf_{y \in W} \left(f(y) + \frac{1}{s\lambda}\|x - y\|^s \right),$$

$$\operatorname{Prox}_{\lambda,[s],W} f(x) := \operatorname*{Argmin}_{y \in W} \left(f(y) + \frac{1}{s\lambda}\|x - y\|^s \right).$$

When $W = B[0, \varepsilon]$ we will just write $e_{\lambda,[s],\varepsilon} f$ and $\operatorname{Prox}_{\lambda,[s],\varepsilon} f$ and we will say that they are the *local s-Moreau envelope and local s-proximal multimapping of* f *with index* λ *associated with* $\varepsilon > 0$.

If $e_{\lambda,[s]}f$ is finite at $\overline{x} \in X$ and if $\overline{y} \in \operatorname{Prox}_{\lambda,[s]}f(\overline{x})$, by Proposition 4.27 one has

$$\partial_F e_{\lambda,[s]}f(\overline{x}) \subset \partial_F f(\overline{y}) \cap \partial_F \left(\frac{1}{s\lambda}\|\cdot\|^s\right)(\overline{x}-\overline{y}). \tag{11.20}$$

The subdifferential $\partial(\frac{1}{s}\|\cdot\|^s)$ is usually denoted by $J_{s,\|\cdot\|}$ and called the *s-duality multimapping* of $(X,\|\cdot\|)$; for $s=2$ one just says the *duality multimapping* and one generally writes $J_{\|\cdot\|}$, as already mentioned in Example 2.66(a). When there is no risk of ambiguity for the norm $\|\cdot\|$, we write J_s and J, omitting the subscript $\|\cdot\|$. By (2.24) we notice that

$$J_s(x) = \{x^* \in X^* : \langle x^*, x\rangle = \|x^*\|_*\|x\|,\ \|x^*\|_* = \|x\|^{s-1}\}, \tag{11.21}$$

from which we see that for each $x \in X$

$$J_s(-x) = -J_s(x) \quad \text{and} \quad J_s(\alpha x) = \alpha^{s-1}J_s(x) \text{ for all } \alpha > 0. \tag{11.22}$$

When $\|\cdot\|$ is Gâteaux differentiable off zero, or equivalently $\frac{1}{s}\|\cdot\|^s$ is Gâteaux differentiable on X, then J_s is single-valued; in such a case one generally identifies $J_s(x)$ with its unique element $D(\frac{1}{s}\|\cdot\|^s)(x)$ and one calls J_s the *s-duality mapping* of $(X,\|\cdot\|)$, and just the *duality mapping* if $s=2$. More will be said later in Subsection 18.1.4 of Chapter 18 with the development and consequences of (18.19).

The following theorem provides some first basic general properties of Moreau s-envelopes. Its assertions (a), (b) and (c) extend the results in Theorem 3.265 related to the standard Moreau envelope $e_\lambda f$ (that is, $e_{\lambda,[s]}f$ with $s=2$).

THEOREM 11.23 (basic properties of Moreau s-envelopes). Let $(X,\|\cdot\|)$ be a normed space. Let a real $s>1$ and $f : X \to \mathbb{R}\cup\{+\infty\}$ be a proper function. Assume that there are reals $\alpha,\beta \geq 0$, $\gamma \in \mathbb{R}$ such that

$$f(x) \geq -\alpha\|x\|^s - \beta\|x\| + \gamma, \text{ for all } x \in X. \tag{11.23}$$

(I) For any real $\lambda > 0$ one has the inclusion $\operatorname{Argmin}_X f \subset \operatorname{Argmin}_X e_{\lambda,[s]}f$ along with the inequality $e_{\lambda,[s]}f(x) \leq f(x)$ for every $x \in X$ and the equality

$$\inf_{x\in X} f(x) = \inf_{x\in X} e_{\lambda,[s]}f(x).$$

(II) Assume that $\lambda \in]0, \frac{1}{s\alpha}[$ (with the convention $1/\alpha = +\infty$ for $\alpha = 0$). One has the following properties.
(a) The function $e_{\lambda,[s]}f$ is finite valued on the space X, and for any real number $r>0$, there exists a real $\tau>0$ (depending on r and λ) such that for every $x \in B(0,r)$

$$e_{\lambda,[s]}f(x) = \inf_{x'\in B(0,\tau)}\left(f(x') + \frac{1}{s\lambda}\|x-x'\|^s\right).$$

(a') For any reals $r>0$ and $\rho>0$ there exists a real $\kappa>0$ such that for every $x \in B(0,r)$

$$\left\{y \in X : f(y) + \frac{1}{s\lambda}\|x-y\|^s \leq e_{\lambda,[s]}f(x) + \rho\right\} \subset \kappa\mathbb{B}_X.$$

(b) The envelope function $e_{\lambda,[s]}f$ is Lipschitzian on any ball $B(0,r)$ with a Lipschitz constant $L \geq \frac{(r+\tau)^{s-1}}{\lambda}$, where τ is as in (a).

(c) Given $x \in X$ with $\sup_{\lambda \in]0, \frac{1}{s\alpha}[} (e_{\lambda,[s]}f)(x) < +\infty$ and given $(\theta(\lambda))_{0<\lambda<(1/s\alpha)}$ with $\theta(\lambda) > 0$ and $\theta(\lambda) \downarrow 0$ as $\lambda \downarrow 0$, then for any family $(y_\lambda)_{0<\lambda<1/(s\alpha)}$ with

$$f(y_\lambda) + \frac{1}{s\alpha}\|x - y_\lambda\|^s \leq e_{\lambda,[s]}f(x) + \theta(\lambda),$$

one has

$$y_\lambda \to x \quad \text{and} \quad f(y_\lambda) \to \liminf_{u \to x} f(u) \quad \text{as } \lambda \downarrow 0.$$

(d) The non-decreasing family $(e_{\lambda,[s]}f)_\lambda$ pointwise converges to $\overline{f}$ as $\lambda \downarrow 0$, where $\overline{f}$ denotes the lower semicontinuous hull of f.
(e) If f is lower semicontinuous, then the family $(e_{\lambda,[s]}f)_\lambda$ pointwise converges to f as $\lambda \downarrow 0$; further, for every non-increasing net $(\lambda_j)_{j\in J}$ in $]0, \frac{1}{s\alpha}[$ tending to 0 and every net $(x_j)_{j\in J}$ in X converging strongly to $\overline{x} \in X$ one has

$$f(\overline{x}) \leq \liminf_{j\in J} e_{\lambda,[s]}f(x_j).$$

(f) If the normed space $(X, \|\cdot\|)$ is a reflexive Banach space whose norm $\|\cdot\|$ is Fréchet differentiable off zero and has the sequential Kadec-Klee property, and if the function f is lower semicontinuous, then $\operatorname{Dom}\operatorname{Prox}_{\lambda,[s]}f$ is dense in X, or more precisely for any $x \in \operatorname{Dom}\partial_F e_{\lambda,[s]}f$, the set $\operatorname{Prox}_{\lambda,[s]}f(x)$ is nonempty and for every $x' \in \operatorname{Prox}_{\lambda,[s]}f(x)$ one has

$$\partial_F e_{\lambda,[s]}f(x) = \left\{\frac{1}{\lambda}J_s(x - x')\right\} \text{ and } \frac{1}{\lambda}J_s(x - x') \in \partial_F f(x');$$

further one has a better Lipschitz constant L than in (b) for $e_{\lambda,[s]}f$ on $B(0,r)$ with $L \geq \frac{(r+\tau)^{s-1}}{s\lambda}$.

Proof. (I) The arguments for (I) are the same for (a) in Theorem 3.265.
(II) Fix any $\lambda \in]0, \frac{1}{s\alpha}[$.
(a)-(a') Fix for a moment $x \in X$ and consider any $x' \in X$. By the minorization assumption we have

$$\frac{1}{s\lambda}\|x' - x\|^s - \alpha\|x'\|^s - \beta\|x'\| + \gamma \leq f(x') + \frac{1}{s\lambda}\|x' - x\|^s. \tag{11.24}$$

Observe that

$$\begin{aligned}\frac{1}{s\lambda}\|x' - x\|^s - \alpha\|x'\|^s &= \left(\frac{1}{s\lambda} - \alpha\right)\|x'\|^s + \frac{1}{s\lambda}\|x' - x\|^s - \frac{1}{s\lambda}\|x'\|^s \\ &= \frac{1 - s\lambda\alpha}{s\lambda}\|x'\|^s + \frac{1}{\lambda}\left(\frac{1}{s}\|x' - x\|^s - \frac{1}{s}\|x'\|^s\right).\end{aligned}$$

Put $\Delta := \frac{1}{s}\|x' - x\|^s - \frac{1}{s}\|x'\|^s$ and consider the function φ defined on $[0,1]$ by

$$\varphi(t) := \frac{1}{s}\|x' - tx\|^s.$$

Since the norm $\|\cdot\|$ is convex, the continuous convex function φ has a right-hand derivative at every real t, and moreover

$$\varphi'_+(t) = \|x' - tx\|^{s-1}N'(x' - tx; -x),$$

where $N'(y; v) := \lim_{\theta\downarrow 0}\theta^{-1}\big(N(y + \theta v) - N(y)\big)$ denotes the directional derivative of the convex function $N := \|\cdot\|$ at the point y. Note also that

$$N'(x' - tx; -x) \geq -\|-x\| = -\|x\|$$

according to the properties of the norm. It results from this that we have

$$\begin{aligned}\Delta := \varphi(1) - \varphi(0) = \int_0^1 \varphi'_+(t)\,dt &= \int_0^1 \|x' - tx\|^{s-1} N'(x' - tx; -x)\,dt \\ &\geq -\|x\| \int_0^1 \|x' - tx\|^{s-1}\,dt \\ &\geq -\|x\| \int_0^1 (\|x'\| + t\|x\|)^{s-1}\,dt.\end{aligned}$$

Further, one knows (see Lemma 3.25) that for any non-negative numbers μ, ν, σ

$$(\mu + \nu)^\sigma \leq 2^\sigma(\mu^\sigma + \nu^\sigma),$$

then

$$\Delta \geq -2^{s-1}\|x\| \int_0^1 (\|x'\|^{s-1} + t^{s-1}\|x\|^{s-1})\,dt,$$

or equivalently

$$\Delta \geq -2^{s-1}\|x\|\|x'\|^{s-1} - \frac{2^{s-1}}{s}\|x\|^s.$$

It ensures that

$$\frac{1}{s\lambda}\|x' - x\|^s - \alpha\|x'\|^s \geq \frac{1 - s\lambda\alpha}{s\lambda}\|x'\|^s - \frac{2^{s-1}}{\lambda}\|x\|\|x'\|^{s-1} - \frac{2^{s-1}}{s\lambda}\|x\|^s. \tag{11.25}$$

From this and (11.24) we see that

$$\zeta(\|x\|) := \frac{1 - s\lambda\alpha}{s\lambda}\|x'\|^s - \frac{2^{s-1}}{\lambda}\|x\|\|x'\|^{s-1} - \frac{2^{s-1}}{s\lambda}\|x\|^s + \gamma \leq f(x') + \frac{1}{s\lambda}\|x - x'\|^2.$$

Noticing that the continuous function $\zeta : [0, +\infty[\to \mathbb{R}$ satisfies $\zeta(t) \to +\infty$ as $t \to +\infty$, it is bounded from below on $[0, +\infty[$. The function $x' \mapsto f(x') + \frac{1}{s\lambda}\|x - x'\|^s$ is then bounded from below on X, hence $e_{\lambda,[s]}f(x)$ is finite. This confirms that the function $e_{\lambda,[s]}f(\cdot)$ is finite valued on the space X.

Now, fix $\overline{x} \in X$ where f is finite. One has by definition of $e_{\lambda,[s]}f$

$$e_{\lambda,[s]}f(x) \leq \frac{1}{s\lambda}\|x - \overline{x}\|^s + f(\overline{x}).$$

Fix a real $\rho > 0$ and take any $x' \in X$ such that

$$f(x') + \frac{1}{s\lambda}\|x' - x\|^s \leq e_{\lambda,[s]}f(x) + \rho.$$

Then

$$\frac{1}{s\lambda}\|x' - x\|^s - \alpha\|x'\|^s - \beta\|x'\| + \gamma \leq \frac{1}{s\lambda}\|x - \overline{x}\|^s + f(\overline{x}) + \rho, \tag{11.26}$$

thus by (11.25) we have

$$\frac{1 - s\lambda\alpha}{s\lambda}\|x'\|^s - \frac{2^{s-1}}{\lambda}\|x\|\|x'\|^{s-1} - \beta\|x'\| - \frac{2^{s-1}}{s\lambda}\|x\|^s \leq \frac{1}{s\lambda}\|x - \overline{x}\|^s + f(\overline{x}) + \rho - \gamma.$$

Putting $\eta := (1 - s\lambda\alpha)^{-1} > 0$, the previous inequality can be written as

$$\|x'\|^s - 2^{s-1}s\eta\|x\|\|x'\|^{s-1} - s\lambda\eta\beta\|x'\| \leq 2^{s-1}\eta\|x\|^s + \eta\|x - \overline{x}\|^s + s\eta\lambda(f(\overline{x}) + \rho - \gamma).$$

Fix now $x \in B(0, r)$. The latter inequality yields

$$\|x'\|^s - 2^{s-1}sr\eta\|x'\|^{s-1} - s\lambda\eta\beta\|x'\| \leq 2^{s-1}\eta r^s + \eta(r + \|\overline{x}\|)^s + s\eta\lambda(f(\overline{x}) + \rho - \gamma),$$

so putting

$$a := 2^{s-1} s r \eta > 0,\ b := s\lambda\eta\beta > 0,\ c := 2^{s-1}\eta r^s + \eta(r + \|\overline{x}\|)^s + s\eta\lambda(f(\overline{x}) + \rho - \gamma),$$

we obtain

$$\|x'\|^s - a\|x'\|^{s-1} - b\|x'\| \leq c.$$

Clearly, the set

$$\{t \geq 0 : t^s - at^{s-1} - bt \leq c\}$$

is independent of $x \in B(0,r)$, and it is not difficult to see that it is bounded. Indeed, suppose that there exists a sequence $(t_n)_n$ of positive reals tending to $+\infty$ such that $t_n^s - at_n^{s-1} - bt_n \leq c$, hence

$$t_n \leq a + \frac{b}{t_n^{s-2}} + \frac{c}{t_n^{s-1}}.$$

(i) If $s > 2$ then, since $t_n \to \infty$ we obtain that $a \geq +\infty$, which is a contradiction.
(ii) If $1 < s \leq 2$ then, since $t_n \to \infty$, there exists $n_0 \in \mathbb{N}$ such that for every $n \geq n_0$, we have $t_n > 1$, then $t_n \geq t_n^{s-1}$ because $s - 1 \leq 1$, hence

$$t_n^s - (a+b)t_n \leq t_n^s - at_n^{s-1} - bt_n \leq c \Longrightarrow t_n^{s-1} \leq a + b + \frac{c}{t_n},$$

with $s - 1 > 0$. Taking the limit we get $a + b \geq +\infty$, which is a contradiction. Consequently, there exists a real $\tau > 0$ (depending on r and λ) such that $\|x'\| < \tau$, which justifies both assertions (a) and (a').
(b) We know by (a) that $e_{\lambda,[s]}f(\cdot)$ is finite valued on X. Taking $x \in B(0,r)$ and keeping in mind that $\lambda \in]0, \frac{1}{s\alpha}[$, we also have by (a)

$$e_{\lambda,[s]}f(x) = \inf_{x' \in B(0,\tau)} \left(f(x') + \frac{1}{s\lambda}\|x - x'\|^s \right).$$

To prepare the Lipschitz property of $e_{\lambda,[s]}f$, we note by the mean value theorem that for any $0 \leq t_1 \leq t_2$, there exists a real $c \in [t_1, t_2]$ such that

$$t_2^s - t_1^s = \left|t_2^s - t_1^s\right| \leq sc^{s-1}\left|t_2 - t_1\right|,$$

then, for any $x, y \in B(0,r)$ and $z \in B(0,\tau)$, there exists c between $\|y - z\|$ and $\|x - z\|$ such that

$$\begin{aligned} \big|\|x - z\|^s - \|y - z\|^s\big| &\leq sc^{s-1}\big|\|x - z\| - \|y - z\|\big| \\ &\leq s(r+\tau)^{s-1}\|x - y\|. \end{aligned} \tag{11.27}$$

Fixing $x' \in B(0,\tau)$ and putting $g_{x'}(x) := f(x') + \frac{1}{s\lambda}\|x - x'\|^s$, by (11.27) we get that for any $x, y \in B(0,r)$

$$g_{x'}(x) - g_{x'}(y) = \frac{1}{s\lambda}(\|x - x'\|^s - \|y - x'\|^s) \leq \frac{1}{\lambda}(r+\tau)^{s-1}\|x - y\|,$$

which implies that

$$\inf_{x' \in \tau\mathbb{B}} g_{x'}(x) \leq \inf_{x' \in \tau\mathbb{B}} g_{x'}(y) + \frac{(r+\tau)^{s-1}}{\lambda}\|x - y\|,$$

then $e_{\lambda,[s]}f(x) - e_{\lambda,[s]}f(y) \leq \frac{(r+\tau)^{s-1}}{\lambda}\|x - y\|$. This means that the function $e_{\lambda,[s]}f$ is Lipschitzian on $B(0,r)$ with the constant $L \geq \frac{(r+\tau)^{s-1}}{\lambda}$.

(c) Consider x, $(\theta(\lambda))_{0<\lambda<(1/s\alpha)}$ and $(y_\lambda)_\lambda$ as in the statement of (c) (and observe that such families $(y_\lambda)_\lambda$ exist since $e_{\lambda,[s]}f(x)$ is finite by (b) for all $\lambda \in]0, \frac{1}{s\alpha}[$). Set

$$g(u) := \sup_{\lambda \in]0, \frac{1}{s\alpha}[} e_{\lambda,[s]}f(u) \quad \text{for all } u \in X$$

and notice that g is lower semicontinuous since each function $e_{\lambda,[s]}f$ is continuous by (b) again. By assumption we have

$$e_{\lambda,[s]}f(x) \le f(y_\lambda) + \frac{1}{s\lambda}\|x - y_\lambda\|^s \le e_{\lambda,[s]}f(x) + \theta(\lambda). \tag{11.28}$$

On the other hand, the family $\big(e_{\lambda,[s]}f(x)\big)_\lambda$ is nonincreasing in λ, hence and it converges to $\sup_{\lambda \in]0, \frac{1}{s\alpha}[} e_{\lambda,[s]}f(x) = g(x)$ as $\lambda \downarrow 0$. For each $\lambda \in]0, \frac{1}{s\alpha}[$ we have by the above inequality (11.28) that

$$\frac{1}{s\lambda}\|x - y_\lambda\|^s - \alpha\|y_\lambda\|^s - \beta\|y_\lambda\| + \gamma \le g(x) + \theta(\lambda), \tag{11.29}$$

hence by (11.25)

$$\left(\frac{1}{s\lambda} - \alpha\right)\|y_\lambda\|^s - \frac{2^{s-1}}{\lambda}\|x\|\|y_\lambda\|^{s-1} - \beta\|y_\lambda\| + \frac{2^{s-1}}{s\lambda}\|x\|^s \le g(x) + \theta(\lambda) - \gamma,$$

which entails as in the proof of (a) that there exist real constants $\lambda_0 \in]0, \frac{1}{s\alpha}[$ and $r > 0$ such that

$$\|y_\lambda\| \le r \quad \text{for all} \in \lambda \in]0, \lambda_0[.$$

It results from (11.29) that for all $\lambda \in]0, \frac{1}{s\alpha_0}[$

$$\|x - y_\lambda\|^s \le s\lambda\big(g(x) + \theta(\lambda) + \alpha r^s + r\beta - \gamma\big),$$

which gives that $(y_\lambda)_\lambda$ converges to x as $\lambda \downarrow 0$.

Further, from the assumption on y_λ we also note that $f(y_\lambda) \le g(x) + \theta(\lambda)$ for all $\lambda \in]0, \frac{1}{s\alpha}[$, hence

$$\liminf_{u \to x} f(u) \le \liminf_{\lambda \downarrow 0} f(y_\lambda) \le \limsup_{\lambda \downarrow 0} f(y_\lambda) \le g(x). \tag{11.30}$$

On the other hand, the inequality $e_{\lambda,[s]}f(\cdot) \le f(\cdot)$ (seen above) gives $f(\cdot) \ge g(\cdot)$. This and the lower semicontinuity of g yield

$$\liminf_{u \to x} f(u) \ge \liminf_{u \to x} g(u) = g(x),$$

which combined with (11.30) assures us that $\lim_{\lambda \downarrow 0} f(y_\lambda) = \liminf_{u \to x} f(u)$, finishing the proof of (c).

(d) Since the family $(e_{\lambda,[s]}f)_\lambda$ is nonincreasing in λ, it converges pointwise to $g(\cdot) = \sup_{\lambda \in]0, \frac{1}{s\alpha}[} e_{\lambda,[s]}f(\cdot)$ as $\lambda \downarrow 0$. To justify (d) we then have to show that g coincides with the lower semicontinuous hull $\overline{f}$ of f. Given any $x \in X$ where g is finite, we have for each $\lambda \in]0, \frac{1}{s\alpha}[$ that $e_{\lambda,[s]}f(x)$ is finite, which allows us to choose $y_\lambda \in X$ such that

$$e_{\lambda,[s]}f(x) \le f(y_\lambda) + \frac{1}{s\lambda}\|x - y_\lambda\|^s \le e_{\lambda,[s]}f(x) + \lambda,$$

which implies by (c) that $\overline{f}(x) \le g(x)$ if $g(x)$ is finite. The latter inequality being obvious if $g(x) = +\infty$, it ensues that $\overline{f}(\cdot) \le g(\cdot)$ on X. Regarding the converse inequality, note first that $f(\cdot) \ge g(\cdot)$ since $e_{\lambda,[s]}f(\cdot) \le f(\cdot)$. Keeping in mind that g is lower semicontinuous and that $\overline{f}$ is the greatest lower semicontinuous function

minorizing f, it follows that $\overline{f}(\cdot) \geq g(\cdot)$. This inequality combined with the other above inequality $\overline{f}(\cdot) \leq g(\cdot)$ yields the desired equality $g(\cdot) = \overline{f}(\cdot)$.
(e) The convergence property in (e) obviously follows from (d). Regarding the inequality related to the lower limit involving the net $(\lambda_j)_j$ it suffices to follow with appropriate slight modifications the proof of the similar inequality in (e) of Theorem 3.265.
(f) Assume that f is lower semicontinuous and that $(X, \|\cdot\|)$ is a reflexive Banach space whose norm $\|\cdot\|$ is Fréchet differentiable (off zero) and has the (sequential) Kadec-Klee property. Since $e_{\lambda,[s]}f$ is continuous, we know by the Asplund property of X that $\operatorname{Dom} \partial_F e_{\lambda,[s]}f$ is dense in X (see Corollary 4.53). Let $x \in \operatorname{Dom} \partial_F e_{\lambda,[s]}f$ and let $x^* \in \partial_F e_{\lambda,[s]}f(x)$. Fix a sequence $(t_n)_n$ in $]0,1[$ with $t_n \downarrow 0$ and a sequence $(y_n)_n$ in X such that

$$f(y_n) + \frac{1}{s\lambda}\|x - y_n\|^s \leq e_{\lambda,[s]}f(x) + t_n^2, \tag{11.31}$$

so fixing (as in (a)) some $\overline{x} \in X$ where f is finite it results that

$$\frac{1}{s\lambda}\|x - y_n\|^s - \alpha\|y_n\|^s - \beta\|y_n\| + \gamma \leq \frac{1}{s\lambda}\|x - \overline{x}\|^s + f(\overline{x}) + t_n^2.$$

According to the proof of (a), the sequence $(y_n)_n$ is bounded, then we can extract a subsequence (that we do not relabel) converging weakly to some $\overline{y} \in X$ and such that $\|x - y_n\| \to \eta$ for some real η. We show that $\eta = \|x - \overline{y}\|$. By the weak lower semicontinuity of the norm we have $\eta \geq \|x - \overline{y}\|$. Put now $x_n := x - t_n(x - y_n)$ and note that $x_n \to x$ since $(y_n)_n$ is bounded. As $x^* \in \partial_F e_{\lambda,[s]}f(x)$, for any real $\varepsilon > 0$ we have for n sufficiently large

$$\begin{aligned}
&\langle x^*, y_n - x\rangle \leq t_n^{-1}\Big(e_{\lambda,[s]}f(x - t_n(x - y_n)) - e_{\lambda,[s]}f(x)\Big) + \varepsilon\|x - y_n\| \\
&\leq t_n^{-1}\left(f(y_n) + \frac{1}{s\lambda}\|(1 - t_n)(x - y_n)\|^s - f(y_n) - \frac{1}{s\lambda}\|x - y_n\|^s + t_n^2\right) + \varepsilon\|x - y_n\|.
\end{aligned}$$

Let

$$\tau_n := \frac{1}{s}\|(1 - t_n)(x - y_n)\|^s - \frac{1}{s}\|x - y_n\|^s = \frac{1}{s}\|x - y_n - t_n(x - y_n)\|^s - \frac{1}{s}\|x - y_n\|^s,$$

and

$$\varphi_n(\theta) := \frac{1}{s}\|x - y_n - t_n\theta(x - y_n)\|^s.$$

Then, by (11.22) we obtain

$$\begin{aligned}
\tau_n &= \varphi_n(1) - \varphi_n(0) = \int_0^1 \varphi_n'(\theta)\, d\theta \\
&= -t_n\left\langle \int_0^1 J_s\big(x - y_n - t_n\theta(x - y_n)\big)\, d\theta, x - y_n \right\rangle \\
&= -t_n\left\langle \int_0^1 J_s\big((1 - t_n\theta)(x - y_n)\big)\, d\theta, x - y_n \right\rangle \\
&= -t_n\langle J_s(x - y_n), x - y_n\rangle \int_0^1 (1 - t_n\theta)^{s-1}\, d\theta = \frac{1}{s}\|x - y_n\|^s((1 - t_n)^s - 1),
\end{aligned}$$

so

$$\begin{aligned}\langle x^*, y_n - x\rangle &\le t_n^{-1}\left(\frac{1}{s\lambda}\|x-y_n\|^s((1-t_n)^s-1)+t_n^2\right)+\varepsilon\|x-y_n\|\\ &= \frac{1}{s\lambda}\|x-y_n\|^s t_n^{-1}((1-t_n)^s-1)+t_n+\varepsilon\|x-y_n\|.\end{aligned}$$

Putting $h(t) := (1-t)^s$ for $0 \le t \le 1$, we have $h'(t) = -s(1-t)^{s-1}$, and hence

$$\langle x^*, y_n - x\rangle \le \frac{1}{s\lambda}\|x-y_n\|^s t_n^{-1}(h(t_n)-h(0))+t_n+\varepsilon\|x-y_n\|.$$

Taking the limit as $n \to \infty$, we get that for every $\varepsilon > 0$

$$\langle x^*, \overline{y} - x\rangle \le \frac{1}{s\lambda}\eta^s h'(0) + \varepsilon\eta,$$

which gives $\langle x^*, \overline{y} - x\rangle \le -\frac{1}{\lambda}\eta^s$, then

$$\frac{\eta^s}{\lambda} \le \|x^*\|\|x-\overline{y}\|. \tag{11.32}$$

Now, take any $z \in X$. Considering again any $\varepsilon > 0$ we have, from the inclusion $x^* \in \partial_F e_{\lambda,[s]}f(x)$ and from the convergence $x + t_n z \to x$ as $n \to \infty$, that for n sufficiently large

$$\begin{aligned}\langle x^*, z\rangle &\le t_n^{-1}\Big(e_{\lambda,[s]}f(x+t_nz) - e_{\lambda,[s]}f(x)\Big) + \varepsilon\|z\|\\ &\le t_n^{-1}\left(f(y_n) + \frac{1}{s\lambda}\|x-y_n+t_nz\|^s - f(y_n) - \frac{1}{s\lambda}\|x-y_n\|^s + t_n^2\right) + \varepsilon\|z\|.\end{aligned}$$

Putting $\sigma_n := \frac{1}{s}\|x-y_n+t_nz\|^s - \frac{1}{s}\|x-y_n\|^s$ and $\psi_n(\theta) := \frac{1}{s}\|x-y_n+t_nz\theta\|^s$, we have by (11.21)

$$\begin{aligned}\sigma_n = \psi_n(1) - \psi_n(0) = \int_0^1 \psi_n'(\theta)\,d\theta &= \int_0^1 \langle J_s(x-y_n+\theta t_nz), t_nz\rangle\,d\theta\\ &\le t_n\|z\|\int_0^1 \|x-y_n+\theta t_nz\|^{s-1}\,d\theta,\end{aligned}$$

hence

$$\langle x^*, z\rangle \le \frac{1}{\lambda}\|z\|\int_0^1 \|x-y_n+\theta t_nz\|^{s-1}\,d\theta + t_n + \varepsilon\|z\|.$$

Noting that $\|x-y_n+\theta t_nz\| \to \eta$ since $t_n \downarrow 0$ and $\|x-y_n\| \to \eta$, it follows from the latter inequality that for every $\varepsilon > 0$

$$\langle x^*, z\rangle \le \frac{1}{\lambda}\|z\|\eta^{s-1} + \varepsilon\|z\|,$$

then $\langle x^*, z\rangle \le \frac{1}{\lambda}\|z\|\eta^{s-1}$, so

$$\|x^*\| \le \frac{\eta^{s-1}}{\lambda}.$$

According to (11.32), we obtain that $\eta \le \|x-\overline{y}\|$, and hence $\eta = \|x-\overline{y}\|$. Consequently

$$\|x-y_n\| \to \|x-\overline{y}\|,$$

and since $(y_n)_n$ converges weakly to $\overline{y}$ and the norm has the sequential Kadec-Klee property, we find that $(y_n)_n$ converges strongly to $\overline{y}$. Passing to the limit in (11.31) and using the lower semicontinuity of f, we get

$$f(\overline{y}) + \frac{1}{s\lambda}\|x - \overline{y}\|^s \leq e_{\lambda,[s]}f(x),$$

hence $e_{\lambda,[s]}f(x) = f(\overline{y}) + \frac{1}{s\lambda}\|x - \overline{y}\|^s$, which proves that $\operatorname{Prox}_{\lambda,[s]}f(x) \neq \emptyset$ and $\overline{y} \in \operatorname{Prox}_{\lambda,[s]}f(x)$. From (11.20) we deduce that for any $x' \in \operatorname{Prox}_{\lambda,[s]}f(x)$

$$\partial_F e_{\lambda,[s]}f(x) \subset \partial_F f(x') \cap \partial_F\left(\frac{1}{s\lambda}\|\cdot\|^s\right)(x - x').$$

Since the norm is Fréchet differentiable off zero, then $\partial_F(\frac{1}{s}\|\cdot\|)^s(\cdot) = J_s(\cdot)$ is a singleton, hence

$$\partial_F e_{\lambda,[s]}f(x) = \frac{1}{\lambda}J_s(x - x') \text{ and } \frac{1}{\lambda}J_s(x - x') \in \partial_F f(x').$$

We can also obtain a better constant of Lipschitz via (b). Indeed, one has

$$x^* = \partial_F e_{\lambda,[s]}f(x) = \frac{1}{s\lambda}J_s(x - x'),$$

then

$$\|x^*\| \leq \frac{1}{s\lambda}\|J_s(x - x')\| = \frac{1}{s\lambda}\|x - x'\|^{s-1} \leq \frac{(r+\tau)^{s-1}}{s\lambda}.$$

According to Theorem 6.55 we obtain $L \geq \frac{(r+\tau)^{s-1}}{s\lambda}$.

□

In the case of a Hilbert space, the next lemma and proposition provide some additional properties to the assertion (d) in Theorem 11.23. We state first the lemma. It will also be used in the proof of the proposition.

LEMMA 11.24. Let H be a Hilbert space and $f : H \to \mathbb{R} \cup \{+\infty\}$ be a proper function. The following assertions hold:
(a) For each real $\lambda > 0$ one has the following link between $e_\lambda f$ and the Legendre-Fenchel conjugate of $f + \frac{1}{2\lambda}\|\cdot\|^2$:

$$-\lambda e_\lambda f = -\frac{1}{2}\|\cdot\|^2 + \left(\lambda f + \frac{1}{2}\|\cdot\|^2\right)^*;$$

so the function $e_\lambda f$ is linearly semiconcave with coefficient $1/\lambda$ whenever it does not take the value $-\infty$.
(b) Assume that f is minorized by a quadratic function $-\alpha\|\cdot\|^2 - \beta\|\cdot\| + \gamma$ with $\alpha, \beta \geq 0$ and $\gamma \in \mathbb{R}$. Then for any $\lambda \in]0, 1/(2\alpha)[$ the function $e_\lambda f$ is Fréchet differentiable at x if and only it is Fréchet subdifferentiable at x; further, at any such a point x one has $\partial_P(-e_\lambda f)(x) = \{-\nabla e_\lambda f(x)\}$.

PROOF. (a) By definition of Moreau envelope we have

$$\begin{aligned}
-\lambda e_\lambda f(x) &= \sup_{y\in H}\left(-\lambda f(y) - \frac{1}{2}\|x\|^2 + \langle x, y\rangle - \frac{1}{2}\|y\|^2\right) \\
&= -\frac{1}{2}\|x\|^2 + \left(\lambda f + \frac{1}{2}\|\cdot\|^2\right)^*(x),
\end{aligned}$$

which is the desired equality in (a). The latter ensures that $\frac{1}{2\lambda}\|\cdot\|^2 - e_\lambda f$ is a convex function, so the linear semiconcavity property with coefficient $1/\lambda$ results

from Theorem 10.15(A).
(b) Fix any $\lambda \in]0, 1/(2\alpha)[$ and any $x \in H$. The local Lipschitz property of $-e_\lambda f$ from Theorem 11.23(b) gives $\partial_C(-e_\lambda f)(x) \neq \emptyset$. By the linear semiconvexity of $-e_\lambda f$ obtained in (a) we know by Theorem 10.29 that $\partial_P(-e_\lambda)f = \partial_F(-e_\lambda f) = \partial_C(-e_\lambda f)$, so $\partial_F(-e_\lambda f)(x) \neq \emptyset$. The assertion (b) then follows from Proposition 4.13(a) saying that a function φ is Fréchet differentiable at a point whenever both functions φ and $-\varphi$ are Fréchet subdifferentiable at this point. □

PROPOSITION 11.25. Let H be a Hilbert space and $f : H \to \mathbb{R} \cup \{+\infty\}$ be a proper function. Let also a real $\lambda > 0$. The following hold:
(a) One has $y \in (\operatorname{Prox}_\lambda f)^{-1}(x)$ if and only if

$$\langle y, x' - x\rangle \leq \left(\frac{1}{2}\|\cdot\|^2 + \lambda f\right)(x') - \left(\frac{1}{2}\|\cdot\|^2 + \lambda f\right)(x), \ \forall x' \in H;$$

in particular, $\operatorname{Prox}_\lambda f \subset (I + \lambda \partial_P f)^{-1}$.
(b) The multimapping $\operatorname{Prox}_\lambda f$ is monotone.
(c) If the function f is minorized by a quadratic function $-\alpha\|\cdot\|^2 - \beta\|\cdot\| + \gamma$ with $\alpha, \beta \geq 0$, $\gamma \in \mathbb{R}$ and if $\lambda < 1/2\alpha$, then $e_\lambda f$ is Fréchet differentiable at any point in $\operatorname{Dom} \partial_F e_\lambda f$, and $\operatorname{Prox}_\lambda f$ is single-valued on $\operatorname{Dom} \partial_F e_\lambda f$ with

$$\{P_\lambda f(x)\} = \operatorname{Prox}_\lambda f(x) = (I + \partial_F f)^{-1}(x) = (I + \partial_P f)^{-1}(x) \text{ for all } x \in \operatorname{Dom} \partial_F e_\lambda f.$$

(d) Assume as in (c) that f is minorized by a quadratic function $-\alpha\|\cdot\|^2 - \beta\|\cdot\| + \gamma$ with $\alpha, \beta \geq 0$ and assume that $\lambda < 1/2\alpha$. Then there exists a dense G_δ set in H at each point of which $e_\lambda f$ is Fréchet differentiable and $\operatorname{Prox}_\lambda f$ is a singleton.

PROOF. (a) The inclusion $x \in \operatorname{Prox}_\lambda f(y)$ means that

$$\frac{1}{2}\|x - y\|^2 + \lambda f(x) \leq \frac{1}{2}\|x' - y\|^2 + \lambda f(x'), \ \forall x' \in H,$$

which, after using the equality $\|u - y\|^2 = \|u\|^2 - 2\langle y, u\rangle + \|y\|^2$ in both members with $u = x$ and $u = x'$ and after simplification, is equivalent to the inequality stated in (a).
(b) Take $y_i \in \operatorname{Prox}_\lambda f(x_i)$ for $i = 1, 2$ and put $\varphi := (1/2)\|\cdot\|^2 + \lambda f$. By (a) we see that $f(x_i)$ is finite (since f is proper) and

$$\langle y_1, x_2 - x_1\rangle \leq \varphi(x_2) - \varphi(x_1) \quad \text{and} \quad \langle y_2, x_1 - x_2\rangle \leq \varphi(x_1) - \varphi(x_2),$$

so adding both inequalities yields $\langle y_1 - y_2, x_1 - x_2 \geq 0$ as desired.
(c) From Theorem 11.23(f) we derive that $\operatorname{Prox}_\lambda f$ is single-valued on $\operatorname{Dom} \partial_F e_\lambda f$ and $\operatorname{Prox}_\lambda f = (I + \lambda \partial_F f)^{-1}$ therein. By (a) we also know that $\operatorname{Prox}_\lambda f \subset (I + \partial_P f)^{-1}$. Since $\partial_P f \subset \partial_F f$, altogether and lemma 11.24 justify the assertion (c).
(d) Under the assumptions in (c), the equality in Lemma 11.24(a) ensures that the convex function $\left(\lambda f + \frac{1}{2\lambda}\|\cdot\|^2\right)^*$ is finite and continuous. Using this and using the Fréchet differentiability of $\|\cdot\|^2$ along with the equality in Lemma 11.24 again, we obtain that $e_\lambda f$ is Fréchet differentiable on a dense G_δ set in the Hilbert space H (since H is an Asplund space). The conclusion in (d) then follows from (c). □

Taking for f the indicator function Ψ_S of a closed set yields the following corollary.

COROLLARY 11.26. For any nonempty closed set S in a Hilbert space H the multimapping Proj_S is monotone.

Theorem 11.23(d) clearly ensures that $e_{\lambda,[s]}f$ is Gâteaux differentiable on X whenever, in addition to the hypotheses therein, f is a proper lower semicontinuous convex function and the norm $\|\cdot\|$ is both strictly convex and Fréchet differentiable off zero. In fact, the same result is valid under the Gâteaux differentiability of $\|\cdot\|$ off zero and the reflexivity of X.

THEOREM 11.27 (differentiability of Moreau s-envelope under convexity). Let $(X,\|\cdot\|)$ be a normed space and $f : X \to \mathbb{R}\cup\{+\infty\}$ be a proper lower semicontinuous convex function. Let also $1 < s < +\infty$ and $0 < \lambda < +\infty$.
(a) The Moreau s-envelope $e_{\lambda,[s]}f$ is finite-valued, convex on X and Lipschitzian on bounded sets, and for any $x \in X$ with $\operatorname{Prox}_{\lambda,[s]}f(x) \neq \emptyset$

$$\partial(e_{\lambda,[s]}f)(x) = \partial f(y) \cap \partial\left(\frac{1}{s\lambda}\|\cdot\|^s\right)(x-y) \quad \text{for all } y \in \operatorname{Prox}_{\lambda,[s]}f(x).$$

Further, the family $(e_{\lambda,[s]}f)_{\lambda>0}$ converges pointwise to f as $\lambda \downarrow 0$, and for any non-increasing net $(\lambda_j)_{j\in J}$ in $]0,+\infty[$ tending to 0 and any net $(x_j)_{j\in J}$ converging weakly to $\overline{x} \in X$ one has

$$f(\overline{x}) \le \liminf_{j\in J} e_{\lambda_j,[s]}f(x_j).$$

(b) If X is a reflexive Banach space, for each real $\lambda > 0$ one has

$$\operatorname{Argmin}_X f = \operatorname{Argmin}_X(e_{\lambda,[s]}f),$$

and also for every $x \in X$

$$\operatorname{Prox}_{\lambda,[s]}f \neq \emptyset,$$

along with for every $y \in \operatorname{Prox}_{\lambda,[s]}f$

$$d\big(0,\partial(e_{\lambda,[s]}f)(x)\big) = \frac{1}{\lambda}\|x-y\|^{s-1} = \sup_{y^*\in\partial(e_{\lambda,[s]}f)(x)} \|y^*\| \le d\big(0,\partial f(x)\big).$$

(c) If X is a reflexive Banach space and its norm $\|\cdot\|$ is strictly convex, then for any real $\lambda > 0$ the s-proximal mapping $P_{\lambda,[s]}f$ is well defined and norm-to-weak continuous on X and for any $x \in \operatorname{dom} f$

$$P_{\lambda,[s]}f(x) \to x \quad \text{and} \quad f(P_{\lambda,[s]}f(x)) \to f(x) \quad \text{as } \lambda \downarrow 0;$$

if in addition the norm $\|\cdot\|$ is (sequentially) Kadec-Klee, then $P_{\lambda,[s]}f$ is norm-to-norm continuous on the space X.
(d) If X is a reflexive Banach space and its norm is both strictly convex and Gâteaux differentiable off zero, then for any real $\lambda > 0$ the Moreau s-envelope $e_{\lambda,[s]}f$ is Gâteaux differentiable on X, the s-proximal mapping $P_{\lambda,[s]}f$ is norm-to-weak continuous on X, and for every $x \in X$

$$\{D(e_{\lambda,[s]}f)(x)\} = \partial f\big(P_{\lambda,[s]}f(x)\big) \cap \{\lambda^-1 J_s\big(x - P_{\lambda,[s]}f(x)\big)\},$$

where we recall that $J_s := D_G\left(\frac{1}{s}\|\cdot\|^s\right)$ is the Gâteaux derivative of $\frac{1}{s}\|\cdot\|^s$.
(e) If X is a reflexive Banach space whose norm is strictly convex, (sequentially) Kadec-Klee and Fréchet differentiable off zero, then the Moreau s-envelope $e_{\lambda,[s]}f$ is of class $\mathcal{C}^1$ on X.
(f) If X is a Hilbert space and $\|\cdot\|$ is the norm associated with the inner product, then

$$\operatorname{Prox}_\lambda f(x) = \{P_\lambda f(x)\} \text{ and } \{D(e_\lambda f)(x)\} = \partial f\big(P_\lambda f(x)\big) \cap \{\lambda^{-1}\big(x - P_\lambda f(x)\big)\};$$

further

$$\langle x - y, P_\lambda f(x) - P_\lambda f(y)\rangle \geq \|P_\lambda f(x) - P_\lambda f(y)\|^2,$$

so in particular the mapping $P_\lambda f$ is Lipschitz on H with constant 1.

PROOF. By Proposition 3.77 the function f is bounded from below by a continuous affine function, so the assumption (11.23) in Theorem 11.23 is satisfied with $\alpha = 0$. We will adapt the arguments employed for Theorem 3.267 related to the standard Moreau envelope.

(a) Theorem 11.23 says that for any $\lambda \in]0, +\infty[$ the Moreau s-envelope $e_{\lambda,[s]}f$ is Lipschitzian on bounded sets and that $(e_{\lambda,[s]}f)_\lambda$ pointwise converges to f as $\lambda \downarrow 0$. The other properties in (a) for $\partial e_{\lambda,[s]}f(x)$ follow from Proposition 3.69. Regarding the inequality concerning the net $(x_j)_{j\in J}$ it is enough to repeat with appropriate changes the arguments used for the analogous inequality in (b) of Theorem 3.267.

(b) Assume that X is reflexive. Given $x \in X$, according to the boundedness from below of f by a continuous affine function, the weakly lower semicontinuous function $\varphi := f(\cdot) + \frac{1}{s\lambda}\|x - \cdot\|^s$ satisfies $\varphi(y) \to +\infty$ as $\|y\| \to +\infty$, and hence it attains its infimum on X by weak compactness of closed balls in the reflexive space X. We deduce that $\mathrm{Prox}_{\lambda,[s]}f(x) \neq \emptyset$.

Consider now any any $u \in \mathrm{Argmin}_X(e_{\lambda,[s]}f)$, which is equivalent to $0 \in \partial(e_{\lambda,[s]}f)(u)$ by convexity of $e_{\lambda,[s]}f$. Choosing $v \in \mathrm{Prox}_{\lambda,[s]}f(u)$, the equality in (a) yields that $0 \in \partial f(v) \cap \partial(\frac{1}{s\lambda}\|\cdot\|^s)(u - v)$. The inclusion $0 \in \partial(\frac{1}{s\lambda}\|\cdot\|^s)(u - v)$ tells us that $u - v$ is a minimizer of $\|\cdot\|^s$, hence $v = u$. Therefore, $0 \in \partial f(u)$, so $u \in \mathrm{Argmin}_X f$ by convexity of f. It results that $\mathrm{Argmin}_X(e_{\lambda,[s]}f) \subset \mathrm{Argmin}_X f$, and this inclusion is an equality since the converse has been established in Theorem 11.23(I).

Let us show the inequalities in (b). Fix any $y^* \in \partial(e_{\lambda,[s]}f)(x)$ and $y \in \mathrm{Prox}_{\lambda,[s]}f(x)$. By (a) we have $y^* \in \partial f(y) \cap \partial(\frac{1}{s\lambda}\|\cdot - y\|^s)(x)$. By the description of $\partial(\frac{1}{s}\|\cdot\|^s)$ in (2.24), we have

$$\|y^*\| = \frac{1}{\lambda}\|x - y\|^{s-1} \quad \text{and} \quad \langle y^*, x - y\rangle = \frac{1}{\lambda}\|x - y\|^s.$$

From this and from the monotonicity of ∂f we derive that for any $x^* \in \partial f(x)$ (if this set is nonempty)

$$\|x^*\|\,\|x - y\| \geq \langle x^*, x - y\rangle \geq \langle y^*, x - y\rangle = \|y^*\|\,\|x - y\|,$$

hence $\|x^*\| \geq \|y^*\|$ if $y \neq x$. Further, if $y = x$, we have $y^* \in \partial(\frac{1}{s\lambda}\|\cdot\|^s)(0)$, hence $y^* = 0$. It follows that in any case

$$\|x^*\| \geq \|y^*\| = \frac{1}{\lambda}\|x - y\|^{s-1} = d\big(0, \partial(e_{\lambda,[s]}f)(x)\big).$$

This being true for every $x^* \in \partial f(x)$, it ensues that

$$d\big(0, \partial f(x)\big) \geq \frac{1}{\lambda}\|x - y\|^{s-1} = \sup_{y^*\in\partial(e_{\lambda,[s]}f)(x)} \|y^*\| = d\big(0, \partial(e_{\lambda,[s]}f)(x)\big),$$

and this translates the relations in (b) whenever $\partial f(x) \neq \emptyset$. It is also clear that the latter relations still hold true when $\partial f(x) = \emptyset$, since in this situation we know that $d\big(0, \partial f(x)\big) = +\infty$.

(c) Under the strict convexity of the norm $\|\cdot\|$ of the reflexive Banach space X the function $\varphi := f(\cdot) + \frac{1}{s\lambda}\|x - \cdot\|^s$ in the proof of (a) is then strictly convex, so its minimizer is unique, which means that $P_{\lambda,[s]}f(x)$ exists for every $x \in X$. Regarding the convergence properties in (c), they directly follow from Theorem 11.23(c) since we know that $\sup_{\lambda>0} e_{\lambda,[s]}f(x) \le f(x)$.

Let us show the norm-to-weak continuity of $P_{\lambda,[s]}f$. Consider any $x \in X$ and any net $(x_i)_{i\in I}$ in X converging strongly to x. Without loss of generality we may and do suppose that this net $(x_i)_{i\in I}$ is bounded. Let us put $y_i := P_{\lambda,[s]}f(x_i)$ and let us choose $y_i^* \in \lambda\partial(e_{\lambda,s}f)(x_i)$ (since $\partial(e_{\lambda,s}f)(x_i) \neq \emptyset$ by continuity of the convex function $e_{\lambda,s}f$). By (a) we have $y_i^* \in \partial(\frac{1}{s}\|\cdot\|^s)(x_i - y_i)$, so (see (11.21))

$$\langle y_i^*, x_i - y_i\rangle = \|x_i - y_i\|^s \quad \text{and} \quad \|y_i^*\| = \|x_i - y_i\|^{s-1}.$$

Since $\lambda^{-1}y_i^* \in \partial(e_{\lambda,[s]}f)(x_i)$ and $e_{\lambda,[s]}f$ is Lipschitz on bounded sets, the net $(y_i^*)_i$ is bounded as well as the net $(y_i)_i$. Take any subnet (that we do not relabel) converging weakly to some $y \in X$. Writing

$$e_{\lambda,[s]}f(x_i) = f(y_i) + \frac{1}{s\lambda}\|x_i - y_i\|^s$$

and using the continuity of $e_\lambda f$ and the weak lower semicontinuity of f and $\|\cdot\|^s$ it ensues that

$$e_{\lambda,[s]}f(x) \ge f(y) + \frac{1}{s\lambda}\limsup_i \|x - y_i\|^s \ge f(y) + \frac{1}{s\lambda}\|x - y\|^s, \tag{11.33}$$

which entails that $y = P_{\lambda,[s]}f(x)$. Then any weakly convergent subnet of the bounded net $(y_i)_i$ in the reflexive Banach space X converges weakly to $P_{\lambda,[s]}f(x)$, hence all the net $(y_i)_i$ converges weakly to $P_{\lambda,[s]}f(x)$. This translates that the mapping $P_{\lambda,[s]}f$ is norm-to-weak continuous.

Assume that the norm $\|\cdot\|$ is in addition (sequentially) Kadec-Klee and let us prove the norm-to-norm continuity of $P_{\lambda,[s]}f$. Consider any $x \in X$ and any sequence $(x_n)_n$ in X with $\|x_n - x\| \to 0$ as $n \to \infty$. Put $y_n := P_{\lambda,[s]}f(x_n)$. The precedent arguments guarantee that the sequence $(y_n)_n$ converges weakly to $y := P_{\lambda,[s]}f(x)$. Further, from the left inequality in (11.33) we have that

$$f(y) + \frac{1}{s\lambda}\|x - y\|^s = e_{\lambda,[s]}f(x) \ge f(y) + \frac{1}{s\lambda}\limsup_{n\to\infty}\|x - y_n\|^s.$$

Noticing that $f(y)$ is finite since we know that $e_\lambda f(x)$ is finite, we derive that $\limsup_{n\to\infty}\|x - y_n\|^s \le \|x - y\|^s$, which combined with the weak lower semicontinuity of $\|\cdot\|^s$ implies that $\|x - y_n\|^s \to \|x - y\|^s$. Since $(x - y_n)_n$ converges weakly to $x - y$, the (sequential) Kadec-Klee property of $\|\cdot\|$ assures us that $\|y_n - y\| \to 0$. This guarantees the norm-to-norm continuity of $P_{\lambda,[s]}f$.

(d) Assume that X is reflexive and the norm $\|\cdot\|$ is both strictly convex and Gâteaux differentiable off zero. We already saw in (c) that $P_{\lambda,[s]}f$ is well defined and norm-to-weak continuous on X. The equality in (a) can then be rewritten in the form

$$\partial(e_{\lambda,[s]}f)(x) = \partial f\big(P_{\lambda,[s]}f(x)\big) \cap \{\lambda^{-1}J_s\big(x - P_{\lambda,[s]}f(x)\big)\}.$$

The set $\partial(e_{\lambda,[s]}f)(x)$ (which is nonempty by continuity and convexity of $e_{\lambda,[s]}f$) is then a singleton for every $x \in X$, thus the continuous convex function $e_{\lambda,[s]}f$ is Gâteaux differentiable on X (see Proposition 3.102) and the formula for its derivative follows from the above equality as well.

(e) Under the assumptions of (e) the Moreau s-envelope $e_{\lambda,[s]}f$ is Gâteaux differentiable with $D(e_{\lambda,[s]}f)(x) = \lambda^{-1}J_s\big(x - P_{\lambda,[s]}f(x)\big)$ by (d), the mapping $P_{\lambda,[s]}f$ is

norm-to-norm continuous by (c) and the duality mapping $J_s := D(\frac{1}{s}\|\cdot\|^s)$ is norm-to-norm* continuous by Fréchet differentiability of $\|\cdot\|^s$ (see Theorem 2.179). Consequently, the Gâteaux derivative $D(e_{\lambda,[s]}f)$ is norm-to-norm* continuous, which means that the function $e_{\lambda,[s]}f$ is of class $\mathcal{C}^1$ on X.
(f) Finally, as regards the properties in the assertion (f), they have been already justified in Theorem 3.267. □

In a Hilbert space the next proposition provides a criterion for the linear semiconvexity of a function.

PROPOSITION 11.28. Let $f : H \to \mathbb{R}\cup\{+\infty\}$ be a proper lower semicontinuous function on a Hilbert space H. Given a real $\lambda > 0$ and a convex subset $C \subset \partial_C f$, the function f is linearly semiconvex with constant $1/\lambda$ on C (that is, $f + \frac{1}{2\lambda}\|\cdot\|^2$ is convex on C) whenever $C \subset \mathrm{Dom}\,(\mathrm{Prox}_{\,\lambda} f)^{-1}$.

PROOF. Consider the function $\varphi := (1/2)\|\cdot\|^2 + \lambda f$. Let $x_1, x_2 \in C$ and $t_1, t_2 \geq 0$ with $t_1 + t_2 = 1$. Put $u := t_1x_1 + t_2x_2$ and choose by assumption some $y \in (\mathrm{Prox}_{\,\lambda} f)^{-1}(u)$. By Proposition 11.25(a) we have $\langle y, x_i - u\rangle \leq \varphi(x_i) - \varphi(u)$, so multiplying with t_i, $i = 1, 2$, and adding both inequalities yield

$$0 = \langle y, t_1x_1 + t_2x_2 - u\rangle \leq t_1\varphi(x_1) + t_2\varphi(x_2) - \varphi(u).$$

Then φ is convex on C, and this clearly ensures the convexity of $f + \frac{1}{2\lambda}\|\cdot\|^2$ on the convex set C. □

11.3. Moreau envelope of primal lower regular functions in Hilbert spaces

In this section we consider X as a Hilbert space H and we will focus the study on various additional regularity properties of Moreau envelopes in such a setting.

In Theorem 11.27 we saw in the setting of Hilbert space that the Moreau envelope of f is differentiable whenever f is a proper lower semicontinuous convex function. The rest of the present section goes beyond convex functions. It is devoted to the differentiability of Moreau envelopes of primal lower regular functions in Hilbert spaces. The study for s-lower regular functions on uniformly convex Banach spaces will be developed later, in Section 18.4 of Chapter 18.

Let us consider first, in addition to the Moreau envelope, the concept of local Moreau envelope. It corresponds to the local Moreau s-envelope $e_{\lambda,[s],\varepsilon}f$ with $s = 2$ in Definition 11.22, and we will write $e_{\lambda,\varepsilon}f$ instead of $e_{\lambda,[2],\varepsilon}f$. Because of its importance we formalize it in a definition.

DEFINITION 11.29. Let X be a normed vector space and $f : X \to \mathbb{R}\cup\{+\infty\}$ be a proper function. Consider two reals $\lambda > 0$ and $\varepsilon > 0$. The *local Moreau envelope* of f with index λ associated with ε is the function from X into $\mathbb{R}\cup\{-\infty, +\infty\}$ defined by

$$e_{\lambda,\varepsilon}f(x) := \inf_{\|y\|\leq\varepsilon}\left(f(y) + \frac{1}{2\lambda}\|x-y\|^2\right), \quad \text{for all } x \in X$$

and the corresponding local proximal mapping is defined by

$$\mathrm{Prox}_{\,\lambda,\varepsilon} f(x) := \operatorname*{Argmin}_{\|y\|\leq\varepsilon}\left(f(y) + \frac{1}{2\lambda}\|x-y\|^2\right), \quad \text{for all } x \in X.$$

When $\mathrm{Prox}_{\,\lambda,\varepsilon} f(x)$ is a singleton, its unique element will be denoted by $P_{\lambda,\varepsilon}f(x)$.

Given any $\lambda, \varepsilon > 0$, it is worth noticing that

$$e_{\lambda,\varepsilon} f(x) = e_\lambda(f + \Psi_{B[0,\varepsilon]})(x) \quad \text{and} \quad \operatorname{Prox}_{\lambda,\varepsilon} f(x) = \operatorname{Prox}_\lambda(f + \Psi_{B[0,\varepsilon]})(x),$$

where we recall that $\Psi_C = \Psi(\cdot, C)$ denotes the indicator function of the subset C.

11.3.1. First properties related to continuity of proximal mapping. Coming back to the Hilbert space H, as above its inner product will be denoted by $\langle \cdot, \cdot \rangle$ and its associated norm will be given by $\|x\| = \sqrt{\langle x, x \rangle}$.

The following proposition provides a basic link between the $\mathcal{C}^1$ property of $e_\lambda f$ and the Lipschitz continuity of $P_\lambda f$.

PROPOSITION 11.30. Let H be a Hilbert space and $f : H \to \mathbb{R} \cup \{+\infty\}$ be a proper lower semicontinuous function minorized by a quadratic function $-\alpha\| \cdot \|^2 - \beta\| \cdot \| + \gamma$ on H, where $\alpha, \beta \geq 0$ and $\gamma \in \mathbb{R}$. Consider $\lambda \in]0, \frac{1}{2\alpha}[$ and let U be an open subset of the Hilbert space H. The following properties are equivalent:
(a) the function $e_\lambda f$ is $\mathcal{C}^1$ on U (resp. $\mathcal{C}^1$ on U with $\nabla e_\lambda f$ locally/globally Lipschitz on U);
(b) the single-valued mapping $P_\lambda f$ is well defined on U and continuous U (resp. Lipschitz continuous locally/globally on U).

When these properties hold, $\nabla e_\lambda f = \lambda^{-1}(I - P_\lambda f)$ on U.

PROOF. Suppose that (a) holds. For each $u \in U$ Theorem 11.23(d) tells us that $\operatorname{Prox}_\lambda f(u) \neq \emptyset$ and $\partial_F e_\lambda f(u) \subset \{\lambda^{-1}(u - y)\}$ for any $y \in \operatorname{Prox}_\lambda f(u)$. Consequently, $\operatorname{Prox}_\lambda f$ is single-valued on U, hence $P_\lambda f$ is well defined on U and $P_\lambda f(u) = u - \lambda \nabla e_\lambda f(u)$ for all $u \in U$. The latter equality also entails the continuity (resp. the corresponding Lipschitz continuity) of $P_\lambda f$ on the open set U.

Suppose now (b). Fix any $u \in U$. As above we know that $\partial_F e_\lambda f(u) \subset \{\lambda^{-1}(u - P_\lambda f(u))\}$. Take any $v \in \partial_L e_\lambda f(u)$. There exist $U \ni u_n \to u$ and $v_n \overset{w}{\to} v$ with $v_n \in \partial_F e_\lambda f(u_n)$, so $v_n = \lambda^{-1}(u_n - x_n)$ with $x_n := P_\lambda f(u_n)$. By continuity of $P_\lambda f$ on U we have $x_n \to x := P_\lambda f(u)$. Since

$$x_n = u_n - \lambda v_n \overset{w}{\to} u - \lambda v = x,$$

we also have $x = P_\lambda f(u) = u - \lambda v$, and hence

$$\partial_L e_\lambda f(u) \subset \{\lambda^{-1}\big(u - P_\lambda f(u)\big)\}.$$

By the local Lipschitz property of $e_\lambda f$ (see Theorem 11.23(b)), the set $\partial_L e_\lambda f(u)$ is nonempty, and hence it is the singleton set $\{\lambda^{-1}\big(u - P_\lambda f(u)\big)\}$ according to the latter inclusion. This singleton property also holds true for the Clarke subdifferential $\partial_C e_\lambda f(u)$ by the equality $\partial_C \varphi(u) = \overline{\text{co}}^*\big(\partial_L \varphi(u)\big)$ for any locally Lipschitz function φ in Theorem 4.120(c). Therefore, $e_\lambda f$ is Gâteaux differentiable on U (see Proposition 2.76) with $\nabla_G e_\lambda f(u) = \lambda^{-1}\big(u - P_\lambda f(u)\big)$ for all $u \in U$. Finally, the continuity (resp. the local/global Lipschitz continuity) of $P_\lambda f$ ensures that $e_\lambda f$ is $\mathcal{C}^1$ on U (resp. $\mathcal{C}^1$ on U with $\nabla e_\lambda f$ locally/globally Lipschitz on U). □

Through the above proposition the convexity of functions on Hilbert spaces can be characterized as follows.

COROLLARY 11.31. Let H be a Hilbert space and $f : H \to \mathbb{R} \cup \{+\infty\}$ be a proper lower semicontinuous function. The following assertions are equivalent:
(a) the function f is convex;
(b) there is some real $\lambda_0 > 0$ such that for any $\lambda \in]0, \lambda_0[$, the single-valued

mapping $P_\lambda f$ is well defined on the space H and κ_λ-Lipschitz therein along with $\liminf_{\lambda\downarrow 0} \lambda^{-1}(\kappa_\lambda - 1) \le 0$.

PROOF. The implication (a)⇒(b) is a direct consequence of Theorem 11.27.

Conversely, suppose that (b) holds. Taking $x_0 \in H$, the non vacuity of $\operatorname{Prox}_\lambda f(x_0)$ clearly entails that f is minorized on H by a quadratic function. Then by Theorem 11.23(d) there is some positive $\lambda_0' \le \lambda_0$ such that for each $\lambda \in]0, \lambda_0'[$ the function $e_\lambda f$ is $\mathcal{C}^1$ on X with $\nabla e_\lambda f = \lambda^{-1}(\mathrm{Id}_H - P_\lambda f)$. For each $\lambda \in]0, , \lambda_0'[$ it results that, for any $x, y \in X$

$$
\begin{aligned}
&\langle \nabla e_\lambda f(x) - \nabla e_\lambda f(y), x - y\rangle \\
&= \frac{1}{\lambda}\big(\|x-y\|^2 - \langle P_\lambda f(x) - P_\lambda f(y), x - y\rangle\big) \\
&\ge \frac{1}{\lambda}\big(\|x-y\|^2 - \|P_\lambda f(x) - P_\lambda f(y)\|\,\|x-y\|\big) \ge -\frac{\kappa_\lambda - 1}{\lambda}\|x-y\|^2.
\end{aligned}
$$

This tells us that the mapping $\nabla e_\lambda f + \frac{1}{\lambda}(\kappa_\lambda - 1)\mathrm{Id}_H$ is monotone, thus the function

$$
\varphi_\lambda := e_\lambda f + \frac{\kappa_\lambda - 1}{2\lambda}\|\cdot\|^2 \tag{11.34}
$$

is convex. By assumption, there is a sequence $(\lambda_n)_n$ in $]0, \lambda_0'|$ tending to 0 and a real $c \ge 0$ such that $(\kappa_{\lambda_n} - 1)/(2\lambda_n) \to -c$ as $n \to \infty$. In addition, we know by Theorem 11.23(c) that $(e_{\lambda_n} f)_n$ converges pointwise to f. We derive from this and (11.34) that $(\varphi_{\lambda_n})_n$ converges pointwise to $f - c\|\cdot\|^2$, which allows us to conclude that f is convex. □

11.3.2. Differentiability properties of Moreau envelope of primal lower regular functions. In this subsection, we will establish in Theorem 11.35 and Theorem 11.40, for a primal lower regular function f, diverse properties of the Moreau envelope and proximal mapping of $f + \Psi(\cdot, B)$ for an appropriate closed ball B: $\mathcal{C}^{1,1}$ property of the Moreau envelope, Lipschitz continuity of the proximal mapping, as well as specific properties near points in the graph of $\partial_C f$. Various preparatory results need to be obtained first.

We start with the following lemma which directly derives from Lemma 11.17 by taking therein $s = 1$, $x^* = 0$ and $\beta = \gamma = \theta = 0$.

LEMMA 11.32. Let X be a normed space and $f : X \to \mathbb{R} \cup \{+\infty\}$ be a function with $f(0) = 0$. Let reals $\alpha > 0$ and $r > 0$ such that $f(x) \ge -\alpha\|x\|^2$ for all $x \in B[0, r]$. Then for any positive real $\lambda < 3/(32\alpha)$ one has for all $u \in B[0, r/4]$

$$
e_{\lambda,r} f(u) = \inf_{\|y\| < 3r/4} \left(f(y) + \frac{1}{2\lambda}\|u - y\|^2 \right)
$$

and

$$
y(u) \in \operatorname{Prox}_{\lambda,r} f(u) \Rightarrow \|y(u)\| < 3r/4.
$$

The second lemma will be derived from the previous one. In this lemma and in the rest of this section $I : H \to H$ denotes the identity mapping from the Hilbert space H into itself, and $T^f_{\varepsilon,\rho}$ is the truncation defined in (11.18) relative to a given subdifferential of f.

LEMMA 11.33. Let $f : H \to \mathbb{R} \cup \{+\infty\}$ be a function on a Hilbert space H which is lower semicontinuous near $0 \in \operatorname{dom} f$. Then for each real $\varepsilon > 0$ with

f lower semicontinuous on $B[0,\varepsilon]$ there exists a real $\lambda_\varepsilon > 0$ such that for every $0 < \lambda \le \lambda_\varepsilon$

$$\mathrm{Prox}_\lambda f(u) \subset \left(I + \lambda T^f_{\varepsilon,\varepsilon/\lambda}\right)^{-1} \quad \text{for all } u \in B[0,\varepsilon/4],$$

where the truncation $T^f_{\varepsilon,\varepsilon/\lambda}$ is relative to a subdifferential ∂ with $\partial_P f \subset \partial f$.

PROOF. Fix any real $\varepsilon > 0$ with f lower semicontinuous on $B[0,\varepsilon]$ and choose a real $\lambda_\varepsilon > 0$ such that

$$\frac{1}{\lambda_\varepsilon}\frac{3}{32}\varepsilon^2 > f(0) - \inf_{\|x\|\le\varepsilon} f(x).$$

Take any real $0 < \lambda \le \lambda_\varepsilon$ and any pair (u,y) with $u \in B[0,\varepsilon/4]$ and $y \in \mathrm{Prox}_{\lambda,\varepsilon} f(u)$. By Lemma 11.32 above we have $y \in B(0,3\varepsilon/4)$, so by definition of $e_{\lambda,\varepsilon}$ we have

$$0 \in \partial_P f(y) + \lambda^{-1}(y-u), \text{ hence } \lambda^{-1}(u-y) \in \partial f(y).$$

Putting $v := \lambda^{-1}(u-y)$ we see that $v \in \partial f(y)$ and

$$\|v\| \le \lambda^{-1}(\|u\| + \|y\|) < \lambda^1(\varepsilon/4 + 3\varepsilon/4) = \varepsilon/\lambda,$$

so $y \in \left(I + \lambda T^f_{\varepsilon,\varepsilon/\lambda}\right)^{-1}(u)$. This finishes the proof. □

The third lemma considers multimappings M whose pertubation $r_0 I + M$ is monotone.

LEMMA 11.34. Let H be a Hilbert space and let $M : H \rightrightarrows H$ be a multimapping such that $r_0 I + M$ is monotone for some real $r_0 \ge 0$, where I denotes the identity mapping on H. Then, for any real $r > r_0$, $(I + r^{-1}M)^{-1}$ is monotone, single-valued and Lipschitz continuous on its domain with $r/(r - r_0)$ as a Lipschitz constant therein.

PROOF. Consider first any monotone multimapping $M : H \rightrightarrows H$ and fix $x_i \in (I+M)^{-1}(v_i)$ for $i = 1,2$. Clearly, $v_i - x_i \in M(x_i)$, hence $\langle (v_1 - x_1) - (v_2 - x_2), x_1 - x_2\rangle \ge 0$, which gives

$$\langle v_1 - v_2, x_1 - x_2 \ge -\|x_1 - x_2\|^2, \text{ hence } \|x_1 - x_2\| \le \|v_1 - v_2\|.$$

This ensures that $(I+M)^{-1}$ is monotone, single valued on its domain, and Lipschitz continuous therein with 1 as a Lipschitz constant.

Now assume that $r_0 I + M$ is monotone. Put $\delta := r - r_0$ and write

$$\begin{aligned}(I + r^{-1}M)^{-1} &= \left(\frac{1}{r}\delta I + \frac{r_0}{r} I + \frac{1}{r} M\right)^{-1} = \left(\frac{\delta}{r}\left(I + \delta^{-1}(r_0 I + M)\right)\right)^{-1} \\ &= (I + M_0)^{-1} \circ \frac{r}{\delta} I,\end{aligned}$$

where $M_0 := \delta^{-1}(r_0 I + M)$. Since M_0 is monotone, we deduce by what precedes that $(I + M_0)^{-1}$ is monotone, single valued on its domain, and Lipschitz therein with 1 as a Lipschitz constant. It results that $(I + r^{-1}M)^{-1}$ is monotone, single valued on its domain, and Lipschitz continuous therein with $\delta^{-1}r$ as a Lipschitz constant. □

We are now in a position to prove the first theorem on differentiability properties of Moreau envelopes of primal lower regular functions.

THEOREM 11.35 (differentiability of Moreau envelopes of plr functions: I). Let $f : H \to \mathbb{R} \cup \{+\infty\}$ be a function on a Hilbert space H which is lower semicontinuous near $0 \in \operatorname{dom} f$ and primal lower regular at 0. Then there exist reals $c > 0$, $0 < \varepsilon_0 < c$ such that for each $0 < \varepsilon \le \varepsilon_0$ there is $\lambda_\varepsilon > 0$ for which the following hold for any $\lambda \in]0, \lambda_\varepsilon]$:
(a) the mapping $P_{\lambda,\varepsilon} f$ is well-defined on $B(0, \varepsilon/4)$ and Lipschitz continuous there with $c/(c - \varepsilon)$ as a Lipschitz constant and

$$P_{\lambda,\varepsilon} f(u) = \left(I + \lambda T^f_{\varepsilon,\varepsilon/\lambda}\right)^{-1}(u) \quad \text{for all } u \in B(0, \varepsilon/4).$$

(b) the function $e_{\lambda,\varepsilon} f$ is differentiable on $B(0, \varepsilon/4)$ and its gradient is Lipschitz continuous there with

$$\nabla(e_{\lambda,\varepsilon} f)(u) = \lambda^{-1}\left(u - P_{\lambda,\varepsilon} f(u)\right) \quad \text{for all } u \in B(0, \varepsilon/4).$$

PROOF. By Corollary 11.19 there are reals $c > 0$ and $0 < \varepsilon_0 < 1/c$ with f lower semicontinuous on $B[0, \varepsilon_0]$ and bounded from below there, and such that for any reals $0 < \varepsilon \le \varepsilon_0$ and $\rho \ge 1$ the multimapping $c\rho I + T^f_{\varepsilon,\rho}$ is monotone. Let any real $\varepsilon \in]0, \varepsilon_0]$. Fix any $\lambda_\varepsilon > 0$ with $\lambda_\varepsilon < \min\{1, \varepsilon\}$ and satisfying the conclusion of Lemma 11.33. Take any $\lambda \in]0, \lambda_\varepsilon]$. The inequality $\varepsilon/\lambda < 1$ ensures that the multimapping $(c\varepsilon/\lambda)I + T^f_{\varepsilon,\varepsilon/\lambda}$ is monotone. Noting that $1/\lambda > c\varepsilon/\lambda$ (since $\varepsilon < 1/c$), Lemma 11.34 yields that $G := \left(I + \lambda T^f_{\varepsilon,\varepsilon/\lambda}\right)^{-1}$ is single valued and Lipschitz continuous on its domain with as Lipschitz constant

$$\frac{1/\lambda}{1/\lambda - c\varepsilon/\lambda} = \frac{1}{1 - c\varepsilon} =: \gamma_\varepsilon.$$

On the other hand, Lemma 11.33 says that

$$\operatorname{Prox}_{\lambda,\varepsilon} f(u) \subset \left(I + \lambda T^f_{\varepsilon,\varepsilon/\lambda}\right)^{-1}(u) \quad \text{for all } u \in B(0, \varepsilon/4). \tag{11.35}$$

Write $\operatorname{Prox}_{\lambda,\varepsilon} f = \operatorname{Prox}_\lambda f_\varepsilon$ and $e_{\lambda,\varepsilon} f = e_\lambda f_\varepsilon$, where $f_\varepsilon := f + \Psi_{\varepsilon\mathbb{B}}$. Take any $u \in B(0, \varepsilon/4)$. By Theorem 11.23(d) there is a sequence $(u_n)_n$ in $B(0, \varepsilon/4)$ converging to u with $\operatorname{Prox}_{\lambda,\varepsilon} f(u_n) \neq \emptyset$, so $\operatorname{Prox}_{\lambda,\varepsilon} f(u_n) = \{G(u_n)\}$. The continuity of G on $B(0, \varepsilon/4)$ yields $G(u_n) \to G(u)$ as $n \to \infty$. Further, for any $y \in B[0, \varepsilon]$

$$f_\varepsilon(G(u_n)) + \frac{1}{2\lambda}\|u_n - G(u_n)\|^2 \le f_\varepsilon(y) + \frac{1}{2\lambda}\|u_n - y\|^2,$$

hence by lower semicontinuity of f_ε

$$f_\varepsilon(G(u)) + \frac{1}{2\lambda}\|u - G(u)\|^2 \le f_\varepsilon(y) + \frac{1}{2\lambda}\|u - y\|^2.$$

It follows that $G(u) \in \operatorname{Prox}_\lambda f_\varepsilon = \operatorname{Prox}_{\lambda,\varepsilon} f(u)$, so $B(0, \varepsilon/4) \subset \operatorname{Dom} \operatorname{Prox}_{\lambda,\varepsilon} f$. From this and (11.35) we deduce that $P_{\lambda,\varepsilon} f$ is well defined on $B(0, \varepsilon/4)$ and γ_ε-Lipschitz therein, which corresponds to the assertion (a).

Finally, the assertion (b) follows from (a) and Proposition 11.30. □

We establish now diverse results in preparation for Theorem 11.40. Some of them have their own interest.

PROPOSITION 11.36. Let $F : H \to \mathbb{R} \cup \{+\infty\}$ be an extended real-valued lower semicontinuous function on the Hilbert space H with $F(0) = 0$ and such that there exists $\sigma > 0$ for which

$$F(x) \geq -\frac{\sigma}{2}\|x\|^2 \quad \text{for all } x \in H. \tag{11.36}$$

Let $\lambda_0 \in]0, \frac{1}{\sigma}[$ and $\eta > 0$. Let $\beta, \gamma > 0$ be positive real numbers such that

$$(1 + 2(1 - \lambda_0\sigma)^{-1})\gamma + \sqrt{2\lambda_0(1 - \lambda_0\sigma)^{-1}\beta} < \eta. \tag{11.37}$$

Then for all $\lambda \in]0, \lambda_0[$, the following hold:
(a) For all $x \in B(0, \gamma)$ and $x' \in \operatorname{Prox}_\lambda F(x)$ one has $\|x'\| < \eta$, $\|x - x'\| < \eta$.
(b) For any $x \in B(0, \gamma)$

$$e_\lambda F(x) = \inf_{x' \in B(0,\eta)} \left\{ F(x') + \frac{1}{2\lambda}\|x - x'\|^2 \right\}.$$

(c) The function $e_\lambda F$ is Lipschitz continuous on $B(0, \gamma)$, with η/λ as a Lipschitz constant.

The proof the proposition requires two lemmas. The first one is concerned with an estimate of approximate proximal points of the function F in the proposition.

LEMMA 11.37. Let $F : H \to \mathbb{R} \cup \{+\infty\}$ be an extended real-valued function defined on the Hilbert space H with $F(0) = 0$ and satisfying (11.36). Let $\lambda \in]0, 1/\sigma[$, $\rho \geq 0$, and $x, x' \in H$. If

$$F(x') + \frac{1}{2\lambda}\|x' - x\|^2 \leq e_\lambda F(x) + \rho, \tag{11.38}$$

we have the estimate

$$\|x'\| \leq 2(1 - \lambda\sigma)^{-1}\|x\| + \sqrt{2\lambda(1 - \lambda\sigma)^{-1}\rho}.$$

PROOF. First with $x' = 0$ we observe for all $x \in H$ the inequality

$$e_\lambda F(x) \leq F(0) + \frac{1}{2\lambda}\|0 - x\|^2 = \frac{1}{2\lambda}\|x\|^2.$$

So for any x, x' satisfying (11.38) we have according to (11.36)

$$-\frac{\sigma}{2}\|x'\|^2 + \frac{1}{2\lambda}\|x' - x\|^2 \leq F(x') + \frac{1}{2\lambda}\|x' - x\|^2 \leq e_\lambda F(x) + \rho \leq \frac{1}{2\lambda}\|x\|^2 + \rho,$$

hence

$$-\frac{\sigma}{2}\|x'\|^2 + \frac{1}{2\lambda}\|x' - x\|^2 \leq \frac{1}{2\lambda}\|x\|^2 + \rho.$$

Computing this, we get

$$\left(\frac{1}{2\lambda} - \frac{\sigma}{2}\right)\|x'\|^2 \leq \perp\frac{1}{\lambda}\langle x', x\rangle + \rho, \text{ that is, } (1 - \lambda\sigma)\|x'\|^2 \leq 2\langle x', x\rangle + 2\lambda\rho.$$

Putting $\alpha := (1 - \lambda\sigma)^{-1}$, we obtain

$$\|x'\|^2 \leq 2\alpha\langle x', x\rangle + 2\lambda\alpha\rho \leq 2\alpha\|x'\|\|x\| + 2\lambda\alpha\rho,$$

hence

$$(\|x'\| - \alpha\|x\|)^2 \leq \alpha^2\|x\|^2 + 2\lambda\alpha\rho,$$

which yields

$$\|x'\| \leq \alpha\|x\| + \sqrt{\alpha^2\|x\|^2 + 2\lambda\alpha\rho} \leq 2\alpha\|x\| + \sqrt{2\lambda\alpha\rho}.$$

□

The second lemma that we need, is related to the proximal subdifferential of the Moreau envelope of a function. It directly follows from Theorem 11.23(d).

LEMMA 11.38. Let $F : H \to \mathbb{R} \cup \{+\infty\}$ be a proper lower semicontinuous function defined on the Hilbert space H. Assume that there exist $\beta, \gamma, \sigma \geq 0$ such that

$$F(y) \geq -\frac{\sigma}{2}\|y\|^2 - \beta\|y\| - \gamma \quad \text{for all } y \in H.$$

Then, for all $\lambda \in]0, \frac{1}{\sigma}[$ (with convention $\frac{1}{\sigma} = +\infty$ when $\sigma = 0$) and all $x \in \operatorname{Dom} \partial_P e_\lambda F$, one has $\operatorname{Prox}_\lambda F(x) \neq \emptyset$ and, for all $x' \in \operatorname{Prox}_\lambda F(x)$, one has

$$\partial_P e_\lambda F(x) = \{\lambda^{-1}(x - x')\} \quad \text{and} \quad \lambda^{-1}(x - x') \in \partial_P F(x').$$

Therefore, $\partial_P e_\lambda F(x)$ and $\operatorname{Prox}_\lambda F(x)$ are singleton sets whenever $x \in \operatorname{Dom} \partial_P e_\lambda F$.

PROOF OF PROPOSITION 11.36. (a) Fix λ_0, η, β and γ as in the statement of Proposition 11.36. Let $\lambda \in]0, \lambda_0[$ and $x \in B(0, \gamma)$. Given $x' \in \operatorname{Prox}_\lambda F(x)$, according to Lemma 11.37 with $\rho = 0$, one obtains

$$\|x'\| \leq 2(1 - \lambda\sigma)^{-1}\|x\| < 2(1 - \lambda_0\sigma)^{-1}\gamma < \eta.$$

Moreover

$$\|x - x'\| \leq \|x\| + \|x'\| \leq \gamma + 2(1 - \lambda\sigma)^{-1}\gamma < \gamma + 2(1 - \lambda_0\sigma)^{-1}\gamma < \eta.$$

(b) Suppose that some $x' \in X$ satisfies the inequality

$$F(x') + \frac{1}{2\lambda}\|x' - x\|^2 \leq e_\lambda F(x) + \beta.$$

Applying Lemma 11.37 we have

$$\|x'\| \leq 2(1-\lambda\sigma)^{-1}\gamma + \sqrt{2\lambda(1 - \lambda\sigma)^{-1}\beta} < 2(1-\lambda_0\sigma)^{-1}\gamma + \sqrt{2\lambda_0(1 - \lambda_0\sigma)^{-1}\beta} < \eta,$$

which justifies our representation in (b).
(c) Observe that for any $x \in B(0, \gamma)$ we have on the one hand

$$e_\lambda F(x) \leq F(0) + \frac{1}{2\lambda}\|x\|^2 = \frac{1}{2\lambda}\|x\|^2$$

and on the other hand the inequalities, for any $y \in B(0, \eta)$,

$$F(y) + \frac{1}{2\lambda}\|x - y\|^2 \geq -\frac{\sigma}{2}\|y\|^2 + \frac{1}{2\lambda}\|x - y\|^2 \geq \frac{-\sigma\eta^2}{2},$$

give through (b) that $e_\lambda F(x) \geq \frac{-\sigma\eta^2}{2}$. This says in particular that $e_\lambda F(\cdot)$ is finite over $B(0, \gamma)$.

So, putting $\Phi_{x'}(x) := F(x') + \frac{1}{2\lambda}\|x' - x\|^2$ for each $x' \in D_{F,\eta} := B(0, \eta) \cap \operatorname{dom} F$ and noting that the family of functions $(\Phi_{x'})_{x' \in D_{F,\eta}}$ is equi-Lipschitzian on $B(0, \gamma)$, we obtain that $e_\lambda F$ is Lipschitz continuous on $B(0, \gamma)$. Let us estimate the Lipschitz constant of $e_\lambda F(\cdot)$ over $B(0, \gamma)$. For any $x \in \operatorname{Dom} \partial_P e_\lambda F$ with $\|x\| < \gamma$ and any $\zeta \in \partial_P e_\lambda F(x)$ we have by Lemma 11.38

$$\zeta = \frac{1}{\lambda}(x - x') \quad \text{for } x' = P_\lambda F(x),$$

hence according to the assertion (a)

$$\|\zeta\| \leq \frac{\eta}{\lambda}.$$

The function $e_\lambda F$ being lower semicontinuous on $B(0,\gamma)$ we deduce from the subdifferential characterization theorem of Lipschitz property (see Theorem 6.55) that $e_\lambda F$ is $\frac{\eta}{\lambda}$-Lipschitz continuous on $B(0,\gamma)$. $\square$

We continue the analysis of a function F satisfying the inequality 11.36 and the equality $F(0)=0$, but by assuming in addition that it is primal lower regular.

PROPOSITION 11.39. Let H be a Hilbert space and $F : H \to \mathbb{R} \cup \{+\infty\}$ be a function which is primal lower regular with constant $c > 0$ on an open convex set containing $B[0,s_0]$, for a constant c satisfying $s_0 < 1/2c$. Assume $F(0)=0$, $0 \in \partial_C F(0)$ and define $\overline{F}(\cdot) = F(\cdot) + \Psi(\cdot, B[0,s_0])$. Let $\lambda_0 \in]0, \frac{1-2cs_0}{2c}[$. Then for any $\beta, \gamma > 0$ satisfying

$$(1 + 2(1-2\lambda_0 c)^{-1})\gamma + \sqrt{2\lambda_0(1-2\lambda_0 c)^{-1}\beta} < s_0, \tag{11.39}$$

the following properties hold.
(a) For all $\lambda \in]0,\lambda_0[$, the function $e_\lambda \overline{F}(\cdot)$ is $\mathcal{C}^{1,1}$ on $B(0,\gamma)$ with $\nabla e_\lambda \overline{F}$ Lipschitz continuous on $B(0,\gamma)$ with $a_\lambda := \frac{1}{\lambda(1-2c\lambda(1+s_0/\lambda))}$ as a Lipschitz constant therein.
(b) $\operatorname{Prox}_\lambda \overline{F}(\cdot)$ is nonempty, single-valued and $(1 + \frac{1}{1-2c\lambda_0(1+s_0)})$-Lipschitz continuous on $B(0,\gamma)$.

PROOF. (a) By assumption, for all $x \in B[0,s_0]$, all $\zeta \in \partial_C F(x)$,

$$F(y) - F(x) \geq \langle \zeta, y - x\rangle - c(1 + \|\zeta\|)\|y-x\|^2$$

for all $y \in B[0,s_0]$. Since $0 \in \partial_C F(0)$ and $F(0)=0$, the latter inequality yields

$$F(x) \geq -c\|x\|^2 \text{ for all } x \in B[0,s_0].$$

Consequently

$$\overline{F}(x) \geq -c\|x\|^2 \text{ for all } x \in H.$$

So $\overline{F}(\cdot)$ satisfies (11.36) with $\sigma = 2c$ and $\overline{F}$ is lower semicontinuous on H since F is by assumption lower semicontinuous at each point of $B[0,s_0]$. Fix $\lambda_0 \in]0, \frac{1-2cs_0}{2c}[$ and $\eta = s_0$. Take, for the function $\overline{F}$, any positive real numbers β, γ satisfying (11.37) in Proposition 11.36 for $\sigma = 2c$, that is, satisfying (11.39). For any $\lambda \in]0,\lambda_0[$, let $x \in B(0,\gamma)$ where $e_\lambda \overline{F}$ is ∂_P-subdifferentiable and let $\zeta \in \partial_P e_\lambda \overline{F}(x)$. We know by Lemma 11.38 that $\operatorname{Prox}_\lambda \overline{F}(x)$ is a singleton and that for $x' = P_\lambda \overline{F}(x)$ we have $\zeta = \lambda^{-1}(x - x')$ and $\zeta \in \partial_P \overline{F}(x')$. Moreover, according to (a) and (c) in Proposition 11.36 we have

$$\|x'\| < s_0, \ \|x - x'\| < s_0 \text{ and } \|\zeta\| < s_0/\lambda,$$

hence in particular $\zeta \in \partial_P \overline{F}(x') = \partial_P F(x')$. Since F is primal lower regular with contant c on an open convex set containing $B[0,s_0]$, we get

$$F(z) - F(x') \geq \langle \zeta, z - x'\rangle - c(1 + \|\zeta\|)\|z - x'\|^2$$

for all $z \in B(0,s_0)$, which yields for any $y \in X$

$$F(z) + \frac{1}{2\lambda}\|z-y\|^2 - \left(F(x') + \frac{1}{2\lambda}\|x'-x\|^2\right)$$

$$\geq \frac{1}{2\lambda}(\|z-y\|^2 - \|x'-x\|^2) + \langle \zeta, z-x'\rangle - c(1+\|\zeta\|)\|z-x'\|^2 =: h(z). \tag{11.40}$$

Note that the equality $x' = P_\lambda \overline{F}(x)$ entails

$$e_\lambda \overline{F}(x) = \overline{F}(x') + \frac{1}{2\lambda}\|x - x'\|^2 = F(x') + \frac{1}{2\lambda}\|x-x'\|^2,$$

and note that by (b) of Proposition 11.36 applied with the function $\overline{F}$ one has

$$e_\lambda \overline{F}(y) = \inf_{z \in B(0,s_0)} \left\{ F(z) + \frac{1}{2\lambda}\|z-y\|^2 \right\} \tag{11.41}$$

for all $y \in B(0,\gamma)$.

On the other hand, from the inequality $\|\zeta\| < s_0/\lambda$ we observe that for any $z \in H$

$$\begin{aligned}\frac{1}{2\lambda}(\|z-y\|^2 - \|x'-x\|^2) + \langle \zeta, z-x'\rangle - c(1+\|\zeta\|)\|z-x'\|^2 \\ \geq \frac{1}{2\lambda}(\|z-y\|^2 - \|x'-x\|^2) + \langle \zeta, z-x'\rangle - c\left(1+\frac{s_0}{\lambda}\right)\|z-x'\|^2.\end{aligned}$$

Further, since $\lambda_0 < \frac{1-2cs_0}{2c}$, or equivalently $(\frac{1}{2c} - s_0)\frac{1}{\lambda_0} > 1$, we also have

$$\frac{1}{2\lambda} - c\left(1+\frac{s_0}{\lambda}\right) = c\left(\left(\frac{1}{2c} - s_0\right)\frac{1}{\lambda} - 1\right) > c\left(\left(\frac{1}{2c} - s_0\right)\frac{1}{\lambda_0} - 1\right) > 0,$$

which implies that

$$\lim_{\|z\|\to+\infty} \frac{1}{2\lambda}(\|z-y\|^2 - \|x'-x\|^2) + \langle \zeta, z-x'\rangle - c\left(1+\frac{s_0}{\lambda}\right)\|z-x'\|^2 = +\infty.$$

The function h in the right hand side in (11.40) is then a coercive lower semicontinuous convex function and hence attains its minimum on the Hilbert space H at some point $\overline{z} \in H$. Writing $\nabla h(\overline{z}) = 0$ and using the equality $\zeta = \lambda^{-1}(x-x')$ yield

$$0 = \frac{1}{\lambda}(\overline{z} - y) + \frac{1}{\lambda}(x-x') - 2c(1+\|\zeta\|)(\overline{z} - x'),$$

which gives

$$\overline{z} = x' - \frac{1}{1-2c\lambda(1+\|\zeta\|)}(x-y).$$

It ensues that

$$\begin{aligned}\inf_H h &= h(\overline{z}) \\ &= \frac{1}{2\lambda}\left(\|x'-y-\frac{1}{1-2c\lambda(1+\|\zeta\|)}(x-y)\|^2 - \|x'-x\|^2\right) \\ &\quad + \langle \zeta, -\frac{1}{1-2c\lambda(1+\|\zeta\|)}(x-y)\rangle - \frac{2c(1+\|\zeta\|)}{2}\left\|\frac{1}{1-2c\lambda(1+\|\zeta\|)}(x-y)\right\|^2,\end{aligned}$$

or equivalently

$$\begin{aligned}\inf_H h = \frac{1}{2\lambda}&\left(\left\|x'-x+(1-\frac{1}{1-2c\lambda(1+\|\zeta\|)})(x-y)\right\|^2 - \|x'-x\|^2\right) \\ &+ \frac{1}{1-2c\lambda(1+\|\zeta\|)}\langle \zeta, y-x\rangle - \frac{2c(1+\|\zeta\|)}{2}\frac{1}{(1-2c\lambda(1+\|\zeta\|))^2}\|(x-y)\|^2.\end{aligned}$$

This can be rewritten as

$$\begin{aligned}\inf_H h = \frac{1}{2\lambda}&\left(-\frac{4c\lambda(1+\|\zeta\|)}{1-2c\lambda(1+\|\zeta\|)}\langle x'-x, x-y\rangle + (\frac{2c\lambda(1+\|\zeta\|)}{1-2c\lambda(1+\|\zeta\|)})^2\|x-y\|^2\right) \\ &+ \frac{1}{1-2c\lambda(1+\|\zeta\|)}\langle \zeta, y-x\rangle - \frac{2c(1+\|\zeta\|)}{2(1-2c\lambda(1+\|\zeta\|))^2}\|x-y\|^2,\end{aligned}$$

and hence

$$\inf_H h = \left(\frac{1}{1-2c\lambda(1+\|\zeta\|)} - \frac{2c\lambda(1+\|\zeta\|)}{1-2c\lambda(1+\|\zeta\|)}\right)\langle\zeta, y-x\rangle$$
$$-\left(\frac{2c(1+\|\zeta\|)}{2(1-2c\lambda(1+\|\zeta\|))^2} - \frac{1}{2\lambda}\Big(\frac{2c\lambda(1+\|\zeta\|)}{1-2c\lambda(1+\|\zeta\|)}\Big)^2\right)\|x-y\|^2$$
$$= \langle\zeta, y-x\rangle - \frac{c(1+\|\zeta\|)}{(1-2c\lambda(1+\|\zeta\|))}\|x-y\|^2.$$

The latter equality combined with the inequality (11.40) and the equality (11.41) gives

$$e_\lambda\overline{F}(y) - e_\lambda\overline{F}(x) \geq \langle\zeta, y-x\rangle - \frac{c(1+\|\zeta\|)}{1-2c\lambda(1+\|\zeta\|)}\|x-y\|^2$$
$$\geq \langle\zeta, y-x\rangle - \frac{c}{1-2c\lambda(1+s_0/\lambda)}(1+s_0/\lambda)\|x-y\|^2$$

for all $y \in B(0,\gamma)$. Setting $k_\lambda := \frac{c}{1-2c\lambda(1+s_0/\lambda)}(1+s_0/\lambda)$, we see through the equality $\|x-y\|^2 = \|y\|^2 - \|x\|^2 - 2\langle x, y-x\rangle$ that, for all $x \in B(0,\gamma) \cap \operatorname{Dom}\partial_P\overline{F}$ and $\zeta \in \partial_P\overline{F}(x)$,

$$e_\lambda\overline{F}(y) + k_\lambda\|y\|^2 - e_\lambda\overline{F}(x) - k_\lambda\|x\|^2 \geq \langle\zeta + 2k_\lambda x, y-x\rangle \text{ for all } y \in B(0,\gamma),$$

which means by the subdifferential characterization of convex functions (see Theorem 6.68) that the function $e_\lambda\overline{F}(\cdot) + k_\lambda\|\cdot\|^2$ is convex on $B(0,\gamma)$. We deduce that $e_\lambda\overline{F}(\cdot) + k_\lambda\|\cdot\|^2 + \frac{1}{2\lambda}\|\cdot\|^2$ is also convex on $B(0,\gamma)$.

Writing, for each $x \in B(0,\gamma)$,

$$-\left(e_\lambda\overline{F}(x) - \frac{1}{2\lambda}\|x\|^2\right) = -\left(\inf_{y\in X}\left\{\overline{F}(y) + \frac{1}{2\lambda}\|x-y\|^2\right\} - \frac{1}{2\lambda}\|x\|^2\right)$$
$$= \sup_{y\in X}\left\{\frac{1}{\lambda}\langle x, y\rangle - \left(\frac{1}{2\lambda}\|y\|^2 + \overline{F}(y)\right)\right\},$$

we see that $-\left(e_\lambda\overline{F}(\cdot) - \frac{1}{2\lambda}\|\cdot\|^2\right)$ is a pointwise supremum of affine functions, and hence it is convex, or equivalently $e_\lambda\overline{F}(\cdot) - \frac{1}{2\lambda}\|\cdot\|^2$ is concave. This ensures the concavity of $e_\lambda\overline{F}(\cdot) - k_\lambda\|\cdot\|^2 - \frac{1}{2\lambda}\|\cdot\|^2$ on $B(0,\gamma)$. We observe that

$$k_\lambda + \frac{1}{2\lambda} = \frac{1}{1-2c\lambda(1+s_0/\lambda)}\left(c(1+s_0/\lambda) + \frac{1}{2\lambda}(1-2c\lambda(1+s_0/\lambda))\right)$$
$$= \frac{1}{2\lambda(1-2c\lambda(1+s_0/\lambda))}.$$

Therefore $e_\lambda\overline{F}(\cdot)$ is simultaneously semiconvex and semiconcave on $B(0,\gamma)$, or more precisely $e_\lambda\overline{F}(\cdot) + (a_\lambda/2)\|\cdot\|^2$ and $(a_\lambda/2)\|\cdot\|^2 - e_\lambda\overline{F}(\cdot)$ are both convex for

$$a_\lambda := \lambda^{-1}(1-2c\lambda(1+s_0/\lambda))^{-1}. \tag{11.42}$$

So, the function $e_\lambda\overline{F}(\cdot)$ being continuous on $B(0,\gamma)$ according to the assertion (c) in Proposition 11.36 we conclude via Theorem 10.37(a) that $e_\lambda\overline{F}(\cdot)$ is $\mathcal{C}^{1,1}$ on $B(0,\gamma)$ with $\nabla e_\lambda\overline{F}(\cdot)$ a_λ-Lipschitz continuous on $B(0,\gamma)$. The assertion (a) of the proposition is then established. It remains to prove the assertion (b).

Since $e_\lambda\overline{F}(\cdot)$ is $\mathcal{C}^{1,1}$ on $B(0,\gamma)$, we have by Proposition 11.30 that

$$\nabla e_\lambda\overline{F}(\cdot) = \lambda^{-1}(I - P_\lambda\overline{F}(\cdot)) \text{ on } B(0,\gamma).$$

Then for all $x, y \in B(0,\gamma)$, we can write

$$\begin{aligned}\|P_\lambda \overline{F}(y) - P_\lambda \overline{F}(x)\| &= \|(I - P_\lambda \overline{F})(x) - (I - P_\lambda \overline{F})(y) - (x-y)\| \\ &\le \|(I - P_\lambda \overline{F})(x) - (I - P_\lambda \overline{F})(y)\| + \|x-y\| \\ &\le (1 + \lambda a_\lambda)\|x-y\|.\end{aligned}$$

According to the definition of a_λ in (11.42) we see that

$$\begin{aligned}1 + \lambda a_\lambda &= 1 + \frac{1}{(1 - 2c\lambda(1 + s_0/\lambda))} \\ &\le 1 + \frac{1}{1 - 2c(\lambda_0 + s_0)},\end{aligned}$$

hence $P_\lambda \overline{F}(\cdot)$ is $\left(1 + \frac{1}{1-2c(\lambda_0+s_0)}\right)$-Lipschitz continuous on $B(0,\gamma)$. This finishes the proof of the proposition. □

We are now in a position to establish the second theorem on differentiability properties for Moreau envelopes of primal lower regular functions. It provides additional properties to Theorem 11.35 when points $(x_0, y_0) \in \operatorname{gph} \partial_C f$ with x_0 suitably close to the reference point are involved.

THEOREM 11.40 (differentiability of Moreau envelopes of plr functions: II). Let $f : H \to \mathbb{R} \cup \{+\infty\}$ be an extended real-valued function defined on the Hilbert space H. Assume that f is primal lower regular with constant $c > 0$ on an open convex set containing $B[u_0, s_0]$ with $s_0 < (1/2c)$. Let $(x_0, y_0) \in \operatorname{gph} \partial_C f$ with $\|x_0 - u_0\| < \frac{s_0}{18}$ and let

$$\overline{f}(\cdot) = f(\cdot) + \Psi\left(\cdot, B\left[x_0, \frac{s_0}{2}\right]\right).$$

Then there exists some threshold $\lambda_0 > 0$ such that for any $\lambda \in]0, \lambda_0[$ the following hold.

(a) The function $e_\lambda \overline{f}$ is $\mathcal{C}^{1,1}$ on $B(u_0, \frac{s_0}{18})$ with $\nabla e_\lambda \overline{f}$ Lipschitz continuous on $B(u_0, \frac{s_0}{18})$ with Lipschitz constant d_λ given by

$$d_\lambda := \frac{1}{\lambda(1 - 2c(\lambda(1 + \|y_0\|) + s_0/2))} + \|y_0\|.$$

(b) The single-valued mapping $P_\lambda \overline{f}$ is well defined on $B(u_0, \frac{s_0}{18})$ and Lipschitz continuous therein with Lipschitz constant k_0 given by

$$k_0 := 1 + \lambda_0 + \frac{1}{1 - 2c(\lambda_0 + s_0/2)}.$$

(c) The equality $P_\lambda \overline{f}(x_0 + \lambda y_0) = x_0$ holds.
(d) One has $\nabla e_\lambda \overline{f} = \lambda^{-1}(I - P_\lambda \overline{f})$ on $B(u_0, \frac{s_0}{18})$.
(e) With k_0 as defined above $\|\nabla e_\lambda \overline{f}(x_0)\| \le k_0 \|y_0\|$.
(f) The inclusion $P_\lambda \overline{f}\big(B(u_0, \frac{s_0}{18})\big) \subset B(u_0, \frac{7}{16}s_0)$ for the proximal mapping $P_\lambda \overline{f}$ is satisfied.

Further, given any $x \in B(u_0, \frac{s_0}{18})$ one has
(g) $\nabla e_\lambda \overline{f}(x) \in \partial_P f(P_\lambda \overline{f}(x))$;
(h) $\|x - x_0\| \ge (1 - c\lambda(2 + \|\nabla e_\lambda \overline{f}(x)\| + k_0\|y_0\|))\| P_\lambda \overline{f}(x) - P_\lambda \overline{f}(x_0)\|$.

PROOF. Let U be an open convex set containing $B[u_0, s_0]$ such that f is primal lower regular on U with constant c.
(a). For each $x \in H$ put

$$F(x) := \frac{1}{1+\|y_0\|}\big(g(x+x_0)-f(x_0)-\langle y_0, x\rangle\big) \quad \text{where} \quad g(x) := f(x)+\Psi(x, B[u_0, s_0])$$

and put also

$$\overline{F}(x) := F(x) + \Psi(x, B[0, \varepsilon]) \quad \text{with} \quad \varepsilon = s_0/2.$$

The function $F(\cdot)$ is lower semicontinuous on H, $F(0) = 0$ and for $\zeta \in \partial_C F(x)$ with $x \in B(u_0 - x_0, \varepsilon)$ we have $(1+\|y_0\|)\zeta + y_0 \in \partial_C f(x+x_0)$. The primal lower regularity property with constant c of f on the open convex set U containing $B[u_0, s_0]$ ensures that F is in turn primal lower regular with constant c on $U_0 := B(0, 17s_0/18)$. Indeed, we have for all $\zeta \in \partial_C F(x)$ with $x \in U_0$

$$f(x'+x_0) - f(x+x_0) \geq \langle (1+\|y_0\|)\zeta + y_0, x'-x\rangle - c(1+(1+\|y_0\|)\|\zeta\| + \|y_0\|)\|x'-x\|^2$$

for all $x' \in U_0 = B(0, 17s_0/18)$, because for such x' we have $x'+x_0 \in B(u_0, s_0) \subset U$. Then for all $x' \in U_0$ we obtain

$$F(x') - F(x) \geq \langle \zeta, x'-x\rangle - \frac{c}{1+\|y_0\|}(1+(1+\|y_0\|)\|\zeta\| + \|y_0\|)\|x'-x\|^2,$$

hence

$$F(x') - F(x) \geq \langle \zeta, x'-x\rangle - c(1+\|\zeta\|)\|x'-x\|^2,$$

which means that F is primal lower regular on $U_0 = B(0, 17s_0/18)$ with constant c. Further, we note that $B(0, 17s_0/18) \supset B[0, \varepsilon]$ and $0 \in \partial_C F(0)$ since $y_0 \in \partial_C f(x_0)$. Consequently, we derive from the latter inequality that $\overline{F}(x) \geq -c\|x\|^2$ for all $x \in H$. As regards the Moreau envelopes, for any $x \in H$ and $\lambda > 0$, we can write

$$\begin{aligned}
& e_{\lambda(1+\|y_0\|)}\overline{F}(x) \\
&= \inf_{u\in H}\left\{F(u) + \Psi(u, B[0,\varepsilon]) + \frac{1}{2\lambda(1+\|y_0\|)}\|x-u\|^2\right\} \\
&= \inf_{u\in H}\left\{\frac{g(u+x_0) - f(x_0) - \langle y_0, u\rangle}{1+\|y_0\|} + \Psi(u, B[0,\varepsilon]) + \frac{1}{2\lambda(1+\|y_0\|)}\|x-u\|^2\right\},
\end{aligned}$$

or equivalently with $B_0 := B[x_0, \varepsilon]$

$$\begin{aligned}
& (1+\|y_0\|)e_{\lambda(1+\|y_0\|)}\overline{F}(x) \\
&= -f(x_0) + \langle y_0, x_0\rangle + \inf_{u\in H}\{g(u+x_0) - \langle y_0, u+x_0\rangle + \Psi(u+x_0, B_0) + (2\lambda)^{-1}\|x-u\|^2\} \\
&= -f(x_0) + \langle y_0, x_0\rangle + \inf_{z\in H}\{g(z) - \langle y_0, z\rangle + \Psi(z, B_0) + (2\lambda)^{-1}\|x+x_0-z\|^2\},
\end{aligned}$$

which gives

$$\begin{aligned}
& (1+\|y_0\|)e_{\lambda(1+\|y_0\|)}\overline{F}(x) = -f(x_0) + \langle y_0, x_0\rangle \\
& + \inf_{z\in H}\{f(z) - \langle y_0, z\rangle + \Psi(z, B[u_0, s_0]\cap B[x_0, \varepsilon]) + (1/(2\lambda))\|x+x_0-z\|^2\}.
\end{aligned}$$

Since $\|x_0 - u_0\| < \frac{s_0}{18}$ and $\varepsilon = \frac{s_0}{2}$, we have $B[u_0, s_0] \cap B[x_0, \varepsilon] = B[x_0, \varepsilon]$ and hence

$$e_{\lambda(1+\|y_0\|)}\overline{F}(x) = \frac{-f(x_0) + \langle y_0, x_0\rangle}{1 + \|y_0\|} + \frac{1}{1 + \|y_0\|}\Bigg(-\frac{\lambda}{2}\|y_0\|^2 - \langle y_0, x\rangle - \langle y_0, x_0\rangle$$
$$+ \inf_{z\in H}\left\{ f(z) + \Psi(z, B[x_0, \varepsilon]) + \frac{1}{2\lambda}\|x + x_0 + \lambda y_0 - z\|^2 \right\}\Bigg).$$

It ensues that

$$e_{\lambda(1+\|y_0\|)}\overline{F}(x) = \frac{-f(x_0) - \langle y_0, x\rangle - \frac{\lambda}{2}\|y_0\|^2}{1 + \|y_0\|} + \frac{1}{1 + \|y_0\|} e_\lambda \overline{f}(x + x_0 + \lambda y_0).$$

So, for any $x \in H$ and any $\lambda > 0$,

(11.43)
$$e_\lambda \overline{f}(x) = (1+\|y_0\|)e_{\lambda(1+\|y_0\|)}\overline{F}(x-(x_0+\lambda y_0))+f(x_0)+\langle y_0, x-(x_0+\lambda y_0)\rangle+\frac{\lambda}{2}\|y_0\|^2.$$

Let $\lambda_0 \in]0, \min\{\frac{1-2cs_0}{2c}, \frac{3}{14c}\}[$. Since the inequality $\lambda_0 < \frac{3}{14c}$ yields

$$1 + \frac{2}{1 - 2c\lambda_0} < \frac{9}{2}, \quad \text{or equivalently} \quad \left(1 + 2(1 - 2c\lambda_0)^{-1}\right)^{-1}\frac{s_0}{2} > \frac{s_0}{9},$$

we can choose $\gamma > 0$ such that

$$\frac{s_0}{9} < \gamma < \frac{s_0}{2}\left(1 + \frac{2}{1 - 2c\lambda_0}\right)^{-1}.$$

Then there exists $\beta > 0$ such that

$$\left(1 + 2(1 - 2c\lambda_0)^{-1}\right)\gamma + \sqrt{2\lambda_0(1 - 2c\lambda_0)^{-1}\beta} < s_0/2.$$

Applying Proposition 11.39 to F, c and $\varepsilon < 1/2c$, we get that for all $\lambda \in]0, \lambda_0[$, the function $e_\lambda\overline{F}(\cdot)$ is $\mathcal{C}^{1,1}$ on $B(0, \gamma)$ with $\nabla e_\lambda\overline{F}$ b_λ-Lipschitz continuous on $B(0, \gamma)$, for $b_\lambda := \frac{1}{\lambda\left(1-2c\lambda(1+\varepsilon/\lambda)\right)}$. Define $\overline{\lambda}_0 = \min\{\lambda_0, \gamma - s_0/9\}$. We deduce according to (11.43) that, for all $\lambda \in]0, \frac{\overline{\lambda}_0}{1+\|y_0\|}[$, $e_\lambda\overline{f}(\cdot)$ is $\mathcal{C}^{1,1}$ on $B(x_0 + \lambda y_0, \gamma)$. Given any $0 < \lambda < \frac{\overline{\lambda}_0}{1+\|y_0\|}$, if x belongs to $B[u_0, s_0/18]$ then

$$\begin{aligned}\|x - (x_0 + \lambda y_0)\| &\le \|x - u_0\| + \|x_0 - u_0\| + \|\lambda y_0\| \\ &\le s_0/18 + s_0/18 + \frac{\overline{\lambda}_0\|y_0\|}{1 + \|y_0\|} \\ &< s_0/9 + \gamma - s_0/9 = \gamma,\end{aligned}$$

so $B[u_0, s_0/18] \subset B(x_0+\lambda y_0, \gamma)$, hence $e_\lambda\overline{f}(\cdot)$ is $\mathcal{C}^{1,1}$ on $B(u_0, s_0/18)$ and according to (11.43), $\nabla e_\lambda\overline{f}$ is d_λ-Lipschitz continuous on $B(u_0, s_0/18)$ for

$$\begin{aligned}d_\lambda &:= (1 + \|y_0\|)b_{\lambda(1+\|y_0\|)} + \|y_0\| \\ &= \frac{(1 + \|y_0\|)}{\lambda(1 + \|y_0\|)(1 - 2c\lambda(1 + \|y_0\|)(1 + \frac{\varepsilon}{\lambda(1+\|y_0\|)}))} + \|y_0\| \\ &= \frac{1}{\lambda(1 - 2c(\lambda(1 + \|y_0\|) + \varepsilon))} + \|y_0\|.\end{aligned}$$

This finishes the proof of the assertion (a).
(b), (c) and (d). As in Proposition 11.39, thanks to Proposition 11.30, we deduce that $\mathrm{Prox}_\lambda \overline{f}$ is nonempty, single-valued and continuous on $B(u_0, \frac{s_0}{18})$, and

$$\nabla e_\lambda \overline{f}(\cdot) = \lambda^{-1}(I - P_\lambda \overline{f})(\cdot) \quad \text{on} \quad B(u_0, s_0/18). \tag{11.44}$$

So, $P_\lambda \overline{f}$ is $(1 + \lambda d_\lambda)$-Lipschitz continuous on $B(u_0, \frac{s_0}{18})$. From the expression of d_λ above we have for every $\lambda \in]0, \frac{\overline{\lambda}_0}{1+\|y_0\|}[$

$$1 + \lambda d_\lambda = 1 + \lambda\|y_0\| + \frac{1}{1 - 2c(\lambda(1 + \|y_0\|) + \varepsilon)},$$

$$1 + \lambda d_\lambda < 1 + \frac{\overline{\lambda}_0\|y_0\|}{1 + \|y_0\|} + \frac{1}{1 - 2c(\overline{\lambda}_0 + \varepsilon)} < 1 + \overline{\lambda}_0 + \frac{1}{1 - 2c(\overline{\lambda}_0 + \varepsilon)},$$

hence $P_\lambda \overline{f}$ is k_0-Lipschitz continuous on $B(u_0, s_0/18)$ for

$$k_0 := 1 + \overline{\lambda}_0 + \big(1 - 2c(\overline{\lambda}_0 + \varepsilon)\big)^{-1}.$$

Furthermore, for any $\lambda \in]0, \overline{\lambda}_0/(1 + \|y_0\|)[$ we claim that

$$x_0 + P_{\lambda(1+\|y_0\|)}\overline{F}(u) = P_\lambda \overline{f}(u + x_0 + \lambda y_0) \tag{11.45}$$

for all $u \in B(0, \varepsilon/4)$, both mappings being nonempty and single valued on this ball. Indeed let $u \in B(0, \varepsilon/4)$ and $p \in H$ such that

$$e_{\lambda(1+\|y_0\|)}\overline{F}(u) = \overline{F}(p) + \frac{1}{2\lambda(1 + \|y_0\|)}\|u - p\|^2.$$

Then, according to (11.43) and the definition of $\overline{F}$, we have

$$\begin{aligned}
&e_\lambda \overline{f}(u + x_0 + \lambda y_0) \\
&= (1 + \|y_0\|)(\overline{F}(p) + \frac{1}{2\lambda(1 + \|y_0\|)}\|u - p\|^2) + f(x_0) \\
&\quad + \langle y_0, u\rangle + \frac{\lambda}{2}\|y_0\|^2 \\
&= f(x_0 + p) + \Psi(x_0 + p, B[u_0, s_0]) + \Psi(p, B[0, \varepsilon]) + \frac{1}{2\lambda}\|u - p\|^2 - \langle y_0, p\rangle \\
&\quad + \langle y_0, u\rangle + \frac{\lambda}{2}\|y_0\|^2 \\
&= f(x_0 + p) + \Psi(x_0 + p, B[u_0, s_0] \cap B[x_0, \varepsilon]) + \frac{1}{2\lambda}\|u + x_0 + \lambda y_0 - (x_0 + p)\|^2,
\end{aligned}$$

and since $B[u_0, s_0] \cap B[x_0, \varepsilon] = B[x_0, \varepsilon]$ we obtain

$$\begin{aligned}
e_\lambda \overline{f}(u + x_0 + \lambda y_0) &= f(x_0 + p) + \Psi(x_0 + p, B[x_0, \varepsilon]) + \frac{1}{2\lambda}\|u + x_0 + \lambda y_0 - (x_0 + p)\|^2 \\
&= \overline{f}(x_0 + p) + \frac{1}{2\lambda}\|u + x_0 + \lambda y_0 - (x_0 + p)\|^2,
\end{aligned}$$

which justifies the desired equality (11.45).

Using the inequality

$$\overline{F}(x) \geq -c\|x\|^2 \text{ for all } x \in H,$$

we see that for all $x \in H$

$$\overline{F}(x) + \frac{1}{2\lambda(1 + \|y_0\|)}\|x\|^2 \geq (\frac{1}{2\lambda(1 + \|y_0\|)} - c)\|x\|^2 \geq 0$$

where the last inequality is due to the fact that $\lambda(1+\|y_0\|) < \overline{\lambda}_0 < \frac{1}{2c}$; so we deduce that $0 \in \operatorname{Prox}_{\lambda(1+\|y_0\|)}\overline{F}(0)$. Since $\operatorname{Prox}_{\lambda(1+\|y_0\|)}\overline{F}(\cdot)$ is single-valued on $B(0,\gamma)$, it follows that $P_{\lambda(1+\|y_0\|)}\overline{F}(0) = 0$. Putting this in (11.45) with $u = 0$ we obtain

$$P_\lambda \overline{f}(x_0 + \lambda y_0) = x_0 \text{ whenever } \lambda \in]0, \overline{\lambda}_0/(1 + \|y_0\|)[. \tag{11.46}$$

So (b), (c) and (d) are established.
(e). To justify (e), it is enough to observe thanks to (11.46), (11.44) and the inequality $1 + \lambda d_\lambda < k_0$ that

$$\begin{aligned}\|\nabla e_\lambda \overline{f}(x_0)\| &= \|\lambda^{-1}(x_0 - P_\lambda \overline{f}(x_0))\| \\ &= \|\lambda^{-1}\big(P_\lambda \overline{f}(x_0 + \lambda y_0) - P_\lambda \overline{f}(x_0)\big)\| \\ &\le \lambda^{-1}(1 + \lambda d_\lambda)\|x_0 + \lambda y_0 - x_0\| = (1 + \lambda d_\lambda)\|y_0\| \le k_0\|y_0\|.\end{aligned}$$

(f),(g). Since f is bounded from below on $B[x_0, \frac{s_0}{2}]$ and $f(x_0) \in \mathbb{R}$, Lemma 11.17 (applied with $\beta = 0$, $\theta = 0$ and $\zeta = 0$) provides some real number $\lambda'_0 > 0$ such that for all $\lambda \in]0, \lambda'_0[$, $\operatorname{Prox}_\lambda \overline{f}(B[x_0, \frac{s_0}{8}]) \subset B[x_0, \frac{3s_0}{8}]$. We set $\overline{\lambda}_1 := \min\{\frac{\overline{\lambda}_0}{1+\|y_0\|}, \lambda'_0\}$. As $B[u_0, \frac{s_0}{18}] \subset B[x_0, \frac{s_0}{8}]$ and $B[x_0, \frac{3s_0}{8}] \subset B(u_0, \frac{7s_0}{16})$, the assertion (f) follows for each $\lambda \in]0, \overline{\lambda}_1[$.

Let $x \in B(u_0, \frac{s_0}{18})$ and $\lambda \in]0, \overline{\lambda}_1[$ be fixed in the remaining of the proof. We first show that $\nabla e_\lambda(x) \in \partial_P f(\operatorname{Prox}_\lambda \overline{f}(x))$. By the assertion (f) proved above we have $P_\lambda \overline{f}(x) \in B(u_0, \frac{7s_0}{16}) \subset B(x_0, \frac{s_0}{2})$. The functions f and $\overline{f}$ are lower semicontinuous and coincide on an open neighborhood of $P_\lambda \overline{f}(x)$. This trivially entails that

$$\partial_P \overline{f}(P_\lambda \overline{f}(x)) = \partial_P f(P_\lambda \overline{f}(x)). \tag{11.47}$$

Combining Lemma 11.38 and the single valuedness of $P_\lambda \overline{f}$ on $B(u_0, \frac{s_0}{18})$, we get $\nabla e_\lambda \overline{f}(x) \in \partial_P f(P_\lambda \overline{f}(x))$, so the assertion (g) holds true.
(h) Finally, let us prove the assertion (h). Since $x, x_0 \in B(u_0, s_0/18) \subset U$, we use the assertion (g) and the property of $\partial_P f$ on U furnished by the inequality (11.2) in Definition 11.1 to see that

$$\begin{aligned}&\langle \nabla e_\lambda \overline{f}(x) - \nabla e_\lambda \overline{f}(x_0), P_\lambda \overline{f}(x) - P_\lambda \overline{f}(x_0)\rangle \\ &\qquad \ge -c(2 + \|\nabla e_\lambda \overline{f}(x)\| + \|\nabla e_\lambda \overline{f}(x_0)\|)\|P_\lambda \overline{f}(x) - P_\lambda \overline{f}(x_0)\|^2 \\ &\qquad \ge -c(2 + \|\nabla e_\lambda \overline{f}(x)\| + k_0\|y_0\|)\|P_\lambda \overline{f}(x) - P_\lambda \overline{f}(x_0)\|^2,\end{aligned}$$

where the latter inequality is due to the assertion (e) shown above. This entails, according to the equality (11.44)

$$\langle x - x_0, P_\lambda \overline{f}(x) - P_\lambda \overline{f}(x_0)\rangle \ge (1 - c\lambda(2 + \|\nabla e_\lambda \overline{f}(x)\| + k_0\|y_0\|))\|P_\lambda \overline{f}(x) - P_\lambda \overline{f}(x_0)\|^2,$$

which confirms the assertion (h) and finishes the proof of the theorem. □

11.4. Subdifferential determination of primal lower regular functions

In addition to Theorem 9.12 and Theorem 9.26 related to subdifferentially and directionally stable functions and to essentially directionally smooth functions, we examine now the subdifferential determination of primal lower regular functions as well as enlarged subdifferential inclusions for functions in that class.

THEOREM 11.41 (subdifferential determination of plr functions). Let H be a Hilbert space and let $f : H \to \mathbb{R} \cup \{+\infty\}$ be lower semicontinuous near a point $\overline{x} \in \operatorname{dom} f$ and $g : H \to \mathbb{R} \cup \{+\infty\}$ be finite at $\overline{x}$ and primal lower regular at $\overline{x}$.

(a) If $\partial_P f(x) \subset \partial_C g(x)$ for all x near $\overline{x}$, then there exist $\eta > 0$ and a real constant C such that

$$f(x) = g(x) + C \quad \text{for all } x \in B(\overline{x}, \eta).$$

(b) If there is a real $\gamma > 0$ satisfying for all x near $\overline{x}$

$$\partial_P f(x) \subset \partial_C g(x) + \gamma \mathbb{B}_H,$$

then for each real $\gamma' > \gamma$ there exist a real $\eta > 0$ such that $B(\overline{x}, \eta) \cap \operatorname{dom} f = B(\overline{x}, \eta) \cap \operatorname{dom} g$ and

$$g(x) - g(y) - \gamma' \|x - y\| \le f(x) - f(y) \le g(x) - g(y) + \gamma' \|x - y\|$$

for all $x \in B(\overline{x}, \eta)$ and $y \in B(\overline{x}, \eta) \cap \operatorname{dom} g$.

PROOF. Below (as defined in (11.18)) $T^f_{\cdot,\cdot}$ will refer to the corresponding truncation of the proximal subdifferential $\partial_P f$, whereas $T^g_{\cdot,\cdot}$ will be the truncation of the C-subdifferential $\partial_C f$. Take a real $\gamma \ge 0$ satisfying the assumption in (b). Considering $f(\cdot + \overline{x}) - f(\overline{x})$ and $g(\cdot + \overline{x}) - g(\overline{x})$, we may and do suppose that $\overline{x} = 0$ and $f(\overline{x}) = g(\overline{x}) = 0$. By lower semicontinuity of f near $\overline{x} = 0$, by primal lower regularity property of g at $\overline{x}$, by Lemma 11.33, by Corollary 11.19, and by Theorem 11.35, there exist reals $c > 0$ and $\varepsilon_0 \in]0, 1/c[$ such that f and g are bounded below on $B[0, \varepsilon_0]$ and such that for each $\varepsilon \in]0, \varepsilon_0]$ the following properties (11.48)-(11.51) hold:

$$(11.48) \qquad (\mathrm{Prox}_\lambda f_\varepsilon)(u) \subset \left(I + \lambda T^f_{\varepsilon,\varepsilon/\lambda}\right)^{-1}(u) \ \forall \lambda > 0, \forall u \in B(0, \varepsilon/4),$$

where $f_\varepsilon := f + \Psi_{\varepsilon\mathbb{B}}$;

$$(11.49) \qquad c\rho I + T^g_{\varepsilon,\rho} \text{ is monotone for all } \rho \ge 1;$$

$$(11.50) \qquad \partial_P f(x) \subset \partial_C g(x) + \gamma \mathbb{B} \quad \text{for all } x \in B(0, \varepsilon);$$

for each $\lambda \in]0, \lambda_\varepsilon]$
(11.51)

$$P_\lambda g_\varepsilon(u) = \left(I + \lambda T^g_{\varepsilon,\varepsilon/\lambda}\right)^{-1}(u) \text{ and } \nabla e_\lambda g_\varepsilon(u) = \lambda^{-1}\left(I - P_\lambda g_\varepsilon\right)(u) \ \forall u \in B(0, \varepsilon),$$

where $g_\varepsilon := g + \Psi_{\varepsilon\mathbb{B}}$ (as defined above for f).

Fix any $\varepsilon \in]0, \varepsilon_0]$. From (11.48) we see that for any $\lambda > 0$ and any $u \in B(0, \varepsilon/4)$

$$(11.52) \quad \partial_P(e_\lambda f_\varepsilon)(u) \subset \lambda^{-1}\left(I - \mathrm{Prox}_\lambda f_\varepsilon\right)(u) \subset \lambda^{-1}\left(u - \left(I + \lambda T^f_{\varepsilon,\varepsilon/\lambda}\right)^{-1}(u)\right).$$

From (11.50) we also see that

$$T^f_{\varepsilon,\varepsilon/\lambda}(\cdot) \subset T^g_{\varepsilon,\varepsilon/\lambda}(\cdot) + \gamma \mathbb{B},$$

so for any $x \in \left(I + \lambda T^f_{\varepsilon,\varepsilon/\gamma}\right)^{-1}(u)$ we have $u \in \left(I + \lambda T^g_{\varepsilon,\varepsilon/\lambda}\right)(x) + \lambda\gamma\mathbb{B}$, hence

$$x \in \left(I + \lambda T^g_{\varepsilon,\gamma+\varepsilon/\lambda}\right)^{-1}(u + \lambda\gamma\mathbb{B}).$$

This and (11.52) give

$$(11.53) \quad \partial_P(e_\lambda f_\varepsilon)(u) \subset \lambda^{-1}\left(u - \left(I + \lambda T^g_{\varepsilon,\gamma+\varepsilon/\lambda}\right)^{-1}(u + \lambda\gamma\mathbb{B})\right) \ \ \forall u \in B(0, \varepsilon/4).$$

Now put $\Lambda_\varepsilon := \min\{\lambda_\varepsilon, \varepsilon, (c\gamma)^{-1}(1 - c\varepsilon), (8\gamma)^{-1}\varepsilon\}$ (where $\gamma^{-1} = +\infty$ if $\gamma = 0$) and fix any $\lambda \in]0, \Lambda_\varepsilon[$. The inequalities $\varepsilon/\lambda > 1$ and $\gamma + \varepsilon/\lambda > 1$ give by (11.49)

$$(c\varepsilon/\lambda)I + T^g_{\varepsilon,\varepsilon/\lambda} \quad \text{and} \quad c(\gamma + \varepsilon/\lambda)I + T^g_{\varepsilon,\gamma+\varepsilon/\lambda} \text{ are monotone.}$$

Since $1/\lambda > c\varepsilon/\lambda$ and $1/\lambda > c(\gamma+\varepsilon/\lambda)$, Lemma 11.34 tells us that $\left(I + \lambda T^g_{\varepsilon,\varepsilon/\lambda}\right)^{-1}$ and $\left(I + \lambda T^g_{\varepsilon,\gamma+\varepsilon/\lambda}\right)^{-1}$ are each one single valued on its domain, and that the mapping $\left(I + \lambda T^g_{\varepsilon,\gamma+\varepsilon/\lambda}\right)^{-1}$ is Lipschitz on its domain with a Lipschitz constant equal to

$$\frac{1/\lambda}{(1/\lambda) - (c\varepsilon/\lambda)} = \frac{1}{1 - c\varepsilon} =: \gamma_\varepsilon \text{ (independent of } \lambda).$$

Since $\left(I + \lambda T^g_{\varepsilon,\varepsilon/\lambda}\right)^{-1}(\cdot) \subset \left(I + \lambda T^g_{\varepsilon,\gamma+\varepsilon/\lambda}\right)^{-1}(\cdot)$ (because $\varepsilon/\lambda < \gamma + \varepsilon/\lambda$), we deduce by (11.51) that

(11.54)

$$\left(I + \lambda T^g_{\varepsilon,\varepsilon/\lambda}\right)^{-1} = \left(I + \lambda T^g_{\varepsilon,\gamma+\varepsilon/\lambda}\right)^{-1} \text{ on Dom } \left(I + \lambda T^g_{\varepsilon,\varepsilon/\lambda}\right)^{-1} \supset B(0, \varepsilon/4)$$

and that $P_\lambda g_\varepsilon$ is γ_ε-Lipschitz on $B(0, \varepsilon/4)$, which yields by (11.51) again

$$\nabla e_\lambda g_\varepsilon(u) = \lambda^{-1} \left(u - \left(I + \lambda T^g_{\varepsilon,\gamma+\varepsilon/\lambda}\right)^{-1}(u)\right) \quad \forall u \in B(0, \varepsilon/4). \tag{11.55}$$

Observing that $\lambda\gamma < \varepsilon/8$ and using (11.54) and the γ_ε-Lipschitz property of $\left(I + \lambda T^g_{\varepsilon,\gamma+\varepsilon/\lambda}\right)^{-1}$ on its domain containing $B(0, \varepsilon/4)$, we also have for every element $u \in B(0, \varepsilon/8)$

$$\left(I + \lambda T^g_{\varepsilon,\gamma+\varepsilon/\lambda}\right)^{-1}(u + \lambda\gamma\mathbb{B}) \subset \left(I + \lambda T^g_{\varepsilon,\gamma+\varepsilon/\lambda}\right)^{-1}(u) + \lambda\gamma_\varepsilon\gamma\mathbb{B},$$

then by (11.53) and (11.55)

$$\partial_P(e_\lambda f_\varepsilon) \subset \nabla(e_\lambda g_\varepsilon)(u) + \gamma_\varepsilon\gamma\mathbb{B}.$$

Since $e_\lambda g_e$ is $\mathcal{C}^{1,1}$ on $B(0, \varepsilon/4)$, it ensues that

$$\partial_P(e_\lambda f_e - e_\lambda g_\varepsilon)(u) \subset \gamma_\varepsilon\gamma\mathbb{B} \quad \text{for all } u \in B(0, \varepsilon/8). \tag{11.56}$$

(a). Under the assumption in (a) we have $\gamma = 0$, so $\partial_P(\varepsilon_\lambda f_\varepsilon - e_\lambda g_\varepsilon)(u) \subset \{0\}$ for all $u \in B(0, \varepsilon/8)$. For each $\lambda \in]0, \Lambda_\varepsilon[$ there is by Corollary 6.57 a real constant C_λ (independent of u) such that $e_\lambda f_\varepsilon(u) = e_\lambda g_\varepsilon(u) + C_\lambda$ for all $u \in B(0, \varepsilon/8)$. Further, by Theorem 11.23(c), as $\lambda \downarrow 0$, the functions $e_\lambda f_\varepsilon(\cdot)$ and $e_\lambda g_\varepsilon(\cdot)$ converge pointwise to f_ε and g_ε respectively, so they converge pointwise to f and g on $B(0, \varepsilon/8)$. It ensues that C_λ converges to the real constant $C := f(0) - g(0)$ (independent of u). Taking the limit as $\lambda \downarrow 0$ gives

$$f(u) = g(u) + C \quad \text{for all } u \in B(0, \varepsilon/8),$$

which translates the assertion (a).

(b). Regarding (b) with $\gamma > 0$, for each fixed $\lambda \in]0, \Lambda_\varepsilon]$ we deduce from (11.56) and from Theorem 6.55 that the function $e_\lambda f_\varepsilon - e_\lambda g_\varepsilon$ is $\gamma_\varepsilon\gamma$-Lipschitz on $B(0, \varepsilon/8)$. Then for any $x, y \in B(0, \varepsilon/8)$ we have

$$e_\lambda g_\varepsilon(x) - e_\lambda g_\varepsilon(y) - \gamma_\varepsilon\gamma\|x-y\| \le e_\lambda f_\varepsilon(x) - e_\lambda f_\varepsilon(y) \le e_\lambda g_\varepsilon(x) - e_\lambda g_\varepsilon(y) + \gamma_\varepsilon\gamma\|x-y\|.$$

Further, it is easy to see that $e_\lambda f_\varepsilon(0)$ and $e_\lambda g_\varepsilon(0)$ are bounded with respect to $\lambda \in]0, \Lambda_\varepsilon[$. Since $e_\lambda f_\varepsilon$ and $e_\lambda g_\varepsilon$ converge pointwise on $B(0, \varepsilon/8)$ to f and g respectively as $\lambda \downarrow 0$, taking $y = 0$ and passing to the limit in the latter inequalities assure us that

$$B(0, \varepsilon/8) \cap \operatorname{dom} f = B(0, \varepsilon/8) \cap \operatorname{dom} g.$$

Therefore, the same inequalities, with any $x \in B(0, \varepsilon/8)$ and any $y \in B(0, \varepsilon/8) \cap \operatorname{dom} g$, furnish after passing to the limit as $\lambda \downarrow 0$

$$g(x) - g(y) - \gamma_\varepsilon \gamma \|x - y\| \le f(x) - f(y) \le g(x) - g(y) + \gamma_\varepsilon \gamma \|x - y\|.$$

To conclude that the assertion (b) holds true, it suffices to notice that $\gamma_\varepsilon \to 1$ as $\varepsilon \downarrow 0$. □

The following finite-dimensional important case of Poliquin deserves to be stated. It directly follows from the previous theorem.

THEOREM 11.42 (R.A. Poliquin). Let $f : \mathbb{R}^n \to \mathbb{R} \cup \{+\infty\}$ be lower semicontinuous near a point $\overline{x} \in \operatorname{dom} f$ and $g : H \to \mathbb{R} \cup \{+\infty\}$ be finite at $\overline{x}$ and primal lower regular at $\overline{x}$. Assume that $\partial_P f(x) \subset \partial_P g(x)$ for all x near $\overline{x}$. Then there exist a neighborhood U of $\overline{x}$ and a real constant C such that

$$f(x) = g(x) + C \quad \text{for all } x \in U.$$

11.5. Prox-regular functions

Continuing with the relaxation of inequality (11.1) and noticing that the subgradient x^* is not localized in the definition of primal lower regular functions in (11.2), one defines prox-regular functions by localizing not only around the reference point $\overline{x}$ but also around a reference subgradient $\overline{x}^*$.

11.5.1. Definition and examples. In order to unify the presentation, we will use a general subdifferential ∂ which is assumed to satisfy the basic properties **Prop.1**···**Prop.4** in Section 6.2 on the class of extended real-valued functions. So, all the analysis will be valid on appropriate spaces for proximal, Fréchet, limiting, approximate and Clarke subdifferentials.

DEFINITION 11.43. Let X be a normed space and $f : X \to \mathbb{R} \cup \{+\infty\}$ be a proper function. Given $\overline{x} \in \operatorname{Dom} \partial f$ and $\overline{x}^* \in \partial f(\overline{x})$, one says that f is *∂-prox-regular at $\overline{x}$ for $\overline{x}^*$* provided there are some reals $\sigma > 0$ and $\varepsilon > 0$ such that for all $x, x' \in B(\overline{x}, \varepsilon)$ with $|f(x) - f(\overline{x})| < \varepsilon$ and all $x^* \in B(\overline{x}^*, \varepsilon)$ with $x^* \in \partial f(x)$ one has

$$f(x') \ge f(x) + \langle x^*, x' - x\rangle - \frac{\sigma}{2}\|x' - x\|^2. \tag{11.57}$$

Such reals $\sigma > 0$ and $\varepsilon > 0$ will be called constants of ∂-prox-regularity of f at $\overline{x}$ for $\overline{x}^*$. If, for every $\overline{x}^* \in \partial f(\overline{x})$, the function f is ∂-prox-regular at $\overline{x}$ for $\overline{x}^*$, the function f is said to be *∂-prox-regular at $\overline{x}$*.

The inequality (11.57) translates that all vectors in $B(\overline{x}^*, \varepsilon) \cap \partial f(B(\overline{x}, \varepsilon))$ are not only variational proximal subgradients at suitable points (see Proposition 4.137(a)), but they are realized as such subgradients in a uniform way with the same constants σ and ε. This *justifies the terminology* in Definition 11.43.

When $f(x) \to f(\overline{x})$ automatically holds as $(x, x^*) \to (\overline{x}, \overline{x}^*)$ with (x, x^*) in the graph of ∂f, the condition $|f(x) - f(\overline{x})| < \varepsilon$ in Definition 11.43 is clearly superfluous. We then formalize this useful property as follows:

DEFINITION 11.44. A function $f : X \to \mathbb{R} \cup \{+\infty\}$ on a normed space X is said to be *∂-subdifferentially continuous at a point $\overline{x}$ for a subgradient $\overline{x}^* \in \partial f(\overline{x})$* when the function from $\operatorname{gph} \partial f$ into $\mathbb{R} \cup \{+\infty\}$, defined by $(x, x^*) \mapsto f(x)$, is continuous at $(\overline{x}, \overline{x}^*)$ with respect to the topology induced on $\operatorname{gph} \partial f$ by the norm topology of $X \times X^*$. If $\partial f(\overline{x})$ is nonempty and if, for every $\overline{x}^* \in \partial f(\overline{x})$, the function f is ∂-subdifferentially continuous at $\overline{x}$ for $\overline{x}^*$, one says that f is *∂-subdifferentially continuous at $\overline{x}$*. When there is no ambiguity, we will omit ∂ and just say subdifferentially continuous at $\overline{x}$ for $\overline{x}^*$.

Lower semicontinuous convex functions f on a normed space X are evidently ∂-prox-regular, and they are also subdifferentially continuous. Indeed, let $(\overline{x}, \overline{x}^*)$ in $\operatorname{gph} \partial f$. Then for any (x, x^*) and (y, y^*) in $\operatorname{gph} \partial f$ with $x^*, y^* \in B(\overline{x}, 1)$, we have with $\gamma := 1 + \|\overline{x}^*\|$

$$f(x) - f(y) \leq \langle x^*, x - y\rangle \leq \gamma \|x - y\|,$$

so $|f(y) - f(x)| \leq \gamma\|y - x\|$. More generally, similar arguments show in the following proposition that primal lower regular functions are prox-regular and subdifferentially continuous.

PROPOSITION 11.45. Let X be a normed space and $f : X \to \mathbb{R} \cup \{+\infty\}$ be a function which is primal lower regular at $\overline{x} \in \operatorname{Dom} \partial_C f$. Then f is ∂_C-prox-regular at $\overline{x}$ and ∂_C-subdifferentially continuous at $\overline{x}$. In fact, there is a neighborhood W of $(\overline{x}, \overline{x}^*)$ such that the restriction of f to $\operatorname{proj}_X(W \cap \operatorname{gph} \partial_C f)$ is Lipschitz continuous therein.

In particular, when X is a Banach space, any $\mathcal{C}^{1,1}$ convexly composite function which is qualified at a point in the domain of its Clarke subdifferential is ∂_C-prox-regular at this point and ∂_C-subdifferentially continuous there.

PROOF. Assume that f is primal lower regular at $\overline{x}$, so in particular it is lower semicontinuous at $\overline{x}$. The prox-regularity of f at $\overline{x}$ being obvious, let us justify that it is subdifferentially continuous at $\overline{x}$. Following the aforementioned arguments, fix any $\overline{x}^* \in \partial_C f(\overline{x})$. There exist reals $c > 0$ and $\delta > 0$ such that, for all $x, y \in B(\overline{x}, \delta) \cap \operatorname{Dom} \partial_C f$ and all $x^*, y^* \in B(\overline{x}^*, 1)$ with $x^* \in \partial_C f(x)$ and $y^* \in \partial_C f(y)$ one has

$$f(x) - f(y) \leq \langle x^*, x - y\rangle + c(1 + \|x^*\|)\|x - y\|^2,$$

so putting $\gamma := 1 + \|\overline{x}^*\| + 2c\delta(2 + \|\overline{x}^*\|)$ we obtain $f(x) - f(y) \leq \gamma\|x - y\|$, and by symmetry $|f(x) - f(y)| \leq \gamma\|x - y\|$. This justifies the Lipschitz property of f on $\operatorname{proj}_X(W \cap \operatorname{gph} \partial_C f)$ for $W := B(\overline{x}, \delta) \times B(\overline{x}^*, 1)$.

The case of $\mathcal{C}^{1,1}$ convexly composite qualified functions follows directly from Theorem 11.13. □

As the following example shows, there are prox-regular continuous functions which are neither subsmooth nor primal lower regular.

EXAMPLE 11.46. Define $f : \mathbb{R} \to \mathbb{R}$ by $f(x) = -x$ if $x \leq 0$ and $f(x) = \sqrt{x}$ if $x \geq 0$. The function f is continuous on $\mathbb{R}$ and prox-regular at 0 for $0 \in \partial_C f(0)$ (see Figure 11.1). Nevertheless, since f is not Lipschitz near 0 it is neither subsmooth at 0 by Proposition 8.18 nor primal lower regular at 0 by Proposition 11.4. As a consequence, f is also not a qualified convexly composite function according to Theorem 11.13 above. □

11.5.2. Subdifferential characterization of prox-regular functions. In the definition of prox-regularity of a function f, not all of $\operatorname{gph} \partial f$ but only a certain localization of this graph is involved. We formalize this in a definition.

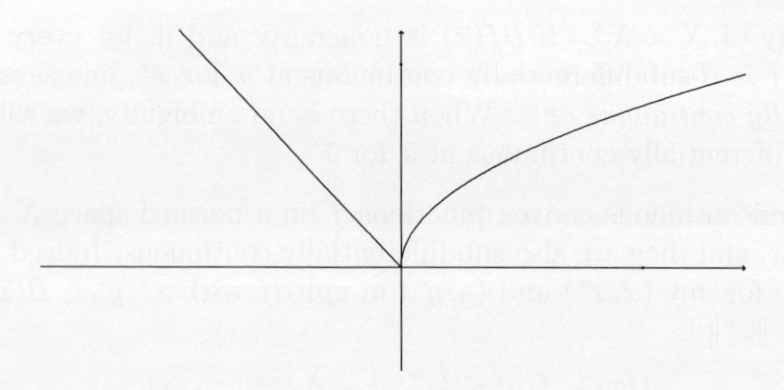

FIGURE 11.1. A continuous prox-regular function failing subsmoothness.

DEFINITION 11.47. Let X be a normed space and $f : X \to \mathbb{R} \cup \{+\infty\}$ be a proper function. For a real $\varepsilon > 0$ and for $(\overline{x}, \overline{x}^*) \in \operatorname{gph} \partial f$ the *f-attentive ε-localization* of ∂f at $\overline{x}$ for $\overline{x}^*$ is the multimapping $\mathcal{L}_\varepsilon^f : X \rightrightarrows X^*$ whose graph is defined by

$$\operatorname{gph} \mathcal{L}_\varepsilon^f := \{(x, x^*) \in \operatorname{gph} \partial f : \|x - \overline{x}\| < \varepsilon, |f(x) - f(\overline{x}| < \varepsilon, \|x^* - \overline{x}^*\| < \varepsilon\}.$$

The same multimapping without the inequality $|f(x) - f(\overline{x})| < \varepsilon$, is called the *simple ε-localization* of ∂f at $\overline{x}$ for $\overline{x}^*$. When there is no risk of ambiguity, we will denote $\mathcal{L}_\varepsilon$ in place of $\mathcal{L}_\varepsilon^f$.

Of course, when f is lower semicontinuous at $\overline{x}$, it is enough to consider a localization with the inequality $f(x) < f(\overline{x}) + \varepsilon$.

Let us establish first the following lemma. It is in the line of Lemma 11.17, but here the linear functional u^* needs to be localized.

LEMMA 11.48. Let X be a normed space and let $f : X \to \mathbb{R} \cup \{+\infty\}$, $\alpha > 0$ and $\varepsilon_1 > 0$ be such that $f(0) = 0$ and $f(x) \geq -\alpha\|x\|^2$ for all $x \in B[0, \varepsilon_1]$. Let a real $\sigma_0 > (32\alpha/3) + 4$. Then given any reals $\sigma \geq \sigma_0$ and $\varepsilon \in]0, \varepsilon_1]$, and given any $u \in B_X(0, \varepsilon/4)$ and $u^* \in B_{X^*}(0, \varepsilon/4)$ one has for every $x \in B[0, \varepsilon] \setminus B(0, 3\varepsilon/4)$

$$f(x) + \langle u^*, u - x\rangle + \frac{\sigma}{2}\|x - u\|^2 > f(0) + \langle u^*, u\rangle + \frac{\sigma}{2}\|u\|^2,$$

so in particular

$$\inf_{\|x\| \leq \varepsilon} \left(f(x) + \langle u^*, u - x\rangle + \frac{\sigma}{2}\|u - x\|^2\right) = \inf_{\|x\| < \frac{3\varepsilon}{4}} \left(f(x) + \langle u^*, u - x\rangle + \frac{\sigma}{2}\|u - x\|^2\right)$$

PROOF. Take any $0 < \varepsilon \leq \varepsilon_1$, $u \in B(0, \varepsilon/4)$, $u^* \in B(0, \varepsilon/4)$ and $x \in B[0, \varepsilon]$ with $\|x\| \geq 3\varepsilon/4$. Consider a real $\sigma > 0$. We notice that

$$\begin{aligned} & f(x) + \langle u^*, u - x\rangle + \frac{\sigma}{2}\|x - u\|^2 \\ \geq & \inf_{\|y\| \leq \varepsilon} f(y) + \frac{\sigma}{2}(\|x\| - \|u\|)^2 - \|u^*\| \, \|x - u\| \\ \geq & \inf_{\|y\| \leq \varepsilon} f(y) + \frac{\sigma}{2} \cdot \frac{\varepsilon^2}{4} - \frac{\varepsilon}{4} \cdot \frac{5\varepsilon}{4}, \end{aligned}$$

and hence

$$f(x) + \langle u^*, u - x\rangle + \frac{\sigma}{2}\|x - u\|^2 \geq \inf_{\|y\| \leq \varepsilon} f(y) + \frac{2\sigma - 5}{4} \cdot \frac{\varepsilon^2}{4}. \tag{11.58}$$

Note also that the inequality

$$\inf_{\|y\|\le\varepsilon} f(y)+\frac{2\sigma-5}{4}\cdot\frac{\varepsilon^2}{4} > \langle u^*,u\rangle+\frac{\sigma}{2}\|u\|^2 \tag{11.59}$$

is implied by

$$\inf_{\|y\|\le\varepsilon} f(y)+\frac{2\sigma-5}{4}\cdot\frac{\varepsilon^2}{4} > \frac{\varepsilon^2}{16}+\frac{\sigma}{2}\cdot\frac{\varepsilon^2}{16},$$

or equivalently by

$$\varepsilon^2\frac{3\sigma-12}{32} > -\inf_{\|y\|\le\varepsilon} f(y).$$

On the other hand, as $f\ge -\alpha\|\cdot\|^2$ on $B[0,\varepsilon]$, we have $-\inf_{\|y\|\le\varepsilon} f(y)\le \alpha\varepsilon^2$, thus to get the inequality (11.59), it is enough to choose $\sigma\ge\sigma_0$, where σ_0 is such that

$$\varepsilon^2\frac{3\sigma_0-12}{32} > \alpha\varepsilon^2,$$

that is,

$$\sigma_0 > \frac{32\alpha}{3}+4.$$

Consequently, for any real $\sigma\ge\sigma_0$, combining (11.58) and (11.59), we obtain

$$f(x)+\langle u^*,u-x\rangle+\frac{\sigma}{2}\|x-u\|^2 > f(0)+\frac{\sigma}{2}\|0-u\|^2+\langle u^*,u-0\rangle,$$

which finishes the proof of the lemma. □

The subdifferential characterization that we will prove next for the behavior of prox-regularity for a function will involve a monotonicity type property of some localization of its subdifferential.

DEFINITION 11.49. Let $M : X \rightrightarrows X^*$ be a multimapping from a normed space X into its topological dual X^*. Given a real $\sigma\ge 0$, one says that M is *σ-hypomonotone* whenever

$$\langle x^*-y^*,x-y\rangle \ge -\sigma\|x-y\|^2$$

for all (x,x^*) and (y,y^*) in $\operatorname{gph} M$.

Clearly, the σ-hypomonotonicity of the multimapping M with $\sigma=0$ corresponds to its monotonicity.

The proof of the following theorem, characterizing the prox-regularity of a function via its subdifferential, makes use of a large part of arguments in the proof of Theorem 11.18. For commodity, in addition to the properties **Prop.1**$\cdots$**Prop.4** in Section 6.2 for the subdifferential ∂, we assume in the rest of this section that $\partial_P f(x)\subset\partial f(x)$ and

$$\partial\big(\langle a^*,\cdot\rangle+\beta+f(\cdot+b)\big)(x)=a^*+\partial f(x+b),$$

for any function $f: X\to\mathbb{R}\cup\{+\infty\}$, any $a^*\in X^*$ and any $\beta\in\mathbb{R}$. Evidently, ∂_{VP}, ∂_F, ∂_L, ∂_A, ∂_C fulfill those properties on appropriate spaces.

THEOREM 11.50 (subdifferential characterization of prox-regular functions). Let X be a Banach space and $f: X\to\mathbb{R}\cup\{+\infty\}$ be a function which is lower semicontinuous near $\overline{x}\in\operatorname{Dom}\partial f$. The following assertions are equivalent:
(a) the function f is ∂-prox-regular at $\overline{x}$ for $\overline{x}^*\in\partial f(\overline{x})$;

(b) $\overline{x}^* \in \partial_{VP} f(\overline{x})$ and there exist reals $\varepsilon_0 > 0$ and $\sigma_0 > 0$ such that the f-attentive ε_0-localization $\mathcal{L}^f_{\varepsilon_0}$ of ∂f at $\overline{x}$ for $\overline{x}^*$ is σ_0-hypomonotone. If in addition f is subdifferentially continuous at $\overline{x}$, then $\mathcal{L}^f_{\varepsilon_0}$ can be taken as the simple ε_0-localization of ∂f at $\overline{x}$ for $\overline{x}^*$.

PROOF. The implication (a)⇒(b) being a direct consequence of the definition of prox-regular functions, let us show the converse implication.

Considering the function $x \mapsto f(x+\overline{x}) - f(\overline{x}) - \langle \overline{x}^*, x\rangle$ in place of f and using the closedness property, we may and do suppose that $\overline{x} = 0$, $\overline{x}^* = 0$, $f(0) = 0$ and $\varepsilon_0 < 1$ with f lower semicontinuous on $B(0, \varepsilon_0)$. Then, since $0 \in \partial_{VP} f(0)$ as easily seen by definition of variational proximal subdifferential, there exist reals $\alpha > 0$ and $\varepsilon_1 > 0$ such that

$$f(x) \geq -\alpha \|x\|^2 \quad \text{for all } x \in B[0, \varepsilon_1].$$

Fix a real $\sigma > \max\{\sigma_0, 4 + (32\alpha/3)\}$ and a positive real

$$\varepsilon < \min\left\{\frac{\varepsilon_0}{1+2\sigma}, \varepsilon_1\right\}. \tag{11.60}$$

Take any $(u, u^*) \in \operatorname{gph} \mathcal{L}^f_{\frac{\varepsilon}{4}}$, where $\mathcal{L}^f_{\frac{\varepsilon}{4}}$ is the f-attentive $(\varepsilon/4)$-localization of ∂f at 0 for 0. Define $\varphi(x) := f(x) + \langle u^*, u - x\rangle + (\sigma/2)\|x - u\|^2$ and

$$\overline{\varphi}(x) = \varphi(x) \text{ if } x \in B[0, \varepsilon] \quad \text{and} \quad \overline{\varphi}(x) = +\infty \text{ elsewhere.}$$

Let $\varepsilon_n \to 0$ with $0 < \varepsilon_n < \varepsilon^2/16$ and, for each $n \in \mathbb{N}$, let $u_n \in X$ be such that

$$\overline{\varphi}(u_n) < \inf_X \overline{\varphi} + \varepsilon_n. \tag{11.61}$$

By Lemma 11.48 we may suppose that $u_n \in B(0, 3\varepsilon/4)$ for all n. The Ekeland variational principle furnishes, for each $n \in \mathbb{N}$, some $x_n \in X$ such that

$$\overline{\varphi}(x_n) < \inf_X \overline{\varphi} + \varepsilon_n,\ \|x_n - u_n\| \leq \sqrt{\varepsilon_n},\ \overline{\varphi}(x_n) = \inf_{x\in X} \left(\overline{\varphi}(x) + \sqrt{\varepsilon_n}\|x - x_n\|\right).$$

For each $n \in \mathbb{N}$, observing that $x_n \in B(0, \varepsilon)$, we have $0 \in \partial(\varphi + \sqrt{\varepsilon_n}\|\cdot - x_n\|)(x_n)$. The fuzzy sum rule yields x'_n, x''_n in X with

$$\|x'_n - x_n\| < \sqrt{\varepsilon_n},\ |f(x'_n) - f(x_n)| < \sqrt{\varepsilon_n},\ \|x''_n - x_n\| < \sqrt{\varepsilon_n}$$

and such that

$$0 \in \partial f(x'_n) - u^* + \sigma\partial\left(\frac{1}{2}\|\cdot\|^2\right)(x''_n - u) + 2\sqrt{\varepsilon_n}\mathbb{B}_{X^*},$$

so there are $x^*_n \in \partial f(x'_n)$ and $y^*_n \in -u^* + \sigma\partial(\frac{1}{2}\|\cdot\|^2)(x''_n - u)$ such that

$$\|x^*_n + y^*_n\| \leq 2\sqrt{\varepsilon_n}. \tag{11.62}$$

For $z^*_n := \sigma^{-1}(y^*_n + u^*)$ we have $z^*_n \in \partial(\frac{1}{2}\|\cdot\|^2)(x''_n)$, hence (see (2.25))

$$\langle z^*_n, x''_n - u\rangle = \|x''_n - u\|^2 \quad \text{and} \quad \|z^*_n\| = \|x''_n - u\|. \tag{11.63}$$

We note that

$$\|z^*_n\| \leq \|x''_n - x_n\| + \|x_n - u_n\| + \|u_n\| + \|u\| < 2\sqrt{\varepsilon_n} + (3\varepsilon/4) + \|u\|$$

along with $(3\varepsilon/4) + \|u\| < \varepsilon$, so for n sufficiently large, say $n \geq n_0$, we have $\|z^*_n\| < \varepsilon$. Fix any $n \geq n_0$. By (11.60) we deduce that

$$\|y^*_n\| \leq \|u^*\| + \sigma\|z^*_n\| < (\varepsilon/4) + \sigma\varepsilon < \varepsilon_0/2,$$

and by (11.62) we get $\|x_n^*\| < \varepsilon_0/2 + \varepsilon_0/2 = \varepsilon_0$. Further, since $\varepsilon_n < \varepsilon^2/16$ we have

$$f(x_n) + \frac{\sigma}{2}\|u - x_n\|^2 + \langle u^*, u - x_n\rangle < f(0) + \frac{\sigma}{2}\|u\|^2 + \langle u^*, u\rangle + \frac{\varepsilon^2}{16},$$

hence

$$f(x_n) \le \frac{\sigma}{2}\|u\|^2 + \langle u^*, x_n\rangle + \frac{\varepsilon^2}{16},$$

which gives by the inequalities $\|u\| < \varepsilon/4$ and $\|u^*\| < \varepsilon/4$ (since $u^* \in T^f_{\frac{\varepsilon}{4}}(u)$) and by (11.60)

$$f(x_n) \le \frac{\sigma}{2}\frac{\varepsilon^2}{16} + \frac{\varepsilon^2}{4} + \frac{\varepsilon^2}{16} < \varepsilon\left(\frac{\sigma}{32} + \frac{5}{16}\right),$$

and hence

$$f(x_n') < f(x_n) + \sqrt{\varepsilon_n} < \varepsilon\left(\frac{\sigma}{32} + \frac{5}{16}\right) + \frac{\varepsilon}{4} = \varepsilon\left(\frac{\sigma}{32} + \frac{9}{16}\right).$$

It ensues that $f(x_n') < \varepsilon_0$. Consequently, $(x_n', x_n^*) \in \mathcal{L}^f_{\varepsilon_0}$ and by σ_0-hypomonotonicity of $\mathcal{L}^f_{\varepsilon_0}$ it results that

$$\langle u^* - x_n^*, u - x_n'\rangle \ge -\sigma_0\|u - x_n'\|^2. \tag{11.64}$$

On the other hand, by (11.63) and (11.62) we also have

$$\begin{aligned}&\langle u^* - x_n^*, u - x_n'\rangle\\ &= \langle \sigma z_n^* - y_n^* - x_n^*, u - x_n''\rangle + \langle \sigma z_n^* - y_n^* - x_n^*, x_n'' - x_n'\rangle\\ &\le -\sigma\|x_n'' - u\|^2 + 2\sqrt{\varepsilon_n}\|u - x_n''\| + (\sigma\varepsilon + 2\sqrt{\varepsilon_n})\|x_n'' - x_n'\|.\end{aligned}$$

Using this, writing $|\,\|x_n' - u\|^2 - \|x_n'' - u\|^2| \le (\|x_n' - u\| + \|x_n'' + u\|)\|x_n'' - x_n'\|$ and noting that $(x_n')_n$ and $(x_n'')_n$ are bounded sequences, we obtain by (11.64)

$$(\sigma - \sigma_0)\|x_n' - u\|^2 \le 2\sqrt{\varepsilon_n}\|u - x_n''\| + c\|x_n'' - x_n'\|,$$

where $c > 0$ is some real constant independent of n. We deduce as $n \to \infty$ that $x_n' \to u$, and hence $x_n \to u$ as well as $u_n \to u$.

Keeping in mind by (11.61) that

$$f(u_n) + \langle u^*, u - u_n\rangle + \frac{\sigma}{2}\|u_n - u\| \le \inf_{\|x\|\le\varepsilon}\left(f(x) + \langle u^*, u - x\rangle + \frac{\sigma}{2}\|u - x\|^2\right) + \varepsilon_n,$$

we derive by the lower semicontinuity of f on $B(0, \varepsilon_0) \ni u$ (because $(u, u^*) \in \operatorname{gph} \mathcal{L}^f_{\frac{\varepsilon}{4}}$) that

$$f(u) \le \liminf_{n\to\infty} f(u_n) \le \inf_{\|x\|\le\varepsilon}\left(f(x) + \langle u^*, u - x\rangle + \frac{\sigma}{2}\|u - x\|^2\right),$$

which implies that

$$f(x) \ge f(u) + \langle u^*, x - u\rangle - \frac{\sigma}{2}\|x - u\|^2 \quad \text{for all } x \in B(0, \varepsilon/4).$$

This inequality translates that f is ∂-prox-regular at $\overline{x}$ for $\overline{x}^*$ (with parameters σ and $\varepsilon/4$), so the proof is complete. □

11.5.3. Differentiability of Moreau envelopes under prox-regularity. A large part of the development in this subsection will be devoted to continuity properties of proximal mappings of Moreau envelopes of prox-regular functions with the aim to derive continuity properties of derivatives of Moreau envelopes of such functions. We know with Theorem 11.27 that, even for convex functions, such properties require a special structure for the norm of the space. Like in Section 11.3, we will then work with X as a Hilbert space H. The situation of uniformly convex Banach spaces will be examined later in Section 18.3.

The arguments in the next lemma are in the lines of those employed for the proof of the assertion (a) in Theorem 11.35.

LEMMA 11.51. Let H be a Hilbert space and $f : H \to \mathbb{R} \cup \{+\infty\}$ be a function which is lower semicontinuous on a ball $B(0, \varepsilon_0)$ along with $f(0) = 0$ and $0 \in \partial f(0)$. Assume that there are reals $\lambda_0 > 0$ and $\alpha > 0$ such that:
(i) $f(x) \geq -\alpha\|x\|^2$ for all $x \in B(0, \varepsilon_0)$;
(ii) $I + \lambda_0 \mathcal{L}^f_{\varepsilon_0}$ is monotone, where $\mathcal{L}^f_{\varepsilon_0}$ denotes the f-attentive ε_0-localization of ∂f at 0 for $0 \in \partial f(0)$.

Then for any positive reals $\lambda < \min\{\lambda_0, 3/(32\alpha)\}$ and $\varepsilon < \min\{\varepsilon_0, \varepsilon_0\lambda, 1\}$, the single-valued mapping $P_{\lambda,\varepsilon} f$ is well defined on $B(0, \varepsilon/4)$ and Lipschitz continuous therein.

PROOF. Put $\overline{f}(x) = f(x)$ if $x \in B[0, \varepsilon]$ and $\overline{f}(x) = +\infty$ elsewhere. We know by Theorem 11.23(b) that $e_\lambda \overline{f}$ is locally Lipschitz continuous on H.

Let us first show that

$$\operatorname{Prox}_{\lambda,\varepsilon} f(u) \subset (I + \lambda\mathcal{L}^f_{\varepsilon_0})^{-1} \quad \text{for all } u \in B(0, \varepsilon/4). \tag{11.65}$$

Fix any $u \in B(0, \varepsilon/4) \cap \operatorname{Dom} \operatorname{Prox}_{\lambda,\varepsilon} f$ and take any $y \in \operatorname{Prox}_{\lambda,\varepsilon} f(u)$. By Lemma 11.32 we know that $\|y\| < 3\varepsilon/4$, hence $0 \in \partial_P f(y) + \lambda^{-1}(y - u)$, that is, $v(u) := \lambda^{-1}(u - y) \in \partial_P f(y) \subset \partial f(y)$. Note that our assumption on ε entails that $\|y\| < \varepsilon_0$ and $\|v(u)\| < \lambda^{-1}(\varepsilon/4 + 3\varepsilon/4) < \varepsilon_0$ and also, by definition of $\operatorname{Prox}_{\lambda,\varepsilon} f$ that

$$f(y) \leq f(0) + \frac{1}{2\lambda}\|u\|^2 < \frac{\varepsilon^2}{32\lambda} < \varepsilon_0.$$

Consequently, $v(u) \in \mathcal{L}f_{\varepsilon_0}(y)$, or equivalently $y \in \left(I + \lambda\mathcal{L}^f_{\varepsilon_0}\right)^{-1}(u)$, which justifies the inclusion (11.65).

Keep u fixed in $B(0, \varepsilon/4)$ and take any $n \in \mathbb{N}$. By the density property of Fréchet subdifferentiability points of $e_\lambda \overline{f}$, there exists $u_n \in B(u, 1/n) \cap B(0, \varepsilon/4)$ with $\partial_F e_\lambda \overline{f}(u_n) \neq \emptyset$. Theorem 11.23(c) guarantees that $\operatorname{Prox}_{\lambda,\varepsilon} f(u_n) \neq \emptyset$. Taking some $x_n \in \operatorname{Prox}_{\lambda,\varepsilon} f(u_n)$ and setting $v_n := \lambda^{-1}(u_n - x_n)$, we have $v_n \in T_{\varepsilon_0}(x_n)$ by (11.65). It results by the monotonicity assumption of $I + \lambda_0 T_{\varepsilon_0}$ that, for any $n, m \in \mathbb{N}$

$$\langle \lambda_0 v_n - \lambda_0 v_m, x_n - x_m \rangle \geq -\|x_n - x_m\|^2,$$

or equivalently

$$(\lambda_0 - \lambda)\|x_n - x_m\|^2 \leq \lambda_0 \langle u_n - u_m, x_n - x_m \rangle.$$

It ensues that $(x_n)_n$ is a Cauchy sequence. Denote by $y(u) \in B[0, \varepsilon]$ its limit.

Fix now any $x \in B[0, \varepsilon]$ and note by definition of x_n that, for every $n \in \mathbb{N}$

$$f(x_n) + \frac{1}{2\lambda}\|u_n - x_n\|^2 \leq f(x) + \frac{1}{2\lambda}\|u_n - x\|^2.$$

By the lower semicontinuity of f on $B(0,\varepsilon_0)$ we deduce that

$$f(y(u)) + \frac{1}{2\lambda}\|u - y(u)\|^2 \le f(x) + \frac{1}{2\lambda}\|u - x\|^2.$$

This means that $y(u) \in \operatorname{Prox}_{\lambda,\varepsilon} f(u)$, and hence $y(u) \in \left(I + \lambda\mathcal{L}^f_{\varepsilon_0}\right)^{-1}(u)$ according to (11.65). We have then proved that

$$B(0,\varepsilon/4) \subset \operatorname{Dom}\left(I + \lambda\mathcal{L}^f_{\varepsilon_0}\right)^{-1} \cap \operatorname{Prox}_{\lambda,\varepsilon} f.$$

Then by Lemma 11.34 and inclusion (11.65) we see that $\left(I + \lambda\mathcal{L}^f_{\varepsilon_0}\right)^{-1}$ is (nonempty) single-valued and Lipschitz continuous on $B(0,\varepsilon/4)$ and for every $u \in B(0,\varepsilon/4)$

$$\left(I + \lambda\mathcal{L}^f_{\varepsilon_0}\right)^{-1}(u) = \operatorname{Prox}_{\lambda,\varepsilon} f(u).$$

This justifies the desired properties for $P_{\lambda,\varepsilon} f$. □

Taking advantage of the Hilbert structure of H we observe given $x, v \in H$ and $\lambda > 0$ we have that, for $u_\lambda := x + \lambda v$

$$(11.66)\quad \begin{cases} f(x') > f(x) + \langle v, x' - x\rangle - \frac{1}{2\lambda}\|x' - x\|^2 \quad \forall x' \in H \setminus \{x\} \\ \Longleftrightarrow \\ f(x') + \frac{1}{2\lambda}\|x' - u_\lambda\|^2 > f(x) + \frac{1}{2\lambda}\|x - u_\lambda\|^2 \quad \forall x' \in H \setminus \{x\}. \end{cases}$$

The next proposition makes use of the latter equivalence.

Proposition 11.52. Let $f : H \to \mathbb{R} \cup \{+\infty\}$ be a function which is lower semicontinuous near $\overline{x} \in \operatorname{Dom}\partial f$ and let $\overline{v} \in \partial f(\overline{x})$. Then f is minorized by a quadratic function and ∂-prox-regular at $\overline{x}$ for $\overline{v}$ if and only if there exist reals $\varepsilon_0 > 0$ and $\lambda_0 > 0$ such that, for any $\lambda \in]0, \lambda_0[$

$$(x, v) \in \operatorname{gph}\mathcal{L}^f_{\varepsilon_0} \ \Rightarrow\ P_\lambda f(x + \lambda v) = x,$$

where we recall that $\mathcal{L}^f_{\varepsilon_0}$ is the f-attentive ε_0-localization of ∂f at $\overline{x}$ for $\overline{v}$. In fact, it suffices that the implication holds for a given $\lambda > 0$.

Proof. Assume that the function f is minorized by a quadratic function and ∂-prox-regular at $\overline{x}$ for $\overline{v}$. By Proposition 4.150(a) there exist some reals $\varepsilon_0 > 0$ and $\lambda_0 > 0$ such that, for any $(x, v) \in \operatorname{gph}\mathcal{L}^f_{\varepsilon_0}$

$$(11.67)\qquad f(x') > f(x) + \langle v, x' - x\rangle - \frac{1}{2\lambda_0}\|x' - x\|^2 \quad \text{for all } x' \in H \setminus \{x\}.$$

By (11.66) we see that, for any $\lambda \in]0, \lambda_0[$

$$(11.68)\ f(x') + \frac{1}{2\lambda}\|x' - (x + \lambda v)\|^2 > f(x) + \frac{1}{2\lambda}\|x - (x + \lambda v)\|^2 \quad \text{for all } x' \in H \setminus \{x\},$$

which means that $P_\lambda f(x + \lambda v) = x$.

Conversely, if for some $\lambda > 0$ we have $P_\lambda f(x + \lambda v) = x$ for any $(x, v) \in \operatorname{gph}\mathcal{L}^f_{\varepsilon_0}$, then f is minorized by a quadratic function (since $e_\lambda f(x + \lambda v) > -\infty$) and by definition of $P_\lambda f$ the inequality (11.68) holds. Consequently, for any $(x, v) \in \operatorname{gph}\partial f$ with $x \in B(\overline{x}, \varepsilon_0)$, $|f(x) - f(\overline{x})| < \varepsilon_0$ and $v \in B(\overline{v}, \varepsilon_0)$, the inequality (11.67) is fulfilled, and this entails that f is ∂-prox-regular at $\overline{x}$ for $\overline{v}$. □

We now establish the $\mathcal{C}^1$ (in fact, $\mathcal{C}^{1,1}$) property for Moreau envelopes of prox-regular functions.

THEOREM 11.53 ($\mathcal{C}^{1,1}$ property of Moreau envelopes of prox-regular functions). Let $f : H \to \mathbb{R}\cup\{+\infty\}$ be function which is lower semicontinuous near $\overline{x} \in \operatorname{Dom} \partial f$ and let $\overline{v} \in \partial f(\overline{x})$. If f is minorized by a quadratic function and ∂-prox-regular at $\overline{x}$ for $\overline{v}$, then there exist reals $\varepsilon_0 > 0$, $\lambda_0 > 0$ and $\delta_0 > 0$ such that each $\lambda \in]0, \lambda_0]$ the following hold:
(a) The single-valued mapping $P_\lambda f$ is well defined and Lipschitz continuous on $B(\overline{x} + \lambda\overline{v}, \delta_0)$ with $P_\lambda f(x) = (I + \lambda\mathcal{L}^f_{\varepsilon_0})^{-1}$ for all $x \in B(\overline{x} + \lambda\overline{v}, \delta_0)$, where $\mathcal{L}^f_{\varepsilon_0}$ is the f-attentive ε_0-localization of ∂f at $\overline{x}$ for $\overline{v}$.
(b) The Moreau envelope $e_\lambda f$ is $\mathcal{C}^{1,1}$ on $B(\overline{x} + \lambda\overline{v}, \delta_0)$ with

$$\nabla e_\lambda f(x) = \lambda^{-1}(x - P_\lambda f(x)) \quad \text{for all } x \in B(\overline{x} + \lambda\overline{v}, \delta_0).$$

PROOF. Since (b) follows from (a) by Proposition 11.30, let us prove (a). Putting

$$F(x) := f(x + \overline{x}) - f(\overline{x}) - \langle \overline{v}, x\rangle,$$

we see that $F(0) = 0$, $0 \in \partial F(0)$ and

$$\begin{aligned} e_\lambda F(x) &= \inf_{y\in H} \left(f(y + \overline{x}) + \frac{1}{2\lambda}(\|x - y\|2 + 2\langle \lambda\overline{v}, x - y\rangle + \|\lambda\overline{v}\|^2) \right) \\ &\quad - \frac{\lambda}{2}\|\overline{v}\|^2 - \langle \overline{v}, x\rangle - f(\overline{x}) \\ &= \inf_{y\in H} \left(f(y + \overline{x}) + \frac{1}{2\lambda}\|(x + \lambda\overline{v}) - y\|2 \right) - \frac{\lambda}{2}\|\overline{v}\|^2 - \langle \overline{v}, x\rangle - f(\overline{x}), \end{aligned}$$

or equivalently

$$e_\lambda F(x) = e_\lambda f(x + \overline{x} + \lambda\overline{v}) - \frac{\lambda}{2}\|\overline{v}\|^2 - \langle \overline{v}, x\rangle - f(\overline{x}),$$

which also ensures that

$$P_\lambda F(x) = P_\lambda f(x + \overline{x} + \lambda\overline{v}). \tag{11.69}$$

By Proposition 4.150(a) and Lemma 11.51 there are reals $\varepsilon_0 > 0$ and $\lambda_1 > 0$ such that

$$I + \lambda_1\mathcal{L}^F_{\varepsilon_0} \text{ is monotone and } F(x) > -\frac{1}{2\lambda_1}\|x\|^2 \text{ for all } x \in X \setminus \{0\}.$$

By Lemma 11.37 we also know that, for any $\lambda \in]0, \lambda_1[$ and any real $\rho > 0$

$$F(y) + \frac{1}{2\lambda}\|x - y\|^2 \le e_\lambda F(y) + \rho \Rightarrow \|y\| \le 2\mu_\lambda\|x\| + \sqrt{2\lambda\mu_\lambda\rho},$$

where $\mu_\lambda := (1 - \lambda_1^{-1}\lambda)^{-1}$, hence putting $\varepsilon_{(\rho,\lambda,\delta)} := 2\mu_\lambda\delta + \sqrt{2\lambda\mu_\lambda\rho}$ we see that, for any $x \in B[0, \delta]$

$$\begin{cases} e_\lambda F(x) = \displaystyle\inf_{\|y\|\le\varepsilon_{(\rho,\lambda,\delta)}} (F(y) + \tfrac{1}{2\lambda}\|x - y\|^2) = e_{\lambda,\varepsilon_{(\rho,\lambda,\delta)}} F(x), \\ \operatorname{Prox}_\lambda F(x) = \displaystyle\operatorname*{Argmin}_{\|y\|\le\varepsilon_{(\rho,\lambda,\delta)}} (F(y) + \tfrac{1}{2\lambda}\|x - y\|^2) = \operatorname{Prox}_{\lambda,\varepsilon_{(\rho,\lambda,\delta)}} F(x). \end{cases} \tag{11.70}$$

Fix positive reals $\lambda_0 < 3\lambda_1/32$ and $\varepsilon_0' < \min\{\varepsilon_0, \lambda_0\varepsilon_0, 1\}$. Choose $\delta, \rho > 0$ small enough that $\varepsilon_{(\rho,\lambda_0,\delta)} < \varepsilon_0'$. By the nondecreasing property of $\lambda \mapsto \varepsilon_{(\rho,\lambda,\delta)}$ we also have $\varepsilon_{(\rho,\lambda,\delta)} < \varepsilon_0'$ for all $\lambda \in]0, \lambda_0]$. Setting $\delta_0 := \min\{\delta, \varepsilon_0'/4\}$, by (11.70) and Lemma 11.32 it results that, for any $\lambda \in]0, \lambda_0]$ and any $x \in B(0, \delta_0)$ we have

$$e_\lambda F(x) = \inf_{\|y\|\le\varepsilon_0'} \left(F(y) + \frac{1}{2\lambda}\|x - y\|^2 \right) = \inf_{\|y\|<3\varepsilon_0'/4} \left(F(y) + \frac{1}{2\lambda}\|x - y\|^2 \right)$$

along with $\operatorname{Prox}_{\lambda}F(x) = \operatorname{Prox}_{\lambda,\varepsilon_0'}F(x)$ and $\|y\| < 3\varepsilon_0'/4$ for every $y \in \operatorname{Prox}_{\lambda,\varepsilon_0'}F(x)$. It ensues from Lemma 11.51 that the single-valued mapping $P_\lambda F$ is well defined on $B(0,\delta_0)$ and Lipschitz continuous therein.

From what precedes and from (11.69) we derive that, for each $\lambda \in]0,\lambda_0]$ the single-valued mapping $P_\lambda f$ is well defined and Lipschitz continuous on the ball $B(\overline{x}+\lambda\overline{v},\delta_0)$. This also ensures by Proposition 11.30, for each $\lambda \in]0,\lambda_0]$, that the Moreau envelope $e_\lambda f$ is of class $\mathcal{C}^{1,1}$ over $B(\overline{x}+\lambda\overline{v},\delta_0)$ and that

$$\nabla e_\lambda f(x) = \lambda^{-1}(x - P_\lambda f(x)) \quad \text{for all } x \in B(\overline{x}+\lambda\overline{v}).$$

The proof is then complete. □

REMARK 11.54. Let f be a function satisfying the assumption of Theorem 11.53 and let

$$\lambda_f := \sup\left\{\lambda > 0 : \exists \beta,\gamma > 0,\ f(\cdot) \geq -\frac{1}{2\lambda}\|\cdot\|^2 - \beta\|\cdot\| - \gamma\right\}.$$

It is worth pointing out that λ_0 in the theorem may not be chosen as any real in $]0,\lambda_f[$. In fact, it may arise that, for some $\lambda \in]0,\lambda_f[$ and some neighborhood of $\overline{x}+\lambda\overline{v}$, the single-valued mapping $P_\lambda f$ is well defined and continuous, but not locally Lipschitz therein. Indeed, let $\varphi : [0,1] \to \mathbb{R}$ be a continuous increasing function with $\varphi(0) = 0$ and with $\varphi'(t) = 0$ for almost every $t \in [0,1]$, and such that there is no open interval of $[0,1]$ over which φ is Lipschitz. The function φ^{-1} is bounded. Define $f : \mathbb{R} \to \mathbb{R}$ by

$$f(x) := -\frac{1}{2}|x|^2 + \int_0^x \varphi^{-1}(t)\,dt \quad \text{for all } x \in \mathbb{R}.$$

The function $f + (1/2)|\cdot|^2$ is convex, and f is prox-regular at 0 for $0 = f'(0)$ (and in fact, f is linearly semiconvex). Further, denoting by μ an upper bound of $|\varphi^{-1}|$ on $\mathbb{R}$ we see that, for every $x \in \mathbb{R}$

$$f(x) \geq -\frac{1}{2}|x|^2 - \mu|x|,$$

so $\lambda_f \geq 1$.

□

11.6. Comments

Functions in the line of s-lower regular functions essentially appeared in the last quarter of the 20th century. R. A. Poliquin in his 1991 paper [**805**, Definition 3.1] defined a lower semicontinuous function $f : \mathbb{R}^n \to \mathbb{R} \cup \{+\infty\}$ to be *primal lower nice* at a point $\overline{x} \in \operatorname{cl}(\operatorname{dom} f)$ through a subdifferential property equivalent to the conditions in Proposition 11.3. The terminology "*primal lower regular*", corresponding to s-lower regular with $s = 1$, seems to be more suitable. In [**805, 806**] Poliquin developed a large part of the theory of primal lower nice functions in finite dimensions with diverse fundamental results showing the interest of such functions in variational analysis. In [**805**, Theorem 4.1] a general subdifferential determination theorem is proved by Poliquin in the form: If $f, g : \mathbb{R}^n \to \mathbb{R} \cup \{+\infty\}$ are two functions which are primal lower nice a point $\overline{x}$ where they are finite and if $\partial_P f(x) = \partial_P g(x)$ for all x near $\overline{x}$, then the there exists a real constant C such that $f(x) = g(x) + C$ for all x near $\overline{x}$. To the best of our knowledge, this was the first subdifferential determination theorem for functions which are neither convex nor locally Lipschitz continuous. In [**806**] Poliquin also extended to the class of primal lower regular functions, the theory of second order epi-differentiability adressed earlier by R. T. Rockafellar for convex functions. According to the equivalence in Proposition 11.3, primal lower regular functions form a subclass of a larger class of functions defined in 1983 by E. De Giorgi, M. Degiovanni, A. Marino and M. Tosques [**320**] as Φ-convex functions. Given an open set U of a Hilbert space H, a function $f : U \to \mathbb{R} \cup \{+\infty\}$ is said in [**320, 321**] to be Φ-convex provided there exists a continuous function $\Phi : (\operatorname{dom} f)^2 \times \mathbb{R}^2 \times \mathbb{R}_+ \to \mathbb{R}_+$ such that

$$f(y) \geq f(x) + \langle v, y - x\rangle - \Phi(x, y, f(x), f(y), \|v\|)\|x - y\|^2$$

for all $x \in \operatorname{Dom} \partial_F f$, $v \in \partial_F f(x)$ and $y \in \operatorname{dom} f$. The primal lower regularity (or the primal lower nice property) of f at $\overline{x}$ corresponds to the Φ-convexity of f on an open neighborhood U_0 of $\overline{x}$ with $\Phi(x, y, f(x), f(y), v) = c(1 + \|v\|)$. In [**321**] Degiovanni, Marino and Tosques developed, for Φ-convex functions, an extensive study of differential evolution problems of the type

$$\begin{cases} \frac{du}{dt} + \partial_F f(u(t)) \ni 0 \\ u(0) = u_0 \in \operatorname{dom} f, \end{cases}$$

proving in this framework many results previously known for subdifferential ∂f of convex functions f or maximal monotone operators $A : H \rightrightarrows H$. The theory of such differential evolution problems was later simplified for qualified $\mathcal{C}^2$ composite convex functions f by S. Guillaume [**450**], and for primal lower regular functions by S. Marcellin and L. Thibault [**689**] and by I. Kecis and L. Thibault [**603**]. The simplifications in [**603, 689**] as well as the versatility of the theory of primal lower regular functions presented in this chapter are due to the form of the rest $c(1 + \|v\|)\|x - y\|^2$ in Definition 11.1.

In Remark 11.15, given a Banach space X and a nondecreasing upper semicontinuous function $\omega : [0, +\infty[\to [0, +\infty[$, we defined a lower semicontinuous function $f : X \to \mathbb{R} \cup \{+\infty\}$ to be *lower* $\omega(\cdot)$*-regular* near $\overline{x} \in \operatorname{cl}(\operatorname{dom} f)$ provided there are an open convex neighborhood U of $\overline{x}$ and reals $c, \gamma > 0$ such that

$$f(y) \geq f(x) + \langle x^*, y - x\rangle - c(1 + \|x^*\|)\|x - y\|\omega(\gamma\|x - y\|)$$

for all $y \in U$, $x \in U \cap \operatorname{Dom} \partial_C f$, and $x^* \in \partial_C f(x)$. The arguments for the theory of that class are essentially (under appropriate conditions on $\omega(\cdot)$) those developed in this chapter for s-lower regular functions. The results are very similar and some can be found in Thibault's paper [**925**]. For example, under the convexity of $t \mapsto t\omega(t)$, Theorem 11.3 in [**925**] ensures that f is lower $\omega(\cdot)$-regular at $\overline{x}$ if and only if there exist reals $\gamma, \delta > 0$ and $c \geq 0$ such that

$$\langle x_1^* - x_2^*, x_1 - x_2\rangle \geq -c(1 + \|x_1^*\| + \|x_2^*\|)\|x_1 - x_2\|\omega(\gamma\|x_1 - x_2\|)$$

for all (x_i, x_i^*) in $\operatorname{gph} \partial_C f$ with $x_i \in B(\overline{x}, \delta)$, for $i = 1, 2$. As said in Remark 11.15, that viewpoint is not developed in the book, since differentiability properties are more versatile with Moreau s-envelope of s-lower regular functions than the Moreau $\omega(\cdot)$-envelope of $\omega(\cdot)$-lower regular ones.

The proofs of Propositions 11.3 and 11.4 use arguments in [**601, 885**]. The interest of examining qualified convexly composite functions $g \circ G$ mainly started in finite dimensions with the 1988 paper [**862**] of R. T. Rockafellar, where g is assumed to be piecewise linear quadratic and where first and second order epi-differentiability of $g \circ G$ were studied. A few years later, R. A. Poliquin showed, in Theorem 5.1 of his 1991 paper [**805**], the primal lower nice property of qualified convexly composite functions $g \circ G$ with general convex functions g on $\mathbb{R}^m$. The interest of qualified convexly composite functions in finite dimensions was further amplified in R. A. Poliquin and R. T. Rockafellar's papers [**807, 808**], where such functions were called "amenable functions". In [**805, 807, 808**] instead of the Robinson-type qualification condition (11.5) in the book, another condition (equivalent to (11.5) in finite dimensions) in terms of the normal cone $N(\operatorname{dom} g; G(\overline{x}))$ is utilized. Theorem 11.7 was first stated in finite dimensions in [**807**, (2.6)] by means of a qualification condition with $N(\operatorname{dom} g; G(\overline{x}))$; the main arguments with g piecewise linear quadratic in [**862**, Theorem 4.5] still work with convex functions g. The proof of the infinite-dimensional extension in Theorem 11.7 in the book involves ideas from C. Combari, R. A. Poliquin and L. Thibault [**273**] and from Combari and Thibault [**274**]. The equality $\partial_C(g \circ G)(\overline{x}) = DG(\overline{x})^* \big(\partial g(G(\overline{x}))\big)$ was previously shown by R. Cominetti [**275**, Theorem 4.2, (23)] under the more stringent assumption (see [**275**, (23), p. 857]) that the restriction of the convex function $g : Y \to \mathbb{R} \cup \{+\infty\}$ to its effective domain $\operatorname{dom} g$ is locally Lipschitz. Theorem 11.13 was proved by I. Kecis and L. Thibault [**601**, Proposition 3.1] as extension of previous results for $\mathcal{C}^{1,1}$ convexly composite functions on $\mathbb{R}^n$ in [**805**, Theorem 5.1] and on Banach spaces in [**268**, Lemma 2.3]; the proof here and in [**601**] are adaptations of the approach by F. Bernard and L. Thibault [**94**] for the prox-regularity of such functions. Qualified convexly composite functions were also considered for diverse objectives in [**184, 656, 806**].

The assertion (a) in Theorem 11.16 with $s = 1$ (that is, for primal lower regular functions) was first shown in finite dimensions by R. A. Poliquin [**805**, Proposition 3.5] and it was extended to Hilbert spaces by A. B. Levy, R. A. Poliquin and L. Thibault [**656**, Theorem 2.4]. As stated, Theorem 11.16 was established by I. Kecis and L. Thibault [**601**, Theorem 3.1]; we also refer to M. Ivanov and N. Zlateva [**545**] for other related results and proofs. A first similar version of Lemma 11.17 was provided by L. Thibault and D. Zagrodny [**926**, Lemma 4.2]; its given statement corresponds to [**601**, Lemma 3.1] and its proof follows principally [**926**, Lemma 3.1] (see also M. Mazade and L. Thibault [**699**, Lemma 3.1]). Theorem 11.18 is taken from [**601**, Theorem 3.2] by Kecis and Thibault; the result was first established by R. A. Poliquin [**805**, Corollary 3.4] in finite dimensions through the concept of *local quadratic conjugate* (see [**805**, p. 390, (3.1)]). The Hilbertian version of Theorem 11.18 appeared, with another method, in the paper [**656**, Proposition 2.2, Corollary 2.3] by A. L. Levy, R. A. Poliquin and L. Thibault. The main arguments in the book for the proof of Theorem 11.18 in Banach spaces follow those of F. Bernard and L. Thibault [**93**] for a similar result for prox-regular functions; see also a characterization of this type by M. Ivanov and N. Zlateva [**547**]. Truncations of subdifferentials of type (11.18) were used by R. A. Poliquin and R. T. Rockafellar [**809**, Definition 3.1] for subdifferential characterizations of prox-regular functions. Proposition 11.20 is an adaptation of [**689**, Proposition 1.6] and [**699**, Proposition 3.4]; previously to [**689, 699**] a similar version was proved by M. Ivanov and N. Zlateva [**547**, Lemma 2.4].

The proof of 11.23 follows, for a very large part, that by I. Kecis and L. Thibault [**602**, Theorem 3.1]. The assertion (a) in Lemma 11.24 was first noticed by E. Asplund [**29**, p. 235, relation (1)] for the square distance function d_C^2 with the indicator function Ψ_C in place of f in Lemma 11.24.

Certain first properties of Moreau envelopes of primal lower regular functions were investigated by L. Thibault and D. Zagrodny [**926**] for applications to the study of subdifferential determination (or integration of subdifferentials) of such functions. The proofs of Proposition 11.30 and its Corollary 11.31 are taken from Bernard and Thibault [**94**]. Regarding the differentiability of Moreau envelopes of primal lower regular functions f at $\overline{x}$, an equivalent form of such a property in the Hilbert setting was first obtained by L. Thibault and D. Zagrodny in their 1995 paper [**926**] (submitted in 1993): Lemma 4.4 in [**926**] showed for suitable values of $\lambda > 0$ that the function $e_\lambda f + \frac{1}{2\lambda}\|\cdot\|^2$ is convex near $\overline{x}$, hence $\partial_F(e_\lambda f)(x) \neq \emptyset$ for x near $\overline{x}$, and on the other hand for x near $\overline{x}$ one also has $\partial_F(-e_\lambda f)(x) \neq \emptyset$ since $e_\lambda f$ is known to be linearly semiconcave (see Lemma 11.24), so altogether $e_\lambda f$ is Fréchet differentiable near $\overline{x}$. Lemma 11.34 is taken from the second paragraph of the proof of Proposition 13.37 in the book of R. T. Rockafellar and R. J-B. Wets [**865**]. Lemma 11.33 and the assertion (a) of Theorem 11.35 correspond to Propositions 3.1 and 3.2 respectively in the paper [**96**] of F. Bernard, L. Thibault and D. Zagrodny.

Lemma 11.37 is due to R. A. Poliquin and R. T. Rockafellar [**809**, Lemma 4.1]. Theorem 11.40 was stated by S. Marcellin and L. Thibault [**689**]; its proof given here follows [**689**] where basic ideas from [**94**] were employed. Another demonstration of Theorem 11.40 can be found in M. Mazade and L. Thibault [**699**, Theorem 4.9].

The arguments for (f) and (g) in Theorem 11.40 follow those in the proof of Proposition 2.8 in [**699**]. Other (finite-dimensional complements) concerning the $\mathcal{C}^1$ property of $e_\lambda f$ and the continuity of $P_\lambda f$ can be found in Example 3.14 of the paper [**969**] by X. Wang.

The subdifferential determination/integration result in Theorem 11.42 was first proved for primal lower regular functions on $\mathbb{R}^n$ by R. A. Poliquin [**805**, Theorem 4.1] (where such functions were called *primal lower nice*). The extension of this result to Hilbert spaces was realized with another method by L. Thibault and D. Zagrodny [**926**, Theorem 4.3]; the statement of the extension corresponds to the assertion (a) of Theorem 11.41. The more general result in the assertion (b) of Theorem 11.41 is due to F. Bernard, L. Thibault and D. Zagrodny [**96**]. In the book the proofs of assertions (a) and (b) of Theorem 11.41 follow the arguments for Theorem 3.1 and Theorem 3.2 respectively in the paper [**96**] of Bernard, Thibault and Zagrodny.

Prox-regular functions were introduced by R. A. Poliquin and R. T. Rockafellar in their 1996 paper [**809**]. One of the main motivations in [**809**] was to show that, for a function f in that class, its twice epi-differentiability at $\overline{x}$ for a vector $\overline{v} \in \partial_L f(\overline{x})$ is equivalent to the proto-differentiability of the multimapping $\partial_L f$ at $\overline{x}$ for $\overline{v}$. Given a function $f : \mathbb{R}^n \to \mathbb{R} \cup \{-\infty, +\infty\}$ and $(\overline{x}, \overline{v})$ in $\operatorname{gph} \partial_L f$, consider for each real $t > 0$ the following second-order difference quotient function $\Delta_t^2(\overline{x}|\overline{v}) : \mathbb{R}^n \to \mathbb{R} \cup \{-\infty, +\infty\}$ given by

$$\Delta_t^2(\overline{x}|\overline{v})(h) := \frac{f(\overline{x} + th) - f(\overline{x}) - t\langle \overline{v}, h\rangle}{\frac{1}{2}t^2} \quad \text{for all } h \in \mathbb{R}^n.$$

When the epigraphs $\operatorname{epi} \Delta_t^2(\overline{x}|\overline{v})(\cdot)$ Painlevé-Peano-Kuratowski converge as $t \downarrow 0$, the limit is the epigraph of some function $d^2 f(\overline{x}|\overline{v}) : \mathbb{R}^n \to \mathbb{R} \cup \{-\infty, +\infty\}$, called the *twice epiderivative* of f at $\overline{x}$ for $\overline{v}$; in such a case one says that f is *twice epi-differentiable* at $\overline{x}$ for $\overline{v}$. One also says that a multimapping $M : \mathbb{R}^n \rightrightarrows \mathbb{R}^m$ is *proto-differentiable* at a point $\overline{x} \in \operatorname{Dom} M$ for a vector $\overline{v} \in M(\overline{x})$ if the graphs of $\operatorname{gph} \Delta_t M(\overline{x}|\overline{v})$ Painlevé-Peano-Kuratowski converge as $t \downarrow 0$, where $\Delta_t M(\overline{x}|\overline{v})$:

$\mathbb{R}^n \rightrightarrows \mathbb{R}^m$ is given for each real $t > 0$ by

$$\Delta_t M(\overline{x}|\overline{v})(h) = \frac{M(\overline{x} + th) - \overline{v}}{t} \quad \text{for all } h \in \mathbb{R}^n.$$

In such a case the limit is the graph of some multimapping $DM(\overline{x}|\overline{v}) : \mathbb{R}^n \rightrightarrows \mathbb{R}^m$, called the *proto-derivative* of M at $\overline{x}$ for $\overline{v}$. Those concepts of twice epi-differentiablity and proto-differentiability were utilized by R. T. Rockafellar [**862, 863, 864**] with convex functions and their subdifferential multimappings. R. A. Poliquin and R. T. Rockafellar [**809**, Theorem 6.1] extended a Rockafellar's result for convex functions in [**864**] to the context of prox-regular functions in showing the following: If a function $f : \mathbb{R}^n \to \mathbb{R} \cup \{-\infty, +\infty\}$ is prox-regular and subdifferentially continuous at $\overline{x}$ for $\overline{v} \in \partial_L f(\overline{x})$, then the multimapping $\partial_L f : \mathbb{R}^n \rightrightarrows \mathbb{R}^n$ is proto-differentiable at $\overline{x}$ for $\overline{v}$ if and only if f is twice epi-differentiable at $\overline{x}$ for $\overline{v}$. It is also shown in [**809**, Theorem 6.1] that in such a case the proto-derivative of the multimapping $\partial_L f$ at $\overline{x}$ for $\overline{v}$ coincides with the L-subdifferential of the function $\frac{1}{2}d^2 f(\overline{x}|\overline{v})(\cdot)$. Other results related to that feature can be found in [**810, 811, 812**].

Definition 11.43 of *prox-regular functions* is due to R. A. Poliquin and R. T. Rockafellar [**809**, Definition 1.1], and Definition 11.44 of subdifferentially continuous functions is also taken from the same paper [**809**] (see Definition 2.1 therein). Theorem 11.50 was first demonstrated by Poliquin and Rockafellar [**809**, Theorem 3.2] in finite dimensions; the proof there used the compactness of closed balls. Its extension to Banach spaces, as stated and proved in the book, was obtained by F. Bernard and L. Thibault [**93**]; a previous different proof in the Hilbert setting was given by Bernard and Thibault [**94**, Theorem 3.4]. Theorem 11.53 on $\mathcal{C}^{1,1}$ property of Moreau envelopes of prox-regular functions was first noticed and established by Poliquin and Rockafellar [**809**, Theorem 4.4] for functions on $\mathbb{R}^n$. The Hilbert version was established by Bernard and Thibault [**94**, Proposition 5.3].

CHAPTER 12

Singular points of nonsmooth functions

12.1. Singular points of nonsmooth mappings

Let X and Y be two normed spaces, U be a nonempty open set of X and $f : U \to Y$ be an arbitrary mapping. The *Dini bilateral directional and one-sided directional derivatives* of f at $x \in U$ in the direction $h \in X$ are defined respectively by

$$d_D f(x;h) := \lim_{t\to 0} t^{-1}[f(x+th)-f(x)] \quad \text{and} \quad d_D^+ f(x;h) := \lim_{t\downarrow 0} t^{-1}[f(x+th)-f(x)].$$

We recall (see (2.41) in Chapter 2) that the Dini one-sided directional derivative $d_D^+ f(x;h)$ (when it exists) is also denoted by $f'(x;h)$.

Similarly, the *Hadamard directional and one-sided directional derivatives* of f at x in the direction h are defined respectively by

$$d_H f(x;h) := \lim_{v\to h, t\to 0} t^{-1}[f(x+tv) - f(x)]$$

$$\text{and} \quad d_H^+ f(x;h) := \lim_{v\to h, t\downarrow 0} t^{-1}[f(x+tv) - f(x)].$$

Obviously, $d_D f(x;h)$ (resp. $d_H f(x;h)$) exists if and only if $d_D^+ f(x;h)$ and $d_D^+ f(x;-h)$ exist (resp. $d_H^+ f(x;h)$ and $d_H^+ f(x;-h)$ exist) and satisfy the equality

$$d_D^+ f(x;-h) = -d_D^+ f(x;h) \quad (\text{resp. } d_H^+ f(x;-h) = -d_H^+ f(x;h)).$$

Whenever $d_H^+ f(x;h)$ exists (resp. $d_H f(x;h)$ exists) in Y for all $h \in X$, it is easily seen that the mapping

$$d_H^+ f(x;\cdot) \text{ (resp. } d_H f(x;\cdot)) \text{ is continuous on } X. \tag{12.1}$$

If f is Lipschitz continuous near x, then the equalities

$$d_D f(x;\cdot) = d_H f(x;\cdot) \quad \text{and} \quad d_D^+ f(x;\cdot) = d_H^+ f(x;\cdot)$$

hold true provided the left-hand sides exist.

DEFINITION 12.1. Given the Lipschitz mapping $f : X \to Y$ between normed spaces X, Y and given $p \in \mathbb{N}$ with $p < \dim X$, the set $\Sigma_{D,p}^Y(f)$ (resp. $\Sigma_{H,p}^Y(f)$) of *Dini* (resp. *Hadamard*) *singular points of mappings* is defined as the set of points $x \in U$ for which there exists a p-dimensional vector subspace F (depending on x) such that for every nonzero $h \in F$ the one-sided directional derivative $d_D^+ f(x;h)$ (resp. $d_H^+ f(x;h)$) exists but the two-sided $d_D f(x;h)$ (resp. $d_H f(x;h)$) does not exist. If f is extended real-valued, we will merely write $\Sigma_{D,p}(f)$ (resp. $\Sigma_{H,p}(f)$).

It is worth noticing that $x \in \Sigma_{D,p}^Y(f)$ if and only if the above properties hold true for some vector subspace F of X satisfying $\dim F \geq p$.

When $f : U \to \mathbb{R} \cup \{-\infty, +\infty\}$ is a function into $\mathbb{R} \cup \{-\infty, +\infty\}$ which is finite at $x \in U$, in addition to the above concepts we need to recall the lower Dini and Hadamard directional derivatives (see (2.40) and (2.41)) and to introduce the similar concepts of upper Dini and upper Hadamard directional derivatives:

$$\underline{d}_D^+(x;h) := \liminf_{t\downarrow 0} t^{-1}[f(x+th)-f(x)]$$

$$\text{and} \quad \overline{d}_D^+(x;h) := \limsup_{t\downarrow 0} t^{-1}[f(x+th)-f(x)],$$

$$\underline{d}_H^+ f(x;h) := \liminf_{v\to h, t\downarrow 0} t^{-1}[f(x+tv)-f(x)]$$

$$\text{and} \quad \overline{d}_H^+ f(x;h) := \limsup_{v\to h, t\downarrow 0} t^{-1}[f(x+tv)-f(x)].$$

As already said just past Definition 2.119, the function $\underline{d}_H^+(x;\cdot)$ is lower semicontinuous on X and the equalities

$$\underline{d}_H^+ f(x;\cdot) = \underline{d}_D^+ f(x;\cdot) \quad \text{and} \quad \overline{d}_H^+ f(x;\cdot) = \overline{d}_D^+ f(x;\cdot)$$

hold true whenever f is Lipschitz near x.

For a nonempty convex set C of X, the dimension of the affine hull of C is the *dimension* of C and is denoted by $\dim C$, so for all $c \in C$

$$\dim C = \dim\big(\mathrm{Vect}\,(C-c)\big) = \dim\big(\mathrm{Vect}(C-C)\big),$$

where $\mathrm{Vect}\,(C-c) := \mathrm{span}\,(C-c)$ denotes the vector subspace of X spanned by the set $C-c$ (which contains zero).

DEFINITION 12.2. For a real-valued locally Lipschitz function $f: U \to \mathbb{R}$ on an open set U of the normed space X, a point $x \in U$ is called a *singular point of order* $p \in \mathbb{N}$ *of* f *relative to the* C*-subdifferential* provided $\dim \partial_C f(x) \geq p$; the set of such points is denoted by $\Sigma_p(f)$, that is,

$$\Sigma_p(f) := \{x \in U : \dim \partial_C f(x) \geq p\}.$$

When there is no risk of ambiguity, we will just say singular point of order p. A singular point of f of order 1 is just called a *singular point* of f, and we will write $\Sigma(f)$ in place of $\Sigma_1(f)$.

Before establishing the relationship between $\Sigma_p(f)$ and $\Sigma_{D,p}(f)$ for real-valued Clarke subregular locally Lipschitz functions, we need two properties concerning the dimension of a vector subspace F, the codimension of another specific subspace and the codimension of the orthogonal $F_\perp$ in the dual space.

Let E be a Hausdorff locally convex space, F, G be closed vector subspaces and $p \in \mathbb{N}$. Recall that the codimension $\mathrm{codim}\, G$ of G in E is the dimension of the quotient vector space E/G. If $F \cap G = \{0\}$, then

$$\dim F \geq p \implies \mathrm{codim}(E/G) \geq p. \tag{12.2}$$

To justify that, assume $\dim F \geq p$ and take $e_1,\cdots,e_p$ in F linearly independent. Denote by $[v]$ the equivalent class in E/G of any $v \in E$. If $\lambda_1,\cdots,\lambda_p$ are reals such that

$$\lambda_1[e_1] + \cdots + \lambda_p[e_p] = 0, \text{ or equivalently } [\lambda_1 e_1 + \cdots + \lambda_p e_p] = 0,$$

this means that $\lambda_1 e_1 + \cdots + \lambda_p e_p \in G$. It follows that

$$\lambda_1 e_1 + \cdots + \lambda_p e_p \in F \cap G,$$

thus $\lambda_1 e_1 + \cdots + \lambda_p e_p = 0$ since $F \cap G = \{0\}$ by assumption. By linear independence of $e_1,\cdots,e_p$ we deduce that $\lambda_i = 0$ for any $i = 1,\cdots,p$. It results that the vectors

$[e_1],\cdots,[e_p]$ in E/G are linearly independent, hence $\dim(E/G) \geq p$, so $\operatorname{codim} G \geq p$ as desired.

Denoting by $F_\perp$ the orthogonal of F in the topological dual E^* of E and considering the codimension $\operatorname{codim}(F_\perp)$ of $F_\perp$ in E^*, one also has

$$\dim F \geq p \iff \operatorname{codim}(F_\perp) \geq p. \tag{12.3}$$

Indeed, assume first $\dim F \geq p$. There exist $e_1,\cdots,e_p$ in F linearly independent, so for each $i \in I := \{1,\cdots,p\}$ denoting $I_i := I \setminus \{i\}$, the element e_i does not belong to the vector subspace in E spanned by all the elements e_j with $j \in I_i$, and this spanned vector subspace is closed. By the Hahn-Banach theorem there exist $e_i^* \in E^*$ such that $\langle e_i^*, e_j\rangle = 1$ if $j = i$ and $\langle e_i^*, e_j\rangle = 0$ if $j \neq i$. Denoting by $[v^*]$ the equivalent class in $E^*/F_\perp$ of any element $v^* \in E^*$, we claim that $[e_1^*],\cdots,[e_p^*]$ are linearly independent. Take $\lambda_1,\cdots,\lambda_p$ in $\mathbb{R}$ with $\lambda_1[e_1^*] + \cdots + \lambda_p[e_p^*] = 0$. The latter is equivalent to $\lambda_1 e_1^* + \cdots + \lambda_p e_p^* \in F_\perp$, so for each $i = 1,\cdots,p$

$$0 = \langle \lambda_1 e_1^* + \cdots + \lambda_p e_p^*, e_i\rangle = \lambda_i,$$

and this justifies the claim that $[e_1],\cdots,[e_p]$ in $E^*/F_\perp$ are linearly independent. This ensures that

$$\operatorname{codim}(F_\perp) = \dim(E^*/F_\perp) \geq p.$$

Conversely, assume that $\dim(E^*/F_\perp) \geq p$ and choose $[e_1^*],\cdots,[e_p^*]$ linearly independent in $E^*/F_\perp$, which means that for $\lambda_i \in \mathbb{R}$

$$\big(\lambda_1 e_1^* + \cdots + \lambda_p e_p^* \in F_\perp\big) \Longrightarrow \lambda_1 = \cdots = \lambda_p = 0. \tag{12.4}$$

For each $i \in I$ (with I as defined above) denoting by G_i the vector subspace in E^* spanned by all e_j^* with $j \in I_i := I \setminus \{i\}$, we have $e_i^* \notin G_i + F_\perp$. Indeed, the inclusion $e_i \in G_i + F_\perp$ would furnish $e^* \in F_\perp$, α and α_j in $\mathbb{R}$ for all $j \in I_i$ such that $e_i^* = \alpha e^* + \sum_{j\in I_i} \alpha_j e_j^*$, and hence $e_i^* - \sum_{j\in I_i} \alpha_j e_j^* \in F_\perp$ which would contradict (12.4). Then, endowing E^* with the $w(E^*, E)$ topology and noting that $G_i + F_\perp$ is $w(E^*, E)$ closed, by the Hahn-Banach theorem again there exists $e_i \in E$ such that

$$\langle e_i^*, e_i\rangle = 1,\ e_i \in (F_\perp)_\perp = F \text{ and } \langle e_i^*, e_j\rangle = 0\ \forall j \in I_i.$$

As above we deduce that $e_1,\cdots,e_p$ are linearly independent, and they belong to F as seen above, hence $\dim F \geq p$.

Now consider any nonempty convex set Q of X^* containing zero, where X^* is the topological dual of a normed space X. Let $Q_\perp$ be its orthogonal in X with respect to the duality pairing $\langle\cdot,\cdot\rangle$, that is, the set of $x \in X$ such that $\langle x^*, x\rangle = 0$ for all $x^* \in Q$. Endow X^* with the $w(X^*, X)$ topology and put $F := \operatorname{cl}_{w^*}(\operatorname{Vect} Q)$. It is readily seen that $\dim Q \geq p$ if and only if $\dim F \geq p$. It then follows from the equivalence (12.3) above (with $E := X^*$ endowed with the $w(X^*, X)$ topology) that

$$\dim Q \geq p \iff \operatorname{codim}(Q_\perp) \geq p. \tag{12.5}$$

PROPOSITION 12.3. *Let U be a nonempty open set of the normed space X and $f : U \to \mathbb{R}$ be a real-valued locally Lipschitz function which is Clarke subregular. For $p \in \mathbb{N}$ with $p < \dim X$, the following hold:*
(a) A point $x \in \Sigma_{D,p}(f)$ if and only if $\operatorname{codim}\big((\partial_C f(x) - \partial_C f(x))_\perp\big) \geq p$.
(b) One has the equalities

$$\Sigma_p(f) = \Sigma_{D,p}(f) = \Sigma_{H,p}(f).$$

PROOF. (a) Fix any $x \in U$ and $h \in X$ and observe that

$$\begin{aligned} d_D^+ f(x;h) + d_D^+ f(x;-h) &= f^\circ(x;h) + f^\circ(x;-h) \\ &= \sup\{\langle u^*, h\rangle : u^* \in \partial_C f(x)\} + \sup\{\langle v^*, -h\rangle : v^* \in \partial_C f(x)\} \\ &= \sup\{\langle u^* - v^*, h\rangle : u^*, v^* \in \partial_C f(x)\} \geq 0. \end{aligned}$$

Consequently, $d_D f(x;h)$ exists in $\mathbb{R}$ if and only if $\langle u^* - v^*, h\rangle \leq 0$ for all $u^*, v^* \in \partial_C f(x)$, or equivalently (according to the symmetry) $\langle u^* - v^*, h\rangle = 0$ for all $u^*, v^* \in \partial_C f(x)$. This means, for $Q := \partial_C f(x) - \partial_C f(x)$, that $d_D f(x;h)$ exists in $\mathbb{R}$ if and only if $h \in Q_\perp$. Therefore, $x \in \Sigma_{D,p}(f)$ if and only if there exists a vector subspace F of X with $\dim F \geq p$ such that $F \cap Q_\perp = \{0\}$, which entails by (12.2) that we have $\operatorname{codim}(Q_\perp) \geq p$.
(b) The Lipschitz property of f ensures that $\Sigma_{D,p}(f) = \Sigma_{H,p}(f)$ and the equality between $\Sigma_{D,p}(f)$ and $\Sigma_p(f)$ follows from (a) above and (12.5). □

The first theorem concerns the smallness of the size of the set $\Sigma^Y_{H,p}(f)$ and uses small sets of the type of Lipschitz surface of finite codimension already introduced in Definition 2.245.

THEOREM 12.4 (L. Zajíček). Let Y be a normed space, X be a separable normed space, U be an open set of X and $f : U \to Y$ be a mapping. Then, for each positive integer $p < \dim X$, the set $\Sigma^Y_{H,p}(f)$ can be covered by countably many Lipschitz surfaces of codimension p.

If X is finite-dimensional and $p = \dim X$, then $\Sigma^Y_{H,p}(f)$ is a countable set.

The proof will be a consequence of the next two lemmas. Before stating the first lemma, recall that, for each positive integer $p \leq \dim X$, the assertion (a) of Proposition 2.253 tells us that the set $\mathrm{Svect}_p(X)$ of all finite p-dimensional vector subspaces of X endowed with the metric d_{Svect} is separable whenever the normed space X is separable; the distance d_{Svect} is defined (see (2.89)) by

$$d_{\mathrm{Svect}}(F,G) = \max\left\{\sup_{u \in F \cap \mathbb{S}_X} d(u, G \cap \mathbb{S}_X),\ \sup_{u \in G \cap \mathbb{S}_X} d(u, F \cap \mathbb{S}_X)\right\},$$

where we recall that $\mathbb{S}_X$ denotes the unit sphere $\{u \in X : \|u\| = 1\}$ of X.

LEMMA 12.5. Let Y be a normed space, X be a separable normed space, and $f : U \to Y$ be a mapping from an open set U of X into Y. For each finite-dimensional vector subspace F of X and for any $n, m \in \mathbb{N}$, consider:
(i) the set Λ_F of points $x \in U$ such that, for every nonzero $h \in F$, the one-sided directional derivative $d_H^+ f(x;h)$ exists but $d_H f(x;h)$ does not exit;
(ii) the set $\Lambda_{F,n,m}$ of points $x \in U$ such that, for all $s, t \in]0, 1/m[$, $h \in F$ with $\|h\| = 1$, and $h', h'' \in B_X(h, 1/m)$, one has $x \pm sh' \in U$ and

$$\left\| \frac{f(x+th'') - f(x)}{t} - \frac{f(x+sh') - f(x)}{s} \right\| < 1/(4n)$$

$$\text{and} \quad \left\| \frac{f(x+th'') - f(x)}{t} + \frac{f(x-sh') - f(x)}{s} \right\| > 1/n.$$

Let also $p \in \mathbb{N}$ with $p \leq \dim X$ and $\mathcal{D}_p$ be a dense countable subset of $\mathrm{Svect}_p(X)$ with respect to the distance d_{Svect}. Then the following inclusions hold:

$$\Lambda_F \subset \bigcup_{(n,m) \in \mathbb{N}^2} \Lambda_{F,n,m} \quad \text{for all } F \in \mathrm{Svect}_p(X)$$

$$\text{and} \qquad \bigcup_{F\in \mathrm{Svect}_p(X)} \Lambda_F \subset \bigcup_{F\in\mathcal{D}_p,(n,m)\in\mathbb{N}^2} \Lambda_{F,n,m}.$$

PROOF. Fix $p \in \mathbb{N}$ with $p \le \dim X$, $F \in \mathrm{Svect}_p(X)$, and $x \in \Lambda_F$. For each $h \in F$ with $\|h\| = 1$, put $\rho(h) := \|d^+_H f(x;h) + d^+_H f(x;-h)\|$ and note that $\rho(h) > 0$. The continuity of $d^+_H f(x;\cdot)$ (see (12.1)) and the finite dimension property of F furnish some integer $n \in \mathbb{N}$ such that $\rho(h) > 2/n$ for all $h \in F$ with $\|h\| = 1$. From the definition of $d^+_H f(x;\cdot)$ we can find, for each $h \in F$ with $\|h\| = 1$, some real $\delta(h) > 0$ such that, for all $s \in]0, \delta(h)[$ and $h' \in X$ with $\|h' - h\| < \delta(h)$, we have $x \pm sh' \in U$ as well as

(12.6)
$$\left\|\frac{f(x+sh')-f(x)}{s} - d^+_H f(x;h)\right\| < \frac{1}{8n} \text{ and } \left\|\frac{f(x-sh')-f(x)}{s} - d^+_H f(x;-h)\right\| < \frac{1}{8n}.$$

Taking again the finite dimension property of F into account, choose $h_1,\cdots,h_k$ in X with $\|h_i\| = 1$, for $i = 1,\cdots,k$, such that

$$\{u \in F : \|u\| = 1\} \subset \bigcup_{1\le i\le k} B_X(h_i, \delta(h_i)/4).$$

Choose also $m \in \mathbb{N}$ such that $1/m < (1/4)\min\{\delta(h_1),\cdots,\delta(h_k)\}$.

To obtain both inclusions of the lemma, it suffices to show that, for any G in $\mathrm{Svect}_p(X)$ with $d_{\mathrm{Svect}}(G,F) < 1/m$, we have $x \in \Lambda_{G,n,m}$.

Take any $G \in \mathrm{Svect}_p(X)$ with $d_{\mathrm{Svect}}(G,F) < 1/m$. Consider any $s,t \in]0,1/m[$, any $v \in G$ with $\|v\| = 1$, and any $h', h'' \in B_X(v, 1/m)$. From the inequality $d_{\mathrm{Svect}}(G,F) < 1/m$ choose $\overline{h} \in F$ with $\|\overline{h}\| = 1$ and $\|\overline{h} - v\| < 1/m$. We can choose some $q \in \{1,\cdots,k\}$ such that $\|\overline{h} - h_q\| < \delta(h_q)/4$. This yields

$$\|h' - h_q\| \le \|h_q - \overline{h}\| + \|\overline{h} - v\| + \|v - h'\| < \delta(h_q),$$

and hence on the one hand $x \pm sh' \in U$ and on the other hand, by (12.6),

$$\left\|\frac{f(x+sh')-f(x)}{s} - d^+_H f(x;h_q)\right\| < \frac{1}{8n}, \ \left\|\frac{f(x-sh')-f(x)}{s} - d^+_H f(x;-h_q)\right\| < \frac{1}{8n}.$$

Similarly, we have $\|h'' - h_q\| < \delta(h_q)$ and

$$\left\|\frac{f(x+th'')-f(x)}{t} - d^+_H f(x;h_q)\right\| < \frac{1}{8n}.$$

Combining those inequalities we obtain

$$\left\|\frac{f(x+th'')-f(x)}{t} - \frac{f(x+sh')-f(x)}{s}\right\| < \frac{1}{4n}$$

and

$$\begin{aligned}\left\|\frac{f(x+th'')-f(x)}{t} + \frac{f(x-sh')-f(x)}{s}\right\| &> \|d^+_H f(x;h_q) + d^+_H f(x;-h_q)\| - \frac{1}{4n} \\ &= \rho(h_q) - 1/(4n) > 1/n.\end{aligned}$$

The desired inclusion $x \in \Lambda_{G,n,m}$ is then justified. □

LEMMA 12.6. Let Y be a normed space, X be a separable normed space, and $f : U \to Y$ be a mapping from an open set U of X into Y. For each finite-dimensional vector subspace F of X with $\dim F < \dim X$, each set Λ_F and each set $\Lambda_{F,n,m}$, as defined in the previous lemma, can be covered by countably many Lipschitz F-surfaces.

If X is a finite-dimensional space and $F = X$, then both sets Λ_F and $\Lambda_{F,n,m}$ are countable.

PROOF. According to the first inclusion of the previous lemma it suffices to prove the result for $\Lambda := \Lambda_{F,n,m}$. To obtain that, it is enough, by Propositions 2.248 and 2.43, to show that $F \cap T^B(\Lambda; u) = \{0\}$ for all $u \in \Lambda$. Suppose by contradiction that there is some $u \in \Lambda$ such that $F \cap T^B(\Lambda; u) \neq \{0\}$, so we can choose some $h \in F \cap T^B(\Lambda; u)$ with $\|h\| = 1$. For $t := 1/(2m)$ and $\gamma := |t^{-1}[f(u + th) - f(u)]|$ there exist $h' \in X$ with $\|h' - h\| < 1/m$ and $0 < s < \big(4nm^2(1+\gamma)\big)^{-1}$ such that $x := u + sh' \in \Lambda$. For $y := u + th$, the inclusion $u \in \Lambda = \Lambda_{F,n,m}$ ensures $x, y \in U$ and

$$\left\| \frac{f(y) - f(u)}{t} - \frac{f(x) - f(u)}{s} \right\| < 1/(4n). \tag{12.7}$$

Put $h'' := 2m(y - x)$ and observe that

$$\|h'' - h\| = \|2m(y - x) - 2m(y - u)\| = 2ms\|h'\| < 2m\left(\frac{1}{4m^2}\right)2 = 1/m,$$

thus the equalities $u = x - sh'$ and $y = x + th''$ along with the inclusion $x \in \Lambda = \Lambda_{F,n,m}$ entail, by the second inequality in the definition of $\Lambda_{F,n,m}$, that

$$\left\| \frac{f(y) - f(x)}{t} + \frac{f(u) - f(x)}{s} \right\| > 1/n.$$

The latter inequality and (12.7) give

$$\left\| \frac{f(u) - f(x)}{t} \right\| \geq \left\| \frac{f(y) - f(x)}{t} + \frac{f(u) - f(x)}{s} \right\| - \left\| \frac{f(y) - f(u)}{t} - \frac{f(x) - f(u)}{s} \right\| > \frac{1}{2n},$$

hence $\|f(u) - f(x)\| > 1/(4nm)$. On the other hand, (12.7) and the definition of γ and s imply

$$\|f(u) - f(x)\| = s\left\| \frac{f(u) - f(x)}{s} \right\| \leq s(\gamma + \frac{1}{4n}) < s(\gamma + 1) < \frac{1}{4nm},$$

which contradicts the preceding inequality $\|f(u) - f(x)\| > 1/(4nm)$ and finishes the proof of the lemma. $\square$

PROOF OF THEOREM 12.4. From the definition of $\Sigma^Y_{H,p}(f)$ and from Lemma 12.5 we have

$$\Sigma^Y_{H,p}(f) = \bigcup_{F \in \mathrm{Svect}_p(X)} \Lambda_F \subset \bigcup_{G \in \mathcal{D}_p, (n,m) \in \mathbb{N}^2} \Lambda_{G,n,m},$$

so the result of the theorem follows from the countability of the set $\mathcal{D}_p$ and from Lemma 12.6. $\square$

The theorem concerning the size of the set of singular points (relative to the C-subdifferential) of C-subregular locally Lipschitz functions is a direct consequence of Proposition 12.3 and Theorem 12.4.

THEOREM 12.7 (L. Zajíček: smallness of sets of singular points of C-subregular Lipschitz functions). Let X be a separable normed space, U be an open set of X and $f : U \to \mathbb{R}$ be a C-subregular locally Lipschitz function. Then, for each positive integer $p < \dim X$, the set $\Sigma_p(f)$ can be covered by countably many Lipschitz surfaces of codimension p.

If X is finite-dimensional and $p = \dim X$, then $\Sigma_p(f)$ is a countable set.

Another theorem related to the lower Hadamard directional derivatives can also be established for nonsubregular real-valued functions.

THEOREM 12.8 (L. Zajíček). Let X be a separable normed space, U be an open set of X and $f : U \to \mathbb{R}$ be a real-valued function. For each positive integer $p \le \dim X$, let $\widehat{\Sigma}_{H,p}$ denote the set of points $x \in U$ for which there exists a p-dimensional vector subspace F of X (depending on x) such that for every nonzero $h \in F$ one has $\underline{d}_H f(x; h) > -\infty$ and

$$\underline{d}_H^+ f(x; h) + \underline{d}_H^+ f(x; -h) > 0.$$

Then, for each positive integer $p < \dim X$, the set $\widehat{\Sigma}_{H,p}(f)$ can be covered by countably many Lipschitz surfaces of codimension p.

If X is finite-dimensional and $p = \dim X$, then $\widehat{\Sigma}_{H,p}(f)$ is a countable set.

As for Theorem 12.4 we proceed with two lemmas.

LEMMA 12.9. Let X be a separable normed space and $f : U \to \mathbb{R} \cup \{+\infty\}$ be an extended real-valued function over an open set U of X. For each finite-dimensional vector subspace F of X and $n \in \mathbb{N}$, consider:
(i) the set $\widehat{\Lambda}_F$ of points $x \in U$ such that, for every nonzero $h \in F$, one has $\underline{d}_H f(x; h) > -\infty$ and

$$\underline{d}_H^+ f(x; h) + \underline{d}_H^+ f(x; -h) > 0;$$

(ii) the set $\widehat{\Lambda}_{F,n}$ of points $x \in \widehat{\Lambda}_F$ such that, for all $s, t \in]0, 1/n[$, $h \in F$ with $\|h\| = 1$, and $h', h'' \in B_X(h, 1/n)$, one has $x \pm sh' \in U$ and

$$\frac{f(x + th'') - f(x)}{t} + \frac{f(x - sh') - f(x)}{s} > 1/n.$$

Let also $p \in \mathbb{N}$ with $p \le \dim X$ and $\mathcal{D}_p$ be a dense countable subset of the metric space $(\mathrm{Svect}_p(X), d_{\mathrm{Svect}})$. Then the following equalities hold

$$\widehat{\Lambda}_F = \bigcup_{n \in \mathbb{N}} \widehat{\Lambda}_{F,n} \quad \text{for all } F \in \mathrm{Svect}_p(X) \quad \text{and} \quad \bigcup_{F \in \mathrm{Svect}_p(X)} \widehat{\Lambda}_F = \bigcup_{F \in \mathcal{D}_p} \widehat{\Lambda}_F.$$

PROOF. Fix $p \in \mathbb{N}$ with $p \le \dim X$, $F \in \mathrm{Svect}_p(X)$, and $x \in \widehat{\Lambda}_F$. For each $h \in F$ with $\|h\| = 1$, note, by definition of $\widehat{\Lambda}_F$, that $\rho(h) := \underline{d}_H f(x; h) + \underline{d}_H f(x; -h)$ is well defined in $\mathbb{R} \cup \{+\infty\}$ and $\rho(h) > 0$. The lower semicontinuity of $\underline{d}_H f(x; \cdot)$ and the finite dimension property of F furnish some $\varepsilon \in]0, 1[$ such that $\rho(h) > \varepsilon$ for all $h \in F$ with $\|h\| = 1$. From the definition of $\underline{d}_H f(x; \cdot)$ we can find, for each $h \in F$ with $\|h\| = 1$, some real $\delta(h) > 0$ such that

$$\underline{d}_H^+ f(x; w) > \min\left\{\underline{d}_H^+ f(x; h) - \frac{\varepsilon}{8}, \frac{1}{\varepsilon}\right\} \text{ and } \underline{d}_H^+ f(x; -w) > \min\left\{\underline{d}_H^+ f(x; -h) - \frac{\varepsilon}{8}, \frac{1}{\varepsilon}\right\} \tag{12.8}$$

for all $w \in B_X(h, \delta(h))$, and

$$(12.9) \qquad \frac{f(x+sh')-f(x)}{s} > \min\left\{\underline{d}_H^+ f(x;h) - \frac{\varepsilon}{8}, \frac{1}{\varepsilon}\right\}$$

$$(12.10) \qquad \text{and} \quad \frac{f(x-sh')-f(x)}{s} > \min\left\{\underline{d}_H^+ f(x;-h) - \frac{\varepsilon}{8}, \frac{1}{\varepsilon}\right\}$$

for all $h' \in B_X(h, \delta(h))$ and $s \in]0, \delta(h)[$.

Taking the finite dimension property of F into account, choose $h_1, \cdots, h_k \in X$ with $\|h_i\| = 1$, for $i = 1, \cdots, k$, such that

$$\{u \in F : \|u\| = 1\} \subset \bigcup_{1 \le i \le k} B_X(h_i, \delta(h_i)/4).$$

Choose also $n \in \mathbb{N}$ such that $1/n < (1/4)\min\{\delta(h_1), \cdots, \delta(h_k)\}$ and $1/n < 3\varepsilon/4$.

To obtain the first equality of the lemma, it suffices to show that $x \in \widehat{\Lambda}_{F,n}$. Consider any $s, t \in]0, 1/n[$, any $v \in F$ with $\|v\| = 1$, and any $h', h'' \in B_X(v, 1/n)$. Choose some $q \in \{1, \cdots, k\}$ such that $\|v - h_q\| < \delta(h_q)/4$. This yields

$$\|h' - h_q\| \le \|h_q - v\| + \|v - h'\| < \delta(h_q) \quad \text{and} \quad \|h'' - h_q\| < \delta(h_q) \text{ as well,}$$

and hence by (12.9)

$$\begin{aligned}
&\frac{f(x+th'')-f(x)}{t} + \frac{f(x-sh')-f(x)}{s} \\
&> \min\left\{\underline{d}_H^+ f(x;h_q) - \frac{\varepsilon}{8}, \frac{1}{\varepsilon}\right\} + \min\left\{\underline{d}_H^+ f(x;-h_q) - \frac{\varepsilon}{8}, \frac{1}{\varepsilon}\right\} \\
&\ge \min\left\{\underline{d}_H^+ f(x;h_q) + \underline{d}_H^+ f(x;h_q) - \frac{\varepsilon}{4}, \frac{1}{\varepsilon}\right\} = \min\left\{\rho(h_q) - \frac{\varepsilon}{4}, \frac{1}{\varepsilon}\right\} > \varepsilon - \frac{\varepsilon}{4} > \frac{1}{n}.
\end{aligned}$$

This proves the desired inclusion $x \in \widehat{\Lambda}_{F,n}$.

Concerning the second equality, taking any $G \in \mathrm{Svect}_p(X)$ with $d_{\mathrm{Svect}}(G, F) < 1/n$, it is sufficient to show that $x \in \widehat{\Lambda}_G$. Consider any $v \in G$ with $\|v\| = 1$ and from the inequality $d_{\mathrm{Svect}}(G, F) < 1/n$ choose $\overline{h} \in F$ with $\|\overline{h}\| = 1$ and $\|\overline{h} - v\| < 1/n$. Choose also some $q \in \{1, \cdots, k\}$ such that $\|\overline{h} - h_q\| < \delta(h_q)/4$. This yields $\|v - h_q\| < \delta(h_q)$, and hence by (12.8) $\underline{d}_H^+ f(x;v) > -\infty$ and

$$\begin{aligned}
\underline{d}_H^+ f(x;v) + \underline{d}_H^+ f(x;-v) &> \min\left\{\underline{d}_H^+ f(x;h_q) - \frac{\varepsilon}{8}, \frac{1}{\varepsilon}\right\} + \min\left\{\underline{d}_H^+ f(x;-h_q) - \frac{\varepsilon}{8}, \frac{1}{\varepsilon}\right\} \\
&\ge \min\left\{\rho(h_q) - \frac{\varepsilon}{4}, \frac{1}{\varepsilon}\right\} > 0,
\end{aligned}$$

which translates the desired inclusion $x \in \widehat{\Lambda}_G$ and finishes the proof. $\square$

LEMMA 12.10. Let X be a separable normed space and $f : U \to \mathbb{R}$ be a real-valued function over an open set U of X. For each finite-dimensional vector subspace F of X with $\dim F < \dim X$, the set $\widehat{\Lambda}_F$ of the previous lemma can be covered by countably many Lipschitz F-surfaces.

If X is finite-dimensional and $F = X$, then the set $\widehat{\Lambda}_F$ is countable.

PROOF. According to the first equality of the previous lemma it suffices to prove the result for $\widehat{\Lambda} := \widehat{\Lambda}_{F,n}$. to show To obtain that, it is enough, by Propositions 2.248 and 2.243, that $F \cap T^B(\widehat{\Lambda}; u) = \{0\}$ for all $u \in \widehat{\Lambda}$. Suppose by contradiction that there is some $u \in \widehat{\Lambda}$ such that $F \cap T^B(\widehat{\Lambda}; u) \neq \{0\}$, so we can choose some $h \in F \cap T^B(\widehat{\Lambda}; u)$ with $\|h\| = 1$.

First suppose that $\underline{d}_H^+ f(u; h) = +\infty$. For $t := 1/(2n)$ we have $y := u + th \in U$ by definition of $\widehat{\Lambda}_{F,n}$. Put $\gamma := |f(y) - f(u)|$. By the inclusion $h \in T^B(\widehat{\Lambda}, u)$ and the equality $\underline{d}_H f(u; h) = +\infty$, choose some $h' \in X$ with $\|h' - h\| < 1/(2n)$ and some $s \in]0, 1/(4n^2)[$ such that $x := u + sh' \in \widehat{\Lambda}$ and $\frac{f(x)-f(u)}{s} > 2n\gamma$. Put $h'' := 2n(y - x)$ and observe that

$$\|h'' - h\| = \|2n(y - x) - 2n(y - u)\| = 2ns\|h'\| < 2n\left(\frac{1}{4n^2}\right)2 = 1/n,$$

thus the equalities $u = x - sh'$ and $y = x + th''$ along with the inclusion $x \in \widehat{\Lambda} = \widehat{\Lambda}_{F,n}$ entail

$$\frac{f(y) - f(x)}{t} > \frac{f(x) - f(u)}{s} + 1/n > 2n\gamma.$$

Consequently, we obtain

$$\gamma \geq f(y) - f(u) = (f(y) - f(x)) + f(x) - f(u) > t\frac{f(y) - f(x)}{t} + 0 \geq \frac{1}{2n}2n\gamma = \gamma,$$

which is a contradiction.

Now suppose $\alpha := \underline{d}_H^+ f(u; h) < +\infty$. Then α is finite by definition of $\widehat{\Lambda} = \widehat{\Lambda}_{F,n}$. According to the definition of $\underline{d}_H^+ f(x; h)$ take some $w \in X$ with $\|w\| < 1/(4n)$ and some positive real $t < 1/(4n)$ such that $y := u + t(h + w) \in U$ and

$$\frac{f(y) - f(u)}{t} < \alpha + \frac{1}{4n}.$$

On the other hand, the inequality $\underline{d}_H^+ f(u; h) > \alpha - 1/(4n)$ ensures, for $s > 0$ small enough and $h' \in X$ close enough to h, that $s^{-1}[f(u + sh') - f(u)] > \alpha - 1/(4n)$. This and the inclusion $h \in T^B(\widehat{\Lambda}; u)$ furnish some $h' \in X$ with $\|h' - h\| < 1/(4n)$ and some positive real $s < \dfrac{t}{4n(|\alpha - \frac{1}{4n}| + 1)}$ such that $x := u + sh' \in \widehat{\Lambda}$ and

$$\frac{f(x) - f(u)}{s} > \alpha - \frac{1}{4n}.$$

For $h'' := t^{-1}(y - x)$, we have

$$\|h'' - h\| \leq \left\|\frac{y - x}{t} - \frac{y - u}{t}\right\| + \|w\| \leq \frac{\|x - u\|}{t} + \frac{1}{4n} \leq \frac{2s}{t} + \frac{1}{4n} < \frac{1}{n}.$$

Writing $u = x - sh'$, $y = x + th''$ and $x \in \widehat{\Lambda} = \widehat{\Lambda}_{F,n}$ ensures by definition of $\widehat{\Lambda}_{F,n}$

$$\frac{f(y) - f(x)}{t} > \frac{f(x) - f(u)}{s} + \frac{1}{n} > \alpha - \frac{1}{4n} + \frac{1}{n} > \alpha + \frac{1}{2n}.$$

We then deduce

$$t\left(\alpha + \frac{1}{4n}\right) > f(y) - f(u) = (f(y) - f(x)) + (f(x) - f(u))$$
$$> t\left(\alpha + \frac{1}{2n}\right) + s\left(\alpha - \frac{1}{4n}\right),$$

hence $s|\alpha - \frac{1}{4n}| > t(\frac{1}{4n})$, which contradicts the choice of $s < \dfrac{t}{4n(|\alpha - \frac{1}{4n}| + 1)}$. the proof of the lemma is then finished. □

PROOF OF THEOREM 12.8. From the definition of $\widehat{\Sigma}_{H,p}(f)$ and from the second equality of Lemma 12.9 we have

$$\widehat{\Sigma}_{H,p}(f) = \bigcup_{F \in \mathrm{Svect}_p(X)} \widehat{\Lambda}_F = \bigcup_{G \in \mathcal{D}_p} \widehat{\Lambda}_G,$$

and hence the result of the theorem follows from the countability of $\mathcal{D}_p$ and from Lemma 12.10. □

12.2. Singular points of convex and semiconvex functions

Theorem 12.7 above established that the set of singular points of a locally Lipschitz function f can be covered by countably many Lipschitz surfaces. Our aim in this section is to show that, if the function f is addition convex (resp. semiconvex), the Lipschitz surfaces in the cover can be required to be defined by differences of convex functions (resp. semiconvex functions as given in Definition 10.1). This is a significant additional property.

DEFINITION 12.11. Let X be a normed space, F be a finite-dimensional vector space, and $\omega : [0, +\infty[\to [0, +\infty[$ be a finite modulus, that is, a nondecreasing function with $\lim_{r \downarrow 0} \omega(r) = 0 = \omega(0)$ (see Definition 10.1). Let also an open convex set U of X.
(i) A function $f : U \to \mathbb{R}$ is said to be a *Lipschitz DC, DSC or DSC_ω function* on U provided f can be represented as $f = f_1 - f_2$ with two Lipschitz functions $f_1, f_2 : U \to \mathbb{R}$ which are convex on U, semiconvex on U, or semiconvex on U with modulus $K\omega(\cdot)$ for some real constant $K > 0$.

So, f is Lipschitz DSC if and only if it is Lipschitz DSC_ω for some finite modulus function ω.

When $U = X$ (or there is no risk of confusion) one says that f is Lipschitz DC, DSC, or DSC_ω.
(ii) A mapping $f : U \to F$ is said to be Lipschitz DC (resp. DSC, DSC_ω) on U, when it is so for all functions $v^* \circ f$ with $v^* \in F^*$.

Some simple stability properties need to be stated:

PROPOSITION 12.12. Let U be an open convex set of the normed space X, F be a finite-dimensional vector space, and $\omega : [0, +\infty[\to [0, +\infty[$ be a nondecreasing (finite) modulus function with $\lim_{r \downarrow 0} \omega(r) = 0 = \omega(0)$. The following hold:
(a) Any linear combination of Lipschitz DC (resp. DSC_ω, DSC) functions from U into $\mathbb{R}$ is Lipschitz DC (resp. DSC_ω, DSC) on U.
(b) Any linear combination of Lipschitz DC (resp. DSC_ω, DSC) mappings from U into F is Lipschitz DC (resp. DSC_ω, DSC) on U.
(c) Given a vector basis $v_1^*, \cdots, v_p^*$ of the dual space F^*, a mapping $f : U \to F$ is Lipschitz DC (resp. DSC_ω, DSC) if and only if $v_i^* \circ f$ is so for $i = 1, \cdots, p$.
(d) A mapping $f : U \to F$ is DSC on U if and only if it is DSC_ω for some finite nondecreasing modulus function ω.
(e) Let U and W be open convex sets of normed spaces X and Y respectively, $A : Y \to X$ be a continuous affine mapping with $A(W) \subset U$, and $f : U \to F$ be a Lipschitz DC (resp. DSC_ω, DSC) mapping. Then the mapping $f \circ A$ has the same property on U.

PROOF. The assertion (a) is obvious and the assertions (b) and (c) follow directly from (a); the assertion (d) is a direct consequence of (c). For a function

$g : U \to \mathbb{R}$, the function $g \circ A$ is convex or $K\omega(\cdot)$-semiconvex on U for some constant $K > 0$ whenever g is convex or $\omega(\cdot)$-semiconvex on U (see Proposition 10.20 for the semiconvex result). Then, this and the assertion (c) guarantee (e). □

Through DC or DSC mappings we define the Lipschitz surfaces which are DC or DSC.

DEFINITION 12.13. Let X be a normed space, F be a nonzero finite-dimensional subspace, and E be a topological vector complement of E in X, so $X = E \oplus F$. When in addition to the properties in Definition 2.245 the Lipschitz mapping $f : E \to F$ is required to be DC, DSC or DSC_ω, one says that the Lipschitz F-surface $C = \{x + f(x) : x \in E\}$ is a Lipschitz DC, DSC or DSC_ω F-surface of the normed space X.

The assertion (c) below provides another description of such surfaces and (b) shows, as in Proposition 2.247, that the property in Definition 12.13 holds for every topological vector complement E of F.

PROPOSITION 12.14. Let X be a normed space, F a finite-dimensional vector subspace of dimension p with $1 \leq p < \dim X$. Let C be a subset of X and $\omega : [0, +\infty[\to [0, +\infty[$ be a nondecreasing modulus function with $\lim_{r \downarrow 0} \omega(r) = 0 = \omega(0)$. The following assertions are equivalent:
(a) the set C is a Lipschitz DC (resp. SDC, SDC_ω) F-surface;
(b) for any topological vector complement E of F in X, there is a Lipschitz DC (resp. SDC, SDC_ω) mapping $f : E \to F$ such that $C = \{x + f(x) : x \in E\}$.
(c) for any topological vector complement E of F in X and for any $x_1^*, \cdots, x_p^* \in X^*$ such that $x_i^* \in E^\perp$ and their restrictions $x_1^*|_F, \cdots, x_p^*|_F$ to F form a basis of F^*, there are Lipschitz DC (resp. DSC, DSC_ω) functions $f_1, \cdots, f_p : E \to \mathbb{R}$ for which

$$C = \{x \in X : x_i^*(x) = f_i(\pi^{E,F}(x)),\ i = 1, \cdots, p\},$$

where $\pi^{E,F}$ denotes the projector on E along F.

PROOF. Let us start with (a) ⇒ (b). Suppose (a), that is, C is a Lipschitz DC (resp. DSC_ω) F-surface. There exist a topological vector complement E_0 of F in X and a Lipschitz DC (resp. DSC_ω) mapping $f_0 : E_0 \to F$ such that $C = \{x + f_0(x) : x \in E_0\}$. Let E be any topological vector complement of F in X. As already seen in the proof of Proposition 2.247 the mapping $A := \pi^{E_0,F}|_E : E \to E_0$ is a topological linear isomorphism. Further, according to the definition of A we have $A(x) - x \in F$ for all $x \in E$. Then we can define a continuous linear mapping $L : E \to F$ with $L(x) := A(x) - x$ for all $x \in E$. Writing, according to the surjectivity of $A : E \to E_0$, the set C as $C = \{A(x) + f_0(A(x)) : x \in E\}$ we see that $C = \{x + L(x) + f_0(A(x)) : x \in E\}$. Since the mapping $f : E \to F$ with $f(x) = L(x) + f_0(A(x))$ for all $x \in E$ is DC (resp. DSC_ω) (see (e) and (b) in Proposition 12.12 for the DSC_ω property), the assertion (b) follows.

Let us prove (b) ⇒ (c). Suppose that (b) holds for C. Let E and $x_1^*, \cdots, x_p^*$ in X^* satisfying the assumptions of (c). By (b) choose a Lipschitz DC (resp. DSC_ω) mapping $f : E \to F$ such that $C = \{u + f(u) : u \in E\}$. For each $i = 1, \cdots, p$, the function $f_i := (x_i^*|_F) \circ f : E \to \mathbb{R}$ is DC (resp. DSC_ω) according to Definition 12.11, and since (by the assumption (c)) $x_1^*|_F, \cdots, x_p^*|_F$ form a basis of the vector space F^* with x_i^* null on E, we have $C = \{x \in X : x_i^*(x) = f_i(\pi^{E,F}(x)), \forall i = 1, \cdots, p\}$, which translates (c).

It remains to show (c) $\Rightarrow$ (a). Let E be a topological vector complement of F in X, so $X = E \oplus F$ and the projectors $\pi^{F,E} : X \to F$ and $\pi^{E,F} : X \to E$ are continuous. Let $u_1, \cdots, u_p$ be a basis of the vector space F and $u_1^*, \cdots, u_p^*$ the dual basis of F^*. The linear functionals $x_i^* := u_i^* \circ \pi^{F,E}$ are continuous, that is, $x_i^* \in X^*$, and they are null on E, that is, $x_i^* \in E^\perp$. So, by (c) there are Lipschitz DC (resp. DSC_ω) functions $f_i : E \to \mathbb{R}$ such that, for $x \in X$, one has $x \in C$ if and only if $x_i^*(x) = f_i(\pi^{E,F}(x))$ for $i = 1, \cdots, p$. The mapping $f : E \to F$ with $f(y) := f_1(y)u_1 + \cdots + f_p(y)u_p$ is DC (resp. DSC_ω) on E by (c) in Proposition 12.12 and by what precedes $C = \{y + f(y) : y \in E\}$, hence (a) holds. $\square$

For sets covered by countably many Lipschitz DSC surfaces, Lemma 12.16 will establish the existence of more suitable coverings with countably many Lipschitz DSC surfaces. To do so, the following lemma is needed.

LEMMA 12.15. Let $(\omega_i)_{i\in\mathbb{N}}$ be a sequence of nondecreasing modulus functions from $[0, +\infty[$ into $[0, +\infty[$ with $\lim_{r\downarrow 0} \omega_i(r) = 0 = \omega_i(0)$ for each $i \in \mathbb{N}$. Then there is a nondecreasing concave modulus function $\omega : [0, +\infty[\to [0, +\infty[$ with $\lim_{r\downarrow 0} \omega(r) = 0 = \omega(0)$ and a decreasing sequence of reals $(\delta_i)_{i\in\mathbb{N}}$ such that, for each $i \in \mathbb{N}$, $\delta_i > 0$ and $\omega(r) \geq \omega_i(r)$ for all $r \in]0, \delta_i[$.

PROOF. According to the property $\lim_{r\downarrow 0} \omega_1(r) = 0$ choose some real $\delta_1 > 0$ such that $\omega_1(r) < 1$ for all $r \in]0, \delta_1[$, and by induction take $\delta_i \in]0, \delta_{i-1}[$ such that $\max\{\omega_1(r), \cdots, \omega_i(r)\} < 1/i$ for all $r \in]0, \delta_i[$. For each $i \in \mathbb{N}$, denote by α_i the increasing affine function on $[0, +\infty[$ with $\alpha_i(0) = 1/i$ and $\alpha_i(\delta_i) = 1$. The function $\omega : [0, +\infty[\to [0, +\infty[$ with $\omega(r) := \inf_{i\in\mathbb{N}} \alpha_i(r)$ is obviously nondecreasing and concave along with $\omega(0) = 0$. Further, given $\varepsilon > 0$ and choosing an integer $i_0 > 1$ with $1/i_0 < \varepsilon$ and choosing also $r_0 > 0$ with $\alpha_{i_0}(r_0) = \varepsilon$, we see that for every $r \in]0, r_0[$ we have $\alpha_{i_0}(r) \leq \varepsilon$, and hence $0 < \omega(r) \leq \varepsilon$, so $\omega(r) \to 0$ as $r \downarrow 0$. Fix any $k \in \mathbb{N}$ and take an arbitrary $r \in]0, \delta_k[$. Consider any $i \in \mathbb{N}$. If $i \leq k$, we have $0 < r < \delta_i$ (because $r < \delta_k \leq \delta_i$), so $\alpha_i(r) > \alpha_i(0) = 1/i \geq 1/k > \omega_k(r)$, the latter inequality being due to the choice of δ_k and the inclusion $r \in]0, \delta_k[$. Suppose now $i > k$. Then $\delta_i < \delta_k$. If $r < \delta_i$, then

$$\alpha_i(r) \geq \alpha_i(0) = 1/i \geq \max\{\omega_1(r), \cdots, \omega_k(r), \cdots, \omega_i(r)\} \geq \omega_k(r).$$

Finally, if $r \geq \delta_i$, we have

$$\alpha_i(r) \geq \alpha_i(\delta_i) = 1 > 1/k > \omega_k(r),$$

the last inequality being due again to the choice of δ_k and the inclusion $r \in]0, \delta_k[$. We then have $\alpha_i(r) \geq \omega_k(r)$, which gives the desired inequality $\omega(r) \geq \omega_k(r)$ for every $r \in]0, \delta_k[$. $\square$

The concepts of uniformly convex norm and uniformly smooth norm, as well as the related moduli of convexity and smoothness and the related power type property, are already recalled in (10.14) and (10.15). More on these concepts and on various properties of uniformly convex Banach spaces will be developed in Section 18.1 in Chapter 18.

LEMMA 12.16. Let X be a separable uniformly convex Banach space which admits an equivalent uniformly smooth norm $\|\cdot\|$ with modulus of smoothness of power type 2. Then, for each integer p with $1 \leq p < \dim X$, a subset C of X can be covered by countably many Lipschitz DSC surfaces of X of codimension p if and only if there exists some nondecreasing (finite) modulus function $\omega : [0, +\infty[\to [0, +\infty[$ with $\lim_{r \downarrow 0} \omega(r) = 0 = \omega(0)$ for which C is covered by countably many Lipschitz DSC_ω surfaces of X of codimension p.

PROOF. The implication $\Leftarrow$ being obvious, let us prove the converse one. Let C be a subset of X for which there exists a sequence $(C_i)_{i \in \mathbb{N}}$ of subsets of X such that $C \subset \bigcup_{i \in \mathbb{N}} C_i$ and each set C_i is a Lipschitz $\mathcal{DSC}_{\omega_i}$ surface of codimension p for some nondecreasing modulus function $\omega_i : [0, +\infty[\to [0, +\infty[$ with $\lim_{r \downarrow 0} \omega_i(r) = 0$. Consider by Lemma 12.15 above some concave nondecreasing modulus function $\omega : [0, +\infty[\to [0, +\infty[$ with $\lim_{r \downarrow 0} \omega(r) = 0 = \omega(0)$, and some decreasing sequence of positive reals $(\delta_i)_{i \in \mathbb{N}}$ such that

$$\omega(r) \geq \omega_i(r) \quad \text{for all } r \in]0, \delta_i[. \tag{12.11}$$

Fix any $i \in \mathbb{N}$ and let us show that $A := C_i$ can be covered by countably many Lipschitz DSC_ω surfaces of X of codimension p. Since A is a Lipschitz DSC_{ω_i} surface of codimension p, there exist a p-dimensional subspace F of X, a topological vector complement E of F in X and a $\mathcal{DSC}_{\omega_i}$ mapping $\varphi : E \to F$ such that $A = \{x + \varphi(x) : x \in E\}$. Let $v_1, \cdots, v_p$ be a vector basis of F and $v_1^*, \cdots, v_p^*$ the dual basis in F^*, that is, $v_k^*(v_k) = 1$ and $v_k^*(v_j) = 0$ for $j \neq k$. For each $k = 1, \cdots, p$ there are $\Lambda_k > 0$ and functions $g_k, h_k : E \to \mathbb{R}$ which are Lipschitz and semiconvex on E both with $\Lambda_k \omega_i(\cdot)$ as a modulus of semiconvexity and such that $v_k^* \circ \varphi = g_k - h_k$.

By the separability of E, choose a sequence $(A_m)_{m \in \mathbb{N}}$ of subsets of E with $\bigcup_{m \in \mathbb{N}} A_m = E$ and $\operatorname{diam} A_m < \delta_i$ for all $m \in \mathbb{N}$. Fix $m \in \mathbb{N}$ and $k \in \{1, \cdots, p\}$, and take for each $a \in A_m$ some $\zeta(a) \in \partial_C g_k(a)$, so by the characterization of subgradients of semiconvex functions (see Theorem 10.29)

$$\langle \zeta(a), x - a \rangle \leq g_k(x) - g_k(a) + \Lambda_k \|x - a\| \omega_i(\|x - a\|) \quad \text{for all } x \in E,$$

hence by (12.11)

$$\langle \zeta(a), x - a \rangle \leq (g_{k|_{A_m}})(x) - (g_{k|_{A_m}})(a) + \Lambda_k \|x - a\| \omega(\|x - a\|) \quad \text{for all } x \in A_m;$$

further, denoting by γ_k a Lipschitz constant of g_k on E we have $\|\zeta(a)\| \leq \gamma_k$ for all $a \in A_m$. Observe that the Banach space E is uniformly convex and the restriction of the norm $\|\cdot\|$ to E is uniformly smooth with modulus of smoothness of power type 2. Considering the bounded set A_m of E, the function $g_{k|_{A_m}}$ and the concave modulus function $\Lambda_k \omega(\cdot)$, Proposition 10.58 yields an extension function $G_{k,m}$ of $g_{k|_{A_m}}$ to E which is semiconvex on E with modulus $\Gamma_{k,m} \omega(\cdot)$ for some $\Gamma_{k,m} > 0$.

Similarly, the function $h_{k|_{A_m}}$ has a Lipschitz extension $H_{k,m} : E \to \mathbb{R}$ to E which is semiconvex on E with modulus $\Theta_{k,m} \omega(\cdot)$ for some $\Theta_{k,m} > 0$. Setting

$$\Phi_m := \sum_{k=1}^{p} (G_{k,m} - H_{k,m}) v_k \quad \text{and} \quad S_m := \{x + \Phi_m(x) : x \in E\},$$

it follows from Proposition 12.12 that S_m is a Lipschitz $\mathcal{DSC}_\omega$ surface of codimension p. On the other hand the equalities $v_k^* \circ \varphi = g_k - h_k$ ensure, for every $x \in E$, that $\varphi(x) = \sum_{k=1}^{p} (g_k(x) - h_k(x)) v_k$, hence for each $x \in A_m$ we have $\varphi(x) = \sum_{k=1}^{p} (G_{k,m}(x) - H_{k,m}(x)) v_k = \Phi_m(x)$. Consequently, since $S_m := \{x + \Phi_m(x) : x \in E\}$, the equalities $C_i =: A = \{x + \varphi(x) : x \in E\}$ give $C_i \subset \bigcup_{m \in \mathbb{N}} S_m$, which entails that C_i, hence C as well, is covered by countably many Lipschitz DSC_ω surfaces of codimension p of X. This translates the desired implication $\Rightarrow$ and finishes the proof. $\square$

After the above results on Lipschitz DC/DSC surfaces and covers with countably many such surfaces, the next step is to show how they contribute to the structures of sets of singular points of convex functions and semiconvex functions, as stated in Theorem 12.22 and Theorem 12.23 below. This will be obtained through a series of lemmas.

LEMMA 12.17. *Let X be a separable normed space, let Q be a countable dense set in X, and let $u_1^*, \cdots, u_p^*$ be linearly independent in X^*. Then there exist $v_1, \cdots, v_p$ in Q linearly independent and $v_1^*, \cdots, v_p^*$ in X^* linearly independent such that*

$$\operatorname{span}\{v_1^*, \cdots, v_p^*\} = \operatorname{span}\{u_1^*, \cdots, u_p^*\} \text{ and } \langle v_i^*, v_j\rangle = \delta_{i,j},$$

where $\delta_{i,j}$ is the Kronecker symbol, $\delta_{i,j} = 1$ if $i = j$ and $\delta_{i,j} = 0$ if $i \neq j$.

PROOF. First suppose that $p = 1$. By density of Q in X there is $v_1 \in Q$ such that $\langle u_1^*, v_1\rangle \neq 0$, so it suffices to put $v_1^* := \langle u_1^*, v_1\rangle^{-1} u_1^*$. Continuing by induction, suppose that the property is true for $p \in \mathbb{N}$. Consider $u_1^*, \cdots, u_p^*, u_{p+1}^*$ linearly independent in X^*. By assumption there are $a_1^*, \cdots, a_p^*$ in X^* linearly independent and $v_1, \cdots, v_p$ in Q linearly independent such that $\langle a_i^*, v_j\rangle = \delta_{i,j}$ for all $i, j \in \{1, \cdots, p\}$ along with $\operatorname{span}\{a_1^*, \cdots, a_p^*\} = \operatorname{span}\{u_1^*, \cdots, u_p^*\}$. Set

$$a^* := u_{p+1}^* - \sum_{k=1}^{p} \langle u_{p+1}^*, v_k\rangle a_k^*$$

and note that $\langle a^*, v_j\rangle = 0$ for all $j \in \{1, \cdots, p\}$ and that $a^* \neq 0$ since

$$u_{p+1}^* \notin \operatorname{span}\{u_1^*, \cdots, u_p^*\} = \operatorname{span}\{a_1^*, \cdots, a_p^*\}.$$

Since $a^* \neq 0$, by density of Q in X again, there is $v_{p+1} \in Q$ such that we have $\langle a^*, v_{p+1}\rangle \neq 0$. Putting $v_{p+1}^* := \langle a^*, v_{p+1}\rangle^{-1} a^*$, we see that $\langle v_{p+1}^*, v_j\rangle = \delta_{i,j}$ for all $j \in \{1, \cdots, p, p+1\}$, which in particular entails that $v_{p+1} \notin \operatorname{span}\{v_1, \cdots, v_p\}$, hence $v_1, \cdots, v_p, v_{p+1}$ are linearly independent. Further, by definition of a^* we have $\operatorname{span}\{a_1^*, \cdots, a_p^*, v_{p+1}^*\} = \operatorname{span}\{u_1^*, \cdots, u_p^*, u_{p+1}^*\}$. Then for each $i \in \{1, \cdots, p\}$ defining $v_i^* := a_i^* - \langle a_i^*, v_{p+1}\rangle v_{p+1}^*$, we obtain

$$\operatorname{span}\{v_1^*, \cdots, v_p^*, v_{p+1}^*\} = \operatorname{span}\{u_1^*, \cdots, u_p^*, u_{p+1}^*\}$$

along with $\langle v_i^*, v_j\rangle = \delta_{i,j}$ for all $j \in \{1, \cdots, p, p+1\}$. This clearly concludes the proof by induction. $\square$

DEFINITION 12.18. Let X be a separable normed space and $p \in \mathbb{N}$ with $p < \dim X$. Let also Q be a countable dense set of X and $\operatorname{Svect}_p(Q)$ be the collection of vector subspaces F of X generated by p linearly independent vectors of Q.
(i) For each $F \in \operatorname{Svect}_p(Q)$, we fix a fixed countable dense set D_F of F^* and we define $\mathcal{A}_F(D_F)$, abbreviated $\mathcal{A}_F$, as the set of all $(v_0^*, v_1^*, \cdots, v_p^*) \in (D_F)^{p+1}$ such that the vectors $v_0^*, v_1^*, \cdots, v_p^*$ are affinely independent in F^*.
(ii) For a continuous semiconvex function $f : U \to \mathbb{R}$ on an open convex set U of X, for $F \in \operatorname{Svect}_p(Q)$ and for $V = (v_0^*, v_1^*, \cdots, v_p^*) \in \mathcal{A}_F$, we define $\Sigma(f, F, V)$ as the set of $x \in U$, for which there are $x_0^*, x_1^*, \cdots, x_p^* \in \partial_C f(x)$ such that $x_i^*|_F = v_i^*$, otherwise stated

$$\Sigma(f, F, V) := \{x \in U : \exists x_i^* \in \partial_C f(x),\, x_i^*|_F = v_i^*, i = 0, 1, \cdots, p\}.$$

We begin by relating the set $\Sigma_p(f)$ of singular points of order p of f to the sets $\Sigma(f, F, V)$.

LEMMA 12.19. Let X be a separable normed space, let Q be a countable dense set of X, and let $p \in \mathbb{N}$ with $p < \dim X$. Let U be an open convex set of X and $f : U \to \mathbb{R}$ be a continuous semiconvex function. Then one has

$$\Sigma_p(f) = \bigcup \{\Sigma(f, F, V) : F \in \mathrm{Svect}_p(Q), V \in \mathcal{A}_F\}.$$

PROOF. By the above definition, for any $x \in \Sigma(f, F, V)$ there exists some $x_0^*, x_1^*, \cdots, x_p^* \in \partial_C f(x)$ which are affinely independent, so the second member of the lemma is included in the first according to Definition 12.2 of $\Sigma_p(f)$.

Let us prove the converse inclusion. Fix any $x \in \Sigma_p(f)$. By definition, there are $x_0^*, x_1^*, \cdots, x_p^* \in \partial_C f(x)$ such that $u_1^* := x_1^* - x_0^*, \cdots, u_p^* := x_p^* - x_0^*$ are linearly independent in X^*. It is clear from Lemma 12.17 that there is some $F \in \mathrm{Svect}_p(Q)$ such that $u_1^*|_F, \cdots, u_p^*|_F$ are linearly independent in F^*. This means that $x_0^*|_F, x_1^*|_F, \cdots, x_p^*|_F$ are affinely independent points in the set $T := \{x^*|_F : x^* \in \partial_C f(x)\}$. We deduce that the set $\mathrm{int}_{F^*} \mathrm{co}\left(\{x_0^*|_F, \cdots, x_p^*|F\}\right)$ is nonempty and included in T (for T is convex). Consequently, $\mathrm{int}_{F^*} T \neq \emptyset$, so the density (assumption) of D_F in F^* furnishes some $V = (v_0^*, v_1^*, \cdots, v_p^*) \in \mathcal{A}_F$ with $v_i^* \in T$. It results that $x \in \Sigma(f, F, V)$, which justifies the desired inclusion. □

LEMMA 12.20. Let X be a separable normed space, let Q be a countable dense set of X, and let $p \in \mathbb{N}$ with $p < \dim X$. Let U be an open convex set of X and $f : U \to \mathbb{R}$ be a continuous semiconvex function. Let $F \in \mathrm{Svect}_p(Q)$ and $V \in \mathcal{A}_F$. Then the set $\Sigma(f, F, V)$ can be covered by countably many Lipschitz F-surfaces.

PROOF. Fix any $x \in \Sigma(f, F, V)$ and any nonzero $h \in F$. By definition of $\Sigma(f, F, V)$ choose $x_0^*, \cdots, x_p^* \in \partial_C f(x)$ such that $V = (x_0^*|_F, \cdots, x_p^*|_F)$. Since the vectors $(x_1^*|_F - x_0^*|_F), \cdots, (x_p^*|_F - x_0^*|_F)$ generate F^*, there exists some integer $1 \le k \le p$ such that

$$(x_k^* - x_0^*)(h) = (x_k^*|_F - x_0^*|_F)(h) \neq 0.$$

This and the Clarke subregularity of f (see Theorem 10.38) entails that $d_H^+ f(x; -h) \neq -d_H^+ f(x; h)$ for any nonzero $h \in F$ since by Proposition 2.74(c)

$$\{\langle x^*, h\rangle : x^* \in \partial_C f(x)\} = [-f^0(x; -h), f^0(x; h)] = [-d_H^+ f(x; -h), d_H^+ f(x; h)].$$

The result of the lemma then follows from Lemma 12.10. □

LEMMA 12.21. Let X be a separable normed space, F be a finite-dimensional vector subspace with $1 \le \dim F < \dim X$, and E be a topological vector complement of F in X, so $X = E \oplus F$. Let C be a Lipschitz F-surface of X, $f : U \to \mathbb{R}$ be a Lipschitz function from an open convex set U of X into $\mathbb{R}$. Let $v^* \in F^*$ and

$$A(v^*) := \{x \in C \cap U : \exists \xi(x) \in \partial_C f(x),\ \xi(x)_{|_F} = v^*\}.$$

(A) If f is convex on U, there exists a Lipschitz convex function $g : E \to \mathbb{R}$ (depending on v^*) such that

$$f(x) - \langle v^*, \pi^{F,E}(x)\rangle = g(\pi^{E,F}(x)) \quad \text{for all } x \in A(v^*). \tag{12.12}$$

(B) If X is a separable uniformly convex Banach space and the function f is semiconvex on U with respect to a nondecreasing (finite) modulus $\omega : [0, +\infty[\to [0, +\infty[$

with $\lim_{r\downarrow 0}\omega(r) = 0 = \omega(0)$, then there exists a Lipschitz semiconvex function $g : E \to \mathbb{R}$ (depending on v^*) such that (12.12) holds.

If in addition X admits an equivalent uniformly smooth norm with modulus of smoothness of power type $1+\beta$ with $0 < \beta \le 1$, then the function g can be taken to have a modulus of semiconvexity of the form $K\omega(\cdot)$ for some constant $K > 0$ whenever either (i) or (ii) below are satisfied
(i) $\beta = 1$,
(ii) $\omega(r) = K_0 r^\alpha$ for some $K_0 > 0$ and $0 < \alpha \le \beta$.

PROOF. Let $v^* \in F^*$ and let $A := A(v^*)$. Since C is a Lipschitz F-surface, the mapping $q := \left(\pi^{E,F}|_C\right)^{-1} : E \to C$ is Lipschitz, so we can choose a Lipschitz constant $\gamma > 1$ for q, which means

$$\|x - x'\| \le \gamma \, \|\pi^{E,F}(x - x')\| \quad \text{for all } x, x' \in C.$$

Consider the function $g_0 : \pi^{E,F}(A) \to \mathbb{R}$ defined by

$$g_0(u) := f(q(u)) - \langle v^*, \pi^{F,E}(q(u))\rangle \quad \text{for all } u \in \pi^{E,F}(A),$$

or equivalently

$$g_0(\pi^{E,F}(x)) = f(x) - \langle v^*, \pi^{F,E}(x)\rangle \quad \text{for all } x \in A.$$

Note that g_0 is Lipschitz on $\pi^{E,F}(A)$ according to the Lipschitz properties of q and f. Further, denoting by L a Lipschitz constant of f, for all $x \in U$, the set $\partial_C f(x)$ is nonempty and $\partial_C f(x) \subset L\mathbb{B}_{X^*}$.
(A) Assume that f is convex. Fix arbitrary $x, a \in A$ and, by definition of A, choose $\zeta(a) \in \partial f(a)$ such that $\zeta(a)|_F = v^*$. Then

$$\langle \zeta(a), x - a\rangle + f(a) \le f(x),$$

hence

$$\langle \zeta(a), \pi^{E,F}(x - a)\rangle + \langle \zeta(a), \pi^{F,E}(x - a)\rangle + f(a) \le f(x),$$

or equivalently

$$\langle \zeta(a), \pi^{E,F}(x - a)\rangle + f(a) - \langle v^*, \pi^{F,E}(a)\rangle \le f(x) - \langle v^*, \pi^{F,E}(x)\rangle.$$

Consequently, for any $y \in \pi^{E,F}(A)$ (thus $y = \pi^{E,F}(x)$ for some $x \in A$),

$$\langle \zeta(a), y - \pi^{E,F}(a)\rangle + g_0(\pi^{E,F}(a)) \le g_0(y); \tag{12.13}$$

further, for each $u \in E$, putting $G_a(u) := \langle \zeta(a), u - \pi^{E,F}(a)\rangle + g_0(\pi^{E,F}(a))$ and $g(u) := \sup_{a\in A} G_a(u)$, we see that the function g is convex, finite and L-Lipschitz on E since $\|\zeta(a)\|_{X^*} \le L$. On the other hand, for any $y \in \pi^{E,F}(A)$ by (12.13) and the definition of g we have $g(y) = g_0(y)$, thus

$$g(\pi^{E,F}(x)) = f(x) - \langle v^*, \pi^{F,E}(x)\rangle \quad \text{for all } x \in A.$$

(B) Assume that X is a uniformly convex Banach space and f is $\omega(\cdot)$-semiconvex on U, where $\omega : [0, +\infty[\to [0, +\infty[$ is nondecreasing with $\lim_{r\downarrow 0}\omega(r) = 0 = \omega(0)$. By Proposition 10.51 there exists a nondecreasing concave modulus $\omega_0 : [0, +\infty[\to [0, +\infty[$ with $\omega_0(\cdot) \le 4\omega(\cdot)$ and such that f is $\omega_0(\cdot)$-semiconvex on U. As above fix arbitrary $x, a \in A$ and, by definition of A, choose $\zeta(a) \in \partial_C f(a)$ such that $\zeta(a)|_F = v^*$. Then by the description of subgradient of semiconvex functions (see Theorem 10.29) one has

$$\langle \zeta(a), x - a\rangle + f(a) \le f(x) + \|x - a\| \, \omega_0(\|x - a\|),$$

hence

$$\begin{aligned}&\langle\zeta(a), \pi^{E,F}(x-a)\rangle + \langle\zeta(a), \pi^{F,E}(x-a)\rangle + f(a)\\ &\leq f(x) + \gamma\|\pi^{E,F}(x-a)\|\,\omega_0(\gamma\|\pi^{E,F}(x-a)\|).\end{aligned}$$

Since $\omega_0(\gamma r) \leq \gamma\omega_0(r)$ according to the concavity of $\omega_0(\cdot)$ and the inequality $\gamma > 1$, it results that

$$\begin{aligned}&\langle\zeta(a), \pi^{E,F}(x-a)\rangle + f(a) - \langle v^*, \pi^{F,E}(a)\rangle\\ &\leq f(x) - \langle v^*, \pi^{F,E}(x)\rangle + \gamma^2\|\pi^{E,F}(x-a)\|\,\omega_0(\|\pi^{E,F}(x-a)\|).\end{aligned}$$

Consequently, for any $y \in \pi^{E,F}(A)$ (thus $y = \pi^{E,F}(x)$ for some $x \in A$),

$$\langle\zeta(a)|_E, y - \pi^{E,F}(a)\rangle \leq g_0(y) - g_0(\pi^{E,F}(a)) + \gamma^2\|y - \pi^{E,F}(a)\|\,\omega_0(\|y - \pi^{E,F}(a)\|);$$

further, $\|\,\zeta(a)|_E\,\| \leq L$. Applying Theorem 10.59 we obtain a Lipschitz semiconvex extension g of g_0 to E. If in addition either the assumption (i) or the assumption (ii) is fulfilled, then the same theorem says that the function g is $K_0\gamma^2\omega_0(\cdot)$-semiconvex for some constant $K_0 > 0$, thus g is also $K\,\omega(\cdot)$-semiconvex on E for some real constant $K > 0$. This finishes the proof of the lemma. □

We can now state and prove the theorems on singular points of convex functions and of semiconvex functions.

THEOREM 12.22 (L. Zajíček). Let X be a separable normed space and $f : U \to \mathbb{R}$ be a continuous convex function from an open convex set U of X into $\mathbb{R}$. For any positive integer p with $p < \dim X$ the set $\Sigma_p(f) := \{x \in U : \dim\big(\partial f(x)\big) \geq p\}$ of singular points of f of order p can be covered by countably many Lipschitz DC surfaces of codimension p.

THEOREM 12.23 (J. Duda and L. Zajíček). Let X be a separable uniformly convex Banach space and $f : U \to \mathbb{R}$ be a continuous semiconvex function on an open convex set U of X. For any positive integer p with $p < \dim X$ the set $\Sigma_p(f) := \{x \in U : \dim\big(\partial_C f(x)\big) \geq p\}$ of singular points of f of order p can be covered by countably many Lipschitz DSC surfaces of codimension p.

If in addition X admits an equivalent uniformly smooth norm with modulus of smoothness of power type $1+\beta$ with $0 < \beta \leq 1$, then the p-codimensional Lipschitz surfaces in the countable cover can be taken to be DSC_ω whenever either (i) or (ii) below is satisfied:

(i) $\beta = 1$,

(ii) $\omega(r) = K_0 r^\alpha$ for some $K_0 > 0$ and $0 < \alpha \leq \beta$.

PROOFS OF THEOREMS 12.22 AND 12.23. Since X is separable and f is locally Lipschitz (the continuous function f is locally Lipschitz under either the convexity or the semiconvexity with respect to a finite modulus), we may write U as a countable union of bounded open convex sets such that f is Lipschitz over each one. So we may suppose that U is bounded and f is Lipschitz over U. Denote by (I) the hypothesis of convexity of f, (II) the hypothesis of semiconvexity of f, and (II+(i)) (resp. (II+(ii))) the hypothesis (II) plus (i) (resp. plus (ii)).

Fix any p-dimensional vector subspace F of X, affinely independent vectors $v_0^*, \cdots, v_p^* \in F^*$, and take any subset A of the set

$$\{x \in U : \exists x_0^*, \cdots, \cdots, x_p^* \in \partial_C f(x) \text{ such that } x_0^*|_F = v_0^*, \cdots, x_p^*|_F = v_p^*\}$$

which is contained in a Lipschitz F-surface. We claim that A is included in a Lipschitz DC F-surface under (I), in a Lipschitz DSC F-surface under (II), in a Lipschitz DSC_ω F-surface under either (II+(i)) or (II+ii) as additional assumption. Indeed, let E be a topological vector complement of F in X, so $X = E \oplus F$. Lemma 12.21 furnishes Lipschitz functions $g_0, g_1, \cdots, g_p : E \to \mathbb{R}$ which are on E convex under (I), semiconvex under (II), and $K\omega(\cdot)$-semiconvex under either (II+(i)) or (II+(ii)) and such that

$$f(x) - \langle v_i^*, \pi^{F,E}(x)\rangle = g_i(\pi^{E,F}(x)) \quad \text{for all } x \in A.$$

Consequently, for each $x = \pi^{E,F}(x) + \pi^{F,E}(x)$ in A, we have

$$\langle v_i^* - v_0^*, \pi^{F,E}(x)\rangle = g_0(\pi^{E,F}(x)) - g_i(\pi^{E,F}(x)), \; i = 1, \cdots, p.$$

Denoting by $v_1, \cdots, v_p$ the vector basis of F whose $v_1^* - v_0^*, \cdots, v_p^* - v_0^*$ is the dual basis, and considering the mapping $g : E \to F$ with

$$g(u) = \big(g_0(u) - g_1(u)\big)v_1 + \cdots + \big(g_0(u) - g_p(u)\big)v_p \quad \text{for all } u \in E,$$

we see that $A \subset \{u + g(u) : u \in E\}$, further the Lipschitz mapping g is DC under (I), DSC under (II), and DSC_ω under either (II+(i)) or (II+(ii)), the semiconvex properties being due to Proposition 12.12(c).

The theorem then follows from Lemma 12.19 and Lemma 12.20. □

We consider now a set C which can be covered by countably many Lipschitz DC or DSC surfaces.

PROPOSITION 12.24. Let X be a separable normed space and p be an integer with $1 \le p < \dim X$.
(A) If a subset C of X can be covered by countably many Lipschitz DC surfaces of X of codimension p, then there exists some Lipschitz convex function $f : X \to \mathbb{R}$ such that $C \subset \Sigma_p(f)$.
(B) Let $\omega : [0, +\infty[\to [0, +\infty[$ be a nondecreasing (finite) modulus function with $\lim_{r\downarrow 0} \omega(r) = 0 = \omega(0)$. If a subset C of X can be covered by countably many Lipschitz DSC_ω surfaces of codimension p of X, then there exists a Lipschitz $\omega(\cdot)$-semiconvex function $f : X \to \mathbb{R}$ such that $C \subset \Sigma_p(f)$.

PROOF. Consider, for each $n \in \mathbb{N}$, a vector subspace F_n of X with $\dim F_n = p$ and a Lipschitz DC (resp. DSC_ω) F_n-surface of X such that $C \subset \bigcup_{n\in N} C_n$.

Fix an arbitrary $n \in \mathbb{N}$ and take a topological vector complement E_n of F_n and continuous linear functionals $x_1^*, \cdots, x_p^* \in X^*$ such that $x_i^* \in E_n^\perp$, for $i = 1, \cdots, p$, and $x^*_{1|F_n}, \cdots, x^*_{p|F_n}$ form a vector basis of F_n^*. By Proposition 2.247 there are functions $\varphi_1, \cdots, \varphi_p, \psi_1, \cdots, \psi_p$ from E_n into $\mathbb{R}$ which are convex (resp. semiconvex with modulus of semiconvexity $\Lambda_n\omega(\cdot)$, for some constant $\Lambda_n > 0$) such that

$$C_n = \{x \in X : x_i^*(x) = \varphi_i(\pi^{E_n,F_n}(x)) - \psi_i(\pi^{E_n,F_n}(x)), \; i = 1, \cdots, p\}. \tag{12.14}$$

Define the functions $g_0, g_1, \cdots, g_p$ on X by putting, for all $x \in X$

$$g_0(x) := \psi_1(\pi^{E_n,F_n}(x)) + \cdots + \psi_p(\pi^{E_n,F_n}(x)) + x_1^*(x) + \cdots + x_p^*(x),$$

and, for $i = 1, \cdots, p$,

$$g_i(x) := \varphi_i(\pi^{E_n,F_n}(x)) + g_0(x) - \psi_i(\pi^{E_n,F_n}(x)) - x_i^*(x).$$

The functions $g_0, g_1, \cdots, g_p$ are obviously Lipschitz on X and they are also convex (resp. semiconvex with modulus of semiconvexity $K_n\omega(\cdot)$, for some constant

$K_n > 0$; see Proposition 12.12). The function $f_n := \max\{g_0, g_1, \cdots, g_p\}$ is clearly Lipschitz on X and convex (resp. $K_n\omega(\cdot)$-semiconvex). Fix any $u \in C_n$ and any nonzero $h \in F$. Then, there exists some $k \in \{1, \cdots, p\}$ such that $x_k^*(h) \neq 0$. Further, on the one hand, from (12.14) and from the definitions of g_i and f_n we have $g_0(u) = g_1(u) = \cdots = g_p(u) = f_n(u)$. On the other hand, the functions f_n, g_0 and g_k (as convex (resp. semiconvex) functions) have usual directional derivative and, since $\pi^{E_n,F_n}(h) = 0$,

$$g_0{}'(u; h) = \sum_{i=1}^{p} \psi_i\left(\pi^{E_n,F_n}(u); \pi^{E_n,F_n}(h)\right) + \sum_{i=1}^{p} x_i^*(h) = \sum_{i=1}^{p} x_i^*(h),$$

$$\begin{aligned} g_k{}'(u; h) &= \varphi_k{}'\left(\pi^{E_n,F_n}(u); \pi^{E_n,F_n}(h)\right) + g_0{}'(u; h) - \psi_k{}'\left(\pi^{E_n,F_n}(u); \pi^{E_n,F_n}(h)\right) - x_k^*(h) \\ &= \sum_{i=1}^{p} x_i^*(h) - x_k^*(h). \end{aligned}$$

Consequently, we obtain (see Danskin's formula in Corollary 2.201)

$$\begin{aligned} f_n{}'(u; h) &= \max\{g_0{}'(u; h), \cdots, g_p{}'(u; h)\} \\ &\geq \max\{g_0{}'(u; h), g_k{}'(u; h)\} = \sum_{i=1}^{p} x_i^*(h) - \min\{0, x_k^*(h)\}, \end{aligned}$$

and also, since $-h \in F$,

$$f_n{}'(u; -h) \geq \sum_{i=1}^{p} x_i^*(-h) - \min\{0, x_k^*(-h)\}.$$

It results that

$$f_n{}'(u; h) + f_n{}'(u; -h) \geq -\min\{0, x_k^*(h)\} - \min\{0, -x_k^*(h)\} = |x_k^*(h)| > 0. \tag{12.15}$$

Now, for each $i \in \mathbb{N}$, choose some constant $c_i > 0$ so small that the function $c_i f_i$ is Lipschitz on X with Lipschitz constant 1 and convex (resp. semiconvex with $\omega(\cdot)$ as modulus of semiconvexity) and such that $c_i|f_i(0)| < 1$. Clearly, the real-valued function $f := \sum_{i\in\mathbb{N}} 2^{-i} c_i f_i$ is Lipschitz on X and convex (resp. semiconvex with $\omega(\cdot)$ as a modulus of semiconvexity). Fix any $u \in C$ and any nonzero $h \in F$ and choose some $n \in \mathbb{N}$ such that $u \in C_n$. Considering the real-valued Lipschitz function $\rho := \sum_{i\in\mathbb{N}\setminus\{n\}} 2^{-i} c_i f_i$, its convexity (resp. semiconvexity) guarantees the existence and the sublinearity of $\rho'(u; \cdot)$, so $\rho'(u; h) + \rho'(u; -h) \geq 0$. This and (12.15) yield

$$f'(u; h) + f'(u; -h) = \rho'(u; h) + \rho'(u; -h) + c_n 2^{-n}\left(f_n{}'(u; h) + f_n{}'(u; -h)\right) > 0,$$

which entails, by definition of $\Sigma_{D,p}(f)$, that $u \in \Sigma_{D,p}(f)$ or equivalently $u \in \Sigma_p(f)$ according to Proposition 12.3. The set C is then included in $\Sigma_p(f)$. □

Lemma 12.16, Theorem 12.22, Theorem 12.23 and Proposition 12.24 lead to the following:

THEOREM 12.25 (L. Zajíček). *Let X be a separable normed space X and p be an integer with $1 \leq p < \dim X$. A subset C of X is included in $\Sigma_p(f)$ for some continuous convex function $f : U \to \mathbb{R}$ from an open convex set U of X if and only if C can be covered by countably many Lipschitz DC surfaces of codimension p.*

THEOREM 12.26. Let X be a separable uniformly convex Banach space which admits an equivalent uniformly smooth norm with modulus of smoothness of power type 2. Then:
(a) For a nondecreasing (finite) modulus function $\omega : [0,+\infty[\to [0,+\infty[$ with $\lim_{r\downarrow 0}\omega(r) = 0 = \omega(0)$, a subset C of X is included in $\Sigma_p(f)$ for some continuous $K\omega(\cdot)$-semiconvex function $f : U \to \mathbb{R}$ (with a constant $K > 0$) from an open convex set U of X if and only if C can be covered by countably many Lipschitz DSC_ω surfaces of codimension p.
(b) A subset C of X is included in $\Sigma_p(f)$ for some continuous semiconvex function $f : U \to \mathbb{R}$ from an open convex set U of X if and only if C can be covered by countably many Lipschitz DSC surfaces of codimension p.

PROOFS OF THEOREMS 12.25 AND 12.26. Concerning the implication $\Rightarrow$ in each case we may suppose that U is bounded and f is Lipschitz on U, since f is locally Lipschitz and X is separable. Consequently, the assertions of Theorem 12.25 and (a) of Theorem 12.26 immediately follow from Theorem 12.22, Theorem 12.23 and Proposition 12.24. The implication $\Rightarrow$ in (b) of Theorem 12.26 also follows from Theorem 12.23. For the converse implication $\Leftarrow$, consider an arbitrary subset C of X which is covered by countably many Lipschitz DSC surfaces of codimension p. By Lemma 12.16 there exists some nondecreasing (modulus) function $\omega : [0,+\infty[\to [0,+\infty[$ such that the set C is covered by countably many Lipschitz DSC_ω surfaces of X. This and Proposition 12.24 guarantee the desired implication $\Leftarrow$ of (b). □

The above result (b) in Theorem 12.26 has a partial extension to general uniformly convex Banach spaces.

PROPOSITION 12.27. Let X be a separable uniformly convex Banach space with an equivalent uniformly smooth norm with modulus of power type $1+\beta$ with $0 < \beta \le 1$. Let $0 < \alpha \le \beta$ and $\omega(r) := K_0 r^\alpha$ for all $r \ge 0$, where K_0 is a positive real constant. Then a subset C of X is included in $\Sigma_p(f)$ for some continuous $K\omega(\cdot)$-semiconvex function $f : U \to \mathbb{R}$ (with a constant $K > 0$) from an open convex set U of X if and only if C can be covered by countably many Lipschitz DSC_ω surfaces of codimension p.

PROOF. As above, we can see that the proposition follows from Theorem 12.23 and Proposition 12.24. □

12.3. Comments

The study of sizes of sets of singular points of different orders of functions probably started with the 1952 article [**19**] by R. D. Anderson and V. L. Klee. In that article [**19**] the authors carried out the study of sizes of such sets for real-valued convex functions on $\mathbb{R}^n$ by means of Hausdorff measures. For the more particular situation of convex functions on $\mathbb{R}^2$, A. S. Besicovitch employed, in his 1963 paper [**100**], rectifiable sets to measure the sizes of such singular sets. The study of smallness of such sets of singular points for functions on infinite-dimensional Banach spaces through Lipschitz surfaces of finite codimensions probably began with the 1979 paper [**986**] (submitted in 1976) by L. Zajíček with convex functions on separable Banach spaces. The main results in both [**19**] and [**100**] are particular cases of [**986**].

All the results in Section 12.1 as well as their proofs are due to L. Zajíček and are taken from his 2012 paper [**996**]. Theorem 12.4, Theorem 12.7, Theorem

12.8 are exactly the results established by L. Zajíček in Theorem 1.4, Theorem 1.1 and Theorem 1.5 respectively in [**996**]. We mention that A.D. Ioffe [**523**, Theorem 1.3(a)] proved previously the case of Theorem 12.4 with extended real-valued functions $f : U \to \mathbb{R} \cup \{-\infty, +\infty\}$ instead of vector-valued mappings $f : U \to Y$. Proposition 12.3 and its proof are taken from [**996**, Lemma 2.4]. Lemma 12.5, Lemma 12.6, Lemma 12.9 and Lemma 12.10 as well as their proofs correspond to Lemma 3.1, Lemma 3.2, Lemma 4.1 and Lemma 4.2 respectively in the same Zajíček's paper [**996**]. Results similar to Theorem 12.7 were previously shown for semiconvex functions (a subclass of C-subregular functions) in finite dimensions in the 1992 paper of G. Alberti, L. Ambrosio and P. Cannarsa [**14**] and in general separable Banach spaces by P. Albano and P. Cannarsa in their 1999 paper [**11**].

Lemma 12.17 and its proof correspond to a part of Lemma 2.3 in Albano and Cannarsa paper [**11**]. Apart Lemma 12.17, the results in Section 12.2 and their proofs in the convex framework are due to L. Zajíček and taken from his 1979 paper [**986**], and in the semiconvex framework the results and their proofs are due to J. Duda and L. Zajíček and taken from their 2013 paper [**364**]. Theorem 12.22 and Theorem 12.25 correspond to Proposition 2 and Proposition 3 in the paper [**986**] by L. Zajíček, while Theorem 12.23 and Theorem 12.26 correspond to Theorem 3.5 and Proposition 3.6 in the paper [**364**] by J. Duda and L. Zajíček. Other results related to singular points of convex functions (or monotone operators) can be found in P. Albano and P. Cannarsa's 1999 paper [**11**], G. Alberti's 1994 paper [**13**], L. Veselý's 1986/87 papers [**954, 955**], L. Zajíček's 1978 paper [**984**].

CHAPTER 13

Non-differentiability points of functions on separable Banach spaces

After the previous chapter devoted to singular points of different orders of some classes of mappings and functions, we will study now the structure of Hadamard non-differentiability points of real-valued functions defined on separable Banach spaces.

13.1. Non-differentiability points of subregular functions

We start with three fundamental cases: Clarke subregular functions, continuous convex functions, and semiconvex functions.

THEOREM 13.1 (Lipschitz hypersurfaces covers of Hadamard nondifferentiability points). Let $f : U \to \mathbb{R}$ be a real-valued function defined on an open set U of a separable normed space X. The following hold:
(a) If f is locally Lipschitz and Clarke subregular, then its set of Hadamard non-differentiability points can be covered by a countable union of Lipschitz hypersurfaces.
(b) If f is continuous and convex, then its set of Hadamard non-differentiability points can be covered by a countable union of Lipschitz DC hypersurfaces.
(c) If f is continuous and semiconvex and if X is a separable uniformly convex Banach space, then its set of Hadamard non-differentiability points can be covered by a countable union of Lipschitz DSC hypersurfaces.

PROOF. Recalling that a locally Lipschitz and Clarke subregular function (resp. continuous convex or semiconvex function) is Hadamard differentiable at a point if and only if its Clarke subdifferential at this point is a singleton, the assertions of the theorem are direct consequences of Theorems 12.4 and 12.23 with $p = 1$. □

The above cover property of non-differentiability points requires a subregularity property of the locally Lipschitz function, like convexity, semiconvexity, Clarke subregularity, etc. Of course, a Lipschitz hypersurface of X is a small set of X (and this will be analyzed in Proposition 13.12 below). When X is finite-dimensional and f is merely locally Lipschitz, the class of Lebesgue null sets may be used in place of countable cover with Lipschitz hypersurfaces. Indeed, in such a framework the classical Rademacher theorem (see Theorem 2.82) says that any locally Lipschitz function is differentiable outside a Lebesgue null set. To consider in infinite-dimensional spaces the case where no subregularity is present, we need to introduce certain concepts of null sets in such spaces.

13.2. Null sets in infinite dimensions

Let us begin with the notion of Aronszajn null sets.

13.2.1. Aronszajn null sets. Recall that a sequence $(v_n)_{n\in\mathbb{N}}$ of a separable normed space X is a *complete sequence* provided that $\overline{\mathrm{span}}\{v_n : n \in \mathbb{N}\} = X$, where we recall that $\overline{\mathrm{span}}(C)$ denotes the closed vector space spanned by a subset C of the space X.

DEFINITION 13.2. Let X be a separable normed space.
(a) Given a nonzero vector $v \in X$ and a Borel set A of X, we write $A \in \mathcal{A}(v)$ if for every $x \in X$ the set $\{t \in \mathbb{R} : x + tv \in A\}$ is Lebesgue negligible in $\mathbb{R}$. We then define $\mathcal{A}(X)$ as the class of Borel sets A of X which, for each complete sequence of nonzero vectors $(v_n)_{n\in\mathbb{N}}$ in X, can be written as $A = \bigcup_{n\in\mathbb{N}} A_n$ with $A_n \in \mathcal{A}(v_n)$ for all $n \in \mathbb{N}$. Each set in $\mathcal{A}(X)$ is said to be *Aronszajn null* in X.
(b) Given a nonzero $v \in X$, a real $\varepsilon \geq 0$ and a Borel set A of X, we write $A \in \widetilde{\mathcal{A}}(v,\varepsilon)$ if the set $\{t \in \mathbb{R} : \varphi(t) \in A\}$ is Lebesgue null in $\mathbb{R}$ for any Lipschitz mapping $\varphi : \mathbb{R} \to X$ such that the Lipschitz constant of $t \mapsto \varphi(t) - tv$ is not greater than ε. (Observe that $\widetilde{\mathcal{A}}(v,\varepsilon)$ with $\varepsilon = 0$ coincides with the set $\mathcal{A}(v)$ defined in (a)). We write $A \in \widetilde{\mathcal{A}}(v)$ if $A = \bigcup_{k\in\mathbb{N}} A_k$ with $A_k \in \widetilde{\mathcal{A}}(v,\varepsilon_k)$ for some $\varepsilon_k > 0$. We then define $\widetilde{\mathcal{A}}(X)$ as the class of Borel sets A of X which, for each complete sequence of nonzero vectors $(v_n)_{n\in\mathbb{N}}$ in X, can be written as $A = \bigcup_{n\in\mathbb{N}} A_n$ with $A_n \in \widetilde{\mathcal{A}}(v_n)$ for all $n \in \mathbb{N}$.

On the one hand, we clearly have $\widetilde{\mathcal{A}}(v) \subset \mathcal{A}(v)$, and on the other hand, for every real $r \neq 0$,

$$\mathcal{A}(rv) = \mathcal{A}(v) \quad \text{and} \quad \widetilde{\mathcal{A}}(rv) = \widetilde{\mathcal{A}}(v),$$

where the first equality follows from the fact that $\{t \in \mathbb{R} : x + tv \in A\} = r\{t \in \mathbb{R} : x + t(rv) \in A\}$. Concerning the second equality, it suffices to observe that given $\varphi : \mathbb{R} \to X$ such that the Lipschitz constant of $t \mapsto \varphi(t) - tv$ is not greater than ε, putting $\varphi_r(t) := \varphi(rt)$ we see that the Lipschitz constant of $t \mapsto \varphi_r(t) - t(rv)$ is not greater than $\varepsilon|r|$ and

$$\{t \in \mathbb{R} : \varphi(t) \in A\} = r\{t \in \mathbb{R} : \varphi_r(t) \in A\}.$$

In addition to the above definition, we will have to consider also other concepts of null sets.

DEFINITION 13.3. Let X be a separable normed space.
(a) We define $\mathcal{C}(X)$ as the class of Borel sets A of X which can be written as $A = \bigcup_{n\in\mathbb{N}} A_n$ where, for each $n \in \mathbb{N}$, the set $A_n \in \mathcal{A}(v_n)$ for some nonzero $v_n \in X$.
(b) We define similarly $\widetilde{\mathcal{C}}(X)$ as the class of Borel subsets A of X which can be written as $A = \bigcup_{n\in\mathbb{N}} A_n$ where, for each $n \in \mathbb{N}$, the set $A_n \in \widetilde{\mathcal{A}}(v_n)$ for some nonzero $v_n \in X$.
(c) More generally, for a non Borel subset A of X, we write $A \in \widetilde{\mathcal{C}}(X)$ whenever $A \subset B$ for some Borel set $B \in \widetilde{\mathcal{C}}(X)$. One says that sets in $\widetilde{\mathcal{C}}(X)$ are $\widetilde{\mathcal{C}}(X)$-null.

Obviously the classes $\mathcal{A}(X)$, $\widetilde{\mathcal{A}}(X)$ and $\widetilde{\mathcal{C}}(X)$ are stable under countable union and

$$\widetilde{\mathcal{A}}(X) \subset \widetilde{\mathcal{C}}(X).$$

It is also easily seen that these three classes are invariant with respect to equivalent norms.

Our next step it to show that $\widetilde{\mathcal{C}}(X) \cap \mathcal{B}(X) \subset \mathcal{A}(X)$. In preparation, let us establish some preliminary results.

LEMMA 13.4. Let E be a finite-dimensional vector subspace with dimension m of a normed space X and let $(v_n)_{n=1}^m$ be a basis of E. Let λ_m be the Lebesgue measure on E. Then, for any Borel set A of X such that $\lambda_m\big(E \cap (x + A)\big) = 0$ for all $x \in X$, there exists $A_1 \in \mathcal{A}(v_1), \cdots, A_m \in \mathcal{A}(v_m)$ such that $A = \bigcup_{n=1}^m A_n$.

PROOF. We proceed by induction on m. For $m = 1$, the result is exactly the definition of $\mathcal{A}(v_1)$. Assume that the property holds true for $m - 1$. Let E be an m-dimensional vector subspace and $v_1, \cdots, v_m$ a basis of E, and let $A \in \mathcal{B}(X)$ be such that $\lambda_m\big(E \cap (x + A)\big) = 0$ for all $x \in X$, where λ_k denotes the k-dimensional Lebesgue measure. Put

$$A_m := \left\{x \in A : \int_{\mathbb{R}} \mathbf{1}_A(x + tv_m)\, dt = 0\right\} = \bigcap_{k \in \mathbb{N}} \left\{x \in A : \int_{-k}^{k} \mathbf{1}_A(x + tv_m)\, dt = 0\right\}.$$

The set A_m is Borelian in X. Given any $x \in X$, if $A_m \cap (x + \mathbb{R}v_m) \neq \emptyset$ taking z in this set, on the one hand we have $z + \mathbb{R}v_m = x + \mathbb{R}v_m$ since $z \in x + \mathbb{R}v_m$, and on the other hand the inclusions $z \in A_m$ and $A_m \subset A$ entail that $\lambda_1\big(A_m \cap (z + \mathbb{R}v_m)\big) = 0$ hence $\lambda_1\big(A_m \cap (x + \mathbb{R}v_m)\big) = 0$. Therefore, in any case we have

$$\lambda_1\big(A_m \cap (x + \mathbb{R}v_m)\big) = 0 \quad \text{for all } x \in X, \tag{13.1}$$

which gives $A_m \in \mathcal{A}(v_m)$.

Now put $A' := A \setminus A_m$ and note by (13.1) that

$$\lambda_1\big(A \cap (x + \mathbb{R}v_m)\big) = \lambda_1\big(A' \cap (x + \mathbb{R}v_m)\big) \quad \text{for all } x \in X,$$

which entails that

$$\lambda_1\big(A' \cap (y + \mathbb{R}v_m)\big) > 0 \quad \text{for all } y \in A' \tag{13.2}$$

since for any $y \in A'$ we have $\lambda_1\big(A \cap (y + \mathbb{R}v_m)\big) > 0$ by the definitions of A' and A_m. Put $F := \operatorname{span}\{v_1, \cdots, v_{m-1}\}$, and fix any $x \in X$. Suppose that $A' \cap (x + \mathbb{R}v_m) \neq \emptyset$ and let $z \in A' \cap (x + \mathbb{R}v_m) = (x + \mathbb{R}v_m) \cap (A \setminus A_m)$. Then $\lambda_1\big(A \cap (z + \mathbb{R}v_m)\big) > 0$ since $z \notin A_m$, hence $\lambda_1\big(A \cap (x + \mathbb{R}v_m)\big) > 0$ since $z + \mathbb{R}v_m = x + \mathbb{R}v_m$ according to the inclusion $z \in x + \mathbb{R}v_m$. Further, using the Fubini theorem and noting that $Q := F \cap (x + A')$ is included in the projection P of $E \cap (x + A')$ into F parallel to $\mathbb{R}v_m$, we also have

$$\begin{aligned} 0 = \lambda_m\big(E \cap (x + A)\big) &\geq \lambda_m\big(E \cap (x + A')\big) \\ &= \int_P \lambda_1\big((u + \mathbb{R}v_m) \cap (x + A')\big)\, d\lambda_{m-1}(u) \\ &\geq \int_Q \lambda_1\big((u + \mathbb{R}v_m) \cap (x + A')\big)\, d\lambda_{m-1}(u). \end{aligned}$$

It follows that $\lambda_{m-1}(Q) = 0$ since $\lambda_1\big((u + \mathbb{R}v_m) \cap (x + A')\big) > 0$ for every $u \in Q$ according to (13.2). $\lambda_{m-1}\big(F \cap (x + A')\big) = 0$ for all $x \in X$. The property for $m - 1$ tells us that there are $A_1 \in \mathcal{A}(v_1), \cdots, A_{m-1} \in \mathcal{A}(v_{m-1})$ such that $A' = A_1 \cup \cdots \cup A_{m-1}$. We then conclude that $A = A_1 \cup \cdots \cup A_{m-1} \cup A_m$. $\square$

LEMMA 13.5. Let $v_1, \cdots, v_m$ be nonzero vectors of a normed space X. Then for any $u \in \operatorname{span}\{v_1, \cdots, v_m\}$ and any $A \in \mathcal{A}(u)$, there exist $A_1 \in \mathcal{A}(v_1), \cdots, A_m \in \mathcal{A}(v_m)$ such that $A = A_1 \cup \cdots \cup A_m$.

PROOF. Let $v_{i_1}, \cdots, v_{i_k}$ be a linearly independent subsystem from $v_1, \cdots, v_m$ spanning the same vector space E than $v_1, \cdots, v_m$. For every $x \in X$, the Fubini theorem ensures that $x + A$ intersects E in a set of k-dimensional Lebesgue measure zero. The above lemma then provides $A_{i_1} \in \mathcal{A}(v_{i_1}), \cdots, A_{i_k} \in \mathcal{A}(v_{i_k})$ such that $A = A_{i_1} \cup \cdots \cup A_{i_m}$, which justifies the statement of the lemma. □

PROPOSITION 13.6. Let X be a separable normed space. Then the following inclusions hold:

$$\widetilde{\mathcal{A}}(X) \subset \widetilde{\mathcal{C}}(X) \cap \mathcal{B}(X) \subset \mathcal{A}(X) \subset \mathcal{C}(X)$$

PROOF. Only the inclusion $\widetilde{\mathcal{C}}(X) \cap \mathcal{B}(X) \subset \mathcal{A}(X)$ needs to be proved, the other ones being obvious. For that inclusion it suffices to show that, given any $v \neq 0$ and $\varepsilon > 0$, we have $\widetilde{\mathcal{A}}(v, \varepsilon) \subset \mathcal{A}(X)$. Fix any $A \in \widetilde{\mathcal{A}}(v, \varepsilon)$ and any complete sequence of nonzero vectors $(v_n)_{n \in \mathbb{N}}$. Consider $u := \sum_{n=1}^{m} \alpha_n v_n$ such that $\|u - v\| < \varepsilon$. Then $A \in \mathcal{A}(u)$ according to the definition of $\mathcal{A}(u)$ and $\widetilde{\mathcal{A}}(v, \varepsilon)$. Lemma 13.5 provides $A_n \in \mathcal{A}(v_n)$ such that $A = \bigcup_{n=1}^{m} A_n$, hence $A \in \mathcal{A}(X)$. □

REMARK 13.7. For the fact that the inclusions $\widetilde{\mathcal{C}}(X) \cap \mathcal{B}(X) \subset \mathcal{A}(X) \subset \mathcal{C}(X)$ are strict, see the comments at the end of the chapter. □

Since every set in $\widetilde{\mathcal{C}}(X)$ is included in some set in $\widetilde{\mathcal{C}}(X) \cap \mathcal{B}(X)$ according to Definition 13.3(c), the second inclusion of the above proposition tells us the following:

PROPOSITION 13.8. Let X be a separable normed space. Then any $\widetilde{\mathcal{C}}(X)$-null set is contained in some Aronszajn null set of X, that is, in some set in $\mathcal{A}(X)$.

13.2.2. Porous sets. In addition to the above concepts of null sets, we will need the important notion of porosity which is another accurate way to see the smallness of sets.

DEFINITION 13.9. Let A be a subset of X.
(a) If X is a metric space, one says that A is *porous at a point* $x \in A$ whenever there exists a constant $0 < c < 1$ such that for every $\varepsilon > 0$ there exists $x' \in B(x, \varepsilon)$ and $r > c\, d(x', x)$ such that $B(x', r) \cap A = \emptyset$. One also says that A is *porous at* x *with constant* c.

In terms of sequences, the set A is porous at $x \in A$ if and only if there exists a constant $0 < c < 1$ and a sequence $(x_n)_{n \in \mathbb{N}}$ converging to x with $x_n \neq x$ and such that

$$B(x_n, c\, d(x_n, x)) \cap A = \emptyset.$$

If X is a normed space and if, given a subset V of X, the element $x' \in B(x, \varepsilon)$ is in the form $x' = x + tv$ for some $t > 0$ and $v \in V$, one says that A is *porous at* x *in the direction-set* V with constant c. Sequentially, the porosity of A at x in the direction-set V means that the above elements x_n are in the form $x_n = x + t_n v_n$ for some $t_n > 0$ and some $v_n \in V$.

If $V = \{v\}$ with $v \neq 0$, one says that A is *porous at* x *in the direction* v with constant c. The set A is said to be *directionally porous at* x (with constant c) if it is porous at x (with constant c) in the direction v for some nonzero $v \in X$.
(b) The set A is called *porous* (resp. *porous in the direction-set* V, *directionally porous*) if it is porous (resp. porous in the direction-set V, directionally porous) at every $x \in A$. If there is a fixed constant $0 < c < 1$ for all $x \in A$, one says that A

is porous (resp. porous in the direction-set V, directionally porous) with constant c.
(c) The set A is called *σ-porous* (resp. *σ-porous in the direction-set* V, *σ-directionally porous*) if it is a countable union of sets which are porous (resp. porous in the direction-set V, directionally porous).

REMARK 13.10. If A is porous at x with constant c, then the above sequential property holds with the same constant c. Conversely, if the sequential property is satisfied with a constant $0 < c < 1$, then it is not difficult to see that A is porous at x with constant c', for every constant $c' \in]0, c[$. □

Clearly the above concepts of porosity are invariant with respect to equivalent norms, and the class of σ-porous (resp. σ-porous in the direction-set V, σ-directionally porous) sets *contains the empty set*, is *hereditary* in the sense that it contains any subset of a set in the class, and is *stable under countable union.* Further, any directionally porous (resp. σ-directionally porous) set is porous (resp. σ-porous). The next proposition provides other first properties.

PROPOSITION 13.11. The following assertions hold:
(a) Any countable set of a normed space is σ-directionally porous.
(b) All porous sets and σ-porous sets in $\mathbb{R}^m$ are Lebesgue negligible.
(c) For any porous set A of a metric space, we have $\operatorname{int}(\operatorname{cl} A) = \emptyset$.
(d) Any σ-porous set in a metric space is a set of first category in this space.
(e) The complement in a complete metric space of any σ-porous set is dense in this space.

PROOF. (a) Consider a point a of a (non-null) normed space X and fix any nonzero vector $v \in X$. Taking $c := 1/2$ and $t_n := 1/n$, and putting $x_n := a + t_n v$, we see that $\{a\} \cap B(x_n, c\|x_n - a\|) = \emptyset$, so $\{a\}$ is directionally porous.
(b) Let A be a porous set of $\mathbb{R}^m$. According to the definition of density point, no point of A is a Lebesgue density point of A, hence A is Lebesgue null in $\mathbb{R}^m$ (see Theorem D.1 in Appendix D).
(c) Let A be a porous set of a metric space X. Suppose that $\operatorname{int}(\operatorname{cl} A) \neq \emptyset$ and fix a point u in this set, so there is a real $\varepsilon > 0$ such that $B(u, 2\varepsilon) \subset \operatorname{cl} A$. We can then choose some $x \in A \cap B(u, \varepsilon)$ and clearly $B(x, \varepsilon) \subset \operatorname{cl} A$. By the porosity of A at x, there are $0 < c < 1$ and a sequence $(x_n)_n$ in $X \setminus \{x\}$ converging to x such that, for every $n \in \mathbb{N}$, the open ball $B(x_n, c\, d(x_n, x))$ does not meet A, so $x_n \notin \operatorname{cl} A$. Choosing n sufficiently large so that $d(x_n, x) < \varepsilon$ we see that $x_n \in B(x, \varepsilon) \subset \operatorname{cl} A$, which contradicts the fact that $x_n \notin \operatorname{cl} A$.
(d) and (e) The assertion (d) follows from (c) and (e) is a direct consequence of the assertion (d). □

Lipschitz hypersurfaces constitute a remarkable subclass of directionally porous sets, as shown in the following proposition.

PROPOSITION 13.12. Any Lipschitz hypersurface of a normed space is directionally porous.

PROOF. Let v be a nonzero vector of a normed space X and E be a topological vector complement, so $X = E \oplus \mathbb{R}v$. Let $f : E \to \mathbb{R}$ be a Lipschitz function and consider the Lipschitz hypersurface $A := \{u + f(u)v : u \in E\}$. Denote by $\|\cdot\|_E$ the restriction to E of the norm of X and denote by $\mathcal{N}$ the equivalent norm of X defined by $\mathcal{N}(u + rv) = \max\{\|u\|_E, |r|\}$ for all $u \in E$ and $r \in \mathbb{R}$; so we will write

$B_{\mathcal{N}}(\cdot,\cdot)$ for the open ball relative to the norm $\mathcal{N}$. Denote by γ a Lipschitz constant of f and choose a real $0 < c < 1/(1+\gamma)$. Take a sequence of reals $(t_n)_n$ tending to 0 with $t_n > 0$. Fixing any $u \in E$ and observing that

$$B_{\mathcal{N}}(u + f(u)v + t_n v, ct_n) \equiv B_{\|\cdot\|_E}(u, ct_n) \times]f(u) + (1-c)t_n, f(u) + (1+c)t_n[$$

and $\mathcal{N}\big(u + f(u)v + t_n v - (u + f(u)v)\big) = t_n$, we see that, for any

$$(w, r) \in B_{\mathcal{N}}\big(u + f(u)v + t_n v, c\mathcal{N}(u + f(u)v + t_n v - (u + f(u)v)\big),$$

we have that

$$|r - f(u)| > (1-c)t_n \geq \gamma c t_n > \gamma \|w - u\|_E \geq |f(w) - f(u)|,$$

so the inequality $|r - f(u)| > |f(w) - f(u)|$ tells us that $r \neq f(w)$. This guarantees that

$$A \cap B_{\mathcal{N}}\big(u + f(u)v + t_n v, c\mathcal{N}(u + f(u)v + t_n v - (u + f(u)v)\big) = \emptyset,$$

hence A is directionally porous at $u + f(u)v \in A$. □

The next proposition examines connections between the porosity of A and the distance function from A. Before stating the results, it is worth noticing the two following easy properties of the distance function.

LEMMA 13.13. Let A be a subset of a normed space X and $x \in A$, and let $f := \operatorname{dist}(\cdot, A)$.
(a) If, for a vector $v \in X$, the bilateral Dini directional derivative $d_D f(x; v)$ exists, then $d_D f(x; v) = 0$.
(b) The function f is Gâteaux differentiable at x if and only if $\overline{d}_D^+ f(x; v) \leq 0$ for all $v \in X$.

PROOF. (a) For the assertion (a), it suffices to observe that

$$t^{-1}\big(f(x + tv) - f(x)\big) \geq 0 \quad \text{and} \quad (-t)^{-1}\big(f(x + (-t)v) - f(x)\big) \leq 0, \text{ for all } t > 0.$$

(b) The implication $\Rightarrow$ follows from (a). Conversely, suppose that $\overline{d}_D^+ f(x; \cdot) \leq 0$. Since the inequality $0 \leq \underline{d}_D^+ f(x; \cdot)$ is obvious, we then have $\lim_{t \downarrow 0} t^{-1} f(x + tv) = 0$ for all $v \in X$. Fixing any $v \in X$, it results that

$$\lim_{t \downarrow 0} (-t)^{-1} f(x - tv) = -\lim_{t \downarrow 0} t^{-1} f(x + t(-v)) = 0.$$

Consequently, $\lim_{t \to 0} t^{-1} f(x + tv) = 0$, which entails that f is Gâteaux differentiable at x (with a null Gâteaux derivative). □

REMARK 13.14. Consider the above function $f := \operatorname{dist}(\cdot, A)$. Since obviously $0 \in \partial_H f(x)$, this function f is Hadamard differentiable, or equivalently Gâteaux differentiable (since f is Lipschitz), at x if and only if $0 \in \partial_H(-f)(x)$ (see Proposition 5.11). The latter is equivalent to $\overline{d}_D^+ f(x; \cdot) = \overline{d}_H^+ f(x; \cdot) \leq 0$. This provides another justification of the equivalence in the above assertion (b). □

PROPOSITION 13.15. Let A be a subset of a normed space X and $x \in A$ and let $f := \operatorname{dist}(\cdot, A)$. The following hold:
(a) The set A is porous at x in a direction $v \neq 0$ if and only if $\overline{d}_D^+ f(x; v) > 0$.
(b) If the set A is porous at x in a direction $v \neq 0$, then the bilateral Dini derivative $d_D f(x; v)$ does not exist.
(c) The set A is directionally porous at x if and only if the distance function

$f = \mathrm{dist}(\cdot, A)$ is not Gâteaux differentiable at x.
(d) The set A is porous at x if and only if the distance function $f = \mathrm{dist}(\cdot, A)$ is not Fréchet differentiable at x.

PROOF. (a) Suppose that A is porous at x in a direction $v \neq 0$. There exists by definition a constant $0 < c < 1$ and a sequence $(t_n)_n$ tending to 0 with $t_n > 0$ such that $B(x + t_n v, ct_n\|v\|) \cap A = \emptyset$, or equivalently $d(x + t_n v, A) \geq ct_n\|v\|$. For $f := \mathrm{dist}(\cdot, A)$, this yields $\overline{d}_D^+ f(x; v) \geq c\|v\| > 0$.

Conversely, if $\overline{d}_D^+ f(x; v) > 0$, then there is some $0 < c < 1$ such that $\overline{d}_D^+ f(x; v) > c\|v\|$, so there exists a sequence $(t_n)_n$ tending to 0 with $t_n > 0$ such that $t_n^{-1} f(x + t_n v) > c\|v\|$. This guarantees that $B(x + t_n v, ct_n\|v\|) \cap A = \emptyset$, which means that A is porous at x in the direction v.
(b) The assertion (b) follows from (a) and from Lemma 13.13(a).
(c) The assertion (c) follows from (a) and from Lemma 13.13(b).
(d) Since $0 \in \partial_F f(x)$ (as easily seen), the function f is Fréchet differentiable at x if and only if $0 \in \partial_F(-f)(x)$ (see Proposition 4.13). Suppose that A is porous at x. There exist a real $0 < c < 1$ and a sequence $(y_n)_n$ converging to x with $y_n \neq x$ such that $B(y_n, c\|y_n - x\|) \cap A = \emptyset$, that is, $f(x + h_n) \geq c\|h_n\|$, where $h_n := y_n - x$. This entails that $0 \notin \partial_F(-f)(x)$, hence f is not Fréchet differentiable at x.

Conversely, if f is not Fréchet differentiable at x, then $0 \notin \partial_F(-f)(x)$, hence there are a real $0 < c < 1$ and a sequence of nonzero vectors $(h_n)_n$ converging to 0 such that $f(x + h_n) > c\|h_n\|$. This ensures that $B(x + h_n, c\|h_n\|) \cap A = \emptyset$, that is, A is porous at x. □

We collect now some properties of sets which are porous in a direction-set. The analysis of these properties yields, among others, to the fact that, in finite-dimensional normed spaces, the concepts of porosity at x and directional porosity at x coincide.

PROPOSITION 13.16. Let A, V and W be subsets of a normed space X and let $x \in A$ and $0 < c < 1$.
(a) If the set A is porous the point x in the direction-set V with constant c, then any subset of $\mathrm{cl}\, A$ containing x is also porous at x in the direction-set V with constant c; thus, if A is porous in the direction-set V with constant c, then so is any of its subsets.
(b) The set A is porous (resp. porous with constant c) at x in the direction-set V if and only if it is porous (resp. porous with constant c) at x in the direction-set $V_\nu := \{v/\|v\| : v \in V, v \neq 0\}$. In particular, A is porous (resp. porous with constant c) at x if and only if it is porous (resp. porous with constant c) at x in the direction-set $\mathbb{S}_X$; further, A is porous (resp. porous with constant c) at x in a direction $v \neq 0$ if and only if it is porous (resp. porous with constant c) at x in the direction-set $\{tv : t \geq 0\}$.
(c) Assume that $V = V_1 \cup \cdots \cup V_m$. If A is porous at x in the direction-set V with constant c, then there exists some $i \in \{1, \cdots, m\}$ such that A is porous at x in the direction-set V_i with constant c. If A is porous (resp. porous with constant c) in the direction-set V, then A can be written as $A = A_1 \cup \cdots \cup A_m$, where each set A_i is porous (resp. porous with constant c) in the direction-set V_i.
(d) Assume that V, W are subsets of the unit sphere $\mathbb{S}_X$ and that there exists $c' \in]0, c[$ such that $\mathrm{dist}(v, W) < c - c'$ for all $v \in V$. If A is porous at x in the direction-set V with constant c, then it is porous at x in the direction-set W with

constant c'.
(e) If V is a compact subset of the unit sphere $\mathbb{S}_X$, then A is porous at x in the direction-set V if and only if there exists some $v \in V$ such that A is porous at x in the direction v. In particular, when X is finite-dimensional, A is porous at x if and only if it is directionally porous at x.

PROOF. The assertions (a) and (b) are obvious as well as the first part of the assertion (c). Concerning the second part, it suffices, for each $i \in \{1, \cdots, m\}$, to take A_i as the set of all $u \in A$ for which A is porous (resp. porous with constant c) at u in the direction-set V_i.

Let us prove (d). Take any $\varepsilon > 0$. By the porosity in the direction-set V with constant c, choose $t > 0$, $v \in V$ (hence $\|v\| = 1$) and $r > 0$ such that $t = t\|v\| < \varepsilon$, $r > ct\|v\| = ct$ and $B(x + tv, r) \cap A = \emptyset$. Choose $w \in W$ such that $\|w - v\| < c - c'$ and put $r' := r - t\|w - v\|$. It is easily seen that $B(x + tw, r') \subset B(x + tv, r)$, so $B(x + tw, r') \cap A = \emptyset$. Further, we note that $t\|w\| = t < \varepsilon$ and

$$c't\|w\| = c't = ct - (c - c')t < r - t\|w - v\| = r'.$$

It ensues that A is porous at x in the direction-set W with constant c'.

Concerning (e), by the compactness of V we can choose $W = \{w_1, \cdots, w_k\} \subset V$ such that $\operatorname{dist}(v, W) < c/3$ for all $v \in V$, where $c \in]0, 1[$ is taken so that A is porous at x in the direction set V with constant c. Then (d) tells us that A is porous at x in the direction-set W with constant $c/3$, so by (c) we obtain that A is porous at x in some direction w_i. Finally, we already observed in (b) that A is porous at $x \in A$ if and only if it is porous at x in the direction-set $\mathbb{S}_X$, so in finite dimensions A is porous at $x \in A$ if and only if it is directionally porous at x according to what precedes. $\square$

Our aim now is to establish a characteristic decomposition of any σ-porous set. We need first to prove two lemmas. In these lemmas, for subsets A and V in the normed space X and for $0 < c < 1$ and $\varepsilon > 0$, we will denote by $P(A, V, c, \varepsilon)$ the set all points $x \in X$ for which there exist $v \in V$, $t \geq 0$ and $r > 0$ such that $t\|v\| < \varepsilon$, $r > ct\|v\|$ and $B(x + tv, r) \cap A = \emptyset$. Since $B(x + tv, r) \cap A = \emptyset$ if and only if $B(x + tv, r) \cap \operatorname{cl} A = \emptyset$, we see that

$$P(A, V, c, \varepsilon) = P(\operatorname{cl} A, V, c, \varepsilon). \tag{13.3}$$

LEMMA 13.17. Let A and V be subsets of a normed space X and let $0 < c < 1$. The following hold:
(a) The set $P(A, V, c, \varepsilon)$ is open for every $\varepsilon > 0$.
(b) The set A is porous at $x \in A$ in the direction-set V with the constant c if and only if

$$x \in \bigcap_{\varepsilon > 0} P(A, V, c, \varepsilon) = \bigcap_{n \in \mathbb{N}} P\left(A, V, c, \frac{1}{n}\right).$$

(c) The set A is porous at $x \in A$ in the direction-set V if and only if

$$x \in \bigcup_{k \in \mathbb{N}} \bigcap_{n \in \mathbb{N}} P\left(A, V, \frac{1}{k+1}, \frac{1}{n}\right).$$

(d) If A is porous in the direction-set V with constant c, then there exists a Borel set $G(A) \supset A$ which is porous in the direction-set V with constant c.
(e) If A is porous (resp. σ-porous) in the direction-set V, then there exists a Borel

set $G'(A) \supset A$ which is porous (resp. σ-porous) in the direction-set V.
(f) If A is porous in the direction-set V (resp. directionally porous), then A can be written in the form $A = \bigcup_{k \in \mathbb{N}} A_k$, where each set A_k is porous in the direction-set V (resp. directionally porous) with constant $\frac{1}{k+1}$.

PROOF. The assertions (a), (b) and (c) are obvious.

Let us prove (d) and (e). If A is porous with constant c in the direction-set V (resp. if A is porous in the direction-set V), we put

$$G := \overline{A} \cap \bigcap_{n \in \mathbb{N}} P\left(A, V, c, \frac{1}{n}\right) \quad \left(\text{resp. } G' := \overline{A} \cap \bigcup_{k \in \mathbb{N}} \bigcap_{n \in \mathbb{N}} P\left(A, V, \frac{1}{k+1}, \frac{1}{n}\right)\right),$$

where $\overline{A} := \operatorname{cl} A$, and by (b) (resp. (c)) we have $G \supset A$ (resp. $G' \supset A$). According to (a) the sets G and G' are Borelian. Since (see (13.3))

$$P\left(A, V, c, \frac{1}{n}\right) = P\left(\overline{A}, V, c, \frac{1}{n}\right) \quad \left(\text{resp. } P\left(A, V, \frac{1}{k+1}, \frac{1}{n}\right) = P\left(\overline{A}, V, \frac{1}{k+1}, \frac{1}{n}\right)\right),$$

we see by (b) (resp. (c)) that $\overline{A}$ is porous with constant c in the direction-set V (resp. porous in the direction-set V) at any point in G (resp. at any point in G'). Proposition 13.16(a) then tells us that G is porous with constant c in the direction set V (resp. G' is porous in the direction-set V). This proves (d) and the first part of (e); the part concerning the σ-porosity easily follows.

Finally, defining A_k as the set of $u \in A$ such that A is porous in the direction-set V (resp. directionally porous) with constant $\frac{1}{k+1}$, the sets A_k fulfill the properties of (f). □

REMARK 13.18. It is worth observing that the set $G(A)$ in (d) above is even a G_δ-set, that is, the intersection of a countable family of open sets, since the closed set $\overline{A}$ of the metric space X involved in the definition of G in the proof above is of course the intersection of a countable family of open sets. □

LEMMA 13.19. Let V and W be two subsets of the normed space X for which there is some real $\alpha > 0$ such that

$$\max\{\|v\|, \|w\|\} \leq \frac{1}{\alpha}\|v + w\| \quad \text{for all } v \in V,\ w \in W.$$

Then any set A which is porous in the direction-set $V + W$ can be written in the form $A = A_1 \cup A_2$, where A_1 is porous in the direction-set V and A_2 is σ-porous in the direction-set W.

PROOF. Putting $A_1 := A \cap \bigcap_{n \in \mathbb{N}} P(A, V, c\alpha/2, 1/n)$, the assertion (b) in the above lemma combined with Proposition 13.16(a) says that A_1 is porous in the direction set V with constant $c\alpha/2$. It then remains to show that $A_2 := A \setminus A_1$ is σ-porous in the direction-set W. Of course, $A_2 = \bigcup_{n \in \mathbb{N}} C_n$, where $C_n := A \setminus P(A, V, c\alpha/2, 1/n)$. Thus it is enough to show that C_n is porous in the direction-set W with constant $c\alpha/2$.. Fix any $x \in C_n$ and consider any $\varepsilon \in]0, 1/n[$. We will prove that $x \in P(C_n, W, \frac{c\alpha}{2}, \varepsilon)$. Since x is in the set A which is porous in the direction-set $V + W$ with constant c, there are $v \in V$, $w \in W$, $t > 0$ and $r > 0$ such that

$$t\|v + w\| < \varepsilon\alpha,\ r > ct\|v + w\|,\ B(x + t(v + w), r) \cap A = \emptyset.$$

Since $t\max\{\|v\|, \|w\|\} \le \alpha^{-1}t\|v+w\| < \varepsilon < 1/n$, we have $t\|w\| < \varepsilon$ and $\frac{r}{2} > \frac{c}{2}t\|v+w\| \ge \frac{c\alpha}{2}t\|w\|$. So, it suffices to show that $B(x+tw, \frac{r}{2}) \cap C_n = \emptyset$ in order to get that $x \in P(C_n, W, \frac{c\alpha}{2}, \varepsilon)$. Suppose by contradiction that there is an element $y \in B(x+tw, \frac{r}{2}) \cap C_n$. By the first inclusion $y \in B(x+tw, \frac{r}{2})$ one would have $B(y+tv, r/2) \subset B(x+t(v+w), r)$, hence $B(y+tv, r/2) \cap A = \emptyset$. This combined with the inequalities

$$t\|v\| < \frac{1}{n} \quad \text{and} \quad \frac{r}{2} > \frac{c}{2}t\|v+w\| \ge \frac{c\alpha}{2}t\|v\|$$

would imply that $y \in P(A, V, \frac{c\alpha}{2}, \frac{1}{n})$, which would contradict the second inclusion $y \in C_n$. Consequently, $B(x+tw, \frac{r}{2}) \cap C_n = \emptyset$, hence $x \in P(C_n, W, \frac{c\alpha}{2}, \varepsilon)$. This being true for all $\varepsilon \in]0, 1/n[$, it results by (b) in the above lemma that C_n is porous at any $x \in C_n$ in the direction-set W with constant $c\alpha/2$, or equivalently C_n is porous in the direction-set W with constant $c\alpha/2$. □

The above lemma directly yields:

PROPOSITION 13.20. *Let V and W be two subsets of the normed space X for which there is some real $\alpha > 0$ such that*

$$\max\{\|v\|, \|w\|\} \le \frac{1}{\alpha}\|v+w\| \quad \text{for all } v \in V,\ w \in W.$$

Then any set A which σ-porous in the direction-set $V+W$ can be written in the form $A = A_1 \cup A_2$, where A_1 is σ-porous in the direction-set V and A_2 is σ-porous in the direction-set W.

The next two propositions which have their own interest will lead to the characteristic decomposition of σ-directionally porous sets.

PROPOSITION 13.21. *Let X be a normed space and V be a finite-dimensional vector subspace spanned by vectors $v_1, \cdots, v_k$. Then any set $A \subset X$ which is σ-porous in the direction-set V can be written in the form $A = \bigcup_{n=1}^{k}(A_n^+ \cup A_n^-)$, where A_n^+ is σ-porous in the direction v_n and A_n^- is σ-porous in the direction $-v_n$.*

PROOF. We will proceed by induction on k. For $k = 1$, the result follows directly from (b) and (c) in Proposition 13.16. Suppose that the result holds true for $k-1 \ge 1$. Let V be the vector subspace spanned by $v_1, \cdots, v_k$, and suppose without loss of generality that $v_1, \cdots, v_k$ are linearly independent. For $V_0 := \operatorname{span}\{v_1, \cdots, v_{k-1}\}$ and $W := \mathbb{R}v_k$, the inequality assumption in Proposition 13.20 above is easily seen to be fulfilled for some real $\alpha > 0$, so the same above proposition and the induction assumption entails the result for k. □

PROPOSITION 13.22. *Let X be a normed space and $V = \overline{\operatorname{span}}\{v_n : n \in \mathbb{N}\}$, and let A be a subset of X. Assume that for each $x \in A$ there exists some $\zeta(x) \in V$ such that A is porous at x in the direction $\zeta(x)$. Then A can be written in the form $A = \bigcup_{n\in\mathbb{N}}(A_n^+ \cup A_n^-)$, where A_n^+ is σ-porous in the direction v_n and A_n^- is σ-porous in the direction $-v_n$.*

PROOF. Without loss of generality, we may suppose that $\|\zeta(x)\| = 1$ for all $x \in A$. Consider the set $A_{k,n}$ of all points $x \in A$ such that A is porous at x in the direction $\zeta(x)$ with constant $\frac{1}{k+1}$ and such that

$$\operatorname{dist}\big(\zeta(x), \mathbb{S}_X \cap \operatorname{span}\{v_1, \cdots, v_n\}\big) < \frac{1}{2k+2}.$$

We observe that $A = \bigcup_{k,n\in\mathbb{N}} A_{k,n}$. Further, applying Proposition 13.16(d) with $V := \{\zeta(x)\}$ and $W := \mathbb{S}_X \cap \operatorname{span}\{v_1, \cdots, v_n\}$, $c = \frac{1}{k+1}$ and $c' := \frac{1}{2k+2}$, ensures that $A_{k,n}$ is porous in the direction-set W, and hence also in the direction-set $\operatorname{span}\{v_1, \cdots, v_n\}$ according to (a) of the same proposition. The result then follows from Proposition 13.21. □

We can now state the desired characteristic decomposition of σ-directionally porous sets. It is just a very interesting particular case of the above proposition.

THEOREM 13.23 (decomposition of σ-porous sets along complete sequence). Let X be a separable normed space and let $(v_n)_{n\in N}$ be a complete sequence of X. Then any σ-directionally porous set A of X can be written in the form $A = \bigcup_{n\in\mathbb{N}}(A_n^+ \cup A_n^-)$, where A_n^+ is σ-porous in the direction v_n and A_n^- is σ-porous in the direction $-v_n$.

If A is Borelian, then the sets A_n^+ and A_n^- can also be required to be Borelian.

PROOF. As said above the first part follows directly from Proposition 13.22 above. If A is Borelian, it it enough to replace A_n^+ and A_n^- by the sets $G'(A_n^+)$ and $G'(A_n^-)$ given by Lemma 13.17(e). □

Suppose that a Borelian set $B \subset X$ is porous in a direction $v \neq 0$. Then for each $u \in X$, the set $B \cap (u + \mathbb{R}v)$ is porous in $u + \mathbb{R}v$, hence Lebesgue negligible in $u + \mathbb{R}v$ (see Proposition 13.11(b)). This combined with Theorem 13.23 and Definition 13.2(a) justifies the following:

PROPOSITION 13.24. Let X be a separable normed space. Then any Borelian σ-directionally porous set of X belongs to $\mathcal{A}(X)$, that is, it is Aronszajn null and hence $\widetilde{\mathcal{C}}(X)$-null too.

13.2.3. Haar null sets. For a locally Lipschitz function $f : U \to \mathbb{R}$, we saw in Proposition 6.6(a) that

$$f^\circ(x; h) = \limsup_{u\to x} \underline{d}_D^+ f(u; h).$$

It is a fundamental question to know properties of classes of subsets A such that

$$f^\circ(x; h) = \limsup_{A\not\ni u\to x} \underline{d}_D^+ f(u; h) \quad \text{for all } x \in U, h \in X.$$

Of course (at the opposite of the problem of non-differentiability points of Lipschitz functions) larger is the set A, more accurate is the formula. In preparation for the introduction of an appropriate class of sets fulfilling the above requirements, known as Haar null sets, let us recall some properties of Radon measures; we refer to [**90**, Chapter 2] for the properties below on Radon measures and for more results.

Let T be a Hausdorff topological space and $\mu : \mathcal{B}(T) \to [0, +\infty]$ be a non-negative measure on the Borel σ-field $\mathcal{B}(T)$ of T such that both following properties (i) and (ii) hold:
(i) $\mu(K) < +\infty$ for every compact set K of T;
(ii) $\mu(B) = \sup\{\mu(K) : K \subset B,\ K\text{compact}\}$ for every $B \in \mathcal{B}(T)$.
The extension of such a measure μ (still denoted by μ) to the μ-completion σ-field of $\mathcal{B}(T)$ is called a *Radon measure* on the topological space T. Recall that Ulam's theorem (see, e.g., [**365**, Theorem 7.1.4]) says that the completion of any finite non-negative measure on the Borel σ-field of any complete separable metric space is a Radon measure.

A subset of T which belongs to the μ-completion of $\mathcal{B}(T)$ for all *finite* Radon measures μ on T, or equivalently for all Radon probability measures μ on T, is said to be a *universally Radon measurable set* in T. The class of such sets will be denoted by $\mathcal{U}_R(T)$. This class is translation invariant whenever T is a topological Abelian group. Indeed, let any $A \in \mathcal{U}_R(T)$ and let any Radon probability measure μ on T. Fix any $t \in T$. Consider the completion (not relabeled) of the probability measure ν on $\mathcal{B}(T)$ defined by $\nu(B) = \mu(t + B)$ for all $B \in \mathcal{B}(T)$. Since ν is clearly a Radon probability measure on T and $A \in \mathcal{U}_R(T)$, we can write $A = A_0 \cup N$ with $A_0 \in \mathcal{B}(T)$ and $N \subset N_0$ for some $N_0 \in \mathcal{B}(T)$ with $\nu(N_0) = 0$. Consequently, $t + A = (t + A_0) \cup (t + N)$ with $t + A_0 \in \mathcal{B}(T)$ and $t + N \subset t + N_0$ along with $\mu(t + N_0) = 0$ since $\mu(t + N_0) = \nu(N_0)$. This means that $t + A$ belongs to the μ-completion of $\mathcal{B}(T)$ for all Radon probability measures μ on T, that is, $(t + A) \in \mathcal{U}_R(T)$.

Further, if $\mu(\cdot) \le \nu(\cdot)$ on $\mathcal{B}(T)$ for two non negative finite Radon measures μ and ν on T, then clearly $\mu(\cdot) \le \nu(\cdot)$ on $\mathcal{U}_R(T)$. If we endow a Borelian subset Q of T with the induced topology, then $\mathcal{U}_R(Q) = \{A \in \mathcal{U}_R(T) : A \subset Q\}$. If ν is a finite Radon measure on Q, then the function which assigns to any $B \in \mathcal{B}(T)$ the non-negative real $\nu\big(B \cap Q)$ is obviously a measure and its completion is a Radon measure on T called the *Radon extension* of ν.

We state in the next theorem some properties relative to the support of a Radon measure, product and convolution of Radon measures (see, e.g., [**106**, Chapter 3] and [**107**, Chapter 7]).

THEOREM 13.25. Let T be a Hausdorff topological space and μ be a non-negative Radon measure on T.
(a) The space T has a smallest closed set S (called the *support of* μ) such that $\mu(T \setminus S) = 0$; so $\mu(U \cap S) > 0$ for every open set U of T satisfying $U \cap S \neq \emptyset$.

If T is metrizable and the measure μ is finite, then its support is separable relative to the induced topology.
(b) Let T_1 and T_2 be two Hausdorff topological spaces and let μ_1 and μ_2 be two non-negative σ-finite Radon measures on T_1 and T_2 respectively. If $A \in \mathcal{U}_R(T_1 \times T_2)$, then the functions $t_1 \mapsto \mu_2\big(A(t_1, \cdot)\big)$ and $t_2 \mapsto \mu_1\big(A(\cdot, t_2)\big)$ are universally Radon measurable on T_1 and T_2 respectively, where $A(t_1, \cdot) := \{t_2 \in T_2 : (t_1, t_2) \in A\}$. Further, there exists a unique Radon measure $\mu_1 \otimes \mu_2$ on $T_1 \times T_2$ such that, for every $A \in \mathcal{U}(T_1 \times T_2)$,

$$(\mu_1 \otimes \mu_2)(A) = \int_{T_1} \mu_2\big(A(t_1, \cdot)\big)\, d\mu_1(t_1) = \int_{T_2} \mu_1\big(A(\cdot, t_2)\big)\, d\mu_2(t_2).$$

(c) If $T_1 = T_2 = T$ and μ_1 and μ_2 are finite, then putting

$$(\mu_1 * \mu_2)(B) = (\mu_1 \otimes \mu_2)\big(\{(t_1, t_2) \in T \times T : t_1 + t_2 \in B\}\big)\ \forall B \in \mathcal{B}(T),$$

the completion of $\mu_1 * \mu_2$ is a Radon measure on T called the *convolution measure* of μ_1 and μ_2. Obviously $\mu_1 * \mu_2 = \mu_2 * \mu_1$.

We can now define the class of Haar null sets.

DEFINITION 13.26. Let $(G, +, d)$ be a metric Abelian additive group which is *complete* and such that the distance d is invariant under translation. A universally Radon measurable subset A of G is said to be *Haar null* whenever there exists a (not necessarily unique) Radon probability measure μ on G such that $\mu(x + A) = 0$ for all $x \in G$. Such a probability is called a *test measure* for the set A.

It is worth observing the following property related to the convolution product. It will be useful in several places below.

PROPOSITION 13.27. Let $(G, +, d)$ be a complete metric Abelian group with a translation invariant distance. If μ is a test measure in G for a Haar null set $A \subset G$, then $\nu * \mu$ is also a test measure for A, for any Radon probability measure ν.

PROOF. Let μ be a test measure for a universally Radon measurable set A in G and let ν be any Radon probability measure on G. Clearly, $\nu * \mu$ is a Radon probability measure on G. Further, for every $x \in G$, we have

$$\begin{aligned}(\nu * \mu)(x + A) &= \int\int \mathbf{1}_{x+A}(y + z)\, d\mu(y) d\nu(z) \\ &= \int\int \mathbf{1}_{x-z+A}(y)\, d\mu(y) d\nu(z) = \int \mu(x - z + A)\, d\nu(z) = 0,\end{aligned}$$

since $\mu(x - z + A) = 0$ for all $z \in G$. This means that $\nu * \mu$ is also a test measure for A. □

If μ_1 and μ_2 are test measures for the sets A_1 and A_2 in G respectively, the above proposition ensures that $\mu_1 * \mu_2$ is a test measure for both A_1 and A_2, so $(\mu_1 * \mu_2)\big(x + (A_1 \cup A_2)\big) = 0$ since

$$(\mu_1 * \mu_2)\big(x + (A_1 \cup A_2)\big) \leq (\mu_1 * \mu_2)(x + A_1) + (\mu_1 * \mu_2)(x + A_2).$$

Then the class of Haar null sets is stable under finite union. We will see below that the stability property even holds for countable union.

The following theorem collects some first properties of Haar null sets.

THEOREM 13.28 (density of complements of Haar null sets). Let $(G, +, d)$ be a complete metric Abelian group with an invariant distance d. The following hold:
(a) Any universally Radon measurable subset of a Haar null set in G is Haar null.
(b) If A is Haar null in G, then $x + A$ is Haar null for every $x \in G$.
(c) If A is Haar null in G, then $G \setminus A$ is dense in G.

PROOF. The assertions (a) and (b) follow directly from the definition.

Regarding the assertion (c), let us proceed by contradiction, in supposing that $G \setminus A$ is not dense in G. Then there is some nonempty open set $U \subset A$, so U is Haar null. Let μ be a test measure for U and let S be the support of μ. Since $G = \bigcup_{x \in G}(x + U)$, there is some $a \in G$ such that $(a + U) \cap S \neq \emptyset$, hence the set $a + U$ being open we have $\mu\big((a + U) \cap S\big) > 0$ according to Theorem 13.25(a). This clearly entails that $\mu(a + U) > 0$, which contradicts that μ is also a test measure for the Haar null set U. □

When the complete metric Abelian group $(G, +, d)$ is *separable and locally compact* relative to the invariant distance d, it is well known (see, e.g.,[**107**, 9.11.4]) that there is a unique (up to multiplication by a positive real) translation invariant measure on G, called the Haar measure in G. Further, in such a case the Haar measure on G is σ-finite, so there exists a Radon probability measure p_G on G equivalent to the Haar measure, that is, for each $B \in \mathcal{U}_R(G)$, the equality $p_G(B) = 0$ holds true if and only if B is a null set for the Haar measure on G. Consequently, in complete separable locally compact metric groups any universally Radon measurable set with Haar measure zero is Haar null. The following proposition shows that the converse also holds true.

PROPOSITION 13.29. Assume that the complete metric Abelian group $(G, +, d)$ is separable and locally compact relative to the invariant distance d. Then a universally Radon measurable set A of G is Haar null in the sense of Definition 13.26 if and only if its Haar measure is zero.

PROOF. Let A be a universally measurable set in G which is Haar null and let μ be a probability test measure for A. Denote by λ_G the Haar measure in G. By Fubini's theorem we see that

$$0 = \int \left[\int \mathbf{1}_A(x+y)\, d\mu(x) \right] d\lambda_G(y) = \int \left[\int \mathbf{1}_A(x+y)\, d\lambda_G(x) \right] d\mu(y) = \lambda_G(A).$$

This confirms the equivalence since the converse implication has been justified above. □

Let T be a complete separable metric space and $C_b(T, \mathbb{R})$ be the space of bounded continuous functions from T into $\mathbb{R}$. With each $f \in C_b(T, \mathbb{R})$ let us associate on the set of Radon probability measures on T the semidistance

$$(\mu, \nu) \mapsto \left| \int_T f(t)\, d\mu(t) - \int_T f(t)\, d\nu(t) \right|.$$

The so-called *weak or narrow topology* on the set of Radon probability measures on T is the topology generated by the family of above semidistances for $f \in C_b(T, \mathbb{R})$; of course, it coincides with the topology induced by the weak* topology of the topological dual of the normed space $C_b(T, \mathbb{R})$ when each Radon measure is identified with the continuous linear functional on $C_b(T, \mathbb{R})$ that it determines. It is known (see, e.g., [**107**, Theoreme 8.9.4]) that the set of Radon probability measures on T equipped with this topology is metrizable and complete.

Assume now that T is a complete separable metric Abelian group $(G, +, d)$ and denote by ε_0^G the Dirac measure concentrated on the zero of G. Let A be a universally Radon measurable subset of G which is Haar null in G and let p be a test measure for A. For each $k \in \mathbb{N}$, putting $U_k := B(0, 1/k)$ there is some $x_k \in G$ such that $0 < r_k := p(x_k + U_k) \leq 1$, hence the completion of the probability measure ν_k, given by $\nu_k(B) := r_k^{-1} p(x_k + B \cap U_k)$ for all $B \in \mathcal{B}(G)$, satisfies $\nu_k(U_k) = 1$. Clearly, $\nu_k(x + A) = 0$ for all $x \in G$, that is, ν_k is a test measure for A. Further, it is easily seen that $\int_G f(x)\, d\nu_k(x) \to f(0)$ as $k \to \infty$ for all $f \in C_b(G, \mathbb{R})$, thus $\nu_k \to \varepsilon_0^G$ weakly as $k \to \infty$, so by Lebesgue's dominated convergence theorem $\nu_k * \mu \to \mu$ weakly, for each Radon probability measure μ on G.

Through those measures ν_k we show first in the next lemma the stability under countable union in the particular case when G is separable.

LEMMA 13.30. Assume that G is a complete separable metric Abelian group with an invariant distance. Let $(A_n)_{n \in \mathbb{N}}$ be a sequence of Haar null sets in G and, for each $n \in \mathbb{N}$, let p_n be a Radon probability test measure for A_n. Then, for each $n \in \mathbb{N}$ there are a real $s_n > 0$, an element $z_n \in G$ and a Radon probability measure μ_n on G such that the following properties (a)-(c) hold:
(a) μ_n is a test measure for A_n in G;
(b) $\mu_n(B) \leq s_n p_n(z_n + B)$ for all $B \in \mathcal{B}(G)$;
(c) the (infinite) convolution probability measure $\mu := \lim_{n \to \infty} \mu_1 * \cdots * \mu_n$ exists and is a test measure for the set $\bigcup_{n \in \mathbb{N}} A_n$; the latter limit is taken with respect to weak topology.

The property (c) entails in particular that $\bigcup_{n \in \mathbb{N}} A_n$ is Haar null in G.

PROOF. On the set of Radon probability measures on G fix a distance ρ whose associated topology coincides with the weak topology. Let $(A_n)_{n\in\mathbb{N}}$ be a sequence of Haar null sets in G. Let $\mu_1 = p_1$ and by induction choose by the analysis preceding the statement of the lemma, for each integer $n \geq 1$, a real $s_n > 0$, an element $z_n \in G$ and a test measure μ_{n+1} for A_{n+1} with $\mu_{n+1}(B) \leq s_{n+1}p_{n+1}(z_{n+1} + B)$ for all $B \in \mathcal{B}(G)$ and such that $\rho(\mu_{n+1} * \nu, \nu) < 2^{-n}$ for every convolution ν of some measures among $\mu_1, \cdots, \mu_n$. The sequence $(\mu_1 * \cdots * \mu_n)_n$ is then a Cauchy sequence, hence by the completeness with respect to the distance ρ its weak limit defines on G the convolution probability measure $\mu := \lim_{n\to\infty} \mu_1 * \cdots * \mu_n$. For each $k \in \mathbb{N}$, observing that $\mu = \mu_k * \nu$ where $\nu := (\mu_1 * \cdots * \mu_{k-1}) * (\lim_{n\to\infty} \mu_{k+1} * \cdots * \mu_{k+n})$ Proposition 13.27 tells us that μ is a test measure for A_k, that is, $\mu(x + A_k) = 0$ for all $x \in G$. Consequently, $\mu(x + \bigcup_{k\in\mathbb{N}} A_k) = 0$ for all $x \in G$, which justifies that $\bigcup_{n\in\mathbb{N}} A_n$ is Haar null in G. □

We can now prove the stability under countable union in its full generality.

THEOREM 13.31 (stability of Haar null property for countable union). Let G be a complete metric Abelian group with an invariant distance. Then any countable union of Haar null sets is Haar null.

PROOF. Let $(A_n)_{n\in\mathbb{N}}$ be a sequence of universally measurable subsets of G which are Haar null. For each $n \in \mathbb{N}$, take a test measure q_n for A_n and note, by Theorem 13.25(a) and the equality $q_n(G) = \sup\{q_n(K) : K \text{ compact in } G\}$ (due to the Radon property of q_n), that the support S_n of q_n is separable. Then the closed subgroup H of G generated by $\bigcup_{n\in\mathbb{N}} S_n$ is separable. Denote by p_n the completion of the restriction of q_n to $\mathcal{B}(H)$. On the one hand, $p_n(H) = q_n(H) = 1$ since the support of q_n is included in H, and on the other hand, for all $h \in H$,

$$p_n\big(h + (A_n \cap H)\big) = p_n\big((h + A_n) \cap H\big) = q_n\big((h + A_n) \cap H\big) \leq q_n(h + A_n) = 0,$$

hence p_n is a test measure for $A_n \cap H$ in H, so $A_n \cap H$ is Haar null in H. By Lemma 13.30 above, for each $n \in \mathbb{N}$ there are a real $s_n > 0$, an element $h_n \in H$ and a Radon probability test measure μ_n for $A_n \cap H$ with $\mu_n(B) \leq s_n p_n(h_n + B)$ for all $B \in \mathcal{B}(H)$ and such that the convolution probability measure $\mu := \lim_{n\to\infty} \mu_1 * \cdots * \mu_n$ exists, where the limit is taken with respect to the weak topology. Let ν be the completion of the extension of μ to $\mathcal{B}(G)$ given by $\nu(Q) := \mu(Q \cap H)$ for all $Q \in \mathcal{B}(G)$. Of course, ν is a Radon probability measure on G. Further, for any $n \in \mathbb{N}$ and $x \in G$, we have for all $h \in H$

$$\begin{aligned}
\mu_n\big(h + (x + A_n) \cap H\big) &\leq s_n p_n\big(h_n + h + (x + A_n) \cap H\big) \\
&= s_n p_n\big((h_n + h + x + A_n) \cap H\big) \\
&= s_n q_n\big((h_n + h + x + A_n) \cap H\big) \\
&\leq s_n q_n(h_n + h + x + A_n) = 0,
\end{aligned}$$

which guarantees that μ_n is a test measure for $(x + A_n) \cap H$ in H. Thus, since $\mu = \mu_n * \gamma$ with $\gamma := (\mu_1 * \cdots * \mu_{n-1}) * (\lim_{k\to\infty} \mu_{n+1} * \cdots * \mu_{n+k})$, the measure μ is also a test measure for $(x+A_n)\cap H$ according to Proposition 13.27. Consequently,

$$\nu\left(x+\bigcup_{n\in\mathbb{N}}A_n\right)=\nu\left(\bigcup_{n\in\mathbb{N}}(x+A_n)\right)=\mu\left(\left(\bigcup_{n\in\mathbb{N}}(x+A_n)\right)\cap H\right)$$
$$=\mu\left(\bigcup_{n\in\mathbb{N}}((x+A_n)\cap H)\right)=0,$$

so ν is a test measure for $\bigcup_{n\in\mathbb{N}}A_n$, and the proof is complete. □

There is no Fubini type full result for Haar null sets; see the comments at the end of the chapter. Fortunately, the next theorem provides a weak result of this type which will be used in the next section.

THEOREM 13.32 (Fubini type property for Haar null sets). *Let $(H,+,d_1)$ be a complete metric Abelian group with an invariant distance, $(T,+,d_2)$ be a locally compact complete separable metric Abelian group, and A be a universally measurable set in $H\times T$. The following are equivalent:*
(a) the set A is Haar null in $H\times T$;
(b) for all h outside a Haar null set in H, the set $A(h,\cdot)$ is Haar null in T.

PROOF. As the Haar measure of a locally compact complete separable Abelian group is σ-finite, there exists a Radon probability measure p_T on T equivalent to the Haar measure on the locally compact Abelian group T, that is, for each $Q\in\mathcal{U}_R(T)$, the equality $p_T(Q)=0$ holds true if and only if Q is a null set for the Haar measure on T. Put

$$B:=\{h\in H:p_T\big(A(h,\cdot)\big)>0\}.$$

Suppose that (b) holds, that is, the set B is Haar null in H. Denote by p_H a test measure in H for B. For any $(\overline{h},\overline{t})\in H\times T$, we have

$$\{h\in H:p_T\big(((\overline{h},\overline{t})+A)(h,\cdot)\big)>0\}=\overline{h}+B,$$

then $(p_H\otimes p_T)\big((\overline{h},\overline{t})+A\big)=0$ according to the Fubini theorem recalled above in Theorem 13.25(b). This means that the set A is Haar null in $H\times T$, so the implication (b)⇒(a) is established.

Conversely, suppose that A is Haar null. Let μ be a test measure for A. Denote by ε_0^H the Dirac measure on H concentrated at the zero of H and put $\nu:=\varepsilon_0^H\otimes p_T$. By Proposition 13.27 we know that $\mu*\nu$ is another test measure for A.

Consider the above set $B:=\{h\in H:p_T\big(A(h,\cdot)\big)>0\}$ which belongs to $\mathcal{U}_R(H)$ according to Theorem 13.25, and consider also the probability measure γ on H defined by $\gamma(F):=\mu(F\times T)$ for all $F\in\mathcal{B}(H)$ and extended (with the same label) to the γ-completion of $\mathcal{B}(H)$. To finish the proof, it suffices to show that γ is a test measure for B. So, fix any $k\in H$ and write first, according to the fact that $\mu*\nu$ is a test measure for A as said above,

$$0=(\mu*\nu)\big((k,0)+A\big)=\int\nu\big((-h_1,-t_1)+(k,0)+A\big)\,d\mu(h_1,t_1)$$
$$=\int\nu\big((k-h_1,-t_1)+A\big)\,d\mu(h_1,t_1).$$

On the other hand,

$$\nu\big((k-h_1,-t_1)+A\big) = \int\left[\int \mathbf{1}_{(k-h_1,-t_1)+A}(h,t)\, d\varepsilon_0^H(h)\right] dp_T(t)$$
$$= \int \mathbf{1}_{(k-h_1,-t_1)+A}(0,t)\, dp_T(t) = p_T\big(-t_1 + A(h_1-k,\cdot)\big),$$

and the function $(h_1,t_1)\mapsto p_T\big(-t_1+A(k-h_1,\cdot)\big)$ is >0 on $(k+B)\times T$ according to the definition of B and $\mathcal{U}_R(H\times T)$-measurable by Theorem 13.25. Consequently, $\gamma(k+B)=\mu\big((k+B)\times T\big)=0$, which finishes the proof. $\square$

The next result, which is quite evident, shows that the class of Haar null sets contains all the previous classes of small sets: $\widetilde{\mathcal{A}}(X), \mathcal{A}(X), \widetilde{\mathcal{C}}(X), \mathcal{C}(X)$.

Proposition 13.33. For any Banach space X, the class $\mathcal{C}(X)$ is included in the class of Haar null sets of X.

Proof. Consider any $v\neq 0$ and take any $A\in\mathcal{A}(v)$, so A is Borelian in X. Then for any $x\in X$, we have $\lambda(\{t\in\mathbb{R} : tv\in x+A\})=0$, where λ denotes the Lebesgue measure on $\mathbb{R}$. Let $g:\mathbb{R}\to]0,+\infty[$ be a positive λ-integrable function and let $\nu(Q)=r\int_Q g(t)\,dt$ for any $Q\in\mathcal{B}(\mathbb{R})$, where $r=(\int_{\mathbb{R}} g(t)\,dt)^{-1}$. Note that $\nu(\{t\in\mathbb{R} : tv\in x+A\})=0$ for all $x\in X$. For the Radon probability measure μ on X defined by $\mu(B):=\nu(\{t\in\mathbb{R} : tv\in B\})$ for every $B\in\mathcal{B}(X)$, it is clear that $\mu(x+A)=0$ for all $x\in X$. Consequently, A is Haar null, and this easily justifies the inclusion of the proposition. $\square$

13.3. Hadamard non-differentiability points of Lipschitz functions

In this section we will develop diverse smallness properties of sets of non-differentiability points of Lipschitz mappings and functions along with various applications of such properties.

13.3.1. Hadamard non-differentiability points of Lipschitz mappings. Given a separable Banach space X and a locally Lipschitz mapping f from X into a Banach space Y satisfying the Radon-Nikodým property, Theorem 13.39 will establish that the set of Hadamard non-differentiability points of f is $\widetilde{\mathcal{A}}(X)$-null. First, we need certain preparatory results.

Lemma 13.34. Let $f:U\to Y$ be a Lipschitz mapping from an open set U of a separable normed space X into a normed space Y and let $g:Y\to\mathbb{R}$ be a Lipschitz function from Y into $\mathbb{R}$. Then there exists a σ-directionally porous set $A\subset U$ such that, for every $x\in U\setminus A$ and every $u\in X$, the set of limit points, as $t\downarrow 0$, of the mapping

$$t\mapsto g\left(\frac{f(x+t(u+v))-f(x+tv)}{t}\right)$$

does not depend on $v\in X$.

Further, that set coincides with the set of limit points, as $t\uparrow 0$, of the same mapping.

Proof. Let $\gamma_f>0$ and $\gamma_g>0$ be two Lipschitz constants of f and g respectively, and observe that the limit points in the statement are finite. Fix $u,v,w\in X$

with $w \neq v$, reals $\delta, \eta > 0$, and open intervals I, J such that $I +]-\eta, +\eta[\subset J$. Denote by C the set (depending on $u, v, w, \delta, \eta, I, J$) of $x \in U$ such that

$$g\left(\frac{f(x + t(u + v)) - f(x + tv)}{t}\right) \notin J \quad \text{for all } 0 < t < \delta, \tag{13.4}$$

and such that there exists a sequence $(t_n)_n$ in $]0, \delta[$ tending to 0 satisfying

$$g\left(\frac{f(x + t_n(u + w)) - f(x + t_n w)}{t_n}\right) \in I \quad \text{for all } n. \tag{13.5}$$

To get the porosity of C in the direction $w - v$ it suffices, for $x \in C$, to check that $B(x + t_n(w - v), r_n) \cap C = \emptyset$, where $r_n := \eta t_n/(2\gamma_f\gamma_g)$. Indeed, for any $y \in B(x + t_n(w - v), r_n)$, the distance between $g\left(\frac{f(y+t_n(u+v))-f(y+t_n v)}{t_n}\right)$ and $g\left(\frac{f(x+t_n(u+w))-f(x+t_n w)}{t_n}\right)$ is not greater than $2\gamma_f\gamma_g t_n^{-1}\|y - (x + t_n(w - v))\| < \eta$. The inclusion $I +]-\eta, \eta[\subset J$ and (13.5) entail that $g\left(\frac{f(y+t_n(u+v))-f(x+t_n v)}{t_n}\right) \in J$, so (13.4) tells us that $y \notin C$. This confirms the porosity of C in the direction $w - v$.

Let A be the union of the sets $C(u, v, w, \delta, \eta, I, J)$ where u, v, w are taken in a fixed countable dense set V of X with $w \neq v$, δ, η are positive rational numbers, and I, J are open intervals with rational end-points and with $I +]-\eta, \eta[\subset J$. By what precedes, the set A is σ-directionally porous. Now suppose that $x \in U$ is such that there are $u_0, v_0, w_0 \in X$ and a real value ρ which is a limit point, as $t \downarrow 0$, of the function $t \mapsto g\left(\frac{f(x+t(u_0+w_0))-f(x+tw_0)}{t}\right)$ but not of the function $t \mapsto g\left(\frac{f(x+t(u_0+v_0))-f(x+tv_0)}{t}\right)$. Then there are $\alpha, \beta \in \mathbb{Q}$ and $\delta, \eta \in]0, +\infty[\cap \mathbb{Q}$ such that $\rho \in]\alpha, \beta[$ and

$$g\left(\frac{f(x + t(u_0 + v_0)) - f(x + tv_0)}{t}\right) \notin]\alpha - 3\eta, \beta + 3\eta[\quad \text{for all } t \in]0, \delta[.$$

Choosing $u, v, w \in V$ with $w \neq v$ such that

$$\max\{\|u - u_0\|, \|v - v_0\|, \|w - w_0\|\} < \eta/(3\gamma_f\gamma_g),$$

and putting $I :=]\alpha - \eta, \beta + \eta[$ and $J :=]\alpha - 2\eta, \beta + 2\eta[$, we see that (13.4) and (13.5) are satisfied. It ensues that $x \in C(u, v, w, \delta, \eta, I, J)$ hence $x \in A$, which proves the first assertion of the lemma.

Finally, to see that the set of limit points coincides with the one as $t \uparrow 0$, it suffices to replace v by $-(u + v)$ in the result of the first assertion. □

In the case of a separable normed space Y with a dense sequence $(y_n)_{n\in\mathbb{N}}$, applying the above lemma with the function g_n given by $g_n(y) := \|y - y_n\|$ we obtain:

LEMMA 13.35. *Let $f : U \to Y$ be a Lipschitz mapping from an open set U of a separable normed space X into a separable normed space Y. Then there exists a σ-directionally porous set $A \subset U$ such that, for all $x \in U \setminus A$, $u, v \in X$ and $y \in Y$,*

$$\liminf_{t\downarrow 0} \left\|\frac{f(x + t(u + v)) - f(x + tv))}{t} - y\right\| = \liminf_{t\downarrow 0} \left\|\frac{f(x + tu) - f(x)}{t} - y\right\|,$$

$$\limsup_{t\downarrow 0} \left\|\frac{f(x + t(u + v)) - f(x + tv)}{t} - y\right\| = \limsup_{t\downarrow 0} \left\|\frac{f(x + tu) - f(x)}{t} - y\right\|.$$

PROPOSITION 13.36. Let $f : U \to Y$ be a Lipschitz mapping from an open set U of a separable normed space X into a normed space Y. Then there exists a σ-directionally porous set $A \subset U$ such that, for every $x \in U \setminus A$, the set E_x of all directions $u \in X$ for which the unilateral Dini directional derivative $d_D^+ f(x;u)$ exists, is a closed linear subspace of X. Further, for each $x \in U \setminus A$, the mapping defined on E_x by $u \mapsto d_D^+ f(x;u)$ is continuous and linear on E_x.

PROOF. Since $f(U)$ is separable, we may suppose that Y is separable. Let A be the set given by Lemma 13.35 and let $x \in U \setminus A$. From the Lipschitz property of f, it is easy to see that the mapping $u \mapsto d_D^+ f(x;u)$ is Lipschitz on E_x and that the set E_x is closed. To obtain that E_x is a vector subspace and $d_D^+ f(x;\cdot)$ is linear on E_x, fix $u, v \in E_x$. Then $d_D^+ f(x;u)$ and $d_D^+ f(x;v)$ exist, so applying the second equality of Lemma 13.35, with v and $y = d_D^+ f(x;u)$, gives

$$\limsup_{t \downarrow 0} \left\| \frac{f(x + t(u+v)) - f(x+tv)}{t} - d_D^+ f(x;u) \right\| = 0,$$

hence writing

$$\frac{f(x+t(u+v)) - f(x)}{t} = \frac{f(x+t(u+v)) - f(x+tv)}{t} + \frac{f(x+tv) - f(x)}{t},$$

we see that $d_D^+ f(x;u+v)$ exists and $d_D^+ f(x;u+v) = d_D^+ f(x;u) + d_D^+ f(x;v)$. Applying the second equality of Lemma 13.35 again but with $v = -u$ and $y = d_D^+ f(x;u)$, it results that

$$\limsup_{t \downarrow 0} \left\| \frac{f(x) - f(x-tu)}{t} - d_D^+ f(x;u) \right\| = 0,$$

hence $d_D^+ f(x;-u)$ exists and $d_D^+ f(x;-u) = -d_D^+ f(x;u)$.

Consequently, E_x is a vector subspace and $d_D^+ f(x;\cdot)$ is linear on E_x. □

Given a class $\mathcal{N}$ of subsets of the normed space X, we recall that it is a Borel *σ-ideal* if $\mathcal{N} \subset \mathcal{B}(X)$ and
(i) $\mathcal{N}$ is stable by countable union;
(ii) for any $B \in \mathcal{B}(X)$ for which there is some $N \in \mathcal{N}$ with $B \subset N$, one has $B \in \mathcal{N}$.

For any Lipschitz mapping $G : U \to Y$ from an open set U of a normed space X into a Banach space Y and for any $v \in X$, the set of $x \in U$ such that $d_D G(x;v)$ exists is easily seen to be Borelian. If in addition X is separable, the set of $x \in U$ such that G is Gâteaux differentiable is also Borelian. This allows us to consider Borel ideals containing sets of the first form.

PROPOSITION 13.37. Let $f : U \to Y$ be a Lipschitz mapping from an open set U of a separable normed space X into a Banach space Y and let $(v_n)_{n \in \mathbb{N}}$ be a complete sequence of nonzero vectors of X. Let $\mathcal{N}$ be a Borel σ-ideal of X which contains, for each $n \in \mathbb{N}$, the set $\{x \in U : d_D f(x;v_n) \text{ does not exist}\}$ and the sets $\{x \in X : d_D g(x;v_n) \text{ does not exist}\}$ for all Lipschitz functions $g : X \to \mathbb{R}$. Then the set

$$\{x \in U : f \text{ is not Gâteaux differentiable}\}$$

belongs to the Borel σ-ideal $\mathcal{N}$.

PROOF. Since each set $\{x \in U : d_D f(x;v_n) \text{ does not exist}\}$ belongs to $\mathcal{N}$ by assumption, it is enough to show that the set N of $x \in U$ where f is not Gâteaux differentiable but $d_D f(x;v_n)$ exists for every $n \in \mathbb{N}$ belongs to $\mathcal{N}$. By what precedes the statement, we know that N is Borelian. By Proposition 13.36, the set N is also σ-directionally porous. Then Theorem 13.23 allows us to write $N = \bigcup_{n\in\mathbb{N}}(N_n^+ \cup N_n^-)$, where the sets N_n^+, N_n^- are Borelian and σ-porous in directions v_n and $-v_n$ respectively. For each $n \in \mathbb{N}$, by definition of σ-porosity there are sequences $(N_{n,k}^+)_{k\in\mathbb{N}}$ and $(N_{n,k}^-)_{k\in\mathbb{N}}$ of porous sets in directions v_n and $-v_n$ respectively such that $N_n^+ = \bigcup_{k\in\mathbb{N}} N_{n,k}^+$ and $N_{n,k}^- = \bigcup_{k\in\mathbb{N}} N_{n,k}^-$. For the real-valued Lipschitz functions $g_{n,k}^+ := d(\cdot, N_{n,k}^+)$ and $g_{n,k}^- := d(\cdot, N_{n,k}^-)$, we know by Proposition 13.15(b) that $N_{n,k}^+ \subset \{x \in X : d_D g_{n,k}^+(x; v_n) \text{ does not exist}\} =: A_{n,k}^+$ and $N_{n,k}^- \subset \{x \in X : d_D g_{n,k}^-(x; v_n) \text{ does not exist}\} =: A_{n,k}^-$. Then we have $N \subset \bigcup_{n,k\in\mathbb{N}}(A_{n,k}^+ \cup A_{n,k}^-)$, so the Borelian set N belongs to the Borel σ-ideal $\mathcal{N}$ since the set $\bigcup_{n,k\in\mathbb{N}}(A_{n,k}^+ \cup A_{n,k}^-)$ itself belongs to $\mathcal{N}$ (due to the assumptions of the proposition). □

In the next lemma, for the Lipschitz mapping $f : U \to Y$ and for $x \in U$ and $v \in X$ we denote

$$\Theta(f, x, v) := \lim_{\varepsilon\downarrow 0}\left(\sup\left\{\left\|\frac{f(x+tv)-f(x)}{t} - \frac{f(x+sv)-f(x)}{s}\right\| : \begin{matrix} 0 < |t| < \varepsilon \\ 0 < |s| < \varepsilon \end{matrix}\right\}\right).$$

LEMMA 13.38. Let $f : U \to Y$ be a mapping from an open set U of a normed space X into a Banach space Y which is Lipschitzian on U with a Lipschitz constant $\gamma > 0$. The following hold:
(a) $d_D f(x; v)$ exists if and only if $\Theta(f, x, v) = 0$.
(b) The function $x \mapsto \Theta(f, x, v)$ is Borelian on U for every $v \in X$.
(c) The function $v \mapsto \Theta(f, x, v)$ is Lipschitz on X for every $x \in U$.
(d) Let $\varphi : \mathbb{R} \to X$ and $r \in \mathbb{R}$ with $\varphi(r) \in U$; if the mapping $t \mapsto \varphi(t) - tv$ is Lipschitzian with Lipschitz constant less than $\Theta(f, \varphi(r), v)/(4\gamma)$, then the mapping $f \circ \varphi : T \to Y$ is not derivable at r, where $T := \varphi^{-1}(U)$.

PROOF. The assertion (a) follows from the definition of $\Theta(f, x, v)$ and the completeness of Y, and (c) is easily obtained with 2γ as a Lipschitz constant on X. Concerning (b) it suffices to note that in the definition of $\Theta(f, x, v)$ we can require ε, t, s to be rational numbers.

To show (d), set $x := \varphi(r)$ and take any real $\varepsilon > 0$. The assumption in (d) ensures that $\Theta(f, x, v) > 0$, so we can choose reals t, s with $0 < |t| < \varepsilon$ and $0 < |s| < \varepsilon$ such that $x + tv$, $x + sv$, $\varphi(r + t)$, $\varphi(r + s)$ belong to U and

$$\left\|\frac{f(x+tv)-f(x)}{t} - \frac{f(x+sv)-f(x)}{s}\right\| > \frac{3}{4}\Theta(f, x, v).$$

Observe, for $\psi(\tau) := \varphi(\tau) - \tau v$, that

$$\begin{aligned}\left\|\frac{f(x+tv)-f(\varphi(r+t))}{t}\right\| &\le \frac{\gamma}{|t|}\|\varphi(r+t)-\varphi(r)-tv\| \\ &= \frac{\gamma}{|t|}\|\psi(r+t)-\psi(r)\| \le \frac{1}{4}\Theta(f, x, v),\end{aligned}$$

since f and ψ are Lipschitz with constants γ and $\Theta(f,x,v)/(4\gamma)$ respectively. A similar inequality also holds with s in place of t. So, writing

$$\begin{aligned}\Delta &:= \left\| \frac{f\circ\varphi(r+t)-f\circ\varphi(r)}{t} - \frac{f\circ\varphi(r+s)-f\circ\varphi(r)}{s} \right\| \\ &\geq \left\| \frac{f(x+tv)-f(x)}{t} - \frac{f(x+sv)-f(x)}{s} \right\| \\ &\quad - \left\| \frac{f(x+tv)-f(\varphi(r+t))}{t} \right\| - \left\| \frac{f(x+sv)-f(\varphi(r+s))}{s} \right\|,\end{aligned}$$

we obtain that

$$\Delta > \frac{3}{4}\Theta(f,x,v) - \frac{2}{4}\Theta(f,x,v) = \frac{1}{4}\Theta(f,x,v),$$

which entails that $\Theta(f\circ\varphi,r,1) \geq \Theta(f,\varphi(r),v)/4 > 0$, hence $f\circ\varphi$ is not derivable at r according to (a). □

THEOREM 13.39 (D. Preiss and L. Zajíček). *Let $f : U \to Y$ be a locally Lipschitz mapping from an open set U of a separable Banach space X into a Banach space Y with the Radon-Nykodým property. Then there exists a set $A \in \widetilde{\mathcal{A}}(X)$ such that f is Gâteaux differentiable, or equivalently Hadamard differentiable, at every point in $U \setminus A$.*

PROOF. We may suppose that f is Lipschitz on U with $\gamma > 0$ as a Lipschitz constant. Let us fix any $v \neq 0$ and prove that the set

$$A := \{x \in U : d_D f(x;v) \text{ does not exists}\}$$

belongs to $\widetilde{\mathcal{A}}(v)$. For each integer k, let us put $A_k := \{x \in U : \Theta(f,x,v) > 1/k\}$ and prove that $A_k \in \widetilde{\mathcal{A}}(v, 1/(4k\gamma))$. From Lemma 13.38(b) we know that A_k is Borelian. Consider any mapping $\varphi : \mathbb{R} \to X$ such that $t \mapsto \varphi(t) - tv$ is Lipschitzian with a Lipschitz constant not greater than $1/(4k\gamma)$. According to Lemma 13.38(d) the mapping $f\circ\varphi$ is not derivable at any $t \in \varphi^{-1}(A_k)$. Since $f\circ\varphi : \varphi^{-1}(U) \to \mathbb{R}$ is Lipschitzian and Y is Radon-Nikodým, the set $\varphi^{-1}(A_k)$ is Lebesgue negligible in $\mathbb{R}$, thus $A_k \in \widetilde{\mathcal{A}}(v, 1/(4k\gamma))$. Since $A = \bigcup_{k\in\mathbb{N}} A_k$ by Lemma 13.38(a), it ensues that $A \in \widetilde{\mathcal{A}}(v)$ as desired.

By what proved above, for any Lipschitz function $g : X \to \mathbb{R}$, the set $\{x \in X : d_D g(x;v) \text{ does not exist}\}$ belongs to $\widetilde{\mathcal{A}}(v)$. Consequently, taking any complete sequence $(v_n)_{n\in\mathbb{N}}$ in X, since the class $\mathcal{N}$ of sets in the form $\bigcup_{n\in\mathbb{N}} N_n$ with $N_n \in \widetilde{\mathcal{A}}(v_n)$ is a Borel σ-ideal of subsets of X, Proposition 13.37 guarantees that the set N of points in U where f is not Gâteaux differentiable can be written as $N = \bigcup_{n\in\mathbb{N}} N_n$, where each $N_n \in \widetilde{\mathcal{A}}(v_n)$. This justifies that $N \in \widetilde{\mathcal{A}}(X)$ as desired. □

Through the above theorem and the concept of Haar null sets, we will extend Theorem 2.83 on the representation of Clarke subdifferential of locally Lipschitz functions in terms of Hadamard derivatives. This requires first the following proposition which answers the question of the previous section concerning the Clarke directional derivative and Haar null sets.

PROPOSITION 13.40. *Let X be a Banach space and $f : U \to \mathbb{R}$ be a real-valued locally Lipschitz function on an open set U of X. Then for any Haar null set A of*

X one has

$$f^\circ(x;v) = \limsup_{A\not\ni u\to x} \underline{d}_D^+ f(u;v) \quad \text{for all } x\in U, v\in X.$$

PROOF. The second member α is obviously not greater than the first. Let us prove the converse inequality. Fix any $x \in U$ and any nonzero v in X. Let X_1 be a topological vector complement of (the one dimensional space) $\mathbb{R}v$, so $X = X_1 \oplus \mathbb{R}v$ and any $u \in X$ can be written as $u = (u_1, u_2 v)$ with $u_1 \in X_1$ and $u_2 \in \mathbb{R}$. Fix any real $\varepsilon > 0$ and by definition of α choose some real $r > 0$ such that $\underline{d}_D^+ f(x+h;v) < \alpha + \varepsilon$ for all $h \in X$ with $x + h \in B_X(x,r) \setminus A$. By Theorem 13.32 there exists a universally measurable set Q_1 in X_1 with $X_1 \setminus Q_1$ Haar null in X_1 and such that, for all $q_1 \in Q_1$, the set $A(q_1,\cdot)$ is Haar Null in $\mathbb{R}v$. Then for any $h \in X$ with $x_1 + h_1 \in Q_1$, the set $-x_2 v - h_2 v + A(x_1 + h_1, \cdot)$ is Haar null in $\mathbb{R}v$, or equivalently the set

$$\{t\in\mathbb{R} : tv \in -x_2 v - h_2 v + A(x_1+h_1,\cdot)\} = \{t \in \mathbb{R} : x+h+tv \in A\}$$

is Lebesgue null in $\mathbb{R}$. It results, for any $\delta \in]0, r/(2\|v\|)[$ and any $h \in B_X(0, r/2)$ with $x_1 + h_1 \in Q_1$, that

$$\delta^{-1}\big(f(x+h+\delta v) - f(x+h)\big) = \delta^{-1}\int_0^\delta \underline{d}_D^+ f(x+h+tv;v)\,dt \le \alpha+\varepsilon.$$

The latter inequality being true for all $h \in (-x_1 + Q_1) \times \mathbb{R}v$ with $\|h\| < r/2$ and the set Q_1 being dense in X_1 by Theorem 13.28(c), it ensues that the inequality holds true for all $h \in B_X(0, r/2)$ and $\delta \in]0, r/(2\|v\|)[$. Consequently,

$$f^\circ(x;v) = \limsup_{\delta\downarrow 0, h\to 0} \delta^{-1}\big(f(x+h+\delta v) - f(x+h)\big) \le \alpha + \varepsilon,$$

which finishes the proof by taking the limits as $\varepsilon \downarrow 0$. □

THEOREM 13.41 (derivative representation of C-subdifferential on separable Banach spaces). Let X be a separable Banach space and let $f : U \to \mathbb{R}$ be a real-valued locally Lipschitz function on an open set U of X. Then, f is Gâteaux differentiable outside a Haar null set in U and for any Haar null set A in X one has for every $x \in U$

$$\partial_C f(x) = \overline{\mathrm{co}}^* \left(\left\{\lim_{n\to\infty} D_G f(x_n) : x_n \to x \text{ with } x_n \in \operatorname{Dom} D_G f \setminus A\right\}\right),$$

and with $N := A \cup (U \setminus \operatorname{Dom} D_G f)$

$$f^\circ(x;h) = \limsup_{N\not\ni u\to x} \langle D_G f(u), h\rangle \quad \text{for all } h \in X.$$

PROOF. Denoting by $E(x)$ the above second member and considering the set $N = A\cup(U\setminus \operatorname{Dom} D_G f)$ in X which is Haar null by Theorem 13.39 and Proposition 13.33, we see that the support function in the direction $v \in X$ of the weak* closed convex set $E(x)$ in X^* is $\limsup_{N\not\ni u\to x} D_G f(u)(v)$. Since the latter coincides with $f^\circ(x;v)$ by Proposition 13.40 and since $f^\circ(x;\cdot)$ is the support function of the weak* closed convex set $\partial_C f(x)$, we conclude that $\partial_C f(x) = E(x)$ as desired. □

In the case of nonseparable Banach spaces we have the following results.

PROPOSITION 13.42. Let $(X, \|\cdot\|)$ be a Banach space and $f : U \to \mathbb{R}$ be a locally Lipschitz real-valued function on an open set U of X. Then, for any $x \in U$ and any $h \in X$, there exists a closed separable subspace E of X containing x and h such that

$$f^\circ(x; h) = \limsup_{\mathrm{Dom}\, D_G f_E \ni u \to x} D_G f_E(u)(h),$$

where f_E denotes the restriction of f to $U \cap E$.

In particular, for any $x \in U$ and $h \in X$, denoting by B the set of $u \in U$ such that the bilateral limit $\lim_{t \to 0} t^{-1}\big(f(u+th) - f(u)\big)$ exists, the following equality holds true

$$f^\circ(x; h) = \limsup_{B \ni u \to x} \lim_{t \to 0} t^{-1}\big(f(u+th) - f(u)\big).$$

PROOF. Choose a sequence $(x_n)_n$ in X converging to x and a sequence $(t_n)_n$ in $]0, +\infty[$ tending to 0 such that $f^\circ(x; h) = \lim_{n \to \infty} t_n^{-1}\big(f(x_n + t_n h) - f(x_n)\big)$. Denote by E the closed vector subspace of X spanned by $\{x, h\} \cup \{x_n : n \in \mathbb{N}\}$, and denote by f_E the restriction of f to $U \cap E$. From the above equality we see that $f^\circ(x; h) = (f_E)^\circ(x; h)$, and on the other hand from Theorem 13.41 we have $(f_E)^\circ(x; h) = \limsup_{\mathrm{Dom}\, D_G f_E \ni u} D_G f_E(u)(h)$ (since E is a separable Banach space). This justifies the first equality of the proposition.

Concerning the second equality of the proposition, denoting its second member by β, the inequality $\beta \le f^\circ(x; h)$ is clear and the converse inequality $f^\circ(x, h) \le \beta$ follows from the first equality of the proposition. $\square$

13.3.2. More on interior tangent property via signed distance function. This section applies Proposition 13.42 above to furnish in the Banach space setting other arguments for an important part of the proof of Theorem 2.154 characterizing the interior tangent property of sets via the signed distance function. For convenience we reproduce in Banach spaces the statement of the theorem.

THEOREM 13.43 (interior tangent property in Banach spaces via signed distance). Let S be a subset of a Banach space $(X, \|\cdot\|)$ and $\overline{x} \in S \cap \mathrm{bdry}\, S$. Then, the following assertions hold:

(a) The set S has the interior tangent property at $\overline{x}$ in a nonzero direction h if and only if

$$(\mathrm{sgd}_S)^\circ(\overline{x}; h) < 0.$$

(b) The set S has the interior tangent property at $\overline{x}$ if and only if $0 \notin \partial_C \mathrm{sgd}_S(\overline{x})$.

PROOF. We keep the arguments in the proof of Theorem 2.154 for the deduction of (b) from (a) and also for the interior tangent property of S at $\overline{x}$ in the direction h under the inequality $(\mathrm{sgd}_S)^\circ(\overline{x}; h) < 0$.

Conversely, suppose that S has the interior tangent property at $\overline{x}$ in the direction h. We keep the arguments in the proof of Theorem 2.154 yielding a real $\varepsilon_0 > 0$ such that for any $0 < \varepsilon < \varepsilon_0$ the following Claims 1-3 in Theorem 2.154 hold true:

Claim 1: $d_S(x) \ge d_S(x + tv)$ for all $t \in]0, \varepsilon[$, $v \in B[h, \varepsilon]$ and $x \in B(\overline{x}, \varepsilon)$.

Claim 2: For all $0 < t < \frac{\varepsilon}{2(1+\|h\|)}$ and $x \in B(\overline{x}, \varepsilon/2) \cap \overline{S^c}$, one has $x - th \notin \overline{S}$ and $d_S(x - th) \ge d_S(x) + \varepsilon t/2$.

Claim 3: For all $0 < t < \frac{\varepsilon}{2(1+\|h\|)}$ and $x \in B(\overline{x}, \varepsilon/2) \cap \overline{S}$, one has $x + th \notin \overline{S^c}$ and $d_{S^c}(x + th) \ge d_{S^c}(x) + \varepsilon t/2$.

Now instead of Claim 4 in the proof of Theorem 2.154 we will invoke Proposition 13.42. Denote by B the set of $u \in B(\overline{x}, \varepsilon/2)$ where the bilateral limit

$$\rho(u;h) := \lim_{\tau \to 0} \tau^{-1}\big(\mathrm{sgd}_S(u+\tau h) - \mathrm{sgd}_S(u)\big)$$

exists. Fix any $x \in B$. If $x \in \overline{S^c}$, we see from Claim 2 that

$$\rho(x;h) = \lim_{\tau \uparrow 0} \tau^{-1}\big(\mathrm{sgd}_S(x+\tau h) - \mathrm{sgd}_S(x)\big) = \lim_{\tau \uparrow 0} \tau^{-1}\big(d_S(x+\tau h) - d_S(x)\big) \le -\frac{1}{2}\varepsilon,$$

and if $x \in \overline{S}$, by Claim 3

$$\rho(x;h) = \lim_{\tau \downarrow 0} \tau^{-1}\big((-d_{S^c}(x+\tau h) - (-d_{S^c}(x))\big) \le -\frac{1}{2}\varepsilon.$$

Consequently, $\rho(x;h) \le -\varepsilon/2$ for all $x \in B$, so the second equality in Proposition 13.42 guarantees that $(\mathrm{sgd}_S)^\circ(\overline{x};h) \le -\varepsilon/2$. It ensues that $(\mathrm{sgd}_S)^\circ(\overline{x};h) \le 0$, so the proof of the theorem is finished. □

13.3.3. Non-differentiability points of one-sided Lipschitz functions. From Theorem 13.39, one can also extend to separable Banach spaces the classical Stepanov theorem concerning the differentiability points of one-sided Lipschitz functions defined on $\mathbb{R}^n$.

Given two normed spaces X, Y and a nonempty open set U of X, recall (see Definition 4.15) that a mapping $f : U \to Y$ is called *one-sided Lipschitz* (or *pointwise Lipschitz*) at $x \in U$ provided that

$$\limsup_{y \to x} \frac{\|f(y) - f(x)\|}{\|y - x\|} < +\infty.$$

THEOREM 13.44 (smallness of sets of points of non-differentiability of one-sided Lipschitz functions). Let $f : U \to \mathbb{R}$ be a real-valued function on an open set U of a separable Banach space X. There exists a set $A \in \widetilde{\mathcal{A}}(X)$ such that f is Hadamard differentiable at each point in $U \setminus A$ where f is one-sided Lipschitz.

PROOF. Since f is bounded around any point at which it is one-sided Lipschitz, we may and do suppose that f is bounded on U. Let $(x_n)_{n \in \mathbb{N}}$ be a dense sequence in U and for each n let $k_n \in \mathbb{N}$ such that $B(x_n, 1/k_n) \subset U$. For each pair of integers (n,k) with $k \ge k_n$ define the functions $f_{n,k}$ and $F_{n,k}$ on $U_{n,k} := B(x_n, 1/k)$ by

$$f_{n,k}(x) := \sup\{\varphi(x) : \varphi \le f \text{ on } U_{n,k} \text{ and } \varphi \text{ is } k\text{−Lipschitz on } U_{n,k}\},$$

$$F_{n,k}(x) := \sup\{\varphi(x) : \varphi \ge f \text{ on } U_{n,k} \text{ and } \varphi \text{ is } k\text{−Lipschitz on } U_{n,k}\}.$$

These functions are clearly k-Lipschitz on $U_{n,k}$ and $f_{n,k} \le f \le F_{n,k}$ on $U_{n,k}$. Further, by Proposition 5.11 the function f is Hadamard differentiable at each point $u \in U$ for which there is some (i,k) with $k \ge k_n$ and $u \in U_{n,k}$ such that both functions $f_{n,k}$ and $F_{n,k}$ are Hadamard differentiable at u with $f_{n,k}(u) = F_{n,k}(u)$. Observe also by Theorem 13.39 that the set

$$A = \bigcup_{n \in \mathbb{N}, k \ge k_n} \{x \in U_{n,k} : \text{either } f_{n,k} \text{ or } F_{n,k} \text{ is not Hadamard differentiable at } x\}$$

belongs to $\widetilde{\mathcal{A}}(X)$. Fix any $u \in U \setminus A$ where f is one-sided Lipschitz. There are $n \in \mathbb{N}$ and $k \ge k_n$ with $u \in U_{n,k}$ such that for every $x \in U_{n,k}$

$$|f(x) - f(u)| \le k\|x - u\|,$$

hence in particular

$$f(u) - k\|u - x\| \le f_{n,k}(x) \le F_{n,k}(x) \le f(u) + k\|u - x\|,$$

so $f_{n,k}(u) = F_{n,k}(u)$. This and what precedes allow us to conclude that f is Hadamard differentiable at the point u. □

Proposition 13.46 below goes further in proving, for any one-sided Lipschitz function f, that $d_H f(x;\cdot)$ exists whenever it is so for $d_D f(x;\cdot)$ except for all x in a σ-directionally porous set. Let us show first the following result which is similar to Proposition 5.2.

PROPOSITION 13.45. Let X and Y be two normed spaces, $f : U \to Y$ be a mapping from an open set U of X into Y, and let $x \in U$. The following assertions are equivalent:
(a) $d_H^+ f(x;0)$ exists;
(b) $d_H f(x;0)$ exists;
(c) $d_H f(x;0) = 0$;
(d) f is one-sided Lipschitz at x.

PROOF. The implications (c) ⇒ (b) and (b) ⇒ (a) are obvious.
(d) ⇒ (c). Suppose (d) and choose a real $\gamma \ge 0$ and a neighborhood V of x such that $\|f(y) - f(x)\| \le \gamma\|y - x\|$ for all $y \in V$. Then for any real $t \neq 0$ and any $v \in X$ with $x + tv \in V$ we have

$$\|t^{-1}\big(f(x + tv) - f(x)\big)\| \le \gamma\|v\| \quad \text{so} \quad \lim_{v\to 0, t\to 0} t^{-1}\big(f(x + tv) - f(x)\big) = 0,$$

which is the property (c).
(a) ⇒ (d). Suppose (a) and choose a real $0 < r < 1$ such that, for all $t \in]0, r]$ and $v \in B[0, r]$, one has

$$\|t^{-1}\big(f(x + tv) - f(x)\big)\| \le 1, \quad \text{that is,} \quad \|f(x + tv) - f(x)\| \le t.$$

For any $y \in B(x, r^2)$ with $y \neq x$, taking $t := r^{-1}\|y - x\|$ and $v := r(y - x)/\|y - x\|$ yields $\|f(y) - f(x)\| \le r^{-1}\|y - x\|$. This finishes the proof. □

PROPOSITION 13.46. Let U be an open set of a separable Banach space X and $f : U \to Y$ be a mapping from U into a normed space Y. Let A be the set of $x \in U$ such that f is one-sided Lipschitz at x and such that there exists some $v \in X$ for which $d_D^+(x; v)$ exists but $d_H^+(x; v)$ does not exist. Then A is σ-directionally porous.

PROOF. Consider, for each integer $k \in \mathbb{N}$, the set

$$A_k := \{u \in A : \|f(y) - f(u)\| \le k\|y - u\|, \ \forall y \in B(u, 1/k)\}.$$

Since $A = \bigcup_{k\in\mathbb{N}} A_k$, it suffices to show that each nonempty set A_k is σ-directionally porous. Fix $k \in \mathbb{N}$ with $A_k \neq \emptyset$ and consider any $x \in A_k$. Choose $v \in X$ such that $d_D^+ f(x; v)$ exists but $d_H^+(x; v)$ does not exist. Proposition 13.45 ensures that $v \neq 0$. We will prove that A_k is porous at x in the direction v. Put $y := d_D^+ f(x; v)$. Since $d_H^+ f(x; v)$ does not exist, there is some $\varepsilon > 0$ such that, for every $n \in \mathbb{N}$, we can choose $h_n \in B(v, 1/n)$ and $t_n \in]0, 1/n[$ with

$$\|t_n^{-1}\big(f(x + th_n) - f(x)\big) - y\| > \varepsilon.$$

Set $x_n := x + t_n v$, $x'_n := x + t_n h_n$, and $\rho := \varepsilon(3k)^{-1}$. It is clearly enough to prove that, for some n_0,

$$B(x_n, \rho t_n) \cap A_k = \emptyset \quad \text{for all } n \geq n_0.$$

Suppose the contrary, that is, there exists an infinite set $N \subset \mathbb{N}$ such that for each $n \in N$ we can choose some $u_n \in B(x_n, \rho t_n) \cap A_k$. Observe that

$$\|u_n - x_n\| < \rho t_n, \quad \|u_n - x'_n\| \leq \|x_n - x'_n\| + \|u_n - x_n\| \leq (t_n/n) + \rho t_n.$$

Since $\|t_n^{-1}(f(x_n) - f(x)) - y\| \to 0$ as $n \to 0$, we can fix some $n \in N$ such that

$$\|t_n^{-1}(f(x_n) - f(x)) - y\| + k/n < \varepsilon/4, \; \|u_n - x_n\| < 1/k, \; \|u_n - x'_n\| < 1/k.$$

Since $u_n \in A_k$, we have

$$\|f(x_n) - f(u_n)\| \leq k\|x_n - u_n\| \quad \text{and} \quad \|f(x'_n) - f(u_n)\| \leq k\|x'_n - u_n\|.$$

It follows from the choice of h_n and the above inequalities that

$$\begin{aligned} \varepsilon &< \left\| \frac{f(x'_n) - f(x)}{t_n} - y \right\| \\ &\leq \left\| \frac{f(x_n) - f(x)}{t_n} - y \right\| + \frac{\|f(x_n) - f(u_n)\|}{t_n} + \frac{\|f(x'_n) - f(u_n)\|}{t_n}, \end{aligned}$$

hence

$$\varepsilon < \left\| \frac{f(x_n) - f(x)}{t_n} - y \right\| + k\rho + k(1/n + \rho) = \left\| \frac{f(x_n) - f(x)}{t_n} - y \right\| + (2/3)\varepsilon + k/n < \varepsilon,$$

which is a contradiction. We then have, as desired, $B(x_n, \rho t_n) \cap A_k = \emptyset$ for n sufficiently large. □

As a direct consequence of Propositions 13.45 and 13.46, we have:

PROPOSITION 13.47. Let U be an open set of a separable Banach space X and $f : U \to Y$ be a mapping from U into a normed space Y. Then the set of $x \in U$ such that f is one-sided Lipschitz at x, Gâteaux differentiable at x but not Hadamard differentiable at x, is σ-directionally porous.

13.4. Zajíček extension of Denjoy-Young-Saks theorem

Our objective in this section is to extend the (one-variable) Denjoy-Young-Saks theorem (see Theorem 6.42) to real-valued functions f defined on an open set U of a separable Banach space X. The first step is the study of points $x \in U$ such that $\liminf\limits_{u \to x} \frac{f(u)-f(x)}{\|u-x\|} = -\infty$ and such that f is bounded from below by $f(x)$ around x on a certain set $K(x, v, \delta)$. This set $K(x, v, \delta)$ is defined, for $v \in X$ with $\|v\| = 1$ and $\delta > 0$, by

$$K(x, v, \delta) = \left\{ y \in X : y \neq x, \; \left\| v - \frac{y - x}{\|y - x\|} \right\| < \delta \right\} = \{x + th : t > 0, \; h \in B(v, \delta) \cap \mathbb{S}_X\},$$

where $\mathbb{S}_X := \{u \in X : \|u\| = 1\}$. We note that $x \notin K(x, v, \delta)$. On the other hand, $K(0, v, \delta)$ is an open cone and its translate by x coincides with $K(x, v, \delta)$, that is,

$$K(x, v, \delta) = x + K(0, v, \delta).$$

LEMMA 13.48. Let $f : U \to \mathbb{R}$ be a function defined on an open set U of a separable Banach space X and let $\delta > 0$ and $v \in X$ with $\|v\| = 1$. Then for the above translated cone $K_x := K(x, v, \delta)$, the set of all $x \in U$ such that

$$f(x) < \liminf_{K_x \ni y \to x} f(y),$$

is σ-directionally porous, and hence $\widetilde{C}(X)$-null.

PROOF. Denote by A the set in the lemma. Write $\mathbb{Q}$ in the form $\mathbb{Q} = \{q_j : j \in \mathbb{N}\}$ and for each $j \in \mathbb{N}$ put

$$A_j := \left\{ x \in A : f(x) < q_j < \liminf_{K_x \ni y \to x} f(y) \right\}.$$

Take j with $A_j \neq \emptyset$ and fix any $x \in A_j$. There exists a real $r > 0$ such that $B(x, r) \subset U$ and such that, for all $y \in K_x \cap B(x, r)$, we have $f(y) > q_j$, and hence $y \notin A_j$. Since K_x is of the form $K_x = x + K_0$ where K_0 is an open cone with $v \in K_0$, there is a real $0 < c < 1$ such that $B(v, c) \subset K_0$. Take a sequence $(t_n)_{n \in \mathbb{N}}$ tending to 0 with $0 < t_n < r/2$, so $B(x + t_n v, ct_n) \subset B(x, r)$. Since $B(t_n v, ct_n) \subset K_0$ (because K_0 is a cone) it results that $B(x + t_n v, ct_n) \subset K_x \cap B(x, r)$, which ensures by what precedes that $B(x + t_n v, ct_n) \cap A_j = \emptyset$. This says that A_j is directionally porous and hence $A = \bigcup_{j \in \mathbb{N}} A_j$ is σ-directionally porous. □

LEMMA 13.49. Let $f : U \to \mathbb{R}$ be a function defined on an open set U of a separable Banach space X and let $0 < \delta < 1$ and $v \in X$ with $\|v\| = 1$. Let $A(f, v, \delta)$ be the set of all $x \in U$ such that the following conditions (i) and (ii) are satisfied:

(i) $f(y) \geq f(x)$ for all $y \in K(x, v, \delta) \cap B_U(x, \delta)$

(ii) $\liminf\limits_{y \to x} \frac{f(y) - f(x)}{\|y - x\|} = -\infty$,

where $B_U(x, \delta) := B(x, \delta) \cap U$.

Then the set $A(f, v, \delta)$ is $\widetilde{\mathcal{C}}(X)$-null.

PROOF. We may suppose that $U = X$. We consider two cases.

Case I. Suppose that $A := A(f, v, \delta)$ is Borelian. For $\varepsilon := \delta/5$, we claim that $A \in \widetilde{\mathcal{A}}(v, \varepsilon)$. Suppose the contrary, so there is a mapping $\varphi : \mathbb{R} \to X$ such that $t \mapsto \varphi(t) - tv$ is Lipschitz on $\mathbb{R}$ with a Lipschitz constant not greater than ε and such that $\lambda(\varphi^{-1}(A)) > 0$, where λ denotes the Lebesgue measure. Clearly we can choose an open interval I of $\mathbb{R}$ such that, for $T := I \cap \varphi^{-1}(A)$,

$$\operatorname{diam} I < \frac{\delta}{1 + \varepsilon} \quad \text{and} \quad \lambda(T) > 0. \tag{13.6}$$

Let us prove that the restriction $g := (f \circ \varphi)_{|T}$ of $f \circ \varphi$ to T is nondecreasing. Consider $t_1, t_2 \in T$ with $t_1 < t_2$. Since $t \mapsto \varphi(t) - tv$ is Lipschitz with a Lipschitz constant $\varepsilon < 1$, we have $\varphi(t_1) \neq \varphi(t_2)$, so by Lemma 10.52

$$\begin{aligned} \left\| v - \frac{\varphi(t_2) - \varphi(t_1)}{\|\varphi(t_2) - \varphi(t_1)\|} \right\| &= \left\| \frac{(t_2 - t_1)v}{\|(t_2 - t_1)v\|} - \frac{\varphi(t_2) - \varphi(t_1)}{\|\varphi(t_2) - \varphi(t_1)\|} \right\| \\ &\leq \frac{2}{|t_2 - t_1|} \left\| \big(t_2 v - t_1 v\big) - \big(\varphi(t_2) - \varphi(t_1)\big) \right\| \leq 2\varepsilon < \frac{\delta}{2}, \end{aligned}$$

thus $\varphi(t_2) \in K\big(\varphi(t_1), v, \delta/2\big)$. Since we also have by (13.6)

$$(1 - \varepsilon)(t_2 - t_1) \leq \|\varphi(t_2) - \varphi(t_1)\| \leq (1 + \varepsilon)(t_2 - t_1) < \delta, \tag{13.7}$$

we see that

$$\varphi(t_2) \in K(\varphi(t_1), v, \delta/2) \cap B(\varphi(t_1), \delta). \tag{13.8}$$

This and the inclusion $\varphi(t_1) \in A$ says by the condition (i) that $f(\varphi(t_2)) \geq f(\varphi(t_1))$, which justifies the nondecreasing property of g.

Now consider the interval $J :=]\inf T, \sup T[$ and put $\tilde{g}(t) := \sup\{g(\tau) : \tau \in T \cap]-\infty, t]\}$ for all $t \in J$, so $\tilde{g}$ is a finite-valued extension of g on J and it is also nondecreasing on J, hence derivable Lebesgue almost everywhere on J. Since $\lambda(T) > 0$, we can choose some $t_0 \in T \cap J$ which is a Lebesgue density point of the set T and where $\tilde{g}$ is derivable. Then putting $\gamma := 1 + |(\tilde{g})'(t_0)|$, there exists some $0 < \alpha < \varepsilon$ such that for any $0 < r < \alpha$ there is some $\theta \in]t_0 - r - \varepsilon r, t_0 - r[\cap T$ such that

$$|f(\varphi(\theta)) - f(\varphi(t_0))| \leq \gamma|\theta - t_0|. \tag{13.9}$$

Setting $x_0 := \varphi(t_0)$ and $\beta := 6\delta^{-1}\gamma(1+\varepsilon)$, the inclusion $\varphi(t_0) \in A$ and the condition (ii) yields some $y \in X \setminus \{x_0\}$ for which

$$r := 6\delta^{-1}\|y - x_0\| < \alpha \quad \text{and} \quad \frac{f(y) - f(x_0)}{\|y - x_0\|} < -\beta, \tag{13.10}$$

and by what precedes we can choose some $\theta \in]t_0 - r - \varepsilon r, t_0 - r[\cap T$ satisfying (13.9). Note that (13.9) says that $f(\varphi(\theta)) - f(x_0) > -\gamma r(1+\varepsilon)$, while the second inequality in (13.10) implies that

$$f(y) - f(x_0) < -\beta\|y - x_0\| = -\gamma r(1+\varepsilon),$$

so $y \neq \varphi(\theta)$. Further, since $\theta \in T$ and $t_0 \in T$, we can take $t_1 = \theta$ and $t_2 = t_0$ in (13.7) and (13.8) to obtain

$$(1-\varepsilon)r \leq \|x_0 - \varphi(\theta)\| \leq (1+\varepsilon)^2 r \quad \text{and} \quad \left\| \frac{x_0 - \varphi(\theta)}{\|x_0 - \varphi(\theta)\|} - v \right\| < \frac{\delta}{2}. \tag{13.11}$$

Keeping in mind that $y \neq \varphi(\theta)$, Lemma 10.52 combined with the first inequality in (13.11) and with the inequality $\varepsilon < 1/3$ entails that

$$\begin{aligned}
\left\| \frac{x_0 - \varphi(\theta)}{\|x_0 - \varphi(\theta)\|} - \frac{y - \varphi(\theta)}{\|y - \varphi(\theta)\|} \right\| &\leq 2\frac{\|y - x_0\|}{\|x_0 - \varphi(\theta)\|} \\
&\leq \frac{2\|y - x_0\|}{(1-\varepsilon)6\delta^{-1}\|y - x_0\|} = \frac{\delta}{3(1-\varepsilon)} < \frac{\delta}{2},
\end{aligned}$$

hence by the last inequality in (13.11) we obtain

$$\left\| \frac{y - \varphi(\theta)}{\|y - \varphi(\theta)\|} - v \right\| < \delta.$$

Observing further by the definition $r = 6\delta^{-1}\|y - x_0\|$ in (13.10) and by (13.11) that

$$\|y - \varphi(\theta)\| \leq \|x_0 - \varphi(\theta)\| + r < 5r < 5\varepsilon = \delta,$$

it ensues that $y \in K(\varphi(\theta), v, \delta) \cap B(\varphi(\theta), \delta)$, which combined with the inclusion $\varphi(\theta) \in A$ yields through the condition (i) that $f(y) \geq f(\varphi(\theta))$. We deduce that

$$f(x_0) - f(\varphi(\theta)) = \big(f(x_0) - f(y)\big) + \big(f(y) - f(\varphi(\theta))\big) \geq f(x_0) - f(y) > \beta\|y - x_0\|,$$

where the last inequality is due to the second inequality in (13.10). On the other hand, by (13.9) we have

$$f(x_0) - f(\varphi(\theta)) \leq \gamma|\theta - t_0| < \gamma(1+\varepsilon)r = \gamma(1+\varepsilon)6\delta^{-1}\|y - x_0\| = \beta\|y - x_0\|,$$

which contradicts the above inequality $f(x_0) - f(\varphi(\theta)) > \beta\|y - x_0\|$. Consequently $A \in \widetilde{\mathcal{A}}(v, \varepsilon)$, and hence A is $\widetilde{\mathcal{C}}(X)$-null.

Case II. Suppose now that $A(f, v, \delta)$ is not necessarily Borelian. We may suppose that f is bounded since it is not difficult to verify that $A(f, v, \delta) \subset \bigcup_{n\in\mathbb{N}} A(f_n, v, \delta)$ with $f_n := \max\{-n, \min\{n, f\}\}$.

The function g defined by

$$g(x) := \min\left\{f(x), \liminf_{y\to x} f(y)\right\} \quad \text{for all } x \in X,$$

is easily seen to be bounded and lower semicontinuous. Further,

$$A(f, v, \delta) \subset A(g, v, \delta) \cup \{x \in A(f, v, \delta) : f(x) > g(x)\}. \tag{13.12}$$

Indeed, let $x \in A(f, v, \delta)$ with $f(x) \le g(x)$. Then $f(x) = g(x)$ since $g(y) \le f(y)$ for all $y \in X$. This and the inclusion $x \in A(f, v, \delta)$ give

$$\liminf_{y\to x} \frac{g(y) - g(x)}{\|y - x\|} \le \liminf_{y\to x} \frac{f(y) - f(x)}{\|y - x\|} = -\infty, \text{ so } \liminf_{y\to x} \frac{g(y) - g(x)}{\|y - x\|} = -\infty.$$

Take any $y \in V(x) := K(x, v, \delta) \cap B(x, \delta)$, hence $f(y) \ge f(x) = g(x)$ since $x \in A(f, v, \delta)$. The same inclusion $x \in A(f, v, \delta)$ also says that $f(u) \ge f(x) = g(x)$ for all $u \in V(x)$, thus $\liminf_{u\to y} f(u) \ge g(x)$ since $V(x)$ is an open set containing y. It ensues that

$$g(y) = \min\left\{f(y), \liminf_{u\to y} f(u)\right\} \ge g(x).$$

Therefore, $x \in A(g, v, \delta)$, hence the inclusion (13.12) is justified.

Write $A(g, v, \delta) = C \setminus L$, where

$$C := \{x \in X : g(y) \ge g(x), \forall y \in V(x)\}, \; L := \left\{x \in X : \liminf_{y\to x} \frac{g(y) - g(x)}{\|y - x\|} > -\infty\right\}.$$

To show that C is closed, consider any sequence $(x_n)_n$ in C converging to some $x \in X$, and take any $y \in V(x)$. Then $\left\|v - \frac{y-x}{\|y-x\|}\right\| < \delta$ and $\|y - x\| < \delta$, so for all n large enough we have $\left\|v - \frac{y-x_n}{\|y-x_n\|}\right\| < \delta$ and $\|y - x_n\| < \delta$, or equivalently $y \in V(x_n)$, which entails $g(x_n) \le g(y)$ since $x_n \in C$. The lower semicontinuity of g ensures that $g(x) \le g(y)$, so $x \in C$, justifying the closedness of C.

Concerning the set L, let us write $L = \bigcup_{n\in\mathbb{N}} L_n$, where

$$L_n := \{x \in X : g(y) - g(x) \ge -n\|y - x\|, \; \forall y \in B(x, 1/n)\},$$

and let us prove that L_n is closed. Fix $n \in \mathbb{N}$ and take any $y \in B(x, 1/n)$. Consider any sequence $(x_k)_k$ in L_n converging to some $x \in X$. For k large enough we have $y \in B(x_k, 1/n)$, hence $g(y) \ge g(x_k) - n\|y - x_k\|$, so the lower semicontinuity of $g - n\|y - \cdot\|$ yields $g(y) \ge g(x) - n\|y - x\|$. This means that $x \in L_n$ and justifies the closedness of L_n.

The set $A(g, v, \delta)$ is then Borelian and Case I guarantees that it is $\widetilde{\mathcal{C}}(X)$-null.

Now fix any $x \in A(f, v, \delta)$ with $f(x) > g(x)$. Take any $y \in V(x)$. Then $f(y) \ge f(x)$ since $x \in A(f, v, \delta)$. The same inclusion $x \in A(f, v, \delta)$ also implies that $f(u) \ge f(x)$ for all $u \in V(x)$, thus $\liminf_{u\to y} f(u) \ge f(x)$ since $V(x)$ is an open set containing y. It ensues that $g(y) = \min\left\{f(y), \liminf_{u\to y} f(u)\right\} \geq f(x)$, thus since $V(x) = K_x \cap B(x,\delta)$ with $K_x := K(x,v,\delta)$, we obtain $\liminf_{K_x \ni y\to x} g(y) \geq f(x) > g(x)$. This means that x belongs to the set in Lemma 13.48 with g in place of f. The latter lemma then ensures that the set $\{x \in A(f,v,\delta) : f(x) > g(x)\}$ is $\widetilde{\mathcal{C}}(X)$-null, which combined with what precedes and (13.12) concludes that $A(f,v,\delta)$ is $\widetilde{\mathcal{C}}(X)$-null. □

LEMMA 13.50. *Let $f : U \to \mathbb{R}$ be a function defined on an open set U of a separable Banach space X and let $\gamma > 0$, $0 < \delta < 1$ and $v \in X$ with $\|v\| = 1$. Let $A_\gamma(f,v,\delta)$ be the set of all $x \in U$ such that the following conditions (i) and (ii) are satisfied:*

(i) $f(y) - f(x) \geq -\gamma\|y - x\|$ *for all* $x \in K(x,v,\delta) \cap B_U(x,\delta)$

(ii) $\liminf_{y\to x} \frac{f(y)-f(x)}{\|y-x\|} = -\infty$,

where $B_U(x,\delta) := B(x,\delta) \cap U$.

Then the set $A_\gamma(f,v,\delta)$ is $\widetilde{\mathcal{C}}(X)$-null.

PROOF. Fix $x^* \in X^*$ with $\langle x^*, v\rangle > 2\gamma$, and set $g = f + x^*$ and $\delta_0 := \min\{\delta, \gamma\|x^*\|^{-1}\}$. Let us show that $A_\gamma(f,v,\delta) \subset A(g,v,\delta_0)$, where the right-hand side set is as in the above lemma. Take any x in $A_\gamma(f,v,\delta)$ and any y in $K(x,v,\delta_0) \cap B_U(x,\delta_0)$. Then $\left\|v - \frac{y-x}{\|y-x\|}\right\| < \delta_0$, thus

$$\left|\langle x^*, v\rangle - \frac{\langle x^*, y - x\rangle}{\|y - x\|}\right| < \delta_0\|x^*\| \leq \gamma,$$

and this gives

$$\langle x^*, y\rangle - \langle x^*, x\rangle \geq (\langle x^*, v\rangle - \gamma)\|y - x\| > \gamma\|y - x\|.$$

Adding the latter with the inequality $f(y) - f(x) \geq -\gamma\|y - x\|$ (due to the condition (i)) we see that $g(y) \geq g(x)$. This and the equality $\liminf_{y\to x} \frac{g(y)-g(x)}{\|y-x\|} = -\infty$ (easily obtained from the condition (ii)) yield that $x \in A(g,v,\delta_0)$ as required. The above lemma then guarantees that $A_\gamma(f,v,\delta)$ is $\widetilde{\mathcal{C}}(X)$-null. □

LEMMA 13.51. *Let $f : U \to \mathbb{R}$ be a function defined on open set U of a separable Banach space X, and let the following conditions (depending on $x \in U$):*

(i) $\underline{d}_H^+(x;v) > -\infty$ *for some* $v \in X$ *(resp. (i')* $\overline{d}_H^+(x;v) < +\infty$ *for some* $v \in X$*);*

(ii) $\liminf_{y\to x} \frac{f(y)-f(x)}{\|y-x\|} = -\infty$ *(resp. (ii')* $\limsup_{y\to x} \frac{f(y)-f(x)}{\|y-x\|} = +\infty$*).*

Then the set of all $x \in U$ satisfying both conditions (i) and (ii) (resp. (i') and (ii')) is $\widetilde{\mathcal{C}}(X)$-null.

PROOF. Denote by P the set of $x \in U$ for which (i) and (ii) hold and choose a dense sequence $(v_j)_{j\in\mathbb{N}}$ in the unit sphere $\mathbb{S}_X$ centered at zero. Fix any $x \in P$. Then there is some $v \in X$ such that (i) and (ii) hold. From these conditions and Proposition 5.2 we obtain $v \neq 0$, so by the positive homogeneity of $\underline{d}_H^+(x;.)$ we may suppose that $\|v\| = 1$. By condition (i) there is some $m \in \mathbb{N}$ such that $\underline{d}_H^+(x;v) > -m$, which furnishes some $\delta > 0$ such that $f(y) - f(x) > -m\|y - x\|$ for all $y \in K(x,v,\delta) \cap B_U(x,\delta)$. Since $\|v\| = 1$ we can find $j \in \mathbb{N}$ satisfying $\|v_j - v\| < \delta$

and we can choose $n \in \mathbb{N}$ satisfying $\|v_j - v\| + 1/n < \delta$. It is not difficult to see that $x \in A_m(f, v_j, 1/n)$, where the latter set is as defined in Lemma 13.50. Then we have the inclusion $P \subset \bigcup_{j,n,m \in \mathbb{N}} A_m(f, v_j, 1/n)$, hence P is $\widetilde{\mathcal{C}}(X)$-null according to Lemma 13.50.

Finally the result concerning the set satisfying conditions (i') and (ii') is obtained by changing f into its opposite $-f$. □

LEMMA 13.52. Let $f : U \to \mathbb{R}$ be a function on an open set U of a separable Banach space X, and let the following conditions (depending on $x \in U$):

(i) $\liminf\limits_{y \to x} \frac{f(y)-f(x)}{\|y-x\|} > -\infty$ and f is not Hadamard subdifferentiable at x;

(ii) $\limsup\limits_{y \to x} \frac{f(y)-f(x)}{\|y-x\|} < +\infty$ and $-f$ is not Hadamard subdifferentiable at x.

Then the set of all $x \in U$ satisfying either (i) or (ii) is $\widetilde{\mathcal{C}}(X)$-null.

PROOF. It is clearly enough to prove that the set A of $x \in U$ fulfilling condition (i) is $\widetilde{\mathcal{C}}(X)$-null. Consider for each $n \in \mathbb{N}$ the set

$$L_n := \{x \in U : f(y) - f(x) \geq -n\|y - x\| \quad \text{for all } y \in B(x, 1/n)\},$$

and a sequence of open sets $(U_k^n)_{k \in \mathbb{N}}$ such that $U = \bigcup_{k \in \mathbb{N}} U_k^n$ and $\operatorname{diam} U_k^n < 1/n$. For $A_k^n := A \cap L_n \cap U_k^n$, we have $A = \bigcup_{n,k \in \mathbb{N}} A_k^n$, so it suffices to show that every nonempty set A_k^n is $\widetilde{\mathcal{C}}(X)$-null.

Fix any $n, k \in \mathbb{N}$ such that $A_k^n \neq \emptyset$. For each $x \in U_k^n$ put

$$g(x) := \sup\{\varphi(x) : \varphi \text{ is Lipschitz with constant } n \text{ on } U_k^n \text{ and } \varphi \leq f \text{ on } U_k^n\}.$$

For each $a \in A_k^n$, the function $\varphi_a : U_k^n \to \mathbb{R}$, with

$$\varphi_a(x) := f(a) - n\|x - a\| \quad \text{for all } x \in U_k^n,$$

is Lipschitz with constant n, and $\varphi_a \leq f$ on U_k^n since $A_k^n \subset L_n$ and $\operatorname{diam} U_k^n < 1/n$. This combined with the definition of $g(x)$ ensures that the function g is finite-valued and Lipschitz on U_k^n with constant n, and also

$$g \leq f \text{ on } U_k^n \quad \text{and} \quad g(a) = f(a) \text{ for all } a \in A_k^n.$$

From this and the definition of Hadamard subdifferential (see Definition 5.3) it is easily seen that f is Hadamard subdifferentiable at a point $a \in A_k^n$ whenever it is so for the function g. From the Preiss-Zajíček Theorem 13.39 we know that there exists a $\widetilde{\mathcal{C}}(X)$-null set $N \subset U_k^n$ such that at each $x \in U_k^n \setminus N$ the Lipschitz function g is Hadamard differentiable, thus we deduce that at any $x \in U_k^n \setminus N$ the function f is Hadamard subdifferentiable. It follows that $A_k^n \subset N$ (since $A_k^n \subset A$), thus A_k^n is $\widetilde{\mathcal{C}}(X)$-null as desired. □

We are now able to prove the theorem:

THEOREM 13.53 (Zajíček extension of Denjoy-Young-Saks theorem). Let $f : U \to \mathbb{R}$ be a function from an open set U of a separable Banach space X. Then there exists a $\widetilde{\mathcal{C}}(X)$-null set $N \subset U$ such that, for each $x \in U \setminus N$, one of the following properties is fulfilled:

(i) f is Hadamard differentiable at x;

(ii) $\overline{d}_H^+(x; \cdot) \equiv +\infty$ on X and $\underline{d}_H^+(x; \cdot) \equiv -\infty$ on X;

(iii) $\overline{d}_H^+(x; \cdot) \equiv +\infty$ on X and $\underline{d}_H^+(x; \cdot)$ is a continuous linear functional on X;

(iv) $\underline{d}_H^+(x;\cdot) \equiv -\infty$ on X and $\overline{d}_H^+(x;\cdot)$ is a continuous linear functional on X.

PROOF. Denote by A the set in Lemma 13.52 and by $A_{-\infty}$ (resp. $A_{+\infty}$) the set of $x \in U$ satisfying conditions (i) and (ii) (resp. (i') and (ii')) in Lemma 13.51. Denote also by A_0 the set of all $x \in U$ such that $\underline{d}_H^+(x;v) + \underline{d}_H^+(x;-v) > 0$ or $\overline{d}_H^+(x;v) + \overline{d}_H^+(x;-v) < 0$ for some $v \in X$, and note that A_0 is $\widetilde{\mathcal{C}}(X)$-null according to Theorem 12.8 (with $p = 1$ therein). This and Lemmas 13.52 and 13.51 tell us that $N := A \cup A_0 \cup A_{-\infty} \cup A_{+\infty}$ is $\widetilde{\mathcal{C}}(X)$-null. Fix any $x \in U \setminus N$ and consider the conditions:

$$\text{(C1)}\ \underline{d}_H^+(x;\cdot) \equiv -\infty \text{ on } X, \quad \text{and} \quad \text{(C2)}\ \overline{d}_H^+(x;\cdot) \equiv +\infty \text{ on } X.$$

We have four cases.

- **Neither (C1) nor (C2) hold:** In this case, Lemma 13.51 gives

$$\liminf_{y\to x} \frac{f(y)-f(x)}{\|y-x\|} > -\infty \quad \text{and} \quad \limsup_{y\to x} \frac{f(y)-f(x)}{\|y-x\|} < +\infty,$$

since $x \notin (A_{-\infty} \cup A_{+\infty})$. This combined with the fact that $x \notin A$ ensures by Lemma 13.52 that f and $-f$ are Hadamard subdifferentiable at x, so f is Hadamard differentiable at x by applying Proposition 5.11(a) (or as easily seen). This is the property (i).

- **Both (C1) and (C2) hold:** This possibility is exactly the property (ii).

- **(C1) holds and not (C2):** Since (C2) does not hold and $x \notin (A_{-\infty} \cup A)$, Lemmas 13.51 and 13.52 entail as in the first case that f is Hadamard subdifferentiable at x, that is, there is some $x^* \in X^*$ such that $\langle x^*,\cdot\rangle \le \underline{d}_H^+(x;\cdot)$. Then for any $v \in X$, since $x \notin A_0$ we have

$$\langle x^*,-v\rangle + \underline{d}_H^+(x;v) \le \underline{d}_H^+(x;-v) + \underline{d}_H^+(x;v) \le 0 \ \text{ hence } \ \underline{d}_H^+(x;v) \le \langle x^*,v\rangle.$$

It results that $\langle x^*,v\rangle = \underline{d}_H^+(x;v)$ for all $v \in X$, which is the property (iii).

- **(C2) holds and not (C1):** From this we obtain as above the property (iv), so the proof is complete. □

13.5. Comments

Concerning Theorem 13.1, its statements (a) and (b) are due to L. Zajíček [**996**] and [**986**] respectively. The statement (c) is due to J. Duda and L. Zajíček [**364**]. The first published result of the three was the statement (b); it was established in the 1979 Zajíček paper [**986**].

It is known that, in infinite-dimensional vector spaces, there is no Radon measure which is invariant under translation of sets. Nevertheless, there are many concepts of null sets in infinite-dimensional spaces. One of the first published concepts is that provided by the 1976 paper of N. Aronszajn [**23**]. Following [**821**] we denoted in Definition 13.2 by $\mathcal{A}(X)$ the class of null sets of a separable normed space X defined in [**23**]. The classes $\widetilde{\mathcal{A}}(X)$, $\mathcal{C}(X)$ and $\widetilde{\mathcal{C}}(X)$ have been introduced as important modifications of $\mathcal{A}(X)$ by D. Preiss and L. Zajíček in their famous 2001 paper [**822**] in order to obtain a refinement of Aronszajn's theorem on the extension to infinite dimensions of Rademacher's theorem on differentiability of Lipschitz functions.

The concept of porosity was probably first conceptualized and developed in the 1967 paper of E. P. Dolzhenko [**345**]. Both surveys [**989**] and [**993**] contain many other results on porosity, σ-porosity and other porosity concepts. Proposition 13.15 is taken from L. Zajíček [**987**]. We have reproduced, from the paper [**821**] of D. Preiss and L. Zajíček, the results and proofs of Lemmas 13.17, 13.19, Propositions 13.16, 13.20, 13.21, 13.22, 13.24, and Theorem 13.23.

Haar null sets have been introduced by J. P. R. Christensen [**227**] in the context of Banach spaces. Through this concept Christensen proved that real-valued locally Lipschitz functions on a separable Banach space is Gâteaux differentiable outside a Haar null set, a version of Rademacher theorem in terms of Haar null sets. Many other results can be found in Christensen's book [**228**]. The first versions of Theorems 13.31 and 13.32 are due to Christensen [**227**] who proved them in separable Banach spaces. The versions presented here for any Banach space are due to J. M. Borwein and W. B. Moors [**137**] and we used many ideas from their proofs. In the proof of Lemma 13.30 we followed the main idea from Y. Benyamini and L. Lindenstrauss [**89**] for the proof of stability by countable union of Haar null sets in separable Banach space.

In Section 13.3 we reproduced, from the 2001 paper by D. Preiss and L. Zajíček [**822**] the results and proofs of Lemmas 13.34, 13.35, 13.38, Propositions 13.36, 13.37, and Theorem 13.39. The latter theorem is the main result in Preiss-Zajíček [**822**]. As said above, it is an extension to separable Banach spaces of the classical Rademacher theorem, and it refined earlier extensions of Rademacher theorem provided by P. Mankiewicz [**687**], J. P. R. Christensen [**228**], N. Aronszajn [**23**], R. R. Phelps [**800**]. The Hadamard differentiability of Lipschitz functions is established in [**687**] outside a set with cube measure null, in [**228**] outside a Haar-null set, in [**23**] outside an Aronszajn null set, in [**800**] outside a Gaussian null set. Let X be a (nonzero) separable Banach space. A measure μ on X is called a *cube measure* whenever, denoting by τ the standard product measure on the Hilbert cube $[0,1]^{\mathbb{N}}$, the measure μ is equal to the image measure τ_L of τ under a continuous affine mapping $L : [0,1]^{\mathbb{N}} \to X$ of the form $L(t_1, \cdots, t_n, \cdots) = a + \sum_{n=1}^{\infty} t_n v_n$, where a is a fixed element of X and $(v_n)_{n\in\mathbb{N}}$ is a fixed linearly independent dense sequence in X such that $\sum_{n=1}^{\infty} \|v_n\| < +\infty$; recall that the image measure τ_L of τ under the mapping L is defined by $\tau_L(B); = \tau\big(L^{-1}(B)\big)$, for all Borelian sets B in X. So, a Borelian subset A of X is said to be *cube measure null* provided that $\tau_L(A) = 0$ for every cube measure τ_L. Regarding Gaussian null sets, recall that a measure μ on X is a *Gaussian measure* if the image measure μ_{x^*} of μ under any continuous linear functional $x^* \in X^*$ is a Gaussian measure on $\mathbb{R}$. The measure μ is said to be *nondegenerate* whenever, for every nonzero $x^* \in X^*$, the Gaussian measure μ_{x^*} on $\mathbb{R}$ is not concentrated at one point; otherwise stated, the Gaussian measure μ is nondegenerate is it is not supported on any closed affine hyperplane of X. This leads to say that a Borel subset A of X is *Gaussian null* provided that $\mu(A) = 0$ for every nondegenerate Gaussian measure μ on X. M. Csörnyei [**300**] showed that the three concepts of Aronszajn null sets, Gaussian null sets and cube measure null sets are equivalent. So, according to Proposition 13.6, Theorem 13.39 of Preiss and Zajíček generalizes all the above results. Examples showing that the inclusions $\widetilde{\mathcal{C}}(X) \cap \mathcal{B}(X) \subset \mathcal{A}(X) \subset \mathcal{C}(X)$ in Proposition 13.6 are strict can be found in the Preiss-Zajíček paper [**822**]. The difference between σ-porosity and Aronsajn nullity is clear through the following result established by D. Preiss and J. Tišer in their 1995 paper [**820**]: Every infinite-dimensional separable Banach space X can be decomposed as $X = A_a \cup A_p$, where A_a is Aronszajn null and A_p is σ-porous. So, there is no comparison in infinite dimensions between the claas of σ-porous sets and that of Aronszajn null sets. Many results concerning Gaussian null or cube measure null sets as well as other concepts of null sets can be found in the book [**89**] of Y. Benyamini and J. Lindenstrauss. Another deep result concerning the richness of points of Fréchet differentiability of Lipschitz functions on Asplund spaces has been proved by D. Preiss [**819**]. Theorem 13.41 has been proved in separable Banach spaces by L. Thibault [**906, 910**] as an extension of the gradient representation established by F. H. Clarke [**233**] in finite dimensions; the proof is based, for a great part, on adaptations of Clarke's finite dimensional arguments. Proposition 13.40 has been observed in [**906, 910**] in the context of separable Banach spaces; its version given for any Banach space is taken from J. M. Borwein and W. B. Moors [**137**].

Theorem 13.44 is an extension to separable Banach space of the classical Stepanov theorem concerning real-valued one-sided Lipschitz functions on $\mathbb{R}^n$. The proof presented here followed the nice method due to J. Malý [**684**]. Other similar statements and proofs for some vector-valued mappings can be found in D. Bongiorno [**110**] and in J. Malý and L. Zajíček [**685**]. General versions of Stepanov theorem for mappings from an open set of a separable Banach space X into a Radon-Nokodým Banach space can be found in the 1998 paper [**109**] of D. Bongiorno with the use of $\mathcal{A}(X)$-null sets and in the 2008 paper [**361**] of J. Duda with the use of the smaller class of $\widetilde{\mathcal{A}}(X)$-null sets. Propositions 13.46 and 13.47 as well as their proofs are taken from the 2012 Zajíček paper [**998**].

For the differentiability of cone-monotone and quasi-convex functions outside certain null sets we refer to: D. Bongiorno [**109, 110**], J. M. Borwein, J. V. Burke and A. S. Lewis [**126**], J. M. Borwein and X. Wang [**149**], Y. Chabrillac and J.-P. Crouzeix [**216**], J.-P. Crouzeix [**299**], J. Duda, P. J. Rabier [**828**], L. Zajíček [**997**], and to references in those papers.

Concerning Section 13.4, Lemmas 13.48, 13.49, 13.50 and 13.51 and their proofs are taken from the 2015 article [**999**] by L. Zajíček. Those lemmas reproduce Zajíček's statements and proofs in [**999**] of Lemmas 3.1-3.4. Lemma 13.52 corresponds to Corollary 4.3 in [**999**]. The Zajíček Theorem 13.53 is the extension by L. Zajíček of the Denjoy-Young-Saks Theorem 6.42 (for functions of one real variable) to real-valued functions on separable Banach spaces. The proof in the manuscript of this Theorem 13.53 is that of Zajíček [**999**, Theorem 5.3]. Previously to [**999**], there were a few articles on that topics. We can cite the 1932 paper [**470**] of U. S. Haslam-Jones and the 1935 paper [**967**] of A. J. Ward for theorems of Denjoy-Young-Saks type for functions of two real variables. In [**470**] the functions of two variables are required to be measurable while no such measurability is assumed in the paper [**967**].

CHAPTER 14

Distance function, metric projection, Moreau envelope

14.1. Distance function and metric projection

The prox-regularity of a set will appear in Chapter 15 as a regularity of its metric projection mapping. The purpose of this section is to establish some important properties of the distance function and the metric projection of a general closed set of a reflexive Banach space. Essentially, we will study in such spaces certain useful conditions for the differentiability of distance functions and for the existence, as well as the genericity, of nearest points. The differentiability of farthest distance functions along with the genericity of farthest points will also be studied.

14.1.1. Density of points with nearest/farthest points. Given a closed nonempty set C of a normed space $(X, \|\cdot\|)$, as said in (2.69) in Chapter 2, a point $y \in C$ is called a *nearest point of* C *to a point* $x \in X$ (or *a nearaest point in* C *of* x) when $\|y - x\| = d_C(x)$, and this is translated as $y \in \operatorname{Proj}_C(x)$. If $y \in C$ and there exists some $x \in X \setminus C$ with $y \in \operatorname{Proj}_C(x)$ we say that y is *a nearest point* in C. The set C is called *proximinal* provided that each point of $X \setminus C$ (or equivalently of X) has at least one nearest point in C, that is, $\operatorname{Proj}_C(x) \neq \emptyset$ for all $x \in X \setminus C$.

The following proposition states some sufficient conditions for proximinality. It also characterizes through that property of proximinality the reflexivity of a Banach space. Recall first that, by the celebrated James theorem (see Theorem C.8 in Appendix C), a Banach space X is reflexive if and only if any one-norm linear functional $x^* \in X^*$ (that is, $\|x^*\|_* = 1$) attains its supremum on the closed unit ball $\mathbb{B}_X$ of X.

PROPOSITION 14.1. Let $(X, \|\cdot\|)$ be a normed space. The following hold:
(a) Every boundedly weakly compact set C of X (that is, the intersection of C with any closed ball is weakly compact) is proximinal.
(b) Every weakly closed set of X is proximinal whenever X is a reflexive Banach space.
(c) The Banach space $(X, \|\cdot\|)$ is reflexive if and only if every closed convex set of X is proximinal.

PROOF. Fix any $x \in X$ and put $r := 1 + d_C(x)$. The $w(X, X^*)$-lower semicontinuous function $\|\cdot - x\|$ attains its infimum on the weak compact set $C \cap B[x, r]$ at some point y. It is then readily seen that y is a nearest point of C to x, which translates (a). Assertion (b) and the implication $\Rightarrow$ of (c) are direct consequences of (a).

To prove the implication $\Leftarrow$ of (c), suppose that the Banach space $(X, \|\cdot\|)$ is not reflexive and by James theorem fix some $u^* \in X^*$ with $\|u^*\| = 1$ such that

$\langle u^*, u\rangle < 1$ for all $u \in \mathbb{B}_X$. Set $C := \{x \in X : \langle u^*, x\rangle = 0\}$ and take any $\overline{x} \in X \setminus C$, that is, $\langle u^*, \overline{x}\rangle \neq 0$. Observe first that the properties of u^* ensure that for all $x \in C$

$$\|\overline{x} - x\| > \langle u^*, \pm(\overline{x} - x)\rangle = \pm\langle u^*, \overline{x}\rangle, \text{ that is, } \|\overline{x} - x\| > |\langle u^*, \overline{x}\rangle|. \tag{14.1}$$

On the other hand, choose a sequence $(u_n)_n$ in X with $\|u_n\| = 1$ such that $\langle u^*, u_n\rangle \to \|u^*\| = 1$ and set $x_n := \overline{x} - \dfrac{\langle u^*, \overline{x}\rangle}{\langle u^*, u_n\rangle} u_n$. Obviously $x_n \in C$ and

$$\|\overline{x} - x_n\| = \frac{|\langle u^*, \overline{x}\rangle|}{|\langle u^*, u_n\rangle|} \to |\langle u^*, \overline{x}\rangle|,$$

which combined with (14.1) says that $d_C(\overline{x}) = |\langle u^*, \overline{x}\rangle|$, and hence it follows from (14.1) that $\overline{x}$ has no nearest points in C. This concludes the proof of the proposition. □

Taking the proposition above into account, for a general closed set C we will study the density and genericity of the set of points of $X \setminus C$ which have nearest points in C. The main characterization theorem of prox-regular sets, namely Theorem 15.6 of Chapter 15, in infinite-dimensional Hilbert or reflexive spaces will use the richness of points which have nearest points in C. Given a nonempty set C in a normed space X, we saw in Proposition 4.32 that

$$\langle x^*, \overline{x} - y_n\rangle \to d_C(\overline{x}) \quad \text{as } n \to \infty, \tag{14.2}$$

whenever $x^* \in \partial_F d_C(\overline{x})$ and $(y_n)_n$ is a minimizing sequence in C for the distance $d_C(\overline{x})$, that is, $y_n \in C$ and $\|y_n - \overline{x}\| \to d_C(\overline{x})$. Through that result the next theorem establishes, under the additional reflexivity of the normed space X and the Kadec-Klee property of the norm, a first richness property of the set of points of $X \setminus C$ admitting nearest points in C. We already recalled in the sentence preceding Theorem 3.267 that a norm $\|\cdot\|$ on X has the (sequential) *Kadec-Klee property* provided that for any sequence $(x_n)_n$ of X converging weakly to x with $\|x_n\| \to \|x\|$ one has $\|x_n - x\| \to 0$; for examples of such norms see Subsection 18.1.2. Any reflexive Banach space X can be renormed with an equivalent norm which is both Kadec-Klee and strictly convex (see Theorem C.11 in Appendix C). The norm $\|\cdot\|$ is *strictly convex* (or *rotund*) whenever

$$\|tx + (1-t)y\| < 1 \quad \text{for } t \in]0,1[\text{ and } \|x\| = \|y\| = 1 \text{ with } x \neq y.$$

The strict convexity of $\|\cdot\|$ is clearly seen to ensure that

$$\Big(\|x\| = \|y\| \text{ and } \|x + y\| \geq \|x\| + \|y\|\Big) \implies (x = y). \tag{14.3}$$

THEOREM 14.2 (nearest points under F-subdifferentiability of distance function). Let C be a nonempty closed set of the normed space X. The following hold:
(a) If $\operatorname{Proj}_C(x) \neq \emptyset$, then, for any $y \in \operatorname{Proj}_C(x)$,

$$\partial_F d_C(x) \subset \partial(\|\cdot\|)(x - y).$$

(b) If $(X, \|\cdot\|)$ is a reflexive Banach space and the norm $\|\cdot\|$ has the (sequential) Kadec-Klee property, then $\operatorname{Proj}_C(x) \neq \emptyset$ for any $x \in X \setminus C$ with $\partial_F d_C(x) \neq \emptyset$, and hence in particular the set of points of $X \setminus C$ which have nearest points in C is dense in $X \setminus C$.

PROOF. Concerning the inclusion property in (a) under the non-emptiness of $\mathrm{Proj}_C(x)$, it follows from Proposition 4.30 related to the Fréchet subdifferential of the infimal convolution since $d_C = \Psi_C \square \|\cdot\|$.

To prove (b) assume that X is a reflexive Banach space and that $\|\cdot\|$ has the Kadec-Klee property. Take $x \in (X \setminus C) \cap \mathrm{Dom}\, \partial_F d_C$. Choose $x^* \in \partial_F d_C(x)$ and take a minimizing sequence $(y_n)_n$ in C for the distance $d_C(x)$, that is, $y_n \in C$ and $\|y_n - x\| \to d_C(x)$ as $n \to \infty$. By the boundedness of $(y_n)_n$ and the reflexivity of X, extracting a subsequence if necessary we may suppose that $(y_n)_n$ converges weakly to some $y \in X$. By the result in Proposition 4.32 recalled in (14.2) above we know that

$$\|x - y\| \geq \langle x^*, x - y\rangle = \lim_n \langle x^*, x - y_n\rangle = d_C(x) = \lim_n \|x - y_n\|.$$

Further, by the lower semicontinuity of the norm we also have $\lim_n \|x - y_n\| \geq \|x - y\|$, and hence $\|x - y\| = \lim_n \|x - y_n\| (= d_C(x))$. Since $(x - y_n)_n$ converges weakly to $x - y$, the Kadec-Klee property of the norm entails that $\|y_n - y\| \to 0$, and this in particular implies that $y \in C$. Since $\|x - y\| = d_C(x)$, we obtain that $y \in \mathrm{Proj}_C(x)$, that is, the first part of the assertion (b) is established.

Finally, the set $\mathrm{Dom}\, \partial_F d_C$ being dense in $\mathrm{dom}\, d_C = X$ according to the reflexivity of X (as seen in Corollary 4.53), the set $(X \setminus C) \cap \mathrm{Dom}\, \partial_F d_C$ is dense in $X \setminus C$ because the latter is open. From this and what precedes, we deduce the desired density property of the second part of the assertion (c). The proof of the theorem is then complete. □

Concerning (b) in the above theorem, a more general density result, say a genericity result for nearest points, will be established later in Theorem 14.29.

Results similar to the assertions in Theorem 14.2 also hold for the farthest distance function.

DEFINITION 14.3. Given a nonempty subset C of a normed space $(X, \|\cdot\|)$, its *farthest distance function* $\mathrm{dfar}_C : X \to [0, +\infty]$ is defined by

$$\mathrm{dfar}_C(x) = \sup\{\|x - y\| : y \in C\} \quad \text{for all } x \in X.$$

The set

$$\mathrm{Far}_C(x) = \{y \in C : \|x - y\| = \mathrm{dfar}_C(x)\}$$

is the *set of farthest points in C to x* (or *set of farthest distance points of x in C*).

Clearly, the function dfar_C is finite at some point if and only if the nonempty set C is bounded. In such a case the function dfar_C is finite on the whole space X. Taking any $x', x'' \in X$ we have for every $y \in C$ that $\|x' - y\| \leq \|x'' - y\| + \|x' - x''\|$, so taking the supremum over $y \in C$ yields $\mathrm{dfar}_C(x') \leq \mathrm{dfar}_C(x'') + \|x' - x''\|$, so by symmetry between x' and x'' we obtain under the boundedness of C

$$(14.4) \qquad |\mathrm{dfar}_C(x') - \mathrm{dfar}_C(x'')| \leq \|x' - x''\| \quad \text{for all } x', x'' \in X.$$

Further, the function dfar_C is convex as the supremum of a family of convex functions. We summarize those properties as follows.

PROPOSITION 14.4. For any nonempty bounded subset C of a normed space $(X, \|\cdot\|)$, the farthest distance function dfar_C is a real-valued Lipschitz convex function with 1 as Lipschitz constant.

Given a nonempty bounded set C in the normed space X, another important feature (easy to verify) is the following:

$$(14.5) \qquad \Big(y \in \mathrm{Far}_C(x)\Big) \Longrightarrow \Big(y \in \mathrm{Far}_C(x + t(x-y)),\ \forall t \in [0,+\infty[\Big).$$

F-subgradients of the farthest distance function dfar_C are connected to subgradients of the norm as follows.

PROPOSITION 14.5. Let C be a nonempty closed bounded set of the normed space $(X, \|\cdot\|)$. If $\mathrm{Far}_C(x) \neq \emptyset$, then, for any $y \in \mathrm{Far}_C(x)$,

$$\partial_F(-\mathrm{dfar}_C)(x) \subset \partial_F(-\|\cdot\|)(x-y).$$

PROOF. Under the non-emptiness of $\mathrm{Far}_C(x)$, the assertion (a) directly follows from Proposition 4. 30 related to the Fréchet subdifferential of the infimal convolution since $-\mathrm{dfar}_C = \Psi_C \square (-\|\cdot\|)$. $\square$

Before showing the properties for dfar_C similar to those of d_C in Theorem 14.2, let us prove a proposition related to F-subgradients of the opposite of dfar_C and to maximizing sequences of dfar_C. The statement and proof are the analogs of those in Proposition 4.32.

PROPOSITION 14.6. Let C be a nonempty bounded subset of a normed space $(X, \|\cdot\|)$ and let $x^* \in \partial_F(-\mathrm{dfar}_C)(\overline{x})$.
(a) One has

$$\limsup_{y\in C, \|y-\overline{x}\|\to \mathrm{dfar}_C(\overline{x})} \langle x^*, y-\overline{x}\rangle = \mathrm{dfar}_C(\overline{x}),$$

that is, for any sequence $(y_n)_n$ in C with $\|y_n - \overline{x}\| \to \mathrm{dfar}_C(\overline{x})$ as $n \to \infty$

$$\langle x^*, y_n - \overline{x}\rangle \to \mathrm{dfar}_C(\overline{x}) \quad \text{as } n\to\infty;$$

in particular $\sup_{y\in C}\langle x^*, y-\overline{x}\rangle = \mathrm{dfar}_C(\overline{x})$.

If in addition $C \neq \{\overline{x}\}$, then one also has $\|x^*\| = 1$.
(b) In particular, for any $\overline{y} \in \mathrm{Far}_C(\overline{x})$ (if any) one has

$$\langle x^*, \overline{y} - \overline{x}\rangle = \mathrm{dfar}_C(\overline{x}).$$

PROOF. We note first that $\|x^*\| \le 1$ by the 1-Lipschitz property of $-\mathrm{dfar}_C(\cdot)$. Let $(y_n)_n$ be any sequence in C with $\|y_n - \overline{x}\| \to \mathrm{dfar}_C(\overline{x})$ and $\mathrm{dfar}_C(\overline{x}) - \|y_n - \overline{x}\| < 1/4$. Set $t_n := 1/(2n) + \sqrt{\mathrm{dfar}_C(u) - \|x_n - u\|}$. Since $t_n(y_n - \overline{x}) \to 0$, by the inclusion $x^* \in \partial_F(-\mathrm{dfar}_C)(\overline{x})$ we can write

$$\liminf_{n\to\infty}\{t_n^{-1}[-\mathrm{dfar}_C(\overline{x} + t_n(\overline{x} - y_n)) + \mathrm{dfar}_C(\overline{x})] - \langle x^*, \overline{x} - y_n\rangle\} \ge 0,$$

or equivalently

$$\limsup_{n\to\infty}\{t_n^{-1}[\mathrm{dfar}_C(\overline{x} + t_n(\overline{x} - y_n)) - \mathrm{dfar}_C(\overline{x})] - \langle x^*, y_n - \overline{x}\rangle\} \le 0.$$

Noting that $t_n < 1$, we also have for every n

$$\begin{aligned}
&\mathrm{dfar}_C(\overline{x} + t_n(\overline{x} - y_n)) - \mathrm{dfar}_C(\overline{x}) \\
&\ge \|\overline{x} + t_n(\overline{x} - y_n) - y_n\| - \mathrm{dfar}_C(\overline{x}) \\
&\ge \|\overline{x} + t_n(\overline{x} - y_n) - y_n\| - \|\overline{x} - y_n\| + \big(\|\overline{x} - y_n\| - \mathrm{dfar}_C(\overline{x})\big) \\
&= t_n\|\overline{x} - y_n\| - \big(\mathrm{dfar}_C(\overline{x}) - \|\overline{x} - y_n\|\big) \ge t_n\|\overline{x} - y_n\| - t_n^2,
\end{aligned}$$

and hence

$$\limsup_{n\to\infty}\Big(\|\overline{x}-y_n\|+\langle x^*,\overline{x}-y_n\rangle-t_n\Big)\le 0.$$

It ensues that

$$\mathrm{dfar}_C(\overline{x})=\lim_{n\to\infty}\|\overline{x}-y_n\|\le\liminf_{n\to\infty}\langle x^*,y_n-\overline{x}\rangle.$$

On the other hand, from the inequality $\|x^*\|\le 1$ we have

$$\limsup_{n\to\infty}\langle x^*,y_n-\overline{x}\rangle\le\lim_{n\to\infty}\|y_n-\overline{x}\|\le\mathrm{dfar}_C(\overline{x}).$$

From the two latter inequalities it follows that $\lim_{n\to\infty}\langle x^*,y_n-\overline{x}\rangle=\mathrm{dfar}_C(\overline{x})$, which confirms the first part of the assertion (a).

Finally, assume that $C\neq\{\overline{x}\}$, so $\mathrm{dfar}_C(\overline{x})>0$. Setting $u_n:=(y_n-\overline{x})/\mathrm{dfar}_C(\overline{x})$, we get $\|u_n\|\to 1$ and $\langle x^*,u_n\rangle\to 1$ as $n\to\infty$. Keeping in mind that $\|x^*\|\le 1$ we conclude that $\|x^*\|=1$. □

Taking the assertion (b) in the above proposition into account and its counterpart for nearest point in Proposition 4.32, the next Proposition 14.8 will provide useful conditions for uniqueness of nearest points and farthest points. Let us establish first a lemma which has its own interest.

LEMMA 14.7. Let C be a nonempty closed set in a normed space $(X,\|\cdot\|)$ and let $\overline{x}\in X$.
(A) Assume that $\mathrm{Proj}_C(\overline{x})\neq\emptyset$.
(a$_1$) For any $y\in\mathrm{Proj}_C(\overline{x})$ and any $t\in\,]0,1[$, one has

$$-d_C\big(\overline{x}+t(y-\overline{x})\big)=-d_C(\overline{x})+t\|y-\overline{x}\|,$$

hence the directional derivative $(-d_C)'(\overline{x};y-\overline{x})$ exists and

$$(-d_C)'(\overline{x};y-\overline{x})=\|y-\overline{x}\|=d_C(\overline{x}).$$

(a$_2$) If the distance function d_C is Gâteaux differentiable at $\overline{x}\notin C$, then for every $y\in\mathrm{Proj}_C(\overline{x})$

$$\langle D(d_C)(\overline{x}),\overline{x}-y\rangle=\|\overline{x}-y\|=d_C(\overline{x}),$$

and $\|D(d_C)(\overline{x})\|=1$.
(B) Assume that the closed set C is bounded and $\mathrm{Far}_C(\overline{x})\neq\emptyset$.
(b$_1$) For any $y\in\mathrm{Far}_C(\overline{x})$ and any real $t>0$

$$\mathrm{dfar}_C\big(\overline{x}+t(\overline{x}-y)\big)=\mathrm{dfar}_C(\overline{x})+t\|\overline{x}-y\|$$

hence

$$(\mathrm{dfar}_C)'(\overline{x};\overline{x}-y)=\|\overline{x}-y\|=\mathrm{dfar}_C(\overline{x}).$$

(b$_2$) If the function dfar_C is Gâteaux differentiable at $\overline{x}$, then $\|D(\mathrm{dfar}_C)(\overline{x})\|=1$ and for any $y\in\mathrm{Far}_C(\overline{x})$

$$\langle D(\mathrm{dfar}_C)(\overline{x}),\overline{x}-y\rangle=\|\overline{x}-y\|=\mathrm{dfar}_C(\overline{x}).$$

PROOF. (A) Assume that $\mathrm{Proj}_C(\overline{x})\neq\emptyset$ and take any $y\in\mathrm{Proj}_C(\overline{x})$. Then for any $t\in\,]0,1[$ we know that $y\in\mathrm{Proj}_C\big(\overline{x}+t(y-\overline{x})\big)$, hence

$$d_C\big(\overline{x}+t(y-\overline{x})\big)=\|\overline{x}+t(y-\overline{x})-y\|=(1-t)\|\overline{x}-y\|,$$

which translates (a$_1$).

Concerning (a$_2$), suppose that $\overline{x}\notin C$ and d_C is Gâteaux differentiable at $\overline{x}$. Then by (a$_1$) we have $\|D(d_C)(\overline{x})\|\ge 1$, and by the Lipschitz property of d_C with

1 as Lipschitz constant we also have $\|D(d_C(\overline{x}))\| \le 1$. Therefore, $\|D(d_C(\overline{x}))\| = 1$, and the proof of (a$_1$) is finished.
(B) Assume now that $\mathrm{Far}_C(\overline{x}) \neq \emptyset$ and take any $y \in \mathrm{Far}_C(\overline{x})$. For any $t > 0$ we have $y \in \mathrm{Far}_C\big(\overline{x} + t(\overline{x} - y)\big)$, which gives

$$\mathrm{dfar}_C\big(\overline{x} + t(\overline{x} - y)\big) = \|\overline{x} + t(\overline{x} - y) - y\| = (1+t)\|\overline{x} - y\|,$$

so (b$_1$) holds true.

Assume in addition that dfar_C is Gâteaux differentiable at $\overline{x}$ and put $\zeta := D(\mathrm{dfar}_C)(\overline{x})$. Then $C \neq \{\overline{x}\}$, hence (b$_1$) entails that $\|\zeta\| \ge 1$. This and the 1-Lipschitz property of dfar_C ensures (as above) that $\|\zeta\| = 1$. The proof of the lemma is complete. □

PROPOSITION 14.8. Let C be a nonempty closed set in a normed space whose norm $\|\cdot\|$ is strictly convex.
(a$_1$) If there exists some $u^* \in X^*$ with $\|u^*\| = 1$ such that $\langle u^*, \overline{x} - y\rangle = d_C(\overline{x})$ for all $y \in \mathrm{Proj}\,_C(\overline{x})$ (if any), then the point $\overline{x}$ possesses at most one nearest point in the set C.
(a$_2$) If the distance function d_C is Gâteaux differentiable at $\overline{x}$, then the point $\overline{x}$ possesses at most one nearest point in the set C.
(b$_1$) If there exists some $u^* \in X^*$ with $\|u^*\| = 1$ such that $\langle u^*, y - \overline{x}\rangle = \mathrm{dfar}_C(\overline{x})$ for all $y \in \mathrm{Far}_C(\overline{x})$ (if any), then the point $\overline{x}$ admits at most one farthest point in the set C.
(b$_2$) If the function dfar_C is Gâteaux differentiable at $\overline{x}$, then the point $\overline{x}$ admits at most one farthest point in the set C.

PROOF. (a) We may suppose $\overline{x} \notin C$ (otherwise both results in (a$_1$) and (a$_2$) are trivial). To prove (a$_1$), consider the case where $\mathrm{Proj}\,_C(\overline{x}) \neq \emptyset$ and take $y_1, y_2 \in \mathrm{Proj}\,_C(\overline{x})$. By the assumption in (a$_1$) we have

$$\begin{aligned}\|(\overline{x} - y_1) + (\overline{x} - y_2)\| &\ge \langle u^*, (\overline{x} - y_1) + (\overline{x} - y_2)\rangle \\ &= \langle u^*, \overline{x} - y_1\rangle + \langle u^*, \overline{x} - y_2\rangle \\ &= 2d_C(\overline{x}) = \|\overline{x} - y_1\| + \|\overline{x} - y_2\|,\end{aligned}$$

hence the strict convexity assumption of the norm $\|\cdot\|$ combined with the equality $\|\overline{x} - y_1\| = \|\overline{x} - y_2\|$ entails that $\overline{x} - y_1 = \overline{x} - y_2$ (see (14.3)), that is, $y_1 = y_2$. This confirms the assertion (a$_1$).

Concerning the second assertion (a$_2$), it clearly follows from (a$_1$) and from the assertion (a$_2$) in Lemma 14.7.
(b) Regarding (b$_1$) and (b$_2$), we may and do assume that $C \neq \{\overline{x}\}$. To prove (b$_1$), suppose that $\mathrm{Far}_C(\overline{x}) \neq \emptyset$, and as above take $y_1, y_2 \in \mathrm{Far}_C(\overline{x})$. By the assumption in (b$_1$) we can write

$$\begin{aligned}\|(y_1 - \overline{x}) + (y_2 - \overline{x})\| &\ge \langle u^*, (y_1 - \overline{x}) + (y_2 - \overline{x})\rangle \\ &= \langle u^*, y_1 - \overline{x}\rangle + \langle u^*, y_2 - \overline{x}\rangle \\ &= 2\mathrm{dfar}_C(\overline{x}) = \|y_1 - \overline{x}\| + \|y_2 - \overline{x}\|,\end{aligned}$$

so (as above) the strict convexity of the norm $\|\cdot\|$ again implies that $y_1 - \overline{x} = y_2 - \overline{x}$. It results that $y_1 = y_2$, justifying (b$_1$).

The final assertion (b$_2$) being a direct consequence of properties (b$_1$) and (b$_2$) in Lemma 14.7, the proof is complete. □

We prove now the analog of Theorem 14.2 for dfar_C and Far_C. The Kadec-Klee property as well as the strict convexity (or rotundity) of a norm have been recalled just before the statement of Theorem 14.2.

THEOREM 14.9 (genericity of points with farthest points). Let $(X, \|\cdot\|)$ be a reflexive Banach space whose norm $\|\cdot\|$ has the (sequential) Kadec-Klee property. Let C be a nonempty closed bounded set in X. The following hold.
(a) One has $\mathrm{Far}_C(x) \neq \emptyset$ for any $x \in X$ with $\partial_F(-\mathrm{dfar}_C)(x) \neq \emptyset$.
(b) The set of points of X which have farthest points in C is of (Baire) second category in X, and hence (in particular) dense in X.
(c) If in addition the norm $\|\cdot\|$ is strictly convex, then the set of points $x \in X$ such that x has a unique farthest point in C is of (Baire) second category in X.

PROOF. (a) Take any $x \in \mathrm{Dom}\,\partial_F(-\mathrm{dfar}_C)$. Choose $x^* \in \partial_F(-\mathrm{dfar}_C)(x)$ and take a sequence $(y_n)_n$ in C such that $\|y_n - x\| \to \mathrm{dfar}_C(x)$ as $n \to \infty$. By the boundedness of $(y_n)_n$ and the reflexivity of X, extracting a subsequence if necessary we may suppose that $(y_n)_n$ converges weakly to some $y \in X$. By Proposition 14.6(a) we can write

$$\|y - x\| \geq \langle x^*, y - x\rangle = \lim_n \langle x^*, y_n - x\rangle = \mathrm{dfar}_C(x) = \lim_n \|y_n - x\|.$$

Further, by the lower semicontinuity of the norm we also have $\lim_n \|y_n - x\| \geq \|y - x\|$, and hence $\|y - x\| = \lim_n \|y_n - x\| (= \mathrm{dfar}_C(x))$. Since $(y_n - x)_n$ converges weakly to $y - x$, the Kadec-Klee property of the norm entails that $\|y_n - y\| \to 0$, and this in particular entails that $y \in C$. Since $\|y - x\| = \mathrm{dfar}_C(x)$, we obtain that $y \in \mathrm{Far}_C(x)$, so the assertion (a) is established.
(b) The reflexive Banach space X being Asplund, the set Q of points where the continuous convex function dfar_C is Fréchet differentiable is a dense G_δ set in X (see Definition 4.50). Clearly, $\partial_F(-\mathrm{dfar}_C)(x) \neq \emptyset$ at each point x where dfar_C is Fréchet differentiable, hence $Q \subset \mathrm{Dom}\,\partial_F(-\mathrm{dfar}_C)$. This and (a) assure us that the set of points $x \in X$ where $\mathrm{Far}_C(x) \neq \emptyset$ is of second category in X.
(c) Now assume in addition that the norm $\|\cdot\|$ is strictly convex. Fix any $x \in X$ where dfar_C is Fréchet differentiable. By (a) we know that $\mathrm{Far}_C(x) \neq \emptyset$, hence using Proposition 14.8(b$_2$) we obtain that $\mathrm{Far}_C(x)$ is a singleton. This finishes the proof of the theorem. □

14.1.2. Differentiability of distance functions and farthest distance functions under differentiable norms. The assertion (a) of Theorem 14.2 can be applied in particular whenever the distance function d_C is Fréchet differentiable at x. Corollary 14.21 below provides a useful criterion for the Fréchet differentiability of the distance function. We study first the case of a general mapping between normed spaces.

DEFINITION 14.10. Let X be a nonzero normed space and $G : X \to Y$ be a mapping from X into a normed space Y. We recall that the *Lipschitz rate* (or *Lipschitz modulus*) $\mathrm{lip}\, G(x)$ of G at $x \in X$ is given by

$$\mathrm{lip}\, G(x) := \lim_{\delta \downarrow 0} \mathrm{lip}_\delta G(x) = \inf_{\delta > 0} \mathrm{lip}_\delta G(x),$$

where

$$\mathrm{lip}_\delta G(x) := \sup\left\{ \frac{\|G(x') - G(x'')\|}{\|x' - x''\|} : \|x' - x\| \leq \delta, \|x'' - x\| \leq \delta, x' \neq x'' \right\}.$$

So, the mapping G is Lipschitz continuous near x if and only if $\operatorname{lip} G(x)$ is finite.

Below we will also use the *Fitzpatrick constant*

$$\operatorname{fitz} G(x) := \lim_{\delta \downarrow 0} \operatorname{fitz}_\delta G(x),$$

where

$$\operatorname{fitz}_\delta G(x) := \sup\left\{ \frac{\|G(x') - G(x'')\|}{\|x' - x''\|} : 2\|x'' - x\| \le \|x' - x\| \le \delta,\ x' \ne x'' \right\}.$$

Obviously $\operatorname{fitz} G(x) \le \operatorname{lip} G(x)$.

PROPOSITION 14.11. Let X be a nonzero normed space, $G : X \to Y$ be a mapping from X into a normed space Y, and let $x \in X$. The following hold:
(a) For any $u \in X$ with $\|u\| = 1$

$$\limsup_{t \downarrow 0} t^{-1}\|G(x + tu) - G(x)\| \le \operatorname{fitz} G(x).$$

(b) If G is Fréchet differentiable at x, then $\|D_F G(x)\| = \operatorname{fitz} G(x)$.

PROOF. (a) Fix $u \in X$ with $\|u\| = 1$. For any real $t > 0$, since $2\|x - x\| \le \|(x + tu) - x\| = t$ we can write

$$t^{-1}\|G(x + tu) - G(x)\| = \frac{\|G(x + tu) - G(x)\|}{\|tu\|} \le \operatorname{fitz}_t G(x),$$

hence taking the limit superior as $t \downarrow 0$ we obtain the desired inequality in (a).
(b) Suppose that G is Fréchet differentiable at x and put $A := D_F G(x)$. By (a) we have $\|A(u)\| \le \operatorname{fitz} G(x)$ for any $u \in X$ with $\|u\| = 1$, and this guarantees that $\|A\| \le \operatorname{fitz} G(x)$. To prove the reverse inequality, fix any $\varepsilon > 0$ and choose $\delta > 0$ such that

$$\|G(x') - G(x) - A(x' - x)\| \le \varepsilon\|x' - x\| \quad \text{for all } x' \in B[x, \delta].$$

Fix any $x', x'' \in X$ satisfying $2\|x'' - x\| \le \|x' - x\| \le \delta$. Then we have

$$\|G(x') - G(x) - A(x' - x) - G(x'') + G(x) + A(x'' - x)\| \le \varepsilon\|x' - x\| + \varepsilon\|x'' - x\|$$

hence

$$\|G(x'') - G(x')\| \le \|A(x'' - x')\| + \varepsilon(\|x' - x\| + \|x'' - x\|).$$

Observe that the inequality $2\|x'' - x\| \le \|x' - x\|$ ensures that

$$2\|x'' - x\| \le \|x' - x''\| + \|x'' - x\|, \quad \text{thus} \quad \|x'' - x\| \le \|x' - x''\|,$$

and the latter entails that

$$\|x' - x\| \le \|x' - x''\| + \|x'' - x\| \le 2\|x' - x''\|.$$

We deduce that

$$\|G(x') - G(x'')\| \le \|A\|\,\|x' - x''\| + 3\varepsilon\|x' - x''\|,$$

which translates that $\operatorname{fitz} G(x) \le \|A\|$. So, $\|D_F G(x)\| = \operatorname{fitz} G(x)$. □

EXAMPLE 14.12. The locally Lipschitz continuous function $f : \mathbb{R} \to \mathbb{R}$, with $f(x) = x^2 \sin(1/x)$ if $x \ne 0$ and $f(0) = 0$, is derivable at 0 with $\operatorname{fitz} f(0) = |Df(0)| = 0$ but $\operatorname{lip} f(0) = 1$. This says on the one hand that the equality in (b) of the above proposition does not hold with $\operatorname{lip} G(x)$ in place of $\operatorname{fitz} G(x)$, and on the other hand that, even for locally Lipschitz continuous mappings which are Fréchet differentiable, $\operatorname{fitz} G(x)$ and $\operatorname{lip} G(x)$ are different constants. □

Consider now the case of real-valued functions.

THEOREM 14.13 (S. Fitzpatrick's theorem for Fréchet differentiability). Let $(X, \|\cdot\|)$ be a normed space, $f : X \to \mathbb{R}$ be a real-valued function, and $x \in X$. Assume that there exists some $u \in X$ with $\|u\| = 1$ such that the norm $\|\cdot\|$ is Fréchet differentiable at u and such that

$$\lim_{t\to 0} t^{-1}\big(f(x+tu) - f(x)\big) = \operatorname{fitz} f(x) \ \Big(\text{resp. } \lim_{t\to 0} t^{-1}\big(f(x+tu) - f(x)\big) = \operatorname{lip} f(x)\Big)$$

with $\operatorname{fitz} f(x) < +\infty$ (resp. $\operatorname{lip} f(x) < +\infty$). Then f is Fréchet differentiable at x and $D_F f(x) = (\operatorname{fitz} f(x))\, u^*$ (resp. $D_F f(x) = (\operatorname{lip} f(x))\, u^*$) where $u^* \in X^*$ denotes the Fréchet derivative of the norm $\|\cdot\|$ at u.

PROOF. Suppose first that the limit $\lambda := \lim_{t\to 0} t^{-1}\big(f(x+tu) - f(x)\big)$ exists and that $\lambda = \operatorname{fitz} f(x)$ along with $\operatorname{fitz} f(x) < +\infty$. Denote $\alpha_f(x) := \operatorname{fitz} f(x)$ and fix any real $0 < \varepsilon \le 1/2$. Take some real $0 < \eta \le \varepsilon$ such that by the Fréchet differentiability of $\|\cdot\|$ at u

$$\|u + y\| - \|u\| \le \langle u^*, y\rangle + \varepsilon\|y\| \quad \text{for all } y \in \eta\mathbb{B}_X.$$

By definitions of λ and $\operatorname{fitz} f(x)$, choose $\delta > 0$ such that

$$|f(x+tu) - f(x) - t\lambda| \le \eta^2|t| \quad \text{for all } t \in [-\delta, \delta]$$

and such that

$$|f(x') - f(x'')| \le (\alpha_f(x) + \eta^2)\|x' - x''\|$$

for all $x', x'' \in X$ satisfying $2\|x'' - x\| \le \|x' - x\| \le \delta$.

Consider any $v \in X$ with $0 < \|v\| \le \eta\delta$ and observe that for $t := \eta^{-1}\|v\|$ we have

$$0 < t \le \delta,\ \|t^{-1}v\| = \eta, \text{ and } 2\|v\| = 2\eta t \le t = \|tu\|.$$

This yields

$$\|u \pm t^{-1}v\| - \|u\| \le \langle u^*, \pm t^{-1}v\rangle + \varepsilon\|t^{-1}v\|$$

and

$$|f(x \pm tu) - f(x+v)| \le (\alpha_f(x) + \eta^2)\| - v \pm tu\|.$$

Consequently, since $\lambda = \alpha_f(x)$ by assumption, we have on the one hand

$$\begin{aligned} f(x+v) - f(x) &= \big(f(x+v) - f(x-tu)\big) + \big(f(x-tu) - f(x)\big) \\ &\le (\alpha_f(x) + \eta^2)\|v + tu\| - t\lambda + \eta^2 t \\ &= \alpha_f(x)(\|v+tu\| - t) + \eta^2(\|v+tu\| + t), \end{aligned}$$

and this entails that

$$\begin{aligned} f(x+v) - f(x) &\le t\alpha_f(x)(\|u + t^{-1}v\| - \|u\|) + 3\eta^2 t \\ &\le t\alpha_f(x)(\langle u^*, t^{-1}v\rangle + \varepsilon\|t^{-1}v\|) + 3\eta^2 t \\ &\le \alpha_f(x)\langle u^*, v\rangle + \varepsilon\alpha_f(x)\|v\| + 3\eta\|v\| \\ &\le \alpha_f(x)\langle u^*, v\rangle + \varepsilon(\alpha_f(x) + 3)\|v\|. \end{aligned}$$

On the other hand, we have similarly

$$\begin{aligned} f(x+v) - f(x) &= \big(f(x+v) - f(x+tu)\big) + \big(f(x+tu) - f(x)\big) \\ &\ge -(\alpha_f(x) + \eta^2)\| - v + tu\| + t\lambda - \eta^2 t \\ &= -\alpha_f(x)(\| - v + tu\| - t) - \eta^2(\| - v + tu\| + t), \end{aligned}$$

which implies that

$$\begin{aligned} f(x+v)-f(x) &\geq -t\alpha_f(x)(\|u-t^{-1}v\|-\|u\|)-3\eta^2 t \\ &\geq -t\alpha_f(x)(\langle u^*,-t^{-1}v\rangle+\varepsilon\|t^{-1}v\|)-3\eta^2 t \\ &\geq \alpha_f(x)\langle u^*,v\rangle-\varepsilon\alpha_f(x)\|v\|-3\eta\|v\| \\ &\geq \alpha_f(x)\langle u^*,v\rangle-\varepsilon(\alpha_f(x)+3)\|v\|. \end{aligned}$$

It follows that

$$|f(x+v)-f(x)-\alpha_f(x)\langle u^*,v\rangle| \leq (\alpha_f(x)+3)\varepsilon\|v\| \quad \text{for all } v\in\eta\delta\mathbb{B}_X,$$

which means that f is Fréchet differentiable at $\overline{x}$ with $D_F f(x)=\alpha_f(x)u^*$. This finishes the proof in the case $\lambda=\alpha_f(x)$.

Suppose now the existence of the limit λ as well as the equality $\lambda=\operatorname{lip} f(x)$. Then (a) in Proposition 14.11 above ensures that $\lambda=\alpha_f(x)=\operatorname{lip} f(x)$, and the result in the present case follows from the previous case. □

As a direct consequence of Theorem 14.13 and Proposition 14.11(b) we have:

COROLLARY 14.14. Let $(X,\|\cdot\|)$ be a reflexive Banach space whose norm $\|\cdot\|$ is Fréchet differentiable (off zero). A real-valued function $f:X\to\mathbb{R}$ is Fréchet differentiable at $x\in X$ if and only if there exists some $u\in X$ with $\|u\|=1$ such that

$$\lim_{t\to 0} t^{-1}\big(f(x+tu)-f(x)\big)=\operatorname{fitz} f(x).$$

The inequality in (a) of Proposition 14.11 leads to introduce, in addition to the constant $\operatorname{fitz} G(x)$, some other constant more connected to the Gâteaux differentiability. For a mapping $G:X\to Y$ from the nonzero normed space X into the normed space Y, it is also of interest to consider the constant

$$\beta_G(x):=\sup_{z\in X}\sup_{\|u\|=1}\limsup_{t\downarrow 0} t^{-1}\|G(x+tz+tu)-G(x+tz)\|. \tag{14.6}$$

Obviously, $\beta_G(x)\leq \operatorname{lip} G(x)$. It is also worth noticing that given any $\varepsilon>0$ and any $v,w\in X$, there exists $\delta>0$ such that for all $0<t<\delta$

$$\|G(x+tw)-G(x+tv)\|\leq t\beta_G(x)\|w-v\|+t\varepsilon. \tag{14.7}$$

There is nothing to prove if $v=w$. Suppose $v\neq w$ and put $\varepsilon':=\varepsilon/\|w-u\|>0$, $z:=v/\|w-v\|$ and $u:=(w-v)/\|w-v\|$. Then

$$\limsup_{\tau\downarrow 0}\|G(x+\tau z+\tau u)-G(x+\tau z)\|<\beta_G(x)+\varepsilon',$$

so there is $\delta'>0$ such that $\|G(x+\tau z+\tau u)-G(x+\tau z)\|<\tau\beta_G(x)+\tau\varepsilon'$ for all $\tau\in]0,\delta'[$. Put $\delta:=\delta'/\|w-v\|$. For each $t\in]0,\delta[$, choosing $\tau:=t\|w-v\|$ yields

$$\|G\big(x+tv+t(w-v)\big)-G(x+tv)\|\leq t\beta_G(x)\|w-v\|+t\varepsilon,$$

which justifies (14.7).

PROPOSITION 14.15. If the mapping $G:X\to Y$ from the nonzero normed space X into the normed space Y is Gâteaux differentiable at $x\in X$, then one has the equality $\|DG(x)\|=\beta_G(x)$.

PROOF. Suppose that G is Gâteaux differentiable at x and denote by A its Gâteaux derivative at x. Then for each $z \in X$ and $u \in X$ with $\|u\| = 1$, we have that

$$t^{-1}\big(G(x+tz+tu)-G(x+tz)\big) = t^{-1}\big(G(x+t(z+u))-G(x)\big)-t^{-1}\big(G(x+tz)-G(x)\big)$$

converges to $A(z+u) - A(z) = Au$ as $t \downarrow 0$, so $\beta_G(x) \le \|A\|$ according to (14.6).

Conversely, for any u with $\|u\| = 1$ we have by (14.6) again

$$\|Au\| = \lim_{t\downarrow 0} t^{-1}\|G(x+tu) - G(x)\| \le \beta_G(x),$$

thus $\|A\| \le \beta_G(x)$. □

In the case of real-valued functions we observe the following:

LEMMA 14.16. Let $f : X \to \mathbb{R}$ be a real-valued function on the nonzero normed space X. Then for any $x \in X$

$$\beta_f(x) = \sup_{z\in X}\ \sup_{\|u\|=1}\ \limsup_{t\downarrow 0} \big(f(x+tz+tu) - f(x+tz)\big).$$

PROOF. Denoting by γ the second member of the equality in the statement, we readily see that $\gamma \le \beta_G(x)$. Now fix any $z, u \in X$ with $\|u\| = 1$ and choose some sequence of positive numbers $(t_n)_n$ tending to zero such that

$$q(z,u) := \limsup_{t\downarrow 0} t^{-1}|f(x+tz+tu)-f(x+tz)| = \lim_{n\to\infty} t_n^{-1}|f(x+t_nz+t_nu)-f(x+t_nz)|.$$

If $|f(x+t_nz+t_nu) - f(x+t_nz)| = f(x+t_nz+t_nu) - f(x+t_nz)$ for large n, then obviously $q(z,u) \le \limsup_{t\downarrow 0} t^{-1}\big(f(x+tz+tu) - f(x+tz)\big) \le \gamma$. If the equality fails for infinitely many n, then for some subsequence that we do not relabel, we have

$$\begin{aligned} q(z,u) &= \lim_{n\to\infty} t_n^{-1}\big(f(x+t_nz) - f(x+t_nz+t_nu)\big) \\ &= \lim_{n\to\infty} t_n^{-1}\big(f(x+t_n(z+u)+t_n(-u)) - f(x+t_n(z+u))\big), \end{aligned}$$

which gives

$$q(z,u) \le \sup_{\zeta\in X}\ \sup_{\|v\|=1}\ \limsup_{t\downarrow 0} t^{-1}\big(f(x+t\zeta+tv) - f(x+t\zeta)\big) = \gamma.$$

By (14.6) and the definition of $q(z,u)$ we can conclude that $\beta_G(x) \le \gamma$. □

THEOREM 14.17 (J.M. Borwein, S. Fitzpatrick and J.R. Giles). Let $(X, \|\cdot\|)$ be a normed space, $f : X \to \mathbb{R}$ be a real-valued function, and $x \in X$. Assume that there exists some $u \in X$ with $\|u\| = 1$ such that the norm $\|\cdot\|$ is Gâteaux differentiable at u and such that

$$\lim_{t\to 0} t^{-1}\big(f(x+tu) - f(x)\big) = \beta_f(x) \quad \Big(\text{resp. } \lim_{t\to 0} t^{-1}\big(f(x+tu) - f(x)\big) = \operatorname{lip} f(x)\Big)$$

with $\beta_f(x) < +\infty$ (resp. $\operatorname{lip} f(x) < +\infty$). Then f is Gâteaux differentiable at x and $D_Gf(x) = (\beta_f(x))\, u^*$ (resp. $D_Gf(x) = (\operatorname{lip} f(x))\, u^*$), where $u^* \in X^*$ denotes the Gâteaux derivative of the norm $\|\cdot\|$ at u.

PROOF. Suppose first that the limit $\lambda := \lim_{t\to 0} t^{-1}\big(f(x+tu) - f(x)\big)$ exists and that $\lambda = \beta_f(x)$ along with $\beta_f(x) < +\infty$. Fix any $y \in X$ with $\|y\| = 1$. Take any real $\varepsilon > 0$ and choose some real $0 < \eta \le \varepsilon$ such that

$$|\, \|u+ty\| - \|u\| - t\langle u^*, y\rangle \,| \le |t|\varepsilon \quad \text{for } |t| \le \eta.$$

By definition of λ there exists some real $\delta > 0$ such that

$$|f(x+tu) - f(x) - t\beta_f(x)| \le \eta^2|t| \quad \text{for } |t| \le \delta.$$

Shrinking δ if necessary, we may suppose according to (14.7) that

$$|f(x+t\eta y) - f(x \pm tu)| \le t\beta_f(x)\|-\eta y \pm u\| + \eta^2 t \quad \text{for all } t \in [0,\delta].$$

Then for all $0 < t < \delta$ we have on the one hand

$$\begin{aligned} f(x+t\eta y) - f(x) &= f(x+t\eta y) - f(x+tu) + f(x+tu) - f(x) \\ &\ge -t\beta_f(x)\|u - \eta y\| - \eta^2 t + t\beta_f(x) - \eta^2 t \\ &= -t\beta_f(x)\big(\|u-\eta y\| - \|u\|\big) - 2\eta^2 t, \end{aligned}$$

hence

$$f(x+t\eta y) - f(x) \ge t\beta_f(x)\big(\langle u^*, \eta y\rangle - \varepsilon\eta\big) - 2\eta^2 t;$$

and on the other hand

$$\begin{aligned} f(x+t\eta y) - f(x) &= f(x+t\eta y) - f(x-tu) + f(x-tu) - f(x) \\ &\le t\beta_f(x)\|u + \eta y\| + \eta^2 t - t\beta_f(x) + \eta^2 t \\ &= t\beta_f(x)\big(\|u+\eta y\| - \|u\|\big) + 2\eta^2 t, \end{aligned}$$

hence

$$f(x+t\eta y) - f(x) \le t\beta_f(x)\big(\langle u^*, \eta y\rangle + \varepsilon\eta\big) + 2\eta^2 t.$$

We deduce that

$$\left|\frac{1}{t\eta}\Big(f(x+t\eta y) - f(x)\Big) - \beta_f(x)\langle u^*, y\rangle\right| \le \varepsilon\big(2 + \beta_f(x)\big),$$

which ensures that $\lim_{t\downarrow 0} t^{-1}\Big(f(x+ty) - f(x) - t\langle \beta_f(x)u^*, y\rangle\Big) = 0$. This translates the Gâteaux differentiability of f at x as well as the equality $D_G f(x) = \beta_f(x)u^*$.

Finally, if we suppose the existence of the limit λ as well as the equality $\lambda = \operatorname{lip} f(x)$ with $\operatorname{lip} f(x) < +\infty$, then we easily see that $\lambda = \beta_f(x) = \operatorname{lip} f(x)$, and the result in the present case follows from the previous case. $\square$

COROLLARY 14.18. Let $(X, \|\cdot\|)$ be a reflexive Banach space whose norm $\|\cdot\|$ is Gâteaux differentiable (off zero). A real-valued function $f : X \to \mathbb{R}$ is Gâteaux differentiable at $x \in X$ if and only if there exists some $u \in X$ with $\|u\| = 1$ such that

$$\lim_{t\to 0} t^{-1}\big(f(x+tu) - f(x)\big) = \beta_f(x).$$

Regarding other corollaries, we start with the farthest distance function. Given a nonempty bounded subset C of a normed space $(X, \|\cdot\|)$, we have seen in Proposition 14.4 that dfar_C is a non-negative real-valued function which is 1-Lipschitz on the space X, that is,

$$|\mathrm{dfar}_C(x') - \mathrm{dfar}_C(x'')| \le \|x' - x''\| \quad \text{for all } x', x'' \in X. \tag{14.8}$$

If in addition X is not reduced to zero, then

$$\operatorname{lip}(\mathrm{dfar}_C)(x) = 1 \quad \text{for all } x \in X. \tag{14.9}$$

Indeed, fix any $x \in X$ and suppose first that $C \ne \{x\}$. Consider a sequence $(t_n)_n$ tending to 0 with $t_n > 0$, and for each $n \in \mathbb{N}$ choose $y_n \in C$ such that

$\mathrm{dfar}_C(x) - t_n^2 \le \|x - y_n\| \le \mathrm{dfar}_C(x)$. Then we note that $x_n := x + t_n(x - y_n) \to x$ and

$$\frac{\mathrm{dfar}_C(x_n) - \mathrm{dfar}_C(x)}{\|x_n - x\|} \ge \frac{(1+t_n)\|x - y_n\| - \|x - y_n\| - t_n^2}{t_n\|x - y_n\|} = \frac{\|x - y_n\| - t_n}{\|x - y_n\|}.$$

This and the convergence $\dfrac{\|x - y_n\| - t_n}{\|x - y_n\|} \to 1$ (as $n \to \infty$) combined with (14.8) give

$$\limsup_{x' \to x, x'' \to x} \frac{|\mathrm{dfar}_C(x'') - \mathrm{dfar}_C(x')|}{\|x'' - x'\|} = 1,$$

which translates the equality $\mathrm{lip}\,(\mathrm{dfar}_C)(x) = 1$. Suppose now that $C = \{x\}$, so $\mathrm{dfar}_C(u) = \|u - x\|$ for all $u \in X$. Fix a nonzero vector $v \in X$. Then for each $t > 0$ putting $x_t := x + tv$ we see that $x_t \to x$ as $t \downarrow 0$ and

$$\frac{\|x_t - x\| - \|x - x\|}{\|x_t - x\|} = \frac{t\|v\|}{t\|v\|} = 1,$$

so as above we deduce that $\mathrm{lip}\,(\mathrm{dfar}_C)(x) = 1$.

Corollary 14.19. *Let C be a nonempty bounded set of a normed vector space $(X, \|\cdot\|)$ and let $x \in X$. If there exists some $u \in X$ with $\|u\| = 1$ such that the norm $\|\cdot\|$ is Fréchet (resp. Gâteaux) differentiable at u and such that*

$$\lim_{t \to 0} t^{-1}\big(\mathrm{dfar}_C(x + tu) - \mathrm{dfar}_C(x)\big) = 1, \tag{14.10}$$

then dfar_C is Fréchet (resp. Gâteaux) differentiable at x and one has the equality $D(\mathrm{dfar}_C)(x) = D(\|\cdot\|)(u)$.

Proof. If there exists some unit vector u (so $X \ne \{0\}$) satisfying the assumptions of the corollary, then (14.9) and Theorem 14.13 (resp. Theorem 14.17) yields the conclusion of the corollary. □

Corollary 14.20. *Let $(X, \|\cdot\|)$ be a nonzero reflexive Banach space whose norm $\|\cdot\|$ is Fréchet differentiable off zero. Let C be a nonempty closed bounded set of X and let $x \in X$. Then the following assertions are equivalent:*
(a) the farthest distance function dfar_C is Fréchet differentiable at x;
(b) there exists some $u \in X$ with $\|u\| = 1$ such that

$$\lim_{t \to 0} t^{-1}\big(\mathrm{dfar}_C(x + tu) - \mathrm{dfar}_C(x)\big) = 1; \tag{14.11}$$

(c) the function dfar_C is Gâteaux differentiable at x and satisfies the equality $\|D_G(\mathrm{dfar}_C)(x)\| = 1$.

Proof. First we note that (b) entails (a) by Corollary 14.19. We also have that (a) implies (c) since $\|D_F \mathrm{dfar}_C(x)\| = 1$ whenever dfar_C is Fréchet differentiable at x by Proposition 14.11(b) and Proposition 14.6(a). Suppose now that X is reflexive and (c) is satisfied. The weak compactness of $\mathbb{B}_X$ gives some $u \in X$ with $\|u\| = 1$ and $\|D_G \mathrm{dfar}_C(x)\| = D_G \mathrm{dfar}_C(x)(u)$. Then,

$$\lim_{t \to 0} t^{-1}\big(\mathrm{dfar}_C(x + tu) - \mathrm{dfar}_C(x)\big) = D_G \mathrm{dfar}_C(x)(u) = 1,$$

so we obtain that (c) entails (b), and the proof is finished. □

Let us now turn to the basic distance functions d_C which arise much more often in several contexts. Given a nonempty closed set C of a normed space $(X, \|\cdot\|)$, it worth noticing that

$$\text{lip}\,(d_C)(x) = 1 \quad \text{for all } x \notin C. \tag{14.12}$$

Indeed, fix $x \notin C$ and take a sequence $(t_n)_n$ tending to 0 with $t_n \in]0,1[$. For each $n \in \mathbb{N}$ choose $y_n \in C$ such that $\|x - y_n\| - t_n^2 \le d_C(x) \le \|x - y_n\|$, so $x_n := x - t_n(x - y_n) \to x$ as $n \to \infty$. Write

$$\frac{d_C(x) - d_C(x_n)}{\|x - x_n\|} \ge \frac{\|x - y_n\| - t_n^2 - \|(1 - t_n)(x - y_n)\|}{t_n\|x - y_n\|} = \frac{\|x - y_n\| - t_n}{\|x - y_n\|}$$

and note that $\dfrac{\|x - y_n\| - t_n}{\|x - y_n\|} \to 1$ (since $\|x - y_n\| \to d_C(x) \neq 0$). Taking also the trivial inequality $|d_C(x') - d_C(x'')| \le \|x' - x''\|$ into account, we obtain the desired equality $\text{lip}\,(d_C)(x) = 1$.

COROLLARY 14.21. Let $(X, \|\cdot\|)$ be a normed vector space and let C be a nonempty closed set of X and $x \notin C$. If there exists some $u \in X$ with $\|u\| = 1$ such that the norm $\|\cdot\|$ is Fréchet (resp. Gâteaux) differentiable at u and such that

$$\lim_{t \to 0} t^{-1}\big(d_C(x + tu) - d_C(x)\big) = 1, \tag{14.13}$$

then d_C is Fréchet (resp. Gâteaux) differentiable at x and $Dd_C(x) = D(\|\cdot\|)(u)$.

PROOF. If there exists some unit vector u satisfying the equality of the corollary, then Theorem 14.13 (resp. Theorem 14.17) and (14.12) give that $f = d_C$ is Fréchet (resp. Gâteaux) differentiable at x. □

When X is a reflexive Banach space, a kind of converse of the above statement relative to the Fréchet differentiability also holds true through the implication (a) $\Rightarrow$ (b) in the following corollary.

COROLLARY 14.22. Let $(X, \|\cdot\|)$ be a nonzero reflexive Banach space whose norm $\|\cdot\|$ is Fréchet differentiable off zero. Let C be a nonempty closed set of X and $x \notin C$. Then the following assertions are equivalent:
(a) the distance function d_C is Fréchet differentiable at x;
(b) there exists some $u \in X$ with $\|u\| = 1$ such that

$$\lim_{t \to 0} t^{-1}\big(d_C(x + tu) - d_C(x)\big) = 1; \tag{14.14}$$

(c) the distance function d_C is Gâteaux differentiable at x and $\|D_G d_C(x)\| = 1$.

PROOF. We already know by the previous corollary that (b) implies (a), and further (a) implies (c) since $\|D_F d_C(x)\| = 1$ whenever d_C is Fréchet differentiable at $x \notin C$ by Proposition 4.31 (which can also be easily checked). Suppose now that X is reflexive and (c) is satisfied. The weak compactness of $\mathbb{B}_X$ gives some $u \in X$ with $\|u\| = 1$ and $\|D_G d_C(x)\| = D_G d_C(x)(u)$. Then,

$$\lim_{t \to 0} t^{-1}\big(d_C(x + tu) - d_C(x)\big) = D_G d_C(x)(u) = 1,$$

so we obtain that (c) entails (b), and the proof is finished. □

The third corollary provides another condition for the differentiability of the distance function when the set of nearest points of x in C is nonempty.

COROLLARY 14.23. Let $(X, \|\cdot\|)$ be a normed vector space and C be a nonempty closed set of X, and let $x \notin C$. If, for some $p(x) \in \mathrm{Proj}_C(x)$, the (bilateral) limit

$$\lim_{t\to 0} t^{-1}\left(d_C\big(x + t(x - p(x))\big) - d_C(x)\right)$$

exists in $\mathbb{R}$ and the norm $\|\cdot\|$ is Fréchet (resp. Gâteaux) differentiable at $x - p(x)$, then the distance function d_C is Fréchet (resp. Gâteaux) differentiable at x and

$$Dd_C(x) = D(\|\cdot\|)\left(\frac{x - p(x)}{\|x - p(x)\|}\right) = D(\|\cdot\|)(x - p(x)).$$

PROOF. First we observe, for each $t \in [-1, 0[$, that $p(x) \in \mathrm{Proj}_C\big(x+t(x-p(x))\big)$, thus

$$t^{-1}\left(d_C\big(x + t(x - p(x))\big) - d_C(x)\right) = t^{-1}\big((1+t)\|x-p(x)\|-\|x-p(x)\|\big) = \|x-p(x)\|.$$

The assumption on the bilateral limit then ensues

$$\lim_{t\to 0} t^{-1}\left(d_C\big(x + t(x - p(x))\big) - d_C(x)\right) = \|x - p(x)\|,$$

which gives, for $u := (x - p(x))/\|x - p(x)\|$,

$$\lim_{t\to 0} t^{-1}\big(d_C(x + tu) - d_C(x)\big) = 1.$$

The result then follows from Corollary 14.21. □

The following theorem continues the study of the differentiability of the distance function at points outside the set. It involves another condition similar to (14.13) but with the unilateral convergence $t \downarrow 0$ in place of the bilateral convergence $t \to 0$.

THEOREM 14.24 (V.S. Balaganski and L.P. Vlasov). Let $(X, \|\cdot\|)$ be a normed space, C be a nonempty closed set of X and $x \in X \setminus C$ such that $\mathrm{Proj}\,_C(x) \neq \emptyset$. Assume that the norm $\|\cdot\|$ is Fréchet (resp. Gâteaux) differentiable off zero and that there exists some $u \in X$ with $\|u\| = 1$ such that

$$\lim_{t\downarrow 0} t^{-1}\big(d_C(x + tu) - d_C(x)\big) = 1.$$

Then d_C is Fréchet (resp. Gâteaux) differentiable at x and, for any $y \in \mathrm{Proj}\,_C(x)$,

$$Dd_C(x) = D(\|\cdot\|)(x - y) = D(\|\cdot\|)(u).$$

PROOF. Fix x and u as above and consider any $y \in \mathrm{Proj}\,_C(x)$, so $d_C(x) = \|x - y\|$. By assumption $\|u\| = 1$ and

$$t^{-1}\big(d_C(x + tu) - d_C(x)\big) \to 1 \quad \text{as } t \downarrow 0. \tag{14.15}$$

Take $y^* = D(\|\cdot\|)(x - y)$, that is, $\langle y^*, x - y\rangle = \|x - y\|$ and $\|y^*\| = 1$. Denote by $\mathcal{B}$ the collection of nonempty bounded sets (resp. singleton sets) of X. Fix any $B \in \mathcal{B}$ and take any sequence of reals $(t_n)_n$ tending to 0 with $0 < t_n < 1$ and any sequence $(b_n)_n$ in B. We observe that, for all $t > 0$,

$$t^{-1}\big(d_C(x + tu) - d_C(x)\big) \le t^{-1}\big(\|(x - y) + tu\| - \|x - y\|\big),$$

which, using (14.15) gives $\langle y^*, u\rangle = \langle D(\|\cdot\|)(x - y), u\rangle \ge 1$, thus $\langle y^*, u\rangle = 1$ since $\langle y^*, u\rangle \le \|y^*\|_*\|u\| = 1$. So, $y^* = D(\|\cdot\|)(u)$. Putting

$$\rho(s) := s + \big(1 - s^{-1}(d_C(x + su) - d_C(x))\big)^{1/2} \quad \text{for all } s > 0,$$

we see that $\rho(1) \geq 1$, and $\rho(s) \to 0$ as $s \downarrow 0$, hence $s\rho(s) \to 0$ as well as $s \downarrow 0$. By continuity of $s \mapsto s\rho(s)$ we may choose, for each $n \in \mathbb{N}$, some $s_n \in]0,1[$ such that $s_n\rho(s_n) = t_n$. Since $s_n^2 \leq s_n\rho(s_n) \leq t_n$, we have $s_n \downarrow 0$. Further, we have

$$\begin{aligned}
&t_n^{-1}\Big(d_C(x+t_nb_n) - d_C(x) - t_n\langle y^*, b_n\rangle\Big)\\
&\geq t_n^{-1}\Big(d_C(x+s_nu) - \|t_nb_n - s_nu\| - d_C(x) - t_n\langle y^*, b_n\rangle\Big)
\end{aligned}$$

(14.16)

$$= \left(\frac{1}{s_n}(d_C(x+s_nu) - d_C(x)) - 1\right)\frac{s_n}{t_n} + \frac{1}{t_n}(s_n - \|t_nb_n - s_nu\| - t_n\langle y^*, b_n\rangle).$$

Since

$$0 \leq R_n := \left(1 - \frac{1}{s_n}(d_C(x+s_nu) - d_C(x))\right)\frac{s_n}{t_n} \leq \rho^2(s_n)\frac{s_n}{t_n} = \rho(s_n),$$

we have $R_n \to 0$ as $n \to \infty$.

On the other hand, for $r_n := \rho(s_n)$ and $Q_n := t_n^{-1}[s_n - \|t_nb_n - s_nu\| - t_n\langle y^*, b_n\rangle]$, we have

$$\begin{aligned}
Q_n &= r_n^{-1}(1 - \|r_nb_n - u\| - r_n\langle y^*, b_n\rangle)\\
&= -r_n^{-1}(\|u + r_nb_n'\| - \|u\| - r_n\langle y^*, b_n'\rangle), \text{ with } b_n' := -b_n,
\end{aligned}$$

so $Q_n \to 0$ as $n \to \infty$ since $y^* = D_F(\|\cdot\|)(u)$. Using this and the equality $\lim_{n\to\infty} R_n = 0$ we obtain from (14.16) that

$$\liminf_{n\to\infty} t_n^{-1}\Big(d_C(x+t_nb_n) - d_C(x) - t_n\langle y^*, b_n\rangle\Big) \geq 0.$$

Further,

$$t_n^{-1}\Big(d_C(x+t_nb_n) - d_C(x) - t_n\langle y^*, b_n\rangle\Big) \leq t_n^{-1}\Big(\|(x-y) + t_nb_n\| - \|x-y\| - t_n\langle y^*, b_n\rangle\Big),$$

and since the right member tends to 0 as $n \to \infty$ according to the equality $y^* = D_F(\|\cdot\|)(x-y)$, we deduce that

$$\limsup_{n\to\infty} t_n^{-1}\Big(d_C(x+t_nb_n) - d_C(x) - t_n\langle y^*, b_n\rangle\Big) \leq 0.$$

Consequently,

$$\lim_{n\to\infty} t_n^{-1}\Big(d_C(x+t_nb_n) - d_C(x) - t_n\langle y^*, b_n\rangle\Big) = 0,$$

and the proof is finished. □

Concerning the conditions in the assumptions of Corollary 14.21 and Theorem 14.24, it is worth pointing out the following.

REMARK 14.25. For a closed set C of a normed space $(X, \|\cdot\|)$ and $x \notin C$, one always has

$$\limsup_{y\to x} \frac{d_C(x) - d_C(y)}{\|x-y\|} = 1.$$

Clearly, the left member is not greater than 1. To verify the reverse inequality, for each $n \in \mathbb{N}$, choose some $u_n \in C$ such that $\|x - u_n\| < d_C(x) + 1/n^2$. For each

$n > 1/d_C(x)$ we can take $y_n \in]x, u_n]$ with $\|x - y_n\| = 1/n$. So, $y_n \to x$ as $n \to \infty$ and, for all $n > 1/d_C(x)$, we have

$$\frac{d_C(x) - d_C(y_n)}{\|x - y_n\|} \geq \frac{\|x - u_n\| - \|u_n - y_n\| - (1/n)^2}{\|x - y_n\|} = 1 - \frac{1}{n}.$$

This combined with the previous inequality justifies the desired equality. □

14.1.3. Genericity of points with nearest points, Lau theorem. Theorem 14.2 above ensures in the Kadec-Klee reflexivity setting the density of the set of points of $X \setminus C$ which have nearest points in C. Given in the same context two closed sets C_1 and C_2, we cannot deduce from that result the density of points of $X \setminus (C_1 \cup C_2)$ which have nearest points both in C_1 and C_2. We then continue the study of richness of points of $X \setminus C$ having nearest points in C with the aim of establishing the genericity of this set, that is, it contains a dense G_δ set of $X \setminus C$. This will be reached with the celebrated Lau's theorem.

Consider thus for the closed set C of the Banach space X and for any $n \in \mathbb{N}$ the set

$$L_n(C) := \{u \in X \setminus C : \exists u^* \in X^* \text{ with } \|u^*\| = 1, \exists \delta > 0, \text{ so that}$$
$$\inf\{\langle u^*, u - x\rangle : x \in C \cap B(u, d_C(u) + \delta)\} > (1 - 2^{-n}) d_C(u)\},$$

and their intersection

$$L(C) := \cap_n L_n(C),$$

(the letter L standing for "Lau"). Consider also

$$\Omega(C) := \{u \in X \setminus C : \exists u^* \in X^* \text{ with } \|u^*\| = 1, \forall \varepsilon > 0, \exists \delta > 0, \text{ so that}$$
$$\inf\{\langle u^*, u - x\rangle : x \in C \cap B(u, d_C(u) + \delta)\} > (1 - \varepsilon) d_C(u)\}.$$

Obviously $\Omega(C) \subset L(C)$.

The proof of Lau's theorem will be achieved through the lemma below.

LEMMA 14.26. Each set $L_n(C)$ is open in X.

PROOF. Fix any $u \in L_n(C)$. By definition of $L_n(C)$, choose some $u^* \in X^*$ with $\|u^*\| = 1$ and some $\delta > 0$ such that

$$0 < t := \inf\{\langle u^*, u - x\rangle : x \in C \cap B(u, d_C(u) + \delta)\} - (1 - 2^{-n}) d_C(u).$$

Take some positive $\eta < \min\{\delta/2, t/2\}$ and fix any $y \in X$ with $\|y - u\| < \eta$. For $\delta_0 := \delta - 2\eta > 0$ we easily verify that

$$C \cap B(u, d_C(u) + \delta) \supset Q := C \cap B(y, d_C(y) + \delta_0).$$

So, for any $x \in Q$ we obtain

$$\langle u^*, u - x\rangle \geq t + (1 - 2^{-n}) d_C(u),$$

and hence

$$\begin{aligned}\langle u^*, y - x\rangle &\geq t + (1 - 2^{-n}) d_C(u) + \langle u^*, y - u\rangle \\ &\geq (1 - 2^{-n}) d_C(y) + t - 2\|u - y\| \\ &\geq (1 - 2^{-n}) d_C(y) + t - 2\eta > (1 - 2^{-n}) d_C(y).\end{aligned}$$

Consequently

$$\inf\{\langle u^*, y - x\rangle : x \in Q\} > (1 - 2^{-n}) d_C(y),$$

which gives $B(u, \eta) \setminus C \subset L_n(C)$. Thus, the openness of $L_n(C)$ is proved. □

The second lemma relates the set $\Omega(C)$ to the Fréchet subdifferential of the distance function d_C.

LEMMA 14.27. For the Banach space X, if $u \in X \setminus C$ and $\partial_F d_C(u) \neq \emptyset$, then $u \in \Omega(C)$.

So, $\Omega(C)$ is dense in $X \setminus C$ whenever the Banach space X is reflexive.

PROOF. Fix $u \in X \setminus C$ with $\partial_F d_C(u) \neq \emptyset$ and fix $u^* \in \partial_F d_C(u)$. We know that $\|u^*\| = 1$ (see Proposition 4.31) and the convergence in Proposition 4.32(a) ensures that for any real $\varepsilon > 0$ there is some $\delta > 0$ such that for all $x \in C$ with $\|x - u\| - d_C(u) < \delta$ we have

$$\langle u^*, u - x \rangle - d_C(u) > -\frac{\varepsilon}{2} d_C(u),$$

which gives

$$\inf\{\langle u^*, u - x \rangle : x \in C \cap B(u, d_C(u) + \delta)\} > (1 - \varepsilon) d_C(u).$$

This implies that $u \in \Omega(C)$.

The second property of the lemma is a direct consequence of what precedes and the fact that the effective domain of $\partial_F d_C$ is dense in $\operatorname{dom} d_C = X$ whenever X is reflexive (see Corollary 4.53). □

The third lemma establishes the equality between $\Omega(C)$ and $L(C)$ when X is reflexive.

LEMMA 14.28. If the Banach space X is reflexive, then $\Omega(C) = L(C)$.

PROOF. It is enough to show that $L(C) \subset \Omega(C)$ since the reverse inclusion is obvious as already said. Fix any $u \in L(C) = \bigcap_n L_n(C)$, so in particular $d_C(u) > 0$. For each $n \in \mathbb{N}$ choose some $u_n^* \in X^*$ with $\|u_n^*\| = 1$ and some $\delta_n > 0$ such that

$$\inf\{\langle u_n^*, u - x_n \rangle : x \in C \cap B(u, d_C(u) + \delta_n)\} > (1 - 2^{-n}) d_C(u)$$

and take a weak-star cluster point u^* of (u_n^*). Without loss of generality, we may suppose $\delta_{n+1} < \delta_n$ along with $\delta_n \downarrow 0$ as $n \to \infty$. For each $n \in \mathbb{N}$ consider the nonempty weakly compact set

$$K_n := \operatorname{cl}_w[C \cap B(u, d_C(u) + \delta_n)]$$

and observe that their intersection $K := \bigcap_n K_n$ is nonempty since $K_{n+1} \subset K_n$. For each $x \in K$ we have on the one hand $\|u - x\| \le d_C(u)$, and on the other hand $\langle u_n^*, u - x \rangle \ge (1 - 2^{-n}) d_C(u)$, which gives $\langle u^*, u - x \rangle \ge d_C(u)$. Since $\|u^*\| \le 1$ we then obtain $\|u^*\| = 1$ and

$$\langle u^*, u - x \rangle = d_C(u) = \|u - x\|.$$

Then for any $\varepsilon > 0$ the set K is contained in the weakly open set

$$V_\varepsilon := \{x \in X : \langle u^*, u - x \rangle > (1 - \frac{\varepsilon}{2}) d_C(u)\}$$

and as the sets K_n are nonincreasing and weakly compact we have some $K_{n_0} \subset V_\varepsilon$. The latter inclusion ensures in particular that

$$\inf\{\langle u^*, u - x \rangle : x \in C \cap B(u, d_C(u) + \delta_{n_0})\} > (1 - \varepsilon) d_C(u).$$

Therefore $u^* \in \Omega(C)$, and hence the desired inclusion $L(C) \subset \Omega(C)$ holds. □

We can now establish Lau's theorem. Recall first that a G_δ set of a topological space is a countable intersection of open sets of the space.

THEOREM 14.29 (K.-S. Lau theorem on genericity of points with nearest points). Let X be a reflexive Banach space endowed with a norm $\|\cdot\|$ satisfying the (sequential) Kadec-Klee property and let C be a nonempty (strongly) closed set of X. Then the set of points of $X \setminus C$ admitting nearest points in C contains a dense G_δ set of $X \setminus C$.

If in addition the (sequential) Kadec-Klee norm is strictly convex, then there exists a dense G_δ set of $X \setminus C$ with unique nearest points in C.

PROOF. Lemmas 14.28 and 14.26 tell us that $\Omega(C)$ is a G_δ set of $X \setminus C$, and by Lemma 14.27 the set $\Omega(C)$ is dense in $X \setminus C$. For the first assertion, we only need to show that the Kadec-Klee property of the norm implies that each point of $\Omega(C)$ admits at least one nearest point. Fix any $u \in \Omega(C)$ and take a sequence $(x_n)_n$ of C such that $\|u - x_n\| \to d_C(u)$. Extracting a subsequence if necessary we may suppose that $(x_n)_n$ converges weakly to some $x \in X$. By definition of $\Omega(C)$ there exists some $u^* \in X^*$ with $\|u^*\| = 1$ such that

$$\|u - x\| \geq \langle u^*, u - x\rangle = \lim_n \langle u^*, u - x_n\rangle \geq d_C(u) = \lim_n \|u - x_n\|.$$

Further, by the lower semicontinuity of the norm we also have $\lim_n \|u - x_n\| \geq \|u - x\|$, and hence $\|u - x\| = \lim_n \|u - x_n\| (= d_C(u))$. Since $(u - x_n)_n$ converges weakly to $u - x$, the Kadec-Klee property of the norm ensures that $\|x - x_n\| \to 0$, and this in particular implies that $x \in C$. Since $\|u - x\| = d_C(u)$, we deduce that x is a nearest point of C to u.

Suppose now that the norm $\|\cdot\|$ is in addition strictly convex. Fix again any $u \in \Omega(C)$. Taking any $y \in C$ such th $\|u - y\| = d_C(u) > 0$ and taking $x_n := y$ for all n in what precedes, the unit element $u^* \in X^*$ obtained above satisfies

$$\langle u^*, u - y\rangle = d_C(u).$$

Consequently, by Proposition 14.8 we obtain that $\mathrm{Proj}_C(u)$ is a singleton. The proof of the theorem is complete. $\square$

That the genericity result in Lau theorem above may fail in the absence of the reflexivity assumption is illustrated in the next example. In fact, it provides in any non-reflexive Banach space closed hyperplanes along with closed bounded convex sets whose set of points with nearest points is not dense in the space.

EXAMPLE 14.30. Let $(X, \|\cdot\|)$ be a non-reflexive Banach space. By James theorem (see Theorem C.8 in Appendix C) there exists some continuous linear functional $a^* \in X^*$ with $\|a^*\| = 1$ which does not attain its supremum on $\mathbb{B}_X$.
(a) Consider the closed vector hyperplane $E := \{v \in X : \langle a^*, v\rangle = 0\}$, so as well-known and easily seen

$$d_E(x)) = |\langle a^*, x\rangle| \quad \text{for all } x \in X.$$

Fix any $x \in X \setminus E$. We claim that the distance $d_E(x)$ is not attained. Suppose by contradiction that there is some $y \in E$ such that $d_E(x) = \|x - y\|$. If $\langle a^*, x\rangle > 0$, putting $u := (x - y)/\|x - y\|$ we have

$$\langle a^*, u\rangle = \frac{1}{\|x - y\|}\langle a^*, x - y\rangle = \frac{1}{d_E(x)}\langle a^*, x\rangle = 1,$$

which contradicts that $a^* \in X^*$ does not attain its supremum on $\mathbb{B}_X$. In the case $\langle a^*, x\rangle < 0$, it suffices to put $u := (y - x)/\|x - y\|$ and to proceed as above.

(b) Define the closed bounded convex set $C := E \cap \mathbb{B}_X$, where E is as in (a). It is readily seen that for any x in the open set $U := \{v \in X : \|x\| < 1 \text{ and } \langle a^*, x\rangle \neq 0\}$ one has $d_C(x) = d_E(x)$, so no point $x \in U$ admits nearest point in C. $\square$

Now, concerning the Kadec-Klee hypothesis of the norm in Theorem 14.29 it is worth pointing out that one can find, for any reflexive Banach space endowed with a non (sequential) Kadec-Klee norm, examples of closed sets C for which there exits a nonempty open set $U \subset X \setminus C$ such that any $u \in U$ has no nearest point in the set C. We turn for that to the Konjagin's construction.

LEMMA 14.31. Assume that the norm $\|\cdot\|$ of the Banach space X does not satisfy the (sequential) Kadec-Klee property. Then there exist a sequence $(x_n)_n$ in X and an element element $x^* \in X^*$ such that

$$(i)\ \langle x^*, x_n\rangle = \|x^*\| = 1 = \lim_n \|x_n\|, \text{ and } (ii)\ \inf_{n\neq m} \|x_n - x_m\| > 0.$$

Conversely, if such x_n and x^* exist and if the Banach space X is reflexive, then the norm $\|\cdot\|$ does not satisfy the (sequential) Kadec-Klee property.

PROOF. Suppose that the norm $\|\cdot\|$ is not (sequentially) Kadec-Klee. Then there is a sequence $(y_n)_n$ in X converging weakly to some y with $\|y_n\| = \|y\| = 1$, but with $\|y_n - y\|$ not converging to 0. Extracting a subsequence we may suppose that for some real $\varepsilon' > 0$ we have $\|y_n - y\| > \varepsilon'$ for all $n \in \mathbb{N}$. Take $x^* \in X^*$ with $\langle x^*, y\rangle = \|x^*\| = 1$. We may suppose that $\langle x^*, y_n\rangle \neq 0$ for all n, and hence we put $z_n := (\langle x^*, y_n\rangle)^{-1} y_n$ for all n. Then $(z_n)_n$ converges weakly to y and

$$\langle x^*, z_n\rangle = 1 = \|x^*\| = \lim_n \|z_n\|,$$

while $\liminf_n \|z_n - y\| \geq \varepsilon'$. Setting $\varepsilon = \varepsilon'/2$ we way also suppose that $\|z_n - y\| > \varepsilon$ for all n. The weak lower semicontinuity property of the norm $\|\cdot\|$ then yields for each n

$$\liminf_k \|z_n - z_k\| \geq \|z_n - y\| > \varepsilon,$$

and hence we have some integer $N(n)$ such that

$$\|z_n - z_k\| > \varepsilon \quad \text{for all } k \geq N(n).$$

So, we can construct an increasing function $s : \mathbb{N} \to \mathbb{N}$ such that $\|z_{s(n)} - z_{s(m)}\| > \varepsilon$ for all integers n, m with $n \neq m$. Then x^* and $x_n := z_{s(n)}$ have the desired properties (i) and (ii).

Conversely, suppose that X is reflexive and that there exist x^* and x_n satisfying (i) and (ii). Extracting a subsequence we may suppose that $(x_n)_n$ converges weakly to some x. Then (i) ensures that

$$\|x\| \geq \langle x^*, x\rangle = 1 = \lim_n \|x_n\|,$$

and hence $\lim_n \|x_n\| = \|x\|$ according to the lower semicontinuity of the norm $\|\cdot\|$. This and (ii) contradict the (sequential) Kadec-Klee property of the norm $\|\cdot\|$ and conclude the proof. $\square$

THEOREM 14.32 (S.V. Konjagin). In any Banach space $(X, \|\cdot\|)$ whose norm is not (sequentially) Kadec-Klee there is a nonempty closed bounded set C for which there exists a nonempty open set $U \subset X \setminus C$ such that any point $u \in U$ has no nearest point in C.

PROOF. Since the norm is not (sequentially) Kadec-Klee, by the preceding lemma there exist $\varepsilon \in]0,1[$, $x^* \in X^*$, and x_n with $\|x_n\| \le 2$ such that

$$\langle x^*, x_n\rangle = 1 = \|x^*\| = \lim_k \|x_k\| \quad \text{and} \quad \inf_{n\neq m} \|x_n - x_m\| \ge \varepsilon.$$

Put $y_n := (1+2^{-n})x_n$ and put also

$$C := \bigcup_n Q_n, \quad \text{where} \quad Q_n := y_n + (B[0,\varepsilon/3] \cap \{x \in X : \langle x^*, x\rangle = 0\}).$$

Observe first that for $n \neq m$ and $p \in Q_n$ and $q \in Q_m$ we have

$$\begin{aligned}\|p-q\| &\ge \|x_n - x_m\| - \|x_m - y_m\| - \|x_n - y_n\| - \|y_m - q\| - \|y_n - p\| \\ &\ge \varepsilon - 2^{1-n} - 2^{1-m} - \varepsilon/3 - \varepsilon/3,\end{aligned}$$

and hence there is some integer N such that for all $m, n \ge N$ with $m \neq n$ we have $\|p-q\| > \varepsilon/9$ for all $p \in Q_n$ and $q \in Q_m$. It easily follows from this that the union $\bigcup_{n\ge N} Q_n$ is closed in X. Since the finite union $\bigcup_{n<N} Q_n$ is also closed, we deduce that C is closed in X.

Consider now the open ball $U := B(0,\varepsilon/9)$ and fix any $u \in U$. Put $v_n := u + y_n - \langle x^*, u\rangle x_n$ and observe that

$$\|v_n - y_n\| \le \|u\| + 2\|u\| < \varepsilon/3 \quad \text{and} \quad \langle x^*, v_n - y_n\rangle = 0.$$

Therefore $v_n \in Q_n$ and

$$\begin{aligned}d_C(u) &\le \liminf_n \|v_n - u\| = \liminf_n \|y_n - \langle x^*, u\rangle x_n\| \\ &= \liminf_n [(1+2^{-n}) - \langle x^*, u\rangle]\|x_n\| \\ &= 1 - \langle x^*, u\rangle,\end{aligned}$$

the second equality using the definition of y_n and the inequality $\langle x^*, u\rangle < 1$. Further, for any $y \in C$ we have $y \in Q_k$ for some integer k, which entails that

$$\langle x^*, y\rangle = \langle x^*, y_k\rangle = (1+2^{-k}) > 1,$$

and this yields

$$\|y-u\| = \|x^*\|\|y-u\| \ge \langle x^*, y\rangle - \langle x^*, u\rangle > 1 - \langle x^*, u\rangle. \tag{14.17}$$

So $d_C(u) \ge 1 - \langle x^*, u\rangle$, which combined with the preceding inequality concerning $d_C(u)$ ensures that $d_C(u) = 1 - \langle x^*, u\rangle > 0$, and hence in particular $U \subset X \setminus C$. On the other hand, since $\|y-u\| > 1 - \langle x^*, u\rangle = d_C(u)$ (the strict inequality being due to (14.17)), we see that no point of U has a nearest point in C. The proof is then complete. □

REMARK 14.33. It is worth pointing out that the above proof also gives that d_C is affine on U since $d_C(u) = 1 - \langle x^*, u\rangle$ for all $u \in U$. □

Despite Konjagin theorem, in any reflexive Banach space there are nonconvex closed sets which satisfy the conclusion of Lau theorem. A large class of such sets is formed by complements of open convex sets (so-called Klee caverns). It is even the case for the class of multiple caverns. A *multiple (Klee) cavern* of X is the complement of the union of a countable (that is, finite or infinite countable) family of pairwise disjoint nonempty open convex sets of X.

PROPOSITION 14.34. Let C be a multiple cavern of a reflexive Banach space $(X, \|\cdot\|)$ (with no property on the norm $\|\cdot\|$). Then there exists a dense G_δ set of $X \setminus C$ whose each point admits a nearest point in C.

PROOF. Suppose first that C is the complement of an open convex set U. We know that the function $-d_C$ is convex (see Lemma 2.165) on the open convex set U and continuous, hence since X is reflexive $-d_C$ (hence also d_C) is Fréchet differentiable (see Theorem 4.49 and the subsequent discussion) on a dense G_δ set Δ of U. We show that any $u \in \Delta$ has a nearest point in C. Denote by u^* the Fréchet derivative of d_C at u and consider a minimizing sequence $(x_n)_n$ of $d_C(u)$. By reflexivity, the bounded sequence $(x_n)_n$ has a subsequence (that we do not relabel) converging weakly to some $x \in X$. We claim that $x \in C$. Otherwise, we have $x, u \in U$ and the convexity of $-d_C$ on U (as seen above) ensures that

$$d_C(x) - d_C(u) \leq \langle u^*, x - u \rangle = \lim_n \langle u^*, x_n - u \rangle = -d_C(u),$$

the last equality being due to Proposition 4.32(a). It follows that $d_C(x) \leq 0$, that is, $x \in C$, which is a contradiction. The claim $x \in C$ is thus true and from the weak lower semicontinuity of the norm we have

$$\|x - u\| \leq \lim_n \|x_n - u\| = d_C(u).$$

This says that x is a nearest point of C to u.

Consider now a countable family $(U_i)_{i \in I}$ of pairwise disjoint open convex sets of X and $C := X \setminus (\bigcup_{i \in I} U_i)$. Fix $i \in I$ and by what precedes take a dense G_δ set Δ_i of U_i whose points possess nearest points in $X \setminus U_i$. Take any $u \in \Delta_i$ and choose a nearest point $x \in X \setminus U_i$ to u, thus $u \in \operatorname{bdry} U_i$. We claim that $x \notin U_j$ for all $j \in I$. Otherwise, there is some $k \neq i$ with $x \in U_k$, which implies that the neighborhood U_k of x must intersect U_i since $u \in \operatorname{bdry} U_i$, and this contradicts the mutual disjoint property of the sets U_j. So, the claim is true and thus $x \in C$. Consequently, x is a nearest point of C to u. We deduce that each point of the set $\Delta := \bigcup_{i \in I} \Delta_i$ has a nearest point in C. Further, for each $i \in I$ the set $D_i := \Delta_i \cup \left(\bigcup_{j \in I, j \neq i} U_j \right)$ is obviously dense in $X \setminus C$ and it is easily seen to be a G_δ set of $X \setminus C$ since Δ_i is a G_δ set of U_i and the sets U_j are open. The obvious equality $\Delta = \bigcap_{i \in I} D_i$ then shows that Δ is a dense G_δ set of $X \setminus C$, and this concludes the proof. □

14.2. Genericity attainment and other properties of Moreau envelopes

A large number of properties of Moreau envelopes and Moreau s-envelopes have been previously established in Subsection 3.21.1 and Section 11.2 respectively. The Moreau envelope $e_\lambda f$ and s-envelope $e_{\lambda,[s]} f$ of a function f and its associated proximal multimapping $\operatorname{Prox}_\lambda f$ and s-proximal multimapping $\operatorname{Prox}_{\lambda,[s]} f$ (all with index $\lambda > 0$) have been defined in those sections (see Definition 3.262 and Definition 11.22). In the present section we will focus on genericity properties of attainment sets of Moreau envelopes of (possibly nonconvex) functions defined on Banach spaces. The genericity of the attainment set of the Moreau envelope of a nonconvex function $f : H \to \mathbb{R} \cup \{+\infty\}$ on a Hilbert space H has been proved in Proposition 11.25(d). In the context of certain general Banach spaces, the density of the attainment set and other basic properties of the Moreau envelope and Moreau s-envelope have been established in Theorem 3.265 and Theorem 11.23.

For the convenience of the reader we begin by restating features from Theorem 3.265 and from Theorem 11.23 for Moreau envelope (that is, in the case when $s = 2$ in Theorem 11.23).

THEOREM 14.35. Let $(X, \|\cdot\|)$ be a normed space and $f : X \to \mathbb{R} \cup \{+\infty\}$ be a proper function. Assume that there are reals $\alpha, \beta \geq 0$, $\gamma \in \mathbb{R}$ such that

$$f(x) \geq -\alpha\|x\|^2 - \beta\|x\| + \gamma, \text{ for all } x \in X.$$

Then for any $\lambda \in]0, \frac{1}{2\alpha}[$ (with the convention $1/\alpha = +\infty$ for $\alpha = 0$), one has the following properties:
(a) For any real number $r > 0$, there exists a real $\tau > 0$ (depending on r and λ) such that for every $x \in B(0, r)$

$$e_\lambda f(x) = \inf_{x' \in B(0,\tau)} \left(f(x') + \frac{1}{2\lambda}\|x - x'\|^2 \right).$$

(a') For any reals $r > 0$ and $\rho > 0$, there exists some real $\kappa > 0$ such that for all $x \in B(0, r)$

$$\left\{ y \in X : f(y) + \frac{1}{2\lambda}\|x - y\|^2 \leq e_\lambda f(x) + \rho \right\} \subset \kappa\mathbb{B}_X.$$

(b) $e_\lambda f$ is Lipschitzian on any ball $B(0, r)$ with a Lipschitz constant $L \geq \frac{r+\tau}{\lambda}$, where τ is as in (a).
(c) $e_\lambda f \uparrow \overline{f}$ as $\lambda \downarrow 0$, where $\overline{f}$ denotes the lower semicontinuous hull of f.
(d) If $(X, \|\cdot\|)$ is a reflexive Banach space whose norm $\|\cdot\|$ is Fréchet differentiable off zero and has the sequential Kadec-Klee property, and if the function f is lower semicontinuous, then $\operatorname{Dom} \operatorname{Prox}_\lambda f$ is dense in X, or more precisely for any $x \in \operatorname{Dom} \partial_F e_\lambda f$, the set $\operatorname{Prox}_\lambda f(x)$ is nonempty and for every $x' \in \operatorname{Prox}_\lambda f(x)$ one has

$$\partial_F e_\lambda f(x) = \left\{ \frac{1}{\lambda} J(x - x') \right\} \quad \text{and} \quad \frac{1}{\lambda} J(x - x') \in \partial_F f(x');$$

further one has a better Lipschitz constant L than in (*b*) for $e_\lambda f$ on $B(0, r)$ with $L \geq \frac{r+\tau}{2\lambda}$.

In addition to the assertion (c) in the above theorem, the following proposition establishes the Attouch-Wets convergence.

PROPOSITION 14.36. Let $(X, \|\cdot\|)$ be a normed space and $f : X \to \mathbb{R} \cup \{+\infty\}$ be a proper function bounded from below by a quadratic function $-\alpha\|\cdot\|^2 - \beta\|\cdot\|^2 + \gamma$, that is

$$f(x) \geq -\alpha\|x\|^2 - \beta\|x\| + \gamma \quad \text{for all } x \in X,$$

with $\alpha, \beta \geq 0$ and $\gamma \in \mathbb{R}$. Then, for any reals $\rho > 0$ and $\varepsilon > 0$ there is $\eta > 0$ such that for all $\lambda \in]0, \eta[$ the following inclusion

$$(\operatorname{epi} e_\lambda f) \cap B[0, \rho] \subset (\operatorname{epi} f) + B(0, \varepsilon) \tag{14.18}$$

holds true; and this combined with the inclusion $\operatorname{epi} f \subset \operatorname{epi} e_\lambda f$ (due to the inequality $\varepsilon_\lambda f \leq f$) ensures the Attouch-Wets convergence of $(e_\lambda f)_\lambda$ to f as $\lambda \downarrow 0$ (see Proposition 1.70).

PROOF. From the assumption it is easy to see that there is some real $\alpha_0 \geq 0$ such that

$$f(x) \geq -\alpha_0(\|x\|^2 + 1) \quad \text{for all } x \in X. \tag{14.19}$$

Endow $X \times \mathbb{R}$ with the sum norm $\|(u,r)\| := \|u\| + |r|$. Let $\rho > 0$ and $\varepsilon > 0$ be arbitrary. Take $\lambda > 0$ and $(x', r) \in \operatorname{epi} e_\lambda f \cap B[0,\rho]$. Then there exists $u \in X$ such that

$$f(u) + \frac{1}{2\lambda}\|u - x'\|^2 \le e_\lambda f(x') + \lambda \le r + \lambda \le \rho + \lambda. \tag{14.20}$$

Since

$$\|u\|^2 \le 2\|u - x'\|^2 + 2\|x'\|^2 \le 2\|u - x'\|^2 + 2\rho^2,$$

we obtain by using (14.19) and (14.20) that

$$\begin{aligned} -2\|u - x'\|^2 &\le -\|u\|^2 + 2\rho^2 \le 1 + \frac{1}{\alpha_0} f(u) + 2\rho^2 \\ &\le 1 + \frac{1}{\alpha_0}\left[\rho + \lambda - \frac{1}{2\lambda}\|u - x'\|^2\right] + 2\rho^2. \end{aligned}$$

Thus $\|u - x'\|^2(\frac{1}{2\lambda\alpha_0} - 2) \le 1 + 2\rho^2 + \frac{\rho+\lambda}{\alpha_0}$, which, for $0 < \lambda < 1/(4\alpha_0)$, is equivalent to the inequality

$$\|u - x'\|^2 \le (\alpha_0 + 2\alpha_0\rho^2 + \rho + \lambda)\frac{2\lambda}{1 - 4\alpha_0\lambda}.$$

Choose a real η with $0 < \eta < \min(\frac{1}{4\alpha_0}, \frac{\varepsilon}{2})$ such that

$$(\alpha_0 + 2\alpha_0\rho^2 + \rho + \lambda)\frac{2\lambda}{1 - 4\alpha_0\lambda} < \varepsilon^2/4 \quad \text{for all } \lambda \in \,]0,\eta[.$$

Then, for any fixed real $\lambda \in \,]0,\eta[$, we have $\|u - x'\|^2 < \varepsilon^2/4$. So, $(x', r) \in (u, r + \varepsilon/2) + B(0,\varepsilon)$, with $(u, r + \varepsilon/2) \in \operatorname{epi} f$ (by relation (14.20)). We then conclude that

$$\operatorname{epi} e_\lambda f \cap B[0,\rho] \subset \operatorname{epi} f + B(0,\varepsilon), \ \forall \lambda \in \,]0,\eta[.$$

□

When the Moreau envelope of a lower semicontinuous function is differentiable on a rich set in X, the convergence of $e_\lambda f$ to f as $\lambda \downarrow 0$ (see (c) in Theorem 14.35) provides a process of approximation of f by functions differentiable on rich subsets of X. Similarly, this process is still valid with $e_{\lambda,[s]}$ for $s > 1$, but the choice $s = 2$ presents the advantage of allowing us in many reasonings and computations to profit from the inner product when $(X, \|\cdot\|)$ is a Hilbert space.

If f is the indicator function of a set $C \subset X$, it is clear that

$$e_\lambda f(x) = \frac{1}{2\lambda} d_C^2(x).$$

Further, $e_\mu f \le e_\lambda f$ whenever $\mu \ge \lambda > 0$.

Assertion (a) in Theorem 14.35 furnishes the Lipschitz property of $e_\lambda f$ over bounded sets. What about the Lipschitz rate of $e_\lambda f$ at a fixed point. The Lipschitz rate of a mapping has been recalled in Definition 14.10.

When the infimum in the definition of $e_\lambda f(u)$ is attained at some $\overline{y}$, that is, $\overline{y} \in \operatorname{Prox}_\lambda f(u)$, the next proposition provides a lower estimate for $\operatorname{lip}(e_\lambda f)(u)$. If the infimum is strongly attained, the lower estimation becomes an equality.

We recall that an infimum $\inf\limits_{y \in Y} \varphi(y)$, for $\varphi : Y \to \mathbb{R} \cup \{-\infty, +\infty\}$, is (see Definition 4.38) *strongly attained at* $\overline{y} \in Y$, if it is attained at $\overline{y}$ and any minimizing sequence $(y_n)_n$ (that is, $\varphi(y_n) \to \inf_{y \in Y} \varphi(y)$) converges to $\overline{y}$. If φ is lower semicontinuous at $\overline{y}$, the requirement that the infimum be attained at $\overline{y}$ can be omitted.

PROPOSITION 14.37. Let $(X, \|\cdot\|)$ be a normed space and $f : X \to \mathbb{R} \cup \{+\infty\}$ be a proper function for which there are real numbers α, β, γ with $\alpha \geq 0$ such that

$$f(x) \geq -\alpha\|x\|^2 - \beta\|x\| + \gamma \quad \text{for all } x \in X.$$

Let also $0 < \lambda < 1/(2\alpha)$, and let $u \in X$. Then the following hold.
(a) There exists a minimizing sequence $(y_n)_n$ of the infimum defining $e_\lambda f(u)$ such that

$$\operatorname{lip}(e_\lambda f)(u) \leq \lambda^{-1} \liminf_{n\to\infty} \|u - y_n\|.$$

(b) If the infimum in the definition of $e_\lambda f(u)$ is attained at $\overline{y}$, then

$$\operatorname{lip}(e_\lambda f)(u) \geq \lambda^{-1}\|u - \overline{y}\|.$$

(c) If the infimum in the definition of $e_\lambda f(u)$ is strongly attained at $\overline{y}$, then the latter inequality is an equality, that is,

$$\operatorname{lip}(e_\lambda f)(u) = \lambda^{-1}\|u - \overline{y}\|.$$

PROOF. (a) By Theorem 14.35 recalled above we know that $e_\lambda f$ is Lipschitz continuous on some open neighborhood U of u. For each integer n put

$$\omega_n := \sup\left\{\frac{e_\lambda f(x') - e_\lambda f(x)}{\|x' - x\|} : x, x' \in U, x' \neq x, \|x - u\| < \frac{1}{n}, \|x' - u\| < \frac{1}{n}\right\}$$

and choose $x_n \neq x'_n \in U$ such that $\|x_n - u\| < 1/n$, $\|x'_n - u\| < 1/n$, and

$$\omega_n - 1/n < \frac{e_\lambda f(x'_n) - e_\lambda f(x_n)}{\|x'_n - x_n\|} \leq \omega_n.$$

Choose also some $y_n \in X$ such that

$$f(y_n) + \frac{1}{2\lambda}\|x_n - y_n\|^2 < e_\lambda f(x_n) + \|x'_n - x_n\|^2.$$

By Theorem 14.35 again, the sequence $(y_n)_n$ is bounded. We then have

$$\begin{aligned}
\operatorname{lip}(e_\lambda f)(u) &= \lim_{n\to\infty} \omega_n = \lim_{n\to\infty} \frac{e_\lambda f(x'_n) - e_\lambda f(x_n)}{\|x'_n - x_n\|} \\
&\leq \liminf_{n\to\infty} \frac{1}{\|x'_n - x_n\|}\Big[f(y_n) + \frac{1}{2\lambda}\|x'_n - y_n\|^2 - f(y_n) \\
&\qquad\qquad - \frac{1}{2\lambda}\|x_n - y_n\|^2 + \|x'_n - x_n\|^2\Big] \\
&= \frac{1}{2\lambda} \liminf_{n\to\infty} \frac{1}{\|x'_n - x_n\|}\Big[\|x'_n - y_n\|^2 - \|x_n - y_n\|^2\Big],
\end{aligned}$$

which entails for $t_n := \|x'_n - x_n\|$

$$\begin{aligned}
\operatorname{lip}(e_\lambda f, u) &\leq \frac{1}{2\lambda} \liminf_{n\to\infty} t_n^{-1} \big|\, \|x'_n - y_n\| - \|x_n - y_n\| \,\big| (t_n + 2\|x_n - y_n\|) \\
&\leq \frac{1}{2\lambda} \liminf_{n\to\infty} (t_n + 2\|x_n - y_n\|) = \lambda^{-1} \liminf_{n\to\infty} \|u - y_n\|,
\end{aligned}$$

where the latter inequality is due to the convergence of $(x_n)_n$ to u.
(b) Suppose that the infimum in the definition of $e_\lambda f(u)$ is attained at $\overline{y}$. If $\overline{y} \neq u$, then

$$\begin{aligned}\operatorname{lip}(e_\lambda f)(u) &\geq \limsup_{t\downarrow 0} \frac{e_\lambda f(u) - e_\lambda f(u - t(u-\overline{y}))}{t\|u-\overline{y}\|} \\ &\geq \limsup_{t\downarrow 0} \frac{1}{t\|u-\overline{y}\|}\left[f(\overline{y}) + \frac{1}{2\lambda}\|u-\overline{y}\|^2 - f(\overline{y}) - \frac{1}{2\lambda}\|u - t(u-\overline{y}) - \overline{y}\|^2\right] \\ &= \|u-\overline{y}\| \lim_{t\downarrow 0} \frac{1-(1-t)^2}{2\lambda t} = \lambda^{-1}\|u-\overline{y}\|.\end{aligned}$$

If $\overline{y} = u$, the inequality $\operatorname{lip}(e_\lambda f)(u) \geq \lambda^{-1}\|u-\overline{y}\|$ still holds since the second member is zero in this case.
(c) Assume now that the infimum in the definition of $e_\lambda f(u)$ is strongly attained at $\overline{y}$. Then the minimizing sequence $(y_n)_n$ in (a) converges to $\overline{y}$, hence

$$\operatorname{lip}(e_\lambda f)(u) \leq \lambda^{-1} \liminf_{n\to\infty} \|u - y_n\| = \lambda^{-1}\|u-\overline{y}\|.$$

This combined with (b) yields the desired equality $\operatorname{lip}(e_\lambda f)(u) = \lambda^{-1}\|u-\overline{y}\|$. □

REMARK 14.38. In addition to proper lower semicontinuous convex functions, it is worth noticing by Theorem 10.29 that the minorization assumption by a quadratic function is satisfied whenever the proper function f is linearly semiconvex on the normed space X and $\operatorname{Dom}\partial_C f \neq \emptyset$; in particular, the minorization is satisfied when X is a Banach space and the proper linearly semiconvex function f is lower semicontinuous on X. □

The following proposition concerning the Fréchet subdifferential and the Bouligand (or lower Hadamard) directional derivative follows directly from Proposition 4.27 concerning the Fréchet subdifferential and Bouligand directional derivative of the infimal convolution.

PROPOSITION 14.39. Assume that the infimum in the definition of $e_\lambda f(u)$ is attained at $\overline{y}$. Then

$$\partial_F(e_\lambda f)(u) \subset \partial_F f(\overline{y}) \cap \partial\left(\frac{1}{2\lambda}\|\cdot\|^2\right)(u-\overline{y})$$

$$\text{and} \quad \underline{d}_H^+(e_\lambda f)(u;h) \leq \min\left\{\underline{d}_H^+ f(u;h), \left(\frac{1}{2\lambda}\|\cdot\|^2\right)'(u-\overline{y};h)\right\},$$

where $(\frac{1}{2\lambda}\|\cdot\|^2)'(u-\overline{y};\cdot)\}$ is the usual directional derivative.

Before studying the Clarke subdifferential of the Moreau envelope $e_\lambda f$ let us investigate certain properties related the non-vacuity of its Fréchet subdifferential.

Like Theorem 14.2, assertion (b) below says that the set of points $u \in X$ such that the infimum defing $e_\lambda f(u)$ is dense in X whenever X is a reflexive Banach space and the norm $\|\cdot\|$ satisfies the Kadec-Klee property.

THEOREM 14.40 (attainment of Moreau envelope under F-subdifferentiability). Let $(X, \|\cdot\|)$ be a normed space and $f : X \to \mathbb{R} \cup \{+\infty\}$ be a function which is proper and bounded from below by a quadratic function

$$f(x) \geq -\alpha\|x\|^2 - \beta\|x\| + \gamma,$$

where $\alpha \geq 0$. Let $0 < \lambda < 1/(2c)$.
(a) If $u^* \in \partial_F e_\lambda f(u)$, one has

$$\lim_{q_{f,u}(y) \to e_\lambda f(u)} \|u - y\| = \lambda \|u^*\| \quad \text{and} \quad \lim_{q_{f,u}(y) \to e_\lambda f(u)} \langle u^*, u - y \rangle = \lambda \|u^*\|^2,$$

where $q_{f,u}(y) := f(y) + \frac{1}{2\lambda}\|u - y\|^2$; otherwise stated, for any minimizing sequence $(y_n)_n$ of the infimum defining $e_\lambda f(u)$

$$\lim_{n\to\infty} \|u - y_n\| = \lambda \|u^*\| \quad \text{and} \quad \lim_{n\to\infty} \langle u^*, u - y_n \rangle = \lambda \|u^*\|^2.$$

(b) Assume that the normed space $(X, \|\cdot\|)$ is a reflexive Banach space whose norm $\|\cdot\|$ has the Kadec-Klee property and assume that in addition to the above minorization hypothesis, the function f is lower semicontinuous. Then

$$\mathrm{Dom}\, \partial_F (e_\lambda f) \subset \mathrm{Dom}\, \mathrm{Prox}_\lambda f,$$

so $\mathrm{Dom}\, \mathrm{Prox}_\lambda f$ is dense in X.

PROOF. A part of the proof of (a) employs arguments similar to the proof of (a) in Proposition 4.32. Fix $0 < \lambda < 1/(2c)$, $u^* \in \partial_F e_\lambda f(u)$, and consider any minimizing sequence $(y_n)_n$ of $e_\lambda f(u)$. Put $t_n = 2^{-n} + \left[f(y_n) + \frac{1}{2\lambda}\|u - y_n\|^2 - e_\lambda f(u)\right]^{1/2}$. Since $t_n \downarrow 0$ and $(y_n - u)_n$ is bounded (see Theorem 14.35), we have by the inclusion $u^* \in \partial_F e_\lambda f(u)$

$$\liminf_{n\to\infty} \left\{ t_n^{-1}[e_\lambda f(u + t_n(y_n - u)) - e_\lambda f(u)] - \langle u^*, y_n - u \rangle \right\} \geq 0.$$

We can estimate

$$\begin{aligned} e_\lambda f(u + t_n(y_n - u)) - e_\lambda f(u) &\leq f(y_n) + \frac{1}{2\lambda}\|u + t_n(y_n - u) - y_n\|^2 - e_\lambda f(u) \\ &= f(y_n) + \frac{1}{2\lambda}(1 - t_n)^2\|u - y_n\|^2 - e_\lambda f(u) \\ &= \left[f(y_n) + \frac{1}{2\lambda}\|u - y_n\|^2 - e_\lambda f(u) \right] - \frac{t_n}{\lambda}\|u - y_n\|^2 \\ &\quad + \frac{t_n^2}{2\lambda}\|u - y_n\|^2, \end{aligned}$$

hence

$$e_\lambda f(u + t_n(y_n - u)) - e_\lambda f(u) \leq t_n^2 - \frac{t_n}{\lambda}\|u - y_n\|^2 + \frac{t_n^2}{2\lambda}\|u - y_n\|^2.$$

This yields

$$\liminf_{n\to\infty} \{-\lambda^{-1}\|u - y_n\|^2 + \langle u^*, u - y_n \rangle\} \geq 0,$$

thus

$$\liminf_{n\to\infty} \lambda^{-1}\|u - y_n\|^2 \leq \liminf_{n\to\infty} \langle u^*, u - y_n \rangle \tag{14.21}$$

$$\text{and} \quad \limsup_{n\to\infty} \lambda^{-1}\|u - y_n\|^2 \leq \limsup_{n\to\infty} \langle u^*, u - y_n \rangle. \tag{14.22}$$

On the other hand, for any fixed $\varepsilon > 0$ and for n large enough we have for all $h \in X$ with $\|h\| = 1$

$$\begin{aligned}
-\varepsilon + \langle u^*, h\rangle &\le \liminf_{n\to\infty} t_n^{-1}[e_\lambda f(u+t_n h) - e_\lambda f(u)] \\
&\le \liminf_{n\to\infty} t_n^{-1}\left[f(y_n) + \frac{1}{2\lambda}\|u+t_n h - y_n\|^2 - f(y_n) - \frac{1}{2\lambda}\|u-y_n\|^2 + t_n^2\right] \\
&\le \liminf_{n\to\infty} t_n^{-1}\left[\frac{t_n}{\lambda}\|u-y_n\| + \left(1+\frac{1}{2\lambda}\right)t_n^2\right] = \liminf_{n\to\infty} \lambda^{-1}\|u-y_n\|,
\end{aligned}$$

hence

$$\|u^*\| \le \lambda^{-1} \liminf_{n\to\infty} \|u-y_n\|, \tag{14.23}$$

which ensures in particular in the case when $\limsup_{n\to\infty} \|u-y_n\| = 0$ that

$$\|u^*\| = \lambda^{-1} \lim_{n\to\infty} \|u-y_n\| = 0 \text{ along with } \|u^*\|^2 = \lambda^{-1} \lim_{n\to\infty} \langle u^*, u-y_n\rangle = 0.$$

Moreover, we have by (14.22)

$$\begin{aligned}
\lambda^{-1}\left(\limsup_{n\to\infty} \|u-y_n\|\right)^2 &= \limsup_{n\to\infty} \lambda^{-1}\|u-y_n\|^2 \\
&\le \limsup_{n\to\infty} \langle u^*, u-y_n\rangle \\
&\le \|u^*\| \limsup_{n\to\infty} \|u-y_n\| \\
&\le \lambda^{-1}(\liminf_{n\to\infty} \|u-y_n\|) \limsup_{n\to\infty} \|u-y_n\|,
\end{aligned}$$

where the last inequality follows from (14.23). Suppose $\limsup_{n\to\infty} \|u-y_n\| > 0$. We then obtain

$$\lambda^{-1} \limsup_{n\to\infty} \|u-y_n\| \le \|u^*\| \le \lambda^{-1} \liminf_{n\to\infty} \|u-y_n\|,$$

or equivalently $\lambda^{-1} \lim_{n\to\infty} \|u-y_n\| = \|u^*\|$. This and (14.21) ensure

$$\begin{aligned}
\lambda\|u^*\|^2 = \lambda^{-1} \lim_{n\to\infty} \|u-y_n\|^2 &\le \liminf_{n\to\infty} \langle u^*, u-y_n\rangle \\
\le \limsup_{n\to\infty} \langle u^*, u-y_n\rangle &\le \|u^*\| \lim_{n\to\infty} \|u-y_n\| = \lambda\|u^*\|^2,
\end{aligned}$$

which means $\lim_{n\to\infty} \langle u^*, u-y_n\rangle = \lambda\|u^*\|^2$. So, in any case the equalities in (a) relative to the sequence $(y_n)_n$ hold.

Suppose now that $(X, \|\cdot\|)$ is a reflexive Banach space and the norm $\|\cdot\|$ has the Kadec-Klee property. Let u, u^* be such that $u^* \in \partial_F e_\lambda f(u)$. Choose a minimizing sequence $(y_n)_n$ of $e_\lambda f(u)$, that is,

$$\lim_{n\to\infty} \left(f(y_n) + \frac{1}{2\lambda}\|u-y_n\|^2\right) = e_\lambda f(u).$$

Theorem 14.35 says that $(y_n)_n$ is bounded, thus by the reflexivity of X we may suppose (without loss of generality) that $(y_n)_n$ converges weakly to some $y \in X$.

Then

$$\|u-y\|\left(\lim_{n\to\infty}\|u-y_n\|\right)=\lambda\|u-y\|\,\|u^*\|\geq\lambda\langle u^*,u-y\rangle=\lambda\lim_{n\to\infty}\langle u^*,u-y_n\rangle$$
$$=\lim_{n\to\infty}\|u-y_n\|^2=\left(\lim_{n\to\infty}\|u-y_n\|\right)^2,$$

where the first and third equalities are due to the equalities in (a). We deduce $\lim_{n\to\infty}\|u-y_n\|\leq\|u-y\|$. On the other hand, the weak lower semicontinuity of the norm ensures $\|u-y\|\leq\lim_{n\to\infty}\|u-y_n\|$. Consequently, $\lim_{n\to\infty}\|u-y_n\|=\|u-y\|$ and this combined with the Kadec-Klee property of the norm entails $\|y_n-y\|\to 0$ as $n\to\infty$. By the lower semicontinuity of f with respect to the norm $\|\cdot\|$ we then obtain

$$f(y)+\frac{1}{2\lambda}\|u-y\|^2\leq\lim_{n\to\infty}\left(f(y_n)+\frac{1}{2\lambda}\|u-y_n\|^2\right)=e_\lambda f(u),$$

so the infimum defining $e_\lambda f(u)$ is attained at y.

It remains to note that the density result follows from the density of $\operatorname{Dom}\partial_F e_\lambda$ in $\operatorname{dom} e_\lambda f=X$ (see Corollary 4.53). □

Before stating a first corollary, let us recall that a norm $\|\cdot\|$ on a vector space X is *locally uniformly convex* (or *locally uniformly rotund*) at a vector $x\in X$ if $\|x_n-x\|\to 0$ whenever $\|x_n\|\to\|x\|$ and $\|\frac{1}{2}(x_n+x)\|\to\|x\|$ as $n\to\infty$. When the norm is locally uniformly convex at any vector in X, one just says that it is *locally uniformly convex* (or *locally uniformly rotund*). It is readily seen that the norm $\|\cdot\|$ is locally uniformly convex if and only if given $(x_n)_n$ in the unit sphere $\mathbb{S}_X$ and $x\in\mathbb{S}_X$ one has $\|x_n-x\|\to 0$ whenever $\|\frac{1}{2}(x_n+x)\|\to 1$ as $n\to\infty$. Results on local uniform convexity of norms are presented in Subsection 18.1.2.

Corollary 14.41. Let $(X,\|\cdot\|)$ be a Banach space and $f:X\to\mathbb{R}\cup\{+\infty\}$ be a proper lower semicontinuous function which is bounded from below by a quadratic function

$$f(x)\geq-\alpha\|x\|^2-\beta\|x\|+\gamma,$$

with $\alpha\geq 0$, and let $0<\lambda<1/(2\alpha)$. Assume that the infimum in the definition of $e_\lambda f(u)$ is attained at $\overline{y}$ and that $\partial_F e_\lambda f(u)\neq\emptyset$. Then for any $u^*\in\partial_F e_\lambda f(u)$ one has

$$\|u^*\|=\lambda^{-1}\|u-\overline{y}\|\quad\text{and}\quad\langle u^*,u-\overline{y}\rangle=\lambda^{-1}\|u-\overline{y}\|^2.$$

If, in addition, the norm is locally uniformly convex at $u-\overline{y}$, then the infimum in the definition of $e_\lambda f(u)$ is strongly attained at $\overline{y}$.

Proof. Since the sequence $(y'_n)_n$ with $y'_n:=\overline{y}$ is a minimizing sequence for the infimum in the definition of $e_\lambda f(u)$, we have by Theorem 14.40 above that $\|u^*\|=\lambda^{-1}\|u-\overline{y}\|$ and $\langle u^*,u-\overline{y}\rangle=\lambda^{-1}\|u-\overline{y}\|^2$.

Suppose in addition that the norm is uniformly convex at $u-\overline{y}$. Take any minimizing sequence $(y_n)_n$ of the infimum in the definition of $e_\lambda f(u)$. According to Theorem 14.40 above again we have

(14.24)
$$\lim_{n\to\infty}\|u-y_n\|=\lambda\|u^*\|=\|u-\overline{y}\|\text{ and }\lim_{n\to\infty}\langle u^*,u-y_n\rangle=\lambda\|u^*\|^2=\lambda^{-1}\|u-\overline{y}\|^2.$$

If $\overline{y}=u$, then $0=\|u-\overline{y}\|=\lim_{n\to\infty}\|u-y_n\|$, hence $y_n\to u=\overline{y}$. Suppose that $\overline{y}\neq u$. In this case the above equalities $\lambda\|u^*\|=\|u-\overline{y}\|$ and $\langle u^*,u-\overline{y}\rangle=\lambda^{-1}\|u-\overline{y}\|^2$

allow us to write

$$\|u-\overline{y}\| + \|u-y_n\| \geq \|(u-\overline{y})+(u-y_n)\| \geq \left\langle \frac{\lambda u^*}{\|u-\overline{y}\|}, (u-\overline{y})+(u-y_n) \right\rangle$$
$$= \|u-\overline{y}\| + \lambda(\|u-\overline{y}\|^{-1})\langle u^*, u-y_n\rangle,$$

thus $\|(u-\overline{y})+(u-y_n)\| \to 2\|u-\overline{y}\|$ since both extreme members of the latter inequalities converge to $2\|u-\overline{y}\|$ according to (14.24). This guarantees by the locally uniform convexity of $\|\cdot\|$ at $u-\overline{x}$ that $\|y_n-\overline{y}\| \to 0$ as $n\to\infty$, so the infimum in the definition of $e_\lambda f(u)$ is strongly attained at $\overline{y}$. □

We can now deduce the following theorem concerning the residual density of the points $u \in X$ such that the infimum defining $e_\lambda f(u)$ is strongly attained.

THEOREM 14.42 (R. Cibulka and M. Fabian). Assume that $(X, \|\cdot\|)$ is a reflexive Banach space and the norm $\|\cdot\|$ is locally uniformly convex. Let $f : X \to \mathbb{R}\cup\{+\infty\}$ be a proper lower semicontinuous function which is bounded from below by a quadratic function

$$f(x) \geq -\alpha\|x\|^2 - \beta\|x\| + \gamma,$$

where $\alpha \geq 0$, and let $0 < \lambda < 1/(2\alpha)$. Then for any $u \in \operatorname{Dom}\partial_F e_\lambda f$ the infimum in the definition of $e_\lambda f(u)$ is strongly attained.

In particular, the set of points $u \in X$ such that the infimum in the definition of $e_\lambda f(u)$ is strongly attained is a dense G_δ set in X.

PROOF. The theorem follows from the latter corollary and Theorem 14.40(b) above, and also from the lemma below. □

LEMMA 14.43. Let $(X, \|\cdot\|)$ be a normed space and $f : X \to \mathbb{R}\cup\{+\infty\}$ be a proper lower semicontinuous function which is bounded from below by a quadratic function

$$f(x) \geq -\alpha\|x\|^2 - \beta\|x\| + \gamma,$$

where $\alpha \geq 0$, and let $0 < \lambda < 1/(2\alpha)$. Then the set of points $u \in X$ where the infimum in the definition of $e_\lambda f(u)$ is strongly attained is a G_δ set in X.

PROOF. For each $x \in X$ and each $\varepsilon > 0$ put

$$P_\varepsilon(x) := \left\{ y \in X : f(y) + \frac{1}{2\lambda}\|x-y\|^2 \leq e_\lambda f(x) + \varepsilon \right\}.$$

We know by (4.24) (or as it can be seen from Definition 4.38) that the set of points u where the infimum defining $e_\lambda f(u)$ is strongly attained coincide with $\bigcap_{n\in\mathbb{N}} \Omega_n$ where

$$\Omega_n := \bigcup_{\varepsilon>0} \{x \in X : \operatorname{diam} P_\varepsilon(x) < 1/n\},$$

so the result will be established once we prove that each Ω_n is open in X. Fix $u \in \Omega_n$ and choose $0 < \varepsilon_0 < 1$ such that $\operatorname{diam} P_{\varepsilon_0}(u) < 1/n$. By Theorem 14.35(a') there exists some real constant $\kappa \geq 0$ so that $\|y\| \leq \kappa$ whenever $f(y) + \frac{1}{2\lambda}\|x - y\|^2 \leq e_\lambda f(x) + \varepsilon_0$ for some $x \in B[u, \varepsilon_0]$. Take some $0 < \delta_0 < \varepsilon_0$ such that $e_\lambda f(x) < e_\lambda f(u) + \varepsilon_0/3$ for all $x \in B(u, \delta_0)$, and choose $0 < \delta < \delta_0$ such that

$$\frac{1}{2\lambda}(\delta^2 + 2\delta(\delta + \kappa + \|u\|)) < \varepsilon_0/3 =: \varepsilon.$$

Consider any $x \in B(u, \delta)$ and any $y \in P_\varepsilon(x)$, and write by definition of $P_\varepsilon(x)$

$$f(y) + \frac{1}{2\lambda}\|x - y\|^2 \le e_\lambda f(x) + \varepsilon < e_\lambda f(x) + 1.$$

We then have

$$\begin{aligned} f(y) + \frac{1}{2\lambda}\|u - y\|^2 &\le f(y) + \frac{1}{2\lambda}\{\|u - x\|^2 + 2\|u - x\|\,\|x - y\| + \|x - y\|^2\} \\ &\le \frac{1}{2\lambda}\{\delta^2 + 2\delta(\|u - x\| + \|u\| + \|y\|)\} + e_\lambda f(x) + \varepsilon \\ &\le \frac{1}{2\lambda}\{\delta^2 + 2\delta(\delta + \kappa + \|u\|)\} + e_\lambda f(u) + \varepsilon + \frac{\varepsilon_0}{3}, \end{aligned}$$

so

$$f(y) + \frac{1}{2\lambda}\|u - y\|^2 \le e_\lambda f(u) + \varepsilon_0.$$

This ensures the inclusion $P_\varepsilon(x) \subset P_{\varepsilon_0}(u)$ hence $\operatorname{diam} P_\varepsilon(x) < 1/n$, which entails $B(u, \delta) \subset \Omega_n$. We conclude that Ω_n is open in X. □

In the case of a closed set we have the following result in complement to Lau Theorem 14.29. Here, under the more stringent requirement of the local uniform convexity of the norm of the reflexive Banach space instead of its Kadec-Klee property (as in Lau theorem) we obtain the stronger result of genericity of points whose distances are strongly attained.

Corollary 14.44. Let C be a nonempty closed set of a Banach space $(X, \|\cdot\|)$ and let $u \in X \setminus C$.
(a) If the distance $d_C(u)$ of u from the set C is attained at $\overline{y} \in C$, if $\partial_F d_C(u) \neq \emptyset$, and if the norm $\|\cdot\|$ is locally uniformly convex at $u - \overline{y}$, then the distance $d_C(u)$ of u from C is strongly attained at $\overline{y}$.
(b) Assume that $(X, \|\cdot\|)$ is reflexive and the norm $\|\cdot\|$ is locally uniformly convex. Then, the distance $d_C(u)$ of u from the set C is strongly attained whenever $\partial_F d_C(u) \neq \emptyset$. Thus the set of points $u \in X$ where the distance $d_C(u)$ of u from C is strongly attained is a dense G_δ set of X.

Proof. We first observe that at a point x, the distance $d_C(x)$ is strongly attained at $\overline{y} \in C$ if and only if the infimum in the equality $\frac{1}{2}d_C^2(x) = \inf_{y \in X}[\Psi_C(y) + \frac{1}{2}\|x - y\|^2$ is strongly attained at $\overline{y}$.

Since $d_C(u) > 0$, the nonemptiness of $\partial_F d_C(u)$ ensures $\partial_F(\frac{1}{2}d_C^2)(u) \neq \emptyset$ by Corollary 4.17. Taking the observation above into account, assertion (a) and assertion (b) then follow from Corollary 14.41 and from Theorem 14.42 respectively. □

Corollary 14.41 and Theorem 14.42 provide conditions in order that the Fréchet subdifferentiability of $e_\lambda f$ at u ensures that the infimum defining $e_\lambda f(u)$ is strongly attained. Conversely, we now study the Fréchet (sub)differentiability of $e_\lambda f$ at u when the infimum in the definition of $e_\lambda f(u)$ is strongly attained.

Proposition 14.45. Let $(X, \|\cdot\|)$ be a normed space, $f : X \to \mathbb{R} \cup \{+\infty\}$ be a proper lower semicontinuous function which is bounded from below by a quadratic function

$$f(x) \ge -\alpha\|x\|^2 - \beta\|x\| + \gamma$$

with $\alpha \geq 0$, and let $0 < \lambda < 1/(2\alpha)$. Let $u \in X$ be such that the infimum in the definition of $e_\lambda f(u)$ is strongly attained at $\overline{y}$. Then

$$\partial_C(e_\lambda f)(u) \subset \partial\left(\frac{1}{2\lambda}\|\cdot\|^2\right)(u-\overline{y}).$$

If in addition the square norm $\|\cdot\|^2$ is Fréchet (resp. Gâteaux) differentiable at $u-\overline{y}$, then $e_\lambda f$ is Fréchet (resp. strictly Gâteaux) differentiable at u with

$$D(e_\lambda f)(u) = \frac{1}{2\lambda}D(\|\cdot\|^2)(u-\overline{y}).$$

PROOF. We know by Theorem 14.35 that $e_\lambda f$ is locally Lipschitz continuous on X. Fix any $u^* \in \partial_C(e_\lambda f)(u)$. Choose a sequence $(t_n)_n$ tending to 0 with $t_n > 0$ and a sequence $(u_n)_n$ converging to u such that

$$(e_\lambda f)^\circ(u;\overline{y}-u) = \lim_{n\to\infty} t_n^{-1}[e_\lambda f(u_n + t_n(\overline{y}-u)) - e_\lambda f(u_n)].$$

For each integer n choose some y_n such that $f(y_n) + \frac{1}{2\lambda}\|u_n - y_n\|^2 < e_\lambda f(u_n) + t_n^2$ and choose also some $y_n^* \in \partial(\frac{1}{2\lambda}\|\cdot\|^2)(u_n + t_n(\overline{y}-u) - y_n)$. Write

$$\begin{aligned}\langle u^*, \overline{y}-u\rangle &\leq \liminf_{n\to\infty} t_n^{-1}\left[f(y_n) + \frac{1}{2\lambda}\|u_n + t_n(\overline{y}-u) - y_n\|^2 - f(y_n)\right.\\ &\qquad \left. -\frac{1}{2\lambda}\|u_n - y_n\|^2 + t_n^2\right]\\ &= (2\lambda)^{-1}\liminf_{n\to\infty} t_n^{-1}[\|u_n + t_n(\overline{y}-u) - y_n\|^2 - \|u_n - y_n\|^2]\\ &\leq \liminf_{n\to\infty}\langle y_n^*, \overline{y}-u\rangle.\end{aligned}$$

Observe also by the boundedness of $(y_n)_n$ (see Theorem 14.35) that

$$\begin{aligned}e_\lambda f(u) &\leq \liminf_{n\to\infty}\left[f(y_n) + \frac{1}{2\lambda}\|u-y_n\|^2\right] \leq \limsup_{n\to\infty}\left[f(y_n) + \frac{1}{2\lambda}\|u-y_n\|^2\right]\\ &\leq \limsup_{n\to\infty}\left[f(y_n) + \frac{1}{2\lambda}\left(\|u_n - y_n\|^2 + 2\|u-u_n\|\,\|u_n-y_n\| + \|u-u_n\|^2\right)\right]\\ &= \limsup_{n\to\infty}\left[f(y_n) + \frac{1}{2\lambda}\|u_n-y_n\|^2\right] \leq \lim_{n\to\infty}[e_\lambda f(u_n) + t_n^2] = e_\lambda f(u),\end{aligned}$$

where the latter equality is due to the continuity of $e_\lambda f$. This says that $\lim_{n\to\infty}[f(y_n) + \frac{1}{2\lambda}\|u-y_n\|^2] = e_\lambda f(u)$, hence $(y_n)_n$ is a minimizing sequence of the infimum defining $e_\lambda f(u)$. The strong attainment assumption then ensures that $y_n \to \overline{y}$ as $n \to \infty$, which entails in particular (according to the choice of y_n^*) that the sequence $(y_n^*)_n$ is bounded. Take a weak star cluster point y^* of $(y_n^*)_n$ and note, by the norm-to-weak* outer semicontinuity of subdifferential of continuous convex functions (see Proposition 3.32(c)), that $y^* \in \partial(\frac{1}{2\lambda}\|\cdot\|^2)(u-\overline{y})$. We deduce

$$\langle u^*, \overline{y}-u\rangle \leq \langle y^*, \overline{y}-u\rangle = -\lambda^{-1}\|u-\overline{y}\|^2.$$

On the other hand, by the inclusion $u^* \in \partial_C e_\lambda f(u)$ we have $\|u^*\| \leq \mathrm{lip}(e_\lambda f)(u)$, and by Proposition 14.37(c) we also have $\mathrm{lip}(e_\lambda f)(u) = \lambda^{-1}\|u-\overline{y}\|$, so $\|u^*\| \leq \lambda^{-1}\|u-\overline{y}\|$. In the case when $\overline{y} = u$ we obtain $u^* = 0 \in \partial(\frac{1}{2\lambda}\|\cdot\|^2)(u-\overline{y})$ since $u-\overline{y} = 0$. In the case when $\overline{y} \neq u$, from the inequalities $\langle u^*, \overline{y}-u\rangle \leq -\lambda^{-1}\|u-\overline{y}\|^2$ and $\|u^*\| \leq \lambda^{-1}\|u-\overline{y}\|$ we derive

$$\lambda^{-1}\|u-\overline{y}\|^2 \leq \langle u^*, u-\overline{y}\rangle \leq \|u^*\|\,\|u-\overline{y}\| \leq \lambda^{-1}\|u-\overline{y}\|^2, \tag{14.25}$$

which gives $\langle \lambda u^*, u-\overline{y}\rangle = \|u-\overline{y}\|^2$ and $\|\lambda u^*\| = \|u-\overline{y}\|$. This implies $\lambda u^* \in \partial(\frac{1}{2}\|\cdot\|^2)(u-\overline{y})$ since we know by (2.25) that $2x^* \in \partial(\|\cdot\|^2)$ whenever $\langle x^*, x\rangle = \|x\|^2$ and $\|x^*\| = \|x\|$. This establishes the inclusion of the proposition.

Assume now that in addition $\|\cdot\|^2$ is Gâteaux differentiable at $u-\overline{y}$. Then the inclusion above says that $\partial_C e_\lambda f(u)$ is a singleton, hence by Proposition 2.76 the function $e_\lambda f$ is strictly Gâteaux differentiable at u and $D_G(e_\lambda f)(u) = \frac{1}{2\lambda}D(\|\cdot\|^2)(u-\overline{y})$. Assume instead that f is Fréchet differentiable at $u-\overline{y}$. The discussion above concerning the case $\overline{y} = u$ guarantees that we may suppose $\overline{y} \neq u$. Then using (14.25) it follows that

$$\left\langle D_G(e_\lambda f)(u), \frac{u-\overline{y}}{\|u-\overline{y}\|}\right\rangle = \lambda^{-1}\|u-\overline{y}\| = \mathrm{lip}(e_\lambda f)(u).$$

Therefore, Theorem 14.13 tells us that $e_\lambda f$ is Fréchet differentiable at u. □

REMARK 14.46. Let $(X, \|\cdot\|)$ be a normed space and let $f : X \to \mathbb{R}\cup\{+\infty\}$ be as in Proposition 14.45. Let $0 < \lambda < 1/(2\alpha)$. Then for any pair (u, y) with $y \in \operatorname{Prox}_\lambda f(u)$ one has

$$\partial_C(e_\lambda f)(u) \cap \partial\left(\frac{1}{2\lambda}\|\cdot\|^2)(u-y\right) \neq \emptyset.$$

Indeed, take a sequence $(t_n)_n$ tending to zero with $t_n > 0$ we can write for any $h \in X$

$$\begin{aligned}(e_\lambda f)^\circ(x; h) &\geq \limsup_{n\to\infty} t_n^{-1}\big[e_\lambda f(x - t_n h + t_n h) - e_\lambda f(x - t_n h)\big]\\ &= \limsup_{n\to\infty} t_n^{-1}\big[e_\lambda f(x) - e_\lambda f(x - t_n h)\big]\\ &\geq \limsup_{n\to\infty} t_n^{-1}\left[f(y) + \frac{1}{2\lambda}\|x-y\|^2 - f(y) - \frac{1}{2\lambda}\|x - t_n h - y\|^2\right],\end{aligned}$$

hence $(e_\lambda f)^\circ(x; h) \geq -q(-h)$ for all $h \in X$, where $q(h) := (\frac{1}{2\lambda}\|\,\|^2)'(x-y; h)$. Therefore, 0 is a minimizer of $h \mapsto (e_\lambda f)^\circ(x;\cdot) + q_0(\cdot)$, where $q_0(h) := q(-h)$, which entails by the Moreau-Rockafellar subdifferential sum rule that

$$0 \in \partial(e_\lambda f)^\circ(x;\cdot)(0) + \partial q_0(0) = \partial_C(e_\lambda f)(x) - \partial q(0).$$

To conclude, it suffices to note that $\partial q(0) = \partial(\frac{1}{2\lambda}\|\,\|^2)(x-y)$. □

Using Proposition 14.45 in taking therein for f the indicator function Ψ_C of C, we deduce the following corollary.

COROLLARY 14.47. Let C be a closed set of a normed space $(X, \|\cdot\|)$ and let $u \in X$ be a point such that the square distance $d_C^2(u)$ of u from C is strongly attained at $\overline{y} \in C$. Then one has

$$\partial_C(d_C^2)(u) \subset \partial(\|\cdot\|^2)(u-\overline{y}).$$

If in addition the norm $\|\cdot\|$ is Fréchet (resp. Gâteaux) differentiable at $u-\overline{y}$, then d_C^2 is Fréchet (resp. strictly Gâteaux) differentiable at u with

$$D(d_C^2)(u) = D(\|\cdot\|^2)(u-\overline{y}).$$

Given a closed set C in a normed space $(X, \|\cdot\|)$ and $u \in X\setminus C$ and $\overline{y} \in C$ with $\|u-\overline{y}\| = d_C(u)$, we have $\|u-\overline{y}\| = d_C(u) > 0$. Then, from Corollary 14.47 we immediately obtain:

COROLLARY 14.48. Let C be a closed set of a normed space $(X, \|\cdot\|)$ and let $u \in X \setminus C$ be a point such that the distance $d_C(u)$ of u from C is strongly attained at $\overline{y} \in C$. Then

$$\partial_C(d_C)(u) \subset \partial(\|\cdot\|)(u - \overline{y}).$$

If in addition the norm $\|\cdot\|$ is Fréchet (resp. Gâteaux) differentiable at $u - \overline{y}$, then d_C is Fréchet (resp. strictly Gâteaux) differentiable at u with

$$D(d_C)(u) = D(\|\cdot\|)(u - \overline{y}).$$

14.3. L-subdifferential by means of Moreau envelopes

The previous section provides, among diverse results, some expressions for (sub)gradients of Moreau envelopes. In the present section we turn to the description of the limiting subsdifferential of a function by means of Fréchet subgradients of its Moreau envelopes.

THEOREM 14.49 (L-subdifferential via Moreau envelopes). Assume that $(X, \|\cdot\|)$ is an Asplund space and that $f : X \to \mathbb{R} \cup \{+\infty\}$ is a proper lower semicontinuous function which is bounded from below by a negative quadratic function. Then for any $x \in \operatorname{dom} f$ one has

$$\partial_L f(x) = {}^{\text{seq}}\operatorname*{Lim\,sup}_{\substack{\lambda \downarrow 0 \\ (u, e_\lambda f(u)) \to (x, f(x))}} \partial_F e_\lambda f(u),$$

where we recall that ${}^{\text{seq}}\operatorname{Lim\,sup}$ means the weak* sequential outer (superior, upper) limit of sets in X^*.

PROOF. Endow $X \times \mathbb{R}$ with the norm $\|(u, r)\| := \|u\| + |r|$, so the dual norm on $X^* \times \mathbb{R}$ is $\|(u^*, r^*)\|_* = \max\{\|u^*\|_*, |r^*|\}$. Fix any $x^* \in \partial_L f(x)$. Since $N_L(S; z) = \mathbb{R}_+ \partial_L d_S(z)$ for any set S with $z \in S$ (see Theorem 4.85(b)), there exists some real $t > 0$ such that

$$\frac{1}{t}(x^*, -1) \in \partial_L d\big(\operatorname{epi} f, (x, f(x))\big).$$

Put $u^* := (1/t)x^*$ and $\alpha^* = -(1/t)$. Then we know by Theorem 4.85(a) that there are sequences $(x_n, r_n)_{n \in \mathbb{N}}$ in $\operatorname{epi} f$ with $(x_n, r_n) \to (x, f(x))$, $(x_n^*, \alpha_n^*)_{n \in \mathbb{N}}$ in $X^* \times \mathbb{R}$ with $(x_n^*, \alpha_n^*) \overset{w^*}{\to} (u^*, \alpha^*)$, $(\varepsilon_n)_{n \in \mathbb{N}}$ with $\varepsilon_n > 0$ and $\varepsilon_n \downarrow 0$, and $(\delta_n)_{n \in \mathbb{N}}$ with $0 < \delta_n < 1/4$ and $\delta_n \downarrow 0$ such that $\|(x_n^*, \alpha_n^*)\|_* \le 1 + \varepsilon_n$ and such that for all $(u, r) \in \operatorname{epi} f \cap B[(x_n, r_n), \delta_n]$

$$-\langle x_n^*, u - x_n \rangle - \alpha_n^*(r - r_n) + \varepsilon_n \|(u, r) - (x_n, r_n)\| \ge 0. \tag{14.26}$$

This ensures in particular $(x_n^*, \alpha_n^*) \in \partial_{F, \varepsilon_n} \Psi_{\operatorname{epi} f}(x_n, r_n)$. Since $-\alpha_n - \varepsilon_n \to -\alpha$ and $-\alpha > 0$, we have $-\alpha_n > \varepsilon_n$ for n large enough, say $n \ge n_0$. For each integer $n \ge n_0$ we get $-\alpha_n > \varepsilon_n$, hence $r_n = f(x_n)$ according to Proposition 4.94(a). It follows by Proposition 14.36 that there are $0 < \lambda_n < 1$ such that $\lambda_n \downarrow 0$ and such that for each integer $n \ge n_0$

$$C_n := \operatorname{epi} e_{\lambda_n} f \cap B\left[(x_n, f(x_n)), \delta_n/2\right] \subset \operatorname{epi} f + B(0, \delta_n^3). \tag{14.27}$$

We also observe that $C_n \neq \emptyset$ since $(x_n, f(x_n)) \in \operatorname{epi} e_{\lambda_n} f$ (keeping in mind the inclusion $\operatorname{epi} f \subset \operatorname{epi} e_\lambda f$ for any $\lambda > 0$). Combining relations (14.26) and (14.27) and putting $c_n = 1 + 2\varepsilon_n$, we get for any $n \ge n_0$ and any $(u, r) \in C_n$

$$c_n \delta_n^3 - \langle x_n^*, u - x_n \rangle - \alpha_n^*(r - f(x_n)) + \varepsilon_n \|(u, r) - (x_n, f(x_n))\| \ge 0.$$

Fix any integer $n \ge n_0$ and consider the function φ_n defined by

$$\varphi_n(u,r) = -\langle x_n^*, u - x_n\rangle - \alpha_n^*(r - f(x_n)) + \varepsilon_n\|(u,r) - (x_n, f(x_n))\|.$$

We note that

$$\varphi_n(x_n, f(x_n)) = 0 \le \varphi_n(u,r) + c_n\delta_n^3, \quad \forall (u,r) \in C_n$$

and $(x_n, f(x_n)) \in C_n$. By the Ekeland variational principle, there are $(x_n', r_n') \in C_n$ such that

$$\|(x_n', r_n') - (x_n, f(x_n))\| \le \delta_n^2 \tag{14.28}$$

and

$$\varphi_n(x_n', r_n') \le \varphi_n(u,r) + c_n\delta_n\|(u,r) - (x_n', r_n')\|, \quad \forall (u,r) \in C_n. \tag{14.29}$$

Thus for $(u,r) \in \operatorname{epi} e_{\lambda_n} f$ with $\|(u,r) - (x_n', r_n')\| < \delta_n^2$ we have

$$\begin{aligned} &- \langle x_n^*, x_n' - x_n\rangle - \alpha_n^*(r_n' - f(x_n)) + \varepsilon_n\|(x_n', r_n') - (x_n, f(x_n))\| \\ &\le -\langle x_n^*, u - x_n\rangle - \alpha_n^*(r - f(x_n)) + \varepsilon_n\|(u,r) - (x_n, f(x_n))\| \\ &\quad + c_n\delta_n\|(u,r) - (x_n', r_n')\|, \end{aligned}$$

that is,

$$\begin{aligned} &\langle x_n^*, u - x_n'\rangle + \alpha_n^*(r - r_n') \\ &\le \varepsilon_n\|(u,r) - (x_n, f(x_n))\| - \varepsilon_n\|(x_n', r_n') - (x_n, f(x_n))\| \\ &\quad + c_n\delta_n\|(u,r) - (x_n', r_n')\|, \end{aligned}$$

hence

$$\langle x_n^*, u - x_n'\rangle + \alpha_n^*(r - r_n') \le (\varepsilon_n + c_n\delta_n)\|(u,r) - (x_n', r_n')\|.$$

This entails that

$$(x_n^*, \alpha_n^*) \in \partial_{F, \varepsilon_n + c_n\delta_n}\Psi_{\operatorname{epi} e_{\lambda_n} f}(x_n', r_n').$$

Since $-\alpha_n - \varepsilon_n - c_n\delta_n \to -\alpha$ with $-\alpha > 0$, as above it ensues that $-\alpha_n > \varepsilon_n + c_n$ for n large enough, say $n \ge n_1$ for some $n_1 \ge n_0$. For each integer $n \ge n_1$ we then have $-\alpha_n > \varepsilon_n + c_n$, hence $r_n' = e_{\lambda_n} f(x_n')$ according to Proposition 4.94(a) again. Further, by the Asplund property of the space X there are (see Definition 4.65) sequences $(u_n, s_n)_{n\in\mathbb{N}}$ in $X \times \mathbb{R}$ and $(u_n^*, s_n^*)_{n\in\mathbb{N}}$ in $X^* \times \mathbb{R}$ with $(u_n^*, s_n^*) \in N^F(\operatorname{epi} e_{\lambda_n} f, (u_n, s_n))$ such that (see Proposition 4.96(a))

$$\|(u_n, s_n) - (x_n', e_{\lambda_n} f(x_n'))\| \le \delta_n, \quad \|(x_n^*, \alpha_n^*) - (u_n^*, s_n^*)\|_* \le 2(\varepsilon_n + c_n\delta_n).$$

Since $s_n^* \to \alpha^*$ and $\alpha^* < 0$, using Proposition 4.94(a) again yields $s_n = e_{\lambda_n} f(u_n)$ for n large enough and

$$\frac{-1}{s_n^*} u_n^* \in \partial_F e_{\lambda_n} f(u_n),$$

according to Proposition 4.68(b). Since $u_n \to x$, $e_{\lambda_n} f(u_n) \to f(x)$, $s_n^* \to \alpha^*$, and

$$\frac{-1}{s_n^*} u_n^* \xrightarrow{w^*} \frac{-1}{\alpha^*} u^* = x^*,$$

we get

$$x^* \in {}^{\rm seq}\operatorname*{Lim\,sup}_{n\to\infty} \partial_F e_{\lambda_n} f(u_n) \quad \text{and} \quad (u_n, e_{\lambda_n} f(u_n)) \to (x, f(x)),$$

which justifies the inclusion of the first member into the second of the desired equality.

Conversely, let $x^* \in \operatorname{seq\,Lim\,sup}_{\lambda \downarrow 0, (u, e_\lambda f(u)) \to (x, f(x))} \partial_F e_\lambda f(u)$. Then there are sequences $\{\lambda_n\}_{n\in\mathbb{N}}$ with $\lambda_n > 0$ and $\lambda_n \downarrow 0$, $(x_n)_{n\in\mathbb{N}}$ in X with $x_n \to x$ and $e_{\lambda_n} f(x_n) \to f(x)$, and $(x_n^*)_{n\in\mathbb{N}}$ in X^* with $x_n^* \xrightarrow{w^*} x^*$, such that

$$x_n^* \in \partial_F e_{\lambda_n} f(x_n) \quad \text{for all } n \text{ sufficiently large.}$$

So, there are sequences $(\varepsilon_n)_{n\in\mathbb{N}}$ with $\varepsilon_n > 0$ and $\varepsilon_n \downarrow 0$, and $(\delta_n)_{n\in\mathbb{N}}$ with $\delta_n > 0$ and $\delta_n \downarrow 0$ such that for n sufficiently large, say $n \geq n_0$,

$$e_{\lambda_n} f(x_n + h) - e_{\lambda_n} f(x_n) - \langle x_n^*, h\rangle + \varepsilon_n \|h\| \geq 0, \ \forall h \in B[0, \delta_n].$$

For each n choose $v_n \in X$ such that

$$f(v_n) + \frac{1}{2\lambda_n}\|v_n - x_n\|^2 \leq e_{\lambda_n} f(x_n) + \delta_n^3.$$

This inequality and the assumption of boundedness from below by a quadratic function ensure that $v_n \to x$ and $\lim_{n\to\infty} f(v_n) = f(x)$, and also that for each $n \geq n_0$

$$\delta_n^3 + e_{\lambda_n} f(x_n + h) - f(v_n) - \frac{1}{2\lambda_n}\|v_n - x_n\|^2 - \langle x_n^*, h\rangle + \varepsilon_n\|h\| \geq 0, \ \forall h \in B[0, \delta_n].$$

Fix $n \geq n_0$. Taking into account that

$$e_{\lambda_n} f(x_n + h) \leq f(v_n + h) + \frac{1}{2\lambda_n}\|v_n - x_n\|^2$$

we deduce that

$$\delta_n^3 + f(v_n + h) - f(v_n) - \langle x_n^*, h\rangle + \varepsilon_n\|h\| \geq 0, \ \forall h \in B[0, \delta_n].$$

The Ekeland variational principle furnishes the existence of $u_n \in X$ which satisfies

$$\|u_n\| \leq \delta_n^2 \quad \text{and} \quad f(v_n + u_n) - \langle x_n^*, u_n\rangle + \varepsilon_n\|u_n\| \leq f(v_n) \tag{14.30}$$

and for all $h \in B[0, \delta_n]$

$$f(v_n + h) - f(v_n + u_n) - \langle x_n^*, h - u_n\rangle + \varepsilon_n\|h\| - \varepsilon_n\|u_n\| + \delta_n\|h - u_n\| \geq 0,$$

which ensures

$$f(v_n + h) - f(v_n + u_n) - \langle x_n^*, h - u_n\rangle + \varepsilon_n\|h - u_n\| + \delta_n\|h - u_n\| \geq 0, \ \forall h \in B[0, \delta_n].$$

By the first inequality in (14.30), for n large enough, u_n is an interior point of $B[0, \delta_n]$, thus

$$x_n^* \in \partial_{F,\varepsilon_n+\delta_n} f(v_n + \cdot)(u_n) = \partial_{F,\varepsilon_n+\delta_n} f(v_n + u_n),$$

so $x^* \in \partial_L f(x)$ since $(v_n + u_n, f(v_n + u_n)_{n\in\mathbb{N}}$ converges to $(x, f(x))$ according to the inequalities in (14.30), to the lower semicontinuity of f, and to the equality $\lim_{n\to\infty} f(v_n) = f(x)$. The desired equality of the theorem is then proved. □

At this stage we saw that the density of the set of points where the metric projection is nonempty requires either a property on the set C or a property on the space X. The next chapter will be related to more regularities on the metric projection in the Hilbert setting.

14.4. Comments

The study in infinite dimensions of the richness of points possessing *nearest points* in a closed set probably began with the 1963 paper by S. B. Stechkin [**899**] and the independent 1968 paper (submitted in September 1966) by M. Edelstein [**371**]. In [**899**] Stechkin showed that, for any nonempty closed set of a uniformly convex Banach space X, the set $K(C)$ of points in X admitting a unique nearest point in C is a dense G_δ set in X. Under the same assumption of uniform convexity of the Banach space, Edelstein [**371**] proved independently the less general result that the set $K(C)$ is dense in X. The general result in Theorem 14.29 for closed sets in reflexive Banach spaces with Kadec-Klee norm, was provided by K.-S. Lau in his 1978 paper [**648**]. Given any nonempty closed set C of a Banach space $(X, \|\cdot\|)$, let us analyze the approaches of those authors for the richness of the above set $K(C)$ of points $x \in X$ such that $\operatorname{Proj}_C(x)$ is a singleton.

Stechkin theorem[1] says: If $(X, \|\cdot\|)$ is a uniformly convex Banach space, then $K(C)$ is a dense G_δ set in X. To present the main ideas of Stechkin's proof, suppose that the norm $\|\cdot\|$ is *uniformly convex*. Putting

$$\rho(x) := \lim_{\varepsilon \downarrow 0} \operatorname{diam}\left(C \cap B[x, d_C(x) + \varepsilon]\right)$$

and putting $F_n := \{x \in X : \rho(x) \geq 1/n\}$ for each $n \in \mathbb{N}$, one sees by completeness of X that $\operatorname{Proj}_C(x)$ is a singleton if and only if $\rho(x) = 0$. Then, one has $X \setminus K(C) = \bigcup_{n\in\mathbb{N}} F_n$. Under the uniform convexity of the norm $\|\cdot\|$, Stechkin proved that F_n is closed and $\operatorname{int} F_n = \emptyset$, so $X \setminus K(C)$ is a countable union of closed sets with empty interior, or equivalently $K(C)$ is a dense G_δ set in X. Stechkin's method to prove that $\operatorname{int} F_n = \emptyset$ in [**899**, Theorem] is essentially based on the following property of uniformly convex norm:

LEMMA 14.50 (Stechkin: [**899**], Lemma 4). Let $(X, \|\cdot\|)$ be a uniformly convex Banach space, and for each $\alpha \in]0,1[$, each $\varepsilon > 0$ and each $x \in \alpha\mathbb{S}_X$ let

$$Q(x, \alpha, \varepsilon) := B[x, 1 - \alpha - \varepsilon] \cap (X \setminus B[0,1]).$$

Then for each $\alpha \in]0,1[$ one has

$$\lim_{\varepsilon \downarrow 0} \sup_{x \in \alpha\mathbb{S}_X} \big(\operatorname{diam} Q(x, \alpha, \varepsilon)\big) = 0,$$

that is, $\operatorname{diam} Q(x, \alpha, \varepsilon) \to 0$ uniformly with respect to $x \in \alpha\mathbb{S}_X$ as $\varepsilon \downarrow 0$.

Concerning the closedness of the set F_n, it holds true even under the weaker property of local uniform convexity of the norm $\|\cdot\|$, see [**899**, Lemma 3]. Note that a lemma (see [**899**, Lemma 2]) similar to Lemma 14.50 allowed Stechkin to show in Theorem 3 of the same paper [**899**] that, if the norm $\|\cdot\|$ of the Banach space is merely *locally uniformly convex*, then the set $\{x \in X : \operatorname{card}(\operatorname{Proj}_C x) \leq 1\}$ is a dense G_δ set in X.

The method employed by Edelstein in the proof of his theorem [**371**, p. 376] is based on specific properties of uniformly convex norms different than the ones used by Stechkin [**899**]. Let $\delta(\cdot)$ denote the modulus of continuity of the norm $\|\cdot\|$ and take any $x \in X \setminus C$. Choosing an arbitrary real ρ with $0 < 2\rho < d_C(x)$, one notes that, for the desired aforementioned density property, it suffices to find

---

[1] An english translation of [**899**] has been kindly communicated to us by S. Cobzas (see [**899**, Theorem 5])

some $\overline{x} \in B[x, \rho]$ with $\operatorname{Proj}_C(\overline{x}) \neq \emptyset$. Putting $x_0 := x$, a sequence $(x_n, y_n)_{n \in \mathbb{N}}$ is constructed with $y_n \in C$ so that

$$\|y_n - x_{n-1}\| \leq r_n \left(1 + \frac{1}{4}\delta\left(\frac{\rho}{2^n r_n}\right)\delta^2\left(\frac{\rho}{2^n r_n}\right)\right)$$

and

$$x_n = x_{n-1} + \frac{r_n}{2}\delta\left(\frac{\rho}{2^n r_n}\right)\frac{y_n - x_{n-1}}{\|y_n - x_{n-1}\|},$$

where $r_n := d_C(x_{n-1})$. From the latter equality and from the uniform convexity of the norm it is not difficult to see that $(x_n)_n$ is a Cauchy sequence converging to some $\overline{x}$ and that $\|x - \overline{x}\| \leq \rho$. Once it is proved that $(y_n)_n$ converges to some $\overline{y}$, it is clear from the construction that $\overline{y} \in C$ and

$$\|\overline{x} - \overline{y}\| = \lim \|x_n - y_{n+1}\| = \lim d_C(x_n) = d_C(\overline{x});$$

so $\overline{y} \in \operatorname{Proj}_C(\overline{x})$. The proof by Edelstein in [**371**, p. 377] that $(y_n)_n$ is a Cauchy sequence is based on the lemma:

LEMMA 14.51 (Edelstein: [**371**], Lemma 2). *Let $(X, \|\cdot\|)$ be a uniformly convex normed space and let $\delta(\cdot)$ denote the modulus of convexity of $\|\cdot\|$. For any $t \in]0, 1]$ and any $x, y \in X$ with $\|x\| \geq \|y\| = 1$ and such that*

$$\left\|x - \frac{1}{2}\delta(t)y\right\| < 1 - \frac{1}{2}\delta(t) + \delta(t)\delta^2(t),$$

one has $\|x - y\| \leq t$.

Now, we state Lau theorem (see [**648**, Theorem 4]) as follows: If $(X, \|\cdot\|)$ is a reflexive Banach space whose norm $\|\cdot\|$ is Kadec-Klee, then $K(C)$ contains a dense G_δ set in X. In order to outline Lau's arguments, consider for each real $\varepsilon \in]0, 1/2[$ the set

$$A_\varepsilon(C) := \{x \in X : \exists(\eta, x^*) \in]0, +\infty[\text{ with } |\,\|x^*\| - 1| < \varepsilon,$$
$$\langle x^*, y - x\rangle \leq -d_C(x)(1 - \varepsilon)\ \forall y \in C \cap B(x, d_C(x) + \eta)\}$$

and put $A(C) := \bigcap_{n>2} A_{1/n}(C)$. Lau's method consists in proving that $A(C)$ is a dense G_δ set in $X \setminus C$. To achieve this, the main part of Lau's development is the lemma:

LEMMA 14.52 (Lau: [**648**], Lemma 3). *If $(X, \|\cdot\|)$ is a Banach space admitting an equivalent norm which is Fréchet differentiable (off zero), then for any $\varepsilon \in]0, 1/2[$ the set $A_\varepsilon(C)$ is an open dense subset in $X \setminus C$ relative to the induced topology in $X \setminus C$.*

For the proof of the density of $A_\varepsilon(C)$ in $X \setminus C$, Lau utilized (with $f = \Psi_C$) a 1976 result of I. Ekeland and G. Lebourg [**383**] saying that, for any real $\varepsilon > 0$ and any proper lower semicontinuous function $f : X \to \mathbb{R} \cup \{+\infty\}$ on a Banach space X admitting an equivalent Fréchet differentiable (off zero) norm, the set[2] $\operatorname{Dom} \partial_{F,\varepsilon} f$ is dense in $\operatorname{dom} f$; it is worth mentioning that in 1978 it was not yet known that $\operatorname{Dom} \partial_F f$ is dense in $\operatorname{dom} f$ under those hypotheses. We also emphasize that the set $L_n(C)$ in Lemma 14.26 in the book corresponds to $A_\varepsilon(C)$ for some ε.

[2] Instead of Fréchet ε-subdifferential, the authors of [**383**] used almost the same concept that they called *local ε-support* in their Definition 1.1.

Under the more stringent assumption that both the Banach space $(X, \|\cdot\|)$ and its dual are uniformly convex, another approach was followed by I. Ekeland and G. Lebourg in their 1976 paper [**383**, Corollary 3.15]; there, the result is deduced from the study of the problem $\inf_{y \in X} \big(g(y) + \|x - y\|^p\big)$. Several results were provided by S. Fitzpatrick [**410, 411**] by means of a subtle constant related to the Fréchet differentiability; this constant is presented in Definition 14.10 in the book and called there the *Fitzpatrick constant.* For other results related to density and genericity of points with nearest points we refer to a survey [**253**] by S. Cobzas and the references therein.

Concerning the first works in infinite dimensions for the density and genericity of points with farthest points, we will cite the 1966 papers [**370**] by M. Edelstein and [**25**] by Asplund along with the 1975 paper [**647**] by K.-S. Lau.

Let C be a nonempty closed bounded subset of a Banach space $(X, \|\cdot\|)$ and let $F(C)$ be the set of points in X which have farthest point in C. If the norm $\|\cdot\|$ is uniformly convex, Edelstein [**370**, Theorem 1] showed that the set $F(C)$ is dense in X. Asplund [**25**] extended the result to the case where $(X, \|\cdot\|)$ is reflexive and locally uniformly convex; in fact, Asplund showed in such a context that $F(C)$ contains a dense G_δ set in X. A more general result was achieved by Lau who proved in [**647**, Theorem 2.3] that in any Banach space $(X, \|\cdot\|)$ the set $F(C)$ contains a dense G_δ set in X whenever the set C is weakly compact. The survey [**253**] contains a long list of references of works related to the density and genericity property of points having farthest points.

The assertion (b) in Proposition 14.6 was probably first noticed by J. Blatter in his 1969 paper [**105**] under the assumption that dfar_C is Gâteaux differentiable at $\overline{x}$; this was a counterpart of the previous 1967 result of L.P. Vlasov [**960**] saying that $\langle D(d_C)(x), x - y\rangle = \|x - y\|$ for every $y \in \mathrm{Proj}\,_C(x)$ whenever the distance function d_C is Gâteaux differentiable at x. The assertion (a) in that Proposition 14.6 is the adaptation of Borwein and Fitzpatrick arguments [**128**, Proposition] to such a context (see also Proposition 4.32 and the reference [**758**]). Regarding the reasoning for Proposition 14.8, it was probably first observed in 1978 by N.V. Zhivkov [**1010**] that the set $\mathrm{Far}_C(x)$ (resp. $\mathrm{Proj}\,_C(x)$) is at most a singleton whenever the related norm $\|\cdot\|$ is *strictly convex* and the function dfar_C (resp. d_C) is Gâteaux differentiable at the point x.

As said above, Theorem 14.29 was established in 1978 by K.-S. Lau [**648**]. The presentation of Theorem 14.29 in the book follows the proof by J. M. Borwein and S. Fitzpatrick [**128**]. Lemma 14.26, Lemma 14.27, and Lemma 14.28 and their proofs correspond to Lemma 5.2, Lemma 5.3 and Lemma 5.6 respectively by Borwein and Fitzpatrick [**128**]; the approach in [**128**] is a modification of K.-S. Lau's proof [**648**] by a systematic use of Fréchet subgradients and their known density for any locally Lipschitz function. The fact that in any non-reflexive Banach space the sets in (a) and (b) of Example 14.30 are such that the set of points having nearest point is not dense in X, was noticed by M. Edelstein in Subsections 1.2 and 5.1 in [**373**]. The result in Theorem 14.32 is due to S.V. Konjagin [**613**] in 1980; for its proof we followed [**613**] and [**128**]. Proposition 14.34 was established by Borwein and Fitzpatrick in Theorem 3.2 and Corollary 3.3 in [**128**].

Theorem 14.13 translates two results by S. Fitzpatrick. Its version with the constant $\mathrm{fitz}\, f(x)$ is from [**411**, Theorem 2.4] while the version with the other constant $\mathrm{lip}\, f(x)$ is from [**411**, Corollary 2.6]; the arguments in the manuscript are taken from Fitzpatrick's article [**411**]. Corollary 14.14 reflects [**411**, Corollary 2.5]. Theorem 14.17 was established by J. M. Borwein, S. Fitzpatrick and J. R. Giles in their 1987 paper [**131**]. The proof in the book corresponds to their proof of Theorem 1 in [**131**]. The constant $\beta_G(x)$ in (14.6) used by Borwein, Fitzpatrick and Giles [**131**] was inspired by a generalized directional derivative introduced by P. Michel and J.-P. Penot [**708**]. The equivalences (a) $\Leftrightarrow$ (c) in Corollary 14.22 and Corollary 14.20 were obtained by Fitzpatrick [**411**, Corollary 3.6] when the norm $\|\cdot\|$ is both Fréchet differentiable and uniformly Gâteaux differentiable and the dual norm $\|\cdot\|_*$ is Fréchet differentiable. The Gâteaux differentiability version of Corollary 14.23 is present in Borwein, Fitzpatrick and Giles [**131**, Corollary 2]. Theorem 14.24 and its proof are taken from the 1996 dense survey on Chebyshev sets by V.S. Balaganski and L.P. Vlasov [**63**, Theorem 1.4].

Proposition 14.37 was established by R. Cibulka and M. Fabian [**229**, Proposition 1]; its proof in the book follows that in [**229**]. The assertion (a) in Theorem 14.40 (the main part of the theorem) is taken from R. Cibulka and M. Fabian [**229**, Proposition 5]; its proof uses ideas of Proposition 4.32(a) whose statement and proof correspond to Proposition 1.4 in the paper by J. M. Borwein and S. Fitzpatrick [**128**] (as said in comments in Section 4.8). The statement and proof of Lemma 14.43 correspond to Lemma 8 in Cibulka and Fabian's paper [**229**]. Corollary 14.47 was first proved by S. Dutta [**369**, Lemma 1]. Its extension 14.45 to Moreau envelopes was obtained by R. Cibulka and M. Fabian [**229**, Proposition 3]. The proof of Proposition 14.45 follows the arguments in [**229, 369**]. The intersection property in Remark 14.46 was noticed in the first conclusion in [**229**, Proposition 3]. Previously to [**229**] various results on density and genericity of $\mathrm{Prox}_\lambda f$ were provided in [**133, 392, 399**]. Differential and subdifferential properties of Moreau envelopes and distance functions in the framework of Banach spaces with uniformly Gâteaux differentiable norms can be found in [**590**]. Theorem 14.49 as stated in the book is taken from [**590**].

CHAPTER 15

Prox-regularity of sets in Hilbert spaces

After the general properties of metric projection established in the previous Chapter 14, we turn our attention in this chapter to the study of prox-regular sets in a *Hilbert space* H, first with varying thickness $\rho(\cdot)$ and then with extended-real constant thickness r. The local property around a fixed point of the set is also developed. The case of more general Banach spaces will be considered later in Chapter 18.

15.1. $\rho(\cdot)$-prox-regularity of sets

Recall (see Definition 4.131) that a vector v in a Hilbert space H is a proximal normal vector of a closed set $C \subset H$ at $x \in C$ if there exists some real $\sigma > 0$ such that

$$x \in \operatorname{Proj}_C(x + \sigma v).$$

Quantifying this in requiring to the positive number σ in the inclusion (for the unit proximal normal vectors of C) to depend in a specific way on the point $x \in C$ (continuously or semicontinuously) leads to the concept of $\rho(\cdot)$-prox-regularity.

DEFINITION 15.1. Let $C \subset H$ be a nonempty closed set of a Hilbert space H and let $\rho : C \to]0, +\infty]$ be a *lower semicontinuous function* with positive extended-real values. The set C is $\rho(\cdot)$-*prox-regular* provided that, for any $x \in C$, any $v \in N^P(C; x)$ with $\|v\| \leq 1$, and any positive real number $t \leq \rho(x)$, one has

$$x \in \operatorname{Proj}_C(x + tv),$$

or equivalently $C \cap B(x + tv, t) = \emptyset$ for any $v \in N^P(C; x)$ with $\|v\| = 1$; the latter can be translated (see the comment after Definition 4.131) into the fact that every proximal normal vector v to C at x with $\|v\| = 1$ can be realized by a t-ball. Such a function ρ, when it exists, is called a *prox-regularity radius function* of C.

The requirement of the lower semicontinuity for the function $\rho(\cdot)$ is natural in the sense that, for each $\overline{x} \in C$, it provides some $0 < r \leq \rho(\overline{x})$ such that the property in Definition 15.1 holds true for all $x \in C$ close enough to $\overline{x}$ and all reals $t \in]0, r]$. Indeed, given $\overline{x} \in C$, by the lower semicontinuity of $\rho(\cdot)$ there is a neighborhood U of $\overline{x}$ and a common constant $0 < r \leq \inf_{u \in U} \rho(u)$ such that $x \in \operatorname{Proj}_C(x + tv)$ for any $x \in U \cap C$, any $v \in N^P(C; x) \cap \mathbb{B}_H$ and any real $t \in]0, r]$.

REMARK 15.2. Consider on the Hilbert space $(H, \langle \cdot, \cdot \rangle, \| \cdot \|)$ another (non necessarily Hilbertian) norm $\| \! | \cdot \| \! |$ which one is also interested in. Denoting by $\mathbb{B}_{\| \! | \cdot \| \! |}$ the closed unit ball relative to this second norm $\| \! | \cdot \| \! |$, one can define the $(\| \! | \cdot \| \! |, \rho(\cdot))$-prox-regularity of C by replacing the initial ball $\mathbb{B}_{\| \cdot \|}$ by $\mathbb{B}_{\| \! | \cdot \| \! |}$, that is, by requiring that $x \in \operatorname{Proj}_C(x + tv)$ for any $x \in C$, any $v \in N^P(C; x) \cap \mathbb{B}_{\| \! | \cdot \| \! |}$ and any real $t \in]0, \rho(x)]$. This can even be done for only the points $x \in C \cap U$, where U is a specific open set with $C \cap U \neq \emptyset$. Of course, $N^P(C;\cdot)$ and $\mathrm{Proj}_C(\cdot)$ are still taken with respect to the initial Hilbertian norm $\|\cdot\|$. □

As in the previous chapters, $\mathrm{Proj}_C(u)$ denotes the (possibly empty) set of nearest points of u in C. Sometimes it will be convenient to write $\mathrm{Proj}\,(u, C)$ or $\mathrm{Proj}\,(C, u)$ instead of $\mathrm{Proj}_C(u)$. When the set $\mathrm{Proj}_C(u)$ is a singleton, we recall that we denote its unique element by $P_C(u)$, $P(u,C)$, $P(C,u)$ or $\mathrm{proj}_C(u)$. So, the notation $P_C(u)$, $P(u,C)$, $P(C;u)$, or $\mathrm{proj}_C(u)$ means that $\mathrm{Proj}_C(u)$ is a nonempty set with a unique element and $\mathrm{Proj}_C(u) = \{P_C(u)\}$.

Throughout this section unless otherwise stated, H will be a real *Hilbert space* endowed with the *inner product* $\langle\cdot,\cdot\rangle$, and $\|\cdot\|$ will be the *associated norm.* Given a nonempty subset C, the proximal normal cone $N^P(C;\cdot)$, the set of nearest points $\mathrm{Proj}_C(x)$ and the metric projection P_C will be taken with respect to this Hilbert norm $\|\cdot\|$. The function $\rho : C \to]0,+\infty]$ will be at least *lower semicontinuous.* However, as we will see later on, many fundamental properties will lead us to focus also our attention on situations when ρ is continuous (instead of being lower semicontinuous). We also recall the convention $1/r = 0$ for $r = +\infty$.

EXAMPLE 15.3. (a) Any nonempty closed convex set C of H is $\rho(\cdot)$-prox-regular with $\rho(x) = +\infty$ for all $x \in C$.
(b) The set $C := H \setminus B(0,r)$ with $r \in]0,+\infty[$ is $\rho(\cdot)$-prox-regular, with $\rho(x) := r$ for all $x \in C$.
(c) In the Euclidean space $\mathbb{R}^2$, let

$$C := \{(x,y) \in \mathbb{R}^2 : y \geq 0,\ (x,y) \notin B(0_{\mathbb{R}^2}, r)\} \text{ with } r \in]0,+\infty[$$

(see Figure 15.1). Put $D := \{(x,y) \in \mathbb{R}\times\mathbb{R} : y \geq 0, x^2+y^2 = r^2\}$. Then, for any continuous function $\rho : C \to]0,+\infty]$ with $\rho(x,y) \leq r$ if $(x,y) \in D$ and $\rho(x,y) \leq +\infty$ if $(x,y) \in C\setminus D$, the set C is $\rho(\cdot)$-prox-regular. It is also $\rho_0(\cdot)$-prox-regular with the lower semicontinuous function $\rho_0 : C \to]0,+\infty]$ defined by $\rho_0(x,y) = r$ if $(x,y) \in D$ and $\rho_0(x,y) = +\infty$ if $(x,y) \in C \setminus D$.

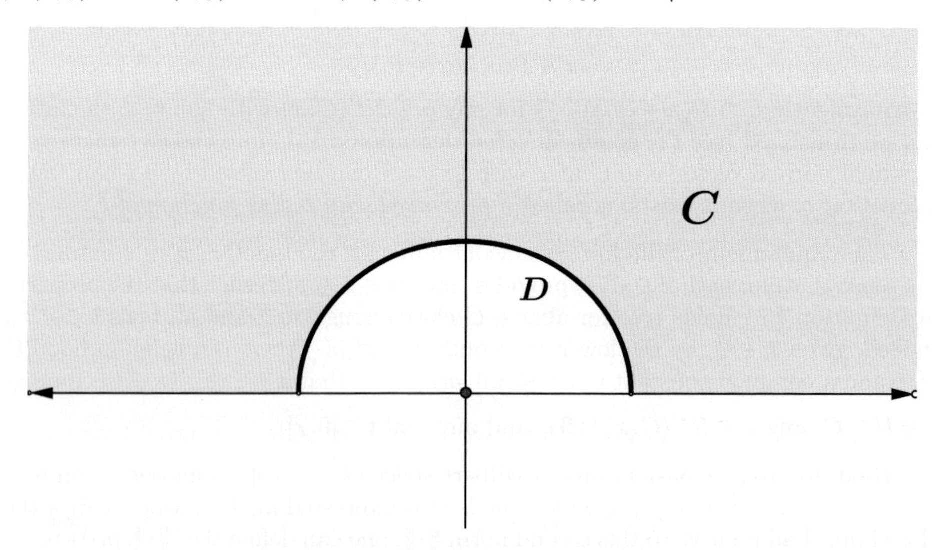

FIGURE 15.1. $\rho(\cdot)$-prox-regular set.

(d) Consider $C := \{(x,y) \in \mathbb{R}\times\mathbb{R} : |y| \geq \exp(-x)\}$ (see Figure 15.2). For any continuous function $\rho : C \to]0, +\infty[$ with $\rho(x, y) \le \big(1 + \exp(-x)\big)^{1/2} \exp(-2x)$ if $y = \pm \exp(-x)$ and $\rho(x, y) \le +\infty$ if $|y| > \exp(-x)$, the set C is $\rho(\cdot)$-prox-regular. It is also $\rho_0(\cdot)$-prox-regular with the lower semicontinuous function $\rho_0 : C \to]0, +\infty]$ defined by $\rho_0(x, y) = \big(1 + \exp(-x)\big)^{1/2} \exp(-2x)$ if $y = \pm \exp(-x)$ and $\rho_0(x, y) = +\infty$ if $|y| > \exp(-x)$.

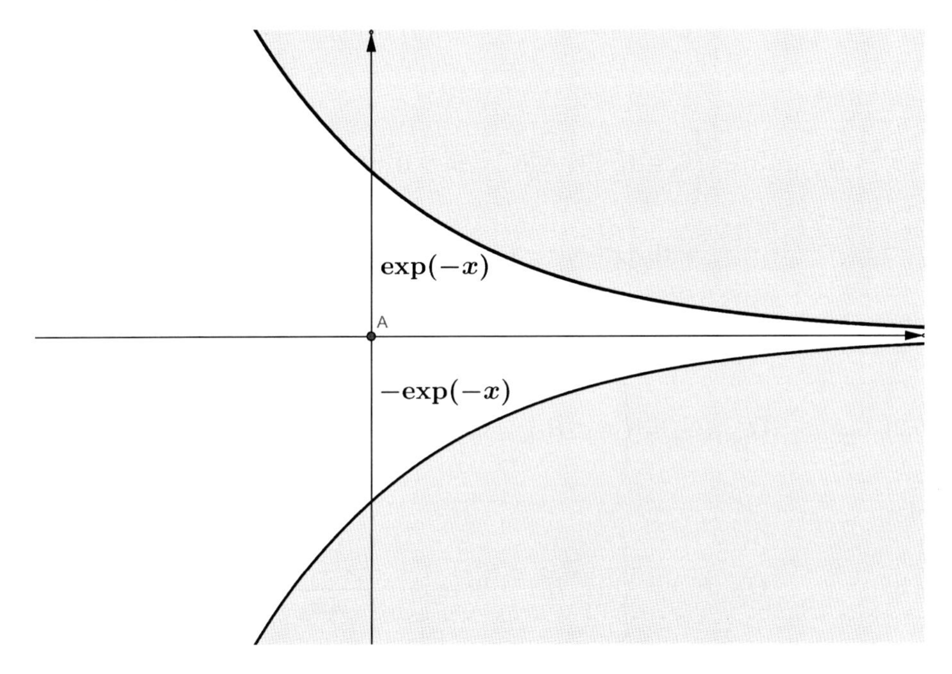

FIGURE 15.2. $\rho(\cdot)$-prox-regular set with $1/\rho(\cdot)$ not bounded above.

(e) For any lower semicontinuous function $\rho : C \to]0, +\infty]$ the set $C := \{(x, y) \in \mathbb{R} \times \mathbb{R} : y \le |x|\}$ fails to be $\rho(\cdot)$-prox-regular. Indeed, at any $(x, |x|) \in C$ with $x \neq 0$, the $\rho(\cdot)$-prox-regularity would entail $\rho(x, |x|) \le |x|\sqrt{2}$, which would contradict by lower semicontinuity the inequality $\rho(0, 0) > 0$.

More examples will be given later. □

In the context of Hilbert space, it is worth mentioning the following simple description of the set of nearest points. It will be used in many places of this chapter.

LEMMA 15.4. Let C be a nonempty set of a Hilbert space H. For $x \in H$, one has the equivalence

$$y \in \mathrm{Proj}_C(x) \Longleftrightarrow y \in C \text{ and } \langle x - y, x' - y \rangle \le \frac{1}{2} \|x' - y\|^2, \ \forall x' \in C. \tag{15.1}$$

PROOF. Clearly, $y \in \mathrm{Proj}_C(x)$ if and only if $y \in C$ and for all $x' \in C$

$$\|x - y\|^2 \le \|x - x'\|^2, \text{ that is, } \|x - y\|^2 \le \|(x - y) - (x' - y)\|^2.$$

In the Hilbert space H, the latter inequality is equivalent to

$$0 \le -2\langle x - y, x' - y \rangle + \|x' - y\|^2, \text{ that is, } \langle x - y, x' - y \rangle \le (1/2)\|x' - y\|^2,$$

which justifies the equivalence of the lemma. □

We start our study of $\rho(\cdot)$-prox-regular sets with Theorem 15.6. In its statement we use for any real $\gamma \in]0,1[$ the notation

$$O^{\gamma}_{\rho(\cdot)}(C) := \Big\{ u \in H : \exists \gamma' \in]0,\gamma[, \exists \eta > 0 \text{ such that } \forall x \in C$$

$$\text{with } \|x-u\| < \eta + d_C(u) \text{ one has } \frac{\|x-u\|}{\rho(x)} < \gamma' \Big\},$$

along with

$$O_{\rho(\cdot)}(C) := \Big\{ u \in H : \exists \gamma \in]0,1[, \exists \eta > 0 \text{ such that } \forall x \in C$$

$$\text{with } \|x-u\| < \eta + d_C(u), \text{ one has } \frac{\|x-u\|}{\rho(x)} < \gamma \Big\},$$

that is, $O_{\rho(\cdot)}(C) = \bigcup\limits_{0<\gamma<1} O^{\gamma}_{\rho(\cdot)}(C)$. It is not difficult to see that

$$O^{\gamma}_{\rho(\cdot)}(C) = \left\{ u \in H : \limsup_{\substack{\|x-u\| \to d_C(u) \\ x \in C}} \frac{\|x-u\|}{\rho(x)} < \gamma \right\}, \tag{15.2}$$

and

$$O_{\rho(\cdot)}(C) = \left\{ u \in H : \limsup_{\substack{\|x-u\| \to d_C(u) \\ x \in C}} \frac{\|x-u\|}{\rho(x)} < 1 \right\}. \tag{15.3}$$

We observe that the inclusion $u \in O^{\gamma}_{\rho(\cdot)}(C)$ amounts to saying that, for any minimizing sequence $(x_n)_n$ in C of $\inf_{x\in C} \|x-u\| = d_C(u)$ (that is, $\|x_n - u\| \to d_C(u)$),

$$\limsup_{n\to\infty} \frac{\|x_n - u\|}{\rho(x_n)} < \gamma.$$

So, keeping in mind the lower semicontinuity of $\rho(\cdot)$, we can notice in particular the important inclusion

$$C \subset O^{\gamma}_{\rho(\cdot)}(C) \quad \text{for all positive } \gamma < 1; \tag{15.4}$$

indeed, fixing any $u \in C$, for every sequence $(x_n)_n$ in C with $\|x_n - u\| \to d_C(u) = 0$, the lower semicontinuity of $\rho(\cdot)$ guarantees that $\liminf\limits_{n\to\infty} \rho(x_n) \geq \rho(u) > 0$ (since $u \in C$), and hence $\lim_{n\to\infty} \frac{\|x_n-u\|}{\rho(x_n)} = 0$. We will also see later that the sets $O^{\gamma}_{\rho(\cdot)}(C)$ and $O_{\rho(\cdot)}(C)$ are open whenever C is $\rho(\cdot)$-prox-regular.

This section will be mainly devoted to establishing several necessary and sufficient conditions for prox-regularity. This will be divided into several steps. Before stating the first theorem on properties of prox-regularity, let us establish the following lemma which provides a general result which has its own interest. It concerns the set of nearest points and in its statement I denotes the *identity mapping* on H.

LEMMA 15.5. Let C be a nonempty closed set of a Hilbert space H. For any point $u \in H$ one has

$$\operatorname{Proj}_C(u) \subset \big(I + d_C(u)\mathbb{B}_H \cap N^P(C;\cdot)\big)^{-1}(u).$$

PROOF. Suppose that $\operatorname{Proj}_C(u)$ is nonempty and fix any $y \in \operatorname{Proj}_C(u)$. Then by definition of proximal normal we have $u - y \in N^P(C; y)$. Since $\|u - y\| = d_C(u)$ the latter inclusion ensures that

$$\frac{1}{d_C(u)}(u - y) \in \mathbb{B}_H \cap N^P(C; y) \quad \text{if } d_C(u) > 0,$$

and hence $u \in y + d_C(u)\mathbb{B}_H \cap N^P(C; y)$ even if $d_C(u) = 0$ (since $\|u - y\| = d_C(u)$). This means that in any case we have

$$y \in \left(I + d_C(u)\mathbb{B}_H \cap N^P(C; \cdot)\right)^{-1}(u),$$

which proves the desired inclusion. □

We can now establish the first theorem providing some first results concerning $\rho(\cdot)$-prox-regular sets.

THEOREM 15.6 (normal properties of $\rho(\cdot)$-prox-regular sets). Let C be a nonempty closed subset of a Hilbert space H and $\rho(\cdot) : C \to]0, +\infty]$ be a lower semicontinuous function. Then anyone of the following assertions (a)-(e) is a characterization of the $\rho(\cdot)$-prox-regularity of C:
(a) For any $x, x' \in C$ and $v \in N^P(C; x)$ one has

$$\langle v, x' - x\rangle \le \frac{1}{2\rho(x)}\|v\|\,\|x' - x\|^2;$$

(b) for any $x \in C$, $v \in N^P(C; x) \cap \mathbb{B}_H$ and any nonnegative real number $t < \rho(x)$, one has $x = P_C(x + tv)$ (we recall that $P_C(y)$ denotes the unique point of $\operatorname{Proj}_C(y)$ when this latter is a singleton);
(c) for any $x_i \in C$, $v_i \in N^P(C; x_i)$ with $i = 1, 2$ one has

$$\langle v_1 - v_2, x_1 - x_2\rangle \ge -\frac{1}{2}\left(\frac{\|v_1\|}{\rho(x_1)} + \frac{\|v_2\|}{\rho(x_2)}\right)\|x_1 - x_2\|^2;$$

(d) for any $u_i \in H$ and for any $y_i \in (I + N^P(C; \cdot))^{-1}(u_i)$ with $i = 1, 2$, one has

$$\langle y_1 - y_2, u_1 - u_2\rangle \ge \left(1 - \frac{\|u_1 - y_1\|}{2\rho(y_1)} - \frac{\|u_2 - y_2\|}{2\rho(y_2)}\right)\|y_1 - y_2\|^2;$$

(e) for any $x \in C$ and any $u \in \operatorname{Dom} P_C$ one has

$$\|x - P_C(u)\| \le \left(1 - \frac{\|u - P_C(u)\|}{\rho(P_C(u))}\right)^{-1}\|x - u\|.$$

PROOF. Denote by (A) the $\rho(\cdot)$-prox-regularity of C.
(A) $\Leftrightarrow$ (a). Fix any $x \in C$ and $v \in N^P(C; x) \cap \mathbb{B}_H$. Fix also any positive real number $t \le \rho(x)$. By Lemma 15.4, the inclusion $x \in \operatorname{Proj}_C(x + tv)$ is equivalent to

$$\langle x + tv - x, x' - x\rangle \le \frac{1}{2}\|x' - x\|^2, \quad \text{for all } x' \in C,$$

which in turn is equivalent to

$$\langle v, x' - x\rangle \le \frac{1}{2t}\|x' - x\|^2, \quad \text{for all } x' \in C.$$

This means that the equivalence between (A) and (a) holds true.

(A) $\Leftrightarrow$ (b). This equivalence is easily verified because we know by Lemma 4.123 that $x = P_C(x + tv)$ whenever $x \in \operatorname{Proj}_C(x + t'v)$ for some $t' > t$.

(a) $\Rightarrow$ (c). Let $v_i \in N^P(C; x_i)$ with $i = 1, 2$. By (b) we have

$$\langle v_1, x_1 - x_2\rangle \geq -\frac{\|v_1\|}{2\rho(x_1)}\|x_1 - x_2\|^2 \text{ and } \langle -v_2, x_1 - x_2\rangle \geq -\frac{\|v_2\|}{2\rho(x_2)}\|x_2 - x_1\|^2,$$

and hence adding the two inequalities yields the desired inequality of (c).

(c) $\Rightarrow$ (a). For $x, x' \in C$ and $v \in N^P(C; x)$ it is enough to take $v_1 = v$, $x_1 = x$, $x_2 = x'$ and $v_2 = 0$ in the inequality of (c) to obtain that of (a).

(c) $\Rightarrow$ (d). Suppose that (c) holds and fix any $u_i \in H$ and any $y_i \in (I + N^P(C; \cdot))^{-1}(u_i)$ for $i = 1, 2$. Then (d) allows us to write

$$\begin{aligned}\langle y_1 - y_2, u_1 - u_2\rangle &= \langle y_1 - y_2, (u_1 - y_1) - (u_2 - y_2)\rangle + \|y_1 - y_2\|^2 \\ &\geq -\frac{1}{2}\left(\frac{\|u_1 - y_1\|}{\rho(y_1)} + \frac{\|u_2 - y_2\|}{\rho(y_2)}\right)\|y_1 - y_2\|^2 + \|y_1 - y_2\|^2,\end{aligned}$$

that is,

$$\langle y_1 - y_2, u_1 - u_2\rangle \geq \left(1 - \frac{\|u_1 - y_1\|}{2\rho(y_1)} - \frac{\|u_2 - y_2\|}{2\rho(y_2)}\right)\|y_1 - y_2\|^2.$$

So, (d) is fulfilled.

(d) $\Rightarrow$ (c). Suppose (d) and take any $x_i \in C$, $v_i \in N^P(C; x_i)$ with $i = 1, 2$. Putting $u_i := x_i + v_i$, we have $x_i \in \left(I + N^P(C; \cdot)\right)^{-1}(u_i)$, which gives by the assumption (d)

$$\langle x_1 - x_2, u_1 - u_2\rangle \geq \left(1 - \frac{\|u_1 - x_1\|}{2\rho(x_1)} - \frac{\|u_2 - x_2\|}{2\rho(x_2)}\right)\|x_1 - x_2\|^2,$$

which is equivalent to

$$\langle x_1 - x_2, v_1 - v_2\rangle \geq -\frac{1}{2}\left(\frac{\|v_1\|}{\rho(x_1)} + \frac{\|v_2\|}{\rho(x_2)}\right)\|x_1 - x_2\|^2,$$

which is (c).

(d) $\Rightarrow$ (e). This implication is trivial.

(e) $\Rightarrow$ (a). Suppose (e) and take any $x, x' \in C$ and $v \in N^P(C; x)$. There are $u \in H$ with $P_C(u) = x$ and $\lambda > 0$ such that $v = \lambda(u - P_C(u))$. Take any $t \in]0, 1]$ sufficiently small so that $t\|x - u\|/(2\rho(x)) < 1$, put $x_t := (1 - t)x + tu$ and note that $x = P_C(x_t)$. Putting $\alpha := 1/(2\rho(x))$ and noting that $1 - \alpha\|x - x_t\| > 0$, the assumption (e) gives that $(1 - \alpha\|x - x_t\|)^2\|x' - x\|^2 \leq \|x' - x_t\|^2$, which by the equality $x' - x_t = (x' - x) + t(x - u)$ is equivalent to

$$(1 - 2\alpha t\|x - u\| + \alpha^2 t^2\|x - u\|^2)\|x' - x\|^2 \leq \|x' - x\|^2 + 2t\langle x' - x, x - u\rangle + t^2\|x - u\|^2.$$

The latter inequality is in turn equivalent to

$$2\langle u - x, x' - x\rangle \leq 2\alpha\|u - x\|\|x' - x\|^2 + t\beta,$$

where $\beta := (1 - \alpha^2\|x' - x\|^2)\|x - u\|^2$. Multiplying by $\lambda/2 > 0$ and making $t \downarrow 0$, we obtain the desired inequality in (a). This finishes the proof of the theorem. $\square$

REMARK 15.7. Continuing with Remark 15.2, let H be a Hilbert space with inner product $\langle \cdot, \cdot\rangle$ and associated norm $\|\cdot\|$, and let $|||\cdot|||$ be another (non necessarily Hilbertian) norm on H. Let C be a nonempty closed set in H and let $N^P(C; \cdot)$ and $\mathrm{Proj}_C(\cdot)$ still denote the P-normal cone and metric projection, both taken with respect to the initial Hilbertian norm $\|\cdot\|$. The proof of Theorem 15.6 also

guarantees that anyone of the following properties characterizes the $(\|\!| \cdot \|\!|, \rho(\cdot))$-prox-regularity of C:
(a) For any $x \in C$, $v \in N^P(C;x)$ one has

$$\langle v, x' - x \rangle \leq \frac{1}{2\rho(x)} \|\!| v \|\!| \, \|x' - x\|^2;$$

(b) for any $x \in C$, $v \in N^P(C;x) \cap \mathbb{B}_{\|\!| \, \|\!|}$ and any non-negative real number $t < \rho(x)$, one has $x = P_C(x + tv)$ (where $P_C(y)$ denotes the unique nearest point with respect to the initial Hilbertian norm $\| \cdot \|$);
(c) for any $x_i \in C$, $v_i \in N^P(C;x_i)$ with $i = 1, 2$ one has

$$\langle v_1 - v_2, x_1 - x_2 \rangle \geq -\frac{1}{2} \left(\frac{\|\!| v_1 \|\!|}{\rho(x_1)} + \frac{\|\!| v_2 \|\!|}{\rho(x_2)} \right) \|x_1 - x_2\|^2;$$

(d) for any $u_i \in H$ and for any $y_i \in (I + N^P(C;\cdot))^{-1}(u_i)$ with $i = 1, 2$, one has

$$\langle y_1 - y_2, u_1 - u_2 \rangle \geq \left(1 - \frac{\|\!| u_1 - y_1 \|\!|}{2\rho(y_1)} - \frac{\|\!| u_2 - y_2 \|\!|}{2\rho(y_2)} \right) \|y_1 - y_2\|^2;$$

(e) for any $x \in C$ and any $u \in \operatorname{Dom} P_C$ the inequality

$$\|x - P_C(u)\| \leq \left(1 - \frac{\|\!| u - P_C(u) \|\!|}{\rho(P_C(u))} \right)^{-1} \|x - u\|$$

is satisfied. □

The next theorem continues the study of properties under the continuity of the function $\rho(\cdot)$.

THEOREM 15.8 (metric projection properties of $\rho(\cdot)$-prox-regular sets). Let C be a nonempty closed subset of a Hilbert space H and $\rho(\cdot) : C \to]0, +\infty]$ be a continuous function. Then one has the following properties.
(f) For any positive real $\gamma < 1$ and any $u \in O^{\gamma}_{\rho(\cdot)}(C)$ the set $\operatorname{Proj}_C(u)$ is a singleton, that is, $P_C(u)$ exists, the equality

$$P_C(u) = \left(I + \gamma\rho(\cdot)\mathbb{B}_H \cap N^P(C;\cdot) \right)^{-1}(u)$$

holds, and the mapping P_C is Lipschitz continuous on $O^{\gamma}_{\rho(\cdot)}(C)$ with $(1-\gamma)^{-1}$ as a Lipschitz constant, that is,

$$\|P_C(u_1) - P_C(u_2)\| \leq (1-\gamma)^{-1} \|u_1 - u_2\| \quad \text{for all } u_1, u_2 \in O^{\gamma}_{\rho(\cdot)}(C).$$

(g) For any positive real $\gamma < 1$ and any $u_i \in O^{\gamma}_{\rho(\cdot)}(C)$, $P_C(u_i)$ exist and

$$\|(I - P_C)(u_1) - (I - P_C)(u_2)\|^2 + (1 - 2\gamma)\|P_C(u_1) - P_C(u_2)\|^2 \leq \|u_1 - u_2\|^2.$$

In particular, for $\gamma \leq 1/2$ the mapping $I - P_C$ is Lipschitz continuous on $O^{\gamma}_{\rho(\cdot)}(C)$ with 1 as a Lipschitz constant.
(h) The mapping P_C is well defined on $O_{\rho(\cdot)}(C)$ and for all $u_1, u_2 \in O_{\rho(\cdot)}(C)$ one has

$$\|P_C(u_1) - P_C(u_2)\| \leq \left(1 - \frac{d_C(u_1)}{2\rho(P_C(u_1))} - \frac{d_C(u_2)}{2\rho(P_C(u_2))} \right)^{-1} \|u_1 - u_2\|.$$

PROOF. (f) and (h). Fix any positive real $\gamma < 1$. For any $u_i \in H$ and any

$$y_i \in (I + \gamma\rho(\cdot)\mathbb{B}_H \cap N^P(C;\cdot))^{-1}(u_i) \quad \text{with } i = 1, 2,$$

the inclusion $u_i \in y_i + \gamma\rho(y_i)\mathbb{B}_H \cap N^P(C; y_i)$ ensures that $\|u_i - y_i\| \le \gamma\rho(y_i)$, and hence it follows from the inequality of (d) in Theorem 15.6 that

$$\|y_1 - y_2\| \le (1-\gamma)^{-1}\|u_1 - u_2\|, \tag{15.5}$$

which entails that the multimapping $(I+\gamma\rho(\cdot)\mathbb{B}_H \cap N^P(C;\cdot))^{-1}(\cdot)$ is, on its domain, single-valued and Lipschitz continuous with $(1-\gamma)^{-1}$ as a Lipschitz constant.

First Proof for the remaining of (f).

Consider any point $u \in O^{\gamma}_{\rho(\cdot)}(C) \setminus C$. On the one hand, we have by (15.2)

$$\limsup_{\substack{\|u-x\| \to d_C(u) \\ x \in C}} \frac{\|x - u\|}{\rho(x)} < \gamma. \tag{15.6}$$

On the other hand, Lau's theorem (see Theorem 14.29) says that there is a sequence $(u_n)_{n\in\mathbb{N}}$ converging to u and such that $P_C(u_n)$ exists for each integer n. Then, $\|u_n - P_C(u_n)\| = d_C(u_n) \to d_C(u)$, and hence $\|u - P_C(u_n)\| \to d_C(u)$. Therefore, (15.6) ensures that $\limsup\limits_{n\to\infty} \frac{\|u - P_C(u_n)\|}{\rho(P_C(u_n))} < \gamma$. Observing that

$$\|u - P_C(u_n)\|/\|u_n - P_C(u_n)\| \underset{n}{\longrightarrow} d_C(u)/d_C(u) = 1,$$

we see that

$$\limsup_n \frac{\|u_n - P_C(u_n)\|}{\rho(P_C(u_n))} = \limsup_n \frac{\|u - P_C(u_n)\|}{\rho(P_C(u_n))} < \gamma.$$

We may then suppose that $\|u_n - P_C(u_n)\| < \gamma\rho(P_C(u_n))$ for all n. This entails that $u_n \in P_C(u_n) + \gamma\rho(P_C(u_n))\mathbb{B}_H \cap N^P(C; P_C(u_n))$, and we deduce from (15.5) that, for all $n, m \in \mathbb{N}$,

$$\|P_C(u_n) - P_C(u_m)\| \le (1-\gamma)^{-1}\|u_n - u_m\|,$$

and hence, for $y_n := P_C(u_n)$, the sequence $(y_n)_n$ is a Cauchy sequence. Denoting its limit by y, we see that for each n

$$\|u_n - y_n\| \le \|u_n - c\| \quad \text{for all } c \in C,$$

and hence

$$\|u - y\| \le \|u - c\| \quad \text{for all } c \in C,$$

that is, $y \in \operatorname{Proj}_C(u)$. Further, the inequality $\|u_n - y_n\| < \gamma\rho(y_n)$ and the continuity of $\rho(\cdot)$ yield $\|u - y\| \le \gamma\rho(y)$. Therefore, for any $u \in O^{\gamma}_{\rho(\cdot)}(C)$ there exists some $y \in \operatorname{Proj}_C(u)$ with $d_C(u) = \|u - y\| \le \gamma\rho(y)$ (because for $u \in C$ it is enough to take $y = u$), which gives $u - y \in N^P(C; y)$, thus we have

$$y \in (I + \gamma\rho(\cdot)\mathbb{B}_H \cap N^P(C;\cdot))^{-1}(u).$$

So, we have

$$O^{\gamma}_{\rho(\cdot)}(C) \subset \operatorname{Dom}\operatorname{Proj}_C \text{ and } O^{\gamma}_{\rho(\cdot)}(C) \subset \operatorname{Dom}(I + \gamma\rho(\cdot)\mathbb{B}_H \cap N^P(C;\cdot))^{-1}.$$

Because of the single valuedness of $(I + \gamma\rho(\cdot)\mathbb{B}_H \cap N^P(C;\cdot))^{-1}(\cdot)$ on its domain and of Lemma 15.5, it follows that the mapping P_C is well defined on $O^{\gamma}_{\rho(\cdot)}(C)$,

$$\{P_C(u)\} = \operatorname{Proj}_C(u) = (I + \gamma\rho(\cdot)\mathbb{B}_H \cap N^P(C;\cdot))^{-1}(u) \quad \text{for all } u \in O^{\gamma}_{\rho(\cdot)}(C),$$

and by (15.5)

$$\|P_C(u) - P_C(u')\| \le (1-\gamma)^{-1}\|u - u'\| \quad \text{for all } u, u' \in O^{\gamma}_{\rho(\cdot)}(C),$$

that is, P_C is Lipschitz continuous on $O^{\gamma}_{\rho(\cdot)}(C)$ with $(1-\gamma)^{-1}$ as a Lipschitz constant. The implication (e) $\Rightarrow$ (f) is then established.

Second proof for the remaining of (f).

Now we claim that for any $u \in O^{\gamma}_{\rho(\cdot)}(C)$ we have

$$\operatorname{Proj}_C(u) \neq \emptyset \quad \text{and} \quad d_C(u) \le \gamma\rho(y) \quad \text{for all } y \in \operatorname{Proj}_C(u). \tag{15.7}$$

The desired properties being obvious for $u \in C$, fix any $u \in O^{\gamma}_{\rho(\cdot)}(C) \setminus C$. Choose a sequence of positive numbers $(\varepsilon_n)_{n\in\mathbb{N}}$ with $\varepsilon_n \downarrow 0$ and a sequence $(z_n)_{n\in\mathbb{N}}$ in C such that $f(z_n) < \inf\limits_C f + \varepsilon_n^2$, where $f(x) := \frac{1}{2}\|x-u\|^2$ for all $x \in H$, and observe that such a sequence $(z_n)_n$ is bounded (because of the definition of f). According to the Ekeland variational principle (see Theorem 2.221), for each integer n there exists some $y_n \in C$ such that

$$\|y_n - z_n\| < \varepsilon_n \quad \text{and} \quad f(y_n) = \inf_{x\in C}\big(f(x) + \varepsilon_n\|x - y_n\|\big). \tag{15.8}$$

On the one hand, the sequence $(y_n)_n$ is bounded and by the equality in (15.8) we have for n large enough

$$\frac{1}{2}\|y_n - u\|^2 \le \frac{1}{2}\|z_n - u\|^2 + \varepsilon_n\|z_n - y_n\|$$

and this ensures that

$$\|y_n - u\| \to d_C(u). \tag{15.9}$$

On the other hand, by the equality in (15.8) again and by the fuzzy sum rule for the proximal subdifferential (see Theorem 4.155(a)), there are

$$y_n' \in C \text{ with } \|y_n' - y_n\| < \varepsilon_n \text{ and } y_n'' \in H \text{ with } \|y_n'' - y_n\| < \varepsilon_n \tag{15.10}$$

such that

$$0 \in \partial f(y_n'') + \varepsilon_n\mathbb{B}_H + N^P(C; y_n'),$$

that is, for some $b_n \in \mathbb{B}_H$ we have $u - y_n'' + \varepsilon_n b_n \in N^P(C; y_n')$. Set $u_n := u + y_n' - y_n'' + \varepsilon_n b_n$ and observe that, by (15.9) and (15.10), $\lim_{n\to\infty}\frac{\|u_n - y_n'\|}{\|u - y_n'\|} = 1$. Since we see by definition of $O^{\gamma}_{\rho(\cdot)}(C)$ (see (15.2)) that $\limsup\limits_{n\to\infty}\frac{\|u - y_n'\|}{\rho(y_n')} < \gamma$, we obtain $\limsup\limits_{n\to\infty}\frac{\|u_n - y_n'\|}{\rho(y_n')} < \gamma$. Deleting a finite number of integers if necessary, we may then suppose that for all n we have $\|u_n - y_n'\| \le \gamma\rho(y_n')$. As $u_n - y_n' \in N^P(C; y_n')$ we obtain $y_n' \in (I + \gamma\rho(\cdot)\mathbb{B}_H \cap N^P(C;\cdot))^{-1}(u_n)$. It follows from (15.5) that

$$\|y_n' - y_m'\| \le (1-\gamma)^{-1}\|u_n - u_m\|,$$

and this implies that $(y_n')_n$ is a Cauchy sequence because $u_n \to u$. Denoting by y the limit of the sequence $(y_n')_n$, we see that $y \in C$ and $(y_n)_n$ also converges to y. From this and (15.9) we derive that $y \in \operatorname{Proj}_C(u)$. Further, the inequality $\|u_n - y_n'\| < \gamma\rho(y_n')$ and the continuity of $\rho(\cdot)$ yield $d_C(u) = \|u - y\| \le \gamma\rho(y)$. The properties in (15.7) are then proved.

Fixing any $u \in O^\gamma_{\rho(\cdot)}(C)$ and $y \in \mathrm{Proj}_C(u)$ we have $u - y \in N^P(C; y)$, and combining this with (15.7) gives $d_C(u) \le \gamma\rho(y)$. So, we have

$$u \in y + d_C(u)\mathbb{B}_H \cap N^P(C; y) \subset y + \gamma\rho(y)\mathbb{B}_H \cap N^P(C; y)$$

and according to (15.7) again

$$O^\gamma_{\rho(\cdot)}(C) \subset \mathrm{Dom}\,\mathrm{Proj}_C \text{ and } O^\gamma_{\rho(\cdot)}(C) \subset \mathrm{Dom}\,(I + \gamma\rho(\cdot)\mathbb{B} \cap N^P(C; \cdot))^{-1}.$$

Because of the single valuedness of $(I + \gamma\rho(\cdot)\mathbb{B}_H \cap N^P(C; \cdot))^{-1}(\cdot)$ on its domain and of Lemma 15.5, it follows that the mapping P_C is well defined on $O^\gamma_{\rho(\cdot)}(C)$,

$$\{P_C(u)\} = \mathrm{Proj}_C(u) = (I + \gamma\rho(\cdot)\mathbb{B}_H \cap N^P(C; \cdot))^{-1}(u) \quad \text{for all } u \in O^\gamma_{\rho(\cdot)}(C),$$

and by (15.5)

$$\|P_C(u) - P_C(u')\| \le (1-\gamma)^{-1}\|u - u'\| \quad \text{for all } u, u' \in O^\gamma_{\rho(\cdot)}(C),$$

that is, P_C is Lipschitz continuous on $O^\gamma_{\rho(\cdot)}(C)$ with $(1-\gamma)^{-1}$ as a Lipschitz constant. All the properties in (f) are then established.

To obtain (h), we see that for $u_i \in O_{\rho(\cdot)}(C)$ with $i = 1, 2$ we have by (f) that $P_C(u_i)$ exists. Then the inequality of (h) easily follows from (d) in Theorem 15.6 by taking $y_i = P_C(u_i)$ therein.

(g) Take $u_i \in O^\gamma_{\rho(\cdot)}(C)$, with $i = 1, 2$. Since (f) above holds, $P_C(u_i)$ exist, and hence $\|u_i - P_C(u_i)\| < \gamma\rho(P_C(u_i))$ according to the definition of $O^\gamma_{\rho(\cdot)}(C)$. Therefore by (d) in Theorem 15.6 we have

$$\langle P_C(u_1) - P_C(u_2), u_1 - u_2\rangle \ge (1-\gamma)\|P_C(u_1) - P_C(u_2)\|^2.$$

This yields

$$\begin{aligned} &\|(I - P_C)(u_1) - (I - P_C)(u_2)\|^2 \\ &= \|u_1 - u_2\|^2 + \|P_C(u_1) - P_C(u_2)\|^2 - 2\langle u_1 - u_2, P_C(u_1) - P_C(u_2)\rangle \\ &\le \|u_1 - u_2\|^2 + (2\gamma - 1)\|P_C(u_1) - P_C(u_2)\|^2, \end{aligned}$$

which translates the property (g). So, the proof of the theorem is complete. □

Before going further we will establish the openness property of the sets $O_{\rho(\cdot)}(C)$ and $O^\gamma_{\rho(\cdot)}(C)$. It will also be of great importance for the sequel to compare the two latter sets with the similar sets

$$U_{\rho(\cdot)}(C) = \left\{ u \in H : \liminf_{\substack{\mathrm{Dom}\,\mathrm{Proj}_C \ni u' \to u \\ x \in \mathrm{Proj}_C(u')}} \frac{\|x - u'\|}{\rho(x)} < 1 \right\} \tag{15.11}$$

and for positive $\gamma < 1$

$$U^\gamma_{\rho(\cdot)}(C) = \left\{ u \in H : \liminf_{\substack{\mathrm{Dom}\,\mathrm{Proj}_C \ni u' \to u \\ x \in \mathrm{Proj}_C(u')}} \frac{\|x - u'\|}{\rho(x)} < \gamma \right\}. \tag{15.12}$$

Observe that the conditions under which the limit inferior is taken in (15.11) and (15.12) make sense because Lau's theorem (see Theorem 14.29) tells us that $\mathrm{Dom}\,\mathrm{Proj}_C$ is dense in the space H.

As easily seen, one always has

(15.13)
$$O^\gamma_{\rho(\cdot)}(C) \subset U^\gamma_{\rho(\cdot)}(C) \text{ and } \{u \in H : \exists y \in \operatorname{Proj}_C(u),\, d_C(u) < \gamma\rho(y)\} \subset U^\gamma_{\rho(\cdot)}(C),$$

and similar inclusions for $U_{\rho(\cdot)}(C)$. The next proposition shows that the three sets in (15.13) coincide whenever C is $\rho(\cdot)$-prox-regular.

PROPOSITION 15.9. Let C be a nonempty closed set C of a Hilbert space H which is $\rho(\cdot)$-prox regular for a continuous function $\rho : C \to]0, +\infty]$. Then the following properties hold.
(a) One has

$$O_{\rho(\cdot)}(C) = \{u \in H : \exists y \in \operatorname{Proj}_C(u),\, d_C(u) < \rho(y)\} = U_{\rho(\cdot)}(C)$$

and for all positive $\gamma < 1$

$$O^\gamma_{\rho(\cdot)}(C) = \{u \in H : \exists y \in \operatorname{Proj}_C(u),\, d_C(u) < \gamma\rho(y)\} = U^\gamma_{\rho(\cdot)}(C).$$

(b) The sets $O_{\rho(\cdot)}(C)$ and $O^\gamma_{\rho(\cdot)}(C)$ for all positive $\gamma < 1$ are open.
(c) The mapping P_C is well defined on the open set $O_{\rho(\cdot)}(C)$ and locally Lipschitz continuous on this open set.

PROOF. Observe first that for any $u' \in \operatorname{Dom}\operatorname{Proj}_C$, any $y' \in \operatorname{Proj}_C(u')$, and any $z \in C$, writing

$$\|u' - z\|^2 = \|u' - y'\|^2 + \|y' - z\|^2 - 2\langle u' - y', z - y'\rangle,$$

the inclusion $u' - y' \in N^P(C; y')$ and (a) of Theorem 15.6 yield

$$\|u' - z\|^2 \geq \|u' - y'\|^2 + \left(1 - \frac{\|u' - y'\|}{\rho(y')}\right) \|z - y'\|^2. \tag{15.14}$$

Fixing now any positive $\gamma < 1$ and any $u \in O^\gamma_{\rho(\cdot)}(C)$ we know by Theorem 15.8(f) that $y := P_C(u)$ exists and the definition of $O^\gamma_{\rho(\cdot)}(C)$ ensures that $\frac{\|u-y\|}{\rho(y)} < \gamma$, that is, $d_C(u) < \gamma\rho_C(y)$, establishing the inclusion of $O^\gamma_{\rho(\cdot)}(C)$ into the second member of the equalities of the proposition concerning $U^\gamma_{\rho(\cdot)}(C)$. The second member of those equalities being always included in the third member (as observed above in (15.13)), let us show the inclusion of the third member into the first one. Take any u in that third member $U^\gamma_{\rho(\cdot)}(C)$ with $u \notin C$. Choose a positive $\gamma' < \gamma$, a sequence $(u_n)_n$ of $\operatorname{Dom}\operatorname{Proj}_C$ converging to u, and $x_n \in \operatorname{Proj}_C(u_n)$ such that

$$\lim_n \frac{\|u_n - x_n\|}{\rho(x_n)} = \liminf_{\substack{\operatorname{Dom}\operatorname{Proj}_C \ni u' \to u \\ x \in \operatorname{Proj}_C(u')}} \frac{\|x - u'\|}{\rho(x)} < \gamma' < \gamma.$$

Deleting a finite number of n if necessary, we may suppose that $\|u_n - x_n\|/\rho(x_n) < \gamma'$ for all n. By (15.14) we then have for all integers m, n

$$\begin{aligned}
(1 - \gamma')\|x_m - x_n\|^2 &\leq \|u_m - x_n\|^2 - \|u_m - x_m\|^2 \\
&\leq (\|u_m - u_n\| + \|u_n - x_n\|)^2 - \|u_m - x_m\|^2 \\
&= (\|u_m - u_n\| + d_C(u_n))^2 - d_C^2(u_m),
\end{aligned}$$

and hence $\|x_m - x_n\|^2 \to 0$ as $m \to \infty$ and $n \to \infty$. Let $x \in C$ be the limit of the Cauchy sequence $(x_n)_n$. Since $\|u_n - x_n\| \to \|u - x\|$ and $\|u_n - x_n\| = d_C(u_n) \to d_C(u)$, we see that $x \in \operatorname{Proj}_C(u)$ and $\|u - x\|/\rho(x) \leq \gamma'$ according to the continuity of the function $\rho(\cdot)$.

Consider now any sequence $(y_n)_n$ in C with $\|u - y_n\| \to d_C(u)$ as $n \to \infty$. According to (15.14) again we have

$$\left(1 - \frac{\|u - x\|}{\rho(x)}\right) \|x - y_n\|^2 \leq \|u - y_n\|^2 - d_C^2(u).$$

Therefore, $y_n \to x$ (because $\|u - x\|/\rho(x) < 1$) and thanks to the continuity of $\rho(\cdot)$ we obtain

$$\lim_n \frac{\|u - y_n\|}{\rho(y_n)} = \frac{\|u - x\|}{\rho(x)} \leq \gamma'.$$

It then follows that $u \in O^\gamma_{\rho(\cdot)}(C)$ according to (15.2), proving the desired inclusion. The equalities concerning $O^\gamma_{\rho(\cdot)}(C)$ are then established and the ones concerning $O_{\rho(\cdot)}(C)$ are direct consequences of those equalities.

The openness of $O_{\rho(\cdot)}(C)$ being a direct consequence of that of $O^\gamma_{\rho(\cdot)}(C)$ for $0 < \gamma < 1$, it remains to show that $O^\gamma_\rho(C)$ is open for $0 < \gamma < 1$. Fix $\overline{u} \in O^\gamma_{\rho(\cdot)}(C)$. By (f) in Theorem 15.8 the point $\overline{y} := P_C(\overline{u})$ exists, and for some positive $\gamma' < \gamma$ we have $\|\overline{u} - \overline{y}\| < \gamma' \rho(\overline{y})$ according to the definition of $O^\gamma_{\rho(\cdot)}(C)$. By continuity of $\rho(\cdot)$ we may choose some $\eta > 0$ such that

$$\text{(15.15)} \qquad \frac{\|u - z\|}{\rho(y)} < \gamma' \quad \text{for all } u \in B(\overline{u}, \eta),\ z \in B(\overline{y}, \eta),\ y \in C \cap B(\overline{y}, \eta).$$

Choose a positive $\varepsilon < \eta$ such that for all $u \in B(\overline{u}, \varepsilon)$

$$\left(1 - \frac{\|\overline{u} - \overline{y}\|}{\rho(\overline{y})}\right)^{-1} \left((d_C(u) + \varepsilon)^2 - d_C^2(\overline{u})\right) < \eta^2.$$

Fix any $u \in B(\overline{u}, \varepsilon)$ and consider a sequence $(y_n)_n$ in C such that $\lim\limits_n \|y_n - u\| = d_C(u)$ and

$$\text{(15.16)} \qquad \limsup_{\substack{\|y - u\| \to d_C(u) \\ y \in C}} \frac{\|y - u\|}{\rho(y)} = \lim_n \frac{\|y_n - u\|}{\rho(y_n)}.$$

For each n, the equality

$$\|y_n - \overline{y}\|^2 = \|y_n - \overline{u}\|^2 - \|\overline{u} - \overline{y}\|^2 + 2\langle \overline{u} - \overline{y}, y_n - \overline{y}\rangle$$

and the inclusion $\overline{u} - \overline{y} \in N^P(C; \overline{y})$ combined with (a) of Theorem 15.6 ensure that

$$\|y_n - \overline{y}\|^2 \leq \|y_n - \overline{u}\|^2 - \|\overline{u} - \overline{y}\|^2 + \frac{\|\overline{u} - \overline{y}\|}{\rho(\overline{y})} \|y_n - \overline{y}\|^2,$$

which implies

$$\left(1 - \frac{\|\overline{u} - \overline{y}\|}{\rho(\overline{y})}\right) \|y_n - \overline{y}\|^2 \leq (\|y_n - u\| + \|u - \overline{u}\|)^2 - \|\overline{u} - \overline{y}\|^2.$$

The second member of the latter inequality converging to $(d_C(u) + \|u - \overline{u}\|)^2 - d_C^2(\overline{u})$ as $n \to \infty$, we obtain for n large enough that $\|y_n - \overline{y}\| < \eta$. Therefore, for n sufficiently large, we have by (15.15) that $\|u - y_n\|/\rho(y_n) < \gamma'$, and hence $\lim\limits_n \frac{\|y_n - u\|}{\rho(y_n)} \leq \gamma' < \gamma$. According to (15.16) and to (15.2), we see that $u \in O^\gamma_{\rho(\cdot)}(C)$, and hence $B(\overline{u}, \varepsilon) \subset O^\gamma_{\rho(\cdot)}(C)$. This concludes the openness of $O^\gamma_{\rho(\cdot)}(C)$.

Finally, the local Lipschitz continuity of P_C is a direct consequence of (f) in Theorem 15.8 and of the openness of $O^\gamma_{\rho(\cdot)}(C)$ for any positive $\gamma < 1$. $\square$

In the definition of prox-regularity as well as in its characterizations (a) – (d) in Theorem 15.6 only points inside the closed set C are considered. The next proposition provides two additional characterizations taking into account points which are outside C. Lemma 15.11 below will be needed. Let us observe first the following feature of the distance function which is very often useful.

LEMMA 15.10. *Let C be a set of a normed space $(X, \|\cdot\|)$ and let $\varphi : [0, \tau[\to \mathbb{R}$ be a derivable function at 0 with $\varphi'(0) = 0 = \varphi(0)$. Then, the function $\varphi \circ d_C$ is Fréchet differentiable at any $u \in C$ with zero as derivative at u; in particular the property holds true with the function d_C^p, for any real $p > 1$.*

PROOF. Write $\varphi(t) = t\varepsilon(t)$ with $\varepsilon(t) \to 0$ as $[0, \tau[\ni t \to 0$. Fix $u \in C$ and note that $\varphi \circ d_C$ is well defined near u. Then for every non-zero $h \in X$ small enough, we obtain with $\eta(h) := \varepsilon(d_C(u + h))$

$$0 \le \left| \|h\|^{-1} \big(\varphi(d_C(u + h)) - \varphi(d_C(u))\big) \right| = \|h\|^{-1} d_C(u + h) |\eta(h)| \le |\eta(h)|,$$

which justifies the Fréchet differentiability result. □

LEMMA 15.11. *Let u be any point of a Hilbert space H and C be any nonempty closed subset of H.*
(a) One has the equalities (even if C is not closed)

$$d_C(u)\partial_P d_C(u) = \partial_P \left(\frac{1}{2} d_C^2\right)(u) \text{ and } d_C(u)\partial_F d_C(u) = \partial_F \left(\frac{1}{2} d_C^2\right)(u) \text{ for all } u \in H.$$

(b) If d_C is ∂_P-subdifferentiable at $u \in H$, then $P_C(u)$ exists and

$$d_C(u)\partial_P d_C(u) = \partial_P \left(\frac{1}{2} d_C^2\right)(u) = \{u - P_C(u)\}.$$

(c) If d_C is ∂_F-subdifferentiable at $u \in H$, then $P_C(u)$ exists and

$$d_C(u)\partial_F d_C(u) = \partial_F \left(\frac{1}{2} d_C^2\right)(u) = \{u - P_C(u)\}.$$

(d) The function d_C is Fréchet differentiable at a point $u \in H \setminus C$ if and only if it is ∂_F-subdifferentiable at that point. In that case, $P_C(u)$ exists and

$$\nabla^F d_C(u) = \frac{1}{d_C(u)}(u - P_C(u)).$$

(e) If the square distance function d_C^2 is Fréchet differentiable at a point $u \in H$, then $P_C(u)$ exists and $\nabla^F(\frac{1}{2} d_C^2)(u) = u - P_C(u)$.

PROOF. Suppose that the left member of the first relation of (a) is nonempty and fix any $v \in \partial_P d_C(u)$. By definition of proximal subgradient there exist some real number $\sigma > 0$ and some neighborhood U of u such that for all $y \in U$

$$\langle v, y - u \rangle \le d_C(y) - d_C(u) + \sigma \|y - u\|^2,$$

and hence

$$\langle d_C(u)v, y - u \rangle \le d_C(u)d_C(y) - d_C^2(u) + \sigma d_C(u)\|y - u\|^2.$$

Observing that the second member of the latter inequality is equal to

$$\frac{1}{2} d_C^2(y) - \frac{1}{2} d_C^2(u) - \frac{1}{2}(d_C(y) - d_C(u))^2 + \sigma d_C(u)\|y - u\|^2,$$

we obtain for all $y \in U$

$$\langle d_C(u)v, y-u\rangle \le \frac{1}{2}d_C^2(y) - \frac{1}{2}d_C^2(u) + \sigma d_C(u)\|y-u\|^2,$$

which says that $d_C(u)v \in \partial_P(\frac{1}{2}d_C^2)(u)$. So the inclusion of the first member of (a) into the second one is established.

Take now any $v \in \partial_P(\frac{1}{2}d_C^2)(u)$. There exists some real number $\sigma > 0$ and some neighborhood U of u such that for all $y \in U$

$$\langle v, y-u\rangle \le \frac{1}{2}d_C^2(y) - \frac{1}{2}d_C^2(u) + \sigma\|y-u\|^2,$$

which yields for $\gamma := d_C(u)$

$$\begin{aligned}\langle v, y-u\rangle &\le \frac{1}{2}d_C^2(y) - \frac{1}{2}d_C^2(u) + \sigma\|y-u\|^2 \\ &= \gamma d_C(y) - \gamma d_C(u) + \frac{1}{2}(d_C(y) - d_C(u))^2 + \sigma\|y-u\|^2 \\ &\le \gamma d_C(y) - \gamma d_C(u) + \frac{1+2\sigma}{2}\|y-u\|^2. \end{aligned} \tag{15.17}$$

So, $\partial_P(\frac{1}{2}d_C^2)(u) \subset \partial_P(\gamma d_C)(u)$. Further, one has $\partial_P(\gamma d_C)(u) = \gamma\partial_P d_C(u)$ if $\gamma > 0$, and this equality still holds if $\gamma = 0$ since in this case $\partial_P d_C(u) \neq \emptyset$ (because $0 \in \partial_P d_C(u)$ whenever $d_C(u) = 0$, as seen from the definition of $\partial_P d_C(u)$). The first equality of the assertion (a) is then established, and the second one is obtained in a similar way.

(b). Assume that $\partial_P d_C(u)$ is nonempty, so $\partial_P(\frac{1}{2}d_C^2)(u)$ is also nonempty by (a). Fix any $v \in \partial_P(\frac{1}{2}d_C^2)(u)$. By the non emptiness assumption of $\partial_P d_C(u)$, Theorem 14.2 ensures that $\mathrm{Proj}_C(u) \neq \emptyset$ (since the case $u \in C$ is trivial). Choosing some $\sigma > 0$ and some neighborhood U of u such that for all $y \in U$

$$\langle v, y-u\rangle \le \frac{1}{2}d_C^2(y) - \frac{1}{2}d_C^2(u) + \sigma\|y-u\|^2,$$

we see that for each $w \in \mathrm{Proj}_C(u)$

$$\langle v, y-u\rangle \le \frac{1}{2}\|y-w\|^2 - \frac{1}{2}\|u-w\|^2 + \sigma\|y-u\|^2$$

for all $y \in U$, and hence $v \in \partial_P(\frac{1}{2}\|\cdot - w\|^2)(u) = \{u - w\}$ (the equality being due to the C^2 property of $\frac{1}{2}\|\cdot - w\|^2$, see Proposition 4.142(b)). This yields that the set $u - \mathrm{Proj}_C(u)$ is a singleton which coincides with $\partial_P(\frac{1}{2}d_C^2)(u)$. Therefore, taking into account the first equality of (a), we obtain that $P_C(u)$ exists and

$$d_C(u)\partial_P d_C(u) = \partial_P\left(\frac{1}{2}d_C^2\right)(u) = \{u - P_C(u)\}$$

whenever $\partial_P d_C(u) \neq \emptyset$.

(c). The proof of (c) is similar because, as said above, Theorem 14.2 also tells us that $\mathrm{Proj}_C(u)$ is nonempty whenever $\partial_F d_C(u) \neq \emptyset$.

(d) and (e). Fix any $u \in H \setminus C$. Obviously, d_C is ∂_F-subdifferentiable at u whenever it is Fréchet differentiable at u. Suppose now that $\partial_F d_C(u) \neq \emptyset$. To obtain the Fréchet differentiability of d_C at u, it suffices to prove the Fréchet differentiability of $f := \frac{1}{2}d_C^2$ at u. By (c) we know that $P_C(u)$ exists and that, for

$v := u - P_C(u)$, we have $\partial_F f(u) = \{v\}$. Fix any $\varepsilon > 0$. By definition of Fréchet subgradient, there exists some $\delta > 0$ such that for any $t \in]0, \delta[$ and any $y \in \mathbb{B}_H$

$$\langle v, ty \rangle \le f(u + ty) - f(u) + \varepsilon t,$$

that is,

$$t^{-1}[f(u + ty) - f(u)] - \langle v, y \rangle \ge -\varepsilon. \tag{15.18}$$

On the other hand, taking the above δ satisfying $\delta < 2\varepsilon$ we also have for all $y \in \mathbb{B}_H$ and $t \in]0, \delta[$

$$\begin{aligned} f(u + ty) - f(u) &\le \frac{1}{2}\|u + ty - P_C(u)\|^2 - \frac{1}{2}\|u - P_C(u)\|^2 \\ &= \langle u - P_C(u), ty \rangle + \frac{1}{2} t^2 \|y\|^2, \end{aligned}$$

and hence

$$t^{-1}[f(u + ty) - f(u)] - \langle v, y \rangle \le \varepsilon.$$

Combining the latter with (15.18) yields the Fréchet differentiability of f at u with $\nabla^F f(u) = v$.

Finally, Lemma 15.10 says that at any point $u \in C$ the function f is Fréchet differentiable with $\nabla^F f(u) = 0$, that is, the equality $\nabla^F f(u) = u - P_C(u)$ still holds. The proof of the lemma is then complete. $\square$

PROPOSITION 15.12. Let C be a nonempty closed subset of a Hilbert space H and $\rho(\cdot) : C \to]0, +\infty]$ be a lower semicontinuous function. Anyone of the following properties can be added to the list of equivalences (a) – (e) in Theorem 15.6:

(α) For any $u \in \operatorname{Dom} \partial_P d_C$ and any $v \in \partial_P d_C(u)$ one has

$$d_C(u) + \langle v, x - u \rangle \le \frac{1}{2\rho(P_C(u))}\|x - P_C(u)\|^2 \quad \forall x \in C;$$

(β) for any $u \in \operatorname{Dom} \partial_P d_C$ and any $v \in \partial_P d_C(u)$ one has

$$\langle v, x - u \rangle \le \frac{1}{2\rho(P_C(u))}\|x - P_C(u)\|^2 \quad \forall x \in C;$$

(γ) for any $x \in C$ and any $v \in N^P(C; x) \cap \mathbb{B}_H$ one has

$$\langle v, u - x \rangle \le d_C(u) + \frac{1}{2\rho(x)}\|P_C(u) - x\|^2 \quad \forall u \in \operatorname{Dom} P_C;$$

(δ) for any $x \in C$ and any $v \in N^P(C; x) \cap \mathbb{B}_H$ one has

$$\langle v, u - x \rangle \le d_C(u) + \frac{1}{2\rho(x)}\big(d_C(u) + \|u - x\|\big)^2 \quad \forall u \in H.$$

PROOF. Recalling by Proposition 4.153 that $\partial_P d_C(u) = N^P(C; u) \cap \mathbb{B}_H$ whenever $u \in C$, we see that (a) of Theorem 15.6 holds under anyone of the assumptions of (α), (β), (γ), (δ).

Now assume that (a) in Theorem 15.6 holds. On the one hand, fix any $u \in \operatorname{Dom} \partial_P d_C$ and $v \in \partial_P d_C(u)$. By (b) in Lemma 15.11 we have that $P_C(u)$ exists. If we suppose $u \notin C$, by (b) in Lemma 15.11 again we have $v = \frac{1}{\|u - P_C(u)\|}(u - P_C(u))$, and hence

$$d_C(u) + \langle v, x - u \rangle = \langle v, u - P_C(u) \rangle + \langle v, x - u \rangle = \langle v, x - P_C(u) \rangle \quad \forall x \in H.$$

The same equality $v = \frac{1}{\|u-P_C(u)\|}(u - P_C(u))$ saying that $v \in N^P(C; P_C(u)) \cap \mathbb{B}_H$, it follows according to (a) of Theorem 15.6 that, for all $x \in C$,

$$d_C(u) + \langle v, x - u\rangle \le \frac{1}{2\rho(P_C(u))}\|x - P_C(u)\|^2,$$

that is, (a) in Theorem 15.6 implies (α) holds. Further, (α) obviously implies (β).

On the other hand, fix in a similar way any $x \in C$ and $v \in N^P(C; x) \cap \mathbb{B}_H$. Then for any $u \in \operatorname{Dom} P_C$ we have by (a) in Theorem 15.6 and the inequality $\|v\| \le 1$ that

$$\langle v, u - x\rangle = \langle v, u - P_C(u)\rangle + \langle v, P_C(u) - x\rangle \le d_C(u) + \frac{1}{2\rho(x)}\|P_C(u) - x\|^2,$$

which ensures (γ).

Finally, suppose that (γ) is fulfilled and fix any $x \in C$ and $v \in N^P(C; x) \cap \mathbb{B}_H$. For every $u \in \operatorname{Dom} P_C$, writing

$$\|P_C(u) - x\| \le \|P_C(u) - u\| + \|u - x\| = d_C(u) + \|x - u\|$$

we obtain by (γ) that

$$\langle v, u - x\rangle \le d_C(u) + \frac{1}{2\rho(x)}\big(d_C(u) + \|u - x\|\big)^2,$$

and this inequality still holds for any $u \in H$ by continuity because $\operatorname{Dom} P_C$ is dense in H according to Lau's theorem (see Theorem 14.29). So, the implication (γ) $\Rightarrow$ (δ) holds true.

So, each one of (α), (β), (γ), (δ) is equivalent to (a) in Theorem 15.6. $\square$

Before stating other characterizations (in terms of the distance function) of $\rho(\cdot)$-prox regularity, let us observe the following tangential and normal regularities of such sets. Furthermore, we also show for such sets that the graph of the normal cone $N^P(C; \cdot)$ is sequentially $\|\cdot\| \times w(H, H)$-closed. This is a significant property since Proposition 4.74(b) provides a closed set in $l^2(\mathbb{N})$ whose limiting normal cone at some point is not weakly sequentially closed and even not strongly closed.

PROPOSITION 15.13. Let C be a nonempty closed set in a Hilbert space H and let $\rho(\cdot) : C \to]0, +\infty]$ be a lower semicontinuous function. The following assertions hold.

(a) If C is $\rho(\cdot)$-prox-regular, then the graph of the normal cone $N^P(C; \cdot)$ is strongly$\times$ weakly sequentially closed.

(b) If C is $\rho(\cdot)$-prox-regular, then the normal regularity

$$N^P(C; x) = N^F(C; x) = N^L(C; x) = N^A(C; x) = N^C(C; x) \quad \text{for all } x \in C$$

is fulfilled along with the tangential regularity

$$T^B(C; x) = T^C(C; x) \quad \text{for all } x \in C.$$

(c) The set C is $\rho(\cdot)$-prox-regular if and only if anyone of the properties in (a)$-$(d) of Theorem 15.6 holds with anyone of the normal cones $N^F(C; \cdot)$, $N^L(C; \cdot)$, $N^A(C; \cdot)$, $N^C(C; \cdot)$ in place of $N^P(C; \cdot)$.

PROOF. Suppose that C is $\rho(\cdot)$-prox-regular and fix any $x, x' \in C$. Take any sequences $(x_n)_n$ in C converging to x and $(v_n)_n$ converging weakly to v with v_n in $N^P(C; x_n)$. Then, by (b) of Theorem 15.6 we have for $\beta := \sup_{n\in\mathbb{N}} \|v_n\| < +\infty$

$$\langle v_n, x' - x_n\rangle \leq \frac{\|v_n\|}{2\rho(x_n)}\|x' - x_n\|^2 \leq \frac{\beta}{2\rho(x_n)}\|x' - x_n\|^2,$$

and hence by the lower semicontinuity of $\rho(\cdot)$ we obtain that

$$\langle v, x' - x\rangle \leq \frac{\beta}{2\rho(x)}\|x' - x\|^2.$$

This implies that $v \in N^P(C; x)$, so the assertion (a) is established.

Recall now that for any $x \in C$

$$N^P(C; x) \subset N^F(C; x) \subset N^L(C; x) = N^A(C; \cdot) \subset N^C(C; x), \tag{15.19}$$

where the latter equality is due to the reflexivity of H (see Theorem 5.68(a)). Suppose that C is $\rho(\cdot)$-prox-regular. The arguments establishing (a) above along with Theorem 4.156(d) show that $N^L(C; x) \subset N^P(C; x)$. This and (15.19) tell us that

$$N^P(C; x) = N^F(C; x) = N^L(C; x).$$

The convexity and closedness of $N^F(C; x)$ (see Proposition 4.8(b)) entails that $N^L(C; x)$ is closed and convex, and hence it is also equal to $N^C(C; x)$ thanks to the equality $N^C(C; x) = \overline{\mathrm{co}} N^L(C; x)$ according to Theorem 4.120. We have then proved the equalities of the assertion (b) concerning the normal cones.

Further, the equality $N^F(C; x) = N^C(C; x)$ ensures that

$$(N^F(C; x))^\circ = (N^C(C; x))^\circ,$$

thus $T^B(C; x) \subset T^C(C; x)$ because of the equality $(N^C(C; x))^\circ = T^C(C; x)$ following from the definition of the Clarke normal cone and because of the relation $T^B(C; x) \subset \big(N^F(C; x)\big)^\circ$ which is due to the inclusion $N^F(C; x) \subset \big(T^B(C; x)\big)^\circ$ observed in Proposition 4.8 (and which is not difficult to check). Therefore, the tangential regularity $T^B(C; x) = T^C(C; x)$ for all $x \in C$ is fulfilled because the inclusion $T^C(C; x) \subset T^B(C; x)$ always holds.

It remains to establish (c). By (15.19) it is obvious that anyone of the properties in (a) – (d) of Theorem 15.6 holds with $N^P(C; \cdot)$ whenever it is so for $N^F(C; \cdot)$, $N^L(C; \cdot)$, $N^A(C; \cdot)$ or $N^C(C; \cdot)$, that is, the implication $\Leftarrow$ is true according to (a)– (d) of Theorem 15.6 again. Finally, the reverse implication is a direct consequence of the assertion (b) above of the proposition. □

The above regularity result allows us to give a characterization of prox-regularity through the Shapiro second order contact property with the Bouligand-Peano tangent cone (see Definition 8.78 for the Shapiro contact property).

The tangential characterization holds both with the Bouligand-Peano tangent cone and the Clarke tangent cone. In the proof we will strongly use Lemma 2.249 (recalled also in (8.26)).

THEOREM 15.14 (Shapiro's second order contact property). Let C be a nonempty closed set of a Hilbert space H and let $\rho(\cdot) : C \to]0, +\infty]$ be a lower semicontinuous function. The following assertions are pairwise equivalent:

(a) The set C is $\rho(\cdot)$-prox-regular;
(b) for all $x, x' \in C$

$$d(x'-x, T^C(C;x)) \le \frac{1}{2\rho(x)}\|x'-x\|^2;$$

(c) for all $x, x' \in C$

$$d(x'-x, T^B(C;x)) \le \frac{1}{2\rho(x)}\|x'-x\|^2.$$

PROOF. Assume first the $\rho(\cdot)$-prox-regularity of C. For $x, x' \in C$, taking as Q the closed convex cone $T^C(C;x)$ in Lemma 2.249 yields

$$d(x'-x, T^C(C;x)) = \sup_{v \in N^C(C;x)\cap \mathbb{B}_H} \langle v, x'-x\rangle.$$

Using the normal regularity $N^P(C;x) = N^C(C;x)$ proved in Proposition 15.13 and using (a) of Theorem 15.6 we obtain

$$d(x'-x, T^C(C;x)) \le \frac{1}{2\rho(x)}\|x'-x\|^2,$$

that is, the implication (a) $\Rightarrow$ (b) is justified.

The second implication (b) $\Rightarrow$ (c) follows directly from the inclusion $T^C(C;x) \subset T^B(C;x)$.

Now assume that the inequality in (c) holds. Fix $x, x' \in C$ and $v \in N^P(C;x) \cap \mathbb{B}_H$. Since $N^P(C;x) \subset N^F(C;x)$, the inclusion $N^F(C;x) \subset \big(T^B(C;x)\big)^\circ$ in (a) of Proposition 4.8 entails $\langle v, h\rangle \le 0$ for all $h \in T^B(C;x)$, hence $v \in K^\circ$ for $K := \overline{\mathrm{co}}\big(T^B(C;x)\big)$ (note also that the latter inclusion can be easily directly verified). Therefore, by Lemma 2.249 again we have that

$$\langle v, x'-x\rangle \le d(x'-x, K) \le d(x'-x, T^B(C;x)) \le \frac{1}{2\rho(x)}\|x'-x\|^2,$$

and this ensures the $\rho(\cdot)$-prox-regularity of C according to (a) of Theorem 15.6, finishing the proof. □

The next proposition measures in some way the lack of convexity of prox-regular sets. It will use the following lemma.

LEMMA 15.15. Let C be a subset of a Hilbert space H and let $x, y \in C$ and $t \in [0,1]$. Assume that for some real $\sigma \ge 0$

$$d_C((1-t)x + ty) - \sigma d_C^2((1-t)x+ty) \le \sigma t(1-t)\|x-y\|^2.$$

Then one has

$$d_C((1-t)x+ty) \le \sigma \min\{t, 1-t\}\|x-y\|^2.$$

PROOF. Put $z_t := (1-t)x + ty$. Observing that

$$d_C^2(z_t) \le \|z_t - x\|^2 = t^2\|y-x\|^2$$

because $x \in C$, we see that

$$\delta_t := d_C^2(z_t) + t(1-t)\|y-x\|^2 \le t^2\|y-x\|^2 + t(1-t)\|y-x\|^2 = t\|y-x\|^2.$$

This and the assumption of the lemma give

$$d_C(z_t) \le \sigma t\|y-x\|^2. \tag{15.20}$$

Using in a similar way that $y \in C$, we also obtain $\delta_t \le (1-t)\|y-x\|^2$, and hence

$$d_C(z_t) \le \sigma(1-t)\|y-x\|^2$$

according to the assumption of the lemma again. Combining this with (15.20) we obtain

$$d_C(z_t) \le \sigma \min\{t, 1-t\}\|y-x\|^2,$$

which is the desired inequality of the lemma. □

PROPOSITION 15.16. Let C be a closed subset of a Hilbert space H and $\rho : C \to]0,+\infty]$ be a continuous function.

(A) If the closed set C of H is $\rho(\cdot)$-prox-regular, then the following hold:
(a) For $z \in (\operatorname{co} C) \cap O_{\rho(\cdot)}(C)$, one has

$$d_C(z) \le \frac{1}{2\rho(P_C(z))} d_C^2(z) + \frac{1}{4\rho(P_C(z))} \sum_{i,j=1}^{n} t_i t_j \left\|x_i - x_j\right\|^2,$$

hence

$$d_C(z) = \|z - P_C(z)\| \le \frac{1}{2\rho(P_C(z))} \sum_{i,j=1}^{n} t_i t_j \left\|x_i - x_j\right\|^2,$$

where $t_i \ge 0$, $\sum_{i=1}^n t_i = 1$, $x_i \in C$, and $z = \sum_{i=1}^n t_i x_i$.
(b) For $n = 2$ the first inequality of (a) means:
For any $x, y \in C$ and $t \in [0,1]$ such that $z_t := (1-t)x + ty \in O_{\rho(\cdot)}(C)$

$$d_C(z_t) - \frac{1}{2\rho(P_C(z_t))} d_C^2(z_t) \le \frac{1}{2\rho(P_C(z_t))} t(1-t)\|y-x\|^2.$$

Concerning the second inequality of (a) for $n = 2$, it can be refined as follows:
(c) For any $x, y \in C$ and $t \in [0,1]$ such that $z_t := (1-t)x + ty \in O_{\rho(\cdot)}(C)$ one has

$$d_C(z_t) \le \frac{1}{2\rho(P_C(z_t))} \min\{t, 1-t\}\|y-x\|^2.$$

In fact, the $\rho(\cdot)$-prox-regularity implies (a) and (a) $\Rightarrow$ (b) $\Rightarrow$ (c).

(B) If there is an open set $V \supset C$ such that P_C is well defined on V and the inequality in (c) is fulfilled for given $x, y \in C$ and $t \in [0,1]$ with $z_t := (1-t)x + ty \in V$, then (d) holds:
(d) There exists a positive $\varepsilon \le 1$ and a function $\alpha(\cdot)$ from $[0,\varepsilon]$ into $]0,+\infty]$ continuous at 0 with $\alpha(0) = \rho(x)$ and such that for all $t \in [0,\varepsilon]$

$$d_C(z_t) \le \frac{1}{2\alpha(t)} \min\{t, 1-t\}\|y-x\|^2.$$

PROOF. (a). Let t_i, x_i and z as in (a). By (f) in Theorem 15.8, $P_C(z)$ exists. Moreover, by (a) in Theorem 15.6 we have for each $i = 1, \dots, n$

$$\langle z - P_C(z), x_i - P_C(z)\rangle \le \frac{1}{2\rho(P_C(z))} \|z - P_C(z)\| \left\|x_i - P_C(z)\right\|^2,$$

so that

$$\left\langle z - P_C(z), \sum_{i=1}^{n} t_i x_i - P_C(z) \right\rangle \le \frac{1}{2\rho(P_C(z))} \|z - P_C(z)\| \sum_{i=1}^{n} t_i \left\|x_i - P_C(z)\right\|^2.$$

Recalling that $z = \sum_{i=1}^n t_i x_i$, we obtain

$$\|z - P_C(z)\| \leq \frac{1}{2\rho(P_C(z))} \sum_{i=1}^n t_i \|x_i - P_C(z)\|^2 . \tag{15.21}$$

Putting $I := \sum_{i=1}^n t_i \|x_i - z\|^2$, an elementary computation taking into account the condition $\sum_{i=1}^n t_i(z - x_i) = 0$ entails, for all $v \in H$,

$$\sum_{i=1}^n t_i \|x_i - v\|^2 = \|z - v\|^2 + I. \tag{15.22}$$

Now we compute I. Taking $v = x_j$ in (15.22), we have

$$\sum_{i=1}^n t_i \|x_i - x_j\|^2 = \|z - x_j\|^2 + I.$$

Thus we obtain both

$$t_j \sum_{i=1}^n t_i \|x_i - x_j\|^2 = t_j \|z - x_j\|^2 + t_j I$$

and

$$\sum_{j=1}^n \sum_{i=1}^n t_j t_i \|x_i - x_j\|^2 = \sum_{j=1}^n t_j \|z - x_j\|^2 + \sum_{j=1}^n t_j I.$$

From this and the equalities $\sum_{j=1}^n t_j = 1$ and $I = \sum_{j=1}^n t_j \|z - x_j\|^2$, we see that

$$I = \frac{1}{2} \sum_{j=1}^n \sum_{i=1}^n t_j t_i \|x_i - x_j\|^2 .$$

Using this expression in (15.22) with $P_C(z)$ in place of v, it follows that

$$\sum_{i=1}^n t_i \|x_i - P_C(z)\|^2 = \|z - P_C(z)\|^2 + \frac{1}{2} \sum_{j=1}^n \sum_{i=1}^n t_j t_i \|x_i - x_j\|^2 .$$

Thus, recalling (15.21), we deduce that

$$\|z - P_C(z)\| \leq \frac{1}{2\rho(P_C(z))} \|z - P_C(z)\|^2 + \frac{\frac{1}{2\rho(P_C(z))}}{2} \sum_{j=1}^n \sum_{i=1}^n t_j t_i \|x_i - x_j\|^2 ,$$

which is obviously equivalent to the first inequality of (a).

Since $\frac{1}{2\rho(P_C(z))} \|z - P_C(z)\| = \frac{1}{2\rho(P_C(z))} d(z, C) < \frac{1}{2}$ (because $z \in O_{\rho(\cdot)}(C)$), the latter inequality entails the second inequality of (a).

(b) and (c). We use notation and assumptions of (b). By the first inequality of (a) we have

$$\|z_t - P_C(z_t)\| \leq \frac{1}{2\rho(P_C(z_t))} (d_C^2(z_t) + t(1-t)\|y - x\|^2), \tag{15.23}$$

which is obviously equivalent to (b). The other implication (b) $\Rightarrow$ (c) follows from Lemma 15.15.

(d). To see (d), note that since $z_t \to x \in V$ as $t \downarrow 0$, the openness of V ensures the existence of some positive $\varepsilon \leq 1$ such that $z_t \in V$ for all $t \in [0, \varepsilon]$. We may put $\alpha(t) := \rho(P_C(z_t))$ for all $t \in [0, \varepsilon]$ and we see that $\alpha(0) = \rho(x)$ and the function $\alpha(\cdot)$

is continuous at 0 thanks to the continuity of P_C at $x \in C$ according to Lemma 15.18 below. The inequality in the assumption of (B) then yields that the inequality in (d) is satisfied with the function $\alpha(\cdot)$. □

REMARK 15.17. Property (b) (resp. (d)) in the above proposition combined with an appropriate additional condition is actually equivalent to the $\rho(\cdot)$-prox-regularity of C, as it will be shown in (q) (resp. (s)) of Theorem 15.22. □

LEMMA 15.18. Let X be a normed space and C be a nonempty subset of X. Then for every sequence $(u_n)_n$ in $\operatorname{Dom} \operatorname{Proj}_C$ converging to some $x \in C$ as $n \to \infty$, one has $x_n \to x$ for any sequence $(x_n)_n$ satisfying $x_n \in \operatorname{Proj}_C(u_n)$.

PROOF. It suffices to note that

$$\|x_n - x\| \le \|x_n - u_n\| + \|u_n - x\| = d_C(u_n) + \|u_n - x\|$$

and $d_C(u_n) \to d_C(x) = 0$. □

Our next step is the characterization of the $\rho(\cdot)$-prox-regularity of the closed set C through the Fréchet differentiability of its square distance function.

We start with the following proposition which has its own interest. Recall first by Proposition 2.74 (or Proposition 3.31) that, for a convex continuous function f which is Gâteaux differentiable on an open set U of H, its Gâteaux gradient $\nabla^G f$ is norm-to-weak continuous on U. This can also be seen by the fact that (for each $h \in H$) both the function $x \mapsto \langle \nabla^G f(x), h\rangle$ and its opposite $x \mapsto \langle \nabla^G f(x), -h\rangle$ are upper semicontinuous on U according to the equality $\langle \nabla^G f(x), h\rangle = \inf_{t>0} t^{-1}\big(f(x+th) - f(x)\big)$.

PROPOSITION 15.19. For a nonempty closed set C and a nonempty open set U of a Hilbert space H the following assertions are equivalent:
(a) the function d_C^2 is Fréchet subregular on U in the sense that $\partial_F(d_C^2)(x) = \partial_C(d_C^2)(x)$ for all $x \in U$;
(b) d_C^2 is Fréchet differentiable on U, or equivalently d_C is Fréchet differentiable on the open set $U \setminus C$;
(c) d_C^2 is of class $\mathcal{C}^1$ on U;
(d) the mapping P_C is well defined on U and norm-to-norm continuous on U;
(e) the mapping P_C is well defined on U and norm-to-weak continuous on U;
(f) d_C^2 is Gâteaux differentiable on U and $\operatorname{Proj}_C(u) \neq \emptyset$ for all $u \in U$;
(g) d_C is Gâteaux differentiable on $U \setminus C$ and $\|\nabla^G d_C(u)\| = 1$ for all $u \in U \setminus C$.

If C is weakly closed (more generally if any weak cluster point of $\operatorname{Proj}_C(U)$ belongs to C), then one can add the following property to the list of equivalences:
(h) The mapping P_C is well defined on U.

Under anyone of the above conditions one has

$$\nabla\left(\frac{1}{2}d_C^2\right)(x) = x - P_C(x) \quad \text{for every } x \in U.$$

PROOF. We note by Lemma 15.10 that the Fréchet differentiability of d_C^2 on U is equivalent to that of d_C on $U \setminus C$. So, the implication (a) ⇒ (b) follows from (e) in Lemma 15.11 and from the non-vacuity of the Clarke subdifferential of the locally Lipschitz function d_C^2. Suppose now that (e) holds. For any $x \in U$ and any sequence $(x_n)_n$ of U with $\|x_n - x\| \to 0$, we have that $P_C(x_n) \xrightarrow[n]{w} P_C(x)$ and

$$\|x_n - P_C(x_n)\| = d_C(x_n) \xrightarrow[n]{} d_C(x) = \|x - P_C(x)\|,$$

and hence $P_C(x_n) \xrightarrow[n]{\|\cdot\|} P_C(x)$. So, (e) $\Rightarrow$ (d), that is, (e) $\Leftrightarrow$ (d) because the converse implication is obvious.

Suppose now that (b) holds. We know by (e) in Lemma 15.11 that P_C is well defined on U and that $\nabla^F d_C^2(u) = 2(u - P_C(u))$ for all $u \in U$. Observe that

$$-d_C^2(x) = \sup_{y \in C}(-\|x - y\|^2) = -\|x\|^2 + \sup_{y \in C}(2\langle x, y\rangle - \|y\|^2) \quad \text{for all } x \in H \tag{15.24}$$

and that, for $f(x) := \sup_{y \in C}(2\langle x, y\rangle - \|y\|^2)$, the function f is convex and continuous on H (the continuity being due to the equality between the first and third member of (15.24)). Further, the Fréchet differentiability of d_C^2 on U entails that of f. The function f being convex continuous and Fréchet differentiable on U, we obtain that its derivative is norm-to-weak continuous on U according to the result justified before the statement of the proposition. So, P_C is norm-to-weak continuous on U, and hence (e) holds. Moreover, by the above equivalence between (d) and (e) we have the norm-to-norm continuity of P_C on U, which ensures the norm-to-norm continuity of $\nabla^F d_C^2$ on U, and this clearly entails (a). Thus, we have established the implications (b) $\Rightarrow$ (c), (b) $\Rightarrow$ (d), and (b) $\Rightarrow$ (a).

Since the implication (c) $\Rightarrow$ (b) is obvious, let us prove (d) $\Rightarrow$ (c). We know by (e) in Lemma 15.11 that at each point $u \in H$ where d_C^2 is Fréchet subdifferentiable we have $\partial_F(d_C^2)(u) = \{2(u - P_C(u))\}$. This and the equality $\partial_C f(x) = \overline{\mathrm{co}}^*\big({}^{\mathrm{seq}}\operatorname{Lim\,sup}_{u \to x} \partial_F f(u)\big)$ for locally Lipschitz function f (as seen in Theorem 4.120(c)), guarantee under (d) (that is, under the condition P_C is well defined and continuous on U) that $\partial_C(d_C^2)(x) = \{2(x - P_C(x))\}$ for all $x \in U$. This says in particular that d_C^2 is Gâteaux differentiable on U with $\nabla^G(d_C^2)(x) = 2(x - P_C(x))$ for all $x \in U$. The norm-to-norm continuity of P_C again yields the existence of $\nabla^F(d_C^2)$ along with its norm-to-norm continuity on U. Consequently, the implication $(d) \Rightarrow (c)$ is true and the equivalence between the properties (a) – (e) holds.

The property (b) implies (f) since (f) is true under (b) by (c) in Lemma 15.11. Suppose (f) and fix $u \in U$. Choosing any y in the nonempty set $\operatorname{Proj}_C(u)$ and taking any $h \in H$, the Gâteaux differentiability of d_C^2 at u tells us that for some $\varepsilon(t)$ converging to 0 as $t \to 0$ we have

$$\begin{aligned} t\langle \nabla^G d_C^2(u), h\rangle &= d_C^2(u + th) - d_C^2(u) + t\varepsilon(t) \\ &\le \|u + th - y\|^2 - \|u - y\|^2 + t\varepsilon(t) \\ &= 2t\langle u - y, h\rangle + t^2\|h\|^2 + t\varepsilon(t), \end{aligned} \tag{15.25}$$

and hence dividing by $t > 0$ and making $t \downarrow 0$ gives $\langle \nabla^G d_C^2(u), h\rangle \le 2\langle u - y, h\rangle$. This being true for all $h \in H$, it follows that $\nabla^G d_C^2(u) = 2(u - y)$, and hence $\operatorname{Proj}_C(u) = \{y\}$, that is, $P_C(u)$ exists and $\nabla^G d_C^2(u) = 2(u - P_C(u))$. Using the equality $f + d_C^2 = \|\cdot\|^2$ given by (15.24) we see that the convex function f is Gâteaux differentiable on U and $\nabla^G f(u) = 2P_C(u)$. The function f being convex and continuous on U, its Gâteaux derivative on the open set U is norm-to-weak continuous (see the result justified before the statement of the proposition). Consequently, P_C is well defined and norm-to-weak continuous on U, that is, (e) is fulfilled. The equivalence between (b) and (f) is then established.

The equivalence (b) $\Leftrightarrow$ (g) is a direct consequence of Corollary 14.22.

Finally, under the weak closedness of C along with the existence of $P_C(u)$ for all $u \in U$, we have for any net $(u_i)_i$ in U converging strongly to $u \in U$

$$\|u_i - P_C(u_i)\| = d_C(u_i) \to d_C(u) = \|u - P_C(u)\|. \tag{15.26}$$

This tells us in particular that the net $(P_C(u_i))_i$ is eventually bounded, and hence its set of weak cluster points is nonempty. For any weak cluster point y of $(P_C(u_i))_i$ we have by the weak lower semicontinuity of $\|\cdot\|$ and by (15.26) that

$$\|u - y\| \le \|u - P_C(u)\|.$$

Since $y \in C$ according to the weak closedness of C (or to the more general assumption on C), we obtain that $y = P_C(u)$. Consequently, P_C is norm-to-weak continuous on U. Therefore, (h) is equivalent to (e) whenever C is weakly closed. The proof of the proposition is then complete. □

The following crucial lemma will also be used in our characterization with the square distance function.

LEMMA 15.20. Assume that the distance function d_C to a nonempty closed set C of a Hilbert space H is Fréchet differentiable on a neighborhood of a point $\overline{u} \notin C$. Then there exists some $\delta > 0$ such that whenever $u \in B(\overline{u}, \delta)$ and $P_C(u) = x$, there exists some $\tau > 0$ such that the point $u_t := u + t(u - x)$ also has $P_C(u_t) = x$ for every $t \in [0, \tau]$.

PROOF. By Proposition 15.19, we may choose some $\varepsilon > 0$ such that $B(\overline{u}, 2\varepsilon) \subset H \setminus C$ and P_C is well defined and continuous on $B(\overline{u}, 2\varepsilon)$, with d_C Fréchet differentiable on $B(\overline{u}, 2\varepsilon)$. Let $\sigma := \sup\{d_C(u) : u \in B(\overline{u}, \varepsilon)\}$. For all $u \in B(\overline{u}, \varepsilon)$ we have on the one hand

$$d_C(u) \ge d_C(\overline{u}) - \|u - \overline{u}\| > 2\varepsilon - \varepsilon = \varepsilon,$$

hence $\varepsilon < d_C(u) \le \sigma$, and on the other hand, for any positive $t < \varepsilon/\sigma$ and for $u_t := u + t(u - P_C(u))$,

$$\begin{aligned}\|u_t - \overline{u}\| &= \|(u - \overline{u}) + t(u - P_C(u))\| \le \|u - \overline{u}\| + t\|u - P_C(u)\| \\ &\le \|u - \overline{u}\| + t d_C(u) < \varepsilon + (\varepsilon/\sigma)\sigma = 2\varepsilon,\end{aligned}$$

thus $u_t \in B(\overline{u}, 2\varepsilon)$. Fix positive numbers $\delta < \varepsilon$ and $s < \delta/\sigma$ (thus $s < 1$) such that

$$\|P_C(u_s) - P_C(u)\| < d_C(u) \quad \text{for all } u \in B(\overline{u}, \delta),$$

which is possible thanks to the continuity of d_C and P_C since $d_C(u) - \|P_C(u_s) - P_C(u)\| \to d_C(\overline{u}) > 0$ as $s \downarrow 0$ and $u \to \overline{u}$. This inequality ensures that for all $u \in B(\overline{u}, \delta)$ we have

$$\begin{aligned}d_C(u_s) &= \|u_s - P_C(u_s)\| = \|u + s(u - P_C(u)) - P_C(u_s)\| \\ &= \|(1 + s)(u - P_C(u_s)) + s(P_C(u_s) - P_C(u))\| \\ &\ge (1 + s)\|u - P_C(u_s)\| - s\|P_C(u_s) - P_C(u)\| \\ &> (1 + s)d_C(u) - s d_C(u) = d_C(u).\end{aligned}$$

Then for all $u \in B(\overline{u}, \delta)$ we have $d_C(u_s) > d_C(u)$ and also $\delta < d_C(u) < \delta/s$ because $\delta < \varepsilon$ and $s < \delta/\sigma$.

Take now any $u \in B(\overline{u},\delta)$ and consider the closed set $D := \{w \in H : d_C(w) \geq d_C(u_s)\}$. As $u \notin D$, according to Lau's theorem (see Theorem 14.29) there is a sequence $D \not\ni y_n \xrightarrow[n]{\|\cdot\|} u$ with $\mathrm{Proj}\,_D(y_n) \neq \emptyset$. Choosing $w_n \in \mathrm{Proj}\,_D(y_n)$ we have $d_C(w_n) = d_C(u_s)$ (because w_n is a boundary point of D). For all n large enough, say $n \geq N$, we have $w_n \in B(\overline{u},2\delta)$ since $y_n \in B(\overline{u},\delta)$ eventually and

$$(15.27)\quad \|y_n - w_n\| = d_D(y_n) \leq \|y_n - u_s\| \xrightarrow[n]{} \|u - u_s\| = s\|u - P_C(u)\| = s d_C(u) < \delta.$$

Fix any $n \in \mathbb{N}$ with $n \geq N$. It ensues that d_C is Fréchet differentiable at w_n and by (d) in Lemma 15.11

$$\nabla^F d_C(w_n) = (w_n - P_C(w_n))/d_C(w_n) \quad \text{and} \quad \|\nabla^F d_C(w_n)\| = 1.$$

Therefore, by Proposition 2.194 the half-space $E := \{v \in H : \langle -\nabla^F d_C(w_n), v\rangle \leq 0\}$ gives the Bouligand-Peano tangent cone to D at w_n, and since proximal normals must belong to the negative polar of this tangent cone, we obtain that $y_n - w_n \in E^\circ = -[0,+\infty[\nabla^F d_C(w_n)$, that is, there exists some $\lambda_n > 0$ such that $y_n - w_n = -\lambda_n \nabla^F d_C(w_n)$, and hence

$$y_n - w_n = -\lambda_n(w_n - P_C(w_n))/d_C(w_n) \quad \text{and} \quad \lambda_n = \|y_n - w_n\|.$$

We have by (15.27) that $\lambda_n < \delta$, and hence

$$\lambda_n < \delta \leq d_C(u) < d_C(u_s) = d_C(w_n).$$

It follows that for $\alpha_n := \lambda_n/d_C(w_n)$ we have $\alpha_n \in]0,1[$ and $y_n = (1-\alpha_n)w_n + \alpha_n P_C(w_n)$. Hence, $P_C(y_n) = P_C(w_n)$ and by Lemma 4.123

$$\begin{aligned}\lambda_n &= \|w_n - y_n\| = \|w_n - P_C(w_n)\| - \|y_n - P_C(w_n)\| \\ &= d_C(w_n) - d_C(y_n) = d_C(u_s) - d_C(y_n).\end{aligned}$$

Putting $t_n := \frac{\alpha_n}{1-\alpha_n} = \frac{d_C(u_s)-d_C(y_n)}{d_C(y_n)}$, we obtain $w_n = y_n + t_n(y_n - P_C(y_n))$. As $(t_n)_n$ converges to $\tau := (d_C(u_s) - d_C(u))/d_C(u) > 0$, we have $w_n \xrightarrow[n]{} u_\tau$ and $u_\tau \in B(\overline{u},2\delta)$ by (15.27) and by the inclusion $y_n \in B(\overline{u},\delta)$. So, by continuity of P_C over $B(\overline{u},2\delta)$ we get $P_C(w_n) \xrightarrow[n]{\|\cdot\|} P_C(u_\tau)$. But we also have $P_C(w_n) \xrightarrow[n]{\|\cdot\|} P_C(u)$ because $P_C(w_n) = P_C(y_n)$ and $y_n \xrightarrow[n]{} u$. Finally, for this number $\tau > 0$ we have $P_C(u_\tau) = P_C(u)$ and Lemma 4.123 guarantees (or as easily checked) that, for all $t \in [0,\tau]$, we have $P_C(u_t) = P_C(u)$, which finishes the proof. □

The next lemma follows from the preceeding one.

LEMMA 15.21. Let C be a closed set in a Hilbert space H and let $\overline{x} \in C$. If the mapping P_C is well-defined and norm-to-norm continuous on a neighborhood U of $\overline{x}$, then there exists some $\varepsilon > 0$ such that for all $x \in C \cap B(\overline{x},\varepsilon)$ and all nonzero $v \in N^P(C;x)$ the equality $P_C\left(x + \varepsilon\frac{v}{\|v\|}\right) = x$ holds.

PROOF. Let $B(\overline{x},\delta)$ be a ball on which P_C is well-defined and norm-to-norm continuous. Take $0 < \varepsilon < \delta/2$ and $x \in C$ with $\|x - \overline{x}\| < \varepsilon$, and consider any nonzero $v \in N^P(C;x)$. By Proposition 4.132(a), there exists $\lambda > 0$ such that $P_C(x+\lambda v) = x$. Set $\lambda_s := \sup\{\lambda \in]0,\varepsilon] : P_C(x + \lambda\frac{v}{\|v\|}) = x\}$. By the continuity of P_C on $B(\overline{x},\delta)$ we have that $P_C(x + \lambda_s\frac{v}{\|v\|}) = x$. Suppose that $\lambda_s < \varepsilon$. As $x + \lambda_s\frac{v}{\|v\|}$ belongs to the open set $B(\overline{x},\delta) \setminus C$ where d_C is Fréchet differentiable according to Proposition 15.19, by Lemma 15.20 there exists some $\eta > 0$ with $\lambda_s + \eta \leq \varepsilon$ such that $P_C(x + (\lambda_s + \eta)\frac{v}{\|v\|}) = x$. This gives a contradiction with the definition of λ_s. Then we can conclude that $\lambda_s = \varepsilon$. $\square$

The following theorem provides the characterization of the $\rho(\cdot)$-prox-regularity of C through the Fréchet differentiability of its square distance function, together with other fundamental characterizations.The statement and proof of the theorem make use of the sets $U_{\rho(\cdot)}(C)$ and $U^{\gamma}_{\rho(\cdot)}(C)$ in (15.11) and (15.12).

THEOREM 15.22 (various characterizations of $\rho(\cdot)$-prox-regularity). Let C be a nonempty closed set C of a Hilbert space H and let $\rho : C \to]0, +\infty]$ be a continuous function. Then the set C is $\rho(\cdot)$-prox-regular if and only if anyone of the following assertions holds:
(f) for any positive $\gamma < 1$ the set $U^{\gamma}_{\rho(\cdot)}(C)$ is open and the mapping P_C is well-defined on $U^{\gamma}_{\rho(\cdot)}(C)$ and Lipschitz continuous there with $(1-\gamma)^{-1}$ as a Lipschitz constant, that is,

$$\|P_C(u_1) - P_C(u_2)\| \le (1-\gamma)^{-1}\|u_1 - u_2\| \quad \text{for all } u_1, u_2 \in U^{\gamma}_{\rho(\cdot)}(C);$$

(g) the set $U_{\rho(\cdot)}(C)$ is open and coincides with $O_{\rho(\cdot)}(C)$, and the mapping P_C is well-defined on $U_{\rho(\cdot)}(C)$ with, for all $u_1, u_2 \in U_{\rho(\cdot)}(C)$,

$$\|P_C(u_1) - P_C(u_2)\| \le \left(1 - \frac{d_C(u_1)}{2\rho(P_C(u_1))} - \frac{d_C(u_2)}{2\rho(P_C(u_2))}\right)^{-1} \|u_1 - u_2\|;$$

(h) the set $U_{\rho(\cdot)}(C)$ is open and the mapping P_C is well-defined on $U_{\rho(\cdot)}(C)$ and locally Lipschitz continuous on $U_{\rho(\cdot)}(C)$;
(i) the set $U_{\rho(\cdot)}(C)$ is open and the mapping P_C is well-defined on $U_{\rho(\cdot)}(C)$ and norm-to-norm continuous on $U_{\rho(\cdot)}(C)$;
(j) the set $U_{\rho(\cdot)}(C)$ is open and the mapping P_C is well-defined on $U_{\rho(\cdot)}(C)$ and norm-to-weak continuous on $U_{\rho(\cdot)}(C)$;
(k) the set $U_{\rho(\cdot)}(C)$ is open and d_C^2 is of class $\mathcal{C}^{1,1}$ on $U_{\rho(\cdot)}(C)$ and $\nabla d_C^2(u) = 2(u - P_C(u))$ for all $u \in U_{\rho(\cdot)}(C)$;
(l) the set $U_{\rho(\cdot)}(C)$ is open and d_C^2 is of class $\mathcal{C}^1$ on $U_{\rho(\cdot)}(C)$;
(m) the set $U_{\rho(\cdot)}(C)$ is open and d_C^2 is Fréchet differentiable on $U_{\rho(\cdot)}(C)$;
(n) the set $U_{\rho(\cdot)}(C)$ is open, Proj_C takes nonempty values on $U_{\rho(\cdot)}(C)$, and d_C^2 is Gâteaux differentiable on $U_{\rho(\cdot)}(C)$;
(o) the set $U_{\rho(\cdot)}(C)$ is open and d_C is Gâteaux differentiable on $U_{\rho(\cdot)}(C) \setminus C$ with $\nabla d_C(u) = (u - P_C(u))/d_C(u)$ for all $u \in U_{\rho(\cdot)}(C) \setminus C$;
(o') the set $U_{\rho(\cdot)}(C)$ is open and d_C is Gâteaux differentiable on $U_{\rho(\cdot)}(C) \setminus C$ with $\|\nabla d_C(u)\| = 1$ for all $u \in U_{\rho(\cdot)}(C) \setminus C$;
(p) the set $U_{\rho(\cdot)}(C)$ is open and, for any $u \in U_{\rho(\cdot)}(C) \setminus C$ such that $x = P_C(u)$ exists, one has

$$x = P_C\left(x + t\frac{u-x}{\|u-x\|}\right) \quad \text{for all } t \in [0, \rho(x)[;$$

(q) the set $U_{\rho(\cdot)}(C)$ is open, P_C is well-defined on $U_{\rho(\cdot)}(C)$ and, for all $x, y \in C$ and $t \in [0,1]$ such that $z_t := (1-t)x + ty \in U_{\rho(\cdot)}(C)$, one has

$$d_C(z_t) - \frac{1}{2\rho(P_C(z_t))} d_C^2(z_t) \le \frac{1}{2\rho(P_C(z_t))} t(1-t)\|y - x\|^2;$$

(r) the set $U_{\rho(\cdot)}(C)$ is open, P_C is well-defined on $U_{\rho(\cdot)}(C)$ and, for all $x, y \in C$ and $t \in [0,1]$ such that $z_t := (1-t)x + ty \in U_{\rho(\cdot)}(C)$, one has

$$d_C(z_t) \le \frac{1}{2\rho(P_C(z_t))} \min\{t, 1-t\}\|y-x\|^2.$$

(s) tor each pair of points $x, y \in C$, there exist a positive $\varepsilon \le 1$ and a continuous function $\alpha : [0,\varepsilon] \to]0,+\infty]$ with $\alpha(0) = \rho(x)$ and such that, for all $t \in [0,\varepsilon]$,

$$d_C(z_t) \le \frac{1}{2\alpha(t)} \min\{t, 1-t\}\|y-x\|^2,$$

where $z_t := (1-t)x + ty$.

If C is weakly closed (which holds whenever H is finite-dimensional), then one can add the following property to the list of equivalences:
(t) the set $U_{\rho(\cdot)}(C)$ is open and P_C is well-defined therein.

PROOF. Denoting by (A) the assertion that C is $\rho(\cdot)$-prox-regular, the equivalences will be established following the array:

$$\begin{array}{ccccccccccc}
(r) & \Rightarrow & (s) & & & & (n) & & & & \\
\Uparrow & & \Downarrow & & & & \Updownarrow & & & & \\
(q) & \Leftarrow & (A) & \Leftarrow & (p) & \Leftarrow & (m) & \Leftrightarrow & (o) & \Leftrightarrow & (o') \\
 & \swarrow & & \searrow & & & \Updownarrow & & & & \\
(f) & \Rightarrow & (h) & & (g) & & (l) & & & & \\
 & & \Updownarrow & \searrow & \Downarrow & & \Updownarrow & & & & \\
 & & (k) & & (i) & \Leftrightarrow & (j) & & & &
\end{array}$$

Theorem 15.8 and Proposition 15.9 ensure that the $\rho(\cdot)$-prox-regularity of C entails (f) and (g). On the other hand, (f) obviously implies (h) and (h) is equivalent to (k) by Proposition 15.19 and Lemma 15.11. Note also that (h) obviously implies (i). To see that (g) also implies (i), observe first that the assumption $U_{\rho(\cdot)}(C) = O_{\rho(\cdot)}(C)$ and (15.3) ensure that, for all $u \in U_{\rho(\cdot)}(C)$, we have $\|u - P_C(u)\|/\rho(P_C(u)) < 1$, or equivalently $d_C(u)/\rho(P_C(u)) < 1$. So, fixing any $\overline{u} \in U_{\rho(\cdot)}(C)$, the inequality of (g) yields, for all $u \in U_{\rho(\cdot)}(C)$,

$$\|P_C(u) - P_C(\overline{u})\| \le 2\left(1 - \frac{d_C(\overline{u})}{\rho(P_C(\overline{u}))}\right)^{-1} \|u - \overline{u}\|,$$

which entails the continuity of P_C at $\overline{u}$ according to the openness assumption of $U_{\rho(\cdot)}(C)$, that is, the implication (g) $\Rightarrow$ (i) is true. Further, according to Proposition 15.19 and Lemma 15.11 again the assertions (i), (j), (l), (m), (n), (o) and (o') are pairwise equivalent.

Let us show (m) $\Rightarrow$ (p). Assuming (m), by Proposition 15.19 we have that P_C is well-defined on $U_{\rho(\cdot)}(C)$ and continuous there. Take any $u \in U_{\rho(\cdot)}(C) \setminus C$ and put $x := P_C(u)$. Choose some $0 < \gamma < 1$ such that $u \in U^\gamma_{\rho(\cdot)}(C)$. By definition of $U^\gamma_{\rho(\cdot)}(C)$ (see (15.12)) and the fact that $U^\gamma_{\rho(\cdot)}(C)$ is open and P_C is well-defined therein, there exists a sequence $(u_n)_n$ in $U^\gamma_{\rho(\cdot)}(C)$ converging to u such that $\|x_n - u_n\|/\rho(x_n) < \gamma$, where $x_n := P_C(u_n)$. The continuity of P_C on $U_{\rho(\cdot)}(C)$ guarantees that $\|x - u\|/\rho(x) \le \gamma$, hence $d_C(u) < \rho(x)$. Take any $t \in [d_C(u), \rho(x)[$. According to the Fréchet differentiability of d_C^2 on the open neighborhood $U_{\rho(\cdot)}(C)$ of u, we have by Lemma 15.20 the existence of some $\tau_0 > 0$ such that $u_\tau \in U_{\rho(\cdot)}(C)$ and $P_C(u_\tau) = x$ for all nonnegative $\tau < \tau_0$, where $u_\tau := u + \tau(u-x)/\|u-x\|$. This leads to consider the nonempty set of all $\tau \in [0, t - d_C(u)]$ such that $u_\tau \in U_{\rho(\cdot)}(C)$

and $P_C(u_\tau) = x$. Let s be its supremum and $(\tau_n)_n$ be a sequence of that set converging to s. For each n we have $d_C(u_{\tau_n}) = \tau_n + d_C(u)$ and $\|u_{\tau_n} - x\| = d_C(u_{\tau_n})$. Combining this with $u_{\tau_n} \xrightarrow[n]{} u_s$, we obtain $d_C(u_s) = s + d_C(u) \le t < \rho(x)$ and $\|u_s - x\| = d_C(u_s)$, and the latter equality is equivalent to $x \in \operatorname{Proj}_C(u_s)$. Consequently,

$$\liminf_{\substack{\operatorname{Dom}\operatorname{Proj}_C \ni u' \to u_s \\ z \in \operatorname{Proj}_C(u')}} \frac{\|z - u'\|}{\rho(z)} \le \frac{\|x - u_s\|}{\rho(x)} < 1,$$

which entails $u_s \in U_{\rho(\cdot)}(C)$ according to the definition of $U_{\rho(\cdot)}(C)$ (see (15.11)), and hence $x = P_C(u_s)$ since P_C is well-defined on $U_\rho(C)$. Further $u_s \notin C$ since $d_C(u_s) = s + d_C(u) > 0$. The function d_C being Fréchet differentiable on the open set $U_{\rho(\cdot)}(C)$, Lemma 15.20 assures us that $s = t - d_C(u)$ (otherwise, we would get a contradiction). Since $u = x + d_C(u)\frac{u-x}{\|u-x\|}$ (thanks to the equality $d_C(u) = \|u - x\|$), we have

$$u_s = x + (s + d_C(u))(u - x)/\|u - x\| = x + t(u - x)/\|u - x\|,$$

so it results that $x = P_C(x + t(u - x)/\|u - x\|)$ for all $t \in [d_C(u), \rho(x)[$. The same equality being still true for all $t \in [0, d_C(u)]$ according to Lemma 4.123 (or as easily seen), it follows that it holds for all $t \in [0, \rho(x)[$, that is, (p) is fulfilled.

Let us prove that (p) entails the $\rho(\cdot)$-prox-regularity of C. Assume then (p) and fix any $x \in C$ and $v \in N^P(C; x) \cap \mathbb{B}_H$. By Definition 15.1 it is enough to show that $x = P_C(x + tv)$ for all positive $t < \rho(x)$. This is obviously true if $v = 0$. Suppose that v is nonzero and put $w := v/\|v\|$. By Proposition 4.132(a) there exists some $s > 0$ such that $x = P_C(x + sw)$. Since $x \in C \subset U_{\rho(\cdot)}(C)$ and $U_\rho(C)$ is open by assumption, we may (by Lemma 4.123) take $s > 0$ small enough that $x + sw \in U_{\rho(\cdot)}(C)$, and hence $u := x + sw \in U_{\rho(\cdot)}(C) \setminus C$. Then by (p), for all $t \in [0, \rho(x)[$, we have

$$x = P_C\left(x + t\frac{u - x}{\|u - x\|}\right), \quad \text{that is, } x = P_C\left(x + t\frac{v}{\|v\|}\right),$$

and since $t\|v\| < \rho(x)$ (because $\|v\| \le 1$) we deduce that x is also the unique nearest point of $x + t\|v\|\frac{v}{\|v\|} = x + tv$ in C, that is, $x = P_C(x + tv)$ for all positive $t < \rho(x)$. This means that C is $\rho(\cdot)$-prox-regular.

The $\rho(\cdot)$-prox-regularity of C ensures (q) by Proposition 15.9 and by Proposition 15.16(b), and the implication (q) $\Rightarrow$ (r) follows from Lemma 15.15. We also have that (r) entails (s) by Proposition 15.16(B) with $V := U_{\rho(\cdot)}(C)$. Let us prove that (s) entails (a) in Theorem 15.6. Fix any $x, y \in C$ and any $v \in N^P(C; x) \cap \mathbb{B}_H$. Take ε and $\alpha(\cdot)$ as given by (s) and put $z_t := (1 - t)x + ty$ for all $t \in]0, \varepsilon]$. By Proposition 4.153 we know that $v \in \partial_P d_C(x)$, and hence by Proposition 4.137(a) there exists some $\sigma > 0$ and some positive $\varepsilon_0 \le \varepsilon$ such that for all $t \in]0, \varepsilon_0]$

$$\langle v, t(y - x)\rangle = \langle v, z_t - x\rangle \le d_C(z_t) + \sigma\|z_t - x\|^2 = d_C(z_t) + \sigma t^2\|y - x\|^2.$$

Therefore, for all $t \in]0, \varepsilon_0]$ we have according to the assumption (s)

$$\langle v, y - x\rangle \le \frac{1}{2\alpha(t)}\|y - x\|^2 + \sigma t\|y - x\|^2,$$

and making $t \downarrow 0$ we obtain $\langle v, y - x\rangle \le \frac{1}{2\rho(x)}\|y - x\|^2$ because $\lim\limits_{t \downarrow 0} \alpha(t) = \rho(x)$ by the assumption (s). This is the desired property (a) in Theorem 15.6.

Finally, under the weak closedness of C the equivalence between (t) and (j) follows from Proposition 15.19. The proof of the theorem in then complete. □

Instead of the differentiability of the distance function d_C, its Fréchet or proximal subdifferentiability can be used to characterize the $\rho(\cdot)$-prox-regularity of C.

PROPOSITION 15.23. Let C be a nonempty closed set C of a Hilbert space H and let $\rho : C \to]0, +\infty]$ be a continuous function. Then the set C is $\rho(\cdot)$-prox-regular if and only if anyone of the two following properties holds:
(a) $U_{\rho(\cdot)}(C)$ is open and $\partial_P d_C(u) \neq \emptyset$ for all $u \in U_{\rho(\cdot)}(C)$;
(b) $U_{\rho(\cdot)}(C)$ is open and $\partial_F d_C(u) \neq \emptyset$ for all $u \in U_{\rho(\cdot)}(C)$.

PROOF. The $\rho(\cdot)$-prox-regularity of C ensures by (k) in Theorem 15.22 the openness of $U_{\rho(\cdot)}(C)$ along with the $\mathcal{C}^{1,1}$-property of d_C on $U_{\rho(\cdot)}(C) \setminus C$, and the latter entails that $\partial_P d_C(u) = \{Dd_C(u)\}$ for all $u \in U_{\rho(\cdot)}(C) \setminus C$ (as seen in Proposition 4.142), thus in particular d_C is ∂_P-subdifferentiable on $U_{\rho(\cdot)}(C)$ (note that $0 \in \partial_P d_C(x)$ for all $x \in C$). So, the $\rho(\cdot)$-prox-regularity of C implies (a). On the other hand, the implication (a) $\Rightarrow$ (b) is obvious because $\partial_P f(\cdot) \subset \partial_F f(\cdot)$ for any function f. Finally, the fact that (b) entails the $\rho(\cdot)$-prox-regularity of C is a direct consequence of (a) and (e) in Lemma 15.11 and of (m) in Theorem 15.22. □

15.2. Uniform and local prox-regularity

15.2.1. Uniform prox-regularity. Let us consider now the case when $\rho(\cdot)$ is a constant function, that is, $\rho(x) = r$ for all $x \in C$, where $r \in]0, +\infty]$. We will write $O_r(C)$, $U_r(C)$, $O_r^\gamma(C)$, $U_r^\gamma(C)$ for the sets corresponding to the constant function associated with the value $r > 0$.

For any closed set C and for the constant r, clearly the equalities

$$\limsup_{\substack{\|x-u\| \to d_C(u) \\ x \in C}} \frac{\|x-u\|}{\rho(x)} = \frac{d_C(u)}{r} = \liminf_{\substack{\mathrm{Dom\,Proj}_C \ni u' \to u \\ x \in \mathrm{Proj}_C(u')}} \frac{\|x-u'\|}{\rho(x)},$$

hold true, and they obviously ensure that

$$O_r(C) = \{u \in H : d_C(u) < r\} = U_r(C) \tag{15.28}$$

and

$$O_r^\gamma(C) = \{u \in H : d_C(u) < \gamma r\} = U_r^\gamma(C) = U_{\gamma r}(C), \tag{15.29}$$

where $0 < \gamma < 1$. The middle sets in (15.28) and (15.29) being trivially open, the sets $U_r(C)$ and $U_r^\gamma(C)$, for any positive $\gamma < 1$, are then always open in H for any closed set C (at the opposite of $U_{\rho(\cdot)}(C)$ and $U_{\rho(\cdot)}^\gamma(C)$ for which we had to require the $\rho(\cdot)$-prox-regularity for their openness).

DEFINITION 15.24. Given a constant $r \in]0, +\infty]$, a nonempty closed set C in a Hilbert space H will be declared to be *uniformly r-prox-regular*, or just *r-prox-regular*, when it is $\rho(\cdot)$-prox-regular for the constant function $\rho(\cdot) \equiv r$. Thus, this means that, for any $x \in C$, any $v \in N^P(C; x)$ with $\|v\| \leq 1$, any positive real number $t \leq r$, one has

$$x \in \mathrm{Proj}_C(x + tv), \quad \text{or equivalently } C \cap B(x + tv, t) = \emptyset.$$

The closed set C is called *uniformly prox-regular*, when it r-prox-regular for some constant $r \in]0, +\infty]$.

Two first remarks deserve to be noticed.

REMARK 15.25. Let C be a nonempty closed set of a Hilbert space H and let $r \in]0, +\infty]$
(a) If C is r-prox-regular, it is obvious that it r'-prox-regular for any extended real $r' \in]0, r]$.
(b) If $r < +\infty$, then C is r-prox-regular if and only if for any $x \in C$ and any $v \in N^P(C; x)$ with $\|v\| = 1$, one has $x \in \mathrm{Proj}_C(x + rv)$. □

Let us give some simple examples of uniformly (resp. non-uniformly) prox-regular sets.

EXAMPLE 15.26. (a) The sets $\mathbb{N}$ and $\mathbb{Z}$ in $\mathbb{R}$ are clearly r-prox-regular with $r = 1/2$.
(b) The set C in Example 15.3(b) is r-prox-regular; more generally the complement $H \setminus B(a, r)$ of any open r-ball $B(a, r)$ in a Hilbert space is r-prox-regular.
(c) The set C in Example 15.3(c) is r-prox-regular.
(d) An r-prox-regular set and its open r-enlargement as well as the properties (c) and (y) in Theorem 15.28 below are illustrated in Figure 15.3. This set (represented

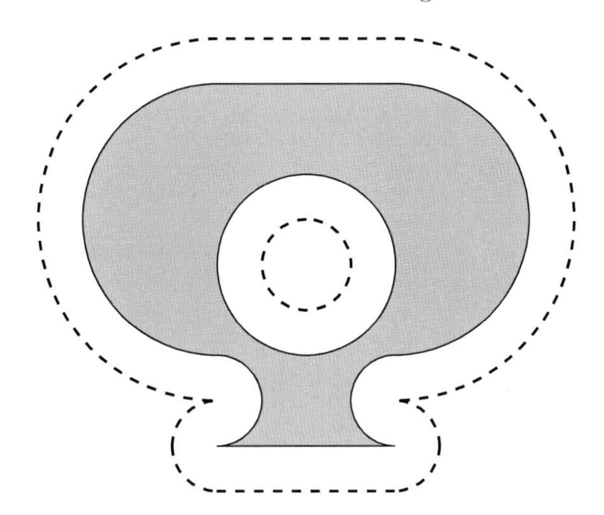

FIGURE 15.3. An r-prox-regular set and its open r-enlargement.

in grey color) looks like a "*tree*".
(e) The set C in Example 15.3(d) is not uniformly prox-regular while it is $\rho(\cdot)$-prox-regular for some continuous function $\rho : C \to]0, +\infty[$. □

Certain characterizations below are formulated as either the hypomonotonicity property or the cocoercivity of a multimapping involving a truncated normal cone. For a normed space X and an extended real $r \in]0, +\infty]$, we recall (see Definition 11.49) that a multimapping $M : X \rightrightarrows X^*$ is said to be $1/r$-*hypomonotone on a subset* U of X provided

$$(15.30)\quad \langle x_1^* - x_2^*, x_1 - x_2 \rangle \geq -\frac{1}{r}\|x_1 - x_2\|^2 \quad \text{for all } x_i \in U \cap \mathrm{Dom}\, M,\ x_i^* \in M(x_i).$$

When $U = X$ one just says that M is $1/r$-hypomonotone. The $1/r$-hypomonotonicity for $r = +\infty$ amounts to the monotonicity of the multimapping M.

While the hypomonotonicity is clearly a weakening of the usual monotonicity, the c-monotonicity or c-strong monotonicity considered in (3.209) is a strengthening of the monotonicity. Here it deserves to be stated in a definition as well as the other concept of c-strong co-monotonicity already utilized just before Lemma 3.274.

DEFINITION 15.27. Given a real $c > 0$, a multimapping $M : X \rightrightarrows X^*$ is *c-strongly monotone* (for short, *c-monotone* or *c-coercive*) on a subset U of a normed space X when

$$\langle x_1^* - x_2^*, x_1 - x_2 \rangle \geq c\|x_1 - x_2\|^2 \quad \text{for all } x_i \in U \cap \operatorname{Dom} M,\ x_i^* \in M(x_i).$$

If instead

$$\langle x_1^* - x_2^*, x_1 - x_2 \rangle \geq c\|x_1^* - x_2^*\|^2 \quad \text{for all } x_i \in U \cap \operatorname{Dom} M,\ x_i^* \in M(x_i),$$

one says that M is *c-strongly co-monotone* (for short, *c-comonotone* (or *c-cocoercive*) on U.

The following theorem provides diverse fundamental characterizations of r-prox-regular sets in a Hilbert space.

THEOREM 15.28 (characterizations of r-prox-regular sets). Let C be a closed subset of a Hilbert space H and $r \in]0, +\infty]$. Then the following assertions are equivalent:
(a) the set C is r-prox-regular;
(b) for any $x, x' \in C$ and $v \in N^P(C; x)$ one has

$$\langle v, x' - x \rangle \leq \frac{1}{2r}\|v\|\,\|x' - x\|^2;$$

(c) for any $x \in C$, any $v \in N^P(C; x) \cap \mathbb{B}_H$, and any non-negative real number $t < r$, one has $x = P_C(x + tv)$;
(d) for any $x_i \in C$, $v_i \in N^P(C; x_i)$ with $i = 1, 2$, one has

$$\langle v_1 - v_2, x_1 - x_2 \rangle \geq -\frac{1}{2}\left(\frac{\|v_1\|}{r} + \frac{\|v_2\|}{r}\right)\|x_1 - x_2\|^2;$$

(e) for any $x_i \in C$, $v_i \in N^P(C; x_i) \cap \mathbb{B}_H$ with $i = 1, 2$, one has

$$\langle v_1 - v_2, x_1 - x_2 \rangle \geq -\frac{1}{r}\|x_1 - x_2\|^2,$$

that is, the multimapping $N^P(C; \cdot) \cap \mathbb{B}_H$ is $1/r$-hypomonotone;
(e') for any (x_i, v_i), $i = 1, 2$, with $x_i \in \operatorname{bdry} C$ and $v_i \in N^P(C; x_i)$ with $\|v_i\| = 1$, one has

$$\langle v_1 - v_2, x_1 - x_2 \rangle \geq -\frac{1}{r}\|x_1 - x_2\|^2;$$

(f) for any $u_i \in H$ and for any $y_i \in (I + N^P(C; \cdot))^{-1}(u_i)$ with $i = 1, 2$, one has

$$\langle y_1 - y_2, u_1 - u_2 \rangle \geq \left(1 - \frac{\|u_1 - y_1\|}{2r} - \frac{\|u_2 - y_2\|}{2r}\right)\|y_1 - y_2\|^2;$$

(g) for any positive $\gamma < 1$, for $u_i \in U_r^\gamma(C)$, and for $y_i \in (I + \gamma r\mathbb{B}_H \cap N^P(C; \cdot))^{-1}(u_i)$ with $i = 1, 2$, one has

$$\langle y_1 - y_2, u_1 - u_2 \rangle \geq (1 - \gamma)\|y_1 - y_2\|^2,$$

that is, the set-valued operator $\left(I + \gamma r\mathbb{B}_H \cap N^P(C; \cdot)\right)^{-1}$ is $(1 - \gamma)$-comonotone on the set $U_r^\gamma(C)$;

(h) For any positive $\gamma < 1$ the mapping P_C is well defined on $U_r^\gamma(C)$ and Lipsfhitz continuous on $U_r^\gamma(C)$ with $(1-\gamma)^{-1}$ as a Lipschitz constant, that is,

$$\|P_C(u_1) - P_C(u_2)\| \le (1-\gamma)^{-1}\|u_1 - u_2\| \quad \text{for all } u_1, u_2 \in U_r^\gamma(C);$$

(i) for any positive $\gamma < 1$ the mapping P_C is well defined on $U_r^\gamma(C)$ and

$$P_C(u) = (I + \gamma r \mathbb{B}_H \cap N^P(C;\cdot))^{-1}(u) \quad \text{for all } u \in U_r^\gamma(C);$$

(j) for any positive $\gamma < 1$ and any $u \in U_r^\gamma(C)$, one has

$$\operatorname{Proj}_C(u) = (I + \gamma r \mathbb{B}_H \cap N^P(C;\cdot))^{-1}(u);$$

(k) the mapping P_C is well defined on $U_r(C)$ with, for all $u_1, u_2 \in U_r(C)$,

$$\|P_C(u_1) - P_C(u_2)\| \le \left(1 - \frac{d_C(u_1)}{2r} - \frac{d_C(u_2)}{2r}\right)^{-1} \|u_1 - u_2\|;$$

(l) the mapping P_C is well defined on $U_r(C)$ and locally Lipschitz continuous there;
(m) the mapping P_C is well defined on $U_r(C)$ and norm-to-norm continuous there;
(n) the mapping P_C is well defined on $U_r(C)$ and norm-to-weak continuous there;
(o) the function d_C^2 is of class $\mathcal{C}^{1,1}$ on $U_r(C)$ and $\nabla d_C^2(u) = 2(u - P_C(u))$ for all $u \in U_r(C)$;
(p) the function d_C^2 is of class $\mathcal{C}^1$ on $U_r(C)$;
(q) the function d_C^2 is Fréchet differentiable on $U_r(C)$, or equivalently d_C is Fréchet differentiable on $U_r(C) \setminus C$;
(r) the multimapping Proj_C takes nonempty values on $U_r(C)$, and d_C^2 is Gâteaux differentiable on $U_r(C)$;
(s) the function d_C is Gâteaux differentiable on $U_r(C) \setminus C$ and $\nabla d_C(u) = (u - P_C(u))/d_C(u)$ for all $u \in U_r(C) \setminus C$;
(s') the function d_C is Gâteaux differentiable on $U_r(C) \setminus C$ and $\|\nabla d_C(u)\| = 1$ for all $u \in U_r(C) \setminus C$;
(t) for any $u \in U_r(C) \setminus C$ such that $x = P_C(u)$ exists, one has

$$x = P_C\left(x + t\frac{u - x}{\|u - x\|}\right) \quad \text{for all } t \in [0, r[;$$

(u) for any $x, y \in C$ and $t \in [0,1]$ such that $z_t := (1-t)x + ty \in U_r(C)$ one has

$$d_C(z_t) - \frac{1}{2r} d_C^2(z_t) \le \frac{1}{2r} t(1-t)\|y - x\|^2;$$

(v) for all $x, y \in C$ and for all $t \in [0,1]$ such that $z_t \in U_r(C)$ (with z_t defined as in (u)) one has

$$d_C(z_t) \le \frac{1}{2r} \min\{t, 1-t\}\|y - x\|^2;$$

(w) $\partial_P d_C(u) \neq \emptyset$ for all $u \in U_r(C)$;
(x) $\partial_F d_C(u) \neq \emptyset$ for all $u \in U_r(C)$.

If C is weakly closed (which holds whenever H is finite-dimensional), then one can add the following to the list of equivalences:
(y) P_C is well defined on $U_r(C)$.

PROOF. The theorem will be established according to the following array:

$$\begin{array}{ccccccccccc}
(x) & \Leftrightarrow & (w) & \Leftrightarrow & (k) & \Leftrightarrow & \cdots & \Leftrightarrow & (t) & & \\
 & & & & \Updownarrow & & & & & & \\
 & & & & (h) & \Leftarrow & & & (g) & \Rightarrow & (i) \\
 & & & & & & & \nearrow & \Uparrow & & \Downarrow \\
 & & & & \Updownarrow & & (e) & & (f) & & (j) \\
 & & & & & \nearrow & & & \Uparrow & \swarrow & \\
 & & & & (d) & \Leftrightarrow & \cdots & \Leftrightarrow & (a) & \Rightarrow & (u) \\
 & & & & & & & & \Updownarrow & \nwarrow & \Downarrow \\
 & & & & & & & & (e') & & (v)
\end{array}$$

By Theorem 15.6 all the properties (a) – (d) are equivalent, and by the comments immediately before the statement of the theorem and by Theorem 15.22 and Proposition 15.23 they are also equivalent to anyone of (h), (k) – (t), (w), and (x); the equivalence between the two properties in (q) is due to Lemma 15.10. The implication (a) $\Rightarrow$ (u) follows from (q) in Theorem 15.22 and the implication (u) $\Rightarrow$ (v) is due to Lemma 15.15. The property (v) also entails (a) by (s) in Theorem 15.22. So, (a), (u), and (v) are equivalent.

The r-prox-regularity of C implies (f) by the assertion (d) in Theorem 15.6, and (f) obviously implies (g). The property (d) trivially entails (e) and (e) implies (g) because, if $r < +\infty$ for $u_i \in U_r^\gamma(C)$ and $y_i \in (I + \gamma r\mathbb{B}_H \cap N^P(C;\cdot))^{-1}(u_i)$ we have $\|(u_i - y_i)/(\gamma r)\| \le 1$ and hence by (e)

$$(\gamma r)^{-1}\langle (u_1 - y_1) - (u_2 - y_2), y_1 - y_2\rangle \ge -r^{-1}\|y_1 - y_2\|^2,$$

which gives (g) (note that the case $r = +\infty$ is obvious). In the proof of Theorem 15.6, the arguments for the implication that (d) in Theorem 15.6 implies (f) in Theorem 15.8 show that (g) implies (h) and (i). Then (d), (e), (f), (g) are each one equivalent to (a) because we already proved that (h) and (a) are equivalent. Taking into account the above implication (g) $\Rightarrow$ (i) and the trivial implication (i) $\Rightarrow$ (j), we now show (j) $\Rightarrow$ (a). Suppose (j) and fix any $x \in C$ and $v \in N^P(C;x) \cap \mathbb{B}_H$. Consider any positive $t < r$ and take $t' \in]t, r[$. For the positive number $\gamma := t'/r < 1$, we have on the one hand $x + tv \in U_r^\gamma(C)$ according to (15.29), and on the other hand $x+tv \in (I+\gamma r\mathbb{B}_H \cap N^P(C;\cdot))(x)$, that is, $x \in (I+\gamma r\mathbb{B}_H \cap N^P(C;\cdot))^{-1}(x+tv)$. It follows from (j) that $x \in \operatorname{Proj}_C(x+tv)$ and this means that (a) holds as desired.

Since (e) obviously entails (e'), it remains to prove (e')$\Rightarrow$(b). Assume the property (e'). Fix any $(\overline{x}, v)$ with $\overline{x} \in \operatorname{bdry} C$ and $v \in N^P(C;\overline{x})$ with $\|v\| = 1$. Put

$$s_0 := \sup\left\{t \in]0, r] : \langle v, x - \overline{x}\rangle \le \frac{1}{2t}\|x - \overline{x}\|^2,\ \forall x \in C\right\},$$

and note that the above set whose supremum is taken is nonempty since $v \in N^P(C;\overline{x})$, so $0 < s_0 \le r$ and $\overline{x} \notin \operatorname{Proj}_C(\overline{x} + tv)$ for any real $t \in]s_0, r]$ if any. We also note that if $s_0 < +\infty$, we have $\overline{x} \in \operatorname{Proj}_C(\overline{x} + s_0 v)$, that is, $d_C(\overline{x} + s_0 v) = s_0$. We claim that $s_0 = r$. By contradiction, suppose $s_0 < r$ and fix a real s such that $s_0 < s < \min\{r, 2s_0\}$. Put $z := \overline{x} + sv$ and $\delta := d_C(z)$. Since $s > s_0$, we have $\overline{x} \notin \operatorname{Proj}_C(\overline{x} + sv)$, or equivalently $\delta = d_C(z) < s$ because $\overline{x} \in C$. Further, writing $\|\overline{x} + s_0 v - z\| = s - s_0 < s_0$, we get that $z \notin C$ since $d_C(\overline{x} + s_0 v) = s_0$ as seen above. By Lau theorem choose a sequence $(z_n)_{n\in\mathbb{N}}$ in $H \setminus C$ converging to z and such that $d_C(z_n) < s$ and $x_n := P_C(z_n)$ exists for every $n \in \mathbb{N}$. For each $n \in \mathbb{N}$ put $v_n := (z_n - x_n)/\|z_n - x_n\|$, $w_n := x_n - \overline{x}$ and $\delta_n := d_C(z_n)$. For each $n \in \mathbb{N}$

the inclusions of the unit vectors v and v_n in $N^P(C;\overline{x})$ and $N^P(C;x_n)$ respectively yield by the assumption (e') that $\langle v_n - v, x_n - \overline{x}\rangle \geq -(1/r)\|x_n - \overline{x}\|^2$, which (by multiplication by $r\delta_n$ and rearrangement) is equivalent to

$$r\langle z_n - z - w_n + (s - \delta_n)v, w_n\rangle + \delta_n\|w_n\|^2 \geq 0. \tag{15.31}$$

Since $x_n \in C$, we also have by definition of s_0 that $\langle v, x_n - \overline{x}\rangle \leq \frac{1}{2s_0}\|x_n - \overline{x}\|^2$, so $\langle v, w_n\rangle \leq \frac{1}{2s_0}\|w_n\|^2 < \frac{1}{s}\|w_n\|^2$. This and (15.31) give for each $n \in \mathbb{N}$

$$r\langle z_n - z, w_n\rangle \geq \frac{\delta_n(r-s)}{s}\|w_n\|^2. \tag{15.32}$$

Write $\|w_n\| = \|\overline{x} - x_n\| \geq \|\overline{x} - z\| - \|z - z_n\| - \|z_n - x_n\| = s - \delta_n - \|z - z_n\|$ and note for some $N \in \mathbb{N}$ that $\lambda_n := s - \delta_n - \|z - z_n\| > 0$ for all $n \geq N$ since $\lambda_n \to s - \delta > 0$. Then we deduce from (15.32) that

$$r\langle z_n - z, w_n\rangle \geq s^{-1}\delta_n(r-s)(s - \delta_n - \|z - z_n\|)^2 \text{ for all } n \geq N.$$

Using the boundedness of $(w_n)_n$ and passing to the limit as $n \to \infty$, we obtain the contradiction $0 \geq s^{-1}\delta(r-s)(s-\delta)^2 > 0$. It results that $s_0 = r$, which gives that $\langle v, x - \overline{x}\rangle \leq \frac{1}{2r}\|x - \overline{x}\|^2$ for all $x \in C$. The property (b) easily follows, and the proof of the theorem is finished. □

REMARK 15.29. Observe with $\|v\| = 1$ that the inequality in (b) of the above theorem is equivalent to

$$\left\langle x - \frac{1}{2}(x + x'), x + rv - \frac{1}{2}(x + x')\right\rangle \geq 0.$$

So, keeping $\|v\| = 1$ and considering the Euclidean space $H = \mathbb{R}^3$ and the point $M(x)$ (that is, the point M with coordinates $x := (x_1, x_2, x_3) \in \mathbb{R}^3$) and the points $M'(x')$, $P(x + rv)$ and $K(\frac{1}{2}(x + x'))$ (that is, K is the middle of the line segment with endpoints M and M'), the above inequality implies $\overrightarrow{KP} \cdot \overrightarrow{KM} \geq 0$, and the latter means geometrically that the point P is always on the same side than the point M with respect to the median plane of the line segment $[M, M']$. This is known as the "*lemma of two points*" of G. Durand for sets that he called "*convex surfaces*" (see the comments in the last section of the chapter). □

The following first corollary states a characterization of r-prox-regularity by a co-monotonicity (or co-coerciveness) property of the metric projection.

COROLLARY 15.30. Let C be a nonempty closed set in a Hilbert space H and let $r \in]0, +\infty]$. The set C is r-prox-regular if and only if for any real $\gamma \in]0, 1[$ the mapping P_C is well-defined on $U_r^\gamma(C)$ and for all $u_1, u_2 \in U_r^\gamma(C)$

$$\langle P_C(u_1) - P_C(u_2), u_1 - u_2\rangle \geq (1 - \gamma)\|P_C(u_1) - P_C(u_2)\|^2.$$

PROOF. If the property of the corollary holds, clearly P_C is $(1-\gamma)^{-1}$-Lipschitz on $U_r^\gamma(C)$, so (h) in Theorem 15.28 gives that C is r-prox-regular. The converse implication directly follows from (g) and (i) in Theorem 15.28. □

Taking $r = +\infty$ in Theorem 15.28 yields the following corollary of characterizations of convexity of closed subsets of H.

COROLLARY 15.31 (various characterizations of convex sets in Hilbert space). For a nonempty closed set C of a Hilbert space H, the following properties are equivalent:

(a) the set C is convex;
(b) the set C is r-prox-regular for all reals $r > 0$;
(c) for any $x_i \in C$, $v_i \in N^P(C; x_i)$ (resp. $v_i \in N^C(C; x_i)$) with $i = 1, 2$ one has

$$\langle v_1 - v_2, x_1 - x_2 \rangle \geq 0,$$

that is, the multimapping $N^P(C; \cdot)$ is monotone;
(d) the mapping P_C is well defined on H and 1-Lipschitz continuous there;
(e) the mapping P_C is well defined on H and norm-to-norm continuous there;
(f) the mapping P_C is well defined on H and norm-to-weak continuous there;
(g) the function d_C^2 is of class $\mathcal{C}^{1,1}$ on H;
(h) the function d_C^2 is of class $\mathcal{C}^1$ on H;
(i) the function d_C^2 is Fréchet differentiable on H, or equivalently d_C is Fréchet differentiable on $H \setminus C$;
(j) the multimapping Proj_C takes nonempty values on H, and d_C^2 is Gâteaux differentiable on H;
(k) the function d_C is Gâteaux differentiable on $H \setminus C$ with $\|\nabla^G d_C(u)\| = 1$ for all $u \in H \setminus C$;
(k') the function d_C is Gâteaux differentiable on $H \setminus C$ with $\nabla^G d_C(u) = (u - P_C(u))/d_C(u)$ for all $u \in H \setminus C$;
(l) $\partial_P d_C(u) \neq \emptyset$ for all $u \in H$;
(m) $\partial_F d_C(u) \neq \emptyset$ for all $u \in H$.

If C is weakly closed (which holds whenever H is finite-dimensional), then one can add the following to the list of equivalences:
(n) P_C is well defined on H.

PROOF. By Proposition 15.19 the assertions (i) and (k) are equivalent. According to Theorem 15.28 all the properties (c) – (j) of the corollary as well as (l) and (m) are equivalent to the r-prox-regularity of C for $r = +\infty$, and assertions (b) and (i) are equivalent by (q) of Theorem 15.28. From (v) of Theorem 15.28, the above properties are also equivalent to $d_C((1-t)x + ty) \leq 0$ for all $x, y \in C$, which corresponds to the convexity of the closed set C. The corollary then holds. □

Another corollary characterizing convex sets is the following:

COROLLARY 15.32. Let C be a closed subset of a Hilbert space H which is r-prox-regular for some real $r \in]0, +\infty[$. If $\operatorname{Enl}_{r_0} C := \{x \in H : d_C(x) \leq r_0\}$ is convex for some $0 < r_0 < r$, then C is convex.

PROOF. Put $C_s := \operatorname{Enl}_s C$. By Theorem 15.28(q) we know that

$$d_C \text{ is Fréchet differentiable on } U_r(C) \setminus C = \{x \in H : 0 < d_C(x) < r\} \tag{15.33}$$

according to the r-prox-regularity of C. On the other hand, since C_{r_0} is convex, by Corollary 15.31(i) above the function $d_{C_{r_0}}$ is Fréchet differentiable at any point $x \notin C_{r_0}$, that is,

$$d_{C_{r_0}} \text{ is Fréchet differentiable at any } x \in H \text{ with } d_{C_{r_0}}(x) > 0. \tag{15.34}$$

Now observe that, for any $x \in H$ with $d_C(x) > r_0$, one has by Lemma 3.260

$$d_C(x) = d_{C_{r_0}}(x) + r_0, \tag{15.35}$$

and hence also $d_{C_{r_0}}(x) = d_C(x) - r_0 > 0$. By (15.34) the function $d_{C_{r_0}}$ is then Fréchet differentiable on the open set $\{x \in H : d_C(x) > r_0\}$, hence according to (15.35) the function d_C is Fréchet differentiable on $\{x \in H : d_C(x) > r_0\}$. This combined with (15.33) guarantees that d_C is Fréchet differentiable outside of C, and the latter property is known by Corollary 15.31(i) again to characterize the convexity of the closed set C of the Hilbert space. □

REMARK 15.33. The result in the corollary fails without the restriction $r_0 < r$. Indeed, $C := \{t \in \mathbb{R} : |t| \geq 1\}$ is a closed nonconvex r-prox-regular set of $\mathbb{R}$ with $r = 1$, but with $r_0 \geq 1$ its r_0-enlargement $\mathrm{Enl}_{r_0} C = \mathbb{R}$ is convex. □

The next theorem is concerned with the situation of epi-Lipschitz sets.

THEOREM 15.34 (G.E. Ivanov; J.P. Vial). Let C be a nonempty closed epi-Lipschitz set of a Hilbert space H with $C \neq H$ and let $r \in]0, +\infty[$. The following are equivalent:
(a) both sets C and $\mathrm{cl}\,(X \setminus C)$ are r-prox-regular;
(b) there exists a mapping $\nu : \mathrm{bdry}\, C \to \mathbb{S}_H$ such that $N^P(C;x) \cap \mathbb{S}_H = \{\nu(x)\}$ for all $x \in \mathrm{bdry}\, C$ and such that for all $x_i \in \mathrm{bdry}\, C$, $i = 1, 2$,

$$\|\nu(x_1) - \nu(x_2)\| \leq \frac{1}{r}\|x_1 - x_2\|;$$

(c) there exists a mapping $\nu : \mathrm{bdry}\, C \to \mathbb{S}_H$ such that $N^P(C;x) \cap \mathbb{S}_H = \{\nu(x)\}$ for all $x \in \mathrm{bdry}\, C$ and such that for all $x_i \in \mathrm{bdry}\, C$, $i = 1, 2$,

$$|\langle \nu(x_1) - \nu(x_2), x_1 - x_2 \rangle| \leq \frac{1}{r}\|x_1 - x_2\|^2.$$

PROOF. Put $C' := \mathrm{cl}\,(X \setminus C)$ and note by Corollary 2.31 that C' is epi-Lipschitz, $C = \mathrm{cl}(\mathrm{int}\, C)$, $C' = \mathrm{cl}(\mathrm{int}\,(C'))$, $\mathrm{bdry}\, C = \mathrm{bdry}\,(C')$ and $N^C(C';x) = -N^C(C;x) \neq \{0\}$ for all $x \in \mathrm{bdry}\, C$, where the non-nullity property is due to Corollary 2.26.

Let us show (a)⇒(b). Assume (a) and take any $x \in \mathrm{bdry}\, C$. Since $N^C(C;x)$ is non-null by what precedes, we may choose a unit vector $v \in N^C(C;x)$, and its opposite belongs to $N^C(C';x)$. Then $v \in N^P(C;x)$ and $-v \in N^P(C';x)$ since $N^P(S;\cdot) = N^C(S;\cdot)$ for any prox-regular set S (see Proposition 15.13(b)), hence Definition 15.24 gives

$$B(x + rv, r) \subset X \setminus C \quad \text{and} \quad B(x - rv, r) \subset X \setminus C' = \mathrm{int}\, C. \tag{15.36}$$

It ensues that $B(x + rv, r) \cap B(x - rv, r) = \emptyset$ as well as

$$B[x + rv, r] \subset C' \quad \text{and} \quad B[x - rv, r] \subset C.$$

The two latter inclusions ensure that

$$N^P(C';x) \cap \mathbb{S}_H \subset N(B[x + rv, r]; x) \cap \mathbb{S}_H = \{-v\}$$

and

$$N^P(C;x) \cap \mathbb{S}_H \subset N(B[x - rv, r]; x) \cap \mathbb{S}_H = \{v\},$$

which yields that both $N^P(C';x) \cap \mathbb{S}_H$ and $N^P(C;x) \cap \mathbb{S}_H$ are singleton sets. Denoting by $\nu(x)$ the unique element in $N^P(C;x) \cap \mathbb{S}_H$, we obtain a mapping $\nu : \mathrm{bdry}\, C \to \mathbb{S}_H$ with $\nu(x) \in N^P(C;x)$ for all $x \in \mathrm{bdry}\, C$.

Now take any $x_i \in \mathrm{bdry}\, C$, $i = 1, 2$, and put $z_i = x_i - r\nu(x_i)$ and $z_i' = x_i + r\nu(x_i)$. By (15.36) we have

$$B(z_1, r) \cap B(z_2', r) = \emptyset \quad \text{and} \quad B(z_1', r) \cap B(z_2, r) = \emptyset,$$

and these equalities entail that $\|z_1 - z_2'\| \geq 2r$ and $\|z_1' - z_2\| \geq 2r$, or equivalently $\|(x_1 - x_2) - r\big(\nu(x_1) + \nu(x_2)\big)\|^2 \geq 4r^2$ and $\|(x_1 - x_2) + r\big(\nu(x_1) + \nu(x_2)\big)\|^2 \geq 4r^2$.

Using the equality $\|a-b\|^2+\|a+b\|^2 = 2(\|a\|^2+\|b\|^2)$ we obtain
$2\|x_1-x_2\|^2+2r^2\|\nu(x_1)+\nu(x_2)\|^2 \geq 8r^2$, hence $\|x_1-x_2\|^2 \geq r^2(4-\|\nu(x_1)+\nu(x_2)\|^2)$.

On the other hand, since $\|\nu(x_i)\| = 1$ we have

$$\|\nu(x_1)+\nu(x_2)\|^2 + \|\nu(x_1)-\nu(x_2)\|^2 = 2(\|\nu(x_1)\|^2+\|\nu(x_2)\|^2) = 4,$$

which gives $4-\|\nu(x_1)+\nu(x_2)\|^2 = \|\nu(x_1)-\nu(x_2)\|^2$. We then derive that we have $\|x_1-x_2\|^2 \geq r^2\|\nu(x_1)-\nu(x_2)\|^2$, which gives the desired inequality in (b). The implication (a)$\Rightarrow$(b) is proved.

The second implication (b)$\Rightarrow$(c) being evident, it remains to show the last one (c)$\Rightarrow$(a). Assuming (c) we have

$$\langle v_1 - v_2, x_1 - x_2\rangle \geq -\frac{1}{r}\|x_1-x_2\|^2$$

for all (x_i, v_i) with $x_i \in \operatorname{bdry} C$ and $v_i \in N^P(C;x_i)\cap \mathbb{S}_H$, $i=1,2$. The equivalence between (a) and (e') in Theorem 15.28 tells us that the set C is r-prox-regular. For each $x \in \operatorname{bdry} C$ we deduce that $N^C(C;x) = N^P(C;x) = \mathbb{R}_+\nu(x)$, hence by epi-Lipschitz property of C we have $N^C(C';x)\cap \mathbb{S}_H = \{-\nu(x)\}$. This and the inclusion $N^P(C';\cdot) \subset N^C(C';\cdot)$ entail for all (x_i,ζ_i), $i=1,2$, with $x_i \in \operatorname{bdry} C'$ and $\zeta_i \in N^P(C';x_i)\cap \mathbb{S}_H$ that $\zeta_i = -\nu(x_i)$, which combined with the assumption (c) gives that

$$\langle \zeta_1 - \zeta_2, x_1 - x_2\rangle \geq -\frac{1}{r}\|x_1-x_2\|^2.$$

The set C' is then r-prox-regular by the equivalence between (a) and (e') in Theorem 15.28 again. This finishes the proof of the theorem. □

Now we provide sufficient conditions for the prox-regularity of epi-Lipschitz sets given by epigraphs of differentiable functions and for the prox-regularity of graphs of differentiable mappings.

PROPOSITION 15.35. Let H_1 and H_2 be two Hilbert spaces and let $F : H_1 \to H_2$ and $f : H_1 \to \mathbb{R}$ be two differentiable mappings whose derivatives are L-Lipschitz on H_1. Then, for $H_1 \times H_2$ and $H_1 \times \mathbb{R}$ endowed with their canonical inner products, the graph $\operatorname{gph} F$ of F and the epigraph $\operatorname{epi} f$ of f are both $1/L$-prox-regular in $H_1 \times H_2$ and $H_1 \times \mathbb{R}$ respectively.

PROOF. We know by Proposition 2.129(b) that

$$T^C(\operatorname{gph} F;(x,F(x))) = \{(u, DF(x)u) \in H_1\times H_2 : u \in H_1\},$$

hence

$$N^C(\operatorname{gph} F;(x,F(x)) = \{(v_1,v_2)\in H_1\times H_2 : \langle v_1,u\rangle + \langle v_2, DF(x)u\rangle = 0,\ \forall u \in H_1\}.$$

Take any $(v_1,v_2) \in N^P(\operatorname{gph} F;(x,F(x))) \subset N^C(\operatorname{gph} F;(x,F(x)))$ and any $x' \in X$. Noting that

$$\begin{aligned}&\langle (v_1,v_2),(x'-x, F(x')-F(x)\rangle\\ &= \left\langle v_1, x'-x\rangle + \langle v_2, \int_0^1 DF(x+t(x'-x)), x'-x\right\rangle dt\end{aligned}$$

we can write

$$\begin{aligned}&\langle (v_1,v_2),(x'-x,F(x')-F(x)\rangle\\&=-\left\langle v_2, DF(x)(x'-x)\rangle+\langle v_2,\int_0^1 DF(x+t(x'-x),x'-x\right\rangle dt\\&=\int_0^1\langle v_2,[-DF(x)+DF(x+t(x'-x))](x'-x)\rangle\,dt,\end{aligned}$$

which gives

$$\begin{aligned}\langle (v_1,v_2),(x'-x,F(x')-F(x)\rangle &\le L\|v_2\|\,\|x'-x\|^2\int_0^1 t\,dt\\&\le \frac{L}{2}\|(v_1,v_2)\|\,\|(x'-x,F(x')-F(x))\|^2.\end{aligned}$$

This and (b) in Theorem 15.28 ensure that $\operatorname{gph} F$ is $1/L$-prox-regular.

In the case of the function f we know that $N^C(\operatorname{epi} f;(x,f(x)))$ coincides with $\mathbb{R}_+(\nabla f(x),-1)$. Take any $(x',s')\in\operatorname{epi} f$. We note that

$$\begin{aligned}\langle(\nabla f(x),-1),(x'-x,s'-f(x))\rangle &\le \langle\nabla f(x),x'-x\rangle-f(x')+f(x)\\&=\int_0^1\langle\nabla f(x)-\nabla f(x+t(x'-x)),x'-x\rangle\,dt\\&\le L\|x'-x\|^2\int_0^1 t\,dt\\&\le\frac{L}{2}\|(\nabla f(x),-1)\|\,\|(x'-x,s'-f(x))\|^2.\end{aligned}$$

From this we see that $\langle\zeta,(x'-x,s'-s)\rangle\le\frac{L}{2}\|\zeta\|\,\|(x'-x,s'-s)\|^2$ for any $(x,s)\in\operatorname{epi} f$ and $\zeta\in N^P(\operatorname{epi} f;(x,s))$ since $N^P(\operatorname{epi} f;(x,s))$ is null if $s>f(x)$. This and (b) in Theorem 15.28 translate the $1/L$-prox-regularity □

REMARK 15.36. Putting $\mu_F:=\inf\limits_{x\in H_1,v_2\in\mathbb{S}_{H_2}}\|DF(x)^*v_2\|$ (where $DF(x)^*$ is the adjoint of $DF(x)$), the above proof shows, for F and f as in the theorem, that $\operatorname{gph} F$ is $(1+\mu_F^2)^{1/2}/L$-prox-regular and $\operatorname{epi} f$ are $(1+\mu_f^2)^{1/2}/L$-prox-regular. Notice also that $\mu_f=\inf_{x\in H_1}\|\nabla f(x)\|$. □

In addition to the properties of Theorem 15.28, the r-prox-regularity can also be characterized by the semiconvexity of the square of the distance function.

PROPOSITION 15.37. Let C be a nonempty closed set of a Hilbert space H and let $r\in\,]0,+\infty]$. The following assertions are equivalent:
(a) the set C is r-prox-regular;
(b) for any real $\gamma\in\,]0,1[$ the function $d_C^2+\frac{\gamma}{1-\gamma}\|\cdot\|^2$ is convex on any open convex subset V of $U_r^\gamma(C)$, otherwise stated d_C^2 is linearly semiconvex on V with $2\gamma/(1-\gamma)$ as coefficient;
(c) the function d_C^2 is locally linearly semiconvex on $U_r(C)$, that is, linearly semiconvex near each point in $U_r(C)$.

PROOF. Suppose that C is r-prox-regular. For any positive $\gamma<1$ and for $\sigma:=\gamma/(1-\gamma)$ the function $f:=d_C^2+\sigma\|\cdot\|^2$ is $\mathcal{C}^1$ on the open set $U_r^\gamma(C)$ according to (p) in Theorem 15.28, and for all $u_i \in U_r^\gamma(C)$ with $i = 1, 2$ we have by (e) in Lemma 15.11

$$\begin{aligned} &\langle \nabla^F f(u_1) - \nabla^F f(u_2), u_1 - u_2 \rangle \\ &= 2\langle (u_1 - P_C(u_1) + \sigma u_1) - (u_2 - P_C(u_2) + \sigma u_2), u_1 - u_2 \rangle \\ &= 2(1+\sigma)\|u_1 - u_2\|^2 - 2\langle P_C(u_1) - P_C(u_2), u_1 - u_2 \rangle \\ &\geq 2(1 + \sigma - (1-\gamma)^{-1})\|u_1 - u_2\|^2 = 0, \end{aligned} \tag{15.37}$$

the last inequality being due to (h) of Theorem 15.28. This ensures that f is convex on any open convex subset of $U_r^\gamma(C)$, as it is well known (see also Theorem 6.68(e)). This justifies the implication (a)$\Rightarrow$(b).

The implication (b)$\Rightarrow$(c) being obvious, it remains to show (c)$\Rightarrow$(a). Fix any $u \in U_r(C)$ and choose by (c) some reals $\delta > 0$ and $\rho \geq 0$ such that $B(u, \delta) \subset U_r(C)$ and such that the function $f := d_C^2 + \rho\|\cdot\|^2$ is convex on $B(u, \delta)$. This convexity property of the function f along with its continuity assures us that $\partial_F f(u) \neq \emptyset$, which yields that $\partial_F d_C^2(u) \neq \emptyset$ since $\rho\|\cdot\|^2$ is Fréchet differentiable at u. Therefore, the function d_C^2 is ∂_F-subdifferentiable on $U_r(C)$, and hence by (x) of Theorem 15.28 and by Lemma 15.11(a) (note that, for $u \in C$, one had $\partial_F d_C(u) \neq \emptyset$ since $0 \in \partial_F d_C(u)$) the set C is r-prox-regular. $\square$

When the function $\rho(\cdot)$ is constant, the tangential condition in Theorem 15.14 can be immediately formulated as follows:

PROPOSITION 15.38. A nonempty closed set C of a Hilbert space H is r-prox-regular if and only if the following Shapiro second order contact property holds:

$$d(x' - x, T^B(C; x)) \leq \frac{1}{2r}\|x' - x\|^2 \quad \text{for all } x, x' \in C.$$

The tangential condition above involves all the points $x, x' \in C$. The corollary below involves only points x, x' in C with $\|x' - x\| < 2s$.

COROLLARY 15.39. Let C be a nonempty closed set of a Hilbert space H and let $r, s > 0$.

(a) If for all $x, x' \in C$ with $\|x' - x\| < 2s$ one has

$$d(x' - x, T^B(C; x)) \leq \frac{1}{2r}\|x' - x\|^2,$$

then the set C is $\min\{r, s\}$-prox-regular.

(b) The set C is then r-prox-regular if and only if the tangential inequality in (a) is fulfilled for all $x, x' \in C$ satisfying $\|x' - x\| < 2r$.

PROOF. Consider $x, x' \in C$ with $\|x' - x\| \geq 2s$. Since zero belongs to $T^B(C; x)$ we have

$$d(x' - x, T^B(C; x)) \leq \|x' - x\| = \frac{1}{\|x' - x\|}\|x' - x\|^2 \leq \frac{1}{2s}\|x' - x\|^2.$$

Combining this with the inequality in the assumption of the corollary, we obtain

$$d(x' - x, T^B(C; x)) \leq \frac{1}{2\min\{r, s\}}\|x' - x\|^2$$

for all $x, x' \in C$. Proposition 15.38 then says that C is $\min\{r, s\}$-prox-regular. This justifies (a), and the other assertion (b) follows directly from (a) and from Proposition 15.38 again. $\square$

A result similar to the corollary above also holds with the normal cone.

PROPOSITION 15.40. Let C be a nonempty closed set of a Hilbert space H and let $r, s > 0$.
(a) If for all $x, x' \in C$ with $\|x' - x\| < 2s$ and for all $v \in N^P(C; x) \cap \mathbb{B}_H$ one has

$$\langle v, x' - x \rangle \le \frac{1}{2r} \|x' - x\|^2,$$

then the set C is $\min\{r, s\}$-prox-regular.
(a') If for all $x, x' \in C$ with $\|x' - x\| < 2s$ and for all $v \in N^P(C; x) \cap \mathbb{B}_H$ and $v' \in N^P(C; x') \cap \mathbb{B}_H$ (resp. for all $(x, v), (x', v')$ with $x, x' \in \operatorname{bdry} C$, $\|x' - x\| < 2s$, $v \in N^P(C; x) \cap \mathbb{S}_H$ and $v' \in N^P(C; x') \cap \mathbb{S}_H$) one has

$$\langle v - v', x - x' \rangle \ge -\frac{1}{r} \|x' - x\|^2,$$

then the set C is $\min\{r, s\}$-prox-regular.
(b) The set C is then r-prox-regular if and only if the inequality in (a) (resp. (a')) is fulfilled for all $x, x' \in C$ (resp. $x, x' \in \operatorname{bdry} C$) satisfying $\|x' - x\| < 2r$.

PROOF. As above take first $x, x' \in C$ with $\|x' - x\| \ge 2s$ and take any v in $N^P(C; x) \cap \mathbb{B}_H$. Since $v \in \mathbb{B}_H$ we have

$$\langle v, x' - x \rangle \le \|x' - x\| = \frac{1}{\|x' - x\|} \|x' - x\|^2 \le \frac{1}{2s} \|x' - x\|^2. \tag{15.38}$$

Combining (15.38) with the inequality in the assumption in (a) of the proposition yields

$$\langle v, x' - x \rangle \le \frac{1}{2 \min\{r, s\}} \|x' - x\|^2$$

for all $x, x' \in C$ and $v \in N^P(C; x) \cap \mathbb{B}_H$. This guarantees the assertion (a) according to (b) of Theorem 15.28.

Now, for any $x, x' \in C$ with $\|x - x'\| \ge 2s$, any $v \in N^P(C; x) \cap \mathbb{B}_H$ and any $v' \in N^P(C; x') \cap \mathbb{B}_H$ (resp. $v \in N^P(C; x) \cap \mathbb{S}_H$ and $v' \in N^P(C; x') \cap \mathbb{S}_H$), the inequality in (15.38) gives

$$\langle v, x' - x \rangle \le \frac{1}{2s} \|x' - x\|^2 \quad \text{and} \quad \langle v', x - x' \rangle \le \frac{1}{2s} \|x' - x\|^2,$$

hence

$$\langle v - v', x - x' \rangle \ge -\frac{1}{s} \|x' - x\|^2.$$

This and the inequality in the assumption in (a') of the proposition assure us that

$$\langle v - v', x - x' \rangle \ge -\frac{1}{\min\{r, s\}} \|x' - x\|^2$$

for any $x, x' \in C$, any $v \in N^P(C; x) \cap \mathbb{B}_H$ and any $v' \in N^P(C; x') \cap \mathbb{B}_H$ (resp. $v \in N^P(C; x) \cap \mathbb{S}_H$ and $v' \in N^P(C; x') \cap \mathbb{S}_H$), which translates the conclusion of the assertion (a').

Finally, the assertion (a) (resp. (a')) in the proposition and the assertion (b) (resp. (e), (e')) in Theorem 15.28 ensure that the assertion (b) in the proposition also holds true. □

We also establish three other useful characterizations of r-prox-regularity under the restriction $\|x - y\| < 2r$ for points x, y in the set.

PROPOSITION 15.41. Let C be a nonempty closed set in a Hilbert space H. For any real $r > 0$ the following assertions are equivalent:
(a) the set C is r-prox-regular;
(b) for any $x, y \in C$ with $\|x - y\| < 2r$ and for any $t \in [0, 1]$

$$B[(1-t)x + ty, \delta_t] \cap C \neq \emptyset,$$

where $\delta_t := r - \sqrt{r^2 - t(1-t)\|x-y\|^2}$;
(c) for any $x, y \in C$ with $\|x - y\| < 2r$ and for any $t \in [0, 1]$

$$d_C((1-t)x + ty) \leq r - \sqrt{r^2 - t(1-t)\|x-y\|^2};$$

(d) for any $x, y \in C$ with $\|x - y\| < 2r$ and for any $t \in [0, 1]$

$$d_C((1-t)x + ty) \leq \frac{1}{2r} \min\{t, 1-t\}\|x-y\|^2.$$

PROOF. Given any $t \in [0, 1]$ and $x, y \in C$ with $\|x - y\| < 2r$, we note that $r^2 - t(1-t)\|x - y\|^2 \geq 0$ and we put $\delta_t := r - \sqrt{r^2 - t(1-t)\|x-y\|^2}$ and $z_t := (1-t)x + ty$ (both depending on x, y). We also note that

$$d_C(z_t) \leq \min\{\|z_t - x\|, \|z_t - y\|\} = \min\{t, 1-t\}\|x - y\| < r.$$

(a)⇒(b). Assume (a) and fix any $t \in [0, 1]$ and $x, y \in C$ with $\|x - y\| < 2r$. Note that the assertion (u) in Theorem 15.28 ensures that $-t(1-t)\|x - y\|^2 \leq -2rd_C(z_t) + d_C^2(z_t)$, that is, $r^2 - t(1-t)\|x-y\|^2 \leq (r - d_C(z_t))^2$, which in turn is equivalent to $d_C(z_t) \leq \delta_t$. Since $d_C(z_t) < r$, we see that $P_C(z_t)$ exists by the assertion (l) in Theorem 15.28, hence $P_C(z_t) \in C \cap B[z_t, \delta_t]$, so $C \cap B[z_t, \delta_t] \neq \emptyset$, as desired.
(b)⇒(c). This implication is trivial.
(c)⇒(d). Assume (c) and take any $t \in [0, 1]$ and $x, y \in C$ with $\|x - y\| < 2r$. As above the assumption $d_C(z_t) \leq \delta_t$ is equivalent to $2rd_C(z_t) - d_C^2(z_t) \leq t(1-t)\|x - y\|^2$, so Lemma 15.15 entails the inequality in (d).
(d)⇒(a). Assume (d) and fix any x, y in C with $\|x - y\| < 2r$ and any v in $N^P(C; x) \cap \mathbb{B}_H$. By Proposition 4.153 we know that $v \in \partial_P d_C(x)$, and hence by Proposition 4.17(a) there exists some $\sigma > 0$ and some real $0 < \varepsilon < 1$ such that for every $t \in]0, \varepsilon]$

$$\langle v, t(y-x)\rangle = \langle v, z_t - x\rangle \leq d_C(z_t) + \sigma\|z_t - x\|^2 = d_C(z_t) + \sigma t^2\|y - x\|^2.$$

Therefore, for all $t \in]0, \varepsilon]$ we have by the assumption (d)

$$\langle v, y - x\rangle \leq \frac{1}{2r}\|y - x\|^2 + \sigma t\|y - x\|^2,$$

and making $t \downarrow 0$ we obtain $\langle v, y-x\rangle \leq \frac{1}{2r}\|y-x\|^2$. This gives the r-prox-regularity of C by Proposition 15.40(b). □

The following important result is a direct consequence of Theorem 15.6(g).

PROPOSITION 15.42. Let C be a (closed) r-prox-regular set of a Hilbert space H. Then for any positive real $\gamma < 1$ and any $u_i \in U_r^\gamma(C)$ with $i = 1, 2$, $P_C(u_i)$ exist and

$$\|(I - P_C)(u_1) - (I - P_C)(u_2)\|^2 + (1 - 2\gamma)\|P_C(u_1) - P_C(u_2)\|^2 \leq \|u_1 - u_2\|^2.$$

Consequently, for $0 < \gamma \leq 1/2$ the mapping $I - P_C$ is Lipschitz on $U_r^\gamma(C)$ with 1 as a Lipschitz constant.

With the normal cone, we can characterize the nearest points of an r-prox-regular set of a Hilbert space.

PROPOSITION 15.43. Let C be an r-prox-regular subset of a Hilbert space H and $x, y \in H$. If $x - y \in N^C(C; y)$ and $\|x - y\| \le r$ (resp. $\|x - y\| < r$), then $y \in \mathrm{Proj}_C(x)$ (resp. $y = P_C(x)$).

PROOF. By Proposition 15.13 we know that $N^C(C; \cdot) = N^P(C; \cdot)$. Assume that $x - y \in N^C(C; y)$ and $\|x - y\| \le r$. The non-vacuity of $N^C(C; y)$ gives us $y \in C$. Combining the r-prox-regularity of C and Theorem 15.28(b), we obtain

$$\langle x - y, x' - y\rangle \le \frac{1}{2r} \|x - y\| \|x' - y\|^2 \quad \text{for all } x' \in C.$$

So, the inequality $\|x - y\| \le r$ yields

$$\langle x - y, x' - y\rangle \le \frac{1}{2} \|x' - y\|^2 \quad \text{for all } x' \in C.$$

Since $y \in C$, the latter inequality and Lemma 15.4 entail that

$$y \in \mathrm{Proj}_C(x).$$

If in addition, $\|x - y\| < r$, then

$$d_C(x) \le \|x - y\| < r.$$

According to Theorem 15.28(k), $\mathrm{Proj}_C(x)$ is a singleton, that is, $y = P_C(x)$. □

15.2.2. Uniform prox-regularity of r-enlargement and r-exterior set. In addition to all the above properties of r-prox-regular sets, our next aim is to study properties of r-enlarged sets and sets of r-exterior points of prox-regular sets. For a set C of a normed space $(X, \|\cdot\|)$ and $r \ge 0$, in addition to the r-enlarged set $\mathrm{Enl}_r\, C$ of C introduced in Definition 3.258 we define the (closed) set of *r-exterior points* of C by

$$\mathrm{Exte}_r\, C := \{u \in X : d_C(u) \ge r\}.$$

We recall that the set of (exact) *r-distance points* to C has been defined in a similar way in Definition 3.258 by

$$D_r(C) := \{u \in X : d_C(u) = r\};$$

so $D_r(C) = (\mathrm{Exte}_r\, C) \cap (\mathrm{Enl}_r\, C)$. Sometimes, it will be convenient to denote, for $r > 0$, by $\mathrm{Tube}_r\, C$ the *open r-tube* $U_r(C) \setminus C$, that is,

$$\mathrm{Tube}_r\, C := \{u \in X : 0 < d_C(u) < r\}.$$

A first general property of enlarged sets has been established in Lemma 3.260, say: For any set C of a normed space X and any $r \ge 0$

$$d(u, C) = r + d(u, \mathrm{Enl}_r\, C) \quad \text{whenever } d(u, C) \ge r.$$

From this result and the next lemma we can prove the prox-regularity of enlarged prox-regular sets.

LEMMA 15.44. Let C be a set in a normed space X and let $0 \le s < r$. Then

$$\mathrm{Tube}_{r-s}(\mathrm{Enl}_s\, C) = \{u \in X : s < d_C(u) < r\} = (\mathrm{Tube}_r\, C) \setminus \mathrm{Enl}_s\, C.$$

PROOF. Let $u \in X$ with $s < d_C(u) < r$. The inequality $d_C(u) > s$ yields by the result of Lemma 3.260 recalled above that $d(u, \mathrm{Enl}_s C) = d(u, C) - s$, hence $0 < d(u, \mathrm{Enl}_s C) < r - s$, that is, $u \in \mathrm{Tube}_{r-s}(\mathrm{Enl}_s C)$.

Conversely, take any $u \in \mathrm{Tube}_{r-s}(\mathrm{Enl}_s C)$, that is, $0 < d(u, \mathrm{Enl}_s C) < r - s$. The inequality $0 < d(u, \mathrm{Enl}_s C)$ means $u \notin \mathrm{Enl}_s C$, or equivalently $d_C(u) > s$. The result recalled above again ensues that $d_C(u) = s + d(u, \mathrm{Enl}_s C)$. Therefore, the inequalities $0 < d(u, \mathrm{Enl}_s C) < r - s$ give $s < d_C(u) < r$ as desired. □

THEOREM 15.45 (prox-regularity of enlarged sets). Let C be a closed set of a Hilbert space H which is r-prox-regular. The following hold:
(a) The enlarged set $\mathrm{Enl}_s C$ is $(r - s)$-prox-regular, for every $0 \le s < r$.
(b) For every $0 < s < r$ the set $D_s(C)$ is a $\mathcal{C}^{1,1}$-submanifold of H at any of its points.

PROOF. (a) The above lemma and Lemma 3.260 give $d(\cdot, \mathrm{Enl}_s C) = d(\cdot, C) - s$ on $\mathrm{Tube}_{r-s}(\mathrm{Enl}_s C) = (\mathrm{Tube}_r C) \setminus \mathrm{Enl}_s C$, and the r-prox-regularity of C entails by Theorem 15.28(q) the Fréchet differentiability of $d(\cdot, C)$ over $\mathrm{Tube}_r C$. Consequently, $d(\cdot, \mathrm{Enl}_s C)$ is Fréchet differentiable on $\mathrm{Tube}_{r-s}(\mathrm{Enl}_s C)$, which translates by Theorem 15.28(q) again the $(r - s)$-prox-regularity of $\mathrm{Enl}_s C$.
(b) By Theorem 15.28(o) the function d_C is of class $\mathcal{C}^{1,1}$ on $U_r(C) \setminus C$ and for any $u \in H$ with $d_C(u) = s$ the gradient $\nabla d_C(u)$ is nonzero since $\|\nabla d_C(u)\| = 1$; so the assertion (b) follows (see Proposition E.9 in Appendix E). □

EXAMPLE 15.46. If C is r-prox-regular and $s = r$, the enlarged set $\mathrm{Enl}_r C$ can fail to be prox-regular. Indeed considering in $\mathbb{R}^2$ the set $C := \{(-1, 0), (1, 0)\}$ which is r-prox-regular for $r = 1$, we see that $\mathrm{Enl}_r C = B[(-1, 0), 1] \cup B[(1, 0), 1]$, and hence $\mathrm{Enl}_r C$ is not σ-prox-regular for any $\sigma > 0$. □

In addition to Lemma 3.260 involved above the following lemma provides some connections between nearest points in C and nearest points in closed enlarged sets of C.

LEMMA 15.47. Let C be a nonempty closed set of a normed space $(X, \|\cdot\|)$. Let $r \ge 0$ and $u \in X$ with $d(u, C) \ge r$. The following hold:
(a) If $u_0 \in \mathrm{Proj}_C(u)$ and $y_0 \in [u_0, u] \cap D_r(C)$, then $y_0 \in \mathrm{Proj}(u, \mathrm{Enl}_r C)$.
(b) If $y \in \mathrm{Proj}(u, \mathrm{Enl}_r C)$ and $z \in \mathrm{Proj}_C(y)$, then $z \in \mathrm{Proj}_C(u)$. Further, if $P_{\mathrm{Enl}_r C}(u) = y$ and $z \in \mathrm{Proj}_C(y)$, then $y \in [z, u]$ and $P_C(u) = z$.

PROOF. Putting $C_r := \mathrm{Enl}_r C$ we know by the result of Lemma 3.260 recalled above that

$$d_C(u) = r + d_{C_r}(u). \tag{15.39}$$

(a) Assume $u_0 \in \mathrm{Proj}_C(u)$ and $y_0 \in [u_0, u] \cap D_r(C)$. Then $u_0 \in \mathrm{Proj}_C(y_0)$ by Lemma 4.123, and from this and (15.39) we obtain

$$\|u - y_0\| + r = \|u - y_0\| + d_C(y_0) = \|u - y_0\| + \|y_0 - u_0\| = \|u - u_0\| = r + d_{C_r}(u)$$

and hence $\|u - y_0\| = d_{C_r}(u)$, that is, $y_0 \in \mathrm{Proj}_{C_r}(u)$ (since $y_0 \in D_r(C) \subset C_r$).
(b) Suppose $y \in \mathrm{Proj}_{C_r}(u)$ and $z \in \mathrm{Proj}_C(y)$. Then

$$\|u - z\| \le \|u - y\| + \|y - z\| = d_{C_r}(u) + d_C(y) \le d_{C_r}(u) + r = d_C(u)$$

(the last equality being due again to (15.39)), hence $z \in \mathrm{Proj}_C(u)$ (since $z \in C$).

Now assume in addition $P_{C_r}(u) = y$. Since $d_C(z) = 0$ and $d_C(u) \ge r$, we observe by the intermediate value theorem that $[z, u] \cap D_r(C) \neq \emptyset$. Taking y' in

$[z,u] \cap D_r(C)$, we have by (a) that $y' \in \mathrm{Proj}_{C_r}(u)$, and hence $y = y' \in [z,u]$. If there exists $z' \neq z$ with $z' \in \mathrm{Proj}_C(u)$, then we see that $z' \notin u + [0,+\infty[(z-u)$ and by (a), for $y'' \in [z',u] \cap D_r(C) \neq \emptyset$ (the non-emptiness is due to the intermediate value theorem again), we have $y'' \in \mathrm{Proj}_{C_r}(u)$, and hence $y'' = y \in [z,u]$, which contradicts the fact that $z' \notin u + [0,+\infty[(z-u)$. ☐

The next lemma shows that the closure of the r-open enlargement of a set coincides with the r-closed enlargement of the set.

LEMMA 15.48. Let C be a nonempty set of a normed space $(X, \|\cdot\|)$. Then for any real $r > 0$ one has

$$\mathrm{cl}_X\,(U_r(C)) = \mathrm{cl}_X\,(\{u \in X : d_C(u) < r\}) = \{u \in X : d_C(u) \le r\} = \mathrm{Enl}_r\, C,$$

or equivalently

$$\mathrm{int}_X(\mathrm{Exte}_r\, C) = \{u \in X : d_C(u) > r\}.$$

PROOF. The inclusion of the first member into the last one is obvious. Concerning the converse inclusion, let $u \in X$ with $d_C(u) = r$. Let $(y_n)_n$ in C be a minimizing sequence of $d_C(u)$, that is,

$$r_n := \|u - y_n\| \to d_C(u) = r \quad \text{as } n \to \infty.$$

Fix any $\varepsilon \in]0,r[$ and choose $N \in \mathbb{N}$ such that $r_N > \varepsilon$ and $r_N - r < \varepsilon$, so $1 - \frac{r}{r_N} < \frac{\varepsilon}{r_N}$, hence $\left(1 - \frac{r}{r_N}\right)^+ < \frac{\varepsilon}{r_N}$. Choose a real t such that

$$1 - \frac{r}{r_N} \le \left(1 - \frac{r}{r_N}\right)^+ < t < \frac{\varepsilon}{r_N} < 1. \tag{15.40}$$

The third inequality in (15.40) allows us to write with $z_N := (1-t)u + ty_N$

$$\|z_N - u\| = t\|u - y_N\| = tr_N < \frac{\varepsilon}{r_N} r_N = \varepsilon,$$

so $z_N \in B(u,\varepsilon)$. Also, from the strict inequality in (15.40) between the first and third member we have

$$d(z_N, C) \le \|z_N - y_N\| = (1-t)\|u - y_N\| = (1-t)r_N < \frac{r}{r_N} r_N = r,$$

thus $z_N \in U_r(C)$.

Consequently, we have $B(u,\varepsilon) \cap U_r(C) \neq \emptyset$. This being true for any $\varepsilon \in]0,r[$, it ensues that $u \in \mathrm{cl}_X(U_r(C))$. The proof is finished. ☐

We continue with some general properties of sets of r-exterior points which will be used in the main results concerning the prox-regularity of $\mathrm{Exte}_r\, C$.

LEMMA 15.49. Let C be a nonempty closed set of a normed space $(X, \|\cdot\|)$. For any reals r, s with $0 < s < r$, one has

$$\mathrm{Exte}_{r-s}\,(\mathrm{Enl}_s\, C) = \mathrm{Exte}_r\, C$$

and

$$D_{r-s}(\mathrm{Enl}_s\, C) = D_r(C).$$

PROOF. Set $C_s := \mathrm{Enl}_s\, C$. Let x in $\mathrm{Exte}_r\, C$ (resp. in $D_r(C)$), that is, $d_C(x) \ge r$ (resp. $d_C(x) = r$), so $d_C(x) \ge s$. Lemma 3.260 then says that $d_C(x) = s + d(x, C_s)$ hence $d(x, C_s) \ge r - s$ (resp. $d(x, D_s(C)) = d(x, C_s) = r - s$), proving the inclusion of the right-hand side into the left-hand (for each one of the required equalities).

Now fix any x in $\mathrm{Exte}_{r-s}(C_s)$ (resp. in $D_{r-s}(C_s)$), that is, $d(x, C_s) \geq r - s$ (resp. $d(x, C_s) = r - s$). Since $r - s > 0$, it results that $x \notin C_s$, otherwise stated $d(x, C) > s$. Lemma 3.260 again ensues that $d(x, C) = s + d(x, C_s)$, hence $d(x, C) \geq r$ since $d(x, C_s) \geq r - s$ (resp. $d(x, C) = r$ since $d(x, C_s) = r - s$). This proves the inclusion of the left-hand side into the right-hand, hence the proof is finished. □

We will also need the properties in the following lemma for the r-exterior set $\mathrm{Exte}_r S$ and the open r-tube $\mathrm{Tube}_r(S) = \{x \in X : 0 < d_S(x) < r\}$ of a set S in a normed space X.

LEMMA 15.50. Let X be a normed space and H be a Hilbert space.
(a) For any real $s > 0$, any nonempty subset S of X and any $u \in X$ one has

$$d(u, S) + d(u, \mathrm{Exte}_s\, S) \geq s.$$

(b) If a nonempty closed set C of the Hilbert space H is r-prox-regular for a real $r > 0$, then for any positive real number $s \leq r$ and any $u \in \mathrm{Tube}_s(C) = U_s(C) \setminus C$, one has

$$d(u, C) + d(u, \mathrm{Exte}_s\, C) = d(u, C) + d(u, D_s(C)) = s.$$

PROOF. (a) Fix any $u \in X$. For every $x \in S$ and every $y \in \mathrm{Exte}_s\, S$

$$\|u - x\| + \|u - y\| \geq \|x - y\| \geq d_S(y) \geq s,$$

which gives

$$\inf_{x \in S} \|u - x\| + \inf_{y \in \mathrm{Exte}_s\, S} \|u - y\| = \inf_{(x,y) \in S \times \mathrm{Exte}_s\, S} (\|u - x\| + \|u - y\|) \geq s,$$

hence $d(u, S) + d(u, \mathrm{Exte}_s\, S) \geq s$.
(b) Fix any positive real number $s \leq r$ and any $u \in U_s(C) \setminus C$, and put $x := P_C(u)$ according to (l) of Theorem 15.28. Then for $u' := x + s(u - x)/\|u - x\|$ one has $x \in \mathrm{Proj}_C(u')$ by (t) of Theorem 15.28, and hence $d_C(u') = s$. Consequently, $u' \in D_s C$ and

$$\begin{aligned} d(u, C) + d(u, \mathrm{Exte}_s\, C) &\leq d(u, C) + d(u, D_s(C)) \\ &\leq \|u - x\| + \|u - u'\| = \|u' - x\| = s. \end{aligned} \tag{15.41}$$

This combines with (a) justifies (b). □

We are now in a position to establish two geometric characterizations of r-prox-regularity in terms of r-exterior points.

THEOREM 15.51 (metric characterization of r-prox-regularity via r-exterior points). Assume that r is a positive real constant. Then, a nonempty closed set C of a Hilbert space H with $C \neq H$ is r-prox-regular if and only if for all $u \in H$ with $0 < d_C(u) < r$ one has

$$d(u, C) + d(u, \mathrm{Exte}_r\, C) = r\,.$$

The latter is also equivalent to

$$d(u, C) + d(u, \mathrm{Exte}_r\, C) = r \quad \text{for all } u \in \{x \in H : 0 < d_C(x) \leq r\} \cup \mathrm{bdry}\, C.$$

PROOF. Lemma 15.50 guarantees that the first equality of the theorem holds whenever C is r-prox-regular. Suppose now that equality and consider any $u \in$

$U_r(C)\setminus C$ for which $\mathrm{Proj}_C(u)$ is a singleton, say $P_C(u) =: x$. Then, for any $y \in]x, u[$ we have $P_C(y) = x$ (see Lemma 4.123). This yields for any $\tau \in]-1, \|u-y\|/\|y-x\|[$

$$\tau^{-1}\big(d_C(y+\tau(y-x)) - d_C(y)\big) = \tau^{-1}\big(\|y+\tau(y-x)-x\| - \|y-x\|\big) = \|y-x\|,$$

which entails the existence of the (bilateral) limit

$$\lim_{\tau\to 0} \tau^{-1}\big(d_C(y+\tau(y-x)) - d_C(y)\big) \quad \text{with } x = P_C(y),$$

and hence by Corollary 14.23 of the Fitzpatrick theorem the distance function d_C is Fréchet differentiable at y. Further, on the one hand $\|u-y\| < \|u-x\| = d_C(u)$ hence $y \notin C$, and on the other hand

$$d_C(y) = \|y-x\| < \|u-x\| = d_C(u) < r,$$

so $y \in U_r(C) \setminus C$. By the equality assumption of the theorem, the function $d\big(\cdot, \mathrm{Exte}_r(C)\big)$ is then Fréchet differentiable at y, which ensues by Lemma 15.11(c) that $P_{\mathrm{Exte}_r(C)}(y) =: u'$ exists. Using successively the equality $d_C(u') = r$ and the equality assumption of the theorem we obtain

$$r \le \|x-u'\| \le \|y-x\| + \|y-u'\| = d(y, C) + d\big(y, \mathrm{Exte}_r(C)\big) = r,$$

so $y \in]x, u'[$ and $x \in \mathrm{Proj}_C(u')$, the second inclusion being due to the fact that $\|x-u'\| = r = d_C(u')$. The inclusions $y \in]x, u'[$ and $y \in]x, u[$ combined with the equality $\|x-u'\| = r$ imply $u' = x + r\frac{u-x}{\|u-x\|}$. This and the inclusion $x \in \mathrm{Proj}_C(u')$ allow us, through (t) of Theorem 15.28 and Lemma 4.123, to conclude that C is r-prox-regular. So, the first desired equivalence is proved.

Regarding the other equivalence, of course the second equality implies the first. Now suppose that $d(u, C) + d(u, \mathrm{Exte}_r C) = r$ for all $u \in U_r(C) \setminus C$. Obviously the equality still holds fo any $u \in H$ with $d_C(u) = r$ since such $u \in \mathrm{Exte}_r C$. Finally, taking any $u \in \mathrm{bdry}\, C$, such u is the limit of a sequence of elements of $U_r(C) \setminus C$, so the continuity of $d(\cdot, C)$ and $d(\cdot, \mathrm{Exte}_r C)$ allows us to conclude the proof. □

To deduce from the above theorem the uniform prox-regularity of sets of r-exterior points, we need two lemmas.

LEMMA 15.52. Let C be a closed set of a normed space $(X, \|\cdot\|)$ and let $r \in]0, +\infty]$. For any $x \in C$, one has

$$d(x, \mathrm{Exte}_r C) \ge r;$$

if $(X, \|\cdot\|)$ is a reflexive Banach space and the norm $\|\cdot\|$ has the Kadec-Klee property, the refined inequality

$$d(x, \mathrm{Exte}_r C) \ge r + d(x, \mathrm{bdry}\, C)$$

holds true for any $x \in C$.

PROOF. We may suppose $\mathrm{Exte}_r C \neq \emptyset$. Fix any $x \in C$. For any $y \in \mathrm{Exte}_r C$ we have $d_C(y) \ge r$, hence $\|x-y\| \ge d_C(y) \ge r$. Therefore, $d(x, \mathrm{Exte}_r C) \ge r$, so the first inequality is proved.

Now suppose that X is a reflexive Banach space and the norm $\|\cdot\|$ has the Kadec-Klee property. The second inequality being equivalent to the first if $x \in \mathrm{bdry}\, C$, fix any $x \in \mathrm{int}\, C$. By Lau theorem (see Theorem 14.29) consider a sequence $(x_n)_n$ in $\mathrm{int}\, C$ converging to x and such that, for each $n \in \mathbb{N}$, we can choose $p_n \in \mathrm{Proj}_{\mathrm{Exte}_r C}(x_n)$, so $p_n \notin C$. Thus, for each $n \in \mathbb{N}$, we have $b_n := x_n + t_n(p_n - x_n) \in$

bdry C for some $t_n \in [0,1[$, so by what precedes $d(b_n, \mathrm{Exte}_r\, C) \geq r$. Further, $p_n \in \mathrm{Proj}_{\,\mathrm{Exte}_r\, C}(b_n)$ (see Lemma 4.123), hence

$$\|b_n - p_n\| = d(b_n, \mathrm{Exte}_r\, C) \geq r,$$

and this gives

$$d(x_n, \mathrm{Exte}_r\, C) = \|x_n - p_n\| = \|p_n - b_n\| + \|b_n - x_n\| \geq r + d(x_n, \mathrm{bdry}\, C).$$

From the continuity of $d(\cdot, \mathrm{Exte}_r\, C)$ and $d(\cdot, \mathrm{bdry}\, C)$, it results $d(x, \mathrm{Exte}_r\, C) \geq r + d(x, \mathrm{bdry}\, C)$, and this finishes the proof. □

LEMMA 15.53. Let C be a closed set of a Hilbert space H which is r-prox-regular. Then, for every positive real $s \leq r$,

$$0 < d(u, C) < s \quad \text{if and only if} \quad 0 < d(u, \mathrm{Exte}_s\, C) < s,$$

that is, the open s-tube $\mathrm{Tube}_s(\mathrm{Exte}_s\, C)$ around $\mathrm{Exte}_s\, C$ coincides with the open s-tube $\mathrm{Tube}_s\, C$ around C.

PROOF. Suppose first $0 < d(u, \mathrm{Exte}_s\, C) < s$. The inequality $d(u, \mathrm{Exte}_s\, C) > 0$ means $u \notin \mathrm{Exte}_s\, C$, that is, $d(u, C) < s$. On the other hand, the inequality $d(u, \mathrm{Exte}_s\, C) < s$ combined with the first inequality of the above lemma gives $u \notin C$, that is, $d(u, C) > 0$.

Conversely suppose $0 < d(u, C) < s$. Then Lemma 15.50 yields $d(u, C) + d(u, \mathrm{Exte}_s\, C) = s$, and hence $0 < d(u, \mathrm{Exte}_s\, C) < s$, which finishes the proof. □

THEOREM 15.54 (prox-regularity of the set of r-exterior points). Let C be a nonempty closed set of a Hilbert space H with $C \neq H$ and let $r \in]0, +\infty[$.
(a) If C is r-prox-regular, then $\mathrm{Exte}_r\, C$ is also r-prox-regular.
(b) The set C is r-prox-regular if and only if for any $u \in \mathrm{Tube}_r(C)$ there exist $u' \in C$ and $u'' \in \mathrm{Exte}_r\, C$ such that

$$\|u - u'\| + \|u - u''\| = r.$$

PROOF. By convenience denote $S := \mathrm{Exte}_r\, C$.
(a) Assume that C is r-prox-regular. The above lemma tells us that the open r-tube around S coincides with the open r-tube around C. Further, for all u in this common open r-tube, $d_C(u) + d_S(u) = r$ by Theorem 15.51, and d_C is Fréchet differentiable at u by (q) in Theorem 15.28. Consequently, the distance function d_S is Fréchet differentiable at any u in the open r-tube $U_r(S) \setminus S$. Theorem 15.28(q) again then guarantees that S is r-prox-regular.
(b) Suppose first that C is r-prox-regular and fix any $u \in \mathrm{Tube}_r(C)$. By (a) the set S is also r-prox-regular, so $P_C(u)$ and $P_S(u)$ exist by (l) in Theorem 15.28. The r-prox-regularity of C also entails that $d_C(u) + d_S(u) = r$ by (b) in Theorem 15.51. Therefore, the equality in (b) is fulfilled by choosing $u' = P_C(u)$ and $u'' = P_S(u)$.

Conversely, suppose that the equality in (b) holds and fix any $u \in \mathrm{Tube}_r(C)$. By the equality assumption there are $u' \in C$ and $u'' \in S$ such that $\|u - u'\| + \|u - u''\| = r$, so $d_C(u) + d_S(u) \leq r$. The converse inequality being always true by Lemma 15.50(a), we deduce that $d_C(u) + d_S(u) = r$ for all $u \in \mathrm{Tube}_r(C)$. Then Theorem 15.51 ensures that the set C is r-prox-regular. This finishes the proof of the theorem. □

15.2.3. Linear semiconvexity of distance function to a prox-regular set. For $0 < s < r$, taking $\gamma := s/r$ in Proposition 15.37 ensures through Theorem 10.15 that the closed set C is r-prox-regular if and only if, for every $s \in]0, r[$, the square distance function d_C^2 is linearly semiconvex with coefficient $\frac{2s}{r-s}$ on any convex set included in $U_s(C)$. The next proposition provides another characterization as the semiconvexity of the distance function itself. The proof requires first two lemmas, each one having its own interest.

In addition to Lemma 15.48, the first lemma relates, for an r-prox-regular set C and for a real s with $0 < s < r$, the set of s-exterior points of C to the s-enlargement set of C.

LEMMA 15.55. Let C be a closed set of a Hilbert space H which is r-prox-regular and let $0 < s < r$. Then one has

$$\mathrm{cl}_H(H \setminus \mathrm{Enl}_s C) = \{u \in H : d_C(u) \geq s\} = \mathrm{Exte}_s C,$$

or equivalently

$$\mathrm{int}_H(\mathrm{Enl}_s C) = \{u \in H : d_C(u) < s\} = U_s(C).$$

PROOF. Since $H \setminus \mathrm{Enl}_s C = \{u \in H : d_C(u) > s\}$, the continuity of d_C gives

$$\mathrm{cl}_H(H \setminus \mathrm{Enl}_s C) \subset \{u \in H : d_C(u) \geq s\}.$$

Fix now any $u \in H$ with $d_C(u) = s$. Choose a sequence $(s_n)_n$ in $]s, r[$ in $]0, s[$ tending to s. Since C is r-prox-regular, $p = P_C(u)$ exists and, for $u_n := p + s_n \frac{u-p}{\|u-p\|}$, we have $p = P_C(u_n)$ thanks to Theorem 15.28(t). We also see that $d_C(u_n) = s_n > s$, hence $u_n \in H \setminus \mathrm{Enl}_s C$. Further, $(u_n)_n$ converges to $p + (s/\|u-p\|)(u-p) = u$ as $n \to \infty$ since $s = d_C(u) = \|u - p\|$. Consequently, $u \in \mathrm{cl}_H(H \setminus \mathrm{Enl}_s C)$, so the desired equality is proved. □

The second lemma shows that the complement of an r-prox-regular set is a union of closed balls with common radius $s < r$.

LEMMA 15.56. Let C be a closed set of a Hilbert space H which is r-prox-regular. Then, for each $0 < s < r$, the set $H \setminus C$ is the union of a family of closed balls with the common radius s.

PROOF. We may suppose that $C \neq H$. Let $y \in H \setminus C$. If $d_C(y) \geq r$, then $B(y, r) \cap C = \emptyset$, hence $B[y, s] \subset H \setminus C$. Suppose now that $0 < d_C(y) < r$. According to the r-prox-regularity of C, Theorem 15.28(l) says that $p := P_C(y)$ exists. For $v := (y-p)/\|y-p\|$, by the r-prox-regularity of C Theorem 15.28(t) gives that we have $p \in \mathrm{Proj}_C(p + rv)$, so $B(p + rv, r) \cap C = \emptyset$. Observe also that

$$\|y - p - rv\| = \left\| \left(1 - \frac{r}{\|y-p\|}\right)(y-p) \right\| = |\|y-p\| - r| = r - d_C(y).$$

If $s \geq r - d_C(y)$, then $y \in B[p+rv, s]$ and $B[p+rv, s] \subset H \setminus C$ since $B[p+rv, s] \subset B(p + rv, r)$.

Suppose that $s < r - d_C(y)$, which ensures in particular $y \neq p + rv$. Set

$$z := y - \|y - p - rv\|^{-1} s(y - p - rv),$$

hence obviously $y \in B[z, s]$. We claim in that case that $B[z, s] \subset B(p + rv, r)$. Indeed, for every $u \in B[z, s]$ we have

$$\begin{aligned}\|u - p - rv\| &\le \|u - z\| + \|z - p - rv\| \\ &= \|u - z\| + \left\| \left(1 - \frac{s}{\|y - p - rv\|}\right)(y - p - rv) \right\| \\ &= \|u - z\| + \big|\|y - p - rv\| - s\big| = \|u - z\| + |r - d_C(y) - s|,\end{aligned}$$

which combined with the inequality $s < r - d_C(y)$ yields

$$\|u - p - rv\| \le \|u - z\| + r - d_C(y) - s \le r - d_C(y)$$

since $\|u - z\| \le s$. This confirms the inclusion $B[z, s] \subset B(p + rv, r)$ of the claim. Therefore, $y \in B[z, s] \subset H \setminus C$.

In conclusion, any point of $H \setminus C$ belongs to some closed ball of radius s included in $H \setminus C$. □

THEOREM 15.57 (linear semiconvexity of distance function as characterization of prox-regularity). Let C be a closed set in a Hilbert space H and let $r \in]0, +\infty]$. The following assertions are equivalent:
(a) the set C is r-prox-regular;
(b) for any real $0 < s < r$ the distance function d_C is linearly semiconvex with constant $1/(r - s)$ on any convex set included in the open s-enlargement $U_s(C)$. (This can also be seen as the linear semiconvexity with coefficient $1/(r - s)$ of d_C on any convex set in the open tube $U_s(C) \setminus C$);
(c) the distance function d_C is locally linearly semiconvex on $U_r(C)$.

PROOF. Assume that C is r-prox-regular and let $0 < s < r$. Fix any real $t \in]s, r[$. We know by Theorem 15.45 that the set $\mathrm{Enl}_s C$ is $(r - s)$-prox-regular, hence Lemma 15.56 above says that $H \setminus \mathrm{Enl}_s C$ is the union of a family of closed balls with radius $r - t$. It follows from Proposition 10.22 that the distance function

$$d(\cdot, \mathrm{cl}_H(H \setminus \mathrm{Enl}_s C)) = d(\cdot, H \setminus \mathrm{Enl}_s C)$$

is linearly semiconcave with coefficient $1/(r - t)$ on any convex set included in $\complement_H(H \setminus \mathrm{Enl}_s C)$. Further, by Lemma 15.55, we have

$$\mathrm{cl}_H(H \setminus \mathrm{Enl}_s C) = \mathrm{Exte}_s C.$$

Then the function $d(\cdot, \mathrm{Exte}_s C)$ is linearly semiconcave with coefficient $1/(r - t)$ on any convex set included in $U_s(C) \subset \mathrm{Enl}_s C$. This being true for any real $t \in]s, r[$, it is easily seen from the definition of semiconcavity that $d(\cdot, \mathrm{Exte}_s C)$ is linearly semiconcave with coefficient $1/(r - s)$ on any convex set included in $U_s(C)$.

Further, for any $x \in \{u \in H : 0 < d_C(u) < s\}$, we have by Theorem 15.51

$$d(x, \mathrm{Exte}_s C) + d(x, C) = s, \quad \text{or equivalently} \quad d(x, C) = s - d(x, \mathrm{Exte}_s C).$$

By Lemma 15.52, for any $x \in C$, we also have $s - d(x, \mathrm{Exte}_s C) \le 0$. Consequently

$$d(x, C) = \max\{0, s - d(x, \mathrm{Exte}_s C)\} \quad \text{for all } x \in U_s(C).$$

This ensures that the distance function d_C is linearly semiconvex with coefficient $1/(r - s)$ on any convex set V included in $U_s(C)$ as the maximum of two functions which are linearly semiconvex on V with $1/(r - s)$ as coefficient. The implication (a)⇒(b) is then justified.

The implication (b)⇒(c) is evident. Finally, for the implication (c)⇒(a), it is enough to proceed as in the proof of the similar implication in the proof of Proposition 15.37. □

The property (b) in Theorem 15.57 characterizes the r-prox-regularity of a set C as the semiconvexity with coefficient $1/(r-s)$ of d_C over any convex set included in $U_s(C)$ with $0 < s < r$. From this we see that for any open convex set $V \subset U_s(C)$ one has

$$\langle \zeta, x' - x\rangle \le d_C(x') - d_C(x) + \frac{1}{2(r-s)}\|x' - x\|^2, \ \forall x, x' \in V, \forall \zeta \in \partial_P d_C(x).$$

In fact, we show below that a similar global inequality holds true, without the restriction to the convex set V, but for all x, x' in the global open set $U_s(C)$.

Let us begin with the following lemma.

LEMMA 15.58. Let C be an r-prox-regular set in a Hilbert space H for some extended real $r \in]0, +\infty]$. Then, for all $x \in U_r(C)$ and $x' \in C$, one has

$$\left(1 - \frac{d_C(x)}{r}\right) \|\mathrm{proj}_C(x) - x'\|^2 \le \|x - x'\|^2 - d_C^2(x),$$

and in particular

$$\left(1 - \frac{d_C(x)}{r}\right) \|\mathrm{proj}_C(x) - x'\|^2 \le \|x - x'\|^2.$$

PROOF. Fix any $x \in U_r(C)$ and any $x' \in C$. We know that $y := \mathrm{proj}\,_C(x)$ exists by the r-prox-regularity of C (see (m) in Theorem 15.28). Notice that $x - y \in N^P(C; y)$. By the characterization (b) in Theorem 15.28 of r-prox-regularity we see that

$$\langle x - y, x' - y\rangle \le \frac{\|x - y\|}{2r}\|x' - y\|^2.$$

Using the equality

$$\|y - x'\|^2 - \|x - x'\|^2 = 2\langle x - y, x' - y\rangle - \|y - x\|^2,$$

we obtain

$$\|y - x'\|^2 - \|x - x'\|^2 \le \frac{d_C(x)}{r}\|y - x'\|^2 - d_C^2(x),$$

which corresponds to the first inequality of the lemma. The second inequality directly follows. □

PROPOSITION 15.59. Let C be a nonempty closed set in a Hilbert space H and let $r \in]0, +\infty]$. The following are equivalent:
(a) the set C is r-prox-regular;
(b) for any $x, x' \in U_r(C)$ and any $\zeta \in \partial_P d_C(x)$ one has

$$\langle \zeta, x' - x\rangle \le d_C(x') - d_C(x) + \frac{1}{2r(1 - r^{-1}d_C(x'))(1 - r^{-1}d_C(x))}\|x' - x\|^2;$$

(c) for any $s \in]0, r[$, any $x, x' \in U_s(C)$ and any $\zeta \in \partial_P d_C(x)$ one has

$$\langle \zeta, x' - x\rangle \le d_C(x') - d_C(x) + \frac{1}{2r(1 - s/r)^2}\|x' - x\|^2.$$

PROOF. Assume that C is r-prox-regular. Fix any $x, x' \in U_r(C)$ and $\zeta \in \partial_P d_C(x)$. We consider two cases.

Case 1: $x \in C$. By the r-prox-regularity of C we know that $y' := \mathrm{proj}_C(x')$ exists (see (m) in Theorem 15.28). Note also that $\zeta \in N^P(C;x) \cap \mathbb{B}_H$ according to the equality $\partial_P d_C(x) = N^P(C;x) \cap \mathbb{B}_H$ in Proposition 4.153 (keep in mind the assumption $x \in C$ in this case 1). Then, the r-prox-regularity of C again ensures (see (b) in Theorem 15.28) that

$$\begin{aligned}\langle \zeta, x' - x\rangle &= \langle \zeta, y' - x\rangle + \langle \zeta, x' - y'\rangle \\ &\le \frac{1}{2r}\|y' - x\|^2 + \|x' - y'\| = \frac{1}{2r}\|y' - x\|^2 + d_C(x').\end{aligned}$$

On the other hand, Lemma 15.58 says that

$$\left(1 - \frac{d_C(x')}{r}\right)\|y' - x\|^2 \le \|x' - x\|^2.$$

Combining the two latter inequalities, we obtain

$$\langle \zeta, x' - x\rangle \le \frac{1}{2(r - d_C(x'))}\|x' - x\|^2 + d_C(x'),$$

or equivalently

$$\langle \zeta, x' - x\rangle \le d_C(x') - d_C(x) + \frac{1}{2r(1 - r^{-1}d_C(x'))(1 - r^{-1}d_C(x))}\|x' - x\|^2.$$

Case 2: $x \notin C$. In this case, by Lemma 15.11(b) we have

$$\zeta = \frac{x - \mathrm{proj}_C(x)}{d_C(x)},$$

which entails that $\zeta \in N^P\big(C; \mathrm{proj}_C(x)\big) \cap \mathbb{S}_H$. Then we can write

$$\begin{aligned}\langle \zeta, x' - x\rangle &= \langle \zeta, x' - \mathrm{proj}_C(x)\rangle + \langle \zeta, \mathrm{proj}_C(x) - x\rangle \\ &= \langle \zeta, x' - \mathrm{proj}_C(x)\rangle - d_C(x) \\ &\le \frac{1}{2(r - d_C(x'))}\|x' - \mathrm{proj}_C(x)\|^2 + d_C(x') - d_C(x),\end{aligned} \tag{15.42}$$

where the latter inequality is due to the result in the case 1. Further, Lemma 15.58 gives that

$$\left(1 - \frac{d_C(x)}{r}\right)\|\mathrm{proj}(x) - x'\|^2 \le \|x - x'\|^2,$$

or equivalently

$$\frac{1}{2(r - d_C(x'))}\|x' - \mathrm{proj}_C(x)\|^2 \le \frac{1}{2r(1 - r^{-1}d_C(x'))(1 - r^{-1}d_C(x))}\|x' - x\|^2.$$

Putting this and (15.42) together we deduce that

$$\langle \zeta, x' - x\rangle \le \frac{1}{2r(1 - r^{-1}d_C(x'))(1 - r^{-1}d_C(x))}\|x' - x\|^2 + d_C(x') - d_C(x),$$

so the implication (a)⇒(b) is justified.

The implication (b)⇒(c) is direct, while the final implication (c)⇒(a) follows from Theorem 15.57. The proof is finished. □

Recalling that $\partial_C d_C(x) = \overline{\text{co}}\left({}^{\text{seq}}\operatorname*{Lim\,sup}_{u\to x} \partial_P d_C(u)\right)$ (see Corollary 4.157), we derive the following corollary.

COROLLARY 15.60. Let C be a nonempty closed set in a Hilbert space H and let $r \in]0, +\infty]$. The following are equivalent:
(a) the set C is r-prox-regular;
(b) for any $x, x' \in U_r(C)$ and any $\zeta \in \partial_C d_C(x)$ one has

$$\langle \zeta, x' - x\rangle \le d_C(x') - d_C(x) + \frac{1}{2r(1 - r^{-1}d_C(x'))(1 - r^{-1}d_C(x))}\|x' - x\|^2;$$

(c) for any $s \in]0, r[$, any $x, x' \in U_s(C)$ and any $\zeta \in \partial_C d_C(x)$ one has

$$\langle \zeta, x' - x\rangle \le d_C(x') - d_C(x) + \frac{1}{2r(1 - s/r)^2}\|x' - x\|^2.$$

15.2.4. Uniform prox-regularity of connected components. We discuss in the present subsection the uniform prox-regularity of union and of connected components. The first proposition is concerned with the connectedness of r-prox-regular sets with diameter less than r. Let us establish first a lemma.

LEMMA 15.61. Let C be an r-prox-regular set of a Hilbert space H with $r \in]0, +\infty]$. Then for any $x_0, x_1 \in C$ with $\|x_0 - x_1\| < 2r$ the mapping $\varphi : [0, 1] \to C$ with $\varphi(t) := P_C((1 - t)x_0 + tx_1)$ is well-defined and continuous, and it satisfies $\varphi(0) = x_0$ and $\varphi(1) = x_1$.

PROOF. By Theorem 15.28(m) we know that the mapping P_C is well-defined on $U_r(C)$ and norm-norm continuous therein. Let $x_0, x_1 \in C$ with $x_0 \neq x_1$ and such that $\|x_0 - x_1\| < 2r$. Putting $x_t := (1 - t)x_0 + tx_t$, we see that

$$d_C(x_t) \le \|x_t - x_0\| < r \text{ if } t \in [0, 1/2] \text{ and } d_C(x_t) \le \|x_t - x_1\| < r \text{ if } t \in [1/2, 1],$$

so $[x_0, x_1] \subset U_r(C)$. The mapping $\varphi : [0, 1] \to C$, given by $\varphi(t) := P_C\big((1 - t)x_0 + tx_1\big)$, is well-defined and continuous on $[0, 1]$ along with $\varphi(0) = x_0$ and $\varphi(1) = x_1$. $\square$

The desired proposition is a direct consequence of the previous lemma.

PROPOSITION 15.62. Let C be a nonempty r-prox-regular set of a Hilbert space H with $\operatorname{diam} C < 2r$, where $r \in]0, +\infty]$. Then C is arcwise connected.

Theorem 15.28(m) also allows us to derive the following condition for the prox-regularity of a union of sets.

PROPOSITION 15.63. Let $(C_i)_{i\in I}$ be a family of nonempty closed sets of a Hilbert space H such that for a real $r > 0$ the inequality $\operatorname{gap}(C_i, C_j) \ge 2r$ holds for all $i \neq j$ in I. Then the set $C := \bigcup_{i\in I} C_i$ is r-prox-regular if and only if each set C_i is r-prox-regular.

PROOF. Since $d_C(x) = \inf_{i\in I} d_{C_i}(x)$, it is clear that $U_r(C) = \bigcup_{i\in I} U_r(C_i)$. Since $\operatorname{gap}(C_i, C_j) \ge 2r$ for $i \neq j$, we see that $d_C(u) < r$ if and only if there is one and only one $i_u \in I$ such that $d(u, C_{i_u}) < r$. We deduce first that $d_C(u) = 0$ entails that $d(u, C_{i_u}) = 0$, hence $u \in C_{i_u} \subset C$. This tells us that C is closed. Secondly, we deduce also from the above equivalence that $U_r(C_i) \cap U_r(C_j) = \emptyset$ for $i \neq j$ and that for any $x \in U_r(C_i)$ one has $d_C(x) = d_{C_i}(x)$ and $\operatorname{Proj}_C(x) = \operatorname{Proj}_{C_i}(x)$ as well.

Therefore, the mapping P_C is well-defined on $U_r(C)$ and continuous therein if and only if for each $i \in I$ the mapping P_{C_i} is well-defined on $U_r(C_i)$ and continuous on $U_r(C_i)$. The equivalence in the proposition then results from Theorem 15.28(m). □

A complete criteria for the r-prox-regularity of a closed set via its connected components is the following:

PROPOSITION 15.64. Let C be a nonempty closed set of a Hilbert space H and let $r \in]0,+\infty[$. The set C is r-prox-regular if and only if all of its connected components are r-prox-regular and $\mathrm{gap}(C', C'') \geq 2r$ for any pair of its distinct connected components C', C''.

PROOF. The implication $\Longleftarrow$ follows directly from Proposition 15.63. Conversely, suppose that C is r-prox-regular and let $(C_i)_{i \in I}$ be the family of its distinct connected components (if any). By Lemma 15.61 it is clear that $\mathrm{gap}(C_i, C_j) \geq 2r$ for any $i \neq j$ in I, and hence each set C_i is r-prox-regular according to Proposition 15.63 again. □

15.2.5. Local (r,α)-prox-regularity. Our next step is to considering the local concept of r-prox-regularity at a point of C with α-latitude.

DEFINITION 15.65. Let C be a set in a Hilbert space H and let $\overline{x} \in C$. Given $r \in]0,+\infty]$ and $\alpha \in]0,+\infty]$, the set C is said to be *(r,α)-prox-regular at $\overline{x}$*, or *r-prox-regular at $\overline{x}$ with α-latitude*, if $C \cap B(\overline{x},\alpha)$ is closed relative to $B(\overline{x},\alpha)$ and if for any $x \in C \cap B(\overline{x},\alpha)$ and $v \in N^P(C;x) \cap \mathbb{B}_H$ one has

$$x \in \mathrm{Proj}_C(x+tv) \quad \text{for every positive real number } t \leq r. \tag{15.43}$$

We say that C is *r-prox-regular at $\overline{x}$* when it is (r,α)-prox-regular at $\overline{x}$ for some $\alpha > 0$. The set C is called *prox-regular at $\overline{x} \in C$* if it is r-prox-regular at $\overline{x}$ for some $r > 0$.

Obviously, for $\alpha = +\infty$ the above property for some $\overline{x} \in C$ corresponds to the (uniform) r-prox-regularity of C. It is also worth noticing that the closedness of C relative to an open set U of H (that is, with respect to the induced topology on U) is equivalent to the equality $U \cap C = U \cap \mathrm{cl}_H C$.

We also note that if $C \cap B(\overline{x},\alpha)$ is open in H for some real $\alpha > 0$, then the set C is r-prox-regular at $\overline{x}$ with $r = +\infty$ (that is, $(+\infty,\alpha')$-prox-regular at $\overline{x}$ for some $\alpha' \in]0,+\infty]$). So, any nonempty open set C of H is r-prox-regular at any of its points with $r = +\infty$.

The following feature, which deserves to be pointed out, is also clear from Definition 15.65, since the proximal normal cone is a local notion.

PROPOSITION 15.66. If a set C in a Hilbert space H is r-prox-regular at $\overline{x} \in C$ with $r \in]0,+\infty]$, then $C \cap U$ is also r-prox-regular at $\overline{x}$ for any neighborhood U of the point $\overline{x}$.

In the case of a closed subset C, the next proposition shows that the prox-regularity of C at any of its points is linked with the $\rho(\cdot)$-prox-regularity.

PROPOSITION 15.67. A nonempty closed set C of a Hilbert space H is prox-regular at any of its points if and only if it is $\rho(\cdot)$-prox-regular for some continuous function $\rho(\cdot)$ from C into $]0,+\infty]$.

PROOF. Suppose that C is $\rho(\cdot)$-prox-regular for some continuous function $\rho(\cdot)$ from C into $]0,+\infty]$ and fix any $\overline{x} \in C$. Taking any positive number $r < \rho(\overline{x})$, the continuity of $\rho(\cdot)$ yields some positive α such that $r < \rho(x)$ for all $x \in C \cap B(\overline{x}, \alpha)$. For any positive $t \leq r$, any $x \in C \cap B(\overline{x}, \alpha)$, and any $v \in N^P(C;x) \cap \mathbb{B}_H$ we then have $t < \rho(x)$, and hence by Definition 15.1 we obtain $x \in \operatorname{Proj}_C(x+tv)$. This justifies that C is prox-regular at $\overline{x}$.

Suppose now that C is prox-regular at any of its points, that is, for each $x \in C$ we have the nonemptiness of the set of positive r for which there exists some positive α such that for all $x' \in C \cap B(x,\alpha)$, $v \in N^P(C;x') \cap \mathbb{B}_H$, and $t \in]0,r[$ the inclusion $x' \in \operatorname{Proj}_C(x'+tv)$ holds. Denote by $s(x)$ the supremum of that set. It is easily checked that the function $s(\cdot)$ is lower semicontinuous on C relative to C with $s(x) \in]0,+\infty]$ for all $x \in C$, and hence by the paracompactness of the metric space C we may choose some continuous function $\rho(\cdot)$ on C with $0 < \rho(x) < s(x)$ for all $x \in C$. For each $x \in C$, the inequality $\rho(x) < s(x)$ guarantees the existence of some r_x with $\rho(x) < r_x < s(x)$ and some $\alpha_x > 0$ such that for all $x' \in C \cap B(x, \alpha_x)$, $v \in N^P(C;x') \cap \mathbb{B}_H$, and $t \in]0, r_x[$, we have $x' \in \operatorname{Proj}_C(x'+tv)$. Consequently, for any $x \in C$ and any $v \in N^P(C;x) \cap \mathbb{B}_H$ we have $x \in \operatorname{Proj}_C(x+tv)$ for all positive $t \leq \rho(x)$, that is, C is $\rho(\cdot)$-prox-regular according to Definition 15.1. □

The local r-prox-regularity is also related to the global r-prox-regularity as follows.

PROPOSITION 15.68. For an extended real $r \in]0,+\infty]$ a closed set C of a Hilbert space H is r-prox-regular if and only if it is r-prox-regular at any of its points.

PROOF. The r-prox-regularity of C obviously ensures the local property. Suppose that C is r-prox-regular at any of its points. Fix any $x \in C$ and $v \in N^P(C;x) \cap \mathbb{B}_H$. The r-prox-regularity of C at x gives by Definition 15.65 some $\alpha > 0$ such that, for any $x' \in C \cap B(x,\alpha)$ and any $v' \in N^P(C;x') \cap \mathbb{B}_H$, one has $x' \in \operatorname{Proj}_C(x'+tv')$ for any positive real number $t \leq r$, and hence in particular for $x' = x$ and $v' = v$ we obtain $x \in \operatorname{Proj}_C(x+tv)$ for any positive real number $t \leq r$. The set C is then r-prox-regular according to Definition 15.1. □

Despite Propositions 15.67 and 15.68, there are closed sets which are prox-regular at some points and not at some others. The closed set $C := \{(x,y) \in \mathbb{R}^2 : x \leq 0 \text{ or } y \leq 0\}$ is prox-regular at any nonzero point but it is not prox-regular at the origin, since otherwise it would be according to Proposition 15.67 $\rho(\cdot)$-prox-regular for some continuous function $\rho(\cdot)$ with positive values and Theorem 15.22 would ensure that there is some open set U containing C whose each point has a unique nearest point in C. This would contradict the fact that any point of $\Delta := \{(x,y) : x = y,\ x > 0\}$ has two nearest points in C. This leads us to study various local properties characterizing the local prox-regularity. To do so, let us introduce for $\overline{x} \in C$ and for positive r, α and $\gamma < 1$ the open sets

$$U_{r,\alpha}(\overline{x}, C) := \{u \in H : d_C(u) < r \text{ and } d_C(u) + \|u - \overline{x}\| < \alpha\}, \tag{15.44}$$

$$U^{\gamma}_{r,\alpha}(\overline{x}, C) := \{u \in H : d_C(u) < \gamma r \text{ and } d_C(u) + \|u - \overline{x}\| < \alpha\}. \tag{15.45}$$

Observe that both sets are contained in $B(\overline{x}, \alpha)$ and $U_{r,\alpha}(\overline{x}, C) = \bigcup_{\gamma \in]0,1[} U^{\gamma}_{r,\alpha}(\overline{x}, C)$.

LEMMA 15.69. Let $\alpha, r \in]0, +\infty]$ and let C be a set in a Hilbert space H such that $C \cap B(\overline{x}, \alpha)$ is closed relative to $B(\overline{x}, \alpha)$ with $\overline{x} \in C$. Let $U_{r,\alpha}(C)$ be defined as above and let $S = \mathrm{cl}_H(C)$.
(a) One has the equalities

$$U_{r,\alpha}(\overline{x}, C) = U_{r,\alpha}(\overline{x}, S) \quad \text{and} \quad U_{r,\alpha}(\overline{x}, C) \setminus C = U_{r,\alpha}(\overline{x}, C) \setminus S,$$

hence in particular, $U_{r,\alpha}(\overline{x}, C) \setminus C$ is open.
(b) One also has the implication

$$u \in U_{r,\alpha}(\overline{x}, C) \Longrightarrow \mathrm{Proj}_C(u) = \mathrm{Proj}_S(u).$$

PROOF. Since $d_S(\cdot) = d_C(\cdot)$, the equality $U_{r,\alpha}(\overline{x}, C) = U_{r,\alpha}(\overline{x}, S)$ directly follows from the definition in (15.44). By the inclusion $U_{r,\alpha}(\overline{x}, C) \subset B(\overline{x}, \alpha)$ and by the equality $S \cap B(\overline{x}, \alpha) = C \cap B(\overline{x}, \alpha)$ due to the assumption that $C \cap B(\overline{x}, \alpha)$ is closed relative to $B(\overline{x}, \alpha)$, we also see that $U_{r,\alpha}(\overline{x}, C) \setminus C = U_{r,\alpha}(\overline{x}, C) \setminus S$, so we have justified (a). Regarding the assertion (b), fix any $u \in U_{r,\alpha}(\overline{x}, C)$. It is clear that $\mathrm{Proj}_C(u) \subset \mathrm{Proj}_S(u)$ (in fact, this inclusion holds for any point in H). Conversely, take any $x \in \mathrm{Proj}_S(u)$ (if any), so $d_C(u) = \|u - x\|$. Then the inclusion $u \in U_{r,\alpha}(\overline{x}, C)$ gives

$$\|x - \overline{x}\| \le \|x - u\| + \|u - \overline{x}\| = d_C(u) + \|u - \overline{x}\| < \alpha,$$

which entails $x \in S \cap B(\overline{x}, \alpha)$. Since $S \cap B(\overline{x}, \alpha) = C \cap B(\overline{x}, \alpha)$, it ensues that $x \in C$, hence $x \in \mathrm{Proj}_C(u)$. The proof is finished. □

The following proposition provides some characterizations and properties and (r, α)-prox regular sets. Various other characterizations and properties of such sets will be established in Section 17.1 of Chapter 17.

THEOREM 15.70 (properties and characterizations of r-prox-regular sets with α-latitude). Let C be a subset of a Hilbert space H and let $\overline{x} \in C$ and $\alpha \in]0, +\infty]$ be such that $C \cap B(\overline{x}, \alpha)$ is closed relative to $B(\overline{x}, \alpha)$. For $r \in]0, +\infty]$ and $S := \mathrm{cl}_H(C)$ consider the following properties:
(a) C is r-prox-regular at $\overline{x}$ with α-latitude.
(b) For any $x \in C \cap B(\overline{x}, \alpha)$ and $v \in N^P(C; x) \cap \mathbb{B}_H$ one has

$$\langle v, x' - x \rangle \le \frac{1}{2r} \|x' - x\|^2 \quad \text{for all } x' \in C.$$

(b') For any $x \in C \cap B(\overline{x}, \alpha)$ and $v \in N^P(C; x) \cap \mathbb{B}_H$ one has

$$\langle v, x' - x \rangle \le \frac{1}{2r} \|x' - x\|^2 \quad \text{for all } x' \in S.$$

(c) For any $x \in C \cap B(\overline{x}, \alpha)$ and $v \in N^P(C; x)$ one has

$$\langle v, x' - x \rangle \le \frac{1}{2r} \|v\| \, \|x' - x\|^2 \quad \text{for all } x' \in C.$$

(c') For any $x \in C \cap B(\overline{x}, \alpha)$ and $v \in N^P(C; x)$ one has

$$\langle v, x' - x \rangle \le \frac{1}{2r} \|v\| \, \|x' - x\|^2 \quad \text{for all } x' \in S.$$

(d) For any $x \in C \cap B(\overline{x}, \alpha)$, any $v \in N^P(C; x) \cap \mathbb{B}_H$, and any nonnegative real number $t < r$ one has $x = P_C(x + tv)$.

(e) For any $x_i \in C \cap B(\overline{x}, \alpha)$ and $v_i \in N^P(C; x_i)$ with $i = 1, 2$ one has

$$\langle v_1 - v_2, x_1 - x_2 \rangle \geq -\frac{1}{2}\left(\frac{\|v_1\|}{r} + \frac{\|v_2\|}{r}\right)\|x_1 - x_2\|^2.$$

(f) For any $x_i \in C \cap B(\overline{x}, \alpha)$ and $v_i \in N^P(C; x_i) \cap \mathbb{B}_H$ with $i = 1, 2$ one has

$$\langle v_1 - v_2, x_1 - x_2 \rangle \geq -\frac{1}{r}\|x_1 - x_2\|^2,$$

that is, the multimapping $N^P(C; \cdot) \cap \mathbb{B}_H$ is $1/r$-hypomonotone on $B(\overline{x}, \alpha)$.
(g) For any $u_i \in H$ and for any $y_i \in (I + N^P(C; \cdot))^{-1}(u_i) \cap B(\overline{x}, \alpha)$ with $i = 1, 2$, one has

$$\langle y_1 - y_2, u_1 - u_2 \rangle \geq \left(1 - \frac{\|u_1 - y_1\|}{2r} - \frac{\|u_2 - y_2\|}{2r}\right)\|y_1 - y_2\|^2.$$

(h) For any positive $\gamma < 1$, any $u_i \in H$, and any $y_i \in (I + \gamma r\mathbb{B}_H \cap N^P(C; \cdot))^{-1}(u_i) \cap B(\overline{x}, \alpha)$ one has

$$\langle y_1 - y_2, u_1 - u_2 \rangle \geq (1 - \gamma)\|y_1 - y_2\|^2,$$

so in particular the multimapping $(I + \gamma r\mathbb{B}_H \cap N^P(C; \cdot))^{-1} \cap B(\overline{x}, \alpha)$ is single-valued on its effective domain.
(i) For any positive $\gamma < 1$ and any $u \in U^{\gamma}_{r,\alpha}(\overline{x}, C)$ one has that $P_C(u)$ exists, the equality

$$P_C(u) = (I + \gamma r\mathbb{B}_H \cap N^P(C; \cdot))^{-1}(u) \cap B(\overline{x}, \alpha)$$

holds, and the mapping P_C satisfies

$$\|P_C(u_1) - P_C(u_2)\| \leq (1 - \gamma)^{-1}\|u_1 - u_2\| \quad \text{for all } u_1, u_2 \in U^{\gamma}_{r,\alpha}(\overline{x}, C).$$

(j) For any positive $\gamma < 1$ the mapping P_C is well defined on $U^{\gamma}_{r,\alpha}(\overline{x}, C)$ and

$$P_C(u) = (I + \gamma r\mathbb{B}_H \cap N^P(C; \cdot))^{-1}(u) \cap B(\overline{x}, \alpha) \quad \text{for all } u \in U^{\gamma}_{r,\alpha}(\overline{x}, \alpha).$$

(k) For any positive $\gamma < 1$ and any $u \in U^{\gamma}_{r,\alpha}(\overline{x}, C)$ one has

$$\operatorname{Proj}_C(u) = (I + \gamma r\mathbb{B}_H \cap N^P(C; \cdot))^{-1}(u) \cap B(\overline{x}, \alpha).$$

(l) The mapping P_C is well defined on $U_{r,\alpha}(\overline{x}, C)$ with for all $u_1, u_2 \in U_{r,\alpha}(\overline{x}, C)$

$$\|P_C(u_1) - P_C(u_2)\| \leq \left(1 - \frac{d_C(u_1)}{2r} - \frac{d_C(u_2)}{2r}\right)^{-1}\|u_1 - u_2\|.$$

(m) The mapping P_C is well defined on $U_{r,\alpha}(\overline{x}, C)$ and locally Lipschitz continuous therein.
(n) The mapping P_C is well defined on $U_{r,\alpha}(\overline{x}, C)$ and norm-to-norm continuous therein.
(o) The mapping P_C is well defined on $U_{r,\alpha}(\overline{x}, C)$ and norm-to-weak continuous therein.
(p) The function d_C^2 is of class $\mathcal{C}^{1,1}$ on $U_{r,\alpha}(\overline{x}, C)$.
(q) The function d_C^2 is of class $\mathcal{C}^1$ on $U_{r,\alpha}(\overline{x}, C)$.
(r) The function d_C^2 is Fréchet differentiable on $U_{r,\alpha}(\overline{x}, C)$.
(s) The multimapping Proj_C takes nonempty values on $U_{r,\alpha}(\overline{x}, C)$, and d_C^2 is Gâteaux differentiable on $U_{r,\alpha}(\overline{x}, C)$.
(t) The function d_C is Gâteaux differentiable on the open set $U_{r,\alpha}(\overline{x}, C) \setminus S =$

$U_{r,\alpha}(\overline{x}, C) \setminus C$ with $\|\nabla^G d_C(u)\| = 1$ for all $u \in U_{r,\alpha}(\overline{x}, C) \setminus S = U_{r,\alpha}(\overline{x}, C) \setminus C$.
(u) For any $u \in U_{r,\alpha}(\overline{x}, C) \setminus C$ and $x = P_C(u)$ one has

$$x = P_C\left(x + t\frac{u - x}{\|u - x\|}\right) \quad \text{for all } t \in [0, r[.$$

(v) For all $x, y \in C \cap B(\overline{x}, \alpha)$ and for all $t \in [0, 1]$ such that $z_t \in U_{r,\alpha}(\overline{x}, C)$ one has

$$d_C(z_t) \leq \frac{1}{2r} \min\{t, 1 - t\}\|y - x\|^2,$$

where $z_t := (1 - t)x + ty$.
(w) $\partial_P d_C(u) \neq \emptyset$ for all $u \in U_{r,\alpha}(\overline{x}, C)$.
(x) $\partial_F d_C(u) \neq \emptyset$ for all $u \in U_{r,\alpha}(\overline{x}, C)$.

Then the following equivalences hold:

$$\text{(a)} \Leftrightarrow \text{(b)} \Leftrightarrow \text{(b')} \Leftrightarrow \text{(c)} \Leftrightarrow \text{(c')} \Leftrightarrow \text{(d)}, \quad \text{(e)} \Leftrightarrow \text{(g)}$$

$$\text{and} \quad \Leftrightarrow \text{(m)} \Leftrightarrow \cdots \Leftrightarrow \text{(u)} \Leftrightarrow \text{(w)} \Leftrightarrow \text{(x)}$$

as well as the implications:

$$\text{(c)} \Rightarrow \text{(e)}, \quad \text{(a)} \Rightarrow \text{(v)} \quad \text{and} \quad \text{(g)} \Rightarrow \text{(l)} \Rightarrow \text{(m)}.$$

In addition:
(α) If (k) holds, then for any positive real numbers $r' < r$ and α' such that $2r' + \alpha' \leq \alpha$ the set C is r'-prox-regular at $\overline{x}$ with α'-latitude.
(β) If (v) holds, then for each $\alpha' \in]0, \alpha[$ and $r' := \min\{r, (\alpha - \alpha')/2\}$ the set C is r'-prox-regular at $\overline{x}$ with α'-latitude.

PROOF. In addition to the evident equivalences (b)⇔(b') and (c)⇔(c'), the sequence of equivalences and implications will be proved as follows:

$$\begin{array}{ccccccccccccc}
\text{(g)} & \Leftrightarrow & \text{(e)} & \Leftarrow & \text{(c)} & \Leftrightarrow & \text{(a)} & \Leftrightarrow & \text{(b)} & \Leftrightarrow & \text{(d)} & \Leftarrow & \text{(u)} \\
\Downarrow & & & & & & & & & & & & \\
\text{(l)} & \Rightarrow & \text{(m)} & \Leftrightarrow & \text{(n)} & \Leftrightarrow & \text{(o)} & \Leftrightarrow & \text{(p)} & \Leftrightarrow & \text{(q)} & \Leftrightarrow & \text{(r)} & \Leftrightarrow & \text{(s)} \\
 & & & & & & & & \Downarrow & & & \nearrow & & & \Updownarrow \\
 & & & & & & & & \text{(w)} & \Rightarrow & \text{(x)} & & & & \text{(t)}
\end{array}$$

Denote as in the statement $S := \mathrm{cl}_H(C)$. The equivalence between the properties (a) – (d) is obtained like in Theorem 15.6 and it is evident that (c) entails (e). The property (e) obviously implies (f) and to see that (f) implies (h) it is enough to observe that, for $u_i \in H$ and $y_i \in (I + \gamma r\mathbb{B}_H \cap N^P(C; \cdot))^{-1}(u_i) \cap B(\overline{x}, \alpha)$ with $i = 1, 2$, we have $y_i \in C \cap B(\overline{x}, \alpha)$ and $(u_i - y_i)/(\gamma t) \in N^P(C; y_i) \cap \mathbb{B}_H$ for some positive real number $t \leq r$, which ensures by (f)

$$(\gamma t)^{-1}\langle (u_1 - y_1) - (u_2 - y_2), y_1 - y_2 \rangle \geq -r^{-1}\|y_1 - y_2\|^2 \geq -t^{-1}\|y_1 - y_2\|^2,$$

and this yields (h). In a similar way, we have (e) ⇒ (g), and obviously we have (g) ⇒ (e).

Let us show (h) ⇒ (i). Fix $\gamma \in [0, 1[$. Consider the multimapping

$$Q(\cdot) := (I + \gamma r\mathbb{B}_H \cap N^P(C; \cdot))^{-1}(\cdot) \cap B(\overline{x}, \alpha).$$

Take $u_i \in U^\gamma_{r,\alpha}(\overline{x}, C) \cap \mathrm{Dom}\, Q$ and $y_i \in Q(u_i)$, $i = 1, 2$. It follows from (h) that

$$\langle y_1 - y_2, u_1 - u_2 \rangle \geq (1 - \gamma)\|y_1 - y_2\|^2,$$

which entails that

$$\|y_1 - y_2\| \le (1-\gamma)^{-1}\|u_1 - u_2\|, \tag{15.46}$$

and hence in particular the multimapping Q is single-valued on the set $U^{\gamma}_{r,\alpha}(\overline{x}, C) \cap \operatorname{Dom} Q$.

We continue by taking any $u \in U^{\gamma}_{r,\alpha}(\overline{x}, C) \cap \operatorname{Dom} \operatorname{Proj}_C$ and any $y \in \operatorname{Proj}_C(u)$. We have $\|y - u\| = d_C(u) < \gamma r$ and $y \in (I + \gamma r\mathbb{B}_H \cap N^P(C;\cdot))^{-1}(u)$, so

$$\|y - \overline{x}\| \le \|y - u\| + \|u - \overline{x}\| = d_C(u) + \|u - \overline{x}\| < \alpha$$

because $u \in U^{\gamma}_{r,\alpha}(\overline{x}, C)$. It ensues that $y \in Q(u)$, thus $\operatorname{Proj}_C(u) \subset Q(u)$. It results that the multimapping Proj_C is single-valued on the set $U^{\gamma}_{r,\alpha}(\overline{x}, C) \cap \operatorname{Dom} \operatorname{Proj}_C$, coincides with Q there and is $(1-\gamma)^{-1}$-Lipschitz continuous there as well. Now consider the general case of any $u \in U^{\gamma}_{r,\alpha}(\overline{x}, C)$. Like in the proof of Theorem 15.6 there is some sequence $(u_n)_n$ of $U^{\gamma}_{r,\alpha}(\overline{x}, C) \cap \operatorname{Dom} \operatorname{Proj}_S$ converging to u according to the Lau theorem (see Theorem 14.29). Note by Lemma 15.69 that $\operatorname{Proj}_S(u_n) = \operatorname{Proj}_C(u_n)$. Then (15.46) ensures that the sequence $(y_n)_n$ with $y_n \in \operatorname{Proj}_S(u_n) = \operatorname{Proj}_C(u_n)$ is a Cauchy sequence, and one sees that its limit $y \in \operatorname{Proj}_S(u)$. By Lemma 15.69 again we have $y \in \operatorname{Proj}_C(u)$. It results that $U^{\gamma}_{r,\alpha}(\overline{x}, C) \cap \operatorname{Dom} \operatorname{Proj}_C = U^{\gamma}_{r,\alpha}(\overline{x}, C)$, and hence according to what precedes P_C is well defined on $U^{\gamma}_{r,\alpha}(\overline{x}, C)$ and $(1-\gamma)^{-1}$-Lipschitz continuous there, and $P_C(u) = Q(u)$ for all $u \in U^{\gamma}_{r,\alpha}(\overline{x}, C)$. The properties in (i) then hold.

Assume (g). By the above implications (g)$\Rightarrow$(e)$\Rightarrow$(f)$\Rightarrow$(h)$\Rightarrow$(i) the assertion (i) holds. We then have by (i) that $P_C(u_i)$ is well defined for $u_i \in U_{r,\alpha}(\overline{x}, C)$ and that

$$P_C(u_i) \in (I + N^P(C;\cdot))^{-1}(u_i) \cap B(\overline{x}, \alpha),$$

for $i = 1, 2$. We may then take $y_i = P_C(u_i)$ in (g) and we obtain

$$\|P_C(u_1) - P_C(u_2)\|^2 \le \left(1 - \frac{d_C(u_1)}{2r} - \frac{d_C(u_2)}{2r}\right)^{-1} \langle P_C(u_1) - P_C(u_2), u_1 - u_2\rangle,$$

and this obviously implies the inequality in (l). So we have proved (g) $\Rightarrow$ (l).

It is evident that the implications (i) $\Rightarrow$ (j) $\Rightarrow$ (k) and (l) $\Rightarrow$ (m) $\Rightarrow$ (n) are fulfilled. We also know by Lemma 15.69 that $U_{r,\alpha}(C) \setminus C$ is open. Further, on the one hand by Proposition 15.19 and Lemmas 15.11 and 15.69 the properties (m) – (t) are pairwise equivalent and on the other hand (p) $\Rightarrow$ (w) $\Rightarrow$ (x) $\Rightarrow$ (r), where the last implication follows from Lemma 15.11 (notice that the equality $U_{r,\alpha}(\overline{x}, C) \setminus S = U_{r,\alpha}(\overline{x}, C) \setminus C$ in (t) follows from Lemma 15.69).

Let us show (u) $\Rightarrow$ (d). Suppose (u) and take any $x \in C \cap B(\overline{x}, \alpha)$, any nonzero $v \in N^P(C;x) \cap \mathbb{B}_H$, and any nonzero $t < r$. Put $w := v/\|v\|$. By definition of $N^P(C;x)$ there is some $s > 0$ such that $x = P_C(x + sw)$. Since $x \in U_{r,\alpha}(\overline{x}, C)$, by Lemma 4.123 we may also suppose $x + sw \in U_{r,\alpha}(\overline{x}, C)$, thus $u := x + sw \in U_{r,\alpha}(\overline{x}, C) \setminus C$. Then for all non-negative $t < r$ the assumption (u) yields $x = P_C(x + t(u-x)/\|u-x\|)$, that is, $x = P_C(x + tv/\|v\|)$, hence $x = P_C(x + tv)$ thanks to the inequality $t \le t/\|v\|$ since $\|v\| \le 1$ and thanks to Lemma 4.123. This means that (d) is fulfilled.

We prove now that assertion (a) implies (v). Suppose (a) holds. Take $x, y \in C \cap B(\overline{x}, \alpha)$ and $t \in [0, 1]$ with $z_t := (1-t)x + ty \in U_{r,\alpha}(\overline{x}, C)$. Since (a) $\Rightarrow$ (i) follows from implications established above, by (i) we have that $P_C(z_t)$ exists and that $P_C(z_t) \in C \cap B(\overline{x}, \alpha)$. The assertion (c) (which is equivalent to (a)) entails that

$$\langle z_t - P_C(z_t), z - P_C(z_t)\rangle \le \frac{1}{2r}\|z_t - P_C(z_t)\|\,\|z - P_C(z_t)\|^2$$

for all $z \in C$. Taking separately $z = x$ and $z = y$ and proceeding like in Proposition 15.16 yields (v), so (a) $\Rightarrow$ (v).

Finally, we establish assertions (α) and (β). Let us start with (β). Suppose (v) and take any $\alpha' \in]0, \alpha[$. We note first that $C \cap B(\overline{x}, \alpha')$ is closed relative to $B(\overline{x}, \alpha')$. Fix $x \in C \cap B(\overline{x}, \alpha')$ and $v \in N^P(C; x) \cap \mathbb{B}_H$. Take any $y \in C$ with $\|x - y\| < \alpha - \alpha'$, hence $y \in C \cap B(\overline{x}, \alpha)$. Put $z_t := (1 - t)x + ty$ for all $t \in [0, 1]$. Since $z_t \to x$ as $t \to 0$, there exists some $\varepsilon > 0$ such that for all $t \in [0, \varepsilon]$ we have $d_C(z_t) < r$ and $d_C(z_t) + \|z_t - \overline{x}\| < \alpha$, that is, $z_t \in U_{r,\alpha}(C)$. On the other hand, $v \in \partial_P d_C(x)$ according to Proposition 4.153, and hence, by the variational characterization of proximal subgradient in Proposition 4.137(a), there is some positive σ and $\varepsilon_0 \le \varepsilon$ such that for all $t \in]0, \varepsilon_0]$

$$\langle v, t(y - x)\rangle = \langle v, z_t - x\rangle \le d_C(z_t) + \sigma\|z_t - x\|^2 = d_C(z_t) + \sigma t^2\|y - x\|^2.$$

We may deduce from what precedes and from assumption (v) that for all positive $t \le \varepsilon_0$

$$\langle v, y - x\rangle \le \frac{1}{2r}\|y - x\|^2 + \sigma t\|y - x\|^2,$$

and this gives as $t \downarrow 0$

$$\langle v, y - x\rangle \le \frac{1}{2r}\|y - x\|^2.$$

On the other hand, with $r' := \min\{r, (\alpha - \alpha')/2\}$ we see that, for any $y \in C$ with $\|y - x\| \ge \alpha - \alpha'$

$$\langle v, y - x\rangle \le \|y - x\| = \frac{2r'}{\|y - x\|}\frac{1}{2r'}\|x - y\|^2 \le \frac{1}{2r'}\|x - y\|^2.$$

It ensues that $\langle v, y - x\rangle \le \frac{1}{2r'}\|y - x\|^2$ for all $y \in C$. Then, by (b) of the theorem we obtain that C is r'-prox-regular at $\overline{x}$ with α'-latitude.

Suppose (k) and let us prove the assertion (α). Fix any positive real numbers $r' < r$ and α' such that $2r' + \alpha' \le \alpha$. Clearly, $C \cap B(\overline{x}, \alpha')$ is closed relative to $B(\overline{x}, \alpha')$. Take any $x \in C \cap B(\overline{x}, \alpha')$ and any $v \in N^P(C; x) \cap \mathbb{B}_H$. Consider any positive $t < r'$ and put $\gamma := t/r' < 1$. Observing that $d_C(x + tv) \le t < \gamma r$ and that

$$d_C(x + tv) + \|x + tv - \overline{x}\| \le t + \|x + tv - \overline{x}\| < 2t + \alpha' \le \alpha,$$

we see that $x + tv \in U^{\gamma}_{r,\alpha}(\overline{x}, C) \subset B(\overline{x}, \alpha)$. Further, we also have

$$x+tv \in \big(I+\gamma r\mathbb{B}_H \cap N^P(C;\cdot)\big)(x), \text{ hence } x \in \big(I+\gamma r\mathbb{B}_H \cap N^P(C;\cdot)\big)^{-1}(x+tv) \cap B(\overline{x}, \alpha).$$

By the assumption (k) we then obtain $x \in \mathrm{Proj}_C(x + tv)$ and the latter inclusion evidently still holds for $t = r'$. So C is r'-prox-regular with α'-latitude. This finishes the proof. $\square$

In the proof of Proposition 15.35, using (b) in Theorem 15.70 in place of (b) in Theorem 15.28 we immediately obtain the following result.

PROPOSITION 15.71. Let H_1 and H_2 be two Hilbert spaces, let U be a nonempty open convex set in H_1, and let $F : U \to H_2$ and $f : U \to \mathbb{R}$ be two differentiable mappings whose derivatives are L-Lipschitz on U. Then, for $H_1 \times H_2$ and $H_1 \times \mathbb{R}$

endowed with their canonical inner products, the graph gph F of F and the epigraph epi f of f are $1/L$-prox-regular at any of their points.

The next theorem establishes a long series of equivalent properties characterizing the local prox-regularity without any mention of α-latitude. We need first the following lemma.

LEMMA 15.72. Let H be a Hilbert space and $\delta, \eta \in]0, +\infty]$. Let C be a subset of H and let $\overline{x} \in C$ be such that $C \cap B(\overline{x}, \delta + \eta)$ is closed relative to $B(\overline{x}, \delta + \eta)$. Assume that the square distance d_C^2 is Fréchet differentiable on $B(\overline{x}, \delta + \eta)$. Then for any pair (u, x) with $u \in B(\overline{x}, \delta) \setminus C$ and $x = P_C(u)$ one has

$$x = P_C(u_t) \quad \text{for all } t \in [0, \eta[,$$

where $u_t := u + t\frac{u-x}{\|u-x\|} = x + (\|u - x\| + t)\frac{u-x}{\|u-x\|}$.

PROOF. Denote $S := \mathrm{cl}_H(C)$ and $U := B(\overline{x}, \delta + \eta)$. According to the Fréchet differentiability of d_C^2 on the open set U, Proposition 15.19 tells us that the mapping P_S is well defined on U and continuous therein. Fix any pair (u, x) with $u \in B(\overline{x}, \delta) \setminus C$ and $x = P_C(u)$, and note that $x = P_S(u)$ (since $P_S(u)$ exists) and that

$$B(\overline{x}, \delta) \setminus C = B(\overline{x}, \delta) \setminus S$$

by the closedness of $C \cap B(\overline{x}, \delta)$ relative to $B(\overline{x}, \delta)$. Take any real $t \in]0, \eta]$. By Lemma 15.20 there exists some $\tau_0 > 0$ such that $P_S(u_\tau) = x$ for all non-negative $\tau < \tau_0$, where $u_\tau := u + \tau(u - x)/\|u - x\|$. The set of all $\tau \in]0, t]$ such that $P_S(u_\tau) = x$ is then nonempty. Let s be its supremum and let $(\tau_n)_n$ be a sequence of that set tending to s. For each $n \in \mathbb{N}$ we have $d_C(u_{\tau_n}) = \tau_n + d_C(u)$ and $\|u_{\tau_n} - x\| = d_C(u_{\tau_n})$. Combining this with $u_{\tau_n} \to u_s$ as $n \to \infty$, we obtain $s + d_C(u) = d_C(u_s) = \|x - u_s\|$, and

$$\|u_s - \overline{x}\| \le \|u_s - u\| + \|u - \overline{x}\| = s + \|u - \overline{x}\| \le t + \|u - \overline{x}\|,$$

so $\|u_s - \overline{x}\| < \delta + \eta$, or equivalently $u_s \in U$. By the existence of P_S on U the equality $d_C(u_s) = \|x - u_s\|$ and the inclusion $u_s \in U$ ensure that $x = P_S(u_s)$. From this we see that $s = t$, since otherwise applying again Lemma 15.20 would furnish a contradiction with the supremum property of s. To conclude it remains to notice that $P_S(u_s) = P_C(u_s)$ since $u_s \in S \cap U = C \cap U$, where the equality is due to closeness property of $C \cap U$ relative to U. □

THEOREM 15.73 (characterizations of local prox-regularity for sets). Let C be a subset of a Hilbert space H which is closed near $\overline{x} \in C$. The following properties are equivalent:
(a) the set C is prox-regular at $\overline{x}$;
(b) there exist a neighborhood U of $\overline{x}$ and a real number $r > 0$ such that for all $x \in C \cap U$ and $v \in N^P(C; x) \cap \mathbb{B}_H$ one has

$$\langle v, x' - x\rangle \le \frac{1}{2r}\|x' - x\|^2 \quad \text{for all } x' \in C;$$

(b') there exist a neighborhood U of $\overline{x}$ and a real number $r > 0$ such that for all $x \in C \cap U$ and $v \in N^P(C; x) \cap \mathbb{B}_H$ one has

$$\langle v, x' - x\rangle \le \frac{1}{2r}\|x' - x\|^2 \quad \text{for all } x' \in C \cap U;$$

(c) there exist a neighborhood U of $\overline{x}$ and a real number $r > 0$ such that for all $x \in C \cap U$ and $v \in N^P(C;x)$ one has

$$\langle v, x' - x\rangle \le \frac{1}{2r}\|v\|\,\|x' - x\|^2 \quad \text{for all } x' \in C;$$

(c') there exist a neighborhood U of $\overline{x}$ and a real number $r > 0$ such that for all $x \in C \cap U$ and $v \in N^P(C;x)$ one has

$$\langle v, x' - x\rangle \le \frac{1}{2r}\|v\|\,\|x' - x\|^2 \quad \text{for all } x' \in C \cap U;$$

(d) there exist a neighborhood U of $\overline{x}$ and a real number $r > 0$ such that for any $x \in C \cap U$, any $v \in N^P(C;x) \cap \mathbb{B}_H$, and any nonnegative real number $t < r$ one has $x = P_C(x + tv)$;
(e) there exist a neighborhood U of $\overline{x}$ and a real number $r > 0$ such that for all $x_i \in C \cap U$ and $v_i \in N^P(C;x_i)$ with $i = 1, 2$ one has

$$\langle v_1 - v_2, x_1 - x_2\rangle \ge -\frac{1}{2}\left(\frac{\|v_1\|}{r} + \frac{\|v_2\|}{r}\right)\|x_1 - x_2\|^2;$$

(f) there exist a neighborhood U of $\overline{x}$ and a real number $r > 0$ such that for all $x_i \in C \cap U$ and $v_i \in N^P(C;x_i) \cap \mathbb{B}_H$ with $i = 1, 2$ one has

$$\langle v_1 - v_2, x_1 - x_2\rangle \ge -\frac{1}{r}\|x_1 - x_2\|^2,$$

that is, the multimapping $N^P(C;\cdot) \cap \mathbb{B}_H$ is $1/r$-hypomonotone on U;
(g) there exist a neighborhood U of $\overline{x}$ and a real number $r > 0$ such that for all $u_i \in H$ and $y_i \in (I + N^P(C;\cdot))^{-1}(u_i) \cap U$ with $i = 1, 2$, one has

$$\langle y_1 - y_2, u_1 - u_2\rangle \ge \left(1 - \frac{\|u_1 - y_1\|}{2r} - \frac{\|u_2 - y_2\|}{2r}\right)\|y_1 - y_2\|^2;$$

(h) there exist a neighborhood U of $\overline{x}$ and a real number $\beta > 0$ such that P_C is well defined on U and β-comonotone or β-cocoercive (hence monotone) there, that is,

$$\langle P_C(u_1), -P_C(u_2), u_1 - u_2\rangle \ge \beta\|P_C(u_1) - P_C(u_2)\|^2 \quad \text{for all } u_1, u_2 \in U;$$

(i) there exists a neighborhood U of $\overline{x}$ such that P_C is well defined on U and Lipschitz continuous on U;
(j) there exists a neighborhood U of $\overline{x}$ such that the mapping P_C is well defined on U and norm-to-norm continuous there;
(k) there exists a neighborhood U of $\overline{x}$ such that the mapping P_C is well defined on U and norm-to-weak continuous there;
(l) the function d_C^2 is of class $\mathcal{C}^{1,1}$ on some neighborhood U of $\overline{x}$;
(m) the function d_C^2 is of class $\mathcal{C}^1$ on some neighborhood U of $\overline{x}$;
(n) the function d_C^2 is Fréchet differentiable on some neighborhood U of $\overline{x}$;
(o) there exists a neighborhood U of $\overline{x}$ such that the multimapping Proj_C takes nonempty values on U and d_C^2 is Gâteaux differentiable on U;
(p) there exists a neighborhood U of $\overline{x}$ such that the function d_C is Gâteaux differentiable on $U \setminus C$ with $\|\nabla^G d_C(u)\| = 1$ for all $u \in U \setminus C$;
(q) there exist a neighborhood U of $\overline{x}$ and a real $r > 0$ such that for any $u \in U \setminus C$ and $x = P_C(u)$ one has

$$x = P_C\left(x + t\frac{u - x}{\|u - x\|}\right) \quad \text{for all } t \in [0, r[;$$

(r) there exist a neighborhood U of $\overline{x}$ and two real numbers $r > 0$ and $\varepsilon \in]0, 1]$ such that for all $x, y \in C \cap U$ and for all $t \in [0, \varepsilon]$ one has

$$d_C(z_t) \leq \frac{1}{2r} \min\{t, 1-t\} \|y - x\|^2,$$

where $z_t := (1-t)x + ty$;
(s) there exists a neighorhood U of $\overline{x}$ such that $\partial_P d_C(u) \neq \emptyset$ for all $u \in U$;
(t) there exists a neighborhood U of $\overline{x}$ such that $\partial_F d_C(u) \neq \emptyset$ for all $u \in U$.

If C is weakly closed, then one can add the following to the list of equivalences:
(u) P_C is well defined on some neighborhood U of $\overline{x}$.

PROOF. After noticing some evident facts, we will begin with the equivalences (b)⇔(b') and (c)⇔(c'), and then we will proceed with the following array.

$$\begin{array}{ccccccccccccccc}
(g) & \Leftrightarrow & (e) & \Leftarrow & (c) & \Leftrightarrow & (a) & \Leftrightarrow & (b) & \Leftrightarrow & (d) & \Leftarrow & (q) & & \\
\Downarrow & & & & & & & & & & & & \Uparrow & & \\
(h) & \Rightarrow & (i) & \Leftrightarrow & (j) & \Leftrightarrow & (k) & \Leftrightarrow & (l) & \Leftrightarrow & (m) & \Leftrightarrow & (n) & \Leftrightarrow & (o) \\
 & & & & & & & & \Downarrow & & & \nearrow & & & \Updownarrow \\
 & & & & & & & & (s) & \Rightarrow & (t) & & & & (p)
\end{array}$$

The equivalences (c) ⇔ (a) ⇔ (b) ⇔ (d) and (e) ⇔ (g) as well as the implication c ⇒ (e) are quite evident.

Assume (b') and let $r > 0$ and U as given therein. Choose a real $\delta > 0$ with $B(\overline{x}, 2\delta) \subset U$. Fix any $x \in C \cap B(\overline{x}, \delta)$ and any $v \in N^P(C; \overline{x}) \cap \mathbb{B}$. Take any $y \in C$. if $\|x - y\| < \delta$, then $y \in C \cap U$, so $\langle v, y - x\rangle \leq \frac{1}{2r}\|x - y\|^2$. If $\|x - y\| \geq \delta$, then

$$\langle v, y - x\rangle \leq \|x - y\| \leq \frac{1}{\delta}\|x - y\|^2.$$

For $r' := \min\{r, \delta/2\}$ we obtain $\langle v, y - x\rangle \leq \frac{1}{2r'}\|y - x\|^2$ for all $y \in C$. This yields (b')⇒(b), hence the equivalence (b)⇔(b') holds true.

Similarly, the equivalence (c)⇔(c') holds.

The implications (g)⇒(h)⇒(i) are direct.

The equivalences (i)⇔(j)⇔$\cdots$⇔(o) ⇔(p) follow easily from the closedness of C near $\overline{x} \in C$ and from Proposition 15.71 while the implications (l)⇒(s)⇒(t) are obvious from properties and definition of proximal subdifferential. The implication (t)⇒(n) can be seen by Lemma 15.11.

The implication (n)⇒(q) being a consequence of Lemma 15.72, it remains to show that (q)⇒(d). Assume (q) and let U and $r > 0$ as given by (q). Choose a real $\delta > 0$ such that $B(\overline{x}, \delta) \subset U$. Take any pair (x, v) with $x \in C \cap B(\overline{x}, \delta)$ and $v \in N^P(C; x) \cap \mathbb{B}$. Suppose $v \neq 0$ and set $w := v/\|v\|$. The definition of proximal normal gives some real $s > 0$ such that $x = P_C(x + sw)$ and $x + sw \in B(\overline{x}, \delta) \setminus C$. Putting $u := x + sw$, by the assumption (q) for every non-negative $t < r$ we have

$$x = P_C\left(x + t\frac{u - x}{\|u - x\|}\right), \quad \text{or equivalently } x = P_C\left(x + t\frac{v}{\|v\|}\right),$$

thus $x = P_C(x + tv)$ thanks to the inequality $t \leq t/\|v\|$ (since $\|v\| \leq 1$) and thanks to Lemma 4.123. The equality $x = P_C(x + tv)$ being still true if $v = 0$, we have then shown (q)⇒(d). The proof of the theorem is finished. □

The case of submanifold immediately deserves to be considered. We recall (see Definition E.8 in Appendix E) that a subset M of a Banach space X is a $\mathcal{C}^{1,1}$-submanifold at $m_0 \in M$ if there exist a closed vector subspace E in X, an open neighborhood U of m_0 in X, an open neighborhood V of zero in X, and a $\mathcal{C}^{1,1}$-diffeomorphism from U onto V with $\varphi(m_0) = 0$ such that $\varphi(M \cap U) = E \cap V$.

PROPOSITION 15.74. (a) If a subset M of a Hilbert space H is a $\mathcal{C}^{1,1}$-submanifold at $m_0 \in M$, then it is prox-regular at m_0.
(b) If $G : H \to Y$ is a mapping from a Hilbert space H into a Banach space Y which is of class $\mathcal{C}^{1,1}$ near a point $\overline{x} \in H$ with $DG(\overline{x})$ surjective and if C is a closed convex set of Y containing $G(\overline{x})$, then the set $G^{-1}(C)$ is prox-regular at $\overline{x}$.

PROOF. We follow the proof of Proposition 8.70.
(b) Let us prove (b) first. Clearly, $G^{-1}(C)$ is closed around $\overline{x}$. On the other hand, by the Banach-Schauder open mapping theorem (see Theorem C.3 in Appendix C) there exists a real $s > 0$ such that $s\mathbb{B}_Y \subset DG(\overline{x})(\mathbb{B}_Y)$. Then Lemma 8.68 furnishes an open neighborhood U of $\overline{x}$ and a real $\gamma > 0$ such that for each $x \in U$ the continuous linear mapping $DG(x)$ is open and $\|y^*\| \le \gamma \|x^*\|$ for all $x^* \in \mathcal{L}(H, \mathbb{R})$ and $y^* \in Y^*$ satisfying $x^* = y^* \circ DG(x)$. Choose an open convex neighborhood $U_0 \subset U$ of $\overline{x}$ such that DG is L-Lipschitz on U_0 for some real $L \ge 0$. Consider any $x, u \in U_0 \cap G^{-1}(C)$ and any $v \in N^C\left(G^{-1}(C); x\right) \cap \mathbb{B}_H$. By Theorem 3.174 there is $y^* \in N^C(C; G(x))$ such that $\langle v, \cdot \rangle = y^* \circ DG(x)$, so $\|y^*\| \le \gamma$ by the choice of γ. Since $\langle y^*, G(u) - G(x) \rangle \le 0$ by convexity of C, we deduce that

$$\begin{aligned}
\langle v, u - x \rangle &= \langle y^*, DG(x)(u - x) \rangle \\
&= \langle y^*, G(u) - G(x) \rangle - \left\langle y^*, \int_0^1 \big(DG(x + t(u - x)) - DG(x)\big)(u - x)\, dt \right\rangle \\
&\le \|y^*\| L \|u - x\|^2 \int_0^1 t\, dt \le \frac{\gamma L}{2} \|u - x\|^2.
\end{aligned}$$

Consequently, (b) in Theorem 15.73 guarantees the prox-regularity of $G^{-1}(C)$ at the point $\overline{x}$.
(a) By definition of $\mathcal{C}^{1,1}$-submanifold (recalled above) there are a closed vector subspace E of H, an open neighborhood U of m_0 in H, a convex neighborhood V of zero in X, and a $\mathcal{C}^{1,1}$ diffeomorphism $\varphi : U \to V$ with $\varphi(m_0) = 0$ such that $M \cap U = \varphi^{-1}(E \cap V)$. Then by (b) proved above, the set M is prox-regular at m_0, which finishes the proof. □

The following local properties corresponding to Proposition 15.13 and Theorem 15.14 also hold with the same proofs.

PROPOSITION 15.75. Let C be a subset of a Hilbert space H and $\overline{x} \in C$. Let $\alpha \in]0, +\infty]$ be such that $C \cap B(\overline{x}, \alpha)$ is closed relative to $B(\overline{x}, \alpha)$. The following hold:
(a) If C is r-prox-regular at $\overline{x}$ with α-latitude, then the normal cone multimapping $N^P(C; \cdot)$ is sequentially norm-to-weak closed relative to $B(\overline{x}, \alpha)$, that is, for any sequence $(x_n)_n$ in $C \cap B(\overline{x}, \alpha)$ converging to $x \in C \cap B(\overline{x}, \alpha)$ and any sequence $(v_n)_n$ in H converging weakly to v with $v_n \in N^P(C; x_n)$, one has $v \in N^P(C; x)$.
(b) If C is r-prox-regular with α-latitude (resp. prox-regular) at $\overline{x}$, then for all $x \in C \cap B(\overline{x}, \alpha)$ (resp. in some neighborhood of $\overline{x}$ relative to C) the normal regularity

$$N^P(C; x) = N^F(C; x) = N^L(C; x) = N^A(C; x) = N^C(C; x)$$

is fulfilled along with the tangential regularity

$$T^B(C;x) = T^C(C;x).$$

(c) The set C is r-prox-regular with α-latitude (resp. prox-regular) at $\overline{x}$ if and only if anyone of the properties (b) – (d) of Theorem 15.70 (resp. anyone of the properties (b) – (g) of Theorem 15.73) holds with anyone of the normal cones $N^F(C;\cdot)$, $N^L(C;\cdot)$,$N^A(C;\cdot)$, $N^C(C;\cdot)$ in place of $N^P(C;\cdot)$.
(d) The set C is r-prox-regular at $\overline{x}$ with α-latitude if and only if

$$d(x'-x, T^B(C;x)) \leq \frac{1}{2r}\|x'-x\|^2 \quad \text{for all } (x,x') \in C \cap B(\overline{x},\alpha) \times C.$$

(e) The set C is prox-regular at $\overline{x}$ if and only if there exist a neighborhood U of $\overline{x}$ and a real number $r > 0$ such that

$$d(x'-x, T^B(C;x)) \leq \frac{1}{2r}\|x'-x\|^2 \quad \text{for all } x, x' \in C \cap U.$$

REMARK 15.76. (a) For a set C of a normed space X which is epi-Lipschitz at $\overline{x} \in C$, we saw in Chapter 2 that the intersection with $U \times X^*$ of the graph of the Clarke normal cone $N^C(C;\cdot)$ is strongly$\times$weakly* sequentially closed relative to $U \times X^*$, for some neighborhood U of $\overline{x}$. From (b) in the above proposition and from (c) in Theorem 15.73 this closedness property also holds true for any set of a Hilbert space H which is prox-regular at $\overline{x}$.

Assume now that H is a *finite-dimensional* Euclidean space.
(b) We know that the Mordukhovich limiting normal cone at any boundary point of a closed set of the finite-dimensional space H is nonzero (see Corollary 4.90). So, for $x \in \operatorname{bdry} C$, by (b) in the above proposition one has $N^P(C;x) \neq \{0\}$ whenever C is prox-regular at x. □

We will see next that the converse implications of (a) and (b) in Proposition 15.75 and of (a) in Remark 15.76 fail. Before that we notice that Definition 8.57, Theorem 15.73(f) and Proposition 15.75 (resp. Definition 8.57, Theorem 15.28(e) and Proposition 15.13(b)) yield the following comparisons with the subsmoothness property for sets.

PROPOSITION 15.77. Let C be a nonempty set of a Hilbert space H.
(a) If C is closed near $\overline{x} \in C$ and prox-regular at $\overline{x}$, then C is subsmooth at $\overline{x}$.
(b) If C is closed and (uniformly) r-prox-regular for some $r \in]0,+\infty]$, then C is uniformly subsmooth.

EXAMPLE 15.78. The converse properties of (a) and (b) in Proposition 15.77 fail. Consider the $\mathcal{C}^1$ function $f : \mathbb{R} \to \mathbb{R}$ in Example 4.138(b) defined by $f(x) = |x|^{3/2}$ if $x \leq 0$ and $f(x) = -|x|^{3/2}$ if $x > 0$. The set $C := \operatorname{epi} f$ in $\mathbb{R}^2$ is subsmooth at the origin by Corollary 8.86. However, since $N^P(C;(0,0)) = 0$ (because $\partial_P f(0) = \emptyset$ for this locally Lipschitz function f, as seen in Example 4.138(b)), the set C is not prox-regular at the origin according to Remark 15.76(b). □

EXAMPLE 15.79. Example 8.87 furnished an epi-Lipschitz set in $\mathbb{R}^2$ which is tangentially regular at any of its points but not subsmooth at $(0, f(0))$. By Proposition 15.77 this tangentially regular set is neither prox-regular at $(0, f(0))$. It results that, relative to the prox-regularity, the converse implications of (a) and (b) in Proposition 15.75 and of (a) in Remark 15.76 fail. □

Example 15.78 and Example 15.79 lead to ask whether, for an epi-Lipschitz set C, the stronger normal regularity $N^C(C;x) = N^P(C;x)$ for all $x \in C$ near $\overline{x}$ entails the prox-regularity of C at $\overline{x}$. The answer is negative as shown in the following example.

EXAMPLE 15.80 (F.H. Clarke, R.J. Stern and P.R. Wolenski). Consider a decreasing sequence $(r_n)_n$ in $]0,1[$ tending to 0 and a decreasing sequence $(t_n)_n$ in $]0,1[$ such that $0 < t_n - r_n$, $t_n + r_n < 1$ and $t_{n+1} + r_{n+1} < t_n - r_n$. Choose a decreasing sequence $(\varepsilon_n)_n$ of $]0,+\infty[$ tending to zero and put $\alpha_n = \varepsilon_n r_n/\sqrt{1+\varepsilon_n^2}$. The open intervals $]t_n, -\alpha_n, t_n + \alpha_n[$ are all pairwise disjoint and included in $]0,1[$ Consider the epigraph C of the function f defined on $\mathbb{R}$ by

$$f(t) = \begin{cases} \sqrt{r_n^2 - (t-t_n)^2} - \sqrt{r_n^2 - \alpha_n^2} & \text{if } |t - t_n| \le \alpha_n \\ 0 & \text{otherwise.} \end{cases}$$

For any $t \in]t_n - \alpha_n, t_n + \alpha_n[$ we have

$$f'(t) = \frac{-(t-t_n)}{\sqrt{r_n^2 - (t-t_n)^2}},$$

so $|f'(t)| \le \varepsilon_n$, $|f'(\sigma_n + 0)| \le \varepsilon_n$ and $|f'(\tau_n - 0)| \le \varepsilon_n$ for $\sigma_n := t_n - \alpha_n$ and $\tau_n := t_n + \alpha_n$. The function f is then Lipschitz continuous. Furthermore, it is easily seen that the proximal and Clarke normal cones of the set C coincide. However, the circles centered at $(t_n, -\sqrt{r_n^2 - \alpha_n^2})$ with radius r_n allow us to see that C is not prox-regular at $(0,0)$ since each point $(t_n, -\sqrt{r_n^2 - \alpha_n^2})$ has many nearest points in C. □

The next result concerning the Hausdorff-Pompeiu excess $\mathrm{exc}(\cdot,\cdot)$ (see Definition 1.61) between the truncated normal cones is a direct consequence of Proposition 1.76 and Proposition 15.75(a).

PROPOSITION 15.81. Assume H is a finite-dimensional Euclidean space that $C \subset H$ is prox-regular at $\overline{x} \in C$. Then, for any real $s > 0$ one has the convergence property

$$\mathrm{exc}\Big(N^P(C;x) \cap s\mathbb{B}, N^P(C;\overline{x}) \cap s\mathbb{B}\Big) \to 0 \quad \text{as } C \ni x \to \overline{x}.$$

REMARK 15.82. The convergence $\mathrm{exc}\Big(N^P(C;\overline{x}) \cap s\mathbb{B}, N^P(C;x) \cap s\mathbb{B}\Big) \to 0$ as $C \ni x \to \overline{x}$ fails even for convex sets. In the Euclidean space $\mathbb{R}^2$, it suffices to take as C a triangle and as $\overline{x}$ anyone of its vertices. □

15.2.6. Directional derivability of the metric projection. In this subsection we show that the metric projection on a prox-regular set C is directionally derivable at boundary points of C. The result can also be seen as derivability in a conical sense. In the statement, it will be convenient to use in some places the notation proj_C instead of P_C.

THEOREM 15.83 (directional derivative of metric projection). Let C be a subset of a Hilbert space H which is closed near $\overline{x} \in C$ and prox-regular at $\overline{x}$. Then there exists some open neighborhood U of $\overline{x}$ such that for any $x \in C \cap U$ and any $y \in H$ one has

$$(\mathrm{proj}_C)'(x;y) := \lim_{t \downarrow 0} \frac{1}{t}\Big(\mathrm{proj}_C(x+ty) - x\Big) = \mathrm{proj}_{T^B(C;x)}(y)$$

and the directional derivative $d'_C(x;y) := \lim_{t\downarrow 0} \frac{1}{t}\big(d_C(x+ty) - d_C(x)\big)$ of d_C at x in the direction y exists and

$$d'_C(x;y) = d\big(y, T^B(C;x)\big),$$

which translates the fact that $d'_C(x;\cdot)$ is a conical derivative.

PROOF. Without loss of generality we may suppose that the set C is closed. By Theorem 15.73 there exist $\varepsilon, r > 0$ such that the mapping P_C is well defined on $B(\overline{x}, \varepsilon)$ and Lipschitz continuous therein, with $\beta \geq 0$ as a Lipschitz constant, and such that

$$\langle v' - v, x - x'\rangle \leq \frac{1}{r}\|x' - x\|^2 \tag{15.47}$$

for all $x, x' \in C \cap B(\overline{x}, \varepsilon)$ and all $v \in N^P(C;x)$, $v' \in N^P(C;x')$ with $\|v\| \leq 1$ and $\|v'\| \leq 1$, and this says in particular that C is prox-regular at each $x \in C \cap B(\overline{x}, \varepsilon)$. Fix any $x \in C \cap B(\overline{x}, \varepsilon)$ and any $y \in H$. Choose $t_0 > 0$ such that for all $t \in]0, t_0]$ we have $x + ty \in B(\overline{x}, \varepsilon)$ and $P_C(x + ty) \in B(\overline{x}, \varepsilon)$. For any $t \in]0, t_0]$ putting $h'_t := \frac{1}{t}\big(P_C(x+ty) - x\big)$, we see that

$$y - h'_t = \frac{1}{t}\big(x + ty - P_C(x+ty)\big),$$

and hence $y - h'_t \in N^P\big(C; P_C(x+ty)\big)$.

Take now any sequence $(t_n)_n$ of $]0, t_0]$ converging to 0 and put $h''_n := h'_{t_n}$. The sequence $(y - h''_n)_n$ being bounded (because of the Lipschitz property of P_C), it has a subsequence $(y - h''_{\sigma(n)})_n$ converging weakly to $y - h$ for some vector $h \in H$. As $N^P(C;x) = N^L(C;x)$ (see Proposition 15.75(b)), we have $y - h \in N^P(C;x)$. Put $h_n := h''_{\sigma(n)}$ for all $n \in \mathbb{N}$. By (15.47) there exists some constant $\gamma > 0$ independent of n such that for $\lambda_n := t_{\sigma(n)}$

$$\langle (y - h_n) - (y - h), x - P_C(x + \lambda_n y)\rangle \leq \frac{\gamma}{r}\|P_C(x + \lambda_n y) - x\|^2.$$

The latter inequality is equivalent to

$$\langle h - h_n, -\lambda_n h_n\rangle \leq \frac{\gamma}{r}\|P_C(x + \lambda_n y) - x\|^2,$$

that is,

$$\|h - h_n\|^2 - \langle h, h - h_n\rangle \leq \frac{\gamma}{r}\frac{1}{\lambda_n}\|P_C(x + \lambda_n y) - x\|^2.$$

Taking the Lipschitz property of P_C into account, we obtain

$$\|h - h_n\|^2 \leq \langle h, h - h_n\rangle + \frac{\beta^2 \lambda_n \gamma}{r}\|y\|^2.$$

So, the sequence $(h_n)_n$ converges strongly to h and by definition of h_n we have $h \in T^B(C;x)$ (because $x + \lambda_n h_n = P_C(x + \lambda_n y) \in C$ for all n).

Take now any $z \in T^B(C;x)$. Since $T^B(C;x) = T^C(C;x)$ according to Proposition 15.75(b) again, there is a sequence $(z_n)_n$ converging to z such that $x + \lambda_n z_n \in C$ for all n. This entails that for all n

$$\|x + \lambda_n y - (x + \lambda_n z_n)\| \geq \|x + \lambda_n y - P_C(x + \lambda_n y)\|,$$

which is equivalent to

$$\|y - z_n\| \geq \left\| y + \frac{1}{\lambda_n}\big(x - P_C(x + \lambda_n y)\big)\right\|.$$

Then, passing to the limit as $n \to \infty$, we get $\|y - z\| \geq \|y - h\|$ for all z in the closed convex cone $T^B(C;x) = T^C(C;x)$, which means that $h = \operatorname{proj}\big(y, T^B(C;x)\big)$. So, for any sequence of positive numbers $(t_n)_n$ converging to 0 there exists a subsequence $(t_{\sigma(n)})_n$ such that

$$\lim_{n\to\infty} \frac{1}{t_{\sigma(n)}} \big(P_C(x + t_{\sigma(n)}y) - x\big) = \operatorname{proj}\big(y, T^B(C;x)\big).$$

Consequently,

$$\lim_{t\downarrow 0} \frac{1}{t}\big(P_C(x+ty) - x\big) = \operatorname{proj}\big(y, T^B(C;x)\big), \tag{15.48}$$

which is the first equality of the theorem.

Regarding the equality relative to the directional derivative of the distance function d_C, observe first that for any positive number t small enough

$$\begin{aligned}\frac{1}{t}\big(d_C(x+ty) - d_C(x)\big) = \frac{1}{t} d_C(x+ty) &= \frac{1}{t}\|P_C(x+ty) - (x+ty)\| \\ &= \left\|\frac{1}{t}\big(P_C(x+ty) - x\big) - y\right\|.\end{aligned}$$

Then, taking (15.48) into account and passing to the limit as $t \downarrow 0$ give that the directional derivative $d'_C(x;y)$ exists and

$$d'_C(x;y) = \|\operatorname{proj}\big(y, T^B(C;x)\big) - y\| = d\big(y, T^B(C;x)\big),$$

which completes the proof. □

The immediate case of a convex set deserves to be stated.

COROLLARY 15.84 (E.H. Zarontonello). *Let C be a closed convex set in a Hilbert space. Then for any $x \in C$ and any $y \in H$ one has*

$$(\operatorname{proj}_C)'(x;y) := \lim_{t\downarrow 0} \frac{1}{t}\big(\operatorname{proj}_C(x+ty) - x\big) = \operatorname{proj}_{T^B(C;x)}(y)$$

and the directional derivative $d'_C(x;y) := \lim_{t\downarrow 0} \frac{1}{t}\big(d_C(x+ty) - d_C(x)\big)$ of d_C at x in the direction y exists and

$$d'_C(x;y) = d\big(y, T^B(C;x)\big).$$

As another consequence of Theorem 15.83 we compute the second order directional derivatives of the square distance function at boundary points.

COROLLARY 15.85. *Let C be a subset of a Hilbert space H which is closed near $\overline{x} \in C$ and prox-regular at $\overline{x}$. Then there exists some open neighborhood U of $\overline{x}$ such that for any $x \in C \cap U$ and any $y \in H$ one has*

$$\left(\frac{1}{2}\nabla d_C^2\right)'(x;y) := \lim_{t\downarrow 0} \frac{1}{t}\left(\frac{1}{2}\nabla d_C^2(x+ty) - \frac{1}{2}\nabla d_C^2(x)\right) = y - \operatorname{proj}\big(y, T^B(C;x)\big).$$

PROOF. It is an immediate consequence of the above theorem and of the explicit computation of the gradient of $d_C^2(\cdot)$ (see (k) in Theorem 15.22). □

The directional derivative of P_C can also be written in terms of the normal cone.

COROLLARY 15.86. Let C be a subset of a Hilbert space H which is closed near $\overline{x} \in C$ and prox-regular at $\overline{x}$. Then there exists some open neighborhood U of $\overline{x}$ such that for any $x \in C \cap U$ and any $y \in H$ one has

$$(\mathrm{proj}_C)'(x; y) = \mathrm{proj}\big(0, y - N(C; x)\big),$$

where $N(C; \cdot)$ denotes any normal cone $N^P(C; \cdot)$, $N^F(C; \cdot)$, $N^L(C; \cdot)$, $N^C(C; \cdot)$.

PROOF. We know that there is an open neighborhood U of $\overline{x}$ such that C is prox-regular at any point in $C \cap U$. Since $N(C; x)$ is the negative polar of the closed convex cone $T^B(C; x)$, the desired equality follows from Theorem 15.83 above and from the following Moreau decomposition theorem for cones. □

In fact, the Moreau decomposition for cones is a consequence of Theorem 3.268 by taking the indicator function Ψ_K of a closed convex cone K as function f there. We also provide below a direct simple proof.

THEOREM 15.87 (Moreau decomposition theorem for polar cones). Let K be a closed convex cone of a Hilbert space H and K° its (negative) polar, that is,

$$K^\circ = \{y \in H : \langle y, x \rangle \le 0, \ \forall x \in K\}.$$

Then, for all $u \in H$

$$u = P_K(u) + P_{K^\circ}(u) \quad \text{with } \langle P_K(u), P_{K^\circ}(u) \rangle = 0.$$

In fact, any $u \in H$ can be decomposed in a unique way as

$$u = v + v' \quad \text{with } v \in K,\ v' \in K^\circ, \text{ and } \langle v', v \rangle = 0.$$

PROOF. Fix any $u \in H$. For any real $t \ge 0$ and any $h \in K$ we can write

$$\langle u - P_K(u), th - P_K(u) \rangle \le 0,$$

so taking $t = 0$ gives $\langle u - P_K(u), -P_K(u) \rangle \le 0$, and on the other hand, dividing by $t > 0$ and making $t \to +\infty$ yields $\langle u - P_K(u), h \rangle \le 0$. From this inequality it follows that $u - P_K(u) \in K^\circ$, and both latter inequalities together with $h = P_K(u)$ entail the equality $\langle u - P_K(u), P_K(u) \rangle = 0$. The equality $u = P_K(u) + (u - P_K(u))$ along with $\langle P_K(u), u - P_K(u) \rangle = 0$ furnishes a certain desired decomposition.

Concerning the uniqueness, suppose that $u = v + v'$ is another decomposition with $v \in K$, $v' \in K^\circ$, and $\langle v', v \rangle = 0$. Then, for any $h \in K$ we have

$$\langle u - v, h - v \rangle = \langle v', h - v \rangle = \langle v', h \rangle \le 0,$$

where the latter inequality is due to the inclusion $v' \in K^\circ$. This guarantees that $v = P_K(u)$. Since $(K^\circ)^\circ = K$, by symmetry we obtain $v' = P_{K^\circ}(u)$. Consequently, the uniqueness of the decomposition follows as well as the equality $u - P_K(u) = P_{K^\circ}(u)$, and the proof is finished. □

Given a closed cone K of a normed space $(X, \|\cdot\|)$, any $v \in X$ and any real $t > 0$, we observe that, for any $y \in K$ and any $z \in \mathrm{Proj}_K(v)$

$$\|tv - tz\| = t\|v - z\| \le t\|v - t^{-1}y\| = \|tv - y\|,$$

hence $tz \in \mathrm{Proj}_K(tv)$. It ensues that $t\mathrm{Proj}_K(v) \subset \mathrm{Proj}_K(tv)$, and this inclusion in turn (with t replaced by t^{-1} and v replaced by tv) gives $t^{-1}\mathrm{Proj}_K(tv) \subset \mathrm{Proj}_K(v)$. It results that

$$t\mathrm{Proj}_K(v) = \mathrm{Proj}_K(tv) \quad \text{for all } t > 0,$$

With closed convex cone in a Hilbert space, taking into account the latter equality which still holds with $t = 0$ in this case, it is worth noticing the following corollary of Moreau decomposition theorem.

COROLLARY 15.88. Let K be a closed convex cone of a Hilbert space H. Then, the function $v \mapsto \|P_K(v)\|$ is sublinear, that is, convex and positively homogeneous; in fact, for all $u, v \in H$,

$$\|P_K(u+v)\| \leq \|P_K(u) + P_K(v)\| \leq \|P_K(u)\| + \|P_K(v)\|.$$

PROOF. We know that the mapping P_K is well defined and it is positively homogeneous by the equality preceding the statement of the corollary. Fixing any $u, v \in H$ and writing, by the Moreau decomposition theorem and by the inclusion $P_{K^\circ}(u) + P_{K^\circ}(v) \in K^\circ$,

$$\begin{aligned}\|P_K(u+v)\| = \|u+v-P_{K^\circ}(u+v)\| &\leq \|u+v-\big(P_{K^\circ}(u)+P_{K^\circ}(v)\big)\| \\ &= \|P_K u) + P_K(v)\| \leq \|P_K(u)\| + \|P_K(v)\|,\end{aligned}$$

the desired inequalities are justified. □

The next other property concerns a stability result for the metric projection to the tangent cone of a prox-regular set. In fact, it will follow from a more general result and from the sequential norm-to-weak closedness property in Proposition 15.75(a).

PROPOSITION 15.89. Let C be a subset of a Hilbert space H which is closed near $\overline{x} \in C$ and whose Clarke normal cone multimapping is sequentially norm-to-weak closed at $\overline{x}$. Then, for any $v \in T^C(C; \overline{x})$, one has

$$\lim_{C \ni x \to \overline{x}} \operatorname{proj}\big(v, T^C(C;x)\big) = v.$$

In particular, the latter equality holds true whenever the set C is either prox-regular at $\overline{x}$, or subsmooth at $\overline{x}$, or epi-Lipschitz at $\overline{x}$.

PROOF. If C is prox-regular at $\overline{x}$, we know by Proposition 15.75 that the Clarke and proximal normal cones coincide near $\overline{x}$ and that the multimapping $N^C(C;\cdot)$ is sequentially norm-to-weak closed at $\overline{x}$; the latter closedness property is also fulfilled if C is either subsmooth at $\overline{x}$ by Proposition 8.63 or epi-Lipschitz at $\overline{x}$ by Proposition 2.94(a).

Now, note first that $v(x) := \operatorname{proj}\big(v, T^C(C;x)\big)$ exists for every $x \in C$, since $T^C(C;x)$ is a closed convex set. Take any sequence $(x_n)_n$ in C converging to $\overline{x}$ and put $v_n := v(x_n)$. Obviously, the sequence $(v_n)_n$ is bounded because $\|v - v_n\| \leq \|v\|$. Fix any weakly convergent subsequence $(v_{s(n)})_{n\in\mathbb{N}}$ tending weakly to some $w \in H$. Observe from the Moreau decomposition theorem above that

$$\langle v_n, v - v_n\rangle = 0 \quad \text{and} \quad v - v_n \in (T^C(C;x_n))^\circ = N^C(C;x_n).$$

The latter inclusion and the sequential norm-to-weak closedness assumption at $\overline{x}$ of $N^C(C;\cdot)$ then ensures that $v - w \in N^C(C;\overline{x})$. According to the assumption $v \in T^C(C;\overline{x})$ it follows that $\langle v - w, v\rangle \leq 0$, and hence $\|v\|^2 \leq \langle w, v\rangle$. On the other hand, the above equality $\langle v_n, v - v_n\rangle = 0$ gives $\|v_{s(n)}\|^2 = \langle v_{s(n)}, v\rangle$, thus $\lim\limits_{n\to\infty} \|v_{s(n)}\|^2$ exists in $\mathbb{R}_+$ and its value is $\langle w, v\rangle$. This and the above inequality

$\|v\|^2 \le \langle w, v\rangle$ entail that $\|v\|^2 \le \lim_{n\to\infty} \|v_{s(n)}\|^2$. Using this and noticing (by the above equality $\langle v_n, v - v_n\rangle = 0$) that

$$\|v - v_{s(n)}\|^2 + \|v_{s(n)}\|^2 = \|v\|^2,$$

we deduce that $\lim_{n\to\infty} \|v - v_{s(n)}\|^2 = 0$. From this we see that the entire sequence $(v_n)_{n\in\mathbb{N}}$ strongly converges to v, which garantees that $\lim_{C\ni x\to\overline{x}} v(x) = v$, as desired. □

In the same line the following lower semicontinuity property holds:

Proposition 15.90. Let C be a subset of a Hilbert space H which is closed near $\overline{x} \in C$. Assume that the Clarke normal cone $N^C(C;\cdot)$ of C is sequentially norm-to-weak closed at $\overline{x}$, which is in particular the case whenever C is either prox-regular at $\overline{x}$, or subsmooth at $\overline{x}$, or epi-Lipschitz at $\overline{x}$. Then, for any $\overline{v} \in H$ the function $(x, v) \mapsto \|\mathrm{proj}\left(v, T^C(C;x)\right)\|$ is lower semicontinuous at $(\overline{x}, \overline{v})$ relative to $C \times H$.

Proof. Denote by $T(x)$ and $N(x)$ the Clarke tangent and normal cones of C at $x \in C$. The family indexed by $x \in C$ of functions $v \mapsto \|P_{T(x)}(v)\|$ being equi-Lipschitz (with 1 as Lipschitz constant), it suffices to show that the function $x \mapsto \|P_{T(x)}(\overline{v})\|$ is lower semicontinuous at $\overline{x}$ relative to C. Choose a sequence $(x_n)_n$ in C converging to $\overline{x}$ and such that

$$\liminf_{C\ni x\to\overline{x}} \|P_{T(x)}(\overline{v})\| = \lim_{n\to\infty} \|P_{T(x_n)}(\overline{v})\|.$$

Notng that the sequence $(\overline{v} - P_{T(x_n)}(\overline{v}))_n$ is bounded since $\|\overline{v} - P_{T(x_n)}(\overline{v})\| \le \|\overline{v}\|$, it admits a subsequence (that we do not relabel) converging weakly to some $\zeta \in H$. Since $\overline{v} - P_{T(x_n)}(\overline{v}) = P_{N(x_n)}(\overline{v}) \in N(x_n)$ by Theorem 15.87 of Moreau decomposition, the assumption of closedness of $N(C;\cdot)$ at $\overline{x}$ yields $\zeta \in N(\overline{x})$. Further, observing that $(P_{T(x_n)})_n$ converges weakly to $\overline{v} - \zeta$, we also have $\lim_{n\to\infty} \|P_{T(x_n)}(\overline{v})\| \ge \|\overline{v} - \zeta\|$ by the weak lower semicontinuity of the norm. It ensues that

$$\lim_{n\to\infty} \|P_{T(x_n)}(\overline{v})\| \ge \|\overline{v} - \zeta\| \ge \|\overline{v} - P_{N(\overline{x})}(\overline{v})\| = \|P_{T(\overline{x})}(\overline{v})\|,$$

where the latter equality is due to the Moreau decomposition theorem again. So, the desired lower semicontinuity is justified.

Finally, as shown in Proposition 15.89 the multimapping $N^C(C;\cdot)$ enjoys the closedness property at $\overline{x}$ under anyone of the particular cases in the statement. □

15.3. Change of metric

Consider a continuous linear mapping $A : H \to H$ which is symmetric in the sense that $\langle Ax, y\rangle = \langle x, Ay\rangle$ for all $x, y \in H$. Suppose that for some real number $\alpha > 0$ the coercivity property

$$\langle Ax, x\rangle \ge \alpha\|x\|^2 \quad \text{for all } x \in H \tag{15.49}$$

holds. Then the real-valued function $(\cdot,\cdot)_A$ defined on $H \times H$ by

$$(x, y)_A := \langle Ax, y\rangle \quad \text{for all } x, y \in H$$

is an inner product. Its associated norm $|\cdot|_A$ has the property

$$\alpha\|x\|^2 \le |x|_A^2 \le \|A\|\,\|x\|^2 \quad \text{for all } x \in H, \tag{15.50}$$

and this property ensures the equivalence of the two norms $\|\cdot\|$ and $|\cdot|_A$. In (15.50) above, $\|A\|$ denotes the norm of the continuous linear mapping A for H endowed with its initial norm $\|\cdot\|$. We also observe that

$$\|Ax\| \geq \alpha\|x\| \quad \text{for all } x \in H \tag{15.51}$$

because $\alpha\|x\|^2 \leq \langle Ax, x\rangle \leq \|Ax\|\,\|x\|$.

PROPOSITION 15.91. For a closed set C of a Hilbert space H and for a continuous linear mapping $A : H \to H$ satisfying (15.49)) the following assertions hold.
(a) If C is r-prox-regular at $\overline{x} \in C$ with respect to the norm $\|\cdot\|$, then it is r'-prox-regular at $\overline{x}$ with respect to the norm $|\cdot|_A$ for $r' := \alpha^{3/2}r/\|A\|$.
(b) If C is r-prox-regular with respect to the norm $\|\cdot\|$, then it is r'-prox-regular with respect to the norm $|\cdot|_A$ for r' as above.

PROOF. Suppose that C is r-prox-regular at $\overline{x}$ with respect to the norm $\|\cdot\|$. Then by Theorem 15.73(c) there exists some neighborhood U of $\overline{x}$ such that for each proximal normal w (with respect to the norm $\|\cdot\|$) to C at $x \in C \cap U$

$$\langle w, x' - x\rangle \leq \frac{1}{2r}\|w\|\,\|x' - x\|^2 \quad \text{for all } x' \in C \cap U. \tag{15.52}$$

Fix any $x, x' \in C \cap U$ and any $v \in N^{P,|\,|_A}(C; x)$ (the proximal normal cone to C at x with respect to the norm $|\,|_A$). It is not difficult to see that $Av \in N^{P,\|\cdot\|}(C; x)$. Then according to (15.52)

$$\begin{aligned}(v, x' - x)_A = \langle Av, x' - x\rangle &\leq \frac{1}{2r}\|Av\|\,\|x' - x\|^2 \\ &\leq \frac{\|A\|}{2r}\|v\|\,\|x' - x\|^2 \leq \frac{\|A\|}{2\alpha^{3/2}r}|v|_A\,|x' - x|_A^2,\end{aligned}$$

the last inequality being due to (15.50). By Theorem 15.73(c) again this means that C is r'-prox-regular at $\overline{x}$ with respect to the norm $|\,|_A$ for $r' := \alpha^{3/2}r/\|A\|$, that is, (a) holds.

The assertion (b) follows from (a) through Proposition 15.68. □

15.4. Prox-regularity in operations

This section is related to the study of prox-regularity of sets under certain operations and transformations. It is essentially concerned with level sets, sublevel sets, intersection of many sets, and inverse images.

Consider first closed subsets C_i of Hilbert spaces H_i, $i = 1, \cdots, m$, and endow the space $H_1 \times \cdots \times H_m$ with its canonic inner product and its associated norm

$$\|(x_1, \cdots, x_m)\| = \left(\|x_1\|_1^2 + \cdots + \|x_m\|_m^2\right)^{1/2}.$$

Clearly, for $C := C_1 \times \cdots \times C_m$ in $H := H_1 \times \cdots \times H_m$ we have for $x := (x_1, \cdots, x_m)$ in H

$$d_C^2(x) = \inf_{(y_1,\cdots,y_m)\in C_1\times\cdots\times C_m} \sum_{i=1}^m \|x_i - y_i\|_i^2 = d_{C_1}^2(x_1) + \cdots + d_{C_m}^2(x_m),$$

so in particular for $r \in]0, +\infty]$ we see that $U_r(C) \subset U_r(C_1) \times \cdots \times U_r(C_m)$. Consequently, the assertion (p) in Theorem 15.28 justifies the following property.

PROPOSITION 15.92. Let $H_1, \cdots, H_m$ be Hilbert spaces and let $H := H_1 \times \cdots \times H_m$ be the Hilbert product space endowed with its canonic inner product and associated norm. If C_i is r_i-prox-regular in H_i with $r_i \in]0, +\infty]$, $i = 1, \cdots, m$, then the set $C_1 \times \cdots \times C_m$ is r-prox-regular in H with $r := \min\{r_1, \cdots, r_m\}$.

Let us continue with some counterexamples showing that that sublevels of smooth functions as well as levels of smooth mappings may fail to be prox-regular. We also give a counterexample for the prox-regularity of intersection of finitely many prox-regular sets.

EXAMPLE 15.93. Endow $\mathbb{R}^2$ with its canonical Euclidean structure. The polynomial function of two variables $g : \mathbb{R}^2 \to \mathbb{R}$ defined by $g(x, y) := xy$ furnishes a first simple example of a smooth function whose sublevel (resp. level) set $\{(x, y) \in \mathbb{R}^2 : g(x, y) \le 0\}$ (resp. $\{(x, y) \in \mathbb{R}^2 : g(x, y) = 0\}$) in $\mathbb{R}^2$ is not prox-regular. For a bounded non-prox-regular sublevel set (resp. level set) of a smooth function we can consider the set $\{(x, y) \in \mathbb{R}^2 : g(x, y) \le 0\}$ (resp. $\{(x, y) \in \mathbb{R}^2 : g(x, y) = 0\}$), where g is the function whose zero sublevel is the union of the closed balls of $\mathbb{R}^2$ of radius 1 centered respectively at $(-1, 0)$ and $(1, 0)$), that is, for all $(x, y) \in \mathbb{R}^2$

$$g(x, y) = \big((x-1)^2 + y^2 - 1\big)\big((x+1)^2 + y^2 - 1\big)).$$

□

The next example illustrates the possibility of non prox-regularity of intersections of prox-regular sets.

EXAMPLE 15.94. Concerning the non preservation of prox-regularity under intersection, consider first the closed set Q of the Euclidean space $\mathbb{R}^2$ defined in the following way. For each $n \in \mathbb{N}$ let D_n denote the closed ball with radius $r = 1/4$ (independent of n) in $\mathbb{R}^2$ with the points $(1/2^{n-1}, 0)$ and $(1/2^n, 0)$ on its boundary and whose ordinate of its center is nonpositive. With $R = 1/2$ define

$$Q := \left\{(x, y) \in \mathbb{R}^2 : y \ge 0,\ \left(x - \frac{1}{2}\right)^2 + y^2 \le R^2\right\} \setminus \bigcup_{n \in \mathbb{N}} \operatorname{int} D_n,$$

and note that it is r-prox-regular, see Figure 15.4. Denote by E the vector subspace given by the axis of abscissa, that is, $E := \mathbb{R} \times \{0\}$. The intersection $Q \cap E$ fails to be prox-regular at $(0, 0)$, in particular $Q \cap E$ is not uniformly prox-regular, that is, there is no extended real $s \in]0, +\infty]$ such that $Q \cap E$ is s-prox-regular.

We also observe, with the linear mapping $A : \mathbb{R} \to \mathbb{R}^2$ defined by $Ax := (x, 0)$ for all $x \in \mathbb{R}$, that the susbset $A^{-1}(Q)$ is not prox-regular in $\mathbb{R}$. □

15.4.1. Uniform prox-regularity in operations. According to the above examples our goal in this first subsection is to provide certain verifiable conditions under which sets of basic structures are uniformly prox-regular. The local prox-regularity counterpart will be studied in the next subsection. The following simple lemma (implicit in the proof of Lemma 15.61 will be employed in several proofs in this subsection.

LEMMA 15.95. Let C be a subset of a normed space X and let $s \in]0, +\infty]$. For any $x, y \in C$ with $\|x - y\| < 2s$ and for any $t \in [0, 1]$ one has

$$x + t(y - x) \in U_s(C).$$

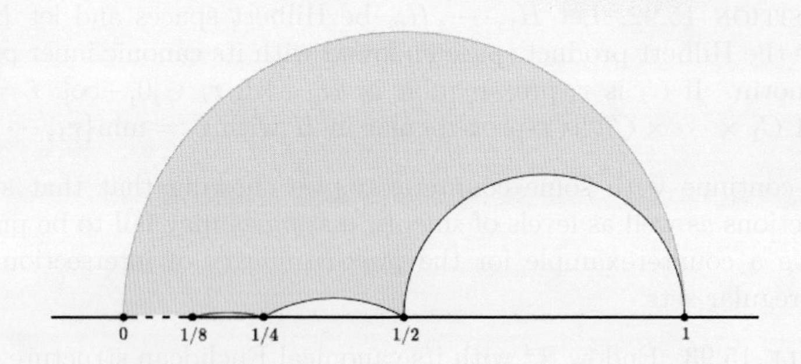

FIGURE 15.4. Intersection of two prox-regular sets which fails to be prox-regular.

PROOF. Fixing any $x, y \in C$ with $\|x - y\| < 2s$ and any $t \in [0, 1]$, it suffices to write with $z_t := x + t(y - x)$

$$d_C(z_t) \le \|z_t - x\| = t\|x - y\| \quad \text{and} \quad d_C(z_t) \le \|z_t - y\| = (1 - t)\|x - y\|,$$

which gives as desired

$$d_C(z_t) \le \min\{t, 1 - t\}\|x - y\| \le \frac{1}{2}\|x - y\| < s.$$

□

The first result is concerned with an easy-to-verify qualification condition for the uniform prox-regularity of constraint sets with infinitely many equalities, or equivalently level sets of nonlinear mappings.

PROPOSITION 15.96. Let $G : H \to Y$ be a mapping from a Hilbert space H into a Banach space Y and let $C := \{x \in H : G(x) = 0\}$. Assume that C is nonempty and that there exists an extended real $s \in]0, +\infty]$ such that:
(*i*) the mapping G is differentiable on $U_s(C)$;
(*ii*) there is a real $\gamma \ge 0$ such that the derivative DG is γ-Lipschitz on $U_s(C)$, that is, for all $x_1, x_2 \in U_s(C)$,

$$\|DG(x_1) - DG(x_2)\| \le \gamma\|x_1 - x_2\|.$$

Assume also that there is some real $\sigma > 0$ such that

$$\sigma\mathbb{B}_Y \subset DG(x)\big(\mathbb{B}_H\big) \quad \text{for every } x \in C. \tag{15.53}$$

Then the set C is r-prox-regular with $r := \min\left\{s, \frac{\sigma}{\gamma}\right\}$.

PROOF. We notice first that the set C is closed by continuity of G on the open set $U_s(C)$. Fix any $x, u \in C$ with $\|u - x\| < 2s$, and note by Lemma 15.95 that $x + t(u - x) \in U_s(C)$ for every $t \in [0, 1]$. By the $\mathcal{C}^1$ property of the mapping G near x along with the surjectivity of $A := DG(x)$ (due to (15.53)) we also have $N^C(C; x) = \{y^* \circ A : y^* \in Y^*\}$ (see Theorem 3.174(a)) Take any $v \in N^C(C; x)$ and for $x^* := \langle v, \cdot\rangle$ choose by the latter equality some $y^* \in Y^*$ such that $x^* = y^* \circ A$. Take also any $y \in \mathbb{B}_Y$. By the inclusion assumption (15.53), there is some $b \in \mathbb{B}_H$ such that $\sigma y = A(b)$, hence

$$\sigma|\langle y^*, y\rangle| = |\langle A^*(y^*), b\rangle| \le \|A^*(y^*)\| = \|x^*\|.$$

This entails that

$$\sigma \|y^*\| \leq \|A^*(y^*)\| = \|x^*\|. \tag{15.54}$$

Writing

$$\begin{aligned} 0 &= \langle y^*, G(u) - G(x) \rangle \\ &= \int_0^1 \langle y^* \circ DG(x + t(u-x)), u - x \rangle \, dt \\ &= \langle y^* \circ A, u - x \rangle + \int_0^1 \langle y^* \circ DG(x + t(u-x)) - y^* \circ A, u - x \rangle \, dt, \end{aligned}$$

we obtain

$$\begin{aligned} \langle x^*, u - x \rangle &= \int_0^1 \langle y^* \circ DG(x) - y^* \circ DG(x + t(u-x)), u - x \rangle \, dt \\ &\leq \gamma \|y^*\| \, \|u - x\|^2 \int_0^1 t \, dt \\ &= \frac{\gamma}{2} \|y^*\| \, \|u - x\|^2. \end{aligned}$$

Combining this with (15.54) and keeping in mind that $x^* = \langle v, \cdot \rangle$, it follows that

$$\langle v, u - x \rangle \leq \frac{\gamma}{2\sigma} \|v\| \, \|u - x\|^2.$$

Putting $r := \min\left\{s, \frac{\sigma}{\gamma}\right\}$, we derive that for any $x, u \in C$ with $\|u - x\| < 2r$ and any $v \in N^C(C; x)$, we have $\langle v, u - x \rangle \leq \frac{1}{2r} \|v\| \, \|u - x\|^2$. Then Proposition 15.40(a) tells us that the set C is r-prox-regular. □

We turn now to the situation of sublevel sets of many nonsmooth functions. The next proposition says in particular that, under a generalized Slater qualification condition, sublevel sets of locally Lipschitz *prox-regular functions* are prox-regular sets. The proof partially follows the approach in Proposition 8.135.

PROPOSITION 15.97. Let $g_1, \ldots, g_m : H \to \mathbb{R}$ be functions on a Hilbert space H such that the set

$$C := \{x \in H : g_1(x) \leq 0, \cdots, g_m(x) \leq 0\}$$

is nonempty. Assume that there is an extended real $s \in]0, +\infty]$ such that:
(i) for each $k \in K := \{1, \ldots, m\}$, the function g_k is locally Lipschitz continuous on the open set $U_s(C)$;
(ii) there is a real $\gamma \geq 0$ such that, for any $k \in K$, for any $x \in \operatorname{bdry} C$, any $\zeta \in \partial_C g_k(x)$, any $x' \in U_s(C)$, and any $\zeta' \in \partial_C g_k(x')$

$$\langle \zeta' - \zeta, x' - x \rangle \geq -\gamma \|x' - x\|^2. \tag{15.55}$$

Assume also that there is a real $\sigma > 0$ such that, for each $x \in \operatorname{bdry} C$, there exists $\overline{v} \in \mathbb{B}_H$ (depending on x) for which for every $k \in K(x) := \{j \in K : g_j(x) = \max_{i \in K} g_i(x)\}$

$$g_k^\circ(x; \overline{v}) \leq -\sigma,$$

or equivalently $\langle \zeta, \overline{v} \rangle \leq -\sigma$ for all $\zeta \in \partial_C g_k(x)$.

Then, the set C is r-prox-regular with $r = \min\{s, \sigma/\gamma\}$.

PROOF. The set C is clearly closed according to the continuity of functions g_k on $U_s(C)$. Define $g : H \to \mathbb{R}$ by $g(x) := \max_{k\in K} g_k(x)$ for all $x \in H$ and note that $C = \{x \in H : g(x) \le 0\}$. By Proposition 2.190 we have

$$\partial_C g(x) \subset \mathrm{co}\left(\bigcup_{k\in K(x)} \partial_C g_k(x)\right) \quad \text{for all } x \in U_s(C). \tag{15.56}$$

For any $x \in \mathrm{bdry}\, C$ we note that any ζ in the right-hand side of (15.56) can be written as $\zeta = \sum_{k\in K(x)} \lambda_k \zeta_k$ with $\zeta_k \in \partial_C g_k(x)$, $\lambda_k \ge 0$ and $\sum_{k\in K(x)} \lambda_k = 1$, so by the inclusion $x \in \mathrm{bdry}\, C$ and the assumption on $\overline{v}$ we have

$$\langle \zeta, \overline{v}\rangle = \sum_{k\in K(x)} \lambda_k \langle \zeta_k, \overline{v}\rangle \le -\sigma, \tag{15.57}$$

hence $\zeta \ne 0$. This and the inclusion (15.56) imply that

$$0 \notin \partial_C g(x) \quad \text{for all } x \in \mathrm{bdry}\, C.$$

Then it follows by Proposition 2.196 that

$$N^C(C; x) \subset \mathbb{R}_+ \partial_C g(x) \quad \text{for all } x \in \mathrm{bdry}\, C. \tag{15.58}$$

Now fix any $x \in \mathrm{bdry}\, C$, any $y \in C$ with $\|x - y\| < 2s$ and any nonzero $v \in N^C(C; x)$. By (15.58) choose a real $\alpha > 0$ and $\zeta \in \partial_C g(x)$ such that $v = \alpha\zeta$, so $\alpha\|\zeta\| = \|v\|$. By (15.56) there are $\lambda_k \ge 0$ with $\lambda_k = 0$ if $k \notin K(x)$ such that $\sum_{k\in K} \lambda_k = 1$ and $\zeta = \sum_{k\in K} \lambda_k \zeta_k$. Take any $k \in K(x)$. Note that for each $t \in [0, 1]$ and $x_t := x + t(y - x)$ we have $x_t \in U_s(C)$ by Lemma 15.95. Then the function $\varphi_k : [0, 1] \to \mathbb{R}$ with $\varphi_k(t) := g_k(x + t(x - y))$ is Lipschitz continuous on $[0, 1]$. Therefore, there is a Lebesgue null set $N_k \subset [0, 1]$ such that φ_k is derivable on $[0, 1] \setminus N_k$ and

$$0 \ge g_k(y) - g_k(x) = \int_0^1 \varphi_k'(t)\, dt.$$

Write $\varphi_k(t) = g_k(G(t))$ with the affine mapping G defined by $G(t) = x + t(y - x)$. For each $t \in]0, 1[\setminus N_k$, the inclusion $\varphi_k'(t) \in \partial_C(g_k \circ G)(t)$ and the Lipschitz property of g_k near $G(t)$ furnishes, by the chain rule in Theorem 2.135, some $\xi_k(t) \in \partial_C g_k(x_t)$ such that $\varphi_k'(t) = \langle \xi_k(t), y - x\rangle$. We deduce that

$$\begin{aligned} 0 \ge \int_0^1 \varphi_k'(t)\, dt &= \int_0^1 \langle \xi_k(t), y - x\rangle\, dt \\ &= \int_0^1 \langle \xi_k(t) - \zeta_k, y - x\rangle\, dt + \langle \zeta_k, y - x\rangle \\ &= \int_0^1 \frac{1}{t}\langle \xi_k(t) - \zeta_k, x_t - x\rangle\, dt + \langle \zeta_k, y - x\rangle, \end{aligned} \tag{15.59}$$

which gives by (15.55)

$$\begin{aligned} 0 &\ge -\int_0^1 \frac{1}{t}\gamma\|x_t - x\|^2\, dt + \langle \zeta_k, y - x\rangle \\ &= -\gamma\|y - x\|^2 \int_0^1 t\, dt + \langle \zeta_k, y - x\rangle = -\frac{\gamma}{2}\|y - x\|^2 + \langle \zeta_k, y - x\rangle, \end{aligned}$$

hence $\langle \zeta_k, y-x\rangle \le \frac{\gamma}{2}\|y-x\|^2$. Keeping in mind that $\lambda_k = 0$ if $k \notin K(x)$, multiplying by λ_k for $k\in K(x)$ and summing over $k\in K(x)$, we see that $\langle \zeta, y-x\rangle \le \frac{\gamma}{2}\|y-x\|^2$. Further, from the inclusion $\overline{v} \in \mathbb{B}_H$ and the inequality $\langle \zeta, \overline{v}\rangle \le -\sigma$ in (15.57), we also have $\|\zeta\| \ge \sigma$. It follows that $\langle \zeta, y-x\rangle \le \frac{\gamma\|\zeta\|}{2\sigma}\|y-x\|^2$. Multiplying both sides by $\alpha \ge 0$, we obtain $\langle v, y-x\rangle \le \frac{\gamma}{2}\|v\|\,\|y-x\|^2$, and this inequality still holds if $x \in \operatorname{int} C$. Proposition 15.40(a) confirms that the set C is r-prox-regular with $r = \min\{s, \sigma/\gamma\}$. □

Strengthening slightly the assumption (ii) in Proposition 15.97 in a symmetric way allows us to weaken the assumption (i).

PROPOSITION 15.98. Let $g_1, \ldots, g_m : H \to \mathbb{R}$ be functions on a Hilbert space H such that the set

$$C := \{x \in H : g_1(x) \le 0, \cdots, g_m(x) \le 0\}$$

is nonempty. Assume that there is an extended real $s \in]0, +\infty]$ such that:
(i) for each $k \in K := \{1, \ldots, m\}$, the function g_k is continuous on the open set $U_s(C)$;
(ii) there is a real $\gamma \ge 0$ such that, for any $k \in K$, for any $x, x' \in U_s(C)$, and for any $\zeta \in \partial_C g_k(x)$ and any $\zeta' \in \partial_C g_k(x')$

$$\langle \zeta' - \zeta, x' - x\rangle \ge -\gamma \|x' - x\|^2 . \tag{15.60}$$

Assume also that there is a real $\sigma > 0$ such that, for each $x \in \operatorname{bdry} C$, there exists $\overline{v} \in \mathbb{B}_H$ (depending on x) for which for every $k \in K(x) := \{j \in K : g_j(x) = \max_{i\in K} g_i(x)\}$

$$g_k^\circ(x; \overline{v}) \le -\sigma,$$

or equivalently $\langle \zeta, \overline{v}\rangle \le -\sigma$ for all $\zeta \in \partial_C g_k(x)$.
Then, the set C is r-prox-regular with $r = \min\{s, \sigma/\gamma\}$.

PROOF. As above the continuity of functions g_k on $U_s(C)$ ensures the set C is closed. Further, using the assumption (15.60) we see by Theorem 10.38 and by the continuity of g_k on $U_s(C)$ that, for each $k \in K$, the function g_k is locally semiconvex on $U_s(C)$, so g_k is locally Lipschitz continuous on $U_s(C)$ by Proposition 10.23(b). Consequently, Proposition 15.97 tells us that the set C is r-prox-regular for r as given in the statement. □

REMARK 15.99. By Proposition 10.39 the condition in (ii) in Proposition 15.98 is fulfilled by any function g_k which is linearly semiconvex with constant γ on an open convex set containing $U_s(C)$. □

Recalling that $\partial_C g(u) = \{Dg(u)\}$ whenever a function $g : H \to \mathbb{R}$ is of class $\mathcal{C}^1$ near u (in fact, if g is strictly Hadamard differentiable at u), we obtain the following first corollary from Proposition 15.97.

COROLLARY 15.100. Let $g_1, \ldots, g_m : H \to \mathbb{R}$ be functions on a Hilbert space H such that the set

$$C := \{x \in H : g_1(x) \le 0, \cdots, g_m(x) \le 0\}$$

is nonempty. Assume that there is an extended real $s \in]0, +\infty]$ such that:
(i) for each $k \in K := \{1, \ldots, m\}$, the function g_k is continuously differentiable on the open set $U_s(C)$;

(ii) there is a real $\gamma \geq 0$ such that, for any $k \in K$, any $x \in \operatorname{bdry} C$ and any $x' \in U_s(C)$

$$\langle \nabla g_k(x') - \nabla g_k(x), x' - x \rangle \geq -\gamma \|x' - x\|^2 . \tag{15.61}$$

Assume also that there is a real $\sigma > 0$ such that, for each $x \in \operatorname{bdry} C$, there exists $\overline{v} \in \mathbb{B}_H$ (depending on x) for which for every $k \in K(x) := \{j \in K : g_j(x) = \max_{i \in K} g_i(x)\}$

$$\langle \nabla g_k(x), \overline{v} \rangle \leq -\sigma.$$

Then, the set C is r-prox-regular with $r = \min\{s, \sigma/\gamma\}$.

Given an open convex neighborhood U of a point $\overline{x}$ in a Banach space X and a function $g : U \to \mathbb{R}$ which is Hadamard differentiable near $\overline{x}$ with

$$\langle Dg(x') - Dg(x), x' - x \rangle \geq -\gamma \|x' - x\|^2 \quad \text{for all } x \text{ near } \overline{x},$$

the tangential characterization in Theorem 10.38 entails that g is semiconvex near $\overline{x}$. Then Corollary 10.30) says that $\partial_C g(x) = \{Dg(x)\}$ for all x near $\overline{x}$. Using this, we see that the next corollary directly follows from Proposition 15.98.

Corollary 15.101. Let $g_1, \dots, g_m : H \to \mathbb{R}$ be functions on a Hilbert space H such that the set

$$C := \{x \in H : g_1(x) \leq 0, \cdots, g_m(x) \leq 0\}$$

is nonempty. Assume that there is an extended real $s \in]0, +\infty]$ such that:
(i) for each $k \in K := \{1, \dots, m\}$, the function g_k is Hadamard differentiable on the open set $U_s(C)$;
(ii) there is a real $\gamma \geq 0$ such that, for any $k \in K$ and any $x, x' \in U_s(C)$

$$\langle \nabla g_k(x') - \nabla g_k(x), x' - x \rangle \geq -\gamma \|x' - x\|^2 . \tag{15.62}$$

Assume also that there is a real $\sigma > 0$ such that, for each $x \in \operatorname{bdry} C$, there exists $\overline{v} \in \mathbb{B}_H$ (depending on x) for which for every $k \in K(x) := \{j \in K : g_j(x) = \max_{i \in K} g_i(x)\}$

$$\langle \nabla g_k(x), \overline{v} \rangle \leq -\sigma.$$

Then, the set C is r-prox-regular with $r = \min\{s, \sigma/\gamma\}$.

If the function $g_k : H \to \mathbb{R}$ is twice differentiable on $U_s(C)$, it is clear that the condition (15.62) amounts to requiring that $\langle D^2 g_k(x)v, v \rangle \geq -\gamma \|v\|^2$ for all $v \in H$. We translate this as follows:

Corollary 15.102. Let $g_1, \dots, g_m : H \to \mathbb{R}$ be functions on a Hilbert space H such that the set

$$C := \{x \in H : g_1(x) \leq 0, \cdots, g_m(x) \leq 0\}$$

is nonempty. Assume that there is an extended real $s \in]0, +\infty]$ such that:
(i) for each $k \in K := \{1, \dots, m\}$, the function g_k is twice differentiable on the open set $U_s(C)$;
(ii) there is a real $\gamma \geq 0$ such that, for any $k \in K$ and any $x \in U_s(C)$

$$\left\langle D^2 g_k(x)v, v \right\rangle \geq -\gamma \|v\|^2 \quad \text{for all } v \in H. \tag{15.63}$$

Assume also that there is a real $\sigma > 0$ such that, for each $x \in \operatorname{bdry} C$, there exists $\overline{v} \in \mathbb{B}_H$ (depending on x) for which for every $k \in K(x) := \{j \in K : g_j(x) = \max_{i \in K} g_i(x)\}$

$$\langle \nabla g_k(x), \overline{v} \rangle \leq -\sigma.$$

Then, the set C is r-prox-regular with $r = \min\{s, \sigma/\gamma\}$.

In addition to the latter corollaries and to Proposition 15.96 the next proposition is concerned with nonlinear inverses of convex sets.

Proposition 15.103. Let $G : H \to Y$ be a mapping from a Hilbert space H into a Banach space Y, let D be a closed convex set in Y and let $C := \{x \in H : G(x) \in D\}$. Assume that C is nonempty and that there exists an extended real $s \in]0, +\infty]$ such that:
(i) the mapping G is differentiable on $U_s(C)$;
(ii) there is a real $\gamma \geq 0$ such that the derivative DG is γ-Lipschitz on $U_s(C)$;
(iii) there is some real $\sigma > 0$ such that

$$\sigma \mathbb{B}_Y \subset DG(x)\big(\mathbb{B}_H\big) - (D - G(x)) \quad \text{for every } x \in \operatorname{bdry} C. \tag{15.64}$$

Then the set C is r-prox-regular with $r := \min\{s, \sigma/\gamma\}$.

Proof. Clearly, the set C is closed by continuity of G on the open set $U_s(C)$. Fix any $x \in \operatorname{bdry} C$ and $u \in C$ with $\|u - x\| < 2s$, and note by Lemma 15.95 that $x + t(u - x) \in U_s(C)$ for every $t \in [0, 1]$. Putting $A := DG(x)$, by the $\mathcal{C}^1$ property of the mapping G near x and by the assumption (15.64) we also have $N^C(C; x) = \{y^* \circ A : y^* \in N(D; G(x))\}$ according to Theorem 7.95. Take any $v \in N^C(C; x)$ and for $x^* := \langle v, \cdot \rangle$ choose by the latter equality some $y^* \in N(D; G(x))$ such that $x^* = y^* \circ A$. Lemma 8.121 implies that $\sigma \|y^*\| \leq \|x^*\|$. By the inclusion $y^* \in N(D; G(x))$ we can write

$$\begin{aligned} 0 &\geq \langle y^*, G(u) - G(x) \rangle \\ &= \int_0^1 \langle y^* \circ DG(x + t(u - x)), u - x \rangle \, dt \\ &= \langle y^* \circ A, u - x \rangle + \int_0^1 \langle y^* \circ DG(x + t(u - x)) - y^* \circ A, u - x \rangle \, dt, \end{aligned}$$

hence we obtain

$$\begin{aligned} \langle x^*, u - x \rangle &\leq \int_0^1 \langle y^* \circ DG(x) - y^* \circ DG(x + t(u - x)), u - x \rangle \, dt \\ &\leq \gamma \|y^*\| \, \|u - x\|^2 \int_0^1 t \, dt \\ &= \frac{\gamma}{2} \|y^*\| \, \|u - x\|^2. \end{aligned}$$

Combining this with the inequality $\sigma \|y^*\| \leq \|x^*\|$ and keeping in mind that $x^* = \langle v, \cdot \rangle$, it follows that

$$\langle v, u - x \rangle \leq \frac{\gamma}{2\sigma} \|v\| \, \|u - x\|^2.$$

Putting $r := \min\left\{s, \frac{\sigma}{\gamma}\right\}$, we deduce that for any $x \in \operatorname{bdry} C$ and any $u \in C$ with $\|u - x\| < 2r$, we have $\langle v, u - x \rangle \leq \frac{1}{2r} \|v\| \, \|u - x\|^2$ for all $v \in N^C(C; x)$. By Proposition 15.40(a) we conclude that the set C is r-prox-regular. $\square$

A similar easy-to-verify result is valid for nonlinear inverse image $G^{-1}(S)$ of a prox-regular set S under the surjectivity of derivative of G. The result extends Proposition 15.96 relative to level sets of mappings.

PROPOSITION 15.104. Let $G : H \to Y$ be a mapping from the Hibert space H into a Hilbert space Y and let S be a subset of Y with $G^{-1}(S) \neq \emptyset$. Assume that S is r-prox-regular for some $r \in]0, +\infty]$, G is γ-Lipschitz on $G^{-1}(S)$ and differentiable on an open enlargement $U_s(G^{-1}(S))$ with DG γ'-Lipschitz on $U_s(G^{-1}(S))$ for some $s \in]0, +\infty]$. Assume also that there is a real $\sigma > 0$ such that

$$\sigma \mathbb{B}_Y \subset DG(x)(\mathbb{B}_X) \quad \text{for all } x \in \operatorname{bdry}\left(G^{-1}(S)\right).$$

Then the set $G^{-1}(S)$ is r'-prox-regular with $r' = \min\left\{s, \frac{\sigma}{\gamma' + r^{-1}\gamma^2}\right\}$.

PROOF. Set $C := G^{-1}(S)$ and fix any $x \in \operatorname{bdry}(C)$ and any $u \in G^{-1}(S)$ with $\|u - x\| < 2s$. By the surjectivity of $DG(x)$ and by Theorem 3.174(a) we know that

$$N^C(C; x) \subset DG(x)^*\big(N^C(S; G(x))\big).$$

Fix $v \in N^C(C; x) \cap \mathbb{B}_H$ and set $x^* := \langle v, \cdot \rangle$. There is an element $y^* \in N^C(S; G(x))$ such that $x^* = DG(x)^*(y^*) = y^* \circ DG(x)$. Lemma 8.68 tells us that $\|y^*\| \le 1/\sigma$. For each $t \in [0, 1]$ put $x_t := x + t(u - x)$ and note by Lemma 15.95 that $x_t \in U_s(C)$. Noticing that $G(u), G(x) \in S$ and keeping in mind the Lipschitz property of G on C and DG on $U_s(C)$, it ensues that

$$\begin{aligned}
\langle x^*, u - x \rangle &= \langle y^*, DG(x)(u - x) \rangle \\
&= \langle y^*, G(u) - G(x) \rangle - \int_0^1 \langle y^*, \big(DG(x_t) - DG(x)\big)(u - x) \rangle \, dt \\
&\le \frac{1}{2r\sigma} \|G(u) - G(x)\|^2 + \frac{\gamma'}{\sigma} \|u - x\|^2 \int_0^1 t \, dt,
\end{aligned}$$

which implies that

$$\langle v, u - x \rangle \le \frac{\gamma^2}{2r\sigma} \|u - x\|^2 + \frac{\gamma'}{2\sigma} \|u - x\|^2.$$

This gives $\langle v, u - x \rangle \le \frac{1}{2\sigma}\left(\gamma' + \frac{\gamma^2}{r}\right)\|u - x\|^2$. Then Proposition 15.40 says that C is r'-prox-regular with $r' := \min\left\{s, \frac{\sigma}{\gamma' + r^{-1}\gamma^2}\right\}$. $\square$

The next proposition furnishes a tangential condition for the prox-regularity of intersections. In the next chapter, we will see in Corollary 16.17 another verifiable condition for the intersection of a prox-regular set and a strongly convex set.

PROPOSITION 15.105. Let C_1, C_2 be two r-prox-regular sets in a Hilbert space H with $C := C_1 \cap C_2 \neq \emptyset$. Assume that there exists a real $\sigma > 0$ such that for each $x \in \operatorname{bdry} C$ there exists a neighborhood U of x such that

$$\sigma \mathbb{B}_H \subset T^B(C_1; x_1) \cap \mathbb{B}_H - T^B(C_2; x_2) \cap \mathbb{B}_H \tag{15.65}$$

for all $x_1 \in C_1 \cap U$ and $x_2 \in C_2 \cap U$. Then the set C is r'-prox-regular with $r' := (r\sigma)/2$.

PROOF. Fix any $x \in \operatorname{bdry} C$, any $y \in C$ and any $v \in N^C(C; x)$. Since C_i is tangentially regular at x, we have $T^B(C_i; x) = T^C(C_i; x)$, $i = 1, 2$. Using this, we obtain from the (full) assumption (15.65) that $N^C(C; x) = N^C(C_1; x) + N^C(C_2; x)$

according to Corollary 7.103. Then we can choose $v_i \in N^C(C_i;x)$, $i=1,2$, such that $v = v_1 + v_2$. Take any $b \in \mathbb{B}_H$ and choose, by the assumption (15.65) with $x_1 = x_2 = x$, some $b_i \in T^C(C_i;x) \cap \mathbb{B}_H$, $i=1,2$, such that $\sigma b = b_1 - b_2$. Then

$$\sigma\langle v_1, b\rangle = \langle v_1, b_1\rangle - \langle v_1, b_2\rangle \le -\langle v_1, b_2\rangle = \langle v_2, b_2\rangle - \langle v, b_2\rangle,$$

which gives $\sigma\langle v_1, b\rangle \le -\langle v, b_2\rangle \le \|v\|$. This being true for every $b \in \mathbb{B}_H$, we obtain $\sigma\|v_1\| \le \|v\|$; and the similar inequality $\sigma\|v_2\| \le \|v\|$ holds true. Therefore, we can write

$$\langle v, y-x\rangle = \langle v_1, y-x\rangle + \langle v_2, y-x\rangle \le \frac{\|v_1\|}{2r}\|y-x\|^2 + \frac{\|v_2\|}{2r}\|y-x\|^2,$$

hence $\langle v, y-x\rangle \le \frac{\|v\|}{r\sigma}\|y-x\|^2$. This justifies the desired prox-regularity of C. □

We turn now to direct images of prox-regular sets.

PROPOSITION 15.106. Let H and Y be Hilbert spaces and C be a (closed) r-prox-regular set in H with $r \in]0,+\infty]$. Let $s \in]0,r]$ and $f: U_s(C) \to Y$ be a differentiable mapping such that Df is γ-Lipschitz over $U_s(C)$ and $\sup\{\|Df(x)\| : x \in \operatorname{bdry} C\} \le \beta$ for some non-negative reals γ, β. Assume that $f(C)$ is closed and there exists a mapping $\sigma : f(C) \to H$ which is L-Lipschitz on $f(C)$ and such that $\sigma(y) \in C \cap f^{-1}(y)$ for every $y \in f(C)$.

Then, for

$$r' := \min\left\{\frac{s}{L}, L^{-2}\left(\gamma + \frac{\beta}{r}\right)^{-1}\right\},$$

the set $f(C)$ is r'-prox-regular.

PROOF. Take $y \in \operatorname{bdry}\big(f(C)\big)$ and $y' \in f(C)$ such that $\|y-y'\| < 2s/L$. Put $x := \sigma(y)$ and $x' := \sigma(y')$, and observe that $\|x'-x\| < 2s$. Note also that $x \in \operatorname{bdry} C$ (otherwise, there is an open set W with $x \in W \subset C$, hence $\sigma^{-1}(W)$ is an open set with $y \in \sigma^{-1}(W) \subset f(C)$, which contradicts that $y \in \operatorname{bdry}\big(f(C)\big)$). Since $T^B(C;x)$ is convex and closed (as seen in (b) of Proposition 15.13), from Theorem 15.14, there exists $u \in T^B(C;x)$ such that

$$\|x'-x-u\| \le \frac{\|x'-x\|^2}{2r}.$$

By Proposition 2.143(a) we have $Df(x)u \in T^B(f(C);f(x))$, and the boundedness property of Df over $\operatorname{bdry} C$ ensures

$$\|Df(x)(x'-x) - Df(x)u\| \le \|Df(x)\|\,\|x'-x-u\| \le \frac{\beta}{2r}\|x'-x\|^2.$$

By Lemma 15.95 we know that $[x,x'] \subset U_s(C)$, hence

$$\begin{aligned}\|f(x')-f(x)-Df(x)(x'-x)\| &= \|\int_0^1 \big(Df(x+t(x'-x)) - Df(x)\big)(x'-x)\,dt \\ &\le \gamma\|x'-x\|^2\int_0^1 t\,dt = \frac{\gamma}{2}\|x'-x\|^2.\end{aligned}$$

Therefore, we have

$$\begin{aligned}d(f(x')-f(x), T^B(f(C);f(x))) &\le \|f(x')-f(x)-Df(x)u\| \\ &\le \|f(x')-f(x)-Df(x)(x'-x)\| \\ &\quad + \|Df(x)(x'-x)-Df(x)u\|,\end{aligned}$$

hence

$$\begin{aligned} d(f(x')-f(x),T^B(f(C);f(x))) &\le \|f(x')-f(x)-Df(x)u\| \\ &\le \left(\frac{\gamma}{2}+\frac{\beta}{2r}\right)\|x'-x\|^2 \\ &\le L^2\left(\frac{\gamma}{2}+\frac{\beta}{2r}\right)\|f(x')-f(x)\|^2. \end{aligned}$$

Recalling Corollary 15.39(a), the proof is complete. □

REMARK 15.107. Another proof of the corollary can be given with similar arguments but with the use of normal vectors through Proposition 15.38 in place of tangent vectors. □

COROLLARY 15.108. Let H and Y be Hilbert spaces and C be a (closed) r-prox-regular set in H with $r \in]0,+\infty]$. Let $s \in]0;r]$ and $f : U_s(C) \to Y$ be a differentiable mapping such that Df is γ-Lipschitz over $U_s(C)$ and $\sup\{\|Df(x)\| : x \in \operatorname{bdry} C\} \le \beta$ for some non-negative reals γ, β. Assume that f is one-to-one over C and such that f^{-1} (the inverse of the restriction $f_C : C \to f(C)$ with $f_C(x) = f(x)$ for all $x \in C$) is L-Lipschitz continuous over C.

Then, for

$$r' := \min\left\{\frac{s}{L}, L^{-2}\left(\gamma+\frac{\beta}{r}\right)^{-1}\right\},$$

the set $f(C)$ is closed and r'-prox-regular.

PROOF. According to the above proposition it suffices to verify that $f(C)$ is closed. Let $(y_n)_n$ be a sequence of $f(C)$ converging to some $y \in Y$. Let $x_n \in C$ given by $f(x_n) = y_n$. We have $\|x_n - x_m\| \le L\|y_n - y_m\|$, hence the Cauchy sequence $(x_n)_n$ converges to some $x \in C$. This ensures $y = f(x) \in f(C)$, justifying the closedness of $f(C)$. □

15.4.2. Local prox-regularity in operations. First we notice as in Proposition 15.92 that the Cartesian product of prox-regular sets S_i at $\overline{x}_i \in S_i$, $i = 1,\cdots,m$, in Hilbert spaces H_i is prox-regular at $(\overline{x}_1,\cdots,\overline{x}_m)$ in the Hilbert space $H_1 \times \cdots \times H_m$. Regarding intersection and inverse image, we saw in Example 15.94 that the sets $Q \cap E$ and $A^{-1}(Q)$ are not prox-regular at $(0,0)$ in $\mathbb{R}^2$ and at 0 in $\mathbb{R}$ respectively, while the sets Q and E are (even uniformly) r-prox-regular and the mapping $A : \mathbb{R} \to \mathbb{R}^2$ is linear.

To provide general sufficient conditions under which the local prox-regularity of intersection or inverse image is preserved, we will utilize the concepts of *truncated normal cone property* for intersection of finitely many sets or for inverse image sets presented in Chapter 8 in Definitions 8.125 and 8.111 respectively. In this way, parallel to the previous approach in Subsection 15.4.1, we present another method following the path taken for subsmooth sets. The reader may also convince himself that certain results in Subsection 15.4.1 for the uniform prox-regularity of sets could have been proved with this method.

If $M \subset H$ is a $\mathcal{C}^{1,1}$-submanifold at a point $m_0 \in M$, we already saw in Proposition 15.74 that M is prox-regular at m_0. We even saw in the same Proposition 15.74 that the inverse image $G^{-1}(D)$ of a closed convex set D of a Banach space Y is prox-regular at $\overline{x} \in G^{-1}(D)$ provided that the mapping $G : H \to Y$ is of class

$\mathcal{C}^{1,1}$ near $\overline{x}$ with $DG(\overline{x})$ surjective. The proposition below shows a similar result under the prox-regularity of D at $G(\overline{x})$. It also establishes the local prox-regularity for intersection of sets under suitable conditions.

PROPOSITION 15.109. Let $(C_k)_{k=1}^m$ be a finite family of closed sets of a Hilbert space H and let D be a closed set of a Hilbert space Y.
(a) If all the sets C_k are prox-regular at a point $\overline{x}$ of their intersection and if they have the (truncated) normal cone intersection property near $\overline{x}$ with respect to the Fréchet normal cone, then their intersection set $\bigcap_{k=1}^m C_k$ is prox-regular at $\overline{x}$.
(b) If a mapping $G : H \to Y$ is of class $\mathcal{C}^{1,1}$ around a point $\overline{x} \in G^{-1}(D)$ and D is prox-regular at $G(\overline{x})$, and if the inverse image set $G^{-1}(D)$ has the (truncated) normal cone inverse image property at $\overline{x}$ with respect to the Fréchet normal cone, then the inverse image set $G^{-1}(D)$ is prox-regular at $\overline{x}$.

PROOF. Assume that each set C_k is prox-regular at $\overline{x}$ and put $C := \bigcap_{k=1}^m C_k$. By Theorem 15.70 it is easily seen that there exist some real number $r > 0$ and some open neighborhood U of $\overline{x}$ such that all the sets C_k are prox-regular at all points of $C \cap U$ and such that for each $k = 1, \cdots, m$ we have for all $x \in C_k \cap U$ and $w \in N^P(C_k; x) \cap \mathbb{B}_H$

$$\langle w, x' - x \rangle \leq \frac{1}{2r} \|x' - x\|^2 \quad \text{for all } x' \in C_k \cap U. \tag{15.66}$$

Restricting the neighborhood U if necessary, we may suppose that for some $\gamma > 0$ the normal cone intersection property (8.67) in Definition 8.125 also holds with the Fréchet normal cone for all points in $U \cap C_1 \cap \cdots \cap C_m$. Fix any $x \in U \cap C$ and $v \in N^P(C; x) \cap \mathbb{B}_H$. By Definition 8.125 there exists for each $i = 1, \cdots, m$ some $w_i \in N^F(C_i; x)$ with $\|w_i\| \leq \gamma$ such that $v = w_1 + \cdots + w_m$. Since $N^F(C_i; x) = N^P(C_i; x)$ (see Proposition 15.75(b)), by (15.66) we obtain

$$\langle v, x' - x \rangle \leq \frac{m\gamma}{2r} \|x' - x\|^2 \quad \text{for all } x' \in C \cap U,$$

which ensures that C is prox-regular at $\overline{x}$, that is, (a) holds.

Let us now establish (b). Fix some open convex neighborhood U of $\overline{x}$ over which (8.57) in Definition 8.111 holds with $N^F(\cdot; \cdot)$ and over which the mapping G as well as its derivative $DG(\cdot)$ are Lipschitz continuous with Lipschitz constants K and K_1 respectively. For $S := G^{-1}(D)$ fix any $x \in U \cap S$ and $v \in N^P(S; x)$ with $\|v\| \leq 1$. Fix by Theorem 15.70 some open neighborhood V of $G(\overline{x})$ and some constant $r > 0$ such that

$$\langle \zeta, y' - y \rangle \leq \frac{1}{2r} \|y' - y\|^2 \quad \text{for all } y, y' \in V \cap D, \text{ and } \zeta \in N^P(D; y) \cap \mathbb{B}_Y. \tag{15.67}$$

Restricting the open convex neighborhood U of $\overline{x}$ if necessary, we may suppose that $G(U) \subset V$. By the normal cone property in Definition 8.111 with $\gamma > 0$, there is some $w \in N^F(D; G(x)) = N^P(D; G(x))$ with $\|w\| \leq \gamma$ such that $v = DG(x)^*(w)$ and hence according to (15.67)

$$\langle w, y' - G(x) \rangle \leq \frac{\gamma}{2r} \|y' - G(x)\|^2 \quad \text{for all } y' \in V \cap D. \tag{15.68}$$

Fix any $x' \in S \cap U$ and write

$$G(x') - G(x) = DG(x)(x' - x) + \int_0^1 \left(DG(x + t(x' - x)) - DG(x)\right)(x' - x)\, dt,$$

and then

$$\langle v, x' - x\rangle = \langle w, G(x') - G(x)\rangle - \int_0^1 \langle w, (DG(x + t(x' - x)) - DG(x))\,(x' - x)\rangle\, dt.$$

Using (15.68) and noticing that

$$\int_0^1 \|(DG(x + t(x' - x)) - DG(x))(x' - x)\|\, dt \le K_1\|x' - x\|^2 \int_0^1 t\, dt,$$

we obtain that

$$\langle v, x' - x\rangle \le \frac{\gamma}{2r}\|G(x') - G(x)\|^2 + \frac{\gamma K_1}{2}\|x' - x\|^2 \le \frac{\gamma}{2}\left(\frac{K^2}{r} + K_1\right)\|x' - x\|^2.$$

The latter being true for all $x, x' \in S \cap U$ and $v \in N^P(S; x) \cap \mathbb{B}_H$, we conclude that the set S is prox-regular at $\overline{x}$. □

The cases with the presence of metric subregularity can be deduced.

PROPOSITION 15.110. Let $(C_k)_{k=1}^m$ be a finite family of closed sets of a Hilbert space H and let D be a closed set of a Hilbert space Y.
(a) If all sets C_k are prox-regular at a point $\overline{x}$ in all the sets C_k and if the sets C_k are metrically subregularly transversal at $\overline{x}$, then their intersection set $\bigcap_{k=1}^m C_k$ is prox-regular at $\overline{x}$.
(b) If a mapping $G : H \to Y$ is of class $\mathcal{C}^{1,1}$ around a point $\overline{x} \in G^{-1}(D)$ and metrically subregularly transversal at $\overline{x}$ to the set D and if D is prox-regular at $G(\overline{x})$, then the set $G^{-1}(D)$ is prox-regular at $\overline{x}$.

PROOF. Put $C := C_1 \cap \cdots \cap C_m$ and fix some open neighborhood U of $\overline{x}$ such that all the sets C_k are prox-regular at all points of $C \cap U$ and such that the metric subregular transversality property in Definition 7.42 holds on U. Fix any $x \in C \cap U$ and take any $v \in N^F(C; x)$ with $\|v\| \le 1$. We have $v \in \partial_F d_C(x)$ since one knows that $\partial_F d_C(x) = N^F(C; x) \cap \mathbb{B}_H$ (see Proposition 4.30). The metric subregular inequality in Definition 7.42 and the definition of Fréchet subgradient easily yields that $v \in \gamma\,\partial_F(d_{C_1} + \cdots + d_{C_m})(x)$ and hence in particular $v \in \gamma\,\partial_L(d_{C_1} + \cdots + d_{C_m})(x)$. The functions d_{C_k} being Lipschitz continuous, the formula of the L-subdifferential of a finite sum of locally Lipschitz continuous functions (see the sum formula (4.41) in Theorem 4.78) assures us that there exist $v_k \in \partial_L d_{C_k}(x)$ such that $\gamma^{-1}v = v_1 + \cdots + v_m$. Since

$$\partial_L d_{C_k}(x) \subset N^L(C_k; x) \cap \mathbb{B}_H = N^F(C_k; x) \cap \mathbb{B}_H$$

(the inclusion following from (4.43) and the equality from Theorem 15.75), we have $v_k \in N^F(C_k; x) \cap \mathbb{B}_H$, and hence we deduce that the normal cone intersection property in Definition 8.125 with respect to the Fréchet normal cone is satisfied near $\overline{x}$ for the sets $C_1, \cdots, C_m$. So, the assertion (a) of the proposition follows from Proposition 15.109(a).

Let us now prove (b). Put $S := G^{-1}(D)$ and fix some open neighborhood U of $\overline{x}$ such that the metric subregular transversality (8.59) holds on U (that is, $d_S \le \gamma\, d_D \circ G$) on U) and such that the mapping G as well as its derivative $DG(\cdot)$ are Lipschitz on U. Fix any $x \in U \cap S$ and $v \in N^F(S; x)$ with $\|v\| \le 1$. As above we have $v \in \partial_F d_S(x)$ and this entails, by the above subregularity property $d_S \le \gamma\, d_D \circ G)$ on U, that $v \in \gamma\,\partial_F(d_D \circ G)(x)$, and hence $v \in \gamma\,\partial_L(d_D \circ G)(x)$. The subdifferential chain rule (see the inclusion formula (4.42)) gives some $w \in \gamma\,\partial_L d_D(G(x))$ such that $v = DG(x)^*(w)$. Since

$$\partial_L d_D(G(x)) \subset N^L(D; G(x)) \cap \mathbb{B}_Y = N^F(D; G(x)) \cap \mathbb{B}_Y,$$

we obtain $w \in N^F(D; G(x)) \cap \gamma\mathbb{B}_Y$, and this assures us that the F-normal cone property in Definition 8.111 holds. This gives the assertion (b) of the proposition through Proposition 15.109(b). □

For the convenience of the reader, in the proof of Proposition 15.109 we made the choice to prove first the prox-regularity of the intersection, and then to give the additional arguments yielding to the prox-regularity of the inverse image. However, the case of the intersection can also be derived from that of the inverse image. Indeed, putting $Y = H^m$, $G(x) = (x, \cdots, x)$ for all $x \in H$, and $D = C_1 \times \cdots \times C_m$ we see that $G^{-1}(D) = \bigcap_{k=1}^{m} C_k$. Therefore, some direct arguments and computation furnish through (b) the result of (a).

Let C and D be subsets of normed spaces X and Y respectively, $g : H \to Y$ be a mapping from X to Y, and $\overline{x} \in C \cap g^{-1}(D)$. Considering $G : X \to X \times Y$ with $G(x) = (x, g(x))$, it is clear (as we saw just after (8.71)) that the metric subregularity qualification condition is satisfied at $\overline{x}$ for $C \cap g^{-1}(D)$ if and only if the mapping G is metrically subregularly transversal at $\overline{x}$ to the set $C \times D$. This and Proposition 15.110 above give the following:

Proposition 15.111. Let C and D be closed sets of Hilbert spaces H and Y respectively and let $g : H \to Y$ be a mapping which is of class $\mathcal{C}^{1,1}$ near a point $\overline{x} \in C \cap g^{-1}(D)$. Assume that C and D are prox-regular at $\overline{x}$ and $g(\overline{x})$ respectively. Assume also that the metric subregularity qualification condition is satisfied at $\overline{x}$ for $C \cap g^{-1}(D)$. Then the set $C \cap g^{-1}(D)$ is prox-regular at $\overline{x}$.

As a first direct corollary, using Corollary 7.75 we derive the local counterpart of Proposition 15.104 relative to inverse images.

Corollary 15.112. Let $G : H \to Y$ be a mapping from a Hilbert space H into a Hilbert space Y, let S be a subset of Y and let $\overline{x} \in G^{-1}(S)$ be such that S is closed at $G(\overline{x})$ and prox-regular at $G(\overline{x})$. Assume that G is of class $\mathcal{C}^{1,1}$ near $\overline{x}$ and that the derivative $DG(\overline{x})$ is surjective. Then the set $G^{-1}(S)$ is prox-regular at $\overline{x}$.

The second corollary of Proposition 15.111 assumes the convexity of C and D. It follows from that proposition and Theorem 7.70(b).

Corollary 15.113. Let $g : H \to Y$ be a mapping from a Hilbert space H into a Hilbert space Y and let C and D be closed convex sets in H and Y respectively. Assume that g is of class $\mathcal{C}^{1,1}$ near a point $\overline{x} \in C \cap g^{-1}(D)$ and assume also that

$$0 \in \operatorname{core}\big(Dg(\overline{x})(C - \overline{x}) - (D - G(\overline{x}))\big).$$

Then the set C is prox-regular at $\overline{x}$.

An analysis of the proofs of Propositions 15.109 and 15.110 reveals that uniform versions also hold under either the truncated normal cone property or the metric subregularity with respective uniform constants.

PROPOSITION 15.114. Assume that all the closed sets C_k, $k = 1, \cdots, m$ in a Hilbert space H are r-prox-regular and that they satisfy the truncated F-normal cone intersection property (8.67) in Definition 8.125 (resp. they are metrically subregularly transversal) at any point of their intersection with the same constant $\gamma > 0$ in (8.67) (resp. the same rate $\gamma > 0$ of metric subregularity in (8.59)). Then the set $\bigcap_{k=1}^{m} C_k$ is r'-prox-regular with $r' := r/(m\gamma)$.

PROPOSITION 15.115. Let D be an r-prox-regular set of a Hilbert space Y and $G : H \to Y$ a mapping from a Hilbert space H into Y. Assume that $G : H \to Y$ is of class $\mathcal{C}^{1,1}$ on an open convex set $U \supset G^{-1}(D)$ and that G and DG are Lipschitz on U with constants K and K_1 respectively. Assume also that the inverse image set $G^{-1}(D)$ has the (truncated) F-normal cone inverse image property (8.57) in Definition 8.111 (resp. the mapping G is metrically subregularly transversal to the set D) at any point of $G^{-1}(D)$ with the same constant $\gamma > 0$ in (8.57) (resp. the same rate $\gamma > 0$ of metric subregularity). Then the inverse image set $G^{-1}(D)$ is r'-prox-regular with $r' := \frac{1}{\gamma(K_1 + r^{-1}K^2)}$.

We examine now the local prox-regularity in the situation when normal transversality conditions are present. Given a subdifferential ∂ and its associated normal cone $N^{\partial}(\cdot;\cdot)$, recall (see Definition 7.52) that a system of sets $C_1 \cdots, C_m$ of a normed space is ∂-normally transversal at a point x in their intersection if

$$\left(v_1 \in N^{\partial}(C_1; x), \cdots, v_m \in N^{\partial}(C_m; x), \sum_{k=1}^{m} v_k = 0\right) \Longrightarrow v_1 = \cdots = v_m = 0.$$

PROPOSITION 15.116. Let $(C_k)_{k=1}^{m}$ be a finite family of closed sets in a Hilbert space H and $\overline{x} \in \bigcap_{k=1}^{m} C_k$. For $C := \bigcap_{k=1}^{m} C_k$, the following properties hold.
(a) If all the sets C_k except at most one are compactly epi-Lipschitzian at $\overline{x}$ and if the sets C_k are L-normally transversal at $\overline{x}$, then the intersection C is prox-regular at $\overline{x}$.
(b) If the Hilbert space H is finite-dimensional and if the sets C_k are L-normally transversal at $\overline{x}$, then the intersection C is prox-regular at $\overline{x}$.

PROOF. (a) First observe that the prox-regularity of the sets C_k at $\overline{x}$ ensures that their F-normal cones coincide near $\overline{x}$ with their L-normal cones according to Proposition 15.75. Then by Proposition 15.110(a) the result follows from the fact that the assumption of compactly epi-Lipschitzian property combined with the L-normal transversality $\overline{x}$ entails (see Theorem 7.58) that the sets C_k are in metric regular, hence also subregular, transversality at $\overline{x}$.
(b) This is a direct consequence of (a). □

Recalling the equality $N^F(S; \overline{x}) = N^L(S; \overline{x})$ for any set S which is prox-regular at $\overline{x}$, Proposition 15.111 combined with the metric subregularity property in Corollary 7.84 furnishes the following other local prox-regularity result for sets of type $C \cap g^{-1}(D)$.

PROPOSITION 15.117. Let C and D be closed sets of a Hilbert space H and the Euclidean space $\mathbb{R}^m$ respectively, and let $g : H \to \mathbb{R}^m$ be a mapping which is of class $\mathcal{C}^{1,1}$ near a point $\overline{x} \in C \cap g^{-1}(D)$. Assume that C and D are prox-regular at $\overline{x}$ and $g(\overline{x})$ respectively. Assume also the following qualification condition: the only vector $\lambda = (\lambda_1, \cdots, \lambda_m)$ in $\mathbb{R}^m$ such that

$$\lambda \in N^F(D; g(\overline{x})) \quad \text{and} \quad \sum_{i=1}^{m} \lambda_i \nabla g_i(\overline{x}) \in -N^F(C; \overline{x})$$

is the null vector $\lambda = (0, \cdots, 0)$. Then the set $C \cap g^{-1}(D)$ is prox-regular at $\overline{x}$.

We derive a first corollary.

COROLLARY 15.118. Let C be a closed set of a Hilbert space H which is prox-regular at $\overline{x} \in C$ and let $g : H \to \mathbb{R}^m$ be a mapping of class $\mathcal{C}^{1,1}$ near $\overline{x} \in M := \{x \in H : g(x) = 0\}$. Assume that the only vector $\lambda = (\lambda_1, \cdots, \lambda_m)$ in $\mathbb{R}^m$ such that

$$\sum_{i=1}^{m} \lambda_i \nabla g_i(\overline{x}) \in N^F(C; \overline{x})$$

is the null vector $\lambda = (0, \cdots, 0)$. Then the set $C \cap M$ is prox-regular at $\overline{x}$.

The case of a real-valued function g (that is, $m = 1$) is of particular interest.

COROLLARY 15.119. Let C be a closed set of a Hilbert space H which is prox-regular at $\overline{x} \in C$ and let $g : H \to \mathbb{R}$ be a real-valued function of class $\mathcal{C}^{1,1}$ near $\overline{x} \in M := \{x \in H : g(x) = 0\}$. If

$$\nabla g(\overline{x}) \notin N^F(C; \overline{x}) \cup (-N^F(C; \overline{x})),$$

then $C \cap M$ is prox-regular at $\overline{x}$.

The next results related to the local prox-regularity of sublevel/level sets are just obvious adaptations of statements and proofs of their counterparts in the previous subsection. So, we omit their proofs. The first two results translate the local point of view of Propositions 15.97 and 15.98 respectively.

PROPOSITION 15.120. Let $g_1, \ldots, g_m : H \to \mathbb{R}$ be functions on a Hilbert space H, let

$$C := \{x \in H : g_1(x) \le 0, \cdots, g_m(x) \le 0\},$$

and let $\overline{x} \in \operatorname{bdry} C$. Assume that:
(i) for each $k \in K := \{1, \ldots, m\}$, the function g_k is locally Lipschitz continuous near the point $\overline{x}$;
(ii) there is a real $\gamma \ge 0$ and a neighborhood U of $\overline{x}$ such that, for any $k \in K$, for any $x \in U \cap \operatorname{bdry} C$, any $\zeta \in \partial_C g_k(x)$, any $x' \in U$, and any $\zeta' \in \partial_C g_k(x')$

$$\langle \zeta' - \zeta, x' - x \rangle \ge -\gamma \|x' - x\|^2 .$$

Assume also that there is a real $\sigma > 0$ such that, for each $x \in \operatorname{bdry} C$ near $\overline{x}$, there exists $\overline{v} \in \mathbb{B}_H$ (depending on x) for which for every $k \in K(x) := \{j \in K : g_j(x) = \max_{i \in K} g_i(x)\}$

$$g_k^\circ(x; \overline{v}) \le -\sigma,$$

or equivalently $\langle \zeta, \overline{v} \rangle \le -\sigma$ for all $\zeta \in \partial_C g_k(x)$.
Then, the set C is prox-regular at $\overline{x}$.

PROPOSITION 15.121. Let $g_1, \ldots, g_m : H \to \mathbb{R}$ be functions on a Hilbert space H, let

$$C := \{x \in H : g_1(x) \le 0, \cdots, g_m(x) \le 0\},$$

and let $\overline{x} \in \operatorname{bdry} C$. Assume that:
(i) for each $k \in K := \{1, \ldots, m\}$, the function g_k is continuous near the point $\overline{x}$;
(ii) there is a real $\gamma \ge 0$ and a neighborhood U of $\overline{x}$ such that, for any $k \in K$, for any $x, x' \in U$, and for any $\zeta \in \partial_C g_k(x)$ and any $\zeta' \in \partial_C g_k(x')$

$$\langle \zeta' - \zeta, x' - x \rangle \ge -\gamma \|x' - x\|^2 .$$

Assume also that there is a real $\sigma > 0$ such that, for each $x \in U \cap \operatorname{bdry} C$ near $\overline{x}$, there exists $\overline{v} \in \mathbb{B}_H$ (depending on x) for which for every $k \in K(x) := \{j \in K : g_j(x) = \max_{i \in K} g_i(x)\}$

$$g_k^\circ(x; \overline{v}) \le -\sigma,$$

or equivalently $\langle \zeta, \overline{v} \rangle \le -\sigma$ for all $\zeta \in \partial_C g_k(x)$.

Then, the set C is prox-regular at $\overline{x}$.

Under the continuous differentiability of the functions g_k, it results the following corollary from Proposition 15.120.

COROLLARY 15.122. Let $g_1, \ldots, g_m : H \to \mathbb{R}$ be functions on a Hilbert space H, let

$$C := \{x \in H : g_1(x) \le 0, \cdots, g_m(x) \le 0\},$$

and let $\overline{x} \in \operatorname{bdry} C$. Assume that:
(i) for each $k \in K := \{1, \ldots, m\}$, the function g_k is continuously differentiable near the point $\overline{x}$;
(ii) there is a real $\gamma \ge 0$ and a neighborhood U of $\overline{x}$ such that, for any $k \in K$, any $x \in U \cap \operatorname{bdry} C$ and any $x' \in U$

$$\langle \nabla g_k(x') - \nabla g_k(x), x' - x \rangle \ge -\gamma \|x' - x\|^2 .$$

Assume also that there is a real $\sigma > 0$ such that, for each $x \in \operatorname{bdry} C$ near $\overline{x}$, there exists $\overline{v} \in \mathbb{B}_H$ (depending on x) for which for every $k \in K(x) := \{j \in K : g_j(x) = \max_{i \in K} g_i(x)\}$

$$\langle \nabla g_k(x), \overline{v} \rangle \le -\sigma.$$

Then, the set C is prox-regular at $\overline{x}$.

15.5. Continuity properties of $C \mapsto P_C(u)$

Consider now the behaviour of $P_C(u)$ with respect to the r-prox-regular set C when we endow the space of r-prox-regular sets with the Hausdorff-Pompeiu distance.

THEOREM 15.123 (continuity of metric projection in prox-regular set-variable). Let C_1 and C_2 be r-prox-regular sets of a Hilbert space H for a constant $r > 0$ such that $\operatorname{haus}(C_1, C_2) \le r$. Then for all $u_1 \in U_r(C_1)$ and $u_2 \in U_r(C_2)$ the following inequalities hold:

$$\begin{aligned} &\left(1 - \frac{1}{r} d_{C_1}(u_1) - \frac{1}{r} d_{C_2}(u_2)\right) \|P_{C_1}(u_1) - P_{C_2}(u_2)\|^2 \\ &\le \|u_1 - u_2\|^2 + 2\big(d_{C_1}(u_1) \operatorname{exc}(C_2, C_1) + d_{C_2}(u_2) \operatorname{exc}(C_1, C_2)\big) \end{aligned} \tag{15.69}$$

and

(15.70)

$$\|P_{C_1}(u_1) - P_{C_2}(u_2)\| \le \frac{2r}{2r - d_{C_1}(u_1) - d_{C_2}(u_2)} \|u_1 - u_2\|$$

$$+ \sqrt{\frac{2r\big(d_{C_1}(u_1)\mathrm{exc}(C_2, C_1) + d_{C_2}(u_2)\mathrm{exc}(C_1, C_2)\big)}{2r - d_{C_1}(u_1) - d_{C_2}(u_2)}}. \tag{15.71}$$

So, for $\gamma \in]0, 1[$ and $u_i \in U_r^\gamma(C_i)$, one has

$$\|P_{C_1}(u_1) - P_{C_2}(u_2)\| \le (1-\gamma)^{-1}\|u_1 - u_2\| + \sqrt{\frac{2\gamma r}{1-\gamma}}\big(\mathrm{haus}(C_1, C_2)\big)^{1/2}. \tag{15.72}$$

Proof. Fix $u_i \in U_r(C_i)$ and put $x_i := P_{C_i}(u_i)$ for $i := 1, 2$ (note that $P_{C_i}(u_i)$ exists according to (k) of Theorem 15.28). Observe that $d_{C_1}(x_2) \le \mathrm{exc}(C_2, C_1)$ since $x_2 \in C_2$. Assume for a moment that $x_i \neq u_i$. By Definition 15.1 of $\rho(\cdot)$-prox-regularity we have $x_i \in \mathrm{Proj}\,_{C_i}\big(x_i + r\frac{u_i - x_i}{\|u_i - x_i\|}\big)$, which entails for all $z \in C_1$

$$\left\| x_1 + r\frac{u_1 - x_1}{\|u_1 - x_1\|} - x_2 \right\| \ge \left\| x_1 + r\frac{u_1 - x_1}{\|u_1 - x_1\|} - z \right\| - \|z - x_2\| \ge r - \|z - x_2\|,$$

and hence, setting $e_{2,1} := \mathrm{exc}(C_2, C_1)$, it results

$$\left\| x_1 + r\frac{u_1 - x_1}{\|u_1 - x_1\|} - x_2 \right\| \ge r - d_{C_1}(x_2) \ge r - e_{2,1}.$$

Since $r - e_{2,1} \ge 0$ according to the assumption $\mathrm{haus}(C_1, C_2) \le r$, we deduce that

$$\begin{aligned}\|x_1 - x_2\|^2 + 2\frac{r}{\|u_1 - x_1\|}\langle u_1 - x_1, x_1 - x_2\rangle + r^2 &\ge r^2 - 2r\, e_{2,1} + (e_{2,1})^2 \\ &\ge r^2 - 2r\, e_{2,1},\end{aligned}$$

which gives without the restriction $x_i \neq u_i$ that

$$\begin{aligned}2r\langle u_1 - x_1, x_2 - x_1\rangle &\le \|u_1 - x_1\|\big(\|x_1 - x_2\|^2 + 2r\, e_{2,1}\big) \\ &= d_{C_1}(u_1)\big(\|x_1 - x_2\|^2 + 2r\, e_{2,1}\big).\end{aligned}$$

Likewise, putting $e_{1,2} := \mathrm{exc}(C_1, C_2)$, we have

$$2r\langle u_2 - x_2, x_1 - x_2\rangle \le d_{C_2}(u_2)\big(\|x_1 - x_2\|^2 + 2r\, e_{1,2}\big),$$

and adding both inequalities we obtain

$$\begin{aligned}&2r\langle u_2 - u_1, x_1 - x_2\rangle + 2r\|x_1 - x_2\|^2 \\ &\le \big(d_{C_1}(u_1) + d_{C_2}(u_2)\big)\|x_1 - x_2\|^2 + 2r\big(d_{C_1}(u_1)e_{2,1} + d_{C_2}(u_2)e_{1,2}\big),\end{aligned} \tag{15.73}$$

which entails

$$\begin{aligned}&- r\|x_1 - x_2\|^2 - r\|u_1 - u_2\|^2 + 2r\|x_1 - x_2\|^2 \\ &\le \big(d_{C_1}(u_1) + d_{C_2}(u_2)\big)\|x_1 - x_2\|^2 + 2r\big(d_{C_1}(u_1)e_{2,1} + d_{C_2}(u_2)e_{1,2}\big),\end{aligned}$$

or equivalently

$$\begin{aligned}&\big(r - d_{C_1}(u_1) - d_{C_2}(u_2)\big)\|x_1 - x_2\|^2 \\ &\le r\|u_1 - u_2\|^2 + 2r\big(d_{C_1}(u_1)e_{2,1} + d_{C_2}(u_2)e_{1,2}\big).\end{aligned}$$

The latter obviously translates the first desired inequality (15.69).

Now let us prove (15.70). Observe that from (15.73) we have

$$\begin{aligned}&\big(2r - d_{C_1}(u_1) - d_{C_2}(u_2)\big)\|x_1 - x_2\|^2 - 2r\|u_1 - u_2\|\,\|x_1 - x_2\|\\&\le 2r\big(d_{C_1}(u_1)e_{2,1} + d_{C_2}(u_2)e_{1,2}\big),\end{aligned}$$

thus

$$\left(\sqrt{2r - d_{C_1}(u_1) - d_{C_2}(u_2)}\|x_1 - x_2\| - \frac{r}{\sqrt{2r - d_{C_1}(u_1) - d_{C_2}(u_2)}}\|u_1 - u_2\|\right)^2$$
$$\le \frac{r^2}{2r - d_{C_1}(u_1) - d_{C_2}(u_2)}\|u_1 - u_2\|^2 + 2r\big(d_{C_1}(u_1)e_{2,1} + d_{C_2}(u_2)e_{1,2}\big).$$

It ensues that

$$\begin{aligned}&\sqrt{2r - d_{C_1}(u_1) - d_{C_2}(u_2)}\|x_1 - x_2\| - \frac{r}{\sqrt{2r - d_{C_1}(u_1) - d_{C_2}(u_2)}}\|u_1 - u_2\|\\&\le \frac{r}{\sqrt{2r - d_{C_1}(u_1) - d_{C_2}(u_2)}}\|u_1 - u_2\| + \sqrt{2r\big(d_{C_1}(u_1)e_{2,1} + d_{C_2}(u_2)e_{1,2}\big)},\end{aligned}$$

and this is equivalent to the desired inequality

$$\|x_1 - x_2\| \le \frac{2r}{2r - d_{C_1}(u_1) - d_{C_2}(u_2)}\|u_1 - u_2\| + \sqrt{\frac{2r\big(d_{C_1}(u_1)e_{2,1} + d_{C_2}(u_2)e_{1,2}\big)}{2r - d_{C_1}(u_1) - d_{C_2}(u_2)}}.$$

From the latter inequality, we readily obtain (15.72), which finishes the proof of the theorem. □

REMARK 15.124. Given any two nonempty closed convex sets C_1, C_2 of a Hilbert space H, taking $r = +\infty$ in (15.69) in the above theorem we derive for all $u_1, u_2 \in H$ the inequality

$$\|P_{C_1}(u_1) - P_{C_2}(u_2)\|^2 \le \|u_1 - u_2\|^2 + 2\big(d_{C_1}(u_1) + d_{C_2}(u_2)\big)\mathrm{haus}(C_1, C_2),$$

already proved in Proposition 1.77(b). □

The next proposition establishes a stability of prox-regularity under the pointwise convergence of distance functions.

PROPOSITION 15.125. *Let $(r_k)_{k\in\mathbb{N}}$ be a sequence of extended reals in $]0, +\infty]$ tending to some $r \in\,]0, +\infty]$. Let $(C_k)_{k\in\mathbb{N}}$ be a sequence of closed sets in a Hilbert space H such that C_k is r_k-prox-regular for each $k \in \mathbb{N}$. If C is a closed set in H such that $(d_{C_k})_{k\in\mathbb{N}}$ pointwise converges to d_C on $U_r(C)$, that is, $d_{C_k}(x) \to d_C(x)$ for every $x \in U_r(C)$, then the set C is r-prox-regular.*

PROOF. Note first that for any sequence $(u_k)_k$ in H converging to $u \in U_r(C)$ we have $d_{C_k}(u_k) \to d_C(u)$ as $k \to \infty$ since

$$|d_{C_k}(u_k) - d_C(u)| \le \|u_k - u\| + |d_{C_k}(u) - d_C(u)|.$$

Fix any $t \in [0,1]$ and any $x, y \in C$ with $\|x - y\| < 2r$. Set $z := (1-t)x + ty$ and note by Lemma 15.95 that $z \in U_r(C)$. Consider $z_k := (1-t)x_k + ty_k$, $\varepsilon_k := d_{C_k}(x) + 1/k$ and $\varepsilon'_k := d_{C_k}(y) + 1/k$, and note that $\varepsilon_k \downarrow 0$ and $\varepsilon'_k \downarrow 0$. For each $k \in \mathbb{N}$ choose $x_k, y_k \in C_k$ such that $\|x_k - x\| < \varepsilon_k$ and $\|y_k - y\| < \varepsilon'_k$, and set $z_k := (1-t)x_k + ty_k$. There exists some $k_0 \in \mathbb{N}$ such that $z_k \in U_r(C)$ and $\|x_k - y_k\| < 2r_k$ for all $k \ge k_0$. For each $k \ge k_0$ we have by (d) in Proposition

15.41 that $d_{C_k}(z_k) \le \frac{1}{2r_k} \min\{t, 1-t\}\|x_k - y_k\|^2$, and hence paasing to the limit as $k \to \infty$ we obtain by what precedes

$$d_C(z) \le \frac{1}{2r} \min\{t, 1-t\}\|x - y\|^2.$$

By (d) in Proposition 15.41 again it results that the set C is r-prox-regular. $\square$

Theorem 15.123 above established the Hölder continuity of the metric projection with respect to the Hausdorff-Pompeiu distance whenever both variable sets are prox-regular. As another result related to the behavior of the metric projection with respect to prox-regular sets, we prove now a theorem on the Attouch-Wets convergence of r_k-prox-regular sets, with $r_k \to r \in]0, +\infty]$. For simplicity we restrict ourselves to the case of a prox-regularity parameter r_k independent of x.

THEOREM 15.126. Let $(r_k)_{k\in\mathbb{N}}$ be a sequence of extended reals in $]0, +\infty]$ tending to some $r \in]0, +\infty]$. Let $(C_k)_{k\in\mathbb{N}}$ be a sequence of closed sets in a Hilbert space H Attouch-Wets converging to a closed set C in H and such that for each $k \in \mathbb{N}$ the set C_k is r_k-prox-regular. Then

(a) the set C is r-prox-regular;
(b) for any real $r' \in]0, r[$ the sequence $\big(P_{C_k}(\cdot)\big)_k$ converges to $P_C(\cdot)$ uniformly on $U_{r'}(C) \cap \sigma\mathbb{B}$ for each real $\sigma > 0$ with $U_{r'}(C) \cap \sigma\mathbb{B} \ne \emptyset$.

PROOF. First, observe that, by Definition 1.66 of the Attouch-Wets convergence, $\big(d_{C_k}\big)_k$ converges to d_C uniformly on each bounded set of H, and hence pointwise on H. Proposition 15.125 says that C is r-prox-regular, so d_C^2 is of class $\mathcal{C}^1$ on $U_r(C)$ by (o) in Theorem 15.28. Let $r' \in]0, r[$ and $s \in]r', r[$. Take any real $\sigma > 0$ with $U_{r'}(C) \cap \sigma\mathbb{B} \ne \emptyset$ and fix two reals $\eta > \beta > \sigma$. There exists $k_0 \in \mathbb{N}$ such that for all $k \ge k_0$ one has $r_k > s$ and $U_s(C) \cap \eta\mathbb{B}_H \subset U_{r_k}(C) \cap \eta\mathbb{B}_H$ according to the uniform convergence of $(d_{C_k})_k$ to d_C on $\eta\mathbb{B}_H$. In particular, $d_{C_k}^2$ is of class $\mathcal{C}^1$ on the open set $U_s(C) \cap \eta\mathbb{U}_H$ (where $\mathbb{U}_H := B_H(0,1)$) for all $k \ge k_0$ according to (o) in Theorem 15.28 again. Since the sequence $(r_k)_k$ converges in $]0, +\infty]$, there is a real $\ell > 0$ such that $2(1 + (1 - s/r_k)^{-1}) < \ell$ for all $k \in \mathbb{N}$. Then for any integer $k \ge k_0$ and for any $x_1, x_2 \in U_s(C) \cap \eta\mathbb{U}_H$ we have by (h) and (o) in Theorem 15.28

$$\|Df_k(x_1) - Df_k(x_2)\| \le \ell\|x_1 - x_2\|, \tag{15.74}$$

where $f_k := d_{C_k}^2$. Put $M := 1 + 2\eta + d_C(0)$. Consider any real $0 < \varepsilon < \min\{1, \beta, \eta - \beta, s - r'\}$ and choose an integer $k_1 \ge k_0$ (depending on ε) such that for $\varepsilon' := \varepsilon/(2M + \ell)$

$$\sup_{x\in 2\eta\mathbb{B}_H} |d_{C_k}(x) - d_C(x)| < (\varepsilon')^2/2 \quad \text{for all } k \ge k_1, \tag{15.75}$$

so we see that $d_{C_k}(x) \le M$ for all $k \ge k_1$ and for all $x \in 2\eta\mathbb{B}_H$. Fix for a moment any $x \in U_{r'}(C) \cap \beta\mathbb{U}_H$ and any $w \in H$ with $\|w\| \le \varepsilon'$. Then for any integer $p \ge k_1$, we have $x + tw \in U_s(C_p) \cap \eta\mathbb{U}_H$ for all $t \in [0, 1]$, so

$$f_p(x + w) - f_p(x) = \langle Df_p(x), w\rangle + \int_0^1 \langle Df_p(x + tw) - Df_p(x), w\rangle \, dt.$$

This yields for any integers $p, q \geq k_1$

$$\begin{aligned}&\langle Df_p(x) - Df_q(x), w\rangle\\ &= (f_p(x+w) - f_p(x)) - (f_q(x+w) - f_q(x))\\ &\quad - \int_0^1 \langle Df_p(x+tw) - Df_p(x), w\rangle\, dt + \int_0^1 \langle Df_q(x+tw) - Df_q(x), w\rangle\, dt.\end{aligned}$$

On the other hand, by (15.74) we have for any integer $k \geq k_1$

$$\left|\int_0^1 \langle Df_k(x+tw) - Df_k(x), w\rangle\, dt\right| \leq \ell\|w\|^2 \int_0^1 t\, dt = \frac{\ell}{2}\|w\|^2;$$

and by (15.75) we also have, with $y = x + w$ or $y = x$, for any integers $p, q \geq k_1$

$$|f_p(y) - f_q(y)| = \left|d_{C_p}(y) + d_{C_q}(y)\right| \left|d_{C_p}(y) - d_{C_q}(y)\right| \leq M(\varepsilon')^2.$$

It ensues that $\langle Df_p(x) - Df_q(x), w\rangle \leq (2M + \ell)(\varepsilon')^2$, hence for every $v \in \mathbb{B}_H$

$$\langle Df_p(x) - Df_q(x), v\rangle \leq (2M + \ell)\varepsilon' = \varepsilon,$$

which gives

$$\|Df_p(x) - Df_q(x)\| \leq \varepsilon. \tag{15.76}$$

This uniform Cauchy property of $(Df_k(\cdot))_k$ on $U_{r'}(C) \cap \beta\mathbb{U}_H$ and all the above features ensure that $(Dd^2_{C_k}(\cdot))_k$ converges uniformly on $U_{r'}(C) \cap \beta\mathbb{U}_H$ to $Dd^2_C(\cdot)$, the Fréchet derivative of d^2_C. Further, by (h) in Theorem 15.28 we have for any $k \geq k_0$

$$\nabla d^2_{C_k}(x) = 2(x - P_{C_k}(x)) \text{ and } \nabla d^2_C(x) = 2(x - P_C(x)) \text{ for all } x \in U_s(C) \cap \beta\mathbb{U}_H.$$

Consequently, the sequence $(P_{C_k})_k$ converges uniformly on $U_{r'}(C) \cap \beta\mathbb{U}_H$ to P_C. This finishes the proof. □

The result has as an immediate consequence the following corollary.

COROLLARY 15.127. *Let U be a nonempty open set of a Hilbert space H and let $(r_k)_{k\in\mathbb{N}}$ be a sequence of extended reals in $]0, +\infty]$ tending to some $r \in]0, +\infty]$. For each $k \in \mathbb{N}$, let $f_k : \operatorname{cl} U \to \mathbb{R} \cup \{+\infty\}$ be lower semicontinuous with r_k-prox-regular epigraph and let $f : \operatorname{cl} U \to \mathbb{R} \cup \{+\infty\}$ be lower semicontinuous. Assume that the sequence $(\operatorname{epi} f_k)_{k\in\mathbb{N}}$ Attouch-Wets converges to $\operatorname{epi} f$. Then $\operatorname{epi} f$ is r-prox-regular in $H \times \mathbb{R}$ equipped with its canonical Hilbert structure.*

15.6. Further results

15.6.1. Representation of multimappings with prox-regular values. Let (S, d) be a metric space and $F : S \rightrightarrows H$ be a multimapping with prox-regular values. The aim of this section is to study under some conditions the existence of a mapping $f : S \times \mathbb{B}_H \to H$ such that $F(s) = f(s, \mathbb{B}_H)$ for all $s \in S$. Such a representation is required to be such that $f(s, \cdot)$ is continuous whenever the multimapping F is continuous with respect to the Hausdorff-Pompeiu distance.

We first establish three lemmae which have their own interest.

LEMMA 15.128. *Let C be a set of a Hilbert space H and $a, b \in H$ with $C \subset B[a, \alpha]$ and $C \cap B(b, \beta) = \emptyset$, where α, β are two positive numbers. Then*

$$\|u - a\|^2 - \|u - b\|^2 \leq \alpha^2 - \beta^2 \quad \text{for all } u \in \overline{\operatorname{co}}\, C.$$

PROOF. It suffices to observe that the closed half-space

$$\{x \in H : 2\langle a-b, x\rangle \le \alpha^2-\beta^2-\|a\|^2-\|b\|^2\} = \{x \in H : \|x-a\|^2-\|x-b\|^2 \le \alpha^2-\beta^2\}$$

contains C (as is easily seen from the second member above according to the assumptions concerning C), and hence it contains $\overline{\text{co}}\, C$. $\square$

The second lemma gives, through the preceding one, an upper estimate of the Hausdorff-Pompeiu distance between a prox-regular set C and its closed convex hull. It will also be used in Proposition 15.136.

LEMMA 15.129. Let C be an r-prox-regular set of a Hilbert space H which is contained in a closed ball $B[a, \alpha]$ with $\alpha > 0$. Then

$$\text{haus}(C, \overline{\text{co}}\, C) \le \alpha^2/r.$$

PROOF. Since $C \subset \overline{\text{co}}\, C$, it suffices to show for any fixed $u \in \overline{\text{co}}\, C$ that one has $d(u, C) \le \alpha^2/r$. We may obviously suppose $d_C(u) > 0$.

If $\alpha \ge r$, then putting $\beta := d_C(u)$ and applying Lemma 15.128 with $b = u$ gives $\|u - a\|^2 \le \alpha^2 - d_C(u)^2$, hence

$$d_C(u) \le \alpha \le \alpha^2/r.$$

Suppose now that $\alpha < r$. The r-prox-regularity of C ensures that $x := P_C(u)$ is well defined and that $C \cap B(x + r\frac{u-x}{\|u-x\|}) = \emptyset$. Applying Lemma 15.128 with $b = x + r\frac{u-x}{\|u-x\|}$ and $\beta = r$ gives for $v := (u - x)/\|u - x\|$

$$\|u - a\|^2 - \|u - x - rv\|^2 \le \alpha^2 - r^2,$$

hence $\|u - x - rv\| \ge \sqrt{r^2 - \alpha^2}$. This yields

$$d_C(u) = \|u - x\| \le r - \|u - x - rv\| \le r - \sqrt{r^2 - \alpha^2} \le \alpha^2/r,$$

completing the proof of the lemma. $\square$

The last lemma is related to the Hausdorff-Pompeiu distance between the convex hulls of two sets.

LEMMA 15.130. Let C_1, C_2 be two nonempty sets of a normed space $(X, \|\cdot\|)$. Then

$$\text{haus}(\overline{\text{co}}\, C_1, \overline{\text{co}}\, C_2) \le \text{haus}(C_1, C_2).$$

PROOF. Putting $h = \text{haus}(C_1, C_2)$ we have for any $x \in C_1$

$$d(x, \overline{\text{co}}\, C_2) \le d(x, C_2) \le h.$$

Since the function $d(\cdot, \overline{\text{co}}\, C_2)$ is convex and continuous, the inequality is preserved for all $x \in \overline{\text{co}}\, C_1$, hence $\sup_{x \in \overline{\text{co}}\, C_1} d(x, \overline{\text{co}}\, C_2) \le h$. Changing C_1 and C_2 gives a similar inequality which combined with the preceding one entails the inequality of the lemma. $\square$

We can now state and prove the representation theorem for multimappings with prox-regular values.

THEOREM 15.131 (representation of continuous multimappings with prox-regular values). Let S be a topological space and $F : S \rightrightarrows H$ be a multimapping with r-prox-regular values in a Hilbert space H. Assume there is a positive number $\alpha < r$ and a mapping $a : S \to H$ such that $F(s) \subset B[a(s), \alpha]$ for all $s \in S$. Assume also that $F(S)$ is bounded in H.

Then there exists a mapping $f : S \times \mathbb{B}_H \to H$ representing F in a Lipschitz way in the sense that
(a) $F(s) = f(s, \mathbb{B}_H)$ for all $s \in S$;
(b) the mapping $f(s, \cdot)$ is Lipschitz continuous on $\mathbb{B}_H$.

The mapping f is also such that $f(\cdot, x)$ is continuous (resp. Hölder continuous with exponent 1/2) for each $x \in \mathbb{B}_H$ whenever the multimapping F is continuous (resp. Lipschitz continuous) with respect to the Hausdorff-Pompeiu distance.

PROOF. Fix a real number $\beta > 0$ such that $F(S) \subset \beta\mathbb{B}_H$. For any $s \in S$ and $x \in \mathbb{B}_H$ put with $P(G(s), \cdot) := P_{G(s)}(\cdot)$

$$G(s) := \overline{\mathrm{co}}\, F(s) \quad \text{and} \quad g(s, x) := P(G(s), \beta x).$$

Fix any real $\sigma > 2\beta$ and take any $s_1, s_2 \in S$ and $x_1, x_2 \in \mathbb{B}_H$. Put $h := \mathrm{haus}(F(s_1), F(s_2))$ and $h' = \mathrm{haus}(G(s_1), G(s_2))$. We know by Lemma 15.130 that $h' \le h$. Observing that $d(\beta x_i, G(s_j)) < \sigma$ for $i, j = 1, 2$ according to the inclusion $G(s_j) \subset \beta\mathbb{B}_H$, we may apply Remark 15.124 (or Proposition 1.77(b)) and we obtain

$$\|P(G(s_1), \beta x_1) - P(G(s_2), \beta x_2)\| \le \beta\|x_1 - x_2\| + \sqrt{2\sigma h'} \le \beta\|x_1 - x_2\| + \sqrt{2\sigma h}.$$

Taking the limit a $\sigma \downarrow 2\beta$ gives

$$\|g(s_1, x_1) - g(s_2, x_2)\| \le \beta\|x_1 - x_2\| + 2\sqrt{\beta h}.$$

Fix now any $s \in S$. By definition of g we have $g(s, \mathbb{B}_H) \subset G(s)$, and for any $y \in G(s)$ putting $x_y := y/\beta \in \mathbb{B}_H$ we see that $y = g(s, x_y)$. So $G(s) = g(s, \mathbb{B}_H)$.
By Lemma 15.129 we have

$$\mathrm{haus}(F(s), G(s)) \le \alpha^2/r < \alpha < r, \tag{15.77}$$

hence $P(F(s), y)$ exists for any $y \in G(s)$ according to the r-prox-regularity of $F(s)$ and $P(F(s), \cdot)$ is a Lipschitz continuous mapping on $U_r^{\alpha/r}(F(s))$ with $\frac{\alpha}{r-\alpha}$ as a Lipschitz constant therein. We may then put $f(s, x) := P(F(s), g(s, x))$ for all $x \in \mathbb{B}_H$ and we easily see from what precedes that $f(s, \mathbb{B}_H) = F(s)$, which is (a) of the theorem.

Further, noting by (15.77) that $g(s_i, x_i) \in U_r^{\alpha/r}(F(x_i))$, Theorem 15.123 allows us to write

$$\begin{aligned}\|f(s_1, x_1) - f(s_2, x_2)\| &\le \frac{r}{r-\alpha}\|g(s_1, x_1) - g(s_2, x_2)\| + \sqrt{\frac{2\alpha r h}{r-\alpha}} \\ &\le \frac{r}{r-\alpha}[\beta\|x_1 - x_2\| + 2\sqrt{\beta h}] + \sqrt{\frac{2\alpha r h}{r-\alpha}},\end{aligned}$$

that is, (b) of the theorem also holds.
The last part of the theorem follows directy from the latter inequality. □

15.6.2. Continuous selections of lower semicontinuous multimappings with prox-regular values. E. Michael introduced a particular class of nonconvex sets in the study of continuous selections of lower (or equivalently, inner) semicontinuous multimappings.

DEFINITION 15.132. A nonempty set C of a normed space $(X, \|\cdot\|)$ is λ-paraconvex (in Michael's sense) for some $\lambda \in [0, 1[$ when for all $s > 0$, $x \in X$ and $y \in \mathrm{co}\big(C \cap B(x, s)\big)$ one has

$$d_C(y) \le \lambda s.$$

When such a $\lambda \in [0,1[$ exists, we say that the set C is paraconvex.

With this concept Michael proved the following theorem in [**707**].

THEOREM 15.133 (E. Michael). Let S be a paracompact Hausdorff topological space, $\lambda \in [0,1[$, and $F : S \rightrightarrows X$ a lower semicontinuous multimapping with nonempty closed λ-paraconvex sets of a Banach space X. Then F admits a continuous selection.

Obviously any convex set is λ-paraconvex for $\lambda = 0$ (hence for any $\lambda \in [0,1[$). Nevertheless there is no implication between paraconvexity in the sense of Michael and prox-regularity of sets, as is seen in the following examples.

EXAMPLE 15.134. The set $\{(x,y) \in \mathbb{R}^2 : x \geq 0 \text{ or } y \geq 0\}$ is obviously not r-prox-regular for any $r > 0$ but it is not difficult to see that it is λ-paraconvex for $\lambda = 1/\sqrt{2}$. □

EXAMPLE 15.135. In $\mathbb{R}^n$ endowed with the usual inner product, the set $C = \{x \in \mathbb{R}^n : \|x\| \geq 1\}$ is obviously 1-prox-regular but it is not paraconvex. Indeed for any $\varepsilon > 0$ we have $\mathrm{co}\big(C \cap B(0, 1+\varepsilon)\big) = B(0, 1+\varepsilon) \ni 0$ and $d_C(0) = 1 \leq 1 + \varepsilon$, so there is no constant $\lambda \in [0,1[$ such that $1 \leq \lambda(1+\varepsilon)$ for all $\varepsilon > 0$. □

Despite the second example above, the inclusion of an r-prox-regular set in some ball of radius less than r ensures the paraconvexity of the set, as shown in the following proposition.

PROPOSITION 15.136. Let C be an r-prox-regular set of a Hilbert space H. Assume that there are some $a \in H$ and some positive $\alpha < r$ such that $C \subset B[a, \alpha]$. Then C is λ-paraconvex for $\lambda = \alpha/r < 1$.

PROOF. Fix any $x \in H$ and $s > 0$ such that $C \cap B(x,s) \neq \emptyset$ and take any $y \in \mathrm{co}\big(C \cap B(x,s)\big)$. There are $t \in [0,1]$ and $u, v \in C \cap B(x,s)$ such that $y = (1-t)u + tv$. We know by Lemma 15.129 that

$$\mathrm{haus}(C, \mathrm{co}C) \leq \alpha^2/r < \alpha < r,$$

hence $d_C(y) < r$, that is, $y \in U_r(C)$. By property (r) in Theorem 15.28 we then have

$$d_C(y) \leq \frac{1}{2r} \min\{1, 1-t\}(\|u - v\|)(\|u - v\|) \leq \frac{1}{2r}\frac{1}{2}(2\alpha)(2s),$$

since on the one hand $\|u - v\| \leq 2\alpha$ for $u, v \in C \subset B[a, \alpha]$ and on the other hand $\|u - v\| \leq 2s$ for $u, v \in B(x,s)$. So we conclude that $d_C(y) \leq \frac{\alpha}{r}s$. □

The following corollary is a direct consequence of the proposition and the Michael theorem above.

COROLLARY 15.137. Let S be a paracompact Hausdorff topological space and $F : S \rightrightarrows H$ be a lower semicontinuous multimapping with nonempty closed r-prox-regular sets of a Hilbert space H. Assume that there exists a positive $\alpha < r$ such that for each $s \in S$ the set $F(s)$ is contained in some closed ball $B[a(s), \alpha]$. Then F admits a continuous selection.

15.7. Comments

The first large study of prox-regular sets under the name of *positive reached sets* in the finite-dimensional space $\mathbb{R}^n$ has been done by H. Federer in his celebrated 1959 paper [**402**]. In that 1959 paper ("Curvature measures") submitted in 1958, Federer observed that there were two classical theories which allowed to compute the volume of an r-enlargement of a closed subset of $\mathbb{R}^n$ as an n-degree polynomial in the positive r-variable, with coefficients having a clear geometrical meaning. One theory was devoted to *convex sets*, while the other to *possibly nonconvex, but sufficiently smooth sets* (actually with $\mathcal{C}^2$-boundary). Then he stated that "the search for a general theory", which must be based neither on convexity nor on differentiability assumptions, "is an obvious challenge". In the above cases of sets, the equality defining the polynomial is known as the *Steiner polynomial formula*, since the result for convex sets was obtained in the nineteen century by J. Steiner in [**901**]. To make concrete what is said above, recall that the Steiner polynomial formula for compact convex sets C in $\mathbb{R}^n$, proved in 1939 by H. Weyl [**971**] for the other type of compact $\mathcal{C}^2$-submanifolds C in $\mathbb{R}^n$, says that

$$\lambda_n(\mathrm{Enl}_r(C)) = \sum_{k=0}^{n} \omega_{n-k}\mathcal{M}_k(C)r^{n-k} \tag{15.78}$$

for $r \in]0,+\infty[$ if C is convex and for r in some interval $]0,r_0[$ if C is a $\mathcal{C}^2$-submanifold. Here λ_n is the n-dimensional Lebesgue measure and ω_j denotes the j-dimensional measure of the closed unit ball of $\mathbb{R}^j$, and we recall that $\mathrm{Enl}_r(C) := \{x \in \mathbb{R}^n : d(x,C) \le r\}$ is the closed r-enlargement of C.

This challenge led to Federer's fundamental paper [**402**] where he introduced the concept of r_0-*positively reached* closed set C in $\mathbb{R}^n$ as a set for which any point $x \in \mathbb{R}^n$ with $d(x,C) < r_0$ has a unique nearest point $P_C(x)$ in C to x, and obtained the desired result of extending the Steiner formula to such sets. In fact, for such a set C Federer's Theorem 5.6 in [**402**] establishes that for each Borel set Q in $\mathbb{R}^n$ and for $r \in]0,r_0[$

$$\lambda_n(Q \cap \mathrm{Enl}_r(C)) = \sum_{k=0}^{n} \omega_{n-k}\Phi_k(C,Q)r^{n-k},$$

where all $\Phi_0(C,\cdot),\cdots,\Phi_n(C,\cdot)$ are countably additive measures, called the *curvature measures associated with* C.

In fact, previously to the Federer's aforementioned paper, given a closed bounded domain C in $\mathbb{R}^n$ whose boundary is a $\mathcal{C}^{1,1}$-submanifold and given the unit exterior normal n_x at each point $x \in \mathrm{bdry}\, C$, N. Aronszajn and K. T. Smith considered in their 1956 paper [**24**, p. 171, 172] the non-negative constants r_0, r_0' and r_0'' defined by

$$\frac{1}{r_0} := \sup\left\{\frac{\|n_x - n_y\|}{\|x-y\|} : x,y \in \mathrm{bdry}\, C,\, x \ne y\right\},$$

$r_0' := \sup\{\rho \in \mathbb{R}_+ : [x-\rho n_x, x+\rho n_x] \cap [y-\rho n_y, y+\rho n_y] = \emptyset, x,y \in \mathrm{bdry}\, C, x \ne y\}$, r_0'' is the supremum of $\rho \in \mathbb{R}_+$ such that for each point $x \in \mathrm{bdry}\, C$ the balls $B[x+\rho n_x,\rho]$ and $B[x-\rho n_x,\rho]$ contain no points in the interior and the exterior of C respectively. In that paper [**24**, p. 172], Aronszajn and Smith claimed that the constants r_0, r_0', r_0'' are all equal and wrote that the proof should be given in a separate note, but it seems that this note was never published. We observe that

$r_0'' > 0$ amounts to saying that bdry C is r_0''-prox-regular according to Definition 15.24. We also notice that the inequality $r_0 > 0$ holds if and only if $x \mapsto n_x$ is Lipschitz continuous on bdry C, which means that bdry C is a $\mathcal{C}^{1,1}_{\rm glo}$-submanifold, where such submanifolds are defined by requiring that the derivatives of φ and φ^{-1} in Definition E.8 in Appendix E be Lipschitz with a uniform Lipschitz constant not depending on m_0. Closed bounded domains C whose boundaries are $\mathcal{C}^{1,1}_{\rm glo}$-submanifolds are employed in [**24**] to valid the Poisson formula for harmonic functions with appropriate integrals over such suitable sets. In the same year 1956 appeared the paper [**835**] of Yu. G. Reshetnyak whose translation of a result (in nowadays terms) is that the epigraph of a real-valued Lipschitz function on $\mathbb{R}^N$ is prox-regular if and only if the function is semiconvex.

We must also mention that, previously to [**402**], [**24**] and [**835**] the class of sets satisfying the property of "*lemma of two points*"in Remark 15.29 was investigated in 1931 by G. Durand [**368**, Lemma 16].

Other different motivations and problems of several independent researches induced diverse authors to introduce many classes of appropriate sets under distinct names. J.-P. Vial used in his 1983 paper [**958**], for a closed set C in $\mathbb{R}^n$, the name of *weakly convex set* with a real constant $r > 0$ whenever for $x, y \in C$ with $0 < \|x - y\| < 2r$ one has $C \cap \Delta_r(x, y) \neq \{x, y\}$, where $\Delta_r(x, y)$ denotes the intersection of all closed balls with radius r containing both x and y. Vial studied this concept as the continuation of his previous 1982 paper [**957**] devoted to the opposite class of *strongly convex sets* C with constant $r > 0$, defined by $\Delta_r(x, y) \subset C$ for any pair $x, y \in C$ with $x \neq y$. Proposition 3.5 in Vial's paper [**958**] showed that the closed set C is weakly convex with a real constant $r > 0$ if and only if

$$C \cap B(x + rv; r\|v\|) = \emptyset \quad \text{for all } x \in C, v \in N^F(C; x) \cap \mathbb{B}_{\mathbb{R}^n}.$$

In fact, Vial considered in $\mathbb{R}^n$ the polar of the Bouligand-Peano tangent cone $T^B(C; x)$, but he noticed in the page 234 that the latter coincides with what is nowadays called the Fréchet normal cone that we denoted by $N^F(C; x)$. It was also used in [**958**, (12), p. 243] the equivalence between the above emptiness property and the inequality

$$\langle v, y - x\rangle \leq \frac{1}{2r}\|x - y\|^2 \quad \text{for all } x, y \in C, v \in N^F(C; x) \cap \mathbb{B}_{\mathbb{R}^n}.$$

For more and other fundamental results established by Vial for weakly convex sets in $\mathbb{R}^n$, we refer the reader to his paper [**958**] as well as to Section 16.2 in the manuscript and the section of comments in Chapter 16.

Translating the notion of φ-convexity of a function f, introduced by E. De Giorgi, M. Degiovanni, A. Marino and M. Tosques in [**320**] and by M. Degiovanni, A. Marino and M. Tosques in [**321**], to the case when f is the indicator function of a set in a Hilbert space H, A. Canino considered in her 1988 paper [**192**] (submitted in 1987) the class of sets in H such that for each set C in the class, there exists a continuous function $p : C \to [0, +\infty[$ (depending on C) satisfying the inequality

$$\langle v, y - x\rangle \leq p(x)\|v\| \, \|y - x\|^2 \quad \text{for all } x, y \in C \text{ and } v \in N^F(C; x).$$

Canino developed in [**192**] a large set of results for such sets that she named *p-convex sets*; in the manuscript a closed p-convex set corresponds to a $\rho(\cdot)$-prox-regular set for an appropriate continuous function $\rho : C \to]0, +\infty]$. A Canino also applied her results to the study in [**192, 193, 194**] of existence, multiplicity and properties of geodesics on p-convex sets, which was the first motivation of her analysis of structure of p-convex sets. In [**195**] Canino also investigated the study of periodic solutions of quadratic Lagrangian systems involving p-convex sets.

Following G. Chavent [**221**], given a subset C of a Hilbert space H let us say that a set of mappings $\mathcal{P} \subset W^{2,\infty}([0,1], H)$ is a *family of finite curvature paths* (called *pseudo-segments* in [**219**]) of C when for each $\gamma \in \mathcal{P}$

(i) $\gamma(t) \in C$ for every $t \in [0,1]$ and $\|\gamma'(t)\|$ is a constant independent of t;
(ii) for any $0 \le t_1 < t_2 \le 1$ the mapping $[0,1] \ni t \mapsto \gamma\big(t_1 + t(t_2 - t_1)\big)$ belongs to $\mathcal{P}$.

Motivated by estimation of distributed parameters in partial differential equations, G. Chavent defined in his 1991 paper [**219**] a set C in H to be *quasi-convex* if there are an open set U in H containing C and a lower semicontinuous function $\eta : U \to]0, +\infty]$ such that for each $u \in U$ and each $\varepsilon \in]0, \eta(u)[$

(i) for any $x, y \in \operatorname{Proj}_{C,\varepsilon}(u)$ there is $\gamma \in \mathcal{P}$ satisfying $\gamma(0) = x$ and $\gamma(1) = y$;
(ii) for some real $\alpha > 0$

$$\frac{1}{2}\frac{d^2}{dt^2}\|u - \gamma(t)\|^2 \ge \alpha\|\gamma'(t)\|^2 \quad \text{for all } t \in [0,1].$$

We recall that $\operatorname{Proj}_{C,\varepsilon} := \{z \in C : \|u - z\| \le d_C(u) + \varepsilon$.

If a quasi-convex set C in H is closed, Chavent showed in Theorems 2.12 and 2.15 in [**219**] that the metric projection mapping P_C is well-defined and continuous on the above open set $U \supset C$. Theorem 1.1 in the 2015 Chavent's paper (submitted in 2013) established in particular that a locally closed set C in H is quasi-convex in the above sense if and only if it p-convex in Canino's sense above for some function $p(\cdot)$. Applications of quasiconvex sets to nonlinear least square problems are provided by Chavent in [**219**]. We also mention Chavent's paper [**221**] dealing with strictly quasiconvex sets

Given a closed set C in a Hilbert space H and an integer $m \ge 1$, A. Shapiro declared in Definition 2.1 of his 1994 paper [**887**] (submitted in 1992) that C is *$O(m)$-convex at a point* $\overline{x} \in C$ if the are real constants $\kappa, \delta > 0$ such that

$$\operatorname{dist}(y - x, T^B(C; x)) \le \kappa\|y - x\|^m$$

for all $x, y \in C \cap B(\overline{x}, \delta)$. This condition with $m = 2$ is what we named (following [**813**]) the *Shapiro second order contact property* in Proposition 15.38. Under the $O(2)$-convexity at $\overline{x}$, Shapiro's Theorem 2.2 in [**887**] established that the metric projection mapping P_C is well-defined on a neighborhood of $\overline{x}$ and Lipschitz continuous there. One of the main tools in [**887**] for that result is the Ekeland variational principle. As example, Shapiro [**887**, p. 134-135] noticed that any set C in the form $C = g^{-1}(K)$ with a closed convex set K of a Banach space Y is $O(2)$-convex at $\overline{x} \in C$ whenever g is of class $\mathcal{C}^{1,1}$ near $\overline{x}$ and the Robinson qualification condition

$$0 \in \operatorname{int}\,[g(\overline{x}) - K - Dg(\overline{x})(H)]$$

is satisfied. Shapiro also proved a result in Theorem 3.1 on directional differentiability of P_C at a point $\overline{x} \in C$ where C is $O(2)$-convex, that is, the existence of $\lim_{t \downarrow 0} t^{-1}[P_C(\overline{x} + th) - P_C(\overline{x})]$ for every $h \in H$. To be complete, let us give the precise statement of Theorem 3.1 in [**887**]:
Let C be a closed set in a Hilbert space H such that

(i) the cone $T^B(C; \overline{x})$ is convex;

(ii)

$$\lim_{C \ni x \to \overline{x}} \frac{\operatorname{dist}\big(x - \overline{x}, T^B(C; x)\big)}{\|x - \overline{x}\|} = 0;$$

(iii) for every $v \in T^B(C; \overline{x})$

$$\lim_{t \downarrow 0} \frac{\operatorname{dist}(\overline{x} + tv, C)}{t} = 0.$$

Under these conditions (which are satisfied if C is $O(2)$-convex at $\overline{x}$), for each fixed $h \in H$ and for $p(\overline{x} + th) \in \operatorname{Proj}_C(\overline{x} + th)$ for $t \geq 0$ small enough, one has

$$p(\overline{x} + th) = \overline{x} + t\, P_{T^B(C;\overline{x})}(h) + o(t).$$

This was one of the principal objectives of the paper [**887**] as expressed by its title "Existence and differentiability of metric projections in Hilbert spaces".

Motivated by sets whose distance function is continuously differentiable on an open exterior tube, F. H. Clarke, R. L. Stern and P. R. Wolenski declared in their 1995 paper [**250**, p. 129, 133], a closed C of a Hilbert space H to be *proximally smooth of a radius* $r \in]0, +\infty[$ when the distance function d_C is continuously differentiable on the open tube

$$\operatorname{Tube}_r(C) := \{u \in H : 0 < d_C(u) < r\}.$$

The main result, Theorem 4.1 in [**250**], proved the equivalence of the following assertions:

(a) The closed set C is proximally smooth of radius r;

(b) for every $u \in \operatorname{Tube}_r(C)$ one has that $\operatorname{Proj}_C(u) \neq \emptyset$ and that d_C is Gâteaux differentiable at u;

(c) for every $u \in \operatorname{Tube}_r(C)$ one has that $\operatorname{Proj}_C(u) \neq \emptyset$, and given any $s \in]0, r[$ one has

$$d\,(u, C) + d\,(u, \operatorname{Exte}_s(C)) = s \quad \text{for all } u \in \operatorname{Tube}_s(C), \tag{15.79}$$

where $\operatorname{Exte}_s(C) : \{u \in H : d_C(u) \geq s\}$;

(d) $\operatorname{Proj}_C(u) \neq \emptyset$ for every $u \in \operatorname{Tube}_r(C)$, and $C \cap B(u + rv, r) = \emptyset$ for every pair $(u, v) \in C \times N^P(C; u)$ with $\|v\| = 1$.

(e) for every $s \in]0, r[$ and every $u \in H$ with $d_C(u) = s$, the proximal normal cone $N^P(\operatorname{Enl}_s(C); u)$ is not reduced to zero;

(f) for every $u \in \operatorname{Tube}_r(C)$ one has $\partial_P d_C(u) \neq \emptyset$.

If H is finite-dimensional, it is also shown in [**250**, Corollary 4.15] that $N^P(C; \cdot) = N^C(C; \cdot)$ whenever C is proximally smooth with some radius $r > 0$.

As already said in the section of comments in Chapter 11, prox-regular functions were introduced by R. A. Poliquin and R. T. Rockafellar in their 1996 paper [**809**]. One of the main motivations in [**809**] was to show that, for a function f in that class, its twice epi-differentiability at $\overline{x}$ for a vector $\overline{v} \in \partial_L f(\overline{x})$ is equivalent to the proto-differentiability of the multimapping $\partial_L f$ at $\overline{x}$ for $\overline{v}$; for the definition of both above concepts (epi-differentiability and proto-differentiability) we refer to the section of comments in Chapter 11. If the indicator function Ψ_C of a closed set

C in $\mathbb{R}^n$ is prox-regular at $\overline{x} \in C$ for a vector $\overline{v} \in N^L(C;\overline{x})$, the set C is declared prox-regular at $\overline{x}$ for $\overline{v}$ by Poliquin and Rockafellar in [**809**]. So, a closed set C in $\mathbb{R}^n$ is prox-regular at a point $\overline{x} \in C$ for a vector $\overline{v} \in N^L(C;\overline{x})$ provided there exist $\sigma, \varepsilon > 0$ (both depending on $\overline{x}$) such that $\langle v, y-x \rangle \leq \frac{1}{2\sigma}\|y-x\|^2$ for all $x, y \in C \cap B(\overline{x}, \varepsilon)$ and all $v \in \partial_L \Psi_C(x) = N^L(C;x)$ with $\|v - \overline{v}\| < \varepsilon$. In the paper [**813**] published in 2000 and submitted in June 1997, R. A. Poliquin, R. T. Rockafellar and L. Thibault showed that a closed set C in a Hilbert space H is prox-regular at $\overline{x} \in C$, in the sense that it is prox-regular at $\overline{x}$ for every $\overline{v} \in N^L(C;\overline{x})$, if and only if it is prox-regular at $\overline{x}$ for the zero vector. The latter property can be expressed by the existence of reals $r(\overline{x}), \delta(\overline{x}) > 0$ such that

$$\langle v, y-x \rangle \leq \frac{1}{2r(\overline{x})}\|y-x\|^2 \tag{15.80}$$

for all $x \in C \cap B(\overline{x}, \delta(\overline{x}))$, $v \in N^L(C;x) \cap \mathbb{B}_H$ and $y \in C$; this amounts to saying that any $x \in C \cap B(\overline{x}, \delta(\overline{x}))$ is the unique nearest point in C of $x + tv$ for any $v \in N^L(C;x) \cap \mathbb{B}_H$ and any $t \in [0, r(\overline{x})[$. This inequality (15.80) is one of the basic properties in the development in [**813**] of the study of prox-regularity of a set C at a point $\overline{x} \in C$. Requiring $r(\overline{x})$ to be a uniform constant $r > 0$ independent of the point $\overline{x} \in C$ yields in [**813**] to the concept of (uniformly) r-prox-regular sets in the Hilbert space H. None of the papers [**958, 192, 193, 194, 195, 219, 221, 887, 250**] mentioned Federer's paper [**402**]. It seems that the paper [**813**] by Poliquin, Rockafellar and Thibault was the first to refer to the pioneering work [**402**] by Federer and the finite-dimensional important results therein related to fundamental properties of the projection mapping in variational nonsmooth analysis.

For closed sets of a Hilbert space, all the concepts of proximally smooth sets, weakly convex sets, p-convex sets with a constant function $p(\cdot)$, positively reached sets in finite dimensions, coincide with uniformly prox-regular sets, whereas the local notion of $O(2)$-convexity of a set at a point coincides with that of prox-regularity at this point. We fixed and adopted the name in [**813**] of prox-regular sets because of the regularity of Lipschitz property of the metric projection mapping over suitable open enlargements and also because of links with (or characterization with) the prox-regularity property of indicator functions of these sets.

The sets $O_{\rho(\cdot)}(C)$ in (15.3) and $U_{\rho(\cdot)(C)}$ in (15.11) were introduced by A. Canino in [**192**]. More precisely, the set $O_{\rho(\cdot)}(C)$ appeared in a similar form in pages 123 and 124 of [**192**] while the set $U_{\rho(\cdot)(C)}$ was used in Proposition 2.8 in [**192**] under the notation of A^* there. Properties (a) and (b) in Proposition 15.9 and their proofs are taken from Proposition 2.8 in the same paper [**192**] by Canino. Certain equivalent forms of the $(|\!|\!|\cdot|\!|\!|, \rho(\cdot))$-prox-regularity property in Remark 15.2 first appeared in the 2018 paper (submitted in 2017) by R. Foygel Barber and W. Ha [**71**] through their concepts of coefficients of concavity for sets in $\mathbb{R}^n$. Those coefficients of concavity are related to the conditions in Remark 15.7. Such local constants are demonstrated by the authors [**71**] to be favorable for the analysis of convergence of projected gradient descent methods for low-rank matrix estimation, namely optimization over a rank-constrained set $C := \{X \in \mathbb{R}^{n \times m} : \operatorname{rank}(X) \leq k\}$. The analysis there employed as $|\!|\!|\cdot|\!|\!|$ the spectral norm (given by the greatest singular value of X) whereas the Euclidean norm $\|\cdot\|$ is the canonical Frobenius norm $\|X\| = \left(\sum_{i,j} X_{i,j}^2\right)^{1/2}$.

The arguments in Theorem 15.8 follow mainly R. A. Poliquin, R. T. Rockafellar and L. Thibault's paper [**813**] for the first proof of the assertion (f), while the second proof of that assertion is an adaptation of ideas of F. Bernard and L. Thibault [**93**]. The proof of Proposition 15.12 appeared in [**265**]. The tangential regularity property in Proposition 15.13(b) was first observed independently J.-P. Vial [**958**, Proposition 3.8] in finite dimensions and by A. Canino [**192**, Proposition 2.14] in the Hilbert setting. Note that, for a positively reached set C in $\mathbb{R}^n$ and for every $x \in C$, the equality $N^P(C;x) = N^F(C;x)$ in Proposition 15.13(b) was already established in the form of an equivalent statement by H. Federer in [**402**, Theorem 4.8(12)]. The arguments for the equivalences between (a)-(d) in Theorem 15.6 are quite direct (see [**813, 265**]) while those for the proof of (e)$\Rightarrow$(a) in that theorem are taken from [**71, 265**].

The fundamental Lemma 15.20 and its proof are taken from the paper [**813**] by R. A. Poliquin, R. T. Rockafellar and L. Thibault.

As stated, in Theorem 15.22 we followed G. Colombo and L. Thibault [**265**].

The equivalences between (p), (r) and (w) in Theorem 15.28 first appeared in Theorem 4.1 of the paper [**250**] by F. H. Clarke, R. L. Stern and P. R. Wolenski. The same Theorem 4.1 in [**250**] also furnished that (p) is equivalent to (c) plus the condition $\mathrm{Tube}_r(C) \subset \mathrm{Dom}\,\mathrm{Proj}_C$. The equivalence (p)$\Leftrightarrow$(h) in Theorem 15.28 can be found in Theorem 4.8 in [**250**]. The characterizations (e), (i), (n), (t) (resp. (x)) of (uniform) r-prox-regularity appeared in Theorem 4.1 (resp. Corollary 4.3) in the paper [**813**] by R. A. Poliquin, R. T. Rockafellar and L. Thibault. The characterizations (s) and (s') in Theorem 15.28 were obtained by F. Bernard, L. Thibault and N. Zlateva in [**97**, Theorem 7.2], and the characterization (u) appeared in the survey article [**265**] of G. Colombo and L. Thibault. The characterization (e') and the proof of (e')$\Rightarrow$(a) are taken from V. V. Goncharov and G. E. Ivanov [**444**, p. 281]. In finite dimensions, it was previously established by H. Federer in [**402**, Theorem 4.8] that C positively reached implies (b), (h) and (p) respectively in Theorem 15.28.

The equivalences (concerned with Chebyshev sets) between (a), (d), (e) and (f) in Corollary 15.31 are due to E. Asplund in his 1969 article [**29**], while the equivalence, under the weak closedness of C, between (a) and (n) was proved earlier in 1961 by V. Klee [**609**]. Corollary 15.32 is new. The equivalence (a)$\Leftrightarrow$(b) in Theorem 15.34 was proved with an equivalent statement in finite dimensions by J. P. Vial [**958**, Proposition 3.6] and extended to infinite dimensional Hilbert space by G. E. Ivanov [**531**, Theorem 7.2] and [**532**, Theorem 3]; the arguments for the implication (a)$\Rightarrow$(b) follows those by Vial [**958**, p. 244]. Some statements in the line of the epigraphical one in Proposition 15.35 were provided in finite dimensions by Vial in Propositions 4.17 and 4.18 in [**958**] and by Clarke, Stern and Wolenski in Theorem 5.2 in [**250**]. The property (b) in Proposition 15.37 was shown in Lemma 4.2(iv) in [**813**] to be a consequence of the r-prox-regularity of C. The characterization of r-prox-regularity through the second order contact property in Proposition 15.37 was established by Poliquin, Rockafellar and Thibault in [**813**, Theorem 4.1(j)]. The non-emptiness property in the assertion (b) of Proposition 15.41 appeared in the paper [**958**, Proposition 3.4] of J.-P. Vial as a consequence of sets satisfying the Vial property (see Definition 16.13 in the next chapter).

The $\mathcal{C}^1$-version of the submanifold property in Theorem 15.45(b) was established in finite dimensions by Clarke, Stern and Wolenski in [**250**, Corollary 4.15(4)].

The characterization of r-prox-regularity and its proof in Theorem 15.51 was given in Theorem 7.11 of Bernard, Thibault and Zlateva [**97**]. Earlier, Clarke, Stern and Wolenski [**250**, Theorem 4.1] proved (as already written above) that d_C is of class $\mathcal{C}^1$ on the open tube $\mathrm{Tube}_r(C)$ if and only if for every $s \in]0,r[$ and every $u \in \mathrm{Tube}_s(C)$ one has $d\,(u,C) + d\,(u, \mathrm{Exte}_s(C)) = s$ as well as the additional property $\mathrm{Proj}\,_C(u) \neq \emptyset$ (with respect to Theorem 15.51).

The implication (a)$\Rightarrow$(b) in Theorem 15.57 that the r-prox-regularity of C entails the linear semiconvexity with coefficient $1/(r-s)$ of the distance function d_C on any convex set included in $U_s(C)$ for every $s \in]0,r[$, is due to M. V. Balashov [**64**, Theorem 2.7]. Theorem 2.4 in that paper [**64**] of Balashov showed, under the connectedness of C and the inequality $\mathrm{diam}\, C \leq 2r$, that C is r-prox-regular whenever there is some $\delta > 0$ such that the distance function d_C is linearly semiconvex with coefficient $1/r$ on every convex set included in $U_\delta(C)$. The proof given in the book for the characterizations in Theorem 15.57 of the r-prox-regularity of C, is taken from the paper [**757**] of F. Nacry and L. Thibault. Lemma 15.58, Proposition 15.59 and Corollary 15.60, as well as their proofs, have been established in the same paper [**757**] by Nacry and Thibault. Given any uniformly convex and uniformly smooth Banach space X, one can also find in Theorem 3.5 of Balashov [**64**] that the distance function of any r-prox-regular set C in X is $\omega(\cdot)$-semiconvex (in the sense of Definition 10.1 in Chapter 10) on $U_r(C)$ for a certain modulus function $\omega(\cdot)$; results of this type in the setting of such Banach spaces will be established in Corollary 18.107 in Chapter 18.

All the characterizations of (local) prox-regularity of C at $\overline{x} \in C$ in Theorem 15.73 as (b), (b'), (c), (c'), (d), (f), (i)-(n), (q), (s), (t), along with (e) in Proposition 15.75, are due R. A. Poliquin, R. T. Rockafellar and L. Thibault [**813**]. The characterization (p) was given by F. Bernard, L. Thibault and N. Zlateva [**97**]. Theorem 15.70 appeared in a slight different form in Colombo and Thibault's survey [**265**] with arguments following the main ideas in [**813**].

Example 15.80 was provided by F. H. Clarke, R. J. Stern and P. R. Wolenski in [**250**, Remark 4.16(3)].

Theorem 15.83 was obtained for convex sets in 1971 by E. Zarantonello [**1001**], and the statement in [**1001**] corresponds to Corollary 15.84 in the manuscript. The full statement (with prox-regular sets) of Theorem 15.83 was proved by A. Canino [**192**, Proposition 2.12] and by A. Shapiro in [**887**, p. 135] as a consequence of his Theorem 3.1 as already said above. Theorem 15.87 is due to J. J. Moreau [**741**].

The inequality (15.72) in Theorem 15.123 was provided M. V. Balashov and G. E. Ivanov [**67**] and the proof of this theorem reproduces their arguments. The proof of Theorem 15.126 follows the approach by F. Bernard, L. Thibault and N. Zlateva [**98**, Theorem 9.4]. Such a result was first observed in finite dimensions by H. Federer in [**402**, Theorem 4.13] for the Hausdorff-Pompeiu convergence.

Example 15.94 is taken from [**98**]. Propositions 15.96 and 15.97 were established in the 2016 paper [**5**] by S. Adly, F. Nacry and L. Thibault. Previously to [**5**, Theorem 4.1], the particular case of $C = \{x \in \mathbb{R}^n : g_1(x) \leq 0\}$ with one locally Lipschitz function $g_1 : \mathbb{R}^n \to \mathbb{R}$ satisfying the assumption (15.60) in the whole space $\mathbb{R}^n = U_\infty(C)$ was given with the (equivalent) condition $\inf\{\|\zeta\| : \zeta \in \partial_C g_1(x), x \in \mathrm{bdry}\, C\} > 0$ by J.-P. Vial in Proposition 4.14(ii) of his 1983 paper [**958**]. While the arguments for Proposition 15.96 are those in [**5**], the proof of Proposition 15.97 in the book is different in certain aspects from that in [**5**, Theorem 4.1]. To illustrate this difference, it is worth noticing first a specific feature from the assumption

$$\langle \zeta' - \zeta, x' - x \rangle \geq -\gamma \|x' - x\|^2 \quad \text{for all } x, x' \in U_s(C), \zeta \in \partial_C g_k(x), \zeta' \in \partial_C g_k(x') \tag{15.81}$$

in Proposition 15.97. This ensures for $f_k := g_k + \frac{\gamma}{2}\|\cdot\|^2$ that $\langle v' - v, x' - x \rangle \geq 0$ for all $x, x' \in U_s(C)$, $v \in \partial_C f_k(x)$, $v' \in \partial_C f_k(x')$, so the function f_k is convex on any open convex set $W \subset U_s(C)$. Since $\partial_C f_k(x) = \partial_C g_k(x) + \gamma x$, we obtain for any $\zeta \in \partial_C g_k(x)$ and any $x' \in W$ that $\langle \zeta + \gamma x, x' - x \rangle \leq g_k(x') - g_k(x) + \frac{\gamma}{2}\|x\|^2 - \frac{\gamma}{2}\|x\|^2$. Easy computations and rearrangements give $\langle \zeta, x' - x \rangle \leq \frac{\gamma}{2}\|x' - x\|^2$. Now take x, y in $U_s(C)$ with $\|x - y\| < 2s$ and take $\zeta_k \in \partial_C g_k(x)$ and $\xi_j \in \partial_C g_j(y)$. Since the compact set $[x, y] \subset U_s(C)$, for some $\varepsilon > 0$ small enough we have $W := [x, y] + B(0, \varepsilon) \subset U_s(C)$, hence by what precedes

$$\langle \zeta_k, y - x \rangle \leq \frac{\gamma}{2}\|y - x\|^2 \quad \text{and} \quad \langle \xi_j, x - y \rangle \leq \frac{\gamma}{2}\|x - y\|^2.$$

We deduce that

$$\langle \zeta_k - \xi_j, x - y \rangle \geq -\gamma \|x - y\|^2. \tag{15.82}$$

This consequence (15.82) is implicitly used in the proof of Theorem 4.1 in [**5**]. Compare to the direct approach (15.59) in the book, the difference becomes clear. Corollary 15.100 was established in [**6**, Theorem 9.1].

Proposition 15.106 was first proved in a similar form in $\mathbb{R}^n$ be H. Federer [**402**, Theorem 4.18]. Here we followed [**265**] for a large part.

The study of local prox-regularity in Subsection 15.4.2 for inverse images and for the intersection of finitely many sets by means of the *truncated normal property* was developed by G. Colombo and L. Thibault [**265**]; Propositions 15.109 and 15.110 are taken from there as well as Corollary 15.118. The assertion (b) in Proposition 15.116 was obtained by H. Federer [**402**, Theorem 4.10(5)]. Proposition 15.117 is probably new. The results in Corollary 15.118 and 15.119 appeared in a less general form in [**265**].

We close this section of Comments by mentioning that diverse results on the structure of singular points of boundaries of prox-regular sets in $\mathbb{R}^N$ have been established by J. Rataj and L. Zajíček [**831**]. This paper [**831**] completes and significantly extends diverse previous results by H. Federer [**402**] on such singularities, and it also presents many new results concerning the study of such points.

CHAPTER 16

Compatible parametrization and Vial property of prox-regular sets, exterior sphere condition

16.1. Compatible parametrization of prox-regular sets

In the previous section we studied the convergence as $s \to \overline{s}$ of $P_{C(s)}$ when the sets $C(s)$ are $r(s)$-prox-regular and Attouch-Wets converge to some set along with $r(s)$ converges to $\overline{r} > 0$ as $s \to \overline{s}$. In this section we are interested in a localization of the continuity of the metric projection with respect to the perturbation variable.

DEFINITION 16.1. Let S be a Hausdorff topological space and $C : S \rightrightarrows H$ be a multimapping with nonempty closed images near $\overline{s} \in \operatorname{dom} C$ and let $\overline{x} \in C(\overline{s})$. We say that C is *prox-regular in x at $\overline{x}$ with compatible parametrization by s at $\overline{s}$* when there exist a neighborhood O of $\overline{s}$ and real numbers $\delta > 0$ and $r > 0$ such that

$$\langle v, x' - x \rangle \leq \frac{1}{2r} \|x' - x\|^2,$$

for all $s \in O$, $x, x' \in C(s) \cap B(\overline{x}, \delta)$, and $v \in N^P(C(s); x)$ with $\|v\| \leq \delta$.

In such a case for any $x \in C(s) \cap B(\overline{x}, \delta)$ with $s \in O$, the set $C(s)$ is prox-regular at x. Indeed, putting $\delta_x := \delta - \|x - \overline{x}\| > 0$ we have

$$\langle v', x'' - x' \rangle \leq \frac{1}{2r} \|x'' - x'\|^2$$

for all $x', x'' \in C(s) \cap B(x, \delta_x)$ and $v' \in N^P(C(s); x')$ with $\|v'\| \leq \delta$. The prox-regularity of $C(s)$ at x then follows from (b) of Theorem 15.73.

THEOREM 16.2 (metric projection of prox-regular sets of compatibly parametrized families). Let $C : S \rightrightarrows H$ be a multimapping which is inner (lower) semicontinuous at $\overline{s}$ in $\operatorname{dom} C$ for $\overline{x} \in C(\overline{s})$ and which takes on nonempty closed values on a neighborhood of $\overline{s}$. Suppose also that C is prox-regular in x at $\overline{x}$ with compatible parametrization by s at $\overline{s}$.

Then for any real $\beta > 1$ there exist two open neighborhoods O and U of $\overline{s}$ and $\overline{x}$ respectively with U bounded, and a closed bounded neighborhood X of $\overline{x}$ such that the following properties hold:
(a) for all $s \in O$, the set $C(s)$ is closed and

$$\begin{aligned} T_X(s, \cdot) : H &\rightrightarrows H \\ u &\mapsto (I + N^L(C(s), \cdot))^{-1}(u) \cap X \end{aligned}$$

is single valued and Lipschitz continuous with constant β on U;
(b) for all $s \in O$, $\operatorname{Proj}(C(s), \cdot) = P_{C(s)}(\cdot)$ is non-empty and single valued on U;
(c) for all $s \in O$ and $u \in U$,

$$T_X(s, u) = P_{C(s)}(u);$$

(d) for all $s, s' \in O$, $u \in U$, $x = T_X(s,u) = P_{C(s)}(u)$, $x' = T_X(s',u) = P_{C(s')}(u)$, and $y = P_{C(s)}(x')$, one has

$$\langle u - x, y - x \rangle \le \frac{1 - \beta^{-1}}{2} \|y - x\|^2.$$

PROOF. Set $\gamma := 1 - \beta^{-1} > 0$. By the prox-regularity of C in x at $\overline{x}$ with compatible parametrization by s at $\overline{s}$, take a neighborhood O of $\overline{s}$ and real numbers $\delta > 0$, and $r > 0$, as in Definition 16.1. Choose a positive real

$$\delta' < \min\left\{\frac{\delta}{8}, \frac{\delta}{6(1+\beta)}, \frac{\gamma r \delta}{2(1+\beta)}, \frac{\gamma r}{2\delta(1+\beta)}\right\}.$$

We may suppose that C takes on nonempty closed values on O. By the inner semicontinuity of C at $\overline{s}$ for $\overline{x}$ (shrinking O if necessary) we may also suppose that $C(s) \cap B(\overline{x}, \delta') \neq \emptyset$ for all $s \in O$, and we choose some $y(s) \in C(s) \cap B(\overline{x}, \delta')$ for each $s \in O$. We also observe that for each $s \in O$ the set $C(s)$ is prox-regular at any point of $C(s) \cap B(\overline{x}, \delta')$, and hence $N^P(C(s); \cdot)$ and $N^L(C(s); \cdot)$ coincide at each point of $C(s) \cap B(\overline{x}, \delta')$. Fix now any $s \in O$ and take any $x_i \in C(s) \cap B(y(s), \delta/2)$ and $v_i \in N^P(C(s); x_i) \cap \mathbb{B}_H$ for $i = 1, 2$. Then $x_i \in C(s) \cap B(\overline{x}, \delta)$, and hence the inequality in Definition 16.1 yields

$$\langle v_1 - v_2, x_1 - x_2 \rangle \ge -\frac{1}{\delta^{-1} r} \|x_1 - x_2\|^2.$$

For the positive real number above $\gamma < 1$ the implication $(f) \Rightarrow (i)$ in Theorem 15.70 ensures that on the set

$$U^{\gamma}_{r/\delta,\delta/2}(y(s), C(s)) = \{u \in H : d_C(u) < \gamma r/\delta \text{ and } d_C(u) + \|u - y(s)\| < \delta/2\}$$

the mapping $P_{C(s)}(\cdot)$ is single valued and Lipschitzian there with $(1-\gamma)^{-1} = \beta$ as a Lipschitz constant therein, and for each $u \in U^{\gamma}_{r/\delta,\delta/2}$

$$P_C(u) = (I + \delta^{-1} r \gamma \mathbb{B}_H \cap N^L(C(s); \cdot))^{-1}(u) \cap B(\overline{x}, \delta/2).$$

Observe that $B(\overline{x}, \delta') \subset U^{\gamma}_{r/\delta,\delta/2}(y(s), C(s))$ since for any $u \in B(\overline{x}, \delta')$ we have on the one hand

$$d_{C(s)}(u) \le \|u - y(s)\| \le \|u - \overline{x}\| + \|\overline{x} - y(s)\| < 2\delta' < \gamma r/\delta,$$

and on the other hand

$$d_{C(s)}(u) + \|u - y(s)\| \le 2\|u - y(s)\| \le 2\|u - \overline{x}\| + 2\|y(s) - \overline{x}\| < 4\delta' < \delta/2.$$

So, the properties above concerning $P_{C(s)}$ are fulfilled on $B(\overline{x}, \delta')$.

Put $X := B[\overline{x}, \delta'(1 + 2\beta)]$. For each $u \in B(\overline{x}, \delta')$ we have $P_{C(s)}(u) \in (I + N^L(C(s); \cdot))^{-1}(u)$ and

$$\begin{aligned}
\|P_{C(s)}(u) - \overline{x}\| &\le \|P_{C(s)}(u) - y(s)\| + \|y(s) - \overline{x}\| \\
&= \|P_{C(s)}(u) - P_{C(s)}(y(s))\| + \|y(s) - \overline{x}\| \\
&\le \beta\|u - y(s)\| + \|y(s) - \overline{x}\| \le \beta\|u - \overline{x}\| + (1+\beta)\|y(s) - \overline{x}\| \\
&\le (1 + 2\beta)\delta',
\end{aligned}$$

hence $P_{C(s)}(u) \in (I + N^L(C(s); \cdot))^{-1}(u) \cap X$. Further, for $u \in B(\overline{x}, \delta')$ we have for any $x \in (I + N^L(C(s); \cdot))^{-1}(u) \cap X$ that

$$\|u - x\| \le \|u - \overline{x}\| + \|x - \overline{x}\| \le 2\delta'(1 + \beta) \le \gamma r/\delta,$$

which easily gives $x \in (I + \frac{\gamma r}{\delta}\mathbb{B}_H \cap N^L(C(s);\cdot))^{-1}(u)$. This and the inclusion $X \subset B(\overline{x}, \delta/2)$ (due to the choice of δ') imply that

$$P_{C(s)}(u) \in (I + N^L(C(s);\cdot))^{-1}(u) \cap X$$
$$\subset \left(I + \frac{\gamma r}{\delta}\mathbb{B}_H \cap N^L(C(s);\cdot)\right)^{-1}(u) \cap B(\overline{x}, \delta/2) = \{P_{C(s)}(u)\}.$$

So, the set $U = B(\overline{x}, \delta')$ and the sets X and O above fulfill the required properties of the proposition.

Now fix also $s' \in O$ and consider $u \in U$ and x, x' and y as stated in (d). We observe on the one hand that $y \in C(s)$ and

$$\|y - \overline{x}\| \le \|y - x'\| + \|x' - \overline{x}\| = d_{C(s)}(x') + \|x' - \overline{x}\|$$
$$\le \|x' - x\| + \|x - \overline{x}\| \le 3\delta'(1 + 2\beta) < \delta,$$

where the third inequality is due to the inclusions $x, x' \in X$, and we observe on the other hand that

$$\|u - x\| \le \|u - \overline{x}\| + \|\overline{x} - x\| \le \delta' + \delta'(1 + 2\beta) = 2\delta'(1 + \beta),$$

where the second inequality follows from the inclusion $x \in X$. The latter observation yielding $\frac{1}{\gamma r}\|u - x\| < \delta$ (according to the choice of δ'), we deduce from the inclusion $u - x \in N^P(C(s); x)$ and from Definition 16.1 that

$$\left\langle \frac{1}{\gamma r}(u - x), y - x \right\rangle \le \frac{1}{2r}\|y - x\|^2,$$

that is,

$$\langle u - x, y - x\rangle \le \frac{\gamma}{2}\|y - x\|^2 = \frac{1 - \beta^{-1}}{2}\|y - x\|^2.$$

This completes the proof. □

For points s and s' close to $\overline{s}$ the next theorem provides an upper bound for the distance between the projections of a point u on $C(s)$ and $C(s')$ respectively by a quantitative expression involving the Hausdorff-Pompeiu distance between the truncations of $C(s)$ and $C(s')$ with the neighborhood X of $\overline{x}$ above.

THEOREM 16.3 (continuity of metric projection of prox-regular sets of compatibly parametrized family). Assume the notation and hypotheses of Theorem 16.2. Let $\beta > 1$ and let O, U, X and T_X be as given by that theorem. For the multimapping $C_X : S \rightrightarrows H$ with

$$C_X(s) := C(s) \cap X \quad \text{for all } s \in S,$$

one has for all $u \in U$ and all $s, s' \in O$

$$\|T_X(s,u) - T_X(s',u)\| \le \frac{h\beta}{2} + \left[2h\beta\|u - T_X(s,u)\| + h^2\left(\frac{\beta^2}{4} + \beta - 1\right)\right]^{\frac{1}{2}},$$

where $h = \mathrm{haus}(C_X(s), C_X(s'))$ is the Hausdorff-Pompeiu distance between the sets $C_X(s)$ and $C_X(s')$ of H.

Further, there exists some real constant $\alpha \ge 0$ such that

$$\|T_X(s,u) - T_X(s',u')\| \le \alpha[\mathrm{haus}(C_X(s), C_X(s'))]^{1/2} + \beta\|u - u'\|,$$

for all $s, s' \in O$ and $u, u' \in U$. In particular $T_X(\cdot,\cdot)$ is continuous (resp. Hölder continuous) on $O \times U$ whenever the multimapping C_X is Hausdorff continuous (resp. Hausdorff Hölder-continuous and S is a metric space) on S.

PROOF. Let $u \in U$ and $s, s' \in O$. Set $x = T_X(s,u)$, $x' = T_X(s',u)$, $h = \text{haus}(C_X(s), C_X(s'))$, and $\mu = \|u - T_X(s,u)\|$. Observe that $x \in X$ by definition of T_X and that

$$x = T_X(s,u) = P_{C(s)}(u) \in C(s),$$

the second equality being due to Theorem 16.2. So, $x \in C_X(s)$ and, for the same reason, $x' \in C_X(s')$. Further, Theorem 16.2 again allows us to consider $y := P_{C(s)}(x')$ and $y' := P_{C(s')}(x)$. We note that

$$\|y' - x\| = d(x, C(s')) \le d(x, C_X(s')) \le \text{haus}(C_X(s), C_X(s')) = h,$$

and likewise $\|y - x'\| \le h$.

On the other hand by (d) of Theorem 16.2 above we have the two inequalities

$$\langle u - x, y - x\rangle - \frac{1-\beta^{-1}}{2}\|y - x\|^2 \le 0, \tag{16.1}$$

and

$$\langle u - x', y' - x'\rangle - \frac{1-\beta^{-1}}{2}\|y' - x'\|^2 \le 0. \tag{16.2}$$

Developing $\|y - x\|^2 = \|(y - x') + (x' - x)\|^2$, the inequality (16.1) becomes

$$\langle u - x, y - x\rangle - \frac{1-\beta^{-1}}{2}\left(\|y - x'\|^2 + 2\langle y - x', x' - x\rangle + \|x' - x\|^2\right) \le 0,$$

that is,

$$\begin{aligned}\langle u - x, y - x'\rangle + \langle u - x, x' - x\rangle &- \frac{1-\beta^{-1}}{2}\|y - x'\|^2 \\ &- (1-\beta^{-1})\langle y - x', x' - x\rangle - \frac{1-\beta^{-1}}{2}\|x' - x\|^2 \le 0.\end{aligned}$$

Likewise (16.2) is equivalent to

$$\begin{aligned}\langle u - x', y' - x\rangle + \langle u - x', x - x'\rangle &- \frac{1-\beta^{-1}}{2}\|y' - x\|2 \\ &- (1-\beta^{-1})\langle y' - x, x - x'\rangle - \frac{1-\beta^{-1}}{2}\|x' - x\|^2 \le 0.\end{aligned}$$

Adding the two latter inequalities and observing that

$$\langle u - x, x' - x\rangle + \langle u - x', x - x'\rangle = \|x' - x\|^2$$

we obtain

$$\begin{aligned}0 \ge\ & \langle u - x, y - x'\rangle + \langle u - x', y' - x\rangle + \|x - x'\|^2 \\ & - \frac{1-\beta^{-1}}{2}(\|y - x'\|^2 + \|y' - x\|^2) \\ & - (1-\beta^{-1})\big(\langle y - x' + x - y', x' - x\rangle + \|x - x'\|^2\big).\end{aligned} \tag{16.3}$$

Note now that

$$\begin{aligned}\langle u - x, y - x'\rangle + \langle u - x', y' - x\rangle &= \langle u - x, y - x' + y' - x\rangle - \langle x' - x, y' - x\rangle \\ &\ge -\|u - x\|(\|y - x'\| + \|y' - x\|) - \langle x' - x, y' - x\rangle \\ &\ge -2h\mu - \langle x' - x, y' - x\rangle,\end{aligned}$$

and that

$$-\frac{1-\beta^{-1}}{2}(\|y - x'\|^2 + \|y' - x\|^2) \ge -h^2(1-\beta^{-1}).$$

Combining the latter two inequalities with (16.3) yields

$$\begin{aligned}0 &\geq -2h\mu - \langle x'-x, y'-x\rangle + \|x'-x\|^2\\ &\quad - (1-\beta^{-1})h^2 - (1-\beta^{-1})\big(\langle y-x'+x-y', x'-x\rangle + \|x'-x\|^2\big)\\ &= -h(2\mu + h(1-\beta^{-1})) + \beta^{-1}\|x'-x\|^2\\ &\quad - \langle (1-\beta^{-1})(y-x') + \beta^{-1}(y'-x), x'-x\rangle,\end{aligned}$$

which implies that

$$\begin{aligned}0 &\geq -h(2\mu + h(1-\beta^{-1})) + \beta^{-1}\|x'-x\|^2\\ &\quad - \big((1-\beta^{-1})\|y-x'\| + \beta^{-1}\|y'-x\|\big)\|x'-x\|\\ &\geq -h(2\mu + h(1-\beta^{-1})) + \beta^{-1}\|x'-x\|^2\\ &\quad - \big((1-\beta^{-1})h + \beta^{-1}h\big)\|x'-x\|\\ &= -h\big(2\mu + h(1-\beta^{-1})\big) + \beta^{-1}\|x'-x\|^2 - h\|x'-x\|.\end{aligned}$$

Multiplying by β gives

$$\begin{aligned}0 &\geq \|x'-x\|^2 - h\beta\|x'-x\| - h\beta\big(2\mu + h(1-\beta^{-1})\big)\\ &= \left(\|x'-x\| - \frac{h\beta}{2}\right)^2 - h\left[2\beta\mu + h\left(\frac{\beta^2}{4} + \beta - 1\right)\right].\end{aligned}$$

Consequently

$$\left(\|x'-x\| - \frac{h\beta}{2}\right)^2 \leq h\left[2\beta\mu + h\left(\frac{\beta^2}{4} + \beta - 1\right)\right],$$

that is,

$$\|x'-x\| \leq h\frac{\beta}{2} + h^{1/2}\left[2\beta\mu + h\left(\frac{\beta^2}{4} + \beta - 1\right)\right]^{1/2},$$

which proves the first part of the theorem.

Let us establish the second part. Take two real numbers $\alpha_1, \alpha_2 \geq 0$ such that $U \subset \alpha_1\mathbb{B}_H$ and $X \subset \alpha_2\mathbb{B}_H$ according to the boundedness of these two sets. We note that for all $z \in C_X(s) \subset X$ and $z' \in C_X(s') \subset X$ we have $\|z - z'\| \leq 2\alpha_2$, hence $h = \mathrm{haus}(C_X(s), C_X(s')) \leq 2\alpha_2$. On the other hand, for $u \in U$ we also have $\|u - T_X(s,u)\| \leq \alpha_1 + \alpha_2$. So according to the first part above

$$\|T_X(s,u) - T_X(s',u)\| \leq \alpha[\mathrm{haus}(C_X(s), C_X(s'))]^{1/2}$$

for

$$\alpha := \frac{\beta}{2}\sqrt{2\alpha_2} + \left[2\beta(\alpha_1+\alpha_2) + 2\alpha_2\left(\frac{\beta^2}{4} + \beta - 1\right)\right]^{1/2}.$$

This inequality combined with (a) of Theorem 16.2 finally yields for all $s, s' \in S$ and $u, u' \in U$

$$\begin{aligned}\|T_X(s,u) - T_X(s',u')\| &\leq \|T_X(s,u) - T_X(s',u)\| + \|T_X(s',u) - T_X(s',u')\|\\ &\leq \alpha[\mathrm{haus}(C_X(s), C_X(s'))]^{1/2} + \beta\|u-u'\|,\end{aligned}$$

which finishes the proof of the theorem. $\square$

16.2. Strongly convex sets and Vial property of prox-regular sets

In this section we will show that uniformly prox-regular sets can be characterized with the Vial property. This property is related to the concept of r-strongly convex segments.

16.2.1. Strongly convex sets.

DEFINITION 16.4. Given $r \in]0, +\infty[$, any intersection of a family of closed balls with radius r in a normed space $(X, \|\cdot\|)$ is generally called a *strongly convex set with constant* r or an *r-strongly convex set* relative to the norm $\|\cdot\|$. When there is no ambiguity, the norm $\|\cdot\|$ will be omitted.

Clearly, the intersection of any family of r-strongly convex sets is an r-strongly convex set. So, given any subset Q of X, the intersection of all r-strongly convex sets containing Q is the smallest r-strongly convex set containing Q; it is called the *r-strongly convex hull of Q*. Of course, if $\operatorname{diam} Q > 2r$, there is no closed ball with radius r containing Q, so its r-strongly convex hull is trivially the whole space X in this case. It is also clear that any r-strongly convex set is a closed convex set.

We will be interested in this subsection and the next one to r-strongly convex hulls of sets containing merely two points x_0, x_1.

DEFINITION 16.5. For any real $r > 0$ and x_0, x_1 in a normed space $(X, \|\cdot\|)$ with $\|x_1 - x_0\| \le 2r$, the r-strongly convex hull of $\{x_0, x_1\}$, denoted by $\Delta_r(x_0, x_1)$ is called the *r-strongly convex segment* (or *r-lens*) *in X with extremities* x_0, x_1. It is then defined by the intersection of all closed balls with radius r containing x_0, x_1, that is,

$$\Delta_r(x_0, x_1) := \bigcap_{x \in X; B[x,r] \ni x_0, x_1} B[x, r]. \tag{16.4}$$

The following proposition establishes some first properties of r-strongly convex segments.

PROPOSITION 16.6. Let a real $r > 0$ and x_0, x_1 be two points of a normed space $(X, \|\cdot\|)$ with $\|x_0 - x_1\| \le 2r$. The following hold:
(a) The set $\Delta_r(x_0, x_1)$ is a closed convex set of X.
(b) If $y_0, y_1 \in \Delta_r(x_0, x_1)$, then $\Delta_r(y_0, y_1) \subset \Delta_r(x_0, x_1)$.
(c) If $x_0, x_1 \in B[y, s]$ and $s \le r$, then $\Delta_r(x_0, x_1) \subset B[y, s]$.
(d) If $\|x_0 - x_1\| \le 2s \le 2r$, then $\Delta_r(x_0, x_1) \subset \Delta_s(x_0, x_1)$.
(e) If $\|x_0 - x_1\| = 2s \le 2r$, then $\Delta_r(x_0, x_1) \subset B[\frac{1}{2}(x_0 + x_1), s]$.
(f) If $y_0, y_1 \in \Delta_r(x_0, x_1)$, then $\|y_0 - y_1\| \le \|x_0 - x_1\|$.
(g) One has $x \in \Delta_r(x_0, x_1)$ if and only if $x \in u + r\mathbb{B}$ for every $u \in (x_0 + r\mathbb{B}) \cap (x_1 + r\mathbb{B})$.
(h) For any $a \in X$ one has $\Delta_r(x_0 + a, x_1 + a) = a + \Delta_r(x_0, x_1)$.
(i) For any $a \in X$ with $\|a\| \le r$ one has the equalities

$$\Delta_r(-a, a) \cap \mathbb{R}a = [-a, a] \quad \text{and} \quad \Delta_r(x_0, x_1) \cap \big(x_0 + \mathbb{R}(x_1 - x_0)\big) = [x_0, x_1].$$

PROOF. (a) As already said above, the assertion (a) is evident since it follows directly from the definition of r-strongly convex segments.
(b) For any closed r-ball $B[x, r]$ containing x_0, x_1, we have $y_0, y_1 \in \Delta_r(x_0, x_1) \subset B[x, r]$, and hence the class of closed r-balls containing x_0, x_1 is included in the class of closed r-balls containing y_0, y_1. This ensures that $D_r(y_0, y_1) \subset D_r(x_0, x_1)$.

(c) Let $y \in X$ and $s \in]0, r]$ be such that $x_0, x_1 \in B[y, s]$. Fix any $x \in \Delta_r(x_0, x_1)$ with $x \neq y$ and put $z := y + \frac{r-s}{\|y-x\|}(y - x)$. We have

$$\|x_0 - z\| \leq \|x_0 - y\| + \|y - z\| \leq s + r - s = r$$

and similarly $\|x_1 - z\| \leq r$, thus $x_0, x_1 \in B[z, r]$, which entails $x \in \Delta_r(x_0, x_1) \subset B[z, r]$ and in particular $\|x - z\| \leq r$. Consequently $r \geq \|x - z\| = \|y - x\| + r - s$, that is, $\|x-y\| \leq s$, or equivalently $x \in B[y, s]$. This says that $\Delta_r(x_0, x_1) \subset B[y, s]$.
(d) Assertion (d) is a direct consequence of (c) and the definition of $\Delta_s(x_0, x_1)$.
(e) Observing that $\Delta_s(x_0, x_1) \subset B[\frac{1}{2}(x_0 + x_1), s]$ for $s = \frac{1}{2}\|x_1 - x_0\|$, we deduce easily the assertion (e) from (d).
(f) Assertion (f) follows directly from (e).
(g) Assertion (g) is easily seen from the definition of strongly convex segment.
(h) By (g) we have $x \in \Delta_r(x_0 + a, x_1 + a)$ if and only if $x \in u + r\mathbb{B}$ for all $u \in a + (x_0 + r\mathbb{B}) \cap (x_1 + r\mathbb{B})$, which in turn is equivalent to $x - a \in u' + r\mathbb{B}$ for all $u' \in (x_0 + r\mathbb{B}) \cap (x_1 + r\mathbb{B})$, or equivalently $x - a \in \Delta_r(x_0, x_1)$ by (g) again. This justifies the assertion (h).
(i) By (h) it suffices to verify the first equality in (i). We may suppose that $a \neq 0$. The inclusion $[-a, a] \subset \Delta_r(-a, a) \cap \mathbb{R}a$ is evident. Suppose that there is some $x \in \Delta_r(-a, a) \cap \mathbb{R}a$ with $x \notin [-a, a]$. Then there is $\alpha \in \mathbb{R}$ with $|\alpha| > 1$ such that $x = \alpha a$. Since $\|a\| \leq r$, we may choose some $u \in (-a + r\mathbb{S}_X) \cap (a + r\mathbb{S}_X)$, that is, $r = \|u + a\| = \|u - a\|$, so in particular $\langle u, a\rangle = 0$ and $B[u, r] \supset \{-a, a\}$. The latter inclusion ensures that $x \in B[u, r]$ since $x \in \Delta_r(-a, a)$. It ensues that

$$r^2 \geq \|x - u\|^2 = \alpha^2\|a\|^2 + \|u\|^2 > \|a\|^2 + \|u\|^2 = r^2,$$

which is a contradiction. This justifies the assertion (i). □

Given x_0, x_1 in a normed space X we know for any $z \in [x_0, x_1]$ that $\|x_1 - x_0\| = \|x_1 - z\| + \|z - x_0\|$. In the Hilbert setting we have the following analogue with the r-strongly convex segment $\Delta_r(x_0, x_1)$ in place of the line segment $[x_0, x_1]$.

PROPOSITION 16.7. Let H be a Hilbert space with Hilbert norm $\|\cdot\|$ and let $r \in]0, +\infty[$ and $x_0, x_1 \in H$ with $\|x_1 - x_0\| \leq 2r$. Then for any $z \in \Delta_r(x_0, x_1)$ one has

$$\|x_1 - x_0\|^2 \geq \|x_1 - z\|^2 + \|z - x_0\|^2.$$

PROOF. Take any $z \in \Delta_r(x_0, x_1)$. By the inclusion $\Delta_r(x_0, x_1) \subset B[\frac{x_0+x_1}{2}, s]$ with $s := \|x_1 - x_0\|/2$ (see Proposition 16.6(e)) it ensues that $\left\|z - \frac{x_0+x_1}{2}\right\| \leq \frac{\|x_1-x_0\|}{2}$. It follows that

$$\begin{aligned}
&\|x_0 - z\|^2 + \|x_1 - z\|^2 \\
=&\left\|\left(\frac{x_0 + x_1}{2} - z\right) + \frac{x_0 - x_1}{2}\right\|^2 + \left\|\left(\frac{x_0 + x_1}{2} - z\right) - \frac{x_0 - x_1}{2}\right\|^2 \\
=&2\left\|\frac{x_0 + x_1}{2} - z\right\|^2 + 2\left\|\frac{x_0 - x_1}{2}\right\|^2 \leq \|x_0 - x_1\|^2.
\end{aligned}$$

□

Two additional properties in Hilbert spaces will also be useful.

PROPOSITION 16.8. Let H be a Hilbert space with Hilbert norm $\|\cdot\|$ and let $r \in]0,+\infty[$ and $x_0, x_1 \in H$ with $\|x_1 - x_0\| \le 2r$.
(a) one has the equality

$$\Delta_r(x_0, x_1) = \bigcup_{t\in[0,1]} B[x_t, \delta(t,r)],$$

where

$$x_t := (1-t)x_0 + tx_1 \quad \text{and} \quad \delta(t,r) := r - \sqrt{r^2 - t(1-t)\|x_0 - x_1\|^2}.$$

(b) If $\|x_0 - x_1\| = 2r$, then

$$\Delta_r(x_0, x_1) = B\left[\frac{1}{2}(x_0 + x_1), r\right].$$

PROOF. (a) Put $x_t := (1-t)x_0 + tx_1$, $\zeta(t) := r^2 - t(1-t)\|x_0 - x_1\|^2$ and $\delta(t) := r - \sqrt{\zeta(t)}$.

Let us first show the inclusion $\supset$. Fix any $t \in [0,1]$ and any $x \in B[x_t, \delta(t)]$, that is, $x \in x_t + \delta(t)\mathbb{B}$. Take any $u \in (x_0 + r\mathbb{B}) \cap (x_1 + r\mathbb{B})$, so $\|u - x_i\| \le r$, $i = 0, 1$, which is equivalent to

$$\|u - x_t\|^2 + t^2\|x_1 - x_0\|^2 + 2t\langle u - x_t, x_1 - x_0\rangle \le r^2$$

and

$$\|u - x_t\|^2 + (1-t)^2\|x_0 - x_1\|^2 + 2(1-t)\langle u - x_t, x_0 - x_1\rangle \le r^2.$$

Multiplying the first inequality by $(1-t)$ and the second by t, then adding the two new inequalities we obtain

$$\|u - x_t\|^2 + \big(t^2(1-t) + (1-t)^2 t\big)\|x_0 - x_1\|^2 \le r^2,$$

or equivalently

$$\|u - x_t\|^2 + t(1-t)\|x_0 - x_1\|^2 \le r^2.$$

This means that $\|u - x_t\| \le \sqrt{\zeta(t)}$, hence $x \in u + (\delta(t) + \sqrt{\zeta(t)})\mathbb{B}$, that is, $x \in u + r\mathbb{B}$. This being true for all $u \in (x_0 + r\mathbb{B}) \cap (x_1 + r\mathbb{B})$, it follows from Proposition 16.6(g) that $x \in \Delta_r(x_0, x_1)$, and this translates the desired inclusion $\supset$.

Concerning the converse inclusion, noting by Proposition 16.6(h) the equality $\Delta_r(x_0, x_1) = a + \Delta_r(-v, v)$, where $a := (x_0 + x_1)/2$ and $v := (x_1 - x_0)/2$, it suffices to show that $\Delta_r(-v, v) \subset \bigcup_{t\in[0,1]} B[v_t, \delta(t)]$, where $v_t := tv + (1-t)(-v) = (2t-1)v$. Clearly, we may suppose $v \ne 0$. Take any $x \in \Delta_r(-v, v)$. If $x \in [-v, v]$, then obviously $x = v_t \in B[v_t, \delta(t)]$ for some $t \in [0,1]$. Suppose that $x \notin [-v, v]$, so $x \notin \mathbb{R}v$ by Proposition 16.6(i). Define the unit vectors

$$e_1 := \|v\|^{-1}v \quad \text{and} \quad e_2 := \|x - \langle x, e_1\rangle e_1\|^{-1}(x - \langle x, e_1\rangle e_1); \tag{16.5}$$

note that e_2 is well-defined since $x \notin \mathbb{R}v = \mathbb{R}e_1$ and note also that $\langle e_1, e_2\rangle = 0$. Define also $u := -\sqrt{r^2 - \|v\|^2}e_2$ and observe (since $u \perp v$) that

$$\|u - v\|^2 = \|u + v\|^2 = \|u\|^2 + \|v\|^2 = r^2, \quad \text{hence } B[u, r] \supset \{-v, v\}, \tag{16.6}$$

which entails that $x \in B[u,r]$ (since $x \in \Delta_r(-v,v)$). Put $\sigma := r^2 - \|v\|^2$. For any $\alpha \in \mathbb{R}$ we have by the definition of u and by the second equality in (16.5)

$$\begin{aligned} u + \alpha(x-u) &= (\alpha-1)\sqrt{\sigma}e_2 + \alpha\big(\langle x, e_1\rangle e_1 + \|x - \langle x, e_1\rangle e_1\| e_2\big) \\ &= \big((\alpha-1)\sqrt{\sigma} + \alpha\|x - \langle x, e_1\rangle e_1\|\big)e_2 + \frac{\alpha\langle x, e_1\rangle}{\|v\|}v \\ &= \big(\alpha(\sqrt{\sigma} + \|x - \langle x, e_1\rangle e_1\|) - \sqrt{\sigma}\big)e_2 + \frac{\alpha\langle x, e_1\rangle}{\|v\|}v. \end{aligned}$$

Choose $\alpha = \dfrac{\sqrt{\sigma}}{\sqrt{\sigma} + \|x - \langle x, e_1\rangle e_1\|} \in [0,1[$ and $t \in \mathbb{R}$ with $2t-1 = \dfrac{\alpha\langle x, e_1\rangle}{\|v\|}$, that is, $t = \dfrac{1}{2}\left(1 + \dfrac{\alpha\langle x, e_1\rangle}{\|v\|}\right)$. With such reals α and t we have $u+\alpha(x-u) = (2t-1)v = v_t$. Then, reminding that $\|x-u\| \le r$ and $0 \le \alpha < 1$, we immediately derive that $\|v_t - u\| = \alpha\|x-u\| < r$, so by (16.6)

$$\|v_t - u\|^2 < r^2 = \|u\|^2 + \|v\|^2.$$

On the other hand, since $v_t \perp u$, we also have $\|v_t - u\|^2 = \|v_t\|^2 + \|u\|^2$. This and the previous inequality yield $\|v_t\|^2 < \|v\|^2$, that is, $(2t-1)^2 < 1$, or equivalently $0 < t < 1$.

Now, according to the equalities $u+\alpha(x-u) = v_t$ and $\|v_t - u\|^2 = \|v_t\|^2 + \|u\|^2$, we can write

$$\|x - v_t\| = (1-\alpha)\|x-u\| = \|x-u\| - \|v_t - u\| \le r - \sqrt{\|v_t\|^2 + \|u\|^2}.$$

Reminding that $r^2 = \|u\|^2 + \|v\|^2$ we also note that

$$\|v_t\|^2 + \|u\|^2 = (4t^2 - 4t + 1)\|v\|^2 + \|u\|^2 = r^2 - 4t(1-t)\|v\|^2.$$

We deduce that

$$\|x - v_t\| \le r - \sqrt{r^2 - 4t(1-t)\|v\|^2} = \delta(t), \quad \text{that is, } x \in B[v_t, \delta(t)],$$

which justifies the desired inclusion $\Delta_r(-v,v) \subset \bigcup_{t\in[0,1]} B[v_t, \delta(t)]$.

(b) Assuming that $\|x_1 - x_0\| = 2r$, we see that $\delta(1/2, r) = r$ and by the equality in the above assertion (a) we have $B(\frac{1}{2}(x_0+x_1), r) \subset \Delta_r(x_0, x_1)$. Consider now any $x \in \Delta_r(x_0,x_1)$. By (a) again there exists some $t \in [0,1]$ such that $\|x - x_t\| \le \delta(t,r)$, which implies

$$\begin{aligned} \|x - x_{1/2}\| &\le \|x - x_t\| + \|x_t - x_{1/2}\| \le \delta(t,r) + |t - (1/2)|\,\|x_1 - x_0\| \\ &= r\left(1 - \sqrt{1 - 4t(1-t)} + |2t-1|\right) = r. \end{aligned}$$

This yields $\Delta_r(x_0,x_1) \subset B[\frac{1}{2}(x_0+x_1), r]$, and hence the desired equality in (b) holds. □

COROLLARY 16.9. Let x_0, x_1, x be in a Hilbert space H such that

$$\left\langle x - \frac{x_0 + x_1}{2}, x_1 - x_0 \right\rangle = 0 \quad \text{and} \quad 0 < \|x_1 - x_0\| < 2r$$

with $r \in]0, +\infty[$. Then $x \in \Delta_r(x_0, x_1)$ if and only if

$$\left\| x - \frac{x_0 + x_1}{2} \right\| \le r - \sqrt{r^2 - \frac{\|x_1 - x_0\|^2}{4}}.$$

PROOF. The implication $\Leftarrow$ follows directly from Proposition 16.8(a) with $t = 1/2$. To prove the converse implication suppose that $x \in \Delta_r(x_0, x_1)$. By Proposition 16.8(a) again there exists some $t \in [0,1]$ such that $x \in B[x_t, \delta(t,r)]$, where $x_t := (1-t)x_0 + tx_1$ and $\delta(t,r) := r - \sqrt{r^2 - t(1-t)\|x_0 - x_1\|^2}$. Then the assumption $\langle x - \frac{x_0+x_1}{2}, x_1 - x_0\rangle = 0$ entails $\langle x - \frac{x_0+x_1}{2}, x_t - \frac{x_0+x_1}{2}\rangle = 0$, which ensures that

$$\|x - x_t\|^2 = \left\| x - \frac{x_0 + x_1}{2} \right\|^2 + \left\| \frac{x_0 + x_1}{2} - x_t \right\|^2.$$

From this and the inclusion $x \in B[x_t, \delta(t,r)]$ we deduce that

$$\left\| x - \frac{x_0 + x_1}{2} \right\| \le \|x - x_t\| \le \delta(t,r) \le \delta(1/2, r),$$

which justifies the implication $\Rightarrow$ and finishes the proof. □

The next proposition continues with the Hilbert space H. Given $x_0, x_1 \in H$ it is clear that for any $x \in [x_0, x_1]$ the equality $\langle x - x_0, x_1 - x_0\rangle = \|x - x_0\|\,\|x_1 - x_0\|$ holds. In the same setting of Hilbert space we also have the following property with the r-strongly convex segment $\Delta_r(x_0, x_1)$ in place of the line segment $[x_0, x_1]$.

PROPOSITION 16.10. Let x_0, x_1 be two points of a Hilbert space $(H, \|\cdot\|)$ and let a positive real r with $\|x_0 - x_1\| \le 2r$. Then for any $x \in \Delta_r(x_0, x_1)$ we have

$$\langle x - x_0, x_1 - x_0\rangle \ge \|x - x_0\|\,\|x_1 - x_0\| \sqrt{1 - \frac{1}{4r^2}\|x_0 - x_1\|^2}.$$

PROOF. We may suppose $x_0 \ne x_1$. Take any $x \in \Delta_r(x_0, x_1)$. By Proposition 16.6(f) we have

$$\|x_0 - x_1\|^2 \ge \|x - x_1\|^2 = \|x - x_0\|^2 + \|x_0 - x_1\|^2 - 2\langle x - x_0, x_1 - x_0\rangle,$$

or equivalently $2\langle x - x_0, x_1 - x_0\rangle \ge \|x - x_0\|^2$. Fix a unit vector u which is a linear combination of $x - x_0$ and $x_1 - x_0$ and which is such that

$$\langle u, x_1 - x_0\rangle = 0 \quad \text{and} \quad \langle u, x - x_0\rangle \ge 0. \tag{16.7}$$

There exist $\alpha, \beta \in \mathbb{R}$ such that

$$x - x_0 = \alpha(x_1 - x_0) + \beta u, \tag{16.8}$$

and (16.7) ensures that $\beta \ge 0$. Further, the latter equality and (16.7) also imply that $\langle x - x_0, x_1 - x_0\rangle = \alpha\|x_1 - x_0\|^2$, and this guarantees that $\alpha \ge 0$ since $x_1 \ne x_0$ and $2\langle x - x_0, x_1 - x_0\rangle \ge \|x - x_0\|^2$ by what precedes.

On the other hand, putting $y := \frac{1}{2}(x_0 + x_1) - \left(r^2 - \frac{1}{4}\|x_0 - x_1\|^2\right)^{1/2} u$ and using the equality of (16.7), we have

$$\begin{aligned}\|y - x_0\|^2 &= \left\| \frac{1}{2}(x_1 - x_0) - \left(r^2 - \frac{1}{4}\|x_0 - x_1\|^2\right)^{1/2} u \right\|^2 \\ &= \frac{1}{4}\|x_0 - x_1\|^2 + r^2 - \frac{1}{4}\|x_0 - x_1\|^2 = r^2,\end{aligned}$$

which implies that $x_0 \in B[y,r]$, and similarly $x_1 \in B[y,r]$. We deduce that $\Delta_r(x_0, x_1) \subset B[y,r]$, and hence $\|x - y\| \le r$ because $x \in \Delta_r(x_0, x_1)$. Since

$$y - x = \left(\alpha - \frac{1}{2}\right)(x_0 - x_1) - \left(\beta + \sqrt{r^2 - \frac{1}{4}\|x_0 - x_1\|^2}\right) u$$

according to (16.8), the last inequality $\|x-y\| \le r$ above along with the equality in (16.7) entail that

$$\left(\alpha-\frac{1}{2}\right)^2 \|x_0-x_1\|^2 + \left(\beta+\sqrt{r^2-\frac{1}{4}\|x_0-x_1\|^2}\right)^2 \le r^2.$$

Consequently,

$$\begin{aligned}\left(\beta+\sqrt{r^2-\frac{1}{4}\|x_0-x_1\|^2}\right)^2 &\le r^2-\frac{1}{4}\|x_0-x_1\|^2+\alpha\,\|x_0-x_1\|^2\\ &\le \left(\sqrt{r^2-\frac{1}{4}\|x_0-x_1\|^2}+\frac{\alpha\|x_0-x_1\|^2}{\sqrt{4r^2-\|x_0-x_1\|^2}}\right)^2,\end{aligned}$$

which (since $\alpha \ge 0$ as seen above) is equivalent to $\beta \le \frac{\alpha\|x_0-x_1\|^2}{\sqrt{4r^2-\|x_0-x_1\|^2}}$. This latter inequality means by (16.8) that

$$\|x-x_0\| = \sqrt{\alpha^2\|x_0-x_1\|^2+\beta^2} \le \alpha\|x_0-x_1\| \left(1-\frac{1}{4r^2}\|x_0-x_1\|^2\right)^{-1/2}.$$

Using again (16.8) and the equality in (16.7), we conclude that

$$\langle x-x_0, x_1-x_0\rangle = \alpha\,\|x_0-x_1\|^2 \ge \|x-x_0\|\,\|x_1-x_0\|\sqrt{1-\frac{1}{4r^2}\|x_0-x_1\|^2}.$$

□

THEOREM 16.11 (Characterizations of strongly convex sets). Let C be a nonempty closed bounded subset of a Hilbert space H and let $r \in]0,+\infty[$. The following assertions are equivalent:
(a) the set C is r-strongly convex;
(b) one has the equality

$$C = \bigcap_{x\in \mathrm{bdry}\, C, v\in N^P(C;x)\cap\mathbb{S}} B[x-rv,r];$$

(c) the set C is convex and one has the inclusion

$$C \subset \bigcap_{x\in \mathrm{bdry}\, C, v\in N(C;x)\cap\mathbb{S}} B[x-rv,r];$$

(d) one has $\operatorname{diam} C \le 2r$ and $\Delta_r(x_1,x_2) \subset C$ for any $x_1,x_2 \in C$;
(e) for all $x' \in C$ and for all $(x,v) \in H\times H$ with $x \in \mathrm{bdry}\, C$ and $v \in N^P(C;x)\cap\mathbb{S}$, one has

$$\langle v, x'-x\rangle \le -\frac{1}{2r}\|x'-x\|^2;$$

(f) for all $x,x' \in C$ and for all $v \in N^P(C;x)$, one has

$$\langle v, x'-x\rangle \le -\frac{\|v\|}{2r}\|x'-x\|^2.$$

PROOF. The proof of the theorem will be established along the following scheme:

$$\begin{array}{ccccccc} (b) & \Leftarrow & (c) & \Leftrightarrow & (e) & \Leftrightarrow & (f)\\ \Downarrow & & \Uparrow & & & & \\ (a) & \Rightarrow & (d) & & & & \end{array}$$

First, the implication (b)$\Rightarrow$(a) and the equivalence (e)$\Leftrightarrow$(f) are trivial. Further, (e) is a translation of (c), and the implication (a)$\Rightarrow$(d) is a direct consequence of the definitions of $\Delta_r(x_1, x_2)$ and of strongly convex sets. To prove (c)$\Rightarrow$(b), put $Q := \bigcap_{x \in \operatorname{bdry} C, v \in N(C;x) \cap \mathbb{S}} B[x - rv, r]$ and assume that (c) is satisfied, so in particular C is weakly compact. Take any $v \in \mathbb{S}$. For the support function $\sigma(C;\cdot)$ of C, there exists some $z \in \operatorname{bdry} C$ such that $\sigma(C;v) = \langle v, z \rangle$, that is, $v \in N(C;z) \cap \mathbb{S}$. By definition of Q, for any $y \in Q$ there is some $b \in \mathbb{B}$ such that $y = z - zv + rb$, hence

$$\langle v, y \rangle = \sigma(C;v) - r + r\langle v, b \rangle \leq \sigma(C;v).$$

It ensues that $\sigma(Q;v) \leq \sigma(C;v)$ for all $v \in \mathbb{S}$, thus $\sigma(Q;\cdot) \leq \sigma(C;\cdot)$. The set C being closed and convex, it follows that $Q \subset C$, hence $Q = C$ since the converse inclusion holds by the assumption (c). The implication (c)$\Rightarrow$(b) is established.

It remains to show (d)$\Rightarrow$(c). Assume (d) holds. First, we note that C is convex, since for any $x_1, x_2 \in C$ writing $\|x_1 - x_2\| \leq \operatorname{diam} C \leq 2r$ we have $[x_1, x_2] \subset \Delta_r(x_1, x_2) \subset C$. Now fix any $(x, v) \in \operatorname{bdry} C \times H$ with $v \in N^P(C;x) \cap \mathbb{S}$. Take any $y \in C$ with $y \neq x$ and put $\alpha := \|x - y\|$. Then $\Delta_r(x, y) \subset C$, so we have $v \in N^P(\Delta_r(x,y);x) = N(\Delta_r(x,y);x)$, where the latter equality is due to the convexity of $\Delta_r(x,y)$. According to Proposition 16.8, we know that

$$\Delta_r(x,y) = \bigcup_{t \in [0,1]} B[u_t, \delta(t,r)],$$

where $u_t := (1-t)x + ty$ and $\delta(t,r) := r - \sqrt{r^2 - t(1-t)\alpha^2}$ for each $t \in [0,1]$. It ensues that for every $t \in]0,1]$

$$\langle v, t(y - x) + \delta(t,r)b \rangle \leq 0 \quad \text{for all } b \in \mathbb{B},$$

or equivalently

$$\delta(t,r) \sup_{b \in \mathbb{B}} \langle v, b \rangle = \delta(t,r) \leq t \langle v, x - y \rangle . \tag{16.9}$$

Now, we observe that for every $t \in]0,1]$

$$\frac{\delta(t,r)}{t} = \frac{1}{t} \frac{r^2 - (r^2 - t(1-t)\alpha^2)}{r + \sqrt{r^2 - t(1-t)\alpha^2}} = \frac{(1-t)\alpha^2}{r + \sqrt{r^2 - t(1-t)\alpha^2}},$$

and this obviously entails

$$\lim_{t \downarrow 0} \frac{\delta(t,r)}{t} = \frac{\alpha^2}{2r}.$$

Coming back to (16.9) and passing to the limit as $t \downarrow 0$ yields

$$\frac{\alpha^2}{2r} \leq \langle v, x - y \rangle, \quad \text{or equivalently } \|y - (x - rv)\|^2 \leq r^2.$$

We deduce that $C \subset B[x - rv, r]$, so the inclusion in (c) holds true. All together, we have shown the implication (d)$\Rightarrow$(c), and the proof of the theorem is finished. $\square$

The equivalence (a)$\Leftrightarrow$(e) in Theorem 16.11 above yields the following direct corollary.

COROLLARY 16.12. Let C be a nonempty closed bounded subset of a Hilbert space H. If C is r-strongly convex for some $r \in]0, +\infty[$, then it is r'-strongly convex for any real $r' \geq r$.

16.2.2. Vial property. We saw in Theorem 16.11 above that the r-strong convexity of a nonempty closed set C, with $\operatorname{diam} C \le 2r$, is characterized by the inclusion $\Delta_r(x_0, x_1) \subset C$ for any $x_0, x_1 \in C$ (so, $\|x_0 - x_1\| \le 2r$). Instead of that inclusion, the Vial property for closed sets requires the non emptiness of the intersection of C with the above r-strongly convex segment.

DEFINITION 16.13. We say that a nonempty closed set C of a normed space $(X, \|\cdot\|)$ has the *Vial property* with a real constant $r > 0$ provided that for each pair of points $x_0, x_1 \in C$ with $0 < \|x_0 - x_1\| < 2r$ the r-strongly convex segment $\Delta_r(x_0, x_1)$ contains a point of C different from x_0 and x_1, or equivalently

$$C \cap \Delta_r(x_0, x_1) \neq \{x_0, x_1\}.$$

The next proposition shows that intersections of sets fulfilling the Vial property with appropriate strongly convex segments are connected. Recall that the gap between two subsets A, B of a normed space is defined by

$$\operatorname{gap}(A, B) := \inf_{x \in A, y \in B} \|x - y\|.$$

PROPOSITION 16.14. Let C be a nonempty closed set of a Hilbert space $(H, \|\cdot\|)$ satisfying the Vial property with a real constant $r > 0$. Then for any $x_0, x_1 \in C$ with $\|x_0 - x_1\| < 2r$ the set $C \cap \Delta_r(x_0, x_1)$ is connected.

PROOF. Suppose that $C_0 := C \cap \Delta_r(x_0, x_1)$ is not connected. There exist two nonempty closed sets A_0 and B_0 such that $C_0 = A_0 \cup B_0$ and $A_0 \cap B_0 = \emptyset$. Fix any $\overline{a} \in A_0$ and any $\overline{b} \in B_0$, and fix a decreasing sequence $(\varepsilon_n)_{n\in\mathbb{N}}$ tending to 0 with $0 < \varepsilon_1 < 2r - \|x_0 - x_1\|$. Choose $a_1 \in A_0$ and $b_1 \in B_0$ such that $\|a_1 - b_1\| < \varepsilon_1 + \operatorname{gap}(A_0, B_0)$, and put $A_1 := A_0 \cap \Delta_r(a_1, b_1)$, $B_1 := B_0 \cap \Delta_r(a_1, b_1)$. Note that the closed sets A_1, B_1 are nonempty and contain a_1 and b_1 respectively. Similarly, choose $a_2 \in A_1$ and $b_2 \in B_1$ such that $\|a_2 - b_2\| < \varepsilon_2 + \operatorname{gap}(A_1, B_1)$, put $A_2 := A_0 \cap \Delta_r(a_2, b_2)$, $B_2 := B_0 \cap \Delta_r(a_2, b_2)$, and note that $a_2 \in A_2$ and $b_2 \in B_2$. So, by induction we define two sequences $(a_n)_{n\in\mathbb{N}}$, $(b_n)_{n\in\mathbb{N}}$ in H and two sequences $(A_n)_{n\in\mathbb{N}}$ and $(B_n)_{n\in\mathbb{N}}$ of nonempty closed sets in H such that for each $n \in \mathbb{N}$:

$$a_n \in A_{n-1},\ b_n \in B_{n-1} \text{ with } \|a_n - b_n\| < \varepsilon_n + \operatorname{gap}(A_{n-1}, B_{n-1})$$

and

$$A_n = A_0 \cap \Delta_r(a_n, b_n) \text{ and } B_n = B_0 \cap \Delta_r(a_n, b_n).$$

For each $n \in \mathbb{N}$ put

$$\gamma_n := \|a_n - b_n\| \quad \text{and} \quad \delta_n := \operatorname{gap}(A_n, B_n) = \inf_{a \in A_n, b \in B_n} \|a - b\|.$$

We note that $\Delta_r(a_{n+1}, b_{n+1}) \subset \Delta_r(a_n, b_n)$ by Proposition 16.6(b) because $a_{n+1}, b_{n+1} \in \Delta_r(a_n, b_n)$, hence $A_{n+1} \subset A_n$ and $B_{n+1} \subset B_n$. For each $n \in \mathbb{N}$ we also note that $a_n \in A_n$, $b_n \in B_n$, and that the inclusion $\{a_{n+1}, b_{n+1}\} \subset \Delta_r(a_n, b_n)$ implies by Proposition 16.6(f) the inequality

$$\gamma_{n+1} = \|a_{n+1} - b_{n+1}\| \le \|a_n - b_n\| = \gamma_n.$$

The decreasing sequence $(\gamma_n)_n$ then converges to some non-negative γ, and this combined with the inequalities $\gamma_{n+1} \le \delta_n + \varepsilon_{n+1}$ and $\delta_n \le \gamma_n$ entails that we also have $\delta_n \to \gamma$ as $n \to \infty$. For any integers $m \ge n \ge 1$, according to (e) of Proposition 16.6 and to the inclusion $b_m \in \Delta_r(a_n, b_n)$, we have $b_m \in B[\frac{1}{2}(a_n + b_n), \frac{1}{2}\gamma_n]$, that is, $\|(b_m - b_n) - (a_n - b_m)\| \le \gamma_n$, or equivalently

$$\|b_m - b_n\|^2 + \|a_n - b_m\|^2 - 2\langle b_m - b_n, a_n - b_m\rangle \le \gamma_n^2.$$

This gives for any integers $m \geq n \geq 1$

$$\begin{aligned}\|b_m - b_n\|^2 &\leq \gamma_n^2 - \|a_n - b_m\|^2 + 2\langle b_m - a_n, a_n - b_m\rangle + 2\langle a_n - b_n, a_n - b_m\rangle \\ &= \gamma_n^2 - 3\|a_n - b_m\|^2 + 2\langle a_n - b_n, a_n - b_m\rangle \\ &\leq \gamma_n^2 - 3\|a_n - b_m\|^2 + 2\|a_n - b_n\| \, \|a_n - b_m\|.\end{aligned}$$

Further, for $m \geq n$ in $\mathbb{N}$ we also have $b_m \in B_m \subset B_n \subset \Delta_r(a_n, b_n)$, hence $\|a_n - b_m\| \leq \|a_n - b_n\|$ by Proposition 16.6(f). It ensues that for $m \geq n$ in $\mathbb{N}$

$$\|b_m - b_n\|^2 \leq \gamma_n^2 - 3\|a_n - b_m\|^2 + 2\|a_n - b_n\|^2 = 3\gamma_n^2 - 3\|a_n - b_m\|^2 \leq 3\gamma_n^2 - 3\delta_n^2.$$

Therefore, $(b_n)_n$ is a Cauchy sequence and converges to some b in H and $b \in \bigcap_{n\in\mathbb{N}} B_n$ since $B_{n+1} \subset B_n$. In a similar way, the sequence $(a_n)_n$ converges to some $a \in \bigcap_{n\in\mathbb{N}} A_n$, and clearly $\|a - b\| = \gamma$. By definition of A_n, B_n it ensues that $\{a, b\} \subset \Delta_r(a_n, b_n)$, hence $\Delta_r(a, b) \subset \Delta_r(a_n, b_n)$ by Proposition 16.6(b); similarly $\Delta_r(a, b) \subset \Delta_r(x_0, x_1)$.

Suppose $\gamma > 0$, that is, $\|a - b\| > 0$. Since $a, b \in C$, according to the Vial property of C and to the inequalities $0 < \|a - b\| \leq \|x_0 - x_1\| < 2r$, there exists some $y \in C \cap \Delta_r(a, b)$ with $y \notin \{a, b\}$. From this and the above inclusion $\Delta_r(a, b) \subset \Delta_r(x_0, x_1)$ we see that $y \in C_0$. Since C_0 is the disjoint union of A_0 and B_0, we have either $y \in A_0$ or $y \in B_0$. Without loss of generality me may suppose $y \in A_0$, so for each $n \in \mathbb{N}$ we have $y \in A_0 \cap \Delta_r(a_n, b_n) = A_n$ according to the above inclusion $\Delta_r(a, b) \subset \Delta_r(a_n, b_n)$. By Proposition 16.6(e) we also have $y \in \Delta_r(a, b) \subset B[\frac{1}{2}(a + b), \frac{1}{2}\gamma]$, and hence $\|y - b\| < \gamma$ because $y \neq a$. This combined with the inclusions $y \in A_n$ and $b \in B_n$ gives $\delta_n \leq \|y - b\| < \gamma$ for all n, which contradicts the convergence above of $(\delta_n)_n$ to γ.

Consequently, $\gamma = 0$, that is, $a = b$, and hence $A_0 \cap B_0 \neq \emptyset$, which contradicts that A_0 and B_0 are disjoint. The proof is then complete. □

The next theorem shows that the Vial property with constant $r > 0$ of a closed set of a Hilbert space is a characterization of its r-prox-regularity.

THEOREM 16.15 (G.E. Ivanov; J.-P. Vial). *A closed set C of a Hilbert space $(H, \|\cdot\|)$ is r-prox-regular for a real $r > 0$ if and only if it satisfies the Vial property with constant $r > 0$.*

The proof of the theorem involves both Proposition 16.8 and Proposition 16.14. It also uses the following lemma.

LEMMA 16.16. *Let x_0, x_1 be two points of a Hilbert space $(H, \|\cdot\|)$, $r \in]0, +\infty[$ and $\varepsilon \in]0, 1]$. Let $a \in H$ be such that*

$$\|x_0 - a\| = r \quad \text{and} \quad \|x_1 - a\| \leq r(1 - \varepsilon).$$

Then the following hold:
(a) For any $y \in B[\frac{x_0 - a}{\|x_0 - a\|}, \varepsilon/2]$

$$\Delta_r(x_0, x_1) \subset B\left[a - r(\frac{y}{\|y\|} - \frac{x_0 - a}{\|x_0 - a\|}), r\right].$$

(b) $B\left[\frac{x_0 - a}{\|x_0 - a\|}, \varepsilon/2\right] \subset N(\Delta_r(x_0, x_1); x_0)$.

PROOF. Put $v_0 := (x_0-a)/\|x_0-a\|$ and fix any $v \in B[v_0, \varepsilon/2]$. For $v_1 := v/\|v\|$ we have

$$\begin{aligned}\|v_1 - v_0\| &\le \|v_1 - v\| + \|v - v_0\| \le \|v_1 - v\| + \varepsilon/2 = |1 - \|v\|| + \varepsilon/2 \\ &= |\,\|v_0\| - \|v\|\,| + \varepsilon/2 \le \|v_0 - v\| + \varepsilon/2 \le \varepsilon.\end{aligned}$$

Put $w := r(v_1 - v_0)$, $z_0 := x_0 + w$, and $z_1 := x_1 + w$, and observe that $\|w\| \le r\varepsilon$. Writing

$$z_0 = x_0 + r(v_1 - v_0) = (x_0 - rv_0) + rv_1 = a + rv_1,$$

we see that $z_0 \in B[a, r]$.

On the other hand, by the assumption $\|x_1 - a\| \le r(1 - \varepsilon)$ we have

$$\|z_1 - a\| \le \|x_1 - a\| + \|w\| \le r(1 - \varepsilon) + r\varepsilon = r,$$

hence $z_1 \in B[a, r]$. Combining this with the previous inclusion $z_0 \in B[a, r]$ we obtain $\Delta_r(z_0, z_1) \subset B[a, r]$, which is equivalent to

$$\Delta_r(x_0, x_1) \subset B[a - r(v_1 - v_0), r] \tag{16.10}$$

according to the definition of z_0 and z_1 and to Proposition 16.6(h). This establishes the assertion (a) of the lemma. Finally, it is easily seen that

$$v_1 \in N(B[0, r]; rv_1) = N(B[a - rv_1 + rv_0, r]; a + rv_0) = N(B[a - r(v_1 - v_0), r]; x_0),$$

and hence $v_1 \in N(\Delta_r(x_0, x_1); x_0)$ thanks to (16.10). This finishes the proof of the lemma. □

PROOF OF THEOREM 16.15. Suppose first that C is r-prox-regular. Fix any $x_0, x_1 \in C$ with $0 < \|x_0 - x_1\| < 2r$. For $u := \frac{1}{2}(x_0 + x_1)$ we have $d_C(u) \le \|u - x_0\| < r$, which ensures that $z := P_C(u)$ exists according to the property (h) of Theorem 15.28. Further, by assertion (u) of the same Theorem 15.28 we have

$$d_C(u) \le r - \left(r^2 - \frac{1}{4}\|x_0 - x_1\|^2\right)^{1/2},$$

that is,

$$\|z - u\| \le r - \left(r^2 - \frac{1}{4}\|x_0 - x_1\|^2\right)^{1/2}.$$

This means with notation in (a) of Proposition 16.8 that $z \in B[u, \delta(1/2, r)]$, and hence $z \in C \cap \Delta_r(x_0, x_1)$ according to the same assertion (a) of Proposition 16.8. Consequently, C satisfies the Vial property with constant r.

Conversely, suppose now that the closed set C satisfies the Vial property with constant $r > 0$ but is not r-prox-regular. The definition of r-prox-regular set (see Definition 15.1) gives points $u \in U_r(C) \setminus C$ and $x \in \operatorname{Proj}_C(u)$ such that $x \notin \operatorname{Proj}_C(x + rv)$ where $v := (u - x)/\|u - x\|$, that is, there exists some $z \in C$ with $z \ne x$ such that $\|x + rv - z\| < r$. Then $\|x - z\| < 2r$ and by Proposition 16.14 the set $C \cap \Delta_r(x, z)$ is connected. Thus there exists a sequence $(x_n)_n$ of points of $C \cap \Delta_r(x, z)$ converging to x with $x_n \ne x$. For $v_n := (x_n - u)/\|x_n - u\|$ we have

$$\langle v_n, x - x_n\rangle = \langle v_n, x - u\rangle + \langle v_n, u - x_n\rangle \le \|x - u\| - \|x_n - u\|,$$

which entails (by the inclusions $x \in \operatorname{Proj}_C(u)$ and $x_n \in C$) that

$$\langle v_n, x - x_n\rangle \le 0. \tag{16.11}$$

Applying now Lemma 16.16 above with $x_0 = x$, $x_1 = z$, $a = x + rv$ and $\varepsilon > 0$ given by the equality $\|z - a\| = r(1 - \varepsilon)$ we obtain that $B[-v, \varepsilon/2] \subset N(\Delta_r(x, z); x)$.

Since $x_n \in \Delta_r(x,z)$, the latter inclusion implies that for all $b \in H$ with $\|b\| \le 1$ we have $\langle -v + \frac{\varepsilon}{2} b, x_n - x\rangle \le 0$, or otherwise stated

$$\langle -v, x - x_n\rangle \ge \frac{\varepsilon}{2}\|x - x_n\|. \tag{16.12}$$

Since $v_n \to -v$, we can choose some integer n_0 such that $\|v_{n_0} + v\| < \varepsilon/2$. We deduce from (16.12) that

$$\langle v_{n_0}, x - x_{n_0}\rangle = \langle v_{n_0} + v, x - x_{n_0}\rangle + \langle -v, x - x_{n_0}\rangle > -\frac{\varepsilon}{2}\|x - x_{n_0}\| + \frac{\varepsilon}{2}\|x - x_{n_0}\|,$$

hence $\langle v_{n_0}, x - x_{n_0}\rangle > 0$, which contradicts (16.11). The proof of the theorem is then complete. □

We derive two corollaries from Theorem 16.15. The first corollary establishes, in addition to the results in Section 15.4 on prox-regularity of intersection of sets, the r-prox-regularity of the intersection of any r-prox-regular set and any r-strongly convex set.

COROLLARY 16.17. Let C and D be two closed sets of a Hilbert space H with $C \cap D \neq \emptyset$. Assume that, for a real $r > 0$, the set C is r-prox-regular and D is r-strongly convex. Then $C \cap D$ is r-prox-regular.

PROOF. Let $x_0, x_1 \in C \cap D =: S$ with $0 < \|x_0 - x_1\| < 2r$. By Theorem 16.15 applied to C there exists $y \in C \cap \Delta_r(x_0, x_1)$ with $y \notin \{x_0, x_1\}$. Further, by the r-strong convexity of D there is a nonempty set $A \subset H$ such that $D = \bigcap_{a\in A} B[a, r]$. Then for each $a \in A$ we have $\{x_0, x_1\} \subset B[a, r]$, hence $\Delta_r(x_0, x_1) \subset B[a, r]$ by Proposition 16.6(c). This ensures that $\Delta_r(x_0, x_1) \subset D$, thus $y \in S \cap \Delta_r(x_0, x_1)$. It results by Theorem 16.15 again that S is r-prox-regular. □

The second corollary is concerned with the connectness of the intersection of any r-prox-regular set with any open ball with radius r.

COROLLARY 16.18. Let C be a nonempty closed set of a Hilbert space H which is r-prox-regular for a real $r > 0$. Then for any $x \in H$ the set $C \cap B(x, r)$ is connected.

PROOF. Let $x \in H$ with $S := C \cap B(x, r) \neq \emptyset$. Suppose that S is not connected, that is, there are two subsets S_0, S_1 of S, closed in S relative to the induced topology, such that $S = S_0 \cup S_1$ and $S_0 \cap S_1 = \emptyset$. Choose $x_i \in S_i$, $i = 0, 1$ and note that $\|x_0 - x_1\| < 2r$. By Proposition 16.14 and Theorem 16.15 the closed set $S' := C \cap \Delta_r(x_0, x_1)$ is connected. Putting $s := \max\{\|x_0 - x\|, \|x_1 - x\|\} < r$, by Proposition 16.6(c) we have $\Delta_r(x_0, x_1) \subset B[x, s] \subset B(x, r)$, so $S' \subset S$. Considering the sets S'_1, S'_2 defined by $S'_i := S_i \cap \Delta_r(x_0, x_1)$, $i = 0, 1$, we see that they are closed in H and nonempty (since $x_i \in S'_i$) and that $S' = S'_1 \cup S'_2$ along with $S'_1 \cap S'_2 = \emptyset$. This contradicts the connectedness of S' and finishes the proof. □

REMARK 16.19. It is worth pointing out that, in addition to Proposition 15.62, the connectedness of r-prox-regular sets with diameter less than $2r$ can be seen through Proposition 16.14 and Theorem 16.15. Indeed, let C be a nonempty (closed) r-prox-regular set in a Hilbert space H with $\operatorname{diam} C < 2r$, where $r \in]0, +\infty[$. Fix $x_0 \in C$ and note that $C \cap \Delta_r(x_0, x)$ is connected for every $x \in C$ by Proposition 16.14 and Theorem 16.15. The equality $C = \bigcup_{x\in C} C \cap \Delta_r(x_0, x)$ ensures that C is connected as the union of a collection of connected sets with a common point x_0. $\square$

Given a real $r > 0$, we have seen in 15.41(b) that a nonempty closed set C in a Hilbert space H is r-prox-regular if and only if for any $x, y \in C$ with $\|x - y\| < 2r$ and any $t \in [0, 1]$

$$C \cap B[(1-t)x + ty, r - \sqrt{r^2 - t(1-t)\|x-y\|^2}] \neq \emptyset.$$

The next proposition shows that it is sufficient to take $t = 1/2$ in that characterization.

PROPOSITION 16.20. Given a real $r > 0$, a nonempty closed set C in a Hilbert space H is r-prox-regular if and only if for any $x, y \in C$ with $\|x - y\| < 2r$ there is some $z \in C$ such that

$$\left\|\frac{x+y}{2} - z\right\| \leq r - \sqrt{r^2 - \frac{\|x-y\|^2}{4}}.$$

PROOF. Suppose first that C is r-prox-regular and take any $x, y \in C$ with $\|x - y\| < 2r$. Then Proposition 15.41(c) guarantees the existence of some $z \in C$ satisfying the desired inequality of the theorem.

Conversely, suppose that the property in the proposition holds. Take any $x, y \in C$ with $0 < \|x - y\| < 2r$. By assumption there is $z \in C$ such that

$$\left\|\frac{x+y}{2} - z\right\| \leq r - \sqrt{r^2 - \frac{\|x-y\|^2}{4}}. \tag{16.13}$$

Then Proposition 16.8 with $t = 1/2$ says that $z \in \Delta_r(x, y)$. Further, one can easily verify by (16.13) and by the inequalities $0 < \|x - y\| < 2r$ that $z \notin \{x, y\}$. Therefore, the closed set C enjoys (see Definition 16.13) the Vial property with constant $r > 0$, so Theorem 16.15 ensures that C is r-prox-regular, and the proof is finished. $\square$

REMARK 16.21. The above characterization of prox-regular sets can be translated in terms of the modulus of convexity of Hilbert spaces. Recalling that the modulus of convexity $\delta_X : [0, 2] \to [0, 1]$ of a normed space $(X, \|\cdot\|)$ is defined (as already recalled in (10.14)) by

$$\delta_X(\varepsilon) = \inf\left\{1 - \left\|\frac{1}{2}(x+y)\right\| : \|x\| = 1, \|y\| = 1, \|x-y\| \geq \varepsilon\right\},$$

it is not difficult to verify that the modulus of convexity of any Hilbert space $(H, \|\cdot\|)$ is given by

$$\delta_H(\varepsilon) = 1 - \sqrt{1 - \frac{\varepsilon^2}{4}} \quad \text{for all } \varepsilon \in [0, 2];$$

we refer to (18.13) for more development. Therefore, the above proposition can be rephrased by saying that the closed C is r-prox-regular if and only if for any $x, y \in C$ with $\|x - y\| < 2r$ there exists some $z \in C$ such that the inequality

$$\left\|\frac{x+y}{2} - z\right\| \leq r\delta_H\left(\frac{\|x-y\|}{r}\right)$$

is satisfied. $\square$

16.2.3. Closedness of Minkowski sums and ball separations properties. In general, the (Minkowski) sum of two prox-regular sets is not prox-regular. In fact, given a prox-regular set C, the sum $C+S$ may be non prox-regular even when S is strongly convex. In the Euclidean space $\mathbb{R}^2$, consider the sets $C := \{(-1,0),(1,0)\}$ and $S := B[(0,0),1]$. For $r := 1$, the closed sets C and S are r-prox-regular and r-strongly convex respectively, while their sum $C+S$ fails to be prox-regular. This subsection addresses certain conditions for the prox-regularity of $C+S$ under the prox-regularity of C and the strong convexity of S. The first step is the study of the closedness of $C+S$ for such types of closed sets.

THEOREM 16.22 (G.E. Ivanov). *Let C be an r-prox-regular set in a Hilbert space H and S be a closed s-strongly convex set in H with two reals $0 < s < r$. Then the set $C+S$ is closed.*

PROOF. Take any $z \in \operatorname{bdry}(C+S)$. There are sequences $(x_n)_n$ with $x_n \in C$ and $y_n \in S$ such that $\|z_n - z\| < \varepsilon_n$ for $z_n := x_n + y_n$ and $\varepsilon_n := 1/n$. Fix $N \in \mathbb{N}$ such that for all $m,n \geq N$ we have $\|z_n - z_m\| < 2(r-s)$, hence since $\|y_n - y_m\| \leq 2s$

$$\|x_n - x_m\| \leq \|z_n - z_m\| + \|y_n - y_m\| < 2r.$$

Therefore, by Proposition 16.20 there is some $x_{n,m} \in C$ such that

$$\left\| x_{n,m} - \frac{1}{2}(x_n + x_m) \right\| \leq r_{n,m} := r - \sqrt{r^2 - \frac{1}{4}\|x_n - x_m\|^2}. \tag{16.14}$$

On the other hand, setting $s_{n,m} := s - \sqrt{s^2 - \frac{1}{4}\|x_n - x_m\|^2}$ we also have

$$B\left[\frac{1}{2}(y_n + y_m), s_{n,m}\right] \subset \Delta\left(\frac{1}{2}(y_n + y_m), s\right) \subset S. \tag{16.15}$$

If $s_{n,m} \leq r_{n,m}$, *a fortiori* $s_{n,m} \leq r_{n,m} + \varepsilon_n + \varepsilon_m$. Suppose now $s_{n,m} > r_{n,m}$. By (16.14) and (16.15) we have with $u_{n,m} := (x_n + x_m)/2$ and $v_{n,m} := (y_n + y_m)/2$

$$B[u_{n,m} + v_{n,m}, s_{n,m}] = u_{n,m} + B[v_{n,m}, s_{n,m}] \subset B[x_{n,m}, r_{n,m}] + S.$$

Noting that $\frac{1}{2}(z_n + z_m) = u_{n,m} + v_{n,m}$, we deduce that

$$B\left[\frac{1}{2}(z_n + z_m), s_{n,m} - r_{n,m}\right] + r_{n,m}\mathbb{B} \subset x_{n,m} + S + r_{n,m}\mathbb{B},$$

which gives $B[\frac{1}{2}(z_n + z_m), s_{n,m} - r_{n,m}] \subset x_{n,m} + S \subset C + S$ according to the cancellation law (1.22). Since $z \in \operatorname{bdry}(C+S)$, it ensues that $\|z - \frac{1}{2}(z_n + z_m)\| \geq s_{n,m} - r_{n,m}$, which combined with both inequalities $\|z_n - z\| < \varepsilon_n$ and $\|z_m - z\| < \varepsilon_m$ yields $s_{n,m} < r_{n,m} + \varepsilon_n + \varepsilon_m$. In any case, we obtain $s_{n,m} < r_{n,m} + \varepsilon_n + \varepsilon_m$, or equivalently

$$\sqrt{r^2 - \frac{1}{4}\|x_n - x_m\|^2} - \sqrt{s^2 - \frac{1}{4}\|x_n - x_m\|^2} \leq r - s + \varepsilon_n + \varepsilon_m.$$

Putting $q_{n,m} := (r - s + \varepsilon_n + \varepsilon_m)^{-1}\left(r^2 - \frac{1}{4}\|x_n - x_m\|^2 - s^2 + \frac{1}{4}\|y_n - y_m\|^2\right)$ we get

$$q_{n,m} \leq \sqrt{r^2 - \frac{1}{4}\|x_n - x_m\|^2} + \sqrt{s^2 - \frac{1}{4}\|x_n - x_m\|^2} \leq \sqrt{r^2 - \frac{1}{4}\|x_n - x_m\|^2} + s,$$

so putting $\theta_{n,m} := r + s + q_{n,m}$ gives

$$r - \theta_{n,m} \le \sqrt{r^2 - \frac{1}{4}\|x_n - x_m\|^2} \le r. \tag{16.16}$$

Writing $\|x_n - x_m\| \le \|y_n - y_m\| + \|z_n - z_m\| \le \|y_n - y_m\| + \varepsilon_n + \varepsilon_m$ and recalling that $\|y_n - y_m\| \le 2s$, we see that

$$\|x_n - x_m\|^2 \le \|y_n - y_m\|^2 + 4s(\varepsilon_n + \varepsilon_m) + (\varepsilon_n + \varepsilon_m)^2,$$

thus we also have

$$0 \le \theta_{n,m} \le r + s - \frac{r^2 - s^2 - s(\varepsilon_n + \varepsilon_m) - 4^{-1}(\varepsilon_n + \varepsilon_m)^2}{r - s + \varepsilon_n + \varepsilon_m},$$

which ensures that $\theta_{n,m} \to 0$ as $n \to \infty$ and $m \to \infty$. This and (16.16) imply that $\|x_n - x_m\| \to 0$ as $n \to \infty$ and $m \to \infty$, hence $(x_n)_n$ converges to some $x \in C$. The equality $y_n = z_n - x_n$ tells us that $(y_n)_n$ converges to some $y \in S$ and that $z = x + y$ belongs to $C + S$, which confirms the closedness property of $C + S$ and finishes the proof. □

PROPOSITION 16.23. *Let C be an r-prox-regular set in a Hilbert space H and S be a closed s-strongly convex set in H with real constants $0 < s < r$. Then, the set $C + S$ is $(r - s)$-prox-regular.*

PROOF. We know by Theorem 16.22 that $C + S$ is closed. For convenience of notation, put $S_1 := C$ and $S_2 := S$. Fix any $x \in S_1 + S_2$, so there are $x_i \in S_i$, with $i = 1, 2$, such that $x = x_1 + x_2$. Take any $v \in N^P(S_1 + S_2; x)$ and any $y \in S_1 + S_2$. Choose $y_i \in S_i$, with $i = 1, 2$, such that $y = y_1 + y_2$. We have $v \in N(S_2; x_2) \cap N^P(S_1; x_1)$. Using both properties (f) in Theorem 16.11 and (b) in Theorem 15.28 we obtain

$$\langle v, y_1 + y_2 - x\rangle \le \frac{\|v\|}{2}\left(\frac{1}{r}\|y_1 - x_1\|^2 - \frac{1}{s}\|y_2 - x_2\|^2\right). \tag{16.17}$$

Set $a := y_1 - x_1$ and $b := y_2 - x_2$ and note that

$$-2\left\langle \sqrt{\frac{s}{r}}a, \sqrt{\frac{r}{s}}b\right\rangle \le \frac{s}{r}\|a\|^2 + \frac{r}{s}\|b\|^2, \quad -\frac{s}{r}\|a\|^2 - \frac{r}{s}\|b\|^2 \le 2\langle a, b\rangle,$$

which entails that

$$\left(1 - \frac{s}{r}\right)\|a\|^2 + \left(1 - \frac{r}{s}\right)\|b\|^2 \le \|a\|^2 + \|b\|^2 + 2\langle a, b\rangle = \|a + b\|^2.$$

The latter inequality means that

$$\frac{1}{r}\|a\|^2 - \frac{1}{s}\|b\|^2 \le \frac{1}{r - s}\|a + b\|^2.$$

This combined with (16.17) yields $\langle v, y_1 + y_2 - x\rangle \le \frac{\|v\|}{2(r-s)}\|y_1 + y_2 - x\|^2$, which ensures the $(r - s)$-prox-regularity of $S_1 + S_2$ by (b) in Theorem 15.28 again. □

We can now give two ball separation results in the next two theorems between a prox-regular set and a strongly convex set.

THEOREM 16.24 (ball separation I). *Let S be an r-prox-regular set of a Hilbert space H with $r > 0$ and C be a non-singleton closed set in H which is r-strongly convex with $C \cap S = \{\overline{x}\}$ and $\overline{x} \in \operatorname{bdry} C$. Then one has the ball separation property*

$$C \subset B[\overline{x} - rv, r] \quad \text{and} \quad B(\overline{x} - rv, r) \cap S = \emptyset.$$

PROOF. By (d) in Theorem 16.11 and by Proposition 16.8(a) we know that $\operatorname{int} C \neq \emptyset$. Let us first show that $(-\overline{x}+\operatorname{int} C)\cap T^B(S;\overline{x}) = \emptyset$. Indeed, suppose there is some h in the latter intersection. There are sequences $(t_n)_n$ tending to 0 with $0 < t_n < 1$ and $(h_n)_n$ converging to h in H such that $\overline{x}+t_n h_n \in S$ for all $n \in \mathbb{N}$. For n large enough, say $n \geq N$, we have $h_n \in -\overline{x}+\operatorname{int} C$, hence (since $0 \in \operatorname{cl}(-\overline{x}+\operatorname{int} C)$) we see that $t_n h_n \in -\overline{x} + \operatorname{int} C$. It ensues that $\overline{x} + t_N h_N \in S \cap \operatorname{int} C$, which is a contradiction.

By the emptiness of the above intersection and the Hahn-Banach separation theorem there is some $v \in X$ with $\|v\| = 1$ such that

$$\langle v, -\overline{x} + z\rangle < 0 \ \forall z \in \operatorname{int} C \quad \text{and} \quad \langle v, h\rangle \geq 0 \ \forall h \in T^B(S;\overline{x}) = T^C(S;\overline{x}),$$

keep in mind that $T^C(S;\overline{x})$ is a closed convex cone and $0 \in \operatorname{cl}(-\overline{x} + C)$. Using the equalities $\left(T^C(S;\overline{x})\right)^{\circ} = N^C(S;\overline{x}) = N^P(S;\overline{x})$, it results that $v \in N(C;\bar{x})$ and $-v \in N^P(S;\overline{x})$. Then from (b) in Theorem 16.11 and from the definition of r-prox-regular sets (see Definition 15.24 we deduce that

$$C \subset B[\overline{x} - rv, r] \quad \text{and} \quad S \cap B(\overline{x} - rv, r) = \emptyset,$$

which is the desired separation property. □

THEOREM 16.25 (ball separation II). Let S be an r-prox-regular set of a Hilbert space H with $r > 0$ and C be a closed set in H which is s-strongly convex with $0 < s < r$. For $g := \operatorname{gap}(S, C)$, where $\operatorname{gap}(S, C) := \inf\{\|x - y\| : x \in S, y \in C\}$, assume that $0 < g < r - s$. Then there exists $a \in H$ such that

$$C \subset B[a, s] \quad \text{and} \quad S \cap B(a, s + g) = \emptyset,$$

and the latter equality entails in particular $S \cap B[a, s] = \emptyset$.

PROOF. By Proposition 16.23 the set $S-C$ is $(r-s)$-prox-regular and $0 \notin S-C$ with $d(0, S - C) = g < r - s$. The point 0 then admits a (unique) nearest point $\overline{z} \in S-C$. Write $\overline{z} = \overline{x}-\overline{y}$ with $\overline{x} \in S$ and $\overline{y} \in C$, so $g = \|\overline{x}-\overline{y}\|$. Note that $\overline{x}$ (resp. $\overline{y}$) is a nearest point of $\overline{y}$ (resp. $\overline{x}$) in S (resp. C). Setting $v := (\overline{y} - \overline{x})/\|\overline{y} - \overline{x}\|$, we see that v belongs to $-N(C;\overline{y})$ and $N^P(S;\overline{x})$. By (b) in Theorem 16.11 and by definition of r-prox-regular sets (see Definition 15.24 we have

$$C \subset B[\overline{y} + sv, s] \quad \text{and} \quad S \cap B(\overline{x} + rv, r) = \emptyset.$$

It remains to prove the inclusion $B(\overline{y} + sv, s + g) \subset B(\overline{x} + rv, r)$. Take any $y \in B(\overline{y} + sv, s + g)$ and note that

$$\begin{aligned}\|y - (\overline{x} + rv)\| &= \left\|\left(y - (\overline{y} + sv)\right) + (\overline{y} - \overline{x} + sv - rv)\right\| \\ &< s + g + \left\|(\overline{y} - \overline{x})\left(1 - \frac{r-s}{g}\right)\right\|,\end{aligned}$$

then $\|y-(\overline{x}+rv)\| < s+g+g\left(\frac{r-s}{g} - 1\right) = r$, which confirms that $B(\overline{y}+sv, s+g) \subset B(\overline{x} + rv, r)$. This finishes the proof of the theorem. □

The proof of Theorem 16.25 also justifies the following proposition.

PROPOSITION 16.26. Let C be an r-prox-regular set in a Hilbert space H and S be a closed s-strongly convex set in H such that $0 < s < r$. Assume that $\operatorname{gap}(C, S) < r - s$. Then $\operatorname{gap}(C, S)$ is attained, that is, there exist $\overline{x} \in C$ and $\overline{y} \in S$ such that

$$\operatorname{gap}(C, S) = \|\overline{x} - \overline{y}\|.$$

16.3. Exterior/interior sphere condition

Let C be a closed r-prox-regular set (with $r \in]0, +\infty[$) of a finite-dimensional Euclidean space and let $x \in \operatorname{bdry} C$. Recalling that in finite dimensions the limiting normal cone at any boundary point is not reduced to zero, we have by prox-regularity $N^P(C;x) = N^L(C;x) \neq \{0\}$. Then according to Definition 15.1, for any $x \in \operatorname{bdry} C$ and any $v \in N^P(C;x)$ with $\|v\| = 1$ (such a vector v exists since $N^P(C;x) \neq \{0\}$) one has $x \in \operatorname{Proj}_C(x + rv)$, or equivalently the proximal normal vector v is realized by an r-ball, that is,

$$C \cap B(x + rv, r) = \emptyset,$$

and we know by Lemma 15.4 that this property is equivalent to the inequality

$$\langle v, x' - x\rangle \leq \frac{1}{2r}\|x' - x\|^2 \quad \text{for all } x' \in C.$$

The case where the emptiness of the above intersection is fulfilled for some unit vector $v \in N^P(C;x)$ instead of every unit vector of $N^P(C;x)$ leads to the following definition.

Definition 16.27. Given a real $r > 0$, one says that a closed set C of a Hilbert space H satisfies the *exterior r-sphere condition/property* when for each $x \in \operatorname{bdry} C$ there exists some unit vector $v \in N^P(C;x)$ which is realized by an r-ball. The terminology is justified by the fact that the property is obviously equivalent, for each $x \in \operatorname{bdry} C$, to the existence of some $y_x \notin C$ such that

$$B(y_x, r) \cap C = \emptyset \quad \text{and} \quad \|x - y_x\| = r.$$

Similarly, one says that the closed set C of H satisfies the *interior r-sphere condition/property* when for any $x \in \operatorname{bdry} C$ there exists some $y_x \in C$ such that

$$B(y_x, r) \subset C \quad \text{and} \quad \|x - y_x\| = r,$$

which (for the given $x \in \operatorname{bdry} C$) is equivalent to

$$x \in B[y_x, r] \subset C.$$

Proposition 16.28. Let C be a nonempty closed subset of a Hilbert space H and let $r \in]0, +\infty[$.
(a) If C is the union of a family of closed r-balls (that is, closed balls with radius r), then C satisfies the interior r-sphere condition.
(b) If C satisfies the interior r-sphere condition, then $C = \operatorname{cl}(\operatorname{int} C)$.
(c) The set C is a union of closed r-balls whenever $C = \operatorname{cl}(\operatorname{int} C)$ and $C' := H \setminus \operatorname{int} C$ is r-prox-regular along with $N^L(C';u) \neq \{0\}$ for all $u \in \operatorname{bdry} C'$ (non-nullity normal condition of C' which holds in finite dimensions).
(d) If C is r-prox-regular and $N^L(C;u) \neq \{0\}$ for any $u \in \operatorname{bdry} C$, then C satisfies the exterior r-sphere condition.
(e) If H is finite-dimensional, then the closed set C satisfies the exterior r-sphere condition whenever it is r-prox-regular.

Proof. The assertions (a) and (b) are obvious and the assertion (d) is obtained with the arguments preceding Definition 16.27. The assertion (e) is a direct consequence of (d). Concerning (c), assume that $C = \operatorname{cl}(\operatorname{int} C)$ and the set $C' := H \setminus \operatorname{int} C$ is r-prox-regular with $N^L(C';u) \neq \{0\}$ for all $u \in \operatorname{bdry} C'$. Fix any $x \in C$. Consider three cases.

Case 1. $d(x, \operatorname{bdry} C) \geq r$.
In this case we have $B(x,r) \subset C$. Otherwise, there exist some $y \in H \setminus C$ and some real $s > 0$ such that $\|y - x\| = s < r$, hence there is some $u \in [x, y] \cap \operatorname{bdry} C$. We deduce that

$$d(x, \operatorname{bdry} C) \leq \|x - u\| \leq \|x - y\| = s < r,$$

which is in contradiction with the inequality $d(x, \operatorname{bdry} C) \geq r$, so the inclusion $B(x,r) \subset C$ is justified. This and the closedness of C imply that $B[x,r] \subset C$, thus x is contained in a closed r-ball included in C.

Case 2. $0 < d(x, \operatorname{bdry} C) < r$.
This second case and the equality $C = \operatorname{cl}(\operatorname{int} C)$ give $d(x, C') = d(x, \operatorname{bdry} C) < r$. Then $P_{C'}x =: u$ exists by r-prox-regularity of C', so for $v := (x-u)/\|x-u\|$ we have $v \in N^P(C'; u)$. The r-prox-regularity of C' again ensures that $C' \cap B(u+rv, r) = \emptyset$, or equivalently $B(u + rv, r) \subset \operatorname{int} C$ according to the definition of C'. This implies that $B[u + rv, r] \subset C$. Further, noticing that

$$x = u + \|x - u\|v \quad \text{and} \quad \|x - u\| = d(x, C') < r,$$

we see that $x \in [u, u + rv]$, so the ball $B[u + rv, r]$ contains x.

Case 3. $d(x, \operatorname{bdry} C) = 0$, that is, $x \in \operatorname{bdry} C$.
The inclusion $x \in \operatorname{bdry} C$ and the assumption $C = \operatorname{cl}(\operatorname{int} C)$ entail that $x \in \operatorname{bdry} C'$, hence $N^L(C'; x) \neq \{0\}$ according to our assumption. Then we can choose a unit vector $v \in N^P(C'; x)$ since $N^P(C'; x) = N^L(C'; x)$ by prox-regularity of C'. Then, we obtain that $B[x + rv, r] \subset C$ by r-prox-regularity of C' as in Case 2, and clearly we also have $x \in B[x + rv, r]$ since $\|v\| = 1$.

In any case we found a closed r-ball containing x and included in C. □

The assertion (e) in the above proposition assures us that any closed r-prox-regular set (with $r \in]0, +\infty[$) in a finite-dimensional Euclidean space satisfies the exterior r-sphere condition. A first natural question concerns the reverse implication. The closed set $\mathbb{N} \times \{0\}$ of $\mathbb{R}^2$ satisfies the exterior r-sphere condition for any real $r > 0$, but it is r-prox-regular for no real $r > \frac{1}{2}$. More specifically, the closed set $\big(\{0\} \cup \{\frac{1}{n} : n \in \mathbb{N}\}\big) \times \{0\}$ of $\mathbb{R}^2$ satisfies the exterior r-sphere condition for every real $r > 0$, but for any real $r > 0$ it fails to be r-prox-regular.

Observing that both closed sets above have empty interiors, two questions have then to be addressed when $C = \operatorname{cl}(\operatorname{int} C)$:
(Q_1): Does the exterior r-sphere condition of C plus its r'-prox-regularity for some positive $r' < r$ entail its r-prox-regularity?
(Q_2): Does the exterior r-sphere condition of C ensure its r'-prox-regularity for some positive $r' < r$?

The answer to both questions is negative. Examples 16.29 and 16.30 below provide counterexamples to (Q_1) and (Q_2) respectively.

EXAMPLE 16.29. Let C be the closed set formed by the complement in $\mathbb{R}^2$ of the union of the two open balls with radius 1 in Figure 16.1. This set C is prox-regular (in fact, 1/2-prox-regular), satisfies the equality $C = \operatorname{cl}(\operatorname{int} C)$ along with the exterior 1-sphere condition. However, it fails to be 1-prox-regular, since the unit proximal normal vector $(0, -1)$ to C at the point $(0, \frac{1}{2})$ cannot be realized by a 1-ball. □

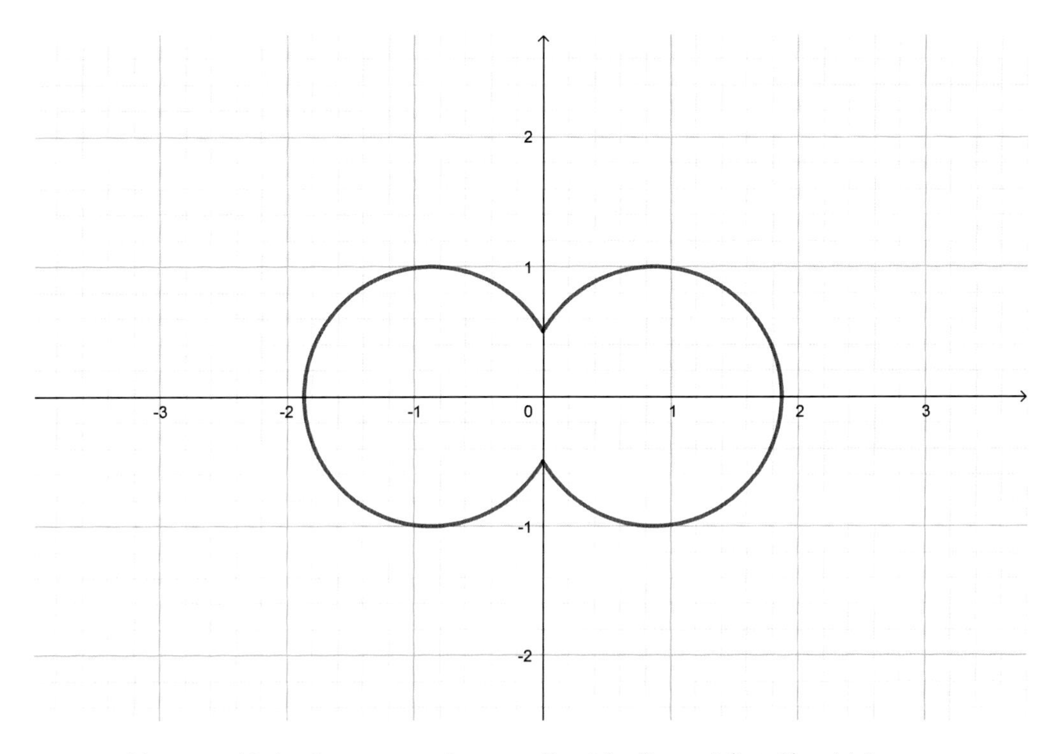

FIGURE 16.1. A prox-regular set C with $C = \mathrm{cl}\,(\mathrm{int}\, C)$ which satisfies the exterior 1-sphere condition, but fails to be 1-prox-regular.

EXAMPLE 16.30 (C. Nour, R.J. Stern and J. Takche). Consider the closed balls $B[c_n, 1]$ with centers $c_n \in \mathbb{R}^2$ and radius 1 in Figure 16.2 such that

$$B[c_n, 1] \cap B[c_{n+1}, 1] = \{(\alpha_n, -1/2^n), (\alpha_n, 1/2^n)\} \quad \text{for every } n \in \mathbb{N}.$$

The closed set C formed by the complement of the union of the corresponding open balls, that is, $C := \mathbb{R}^2 \setminus \big(\bigcup_{n\in\mathbb{N}} B(c_n, 1)\big)$, satisfies the equality $C = \mathrm{cl}(\mathrm{int}\, C)$ along with the exterior 1-sphere condition. However this set C fails to be r-prox-regular for any $r > 0$, since vertical proximal normal vectors at $(\alpha_n, 1/2^n)$ are realized only by balls of radius not greater than $1/2^n$. $\square$

Despite the preceding examples, the next proposition shows that under the additional assumption of the epi-Lipschitz property of the involved set the answer to question (Q_2) is positive. The proof of the proposition utilizes the following lemma.

LEMMA 16.31. Let $f : U \to \mathbb{R}$ be a γ-Lipschitz continuous function on an open set U of a Hilbert space H and let $H \times \mathbb{R}$ be endowed with the canonic inner product and Hilbert norm. Let a point $x \in U$, a real $r > 0$ and a unit vector $(v, -\alpha) \in H \times \mathbb{R}$ with $\alpha \geq 0$ such that

$$B\big((x, f(x)) + r(v, -\alpha), r\big) \cap \mathrm{epi}\, f = \emptyset.$$

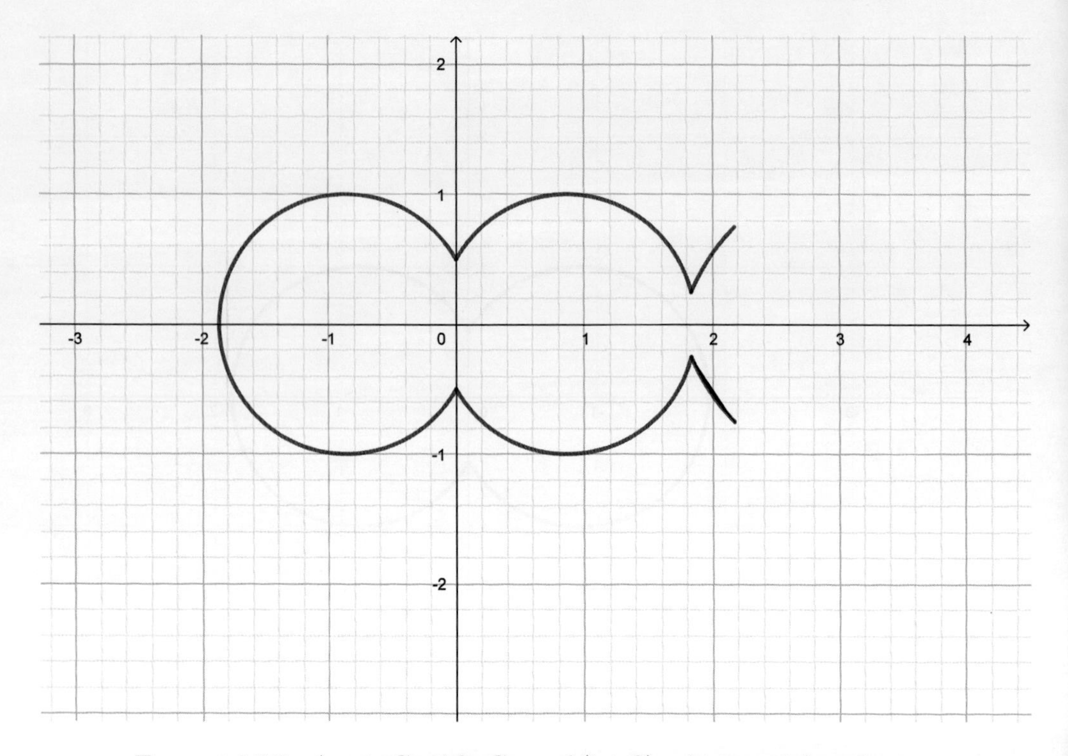

FIGURE 16.2. A set C with $C = \mathrm{cl}\,(\mathrm{int}\, C)$ which satisfies the exterior 1-sphere condition, but fails to be r-prox-regular for any $r > 0$.

Then, for any $(\zeta, -\beta) \in N^P(\mathrm{epi}\, f; (x, f(x)))$

$$\langle (\zeta, -\beta), (x', s) - (x, f(x)) \rangle \leq \frac{\|(\zeta, -\beta)\|}{2r'} \|(x', s) - (x, f(x))\|^2 \quad \forall (x', s) \in \mathrm{epi}\, f,$$

where $r' := \frac{r}{(1+\gamma^2)^{3/2}}$.

PROOF. The assumption $B\big((x, f(x)) + r(v, -\alpha), r\big) \cap \mathrm{epi}\, f = \emptyset$ means that

$$\langle (v, -\alpha), (x', s) - (x, f(x)) \rangle \leq \frac{1}{2r} \|(x', s) - (x, f(x))\|^2$$

for all $x' \in U$ and $s \geq f(x')$, hence $(v, -\alpha) \in N^P(\mathrm{epi}\, f; (x, f(x)))$. The function f being γ-Lipschitz continuous we have $\|v\| \leq \gamma\alpha$, and hence $\alpha > 0$. So the preceding inequality is equivalent for $w := v/\alpha$ to

$$\left\langle \frac{(w, -1)}{\|(w, -1)\|}, (x', s) - (x, f(x)) \right\rangle \leq \frac{1}{2r} \|(x', s) - (x, f(x))\|^2. \tag{16.18}$$

Since $\|w\| \leq \gamma$, we have $\|(w, -1)\| \leq \sqrt{1 + \gamma^2}$. Then taking $s = f(x')$ in (16.18) and using the γ-Lipschitz property of f we easily obtain for $r' := \frac{r}{(1+\gamma^2)^{3/2}}$ that

$$\langle w, x' - x \rangle \leq f(x') - f(x) + \frac{1}{2r'} \|x' - x\|^2$$

for all $x' \in U$. Proposition 10.39 yields that

$$\langle \zeta, x' - x \rangle \leq f(x') - f(x) + \frac{1}{2r'}\|x' - x\|^2$$

for all $\zeta \in \partial_P f(x)$ and $x' \in U$. This readily entails for all $x' \in U$, $s \geq f(x')$, and $\zeta \in \partial_P f(x)$ that

$$\begin{aligned}\langle (\zeta, -1), (x', s) - (x, f(x)) \rangle &\leq \frac{1}{2r'}\|(x', s) - (x, f(x))\|^2 \\ &\leq \frac{\|(\zeta, -1)\|}{2r'}\|(x', s) - (x, f(x))\|^2.\end{aligned}$$

This finishes the proof since the proximal normal cone $N^P\big(\operatorname{epi} f; (x, f(x))\big)$ is generated by $\partial_P f(x) \times \{-1\}$ because of the Lipschitz continuity of the function f. □

PROPOSITION 16.32. Let $f : H \to \mathbb{R}$ be a γ-Lipschitz continuous real-valued function on a Hilbert space H. Assume that the epigraph $\operatorname{epi} f$ of f satisfies the exterior r-sphere condition in $H \times \mathbb{R}$ endowed with its canonic Hilbert structure. Then $\operatorname{epi} f$ is r'-prox-regular for $r' := \frac{r}{(1+\gamma^2)^{3/2}}$.

PROOF. Assume that $\operatorname{epi} f$ satisfies the exterior r-sphere condition. Fix any $x \in H$. There exists some unit vector $(v, -\alpha) \in N^P\big(\operatorname{epi} f; (x, f(x))\big)$ (with $\alpha \geq 0$) such that $B\big((x, f(x)) + r(v, -\alpha), r\big) \cap \operatorname{epi} f = \emptyset$. Putting $r' := \frac{r}{(1+\gamma^2)^{3/2}}$, Lemma 16.31 tells us that, for any $(\zeta, -\beta) \in N^P(\operatorname{epi} f; (x, f(x)))$ we have

$$\langle (\zeta, -\beta), (x', s) - (x, f(x)) \rangle \leq \frac{\|(\zeta, -\beta)\|}{2r'}\|(x', s) - (x, f(x))\|^2 \quad \forall (x', s) \in \operatorname{epi} f.$$

Note also that $N^P(\operatorname{epi} f; (x, \rho))$ is reduced to zero for any real $\rho > f(x)$ (since $(x, \rho) \in \operatorname{int}(\operatorname{epi} f)$). It results (see (b) in Theorem 15.28) that the set $\operatorname{epi} f$ is r'-prox-regular. □

LEMMA 16.33. Let C be a closed set in a Hilbert space H which is epi-Lipschitz at a point $\overline{x} \in \operatorname{bdry} C$. Assume that there are two reals $\eta, \rho > 0$ such that for each $x \in B(\overline{x}, \eta) \cap \operatorname{bdry} C$ there is a unit vector $\zeta_x \in H$ such that

$$B(x + \rho\zeta_x,\, \rho) \cap C = \emptyset.$$

Then C is prox-regular at $\overline{x}$; more precisely there is an open neighborhood W of $\overline{x}$ and a real $r > 0$ such that for all $x \in W \cap C$ and all $v \in N^P(C; x)$

$$\langle v, x' - x \rangle \leq \frac{\|v\|}{2r}\|x' - x\|^2 \quad \text{for all } x' \in C.$$

PROOF. By Remark 2.21, striking $\eta > 0$ if necessary, there are a unit vector $\overline{h} \in H$ and a γ-Lipschitz function $f : E := (\mathbb{R}\overline{h})^{\perp} \to \mathbb{R}$ such that

$$B_H(\overline{x}, \eta) \cap C = B_H(\overline{x}, \eta) \cap A(\operatorname{epi} f), \tag{16.19}$$

where $A : E \times \mathbb{R} \to H$ is given by $A(u, s) = u + s\overline{h}$. Endow $E \times \mathbb{R}$ with the canonic inner product and Hilbert norm. By Remark 2.21 we also know that the linear mapping A is a bijective isometry. Let $\pi_E : H \to E$ and $H \ni x \mapsto \pi_{\overline{h}}(x)\overline{h} \in \mathbb{R}\overline{h}$ be the canonic orthogonal projections, so any $x \in H$ admits the representation $x = \pi_E(x) + \pi_{\overline{h}}(x)\overline{h}$.

Put $\sigma := \min\{\rho, \eta/3\}$. For any $x \in B(\overline{x}, \eta/3) \cap \operatorname{bdry} C$, it is easily verified that

$$B_H(x + \sigma\zeta_x, \sigma) \subset B_H(x + \rho\zeta_x, \rho) \cap B_H(\overline{x}, \eta),$$

hence by (16.19) and the assumption of emptiness in the statement of the lemma, it ensues that $B_H(x+\sigma\zeta_x,\sigma)\cap A(\operatorname{epi} f)=\emptyset$. Consequently, for any $x\in B_H(\overline{x},\eta/3)\cap$ bdry C noting that $\pi_{\overline{h}}(x)=f(x)$, we see that

$$B_{E\times\mathbb{R}}\big((\pi_E(x),f(\pi_E(x))+\sigma(\pi_E(\zeta_x),\pi_{\overline{h}}(\zeta_x)),\sigma\big)\cap\operatorname{epi} f=\emptyset.$$

Taking a neighborhood V of $\pi_E(\overline{x})$ in E such that $u+f(u)\overline{h}\in B(\overline{x},\eta/3)$ for all $u\in V$ (keep in mind that $\overline{x}=\pi_E(\overline{x})+f(\pi_E(\overline{x}))\overline{h}$), we deduce that, for any $u\in V$ and $\xi_u:=\zeta_{u+f(u)\overline{h}}$, that

$$B_{E\times\mathbb{R}}\big((u,f(u)+\sigma(\pi_E(\xi_u),\pi_{\overline{h}}(\xi_u)),\sigma\big)\cap\operatorname{epi} f=\emptyset.$$

Lemma 16.31 implies that there is a real $\rho'>0$ such that for any $u\in V$ and any $(\zeta,-\beta)\in N^P(\operatorname{epi} f;(u,f(u)))$ one has

$$\langle(\zeta,-\beta),(u',s)-(u,f(u))\rangle\le\frac{\|(\zeta,-\beta)\|}{2\rho'}\|(u',s)-(u,f(u))\|^2$$

for all $(u',s)\in(V\times\mathbb{R})\cap\operatorname{epi} f$. From this we derive a real $\rho''>0$ and an open neighborhood W of $\overline{x}$ in H such that, for any $x\in W\cap C$ and any $v\in N^P(C;x)$

$$\langle v,x'-x\rangle\le\frac{\|v\|}{2\rho''}\|x'-x\|^2\quad\text{for all } x'\in W\cap C.$$

Using the equivalence between the properties (b) and (b') in Theorem 15.73 we obtain a real $r>0$ and an open neighborhood U of $\overline{x}$ in H such that, for any $x\in U\cap C$ and any $v\in N^P(C;x)$

$$\langle v,x'-x\rangle\le\frac{\|v\|}{2\rho''}\|x'-x\|^2\quad\text{for all } x'\in C,$$

which finishes the proof of the lemma. □

PROPOSITION 16.34. Let H be a finite-dimensional Euclidean space and C be an epi-Lipschitz set in H whose boundary bdry C is bounded in H. Then the following are equivalent:
(a) the set C is a union of closed balls with constant radius;
(b) the set C satisfies the interior r-sphere condition for some real $r>0$;
(c) the set $H\setminus\operatorname{int} C$ is uniformly prox-regular.

PROOF. Since the set C is closed and epi-Lipschitz at any of its points, we first notice that the closed set $C':=H\setminus\operatorname{int} C$ is epi-Lipschitz at any of its points and that $C=\operatorname{cl}(\operatorname{int} C)$ (see Corollary 2.31). We also note from the latter equality that bdry $C=$ bdry (C'), as already seen also in Corollary 2.31.

The implications (c)⇒(a) and (a)⇒(b) follow respectively from the assertions (c) and (e) in Proposition 16.28. It remains to show (b)⇒(c). Suppose that (b) holds. For any $x\in\operatorname{bdry}(C')=\operatorname{bdry} C$, by the interior r-sphere property of C there exists a unit vector $\zeta_x\in H$ such that $B(x+r\zeta_x,r)\subset C$, hence $B(x+r\zeta_x,r)\subset\operatorname{int} C$, which implies that $B(x+r\zeta_x,r)\cap C'=\emptyset$. Then, for each point $x\in\operatorname{bdry}(C')$ Lemma 16.33 furnishes some real $r_x>0$ and some open neighborhood U_x of x such that for any $y\in U_x\cap\operatorname{bdry}(C')$ and any $v\in N^P(C';y)$

$$\langle v,y'-y\rangle\le\frac{\|v\|}{2r_x}\|y'-y\|^2\quad\text{for all } y'\in C'.$$

By the compactness of $\operatorname{bdry} C = \operatorname{bdry}(C')$ there are $x_1, \cdots, x_k \in \operatorname{bdry} C'$ such that $U_{x_1}, \cdots, U_{x_k}$ cover $\operatorname{bdry} C'$. Setting $r' := \min\{r_{x_1}, \cdots, r_{x_k}\} > 0$, we see that for any $y \in \operatorname{bdry}(C')$ and any $v \in N^P(C'; y)$ we have

$$\langle v, y' - y \rangle \leq \frac{\|v\|}{2r'} \|y' - y\|^2 \quad \text{for all } y' \in C'.$$

It results that the closed set C' is r'-prox-regular according to the property (b) in Theorem 15.28, which justifies the implication (b)⇒(c) and finishes the proof. □

Similarly, we have the following equivalences.

PROPOSITION 16.35. Let H be a finite-dimensional Euclidean space and C be an epi-Lipschitz set in H whose boundary $\operatorname{bdry} C$ is bounded.. Then the following are equivalent:
(a) the set C is uniformly prox-regular;
(b) the set C satisfies the exterior r-sphere condition for some real $r > 0$;
(c) the set $H \setminus \operatorname{int} C$ is a union of closed balls with constant radius.

PROOF. Putting $C' := H \setminus \operatorname{int} C$, we observe from Corollary 2.31 that C' is epi-Lipschitz at any of its points and that $C' = \operatorname{cl}(\operatorname{int}(C'))$ along with $C = H \setminus \operatorname{int}(C')$ and $\operatorname{bdry}(C') = \operatorname{bdry} C$.

The implication (a)⇒(c) follows from the assertion (c) in Proposition 16.28.

Suppose that (c) holds with the constant radius $r > 0$ and take any $x \in \operatorname{bdry} C = \operatorname{bdry}(C')$. There exists a point $z_x \in H$ such that $B[z_x, r] \subset C'$, hence $z_x \neq x$ and $B(z_x, r) \subset \operatorname{int}(C')$. Putting $\zeta_x := (z_x - x)/\|z_x - x\|$, we deduce for any $x \in \operatorname{bdry} C$ that $B(x + r\zeta_x, r) \cap C = \emptyset$ and the unit vector $\zeta_x \in N^P(C, x)$. This justifies the implication (c)⇒(b).

Now, suppose that (b) holds. For any $x \in \operatorname{bdry} C$ there exits a unit vector $\zeta_x \in N^P(C, x)$ such that $B(x + r\zeta_x, r) \cap C = \emptyset$. Using Lemma 16.33 it suffices to proceed as in Proposition 16.34 to see that there exists some real $r' > 0$ such that the set C is r'-prox-regular. Then the implication (b)⇒(a) holds true and the proof is complete. □

Now let us investigate a little more the interior sphere condition. Consider a real $r > 0$ and an r-prox-regular closed set C of a finite-dimensional Euclidean space H such that $C = \operatorname{cl}(\operatorname{int} C)$. The assertions (a) and (d) in Proposition 16.28 imply that the set $C' := H \setminus \operatorname{int} C$ is a union of closed r-balls.

Let C be the closed set of Figure 16.3 formed by the closed region containing the origin and delimited by the three arcs of circle (c), (d), (e) with radius 1. This set C obviously satisfies the interior 1-sphere condition but it fails to be the union of closed 1-balls since the origin is not contained in any closed 1-ball included in C. So, it is interesting to know whether any set satisfying the interior r-sphere condition is the union of closed r'-balls for some positive constant $r' \leq r$. The next theorem provides an affirmative answer.

THEOREM 16.36 (C. Nour, R.J. Stern and J. Takche). Let C be a closed set of a Hilbert space H satisfying the interior r-sphere condition with $r \in]0, +\infty[$. Assume that the set $C' := H \setminus \operatorname{int} C$ is proximinal (which is the case when H is finite-dimensional). Then C is a union of closed $\frac{r}{2}$-balls.

PROOF. The interior sphere condition of C tells us that $C = \operatorname{cl}(\operatorname{int} C)$ according to Proposition 16.28(b). This ensures that $\operatorname{bdry} C = \operatorname{bdry}(C')$, where $C' := H \setminus$

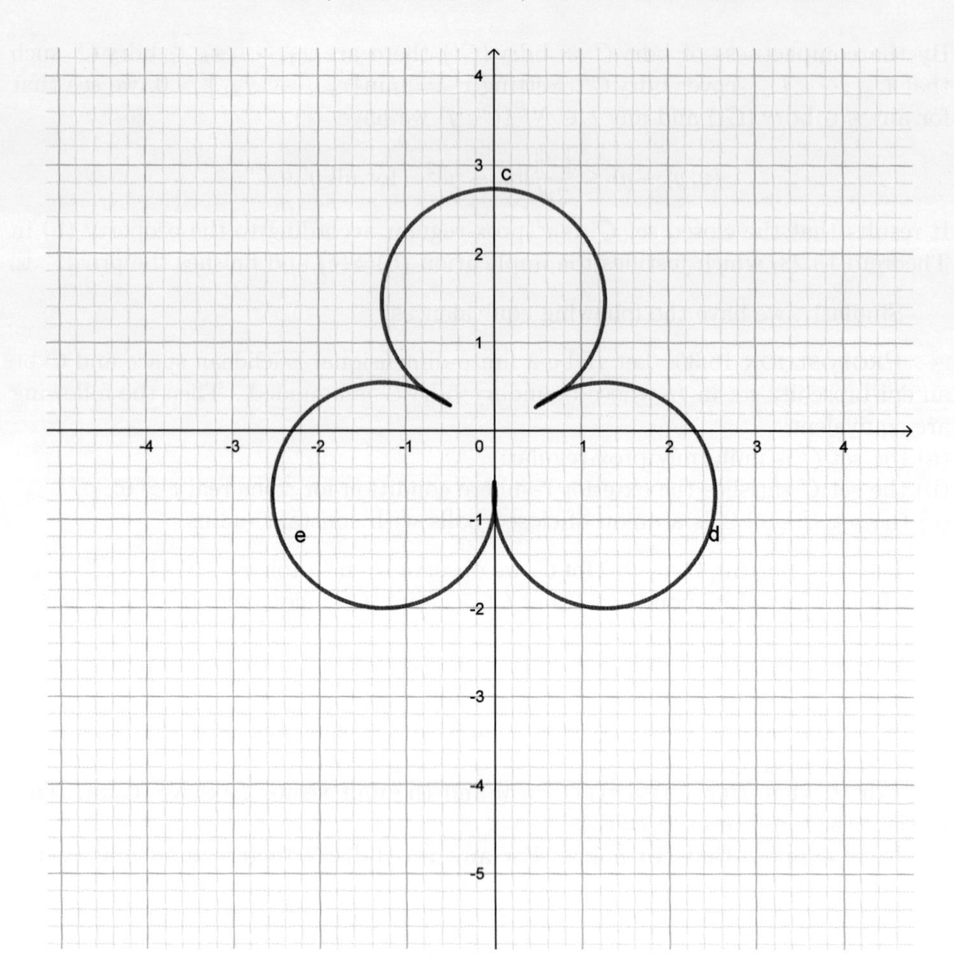

FIGURE 16.3. A closed set C which satisfies the interior r-sphere condition, but fails to be union of closed r-balls.

$\operatorname{int} C$. Further, for each $x \in \operatorname{int} C$ we have $\operatorname{Proj}_{C'}(x) \neq \emptyset$ by the proximinality of C', hence $\operatorname{Proj}_{\operatorname{bdry} C'}(x) \neq \emptyset$. It ensues that $\operatorname{Proj}_{\operatorname{bdry} C}(x) \neq \emptyset$ for any $x \in \operatorname{int} C$, thus also for any $x \in C$. Consequently, the theorem follows from Proposition 16.37 below. $\square$

PROPOSITION 16.37. Let C be a closed set of a Hilbert space H satisfying the interior r-sphere condition with $r \in]0, +\infty[$. Then for any $x \in C$ satisfying the condition $\operatorname{Proj}(x, \operatorname{bdry} C) \neq \emptyset$ there exists some closed $\frac{r}{2}$-ball containing x and included in C.

PROOF. If $x \in \operatorname{bdry} C$, then the interior r-sphere condition gives some closed r-ball containing x and contained in C, and from this one easily obtains a closed $\frac{r}{2}$-ball containing x and contained in C. So suppose that $x \in \operatorname{int} C$ with $\operatorname{Proj}(x, \operatorname{bdry} C) \neq \emptyset$. Choose $p \in \operatorname{Proj}(x, \operatorname{bdry} C)$ and put $r_0 := \|x-p\| = d(x, \operatorname{bdry} C) > 0$. Observing that $B[x, r_0] \subset C$, we see that, if $r_0 \geq \frac{r}{2}$, we have

$$x \in B[x, r/2] \subset C.$$

So, suppose that $0 < r_0 < \frac{r}{2}$. Take any positive real $\varepsilon \le r_0/2$ and choose some $z_\varepsilon \in B[p, \varepsilon] \cap (H \setminus C)$ (which is a nonempty set because $p \in \operatorname{bdry} C$) and put

$$x_\varepsilon := z_\varepsilon + \langle x - z_\varepsilon, v\rangle v = z_\varepsilon + t_\varepsilon v,$$

where $v := \frac{x-p}{\|x-p\|}$ and $t_\varepsilon := \langle x - z_\varepsilon, v\rangle$. We observe that

$$\langle x_\varepsilon - x, v\rangle = \langle z_\varepsilon + t_\varepsilon v - x, v\rangle = t_\varepsilon \langle v, v\rangle - \langle x - z_\varepsilon, v\rangle = t_\varepsilon - t_\varepsilon = 0,$$

that is, $x_\varepsilon - x \perp v$. Further, since $p = x - r_0 v$ we have

$$\begin{aligned}\|z_\varepsilon - p\|^2 &= \|(x_\varepsilon - x) + (r_0 - t_\varepsilon)v\|^2 \\ &= \|x_\varepsilon - x\|^2 + (r_0 - t_\varepsilon)^2 + 2(r_0 - t_\varepsilon)\langle x_\varepsilon - x, v\rangle \\ &= \|x_\varepsilon - x\|^2 + (r_0 - t_\varepsilon)^2.\end{aligned}$$

This and the inequality $\|z_\varepsilon - p\| \le \varepsilon$ entail that

$$\|x_\varepsilon - x\|^2 \le \varepsilon^2 - (r_0 - t_\varepsilon)^2. \tag{16.20}$$

Consequently, $\|x_\varepsilon - x\| \le \varepsilon \le r_0$ and hence $x_\varepsilon \in B[x, r_0]$.We also have from (16.20) $\varepsilon^2 - (r_0 - t_\varepsilon)^2 \ge 0$, or equivalently $r_0 - \varepsilon \le t_\varepsilon \le r_0 + \varepsilon$, which by the inequality $\varepsilon \le r_0/2$ implies that

$$\varepsilon \le t_\varepsilon \le r_0 + \varepsilon. \tag{16.21}$$

Since $z_\varepsilon \notin C$ and $x_\varepsilon \in C$ with $x_\varepsilon = z_\varepsilon + t_\varepsilon v$, we can choose some real $t'_\varepsilon \in]0, t_\varepsilon]$ such that the point

$$p_\varepsilon := z_\varepsilon + t'_\varepsilon v = x_\varepsilon + (t'_\varepsilon - t_\varepsilon)v$$

belongs to the boundary $\operatorname{bdry} C$ of C.

Claim 1. $\|p_\varepsilon - z_\varepsilon\| = t'_\varepsilon \le \varepsilon$.

Writing

$$\|p_\varepsilon - x\|^2 = \|(x_\varepsilon - x) - (t_\varepsilon - t'_\varepsilon)v\|^2 = \|x_\varepsilon - x\|^2 + (t_\varepsilon - t'_\varepsilon)^2$$

and using (16.20) and the inequality $\|x - p_\varepsilon\| \ge d(x, \operatorname{bdry} C) = r_0$, we see that

$$\varepsilon^2 - (r_0 - t_\varepsilon)^2 \ge \|x_\varepsilon - x\|^2 \ge r_0^2 - (t_\varepsilon - t'_\varepsilon)^2.$$

The inequality between the first hand side and the third can be rewritten as

$$2r_0(r_0 - t_\varepsilon) \le \varepsilon^2 - t_\varepsilon^2 + (t_\varepsilon - t'_\varepsilon)^2. \tag{16.22}$$

On the other hand, (16.21) yields $t_\varepsilon(r_0 - \varepsilon) \le r_0^2 - \varepsilon^2$, that is, $\varepsilon(\varepsilon - t_\varepsilon) \le r_0(r_0 - t_\varepsilon)$. Combining the latter inequality with (16.22) gives $2\varepsilon(\varepsilon - t_\varepsilon) \le \varepsilon^2 - t_\varepsilon^2 + (t_\varepsilon - t'_\varepsilon)^2$, or equivalently $(t_\varepsilon - \varepsilon)^2 \le (t_\varepsilon - t'_\varepsilon)^2$, which in turn means $0 \le t_\varepsilon - \varepsilon \le t_\varepsilon - t'_\varepsilon$ since $t_\varepsilon \ge \varepsilon$ by (16.21). It results that $t'_\varepsilon \le \varepsilon$, and this finishes the proof of the claim.

By the interior r-sphere condition of C we can choose, according to Definition 16.27 a unit vector $v_\varepsilon \in H$ such that $B(p_\varepsilon + rv_\varepsilon, r) \subset C$. Setting $C' := H \setminus \operatorname{int} C$, the latter inclusion is equivalent to

$$\langle v_\varepsilon, z - p_\varepsilon\rangle \le \frac{1}{2r}\|z - p_\varepsilon\|^2 \quad \text{for all } z \in C'.$$

From this and the relation $z_\varepsilon \in C'$ (because $z_\varepsilon \notin C$) we have

$$\langle v_\varepsilon, v\rangle = -\langle v_\varepsilon, -v\rangle = -\left\langle v_\varepsilon, \frac{z_\varepsilon - p_\varepsilon}{\|z_\varepsilon - p_\varepsilon\|}\right\rangle \ge -\frac{1}{2r}\|z_\varepsilon - p_\varepsilon\|. \tag{16.23}$$

Put $y_\varepsilon := p_\varepsilon + rv_\varepsilon$. If $\|y_\varepsilon - x\| \le r$, then for $y_x := x + \frac{r}{2}\frac{y_\varepsilon - x}{\|y_\varepsilon - x\|}$ it is easily seen that $\|y_x - y_\varepsilon\| \le r/2$, thus

$$x \in B[y_x, r/2] \subset B[y_\varepsilon, r] \subset C.$$

So we assume that $\|y_\varepsilon - x\| > r$.

Claim 2. *For* $r_\varepsilon := \frac{r_0^2\|y_\varepsilon - x\|}{\|y_\varepsilon - x\|^2 + r_0^2 - r^2}$ *and* $c_\varepsilon := x + r_\varepsilon \frac{y_\varepsilon - x}{\|y_\varepsilon - x\|}$ *we have*

$$B[c_\varepsilon, r_\varepsilon] \subset B[x, r_0] \cup B[y_\varepsilon, r] \subset C.$$

Fix any $u \in B[c_\varepsilon, r_\varepsilon] \setminus B[x, r_0]$. Observing that the inclusion $u \in B[c_\varepsilon, r_\varepsilon]$ is equivalent to

$$\|u - x\|^2 \le 2r_\varepsilon \left\langle u - x, \frac{y_\varepsilon - x}{\|y_\varepsilon - x\|} \right\rangle,$$

we obtain that

$$\begin{aligned}\|u - y_\varepsilon\|^2 &= \|(u - x) - (y_\varepsilon - x)\|^2 \\ &= \|u - x\|^2 + \|y_\varepsilon - x\|^2 - 2\langle u - x, y_\varepsilon - x\rangle \\ &\le \|u - x\|^2 + \|y_\varepsilon - x\|^2 - \frac{\|y_\varepsilon - x\|}{r_\varepsilon}\|u - x\|^2,\end{aligned}$$

which in turn implies that

$$\begin{aligned}\|u - y_\varepsilon\|^2 &\le \|y_\varepsilon - x\|^2 - \left(\frac{\|y_\varepsilon - x\|}{r_\varepsilon} - 1\right)\|u - x\|^2 \\ &= \|y_\varepsilon - x\|^2 - \frac{\|y_\varepsilon - x\|^2 - r^2}{r_0^2}\|u - x\|^2 \\ &\le \|y_\varepsilon - x\|^2 - \frac{\|y_\varepsilon - x\|^2 - r^2}{r_0^2} r_0^2 = r^2,\end{aligned}$$

where the last inequality is due to the fact that $\|u - x\| > r_0$ (and to our assumption $\|y_\varepsilon - x\| > r$). Consequently, the inclusion of Claim 2 is verified.

Claim 3. $\|y_\varepsilon - x\| \le \sqrt{r^2 + r_0^2} + 4\varepsilon$.

Observe first that

$$\begin{aligned}\|y_\varepsilon - x\|^2 &= \|(y_\varepsilon - p_\varepsilon) + (p_\varepsilon - p) + (p - x)\|^2 \\ &= \|rv_\varepsilon + (p_\varepsilon - p) - r_0 v\|^2 \\ &= r^2 + \|p - p_\varepsilon\|^2 + r_0^2 + 2r\langle v_\varepsilon, p_\varepsilon - p\rangle - 2rr_0\langle v_\varepsilon, v\rangle - 2r_0\langle p_\varepsilon - p, v\rangle \\ &\le r^2 + \|p - p_\varepsilon\|^2 + r_0^2 + 2r_0\|p_\varepsilon - p\| + 2r\|p_\varepsilon - p\| - 2rr_0\langle v_\varepsilon, v\rangle.\end{aligned}$$

On the other hand, (16.23) and Claim 1 ensure that

$$-2rr_0\langle v_\varepsilon, v\rangle \le r_0\|z_\varepsilon - p_\varepsilon\| \le \varepsilon r_0,$$

and

$$\|p_\varepsilon - p\| \le \|p_\varepsilon - z_\varepsilon\| + \|z_\varepsilon - p\| \le \varepsilon + \varepsilon = 2\varepsilon.$$

Consequently, we deduce

$$\begin{aligned}
\|y_\varepsilon - x\|^2 &\le r^2 + 4\varepsilon^2 + r_0^2 + 4r_0\varepsilon + 4r\varepsilon + r_0\varepsilon \\
&= r^2 + r_0^2 + 4\varepsilon^2 + \varepsilon(5r_0 + 4r) \\
&\le r^2 + r_0^2 + 4\varepsilon^2 + \frac{13}{2}\varepsilon r \\
&\le r^2 + r_0^2 + 16\varepsilon^2 + 8\varepsilon\sqrt{r^2 + r_0^2} \\
&= \left(\sqrt{r^2 + r_0^2} + 4\varepsilon\right)^2,
\end{aligned}$$

where the first inequality is trivial and the second one is due to the assumption $r_0 < \frac{r}{2}$. This finishes the proof of Claim 3.

Claim 4. *There exists some positive $\overline{\varepsilon} \le \frac{r_0}{2}$ such that $r_{\overline{\varepsilon}} \ge \frac{r}{2}$.*

Using the definition of r_ε it is readily seen that the inequality $r_\varepsilon \ge \frac{r}{2}$ is equivalent to

$$r\|y_\varepsilon - x\|^2 - 2r_0^2\|y_\varepsilon - x\| + r(r_0^2 - r^2) \le 0.$$

It is then sufficient to show the existence of some positive $\varepsilon \le \frac{r_0}{2}$ such that $\|y_\varepsilon - x\|$ is between the two roots $\frac{r_0^2 \pm \sqrt{\Delta'}}{r}$ where $\Delta' := (r^2 - r_0^2)^2 + r^2 r_0^2 > 0$. Noting that

$$\frac{r_0^2 - \sqrt{\Delta'}}{r} < 0 \le \|y_\varepsilon - x\|,$$

it is enough to establish the existence of some positive $\overline{\varepsilon} \le \frac{r_0}{2}$ such that

$$\|y_{\overline{\varepsilon}} - x\| \le \frac{r_0^2 + \sqrt{\Delta'}}{r},$$

and by Claim 3 this is a fortiori possible if we can find a positive $\overline{\varepsilon} \le \frac{r_0}{2}$ such that

$$\sqrt{r^2 + r_0^2} + 4\overline{\varepsilon} \le \frac{r_0^2 + \sqrt{\Delta'}}{r}, \text{ that is, } 4\overline{\varepsilon} \le \frac{r_0^2 + \sqrt{\Delta'}}{r} - \sqrt{r^2 + r_0^2}.$$

Since $\frac{r_0^2 + \sqrt{\Delta'}}{r} - \sqrt{r^2 + r_0^2} > 0$ (as easily checked), we can choose

$$\overline{\varepsilon} := \min\left\{\frac{r_0}{2}, \frac{1}{4}\left(\frac{r_0^2 + \sqrt{\Delta'}}{r} - \sqrt{r^2 + r_0^2}\right)\right\} > 0$$

and this choice gives the conclusion of Claim 4.

Considering $y_x = x + \frac{r}{2}\frac{y_\varepsilon - x}{\|y_\varepsilon - x\|}$ we see by Claim 4 and the definition of $c_{\overline{\varepsilon}}$ that $\|c_{\overline{\varepsilon}} - y_x\| = r_{\overline{\varepsilon}} - \frac{r}{2}$, hence $B[y_x, r/2] \subset B[c_{\overline{\varepsilon}}, r_{\overline{\varepsilon}}]$. It results by the inequality $\overline{\varepsilon} \le r_0/2$ in Claim 4 and by Claim 2 that

$$x \in B[y_x, r/2] \subset B[c_{\overline{\varepsilon}}, r_{\overline{\varepsilon}}] \subset B[x, r_0] \cup B[y_{\overline{\varepsilon}}, r] \subset C.$$

This finishes the proof of the proposition. □

16.4. Comments

A. B. Levy, R. A. Poliquin and R. T. Rockafellar introduced in their 2000 paper [**657**] the compatibility property for a parametrized family $(f_s)_{s\in S}$ of prox-regular functions defined on $\mathbb{R}^n$. In [**657**] several significant results are proved for such families. Considering the situation when f_s is the indicator function of a closed set $C(s)$ in $\mathbb{R}^n$, S. M. Robinson [**843**] established several additional results. In particular, for a family $(C(s))_{s\in S}$ of prox-regular sets in $\mathbb{R}^n$ which is compatibly parametrized at $\overline{s} \in S$ for $\overline{x} \in C(\overline{s})$, Robinson proved in [**843**] the Painlevé-Peano continuity of $(s,u) \mapsto P_{C(s)}(u)$ along with the equality

$$P_{C(s)}(u) = \big(I + N^L(C(s);\cdot)\big)^{-1}(u) \cap X$$

for some neighborhood X of $\overline{x}$ and for all s near $\overline{s}$ and u near $\overline{x}$. Robinson's work allows him to study in the setting of prox-regular sets the behavior of solutions $u(s)$ of variational inequalities in the form

$$F(s,u) + N^L(C(s);u(s)) \ni 0.$$

The results for the above variational inequalities are continuation of results obtained in a series of Robinson's papers (see the references in [**842**]) for polyhedral convex sets $C(s)$ with $s \in S$. Theorem 16.2 and Theorem 16.3 are taken from the paper [**883**] of M. Sebbah and L. Thibault. They correspond to extensions to Hilbert spaces of Robinson's results in $\mathbb{R}^n$.

Strongly convex functions and sets already appeared in the 1966 paper of B. T. Polyak [**817**] and in the 1966 large survey paper of E. S. Levintin and B. T. Polyak [**654**, page 7] respectively. Rephrasing [**817**] and [**654**] respectively, with the point $(1-t)x + ty$ in place of the middle point $(x+y)/2$, a function $f : X \to \mathbb{R}$ and a closed set C of a normed space X can be declared strongly convex in the sense of [**817**] and [**654**] respectively if there is a real constant $\gamma > 0$ such that

$$f((1-t)x+ty) \le (1-t)f(x) + tf(y) - \gamma t(1-t)\|x-y\|^2 \;\; \forall t \in [0,1], \forall x,y \in X,$$

$$(1-t)x + ty + \gamma t(1-t)\|x-y\|^2 \mathbb{B}_X \subset C \;\; \forall t \in [0,1], \forall x,y \in C. \tag{16.24}$$

Strongly convex sets (in the latter sense) are utilized in [**654**, Theorem 6.1] to establish strong convergence results with quadratic rate in reflexive Banach spaces for conditional gradient algorithms.

The first systematic study of strongly convex sets was probably the 1982 paper (in fact submitted in 1978) by J.-P. Vial [**957**] in finite-dimensional Euclidean spaces; the concept of r-strongly convex segment $\Delta_r(x,y)$ in Definition 16.5 seems to have been introduced therein under the notation $D_r(x,y)$ but without any name. Through $\Delta_r(x,y)$, J.-P. Vial [**957**] said that a closed set C of a finite-dimensional Euclidean space E with $\mathrm{diamC} \le 2r$ is strongly convex with real constant $r > 0$ if

$$\Delta_r(x,y) \subset C \text{ for all } x,y \in C.$$

In [**957**, Theorem 1(ii)] Vial showed in the finite-dimensional Euclidean space E that a closed set C is strongly convex with constant $r > 0$ in his above sense if and only if for all $t \in [0,1]$ and all $x,y \in C$

$$(1-t)x + ty + \frac{1}{2r}t(1-t)\|x-y\|^2 \mathbb{B}_E \subset C,$$

which corresponds to the definition of Levintin-Polyak in (16.24) above. Theorem 1(iv) in [**957**] also proved, again in the finite-dimensional Euclidean setting E, that C is strongly convex with constant $r > 0$ if and only if

$$C = \bigcap_{x\in E;\, B[x,r]\supset C} B[x,r]. \tag{16.25}$$

In the book, we followed E. S. Polovinkin and M. V. Balashov [**815**, Definition 3.1.1] in defining strongly convex sets with constant $r > 0$ in a normed space X as sets C satisfying the latter property (16.25). At this step, it is worth recalling that any closed convex set in a Banach space (even in a Hausdorff locally convex space) is an intersection of closed affine half-spaces. Replacing closed affine half-spaces by closed balls, a closed set C of a Banach space $(X, \|\cdot\|)$ is said to satisfy the *Mazur intersection property* whenever it is an intersection of closed balls; if this holds true for all closed convex sets in X, the Banach space $(X, \|\cdot\|)$ is nowadays said to satisfy (see, for example, the extensive survey [**447**]) the *Mazur intersection property* in honour to S. Mazur's 1933 paper [**701**] initiating the study of Banach spaces with such a property. The equality (16.25) then translates the strengthening of Mazur intersection property in requiring that the closed balls, whose intersection coincides with C, must all have the same radius r.

J.-P. Vial utilized the concept of r-strongly convex segment $\Delta_r(x, y)$ to develop, in his 1983 paper [**958**], several properties of sets of finite-dimensional Euclidean spaces satisfying the condition in Definition 16.13; such sets were called *weakly convex sets* in [**958**]. Notice that Vial's paper [**958**] was submitted in July 1981.

Properties (b), (d), (e), (f) in Proposition 16.6 are taken from G. E. Ivanow [**531**] where the term of "strongly convex segment"has been given. Proposition 16.8 and its proof correspond to Theorem 3.3.2 in the book by E. S. Polovinkin and M. V. Balashov [**815**]. Proposition 16.10 corresponds to Property 2.10 in the paper [**531**, p. 1117] by G. E. Ivanov, and the proof is also taken from there. Proposition 16.14 has been first proved by J.-P. Vial [**958**, Proposition 3.3] in finite-dimensional Euclidean spaces and extended by G. E. Ivanov [**531**, Theorem 4.1] to Hilbert spaces; the proof given in the book is the adaptation to the Hilbert setting of the proof provided by M. V. Balashov and G. E. Ivanov in [**67**, Lemma 4.1] in the uniformly convex Banach setting.

J.-P. Vial [**958**, Proposition 3.5] showed in finite-dimensional Euclidean spaces that a closed set C satisfies the Vial property (the weak convexity in his terminology) with a real constant $r > 0$ if and only if for every $x \in C$ and every unit vector $v \in N^F(S; x)$ one has $C \cap B(x + rv; r) = \emptyset$; this feature has been extended in a deep way to Hilbert spaces by G. E. Ivanov [**531**, Theorem 6.1]. Theorem 16.15 is the translation of that result with prox-regularity, since we know (see the previous chapter) that the above emptiness condition corresponds to the r-prox-regularity of the closed set C. The proof presented in Theorem 16.15 that the Vial property with constant $r > 0$ implies its r-prox-regularity, is the adaptation to the Hilbert setting of the very nice proof of M. V. Balashov and G. E. Ivanov [**67**, Theorem 2.1]. Lemma 16.16 is the Hilbert version of [**67**, Lemma 3.19] (proved in uniformly convex Banach spaces in [**67**]). Corollary 16.17 can be found in [**958**, Proposition 3.2] and [**531**, Lemma 3.6]. Proposition 16.20 is due to G. E. Ivanov [**531**, Lemma 4.2]. The proof given in the book for the implication $\Leftarrow$ of this proposition is that of Ivanov in [**531**, p. 1122]; the arguments for the direct implication $\Rightarrow$ are different. In [**531**, Lemma 4.2] the result in Proposition 16.20 is formulated in terms of the modulus of convexity of the Hilbert space as in Remark 16.21. Theorem 16.22 is due to G. E. Ivanov [**533**, Theorem 1.12.3] as well as the proof in the manuscript. Proposition 16.23 is also due to G. E. Ivanov [**533**, Theorem 1.12.4]; the proof that we gave follows the one in the paper by S. Adly, N. Nacry and L. Thibault [**8**]. The proof in [**533**] is based on Lemma 1.2.1 there, establishing that, given x_i, y_i, $i = 1, 2$, with $0 < \|x_i - y_i\| < 2r_i$, one has

$$\Delta_{r_1+r_2,0}(x_1 + x_2, y_1 + y_2) \subset \Delta_{r_1,0}(x_1, y_1) + \Delta_{r_2,0}(x_2, y_2),$$

where $\Delta_{r,0}(x,y) := \Delta_r(x,y) \setminus \{x,y\}$. Theorems 16.24 and 16.25 were essentially established in finite-dimensional Euclidean spaces by J.-P. Vial [**958**, Theorem 5.1] and in Hilbert spaces by G. E. Ivanov [**533**, Theorem 1.18.2].

Sets which are unions of closed r-balls have been utilized in Optimal Control Theory by P. Cannarsa and C. Sinestrari in their 1995 paper [**200**] and in their 2004 book [**201**]. Previously, closed sets whose complements are unions of open r-balls have been considered by N. V. Efimov and S. B. Stechkin in their 1959 paper [**379**]; such sets are studied in details in G. E. Ivanov's 2005 article under the name of *weakly convex sets in the sense of Efimov-Stechkin*. The interior r-ball/sphere condition has been used by O. Alvarez, P. Cardaliaguet and R. Monneau for dislocation dynamics in their 2005 paper [**15**], by P. Cannarsa and P. Cardiaguet [**198**] in their 2006 paper for perimeter estimates of reachable sets in Control theory, and by P. Cannarsa and H. Frankowska in the their 2006 paper [**199**] for the study of certain regularity properties of the minimal time function in Optimal Control Theory. All the results developed in Section 16.3 are due to C. Nour, R. J. Stern and J. Takche. Proposition 16.32 and its proof are taken from Nour, Stern and Takche [**775**] (see Theorem 3.7 and Lemma 3.8 in [**775**]). Theorem 16.36 and Proposition 16.37 are taken from their 2011 paper [**776**]. Their statements and proofs correspond to the statement and proof of Conjecture 1.3 of Nour, Stern and Takche [**776**, p. 592-595]. Example 16.29 and Example 16.30 are due to Nour, Stern and Takche [**775**, Examples 2.2 and 2.3] and Figure 16.3 is also taken from the same paper [**775**] by Nour, Stern and Takche (see Figure 4.1 and Example 4.1 therein); other illustrative examples can also be found in [**775, 774**].

CHAPTER 17

Differentiability of metric projection onto prox-regular sets

Throughout this chapter p is an integer $p \geq 1$ and H is still a Hilbert space as in the previous chapter. Given a set C of the Hilbert space H which is closed near $\overline{x} \in \operatorname{bdry} C$, the aim of this chapter is to study the $\mathcal{C}^p$-differentiability of the metric projection P_C at points in $H \setminus C$ near $\overline{x}$. By the previous chapter, we know that such a set must be prox-regular at $\overline{x}$. In order to put the study in the context of quantifying the set over which the differentiability holds, we will use the concept of (r,α)-prox-regularity. Then, we are drawn to establish first for such sets, in addition to Subsection 15.2.5, diverse other properties which will be used in the development of the chapter. Doing so, we will also establish diverse other fundamental features which illustrate the relevance and the amenability of this quantitative local prox-regularity.

17.1. Further properties of (r,α)-prox-regularity

Given an (r,α)-prox-regular set C at $\overline{x} \in C$ (see Definition 15.65), we begin by examining the latitude of its prox-regularity at points around $\overline{x}$ as well as the latitude of prox-regularity of translated sets of C.

Proposition 17.1. Let C be a set in a Hilbert space H which is (r,α)-prox-regular at $\overline{x} \in C$ with $r,\alpha \in]0,+\infty]$. The following hold:
(a) For any $y \in C \cap B(\overline{x},\alpha)$, the set C is $(r,\alpha - \|y-\overline{x}\|)$-prox-regular at y.
(b) For any $w \in H$ the set $w + C$ is (r,α)-prox-regular at $w + \overline{x}$.

Proof. To prove (a), observe that for $y \in B(\overline{x},\alpha) \cap C$ we have

$$B(y,\alpha - \|y-\overline{x}\|) \subset B(\overline{x},\alpha).$$

From this inclusion we easily see that $C \cap B(y,\alpha - \|y-\overline{x}\|)$ is closed relative to $B(y,\alpha - \|y-\overline{x}\|)$ and that (15.43) still holds for all $x \in B(y,\alpha - \|y-\overline{x}\|) \cap C$ and all $v \in N^P(C;x) \cap \mathbb{B}_H$. This means that the set C is $(r,\alpha - \|y-\overline{x}\|)$-prox-regular at y.

Now we justify the assertion (b). Set $S := w + C$. Noting that $S \cap B(w + \overline{x},\alpha) = w + C \cap B(\overline{x},\alpha)$, we easily see that $S \cap B(w+\overline{x},\alpha)$ is closed relative $B(w+\overline{x},\alpha)$. Let any $y \in S \cap B(w+\overline{x},\alpha)$, that is, $y - w \in C \cap B(\overline{x},\alpha)$. Let any $v \in N^P(S;y) \cap \mathbb{B}_H$. Then there exists a real $s > 0$ such that $y + sv \in \operatorname{Proj}_S(y)$, or equivalently $y - w + sv \in \operatorname{Proj}_C(y-w)$. So, $v \in N^P(C;y-w) \cap \mathbb{B}_H$. Applying the equivalence (a)⇔(b) in Theorem 15.70 and taking $y - w \in C \cap B(\overline{x},\alpha)$ and $v \in N^P(C;y-w) \cap \mathbb{B}_H$ in the property (b) therein, we obtain

$$\langle v, y'-y\rangle = \langle v, y' - w - (y-w)\rangle \leq \frac{1}{2r}\|y'-y\|^2 \text{ for all } y' \in S,$$

hence the set S is (r,α)-prox-regular at $w+\overline{x}$ according to the equivalence (a)$\Leftrightarrow$(b) in Theorem 15.70 again. $\square$

Concerning the quantitative concept of (r,α)-prox-regularity of sets of a Hilbert space H, we have the following property related to the hypomonotonicity of the truncated normal cone.

PROPOSITION 17.2. *Let C be a subset of a Hilbert space H and let $\overline{x} \in C$ and $\alpha \in]0,+\infty]$ be such that $C \cap B(\overline{x},\alpha)$ is closed relative to $B(\overline{x},\alpha)$. Assume that there exists $0 < r < \alpha$ such that the multimapping $N^P(C;\cdot) \cap \mathbb{B}_H$ is $\frac{1}{r}$-hypomonotone on $B(\overline{x},\alpha)$. Then, for the extended real $\sigma := \frac{1}{2}(\alpha - r) > 0$ the set C is $(\frac{r}{2},\sigma)$-prox-regular at $\overline{x}$.*

PROOF. Clearly, the closedness of $C \cap B(\overline{x},\alpha)$ relative to $B(\overline{x},\alpha)$ entails that $C \cap B(\overline{x},\sigma)$ is closed relative to $B(\overline{x},\sigma)$. By the hypomonotonicity assumption we have, for all $x_i \in C \cap B(\overline{x},\alpha)$ and all $v_i \in N^P(C;x_i) \cap \mathbb{B}_H$,

$$\langle v_1 - v_2, x_1 - x_2\rangle \geq -\frac{1}{r}\|x_1 - x_2\|^2.$$

Fix $x \in C \cap B(\overline{x},\alpha)$ and $v \in N^P(C;x) \cap \mathbb{B}_H$. The latter inequality, for $x_1 = x, x_2 = x'$ and $v_1 = v, v_2 = 0$, gives

$$\langle v, x' - x\rangle \leq \frac{1}{r}\|x' - x\|^2 \quad \text{for all } x' \in C \cap B(\overline{x},\alpha).$$

So, for all $x' \in C \cap B(\overline{x},\alpha)$ we have

$$0 \leq \|x' - x\|^2 - 2\left\langle \frac{r}{2}v, x' - x\right\rangle,$$

or equivalently

$$\left\|\frac{r}{2}v\right\|^2 \leq \left\|x' - x - \frac{r}{2}v\right\|^2.$$

This ensures that $x \in \operatorname{Proj}(x + \frac{r}{2}v, C \cap B(\overline{x},\alpha))$.

On the other hand, for the real number $\sigma := \frac{1}{2}(\alpha - r) > 0$ (recall that $r < \alpha$), we have that for all $x \in B(\overline{x},\sigma)$

$$\left\|x + \frac{r}{2}v - \overline{x}\right\| \leq \|x - \overline{x}\| + \frac{r}{2} < \frac{\alpha}{2} - \frac{r}{2} + \frac{r}{2} = \frac{\alpha}{2},$$

hence $\operatorname{Proj}(x + \frac{r}{2}v, C \cap B(\overline{x},\alpha)) = \operatorname{Proj}(x + \frac{r}{2}v, C)$ according to Lemma 2.219.

Altogether, it results that for all $x \in C \cap B(\overline{x},\sigma)$ and all $v \in N^P(C;x) \cap \mathbb{B}_H$

$$x \in \operatorname{Proj}\left(x + \frac{r}{2}v, C\right),$$

which allows us to conclude by (15.43) that C is $(\frac{r}{2},\sigma)$-prox-regular at $\overline{x}$. $\square$

Given a function $f : H \to \mathbb{R} \cup \{+\infty\}$ which is lower semicontinuous on an open convex set $U \subset H$ with $U \cap \operatorname{dom} f \neq \emptyset$, we recall that f is primal lower regular with constant $c \geq 0$ on U (see Definition 11.1) provided that, for all $x \in U \cap \operatorname{Dom} \partial_P f$ and for all $v \in \partial_P f(x)$ one has

$$f(y) \geq f(x) + \langle v, y - x\rangle - c(1 + \|v\|)\|y - x\|^2 \tag{17.1}$$

for every $y \in U$. The next result establishes a relationship between the primal lower regular property of the indicator function of a set and the local quantitative prox-regularity of the set.

PROPOSITION 17.3. Let C be a set in a Hilbert space H and $\overline{x} \in C$. Let $\alpha \in]0,+\infty]$ be such that $C \cap B(\overline{x},\alpha)$ is closed relative to $B(\overline{x},\alpha)$.
(a) If the set C is (r,α)-prox-regular at $\overline{x}$ with $r \in]0,+\infty]$, then the indicator function Ψ_C is primal lower regular on $B(\overline{x},\alpha)$ with the constant $\frac{1}{2r}$.
(b) If the indicator function Ψ_C is primal lower regular with a real constant $c > 0$ on $B(\overline{x},\alpha)$, then the set C is (r,α)-prox-regular at $\overline{x}$ for $r := \min\{(4c)^{-1}, \alpha/2\}$.

PROOF. We note that Ψ_C is lower semicontinuous on $B(\overline{x},\alpha)$.

First, the (r,α)-prox-regularity of C leads to the primal lower regularity of Ψ_C as quantified in (a). Indeed, if C is (r,α)-prox-regular at $\overline{x}$, then according to the equivalence (a)$\Leftrightarrow$(b) in Theorem 15.70 we have for all $x \in B(\overline{x},\alpha) \cap C$,

$$\langle v, x' - x\rangle \le \frac{1}{2r}\|x' - x\|^2 \quad \forall v \in N^P(C;x) \cap \mathbb{B}, \forall x' \in C.$$

Now if $v \in N^P(C;x)$, putting it in the latter inequality, we obtain for all x such that $\|x - \overline{x}\| < \alpha$, all $v \in N^P(C;x) = \partial_P \Psi_C(x)$

$$\begin{aligned}\Psi_C(x') &\ge \Psi_C(x) + \langle v, x' - x\rangle - \frac{\|v\|}{2r}\|x' - x\|^2 \\ &\ge \Psi_C(x) + \langle v, x' - x\rangle - \frac{1 + \|v\|}{2r}\|x' - x\|^2\end{aligned}$$

for all x' with $\|x' - x\| < \alpha$, hence the indicator function Ψ_C is primal lower regular on $B(\overline{x},\alpha)$ with the constant $\frac{1}{2r}$. So, the assertion (a) is established.

To prove the assertion (b), suppose that Ψ_C is primal lower regular on $B(\overline{x},\alpha)$ with a real constant $c > 0$, that is, for all $x \in B(\overline{x},\alpha)$, all $v \in \partial_P \Psi_C(x) = N^P(C;x)$, the inequality

$$\Psi_C(x') \ge \Psi_C(x) + \langle v, x' - x\rangle - c(1 + \|v\|)\|x' - x\|^2$$

is valid for all $x' \in B(\overline{x},\alpha)$. Take any $x \in C \cap B(\bar{x},\alpha)$ and any $v \in N^P(C;x) \cap \mathbb{B}_H$. Then we get

$$\langle v, x' - x\rangle \le c(1 + \|v\|)\|x' - x\|^2 \le 2c\|x' - x\|^2 \quad \forall x' \in C \cap B(\overline{x},\alpha).$$

When $\|x' - \overline{x}\| \ge \alpha$ with $x' \in C$ write

$$\langle v, x' - x\rangle \le \|x' - x\| = \|x' - x\|^2 \frac{1}{\|x' - x\|} \le \frac{1}{\alpha}\|x' - x\|^2.$$

We deduce that

$$\langle v, x' - x\rangle \le \frac{1}{2\min\{(4c)^{-1}, \alpha/2\}}\|x' - x\|^2 \quad \forall x' \in C.$$

Putting $r := \min\{(4c)^{-1}, \alpha/2\}$, we conclude by the equivalence (a)$\Leftrightarrow$(b) in Theorem 15.70 that the set C is (r,α)-prox-regular at $\overline{x}$. $\square$

In the next theorem we use the following restricted local enlargement.

DEFINITION 17.4. Let C be a subset of a Hilbert space H with $\overline{x} \in C$ and let $r, \alpha \in]0,+\infty]$ be two extended reals. We call the set

$$\mathcal{R}_C(\overline{x}, r, \alpha) := \{x + tv : x \in C \cap B(\overline{x},\alpha), t \in [0, r[, v \in N^P(C;x) \cap \mathbb{B}_H\}$$

the *local enlargement of* C *around* $\overline{x}$ *with restricted rays of thickness* r *starting from points in* $C \cap B(\overline{x},\alpha)$ *along proximal normals in* $\mathbb{B}_H$.

We observe that

$$C \cap B(\overline{x}, \alpha) \subset \mathcal{R}_C(\overline{x}, r, \alpha) \quad \text{and} \quad \mathcal{R}_C(\overline{x}, r, \alpha) = \bigcup_{\gamma \in]0,1[} \mathcal{R}_C(\overline{x}, \gamma r, \alpha). \tag{17.2}$$

Denoting by I the identity mapping on H, we also observe that

$$u \in \mathcal{R}_C(\overline{x}, r, \alpha) \Leftrightarrow (I + t\mathbb{B}_H \cap N^P(C; \cdot))^{-1}(u) \cap B(\overline{x}, \alpha) \neq \emptyset \text{ for some } t \in [0, r[.$$

Indeed, if $u \in \mathcal{R}_C(\overline{x}, r, \alpha)$, then there exist $x \in B(\overline{x}, \alpha)$, $v \in N^P(C; x) \cap \mathbb{B}_H$ such that $u = x + tv$ for some $t \in [0, r[$. Then $u \in x + t\mathbb{B}_H \cap N^P(C; x)$. Therefore

$$(I + t\mathbb{B}_H \cap N^P(C; \cdot))^{-1}(u) \cap B(\overline{x}, \alpha) \neq \emptyset.$$

Conversely if the latter intersection is not empty for some $t \in [0, r[$, there exists $x \in B(\overline{x}, \alpha)$ such that $u \in x + t\mathbb{B}_H \cap N^P(C; x)$, that is, $u = x + tv$ for some $v \in N^P(C; x) \cap \mathbb{B}_H$, so $u \in \mathcal{R}_C(\overline{x}, r, \alpha)$.

The next theorem shows in particular that the properties (i) and (l) established over the set $U^{\gamma}_{r,\alpha}(\overline{x}, C)$ in Theorem 15.70 are also valid over the set $\mathcal{R}_C(\overline{x}, \gamma r, \alpha)$

THEOREM 17.5 (metric projection to (r, α)-prox-regular sets). Let C be a set in a Hilbert space H which is (r, α)-prox-regular at $\overline{x} \in C$ for some extended reals $r, \alpha \in]0, +\infty]$. The following properties hold.
(i') For any real $\gamma \in]0, 1[$ and any $u \in \mathcal{R}_C(\overline{x}, \gamma r, \alpha)$ the set $\operatorname{Proj}_C(u)$ is a singleton, that is, $P_C(u)$ exists; further, the equality

$$P_C(u) = (I + \gamma r\mathbb{B}_H \cap N^P(C; \cdot))^{-1}(u) \cap B(\overline{x}, \alpha)$$

holds, and the mapping P_C satisfies

$$\|P_C(u_1) - P_C(u_2)\| \le (1 - \gamma)^{-1}\|u_1 - u_2\| \quad \text{for all } u_1, u_2 \in \mathcal{R}_C(\overline{x}, \gamma r, \alpha).$$

(i") For any real $\gamma \in]0, 1[$ and any $u_i \in \mathcal{R}_C(\overline{x}, \gamma r, \alpha)$, $i = 1, 2$, $P_C(u_i)$ exists and

$$\|(I - P_C)(u_1) - (I - P_C)(u_2)\|^2 + (1 - 2\gamma)\|P_C(u_1) - P_C(u_2)\|^2 \le \|u_1 - u_2\|^2.$$

In particular, for $\gamma \le 1/2$ the mapping $I - P_C$ is Lipschitz continuous on $\mathcal{R}_C(\overline{x}, \gamma r, \alpha)$ with 1 as Lipschitz constant.
(l') The mapping P_C is well defined on $\mathcal{R}_C(\overline{x}, r, \alpha)$ and for all $u_1, u_2 \in \mathcal{R}_C(\overline{x}, r, \alpha)$ one has

$$\|P_C(u_1) - P_C(u_2)\| \le \left(1 - \frac{d_C(u_1)}{2r} - \frac{d_C(u_2)}{2r}\right)^{-1} \|u_1 - u_2\|.$$

PROOF. (i') Fix any $0 < \gamma < 1$ and any $u \in \mathcal{R}_C(\overline{x}, \gamma r, \alpha)$. Then $u = x + tv$ for some $x \in C \cap B(\overline{x}, \alpha)$, $v \in N^P(C; x) \cap \mathbb{B}_H$ and $0 \le t < \gamma r$. On the one hand, according to the equivalence (a)$\Leftrightarrow$(d) in Theorem 15.70 one has $x = P_C(u)$. On the other hand, $x \in (I + \gamma r\mathbb{B}_H \cap N^P(C; \cdot))^{-1}(u) \cap B(\overline{x}, \alpha)$, so this tells us that

$$P_C(u) \in (I + \gamma r\mathbb{B}_H \cap N^P(C; \cdot))^{-1}(u) \cap B(\overline{x}, \alpha).$$

The right-hand side of the latter inclusion being a singleton set according to the implication (a)$\Rightarrow$(h) in Theorem 15.70, this inclusion is in fact an equality. By the implication (a)$\Rightarrow$(h) in Theorem 15.70 again, for $u_i \in \mathcal{R}_C(\overline{x}, \gamma r, \alpha)$ with $i = 1, 2$ we have

$$\langle P_C(u_1) - P_C(u_2), u_1 - u_2 \rangle \ge (1 - \gamma)\|P_C(u_1) - P_C(u_2)\|^2,$$

and hence

$$\|P_C(u_1) - P_C(u_2)\| \le (1 - \gamma)^{-1}\|u_1 - u_2\|.$$

The assertion (i') is justified.
(i") Fix any $0 < \gamma < 1$ and $u_i \in \mathcal{R}_C(\overline{x}, \gamma r, \alpha)$ for $i = 1, 2$. We saw in the above analysis that $P_C(u_i)$ exists and that

$$\langle P_C(u_1) - P_C(u_2), u_1 - u_2\rangle \geq (1-\gamma)\|P_C(u_1) - P_C(u_2)\|^2.$$

This yields

$$\begin{aligned}&\|(I - P_C)(u_1) - (I - P_C)(u_2)\|^2\\ &= \|u_1 - u_2\|^2 + \|P_C(u_1) - P_C(u_2)\|^2 - 2\langle u_1 - u_2, P_C(u_1) - P_C(u_2)\rangle\\ &\leq \|u_1 - u_2\|^2 + (2\gamma - 1)\|P_C(u_1) - P_C(u_2)\|^2,\end{aligned}$$

which translates the desired inequality in (i").
(l') As above, for $u_i \in \mathcal{R}_C(\overline{x}, r, \alpha)$ we have that $P_C(u_i)$ exists and belongs to $B(\overline{x}, \alpha)$, thus the implication (a)$\Leftrightarrow$(g) in Theorem 15.70 entails that with $p_i := P_C(u_i)$

$$\langle p_1 - p_2, u_1 - u_2\rangle \geq \left(1 - \frac{\|u_1 - p_1\|}{2r} - \frac{\|u_2 - p_2\|}{2r}\right)\|p_1 - p_2\|^2.$$

This implies that

$$\|P_C(u_1) - P_C(u_2)\| \leq \left(1 - \frac{d_C(u_1)}{2r} - \frac{d_C(u_2)}{2r}\right)^{-1}\|u_1 - u_2\|.$$

The proof is finished. □

Before going further, given $r, \alpha \in]0, +\infty]$ let us establish some properties of the sets $\mathcal{R}_C(\overline{x}, r, \alpha)$ and $\mathcal{R}_C(\overline{x}, \gamma r, \alpha)$ and compare them with the following similar local enlargements.

Definition 17.6. Let C be a set in a Hilbert space H, let $\overline{x} \in C$ and let $r, \alpha \in]0, +\infty]$ be two extended reals. We define the *local enlargement* $\mathcal{W}_C(\overline{x}, r, \alpha)$ *of C around $\overline{x}$ with latitude r and with α-localization of nearest points* by

$$\mathcal{W}_C(\overline{x}, r, \alpha) := \{u \in H : \operatorname{Proj}_C(u) \cap B(\overline{x}, \alpha) \neq \emptyset, d_C(u) < r\};$$

so for any positive $\gamma < 1$ we have

$$\mathcal{W}_C(\overline{x}, \gamma r, \alpha) := \{u \in H : \operatorname{Proj}_C(u) \cap B(\overline{x}, \alpha) \neq \emptyset, d_C(u) < \gamma r\}.$$

Denoting $S := \mathrm{cl}_H C$, for any $u \in S \cap \mathcal{W}_C(\overline{x}, r, \alpha)$ there is $p \in \operatorname{Proj}_C(u) \cap B(\overline{x}, \alpha)$, so $\|u - p\| = d_C(u) = 0$, hence $u = p$, which in turn gives $u \in C$. It ensues that $S \cap \mathcal{W}_C(\overline{x}, r, \alpha) = C \cap \mathcal{W}_C(\overline{x}, r, \alpha)$, so

$$(\mathrm{cl}_H C) \cap \mathcal{W}_C(\overline{x}, r, \alpha) = C \cap \mathcal{W}_C(\overline{x}, r, \alpha) = C \cap B_H(\overline{x}, \alpha), \tag{17.3}$$

since the second equality is quite direct. Further,

$$\operatorname{Proj}_{\mathrm{cl}_H C}(x) = \operatorname{Proj}_C(x) \quad \text{for all } x \in \mathcal{W}_C(\overline{x}, r, \alpha). \tag{17.4}$$

Indeed, fix $x \in \mathcal{W}_C(\overline{x}, r, \alpha)$ and take any $p \in \operatorname{Proj}_S(x)$ (if any).

Observe that $\mathcal{W}_C(\overline{x}, r, \alpha)$ (resp. $\mathcal{W}_C(\overline{x}, \gamma r, \alpha)$) contains $\overline{x}$, and is always contained in $\mathcal{R}_C(\overline{x}, r, \alpha)$ (resp. $\mathcal{R}_C(\overline{x}, \gamma r, \alpha)$). Indeed, fix $\gamma \in]0, 1]$. Take any $u \in \mathcal{W}_C(\overline{x}, \gamma r, \alpha)$, so $d_C(u) < \gamma r$ and we can choose some $x \in \operatorname{Proj}_C(u) \cap B(\overline{x}, \alpha)$. Then, when $d_C(u) \neq 0$, we can write u as

$$u = x + tv \quad \text{where } t := d_C(u) < \gamma r, \quad \text{and } v := \frac{1}{t}(u - x) \in N^P(C; x) \cap \mathbb{B}_H,$$

which yields $u \in \mathcal{R}_C(\overline{x}, \gamma r, \alpha)$. When $d_C(u) = 0$, the inclusion $x \in \operatorname{Proj}_C(u) \cap B(\overline{x}, \alpha)$ gives $u = x \in C \cap B(\overline{x}, \alpha)$, so choosing $t = 0$ and $v = 0$ the equality $u = u + tv$ leads also to $u \in \mathcal{R}_C(\overline{x}, \gamma r, \alpha)$. It results that $\mathcal{W}_C(\overline{x}, \gamma r, \alpha) \subset \mathcal{R}_C(\overline{x}, \gamma r, \alpha)$.

Figure 17.1 presents a closed set C in $\mathbb{R}^2$ with three points x_i (with $i = 1, 2, 3$) where the local enlargement $W_C(x_i, r_i, \alpha_i)$ is illustrated: point $x_1 = (-1, 1)$, with its local enlargement $W_C(x_1, 1, \sqrt{2})$; point $x_2 = (7/4, 2)$ with its local enlargement $W_C(x_2, +\infty, 1/2)$; and point $x_3 = (3, 1)$ with its local enlargement $W_C(x_3, 1/2, 1/2)$.

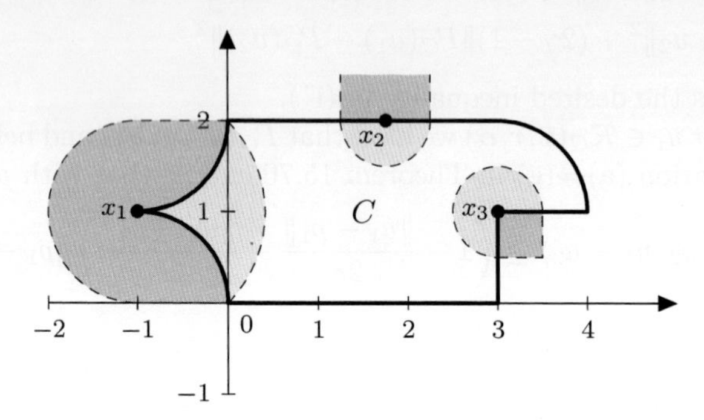

FIGURE 17.1. Local enlargements with localization of nearest points.

Whenever C is (r, α)-prox-regular at $\overline{x}$, we show in Theorem 17.7 below that the two types of local enlargement $\mathcal{R}_C(\overline{x}, r, \alpha)$ and $\mathcal{W}_C(\overline{x}, r, \alpha)$ coincide.

THEOREM 17.7 (openness of local enlargement). Let C be a subset of a Hilbert space H which is (r, α)-prox-regular at $\overline{x} \in C$ for some extended reals $r, \alpha \in]0, +\infty]$. The following properties hold.
(a) For any $0 < \gamma \le 1$ one has

(17.5)
$$\mathcal{R}_C(\overline{x}, \gamma r, \alpha) = \mathcal{W}_C(\overline{x}, \gamma r, \alpha) := \{u \in H : \operatorname{Proj}_C(u) \cap B(\overline{x}, \alpha) \neq \emptyset,\ d_C(u) < \gamma r\}$$

and also

$$\mathcal{R}_C(\overline{x}, \gamma r, \alpha) = \{u \in H : \operatorname{Proj}_C(u) \cap B(\overline{x}, \alpha) = \{P_C(u)\},\ d_C(u) < \gamma r\}. \tag{17.6}$$

(b) For any $0 < \gamma \le 1$ the set $\mathcal{R}_C(\overline{x}, \gamma r, \alpha) = \mathcal{W}_C(\overline{x}, \gamma r, \alpha)$ is an open set containing $\overline{x}$ and $\mathcal{W}_C(\overline{x}, \gamma r, \alpha) \setminus C$ is also an open set.
(c) The mapping P_C is well defined on the open set $\mathcal{R}_C(\overline{x}, r, \alpha)$ and locally Lipschitz continuous on this open set.

PROOF. Denote $S := \operatorname{cl}_H(C)$.
(a) The inclusion $\supset$ has been already justified above. To prove the converse inclusion, take any $u \in \mathcal{R}_C(\overline{x}, \gamma r, \alpha)$. Necessarily $\operatorname{Proj}_C(u) \neq \emptyset$, and $d_C(u) < \gamma r$. Indeed $u = x + tv$ for some $x \in C \cap B(\overline{x}, \alpha)$, $0 < t < \gamma r$, and $v \in N^P(C, x) \cap \mathbb{B}_H$. Then, the (r, α)-prox-regularity of C at $\overline{x}$ entails that $x = P_C(u)$ according to the equivalence (a)$\Leftrightarrow$(d) in Theorem 15.70, hence in particular $\operatorname{Proj}_C(u) \cap B(\overline{x}, \alpha) \neq \emptyset$. Since we also have $d_C(u) = \|u - x\| = \|tv\| < \gamma r$, it ensues that $u \in \mathcal{W}_C(\overline{x}, \gamma r, \alpha)$. The inclusion $\mathcal{R}_C(\overline{x}, \gamma r, \alpha) \subset \mathcal{W}_C(\overline{x}, \gamma r, \alpha)$ then holds true. Therefore, (17.5) is then justified and (17.6) is a direct consequence of this and of the equivalence (a)$\Leftrightarrow$(d) in Theorem 15.70.

(b) We first notice that the set $\mathcal{R}_C(\overline{x}, r, \alpha)$ obviously contains $\overline{x}$. To show its openness, fix any $\overline{u} \in \mathcal{R}_C(\overline{x}, \gamma r, \alpha)$. Then, there exists $\overline{y} \in B(\overline{x}, \alpha) \cap C$ and $\overline{v} \in N^P(C, \overline{y}) \cap \mathbb{B}_H$ such that

$$\overline{u} = \overline{y} + t\overline{v} \quad \text{where } t \in [0, \gamma r[. \tag{17.7}$$

Since $\overline{y} \in B(\overline{x}, \alpha) \cap C$, there exists $\eta > 0$ such that $B[\overline{y}, \eta] \subset B(\overline{x}, \alpha)$. Observe also that $\|\overline{u} - \overline{y}\| < \gamma r$ according to (17.7). By continuity of the norm we may choose $\eta' > 0$ such that

$$\|u - y\| < \gamma r \quad \text{for all } u \in B(\overline{u}, \eta') \text{ and all } y \in B(\overline{y}, \eta'). \tag{17.8}$$

Put $\eta'' := \min\{\eta, \eta'\}$. Since $(d_C(u) + \mu)^2 - d_C^2(\overline{u}) \to 0$ as $u \to \overline{u}$ and $\mu \downarrow 0$, consider a positive real $\varepsilon < \eta''$ such that for all $u \in B(\overline{u}, \varepsilon)$

$$\left(1 - \frac{t}{r}\right)^{-1} \left((d_C(u) + \varepsilon)^2 - d_C^2(\overline{u})\right) < (\eta'')^2. \tag{17.9}$$

Fix any point $u \in B(\overline{u}, \varepsilon)$. Choose by Lau theorem (see Theorem 14.29) a sequence $(u_n)_n$ in $\operatorname{Dom} \operatorname{Proj}_S(\cdot)$ converging to u and for each integer $n \in \mathbb{N}$ take some $y_n \in \operatorname{Proj}_S(u_n)$. Observing (by the equality $d_C = d_S$) that

$$d_C(u) \le \|u - y_n\| \le \|u - u_n\| + \|u_n - y_n\| = \|u - u_n\| + d_C(u_n),$$

we see that $\lim\limits_{n\to\infty} \|u - y_n\| = d_C(u)$. Moreover, for each $n \in \mathbb{N}$ the equality

$$\|y_n - \overline{y}\|^2 = \|y_n - \overline{u}\|^2 - \|\overline{u} - \overline{y}\|^2 + 2\langle \overline{u} - \overline{y}, y_n - \overline{y}\rangle$$

and the inclusion $\overline{u} - \overline{y} \in N^P(C; \overline{y})$ ensure according to Theorem 15.70(c') that

$$\|y_n - \overline{y}\|^2 \le \|y_n - \overline{u}\|^2 - \|\overline{u} - \overline{y}\|^2 + \frac{t}{r}\|y_n - \overline{y}\|^2$$

since C is (r, α)-prox-regular at $\overline{x}$ and since $\|\overline{u} - \overline{y}\| \le t$ by (17.7). It follows that

$$\begin{aligned}
(1 - \frac{t}{r})\|y_n - \overline{y}\|^2 &\le \|y_n - \overline{u}\|^2 - \|\overline{u} - \overline{y}\|^2 \\
&\le (\|y_n - u\| + \|u - \overline{u}\|)^2 - \|\overline{u} - \overline{y}\|^2 \\
&\le (\|y_n - u\| + \|u - \overline{u}\|)^2 - d_C^2(\overline{u}).
\end{aligned}$$

Since $(\|y_n - u\| + \|u - \overline{u}\|)^2 - d_C^2(\overline{u}) \to (d_C(u) + \|u - \overline{u}\|)^2 - d_C^2(\overline{u})$ as $n \to \infty$ and since $\|u - \overline{u}\| < \varepsilon$, we obtain on the one hand by (17.9) that for some integer n_1

$$\|y_n - \overline{y}\| < \eta'' \text{ for all } n \ge n_1.$$

It follows that for every integer $n \ge n_1$

$$y_n \in B(\overline{y}, \eta'') \cap S \subset B(\overline{x}, \alpha) \cap S = B(\overline{x}, \alpha) \cap C,$$

where the latter equality is due to the fact that $B(\overline{x}, \alpha) \cap C$ is closed relative to $B(\overline{x}, \alpha)$ by (r, α)-prox-regularity of C at $\overline{x} \in C$. On the other hand, since $(u_n)_n$ converges to u and $u \in B(\overline{u}, \varepsilon)$, we may suppose that

$$u_n \in B(\overline{u}, \varepsilon) \text{ for all } n \ge n_1.$$

According to (17.8) we also have for all $n \ge n_1$

$$\|u_n - y_n\| = d_C(u_n) \le \|u_n - \overline{y}\| < \gamma r, \tag{17.10}$$

which combined with the inclusion $y_n \in B(\overline{x},\alpha)\cap C$ and the equality $u_n = y_n + \|u_n - y_n\|\frac{(u_n-y_n)}{\|u_n-y_n\|}$ gives

$$y_n \in \left(I + \gamma r\mathbb{B}_H \cap N^P(C;\cdot)\right)^{-1}(u_n)\cap B(\overline{x},\alpha).$$

Applying the implication (a)$\Rightarrow$(h) in Theorem 15.70 we obtain

$$\|y_n - y_m\| \le (1-\gamma)^{-1}\|u_n - u_m\| \text{ for all } n,m \ge n_1,$$

and this implies that $(y_n)_n$ is a Cauchy sequence because $u_n \to u$. Denoting by y the limit of the sequence $(y_n)_n$, we get that

$$y \in B[\overline{y},\eta'']\cap S\subset B(\overline{x},\alpha)\cap S = B(\overline{x},\alpha)\cap C \text{ and } \|u-y\| = \lim_{n\to\infty}\|u-y_n\| = d_C(u),$$

and by (17.10) and (17.8) we have $\|u-y\| \le \|u-\overline{y}\| < \gamma r$. When $d_C(u) > 0$, putting $t := \|u-y\|$ and $v := t^{-1}(u-y)$, we can write

$$u = y + tv \text{ with } y \in C\cap B(\overline{x},\alpha),\ 0 < t < \gamma r \text{ and } v \in N^P(C;y)\cap \mathbb{B}_H,$$

thus $u \in \mathcal{R}_C(\overline{x},\gamma r,\alpha)$. The case $d_C(u) = 0$ gives

$$u = y = y + t'v' \text{ where } t' = 0 \text{ and } v' = 0.$$

In any case, we then have $u \in \mathcal{R}_C(\overline{x},\gamma r,\alpha)$, thus $B(\overline{u},\varepsilon) \subset \mathcal{R}_C(\overline{x},\gamma r,\alpha)$. This concludes the openness of $\mathcal{R}_C(\overline{x},\gamma r,\alpha)$.

Regarding $\mathcal{W}_C(\overline{x},\gamma r,\alpha)\setminus C$, denote $\mathfrak{W} := \mathcal{W}_C(\overline{x},\gamma r,\alpha)$ and keep in mind the notation $S := \mathrm{cl}_H C$. The equality $S\cap\mathfrak{W} = C\cap\mathfrak{W}$ (see (17.3)) assures us that

$$\mathfrak{W}\setminus C = \mathfrak{W}\setminus \mathfrak{W}\cap C = \mathfrak{W}\setminus\mathfrak{W}\cap S = \mathfrak{W}\setminus S,$$

so $\mathfrak{W}\setminus C$ is open by the openness of $\mathfrak{W}$.
(c) Finally, the assertion (c) follows from (i') in Theorem 17.5. □

Assume that C is (r,α)-prox regular at $\overline{x}\in C$. Consider $x \in \mathcal{W}_C(\overline{x},\gamma r,\alpha)$ with $\gamma \in]0,1[$ and consider a real $t \ge 0$ with $(1+t)\gamma \le 1$. Then, $P_C(x)$ exists by Theorem 17.7(c) and $P_C(x) \in B(\overline{x},\alpha)$ by definition of $\mathcal{W}_C(\overline{x},\gamma r,\alpha)$. Further, setting $v := x - P_C(x)$ we have $(1+t)x - tP_C(x) = P_C(x) + (1+t)v$ with $\|(1+t)v\| = (1+t)d_C(x) < (1+t)\gamma r$, hence $\|(1+t)v\| < r$. Then, Definition 15.65 of (r,α)-prox-regularity gives $P_C\big((1+t)x - tP_C(x)\big) = P_C(x)$. We state this property as a lemma.

Lemma 17.8. Let C be a set of a Hilbert space H which is (r,α)-prox-regular at $\overline{x}\in C$. Then for any $\gamma\in]0,1[$ and any $x \in \mathcal{W}_C(\overline{x},\gamma r,\alpha)$ one has

$$P_C\big((1+t)x - tP_C(x)\big) = P_C(x) \quad \text{for every real } t > 0 \text{ with } (1+t)\gamma \le 1.$$

As in the case of uniform prox-regularity, we considered in Definition 15.65 of (r,α)-prox-regularity and in its characterizations in Theorem 15.70 only points inside the set C. The additional assertions in the next proposition will give other characterizations with points outside C.

Proposition 17.9. Let C be a subset of a Hilbert space H and let $\overline{x}\in C$. Let $\alpha \in]0,+\infty]$ be such that $C\cap B(\overline{x},\alpha)$ is closed relative to $B(\overline{x},\alpha)$ and let $r \in]0,+\infty]$. The following other properties can be added to the list of equivalences $(a)-(d)$ in Theorem 15.70.
(α) For any $u \in \mathrm{Dom}\,\partial_P d_C \cap \mathcal{R}_C(\overline{x},r,\alpha)$ and any $v \in \partial_P d_C(u)$ one has

$$d_C(u) + \langle v, x-u\rangle \le \frac{1}{2r}\|x - P_C(u)\|^2 \quad \text{for all } x \in C.$$

(β) For any $u \in \operatorname{Dom} \partial_P d_C \cap \mathcal{R}_C(\overline{x}, r, \alpha)$ and any $v \in \partial_P d_C(u)$ one has

$$\langle v, x - u\rangle \leq \frac{1}{2r}\|x - P_C(u)\|^2 \quad \text{for all } x \in C.$$

(γ) For any $x \in C \cap B(\overline{x}, \alpha)$ and any $v \in N^P(C; x) \cap \mathbb{B}_H$ one has with $S := \operatorname{cl}_H(C)$

$$\langle v, u - x\rangle \leq d_C(u) + \frac{1}{2r}\|y - x\|^2 \quad \text{for all } u \in \operatorname{Dom}\operatorname{Proj}_S \text{ and } y \in \operatorname{Proj}_S(u).$$

(δ) For any $x \in C \cap B(\overline{x}, \alpha)$ and any $v \in N^P(C; x) \cap \mathbb{B}_H$ one has

$$\langle v, u - x\rangle \leq d_C(u) + \frac{1}{2r}(d_C(u) + \|u - x\|)^2 \quad \text{for all } u \in H.$$

PROOF. Observe first that (α) $\Rightarrow$ (β). Next we show that (β) implies that C is (r, α)-prox regular at $\overline{x}$. Indeed, take any $u \in C \cap B(\overline{x}, \alpha)$ and any $v \in N^P(C; u) \cap \mathbb{B}_H$. We obviously have $u \in \mathcal{R}_C(\overline{x}, r, \alpha)$ and $P_C(u) = u$. Further, we know by Proposition 4.153 that $\partial_P d_C(u) = N^P(C; u) \cap \mathbb{B}_H$, thus $v \in \partial_P d_C(u)$. According to the assumption of property (β), we have

$$\langle v, x - u\rangle \leq \frac{1}{2r}\|x - u\|^2 \quad \text{for all } x \in C,$$

which by (b) in Theorem 15.70 ensures that C is (r, α)-prox-regular ar $\overline{x}$.

Let us show that the (r, α)-prox-regularity of C at $\overline{x}$ implies the property (α). Fix any $u \in \operatorname{Dom} \partial_P d_C \cap \mathcal{R}_C(\overline{x}, r, \alpha)$ and any $v \in \partial_P d_C(u)$. According to (17.6) in Theorem 17.7 we have that $P_C(u)$ exists and $P_C(u) \in B(\overline{x}, \alpha)$. If we suppose $u \notin C$, then $u \neq P_C(u)$ and $d_S(u) = d_C(u) = \|u - P_C(u)\| > 0$, so by (b) in Lemma 15.11 we have that $P_S(u)$ exists, so $P_S(u) = P_C(u)$. By Lemma 15.11 again

$$v = \frac{u - P_S(u)}{\|u - P_S(u)\|} = \frac{u - P_C(u)}{\|u - P_C(u)\|},$$

the equa which entails for every $x \in H$

$$d_C(u) + \langle v, x - u\rangle = \langle v, u - P_C(u)\rangle + \langle v, x - u\rangle = \langle v, x - P_C(u)\rangle.$$

We note that the same equality $v = \frac{u - P_C(u)}{\|u - P_C(u)\|}$ ensures that $v \in N^P(C; P_C(u)) \cap \mathbb{B}_H$, so the (r, α)-prox-regularity of C at $\overline{x}$ gives by (b) in Theorem 15.70 that

$$d_C(u) + \langle v, x - u\rangle \leq \frac{1}{2r}\|x - P_C(u)\|^2 \quad \text{for all } x \in C,$$

and the inequality still holds true whenever $u \in C$ according again to (b) in Theorem 15.70. This translates the property (α).

Now we prove that (γ) $\Rightarrow$ (δ). Assume (γ) and fix any $x \in C \cap B(\overline{x}, \alpha)$ and any $v \in N^P(C; x) \cap \mathbb{B}$. For every $u \in \operatorname{Dom}\operatorname{Proj}_S$ and every $y \in \operatorname{Proj}_S(u)$, writing

$$\|y - x\| \leq \|y - u\| + \|u - x\| = d_S(u) + \|x - u\| = d_C(u) + \|x - u\|,$$

we obtain by (γ) that

$$\langle v, u - x\rangle \leq d_C(u) + \frac{1}{2r}(d_C(u) + \|u - x\|)^2,$$

and this inequality still holds for any $u \in H$ by continuity of $d_C(\cdot)$ and density of $\operatorname{Dom}\operatorname{Proj}_S$ in H, according to Lau theorem (see Theorem 14.29). This justifies the implication (γ) $\Rightarrow$ (δ).

We note that (δ) implies (b) in Theorem 15.70 by taking $u \in C$ in the inequality in (δ), so (δ) implies that C is (r, α)-prox-regular at $\overline{x}$.

It remains to show that the (r,α)-prox-regularity of C at $\overline{x}$ implies (γ). Assume C is (r,α)-prox-regular at $\overline{x}$ and fix any $x \in C \cap B(\overline{x},\alpha)$ and any $v \in N^P(C;x) \cap \mathbb{B}_H$. For every $u \in \operatorname{Dom}\operatorname{Proj}_S$ and every $y \in \operatorname{Proj}_S(u)$ we have, by (b') in Theorem 15.70 and by the inequality $\|v\| \le 1$ and the equality $d_S(u) = d_C(u)$, that

$$\langle v, u - x\rangle = \langle v, u - y\rangle + \langle v, y - x\rangle \le d_C(u) + \frac{1}{2r}\|y - x\|^2,$$

which ensures (γ) and finishes the proof. $\square$

The following additional property of (r,α)-prox-regular sets is also of interest.

PROPOSITION 17.10. Let C be a subset of a Hilbert space H which is (r,α)-prox-regular at $\overline{x} \in C$ for some $r, \alpha \in]0,+\infty]$. Then for all $x \in C \cap B(\overline{x},\alpha)$ and all $v \in \partial_P d_C(x) = N^P(C;x) \cap \mathbb{B}$

$$\langle v, u - x\rangle \le \frac{2}{r}\|u - x\|^2 + d_C(u) \quad \text{for all}\, u \in \mathcal{R}_C(\overline{x}, r, \alpha).$$

PROOF. Fix any $x \in C \cap B(\overline{x},\alpha)$ and any $v \in \partial_P d_C(x) = N^P(C;x) \cap \mathbb{B}$, where we recall that the latter equality is due to Proposition 4.153. Fix also any $u \in \mathcal{R}_C(\overline{x}, r, \alpha)$. By Theorem 17.7 we have that $P_C(u)$ exists, so utilizing (γ) in Proposition 17.9 we have

$$\langle v, u - x\rangle \le d_C(u) + \frac{1}{2r}\|P_C(u) - x\|^2.$$

Noticing that $\|P_C(u) - x\| \le \|P_C(u) - u\| + \|u - x\| \le 2\|u - x\|$, the desired inequality of the proposition follows. $\square$

When C is r-prox-regular, we know by Theorem 17.7(a) that for any $\overline{x} \in C$

$$U_r(C) = \mathcal{W}_C(\overline{x}, r, +\infty) = \mathcal{R}_C(\overline{x}, r, +\infty).$$

So, Proposition 17.10 immediately yields the following corollary.

COROLLARY 17.11. Let C be an r-prox-regular set of a Hilbert space H with $r \in]0,+\infty]$. Then for all $x \in C$ and all $v \in \partial_P d_C(x) = N^P(C;x) \cap \mathbb{B}_H$ one has

$$\langle v, u - x\rangle \le \frac{2}{r}\|u - x\|^2 + d_C(u) \quad \text{for all}\, u \in U_r(C).$$

The next theorem provides several other characterizations of the (r,α)-prox-regularity of C. Some of them are in terms of some kinds of differentiability of its square distance function. Recall that

$$\mathcal{W}_C(\overline{x}, r, \alpha) = \{u \in H : \operatorname{Proj}_C(u) \cap B(\overline{x},\alpha) \ne \emptyset \text{ and } d_C(u) < r\}.$$

THEOREM 17.12 (metric characterizations of (r,α)-prox-regularity). Let C be a set in a Hilbert space H, and let $\overline{x} \in C$ and $\alpha \in]0,+\infty]$ be such that $C \cap B(\overline{x},\alpha)$ is closed relative to $B(\overline{x},\alpha)$. Let also $r \in]0,+\infty]$ and $S := \operatorname{cl}_H(C)$. The set C is (r,α)-prox-regular at $\overline{x}$ if and only if anyone of the following assertions holds:
(h) for any positive $\gamma < 1$ the set $\mathcal{W}(\overline{x}, \gamma r, \alpha)$ is open and the mapping P_C is well defined on $\mathcal{W}(\overline{x}, \gamma r, \alpha)$ and $(1-\gamma)^{-1}$-Lipschitz continuous there, that is,

$$\|P_C(u_1) - P_C(u_2)\| \le (1-\gamma)^{-1}\|u_1 - u_2\| \quad \text{for all}\, u_1, u_2 \in \mathcal{W}_C(\overline{x}, \gamma r, \alpha);$$

(i) the set $\mathcal{W}_C(\overline{x}, r, \alpha)$ is open and the mapping P_C is well defined on $\mathcal{W}_C(\overline{x}, r, \alpha)$ with for all $u_1, u_2 \in \mathcal{W}_C(\overline{x}, r, \alpha)$

$$\|P_C(u_1) - P_C(u_2)\| \le \left(1 - \frac{d_C(u_1)}{2r} - \frac{d_C(u_2)}{2r}\right)^{-1}\|u_1 - u_2\|;$$

(j) the set $\mathcal{W}_C(\overline{x},r,\alpha)$ is open and the mapping P_C is well defined on $\mathcal{W}_C(\overline{x},r,\alpha)$ and locally Lipschitz continuous on $\mathcal{W}_C(\overline{x},r,\alpha)$;
(k) the set $\mathcal{W}_C(\overline{x},r,\alpha)$ is open and the mapping P_C is well defined on $\mathcal{W}_C(\overline{x},r,\alpha)$ and norm-to-norm continuous on $\mathcal{W}_C(\overline{x},r,\alpha)$;
(l) the set $\mathcal{W}_C(\overline{x},r,\alpha)$ is open and the mapping P_C is well defined on $\mathcal{W}_C(\overline{x},r,\alpha)$ and norm-to-weak continuous on $\mathcal{W}_C(\overline{x},r,\alpha)$;
(m) the set $\mathcal{W}_C(\overline{x},r,\alpha)$ is open and d_C^2 is of class $\mathcal{C}^{1,1}$ on $\mathcal{W}_C(\overline{x},r,\alpha)$ with

$$\nabla\left(\frac{1}{2}d_C^2\right)(u) = (u - P_C(u)) \subseteq \quad \text{for all } u \in \mathcal{W}_C(\overline{x},r,\alpha);$$

(n) the set $\mathcal{W}_C(\overline{x},r,\alpha)$ is open and d_C^2 is of class $\mathcal{C}^1$ on $\mathcal{W}_C(\overline{x},r,\alpha)$;
(o) the set $\mathcal{W}_C(\overline{x},r,\alpha)$ is open and d_C^2 is Fréchet differentiable on $\mathcal{W}_C(\overline{x},r,\alpha)$;
(p) the set $\mathcal{W}_C(\overline{x},r,\alpha)$ is open and d_C^2 is Gâteaux differentiable on $\mathcal{W}_C(\overline{x},r,\alpha)$ and $\operatorname{Proj}_S(u) \neq \emptyset$ for all $u \in \mathcal{W}_C(\overline{x},r,\alpha)$;
(q) the set $\mathcal{W}_C(\overline{x},r,\alpha)$ is open and d_C is Gâteaux differentiable on $\mathcal{W}_C(\overline{x},r,\alpha)\setminus S = \mathcal{W}_C(\overline{x},r,\alpha)\setminus C$ with $\nabla d_C(u) = (u - P_C(u))/d_C(u)$ for all $u \in \mathcal{W}_C(\overline{x},r,\alpha)\setminus S$;
(r) the set $\mathcal{W}_C(\overline{x},r,\alpha)$ is open and for any pair (u,x) with $u \in \mathcal{W}_C(\overline{x},r,\alpha)\setminus C$ and $x = P_C(u)$ one has

$$x = P_C\left(x + t\frac{u-x}{\|u-x\|}\right) \quad \text{for all}\, t \in [0,r[; \tag{17.11}$$

(s) the set $\mathcal{W}_C(\overline{x},r,\alpha)$ is open and for any pair (u,x) with $u \in \mathcal{W}_C(\overline{x},r,\alpha)\setminus S$ and $x = P_C(u)$ one has

$$x = P_C\left(x + t\frac{u-x}{\|u-x\|}\right) \quad \text{for all}\, t \in [0,r[. \tag{17.12}$$

PROOF. The sequence of equivalences will be proved as follows:

$$\begin{array}{ccccccccccc}
(q) & \Leftrightarrow & (p) & \Leftrightarrow & (o) & & \Longleftrightarrow & (n) & \Leftrightarrow & (m) & \Leftrightarrow & (l) \\
 & & & & \Downarrow & & & & & & & \Updownarrow \\
 & & (r) & \Leftrightarrow & (s) & & & & (i) & \Longrightarrow & & (k) \\
 & & & & \Downarrow & & & \nearrow & & & & \Uparrow \\
 & & & & (r,\alpha) & \text{prox-regular at } \overline{x} & \Longrightarrow & & & (h) & \Rightarrow & (j)
\end{array}$$

Several arguments below are in the line of those in the proof of Theorem 15.22.

Theorem 17.5 and Theorem 17.7 ensure that the (r,α)-prox-regularity of C entails both (h) and (i). Assertion (h) obviously implies (j) and (j) implies (k).

To see that (i) implies (k), assume (i) and fix any $\overline{u} \in \mathcal{W}_C(\overline{x},r,\alpha)$. Observe that the definition of $\mathcal{W}_C(\overline{x},r,\alpha)$ ensures that for every $u \in \mathcal{W}_C(\overline{x},r,\alpha)$ we have

$$1 - \frac{d_C(u)}{2r} - \frac{d_C(\overline{u})}{2r} > 1 - \frac{1}{2} - \frac{d_C(\overline{u})}{2r},$$

so the inequality in (i) yields

$$\|P_C(u) - P_C(\overline{u})\| \le 2\left(1 - \frac{d_C(\overline{u})}{r}\right)^{-1}\|u - \overline{u}\|.$$

This entails the continuity of P_C at $\overline{u}$ according to the openness assumption of $\mathcal{W}_C(\overline{x},r,\alpha)$, hence (i) implies (k).

Further, from Lemma 15.11 and Proposition 15.19 we get that the assertions from (k) to (q) are pairwise equivalent; notice that the equality $\mathcal{W}_C(\overline{x},r,\alpha)\setminus S =$

$\mathcal{W}_C(\overline{x}, r, \alpha) \setminus C$ in (q) follows from the equality $\mathcal{W}_C(\overline{x}, r, \alpha) \cap S = \mathcal{W}_C(\overline{x}, r, \alpha) \cap C$ in (17.3) as already seen in the final part of the proof of (b) in Theorem 17.7.

The equivalence (r)$\Leftrightarrow$(s) is direct since $\mathcal{W}_C(\overline{x}, r, \alpha) \setminus S = \mathcal{W}_C(\overline{x}, r, \alpha) \setminus C$.

Now we show that (o)$\Rightarrow$s). Assume (o). By Proposition 15.19 P_S is well defined on the open set $\mathcal{W}_C(\overline{x}, r, \alpha)$ and continuous there. Take any pair (u, x) such that $u \in \mathcal{W}_C(\overline{x}, r, \alpha) \setminus S$ and $x = P_C(u)$. In particular, $u \in \mathcal{W}_C(\overline{x}, r, \alpha) \setminus S$ and $x = P_S(u)$ (since $P_S(u)$ exists by the fact that P_S is well defined on $\mathcal{W}_C(\overline{x}, r, \alpha)$). Note also that $x = P_C(u) \in B(\overline{x}, \alpha)$ since $\operatorname{Proj}_C(u) \cap B(\overline{x}, \alpha) \neq \emptyset$ by definition of $\mathcal{W}_C(\overline{x}, r, \alpha)$. Consider any $t \in]d_C(u), r[$. According to the Fréchet differentiability of $d_S^2 = d_C^2$ on the open set $\mathcal{W}_C(\overline{x}, r, \alpha)$, by Lemma 15.20 there exists some $\tau_0 > 0$ such that $u_\tau \in \mathcal{W}_C(\overline{x}, r, \alpha)$ and $P_S(u_\tau) = x$ for all nonnegative real $\tau \leq \tau_0$, where $u_\tau = u + \tau(u - x)/\|u - x\|$. The set of all $\tau \in]0, t - d_C(u)]$ such that $u_\tau \in \mathcal{W}_C(\overline{x}, r, \alpha)$ and $P_S(u_\tau) = x$ is then nonempty. Let s be its supremum and $(\tau_n)_n$ be a sequence of that set converging to this real $s \in]0, t - d_C(u)]$. For each n we have $d_C(u_{\tau_n}) = \tau_n + d_C(u)$ and $\|u_{\tau_n} - x\| = d_C(u_{\tau_n})$. Combining this with the fact that $u_{\tau_n} \to u_s$ as $n \to \infty$, we obtain $\|u_s - x\| = d_C(u_s)$ and

$$d_C(u_s) = s + d_C(u) \leq t < r \quad \text{and} \quad x \in \operatorname{Proj}_C(u_s) \cap B(\overline{x}, \alpha),$$

so $u_s \in \mathcal{W}_C(\overline{x}, r, \alpha)$. Further, we have $u_s \notin S$ since $d_C(u_s) = s + d_C(u) > 0$, and we have also $x = P_S(u_s)$ since P_S is well defined on $\mathcal{W}_C(\overline{x}, r, \alpha)$. Therefore, Lemma 15.20 ensures that $s = t - d_C(u)$, and for this value of s we have

$$u_s = x + (s + d_C(u))(u - x)/\|u - x\| = x + t(u - x)/\|u - x\|. \tag{17.13}$$

Note also that the equality $x = P_S(u_s)$ implies $x = P_S(u_s) = P_C(u_s)$ since $x \in C \subset S$. It ensues by (17.13) that $x = P_C(x + t(u - x)/\|u - x\|)$ for all $t \in]d_C(u), r[$, and this also holds (by Lemma 4.123) for $t \in [0, d_C(u)]$ because $x = P_C(u)$. So, the implication (o)$\Rightarrow$(s) holds.

Finally, let us prove that (s) implies the (r, α)-prox-regularity of C. Assume that (s) holds and fix any $t \in [0, r[$ and any pair (u, x) satisfying $x \in B(\overline{x}, \alpha) \cap C$ and $v \in N^P(C; x) \cap \mathbb{B}_H$ with $v \neq 0$. Put $w := v/\|v\|$. Since $w \in N^P(C; x)$, by definition of proximal normal there exists a real $s > 0$ such that $x = P_C(x + sw)$. Since $\mathcal{W}_C(\overline{x}, r, \alpha)$ is open and contains the point x (because $x \in C \cap B(\overline{x}, \alpha)$), we may take $s > 0$ small enough that $x + sw \in \mathcal{W}_C(\overline{x}, r, \alpha)$, and hence $u := x + sw \in \mathcal{W}_C(\overline{x}, r, \alpha) \setminus S$ since $d_C(x + sw) = s > 0$. By (s) we have $x = P_C(x + \theta \frac{u - x}{\|u - x\|})$ for all $\theta \in [0, r[$. Note that $P_C(x + \theta \frac{u-x}{\|u-x\|}) = P_C(x + \theta w)$. Setting $t' := t\|v\|$, we have $t' \in [0, r[$ since $\|v\| \leq 1$, thus recalling that $w = v/\|v\|$ it ensues that

$$x = P_C(x + t'w) = P_C(x + tv).$$

The equality $x = P_C(x + tv)$ being still true for $v = 0$, it follows that $x = P_C(x + tv)$ for all $x \in C \cap B(\overline{x}, \alpha)$, all $v \in N^P(C; x) \cap \mathbb{B}_H$ and all $t \in [0, r[$. Consequently, the set C is (r, α)-prox-regular at $\overline{x}$ by Definition 15.65 of (r, α)-prox-regularity of sets. The proof is complete. $\square$

The next theorem extends to the (r, α)-prox-regularity context, with a similar proof, the geometric characterization established for uniform r-prox-regular sets in Theorem 15.51. The proof of the theorem will use the following lemma whose proof is the adaptation of arguments in Lemma 15.50.

LEMMA 17.13. *Let C be a set in a Hilbert space H which is (r, α)-prox-regular at $\overline{x} \in C$ for $r, \alpha \in]0, +\infty]$. Then for any positive real number $t \leq r$ and any*

$u \in \mathcal{R}_C(\overline{x}, t, \alpha) \setminus C$ one has

$$d(u, C) + d(u, \mathrm{Exte}_t(C)) = d(u, C) + d(u, D_t(C)) = t,$$

where we recall that $D_t(C) = \{x \in H : d_C(x) = t\}$.

PROOF. Note first by Theorem 17.7 that $\mathcal{R}_C(\overline{x}, r, \alpha) = W_C(\overline{x}, r, \alpha)$. Fix any positive real number $t \leq r$ and any $u \in \mathcal{R}_C(\overline{x}, t, \alpha) \setminus C$, and note that $x := P_C(u)$ exists according to (j) of Theorem 17.12. Note also that $x \neq u$ since $x \in C$ and $u \notin C$. Then, for $u' := x + t(u - x)/\|u - x\|$ one has $x \in \mathrm{Proj}_C(u')$ by (r) of Theorem 17.12, and hence $d_C(u') = t$. Consequently, we have $u' \in D_t(C)$, hence we obtain

$$\begin{aligned} d(u, C) + d(u, \mathrm{Exte}_t(C)) &\leq d(u, C) + d(u, D_t(C)) \\ &\leq \|u - x\| + \|u - u'\| \\ &= \|u - x\| + \left|1 - \frac{t}{\|u - x\|}\right| \|u - x\| = t, \end{aligned}$$

where the latter equality is due to the fact that $\|u - x\| = d_C(u) \leq t$. Thus, we have shown that

$$d(u, C) + d(u, \mathrm{Exte}_t(C)) \leq d(u, C) + d(u, D_t(C)) \leq t.$$

On the other hand, for any $y \in \mathrm{Exte}_t(C)$ one has

$$\|y - u\| \geq \|y - P_C(u)\| - \|u - P_C(u)\| \geq d_C(y) - d_C(u) \geq t - d_C(u),$$

which gives $d(u, \mathrm{Exte}_t(C)) \geq t - d_C(u)$. The equalities of the lemma are then justified. □

THEOREM 17.14 ((r, α)-prox-regularity via r-exterior set). Let C be a set in a Hilbert space H and let $\overline{x} \in C$. Let $r \in]0, +\infty[$ and $\alpha \in]0, +\infty]$ be such that $C \cap B(\overline{x}, \alpha)$ is closed relative to $B(\overline{x}, \alpha)$. The set C is (r, α)-prox-regular at $\overline{x}$ if and only if $\mathcal{R}_C(\overline{x}, r, \alpha)$ is open, coincides with $W_C(\overline{x}, r, \alpha)$, and for all $u \in \mathcal{R}_C(\overline{x}, r, \alpha) \setminus C$ one has

$$d(u, C) + d(u, \mathrm{Exte}_r(C)) = r.$$

PROOF. If C is (r, α)-prox-regular at $\overline{x}$, Lemma 17.13 ensures that the equality in the statement holds, and Theorem 17.7 says that $\mathcal{R}_C(\overline{x}, r, \alpha)$ is open and coincides with $W_C(\overline{x}, r, \alpha)$. To prove the converse implication, denote $S := \mathrm{cl}_H(C)$. Assume that the equality in the statement of the theorem as well as the other above properties hold. Consider any $u \in W_C(\overline{x}, r, \alpha)$ for which $x = P_C(u)$ exists, and note that $x \in \mathrm{Proj}_S(u)$ and $x \in C \cap B(\overline{x}, \alpha)$ thanks to the property $\mathrm{Proj}_C(u) \cap B(\overline{x}, \alpha) \neq \emptyset$ by Definition 17.6 of $W_C(\overline{x}, r, \alpha)$. Then for any $y \in]x, u[$ Lemma 4.123 gives $x = P_C(y) = P_S(y)$, which in particular entails $y \notin S$. This yields for any $\tau \in]-1, \|u - y\|/\|y - x\|[$

$$\tau^{-1}\big(d_S(y + \tau(y - x)) - d_S(y)\big) = \tau^{-1}\big(\|y + \tau(y - x) - x\| - \|y - x\|\big) = \|y - x\|,$$

which guarantees the existence of the bilateral limit

$$\lim_{\tau \to 0} \tau^{-1}\big(d_S(y + \tau(y - x)) - d_S(y)\big) \quad \text{with} \quad x = P_S(y).$$

Corollary 14.23 of the Fitzpatrick theorem entails that the function $d_S = d_C$ is Fréchet differentiable at y. Note also that y can be written as

$$y = x + \|y - x\| \frac{y - x}{\|y - x\|}$$

with $x \in C \cap B(\overline{x}, \alpha)$, $\frac{y-x}{\|y-x\|} \in N^P(C;x) \cap \mathbb{B}$, and $\|y-x\| < \|u-x\| = d_C(u) < r$, so $y \in \mathcal{R}_C(\overline{x}, r, \alpha)$, hence $y \in \mathcal{R}_C(\overline{x}, r, \alpha) \setminus S$. Since $d(\cdot, \mathrm{Exte}_r(C)) = r - d(\cdot, C)$ on the open set $\mathcal{R}_C(\overline{x}, r, \alpha) \setminus S$ by assumption, the function $d(\cdot, \mathrm{Exte}_r(C))$ is also Fréchet differentiable at y. From Lemma 15.11 we derive that $P_{\mathrm{Exte}_r(C)}(y) := u'$ exists, and $y \notin \mathrm{Exte}_r(C)$ since $d_C(y) = \|y-x\| < r$ as seen above. Therefore, we have $d_C(u') = r$, then using the inclusion $y \in \mathcal{R}_C(\overline{x}, r, \alpha) \setminus S$ and the equality assumption on $\mathcal{R}_C(\overline{x}, r, \alpha) \setminus S$ we obtain

$$\begin{aligned} r = d_C(u') \le \|x - u'\| &\le \|x - y\| + \|y - u'\| \\ &= d(y, C) + d(y, \mathrm{Exte}_r(C)) = r, \end{aligned}$$

so $y \in]x, u'[$ and $x \in \mathrm{Proj}_C(u')$ since $\|x - u'\| = r = d_C(u')$. The inclusions $y \in]x, u'[$ and $y \in]x, u[$ combined with the equality $\|x - u'\| = r$ entail that $u' = x + r\frac{u-x}{\|u-x\|}$. Using this and the inclusion $x \in \mathrm{Proj}_C(u')$ along with the openness assumption of $\mathcal{W}_C(\overline{x}, r, \alpha)$, we may apply (r) in Theorem 17.12 to conclude that C is (r, α)-prox-regular at $\overline{x} \in C$. □

17.2. Differentiability of metric projection

We already pointed out in the beginning of the chapter that, if the metric projection mapping P_C is Fréchet differentiable on $U \setminus C$ for some open neighborhood U of $\overline{x} \in \mathrm{bdry}\, C$, then the set C must be prox-regular near the point $\overline{x}$. However, the simple example below shows that, even for a convex set C, such a Fréchet differentiability may fail.

EXAMPLE 17.15. Consider the closed convex cone

$$C := \{x \in \mathbb{R}^2 : x_1 \le 0, x_2 \le 0\}$$

in $\mathbb{R}^2$ and note that

$$P_C(x) = \begin{cases} (x_1, 0) & \text{if } x_1 \le 0, x_2 \ge 0 \\ (0, 0) & \text{if } x_1 \ge 0, x_2 \ge 0 \\ (0, x_2) & \text{if } x_1 \ge 0, x_2 \le 0. \end{cases}$$

It is clear that there is no open neighborhood U of $(0,0)$ such that P_C is differentiable on $U \setminus C$. □

The above example suggests to assume some smoothness property of the boundary of the set C.

17.2.1. Variational and prox-regularity properties of submanifolds.

Remind that we consider an integer $p \ge 1$. Given an open set U of a Banach X and a mapping $f : U \to Y$ from U into a Banach space Y, the space of continuous mappings from U into X is usually denoted by $\mathcal{C}(U, Y)$ or $\mathcal{C}^0(U, Y)$. Accordingly, the space of mappings from U into Y which are locally uniformly continuous (resp. locally Lipschitz continuous) will be denoted by $\mathcal{C}^{0,0}(U, Y)$ (resp. $\mathcal{C}^{0,1}(U, Y)$). Similarly, $\mathcal{C}^p(U, Y)$ being the space of mappings $f : U \to Y$ whose derivatives $Df, \cdots, D^p f$ exist and are continuous on U, we will denote by $\mathcal{C}^{p,0}(U, Y)$ (resp. $\mathcal{C}^{p,1}(U, Y)$) the space of mappings $f \in \mathcal{C}^p(U, Y)$ such that the p-th derivative $D^p f$ is locally uniformly continuous (resp. locally Lipschitz continuous) on U. If X is finite-dimensional, by compactness of closed balls it is clear that any continuous mapping on U is locally uniformly continuous there, so $\mathcal{C}^k(U, Y)$ and $\mathcal{C}^{k,0}(U, Y)$ coincide for any $k \in \{0\} \cup \mathbb{N}$. Such a coincidence fails when X is infinite-dimensional; see comments and more in Appendix E.

We recall (see Definition E.8 in Appendix E) that a set M of a Banach space X is said to be a *$\mathcal{C}^p$-submanifold of X at $m_0 \in M$* if there exist a closed vector subspace E in X, an open neighborhood U of m_0 in X, an open neighborhood V of zero in X, and a mapping $\varphi : U \to V$ from U onto V such that

(i) φ is a $\mathcal{C}^p$-diffeomorphism, that is, $\varphi : U \to \varphi(U) = V$ is bijective and φ, φ^{-1} are both mappings of class $\mathcal{C}^p$;

(ii) $\varphi(m_0) = 0$ and $\varphi(U \cap M) = \varphi(U) \cap E$.

If in addition the p-th derivative of φ satisfies the condition

(iii) $D^p\varphi$ is locally uniformly continuous (resp. locally Lipschitz continuous) on U,

one says that M is a $\mathcal{C}^{p,0}$-submanifold (resp. $\mathcal{C}^{p,1}$-submanifold) at m_0.

Given the above description, one calls E a *model subspace* and (U, φ) a *local chart.* One says that M is a *$\mathcal{C}^p$-submanifold (resp. $\mathcal{C}^{p,0}$-submanifold, or $\mathcal{C}^{p,1}$-submanifold) of X* if it is a $\mathcal{C}^p$-submanifold (resp. $\mathcal{C}^{p,0}$-submanifold, or $\mathcal{C}^{p,1}$ submanifold) at every point $m_0 \in M$ with the same model space E.

One says that a set C in X has *$\mathcal{C}^p$-smooth boundary* (resp. *$\mathcal{C}^p$-smooth boundary at $\overline{x} \in \operatorname{bdry} C$*) if $\operatorname{bdry} C$ is a $\mathcal{C}^p$-submanifold of X (resp. $\mathcal{C}^p$-submanifold of X at the point $\overline{x}$). Sets with $\mathcal{C}^{p,0}$-smooth and $\mathcal{C}^{p,1}$-smooth boundaries (at a point) are defined similarly.

For a $\mathcal{C}^p$-submanifold M of X at $m_0 \in M$, the *tangent (vector) space* of M at m_0 is defined as

$$T_{m_0}M := \left\{h \in X : \exists \gamma :]-1, 1[\to M \ C^1\text{-curve with } \gamma(0) = m_0, \gamma'(0) = h\right\}.$$

For the model space E and the local chart (U, φ), with $\varphi(U) = V$, as above, we know by Corollary 2.134 that $T^C(M; m_0) = T^B(M; m_0) = D\varphi^{-1}(0)\big(E\big)$, and clearly $T_{m_0}M \subset T^B(M; m_0)$. On the other hand, choose $\eta > 0$ with $\eta\mathbb{B}_E \subset V$. For any $v \in \eta\mathbb{B}_E$ we can define the mapping $\gamma :]-1, 1[\to M$ by $\gamma(t) = \varphi^{-1}(tv)$ for all $t \in]-1, 1[$. Obviously, $\gamma(0) = m_0$ and $\gamma'(0) = D\varphi^{-1}(0)v$, so $D\varphi^{-1}(0)v \in T_{m_0}M$. It follows that $D\varphi^{-1}(0)(E) \subset T_{m_0}M$. Consequently, we have

$$T^C(M; m_0) = T^B(M; m_0) = T_{m_0}M = D\varphi^{-1}(0)\big(E\big), \tag{17.14}$$

which tells us in particular that $D\varphi^{-1}(0)\big(E\big)$ is independent of the local chart (U, φ) and of the model space E.

We start with a property related to the codimension of the tangent space to the boundary of a set when this boundary is a smooth submanifold. We recall that, in addition to $\operatorname{cl} C$, we also use the notation $\overline{C}$ to denote the closure of a set C when it is convenient.

PROPOSITION 17.16. Let C be a subset of a Hilbert space H and let $\overline{x} \in \operatorname{bdry} C$ be such that $\overline{x} \in \overline{\operatorname{int} C}$ and C is closed near $\overline{x}$. If $\operatorname{bdry} C$ is a $\mathcal{C}^p$-submanifold of H at $\overline{x}$, then the tangent space $T_{\overline{x}}(\operatorname{bdry} C)$ is a closed vector subspace of H of codimension 1; more precisely, there exist a closed vector subspace E of codimension 1, an open neighborhood U of $\overline{x}$ in H and a $\mathcal{C}^p$-diffeormorphism $\varphi : U \to \varphi(U) \subset H$ such that $\varphi(U \cap \operatorname{bdry} C) = \varphi(U) \cap E$.

PROOF. Set $M := \operatorname{bdry} C$. Let U be an open connected neighborhood of $\overline{x}$ such that there exists a $\mathcal{C}^p$-diffeomorphism $\varphi : U \to \varphi(U) \subset H$ and a closed vector subspace E of H such that $\varphi(\overline{x}) = 0$ and

$$\varphi(U \cap M) = \varphi(U) \cap E.$$

Denote $V := \varphi(U)$. As recalled above, we know that $T_{\overline{x}}(\operatorname{bdry} C) = T_{\overline{x}}M = D\varphi^{-1}(0)E$. It is enough to show that E is a subspace of codimension 1. Assume the contrary and consider two distinct vectors $v_1, v_2 \in V \setminus E$. Without loss of generality, we may assume that there exists $\delta > 0$ such that $V = B(0, \delta)$. Denoting by $\mathcal{H}_E$ a Hamel basis of E, we can distinguish two cases:

(I) The set $\mathcal{H}_E \cup \{v_1, v_2\}$ is linearly independent: Then, for each $t \in [0,1]$, putting

$$\gamma(t) = (1-t)v_1 + tv_2 \in V \setminus E,$$

the mapping $\gamma : [0,1] \to V \setminus E$ defines a continuous curve with $\gamma(0) = v_1$ and $\gamma(1) = v_2$.

(II) The set $\mathcal{H}_E \cup \{v_1, v_2\}$ is not linearly independent: Since $\operatorname{codim}(E) \geq 2$, there exists $v_3 \in V \setminus E$ such that both sets $\mathcal{H}_E \cup \{v_1, v_3\}$ and $\mathcal{H}_E \cup \{v_2, v_3\}$ are linearly independent. Then, using the latter part, we can construct two continuous curves $\gamma_1 : [0, 1/2] \to V \setminus E$ and $\gamma_2 : [1/2, 1] \to V \setminus E$ such that $\gamma_1(0) = v_1$, $\gamma_1(1/2) = \gamma_2(1/2) = v_3$ and $\gamma_2(1) = v_2$. Then, considering the mapping $\gamma : [0,1] \to V \setminus E$ given by

$$\gamma(t) = \begin{cases} \gamma_1(t) & t \in [0, 1/2] \\ \gamma_2(t) & t \in]1/2, 1], \end{cases}$$

we arrive at the same conclusion as (I).

Since v_1 and v_2 are two arbitrary distinct points of $V \setminus E$, the existence of such a continuous curve γ entails that $V \setminus E$ is path-connected, and therefore is connected. Then, since $\varphi^{-1} : V \to U$ is continuous, we derive that $U \setminus M = \varphi^{-1}(V \setminus E)$ is connected too. This is clearly a contradiction since the two open sets $U \cap \operatorname{int} C$ and $U \cap (H \setminus \overline{C})$ are nonempty (according to the assumptions $\overline{x} \in \overline{\operatorname{int} C}$ and $\overline{x} \in \operatorname{bdry} C$ with C closed near $\overline{x}$), and they satisfy the equality

$$U \setminus M = (U \cap \operatorname{int} C) \cup (U \cap (H \setminus \widehat{C})).$$

The proof is therefore complete. □

Our next step is to show in Proposition 17.21 that a closed set with the interior tangent cone property and whose boundary is a $\mathcal{C}^p$ submanifold can be represented locally as the epigraph of a $\mathcal{C}^p$-function.

Before proving the proposition we need some features for sets C with the interior tangent cone property at $x \in \operatorname{bdry} C$ and whose boundaries are smooth near x.

Proposition 17.17. Let C be a subset of a Hilbert space H which is closed near a point $x \in \operatorname{bdry} C$ and has the interior tangent cone property at x. Assume that $\operatorname{bdry} C$ is a $\mathcal{C}^1$-submanifold at x. Then $T^C(C; x)$ is a half-space; more precisely there exits a unit vector $\widehat{n}(x) \in H$ such that

$$T^C(C; x) = \{h \in H : \langle \widehat{n}(x), h \rangle \geq 0\},$$

so

$$N^C(C; x) = \{-t\widehat{n}(x) : t \geq 0\}. \tag{17.15}$$

PROOF. By the epi-Lipschitz property of C at x we know by Proposition 2.25 that int $\big(T^C(C;x)\big) \neq \emptyset$ and we also know by Corollary 2.32 that

$$T^C(\mathrm{bdry}\, C;x) = T^C(C;x) \cap -T^C(C;x).$$

Consider the closed vector subspace $E(x)$ of codimension 1, the open neighborhood U of x and the $\mathcal{C}^p$-diffeomorphism $\varphi : U \to \varphi(U) \subset H$, all given by Proposition 17.16, so $\varphi(U \cap \mathrm{bdry}\, C) = E(x) \cap \varphi(U)$. Taking an orthogonal unit vector $\widehat{\nu}_{E(x)}$ of $E(x)$, we see that $H = E(x) \oplus \mathbb{R}\widehat{n}(x)$ and

$$N^C(\mathrm{bdry}\, C;x) = \mathbb{R}\widehat{n}(x),$$

where $\widehat{n}(x) := D\varphi(x)^*\widehat{\nu}_{E(x)}/\|D\varphi(x)^*\widehat{\nu}_{E(x)}\|$. It ensues that
(17.16)

$$T^C(C;x) \cap -T^C(C;x) = T^C(\mathrm{bdry}\, C;x) = \{h \in X : \langle \widehat{n}(x), h\rangle = 0\} = E(x).$$

Consider the continuous linear mappings $\pi_{E(x)} : H \to E(x)$ and $\pi_{\widehat{n}(x)} : H \to \mathbb{R}$ defined by

$$u = \pi_{E(x)}(u) + \pi_{\widehat{n}(x)}(u)\widehat{n}(x) \quad \text{for every } u \in H.$$

Since int $\big(T^C(C;x)\big) \neq \emptyset$ as said above , we have $T^C(C;x) \neq E(x)$, hence there is some $u \in T^C(C;x)$ such that $\pi_{\widehat{n}(x)}(u) \neq 0$. Since $E(x) \subset T^C(C;x)$, we have by convexity of the cone $T^C(C;x)$ that

$$\pi_{\widehat{n}(x)}(u)\widehat{n}(x) = u + \pi_{E(x)}(-u) \in T^C(C;x),$$

so $\mathbb{R}_+\big(\pi_{\widehat{n}(x)}(u)\widehat{n}(x)\big) \subset T^C(C;x)$. Suppose $\pi_{\widehat{n}(x)}(u) > 0$. Then $\mathbb{R}_+\widehat{n}(x) \subset T^C(C;x)$, hence (keeping in mind the convexity of $T^C(C;x)$ along with the inclusion $E(x) \subset T^C(C;x)$ and the equality $h = \pi_{E(x)}(h) + \langle \widehat{n}(x), h\rangle\widehat{n}(x)$) we see that

$$\{h \in H : \langle \widehat{n}(x), h\rangle \geq 0\} \subset T^C(C;x). \tag{17.17}$$

In fact, this inclusion is an equality. Otherwise, the existence of $v \in T^C(C;x)$ with $\langle \widehat{n}(x), v\rangle < 0$ would yield by what precedes $\mathbb{R}_+\big(\pi_{\widehat{n}(x)}(v)\widehat{n}(x)\big) \subset T^C(C;x)$, hence (as above) we would have

$$\{h \in H : \langle \widehat{n}(x), h\rangle < 0\} \subset T^C(C;x).$$

This combined with (17.17) would give $T^C(C;x) = H$, which would contradict (17.16).

It results that

$$\text{either } T^C(C;x) = \{h \in X : \langle \widehat{n}(x), h\rangle \leq 0\} \text{ or } T^C(C;x) = \{h \in X : \langle \widehat{n}(x), h\rangle \geq 0\},$$

which confirms that $T^C(C;x)$ is a half-space. We may suppose that $\widehat{\nu}_{E(x)}$ is chosen so that the second latter equality holds true. We then derive that $N^C(C;x) = \{-t\widehat{n}(x) : t \geq 0\}$ as desired. □

It is worth noting that the vector $\widehat{n}(x)$ does not depend on the diffeomorphism nor the model space chosen to describe bdry C as submanifold, since it is fully determined by $N^C(C;x)$ via (17.15).

DEFINITION 17.18. Let C be a subset of a Hilbert space H which is closed near a point $x \in \mathrm{bdry}\, C$ and has the interior tangent cone property at x. Assume that bdry C is a $\mathcal{C}^1$-submanifold at x. The vector $\widehat{n}(x)$ given by Proposition 17.17 as the unique unit vector $\widehat{n}(x) \in H$ such that

$$N^C(C;x) = \{-t\widehat{n}(x) : t \geq 0\},$$

is called the *unit interior normal vector of* $\operatorname{bdry} C$ *at* x, since it allows to describe the normal cone $N^C(\operatorname{bdry} C; x)$ as above and it "aims" to $\operatorname{int} C$.

In what follows, under the assumptions of Proposition 17.17 we will keep the notation $\widehat{n}(x)$ for the unit interior normal vector of $\operatorname{bdry} C$ at x and we will denote by $Z(x)$ the tangent vector space $T^C(\operatorname{bdry} C; x)$ whenever it will be convenient.

In preparation for the proof of the next Proposition 17.21 we also state and argue two simple lemmas.

LEMMA 17.19. Let C be a subset of a Hilbert space H and U an open subset of H.
(a) The following equalities hold:

$$\operatorname{int}_U(U \cap C) = U \cap \operatorname{int} C, \quad \operatorname{cl}_U(U \cap C) = U \cap \operatorname{cl} C, \quad \operatorname{bdry}_U(U \cap C) = U \cap \operatorname{bdry} C.$$

(b) If $U \cap C = U \cap \overline{\operatorname{int} C}$, then

$$U \cap C = \operatorname{cl}_U\big(\operatorname{int}_U(U \cap C)\big) = \operatorname{cl}_U(U \cap \operatorname{int} C).$$

PROOF. The first two equalities in (a) easily follow from the openness of U and the third is a consequence of the former equalities. Finally, if $U \cap C = U \cap \overline{\operatorname{int} C}$, then we see from (a) that

$$\operatorname{cl}_U\big(\operatorname{int}_U(U \cap C)\big) = \operatorname{cl}_U(U \cap \operatorname{int} C) = U \cap \operatorname{cl}(\operatorname{int} C) = U \cap C.$$

□

The second lemma is concerned with a connectedness property of epigraphs. Recall that the strict epigraph $\operatorname{epi}_s f$ of a function $f : U \to \mathbb{R} \cup \{-\infty, +\infty\}$ is given by

$$\operatorname{epi}_s f := \{(x, r) \in U \times \mathbb{R} : f(x) < r\}.$$

LEMMA 17.20. Let $f : U \to \mathbb{R}$ be a continuous function on an open set U of a normed space X and let $\overline{x} \in U$. Then for any $\eta \in]0, +\infty]$ there exists $\delta_0 > 0$ such that for every $0 < \delta \le \delta_0$ the sets

$$(\operatorname{epi}_s f) \cap \big(B(\overline{x}, \delta) \times]f(\overline{x}) - \eta, f(\overline{x}) + \eta[\big) \text{ and } (\operatorname{epi} f) \cap \big(B(\overline{x}, \delta) \times]f(\overline{x}) - \eta, f(\overline{x}) + \eta[\big)$$

are path-connected.

PROOF. Take any $\eta \in]0, +\infty]$. There exists $\delta_0 > 0$ such that $B(\overline{x}, \delta_0) \subset U$ and such that $|f(x) - f(\overline{x})| < \eta$ for every $x \in B(\overline{x}, \delta_0)$. Consider any $\delta \in]0, \delta_0]$, any (x_0, r_0) and (x_1, r_1) in $P_{\eta,\delta}$, where

$$P_{\eta,\delta} := E_f \cap \big(B(\overline{x}, \delta) \times]f(\overline{x}) - \eta, f(\overline{x}) + \eta[\big),$$

with $E_f := \operatorname{epi}_s f$ (resp. $E_f := \operatorname{epi} f$). Then one has $r_0 = f(x_0) + \theta_0$ and $r_1 = f(x_1) + \theta_1$ with reals $\theta_0 > 0$ and $\theta_1 > 0$ (resp. $\theta_0 \ge 0$ and $\theta_1 \ge 0$). Put

$$\beta_0 := \max\Big\{|r_0 - f(\overline{x})|, |r_1 - f(\overline{x})|, \max_{t \in [0,1]} |f((1-t)x_0 + tx_1)|\Big\} < \eta$$

and choose $\beta \in]\beta_0, \eta[$. Set $\alpha := f(\overline{x}) + \beta$ and note that one has $\alpha \in]f(\overline{x}) - \eta, f(\overline{x}) + \eta[$ along with $\alpha \ge r_0$ and $\alpha \ge r_1$. Then one deduces by definition of $P_{\eta,\delta}$ that

$$[(x_0, r_0), (x_0, \alpha)] \subset P_{\eta,\delta} \quad \text{and} \quad [(x_1, r_1), (x_1, \alpha)] \subset P_{\eta,\delta}.$$

Further, for $z_t := (1-t)x_0 + tx_1$ we see that $\{(z_t, \alpha) : t \in [0,1]\} \subset P_{\eta,\delta}$. It results that $P_{\eta,\delta}$ is pathwise connected as desired. □

The strict epigraph $\operatorname{epi}_s f$ of a function $f : U \to \mathbb{R} \cup \{-\infty, +\infty\}$ has been recalled before the statement of Lemma 17.20. Recall also that the strict hypograph $\operatorname{hypo}_s f$ of the function f is defined as

$$\operatorname{hypo}_s f := \{(x, r) \in U \times \mathbb{R} : f(x) > r\}.$$

Given a closed vector subspace Z of a Hilbert space H and its orthogonal vector space $Z^\perp$ in H, we know that H is the topological direct sum $H = Z \oplus Z^\perp$ of Z and $Z^\perp$. In this case, since there is no ambiguity about the complement $Z^\perp$ of Z, it will be convenient to denote by π_Z (resp. $\pi_{Z^\perp}$) the projector mapping $\pi^{Z,Z^\perp}$ (resp. $\pi^{Z^\perp,Z}$) onto Z parallel to $Z^\perp$ (resp. onto $Z^\perp$ parallel to Z) (recalled in (2.88)), that is, $\pi_Z : H \to Z$ and $\pi_{Z^\perp} : H \to Z^\perp$ are defined by

$$x = \pi_Z(x) + \pi_{Z^\perp}(x) \quad \text{with } \pi_Z(x) \in Z,\ \pi_{Z^\perp}(x) \in Z^\perp \quad \text{for every } x \in H. \tag{17.18}$$

Clearly, $\pi_Z(x)$ coincides with the orthogonal projection of x onto Z, which in turn coincides with the metric projection $P_Z(x)$ of x on Z.

For a Cartesian product $X_1 \times X_2$ of two nonempty sets, we will denote by pr_{X_1}, or pr_1, the first coordinate projector mapping from $X_1 \times X_2$ into X_1 defined by

$$\operatorname{pr}_{X_1}(x_1, x_2) = x_1 \quad \text{for all } (x_1, x_2) \in X_1 \times X_2. \tag{17.19}$$

The mapping pr_{X_2}, or pr_2, is defined in a similar way.

Proposition 17.21. Let C be a subset of a Hilbert space H which is closed near $\overline{x} \in \operatorname{bdry} C$ and has the interior tangent property at $\overline{x}$. Assume that $\operatorname{bdry} C$ is a $\mathcal{C}^p$-submanifold at $\overline{x}$ with $p \geq 1$ and denote by $Z(\overline{x}) := T_{\overline{x}}(\operatorname{bdry} C)$ the tangent vector space to the boundary of C at $\overline{x}$. Let $\widehat{n}(\overline{x})$ denote the unit interior normal vector of $\operatorname{bdry} C$ at $\overline{x}$. The following hold.
(a) There exist a real $\varepsilon > 0$, a connected open neighborhood Q_0 of $z_0 := \pi_{Z(\overline{x})}(\overline{x})$ in $Z(\overline{x})$ (relative to the induced topology on $Z(\overline{x})$), a function $f : Q_0 \subseteq Z(\overline{x}) \to \mathbb{R}$ which is of class $\mathcal{C}^p$ on Q_0 with $\overline{x} = z_0 + f(z_0)\widehat{n}(\overline{x})$ and $\nabla f(z_0) = 0$ and such that for the connected open neighborhood

$$U_0 := \{z + t\widehat{n}(\overline{x}) : z \in Q_0, t \in]f(z_0) - \varepsilon, f(z_0) + \varepsilon[\}$$

of $\overline{x}$ in H one has $U_0 \subset U$ along with

$$U_0 \cap C = \{z + t\widehat{n}(\overline{x}) \in U_0 \ :\ z \in Z(\overline{x}), f(z) \leq t\}.$$

(b) If $p \geq 2$, then C is prox-regular at $\overline{x}$. Furthermore, endowing $Z(\overline{x}) \times \mathbb{R}$ with the canonical inner product

$$\langle (z, t), (z', t') \rangle = \langle z, z' \rangle + tt',$$

if $r \in]0, +\infty]$ is the prox-regularity constant of C at $\overline{x}$ (that is, C is r-prox-regular at $\overline{x}$), then $\operatorname{epi} f$ is also r-prox-regular at $(z_0, f(z_0))$, where $z_0 := \pi_{Z(\overline{x})}(\overline{x})$ as mentioned above.

Proof. (a) By Proposition 17.16 choose an open neighborhood U of $\overline{x}$, a $\mathcal{C}^p$-diffeomorphism $\varphi : U \to \varphi(U) \subset H$, and a closed vector subspace Z of H of codimension 1 such that $\varphi(\overline{x}) = 0$ and

$$\varphi(U \cap \operatorname{bdry} C) = Z \cap \varphi(U).$$

By replacing φ by $D\varphi^{-1}(0) \circ \varphi$, we can choose $Z = Z(\overline{x})$ and $D\varphi(\overline{x}) = \operatorname{id}_H$. Let ν be a unit vector of H orthogonal to Z. We have that $\mathbb{R}\nu$ is a topological vector subspace complement of Z in H, that is, $H = Z \oplus \mathbb{R}\nu$. Noticing that $Z = \{x \in H : \langle \nu, x\rangle = 0\}$, we see that, for $z + t\nu \in U$ with $z \in Z$ and $t \in \mathbb{R}$,

$$z + t\nu \in U \cap \operatorname{bdry} C \Leftrightarrow \langle \varphi(z + t\nu), \nu\rangle = 0.$$

Consider the open set $W := \{(z,t) \in Z \times \mathbb{R} \ : \ z + t\nu \in U\}$ in $Z \times \mathbb{R}$, where Z is equipped with the induced norm, and consider also the $\mathcal{C}^p$ function $F : W \to \mathbb{R}$ defined by

$$F(z,t) := \langle \varphi(z + t\nu), \nu\rangle, \quad \text{for all } (z,t) \in W.$$

Write $\overline{x} = z_0 + t_0\nu$ with $z_0 \in Z$ and $t_0 \in \mathbb{R}$, and note that $F(z_0,t_0) = 0$ and that the derivative with respect to the second variable t at (z_0,t_0) satisfies

$$D_2F(z_0,t_0) = \langle D\varphi(z_0 + t_0\nu)\nu, \nu\rangle = \langle D\varphi(\overline{x})\nu, \nu\rangle = \|\nu\|^2 = 1.$$

We can apply the implicit function theorem (see Theorem E.2 in Appendix E) to obtain an open connected neighborhood Q_0 of z_0 in Z, a real $\varepsilon > 0$ and a $\mathcal{C}^p$ function $f : Q_0 \to]t_0 - \varepsilon, t_0 + \varepsilon[$ such that

$$U_0 := \{z + t\nu : z \in Q_0, \ t \in \,]t_0 - \varepsilon, t_0 + \varepsilon[\, \} \subset U$$

and such that, for $z \in Z$ and $t \in \mathbb{R}$

$$\Big(z + t\nu \in U_0 \cap \operatorname{bdry} C\Big) \Leftrightarrow \Big(z + t\nu \in U_0 \text{ and } F(z,t) = 0\Big) \Leftrightarrow \Big(z + t\nu \in U_0 \text{ and } t = f(z)\Big),$$

so in particular $t_0 = f(z_0)$. The set C being epi-Lipschitz at $\overline{x}$, shrinking Q_0 and ε if necessary we may and do suppose that $U_0 \cap \operatorname{int} C$ is connected and $U_0 \cap C = U_0 \cap \left(\overline{\operatorname{int} C}\right)$ according to Lemma 17.20 and Proposition 2.27(a). Furthermore, for any $h \in Z$ we have

$$\langle \nabla f(z_0), h\rangle = -D_2F(z_0,t_0)^{-1} \circ D_1F(z_0,t_0)h = -D_1F(z_0,t_0)h = -\langle D\varphi(\overline{x})h, \nu\rangle = 0,$$

since $D_2F(z_0,t_0) = \mathrm{id}_{\mathbb{R}}$ and $D\varphi(\overline{x})\big|_Z = \mathrm{id}_Z$. Thus, $\nabla f(z_0) = 0$.

Endow $Z \times \mathbb{R}$ with the canonical inner product

$$\langle (z,r), (z',r')\rangle_{Z\times\mathbb{R}} := \langle z, z'\rangle + rr'$$

and with the associated norm. With the linear isometric isomorphism $L : Z \times \mathbb{R} \to H$ defined by $L(z,t) := z + t\nu$, by Lemma 17.20 we may shrink again (if necessary) Q_0 and $\varepsilon > 0$ so that $(L^{-1}(U_0)) \cap \operatorname{epi}_s f$ and $(L^{-1}(U_0)) \cap \operatorname{hypo}_s f$ are connected, hence we see that they are the two connected components of $L^{-1}(U_0) \setminus \operatorname{gph} f$. It results that $U_0 \cap L(\operatorname{epi}_s f)$ and $U_0 \cap L(\operatorname{hypo}_s f)$ are the two connected components of $U_0 \setminus \operatorname{bdry} C$. Since $U_0 \cap \operatorname{int} C$ is a connected set included $U_0 \setminus \operatorname{bdry} C$, it is easily seen that either $U_0 \cap \operatorname{int} C = U_0 \cap L(\operatorname{epi}_s f)$ or $U_0 \cap \operatorname{int} C = U_0 \cap L(\operatorname{hypo}_s f)$. Noticing that

$$\begin{aligned} U_0 \cap L(\operatorname{hypo}_s f) &= \{z + t\nu : z \in Q_0, \ t \in \,]t_0 - \varepsilon, t_0 + \varepsilon[, \ t < f(z)\} \\ &= \{z + t(-\nu) : z \in Q_0, \ t \in \,]-t_0 - \varepsilon, -t_0 + \varepsilon[, \ (-f)(z) < t\}, \end{aligned}$$

and changing ν by $-\nu$ and t_0 by $-t_0$ if necessary, we may suppose that the equality $U_0 \cap \operatorname{int} C = U_0 \cap L(\operatorname{epi}_s f)$ holds true. By Lemma 17.19 we derive that the equality $U_0 \cap C = U_0 \cap L(\operatorname{epi} f)$ holds.

Let us denote $A := L^{-1}$ and let us keep $Z \times \mathbb{R}$ endowed with the above canonical inner product. Writing any $x \in H$ as $x = \pi_Z(x) + \pi_{\mathbb{R}}(x)\nu$ with $\pi_Z(x) \in Z$ and $\pi_{\mathbb{R}}(x) \in \mathbb{R}$, the bijective linear mapping $A : H \to Z \times \mathbb{R}$ satisfies $A(x) := (\pi_Z(x), \pi_{\mathbb{R}}(x))$ and it is an isomorphism such that $A(U_0 \cap C) = A(U_0) \cap (\operatorname{epi} f)$.

Since f is of class $\mathcal{C}^1$, at any $z \in \pi_Z(U_0)$ we have $\partial_C f(z) = \{\nabla f(z)\}$, and therefore (see Proposition 2.87)

$$N^C\big(\mathrm{epi}\, f; (z, f(z))\big) = \{\lambda(\nabla f(z), -1) \,:\, \lambda \geq 0\}.$$

Further, taking the linear isomorphism A into account, we have for any $x \in C \cap U_0$ (see Proposition 2.130)

$$\begin{aligned} N^C(C; x) = N^C(U_0 \cap C; x) &= A^*\big(N^C(A(U_0) \cap (\mathrm{epi}\, f); A(x))\big) \\ &= A^*\big(N^C(\mathrm{epi}\, f; A(x))\big), \end{aligned}$$

where A^* denotes the adjoint of A. This yields by (17.15)

$$\begin{aligned} \{-\lambda\widehat{n}(\overline{x}) \,:\, \lambda \geq 0\} = N^C(C; \overline{x}) &= \{\lambda A^*(\nabla f(z_0), -1) \,:\, \lambda \geq 0\} \\ &= \{A^*(0, -\lambda) \,:\, \lambda \geq 0\}. \end{aligned}$$

Observing that $A^* = L$, we get that $\widehat{n}(\overline{x}) = \nu$, which finishes the proof of (a).
(b) Assume that $p \geq 2$. By (a) the above function f is of class $\mathcal{C}^{1,1}$, hence $\mathrm{epi}\, f$ is prox-regular at z_0 by Proposition 15.71. The set C is then prox-regular at $\overline{x}$ since $U_0 \cap C = U_0 \cap A^{-1}(\mathrm{epi}\, f)$ and A is a linear isomorphism.

Let $r \in]0, +\infty]$ be a constant of prox-regularity of C at $\overline{x}$. Then, by the equivalence (a)$\Leftrightarrow$(c) in Theorem 15.70 there exists $\alpha > 0$ such that for every $x \in C \cap B_H(\overline{x}, \alpha)$ and every $\xi \in N^P(C; x)$, one has

$$\langle \xi, x' - x \rangle \leq \frac{1}{2r} \|\xi\| \|x' - x\|^2, \ \forall x' \in C.$$

Coming back to the arguments for (a), by Lemma 17.20 and Proposition 2.27(a) we may and do suppose that $U_0 \subseteq B(\overline{x}, \alpha)$. Now, fix $(z, t) \in A(U_0) \cap (\mathrm{epi}\, f)$ and $\zeta \in N^P\big(\mathrm{epi}\, f; (z, t)\big) \cap \mathbb{B}_{Z \times \mathbb{R}}$. For every $(z', t') \in A(U_0) \cap (\mathrm{epi}\, f)$, we have by Proposition 4.135 that

$$\begin{aligned} \langle \zeta, (z', t') - (z, t) \rangle &= \langle (A^*)^{-1} A^* \zeta, (z', t') - (z, t) \rangle \\ &= \langle A^* \zeta, A^{-1}(z', t') - A^{-1}(z, t) \rangle \\ &\leq \frac{1}{2r} \|A^* \zeta\| \|A^{-1}\big((z', t') - (z, t)\big)\|^2 \\ &\leq \frac{1}{2r} \|(z', t') - (z, t)\|^2, \end{aligned}$$

where the last inequality follows from the equalities $A^{-1} = A^*$ and $\|A^*\| = 1$. Now, consider $(z', t') \in (\mathrm{epi}\, f) \setminus A(U_0)$. Since $z' \in Q_0$ (keep in mind that f is defined only on Q_0) and since

$$A(U_0) = Q_0 \times \,]t_0 - \varepsilon, t_0 + \varepsilon[,$$

we have necessarily that $t' \notin]t_0 - \varepsilon, t_0 + \varepsilon[$, and in fact $t' \geq t_0 + \varepsilon > t$ because $t' \geq f(z') > t_0 - \varepsilon$. Since $\max\{t, f(z')\} < t_0 + \varepsilon \leq t'$, we can define $t'' = \max\{t, f(z')\}$ and, noting that $\mathrm{pr}_{\mathbb{R}}(\zeta) \leq 0$ and $(z', t'') \in A(U_0) \cap \mathrm{epi}\, f$, we can write by what precedes

$$\begin{aligned}\langle \zeta, (z', t') - (z, t) \rangle &= \langle \zeta, (z', t'') - (z, t) \rangle + \langle \zeta, (0, t' - t'') \rangle \\ &\leq \langle \zeta, (z', t'') - (z, t) \rangle \leq \frac{1}{2r} \|(z', t'') - (z, t)\|^2 \\ &= \frac{1}{2r} (\langle z' - z, z' - z \rangle + (t'' - t)^2) \\ &\leq \frac{1}{2r} (\langle z' - z, z' - z \rangle + (t' - t)^2) = \frac{1}{2r} \|(z', t') - (z, t)\|^2,\end{aligned}$$

where the last inequality is due to the fact that $t \leq t'' < t'$. We then obtain that, for all $(z', t') \in \operatorname{epi} f$

$$\langle \zeta, (z', t') - (z, t) \rangle \leq \frac{1}{2r} \|(z', t') - (z, t)\|^2.$$

The set $\operatorname{epi} f$ being also clearly closed near $(z_0, f(z_0))$, the r-prox-regularity of $\operatorname{epi} f$ at $(z_0, f(z_0))$ is justified by the equivalence (a)$\Leftrightarrow$(c) in Theorem 15.70 again. This finishes the proof. □

17.2.2. Smoothness of metric projection onto prox-regular sets with smooth boundary. Throughout this subsection we will use the open normal ray and the λ-truncated open normal ray of a set at suitable points.

DEFINITION 17.22. Let C be a subset of a Hilbert space H which is closed near $\overline{x} \in C$. When the proximal normal cone of the set C at $\overline{x}$ has the form

$$N^P(C; \overline{x}) = \{t\nu \ : \ t \geq 0\},$$

for some unit vector $\nu \in \mathbb{S}_H$, we define its *open normal ray* $\operatorname{Ray}_{\overline{x}}(C)$ and its λ-*truncated open normal ray* (for $\lambda > 0$) at $\overline{x}$ by

$$\begin{aligned}\operatorname{Ray}_{\overline{x}}(C) &= \{\overline{x} + t\nu \ : \ t > 0\} \\ \operatorname{Ray}_{\overline{x},\lambda}(C) &= \{\overline{x} + t\nu \ : \ t \in \,]0, \lambda[\,\}.\end{aligned}$$

We already know (see Proposition 15.71) that the epigraph of a $\mathcal{C}^2$-function on an open set of a Hilbert space is prox-regular. The first result of this subsection shows the smoothness of the metric projection of such an epigraph on neighborhoods of λ-truncated open rays, for suitable $\lambda > 0$.

THEOREM 17.23 (smoothness of metric projection to epigraphs). Let H be a Hilbert space, O_0 be an open ball $B(\overline{x}, \eta)$ in H with $\eta \in]0, +\infty]$, and $f : O_0 \subset H \to \mathbb{R}$ be a function of class $\mathcal{C}^{p+1}$ (with $p \geq 1$) on O_0 such that $\nabla f(\overline{x}) = 0$. Let any $r, s \in]0, +\infty]$ with $r + s \leq \eta$ such that r is a constant of prox-regularity of $\operatorname{epi} f$ at $(\overline{x}, f(\overline{x}))$ and let

$$\lambda = \min \left\{ r, \ \left(-2 \inf \left\{ \langle u, D^2 f(\overline{x}) u \rangle \ : \ u \in \mathbb{B}_H \right\} \right)^{-1} \right\}.$$

Then there exists an open set W in H containing $\operatorname{Ray}_{(\overline{x}, f(\overline{x})),\lambda}(\operatorname{epi} f)$ such that

(a) $d_{\operatorname{epi} f}$ is of class $\mathcal{C}^{p+1}$ on W;
(b) $P_{\operatorname{epi} f}$ is of class $\mathcal{C}^p$ on W.

PROOF. Let us denote $C := \operatorname{epi} f$, and $\operatorname{pr}_H : H \times \mathbb{R} \to H$, $\operatorname{pr}_{\mathbb{R}} : H \times \mathbb{R} \to \mathbb{R}$ the coordinate projectors associated with the Cartesian product $H \times \mathbb{R}$ as defined in (17.19). Also, for simplicity, we will write $u = (u_1, u_2)$ for each $u \in H \times \mathbb{R}$. According to the convention $1/0 = +\infty$ and noting that $\inf_{u \in \mathbb{B}_H} \langle u, D^2 f(\overline{x}) u \rangle \leq 0$, one sees that $\lambda > 0$. For each $x \in O_0$, keeping in mind that f is at least $\mathcal{C}^2$ on O_0, we have $\partial_P f(x) = \{\nabla f(x)\}$ (see Proposition 4.142), hence

$$N^P(C;(x, f(x))) = \{t(\nabla f(x), -1) : t \in [0, +\infty[\}. \tag{17.20}$$

We know by Proposition 15.71 that C is prox-regular at $v_0 := (\overline{x}, f(\overline{x}))$. Take as in the statement any $r, s \in]0, +\infty]$ with $r + s \leq \eta$ so that C is r-prox-regular at v_0. By Theorem 17.12 there exists $\alpha \in]0, s]$ small enough such that C is (r, α)-prox-regular at v_0 and such that, denoting $O := \mathcal{W}_C(v_0, r, \alpha)$, we have that O and $O \setminus C$ are open in H (see Theorem 17.7(b)), the mapping P_C is well defined on O, the distance function d_C is continuously differentiable in $O \setminus C$ and

$$\nabla d_C(v) = \frac{v - P_C(v)}{d_C(v)}, \ \forall v \in O \setminus C. \tag{17.21}$$

Further, noticing that

$$O := \mathcal{W}_C(v_0, r, \alpha) \subset B_{H\times\mathbb{R}}(v_0, \alpha + r) \subset B_{H\times\mathbb{R}}(v_0, \eta) \subset B_H(\overline{x}, \eta) \times B_{\mathbb{R}}(f(\overline{x}), \eta),$$

we have $\mathrm{pr}_H(O) \subset B_H(\overline{x}, \eta)$, hence f is of class $\mathcal{C}^{p+1}$ on $\mathrm{pr}_H(O)$, which in turn ensures (see Proposition 4.142) that $\partial f(x) = \{\nabla f(x)\}$ for all $x \in \mathrm{pr}_H(O)$. Then by Proposition 4.149(b) we obtain for every $x \in \mathrm{pr}_H(O)$

$$N^P(C;(x, f(x)) = \{t(\nabla f(x), -1) : t \geq 0\}. \tag{17.22}$$

On the other hand, since O coincides with $\mathcal{R}_C(v_0, r, \alpha)$ by Theorem 17.7, we have $\mathrm{Ray}_{\,v_0,\lambda}(C) \subset O$. Fix any $u_0 \in \mathrm{Ray}_{\,v_0,\lambda}(C)$. We can then choose a convex neighborhood U of u_0 in $H \times \mathbb{R}$ and two convex neighborhoods V, V' of v_0 in $H \times \mathbb{R}$ such that

- $U \subset O \setminus C$, $V' \subset O$, $V \subset V'$;
- $(v_1, f(v_1)) \in V'$ for every $v_1 \in \mathrm{pr}_H(V)$;
- there exists $\delta > 0$ such that $U + (\{0\} \times]-\delta, \delta[\,) \subset O \setminus C$; and
- $\mathrm{diam}\,(\mathrm{pr}_{\mathbb{R}}(V')) < \delta$.

From those properties, we have that for each $v \in V$

$$(v_1, f(v_1)) \in V' \quad \text{and} \quad U - (0, v_2 - f(v_1)) \subset O \setminus C.$$

Let us define the mapping

$$\begin{aligned} F : U \times V &\to H \times \mathbb{R} \\ (u, v) &\mapsto u - v - d_C(u)\varphi(v), \end{aligned}$$

where

$$\varphi(v) = \frac{(\nabla f(v_1), -1)}{\|(\nabla f(v_1), -1)\|} \quad \text{for all } v \in V.$$

We claim that

$$F(u, v) = 0 \ \Leftrightarrow \ v = P_C(u). \tag{17.23}$$

For the implication $\Leftarrow$, let us suppose that $v = P_C(u)$. Then, $u - v \in N^P(C; v)$ and by (17.20) and the definition of φ, there exists $t \geq 0$ such that

$$u = v + t\varphi(v).$$

Thus, noting that $d_C(u) = \|u - P_C(u)\| = t\|\varphi(v)\| = t$, we conclude that $F(u, v) = 0$. On the other hand, to prove the converse implication $\Rightarrow$, let us suppose that

$F(u,v) = 0$, so $\|u - v\| = d_C(u)$. Putting $v' = (v_1, f(v_1))$ and noting that $\varphi(v') = \varphi(v)$, we can write

$$u = v + d_C(u)\varphi(v) = v' + d_C(u)\varphi(v') + (0, v_2 - f(v_1)).$$

Therefore, with $u' := u - (0, v_2 - f(v_1))$, we have $u' = v' + d_C(u)\varphi(v')$ with $\varphi(v') \in N^P(C; v') \cap \mathbb{B}_H$ and $d_C(u) < r$ since $u \in \mathcal{W}_C(v_0, r, \alpha)$, hence $P_C(u') = v'$ according to the equivalence (a)$\Leftrightarrow$(d) in Theorem 15.70. From this we also see that

$$d_C(u') = \|u' - v'\| = \|u - v\| = d_C(u).$$

Define the mapping

$$g :]-1, 1+\delta'[\to O \setminus C, \quad \text{given by} \quad g(t) = u - \big(0, t(v_2 - f(v_1))\big),$$

with some $\delta' > 0$ for which g is well-defined. Then, for each $t \in]-1, 1+\delta'[$ we have by (17.21)

$$(d_C \circ g)'(t) = -Dd_C(g(t))\big(0, v_2 - f(v_1)\big) = -\mathrm{pr}_{\mathbb{R}}\left(\frac{g(t) - P_C(g(t))}{d_C(g(t))}\right)(v_2 - f(v_1)).$$

Noting that $g(t) - P_C(g(t)) \in N^P\big(C; P_C(g(t))\big)$ and recalling that $g(t) \notin C$, by (17.20) we obtain that $\mathrm{pr}_{\mathbb{R}}\left(\frac{g(t)-P_C(g(t))}{d_C(g(t))}\right) < 0$. Thus, $\mathrm{sgn}\,((d_C \circ g)'(t)) = \mathrm{sgn}\,(v_2 - f(v_1))$ for all $t \in]-1, 1+\delta'[$ (where $\mathrm{sgn}\,(\cdot)$ denotes the sign function on $\mathbb{R} \setminus \{0\}$), and we get that if $v_2 \neq f(v_1)$, then

$$(d_C \circ g)(1) \neq (d_C \circ g)(0), \quad \text{that is,} \ d_C(u') \neq d_C(u),$$

since $d_C \circ g$ is strictly monotone. Since $d_C(u) = d_C(u')$, we conclude that $v_2 = f(v_1)$ and therefore $u = u'$ and $v = v'$. In particular, $P_C(u) = v$, which proves the implication $\Rightarrow$ in (17.23). Our claim of equivalence in (17.23) is justified.

Now from the inclusion $u_0 \in \mathrm{Ray}_{\,v_0,\lambda}(C)$ and from (17.20) with $\nabla f(\overline{x}) = 0$, we have $u_0 = v_0 + t(0,-1)$ for some $t \geq 0$ with $t < \lambda < r$. This and the inclusion $(0,-1) \in N^P(C; v_0) \cap \mathbb{B}_{H\times\mathbb{R}}$ along with the (r,α)-prox-regularity of C at v_0 entail, by the equivalence (a)$\Leftrightarrow$(d) in Theorem 15.70 that $P_C(u_0) = v_0$. Then $F(u_0, v_0) = 0$ by (17.23). To be able to apply the implicit function theorem to F at (u_0, v_0), we need to verify that (see Theorem E.2 in Appendix E)

$$D_2F(u_0, v_0) = -\mathrm{id}_{H\times\mathbb{R}} - d_C(u_0) \cdot D\varphi(v_0)$$

is an isomorphism. Let us define the mappings $\varphi_1 : (H \times \mathbb{R}) \setminus \{0\} \to H \times \mathbb{R}$ and $\varphi_2 : H \to H \times \mathbb{R}$ given by

$$\varphi_1(y) = \frac{y}{\|y\|} \quad \text{and} \quad \varphi_2(x) = (x, -1).$$

We can write $\varphi = \varphi_1 \circ \varphi_2 \circ \nabla f \circ \mathrm{pr}_H$. Recalling that for all $h \in H \times \mathbb{R}$

$$D\varphi_1(y)h = \frac{\|y\|h - \langle \varphi_1(y), h\rangle y}{\|y\|^2},$$

we have that

$$
\begin{aligned}
&D\varphi(v_0)h = D\varphi_1((\nabla f(\overline{x}),-1))\circ D\varphi_2(\nabla f(\overline{x}))\circ D(\nabla f)(\overline{x})\circ \mathrm{pr}_H(h)\\
&= D\varphi_1((0,-1))(D(\nabla f)(\overline{x})h_1,0)\\
&= \|(0,-1)\|^{-2}\left(\|(0,-1)\|(D(\nabla f)(\overline{x})h_1,0)-\left\langle \frac{(0,-1)}{\|(0,-1)\|},(D(\nabla f)(\overline{x})h_1,0)\right\rangle(0,-1)\right)\\
&= (D(\nabla f)(\overline{x})h_1,0).
\end{aligned}
$$

Thus, $D\varphi(v_0) = (D(\nabla f)(\overline{x})\circ\mathrm{pr}_H,0)$. With this equality at hands, let us then show that $\mathrm{id}_{H\times\mathbb{R}} + d_C(u_0)D\varphi(v_0)$ is bijective. We may suppose that $D^2f(\overline{x}) \neq 0$, since otherwise the bijectivity is trivial.

surjectivity: Let us consider $h \in X\times\mathbb{R}$ with $h\neq 0$. Since

$$(\mathrm{id}_{H\times\mathbb{R}} + d_C(u_0)D\varphi(v_0))^*h = \mathrm{id}_{H\times\mathbb{R}}(h) + d_C(u_0)(D\varphi(v_0))^*h,$$

it follows that

$$
\begin{aligned}
&\|(\mathrm{id}_{H\times\mathbb{R}} + d_C(u_0)D\varphi(v_0))^*h\|^2\\
&= \|h\|^2 + 2d_C(u_0)\langle(D\varphi(v_0))^*h,h\rangle + d_C(u_0)^2\|(D\varphi(v_0))^*h\|^2\\
&= \|h\|^2 + 2d_C(u_0)\langle h_1, D(\nabla f)(\overline{x})h_1\rangle + d_C(u_0)^2\|(D\varphi(v_0))^*h\|^2\\
&\geq \|h\|^2 + 2d_C(u_0)\left\langle \frac{h_1}{\|h\|}, D^2f(\overline{x})\frac{h_1}{\|h\|}\right\rangle\|h\|^2,
\end{aligned}
$$

which ensures that

$$
\begin{aligned}
\|(\mathrm{id}_{H\times\mathbb{R}} + d_C(u_0)D\varphi(v_0))^*h\|^2 &\geq \left(1+2\inf_{x\in\mathbb{B}_H}\{\langle x,D^2f(\overline{x})x\rangle\}d_C(u_0)\right)\cdot\|h\|^2\\
&\geq \left(1-\frac{1}{\lambda}d_C(u_0)\right)\cdot\|h\|^2, \qquad (17.24)
\end{aligned}
$$

where the last inequality is due to the definition of λ. Since $u_0 \in \mathrm{Ray}_{\,v_0,\lambda}(C) \subset \mathcal{W}_C(v_0,\lambda,\alpha)$, we have that $c := 1-\lambda^{-1}d_C(u_0) > 0$, and so, by Theorem C.6 in Appendix C, the desired surjectivity follows.

injectivity: Let $h \in X\times\mathbb{R}$ be such that $\big(\mathrm{id}_{H\times\mathbb{R}} + d_C(u_0)D\varphi(v_0)\big)h = 0$. Then necessarily $h_2 = 0$ since $\mathrm{pr}_{\mathbb{R}}\,(D\varphi(v_0)h) = 0$, and hence taking account the inequality $\inf_{x\in\mathbb{B}_H}\{\langle x,D^2f(\overline{x})x\rangle\}\leq 0$, we can write

$$
\begin{aligned}
2\inf_{x\in\mathbb{B}_H}\{\langle x,D^2f(\overline{x})x\rangle\}\|h\|^2 &\leq \inf_{x\in\mathbb{B}_H}\{\langle x,D^2f(\overline{x})x\rangle\}\|h\|^2 \leq \left\langle \frac{h_1}{\|h\|}, D^2f(\overline{x})\frac{h_1}{\|h\|}\right\rangle\|h\|^2\\
&= \langle h_1, D^2f(\overline{x})h_1\rangle = \langle h,(D(\nabla f)(\overline{x})h_1,0)\rangle = \langle h, D\varphi(v_0)h\rangle\\
&= d_C(u_0)^{-1}\langle h, d_C(u_0)D\varphi(v_0)h\rangle = -d_C(u_0)^{-1}\|h\|^2,
\end{aligned}
$$

where the last equality is due to the fact that we have supposed that $\big(\mathrm{id}_{H\times\mathbb{R}} + d_C(u_0)D\varphi(v_0)\big)h = 0$. But since $-d_C(u_0)^{-1} < -\lambda^{-1} \leq 2\inf_{x\in\mathbb{B}_H}\{\langle x,D^2f(\overline{x})x\rangle\}$ according to the definition of λ, we have that necessarily $h=0$, which proves the desired injectivity.

Now, we can apply the implicit function theorem (see Theorem E.2 in Appendix E) in the following way. Since d_C is of class $\mathcal{C}^1$ in U, we have that F is of class $\mathcal{C}^1$ on $U\times V$. Therefore, there exist two neighborhoods $U_1\subset U$ of u_0 and $V_1\subset V$ of v_0 and a mapping $\phi : U_1 \to V_1$ such that

(i) ϕ is of class $\mathcal{C}^1$;
(ii) For each $u' \in U_1$, $F(u', \phi(u')) = 0$;
(iii) For each $(u', v') \in U_1 \times V_1$, $F(u', v') = 0 \Rightarrow v' = \phi(u')$.

Then, by (ii) and (iii) and by (17.23) we get that $P_C = \phi$ in U_1, and therefore P_C is of class $\mathcal{C}^1$ on U_1 according to (i). Now, looking at the formula (17.21), we get that d_C is of class $\mathcal{C}^2$ on U_1, so F is of class $\mathcal{C}^2$ on $U_1 \times V_1$. We can apply recursively this argument as follows (using, for short, the notation IFT for the "implicit function theorem"):

$$
\begin{aligned}
d_C \text{ is of class } \mathcal{C}^2 \text{ in } U_1 \Rightarrow F &\text{ is of class } \mathcal{C}^2 \text{ on } U_1 \times V_1 \\
&\underset{\text{IFT}}{\Rightarrow} \exists U_2 \in \mathcal{N}(u_0),\ P_C \text{ is of class } \mathcal{C}^2 \text{ on } U_2 \\
&\ \ \vdots \\
&\Rightarrow F \text{ is of class } \mathcal{C}^p \text{ on } U_{p-1} \times V_{p-1} \\
&\underset{\text{IFT}}{\Rightarrow} \exists U_p \in \mathcal{N}(u_0),\ P_C \text{ is of class } \mathcal{C}^p \text{ on } U_p \\
&\Rightarrow d_C \text{ is of class } \mathcal{C}^{p+1} \text{ on } U_p.
\end{aligned}
$$

Since ∇f is assumed to be of class $\mathcal{C}^p$ and not of class $\mathcal{C}^{p+1}$, the argument ends at this iteration, because we cannot ensure that F is of class $\mathcal{C}^{p+1}$. The proof is finished by defining W as the union of the U_p obtained by this way for each $u_0 \in \mathrm{Ray}_{v_0,\lambda}(C)$, and keeping in mind that $C = \operatorname{epi} f$. □

Given a closed vector subspace E of a Hilbert space H and its orthogonal vector space $E^\perp$, we already saw through (17.18) that the projector mapping π_E onto E parallel to $E^\perp$ coincides with the orthogonal projector mapping Π_E on E, where for any $x \in H$ the element $\Pi_E(x)$ is the unique one in E such that $\langle x - \pi_E(x), y\rangle = 0$ for all $y \in E$.

REMARK 17.24. Keep the assumptions of Theorem 17.23 above and notation $C := \operatorname{epi} f$. Since $\mathrm{Ray}_{v_0,\lambda}(C) \subset O$, for the point $u_0 \in \mathrm{Ray}_{v_0,\lambda}(C)$ we have seen in the proof of the theorem that $P_C(u_0) = v_0$, so

$$u_0 - P_C(u_0) = u_0 - v_0 = t(\nabla f(\overline{x}), -1) = t(0, -1)$$

for some $t \in]0, \lambda[$. Consequently,

$$\nabla d_C(u_0) = \frac{u_0 - P_C(u_0)}{\|u_0 - P_C(u_0)\|} = (0, -1) = \frac{(\nabla f(\overline{x}), -1)}{\|(\nabla f(\overline{x}), -1)\|} = \varphi(v_0),$$

where the latter equality is due to the definition of φ. This allows us to point out that

$$
\begin{aligned}
DP_C(u_0) &= -[D_2F(u_0, v_0)]^{-1} \circ D_1F(u_0, v_0) \\
&= -[D_2F(u_0, v_0)]^{-1} \circ \left(\mathrm{id}_{H\times\mathbb{R}} - \left\langle \frac{u_0 - P_C(u_0)}{d_C(u_0)}, \cdot \right\rangle \frac{u_0 - P_C(u_0)}{d_C(u_0)} \right) \\
&= -[D_2F(u_0, v_0)]^{-1} \circ \Pi_{H\times\{0\}},
\end{aligned}
$$

where the latter equality follows from the above equality $(u_0 - P_C(u_0))/d_C(u_0) = (0, -1)$ and where according to the above notation, $\Pi_{H\times\{0\}}$ is the orthogonal projection on the closed vector subspace $H \times \{0\}$ of $H \times \mathbb{R}$.

We also note that $-D_2F(u_0, v_0)$ maps $H \times \{0\}$ onto $H \times \{0\}$. In particular, we have that $DP_C(u_0)$ restricted to $H \times \{0\}$ is invertible as a mapping from $H \times \{0\}$ to $H \times \{0\}$. □

The following lemma is motivated by the definition of λ in Theorem 17.23. Before stating the lemma, let us recall (see Proposition 15.71) that given a $\mathcal{C}^{1,1}$-function $f : U \to \mathbb{R}$ on an open set U of a Hilbert space H, its epigraph $\operatorname{epi} f$ is prox-regular at any of its points.

LEMMA 17.25. Let U be an open set of a Hilbert space H, let $p \in \mathbb{N}$ and let $f : U \to \mathbb{R}$ be a function of class $\mathcal{C}^{p+1}$ near $\overline{x} \in U$ such that $\nabla f(\overline{x}) = 0$. Let $r \in]0, +\infty]$ be a constant of prox-regularity of $\operatorname{epi} f$ at $(\overline{x}, f(\overline{x}))$. Then, one has

$$\inf \left\{ \langle u, D^2 f(\overline{x}) u \rangle \ : \ u \in \mathbb{B}_H \right\} \geq -\frac{1}{r}.$$

PROOF. Let us denote $O := B_{X\times\mathbb{R}}((\overline{x}, f(\overline{x})), \alpha)$ with $\alpha > 0$ small enough such that $\operatorname{pr}_H(O) \subset U$, f is of class $\mathcal{C}^{p+1}$ on $\operatorname{pr}_H(O)$ and $\operatorname{epi} f$ is (r, α)-prox-regular at $(\overline{x}, f(\overline{x}))$. Then, for every $(x, s) \in O \cap \operatorname{epi} f$ and every $\xi \in N^P(\operatorname{epi} f; (x, s))$, we have by the equivalence (a)⇔(c) in Theorem 15.70 that

$$\langle \xi, (x', s') - (x, s) \rangle \leq \frac{1}{2r} \|\xi\| \|(x', s') - (x, s)\|^2, \ \forall (x', s') \in \operatorname{epi} f. \tag{17.25}$$

Fix $h \in H$. Since for every $x \in \operatorname{pr}_H(O)$ we have (as already justified in the proof of Theorem 17.23) that

$$N^P\big(\operatorname{epi} f; (x, f(x))\big) = \{t(\nabla f(x), -1) \ : \ t \geq 0\},$$

using the equality $\nabla f(\overline{x}) = 0$ we can write

$$\begin{aligned}
&\langle h, D^2 f(\overline{x}) h \rangle = \lim_{t \downarrow 0} \left\langle th, \frac{\nabla f(\overline{x} + th) - \nabla f(\overline{x})}{t^2} \right\rangle \\
&= \lim_{t \downarrow 0} \left\langle \big(th, f(\overline{x} + th) - f(\overline{x})\big), \frac{\big(\nabla f(\overline{x} + th), -1\big) - (0, -1)}{t^2} \right\rangle \\
&= \lim_{t \downarrow 0} \left\langle \big(\overline{x} + th, f(\overline{x} + th)\big) - \big(\overline{x}, f(\overline{x})\big), \frac{\big(\nabla f(\overline{x} + th), -1\big)}{t^2} \right\rangle + \frac{f(\overline{x} + th) - f(\overline{x})}{t^2},
\end{aligned}$$

thus, according to (17.25), we obtain

$$\begin{aligned}
&\langle h, D^2 f(\overline{x}) h \rangle \\
&\geq \lim_{t \downarrow 0} -\frac{1}{2rt^2} \|\big(\nabla f(\overline{x} + th), -1\big)\| \left\|\big(th, f(\overline{x} + th) - f(\overline{x})\big)\right\|^2 + \frac{f(\overline{x} + th) - f(\overline{x})}{t^2} \\
&= \lim_{t \downarrow 0} -\frac{1}{2r} \|\big(\nabla f(\overline{x} + th), -1\big)\| \left\| \left(h, \frac{f(\overline{x} + th) - f(\overline{x})}{t} \right) \right\|^2 + \frac{f(\overline{x} + th) - f(\overline{x})}{t^2} \\
&= -\frac{1}{2r} \|\big(\nabla f(\overline{x}), -1\big)\| \left\|\big(h, Df(\overline{x})h\big)\right\|^2 + \lim_{t \downarrow 0} \frac{f(\overline{x} + th) - f(\overline{x}) - tDf(\overline{x})h}{t^2} \\
&= -\frac{1}{2r} \|(h, 0)\|^2 + \frac{1}{2} \langle h, D^2 f(\overline{x}) h \rangle = -\frac{1}{2r} \|h\|^2 + \frac{1}{2} \langle h, D^2 f(\overline{x}) h \rangle,
\end{aligned}$$

where the penultimate equality is due to the nullity of both $\nabla f(\overline{x})$ and $Df(\overline{x})$. The conclusion follows. □

With the help of Lemma 17.25 we can prove the following lemma which is one the steps preparing Theorem 17.30 below.

LEMMA 17.26. Let C be a set in a Hilbert space H which is closed near $\overline{x} \in \operatorname{bdry} C$. Assume that C has the interior tangent cone property at $\overline{x}$ and that $\operatorname{bdry} C$ is a $\mathcal{C}^{p+1}$-submanifold (with $p \geq 1$) at $\overline{x}$. Then C is prox-regular at $\overline{x}$ and there exist a real $\lambda > 0$ and an open set V containing $\operatorname{Ray}_{\overline{x},\lambda}(C)$ such that

- d_C is of class $\mathcal{C}^{p+1}$ on V;
- P_C is of class $\mathcal{C}^p$ on V.

PROOF. We already know by Proposition 17.21(b) that C is prox-regular at $\overline{x}$. Recalling that $Z(\overline{x}) := T_{\overline{x}}(\operatorname{bdry} C)$ and applying Proposition 17.21, there exist an open neighborhood U of $\overline{x}$ in H and a function $f : \pi_{Z(\overline{x})}(U) \subset Z(\overline{x}) \to \mathbb{R}$ such that, denoting $z := \pi_{Z(\overline{x})}(\overline{x})$, f is of class $\mathcal{C}^{p+1}$ in $\pi_{Z(\overline{x})}(U)$, $\nabla f(z) = 0$,

$$U \cap C = \{z' + t\widehat{n}(\overline{x}) \in U \ : z' \in \pi_{Z(\overline{x})}(U),\ f(z') \leq t\},$$

and also $\operatorname{epi} f$ is prox-regular at $(z, f(z))$; keep in mind that $\widehat{n}(\overline{x})$ denotes the unit interior normal of $\operatorname{bdry} C$ at $\overline{x}$. Choose $\eta > 0$ such that $B(\overline{x}, \eta) \subset U$ and choose also $r_0, \alpha_0 > 0$ with $r_0 + \alpha_0 < \eta$ such that C and $\operatorname{epi} f$ are (r_0, α_0)-prox-regular at $\overline{x}$ and $(z, f(z))$ respectively. Note by Theorem 17.7(b) that $\mathcal{W}_C(\overline{x}, r_0, \alpha_0)$ is open. Let $L : Z(\overline{x}) \times \mathbb{R} \to H$ be the canonical isomorphism given by $L(z', t) = z' + t\widehat{n}(\overline{x})$ for all $(z', t) \in Z(\overline{x}) \times \mathbb{R}$, so

$$U \cap C = U \cap L(\operatorname{epi} f).$$

By Theorem 17.23 and the inequality of Lemma 17.25, we have that $P_{\operatorname{epi} f}$ is of class $\mathcal{C}^p$ on an open neighborhood W of $\operatorname{Ray}_{(z,f(z)),r_0/2}(\operatorname{epi} f)$. By continuity of $P_{\operatorname{epi} f}$ near $(z, f(z)) \in L^{-1}(U)$ (due to the prox-regularity of $\operatorname{epi} f$ at $(z, f(z))$) and by the openness of $L^{-1}(\mathcal{W}_C(\overline{x}, r_0, \alpha_0)) \ni (z, f(z))$ (due to the openness of $\mathcal{W}_C(\overline{x}, r_0, \alpha_0)$), choose $0 < r < r_0$ and $0 < \alpha < \alpha_0$ such that $P_{\operatorname{epi} f}(w) \in L^{-1}(U)$ for all $w \in L^{-1}(\mathcal{W}_C(\overline{x}, r, \alpha))$. Put $r' := r/2$ and note that $\mathcal{W}_C(\overline{x}, r', \alpha)$ is open by (r', α)-prox-regularity of C at $\overline{x}$ and by Theorem 17.7(b) again. Observing by (17.15) that

$$\begin{aligned}\operatorname{Ray}_{\overline{x},r'}(C) &= \{\overline{x} - t\widehat{n}(\overline{x}) : t \in]0, r'[\} = L\big(\{(z, f(z)) + t(0, -1) : t \in]0, r'[\}\big) \\ &= L\big(\operatorname{Ray}_{(z,f(z)),r'}(\operatorname{epi} f)\big),\end{aligned}$$

we have that $W' := W \cap L^{-1}(\mathcal{W}_C(\overline{x}, r', \alpha))$ is an open neighborhood of the truncated ray $\operatorname{Ray}_{(z,f(z)),r'}(\operatorname{epi} f)$. For each $w \in W'$ we have $L(w) \in \mathcal{W}_C(\overline{x}, r', \alpha)$, so the (r', α)-prox-regularity of C at $\overline{x}$ ensures the existence of $P_C(L(w))$ and the definition of $\mathcal{W}_C(\overline{x}, r', \alpha)$ gives $P_C(L(w)) \in B_H(\overline{x}, \alpha)$, hence $P_C(L(w)) \in U \cap C$, which in turn entails that

$$L^{-1}\big(P_C(L(w))\big) \in L^{-1}(U \cap C) = \operatorname{epi} f \cap L^{-1}(U).$$

Since L is an isometric isomorphism, we obtain for each $w \in W'$

$$\begin{aligned}&\|L(w) - P_C(L(w))\| = \|w - L^{-1}\big(P_C(L(w))\big)\| \\ &\geq \|w - P_{\operatorname{epi} f}(w)\| = \|L(w) - L(P_{\operatorname{epi} f}(w))\| \geq \|L(w) - P_C(L(w))\|,\end{aligned}$$

where the latter inequality is due to the inclusion $L(P_{\operatorname{epi} f}(w)) \in C$ since

$$L(P_{\operatorname{epi} f}(w)) \in U \cap L(\operatorname{epi} f) = U \cap C.$$

Therefore, for each $v \in V := L(W')$, we have

$$P_C(v) = (L \circ P_{\operatorname{epi} f} \circ L^{-1})(v), \tag{17.26}$$

hence P_C (which is well-defined on V) is of class $\mathcal{C}^p$ on V. Further, since W is open, the set V is an open neighborhood of $\operatorname{Ray}_{x,r'}(C)$. This finishes the proof of the lemma. □

From Remark 17.24, we see that, in the proof of the preceding lemma, for each $u_0 \in \operatorname{Ray}_{\bar{x},r'}(C)$, the operator $DP_{\operatorname{epi} f}(L^{-1}(u_0))$ restricted to $Z(\bar{x}) \times \{0\}$ is invertible as a mapping from $Z(\bar{x}) \times \{0\}$ onto $Z(\bar{x}) \times \{0\}$. From this observation, we can conclude that the operator

$$DP_C(u_0) = L \circ DP_{\operatorname{epi} f}(L^{-1}(u_0)) \circ L^{-1}$$

restricted to $Z(\bar{x})$ also is invertible as a mapping from $Z(\bar{x})$ onto $Z(\bar{x})$. This yields the following proposition.

Proposition 17.27. Let C be a set in a Hilbert space H which is closed near $\overline{x} \in \operatorname{bdry} C$. Assume that C has the interior tangent cone property at $\overline{x}$ and that its boundary is a $\mathcal{C}^{p+1}$-submanifold (with $p \geq 1$) at $\overline{x}$. Then there exist $r, \alpha > 0$ such that C is (r,α)-prox-regular at $\overline{x}$ and such that $B_H(\overline{x},\alpha) \cap \operatorname{bdry} C$ is a $\mathcal{C}^{p+1}$-submanifold and for each $u_0 \in \operatorname{Ray}_{\overline{x},r'}(C)$ the operator $DP_C(u_0)$ is invertible as a mapping from $Z(\overline{x})$ onto $Z(\overline{x})$, where $r' := r/2$ and $Z(\overline{x}) := T_{\overline{x}}(\operatorname{bdry} C)$.

Given an extended real $\lambda > 0$, a nonempty set C of a Hilbert space H and a point $x \in H \setminus C$ such that $P_C(x)$ exists, we extend Definition 17.22 by defining the *λ-truncated open normal ray* starting from $P_C(x)$ in the direction $x - P_C(x)$ as the set

$$\operatorname{Ray}^{x}_{P_C(x),\lambda}(C) := \left\{ P_C(x) + t\frac{x - P_C(x)}{\|x - P_C(x)\|} : t \in]0, \lambda[\right\}. \tag{17.27}$$

Of course, $\operatorname{Ray}^{x}_{P_C(x),\lambda}(C)$ coincides with $\operatorname{Ray}_{P_C(x),\lambda}(C)$ whenever there is a unique unit proximal normal to C at $P_C(x)$.

Suppose that the point $x \in H$ outside the set C is such that $P_C(x)$ exists and C is (r,α)-prox-regular at $P_C(x)$. By definition we have $\operatorname{Ray}^{x}_{P_C(x),\lambda}(C) \subset \mathcal{R}_C(P_C(x),\lambda,\alpha)$. On the other hand, with $\zeta := (x - P_C(x))/\|x - P_C(x)\|$, any $u \in \operatorname{Ray}^{x}_{P_C(x),r}(C)$ can be written as $u = P_C(x) + t\zeta$ with $t \in]0, r[$, so

$$u \in \big(I + t\mathbb{B}_H \cap N^P(C;\cdot)\big)(P_C(x)),$$

or equivalently

$$P_C(x) \in \big(I + t\mathbb{B}_H \cap N^P(C;\cdot)\big)^{-1}(u) \cap B(P_C(x),\alpha).$$

Then, the assertion (i') in Theorem 17.5 yields that $P_C(u)$ exists and that the equality $P_C(u) = P_C(x)$ holds.

We summarize this feature in the following lemma.

Lemma 17.28. Let C be a nonempty set in a Hilbert space H and let $x \in H \setminus C$ be such that $P_C(x)$ exists. Assume that C is (r,α)-prox-regular at $P_C(x)$ with $r, \alpha \in]0, +\infty]$. Then, for every $u \in \operatorname{Ray}^{x}_{P_C(x),r}(C)$ one has that $P_C(u)$ exists and $P_C(u) = P_C(x)$.

Now we show in particular that the differentiability of P_C at $x \in H \setminus C$ entails its differentiability at points in the truncated open ray $\mathrm{Ray}\,^{x}_{P_C(x),r}(C)$ whenever C is r-prox-regular at $P_C(x)$.

PROPOSITION 17.29. Let C be a nonempty set of a Hilbert space H and let $x \in H \setminus C$. Assume that $P_C(x)$ exists and that C is r-prox-regular at $P_C(x)$ for some $r \in]0, +\infty]$ with $d_C(x) < r$. Then, for each $y \in \mathrm{Ray}\,^{x}_{P_C(x),r}(C)$ the following hold.
(a) If $D_G P_C(x)$ (resp. $D_F P_C(x)$) exists, then $D_G P_C(y)$ (resp. $D_F P_C(y)$) also exists.
(b) If $P_C(\cdot)$ is of class $\mathcal{C}^p$ (with $p \geq 1$) near x, then it is also of class $\mathcal{C}^p$ near y.

PROOF. Without loss of generality, we may and do suppose that $r < +\infty$. Choose $\alpha > 0$ such that C is (r, α)-prox-regular at $P_C(x)$. By Definition of $\mathrm{Ray}\,^{x}_{P_C(x),r}(C)$ in (17.27) we may write $y = P_C(x) + \mu(x - P_C(x))$ with a real $\mu > 0$ satisfying $\mu\|x - P_C(x)\| < r$. Choose $\gamma \in]0, 1[$ such that $\max\{\mu d_C(x), d_C(x)\} < \gamma r$. The above features concerning y and the equality $x = P_C(x) + (x - P_C(x))$ along with the inequality $\|x - P_C(x)\| < \gamma r$ (according to the choice of γ) guarantee that both y, x belong to $\mathrm{Ray}\,^{x}_{P_C(x),\gamma r}(C)$. It ensues by definition of $\mathrm{Ray}\,^{x}_{P_C(x),\gamma r}(C)$ that $[x, y] \subset \mathrm{Ray}\,^{x}_{P_C(x),\gamma r}(C)$ and $d_C(w) \leq \|w - P_C(x)\| < \gamma r$ for all $w \in [x, y]$. Taking into account the (r, α)-prox-regularity of C at $P_C(x)$, Lemma 17.28 says that $P_C(w) = P_C(x)$ for all $w \in [x, y]$, and Theorem 17.7 ensures that

$$[x, y] \subset \mathrm{Ray}\,^{x}_{P_C(x),\gamma r}(C) \subset \mathcal{R}_C(P_C(x), \gamma r, \alpha) = \mathcal{W}_C(P_C(x), \gamma r, \alpha).$$

Keeping in mind that the set $\mathcal{W}_C(P_C(x), \gamma r, \alpha)$ is open by the (r, α)-prox-regularity of C at $P_C(x)$ (see Theorem 17.7(b)), the inclusion $[x, y] \subset \mathcal{W}_C(P_C(x), \gamma r, \alpha)$ combined with the compactness of $[x, y]$ furnishes some real $\delta > 0$ such that

$$[x, y] + B[0, 2\delta] \subset \mathcal{W}_C(P_C(x), \gamma r, \alpha). \tag{17.28}$$

On the other hand, we know by (h) in Theorem 17.12 that $P_C(\cdot)$ is Lipschitz continuous on $\mathcal{W}_C(P_C(x), \gamma r, \alpha)$ with Lipschitz constant $K := (1 - \gamma)^{-1}$, hence in particular, $\|D_G P_C(w)\| \leq K$ for all $w \in [x, y]$ such that $D_G P_C(w)$ exists.

Now, choose a sequence $\{y_i\}_{i=0}^{n} \subset [x, y]$ such that $y_0 = x$, $y_n = y$ and

$$y_{i+1} = y_i + t_i(y_i - P_C(y_i))$$

with $t_i \in \mathbb{R}$ small enough such that $|t_i| < \min\{(1+K)^{-1}, \gamma^{-1}(1-\gamma)\}$ and $\gamma r|t_i| < \delta$. Then for each $i \in \{0, \cdots, n\}$ we have by (17.28)

$$y_i + h + t_i(y_i + h - P_C(y_i + h)) \in \mathcal{W}_C(P_C(x), \gamma r, \alpha), \ \forall h \in B_H(0, \delta). \tag{17.29}$$

As $P_C(y_i) = P_C(x)$ for all $i \in \{0, \cdots, n\}$, it is enough to show that both statements (a) and (b) hold replacing y by y_1. The general case is obtained inductively, replacing in the ith step the roles of x and y by y_{i-1} and y_i respectively.

Assume then $n = 1$ and denote $t := t_1$. Note first that the operator

$$A := \mathrm{id}_H + t(\mathrm{id}_H - D_G P_C(x))$$

is invertible, since $\|\mathrm{id}_H - A\| = |t|\|\mathrm{id}_H - D_G P_C(x)\| \leq |t|(1+K) < 1$ (see Proposition C.1 in Appendix C). We will show that P_C is Gâteaux differentiable at y and $D_G P_C(y) = D_G P_C(x) \circ A^{-1}$. Fix $h \in H$, $s \in]0, +\infty[$ and denote $u = A^{-1}h$. By definition of A we can write

$$y + sAu = x + t(x - P_C(x)) + su + st(u - D_G P_C(x)u)$$
$$= (1 + t)(x + su) - t(P_C(x) + sD_G P_C(x)u).$$

Also, we know that $P_C(x + su) = P_C(x) + sD_G P_C(x)u + o(s)$ with $o(s)/s \to 0$ as $s \to 0$, and hence combining both equalities we obtain that

$$y + sAu = (1 + t)(x + su) - tP_C(x + su) + o(s). \tag{17.30}$$

Taking s small enough, we may suppose that $x_s := x + su \in \mathcal{W}_C(P_C(x), \gamma r, \alpha)$ according to the openness of $\mathcal{W}_C(P_C(x), \gamma r, \alpha)$. Then, $P_C(x_s)$ exists and putting

$$z_s := x_s + t(x_s - P_C(x_s)) = (1 + t)x_s - tP_C(x_s)$$

and noting that $\gamma(1+t) < 1$, Lemma 17.8 gives $P_C(z_s) = P_C(x_s)$, hence by (17.30)

$$P_C\big(y + sAu + o(s)\big) = P_C(x + su).$$

Since $P_C(\cdot)$ is Lipschitz continuous on $\mathcal{W}_C(P_C(x), \gamma r, \alpha)$ as seen above, we can write for s small enough

$$P_C(y+sh) = P_C(y+sAu) = P_C(x+su)+o(s) = P_C(y)+sD_G P_C(x)\circ A^{-1}(h)+o(s),$$

where the last equality follows from the equalities $P_C(y) = P_C(x)$ and $u = A^{-1}h$. Since h is arbitrary, we conclude that $D_G P_C(y)$ exists and it coincides with the operator $D_G P_C(x)\circ A^{-1}$, as we claimed. The case when P_C is Fréchet differentiable is obtained with the same arguments through the fact that $o(A^{-1}h) = o(h)$.

Finally, to prove (b) suppose that $P_C(\cdot)$ is of class $\mathcal{C}^p$ near x. It is direct that the mapping $u \mapsto \Phi(u) := (1+t)u - tP_C(u)$ is also of class $\mathcal{C}^p$ near x and $D_F\Phi(x) = A$. Recall also that both x and $(1+t)x - tP_C(x)$ are in the open set $\mathcal{W}_C(P_C(x), \gamma r, \alpha)$. Using the inverse mapping theorem (see Theorem E.1 in Appendix E), there exists an open neighborhood $U \subset \mathcal{W}_C(P_C(x), \gamma r, \alpha)$ of x and an open neighborhood V of $y = \Phi(x)$ also included in $\mathcal{W}_C(P_C(x), \gamma r, \alpha)$ such that P_C is of class $\mathcal{C}^p$ on U and such that the restriction $\Phi_{|U} : U \to V$ (of Φ to U with images in V) is invertible and $\Phi_{|U}^{-1}$ is of class $\mathcal{C}^p$ on V. For each $u \in U$ by Lemma 17.8 we have $P_C(\Phi(u)) = P_C(u)$, which yields

$$P_C(v) = P_C \circ \Phi \circ (\Phi_{|U}^{-1})(v) = P_C \circ (\Phi_{|U}^{-1})(v) \quad \text{for all } v \in V,$$

and so the conclusion follows by chain rule. The proof of the proposition is complete. □

From the development carried out, in fact from Proposition 17.21(b), Lemma 17.26, Proposition 17.27 and Proposition 17.29, the following theorem directly follows. The theorem furnishes conditions for the smoothness of metric projection in a very large quantitative way involving the threshold (r, α) of prox-regularity.

THEOREM 17.30 (smoothness of metric projection to sets with smooth boundary and interior tangent property). *Let C be a set in a Hilbert space H which is closed near $\overline{x} \in \operatorname{bdry} C$ and has the interior tangent cone property at $\overline{x}$. Assume that $\operatorname{bdry} C$ is a $\mathcal{C}^{p+1}$-submanifold near $\overline{x}$ with an integer $p \geq 1$. Then C is prox-regular at $\overline{x}$, and given extended reals $r, \alpha > 0$ such that $B_H(\overline{x}, \alpha) \cap \operatorname{bdry} C$ is a $\mathcal{C}^{p+1}$-submanifold and C is r-prox-regular at $\overline{x}$, there exists an open neighborhood V of the r-truncated open ray $\operatorname{Ray}_{\overline{x},r}(C)$ such that*

- *d_C is of class $\mathcal{C}^{p+1}$ on V;*

- P_C is of class $\mathcal{C}^p$ on V.

Furthermore, if the set C is (r,α)-prox-regular at $\overline{x}$ and has the interior tangent cone property at each point in $B_H(\overline{x},\alpha)\cap C$, then $\mathcal{W}_C(\overline{x},r,\alpha)\setminus C$ is open and

- d_C is of class $\mathcal{C}^{p+1}$ on the open set $\mathcal{W}_C(\overline{x},r,\alpha)\setminus C$;
- P_C is of class $\mathcal{C}^p$ on the open set $\mathcal{W}_C(\overline{x},r,\alpha)\setminus C$.

A first corollary is concerned with $\rho(\cdot)$-prox-regular sets.

COROLLARY 17.31. Let C be a $\rho(\cdot)$-prox-regular set in a Hilbert space H which satisfies the interior tangent cone property at any of its points. Assume that its boundary $\operatorname{bdry} C$ is, for some integer $p\geq 1$, a $\mathcal{C}^{p+1}$-submanifold at any of its points. Then the following hold:

- d_C is of class $\mathcal{C}^{p+1}$ on $U_{\rho(\cdot)}(C)\setminus C$;
- P_C is of class $\mathcal{C}^p$ on $U_{\rho(\cdot)}(C)\setminus C$.

PROOF. Fix $u\in U:=U_{\rho(\cdot)}(C)\setminus C$. Since C is $\rho(\cdot)$-prox-regular, we have that there exists $y\in \operatorname{Proj}_C(u)$ such that $d_C(u)<\rho(y)$. Let us fix a real r with $d_C(u)<r<\rho(y)$. Since ρ is continuous, there exists a neighborhood V of y on which $\rho(v)>r$ for each $v\in C\cap V$. Therefore, by properties related to $U_{\rho(\cdot)}(C)$ in Theorem 15.22 and by Theorem 15.70 the set C is r-prox-regular at y. Then, by Theorem 17.30 there exists $\alpha>0$ small enough such that P_C is well-defined on $\mathcal{W}_C(y,r,\alpha)\setminus C$ and of class $\mathcal{C}^p$ on this open set. Noting that

$$u\in(\mathcal{W}_C(y,r,\alpha)\setminus C)\cap U\subseteq U,$$

and that both sets $\mathcal{W}_C(y,r,\alpha)\setminus C$ and U are open, we conclude that P_C is well-defined near u and of class $\mathcal{C}^p$ near u. Since u is arbitrary in the open set U, the conclusion follows. □

Taking $\rho(\cdot)$ constant immediately yields the next corollary related to the important case of r-prox-regularity.

COROLLARY 17.32. Let C be an r-prox-regular set (with $r\in\,]0,+\infty]$) in a Hilbert space H which satisfies the interior tangent cone property at any of its points. Assume that its boundary $\operatorname{bdry} C$ is, for some integer $p\geq 1$, a $\mathcal{C}^{p+1}$-submanifold at any of its points. Then the following hold:

- d_C is of class $\mathcal{C}^{p+1}$ on $U_r(C)\setminus C$;
- P_C is of class $\mathcal{C}^p$ on $U_r(C)\setminus C$.

Now recall that a *convex body* in a normed space is a closed convex set with nonempty topological interior. In the case when C is a *convex body*, recalling that all convex closed sets are $(+\infty)$-prox-regular, we directly derive the following other corollary of Theorem 17.30.

COROLLARY 17.33 (R.B. Holmes). Let C be a convex body of a Hilbert space H. Assume that $\operatorname{bdry} C$ is a $\mathcal{C}^{p+1}$-submanifold at a point $\overline{x}\in\operatorname{bdry} C$, with $p\geq 1$. Then there exists an open neighborhood W of $\operatorname{Ray}_{\overline{x}}(C)$ such that

- d_C is of class $\mathcal{C}^{p+1}$ on W;
- P_C is of class $\mathcal{C}^p$ on W.

The next theorem directly follows.

THEOREM 17.34 (R.B. Holmes: smoothness of metric projection to convex bodies). Let C be a convex body of a Hilbert space H whose boundary $\operatorname{bdry} C$ is, for some integer $p \geq 1$, a $\mathcal{C}^{p+1}$-submanifold at any of its points. Then

- d_C is of class $\mathcal{C}^{p+1}$ on $H \setminus C$;
- P_C is of class $\mathcal{C}^p$ on $H \setminus C$.

17.3. Characterization of epi-Lipschitz sets with smooth boundary

For a set C which is closed near $\overline{x} \in \operatorname{bdry} C$ and satisfies the interior tangent condition at $\overline{x}$, we saw in Theorem 17.30 that the metric projection is $\mathcal{C}^p$-smooth around a truncated ray $\mathrm{Ray}_{\overline{x},\lambda}(C)$ whenever $\operatorname{bdry} C$ is a $\mathcal{C}^{p+1}$-submanifold at $\overline{x}$. The present section is devoted to the study of an additional condition for the converse in order to characterize the $\mathcal{C}^{p+1}$-submanifold property of $\operatorname{bdry} C$ at $\overline{x}$ in terms of the above $\mathcal{C}^p$-smoothness of P_C plus this condition.

17.3.1. Properties of derivatives of metric projection. In this first subsection we will establish some basic properties of the derivative of the metric projection when it exists near the point of interest.

LEMMA 17.35. Let C be a nonempty set in a Hilbert space H, let $x \in H \setminus C$ and let $S := \operatorname{cl} C$. Assume that $P_C(\cdot)$ is well-defined and continuous on a neighborhood U of x with $P_C(u) = P_S(u)$ for all $u \in U$; If $D_G P_C(x)$ exists, then it coincides with the second Gâteaux derivative at x of the convex function $\varphi_C : H \to \mathbb{R}$ given by

$$\varphi_C(h) := \frac{1}{2}\|h\|^2 - \frac{1}{2}d_C^2(h) = \frac{1}{2}\|h\|^2 - \frac{1}{2}d_S^2(h) \quad \text{for all } h \in X. \tag{17.31}$$

Consequently, $D_G P_C(x)$ is a symmetric and positive operator.

PROOF. We already saw that the continuous function φ_C is convex according to the equality $\varphi_C(h) = \sup_{y \in C}\{\langle h, y\rangle - \frac{1}{2}\|y\|^2\}$ in (2.71). Since $P_S(\cdot) = P_C(\cdot)$ is well-defined and continuous on U, the equivalence (d)$\Leftrightarrow$(b) in Proposition 15.19 says that $d_S^2 = d_C^2$ is Fréchet differentiable on U, hence by (17.31) the function φ_C is Fréchet differentiable on U. Using this Snd Lemma 15.11(e) yields $\nabla_F \varphi_C(u) = P_C(u)$ for all $u \in U$. In particular, since P_S is continuous on U, we get that φ_C is of class $\mathcal{C}^1$ on U with $D_F\varphi_C(u) = \langle P_S(u), \cdot\rangle = \langle P_C(u), \cdot\rangle$ for all $u \in U$. Since $x \in U$, we conclude that $D_G P_C(x)$, when it exists, is the second (Gâteaux) derivative of φ_C at x, and so, by convexity of φ_C, it ensues that $D_G P_C(x)$ is a symmetric and positive operator. □

For a nonempty set C of a Hilbert space H and a point $x \in H \setminus C$ we define, if $P_C(x)$ exists, the set

$$H_C[x] := \{h \in H \ : \ \langle h, x - P_C(x)\rangle = 0\}, \tag{17.32}$$

that is, $H_C[x]$ is the vector hyperplane orthogonal to $x - P_C(x)$. If there is no confusion, we will simply write $H[x]$ instead of $H_C[x]$. It is worth noticing that the (orthogonal projection) operator $\Pi_{H[x]}$ is symmetric since (as known) Π_E is symmetric for any closed vector subspace of H: indeed, for any $h, h' \in H$ the equalities $\langle \Pi_E h, h' - \Pi_E h'\rangle = 0 = \langle h - \Pi_E h, \Pi_E h'\rangle$ yield

$$\langle \Pi_E h, h'\rangle = \langle \Pi_E h, \Pi_E h'\rangle = \langle h, \Pi_E h'\rangle. \tag{17.33}$$

The following proposition considers the situation of composition of operators $D_G P_C(x)$ and $\Pi_{H[x]}$.

PROPOSITION 17.36. Let C be a nonempty subset of a Hilbert space H, let $x \in H \setminus C$ and let $S := \operatorname{cl} C$. Assume that P_C is well defined and continuous on a neighborhood U of x with $P_C(u) = P_S(u)$ for all $u \in U$ and that $D_G P_C(x)$ exists. Then the composition operator $D_G P_C(x) \circ \Pi_{H[x]}$ is symmetric and

$$D_G P_C(x) \circ \Pi_{H[x]} = D_G P_C(x) = \Pi_{H[x]} \circ D_G P_C(x).$$

In particular, $D_G P_C(x) H \subseteq H[x]$.

PROOF. First, noting that $P_C(z) = P_C(x)$ for every $z \in [P_C(x), x]$, we see that

$$D_G P_C(x)(P_C(x) - x) = \lim_{t \downarrow 0} \frac{P_C(x + t(P_C(x) - x)) - P_C(x)}{t} = 0.$$

For any $y \in H$, the previous equality and the inclusion $y - \Pi_{H[x]}(y) \in \mathbb{R}(x - P_C(x))$ ensure that $D_G P_C(x)(y - \Pi_{H[x]}(y)) = 0$. This tells us that

$$D_G P_C(x) \circ \Pi_{H[x]} = D_G P_C(x).$$

On the other hand, by Lemma 17.35 and by (17.33) the operators $D_G P_C(x)$ and $\Pi_{H[x]}$ are both symmetric. Applying twice the above equality $D_G P_C(x) \circ \Pi_{H[x]} = D_G P_C(x))$, it follows that

$$\begin{aligned} \langle \Pi_{H[x]} \circ D_G P_C(x) h, h' \rangle &= \langle h, D_G P_C(x) \circ \Pi_{H[x]} h' \rangle \\ &= \langle h, D_G P_C(x) h' \rangle = \langle D_G P_C(x) h, h' \rangle \\ &= \langle D_G P_C(x) \circ \Pi_{H[x]} h, h' \rangle, \end{aligned}$$

for all $h, h' \in X$. We conclude that $D_G P_C(x)$ and $\Pi_{H[x]}$ commute and that the desired equalities hold true. □

The following proposition provides a suitable upper bound for the norm of $D_G P_C(x)$ under the prox-regularity of C at $P_C(x)$.

PROPOSITION 17.37. Let C be a nonempty set of a Hilbert space H and let $x \in H \setminus C$. Assume that $P_C(x)$ exists and that C is r-prox-regular at $P_C(x)$ for some $r \in]0, +\infty]$ with $d_C(x) < r$. If $D_G P_C(x)$ exists, then

$$\|D_G P_C(x)\| \le \frac{1}{1 - r^{-1} d_C(x)}.$$

In particular, for any $\delta \in \left]0, 1 - \frac{d_C(x)}{r}\right[$, one has that the operator $\mathrm{id}_H - \delta D_G P_C(x)$ is positive and invertible.

PROOF. If there is some real $r_0 > 0$ such that the result holds for every $r \in]r_0, +\infty[$, it is clear that it still valid for $r = +\infty$ by making $r \to +\infty$. It then suffices to give the proof for $r < +\infty$. So, fix $r \in]0, +\infty[$.

Choose $\alpha > 0$ such that C is (r, α)-prox-regular at $P_C(x)$. Take any $\gamma \in]0, 1[$ such that $d_C(x) < \gamma r$. Then $x \in \mathcal{W}_C(P_C(x), \gamma r, \alpha)$, and (h) in Theorem 17.12 tells us that $P_C(\cdot)$ is $(1-\gamma)^{-1}$-Lipschitz continuous on this set $\mathcal{W}_C(P_C(x), \gamma r, \alpha)$ which is known to be open by Theorem 17.7(b). It follows that for every $h \in H$ there is a function $\varepsilon :]0, +\infty[\to]0, +\infty[$ with $\varepsilon(t) \to 0$ as $t \downarrow 0$ such that

$$\begin{aligned} \|D_G P_C(x) h\| &\le t^{-1} \big(\|P_C(x + th) - P_C(x)\| + t\varepsilon(t) \big) \\ &\le (1-\gamma)^{-1} \|h\| + \varepsilon(t), \end{aligned}$$

thus $\|D_G P_C(x)\| \leq (1-\gamma)^{-1}$. Passing to the limit as $\gamma \downarrow r^{-1} d_C(x)$ gives

$$\|D_G P_C(x)\| \leq (1 - r^{-1} d_C(x))^{-1} = \frac{r}{r - d_C(x)}.$$

Finally, for each $\delta \in \left]0, \frac{r-d_C(x)}{r}\right[$ we obtain

$$\|\mathrm{id}_H - (\mathrm{id}_H - \delta D_G P_C(x))\| = \delta \|D_G P_C(x)\| < 1,$$

and hence $\mathrm{id}_H - \delta D_G P_C(x)$ is positive and invertible, since every linear operator from H into H with norm less than 1 is known to enjoy both these properties (see Propositions C.1 and C.2 in Appendix C). □

Let us come back to the extended truncated open normal ray

$$\mathrm{Ray}\,^{x}_{P_C(x),\lambda}(C) := \left\{ P_C(x) + t \frac{x - P_C(x)}{\|x - P_C(x)\|} : t \in]0, \lambda[\right\}$$

defined in (17.27), where $\lambda > 0$ and $x \in H \setminus C$ is such that $P_C(x)$ exists. With this and the above proposition we complement Proposition 17.29 as follows.

PROPOSITION 17.38. *Let C be a nonempty set of a Hilbert space H, let $x \in H \setminus C$ and let $S := \mathrm{cl}\, C$. Assume that P_C is well-defined and continuous on a neighborhood U of x with $P_C(u) = P_S(u)$ for all u in this neighborhood. Assume that C is r-prox-regular at $P_C(x)$ for some $r \in]0, +\infty]$ with $d_C(x) < r$. Assume also that $D_G P_C(x)$ exists and is injective over $H[x]$ (resp. surjective onto $H[x]$). Then for each $y \in \mathrm{Ray}\,^{x}_{P_C(x),r}(C)$ one has that $D_G P_S(y)|_{H[y]}$ exists and is injective over $H[y] = H[x]$ (resp. surjective onto $H[y] = H[x]$).*

PROOF. Without loss of generality, we may and do suppose that $r < +\infty$. By Proposition 17.29 $D_G P_C(y)$ exists for every $y \in \mathrm{Ray}\,^{x}_{P_C(x),r}(C)$. Fix any $y \in \mathrm{Ray}\,^{x}_{P_C(x),r}(C)$. The proof of Proposition 17.29 reveals that there exists a constant $K > 0$ such that $\|D_G P_C(w)\| \leq K$ for all $w \in [x, y]$. The proof of the same Proposition 17.29 also reveals that there is a real t with $|t|(1 + K) < 1$ and some $u \in [x, y]$ such that for the continuous linear operator

$$A := \mathrm{id}_H + t(\mathrm{id}_H - D_G P_C(u))$$

from H into itself, we have $D_G P_C(y) = D_G P_C(u) \circ A^{-1}$, where the invertibility of A is due to the inequality

$$\|\mathrm{id}_H + t(\mathrm{id}_H - D_G P_C(u))\| = |t| \, \|\mathrm{id}_H - D_G P_C(u)\| \leq |t|(1 + K) < 1.$$

The proof will follow from the bijectivity of A, once we will show that $A(H[x]) = H[x]$. Let us argue that equality. First, by continuity of P_C near x we may use the definition of A and Proposition 17.36 to write for any $h \in H[x]$

$$\begin{aligned}
\langle Ah, x - P_C(x)\rangle &= (1+t)\langle h, x - P_C(x)\rangle - t\langle D_G P_C(x)h, x - P_C(x)\rangle \\
&= (1+t)\langle h, x - P_C(x)\rangle - t\langle \Pi_{H[x]} \circ D_G P_C(x)h, x - P_C(x)\rangle = 0,
\end{aligned}$$

and so the inclusion $A(H[x]) \subset H[x]$ holds true. On the other hand, the inequality $\|\mathrm{id}_H - A\| < 1$ entails that $A^{-1} = \mathrm{id}_H + \sum_{k=1}^{\infty} A^k$ (according to Proposition C.1 in Appendix C). Using this and the preceding inclusion $A(H[x]) \subset H[x]$ along with the feature that $H[x]$ is a closed vector space, we also see that $A^{-1}(H[x]) \subset H[x]$, hence $H[x] \subset A(H[x])$. The equality $A(H[x]) = H[x]$ is then justified as desired. □

Now, in addition to the role of Proposition 17.36 in the proof of Proposition 17.38, we prove, through Proposition 17.36 again, a lemma which will be used in the next section. The lemma will involve the concept of partial derivative with respect to a subspace.

DEFINITION 17.39. Let X and Y be two normed spaces, U be a nonempty open set in X and $f : U \to Y$ be a mapping. Let also E be a closed vector subspace of X and $\overline{x} \in U$. One says that f is *partially Gâteaux-differentiable (partially G-differentiable, for short) at $\overline{x}$ with respect to E* if there exists a continuous linear operator $A : E \to Y$ such that, for every $h \in E$ one can write

$$f(\overline{x} + th) = f(\overline{x}) + tAh + o(t) \quad \text{for all } t \text{ with } \overline{x} + th \in U.$$

In such a case A is unique, and it is called the *partial G-derivative of f at $\overline{x}$ with respect to E* and we denote it by $D_{G,E}f(\overline{x})$. Analogously, we say that f is *partially Fréchet-differentiable (partially F-differentiable, for short) at $\overline{x}$ with* respect to E if there exists a continuous linear operator $A : E \to Y$ such that

$$f(\overline{x} + h) = f(\overline{x}) + Ah + o(h), \quad \text{for all } h \in E \cap (-\overline{x} + U).$$

In such a case A is unique, and it is called the *partial F-derivative of f at $\overline{x}$ with respect to E* and we denote it by $D_{F,E}f(\overline{x})$. If $E = X$, it is clear that $D_{G,X}f(\overline{x})$ (resp. $D_{F,X}f(\overline{x})$) coincides with the usual G-derivative (resp. F-derivarive). We recall that the above notation $o(h)$ means that $\|h\|^{-1}o(h) \to 0$ as $h \to 0$.

For a vector subspace E of H it will be convenient in the rest of the section to denote by $\widehat{\imath}_E : E \to H$ the canonical injection of E into H defined by $\widehat{\imath}_E(x) = x$ for all $x \in E$.

LEMMA 17.40. Let C be a nonempty closed set in a Hilbert space H, let $x \in H \setminus C$ and let $S := \operatorname{cl} C$. Assume that $P_C(x)$ is well defined and continuous on a neighborhood of x with $P_C(u) = P_S(u)$ for all u in this neighborhood. Assume also that C is r-prox-regular at $P_C(x)$ for some $r \in]0, +\infty]$ with $d_C(x) < r$. If $D_G P_C(x)$ exists and $D_G P_C(x)(H) = H[x]$, then the following hold.
(a) The operator $D_G P_C(x)|_{H[x]}$ is invertible as a mapping from $H[x]$ onto $H[x]$.
(b) The partial G-derivative of P_C at $P_C(x)$ with respect to $H[x]$ exists and it coincides with $\widehat{\imath}_{H[x]}$.

PROOF. By Proposition 17.36 we can define the continuous linear mapping $A : H[x] \to H[x]$ by $A(h) := \left(D_G P_C(x)|_{H[x]}\right)(h)$ for all $h \in H[x]$. Using the symmetry of $D_G P_C(x)$ (see Lemma 17.35) and denoting as above by $\widehat{\imath}_{H[x]}$ the canonical injection from $H[x]$ into H, we obtain

$$A^* = \left(\Pi_{H[x]} \circ D_G P_C(x) \circ \widehat{\imath}_{H[x]}\right)^* = \Pi_{H[x]} \circ D_G P_C(x) \circ \widehat{\imath}_{H[x]} = A,$$

where the second equality follows from the equality $\Pi^*_{H[x]} = \widehat{\imath}_{H[x]}$, when we consider $\Pi_{H[x]}$ as a mapping from H into $H[x]$. This symmetry of A and the surjectivity of A (due to the assumption $D_G P_C(x)(H) = H[x]$ and to the equality $D_G P_C(x) = D_G P_C(x) \circ \Pi_{H[x]}$ in Proposition 17.36) imply (see Theorem C.6 in Appendix C) that A is invertible as a mapping from $H[x]$ onto $H[x]$. The assertion (a) is then justified.

For (b), fix $h \in H[x]$ and denote $u = A^{-1}h$. By hypothesis, we know that

$$P_C(x + tu) = P_C(x) + tD_G P_C(x)u + o(t).$$

Since P_C is Lipschitz continuous near $P_C(x)$ (by prox-regularity of C at $P_C(x)$), for $t > 0$ small enough we deduce that

$$P_C(x + tu) = P_C\big(P_C(x + tu)\big) = P_C\big(P_C(x) + tD_G P_C(x)u\big) + o(t).$$

Combining both equalities and recalling that $D_G P_C(x)u = h$, it results that for $t > 0$ small enough

$$P_C(P_C(x) + th) = P_C(x) + th + o(t),$$

which entails, by arbitrariness of h, that the partial G-derivative of P_C at $P_C(x)$ with respect to $H[x]$ exists and it coincides with $\widehat{i}_{H[x]}$. This finishes the proof. $\square$

17.3.2. Smoothness of the boundary of a set via the metric projection. The study of additional conditions to the smoothness of the metric projection in order to obtain the smoothness of the boundary is the objective of this subsection. Such conditions will make use of the concept of partial differentiability with respect to a vector subspace defined in Definition 17.39 above.

Let us start with the case when the set C of interest is an epigraph. Consider a Lipschitz function $f : H \to \mathbb{R}$ on a Hilbert space H with Lipschitz constant $\gamma > 0$ and a point $\overline{x} \in H$. Endowing $H \times \mathbb{R}$ with its natural inner product $\langle (x, r), (x', r') \rangle := \langle x, x' \rangle_H + rr'$, put $C := \operatorname{epi} f \subset H \times \mathbb{R}$ and assume that C *is* (r, α)*-prox-regular at* $(\overline{x}, f(\overline{x}))$ with extended reals $r > 0$ and $\alpha > 0$. On $\operatorname{Dom} P_C$ define

$$\Lambda_1 := \operatorname{pr}_H \circ P_C \quad \text{and} \quad \Lambda_2 := \operatorname{pr}_{\mathbb{R}} \circ P_C.$$

We will study the relationship between the smoothness of P_C at (x, λ) in the set $\mathcal{W}_C((\overline{x}, f(\overline{x})), r, \alpha) \setminus C$ and the smoothness of f at $\Lambda_1(x, \lambda)$. Recall that by Proposition 4.147(a)

$$(x^*, r) \in N^{VP}\big(\operatorname{epi} f; (\overline{x}, f(\overline{x}))\big) \implies r \le 0$$

and that by Proposition 4.149(a) the Lipschitz property of f entails the implication

$$(x^*, 0) \in N^P\big(\operatorname{epi} f; (\overline{x}, f(\overline{x}))\big) \implies x^* = 0.$$

This entails that for all $(x, \lambda) \in (\operatorname{Dom} P_C) \setminus C$, we have

$$\lambda - \Lambda_2(x, \lambda) = \operatorname{pr}_{\mathbb{R}}((x, \lambda) - P_C(x, \lambda)) < 0. \tag{17.34}$$

LEMMA 17.41. Let $f : H \to R$ be a Lipschitz function such that $C := \operatorname{epi} f$ is (r, α)-prox-regular at $(\overline{x}, f(\overline{x}))$ with $r, \alpha \in]0, +\infty]$. Let $(x, \lambda) \in \mathcal{W}_C((\overline{x}, f(\overline{x})), r, \alpha) \setminus C$ be such that the partial G-derivative of P_C with respect to $H[x, \lambda]$ exists at $P_C(x, \lambda)$ and it coincides with the canonical injection $\widehat{i}_{H[x,\lambda]}$ of $H[x, \lambda]$ into $H \times \mathbb{R}$. Then f is G-differentiable at $\Lambda_1(x, \lambda)$.

PROOF. Let $(x, \lambda) \in \mathcal{W}_C((\overline{x}, f(\overline{x})), r, \alpha) \setminus C$ be such that the partial G-derivative of P_C with respect to $H[x, \lambda]$ exists at $P_C(x, \lambda)$ and coincides with $\widehat{i}_{H[x,\lambda]}$. Set $x_1 := \Lambda_1(x, \lambda)$ and $\widehat{n} := d_C(x, \lambda)^{-1}(P_C(x, \lambda) - (x, \lambda))$. By hypothesis, on the one hand we have that $P_C((x_1, f(x_1)) + th) = (x_1, f(x_1)) + th + o(t)$, for each $h \in H[x, \lambda]$. On the other hand, since C is prox-regular at $(x_1, f(x_1))$, we have by Theorem 15.83 that for every $(y, s) \in H \times \mathbb{R}$,

$$P_C((x_1, f(x_1)) + t(y, s)) = (x_1, f(x_1)) + tP_{T^B(C;(x_1, f(x_1)))}(y, s) + o(t).$$

Combining both expansions of $P_C((x_1, f(x_1)) + th)$, it ensues that

$$P_{T^B(C;(x_1, f(x_1)))}(h) = h$$

for every $h \in H[x,\lambda]$, which implies that $H[x,\lambda] \subset T^B(C;(x_1, f(x_1)))$. Further, since $C = \operatorname{epi} f$ is tangentially regular at $(x_1, f(x_1))$ (see Proposition 15.75), it follows that the directional derivative $f'(x_1;\cdot)$ exists and coincides with the Clarke directional derivative $f^\circ(x_1;\cdot)$. Then $H[x,\lambda] \subset \operatorname{epi} f'(x_1;\cdot)$ and $f'(x_1;\cdot) : H \to \mathbb{R}$ is a positively homogeneous continuous convex function. Lemma 17.42 below tells us that $f'(x_1;\cdot)$ is a continuous linear functional on H, so f is G-differentiable at $\Lambda_1(x,\lambda)$, finishing the proof. □

LEMMA 17.42. Let E be a vector space and $\varphi : E \to \mathbb{R} \cup \{+\infty\}$ be a positively homogeneous convex function. If $\operatorname{epi} \varphi$ contains a vector hyperplane of $E \times \mathbb{R}$, then φ is a linear functional.

PROOF. Let $\mathcal{H} \subset E \times \mathbb{R}$ be the vector hyperplane included in $\operatorname{epi} \varphi$. We can write $\mathcal{H} = \{(x,r) \in E \times \mathbb{R} : \zeta(x) - \beta r = 0\}$, where $\beta \in \mathbb{R}$ and $\zeta : E \to \mathbb{R}$ is a linear functional. We claim that $\beta \neq 0$. If $\beta = 0$, taking 0_E as the zero vector in E one would have for every $r \in \mathbb{R}$ that $(0_E, r) \in \mathcal{H}$, hence $(0_E, r) \in \operatorname{epi} \varphi$, that is, $\varphi(0_E) \leq r$ for all $r \in \mathbb{R}$, which would contradict the assumption $\varphi(E) \subset \mathbb{R} \cup \{+\infty\}$. Consequently, $\beta \neq 0$ and we can write $\mathcal{H}$ as $\mathcal{H} = \{(x,r) \in E \times \mathbb{R} : \xi(x) - r = 0\}$, where $\xi := \beta^{-1}\zeta$. Then for any $x \in E$, we see that $(x, \xi(x)) \in \mathcal{H}$, which implies that $\varphi(x) \leq \xi(x)$ according to the inclusion assumption $\mathcal{H} \subset \operatorname{epi} \varphi$. It ensues that the positively homogeneous convex function φ is finite on E, so $\varphi(0_E) = 0$ and $0 = \varphi(x - x) \leq \varphi(x) + \varphi(-x)$. We derive that $-\varphi(x) \leq \varphi(-x) \leq \xi(-x)$, which yields $\xi(x) \leq \varphi(x)$. This combined with the previous converse inequality gives $\varphi(x) = \xi(x)$ for all $x \in E$, which justifies the linearity of the function φ. □

The next proposition provides an expression for the derivative of f at $\Lambda_1(x,\lambda)$.

PROPOSITION 17.43. Let $f : H \to R$ be a Lipschitz function such that $C := \operatorname{epi} f$ is (r,α)-prox-regular at $(\overline{x}, f(\overline{x}))$ with $r, \alpha \in]0,+\infty]$. Let (x,λ) be a point in $\mathcal{W}_C((\overline{x}, f(\overline{x})), r, \alpha) \setminus C$ such that f is G-differentiable at $\Lambda_1(x,\lambda)$. Then the following hold:

(a) $\nabla f(\Lambda_1(x,\lambda)) = -\frac{x-\Lambda_1(x,\lambda)}{\lambda-\Lambda_2(x,\lambda)} = -\frac{\mathrm{pr}_H\big((x,\lambda)-P_C(x,\lambda)\big)}{\mathrm{pr}_{\mathbb{R}}\big((x,\lambda)-P_C(x,\lambda)\big)}$;

(b) the partial G-derivative of $P_C(\cdot)$ at $P_C(x,\lambda)$ with respect to

$$T[x,\lambda] := \{(h,s) \in X \times \mathbb{R} \;:\; \langle (h,s), (\nabla f(\Lambda_1(x,\lambda)), -1)\rangle = 0\}$$

exists and it coincides with the canonical injection $\widehat{\imath}_{T[x,\lambda]}$ of $T[x,\lambda]$ into $H \times \mathbb{R}$.

PROOF. For part (a), since f is G-differentiable at $\Lambda_1(x,\lambda)$ we have, by Proposition 4.142, that $\partial_P f(\Lambda_1(x,\lambda))$ is contained in $\{\nabla f(\Lambda_1(x,\lambda))\}$. Further, by equation (17.34), we know that $\lambda - \Lambda_2(x,\lambda) < 0$, and so we can write

$$\left(-\frac{x-\Lambda_1(x,\lambda)}{\lambda-\Lambda_2(x,\lambda)}, -1\right) = \frac{(x,\lambda) - P_C(x,\lambda)}{|\mathrm{pr}_{\mathbb{R}}((x,\lambda) - P_C(x,\lambda))|} \in N^P(C; P_C(x,\lambda)).$$

Since by construction $P_C(x,\lambda) = (\Lambda_1(x,\lambda), f(\Lambda_1(x,\lambda)))$, we deduce that

$$-\frac{x-\Lambda_1(x,\lambda)}{\lambda-\Lambda_2(x,\lambda)} \in \partial_P f(\Lambda_1(x,\lambda)),$$

which proves the desired equality.

For part (b), as in Lemma 17.41 we note by prox-regularity that $C = \operatorname{epi} f$ is tangentially regular at $(x_1, f(x_1))$ with $x_1 := \Lambda_1(x,\lambda)$, hence we get that $f'(x_1;\cdot)$

exists and that we can write

$$T^B(C;P_C(x,\lambda)) = T^C(C;P_C(x,\lambda)) = T^C(\operatorname{epi} f;P_C(x,\lambda)) = \operatorname{epi} f'(\Lambda_1(x,\lambda);\cdot).$$

Since the Lipschitz function f is G-differentiable at $\Lambda_1(x,\lambda)$, it results that

$$T^B(C;P_C(x,\lambda)) = \{(h,s) \in H \times \mathbb{R} \,:\, D_G f(\Lambda_1(x,\lambda))h \le s\},$$

and hence $T[x,\lambda] \subset T^B(C;P_C(x,\lambda))$. Further, by Theorem 15.83, we know that $P_C(\cdot)$ is directionally differentiable at $P_C(x,\lambda)$ and that for every $(h,s) \in X \times \mathbb{R}$,

$$P_C(P_C(x,\lambda) + t(h,s)) = P_C(x,\lambda) + tP_{T^B(C;P_C(x,\lambda))}(h,s) + o(t).$$

This and the above inclusion $T[x,\lambda] \subseteq T^B(C;P_C(x,\lambda))$, entail that for all $(h,s) \in T[x,\lambda]$,

$$P_C(P_C(x,\lambda) + t(h,s)) = P_C(x,\lambda) + t(h,s) + o(t),$$

and so, the partial G-derivative of P_C at $P_C(x,\lambda)$ with respect to $T[x,\lambda]$ exists and it coincides with the canonical injection $\widehat{i}_{T[x,\lambda]}$. The proof is now complete. □

We continue the study with the epigraph of f by assuming now that its metric projection is of class $\mathcal{C}^p$ around a suitable point.

PROPOSITION 17.44. Let $f : H \to \mathbb{R}$ be a Lipschitz function such that $C := \operatorname{epi} f$ is (r,α)-prox-regular at $(\overline{x}, f(\overline{x}))$ with $r,\alpha \in]0,+\infty]$. Assume that f is G-differentiable near $\overline{x}$ and that $P_C(\cdot)$ is of class $\mathcal{C}^p$ (with $p \ge 1$) on an open neighborhood U of some point $(x_0,\lambda_0) \in \mathcal{W}_C\big((\overline{x}, f(\overline{x})), r, \alpha\big) \setminus C$ satisfying $P_C\big((x_0,\lambda_0)\big) = (\overline{x}, f(\overline{x}))$. Assume also that $\nabla f(\overline{x}) = 0$ and that $DP_C(u)|_{H[u]}$ is invertible for each $u \in U$ as a mapping from $H[u]$ into $H[u]$. Then, f is of class $\mathcal{C}^{p+1}$ near $\overline{x}$.

PROOF. Let us set $\overline{v} := (\overline{x}, f(\overline{x}))$ and $u_0 := (x_0,\lambda_0)$, so $P_C(u_0) = \overline{v}$. For simplicity, we will write $u = (u_1,u_2)$ for each $u \in H \times \mathbb{R}$. Without loss of generality, we may and do suppose that $f(\overline{x}) = 0$. Put $\widehat{n} := d_C(u_0)^{-1}(u_0 - P_C(u_0))$, and consider the mapping $F : U \to H \times \mathbb{R}$ defined by

$$F(u) = P_C(u) + \langle \widehat{n}, u - u_0\rangle\, \widehat{n} \quad \text{for all } u \in U.$$

Clearly, F is of class $\mathcal{C}^p$ on U. By Proposition 17.36 we also have for all $h \in H \times \mathbb{R}$

$$DF(u_0)h = DP_C(u_0)h + \langle \widehat{n}, h\rangle \widehat{n} = DP_C(u_0) \circ \Pi_{H[u_0]} h + \langle \widehat{n}, h\rangle \widehat{n},$$

and so $DF(u_0)$ is invertible, since $DP_C(u_0)$ restricted to $H[u_0]$ is invertible as a mapping from $H[u_0]$ to $H[u_0]$ and since $\widehat{n} \perp H[u_0]$ by definition of $H[u_0]$. Then, the inverse mapping theorem (see Theorem E.1 in Appendix E) furnishes an open neighborhood $U_0 \subset U$ of u_0 and an open neighborhood V of $F(u_0) = P_C(u_0) = \overline{v}$ such that $F|_{U_0} : U_0 \to V$ is bijective (as a mapping from U_0 into V) and its inverse $F|_{U_0}^{-1}$ is of class $\mathcal{C}^p$ on V.

Since $\Lambda_1(u_0) = \mathrm{pr}_H(P_C(u_0)) = \mathrm{pr}_H(\overline{v}) = \mathrm{pr}_H\big((\overline{x}, f(\overline{x}))\big) = \overline{x}$ and $\nabla f(\overline{x}) = 0$, we have by Proposition 17.43(a) that in fact $\widehat{n} = (0,-1)$ and $H[u_0] = H \times \{0\}$. Choose then an open neighborhood V_H of $\overline{x}$ in H such that $V_H \times \{0\} \subset V$ (which can be done since $f(\overline{x}) = 0$ and V is a neighborhood of $(\overline{x}, f(\overline{x}))$) and such that f is G-differentiable on V_H. Recalling that$\mathcal{W}_C((\overline{x}, f(\overline{x})), r, \alpha) \setminus C$ is open in H by (r,α)-prox-regularity of C at $(\overline{x}, f(\overline{x}))$ (see Theorem 17.7(b)), shrinking U_0 and V if necessary, we may suppose that $U_0 \subset \mathcal{W}_C((\overline{x}, f(\overline{x})), r, \alpha) \setminus C$, and so $U_0 \subset \operatorname{Dom} P_C \setminus C$. By (17.34) it ensues that $u_2 - \mathrm{pr}_{\mathbb{R}}(P_C(u)) < 0$ for each $u \in U_0$. Considering the mapping $G : V_H \to U_0$, defined by

$$G(x) := F|_{U_0}^{-1}(x,0) \quad \text{for all}\, x \in V_H,$$

and using Proposition 17.43(a), for every $x \in V_H$ we can write

$$\nabla f(x) = \nabla f(\mathrm{pr}_H(x,0)) = (\nabla f) \circ \mathrm{pr}_H \circ F(G(x))$$
$$= (\nabla f) \circ \Lambda_1(G(x)) = -\frac{\mathrm{pr}_H \circ G(x) - \Lambda_1 \circ G(x)}{\mathrm{pr}_{\mathbb{R}} \circ G(x) - \Lambda_2 \circ G(x)},$$

where (keeping the equality $\widehat{n} = (0,-1)$ in mind) the third equality is due to the fact that

$$\mathrm{pr}_H \circ F(u) = \mathrm{pr}_H \circ P_C(u) + \langle \widehat{n}, u - u_0 \rangle \mathrm{pr}_H(\widehat{n}) = \mathrm{pr}_H \circ P_C(u) = \Lambda_1(u).$$

Since G, Λ_1 and Λ_2 are of class $\mathcal{C}^p$, we conclude that ∇f is of class $\mathcal{C}^p$ on V_H. This yields that f is of class $\mathcal{C}^{p+1}$ on V_H, so the proof is finished. □

Now, after the development for the epigraph of Lipschitz functions, we can come back to the general case of study, that is, the case when C is a closed subset of a Hilbert space H.

PROPOSITION 17.45. Let C be a nonempty set of a Hilbert space H, let $x \in H \setminus C$ and let $S := \mathrm{Cl}\, C$. Assume that P_C is well defined on a neighborhood U of x with $P_C(u) = P_S(u)$ for all $u \in U$ and that C is (r,α)-prox-regular at $P_C(x)$ for some $\alpha \in]0,+\infty]$ and some $r \in]0,+\infty]$ with $d_C(x) < r$. Assume also that P_C is G-differentiable at x and that $D_G P_C(x)|_{H[x]}$ is injective over $H[x]$. Then

$$\mathbb{R}_+\{x - P_C(x)\} \subset N^P(C; P_C(x)) \subset \mathbb{R}\{x - P_C(x)\},$$

where $\mathbb{R}_+ := [0,+\infty[$.

If in addition C has the interior tangent cone property (or equivalently the epi-Lipschitz property) at $P_C(x)$, then

$$\mathbb{R}_+\{x - P_C(x)\} = N^P(C; P_C(x)).$$

PROOF. Since $P_C(x)$ exists, the inclusion $\mathbb{R}_+\{x - P_C(x)\} \subset N^P(C; P_C(x))$ is direct from the definition of proximal normal cone. Suppose that the second inclusion fails, that is, there exists $\xi \in N^P(C; P_C(x)) \setminus \mathbb{R}\{x - P_C(x)\}$. By definition of proximal normal cone and by Definition 17.6 of $\mathcal{W}_C(P_C(x), r, \alpha)$, that means that there exists $y \in \mathcal{W}_C(P_C(x), r, \alpha) \setminus C$ and $\beta > 0$ such that $P_C(y) = P_C(x)$ and $\beta\xi = y - P_C(y) \notin \mathbb{R}\{x - P_C(x)\}$. For each $t \in [0,1]$, the convexity of $N^P(C; P_C(x))$ and the equality $P_C(x) = P_C(y)$ yield that $(x + t(y-x)) - P_C(x) \in N^P(C; P_C(x))$. Using this we can write

$$\|x + t(y-x) - P_C(x)\| = \|t(y - P_C(y)) + (1-t)(x - P_C(x))\| \le t d_C(y) + (1-t) d_C(x) < r,$$

where the strict inequality is due to the inclusion of y and x in $\mathcal{W}_C(P_C(x), r, \alpha)$. We derive from this and from Definition 15.65 of (r,α)-prox-regularity that $P_C(x + t(y-x)) = P_C(x)$ for all $t \in [0,1]$. By Proposition 17.36, the latter equality and the G-differentiability assumption of P_C at x entail

$$D_G P_C(x)\left(\Pi_{H[x]}(y-x)\right) = D_G P_C(x)(y-x) = \lim_{t \downarrow 0} \frac{P_C(x + t(y-x)) - P_C(x)}{t} = 0.$$

By the injectivity assumption of $D_G P_C(x)|_{H[x]}$, we deduce that $\Pi_{H[x]}(y-x) = 0$. But, since $y - P_C(x) = y - P_C(y) \notin \mathbb{R}\{x - P_C(x)\} = H[x]^\perp$, we know that $\Pi_{H[x]}(y - P_C(x)) \neq 0$, and so

$$0 = \Pi_{H[x]}(y-x) = \Pi_{H[x]}(y - P_C(x)) - \Pi_{H[x]}(x - P_C(x)) = \Pi_{H[x]}(y - P_C(x)) \neq 0,$$

which is a contradiction. It results that $N^P(C; P_C(x)) \subset \mathbb{R}\{x - P_C(x)\}$, which proves the first part of the proposition.

For the second part, assume in addition that C has the interior cone property at $P_C(x)$ (or equivalently the epi-Lipschitz property at $P_C(x)$ since C is closed near $P_C(x)$ by proox-regularity at $P_C(x)$). Then, $T^C(C; P_C(x))$ has nonempty interior (see Proposition 2.25). Suppose by contradiction that $N^P(C; P_C(x))$ is not equal to $\mathbb{R}_+\{x - P_C(x)\}$. Then, by the first part of the present proposition and the fact that $N^P(C; P_C(x))$ is a cone, we have that $N^P(C; P_C(x)) = \mathbb{R}\{x - P_C(x)\}$. Since C is prox-regular at $P_C(x)$, we know that $N^C(C; P_C(x))$ coincides with $N^P(C; P_C(x))$, and so, $T^C(C; P_C(x))$ being a closed convex cone, it ensues that

$$T^C(C; P_C(x)) = \left[N^P(C; P_C(x))\right]^\circ = \{h \in X \,:\, \langle h, x - P_C(x)\rangle = 0\}.$$

Since the last set in the above equality has empty interior, this contradicts that $\operatorname{int}\left(T^C(C; P_C(x))\right) \neq \emptyset$. Thus, $N^P(C; P_C(x))$ coincides $\mathbb{R}_+\{x - P_C(x)\}$, which finishes the proof. □

We already know (see Proposition 17.21) that, when a set C of H is closed near $\overline{x} \in \operatorname{bdry} C$ with the interior tangent cone property at $\overline{x}$ and its boundary is a $\mathcal{C}^2$-submanifold at $\overline{x} \in \operatorname{bdry} C$, the set C is both epi-Lipschitz at $\overline{x}$ and prox-regular at $\overline{x}$. The local characterization of the property of $\mathcal{C}^{p+1}$-submanifold of $\operatorname{bdry} C$ at $\overline{x} \in \operatorname{bdry} C$ is the following.

THEOREM 17.46 (D. Salas and L. Thibault). Let C be a subset of a Hilbert space H which is closed near $\overline{x} \in \operatorname{bdry} C$ and has the interior tangent cone property at $\overline{x}$. The following assertions are equivalent:
(a) the boundary $\operatorname{bdry} C$ is a $\mathcal{C}^{p+1}$-submanifold at $\overline{x}$;
(b) the set C is prox-regular at $\overline{x}$, and for any $r, \alpha > 0$ for which C is (r, α)-prox-regular at $\overline{x}$ as well as epi-Lipschitz at each point in $C \cap B_H(\overline{x}, \alpha)$ and for which $B_H(\overline{x}, \alpha) \cap \operatorname{bdry} C$ is a $\mathcal{C}^{p+1}$-submanifold, one has that P_C is of class $\mathcal{C}^p$ on the open set $\mathcal{W}_C(\overline{x}, r, \alpha) \setminus C$ and that the mapping $DP_C(u)$ restricted to $H[u]$ is invertible as a mapping from $H[u]$ onto $H[u]$ for every $u \in \mathcal{W}_C(\overline{x}, r, \alpha) \setminus C$;
(c) the set C is prox-regular at $\overline{x}$, and for any prox-regularity constant $r > 0$ of C at $\overline{x}$ there exits $\alpha > 0$ such that $\mathcal{W}_C(\overline{x}, r, \alpha) \setminus C$ is open and P_C is of class $\mathcal{C}^p$ therein and such that, for every $u \in \mathcal{W}_C(\overline{x}, r, \alpha) \setminus C$ the operator $DP_C(u)$ restricted to $H[u]$ is invertible as an operator from $H[u]$ onto $H[u]$;
(d) there exists an open neighborhood U of $\overline{x}$ such that $U \setminus C$ is open, P_C is of class $\mathcal{C}^p$ on $U \setminus C$, and for every $u \in U \setminus C$ the operator $DP_C(u)$ restricted to $H[u]$ is surjective onto $H[u]$.

PROOF. The implications (b)⇒(c) and (c)⇒(d) are evident. Concerning the implication (a)⇒(b), suppose that (a) is satisfied. By the comments preceding the statement of the theorem, the set C is prox-regular at $\overline{x}$. So, consider any $r, \alpha > 0$ for which C is (r, α)-prox-regular at $\overline{x}$ and epi-Lipschitz at each point in $C \cap B_H(C, \alpha)$ and for which $B_H(\overline{x}, \alpha) \cap \operatorname{bdry} C$ is a $\mathcal{C}^{p+1}$-submanifold. The sets $\mathcal{W}_C(\overline{x}, r, \alpha)$ and $\mathcal{W}_C(\overline{x}, r, \alpha) \setminus C$ are open by Theorem 17.7(b), and Theorem 17.30 tells us that P_C is well-defined and of class $\mathcal{C}^p$ on $\mathcal{W}_C(\overline{x}, r, \alpha) \setminus C$, while for $u \in \mathcal{W}_C(\overline{x}, r, \alpha) \setminus C$ the invertibility of $DP_C(u)|_{H[u]} : H[u] \to H[u]$ follows from Proposition 17.27, Lemma 17.28 and Proposition 17.38. The implication (a)⇒(b) is then justified.

It remains to prove (d)⇒(a). Under the hypothesis of (d) we know by the continuity of P_C near $\overline{x}$ that C is prox-regular at $\overline{x}$ (see the equivalence (j)⇔(a) in Theorem 15.73). Denote $S := \mathrm{cl}\, C$. Choose $r_0, \alpha_0 > 0$ such that C is (r_0, α_0)-prox-regular at $\overline{x}$ and choose $r \in]0, r_0[$ and $\alpha \in]0, \alpha_0[$ such that $2r + \alpha < \alpha_0$ and such that

$$U_0 := \mathcal{W}(C, r, \alpha) \subset U.$$

Given any $x \in U_0$ and any $y \in \mathrm{Proj}_S(x)$ we have

$$\|y - \overline{x}\| \leq \|x - y\| + \|x - \overline{x}\| < 2r + \alpha < \alpha_0,$$

so $y \in B(\overline{x}, \alpha_0)$, and hence by (17.3)

$$y \in S \cap B(\overline{x}, \alpha_0) \subset S \cap \mathcal{W}(C, r_0, \alpha_0) = C \cap \mathcal{W}(C, r_0, \alpha_0),$$

which in turn entails that $y = P_C(x)$ since $P_C(x)$ exists. Then $P_S(x) = P_C(x)$ for all $x \in U_0$, so in particular P_C is (well defined and) continuous on U_0 with $P_C(x) = P_S(x)$ for all $u \in U_0$. Further, for any $u \in U_0$ we have by assumption $DP_C(u)\big(H[u]\big) = H[u]$, which ensures by Proposition 17.36 that $Dp_C(u)(H) = H[u]$. Thus, by Lemma 17.40, we get that for every $u \in U_0 \setminus C$, $DP_C(u)|_{H[u]}$ is invertible as an operator from $H[u]$ to $H[u]$.

We know that the interior tangent cone property assumption is equivalent to the epi-Lipschitz property of C at $\overline{x}$ and that it assures us (see Corollary 2.26) that there exists a nonzero vector $\nu \in N^C(C; \overline{x})$. Further, the prox-regularity of C at $\overline{x}$ gives that $N^C(C; \overline{x}) = N^P(C; \overline{x})$. By shrinking the norm of ν if necessary, we may suppose that $x_0 := \overline{x} + \nu \in U_0$ and $P_C(x_0) = \overline{x}$. Further, since $d_C(x_0) < r$, we have by Proposition 17.45

$$N^C(C; \overline{x}) = N^P(C; \overline{x}) = \mathbb{R}_+(x_0 - P_C(x_0)) = \mathbb{R}_+\nu. \tag{17.35}$$

In particular, it ensues that $T^C(C; \overline{x}) = \{h \in H \ : \ \langle h, \nu \rangle \leq 0\}$ and that $\widehat{n} := -\|\nu\|^{-1}\nu \in I(C; \overline{x})$ since $I(C; \overline{x})$ is the topological interior of $T^C(C; \overline{x})$ by the interior tangent cone property of C at $\overline{x}$ (see Proposition 2.25). Denote $Z := \{\widehat{n}\}^\perp$ and $\overline{z} := \pi_Z(\overline{x})$. By Remark 2.21 and by the interior tangent cone property of C, there exist an open neighborhood O of $\overline{z}$ in Z, a real $\varepsilon > 0$ and a Lipschitz function $f : Z \to \mathbb{R}$ such that $\overline{x} = \overline{z} + f(\overline{z})\widehat{n}$ and such that the set

$$U_1 := L\left(O \times \left]f(\overline{z}) - \varepsilon, f(\overline{z}) + \varepsilon\right[\right)$$

is contained in U_0 and $U_1 \cap C = U_1 \cap L(\mathrm{epi}\, f)$, where $L : Z \times \mathbb{R} \to H$ is the canonic isomorphism given by $L(z, \lambda) = z + \lambda\widehat{n}$. Denote $C' := \mathrm{epi}\, f$ and note that $L^{-1}(\overline{x}) = (\overline{z}, f(\overline{z}))$ and C' is prox-regular at $(\overline{z}, f(\overline{z}))$ as directly seen through the equivalence (c)⇔(a) in Theorem 15.73. Since L is an isometric isomorphism and P_C is continuous on U_0, it is easy to verify (as for (17.26)) that there exists an open neighborhood $U_2 \subset U_1$ of $\overline{x}$ in H such that, denoting $W := L^{-1}(U_2)$,

$$P_{C'}|_W = P_{\mathrm{epi}\, f}|_W = L^{-1} \circ P_C \circ L|_W.$$

Therefore, $P_{\mathrm{epi}\, f}$ is well-defined on W and it is of class $\mathcal{C}^p$ on the open set $W \setminus C'$. Also, for each $(z, \lambda) \in W \setminus C'$, it is clear that

$$DP_{\mathrm{epi}\, f}(z, \lambda) = L^{-1} \circ DP_C(L(z, \lambda)) \circ L$$

and that

$$\begin{aligned} H_{C'}[z, \lambda] &= \{(z', \lambda') \in H \times \mathbb{R} \ : \ \langle L^{-1}(L(z, \lambda) - P_C(L(z, \lambda))), (z', \lambda') \rangle = 0\} \\ &= L^{-1}(H_C[L(z, \lambda)]), \end{aligned}$$

where the latter equality is due to the equality $L^* = L^{-1}$ (easily verified).

For each $(z,\lambda) \in W \setminus C'$ it results that $DP_{\mathrm{epi}\, f}(z,\lambda)|_{H_{C'}[z,\lambda]}$ is invertible as a mapping from $H_{C'}[z,\lambda]$ onto itself. Choose $r_1, \alpha_1 > 0$ such that C' is (r_1,α_1)-prox-regular at $(\overline{z}, f(\overline{z}))$ and shrink W (by shrinking U_2) if necessary so that $W \subset \mathcal{W}_{C'}((\overline{z}, f(\overline{z})), r_1, \alpha_1)$. Using Lemma 17.40 again, we obtain that for every $(z,\lambda) \in W \setminus C'$ the partial G-derivative of $P_{\mathrm{epi}\, f}$ with respect to $H_{C'}[z,\lambda]$ exists at $P_{\mathrm{epi}\, f}(z,\lambda)$ and it coincides with the identity $\mathrm{id}_{H_{C'}[z,\lambda]}$.

We can now apply Lemma 17.41 to deduce that for every $(z,\lambda) \in W \setminus C'$, the function f is G-differentiable at $\Lambda_1(z,\lambda)$, where $\Lambda_1 := \mathrm{pr}_Z \circ P_{\mathrm{epi}\, f}$. We claim that $\Lambda_1(W \setminus C')$ is a neighborhood of $\overline{z}$. Indeed, fix a positive real $\alpha' < \alpha_1$ small enough such that $B_{Z\times\mathbb{R}}((\overline{z}, f(\overline{z})), \alpha') \subset W$. Since f is continuous at $\overline{z}$, we can fix $\alpha'' > 0$ such that $(z, f(z)) \in B_{Z\times\mathbb{R}}((\overline{z}, f(\overline{z})), \alpha')$ for each $z \in B_Z(\overline{z}, \alpha'')$. Then, for any such z, since $N^P(C'; (z, f(z)))$ is not reduced to zero (according to the normal regularity $N^P(C'; (z, f(z))) = N^C(C'; (z, f(z)))$ and to the epi-Lipschitz property of C') at $(z, f(z))$, we get that there exists a nonzero $\xi \in N^P(C'; (z, f(z)))$ such that

$$\|(z, f(z)) + \xi - (\overline{z}, f(\overline{z}))\| < \alpha' \quad \text{and} \quad (z, f(z)) = P_{C'}((z, f(z)) + \xi).$$

Hence, $(z, f(z)) + \xi \in B_{Z\times\mathbb{R}}((\overline{z}, f(\overline{z})), \alpha') \setminus C'$ and $\Lambda_1((z, f(z)) + \xi) = z$. This entails the inclusion

$$\Lambda_1(W \setminus C') \supset \Lambda_1(B_{Z\times\mathbb{R}}((\overline{z}, f(\overline{z})), \alpha') \setminus C') \supset B_Z(\overline{z}, \alpha''),$$

proving our claim. It then follows that f is G-differentiable near $\overline{z}$. Furthermore, keeping in mind that $L^* = L^{-1}$, we can write (see Proposition 2.130)

$$\begin{aligned} N^P(\mathrm{epi}\, f; (\overline{z}, f(\overline{z}))) &= N^C(\mathrm{epi}\, f \cap L^{-1}(U_0); (\overline{z}, f(\overline{z}))) = N^C(L^{-1}(C \cap U_0); (\overline{z}, f(\overline{z}))) \\ &= L^* \left[N^C(C \cap U_0; \overline{x})\right] = L^{-1}\left[N^C(C \cap U_0; \overline{x})\right] \\ &= L^{-1}\left[N^P(C; \overline{x})\right] = \{(\pi_Z(\xi), \langle \xi, \widehat{n}\rangle) \ : \xi \in N^P(C; \overline{x})\}. \end{aligned}$$

So using the equalities $N^P(C; \overline{x}) = \mathbb{R}_+\nu = -\mathbb{R}_+\widehat{n}$ due to (17.35) and to the definition of $\widehat{n}$, we see that

$$N^P(\mathrm{epi}\, f; (\overline{z}, f(\overline{z}))) = \mathbb{R}_+\{(-\pi_Z(\widehat{n}), -1)\} = \mathbb{R}_+\{(0, -1)\},$$

which tells us that $\nabla f(\overline{z}) = 0$. All the hypotheses of Proposition 17.44 are then fulfilled, which guarantees that f is of class $\mathcal{C}^{p+1}$ near $\overline{z}$. Finally, noting that $U_1 \cap \mathrm{bdry}\, C = U_1 \cap L(\mathrm{gph}\, f)$, where we recall that $\mathrm{gph}\, f$ denotes the graph of f, we conclude that $\mathrm{bdry}\, C$ is a $\mathcal{C}^{p+1}$-submanifold at $\overline{x}$ (according to Theorem E.10 in Appendix E). The proof of the theorem is then complete. □

We know by Proposition 15.67 that if a closed set C is prox-regular, then there exists a continuous function $\rho : C \to]0, +\infty]$ (called prox-regularity function) such that it is $\rho(\cdot)$-prox-regular, that is, for every $x \in C$ and every $\zeta \in N^P(C; x) \cap \mathbb{B}_H$ one has that

$$x \in \mathrm{Proj}\,_C(x + t\zeta), \qquad \text{for every real } t \in [0, \rho(x)].$$

Then, for every $x \in C$, and every $r \in]0, \rho(x)[$, by continuity of ρ at x there exists $\alpha > 0$ such that C is (r,α)-prox-regular at x and such that

$$\mathcal{W}_C(x, \alpha, r) \subset U_{\rho(\cdot)}(C) := \{u \in X \ : \ \exists y \in \mathrm{Proj}\,_C(u),\ d_C(u) < \rho(y)\}.$$

On the other hand, if the closed set C is $\rho(\cdot)$-prox-regular, then $U_{\rho(\cdot)}(C) \setminus C$ is the union of the collection of sets $\mathcal{W}_C(x, \alpha, r) \setminus C$ over $x \in \mathrm{bdry}\, C$ and $r < \rho(x)$, for which C is (r, α)-prox-regular. This combined with Theorem 17.46 guarantees the following theorem which provides a fundamental complete quantitative characterization of epi-Lipschitz sets in Hilbert spaces with smooth boundaries.

THEOREM 17.47 (characterizations of epi-Lipschitz sets with smooth boundary). Let C be a closed set of a Hilbert space H with the interior tangent cone property (or equivalently with the epi-Lipschitz property) at any boundary point. The following assertions are equivalent:
(a) the boundary $\operatorname{bdry} C$ is a $\mathcal{C}^{p+1}$-submanifold at each of its points;
(b) the set C is prox-regular, and for any continuous function $\rho : C \to]0, +\infty]$ such that C is $\rho(\cdot)$-prox-regular, P_C is of class $\mathcal{C}^p$ on $U_{\rho(\cdot)}(C) \setminus C$ and for every $u \in U_{\rho(\cdot)}(C) \setminus C$ the operator $DP_C(u)$ restricted to $H[u]$ is invertible as an operator from $H[u]$ onto $H[u]$;
(c) there exists an open set U containing C such that P_C is of class $\mathcal{C}^p$ on $U \setminus C$ and for every $u \in U \setminus C$ the operator $DP_C(u)$ restricted to $H[u]$ is surjective onto $H[u]$.

The following corollary furnishes a relationship between the thresholds of prox-regularity and the smoothness of the metric projection.

COROLLARY 17.48. Let C be a closed set of a Hilbert space H with the interior tangent cone property (or equivalently with the epi-Lipschitz property) at any boundary point.
(I) For $\overline{x} \in \operatorname{bdry} C$ and for $r, \alpha \in]0, +\infty]$, the following assertions are equivalent:
(a) C is (r, α)-prox-regular at the point $\overline{x} \in \operatorname{bdry} C$ and $(\operatorname{bdry} C) \cap B(\overline{x}, \alpha)$ is a $\mathcal{C}^{p+1}$-submanifold;
(b) the set $\mathcal{W}_C(\overline{x}, r, \alpha)$ is open, P_C is of class $\mathcal{C}^p$ on $\mathcal{W}_C(\overline{x}, r, \alpha) \setminus C$, and for every $u \in \mathcal{W}_C(\overline{x}, r, \alpha) \setminus C$ the operator $DP_C(u)$ restricted to $H[u]$ is invertible as an operator from $H[u]$ onto $H[u]$.

(II) For every continuous function $\rho : C \to]0, +\infty]$ the following assertions are also equivalent:
(a) the set C is $\rho(\cdot)$-prox-regular and its boundary $\operatorname{bdry} C$ is a $\mathcal{C}^{p+1}$-submanifold at each of its points;
(b) the set $U_{\rho(\cdot)}(C)$ is open, P_C is of class $\mathcal{C}^p$ on $U_{\rho(\cdot)}(C) \setminus C$, and for every point $u \in U_{\rho(\cdot)}(C) \setminus C$ the operator $DP_C(u)$ restricted to $H[u]$ is invertible as an operator from $H[u]$ onto $H[u]$.

The situation of a convex body directly follows.

THEOREM 17.49 (S. Fitzpatrick and R.R. Phelps). Let C be a convex body of a Hilbert space H. The following assertions are equivalent:
(a) the boundary $\operatorname{bdry} C$ is a $\mathcal{C}^{p+1}$-submanifold at any of its points;
(b) the mapping P_C is of class $\mathcal{C}^p$ on $H \setminus C$ and for every $u \in H \setminus C$ the operator $DP_C(u)$ restricted to $H[u]$ is invertible as an operator from $H[u]$ onto $H[u]$;
(c) there exists an open set U containing C such that P_C is of class $\mathcal{C}^p$ on $U \setminus C$ and for every point $u \in U \setminus C$ the operator $DP_C(u)$ restricted to $H[u]$ is surjective onto $H[u]$.

17.4. Metric projection onto submanifold

We analyze in this section the situation of submanifolds in Hilbert space.

17.4.1. Differentiability of metric projection onto submanifold. We will begin by establishing around any $\mathcal{C}^{p+1}$-submanifold (in a Hilbert space) the $\mathcal{C}^p$ property of its metric projection. To do so, we will employ the next lemma. Before stating the lemma, recall (see Proposition 15.74) that any $\mathcal{C}^2$-submanifold M at $m_0 \in M$ in a Hilbert space is prox-regular at m_0, so by Theorem 17.4.1 the metric projection mapping P_M is well-defined near m_0 and continuous therein. Recall also the following theorem (see Theorem E.10 in Appendix E) on the representation of a submanifold as graph of a suitable mapping.

THEOREM 17.50. Let M be a subset of a Hilbert space H and $m_0 \in M$, and let Z be a closed vector subspace in H. Given an integer $p \in \mathbb{N}$, the set M is a $\mathcal{C}^p$-submanifold (resp. $\mathcal{C}^{p,0}$-submanifold, or $\mathcal{C}^{p,1}$-submanifold) at m_0 with Z as model space if and only if there exist a neighborhood U of m_0 in H, a neighborhood $V_Z \subset Z$ of zero in Z, and a mapping $\theta : V_Z \to Z^\perp$ such that

(a) $\theta \in \mathcal{C}^p(V_Z, Z^\perp)$ (resp. $\theta \in \mathcal{C}^{p,0}(V_Z, Z^\perp)$, or $\theta \in \mathcal{C}^{p,1}(V_Z, Z^\perp)$);
(b) $\theta(0) = 0$ and $D\theta(0) = 0$;
(c) $M \cap U = \big(m_0 + L^{-1}(\operatorname{gph}\theta)\big) \cap U$,

where $L : H \to Z \times Z^\perp$ is the canonic isomorphism given by the equality $L(x) = (\pi_Z(x), \pi_{Z^\perp}(x))$.

Further, under (a), (b) (c) one has $T_{m_0}M = Z$.

LEMMA 17.51. Let an integer $p \geq 1$ and M be a $\mathcal{C}^{p+1}$-submanifold (resp. $\mathcal{C}^{p+1,0}$-submanifold, or $\mathcal{C}^{p+1,1}$-submanifold) at zero in a Hilbert space H.
(a) There exists an open neighborhood U of zero in H such that P_M is well-defined on U and such that, for the closed vector subspace Z of H, for the canonical isomorphism $L : H \to Z \times Z^\perp$ with $L(x) := (\pi_Z(x), \pi_{Z^\perp}(x))$, for the open neighborhood V_Z of zero in Z and for the mapping $\theta : V_Z \to Z^\perp$, all given by Theorem 17.50 above, the properties (a), (b), (c) in this theorem are satisfied.
(b) With Z, V_Z, L and θ as in (a) let $\Phi : V_Z \times Z^\perp \to H$ be defined by

$$\Phi(v, z_2) := L^{-1}(v - D\theta(v)^* z_2, \theta(v) + z_2) \quad \text{for all } (v, z_2) \in V_Z \times Z^\perp.$$

Then, there exists an open neighborhood W of $(0,0)$ in $Z \times Z^\perp$ such that the restriction $\Phi_W : W \to \Phi(W)$ of Φ to W is a $\mathcal{C}^p$-diffeomorphism (resp. with in addition $\Phi_W \in \mathcal{C}^{p,0}(W, H)$, or $\Phi_W \in \mathcal{C}^{p,1}(W, H)$) and an open neighborhood $O \subset \Phi(W)$ of zero in H such that

$$P_M(u) = L^{-1}(\operatorname{pr}_Z \circ \Phi_W^{-1}(u), \theta \circ \operatorname{pr}_Z \circ \Phi_W^{-1}(u)) \quad \text{for all } u \in O.$$

PROOF. (a) By Proposition 15.74 we know that M is prox-regular at zero, so there exists an open neighborhood U_0 of zero in H over which P_M is well defined (and continuous) (see Theorem 15.73(j)). Theorem 17.50 then yields an open neighborhood $U \subset U_0$ of zero in H, a closed subspace Z of H, a neighborhood $V_Z \subset Z$ of zero in Z, and mappings $\theta : V_Z \to Z^\perp$ and $L : H \to Z \times Z^\perp$ such that the properties in that Theorem 17.50 hold true.
(b) Let U, Z, V_Z, L and θ as above. We notice first that for every $m \in M \cap U$, choosing $v \in V_Z$ with $m = L^{-1}(v, \theta(v))$ and noting that $L^* = L^{-1}$, we have by Proposition 2.130 and Proposition 2.129

$$\begin{aligned} N^C(M; m) &= L^{-1}\left(N^C(\operatorname{gph}\theta; (v, \theta(v)))\right) \\ &= L^{-1}\left(\{(-D\theta(v)^* z_2, z_2) \,:\, z_2 \in Z^\perp\}\right). \end{aligned}$$

We note also by Theorem 17.50 that the mapping Φ (as defined in the statement) is of class $\mathcal{C}^p$ (resp. with in addition $\Phi \in \mathcal{C}^{p,0}(V_Z \times Z^\perp, H)$, or $\Phi \in \mathcal{C}^{p,0}(V_Z \times Z^\perp, H)$), and that $\Phi(0) = 0$ along with $D\Phi(0) = L^{-1} \circ \mathrm{id}_{Z\times Z^\perp} = L^{-1}$ (as easily verified). Therefore, the local inversion mapping theorem (resp. combined with Proposition E.6 in Appendix E) furnishes an open neighborhood W of $(0,0)$ in $Z \times Z^\perp$ such that $W \subset L(U)$ and the restriction $\Phi_W : W \to \Phi(W)$ of Φ to W is a $\mathcal{C}^p$-diffeomorphism on W (resp. with in addition $\Phi_W \in \mathcal{C}^{p,0}(W, H)$, or $\Phi_W \in \mathcal{C}^{p,1}(W, H)$).

Choose $\delta > 0$ such that $B_{Z\times Z^\perp}((0,0), 3\delta) \subset W$ and $B_H(0, 2\delta) \subset U$. Put $O = B_H(0, \delta)$. Note that P_M is well-defined on $O \subset U$. Fix any $u \in O$. Writing (since $0 \in M$)

$$\|P_M(u)\| \le \|u - P_M(u)\| + \|u\| \le \|u - 0\| + \|u\| < 2\delta,$$

we see that $P_M(u) \in B_H(0, 2\delta)$, and hence there exists $v \in V_Z$ such that

$$P_M(u) = L^{-1}(v, \theta(v)), \text{ hence } \|v\| = \|\pi_Z(P_M(u))\| < 2\delta. \tag{17.36}$$

Noticing also that

$$u - P_M(u) \in N^C(M; P_M(u)) = L^{-1}\left(\{(-D\theta(v)^* z_2, z_2) \,:\, z_2 \in Z^\perp\}\right),$$

we obtain some $z_2 \in Z^\perp$ such that

$$z_2 = \pi_{Z^\perp}(u - P_M(u)) \quad \text{and} \quad -D\theta(v)^*(v) z_2 = \pi_Z(u - P_M(u)).$$

Observing by the first equality in (17.36) that

$$v = \pi_Z(P_M(u)) \quad \text{and} \quad \theta(v) = \pi_{Z^\perp}(P_m(u)),$$

we also have that

$$\begin{aligned}\Phi(v, z_2) &= L^{-1}\big(v - D\theta(v)^* z_2, \theta(v) + z_2\big) = v - D\theta(v)^* z_2 + \theta(v) + z_2 \\ &= \pi_Z(P_M(u)) + \pi_Z(u - P_M(u)) + \pi_{Z^\perp}(P_M(u)) + \pi_{Z^\perp}(u - P_M(u)) \\ &= \pi_Z(u) + \pi_{Z^\perp}(u) = u,\end{aligned}$$

so

$$u = \Phi(v, z_2). \tag{17.37}$$

The above equality $z_2 = \pi_{Z^\perp}(u - P_M(u))$ and the inclusion $0 \in M$ yields

$$\|z_2\| = \|\pi_{Z^\perp}(u - P_M(u))\| \le \|u - P_M(u)\| \le \|u\| \le \delta,$$

which combined with the inequality $\|v\| < 2\delta$ implies that $(v, z_2) \in W$. This and (17.37) ensure, by bijectivity of Φ_W, that we can write $(v, z_2) = \Phi_W^{-1}(u)$. It follows that $v = \mathrm{pr}_Z \circ \Phi_W^{-1}(u)$, and hence by the first equality in (17.36)

$$P_M(u) = L^{-1}(v, \theta(v)) = L^{-1}(\mathrm{pr}_Z \circ \Phi_W^{-1}(u), \theta \circ \mathrm{pr}_Z \circ \Phi_W^{-1}(u)).$$

This being true for all $u \in O$, the proof is then finished. □

Recall (see Proposition 15.19) for a closed set C of a Hilbert space H that P_C is well-defined and continuous on an open set U of H if and only if d_C^2 is continuously differentiable on U, and in such a case $\nabla_F(\frac{1}{2}d_C^2)(u) = u - P_C(u)$ for all $u \in U$. This and Lemma 17.51 just above guarantee the following theorem.

THEOREM 17.52 (smoothness of metric projection onto submanifolds). Let $p \ge 1$ be an integer and M be a $\mathcal{C}^{p+1}$-submanifold at $m_0 \in M$ in a Hilbert space H. Then there exists an open neighborhood U of m_0 such that both properties (a) and (b) hold:

(a) the square distance function $d_M^2(\cdot)$ is of class $\mathcal{C}^{p+1}$ on U;
(b) the metric projection P_M is well-defined on U and it is of class $\mathcal{C}^p$ therein.

REMARK 17.53. Theorem 17.52 can also be utilized in place of Proposition 17.21 and Theorem 17.23 to develop a proof of Theorem 17.30. □

As a direct consequence of Theorem 17.52 and Proposition 17.29(b) we have the following corollary.

COROLLARY 17.54. Let $p \geq 1$ be an integer and M be a closed set of a Hilbert space H which is a $\mathcal{C}^{p+1}$-submanifold of H. Then M is a prox-regular set and for any of its continuous prox-regularity radius function $\rho : M \to]0, +\infty]$ both properties (a) and (b) hold:

(a) the square distance function $d_M^2(\cdot)$ is of class $\mathcal{C}^{p+1}$ on the open set $U_{\rho(\cdot)}(M)$;
(b) the metric projection P_M is well-defined on the open set $U_{\rho(\cdot)}(M)$ and it is of class $\mathcal{C}^p$ therein.

The situation of $\mathcal{C}^{p+1,0}$-submanifold will be analyzed much more in the next subsection in order to obtain a complete characterization for such a case.

17.4.2. Characterization of submanifolds via metric projection. Let us begin with the following lemma.

LEMMA 17.55. Let H, Y and Z be three Hilbert spaces and let U be an open neighborhood of 0 in H. Let $\Lambda : U \subset H \to \mathcal{L}(Y, Z)$ be a continuous mapping, let $f : U \times Y \to Z$ be defined by

$$f(u, y) := \Lambda(u)y \quad \text{for all } (u, y) \in U \times Y,$$

and let $p \geq 1$ be an integer. The following hold.
(a) The mapping Λ is of class $\mathcal{C}^p$, whenever f is of class $\mathcal{C}^p$ and that there exists an open neighborhood V of 0 in Y such that the family $\big(D^p f(\cdot, v)\big)_{v \in V}$ is locally uniformly equi-continuous, that is, for every $u_0 \in U$, there exists $\delta_0 > 0$ with $B_H(u_0, \delta_0) \subset U$ such that for every $\varepsilon > 0$ there is $\delta > 0$ satisfying

$$\forall u, u' \in B_X(u_0, \delta_0),\ \|u - u'\| \leq \delta \Longrightarrow \sup_{v \in V} \|D^p f(u, v) - D^p f(u', v)\| \leq \varepsilon. \tag{17.38}$$

(b) If for some open neighborhood V of zero in Y the restriction $f\big|_V$ of f to $U \times V$ belongs to $\mathcal{C}^{p,0}(U \times V, Z)$ (resp. $f\big|_V \in \mathcal{C}^{p,1}(U \times V, Z)$), then $\Lambda \in \mathcal{C}^{p,0}(U, \mathcal{L}(Y, Z))$ (resp. $\Lambda \in \mathcal{C}^{p,1}(U, \mathcal{L}(Y, Z))$).

PROOF. (a) Assume that f is of class $\mathcal{C}^p$ and that there exists $\eta > 0$ such that for $V := B_Y(0, \eta)$, the family $\big(D^p f(\cdot, v)\big)_{v \in V}$ is locally uniformly equi-continuous. For each $k \in \{1, \cdots, p-1\}$, by continuous differentiability of $D^k f$ and by the definition of f we see that the family $\big(D^k f(\cdot, v)\big)_{v \in V}$ is also locally uniformly equi-continuous.

For each $k \in \{1, \ldots, p\}$ define $F_k : U \to \mathcal{L}(H^k, \mathcal{L}(Y, Z))$ by

$$F_k(u)(x_1, \cdots, x_k)y := D^k f(u, y)((x_1, 0), \cdots, (x_k, 0)),$$

for every $(x_1, \cdots, x_k) \in H^k$ and every $y \in Y$. Let us prove that F_k is continuous for every $k \in \{1, \ldots, p\}$. For every $x \in H^k$ denote $\widetilde{x} := ((x_1, 0), \ldots, (x_k, 0)) \in (H \times Y)^k$.

Fix $u \in U$ and take any $\varepsilon > 0$. For $h \in H$ with $u + h \in U$ writting

$$\begin{aligned}\|F_k(u+h) - F_k(u)\| &= \sup_{x \in \mathbb{B}_{H^k}} \|F_k(u+h)(x) - F_k(u)(x)\|_{\mathcal{L}(Y,Z)} \\ &= \sup_{y \in B_Y(0,1)} \sup_{x \in \mathbb{B}_{H^k}} \|F_k(u+h)(x)y - F_k(u)(x)y\|_Z \\ &= \frac{1}{\eta} \sup_{y \in B_Y(0,\eta)} \sup_{x \in \mathbb{B}_{H^k}} \|D^k f(u+h,y)\widetilde{x} - D^k f(u,y)\widetilde{x}\|_Z,\end{aligned}$$

we see that

$$(17.39)\quad \|F_k(u+h) - F_k(u)\| \le \frac{1}{\eta} \sup_{y \in B_Y(0,\eta)} \|D^k f(u+h,y) - D^k f(u,y)\|_{\mathcal{L}(H^k,Z)}.$$

By (17.38) and by the above local uniform equi-continuity of $(D^k f(\cdot,v))_{v \in V}$, we can choose $\delta > 0$ sufficiently small (depending on u, ε and k) such that

$$(17.40)\qquad \sup_{y \in B_Y(0,\eta)} \|D^k f(u+h,y) - D^k f(u,y)\|_{\mathcal{L}(H^k,Z)} \le \eta\varepsilon, \quad \forall h \in B_H[0,\delta].$$

It results that for every $h \in B_H(0,\delta)$ we have that $\|F_k(u+h) - F_k(u)\| \le \varepsilon$, which justifies the continuity of F_k.

Let us show now by induction on $k \in \{1, \cdots, p\}$ that $D^k\Lambda = F_k$. Let us start with $k = 1$. Fix $u \in U$ and consider any $\varepsilon > 0$. As we did above, we can choose $\delta > 0$ sufficiently small depending on u, ε and k, such that the inequality (17.40) holds true for $k = 1$. For any $h \in B_X(0,\delta)$, noticing that for every $y \in B_Y(0,\eta)$

$$\begin{aligned}\|\Lambda(u+h)y - \Lambda(u)y - F_1(u)(h)y\|_Z &= \|f(u+h,y) - f(u,y) - Df(u,y)(h,0)\|_Z \\ &= \left\| \int_0^1 \big(Df(u+th,y) - Df(u,y)\big)(h,0)\,dt \right\|_Z\end{aligned}$$

and that

$$\begin{aligned}&\left\| \int_0^1 \big(Df(u+th,y) - Df(u,y)\big)(h,0)\,dt \right\|_Z \\ &\le \left(\int_0^1 \|Df(u+th,y) - Df(u,y)\|_{\mathcal{L}(H,Z)}\,dt \right) \|h\| \le \varepsilon\eta\|h\|\end{aligned}$$

(where the last inequality is due to the fact that $\|th\| \le \delta$), it ensues that

$$\begin{aligned}&\|\Lambda(u+h) - \Lambda(u) - F_1(u)(h)\| = \frac{1}{\eta} \sup_{y \in B_Y(0,\eta)} \|\Lambda(u+h)y - \Lambda(u)y - F_1(u)(h)y\|_Z \\ &\le \left(\frac{1}{\eta} \sup_{y \in B_Y(0,\eta)} \int_0^1 \|Df(u+th,y) - Df(u,y)\|_{\mathcal{L}(H,Z)}\,dt \right) \|h\| \le \varepsilon\|h\|.\end{aligned}$$

Since $F_1(u) \in \mathcal{L}(H, \mathcal{L}(Y,Z))$ and $\varepsilon > 0$ is arbitrary, it follows that $D\Lambda(u)$ exists and it coincides with $F_1(u)$. This justifies the desired property for $k = 1$.

Now, suppose that the property holds true for $1, \cdots, k-1$ with $1 < k \le p$. Then $\widetilde{\Lambda} := D^{k-1}\Lambda$ exists and it coincides with F_{k-1}. Fix $u \in U$ and take any $\varepsilon > 0$. As above, we can choose $\delta > 0$ sufficiently small so that the inequality (17.40) holds for $k-1$, hence for every $h \in B_H[0,\delta]$ writing

$$\begin{aligned}
&\|\widetilde{\Lambda}(u+h) - \widetilde{\Lambda}(u) - F_k(u)\| \\
=&\frac{1}{\eta} \sup_{y\in B_Y[0,\eta]} \sup_{x\in \mathbb{B}_{H^{k-1}}} \|\widetilde{\Lambda}(u+h)(x)y - \widetilde{\Lambda}(u)(x) - F_k(u)(x)y\|_Z \\
=&\frac{1}{\eta} \sup_{y\in B_Y[0,\eta]} \sup_{x\in \mathbb{B}_{H^{k-1}}} \left\| \int_0^1 \big(D^k f(u+th,y) - D^k f(u,y)\big)(\widetilde{x},(h,0))\,dt \right\|_Z,
\end{aligned}$$

we obtain that

$$\begin{aligned}
&\|\widetilde{\Lambda}(u+h) - \widetilde{\Lambda}(u) - F_k(u)\| \\
\le &\left(\frac{1}{\eta} \sup_{y\in B_Y[0,\eta]} \int_0^1 \|D^k f(u+th,y) - D^k f(u,y)\|_{\mathcal{L}(H^k,Z)}\,dt \right) \|h\| \le \varepsilon \|h\|,
\end{aligned}$$

where again the last inequality is due to the fact that $\|th\| \le \delta$ and where $\widetilde{x} := ((x_1,0),\cdots,(x_{k-1},0))$ for all $x \in H^{k-1}$. We derive as above that $D^k\Lambda$ exists and it coincides with F_k. This tells us that Λ is of class $\mathcal{C}^p$, which justifies the assertion (a) of the lemma.

(b) Now assume that $f\big|_V \in \mathcal{C}^{p,0}(U\times V, Z)$. By Lemma 17.56 below the local uniform continuity of $D^p\left(f\big|_V\right)$ implies the local uniform equi-continuity of the family $\big(D^p f(\cdot,v)\big)_{v\in V}$. This assures us by (a) that $D^p\Lambda$ exists, is continuous and coincides with F_p. We then have to show the locally uniform continuity of F_p.

Fix any $u_0 \in U$ and choose $\delta_0 > 0$ sufficiently small so that $B_{H\times Y}((u_0,0),\delta_0) \subset U\times V$ and so that $D^p\left(f\big|_V\right)$ is uniformly continuous on $B_{H\times Y}((u_0,0),\delta_0)$. Observe that $D^p\left(f\big|_V\right)$ coincides with $D^p f$ in $B_{H\times Y}((u_0,0),\delta_0)$. Take any $\varepsilon > 0$ and choose $\delta > 0$ such that for all $(u,y),(u',y') \in B_{H\times Y}((u_0,0),\delta_0)$ one has

$$\|(u',y') - (u,y)\| \le \delta \Rightarrow \|D^p f(u',y') - D^p f(u,y)\|_{\mathcal{L}(H^p,Z)} \le \frac{\delta_0\varepsilon}{2}.$$

Consider any $u,u' \in B_H(u_0,\delta_0/2)$. Since $\|(x,y)\| = (\|x\|^2 + \|y\|^2)^{1/2}$ for all $(x,y) \in H\times Y$, we have $B_H(u_0,\delta_0/2)\times B_Y(0,\delta_0/2) \subset B_{H\times Y}((u_0,0),\delta_0)$. Then, if $\|u-u'\| \le \delta$ we have as in (17.39)

$$\|F_p(u) - F_p(u')\| \le \frac{2}{\delta_0} \sup_{y\in B_Y(0,\delta_0/2)} \|D^p f(u,y) - D^p f(u',y)\| \le \varepsilon.$$

This justifies that F_p is uniformly continuous on $B_X(u_0,\delta_0/2)$, which in turn (by the arbitrariness of $u_0 \in U$) confirms the local uniform continuity of F_p. It results that $\Lambda \in \mathcal{C}^{p,0}(U,\mathcal{L}(Y,Z))$. The arguments for the local Lipschitz continuity of $D^p\Lambda$ whenever $f\big|_V \in \mathcal{C}^{p,1}(U\times V, Z)$ are similar. The proof of the lemma is finished. □

LEMMA 17.56. *Let H, Y and Z be three Hilbert spaces, U be an open neighborhood of 0 in H and V be a bounded open neighborhood of 0 in Y. Let $f : U\times Y \to Z$ be such that its restriction to $U\times V$ is class $\mathcal{C}^{p,0}$ with an integer $p \ge 1$. Assume that $f(u,\cdot)$ is linear for every $u \in U$. Then the family $\big(D^p f(\cdot,v)\big)_{v\in V}$ is locally uniformly equicontinuous.*

PROOF. Let a real $\beta > 0$ such that $V \subset \beta\mathbb{B}_Y$. Fix any $u_0 \in U$. Choose a real $\delta_0 > 0$ such that $W := B_X(u_0,\delta_0)\times B_Y(0,\delta_0)$ is included in $U\times V$ and such that

$D^p f|_W$ is uniformly continuous. Take any $\varepsilon > 0$ and choose $\delta > 0$ such that for any $(u, v), (u', v') \in W$

$$\|(u, v) - (u', v')\| < \delta \Longrightarrow \|D^p f(u, v) - D^p f(u', v')\| < \varepsilon\delta_0/(2\beta).$$

Then, for any $u, u' \in B_H(u_0, \delta_0)$ satisfying $\|u - u'\| < \delta$ and for any $v \in V$

$$\begin{aligned}\|D^p f(u, v) - D^p f(u', v)\| &= \left\| D^p \left(\frac{2\beta}{\delta_0} f\left(u, \frac{\delta_0}{2\beta} v\right)\right) - D^p \left(\frac{2\beta}{\delta_0} f\left(u', \frac{\delta_0}{2\beta} v\right)\right)\right\| \\ &= \frac{2\beta}{\delta_0} \left\| D^p f\left(u, \frac{\delta_0}{2\beta} v\right) - D^p f\left(u', \frac{\delta_0}{2\beta} v\right)\right\| < \varepsilon.\end{aligned}$$

This yields that for all $u, u' \in B_H(u_0, \delta_0)$

$$\|u - u'\| < \delta \Longrightarrow \sup_{v \in V} \|D^p f(u, v) - D^p f(u', v)\| \le \varepsilon.$$

This confirms the local uniform equicontinuity of the family $\left(D^p f(\cdot, v)\right)_{v \in V}$. □

We can now establish the theorem characterizing $\mathcal{C}^{p+1,0}$-submanifolds in infinite-dimensional Hilbert spaces via the square distance function or via the metric projection.

THEOREM 17.57 (characterization of $\mathcal{C}^{p+1,0}$-submanifolds via metric projection). *Let M be a set of a Hilbert space H which is closed near $m_0 \in M$.*

(a) *The following assertions (a.o), (a.i) and (a.ii) are pairwise equivalent:*
  (a.o) *M is a $\mathcal{C}^{p+1,0}$-submanifold at m_0;*
  (a.i) *$d_M^2(\cdot)$ is of class $\mathcal{C}^{p+1,0}$ on an open neighborhood U of m_0;*
  (a.ii) *P_M is well-defined on an open neighborhood U of m_0 and it belongs to $\mathcal{C}^{p,0}(U, H)$.*

(b) *Similarly, the following assertions are also equivalent:*
  (b.o) *M is a $\mathcal{C}^{p+1,1}$-submanifold at m_0;*
  (b.i) *$d_M^2(\cdot)$ is of class $\mathcal{C}^{p+1,1}$ on an open neighborhood U of m_0;*
  (b.ii) *P_M is well-defined on an open neighborhood U of m_0 and it belongs to $\mathcal{C}^{p,1}(U, H)$.*

PROOF. We may and do suppose that M is closed and $m_0 = 0$. By Proposition 15.19 the properties (a.i) and (a.ii) (resp. (b.i) and (b.ii)) are equivalent. We also observe by Lemma 17.51 that the implication (a.o)⇒(a.ii) (resp. (b.o)⇒(b.ii)) also holds true.

It remains to show (a.i)⇒(a.o) (the proof of (b.i)⇒(b.o) being similar). Assume (a.i) and fix $\delta > 0$ sufficiently small so that $d_M^2(\cdot)$ is of class $\mathcal{C}^{p+1,0}$ on $B_H(0, 2\delta)$. By Proposition 15.19 again P_M is well-defined on $B_H(0, 2\delta)$ and $\frac{1}{2}\nabla(d_M^2)(u) = u - P_M(u)$ for all $u \in B_H(0, 2\delta)$. It ensues that P_M is of class $\mathcal{C}^{p,0}$ on $B_H(0, 2\delta)$. Further, for $x \in B_H(0, \delta)$ we have that $P_M(x) \in B_H(0, 2\delta)$ and

$$P_M(P_M(x)) = P_M(x) \text{ and } P_M(0) = 0.$$

Putting $A := DP_M(0)$, we deduce from both latter equalities that $A \circ A = A$. Further, we know by Lemma 17.35 that A is the second derivative of a (convex) function, so it is also symmetric. We derive that for each $x \in H$

$$\langle x - Ax, Ax' \rangle = 0, \quad \forall x' \in H.$$

It follows that $DP_M(0) = \Pi_Z$, where Z is the closed vector subspace $Z = \overline{DP_M(0)H}$. Consider the $\mathcal{C}^p$-mapping $\Phi : B_H(0, 2\delta) \to Z \times Z^\perp$ defined by

$$\Phi(x) = \big(\pi_Z(P_M(x)), \pi_{Z^\perp}(x - P_M(x))\big) \quad \text{for all } x \in B_H(0, 2\delta).$$

Observe that $\Phi(0) = 0$ and $D\Phi(0) = L$, where $L : H \to Z \times Z^\perp$ is the above canonic isomorphism given by $L(h) = (\pi_Z(h), \pi_{Z^\perp}(h))$. Note also by Lemma E.5 in Appendix E that $\Phi \in \mathcal{C}^{p,0}(B_H(0, 2\delta), Z \times Z^\perp)$. Since $D\Phi(0)$ is invertible, by the local inverse mapping theorem and by Proposition E.6 in Appendix, there exists $0 < \delta_0 < \delta$ such that, for $\mathcal{U} = B_H(0, 2\delta_0)$ in H, the mapping $\Phi\big|_{\mathcal{U}} : \mathcal{U} \to \Phi(\mathcal{U})$ is a $\mathcal{C}^p$-diffeomorphism and $\big(\Phi\big|_{\mathcal{U}}\big)^{-1} \in \mathcal{C}^{p,0}(\Phi(\mathcal{U}), \mathcal{U})$. For convenience, denote $\Phi_{\mathcal{U}} := \Phi\big|_{\mathcal{U}}$. Choose an open neighborhood $V = V_Z \times V_{Z^\perp} \subset \Phi(\mathcal{U})$ of $(0,0)$ in $Z \times Z^\perp$ such that $U := (\Phi_{\mathcal{U}})^{-1}(V) \subset B_H(0, \delta_0)$. Put also $\Phi_U := \Phi\big|_U$. Define the $\mathcal{C}^p$-mapping $\theta : V_Z \to Z^\perp$ by

$$\theta(v_1) := \pi_{Z^\perp} \circ \Phi_U^{-1}(v_1, 0), \quad \forall v_1 \in V_Z.$$

Observe that $\theta(0) = 0$ and $D\theta(0) = 0$.

We claim that $M \cap U = L^{-1}(\operatorname{gph}\theta) \cap U$. Indeed, for every $u \in U$, from the evident equality $\Phi_{\mathcal{U}}(P_M(u)) = (\operatorname{pr}_Z \circ \Phi(u), 0)$ and from the injectivity of $\Phi_{\mathcal{U}}$ it is easily seen that

$$\Phi_U(M \cap U) = V_Z \times \{0\} = (Z \times \{0\}) \cap \Phi_{\mathcal{U}}(U).$$

It results that we have

$$\begin{aligned} u \in M \cap U &\Leftrightarrow (v_1, v_2) = v := \Phi_U(u) \in V_Z \times \{0\} \text{ and } \Phi_U^{-1}(v) = P_M(\Phi_U^{-1}(v)) \\ &\Leftrightarrow \begin{aligned}[t] (\pi_Z(u), \pi_{Z^\perp}(u)) &= (\pi_Z \circ \Phi_U^{-1}(v_1, 0), \pi_{Z^\perp} \circ \Phi_U^{-1}(v_1, 0)) \\ &= (v_1, \theta(v_1)) \in \operatorname{gph}(\theta) \cap L(U), \end{aligned} \end{aligned}$$

where the last equality follows from the fact that, by the above equality $\Phi_U^{-1}(v) = P_M(\Phi_U^{-1}(v))$ and by the other equality $\pi_Z \circ P_M(x) = \operatorname{pr}_Z \circ \Phi(x)$ for all $x \in U$ (due to the definition of Φ) we have

$$\pi_Z(\Phi_U^{-1}(v_1, 0)) = \pi_Z \circ P_M(\Phi_U^{-1}(v_1, 0)) = \operatorname{pr}_Z \circ (\Phi(\Phi_U^{-1}(v_1, 0))) = v_1.$$

The claim is then justified.

By Theorem 17.50, to obtain that M is a $\mathcal{C}^{p+1,0}$-submanifold at 0, it remains to prove that the $\mathcal{C}^p$-mapping θ in fact belongs to $\mathcal{C}^{p+1,0}(V_Z, Z^\perp)$. For that, fix $x \in U$ and denote $(z_1, z_2) := \Phi(x) \in V$, so $z_1 = \pi_Z(P_M(x))$ by definition of Φ. Since $\Phi(P_M(x)) = (\pi_Z(P_M(x)), 0)$ (by definition of Φ again), it ensues that

$$P_M(x) = \Phi_U^{-1}(\pi_Z(P_M(x)), 0) = \Phi_U^{-1}(z_1, 0),$$

and hence by definition of θ

$$\pi_{Z^\perp}(P_M(x)) = \theta(z_1) \text{ and } L(P_M(x)) = (z_1, \theta(z_1)).$$

Then, recalling that $M \cap U = L^{-1}(\operatorname{gph}\theta) \cap U$ and noticing that $L^* = L^{-1}$, we can write by the inclusion of the proximal normal cone into the Clarke one

$$\begin{aligned} x - P_M(x) &\in N^C(L^{-1}(\operatorname{gph}\theta); P_M(x)) = L^*\big(N^C(\operatorname{gph}\theta; L(P_M(x)))\big) \\ &= L^*\big(N^C(\operatorname{gph}\theta; (z_1, \theta(z_1)))\big) = L^{-1}\big(\{(-D\theta(z_1)^*\xi, \xi) \;:\; \xi \in Z^\perp\}\big), \end{aligned}$$

hence $L(x - P_M(x)) = (-D\theta(z_1)^*\xi, \xi)$ for some $\xi \in Z^\perp$. Since by definition of L and Φ we have $\operatorname{pr}_{Z^\perp}(L(x - P_M(x))) = \pi_{Z^\perp}(x - P_M(x)) = \operatorname{pr}_{Z^\perp} \circ \Phi(x)$, it ensues that $\xi = z_2$, and so by the equality $x - P_M(x) = \frac{1}{2}\nabla d_M^2(x) = \frac{1}{2}\nabla d_M^2(\Phi_U^{-1}(z_1, z_2))$ we get that

$$D\theta(z_1)^* z_2 = -\frac{1}{2}\pi_Z \circ (\nabla d_M^2) \circ \Phi_U^{-1}(z_1, z_2). \tag{17.41}$$

This equality holds for every $(z_1, z_2) \in V$. By Lemma E.5 in Appendix E, it is immediate that $-\frac{1}{2}\pi_Z \circ (\nabla d_M^2) \circ \Phi_U^{-1} \in \mathcal{C}^{p,0}(V, Z)$, and so the mapping $f : V_Z \times Z^\perp \to Z$ given by $f(v, \xi) := D\theta(v)^*\xi$ fulfills the assumptions of Lemma 17.55(b). This ensures that $D\theta(\cdot)^* \in \mathcal{C}^{p,0}(V_Z, \mathcal{L}(Z^\perp, Z))$, which in turn implies (noticing that the adjoint map $* : \mathcal{L}(Z, Z^\perp) \to \mathcal{L}(Z^\perp, Z)$ is of class $\mathcal{C}^\infty$ and invertible by reflexivity of the involved spaces) that $\theta \in \mathcal{C}^{p+1,0}(V_Z, Z^\perp)$. The proof is then finished. □

In finite dimensions we directly derive:

THEOREM 17.58 (J.-B. Poly and G. Raby). *Let M be a set of a finite-dimensional Euclidean space E which is closed near $m_0 \in M$. The following assertions (a.o), (a.i) and (a.ii) are pairwise equivalent:*

(a.o) *M is a $\mathcal{C}^{p+1}$-submanifold at m_0;*
(a.i) *$d_M^2(\cdot)$ is of class $\mathcal{C}^{p+1}$ on an open neighborhood U of m_0;*
(a.ii) *P_M is well-defined on an open neighborhood U of m_0 and it belongs to $\mathcal{C}^p(U, E)$.*

17.4.3. Smoothness property of signed distance function. We will present the result on smoothness of signed distance function by considering closed sets with the interior tangent cone property at its points. We start with the following proposition.

PROPOSITION 17.59. *Let C be a closed set of a Hilbert space H which has the interior tangent cone property at a point $\overline{x} \in \operatorname{bdry} C$. If $\operatorname{bdry} C$ is a $\mathcal{C}^p$-submanifold at $\overline{x}$ with $p \in \mathbb{N}$, then there exists a vector hyperplane Z of H, a unit orthogonal vector ν_Z to Z, a neighborhood U of $\overline{x}$ and a $\mathcal{C}^p$-diffeomorphism $\varphi : U \to \varphi(U)$ with $\varphi(\overline{x}) = 0$ such that*

$$\varphi(U \cap \operatorname{bdry} C) = \varphi(U) \cap Z \quad \text{and} \quad \varphi(U \cap C) = \varphi(U) \cap \{\langle \nu_Z, \cdot\rangle \le 0\}.$$

Further, at each $x \in U \cap \operatorname{bdry} C$ there exists a unique unit C-normal vector $\widehat{\nu}(x)$ to C, and this vector is given by

$$\widehat{\nu}(x) = \frac{D\varphi(x)^*\nu_Z}{\|D\varphi(x)^*\nu_Z\|}.$$

PROOF. We know that the set $\operatorname{cl}_H(C^c)$ (where $C^c := H \setminus C$) satisfies also the interior tangent cone property at $\overline{x}$ by Proposition 2.29. We also notice by Proposition 17.16 that there exists a vector hyperplane Z of X, a neighborhood U of $\overline{x}$ and a $\mathcal{C}^p$-diffeomorphism $\varphi : U \to \varphi(U)$ with $\varphi(\overline{x}) = 0$ such that

$$\varphi(U \cap \operatorname{bdry}_H C) = \varphi(U) \cap Z.$$

Without loss of generality, we may suppose that $\varphi(U) = B_X(0, \delta)$ for some real $\delta > 0$. Let ν_Z be a unit vector in H orthogonal to Z, so $Z^\perp = \mathbb{R}\nu_Z$. As a first step let us prove that $V := \varphi(U \cap \operatorname{int}_H C)$ is connected. Suppose the contrary, that is, there exist two disjoint nonempty open sets V_1 and V_2 such that $V_1 \cup V_2 = V$.

Choose $v_1 \in V_1$ and $v_2 \in V_2$ and note that the line segment $[v_1, v_2]$ is included in $\varphi(U)$ by convexity of $\varphi(U)$. Define

$$t_1 := \sup\{t \in [0,1] \ : \ [v_1, tv_2 + (1-t)v_1] \subset V_1\},$$
$$t_2 := \inf\{t \in [0,1] \ : \ [tv_2 + (1-t)v_1, v_2] \subset V_2\}.$$

It is not difficult to see that $0 < t_1 \le t_2 < 1$. Therefore, there exists at least an element $v_0 \in [v_1, v_2]$ such that $\varphi^{-1}(v_0) \in \mathrm{bdry}_H(\mathrm{int}_H C) \subset \mathrm{bdry}_H C$, and therefore $v_0 \in Z$. Consider the sets

$$B^+ = B_X(0,\delta) \cap \{x \in X \ : \ \langle \nu_Z, x\rangle > 0\} \text{ and } B^- = B_X(0,\delta) \cap \{x \in X \ : \ \langle \nu_Z, x\rangle < 0\}.$$

Clearly, $V \subset B^+ \cup B^-$, and therefore either $v_1 \in B^+$ or $v_1 \in B^-$. By replacing ν_Z by $-\nu_Z$ if necessary, we may suppose that $v_1 \in B^-$. Since $\langle \nu_Z, v_0\rangle = 0$ according to the inclusion $v_0 \in Z$, we obtain by linearity that $v_2 \in B^+$. Moreover, $\varphi(C^c \cap U)$ is also included in $B^+ \cup B^-$. Noticing that $\varphi(C^c \cap U) \neq \emptyset$ because $\overline{x} \in \mathrm{bdry}_H C$, we deduce that at least one of the sets $\varphi(C^c \cap U) \cap B^+$ and $\varphi(C^c \cap U) \cap B^-$ is nonempty. Without losing generality, let us suppose that the first one is nonempty. Then, we can write

$$B^+ = (\varphi(C^c \cap U) \cap B^+) \cup (V \cap B^+),$$

where the sets in the union are nonempty, open and disjoint. This ensures that B^+ is disconnected, which is clearly a contradiction since B^+ is convex. Thus, V is connected which entails that $U \cap \mathrm{int}_H C = \varphi^{-1}(V)$ is also connected.

Let us show now that, replacing ν_Z by $-\nu_Z$ if necessary, we have

$$V = \{x \in \varphi(U) \ : \ \langle \nu_Z, x\rangle < 0\}.$$

Suppose the contrary, and assume that there exist $v_1, v_2 \in V$ such that $\langle \nu_Z, v_1\rangle < 0$ and $\langle \nu_Z, v_2\rangle > 0$. Then, by connectedness of V, the sets

$$\{x \in \varphi(U) \ : \ \langle \widehat{n}_Z, x\rangle < 0\} \text{ and } \{x \in \varphi(U) \ : \ \langle \widehat{n}_Z, x\rangle > 0\}$$

must intersect. This is a contradiction. Then, either V is contained in the first one of the two latter sets or in the second one. If the second case holds, we can replace ν_Z by $-\nu_Z$. This gives

$$V \subset \{x \in \varphi(U) \ : \ \langle \nu_Z, x\rangle < 0\} = \varphi(U) \cap \{\langle \nu_Z, \cdot\rangle < 0\}.$$

If the inclusion is strict, then there exist $y, y' \in \varphi(U)$ such that $\varphi^{-1}(y) \in \mathrm{int}_H C$, $\varphi^{-1}(y') \in H \setminus \mathrm{int}_H C = \mathrm{cl}_H(C^c)$ and $\langle \nu_Z, z\rangle < 0$ for all $z \in [y, y']$. Observe that $[y, y'] \subset \varphi(U)$ since $\varphi(U) = B(0,\delta)$. Define

$$t_0 := \sup\{t \in [0,1] \ : \ ty' + (1-t)y \in \varphi(U \cap \mathrm{int}_H C)\},$$

and $y_0 := t_0 y + (1-t_0)x \subset [y, y'] \subset \varphi(U)$. The inclusion $y_0 \in [y, y']$ gives the inequality $\langle \nu_Z, y_0\rangle < 0$. By definition of t_0 and the openness of $\varphi(U \cap \mathrm{int}_H C)$ in H, we also have $y_0 \in \mathrm{bdry}_H\big(\varphi(U \cap \mathrm{int}_H C)\big)$, hence

$$y_0 \in \varphi(U) \cap \mathrm{bdry}_H\big(\varphi(U \cap \mathrm{int}_H C)\big) = \mathrm{bdry}_{\varphi(U)}\big(\varphi(U \cap \mathrm{int}_H C)\big),$$

which yields by homeomorphism property of $\varphi : U \to \varphi(U)$

$$\varphi^{-1}(y_0) \in \mathrm{bdry}_U(U \cap \mathrm{int}_H C) = U \cap \mathrm{bdry}_H(U \cap \mathrm{int}_H C),$$

where the latter equality is due to Lemma 17.19. From the inclusion $\varphi^{-1}(y_0) \in \mathrm{bdry}_H(U \cap \mathrm{int}_H C)$ we see one hand that $\varphi^{-1}(y_0) \in \mathrm{cl}_H(U \cap \mathrm{int}_H C) \subset \mathrm{cl}_H C$, and one the other hand by the openness of $U \cap \mathrm{int}_H C$ that $\varphi^{-1}(y_0) \notin U \cap \mathrm{int}_H C$, so $\varphi^{-1}(y_0) \notin \mathrm{int}_H C$ since $\varphi^{-1}(y_0) \in U$. It ensues that $\varphi^{-1}(y_0) \in U \cap \mathrm{bdry}_H C$, hence

$$y_0 \in \varphi(U \cap \mathrm{bdry}_H C) = \varphi(U) \cap Z,$$

which in turn entails that $\langle \nu_Z, y_0 \rangle = 0$, and this is in contradiction with the above inequality $\langle \nu_Z, y_0 \rangle < 0$. Consequently, the equality $V = \varphi(U) \cap \{\langle \nu_Z, \cdot \rangle < 0\}$ holds true, that is,

$$\varphi(U \cap \mathrm{int}_H C) = \varphi(U) \cap \{\langle \nu_Z, \cdot \rangle < 0\}.$$

Combining this with the equality $\varphi(U \cap \mathrm{bdry}_H C) = \varphi(U) \cap \{\langle \widehat{n}_Z, \cdot \rangle = 0\}$, we derive that as desired

$$\varphi(U \cap C) = \varphi(U) \cap \{\langle \widehat{n}_Z, \cdot \rangle \le 0\}.$$

Finally, the property and equality concerning $\widehat{\nu}(x)$ in the statement follows from the latter equality. □

Let C be a subset of a Hilbert space H and U_0 be an open set of H such that $U_0 \cap C$ is closed relative to U_0. Assume that C has the interior tangent cone property at each point of $U_0 \cap \mathrm{bdry}\, C$ and that $\mathrm{bdry}\, C$ is a $\mathcal{C}^1$ submanifold of H at each point of $U_0 \cap \mathrm{bdry}\, C$. Then for each $x \in U_0 \cap \mathrm{bdry}\, C$ the unique unit normal vector $\widehat{\nu}_C(x)$ to C at x with $N^C(C; x) = \mathbb{R}_+ \widehat{\nu}_C(x)$ exists according to Proposition 17.59; it is called the *unit exterior normal vector* to C at x. The mapping

$$\widehat{\nu}_C : U_0 \cap \mathrm{bdry}\ C \to \mathbb{S}_H \tag{17.42}$$

is then well-defined. When there is no risk of confusion, one denotes simply $\widehat{\nu}(x)$ in place of $\widehat{\nu}_C(x)$. In the case where the boundary is in addition a $\mathcal{C}^{p+1}$-submanifold, Proposition 17.59 and Theorem 17.52 directly guarantee the following proposition.

PROPOSITION 17.60. Let C be a subset of a Hilbert space H and U_0 be an open set of H such that $U_0 \cap C$ is closed relative to U_0. Assume that C has the interior tangent cone property at each point of $U_0 \cap \mathrm{bdry}\, C$ and $\mathrm{bdry}\, C$ is a $\mathcal{C}^{p+1}$ submanifold of H at each point of $U_0 \cap \mathrm{bdry}\, C$ with $p \in \mathbb{N}$. Then there is an open set U containing $U_0 \cap \mathrm{bdry}\, C$ such that the mapping $x \mapsto \widehat{\nu}(P_{\mathrm{bdry}\, C}(x))$ is well-defined on U and of class $\mathcal{C}^p$ on U.

We can now state and prove the theorem related to the smoothness of the signed distance function as well as to sublevel representations of sets with such a property. The statement involves the concepts of exterior unit normal vector and mapping defined in (17.42).

THEOREM 17.61 (smoothness of signed distance function). Let C be a closed subset of a Hilbert space H with the interior tangent cone property at any of its points and let $p \in \mathbb{N}$. The following assertions are equivalent:
(a) The boundary $\mathrm{bdry}\, C$ of C is a $\mathcal{C}^{p+1}$-submanifold at each of its points;
(b) there exists an open set $U \supset \mathrm{bdry}\, C$ over which the signed distance function sgd_C is of class $\mathcal{C}^{p+1}$ and such that the exterior unit normal mapping $\widehat{\nu}$ is well defined on $U \cap \mathrm{bdry}\, C$ with

$$\nabla \mathrm{sgd}_C(x) = \widehat{\nu}(P_{\mathrm{bdry}\, C}(x)) \quad \text{for all } x \in U,$$

which gives the Eikonal equation

$$\|\nabla \mathrm{sgd}_C(x)\| = 1 \quad \text{for all } x \in U;$$

(c) the signed distance function sgd_C is of class $\mathcal{C}^{p+1}$ in an open set U containing $\mathrm{bdry}\, C$;
(d) there exists a function $g : H \to \mathbb{R}$ of class $\mathcal{C}^{p+1}$ such that

$$C = \{x \in H : g(x) \le 0\} \text{ and } \mathrm{bdry}\, C = \{x \in H : g(x) = 0\},$$

and such that 0 is not a critical value of g, that is, $Dg(x) \neq 0$ for any $x \in g^{-1}(0)$.

PROOF. On the one hand, the implication (d)⇒(a) is direct, since the fact that 0 is not a critical value of g yields directly that $\mathrm{bdry}\, C = \{x \in X \ : \ g(x) = 0\}$ is a $\mathcal{C}^{p+1}$-submanifold of H according to Proposition E.9 in Appendix E.

Let us show the implication (c)⇒(d). Since H is a metric space, there exists an open set V containing $\mathrm{bdry}\, C$ such that $\overline{V} \subset U$. Then, we can consider the open cover $\mathscr{U} = \{U, U_-, U_+\}$ where $U_- =: (\mathrm{int}\, C) \setminus \overline{V}$ and $U_+ := H \setminus (C \cup \overline{V})$. Consider a corresponding locally finite $\mathcal{C}^{\infty}$-partition of unity $\{\varphi, \varphi_-, \varphi_+\}$ subordinated to this cover (see Theorem A.4 in Appendix A). Then, the function g is simply defined by

$$g(x) = -1 \cdot \varphi_-(x) + \varphi(x) \cdot \mathrm{sgd}_C(x) + 1 \cdot \varphi_+(x) \quad \text{for all } x \in H.$$

Noting that $\varphi|_V = 1$, we see that at every point $x \in \mathrm{bdry}\, C$, $\nabla g(x) = \nabla \mathrm{sgd}_C(x)$, and therefore by Theorem 2.154 the regularity property $Dg(x) \neq 0$ also holds, so the implication (c)⇒(d) is justified.

The implication (b)⇒(c) being obvious, it remains to show (a)⇒(b). Suppose that (a) holds. By Proposition 17.59 and Theorem 17.46, we know that the sets C and $S := \mathrm{cl}(H \setminus C)$ are epi-Lipschitz and prox-regular and that there exists an open set U_1 containing $\mathrm{bdry}\, C$ such that d_C and d_S are of class $\mathcal{C}^{p+1}$ on $U_1 \setminus C$ and $U_2 \setminus S$ respectively. Furthermore, by Theorem 2.154, we know that for each $x \in \mathrm{bdry}\, C$, $0 \notin \partial_C \mathrm{sgd}_C(x)$. Choose by Proposition 17.60 an open set U_2 containing $\mathrm{bdry}\, C$ such that $x \mapsto \widehat{\nu}(P_{\mathrm{bdry}\, C}(x))$ is of class C^p on U_2. Putting $U := U_1 \cap U_2$ we only need to show that for every $x \in U$, the function sgd_C is G-differentiable at x with $\nabla \mathrm{sgd}_C(x) = \widehat{\nu}(P_{\mathrm{bdry}\, C}(x))$. Let $x \in U$ and $h \in \mathbb{S}_X$. If $x \in H \setminus C$, then we have by Lemma 15.11(d)

$$\begin{aligned}
\lim_{t \downarrow 0} \frac{\mathrm{sgd}_C(x + th) - \mathrm{sgd}_(x)}{t} &= \lim_{t \downarrow 0} \frac{d_C(x + th) - d_C(x)}{t} \\
&= \left\langle \frac{x - P_C(x)}{\|x - P_C(x)\|}, h \right\rangle = N^C(C; P_C(x)) \cap \mathbb{S}_H \\
&= \langle \widehat{n}_C(P_{\mathrm{bdry}\, C}(x)), h \rangle,
\end{aligned}$$

where the last equality follows noting that $P_C(x) = P_{\mathrm{bdry}\, C}(x)$. Similarly, if $x \in \mathrm{int}\, C$, utilizing the same arguments and noting that $P_S(x) = P_{\mathrm{bdry}\, C}(x)$ and that $N^C(S; P_S(x)) = -N^C(C; P_S(x))$ (since C is epi-Lipschitz), we see that

$$\begin{aligned}
&\lim_{t \downarrow 0} \frac{\mathrm{sgd}_C(x + th) - \mathrm{sgd}_C(x)}{t} = -\lim_{t \downarrow 0} \frac{d_S(x + th) - d_S(x)}{t} \\
&= -\left\langle \frac{x - P_S(x)}{\|x - P_S(x)\|}, h \right\rangle = -N^C(S; P_S(x)) \cap \mathbb{S}_H \\
&= N^C(C; P_S(x)) \cap \mathbb{S}_H = N^C(C; P_{\mathrm{bdry}\, C}(x)) \cap \mathbb{S}_H = \langle \widehat{\nu}_C(P_{\mathrm{bdry}\, C}(x)), h \rangle.
\end{aligned}$$

Finally, for x fixed as above in $U \cap \mathrm{bdry}\, C$ we know that $P_{\mathrm{bdry}\, C}$ is differentiable at x by Theorem 17.52 and that $DP_{\mathrm{bdry}\, C}(x) = \Pi_{T_x(\mathrm{bdry}\, C)}$ by Theorem 15.83, where the latter equality is due to the prox-regularity of $\mathrm{bdry}\, C$ at x. Furthermore, for any $u \in U$ we observe (with some of above arguments) that $\widehat{\nu}_S(P_{\mathrm{bdry}\,C}(x)) = -\widehat{\nu}_C(P_{\mathrm{bdry}\,C}(x))$, and hence

$$\mathrm{sgd}_C(u) = \begin{cases} d_C(u) = \langle \widehat{\nu}_C(P_{\mathrm{bdry}\,C}(u)), u - P_{\mathrm{bdry}\,C}(u) \rangle, & \text{if } u \in H \setminus C, \\ -d_S(u) = \langle \widehat{\nu}_C(P_{\mathrm{bdry}\,C}(u)), u - P_{\mathrm{bdry}\,C}(u) \rangle, & \text{if } u \in S. \end{cases}$$

Therefore, we can write

$$\begin{aligned} &t^{-1}\mathrm{sgd}_C(x+th) \\ &= t^{-1}\langle \widehat{\nu}_C(P_{\mathrm{bdry}\,C}(x+th)), x+th - P_{\mathrm{bdry}\,C}(x+th) \rangle \\ &= t^{-1}\langle \widehat{\nu}_C(P_{\mathrm{bdry}\,C}(x+th)), x+th - (P_{\mathrm{bdry}\,C}(x) + DP_{\mathrm{bdry}\,C}(x)(th) + o(t)) \rangle \\ &= \langle \widehat{\nu}_C(P_{\mathrm{bdry}\,C}(x+th)), h \rangle - \langle \widehat{\nu}_C(P_{\mathrm{bdry}\,C}(x+th)), DP_{\mathrm{bdry}\,C}(x)h \rangle + \frac{o(t)}{t}, \end{aligned}$$

which entails by the equality $\mathrm{sgd}_C(x) = 0$ that

$$t^{-1}\mathrm{sgd}_C(x+th) - \mathrm{sgd}_C(x) \underset{t \downarrow 0}{\longrightarrow} \langle \widehat{\nu}_C(x), h \rangle - \langle \widehat{\nu}_C(x), DP_{\mathrm{bdry}\,C}(x)h \rangle = \langle \widehat{\nu}_C(x), h \rangle$$

where the last equality follows from the fact that $\widehat{\nu}_C(x)$ is orthogonal to $T_x(\mathrm{bdry}\,C)$. So, as desired the function sgd_C is G-differentiable on U with

$$\nabla \mathrm{sgd}_C(x) = \widehat{\nu}_C(P_{\mathrm{bdry}\,C}(x)) \quad \text{for every } x \in U.$$

This finishes the proof of the theorem. □

17.5. Comments

It seems that the differentiability of metric projection started with the fundamental 1952 paper [**761**] of J. Nash where he characterized analytic manifolds in finite dimensions by analyticity of metric projection. Although only analytic submanifolds are examined in [**761**], the proof of Lemma 1 in [**761**] also shows that, given a subset M in $\mathbb{R}^n$ which is a $\mathcal{C}^{p+1}$-submanifold at $m_0 \in M$ then the metric projection P_M is well-defined and of class $\mathcal{C}^p$ near m_0. A second great step arrived with the 1973 paper [**501**] by R. B. Holmes and the 1982 paper [**412**] by S. Fitzpatrick and R. R. Phelps. R. B. Holmes showed in [**501**] that, for any closed convex body C in a Hilbert space H whose boundary is a $\mathcal{C}^{p+1}$-submanifold, all the derivatives $DP_C, \cdots, D^pP_C$ exist on $H \setminus C$ and are continuous there, otherwise stated P_C is of class $\mathcal{C}^p$ on $H \setminus C$. A local version of such a result is also proved in [**501**] when the boundary of the closed convex body C is a $\mathcal{C}^{p+1}$-submanifold only near a point $\overline{x} \in \mathrm{bdry}\,C$. S. Fitzpatrick and R. R. Phelps completed both global and local above results by identifying the non-trivial condition (Cond) allowing them to prove in [**412**] that the boundary of the convex body C in H is a $\mathcal{C}^{p+1}$-submanifold (resp. $\mathcal{C}^{p+1}$-submanifold near $\overline{x} \in \mathrm{bdry}\,C$) if and only P_C is of class $\mathcal{C}^p$ on $H \setminus C$ (resp. on a suitable open set outside C) and condition (cond) holds. The third great step was the 1984 paper [**816**] where J.-B. Poly and G. Raby demonstrated that a set M in $\mathbb{R}^n$ is a $\mathcal{C}^{p+1}$-submanifold at a point $m_0 \in M$ if and only if the metric projection P_M exists on an open neighborhood U of m_0 and is of class $\mathcal{C}^p$ on U. In addition, we must mention three other fundamental papers by E. H. Zarantonello, by A. Canino and by A. Shapiro, already cited in Comments in Chapter 15 for the directional derivative of metric projection. Those papers are related to another direction. Given any closed convex subset C of a Hilbert space H and any point $\overline{x} \in C$, as already discussed in Comments section in Chapter 15,

E. H. Zarantonello showed in his 1971 paper [**1001**] that the directional derivative of P_C exists along with the equality

$$P'_C(\overline{x}; h) = P_{T(C;\overline{x})}(h) \quad \text{for all } h \in H;$$

various applications to spectral theory of conic operators are developed in [**1001**]. We recall that the directional derivative $F'(x; \cdot)$ of a mapping F between Banach spaces X, Y, when it exists, is defined by

$$F'(x; h) = \lim_{t \downarrow 0} t^{-1}\big(F(x + th) - F(x)\big) \quad \text{for all } h \in X.$$

Independently, A. Canino in her 1988 paper [**192**] and A. Shapiro in his 1994 paper [**887**] extended the above Zarantonello result for sets called p-convex in [**192**] and $O(2)$-convex in [**887**]; p-convex sets and $O(2)$-convex sets coincide with prox-regular sets. The directional derivative of the metric projection of a closed convex set C at points outside C was studied a few years later by A. Haraux [**466**] and F. Mignot [**710**]. Following [**466**] let us say that a closed convex set C in a Hilbert space H is *polyhedric* at a point $x \in H$ provided

$$\mathrm{cl}\,\big([v]^{\perp} \cap \mathbb{R}_{+}(C - P_C x)\big) = [v]^{\perp} \cap T(C; P_C x) \quad \text{for all } v \in N(C; P_C x),$$

where $[v]^{\perp}$ denotes the vector subspace orthogonal to the vector v. Assuming that the closed convex set C of the Hilbert space H is polyhedric at a point $x \in H$, it is proved in [**466**] that the directional derivative of P_C at x exists and satisfies the equality

$$P'_C(x; h) = P_{\Gamma(C;x)}(h) \quad \text{for all } h \in H, \tag{17.43}$$

where $\Gamma(C; x) := [x - P_C(x)]^{\perp} \cap T(C; P_C x)$; applications to variational inequalities in Sobolev spaces are given in [**466**]. Given on the Hilbert space H a coercive continuous (not necessarily symmetric) bilinear functional $a : H \times H \to \mathbb{R}$ and defining the a-projection of $x \in H$ onto the closed convex set C as the unique point $P_{C,a}(x) \in C$ such that

$$a\big(x - P_{C,a}(x), y - P_{C,a}(x)\big) \le 0 \quad \text{for all } y \in C,$$

the equality in (17.43) is proved in [**710**] with $P_{C,a}$ in place of P_C whenever C is polyhedric at x; the result is applied in [**710**] to variational inequalities and to optimal control problems governed by certain variational inequalities. We must also mention the book [**111**] of J. F. Bonnans and A. Shapiro as a very good reference with diverse applications of polyhedricity of convex sets related to second order optimality conditions for optimization and optimal control problems. Important complement results for polyhedricity of convex sets can be found in the papers [**965, 966**] of G. Wachsmuth.

Theorem 17.30 was proved by R. Correa, D. Salas and L. Thibault in [**294**] and its converse yielding to Theorem 17.46 is due to D. Salas and L. Thibault [**880**]. All the results in Theorem 17.23, Lemma 17.25, Lemma 17.26 and Proposition 17.27, their proofs along with the proof of Theorem 17.30 are taken from the 2018 paper [**294**] of Correa, Salas and Thibault. Corollary 17.33 and Theorem 17.34 correspond to much earlier results of R. B. Holmes in his 1973 paper [**501**]. The approach by Holmes in [**501**], using the convexity of the set C, is based on the Minkowski functional of a convex body, while the approach in [**294**] employes, for a very large part, variational analysis arguments for the prox-regularity and the epi-Lipschitz property of sets.

Theorem 17.46 and Theorem 17.47 are due to D. Salas and L. Thibault [**880**]. From the same paper [**880**] we have taken the proofs of those two theorems. From [**880**] we have also taken the proofs of Proposition 17.38, Lemma 17.40, Lemma 17.41, Proposition 17.43, and Proposition 17.44.

Corollary 17.48 appeared also in Salas and Thibault's article [**880**], whereas Theorem 17.49 is a result proved many years earlier in 1982 by S. Fitzpatrick and R. R. Phelps in [**412**]. The result of Fitzpatrick and Phelps completed, for a convex body, the one of Holmes [**501**] (that is, the statement of Theorem 17.34) and their approach is based on the Minkowski functional of the convex body as in [**501**]. In fact, the proofs of Lemma 17.35 and Proposition 17.36 in the book are inspired by the similar results for convex sets established in [**412**].

Theorem 17.52 and Theorem 17.57 as well as their proofs are taken from the paper [**878**] by D. Salas and L. Thibault. The finite-dimensional characterization in Theorem 17.58 is due to J.-B. Poly and G. Raby [**816**] and the proofs in the manuscript along with the proofs in [**878**] use various ideas from Poly and Raby [**816**]. We also mention that, given an analytic submanifold M in $\mathbb{R}^n$, the analyticity property of its metric projection near M was established by J. Nash in his 1952 paper [**761**]. As said in the first lines of this section of Comments, the implication (a.o)$\Rightarrow$(a.ii) (that is the finite-dimensional version of Theorem 17.52) can also be seen as a direct consequence of the arguments employed for Lemma 1 in this Nash's paper [**761**].

Concerning Theorem 17.61 on smoothness of signed distance function the finite-dimensional version of the implication (a)$\Rightarrow$(b) was proved by S. G. Krantz and H. R. Parks [**616**, Theorems 2 and 3], and by M. Delfour and J.-P. Zolésio [**324**, Theorem 5.6] with applications to shape optimization problems. The converse implication (c)$\Leftarrow$(a) was proved by Delfour and Zolésio [**324**, Theorem 5.6]. The approach in the setting of Hilbert spaces in the book for Theorem 17.61 as well as for Proposition 17.59 follows the development by D. Salas and L. Thibault [**879**]. Other results on signed distance functions along with many applications to shape optimization problems can be found in the book [**325**] by Delfour and Zolésio.

CHAPTER 18

Prox-regularity of sets in uniformly convex Banach spaces

The theory of prox-regular sets in Hilbert spaces has been extensively studied in Chapter 15. In the present one we consider the setting of uniformly convex Banach spaces.

18.1. Uniformly convex Banach spaces

This section is devoted to recalling some basic properties of uniformly convex Banach spaces which will be utilized in the development of the extension of prox-regular sets to such spaces. Almost all the results will be demonstrated in detail. Doing so, we will prepare some tools and non-Hilbertian arguments which will be encountered in the next sections.

Throughout this chapter, all vector spaces which we will work with, are assumed to be *nonreduced to the trivial null space.*

We start with the strict convexity of a normed space.

18.1.1. Strictly convex normed spaces. Although the concept of strict convexity has been already utilized in the book (see, for example, Theorem 14.29 and the sentence preceding Theorem 14.2), to facilitate the presentation of this section we begin by recalling the definition.

DEFINITION 18.1. A norm $\|\cdot\|$ on a vector space X is called *strictly convex* or *rotund* provided that, for all $t \in]0,1[$ and x, y in X with $x \neq y$ and $\|x\| = \|y\| = 1$, one has

$$\|tx + (1-t)y\| < 1.$$

One also says that the normed space $(X, \|\cdot\|)$ is *strictly convex* or *rotund.*

Some useful characterizations must be established. The assertion (c) below means that the strict convexity of $(X, \|\cdot\|)$ amounts to saying that the unit sphere $\mathbb{S}_X$ of X relative to $\|\cdot\|$ does not contain any line segment.

LEMMA 18.2. Let $\|\cdot\|$ be a norm on a vector space X. The following are equivalent:

(a) the norm $\|\cdot\|$ is strictly convex;

(b) for any $x, y \in \mathbb{S}_X$ with $x \neq y$, the inequality $\|\frac{1}{2}(x+y)\| < 1$ holds;

(c) for any $x, y \in \mathbb{S}_X$ with $x \neq y$, there exists some $t_0 \in]0,1[$ such that $\|t_0 x + (1-t_0)y\| < 1$;

(d) for any nonzero $x, y \in X$, the equality $\|x+y\| = \|x\| + \|y\|$ entails that $y = \lambda x$ with $\lambda = \|y\|/\|x\|$.

PROOF. The implications (a) $\Rightarrow$ (b) and (b) $\Rightarrow$ (c) are obvious. Conversely, assume (c) and fix $0 < t < 1$ and $x, y \in \mathbb{S}_X$ with $x \neq y$. Let $t_0 \in]0,1[$ be given by (c). If $t = t_0$, there is nothing to prove. Put $z_t := tx + (1-t)y$ and $q := t_0 x + (1-t_0)y$. If $0 < t < t_0$, then $z_t \in [q, y]$, hence $z_t = sq + (1-s)y$ for some $0 < s < 1$. Consequently, the assumption (c) entails

$$\|z_t\| \leq s\|q\| + (1-s)\|y\| = s\|q\| + (1-s) < s + (1-s) = 1,$$

so $\|z_t\| < 1$ as required. The case $t_0 < t < 1$ is similar, so (c) $\Rightarrow$ (a) is proved.

Now let us show that (a) $\Rightarrow$ (d). Fix two nonzero $x, y \in X$ such that $\|x+y\| = \|x\| + \|y\|$. The latter equality is equivalent to

$$\left\| \frac{\|x\|}{\|x\|+\|y\|} \frac{x}{\|x\|} + \frac{\|y\|}{\|x\|+\|y\|} \frac{y}{\|y\|} \right\| = 1,$$

hence the assumption (a) of strict convexity of $\|\cdot\|$ guarantees that

$$\frac{x}{\|x\|} = \frac{y}{\|y\|} \text{ or equivalently } y = \frac{\|y\|}{\|x\|} x,$$

which is (d).

Finally, we will prove that (d) $\Rightarrow$ (b). Assume (d). If there are $x, y \in \mathbb{S}_X$ with $x \neq y$ such that $\|(x+y)/2\| = 1$, then $\|x+y\| = 2 = \|x\| + \|y\|$, hence (d) yields that $y = (\|y\|/\|x\|)x = x$, which contradicts that $x \neq y$. So, (b) is satisfied, that is, the implication (d) $\Rightarrow$ (b) holds true, and the proof is finished. □

EXAMPLE 18.3. - Any Euclidean or Hilbert space is clearly strictly convex.
- The sum norm and the max norm on $\mathbb{R}^k$ are easily seen to be not strictly convex.
- The spaces $\ell^1_{\mathbb{R}}(\mathbb{N})$, $c^0_{\mathbb{R}}(\mathbb{N})$, $\ell^\infty_{\mathbb{R}}(\mathbb{N})$, $L^1(T, \mathbb{R})$ and $L^\infty(T, \mathbb{R})$ endowed with their canonical norms are also easily seen to be not strictly convex.
- For any real $p > 1$, the spaces $\ell^p_{\mathbb{R}}(\mathbb{N})$ and $L^p(T, \mathbb{R})$ endowed with their canonical norms are known to be strictly convex, and in fact they are more than strictly convex since we will see later that they are uniformly convex; see the comments at the end of the chapter for references. □

Given $x \in \mathbb{S}_X$ and a nonzero $x^* \in N(\mathbb{B}_X; x)$, the normal cone of the unit ball $\mathbb{B}_X$ at x, it is clear that the latter inclusion means $\langle x^*, y \rangle \leq \langle x^*, x \rangle$ for all $y \in \mathbb{B}_X$, so (recalling the definition of supporting functional in Definition 2.224) we see that x^* is a *supporting functional of* $\mathbb{B}_X$ *at* x. It is also clear that for $x \in \mathbb{S}_X$ (see 2.23)

$$\partial\|\cdot\|(x) = N(\mathbb{B}_X; x) \cap \mathbb{S}_{X^*},$$

that is, the set of unit supporting functionals of $\mathbb{B}_X$ at x coincides with $\partial\|\cdot\|(x)$. Strictly convex spaces can be characterized via unit supporting functionals of unit balls as follows.

PROPOSITION 18.4. Given a normed space $(X, \|\cdot\|)$, the following are equivalent:
(a) the norm $\|\cdot\|$ is strictly convex;
(b) for any $x, y \in \mathbb{S}_X$ with $x \neq y$, one has

$$N(\mathbb{B}_X; x) \cap \mathbb{S}_{X^*} \cap N(\mathbb{B}_X; y) = \emptyset;$$

(c) any supporting functional of $\mathbb{B}_X$ supports it at a unique point of its boundary;
(d) for any $x, y \in \mathbb{S}_X$ with $x \neq y$, one has

$$\partial\|\cdot\|(x) \cap \partial\|\cdot\|(y) = \emptyset.$$

PROOF. By the analysis preceding the proposition, the assertions (b), (c) and (d) are pairwise equivalent. Assume (a) and suppose that (b) does not hold, that is, there are $x_1, x_2 \in \mathbb{S}_X$ with $x_1 \neq x_2$ and some $u^* \in N(\mathbb{B}_X; x_1) \cap N(\mathbb{B}_X; x_2)$ with $\|u^*\| = 1$. Then $\langle u^*, x_i\rangle = 1$, for $i = 1, 2$, so putting $z := (x_1 + x_2)/2$ we have $\langle u^*, z\rangle = 1$. This combined with the inequality $\|z\| \le 1$ gives that $\|z\| = 1$, contradicting the strict convexity of $\|\cdot\|$. This justifies the implication (a)⇒(b).

Conversely, assume (b) and suppose that $\|\cdot\|$ is not strictly convex, or equivalently there are two distinct $x_1, x_2 \in \mathbb{S}_X$ with $z := (x_1 + x_2)/2$ in $\mathbb{S}_X$. Choosing $z^* \in \partial\|\cdot\|(z) \neq \emptyset$, we have (see (2.23)) that $\|z^*\| = 1$ and $\langle z^*, z\rangle = 1$, and the latter equality means that $\langle z^*, x_1 + x_2\rangle = 2$. Since the equality $\|z^*\| = 1$ yields $\langle z^*, x_i\rangle \le 1$, for $i = 1, 2$, we deduce that $\langle z^*, x_i\rangle = 1$. The latter combined with the equality $\|z^*\| = 1$ entails that $z^* \in N(\mathbb{B}_X; x_i)$, contradicting (b). So, (b)⇒(a) holds true. □

If $(X, \|\cdot\|)$ is strictly convex and Y is a vector subspace of X, it is trivial that Y endowed with the induced norm is strictly convex.

Since $\mathbb{R}^2$ endowed with either the sum or the max norm is not strictly convex, appropriate norms on a Cartesian product must be used for the preservation of strict convexity.

PROPOSITION 18.5. Let $(X_i, \|\cdot\|_i)$, $i = 1, \cdots, k$, be strictly convex normed spaces and let $|\!|\!|\cdot|\!|\!|$ be the norm on $X := X_1 \times \cdots \times X_k$ defined for all $x := (x_1, \cdots, x_k) \in X$ by

$$|\!|\!|x|\!|\!| := (\|x_1\|_1^2 + \cdot + \|x_k\|_k^2)^{1/2}.$$

Then, $(X, |\!|\!|\cdot|\!|\!|)$ is strictly convex.

PROOF. Let x, y be two points in X with $|\!|\!|x|\!|\!| = |\!|\!|y|\!|\!| = 1$ and $x \neq y$, so writing $x = (x_1, \cdots, x_k)$ with $x_i \in X_i$, we have $x_j \neq y_j$ for some $j \in \{1, \cdots, k\}$. Denoting by $\mathcal{N}$ the usual Euclidean norm on $\mathbb{R}^k$, we observe that $|\!|\!|x|\!|\!| = \mathcal{N}(\|x_1\|_1, \cdot, \|x_k\|_k)$, so both k-tuples $(\|x_1\|_1, \cdot, \|x_k\|_k)$ and $(\|y\|_1, \cdots, \|y_k\|_k)$ are on the unit sphere of $(\mathbb{R}^k, \mathcal{N})$. In the case when $\|x_j\|_j \neq \|y_j\|_j$, the strict convexity of $\mathcal{N}$ yields

$$\begin{aligned} |\!|\!|x+y|\!|\!| &= \mathcal{N}(\|x_1 + y_1\|_1, \cdots, \|x_k + y_k\|_k) \\ &\le \mathcal{N}(\|x_{n,1}\|_1 + \|y_{n,1}\|_1, \cdots, \|x_{n,k}\|_k + \|y_{n,k}\|_k) < 2. \end{aligned}$$

Suppose now that $\|x_j\|_j = \|y_j\|_j$. From the strict convexity of $\|\cdot\|_j$ and Lemma 18.2(d) it follows that $\|x_j + y_j\|_j < \|x_j\|_j + \|y_j\|_j$, which by the definition of $\mathcal{N}$ entails that

$$\begin{aligned} |\!|\!|x+y|\!|\!| &= \mathcal{N}(\|x_1 + y_1\|_1, \cdots, \|x_k + y_k\|_k) \\ &< \mathcal{N}(\|x_{n,1}\|_1 + \|y_{n,1}\|_1, \cdots, \|x_{n,k}\|_k + \|y_{n,k}\|_k) \\ &\le \mathcal{N}(\|x_1\|_1, \cdots, \|x_k\|_k) + \mathcal{N}(\|y_1\|_1, \cdots, \|y_k\|_k) = 2. \end{aligned}$$

In any case, we obtain $|\!|\!|x+y|\!|\!| < 2$, which justifies the strict convexity of $|\!|\!|\cdot|\!|\!|$ by Lemma 18.2(d) again. □

REMARK 18.6. For any real $p > 1$ and $|\!|\!|x|\!|\!|_p := (\|x_1\|_1^p + \cdots + \|x_k\|_k^p)^{1/p}$ the same arguments show that the normed space $(X, |\!|\!|\cdot|\!|\!|_p)$ is strictly convex whenever each norm $\|\cdot\|_i$ is strictly convex. □

REMARK 18.7. We mention (see the comments at the end of the chapter) that, given a closed vector subspace Y of a strictly convex normed space X, the quotient space X/Y endowed with its canonical norm may fail to be strictly convex. □

From Lemma 18.2 we can derive a useful property for nearest points when the norm is strictly convex. This property extends the similar Hilbert one in Lemma 4.123. We recall that $\mathrm{Proj}_C(u)$ denotes the set of nearest points of u in a set C of a normed space and that $P_C(u)$ is the unique element in $\mathrm{Proj}_C(u)$ when the latter is a singleton.

LEMMA 18.8. *Let $(X, \|\cdot\|)$ be a strictly convex normed vector space, C be a nonempty subset of X and $u \in X$. Assume that $\mathrm{Proj}_C(u) \neq \emptyset$. Then for any $p \in \mathrm{Proj}_C(u)$ and any $t \in]0,1]$, one has*

$$\mathrm{Proj}_C(u + t(p-u)) = \{p\}, \quad \text{that is, } p = P_C(u + t(p-u)).$$

PROOF. The case $t = 1$ being obvious, we may suppose $t \in]0,1[$. Putting $u_t := u + t(p-u)$, we already know by Lemma 4.123 that $p \in \mathrm{Proj}_C(u_t)$.

Take any element $p_t \in \mathrm{Proj}_C(u_t)$. Then, writing

$$\begin{aligned} d_C(u_t) = \|u_t - p_t\| = \|u - p_t + t(p-u)\| &\geq \|u - p_t\| - t\|u-p\| \\ &\geq d_C(u) - t\|u-p\| = \|u_t - p\| = d_C(u_t), \end{aligned}$$

we see that all the latter inequalities are equalities, and hence

$$\|u - p_t\| = d_C(u) \tag{18.1}$$

and

$$\|u - p_t\| = \|u - p_t + t(p-u)\| + \|t(u-p)\|. \tag{18.2}$$

If $t(u-p) = 0$, then $u \in C$ and the result is obvious. If $u - p_t + t(p-u) = 0$, the inequality $\|u - p_t + t(p-u)\| \geq (1-t)d_C(u)$ entails $d_C(u) = 0$, so (18.1) gives $u = p_t$, hence $u \in C$ and the result is again obvious. So, suppose $t(u-p) \neq 0$ and $u - p_t + t(p-u) \neq 0$. By the strict convexity of the norm $\|\cdot\|$, the equality (18.2) entails by Lemma 18.2(d) that there exists $\mu > 0$ with

$$u - p_t + t(p-u) = \mu t(u-p),$$

that is,

$$u - p_t = t(\mu+1)(u-p). \tag{18.3}$$

Using (18.1) and taking the norm of both members of (18.3) yield the equality $d_C(u) = t(\mu+1)d_C(u)$ and since $d_C(u) > 0$ (because $u - p \neq 0$) we have $t(\mu+1) = 1$. Putting this value in (18.3) gives $p_t = p$, which completes the proof. □

The next proposition shows that the norm of a normed space is strictly convex (resp; Gâteaux differentiable off zero) whenever its dual norm is Gâteaux differentiable off zero (resp; Strictly convex).

PROPOSITION 18.9. *Let $(X, \|\cdot\|)$ be a normed space. The following assertions hold.*
(a) If the dual norm $\|\cdot\|_*$ of X^* is strictly convex, then the norm $\|\cdot\|$ is Gâteaux differentiable off zero.
(b) If the dual norm $\|\cdot\|_*$ is Gâteaux differentiable off zero, then the norm $\|\cdot\|$ is strictly convex.

PROOF. (a) Suppose that $\|\cdot\|$ is not Gâteaux differentiable away from zero. Then there is some $x \in \mathbb{S}_X$ with two different elements x^*, y^* in $\partial\|\cdot\|(x)$. Then (see (2.23)) $\|x^*\|_* = \|y^*\|_* = 1$ and $\langle x^*, x\rangle = 1 = \langle y^*, x\rangle$, so we have $\langle (x^*+y^*)/2, x\rangle = 1$. It ensues that $\|(x^*+y^*)/2\|_* = 1$, and hence we see that the dual norm $\|\cdot\|_*$ of

X^* is not strictly convex.
(b) Suppose that $\|\cdot\|$ is not strictly convex. Then there are two different points $x, y \in X$ with $\|x\| = \|y\| = 1$ and $\|(x+y)/2\| = 1$ according to Lemma 18.2(b). The latter equality gives some $x^* \in X^*$ with $\|x^*\| = 1$ and $\langle x^*, (x+y)/2 \rangle = 1$. It results that $\langle x^*, x \rangle + \langle x^*, y \rangle = 2$ along with $\langle x^*, x \rangle \leq 1$ and $\langle x^*, y \rangle \leq 1$. We deduce that $\langle x^*, x \rangle = 1$ and $\langle x^*, y \rangle = 1$, hence both x and y are in $\partial \|\cdot\|_*(x^*)$ (see 2.23). The latter entails that $\|\cdot\|_*$ is not Gâteaux differentiable at x^*. This finishes the proof of the proposition. □

The above proposition provides only a partial duality between the strict convexity of the dual norm and the Gâteaux differentiability of the primal norm. Indeed, for a normed space $(X, \|\cdot\|)$ it is known (see the comments at the end of the chapter) that the strict convexity (resp. the Gâteaux differentiability off zero) of the primal norm $\|\cdot\|$ does not imply the Gâteaux differentiability off zero (resp. the strict convexity) of the dual norm $\|\cdot\|_*$. In contrast, we will see later that a complete duality exists for the uniform convexity.

The differentiability of a norm offers two criteria for weak* and strong convergences. This result, which is often very useful, is usually called Šmulian lemma. The lemma can be deduced from the Asplund-Rockafellar Theorem 3.118. We provide direct arguments below.

LEMMA 18.10 (V.L. Šmulian lemma). Let $(X, \|\cdot\|)$ be a normed space and let $u \in X$ with $\|u\| = 1$. The following hold.
(a) The norm $\|\cdot\|$ is Gâteaux differentiable at u if and only if

$$(\langle x_n^*, u \rangle \to 1 = \langle x^*, u \rangle \text{ and } \|x_n^*\|_* \leq 1 = \|x^*\|_*) \Rightarrow x_n^* \xrightarrow{w^*} x^*,$$

which in turn is equivalent to the property

$$(\langle x_n^*, u \rangle \to 1 \text{ and } \|x_n^*\|_* \leq 1) \Rightarrow (x_n^*)_n \ w^* - \text{converges}.$$

(b) The norm $\|\cdot\|$ is Fréchet differentiable at u if and only if

$$(\langle x_n^*, u \rangle \to 1 = \langle x^*, u \rangle \text{ and } \|x_n^*\|_* \leq 1 = \|x^*\|_*) \Rightarrow \|x_n^* - x^*\|_* \longrightarrow 0,$$

which in turn is equivalent to the property

$$(\langle x_n^*, u \rangle \to 1 \text{ and } \|x_n^*\|_* \leq 1) \Rightarrow (x_n^*)_n \text{ strongly converges}.$$

PROOF. (a) Call (a') and (a") the first and second properties in (a) other than the Gâteaux differentiability. Suppose first that (a') is not satisfied, and let $x^* \in X^*$ and $(x_n^*)_n$ in X^* such that $\langle x_n^*, u \rangle \to 1 = \langle x^*, u \rangle$ and $\|x_n^*\|_* \leq 1 = \|x^*\|_*$ while $(x_n^*)_n$ does not w^*-converge to x^*. There exists some $v \in \mathbb{S}_X$ and some $\varepsilon > 0$ such that $|\langle x_{s(n)}^* - x^*, v \rangle| \geq \varepsilon$ for some subsequence $(x_{s(n)}^*)_n$. Taking a subnet of $(x_{s(n)}^*)_n$ converging weakly* to some $y^* \in X^*$, we see that $|\langle y^* - x^*, v \rangle| \geq \varepsilon$ and we see also by (2.23) that $y^* \in \partial \|\cdot\|(u)$ since $\langle y^*, u \rangle = 1$ with $\|y^*\|_* \leq 1$ (by the weak* lower semicontinuity of $\|\cdot\|_*$). We then obtain two different elements x^*, y^* in $\partial \|\cdot\|(u)$, so $\|\cdot\|$ is not Gâteaux differentiable at u. This means that the Gâteaux differentiability of $\|\cdot\|$ at u implies (a').

Suppose that (a') is satisfied and let any sequence $(x_n^*)_n$ in X^* such that $\langle x_n^*, u \rangle \to 1$ and $\|x_n^*\|_* \leq 1$. Choosing some $x^* \in \mathbb{S}_{X^*}$ with $\langle x^*, u \rangle = \|u\| = 1$, we can apply (a') to obtain that $x_n^* \xrightarrow{w^*} x^*$. This translates that (a') implies (a").

Now suppose that (a") holds and take y_1^*, y_2^* in $\partial\|\cdot\|(u)$. Putting $x_{2n-1}^* := y_1^*$ and $x_{2n}^* := y_2^*$, by (a") the sequence $(x^*)_n$ converges weakly*, hence $y_1^* = y_2^*$. Then $\partial\|\cdot\|(u)$ is a singleton, thus (the continuous convex function) $\|\cdot\|$ is Gâteaux differentiable at u (see Proposition 2.76 or Proposition 3.102).
(b) As above denote by (b') and (b") the first and second property in (b) other than the Fréchet differentiability. Suppose that $\|\cdot\|$ is Fréchet differentiable at u, and let $x^* \in X^*$ and $(x_n^*)_n$ in X^* such that $\langle x_n^*, u\rangle \to 1 = \langle x^*, u\rangle$ and $\|x_n^*\|_* \le 1 = \|x^*\|_*$. Fix any real $\varepsilon > 0$. By Lemma 4.48(b) we know that

$$\|h\|^{-1}(\|u+h\| + \|u-h\| - 2\|u\|) \to 0 \quad \text{as } h \to 0,$$

so there exists a real $\eta > 0$ such that for all $h \in X$ with $\|h\| \le \eta$

$$\|u+h\| + \|u-h\| - 2 \le (\varepsilon/2)\|h\|.$$

Choose an integer N such that $|\langle x_n^*, u\rangle - 1| \le (\varepsilon\eta)/2$. For each integer $n \ge N$ we have for all $h \in X$ with $\|h\| \le \eta$

$$\begin{aligned}\langle x_n^* - x^*, h\rangle &= \langle x_n^*, u+h\rangle + \langle x^*, u-h\rangle - \langle x_n^*, u\rangle - \langle x^*, u\rangle \\ &\le \|u+h\| + \|u-h\| - \langle x_n^*, u\rangle - \langle x^*, u\rangle \\ &\le 2 - \langle x_n^*, u\rangle - \langle x^*, u\rangle + (\varepsilon/2)\|h\| \\ &= 1 - \langle x_n^*, u\rangle + (\varepsilon/2)\|h\| \le \varepsilon\eta,\end{aligned}$$

which gives $\|x_n^* - x^*\|_* \le \varepsilon$. Consequently, the Fréchet differentiability of $\|\cdot\|$ at u implies (b').

The proof that (b') implies (b") is similar to that of the implication (a')⇒(a").

Now suppose that (b") holds but $\|\cdot\|$ is not Fréchet differentiable at u. By Lemma 4.48(b) there are some real $\varepsilon > 0$ and some sequence $h_n \to 0$ such that, for all $n \in \mathbb{N}$

$$\|u+h_n\| + \|u-h_n\| - 2 \ge \varepsilon\|h_n\|.$$

For each $n \in \mathbb{N}$ choose x_n^*, y_n^* in X^* with $\|x_n^*\|_* = \|y_n^*\|_* = 1$ such that $\langle x_n^*, u+h_n\rangle = \|u+h_n\|$ and $\langle y_n^*, u-h_n\rangle = \|u-h_n\|_*$. We note that $\langle x_n^*, u\rangle \to 1$ and $\langle y_n^*, u\rangle \to 1$ as $n \to \infty$, so taking $x^* \in X^*$ with $\|x^*\|_* = 1$ and $\langle x^*, u\rangle = \|u\| = 1$, we get by (b") that $\|x_n^* - x^*\| \to 0$ and $\|y_n^* - x^*\| \to 0$, hence in particular $\|x_n^* - y_n^*\|_* \to 0$. However, for each $n \in \mathbb{N}$ we have $\langle x_n^* + y_n^*, u\rangle \le \|x_n^* + y_n^*\|_* \le 2$, so

$$\begin{aligned}\langle x_n^* - y_n^*, h_n\rangle &= \langle x_n^*, u+h_n\rangle + \langle y_n^*, u-h_n\rangle - \langle x_n^* + y_n^*, u\rangle \\ &\ge \|u+h_n\| + \|u-h_n\| - 2 \ge \varepsilon\|h_n\|,\end{aligned}$$

thus $\|x_n^* - y_n^*\| \ge \varepsilon$. This contradicts the fact that $\|x_n^* - y_n^*\|_* \to 0$ as $n \to \infty$. The property (b") then implies the Fréchet differentiability of $\|\cdot\|$ at u, and the proof is finished. □

REMARK 18.11. Denoting by g the indicator function of $\mathbb{B}_{X^*}$, it is easily seen that the equivalence between (b") and the Fréchet differentiability of $\|\cdot\|$ at u is also a consequence of Theorem 4.58(b). □

18.1.2. Uniformly convex Banach spaces. Let us come back to Lemma 18.2. Specifying the case where in (b) of this lemma the distance from $\frac{1}{2}(x+y)$ to the unit sphere $\mathbb{S}_X$ of $(X, \|\cdot\|)$ can be quantified in a uniform way with respect to both x and y in $\mathbb{S}_X$, one defines the uniform convexity of a norm as follows.

DEFINITION 18.12. A norm $\|\cdot\|$ on a vector space X is called *uniformly convex* (or *uniformly rotund*) provided that for every real $\varepsilon\in]0,2]$ there exists $\eta>0$ (depending on ε) such that $\left\|\frac{1}{2}(x+y)\right\|\le 1-\eta$ for all $x,y\in\mathbb{S}_X$ with $\|x-y\|\ge\varepsilon$. In such a case one also says that $(X,\|\cdot\|)$ is a uniformly convex (or uniformly rotund) normed space.

Obviously, the uniform convexity of a norm implies its strict convexity according to Lemma 18.2(b).

Taking into account the condition in the above definition, it is natural to consider the following function.

DEFINITION 18.13. One defines the *modulus of convexity* (or *modulus of rotundity*) of a norm $\|\cdot\|$ on a vector space X as the function $\delta_{\|\cdot\|}:[0,2]\to[0,1]$ given by

$$\delta_{\|\cdot\|}(\varepsilon):=\inf\left\{1-\left\|\frac{x+y}{2}\right\|:\|x\|=\|y\|=1,\|x-y\|\ge\varepsilon\right\}\quad\text{for all }\varepsilon\in[0,2];$$

when there is no risk of confusion we will also use the notation δ_X in place of $\delta_{\|\cdot\|}$. The norm $\|\cdot\|$ or the normed space $(X,\|\cdot\|)$ is then *uniformly convex* (or *uniformly rotund*) if and only if $\delta_{\|\cdot\|}(\varepsilon)>0$ for all $\varepsilon\in]0,2]$.

It is evident that the function $\delta_{\|\cdot\|}$ is nondecreasing and satisfies the equality

$$\delta_{\|\cdot\|}(0)=0.$$

Given a real $\alpha>0$, for the new norm $\alpha\|\cdot\|$ it is quite clear that, for any real $\varepsilon\in[0,2]$

$$\delta_{\alpha\|\cdot\|}(\varepsilon)=\delta_{\|\cdot\|}(\varepsilon).$$

If the real normed vector space $(X,\|\cdot\|)$ is one-dimensional, then one has $\delta_{\|\cdot\|}(\varepsilon)=\delta_{\mathbb{R}}(\varepsilon)$ for all $\varepsilon\in[0,2]$, where $\delta_{\mathbb{R}}$ stands for the modulus of convexity of the absolute value $|\cdot|$ of $\mathbb{R}$. Further, an easy calculus yields that

$$\delta_{\mathbb{R}}(0)=0\quad\text{and}\quad\delta_{\mathbb{R}}(\varepsilon)=1\ \text{if }\varepsilon\in]0,2].\tag{18.4}$$

If the vector space X is *at least two-dimensional*, in the definition of modulus of convexity of $\|\cdot\|$ the points x,y can be taken in the closed unit ball $\mathbb{B}_X$ instead of the unit sphere $\mathbb{S}_X$ both with respect to the norm $\|\cdot\|$. Otherwise stated, for all $\varepsilon\in[0,2]$ one has the equality

$$\delta_{\|\cdot\|}(\varepsilon)=\inf\left\{1-\left\|\frac{x+y}{2}\right\|:\|x\|\le 1,\|y\|\le 1,\|x-y\|\ge\varepsilon\right\}.\tag{18.5}$$

To justify the equality for a fixed $\varepsilon\in[0,2]$, denote the right member by $\zeta(\varepsilon)$ and note first that the inequality $\zeta(\varepsilon)\le\delta_{\|\cdot\|}(\varepsilon)$ is evident. To sketch the main lines of the converse inequality, take first any $x,y\in X$ with $\|x\|=1$, $\|y\|\le 1$ and $\|x-y\|\ge\varepsilon$. Let E be a two-dimensional vector subspace of X containing both x and y, and let $x',y'\in E$ with $\|x'\|=\|y'\|=1$ such that $x'-y'=x-y$ and such that y, x' and y' lie all together in one of the half-planes in E whose boundary is the straight line passing by x and $-x$ (see Figure 18.1). Choose two reals $\lambda\ge 1$ and $\mu\ge 0$ such that

$$\lambda\left(\frac{x+y}{2}\right)=\mu x'+(1-\mu)x.\tag{18.6}$$

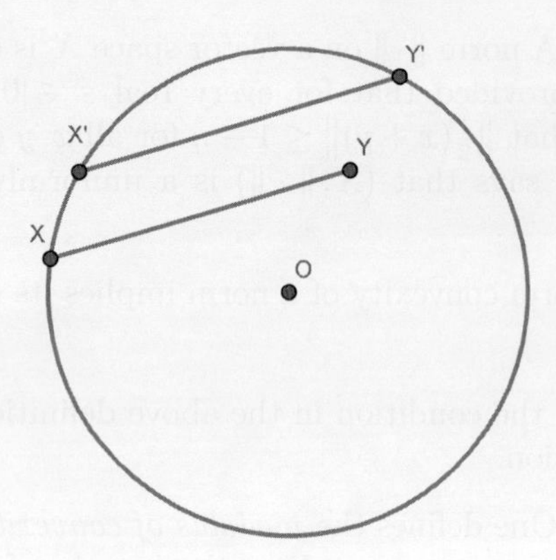

FIGURE 18.1. Illustration for x, y, x', y'.

Since $x - y = x' - y'$, we have

$$\frac{\lambda}{2}x' + \frac{\lambda}{2}y' = \frac{\lambda}{2}x' + \frac{\lambda}{2}(y - x + x') = \frac{\lambda}{2}(y - x) + \lambda x' = \lambda(\frac{x+y}{2}) + \lambda x' - \lambda x,$$

which by (18.6) gives

$$\lambda\left(\frac{x' + y'}{2}\right) = (\lambda + \mu)x' + (1 - \lambda - \mu)x = x + (\lambda + \mu)(x' - x). \tag{18.7}$$

Put $x_1 := \mu x' + (1 - \mu)x = x + \mu(x' - x)$ and $x_2 := x + (\lambda + \mu)(x' - x)$, and note that

$$\|x_2\| \geq (\lambda + \mu)\|x'\| - (\lambda + \mu - 1)\|x\| = \lambda + \mu - (\lambda + \mu - 1) = 1.$$

Therefore, for $\alpha := \frac{\mu}{\lambda+\mu}$ observing that $x_1 = (1 - \alpha)x + \alpha x_2$ along with $0 \leq \alpha \leq 1$ it follows (since $\|x\| = 1 \leq \|x_2\|$) that

$$\|x_1\| \leq (1 - \alpha)\|x\| + \alpha\|x_2\| \leq (1 - \alpha)\|x_2\| + \alpha\|x_2\| = \|x_2\|.$$

Then by (18.6) and (18.7) one arrives at

$$\left\|\frac{x+y}{2}\right\| \leq \left\|\frac{x' + y'}{2}\right\|, \text{ hence } 1 - \left\|\frac{x+y}{2}\right\| \geq 1 - \left\|\frac{x' + y'}{2}\right\| \geq \delta_{\|\cdot\|}(\varepsilon). \tag{18.8}$$

Consider now any $x, y \in X$ with $\|x\| < 1$, $\|y\| < 1$ and $\eta := \|x - y\| \geq \varepsilon$. Put $m := (x + y)/2$. On the straight line passing by x and y there is a line segment $[u, v]$ with $u \in \mathbb{S}_X$, $\|v\| < 1$, $\|u - v\| \geq \eta$ and $\|(u + v)/2\| \geq \|m\|$ (see Figure 18.2). This combined with (18.8) yields

$$\delta_{\|\cdot\|}(\varepsilon) \leq 1 - \left\|\frac{u+v}{2}\right\| \leq 1 - \left\|\frac{x+y}{2}\right\|.$$

The latter inequality and (18.8) entail that, for all $x, y \in \mathbb{B}_X$ with $\|x - y\| \geq \varepsilon$ one has $\delta_{\|\cdot\|}(\varepsilon) \leq 1 - \|(x + y)/2\|$, and hence $\delta(\varepsilon) \leq \zeta(\varepsilon)$. Consequently, the stated equality $\delta_{\|\cdot\|}(\varepsilon) = \zeta(\varepsilon)$ is justified.

If we set $\varphi(x, y) = 1 - \|(x + y)/2\|$ and $G_\varepsilon = \{(x, y) \in \mathbb{B}_X \times \mathbb{B}_X : \|x - y\| \geq \varepsilon\}$, $G'_\varepsilon = \{(x, y) \in \mathbb{B}_X \times \mathbb{B}_X : \|x - y\| = \varepsilon\}$, $G''_\varepsilon = \{(x, y) \in \mathbb{S}_X \times \mathbb{S}_X : \|x - y\| = \varepsilon\}$, the same above arguments ensure that

$$\zeta'(\varepsilon) := \inf_{(x,y)\in G'_\varepsilon} \varphi(x, y) = \inf_{(x,y)\in G''_\varepsilon} \varphi(x, y) =: \zeta''(\varepsilon).$$

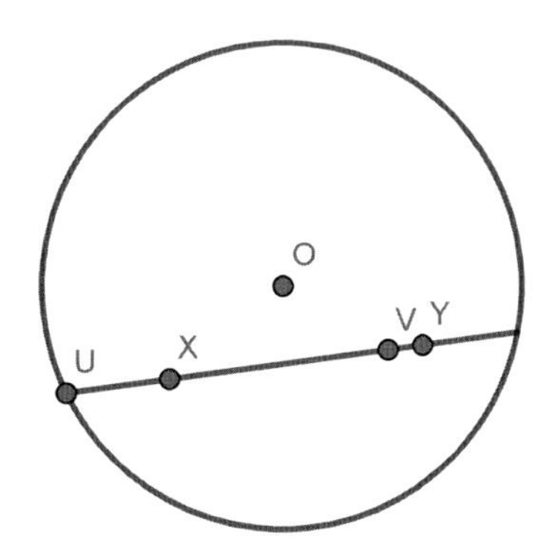

FIGURE 18.2. Illustration for x, y, u, v.

We then clearly see that $\delta_{\|\cdot\|}(\varepsilon) = \zeta(\varepsilon) \leq \zeta'(\varepsilon) = \zeta''(\varepsilon)$. Take now two elements x, y in $\mathbb{S}_X$ with $\|x - y\| \geq \varepsilon$. Obviously, on the line segment $[x, y]$ there are two points x', y' with $\frac{x+y}{2} = \frac{x'+y'}{2}$ and $\|x' - y'\| = \varepsilon$. Noting that x', y' lie in $\mathbb{B}_X$, it results that $(x', y') \in G'_\varepsilon$, and hence $\delta_{\|\cdot\|}(\varepsilon) \geq \zeta'(\varepsilon)$, so $\delta_{\|\cdot\|}(\varepsilon) = \zeta(\varepsilon) = \zeta'(\varepsilon) = \zeta''(\varepsilon)$.

Otherwise stated, when X is *at least two-dimensional* the following equalities hold:

$$\begin{aligned} \delta_{\|\cdot\|}(\varepsilon) &= \inf\left\{1 - \left\|\frac{x+y}{2}\right\| : \|x\| \leq 1, \|y\| \leq 1, \|x-y\| \geq \varepsilon\right\} \\ &= \inf\left\{1 - \left\|\frac{x+y}{2}\right\| : \|x\| \leq 1, \|y\| \leq 1, \|x-y\| = \varepsilon\right\} \qquad (18.9) \\ &= \inf\left\{1 - \left\|\frac{x+y}{2}\right\| : \|x\| = 1, \|y\| = 1, \|x-y\| = \varepsilon\right\}, \end{aligned}$$

providing in this way various formulas for the modulus of convexity.

It is also worth observing that, if $(X, \|\cdot\|)$ is uniformly convex and Y is a vector subspace of X, then Y endowed with the induced norm is trivially uniformly convex. In fact, the inequality $\delta_Y(\varepsilon) \geq \delta_X(\varepsilon)$ obviously holds true for all $0 \leq \varepsilon \leq 2$. It is also quite easy to verify the following characterizations of uniform convexity of a norm.

PROPOSITION 18.14. For a normed space $(X, \|\cdot\|)$ the following assertions are equivalent:
(a) the norm $\|\cdot\|$ is uniformly convex;
(b) for any sequences $(x_n)_n$ and $(y_n)_n$ in X with $\|x_n\| \to 1$, $\|y_n\| \to 1$ and $\|(x_n + y_n)/2\| \to 1$ as $n \to \infty$, one has $\|x_n - y_n\| \to 0$;
(c) for any sequences $(x_n)_n$ and $(y_n)_n$ in the space X such that $\limsup_{n\to\infty} \|x_n\| \leq 1$ and $\limsup_{n\to\infty} \|y_n\| \leq 1$ along with $\|(x_n + y_n)/2\| \to 1$ as $n \to \infty$ as $n \to \infty$, one has that $\|x_n - y_n\| \to 0$.

Furthermore, the equivalences still hold with nets (with a common index set) in place of sequences in (b) and (c).

PROOF. It is first an easy exercise to verify the equivalence (a)$\Leftrightarrow$(b), whereas the implication (c)$\Rightarrow$(b) is obvious. Now assume (b) and let $(x_n)_n$ and $(y_n)_n$ be two sequences in X such that both upper limits are not greater than one and

$\lim_{n\to\infty} \|(x_n + y_n)/2\| = 1$. Since

$$2 = \lim_{n\to\infty} \|x_n + y_n\| \le \liminf_{n\to\infty} \|x_n\| + \limsup_{n\to\infty} \|y_n\| \le \limsup_{n\to\infty} \|x_n\| + \limsup_{n\to\infty} \|y_n\| \le 2,$$

we derive that $\lim_{n\to\infty} \|x_n\| = 1$. Similarly, $\lim_{n\to\infty} \|y_n\| = 1$, hence (b) tells us that $\|x_n - y_n\| \to 0$ as $n \to \infty$.

Finally, we note that the above arguments also work with nets. □

A first corollary examines finite-dimensional spaces.

COROLLARY 18.15. A finite-dimensional normed space $(X, \|\cdot\|)$ is uniformly convex if and only if it is strictly convex.

PROOF. We already know that the uniform convexity of $\|\cdot\|$ implies its strict convexity. Now assume that $\|\cdot\|$ is strictly convex and not uniformly convex. By (b) in the above proposition there are a real $\varepsilon > 0$ and sequences $(x_n)_n$ and $(y_n)_n$ in $\mathbb{S}_X$ with $\|x_n + y_n\| \to 2$ as $n \to \infty$ and $\|x_n - y_n\| \ge \varepsilon$ for all $n \in \mathbb{N}$. By compactness of $\mathbb{S}_X$ we may suppose that $(x_n)_n$ and $(y_n)_n$ converge to x and y in $\mathbb{S}_X$ respectively. We also note that $x \neq y$ according to the inequality $\|x - y\| \ge \varepsilon$, and that $\|x + y\| = 2$, which (according to Lemma 18.2) contradicts the strict convexity of the norm $\|\cdot\|$. □

Another corollary is concerned with quotient spaces. It shows that (unlike the strict convexity, see Remark 18.7) the uniform convexity is preserved on quotient space. We recall that given a closed (strict) vector subspace Y of a normed space $(X, \|\cdot\|)$, the canonical norm over the quotient space X/Y is defined for any equivalence class $[x] := x + Y$ in X/Y by

$$\|[x]\|_{X/Y} := \inf\{\|u\| : u \in x + Y\} = d(x, Y). \tag{18.10}$$

COROLLARY 18.16. Let Y be a closed strict vector subspace of a uniformly convex normed space $(X, \|\cdot\|)$. The quotient normed space $(X/Y, \|\cdot\|_{X/Y})$ is uniformly convex.

PROOF. Let $([x_n])_n$ and $([x'_n])_n)$ in the sphere of X/Y such that $\|[x_n] + [x'_n]\|_{X/Y} \to 2$ as $n \to \infty$. For each $n \in \mathbb{N}$ choose $u_n \in [x_n]$ and $u'_n \in [x'_n]$ such that

$$\|[x_n]\|_{X/Y} \le \|u_n\| \le \|[x_n]\|_{X/Y} + \frac{1}{n} \text{ and } \|[x'_n]\|_{X/Y} \le \|u'_n\| \le \|[x'_n]\|_{X/Y} + \frac{1}{n}.$$

Since $\|[x_n + x'_n]\|_{X/Y} \le \|u_n + u'_n\| \le 2 + (2/n)$, it ensues that $\|u_n\| \to 1$, $\|u'_n\| \to 1$ and $\|u_n + u'_n\| \to 2$, so the uniform convexity of $\|\cdot\|$ entails that $\|u_n - u'_n\| \to 0$. This trivially ensures that $\|[x_n] - [x'_n]\|_{X/Y} \to 0$ as $n \to 0$. □

A uniformly convex norm enjoys the Kadec-Klee property.

PROPOSITION 18.17. If a normed space $(X, \|\cdot\|)$ is uniformly convex, then its norm $\|\cdot\|$ enjoys the Kadec-Klee property, that is, for any net $(x_j)_{j\in J}$ in X one has $\|x_j - x\| \to 0$ if and only if $(x_j)_{j\in J}$ converges weakly to x and $\|x_j\| \to \|x\|$.

PROOF. Let $(x_j)_{j\in J}$ be a net converging weakly to x in X along with $\|x_j\| \to \|x\|$. We may suppose that $x \neq 0$, otherwise we are done. Then, we may also suppose that $x_j \neq 0$ for all $j \in J$, so putting $u_j := x_j/\|x_j\|$ and $u := x/\|x\|$ we have that $(\frac{1}{2}(u_j + u))_{j\in J}$ converges weakly to u. By the weak lower semicontinuity of $\|\cdot\|$ it ensues that $\|\frac{1}{2}(u_j+u)\| \to 1$, hence $\|u_j-u\| \to 0$ by the uniform convexity of $\|\cdot\|$, which in turn gives $\|x_j - x\| \to 0$. The converse implication is obvious. $\square$

When we endow with a suitable norm the product of finitely many uniformly convex normed spaces, it is uniformly convex.

PROPOSITION 18.18. Let $(X_i, \|\cdot\|_i)$, $i = 1, \cdots, k$, be uniformly convex normed spaces and let $\|\!|\cdot|\!\|$ be the norm on $X := X_1 \times \cdots \times X_k$ defined for all $x := (x_1, \cdots, x_k) \in X$ by

$$|\!\|x|\!\| := (\|x_1\|_1^2 + \cdots + \|x_k\|_k^2)^{1/2}.$$

Then, $(X, \|\!|\cdot|\!\|)$ is uniformly convex.

PROOF. Let $(x)_n$ and $(y_n)_n$ be sequences in X with $|\!\|x_n|\!\| = |\!\|y_n|\!\| = 1$ such that $|\!\|x_n + y_n|\!\| \to 2$ as $n \to \infty$. Denoting by $\mathcal{N}$ the usual Euclidean norm on $\mathbb{R}^k$, we notice that $|\!\|x|\!\| = \mathcal{N}(\|x_1\|_1, \cdot, \|x_k\|_k)$. Writing

$$\begin{aligned}
|\!\|x_n + y_n|\!\| &= \mathcal{N}(\|x_{n,1} + y_{n,1}\|_1, \cdots, \|x_{n,k} + y_{n,k}\|_k) \\
&\leq \mathcal{N}(\|x_{n,1}\|_1 + \|y_{n,1}\|_1, \cdots, \|x_{n,k}\|_k + \|y_{n,k}\|_k) \\
&\leq \mathcal{N}(\|x_{n,1}\|_1, \cdots, \|x_{n,k}\|_k) + \mathcal{N}(\|y_{n,1}\|_1, \cdots, \|y_{n,k}\|_k) \\
&= |\!\|x_n|\!\| + |\!\|y_n|\!\| = 2,
\end{aligned}$$

we see that $\mathcal{N}(\|x_{n,1}\|_1 + \|y_{n,1}\|_1, \cdots, \|x_{n,k}\|_k + \|y_{n,k}\|_k) \to 2$ as $n \to \infty$. Since $\mathcal{N}(\|x_{n,1}\|_1, \cdots, \|x_{n,k}\|_k) = \mathcal{N}(\|y_{n,1}\|_1, \cdots, \|y_{n,k}\|_k) = 1$, the parallelogram equality in the Euclidean space $\mathbb{R}^k$ ensures that, for each $i = 1, \cdots, k$ we have $\|x_{n,i}\|_i - \|y_{n,i}\|_i \to 0$ as $n \to \infty$.

Suppose that $|\!\|x_n - y_n|\!\| \not\to 0$ as $n \to \infty$. Extracting a subsequence, we may consider that

$$\lim_{n\to\infty} |\!\|x_n - y_n|\!\| = s > 0 \text{ and } \lim_{n\to\infty} \|x_{n,i}\|_i = \lim_{n\to\infty} \|y_{n,i}\|_i =: r_i, \ \forall i = 1, \cdots, k,$$

and that for some $j \in \{1, \cdots, k\}$

$$\lim_{n\to\infty} \|x_{n,j} - y_{n,j}\|_j =: t > 0 \text{ and } \|x_{n,j}\|_j > 0, \|y_{n,j}\|_j > 0, \ \forall n \in \mathbb{N}.$$

Since

$$\lim_{n\to\infty} \|r_j^{-1}x_{n,j} - r_j^{-1}y_{n,j}\|_j = r_j^{-1}t > 0 \text{ and } \lim_{n\to\infty} \|r_j^{-1}x_{n,j}\|_j = \lim_{n\to\infty} \|r_j^{-1}y_{n,j}\|_j = 1,$$

the uniform convexity of the norm $\|\cdot\|_j$ entails that

$$\limsup_{n\to\infty} \|r_j^{-1}x_{n,j} + r_j^{-1}y_{n,j}\|_j < 2, \text{ that is, } \limsup_{n\to\infty} \|x_{n,j} + y_{n,j}\|_j < 2r_j.$$

On the other hand, for each $i \neq j$ in $\{1, \cdots, k\}$, since $\lim_{n\to\infty} \|x_{n,i}\|_i = \lim_{n\to\infty} \|y_{n,i}\|_i =: r_i$, we also have $\limsup_{n\to\infty} \|x_{n,i} + y_{n,i}\|_i \leq 2r_i$. We deduce that

$$\begin{aligned}
2 = \lim_{n\to\infty} |\!\|x_n + y_n|\!\| &= \lim_{n\to\infty} \mathcal{N}(\|x_{n,1} + y_{n,1}\|_1, \cdots, \|x_{n,k} + y_{n,k}\|_k) \\
&\leq \mathcal{N}(\limsup_{n\to\infty} \|x_{n,1} + y_{n,1}\|_1, \cdots, \limsup_{n\to\infty} \|x_{n,k} + y_{n,k}\|_k) \\
&< \mathcal{N}(2r_1, \cdots, 2r_j, \cdots, 2r_k) = 2.
\end{aligned}$$

This contradiction ensures that $|\!\|x_n - y_n|\!\| \to 0$ as $n \to \infty$, which translates the uniform convexity of the norm $\|\!|\cdot|\!\|$. $\square$

REMARK 18.19. Taking any real $p > 1$ and putting for all $x \in X = X_1 \times \cdots \times X_k$

$$\|x\|_p := (\|x_1\|_1^p + \cdots + \|x_k\|_k^p)^{1/p},$$

slight modifications of the above arguments show that the normed space $(X, \|\cdot\|_p)$ is uniformly convex whenever each $(X_i, \|\cdot\|_i)$ is uniformly convex. □

In fact the Kadec-Klee property in Proposition 18.17 holds true for the more general class of locally uniformly convex norms as we will see in Proposition 18.22 below. The definition of locally uniformly convex norms corresponds to the localization when y is taken as a fixed element $\overline{x} \in \mathbb{S}_X$ in Definition 18.12 as follows.

DEFINITION 18.20. Let $\|\cdot\|$ be a norm on a vector space X and let $\overline{x} \in X$ with $\|\overline{x}\| = 1$. The norm $\|\cdot\|$ is said to be *locally uniformly convex* (or *locally uniformly rotund*) at $\overline{x}$ provided that for every real $\varepsilon \in]0, 2]$ there exists $\eta > 0$ (depending on ε and $\overline{x}$) such that $\left\|\frac{1}{2}(x + \overline{x})\right\| \leq 1 - \eta$ for every $x \in \mathbb{S}_X$ with $\|x - \overline{x}\| \geq \varepsilon$. The norm is *locally uniformly convex* (or *locally uniformly rotund*) if it is locally uniformly convex at any $\overline{x} \in X$ with $\|\overline{x}\| = 1$. In such a case one also says that $(X, \|\cdot\|)$ is a locally uniformly convex (or locally uniformly rotund) normed space.

Clearly, if $(X, \|\cdot\|)$ is locally uniformly convex and Y is a vector subspace of X, then Y endowed with the induced norm is locally uniformly convex. Further, the same proof for Proposition 18.14 also guarantees the following similar characterizations of locally uniformly convex norms.

PROPOSITION 18.21. For a normed space $(X, \|\cdot\|)$ the following assertions are equivalent:
(a) the norm $\|\cdot\|$ is locally uniformly convex;
(b) for any $x \in X$ with $\|x\| = 1$ and any sequence $(x_n)_n$ in X with $\|x_n\| \to 1$ and such that $\|(x_n + x)/2\| \to 1$ as $n \to \infty$, one has $\|x_n - x\| \to 0$;
(c) for any $x \in X$ with $\|x\| = 1$ and any sequence $(x_n)_n$ in X with $\limsup_{n\to\infty} \|x_n\| \leq 1$ and such that $\|(x_n + x)/2\| \to 1$ as $n \to \infty$, one has $\|x_n - x\| \to 0$.

Furthermore, the equivalences still hold with nets (with a common index set) in place of sequences in (b) and (c).

As said above locally uniformly convex norms possess the Kadec-Klee property and this is justified with the same arguments utilized for uniformly convex norms in Proposition 18.17.

PROPOSITION 18.22. If a normed space $(X, \|\cdot\|)$ is locally uniformly convex, then its norm $\|\cdot\|$ possesses the Kadec-Klee property, that is, for any net $(x_j)_{j\in J}$ in X one has $\|x_j - x\| \to 0$ if and only if $(x_j)_{j\in J}$ converges weakly to x and $\|x_j\| \to \|x\|$.

Now we use the Šmulian lemma to establish the Fréchet differentiability of a norm under the local uniform convexity of the dual norm. The result is the locally uniformly convex norm counterpart of Proposition 18.9(a).

PROPOSITION 18.23. Let $(X, \|\cdot\|)$ be a normed space. If the dual norm $\|\cdot\|_*$ is locally uniformly convex, then the norm $\|\cdot\|$ is Fréchet differentiable off zero.

PROOF. Assume that $\|\cdot\|_*$ is locally uniformly convex and fix any $u \in \mathbb{S}_X$ (if any). Consider $x^* \in \mathbb{S}_{X^*}$ with $\langle x^*, u\rangle = 1$ and a sequence $(x_n^*)_n$ in $\mathbb{B}_{X^*}$ with $\langle x_n^*, u\rangle \to 1$. Noting that

$$2 \geq \|x_n^* + x^*\|_* \geq \langle x_n^* + x^*, u\rangle \quad \text{and} \quad \langle x_n^* + x^*, u\rangle \to 2,$$

we see that $\|x_n^* + x^*\|_* \to 2$, so the local uniform convexity of the norm $\|\cdot\|_*$ entails that $\|x_n^* - x^*\|_* \to 0$. Consequently, the Šmulian lemma (see Lemma 18.10) ensures that the primal norm $\|\cdot\|$ is Fréchet differentiable at u. □

18.1.3. Uniformly smooth Banach spaces. Let $(X, \|\cdot\|)$ be a normed space. It is usual to say that the norm $\|\cdot\|$ is *smooth* when it is Gâteaux differentiable away from the origin. Given $\overline{x} \in \mathbb{S}_X$, by Lemma 4.48 one sees that $\|\cdot\|$ is Gâteaux differentiable at $\overline{x}$ if and only if for each $y \in \mathbb{S}_X$

$$\lim_{t \downarrow 0} t^{-1} \left(\frac{1}{2}(\|\overline{x} + ty\| + \|\overline{x} - ty\|) - 1 \right) = 0,$$

and that $\|\cdot\|$ is Fréchet differentiable at $\overline{x}$ if and only if the above limit is uniform with respect to $y \in \mathbb{S}_X$. The same lemma also guarantees that $\|\cdot\|$ is uniformly Fréchet differentiable on the unit sphere $\mathbb{S}_X$ if and only if the above limit is uniform both with respect to $y \in \mathbb{S}_X$ and with respect to $\overline{x} \in \mathbb{S}_X$, or equivalently

$$\frac{1}{t} \sup_{x,y \in \mathbb{S}_X} \left| \frac{1}{2}(\|\overline{x} + ty\| + \|\overline{x} - ty\|) - 1 \right| \to 0 \quad \text{as } t \downarrow 0.$$

When the norm is uniformly Fréchet differentiable on the unit sphere $\mathbb{S}_X$, one merely says that it is *uniformly Fréchet differentiable.*

The following definition is based on the latter characterization of uniform Fréchet differentiability of a norm.

DEFINITION 18.24. Let $(X, \|\cdot\|)$ be a normed space. The *modulus of smoothness* $\varrho_{\|\cdot\|} : [0, +\infty[\to [0, +\infty[$ *of the norm* $\|\cdot\|$ is defined by

$$\varrho_{\|\cdot\|}(\tau) := \sup \left\{ \frac{\|x + \tau y\| + \|x - \tau y\|}{2} - 1 : \|x\| = 1, \|y\| = 1 \right\}.$$

Obviously $\varrho_{\|\cdot\|}(0) = 0$. The norm $\|\cdot\|$ or the space $(X, \|\cdot\|)$ is said to be *uniformly smooth* when

$$\lim_{\tau \downarrow 0} \frac{\varrho_{\|\cdot\|}(\tau)}{\tau} = 0,$$

which is equivalent to say that it is uniformly Fréchet differentiable according to what precedes. When there is no ambiguity on the norm, it will often be convenient (like for the modulus of convexity) to write ϱ_X in place of $\varrho_{\|\cdot\|}$.

Given a real $\alpha > 0$ and considering the new norm $\alpha\|\cdot\|$, it is readily seen that, for any real $\tau \geq 0$

$$\varrho_{\alpha\|\cdot\|}(\tau) = \varrho_{\|\cdot\|}(\tau).$$

If $(X, \|\cdot\|)$ is one-dimensional, then one has $\varrho_{\|\cdot\|}(\tau) = \varrho_{\mathbb{R}}(\tau)$ for all $\tau \geq 0$, where $\varrho_{\mathbb{R}}$ denotes the modulus of smoothness of the absolute value of $\mathbb{R}$. An easy calculus also gives

$$\varrho_{\mathbb{R}}(\tau) = \begin{cases} 0 & \text{if } 0 \leq \tau \leq 1 \\ \tau - 1 & \text{if } \tau > 1. \end{cases} \tag{18.11}$$

If we endow a nonzero closed vector subspace Y of a normed space $(X \|\cdot\|)$ with the induced norm $\|\cdot\|_Y$, it is obvious that $\varrho_Y(\cdot) \leq \varrho_X(\cdot)$, so $(Y, \|\cdot\|_Y)$ is uniformly smooth whenever $(X, \|\cdot\|)$ is uniformly convex.

Let examine some other basic properties. Clearly, $\varrho_{\|\cdot\|}(\tau) \le \tau$ and $\varrho_{\cdot\|}(\cdot)$ is a convex function on $[0,+\infty[$ as the supremum of a collection of convex functions. This convexity property combined with the equality

$$\varrho_{\|\cdot\|}(\tau)/\tau = \tau^{-1}\big(\varrho_{\|\cdot\|}(0+\tau) - \varrho_{\|\cdot\|}(0)\big)$$

ensures that the function $\tau \mapsto \varrho_{\|\cdot\|}(\tau)/\tau$ is nondecreasing on $]0,+\infty[$ (see 2.19).

On the other hand, for each $\tau \ge 0$ we can write

$$\begin{aligned} 2\varrho_{\|\cdot\|}(\tau) &= \sup_{x\in\mathbb{S}_X}\ \sup_{y\in\mathbb{S}_X} \left(\|x+\tau y\| + \|x-\tau y\| - 2\right) \\ &= \sup_{x\in\mathbb{S}_X}\ \sup_{y\in\tau\mathbb{S}_X} \left(\|x+y\| + \|x-y\| - 2\right) \\ &= \sup_{x\in\mathbb{S}_X}\ \sup_{y\in\tau\mathbb{B}_X} \left(\|x+y\| + \|x-y\| - 2\right), \end{aligned}$$

where the latter equality is due to the fact the supremum of a convex function (here $y \mapsto \|x+y\| + \|x-y\| - 2$) on a closed set with nonempty boundary coincides with the supremum over the boundary. So, for all $\tau \ge 0$ the equality

$$\varrho_{\|\cdot\|}(\tau) = \sup\left\{\frac{\|x+y\| + \|x-y\|}{2} - 1 : \|x\| = 1, \|y\| \le \tau\right\} \tag{18.12}$$

holds true, and this obviously entails that the function $\tau \mapsto \varrho_{\|\cdot\|}(\tau)$ is nondecreasing on $[0,+\infty[$.

In the following proposition we state the above properties of the modulus of smoothness along with other classical ones in the assertions (si). In the proposition, before stating the assertions (si) we will establish the assertions (ci) related to the modulus of convexity.

PROPOSITION 18.25. Let $(X, \|\cdot\|)$ be a normed space. The following hold:
(c1) $\delta_{\|\cdot\|}(0) = 0$ and $\delta_{\|\cdot\|}(\varepsilon) \le 1$ for all $\varepsilon \in [0,2]$.
(c2) $\delta_{\|\cdot\|}(\cdot)$ is nondecreasing on $[0,2]$.
(c3) If X is at least two-dimensional, then the function $\varepsilon \mapsto \delta_{\|\cdot\|}(\varepsilon)/\varepsilon$ is nondecreasing on $]0,2]$, hence in particular

$$\delta_{\|\cdot\|}(\varepsilon) \le \varepsilon/2 \quad \text{for all } \varepsilon \in [0,2].$$

(c4) If $\|\cdot\|$ is uniformly convex and X is at least two-dimensional, then the function $\delta_{\|\cdot\|}$ in (c2) is increasing on $[0,2]$.

(s1) $\varrho_{\|\cdot\|}(0) = 0$ and $\varrho_{\|\cdot\|}(\tau) \le \tau$ for all $\tau \ge 0$.
(s2) The function $\varrho_{\|\cdot\|}$ is convex, continuous and nondecreasing on $[0,+\infty[$.
(s3) The function $\tau \mapsto \varrho_{\|\cdot\|}(\tau)/\tau$ is nondecreasing on $]0,+\infty[$.

PROOF. All the properties (s1), (s2) and (s3) have been justified in the lines above preceding the statement of the proposition, whereas (c1) and (c2) directly follow from the definition of $\delta_{\|\cdot\|}$. It remains to show (c3) and (c4), assuming that X is at least two-dimensional.

To show (c3), fix $0 < s < t \le 2$ and consider any $x, y \in X$ with $\|x\| = \|y\| = 1$ and $\|x-y\| = t$. Put $z := (x+y)/\|x+y\|$ and $m := (x+y)/2$. Choose x' and y' such that $x' - z = \dfrac{s}{t}(x-z)$ and $y' - z = \dfrac{s}{t}(y-z)$. It ensues that $x' - y' = \frac{s}{t}(x-y)$, $\|x'-y'\| = s$ and

$$x' + y' = \frac{s}{t}\|x+y\|z + 2\left(1 - \frac{s}{t}\right)z, \text{ so } \frac{1}{2}(x'+y') = \left(1 - \frac{s}{t} + \frac{s}{t}\|\frac{x+y}{2}\|\right)z,$$

and since $\|z\| = 1$, the latter equality gives with $m' := (x' + y')/2$ that

$$z - m' = z - \left(1 - \frac{s}{t} + \frac{s}{t}\|\frac{x+y}{2}\|\right) z = (1 - \|m'\|)z, \text{ hence } \|z - m'\| = 1 - \|m'\|,$$

and we also have

$$z - m' = \frac{s}{t}\left(1 - \frac{\|x+y\|}{2}\right) z = \frac{s}{t}(1 - \|m\|)z, \text{ hence } \|z - m'\| = \frac{s}{t}(1 - \|m\|).$$

Therefore, we obtain $s^{-1}(1 - \|m'\|) = t^{-1}(1 - \|m\|)$, which combined with the inclusions $x', y' \in \mathbb{B}_X$ and with the equality (18.9) yields

$$s^{-1}\delta_{\|\cdot\|}(s) \le s^{-1}(1 - \|m'\|) = t^{-1}(1 - \|m\|) = t^{-1}\left(1 - \|\frac{x+y}{2}\|\right).$$

It results that $s^{-1}\delta_{\|\cdot\|}(s) \le t^{-1}\delta_{\|\cdot\|}(t)$. Further, this and the nondecreasing property of $\varepsilon \mapsto \delta_{\|\cdot\|}(\varepsilon)$ ensures that $s^{-1}\delta_{\|\cdot\|}(s) \le r^{-1}\delta_{\|\cdot\|}(r) \le r^{-1}\delta_{\|\cdot\|}(2)$ for all $r \in]s, 2[$, so taking the limit as $r \uparrow 2$ we obtain $s^{-1}\delta_{\|\cdot\|}(s) \le 2^{-1}\delta_{\|\cdot\|}(2)$. The assertion of non-decreasing property in (c3) is then justified. Further, this and (c1) ensure that for any $\varepsilon \in]0, 2]$

$$\frac{\delta_{\|\cdot\|}(\varepsilon)}{\varepsilon} \le \frac{\delta_{\|\cdot\|}(2)}{2} \le \frac{1}{2},$$

which gives $\delta_{\|\cdot\|}(\varepsilon) \le \varepsilon/2$ for any $\varepsilon \in [0, 2]$. The assertion (c3) is established.

Assume now that $\|\cdot\|$ is uniformly convex. Then, for $0 < s < t \le 2$ the assertion (c3) and the inequality $\delta_{\|\cdot\|}(t) > 0$ (due the uniform convexity of $\|\cdot\|$) allow us to write

$$s^{-1}\delta_{\|\cdot\|}(s) \le t^{-1}\delta_{\|\cdot\|}(t) < s^{-1}\delta_{\|\cdot\|}(t),$$

thus $\delta_{\|\cdot\|}(s) < \delta_{\|\cdot\|}(t)$, which translates (c4). □

REMARK 18.26. Unlike the modulus of smoothness $\varrho_{\|\cdot\|}$, the modulus of convexity $\delta_{\|\cdot\|}$ may fail to be convex, see (18.26) in the comments at the end of the chapter. □

Many classical Banach spaces endowed with their canonical norms are both uniformly convex and uniformly smooth. Consider first any Hilbert space $(H, \|\cdot\|)$ with $\dim H \ge 2$. Let $\varepsilon \in]0, 2]$. Take any $x, y \in \mathbb{S}_H$ with $\|x - y\| \ge \varepsilon$. By the parallelogram law we have

$$\|x + y\|^2 = 2(\|x\|^2 + \|y\|^2) - \|x - y\|^2 \le 4 - \varepsilon^2,$$

and the inequality is an equality if $\|x - y\| = \varepsilon$. It follows that

$$\left\|\frac{x+y}{2}\right\| \le \sqrt{1 - \frac{\varepsilon^2}{4}}, \text{ or equivalently } 1 - \left\|\frac{x+y}{2}\right\| \ge 1 - \sqrt{1 - \frac{\varepsilon^2}{4}},$$

and hence by Definition 18.13 we obtain

$$\delta_H(\varepsilon) = 1 - \sqrt{1 - (1/4)\varepsilon^2} \ge \varepsilon^2/8. \tag{18.13}$$

Note also that the Norlander inequality gives for any (nonzero) vector space $(X, \|\cdot\|)$ and any (nonzero) Hilbert space H (see the Norlander theorem (Theorem 18.122) in the comments at the end of the chapter)

$$\delta_{\|\cdot\|}(\varepsilon) \le \delta_H(\varepsilon) \quad \text{for all } \varepsilon \in [0, 2]. \tag{18.14}$$

Now assume again that H is a Hilbert space with $\dim H \geq 2$ and let $\tau \in]0,+\infty[$. Take any $x \in \mathbb{S}_H$ and $y \in \tau\mathbb{S}_H$. Using the inner product $\langle\cdot,\cdot\rangle$ to which the Hilbertian norm $\|\cdot\|$ is associated, we can write

$$\|x+\tau y\| = \sqrt{1+2\tau\langle x,y\rangle+\tau^2} \quad \text{and} \quad \|x-\tau y\| = \sqrt{1-2\tau\langle x,y\rangle+\tau^2}.$$

It ensues that

$$\begin{aligned}\frac{1}{2}(\|x+\tau y\|+\|x-\tau y\|)-1 &= \frac{1}{2}\sqrt{1+2\tau\langle x,y\rangle+\tau^2}+\frac{1}{2}\sqrt{1-2\tau\langle x,y\rangle+\tau^2}-1\\ &\leq \sqrt{\frac{1}{2}(1+2\tau\langle x,y\rangle+\tau^2)+\frac{1}{2}(1-2\tau\langle x,y\rangle+\tau^2)}-1\\ &= \sqrt{1+\tau^2}-1,\end{aligned}$$

where the inequality is due to the concavity of $t \mapsto \sqrt{t}$ on $[0,+\infty[$; note also that the inequality is an equality if x,y are orthogonal. It then follows that

$$\varrho_H(\tau) = (1+\tau^2)^{1/2}-1 < \min\{\tau,\tau^2/2\}. \tag{18.15}$$

In addition to Hilbert spaces, it is well known (see [**79, 334, 665, 666, 704**]) that all the Banach spaces l^p, L^p, and W^p_m $(1<p<+\infty)$ are (for their usual norms) uniformly convex and uniformly smooth. More precisely, for $\varepsilon \in]0,2]$ and for $\tau > 0$
(18.16)

$$\delta_{l^p}(\varepsilon) = \delta_{L^p}(\varepsilon) = \delta_{W^p_m}(\varepsilon) = \begin{cases} \frac{p-1}{8}\varepsilon^2 + o(\varepsilon^2) > \frac{p-1}{8}\varepsilon^2, & 1<p<2,\\ 1-\left[1-\left(\frac{\varepsilon}{2}\right)^p\right]^{1/p} > \frac{1}{p}\left(\frac{\varepsilon}{2}\right)^p, & p \geq 2,\end{cases}$$

$$\varrho_{l^p}(\tau) = \varrho_{L^p}(\tau) = \varrho_{W^p_m}(\tau) = \begin{cases} (1+\tau^p)^{1/p}-1 < \frac{1}{p}\tau^p, & 1<p<2,\\ \frac{p-1}{2}\tau^2 + o(\tau^2) < \frac{p-1}{2}\tau^2, & p \geq 2.\end{cases}$$

We saw above in (s3) of Proposition 18.25 that $\tau \mapsto \tau^{-1}\varrho_{\|\cdot\|}(\tau)$ is nondecreasing. The next proposition examines the function $\tau \mapsto \tau^{-2}\varrho_{\|\cdot\|}(\tau)$. Let us establish first the following lemma.

LEMMA 18.27. *There exists some real $0 < \tau_0 < (-1+\sqrt{2})/2$ such that for any normed space $(X,\|\cdot\|)$*

$$\varrho_{\|\cdot\|}(2\tau) \leq (4+15\tau)\varrho_{\|\cdot\|}(\tau) \quad \text{for all } \tau \in [0,\tau_0].$$

PROOF. Let $\tau \in]0,1[$, and let $x \in \mathbb{S}_X$ and $y \in X$ with $\|y\| \leq \tau$. By the property (18.12) for modulus of uniform smoothness we have

$$\frac{1}{2}\left(\left\|\frac{x+y}{\|x+y\|}+\frac{y}{\|x+y\|}\right\|+\left\|\frac{x+y}{\|x+y\|}-\frac{y}{\|x+y\|}\right\|\right)-1 \leq \varrho_{\|\cdot\|}\left(\frac{\tau}{\|x+y\|}\right),$$

which, by the equality $\|x\| = 1$, is equivalent to

$$\|x+2y\| \leq 2\|x+y\|\,\varrho_{\|\cdot\|}(\tau/\|x+y\|)+2\|x+y\|-1.$$

Changing y in $-y$ gives also

$$\|x-2y\| \leq 2\|x-y\|\,\varrho_{\|\cdot\|}(\tau/\|x-y\|)+2\|x-y\|-1.$$

We deduce that

$$\begin{aligned}&\frac{1}{2}(\|x+2y\|+\|x-2y\|)-1\\&\leq \|x+y\|\varrho_{\|\cdot\|}(\tau/\|x+y\|)+\|x-y\|\varrho_{\|\cdot\|}(\tau/\|x-y\|)+2\Big(\frac{1}{2}(\|x+y\|+\|x-y\|)-1\Big)\\&\leq \|x+y\|\varrho_{\|\cdot\|}(\tau/\|x+y\|)+\|x-y\|\varrho_{\|\cdot\|}(\tau/\|x-y\|)+2\varrho_{\|\cdot\|}(\tau)\\&\leq 2(1+\tau)\varrho_{\|\cdot\|}\big(\tau/(1-\tau)\big)+2\varrho_{\|\cdot\|}(\tau),\end{aligned}$$

where the latter inequality is due to the non-decreasing property of $\varrho_{\|\cdot\|}$. Taking the supremum over $x \in \mathbb{S}_X$ and $y \in \tau\mathbb{B}_X$ yields by (18.12) again

$$\varrho_{\|\cdot\|}(2\tau) \leq 2(1+\tau)\varrho_{\|\cdot\|}\big(\tau/(1-\tau)\big)+2\varrho_{\|\cdot\|}(\tau).$$

On the other hand, for $\tau \in]0,1/2]$ we notice that $1/(1-\tau) \leq 1+2\tau$, thus by the nondecreasing property of $\varrho_{\|\cdot\|}$ again

$$\varrho_{\|\cdot\|}\big(\tau/(1-\tau)\big) \leq \varrho_{\|\cdot\|}\big(\tau(1+2\tau)\big) \leq (1-2\tau)\varrho_{\|\cdot\|}(\tau)+2\tau\varrho_{\|\cdot\|}(2\tau),$$

where the latter inequality is due the convexity of $\varrho_{\|\cdot\|}$. Therefore, for $\tau \in]0,1/2]$

$$\varrho_{\|\cdot\|}(2\tau) \leq (2+2\tau)(1-2\tau)\varrho_{\|\cdot\|}(\tau)+4\tau(1+\tau)\varrho_{\|\cdot\|}(2\tau)+2\varrho_{\|\cdot\|}(\tau),$$

or equivalently

$$(1-4\tau-4\tau^2)\varrho_{\|\cdot\|}(2\tau) \leq (4-2\tau-4\tau^2)\varrho_{\|\cdot\|}(\tau).$$

Observing for $0<\tau<(-1+\sqrt{2})/2<1/2$ that $1-4\tau-4\tau^2>0$, it follows that

$$\varrho_{\|\cdot\|}(2\tau) \leq \frac{4-2\tau-4\tau^2}{1-4\tau-4\tau^2}\varrho_{\|\cdot\|}(\tau) = \big(4+14\tau+o(\tau)\big)\varrho_{\|\cdot\|}(\tau).$$

Choosing $0<\tau_0<(-1+\sqrt{2})/2$ such that for all $\tau \in]0,\tau_0]$

$$\frac{4-2\tau-4\tau^2}{1-4\tau-4\tau^2} = 4+14\tau+o(\tau) \leq 4+15\tau,$$

we see that $\varrho_{\|\cdot\|}(2\tau) \leq (4+15\tau)\varrho_{\|\cdot\|}(\tau)$ for all $\tau \in [0,\tau_0]$ as desired. □

PROPOSITION 18.28. Let $(X,\|\cdot\|)$ be a normed space. There exists a real constant $c>0$ such that for all reals θ and τ with $0<\tau<\theta$

$$\frac{\rho(\theta)}{\theta^2} \leq c\,\frac{\rho(\tau)}{\tau^2};$$

in fact c can be chosen as $c := \dfrac{4\tau_0}{\varrho_{\|\cdot\|}(\tau_0)}\displaystyle\prod_{k=1}^{\infty}\Big(1+\frac{15\tau_0}{2^k4}\Big)$ with τ_0 as given by Lemma 18.27 above.

PROOF. Let $\tau_0>0$ be given by Lemma 18.27. We distinguish three cases.

Case 1: $0<\tau_0\leq\tau<\theta$**.** Then $1/\theta \leq 1/\tau$, which combined with the inequality $\varrho(\theta) \leq \theta$ due to (s1) of Proposition 18.25 gives

$$\frac{\varrho(\theta)}{\theta} \leq \frac{\varrho(\theta)}{\tau} \leq \frac{\theta}{\tau}, \text{ so } \frac{\varrho(\theta)}{\theta^2} \leq \frac{1}{\tau}.$$

Further, by the nondecreasing property of the function $t \mapsto \varrho_{\|\cdot\|}(t)/t$ seen in (s3) of Proposition 18.25 we have

$$\varrho_{\|\cdot\|}(\tau)/\tau \geq \varrho_{\|\cdot\|}(\tau_0)/\tau_0,$$

so using the inequality $\varrho_{\|\cdot\|}(t) \leq t$ (see (s1) in Proposition 18.25) we obtain

$$\varrho_{\|\cdot\|}(\tau)/\tau^2 \geq \varrho_{\|\cdot\|}(\tau_0)/(\tau_0\tau) \geq (\varrho_{\|\cdot\|}(\tau_0)/\tau_0)(\varrho_{\|\cdot\|}(\theta)/\theta^2).$$

Case 2: $0 < \tau < \theta \leq \tau_0$. Choosing $m \in \mathbb{N}$ satisfying $\theta/(2^m) \leq \tau < \theta/(2^{m-1})$ and noticing by the nondecreasing property of $t \mapsto \varrho_{\|\cdot\|}(t)$

$$\frac{\varrho_{\|\cdot\|}(\theta)}{\theta^2} = \frac{1}{\theta^2}\varrho_{\|\cdot\|}\left(\frac{\theta}{2^m}\right)\prod_{k=1}^{m}\frac{\varrho_{\|\cdot\|}(\theta/2^{k-1})}{\varrho_{\|\cdot\|}(\theta/2^k)} \leq \frac{1}{\theta^2}\varrho_{\|\cdot\|}(\tau)\prod_{k=1}^{m}\frac{\varrho_{\|\cdot\|}(\theta/2^{k-1})}{\varrho_{\|\cdot\|}(\theta/2^k)},$$

we obtain by Lemma 18.27

$$\begin{aligned}\frac{\varrho_{\|\cdot\|}(\theta)}{\theta^2} &\leq \varrho_{\|\cdot\|}(\tau)\frac{4^m}{\theta^2}\prod_{k=1}^{m}\left(1+\frac{15\theta}{2^k4}\right)\\ &\leq \frac{4\varrho_{\|\cdot\|}(\tau)}{\tau^2}\prod_{k=1}^{\infty}\left(1+\frac{15\theta}{2^k4}\right)\\ &\leq \frac{4\varrho_{\|\cdot\|}(\tau)}{\tau^2}\prod_{k=1}^{\infty}\left(1+\frac{15\tau_0}{2^k4}\right).\end{aligned}$$

Case 3: $0 < \tau < \tau_0 < \theta$. In this case we can write

$$\frac{\varrho_{\|\cdot\|}(\theta)}{\theta^2} \leq \frac{\tau_0}{\varrho_{\|\cdot\|}(\tau_0)}\frac{\varrho_{\|\cdot\|}(\tau_0)}{\tau_0^2} \leq \frac{\tau_0}{\varrho_{\|\cdot\|}(\tau_0)}\frac{4\varrho_{\|\cdot\|}(\tau)}{\tau^2}\prod_{k=1}^{\infty}\left(1+\frac{15\tau_0}{2^k4}\right),$$

where the left inequality is due to Case 1 and the right to Case 2.

Since $\tau_0/\varrho_{\|\cdot\|}(\tau_0) \geq 1$ (see (s1) in Proposition 18.25), the choice

$$c := \frac{4\tau_0}{\varrho_{\|\cdot\|}(\tau_0)}\prod_{k=1}^{\infty}\left(1+\frac{15\tau_0}{2^k4}\right)$$

finishes the proof. □

REMARK 18.29. Using the known fact that $\varrho_{\|\cdot\|}(\cdot) \geq \varrho_H(\cdot)$ (see (18.17) in Remark 18.32 below), for the universal constant

$$C := \frac{4\tau_0}{\varrho_H(\tau_0)}\prod_{k=1}^{\infty}\left(1+\frac{15\tau_0}{2^k4}\right)$$

(not depending on the norm $\|\cdot\|$) and for the constant c in Proposition 18.28 we see that $C \geq c$. Consequently, the statement of Proposition 18.28 holds true with the universal constant C in place of c. □

Unlike relationships between strict convexity and smoothness of norms, there is a complete duality concerning the uniform corresponding properties.

THEOREM 18.30 (V.L. Šmulian theorem: uniform smoothness of dual norm versus uniform convexity of primal norm). *Given a normed space $(X, \|\cdot\|)$ and the dual norm $\|\cdot\|_*$ on X^*, the following equivalences hold.*
(a) The normed space $(X, \|\cdot\|)$ is uniformly convex if and only if $(X^*, \|\cdot\|_*)$ is uniformly smooth.
(b) The normed space $(X, \|\cdot\|)$ is uniformly smooth if and only if $(X^*, \|\cdot\|_*)$ is uniformly convex.

The proof of the theorem uses the following lemma providing an expression of the modulus of smoothness in terms of the modulus of convexity.

LEMMA 18.31. Given a normed space $(X, \|\cdot\|)$, the following equalities hold for every real $\tau \geq 0$:

$$\varrho_{\|\cdot\|_*}(\tau) = \sup\left\{ \frac{\tau\varepsilon}{2} - \delta_{\|\cdot\|}(\varepsilon) : \varepsilon \in [0,2]\right\}$$

and

$$\varrho_{\|\cdot\|}(\tau) = \sup\left\{ \frac{\tau\varepsilon}{2} - \delta_{\|\cdot\|_*}(\varepsilon) : \varepsilon \in [0,2]\right\}.$$

PROOF. Fix any real $\tau \geq 0$. We can write

$$\begin{aligned}
2\varrho_{\|\cdot\|_*}(\tau) &= \sup\{ \|x^* + \tau y^*\|_* + \|x^* - \tau y^*\|_* - 2 : x^*, y^* \in \mathbb{S}_{X^*}\} \\
&= \sup\{ x^*(x) + \tau y^*(x) + x^*(y) - \tau y^*(y) - 2 : x, y \in \mathbb{S}_X, x^*, y^* \in \mathbb{S}_{X^*} \} \\
&= \sup\{ \|x+y\| + \tau\|x-y\| - 2 : x, y \in \mathbb{S}_X \} \\
&= \sup\{ \|x+y\| + \tau\varepsilon - 2 : x, y \in \mathbb{S}_X, \|x-y\| \geq \varepsilon, \varepsilon \in [0,2]\},
\end{aligned}$$

from which we see that

$$\begin{aligned}
2\varrho_{\|\cdot\|_*}(\tau) &= \sup_{\varepsilon\in[0,2]} \left(\tau\varepsilon - 2\inf\left\{ 1 - \left\|\frac{x+y}{2}\right\| : x, y \in \mathbb{S}_X, \|x-y\| \geq \varepsilon \right\}\right) \\
&= \sup_{\varepsilon\in[0,2]} (\tau\varepsilon - 2\delta_{\|\cdot\|}(\varepsilon)).
\end{aligned}$$

This justifies the first equality of the lemma and the second is obtained in a similar way. □

REMARK 18.32. (a) Regarding similar features for the modulus of smoothness one has for any $\varepsilon \in [0,2]$ the estimates

$$\delta_{\|\cdot\|_*}(\varepsilon) \geq \sup\left\{ \frac{\tau\varepsilon}{2} - \varrho_{\|\cdot\|}(\tau) : \tau \in [0,+\infty[\right\}$$

and

$$\delta_{\|\cdot\|}(\varepsilon) \geq \sup\left\{ \frac{\tau\varepsilon}{2} - \varrho_{\|\cdot\|_*}(\tau) : \tau \in [0,+\infty[\right\}.$$

Indeed, given any $\varepsilon \in [0,2]$ Lemma 18.31 gives for every real $\tau > 0$ the inequality

$$\varrho_{\|\cdot\|}(\tau) \geq \frac{\tau\varepsilon}{2} - \delta_{\|\cdot\|_*}(\varepsilon), \quad \text{hence } \delta_{\|\cdot\|_*}(\varepsilon) \geq \frac{\tau\varepsilon}{2} - \varrho_{\|\cdot\|}(\tau),$$

and this justifies the inequality $\delta_{\|\cdot\|_*}(\varepsilon) \geq \sup\{ \frac{\tau\varepsilon}{2} - \varrho_{\|\cdot\|}(\tau) : \tau \in [0,+\infty[\}$. The second desired inequality holds true similarly.

Notice that the above inequalities fail to be equalities since moduli of convexity of norms are not in general convex functions according to Remark 18.26, while the right-hand sides of the inequalities are convex functions of ε.

(b) According to (18.14) the above lemma also ensures that for any (nonzero) normed space $(X, \|\cdot\|)$ one has

$$\varrho_H(\tau) \leq \varrho_{\|\cdot\|}(\tau) \quad \text{for every } \tau \in [0,+\infty[, \tag{18.17}$$

where we recall that ϱ_H denotes the modulus of smoothness of any (nonzero) Hilbert space. □

PROOF OF THEOREM 18.30. For each real $\tau > 0$ the above lemma tells us that

$$\frac{\varrho_{\|\cdot\|_*}(\tau)}{\tau} = \sup\left\{\frac{\varepsilon}{2} - \frac{\delta_{\|\cdot\|}(\varepsilon)}{\tau} : \varepsilon \in [0,2]\right\}. \tag{18.18}$$

Assume that the dual norm $\|\cdot\|_*$ is uniformly smooth, that is, $\tau^{-1}\varrho_{\|\cdot\|_*}(\tau) \to 0$ as $\tau \downarrow 0$. Then, fixing any $\varepsilon \in]0,2]$ we can choose some $\tau_\varepsilon > 0$ such that $\tau_\varepsilon^{-1}\varrho_{\|\cdot\|_*}(\tau_\varepsilon) < \varepsilon/4$. We deduce according to 18.18 that

$$\frac{\varepsilon}{2} - \frac{\delta_{\|\cdot\|}(\varepsilon)}{\tau_\varepsilon} \le \frac{\varrho_{\|\cdot\|_*}(\tau_\varepsilon)}{\tau_\varepsilon} < \frac{\varepsilon}{4},$$

which gives $\delta_{\|\cdot\|}(\varepsilon) > \tau_\varepsilon\varepsilon/4 > 0$, so the primal norm $\|\cdot\|$ is uniformly convex. Working with the second formula in Lemma 18.31, the same arguments show that the dual norm $\|\cdot\|_*$ is uniformly convex whenever the primal norm $\|\cdot\|$ is uniformly smooth.

Assume now that $\|\cdot\|$ is uniformly convex, and fix any $\eta \in]0,1[$. Define $\tau_0 := \delta_{\|\cdot\|}(2\eta)$ and note that $\tau_0 > 0$ by uniform convexity of $\|\cdot\|$. On the one hand, for $0 \le \varepsilon \le 2\eta$ we have

$$\frac{\varepsilon}{2} - \frac{\delta_{\|\cdot\|}(\varepsilon)}{\tau_0} \le \frac{\varepsilon}{2} \le \eta,$$

and, on the other hand, for $2\eta < \varepsilon \le 2$ we see by the nondecreasing property of $\varepsilon \mapsto \delta_{\|\cdot\|}(\varepsilon)$ that

$$\frac{\varepsilon}{2} - \frac{\delta_{\|\cdot\|}(\varepsilon)}{\tau_0} \le 1 - \frac{\delta_{\|\cdot\|}(\varepsilon)}{\tau_0} \le 1 - \frac{\delta_{\|\cdot\|}(2\eta)}{\tau_0} = 0 < \eta.$$

By (18.18) it ensues that $\tau_0^{-1}\varrho_{\|\cdot\|_*}(\tau_0) \le \eta$, which combined with the nondecreasing property of $\tau \mapsto \tau^{-1}\varrho_{\|\cdot\|_*}(\tau)$ (see Proposition 18.25(s3)) yields that $\tau^{-1}\varrho_{\|\cdot\|_*}(\tau) \le \eta$ for all $\tau \in]0,\tau_0]$. This says that $\tau^{-1}\varrho_{\|\cdot\|_*}(\tau) \to 0$ as $\tau \downarrow 0$ and justifies the uniform smoothness of the dual norm $\|\cdot\|_*$. Similarly, using the second formula in Lemma 18.31 one can prove that $\|\cdot\|$ is uniformly smooth whenever its dual norm $\|\cdot\|_*$ is uniformly convex. □

A basic result of uniformly convex (resp. uniformly smooth) Banach spaces is their reflexivity.

THEOREM 18.33 (D. Milman; B.J. Pettis). Every uniformly convex (resp. uniformly smooth) Banach space is reflexive.

PROOF. Assume first that $(X,\|\cdot\|)$ is a uniformly convex Banach space and take any $x^{**} \in X^{**}$ with $\|x^{**}\|_{**} = 1$. Identifying X as a subspace of X^{**}, Goldstine's theorem (see Theorem C.7 in Appendix) tells us that there exists a net $(x_j)_{j\in J}$ in X which $w(X^{**},X^*)$-converges to x^{**} in X^{**}. The net $(\frac{1}{2}(x_i + x_j))_{(i,j)\in J\times J}$ $w(X^{**},X^*)$-converges to x^{**}, hence $\lim\limits_{(i,j)\in I\times J} \|\frac{1}{2}(x_i + x_j)\| = 1$ by the lower semicontinuity of $\|\cdot\|_{**}$ with respect to the topology $w(X^{**},X^*)$. The uniform convexity of $\|\cdot\|$ implies that $\lim\limits_{(i,j)\in I\times J} \|x_i - x_j\| = 0$, that is, $(x_j)_{j\in J}$ is a Cauchy net in X. It then converges in $(X,\|\cdot\|)$ to some x, so $x^{**} = x$. This justifies the reflexivity of $(X,\|\cdot\|)$.

Assume now that $(X,\|\cdot\|)$ is uniformly smooth, so $(X^*,\|\cdot\|_*)$ is uniformly convex by Theorem 18.30. According to what precedes, $(X^*,\|\cdot\|_*)$ is reflexive, thus $(X,\|\cdot\|)$ is reflexive too. □

The next proposition is concerned with the stability of uniform smoothness for the product of finitely many spaces endowed with suitable norms.

PROPOSITION 18.34. Let $(X_i, \|\cdot\|_i)$, $i = 1, \cdots, k$, be uniformly smooth normed spaces and let $\|\!|\cdot|\!\|$ be the norm on $X := X_1 \times \cdots \times X_k$ defined for all $x := (x_1, \cdots, x_k) \in X$ by
$$\|\!|x|\!\| := (\|x_1\|_1^2 + \cdots + \|x_k\|_k^2)^{1/2}.$$
Then, $(X, \|\!|\cdot|\!\|)$ is uniformly smooth.

PROOF. We know that the dual norm $\|\!|\cdot|\!\|_*$ (of $\|\!|\cdot|\!\|$) on $X^* = X_1^* \times \cdots \times X_k^*$ is given for all $x^* = (x_1^*, \cdots, x_k^*)$ by
$$\|\!|x^*|\!\|_* = \big((\|x_1^*\|_{1*})^2 + \cdots + (\|x_1^*\|_{k*})^2\big)^{1/2},$$
where $\|\cdot\|_{i*}$ denotes the dual norm of $\|\cdot\|_i$. Since $\|\cdot\|_{i*}$ is uniformly convex for each $i = 1, \cdots, k$ (see Theorem 18.30(b)), Proposition 18.18 tells us that the space $(X^*, \|\!|\cdot|\!\|_*)$ is uniformly convex, hence $(X, \|\!|\cdot|\!\|)$ is uniformly smooth by Theorem 18.30(b) again. □

REMARK 18.35. Using Remark 18.19 it is clear that the result of the above proposition still holds if the norm $\|\!|\cdot|\!\|$ is replaced by the norm $\|\!|\cdot|\!\|_p$ in Remark 18.19, where p is any real in $]1, +\infty[$. □

Consider now the case of a quotient normed spaces.

PROPOSITION 18.36. Let Y be a closed strict vector subspace of a uniformly smooth normed space $(X, \|\cdot\|)$. The quotient normed space $(X/Y, \|\cdot\|_{X/Y})$ is uniformly smooth, where $\|\cdot\|_{X/Y}$ denotes the canonical norm of X/Y in (18.10).

PROOF. We know that the closed subspace $Y^\perp := \{x^* \in X^* : \langle x^*, y\rangle = 0, \forall y \in Y\}$ in X^* endowed with the induced norm is isometrically identified with the topological dual of $(X/Y, \|\cdot\|_{X/Y})$, this topological dual being endowed with the dual norm of $\|\cdot\|_{X/Y}$. Then $Y^\perp$ endowed with that norm is uniformly convex according to the uniform convexity of $(X^*, \|\cdot\|_*)$ by Theorem 18.30(b). The same theorem then entails that X/Y endowed with its canonical norm is uniformly smooth. □

The above examples in (18.16) of L^p and W_m^p spaces tell us that the moduli of convexity (resp. smoothness) of their usual norms are equivalent near 0 to simple functions $\varepsilon \mapsto \kappa\varepsilon^q$ with $\kappa, q > 0$ (resp. $\tau \mapsto c\tau^s$ with $c, s > 0$). Given a normed space $(X, \|\cdot\|)$, it is often of interest to consider that the modulus of convexity (resp. smoothness) of the norm $\|\cdot\|$ is controlled by such functions.

A basic situation of such a control yielding the uniform convexity of the norm $\|\cdot\|$ is when there are real constants $\kappa, q > 0$ such that $\kappa\varepsilon^q \le \delta_{\|\cdot\|}(\varepsilon)$ for all $\varepsilon \in [0, 2]$. In such a case, by (18.14) and (18.13) one has for all $\varepsilon \in]0, 2]$
$$\kappa\varepsilon^q \le \delta_{\|\cdot\|}(\varepsilon) \le \delta_H(\varepsilon) = (\varepsilon^2/8)(1 + o(\varepsilon^2)).$$
In particular, one has $8\kappa\varepsilon^{q-2} \le 1 + o(\varepsilon^2)$ for $\varepsilon > 0$ near 0, so necessarily $q \ge 2$.

Similarly, an important control guaranteeing the uniform smoothness of the norm $\|\cdot\|$ is furnished by real constants $c > 0$ and $s > 1$ such that $\varrho_{\|\cdot\|}(\tau) \le c\tau^s$ for all $\tau > 0$. With these constants, by (18.15) and Remark 18.32(b)
$$(\tau^2/2)(1 + o(\varepsilon^2)) = \varrho_H(\tau) \le \varrho_{\|\cdot\|}(\tau) \le c\tau^s,$$
which gives $\tau^{2-s}(1 + o(\varepsilon^2)) \le 2c$. Considering the latter inequality for $\tau > 0$ near 0, we see that necessarily $s \le 2$, thus $1 < s \le 2$.

DEFINITION 18.37. Given a normed space $(X, \|\cdot\|)$, the modulus of convexity of the norm $\|\cdot\|$ is said to be of *power type* when there exist real constants $\kappa > 0$ and $q \geq 2$ such that

$$\kappa\varepsilon^q \leq \delta_{\|\cdot\|}(\varepsilon) \quad \text{for all } \varepsilon \in [0,2].$$

In such a case, one says that the modulus of convexity of $\|\cdot\|$ (or the uniformly convex space $(X, \|\cdot\|)$) is of *power type* $q \geq 2$.

The modulus of smoothness of the norm $\|\cdot\|$ is of *power type* whenever there are real constants $c > 0$ and $s \in]1,2]$ such that

$$\varrho_{\|\cdot\|}(\tau) \leq c\tau^s \quad \text{for all } \tau \geq 0.$$

In a such a case, the modulus of smoothness of $\|\cdot\|$ (or the uniform smooth space $(X, \|\cdot\|)$) is then said to be of *power type* $s \in]1,2]$.

For any real $q \geq 2$ there are Banach spaces whose norms are uniformly convex of power type q. Indeed, endowed with its usual norm any L^p (or W^p_m) space is by (18.16) uniformly convex of power type 2 if $1 < p \leq 2$, and of power type p if $p \in [2, +\infty[$. By (18.16) again, we see that, for any $s \in]1,2]$ there are Banach spaces whose norms are uniformly smooth of power type s, since any L^p (or W^p_m) space is uniformly smooth of power type p if $1 < p < 2$, and of power type 2 if $p \in [2, +\infty[$.

Lemma 18.31 allows us to prove the following result related to moduli of convexity and smoothness of power type.

PROPOSITION 18.38. Let $(X, \|\cdot\|)$ be a normed space and let $q \in [2, +\infty[$ and $s \in]1,2]$ be conjugate in the sense that $\frac{1}{q} + \frac{1}{s} = 1$. Then the modulus of convexity of the norm $\|\cdot\|$ (resp. $\|\cdot\|_*$) is of power type q if and only if the modulus of smoothness of $\|\cdot\|_*$ (resp. $\|\cdot\|$) is of power type s.

PROOF. Assume first that $\delta_{\|\cdot\|}$ is of power type q, that is, there is a real constant $\kappa > 0$ such that $\kappa\varepsilon^q \leq \delta_{\|\cdot\|}(\varepsilon)$ for all $\varepsilon \in [0,2]$. Fix any real $\tau \geq 0$. By Lemma 18.31 it ensues that

$$\varrho_{\|\cdot\|_*}(\tau) = \sup_{\varepsilon \in]0,2[} \left(\frac{\tau\varepsilon}{2} - \delta_{\|\cdot\|}(\varepsilon)\right) \leq \sup_{\varepsilon \in]0,2[} \left(\frac{\tau\varepsilon}{2} - \kappa\varepsilon^q\right).$$

Putting $g(\varepsilon) = \tau\varepsilon/2 - \kappa\varepsilon^q$ and noting that $g'(\varepsilon) = (\tau/2) - \kappa q\varepsilon^{q-1}$, we see that the derivative g' is null at $\varepsilon_\tau := \left((2\kappa q)^{-1}\tau\right)^{1/(q-1)}$ with $\varepsilon_\tau \in]0,2[$, and that $g(\varepsilon_\tau)$ is the supremum of g on $]0,\varepsilon[$. The value

$$g(\varepsilon_\tau) = \frac{\tau}{2}\left(\frac{\tau}{2\kappa q}\right)^{1/(q-1)} - \left(\frac{\kappa}{(2\kappa q)^{q/(q-1)}}\right)\tau^{q/(q-1)}$$

and the equality $s = q/(q-1)$ easily give $\varrho_{\|\cdot\|_*}(\tau) \leq c\tau^s$ for some real constant $c > 0$ (independent of τ), which translates that the modulus of smoothness of $\|\cdot\|_*$ is of power type s.

Conversely, suppose that there is a real constant $c > 0$ such that $\varrho_{\|\cdot\|_*}(\tau) \leq c\tau^s$ for all $\tau \geq 0$. Fix any $\varepsilon \in [0,2]$ and note by Remark 18.32 that

$$\delta_{\|\cdot\|}(\varepsilon) \geq \sup_{\tau > 0}\left(\frac{\tau\varepsilon}{2} - \varrho_{\|\cdot\|_*}(\tau)\right) \geq \sup_{\tau > 0}\left(\frac{\tau\varepsilon}{2} - c\tau^s\right).$$

Proceeding as above one obtains some real constant $\kappa > 0$ (independent of ε) satisfying the inequality $\delta_{\|\cdot\|}(\varepsilon) \geq \kappa\varepsilon^q$. This justifies the converse implication. So the equivalence holds true for $\delta_{\|\cdot\|}$ and $\varrho_{\|\cdot\|_*}$. The equivalence for $\delta_{\|\cdot\|_*}$ and $\varrho_{\|\cdot\|}$ is similar. □

18.1.4. Characterizations of uniformly convex/smooth norms via duality mappings. In addition to the material developed so far, another tool which will be utilized in the next sections is the one of duality mapping. As already seen in (11.21), with the norm $\|\cdot\|$ on a normed space X and with a real $p > 1$ one associates the multimapping $J_{p,\|\cdot\|} : X \rightrightarrows X^*$ defined for every $x \in X$ by

(18.19)
$$J_{p,\|\cdot\|}(x) := \partial\left(\frac{1}{p}\|\cdot\|^p\right)(x) = \{x^* \in X^* : \langle x^*, x\rangle = \|x\|\,\|x^*\|_*,\ \|x^*\|_* = \|x\|^{p-1}\},$$

see (2.24) for the expression of $\partial(\frac{1}{p}\|\cdot\|^p)(x)$. It is immediate (as seen in (11.22)) that

$$J_{p,\|\cdot\|}(-x) = J_{p,\|\cdot\|}(x) \quad \text{and} \quad J_{p,\|\cdot\|}(\lambda x) = \lambda^{p-1}J_{p,\|\cdot\|}(x) \quad \forall\,\lambda \ge 0. \tag{18.20}$$

In the particular case $p = 2$, denoting $J_{\|\cdot\|} := J_{2,\|\cdot\|}$ we obtain for all $x \in X$

$$\langle x^*, x\rangle = \|x^*\|_*\,\|x\| \quad \text{and} \quad \|x^*\|_* = \|x\| \quad \text{for all } x^* \in J_{\|\cdot\|}(x).$$

Then, one calls $J_{\|\cdot\|} := J_{2,\|\cdot\|}$ the *normalized duality multimapping.* It is worth noticing that

$$J_{p,\|\cdot\|}(x) = J_{\|\cdot\|}(x) = \partial\|\cdot\|(x) \quad \text{for any } x \in \mathbb{S}_X \text{ and } p > 1. \tag{18.21}$$

From this and from the second equality in (18.20) we see that

$$J_{p,\|\cdot\|}(x) = \|x\|^{p-2}J_{\|\cdot\|}(x) \quad \text{for all nonzero } x \in X. \tag{18.22}$$

Denote by $q > 1$ the conjugate exponent of p given by $1/p + 1/q = 1$, so (as we know) $\frac{1}{q}\|\cdot\|_*^q$ is the Legendre-Fenchel conjugate of $\frac{1}{p}\|\cdot\|^p$ (see (3.41). It ensues (for $x \in X$ and $x^* \in X^*$) that $x^* \in \partial(\frac{1}{p}\|\cdot\|)(x)$ if and only if $x \in \partial(\frac{1}{q}\|\cdot\|_*)(x^*)$. If X is reflexive, for any $x^* \in X^*$ we can choose some $x \in X = (X^*)^*$ such that $x \in \partial(\frac{1}{q}\|\cdot\|_*)(x^*)$, hence $x^* \in \partial(\frac{1}{p}\|\cdot\|)(x) = J_{p,\|\cdot\|}(x)$. Consequently, if $(X, \|\cdot\|)$ is a reflexive Banach space, both multimappings

$$J_{p,\|\cdot\|} : X \rightrightarrows X^* \text{ and } J_{q,\|\cdot\|_*} : X^* \rightrightarrows X \text{ are surjective.} \tag{18.23}$$

When the norm $\|\cdot\|$ is strictly convex, it is easy to check through Proposition 18.4, (18.21) and (18.22) (or directly through (18.19)) that $J_{p,\|\cdot\|}(x) \cap J_{p,\|\cdot\|}(y) = \emptyset$ for $x \ne y$ in X. If the norm $\|\cdot\|$ is Gâteaux differentiable away from zero, or equivalently $\|\cdot\|^p$ is Gâteaux differentiable on X, we see for each $x \in X$ that $J_{p,\|\cdot\|}(x)$ is a singleton, and hence we can identify $J_{p,\|\cdot\|}(x)$ with its unique element $D_G(\frac{1}{p}\|\cdot\|^p)(x)$. In such a case $J_{p,\|\cdot\|}$ is seen as a mapping, so one has the basic equalities

(18.24)
$$\langle J_{p,\|\cdot\|}(x), x\rangle = \|J_{p,\|\cdot\|}(x)\|_*\,\|x\| \quad \text{and} \quad \|J_{p,\|\cdot\|}(x)\|_* = \|x\|^{p-1} \quad \text{for all } x \in X.$$

In the particular case $p = 2$ and under the Gâteaux differentiability of $\|\cdot\|$, these equalities become

$$\langle J_{\|\cdot\|}(x), x\rangle = \|J_{\|\cdot\|}(x)\|_*\,\|x\|. \quad \text{and} \quad \|J_{\|\cdot\|}(x)\|_* = \|x\| \quad \text{for all } x \in X, \tag{18.25}$$

and one says that $J_{\|\cdot\|}$ is the *normalized duality mapping.*

Let $q > 1$ be as above the conjugate exponent of $p > 1$. When $(X, \|\cdot\|)$ is a reflexive Banach space and both norms $\|\cdot\|$ and $\|\cdot\|_*$ are Gâteaux differentiable off the origin, then both $J_{p,\|\cdot\|} : X \to X^*$ and $J_{q,\|\cdot\|_*} : X^* \to X$ are single-valued mappings; further, for $x \in X$ and $x^* \in X^*$ we have that $x^* = J_{p,\|\cdot\|}(x)$ if and only if $x = J_{q,\|\cdot\|_*}(x^*)$, so $J_{p,\|\cdot\|}$ and $J_{q,\|\cdot\|_*}$ are bijective mappings and anyone is the inverse of the other. Then, using Theorem 18.33 and Theorem 18.30 we see that, for any Banach space $(X, \|\cdot\|)$ which is uniformly convex and uniformly smooth, the duality mapping $J_{p,\|\cdot\|} : X \to X^*$ is bijective and its inverse is the duality mapping $J_{q,\|\cdot\|_*} : X^* \to X$. In particular, under the uniform convexity and uniform smoothness of a Banach space $(X, \|\cdot\|)$, the normalized duality mapping $J_{\|\cdot\|} : X \to X^*$ is bijective and its inverse is the duality mapping $J_{\|\cdot\|_*} : X^* \to X$.

When there is no ambiguity, it will often be convenient to denote $J_{p,X}$ or J_p instead of $J_{p,\|\cdot\|}$, and similarly J_X or J in place of $J_{\|\cdot\|}$. In certain cases, we will also write J_q^* instead of $J_{q,\|\cdot\|_*}$ and J^* instead of $J_{\|\cdot\|_*}$.

The following proposition provides a first characterization of uniform convexity of a norm. The characterization is obtained by means of the duality multimapping. It will be useful for the next subsection.

PROPOSITION 18.39 (S. Reich). Let $(X, \|\cdot\|)$ be a normed space and let $j : \mathbb{S}_X \to X^*$ be a selection of the restriction $J_{\|\cdot\|}|_{\mathbb{S}_X}$ of $J_{\|\cdot\|}$ to $\mathbb{S}_X$. Let $\gamma, \gamma_j : [0, 2] \to \mathbb{R}_+$ be defined for every $\varepsilon \in [0, 2]$ by

$$\gamma(\varepsilon) := \inf\{1 - \langle x^*, y\rangle : \|x\| = \|y\| = 1, \|x - y\| \geq \varepsilon, x^* \in J_{\|\cdot\|}(x)\},$$

$$\gamma_j(\varepsilon) := \inf\{1 - \langle j(x), y\rangle : \|x\| = \|y\| = 1, \|x - y\| \geq \varepsilon\}.$$

Then one has

$$2\delta_{\|\cdot\|}(\varepsilon) \leq \gamma(\varepsilon) \quad \text{and} \quad \min\{\varepsilon/4, \gamma_j(\varepsilon/4)\} \leq \delta_{\|\cdot\|}(\varepsilon).$$

Consequently, the following assertions are equivalent:
(a) the norm $\|\cdot\|$ is uniformly convex;
(b) for every $\varepsilon \in]0, 2]$ one has $\gamma(\varepsilon) > 0$;
(c) there exists a selection $j : \mathbb{S}_X \to X^*$ of $J_{\|\cdot\|}|_{\mathbb{S}_X}$ such that for every $\varepsilon \in]0, 2]$ one has $\gamma_j(\varepsilon) > 0$.

PROOF. Fix any $\varepsilon \in [0, 2]$. Take any $x, y \in \mathbb{S}_X$ with $\|x - y\| \geq \varepsilon$ and any $x^* \in J_{\|\cdot\|}(x)$. By definition of $\delta_{\|\cdot\|}$ we have $\left\|\frac{1}{2}(x + y)\right\| \leq 1 - \delta_{\|\cdot\|}(\varepsilon)$, hence

$$\frac{1}{2} + \frac{1}{2}\langle x^*, y\rangle = \langle x^*, \frac{1}{2}(x + y)\rangle \leq \left\|\frac{1}{2}(x + y)\right\| \leq 1 - \delta_{\|\cdot\|}(\varepsilon),$$

which in turn gives $1 - \langle x^*, y\rangle \geq 2\delta_{\|\cdot\|}(\varepsilon)$. It results that

$$\gamma(\varepsilon) = \inf\{1 - \langle x^*, y\rangle : \|x\| = \|y\| = 1, \|x - y\| \geq \varepsilon, x^* \in J_{\|\cdot\|}(x)\} \geq 2\delta_{\|\cdot\|}(\varepsilon).$$

Now let $j : \mathbb{S}_X \to X^*$ be a selection of $J_{\|\cdot\|}|_{\mathbb{S}_X}$. Take any $x, y \in \mathbb{S}_X$ such that $\|x - y\| \geq \varepsilon$. First, suppose that $x \neq -y$ and put $z := \frac{x+y}{\|x+y\|}$.

Case I: $\|x + y\| < 2 - \varepsilon/2$. Trivially, one has $1 - \left\|\frac{x+y}{2}\right\| > \varepsilon/4$.

Case II: $\|x + y\| \geq 2 - \varepsilon/2$. In this case

$$\|x - z\| = \frac{\|x - y - (2 - \|x + y\|)x\|}{\|x + y\|} \geq \frac{1}{2}(\|x - y\| - (2 - \|x + y\|)) \geq \frac{1}{2}\left(\varepsilon - \frac{\varepsilon}{2}\right) = \frac{\varepsilon}{4},$$

and similarly $\|y - z\| \geq \varepsilon/4$. By definition of $\gamma_j(\cdot)$ we deduce that

$$1 - \left\|\frac{x+y}{2}\right\| = 1 - \frac{1}{2}\langle j(z), x+y\rangle = \frac{1}{2}(1 - \langle j(z), x\rangle) + \frac{1}{2}(1 - \langle j(z), y\rangle)$$
$$\geq \frac{1}{2}\gamma_j(\varepsilon/4) + \frac{1}{2}\gamma_j(\varepsilon/4) = \gamma_j(\varepsilon/4).$$

On the other hand, if $x = -y$, one has $1 - \left\|\frac{x+y}{2}\right\| = 1$. Consequently, we obtain

$$\delta_{\|\cdot\|}(\varepsilon) \geq \min\{\varepsilon/4, \gamma_j(\varepsilon/4)\},$$

which finishes the proof of the inequalities of the lemma.

Recalling that $\|\cdot\|$ is uniformly convex if and only if $\delta_{\|\cdot\|}(\varepsilon) > 0$ for every $\varepsilon \in]0, 2]$, the equivalence between the assertions (a), (b) and (c) follow from the preceding inequalities. □

In addition to the above characterization of uniform convexity via the duality multimapping, the uniform smoothness of a norm can also be characterized via its duality mapping as shown in the next proposition. We need first a lemma.

Lemma 18.40. Let U be an open convex set of a normed space E and $f : U \to \mathbb{R}$ be a convex function which is uniformly Fréchet differentiable on a set Q for which $Q + \eta\mathbb{B} \subset U$ for some $\eta > 0$. Then the derivative Df is uniformly continuous on Q.

Proof. Fix any real $\varepsilon > 0$. By uniform Fréchet differentiability of f on Q choose a positive real $t < \eta/2$ such that for all $x \in Q$ and $y \in x + 2t\mathbb{B}$

$$|f(y) - f(x) - Df(x)(y - x)| \leq \varepsilon\|y - x\|.$$

Then for any $x \in Q$, any $y \in x + t\mathbb{B}$ and any $h \in \mathbb{B}$

$$Df(y)(h) \geq t^{-1}\big(f(y) - f(y - th)\big) = t^{-1}\big(f(y) - f(x)\big) + t^{-1}\big(f(x) - f(y - th)\big)$$
$$\geq t^{-1}Df(x)(y - x) + t^{-1}\big(-Df(x)(y - x - th) - \varepsilon\|y - x - th\|\big)$$
$$\geq t^{-1}Df(x)(y - x) + t^{-1}\big(-Df(x)(y - x - th) - \varepsilon\|y - x\| - \varepsilon\|th\|\big),$$

which gives by the inequality $\|y - x\| \leq t$ that $Df(y)(h) \geq Df(x)(h) - 2\varepsilon$. It ensues that, for any $x \in Q$ and $y \in E$ with $\|y - x\| \leq t$ we have $\|Df(y) - Df(x)\| \leq 2\varepsilon$, and this entails in particular that Df is uniformly continuous on Q. □

Proposition 18.41. Let $(X, \|\cdot\|)$ be a normed space and let $p \in]1, +\infty[$. The following assertions are equivalent:
(a) the norm $\|\cdot\|$ is uniformly smooth;
(b) the normalized duality mapping $J_{\|\cdot\|}$ is single-valued and uniformly continuous on the unit sphere $\mathbb{S}_X$;
(c) the duality mapping $J_{p,\|\cdot\|}$ is single-valued on X and uniformly continuous on any bounded subset of X.

Proof. We show that (a)⇔(b) and (b)⇔(c).
(a)⇒(b). Assume that the norm $\|\cdot\|$ is uniformly smooth and fix any real ε in $]0, 2[$. By Theorem 18.30 the dual norm $\|\cdot\|_*$ is uniformly convex, and hence $\delta_{\|\cdot\|_*}(\varepsilon) > 0$. Noting that $J := J_{\|\cdot\|}$ is single-valued on X according to the (Fréchet) differentiability of $\|\cdot\|^2$ on X (due to that of $\|\cdot\|$ on $\mathbb{S}_X$) and taking any $x, y \in \mathbb{S}_X$ with $\|x - y\| < \delta_{\|\cdot\|_*}(\varepsilon)$, we can write

$$\|J(x) + J(y)\|_* \geq \langle J(x) + J(y), y\rangle = \langle J(x), x\rangle + \langle J(y), y\rangle - \langle J(x), x - y\rangle$$
$$= 2 - \langle J(x), x - y\rangle > 2 - \delta_{\|\cdot\|_*}(\varepsilon),$$

which yields by definition of $\delta_{\|\cdot\|_*}(\varepsilon)$ that $\|J(x) - J(y)\|_* < \varepsilon$. This justifies the implication (a)$\Rightarrow$(b).

(b)$\Rightarrow$(a). Assume that J is single-valued and uniformly continuous on the sphere $\mathbb{S}_X$. Fix any real $\varepsilon > 0$ and choose a real $\eta > 0$ such that

$$\|J(x) - J(y)\| \le \varepsilon \quad \text{for all } x, y \in \mathbb{S}_X \text{ with } \|x - y\| \le \eta. \tag{18.26}$$

Take any $x, h \in \mathbb{S}_X$. Note that for every $t \in]0, 1/2]$ we have $\|x+th\| \ge \|x\| - t\|h\| \ge 1/2$, hence

$$\begin{aligned}\left\|\frac{x+th}{\|x+th\|} - x\right\| &= \frac{\|\, x+th - \|x+th\|x \,\|}{\|x+th\|} \\ &\le 2(\|\, x - \|x+th\|x \,\| + t) = 2(|\, 1 - \|x+th\| \,| + t) \\ &= 2(|\, \|x\| - \|x+th\| \,| + t) \le 2(t+t) = 4t.\end{aligned} \tag{18.27}$$

Take any $t \in]0, \eta/4]$. By (18.25) we can write

$$\begin{aligned}\langle J(x), th\rangle &= \langle J(x), x+th\rangle - 1 \le \|x+th\| - 1 \\ &= \|x+th\| - \|x\| \\ &= \|x+th\| \left\langle J\left(\frac{x+th}{\|x+th\|}\right), \frac{x+th}{\|x+th\|}\right\rangle - \|x\| \\ &= \left\langle J\left(\frac{x+th}{\|x+th\|}\right), x+th\right\rangle - \|x\|,\end{aligned}$$

so using the inequality $\|x\| \ge \left\langle J\left(\frac{x+th}{\|x+th\|}\right), x\right\rangle$ in the latter side we obtain

$$\langle J(x), th\rangle \le \|x+th\| - \|x\| \le \left\langle J\left(\frac{x+th}{\|x+th\|}\right), x+th\right\rangle - \left\langle J\left(\frac{x+th}{\|x+th\|}\right), x\right\rangle,$$

or equivalently

$$\langle J(x), th\rangle \le \|x+th\| - \|x\| \le \left\langle J\left(\frac{x+th}{\|x+th\|}\right), th\right\rangle.$$

It follows that

$$0 \le t^{-1}\big(\|x+th\| - \|x\| - t\langle J(x), h\rangle\big) \le \langle J\left(\frac{x+th}{\|x+th\|}\right) - J(x), h\rangle,$$

which in turn entails that

$$0 \le t^{-1}\big(\|x+th\| - \|x\| - t\langle J(x), h\rangle\big) \le \left\|J\left(\frac{x+th}{\|x+th\|}\right) - J(x)\right\| \le \varepsilon,$$

where the latter inequality is due to (18.26) since $\left\|\frac{x+th}{\|x+th\|} - x\right\| \le \eta$ according to (18.27). This being true for all $t \in]0, \eta/4]$ and all $x, h \in \mathbb{S}_X$, the norm $\|\cdot\|$ is uniformly Fréchet differentiable on the unit sphere, otherwise stated $\|\cdot\|$ is uniformly smooth. This means that (b) implies (a).

(b)$\Leftrightarrow$(c). The implication (c)$\Rightarrow$(b) is obvious. To prove the converse implication, assume that J is single-valued and uniformly continuous on $\mathbb{S}_X$, and fix any real $r > 0$. By the above equivalence (b)$\Leftrightarrow$(a) we know that the norm $\|\cdot\|$ is uniformly Fréchet differentiable on the unit sphere $\mathbb{S}_X$. Let us show that $\|\cdot\|^p$ is uniformly Fréchet differentiable on $r\mathbb{B}_X$.The single-valuedness of J on $\mathbb{S}_X$ entails in particular that $\|\cdot\|$ is Gâteaux differentiable at each point in $\mathbb{S}_X$, hence $\|\cdot\|^p$ is Gâteaux differentiable at any point of X and $J_p := J_{p,\|\cdot\|}$ is single-valued on the whole space X. Fix any real $\varepsilon > 0$. For any $x \in r\mathbb{B}_X$, $h \in \mathbb{S}_X$ and $t > 0$, putting $q(t,x;h) := t^{-1}(\|x+th\|^p - \|x\|^p)$ we have $0 \le q(t,x;h) - p\langle J_p(x), h\rangle$ (since $\|\cdot\|^p$ is convex) and

$$\begin{aligned} & q(t,x;h) - p\langle J_p(x), h\rangle \\ & = t^{-1}\big(\|x+th\| - \|x\|\big) \int_0^1 p\big(\|x\| + \theta(\|x+th\| - \|x\|)\big)^{p-1} d\theta - p\langle J_p(x), h\rangle, \\ & \le p \int_0^1 (\|x\| + \theta t)^{p-1} d\theta + p\|x\|^{p-1}, \end{aligned}$$

so we may choose some positive real $\rho < \min\{r, 1\}$ such that

$$\sup_{x \in \rho\mathbb{B}_X, h \in \mathbb{S}_X} |q(t,x;h) - p\langle J_p(x), h\rangle| \le \varepsilon \quad \text{for all } t \in \,]0, \rho[. \tag{18.28}$$

Now, set $Q_{\rho,r} := \{x \in X : \rho \le \|x\| \le r\}$. Take any $0 < t < 1$ and $x \in Q_{\rho,r}$, and put $u := \|x\|^{-1}x$ and $s := s(t,x) := \|x\|^{-1}t$ (depending on both t and $\|x\|$). Consider any $h \in \mathbb{S}_X$. We note that, for

$$R(t,x;h) := t^{-1}\big(\|x+th\|^p - \|x\|^p - pt\langle J_p(x), h\rangle\big),$$

we have

$$\begin{aligned} & R(t,x;h) \\ & = \|x\|^{p-1}\{s^{-1}\big(\|u+sh\|^p - \|u\|^p - ps\langle J_p(u), h\rangle\big)\} \\ & = \|x\|^{p-1}\left\{s^{-1}\big(\|u+sh\| - \|u\| - s\langle J_p(u), h\rangle\big) \int_0^1 p\big(\|u\| + \theta(\|u+sh\| - \|u\|)\big)^{p-1} d\theta\right\} \\ & \quad + \|x\|^{p-1}\left\{\left(\int_0^1 p\big(\|u\| + \theta(\|u+sh\| - \|u\|)\big)^{p-1} d\theta - p\right)\langle J_p(u), h\rangle\right\} \\ & := R_1(t,x;h) + R_2(t,x:h). \end{aligned}$$

Writing

$$\begin{aligned} \int_0^1 \big(\|u\| + \theta(\|u+sh\| - \|u\|)\big)^{p-1} d\theta & \le \int_0^1 (1 + \theta s\|h\|)^{p-1} d\theta \\ & \le \int_0^1 (1 + \rho^{-1}\theta)^{p-1} d\theta \le (1+\rho^{-1})^{p-1}, \end{aligned}$$

we see that

$$|R_1(t,x;h)| \le pr^{p-1}(1+\rho^{-1})^{p-1}\big|s^{-1}\big(\|u+sh\| - \|u\| - s\langle J_p(u), h\rangle\big)\big|,$$

which ensures that $\sup_{x \in Q_{\rho,r}, h \in \mathbb{S}_X} |R_1(t,x;h)| \to 0$ as $t \downarrow 0$, since (by the equality $J_p(y) = J(y)$ for all $y \in \mathbb{S}_X$ in (18.21))

$$\sup_{v \in \mathbb{S}, h \in \mathbb{S}} \tau^{-1}\big(\|v+\tau h\| - \|v\| - \tau\langle J_p(v), h\rangle\big) = \sup_{v \in \mathbb{S}, h \in \mathbb{S}} \tau^{-1}\big(\|v+\tau h\| - \|v\| - \tau\langle J(v), h\rangle\big)$$

tends to 0 as $\tau \downarrow 0$ (by the uniform Fréchet differentiability of $\|\cdot\|$ on $\mathbb{S}_X$) and since $\sup_{x \in Q_{\rho,r}} s(t,x) \to 0$ as $t \downarrow 0$. Then, there exists a real $\eta_1 \in \,]0,1[$ such that

$$\sup_{x \in Q_{\rho,r}, h \in \mathbb{S}_X} |R_1(t,x;h)| \le \varepsilon/2 \quad \text{for all } t \in \,]0, \eta_1[. \tag{18.29}$$

Regarding $R_2(x,t;h)$, we have

$$|R_2(t,x;h)| \le pr^{p-1} \int_0^1 \big|(1+\theta(\|u+sh\|-\|u\|))^{p-1}-1\big|\,d\theta. \tag{18.30}$$

Choose a real $\delta \in]0,1[$ such that

$$|(1+\zeta)^{p-1}-1| \le \varepsilon/(2pr^{p-1}) \quad \text{for all } \zeta \in [-\delta,\delta].$$

Put $\eta_2 := \rho\delta$ and observe that for any $t \in]0,\eta_2[$ we have for every $\theta \in [0,1]$

$$|\theta(\|u+sh\|-\|u\|)| \le s \le t/\rho < \delta,$$

hence by what precedes $\big|(1+\theta(\|u+sh\|-\|u\|))^{p-1}-1\big| \le \varepsilon/(2pr^{p-1})$, which in turn implies by (18.30) that

$$\sup_{x\in Q_{\rho,r},h\in\mathbb{S}_X} |R_2(t,x;h)| \le \varepsilon/2 \quad \text{for all } t \in]0,\eta_2[.$$

For $\eta := \min\{\eta_1,\eta_2\}$, the latter inequality combined with (18.29) implies that we have $\sup\limits_{x\in Q_{\rho,r},h\in\mathbb{S}_X} R(t,x;h) \le \varepsilon$. From this and (18.28) (noting that $\eta < \rho$ since $\delta < 1$) it results that for all $t \in]0,\eta[$

$$\sup_{x\in r\mathbb{B}_H,h\in\mathbb{S}_X} t^{-1}\big|\|x+th\|^p-\|x\|^p-t\langle J_p(x),h\rangle\big| \le \varepsilon \quad \text{for all } t \in]0,\eta[,$$

which says that the function $(1/p)\|\cdot\|^p$ is uniformly Fréchet differentiable on $r\mathbb{B}_X$. By Lemma 18.40 above we derive that J_p is uniformly continuous on $r\mathbb{B}_X$. Consequently, the implication (b)⇒(c) holds true, and the proof is finished. □

In addition to the above equivalences in Proposition 18.41, we show that a suitable multiple of the modulus of smoothness majorizes the modulus of continuity on the sphere of the derivative of a differentiable norm. This has been utilized in Theorem 10.54 in the form stated in (10.16).

PROPOSITION 18.42. *Let X be a (nonzero) vector space and $\|\cdot\|$ be a norm on X Gâteaux differentiable off zero, and let $\Lambda(x) \in X^*$ be the Gâteaux derivative of $\|\cdot\|$ at each nonzero $x \in X$. The following hold:*
(a) For any nonzero $x,y \in X$ with $x \ne y$, one has

$$\|\Lambda(x)-\Lambda(y)\| \le \frac{2\|x\|}{\|x-y\|}\varrho_{\|\cdot\|}\left(\frac{2\|x-y\|}{\|x\|}\right).$$

(b) Denoting by $\omega^{c}_{\Lambda,\mathbb{S}_X}$ the least modulus of continuity of Λ relative to the sphere $\mathbb{S}_X$, that is,

$$\omega^{c}_{\Lambda,\mathbb{S}_X}(\tau) := \sup\{\|\Lambda(x)-\Lambda(y)\| : x,y \in \mathbb{S}_X, \|x-y\| \le \tau\} \quad \text{for all } \tau \ge 0,$$

one has

$$\omega^{c}_{\Lambda,\mathbb{S}_X}(\tau) \le \frac{2}{\tau}\varrho_{\|\cdot\|}(2\tau) \quad \text{for all } \tau > 0.$$

PROOF. (a) Fix any $u \in \mathbb{S}_X$ and any nonzero $y \in X$ with $y \ne u$, and fix also any vector $h \in X$ with $\|h\| = \|u-y\|$. By the convexity of the norm we have

$$\begin{aligned}\langle\Lambda(y),h\rangle-\langle\Lambda(u),h\rangle &\le \|y+h\|-\|y\|-\langle\Lambda(u),h\rangle\\ &= \|y+h\|-\|u\|+\langle\Lambda(u),u-y-h\rangle+\|u\|-\|y\|+\langle\Lambda(u),y-u\rangle\\ &\le \|y+h\|-\|u\|+\langle\Lambda(u),u-y-h\rangle,\end{aligned}$$

which gives

$$\begin{aligned}\langle \Lambda(y) - \Lambda(u), h\rangle &\le \|y+h\| - \|u\| + \|u + (u-y-h)\| - \|u\| \\ &= \|u+tv\| + \|u-tv\| - 2,\end{aligned}$$

where $v := (y+h-u)/\|y+h-u\|$ and $t := \|y+h-u\|$. It ensues by definition of the modulus of smoothness that

$$\langle \Lambda(y) - \Lambda(u), h\rangle \le 2\varrho_{\|\cdot\|}(t) = 2\varrho(\|y-u+h\|).$$

Since $\varrho_{\|\cdot\|}(\|y-u+h\|) \le \varrho_{\|\cdot\|}(\|y-u\| + \|h\|) = \varrho_{\|\cdot\|}(2\|y-u\|)$, we deduce that $\langle \Lambda(y) - \Lambda(u), h\rangle \le 2\varrho_{\|\cdot\|}(2\|y-u\|)$. The latter inequality being true for all $h \in X$ with $\|h\| = \|y-u\|$, we derive that

$$\|\Lambda(y) - \Lambda(u)\|\, \|y-u\| \le 2\varrho_{\|\cdot\|}(2\|y-u\|).$$

Consider now any nonzero $x, y \in X$ with $x \neq y$. The latter inequality (with $x/\|x\|$ and $y/\|x\|$ in place of u and y respectively) ensures that

$$\left\|\Lambda\left(\frac{y}{\|x\|}\right) - \Lambda\left(\frac{x}{\|x\|}\right)\right\| \frac{\|y-x\|}{\|x\|} \le 2\varrho_{\|\cdot\|}\left(\frac{2\|y-x\|}{\|x\|}\right).$$

Since $\Lambda(tw) = \Lambda(w)$ for every $t > 0$ and every nonzero $w \in X$, we deduce that

$$\|\Lambda(y) - \Lambda(x)\| \le \frac{2\|x\|}{\|y-x\|}\varrho_{\|\cdot\|}\left(\frac{2\|y-x\|}{\|x\|}\right),$$

hence the inequality in (a) is justified.
(b) Take any real $\tau > 0$ and any $x, y \in \mathbb{S}_X$ with $0 < \|y-x\| \le \tau$. By (a) and the nondecreasing property of $t \mapsto \varrho_{\|\cdot\|}(t)/t$ on $]0, +\infty[$ (see the property (s3) in Proposition 18.25), it follows that

$$\|\Lambda(x) - \Lambda(y)\| \le 2\frac{\varrho_{\|\cdot\|}(2\|y-x\|)}{\|y-x\|} \le 2\frac{\varrho_{\|\cdot\|}(2\tau)}{\tau},$$

and hence $\omega^c_{\Lambda,\mathbb{S}_X}(\tau) \le \frac{2}{\tau}\varrho_{\|\cdot\|}(2\tau)$. □

Proposition 18.41(c) characterizes the uniform smoothness of a Gâteaux differentiable norm as uniform continuity of the duality mapping on balls. As a corollary of Proposition 18.42 we obtain another characterization via the uniform continuity of the derivative of the norm outside balls centered at zero.

COROLLARY 18.43. A norm $\|\cdot\|$ on a (nonzero) vector space X is uniformly smooth if and only if it is Gâteaux differentiable (off zero) and the mapping $x \mapsto D\|\cdot\|(x)$ is uniformly continuous outside every ball centered at the origin.

PROOF. The implication $\Leftarrow$ being a direct consequence of Proposition 18.41(b) and of the equality $J(x) = D\|\cdot\|(x)$ for all $x \in \mathbb{S}_X$ (see (18.21)), let us prove the reverse one. Suppose that $\|\cdot\|$ is uniformly smooth and denote $\Lambda(x) := D\|\cdot\|(x)$ for every $x \neq 0$. Fix any real $\alpha > 0$. Take any $\varepsilon > 0$ and (keeping in mind that $\varrho_{\|\cdot\|}(\tau)/\tau \to 0$ as $\tau \downarrow 0$) choose $\eta > 0$ such that $\varrho_{\|\cdot\|}(\tau)/\tau \le \varepsilon/4$ whenever $0 < \tau \le \eta$. For all $x, y \in X \setminus B(0, \alpha)$ with $0 < \|x-y\| \le \alpha\eta/2$, by Proposition 18.42(a) and by the nondecreasing property of $\varrho_{\|\cdot\|}(\tau)/\tau$ (see Proposition 18.25(s3)) we have

$$\|\Lambda(x) - \Lambda(y)\| \le \frac{2\|x\|}{\|x-y\|}\varrho_{\|\cdot\|}\left(\frac{2\|x-y\|}{\|x\|}\right) \le \frac{4}{\eta}\varrho_{\|\cdot\|}(\eta) \le \varepsilon,$$

which justifies the implication $\Rightarrow$. □

18.1.5. Xu-Roach theorems on moduli of convexity and smoothness. For convenience let us set as usual for all $r, s \in \mathbb{R}$

$$r \wedge s := \min\{r, s\} \quad \text{and} \quad r \vee s := \max\{r, s\}.$$

Given a normed space $(X, \|\cdot\|)$, for $u, v \in X$ with $\|u\| = 1$ and $v \neq 0$, we see on the one hand by definition of the modulus of convexity $\delta_{\|\cdot\|}$ that

$$\left\| u + \frac{v}{\|v\|} \right\| \leq 2 - 2\delta_{\|\cdot\|}\left(\left\| u - \frac{v}{\|v\|} \right\| \right),$$

and on the other hand that with $\|v\| \geq 1$

$$\|u + v\| \leq \left\| u + \frac{v}{\|v\|} \right\| + \left\| \left(1 - \frac{1}{\|v\|} \right) v \right\| = \left\| u + \frac{v}{\|v\|} \right\| + \|v\| - 1,$$

so combining those inequalities and using the equality $\|u\| = 1$ give for $\|v\| \geq 1$

$$\|u + v\| \leq \|u\| + \|v\| - 2\delta_{\|\cdot\|}\left(\left\| \frac{u}{\|u\|} - \frac{v}{\|v\|} \right\| \right).$$

Then for any nonzero $x, y \in X$, putting $\alpha := \|x\| \wedge \|y\|$ we deduce that

$$\|\alpha^{-1}x + \alpha^{-1}y\| \leq \|\alpha^{-1}x\| + \|\alpha^{-1}y\| - 2\delta_{\|\cdot\|}\left(\left\| \frac{x}{\|x\|} - \frac{y}{\|y\|} \right\| \right),$$

which is equivalent to

$$\|x + y\| \leq \|x\| + \|y\| - 2(\|x\| \wedge \|y\|)\delta_{\|\cdot\|}\left(\left\| \frac{x}{\|x\|} - \frac{y}{\|y\|} \right\| \right).$$

In this subsection, with the above type of method and with various more technical arguments, much more involved inequalities (Xu-Roach inequalities) will be established for the modulus of convexity and the modulus of smoothness of norms. The inequalities will yield to the celebrated Xu-Roach theorems for such moduli.

Given a real $p > 1$, let us denote by $\mathfrak{S}_{\|\cdot\|}([0, 2], \mathbb{R}_+)$ the set of increasing functions $\phi : [0, 2] \to \mathbb{R}_+$ such that $\phi(0) = 0$ and there exists a real $K > 0$ satisfying

$$\phi(t) \geq K\delta_{\|\cdot\|}(t/2) \quad \text{for all } t \in [0, 2]. \tag{18.31}$$

THEOREM 18.44 (Zong-Ben Xu and G.F. Roach theorem for uniform convexity). *Let $(X, \|\cdot\|)$ be a Banach space which is at least two-dimensional and let a real $p > 1$. The following assertions are equivalent:*
(a) the norm $\|\cdot\|$ is uniformly convex;
(b) there exist a function $\phi_p \in \mathfrak{S}_{\|\cdot\|}([0, 2], \mathbb{R}_+)$ such that, for all $x, y \in X$ not both zero, all $x^* \in J_{p,\|\cdot\|}(x)$ and all $y^* \in J_{p,\|\cdot\|}(y)$,

$$\langle x^* - y^*, x - y \rangle \geq (\|x\| \vee \|y\|)^p \phi_p\left(\frac{\|x - y\|}{\|x\| \vee \|y\|} \right);$$

(b') there exist a function $\phi_p \in \mathfrak{S}_{\|\cdot\|}([0, 2], \mathbb{R}_+)$ and a selection j_p of $J_{p,\|\cdot\|}$ such that, for all $x, y \in X$ not both zero,

$$\langle j_p(x) - j_p(y), x - y \rangle \geq (\|x\| \vee \|y\|)^p \phi_p\left(\frac{\|x - y\|}{\|x\| \vee \|y\|} \right);$$

(c) there exist a function $\phi_p \in \mathfrak{S}_{\|\cdot\|}([0, 2], \mathbb{R}_+)$ such that, for all $x, y \in X$ not both zero and all $x^* \in J_{p,\|\cdot\|}(x)$

$$\|x + y\|^p \geq \|x\|^p + p\langle x^*, y \rangle + \sigma_p(x, y),$$

where

$$\sigma_p(x,y) := p\int_0^1 \frac{(\|x+ty\| \vee \|x\|)^p}{t}\phi_p\left(\frac{t\|y\|}{\|x+ty\| \vee \|x\|}\right)dt; \tag{18.32}$$

(c') there exist a function $\phi_p \in \mathfrak{S}_{\|\cdot\|}([0,2],\mathbb{R}_+)$ and a selection j_p of $J_{p,\|\cdot\|}$ such that, for all $x,y \in X$ not both zero,

$$\|x+y\|^p \geq \|x\|^p + p\langle j_p(x), y\rangle + \sigma_p(x,y),$$

where as defined in (c)

$$\sigma_p(x,y) := p\int_0^1 \frac{(\|x+ty\| \vee \|x\|)^p}{t}\phi_p\left(\frac{t\|y\|}{\|x+ty\| \vee \|x\|}\right)dt.$$

The proof of the theorem goes through several lemmas.

LEMMA 18.45. Let $(X,\|\cdot\|)$ be a normed space, let $x,y \in \mathbb{S}_X$ with $\varepsilon := \|x-y\| > 0$, and let $t \in]0,1[$. Then, for any reals $r,s \geq 0$ with $r+s=1$ one has

$$\|rx+sty\| \leq r + st - 2(r\wedge s)t\delta_{\|\cdot\|}(\varepsilon).$$

PROOF. Put $\alpha := (r+st)^{-1}$ and $\beta := (r+st)^{-1}r(1-t)$, and note that $\alpha > 0$ and $0 \leq \beta \leq 1$. It is easily verified that

$$\alpha(rx+sty) = \beta x + (1-\beta)(rx+sy).$$

We deduce that

$$\begin{aligned}\|rx+sty\| &= (r+st)\|\beta x + (1-\beta)(rx+sy)\| \\ &\leq (r+st)\big(\beta\|x\| + (1-\beta)\|rx+sy\|\big) \\ &= (1-t)r + t\|rx+sy\|. \end{aligned} \tag{18.33}$$

On the other hand, recall that the function $\tau \mapsto q(\tau,h) := \tau^{-1}(\|x+\tau h\| - \|x\|)$ is nondecreasing on $\mathbb{R}\setminus\{0\}$ (by convexity of $\|\cdot\|$). Then, if $s \leq 1/2$, using the latter nondecreasing property along with the definition of modulus of convexity one has

$$\begin{aligned}\|rx+sy\| &= 1 + s\cdot s^{-1}\big(\|x+s(y-x)\| - \|x\|\big) \\ &= 1 + sq(s,y-x) \leq 1 + sq\left(\frac{1}{2}, y-x\right) \\ &= 1 - 2s\left(1 - \left\|\frac{1}{2}(x+y)\right\|\right) \leq 1 - 2s\delta_{\|\cdot\|}(\varepsilon).\end{aligned}$$

By symmetry, if $r \leq 1/2$ one also has

$$\|rx+sy\| \leq 1 - 2r\delta_{\|\cdot\|}(\varepsilon).$$

From the two latter inequalities and from (18.33) we derive that

$$\begin{aligned}\|rx+sty\| &\leq (1-t)r + t\big(1 - 2(r\wedge s)\delta_{\|\cdot\|}(\varepsilon)\big), \\ &= r + st - 2(r\wedge s)t\delta_{\|\cdot\|}(\varepsilon),\end{aligned}$$

which translates the desired inequality. $\square$

LEMMA 18.46. Let $(X,\|\cdot\|)$ be a normed space which is at least two-dimensional and let a real $p > 1$. The norm $\|\cdot\|$ is uniformly convex if and only if for each pair

$(r,s) \in [0,1]^2$ with $r+s=1$, there exist a function $\eth_p(r,s,\cdot) : [0,2] \to \mathbb{R}_+$ with $\eth_p(r,s,0) = 0$ such that $\eth_p(r,s,\cdot)$ is increasing and

$$\|rx+sy\|^p + (\|x\| \vee \|y\|)^p \eth_p\left(r,s,\frac{\|x-y\|}{\|x\| \vee \|y\|}\right) \le r\|x\|^p + s\|y\|^p$$

for all $x, y \in X$ not both zero.

Proof. Assume that the property in the lemma holds. Fix any $\varepsilon \in]0,2]$ and any $x, y \in \mathbb{S}_X$ with $\|x-y\| \ge \varepsilon$. Taking $r = s = 1/2$ in the property in the lemma, we obtain

$$\left\|\frac{x+y}{2}\right\|^p + \eth_p\left(\frac{1}{2},\frac{1}{2},\varepsilon\right) \le 1,$$

hence in particular $0 < \eth_p\left(\frac{1}{2},\frac{1}{2},\varepsilon\right) < 1$ since $\eth_p\left(\frac{1}{2},\frac{1}{2},\varepsilon\right) > \eth_p\left(\frac{1}{2},\frac{1}{2},0\right) = 0$. We deduce by definition of modulus of convexity that

$$\delta_{\|\cdot\|}(\varepsilon) \ge 1 - \left(1 - \eth_p\left(\frac{1}{2},\frac{1}{2},\varepsilon\right)\right)^{1/p} > 0,$$

so $\|\cdot\|$ is uniformly convex.

Conversely, assume that $\|\cdot\|$ is uniformly convex and fix $r, s \ge 0$ with $r+s=1$. Consider the functions $\varphi_1(r,s,\cdot)$, $\varphi_2(r,s,\cdot)$ from $[0,2]$ into $\mathbb{R}_+$ defined by

$$\varphi_1(r,s,\varepsilon) := r + s(1-(\varepsilon/2))^p - (1-s(\varepsilon/2))^p, \tag{18.34}$$

$$\varphi_2(r,s,\varepsilon) := r\{1 - sr^{p-1}/[s^{1/(p-1)} - \left(s - 2(r\wedge s)\delta_{\|\cdot\|}(\varepsilon/2)\right)^{p/(p-1)}]^{p-1}\}. \tag{18.35}$$

Computing the derivative shows that $\varphi_1(r,s,\cdot)$ is increasing on $[0,2]$. The space X being at least two-dimensional, the uniform convexity of $\|\cdot\|$ entails by (c4) in Proposition 18.25 that $\delta_{\|\cdot\|}$ is increasing on $[0,2]$, so the function $\varphi_2(r,s,\cdot)$ is also increasing on $[0,2]$. Fix any $x \in \mathbb{S}_X$ and any nonzero $y \in \mathbb{B}_X$. Put $t_0 := \|y\|$, $y_0 := y/t_0$, $\varepsilon := \|x-y\|$ and $\varepsilon_0 := \|x-y_0\|$. Define $f : [0,1] \to \mathbb{R}_+$ by

$$f(t) := r + st^p - \left(r + st - 2(r\wedge s)t\delta_{\|\cdot\|}(\varepsilon_0)\right)^p \quad \text{for all } t \in [0,1].$$

By Lemma 18.45 we have

$$\begin{aligned} f(t_0) &\le r + st_0^p - \|rx + st_0y_0\|^p \\ &= r + st_0^p - \|rx+sy\|^p = r\|x\|^p + s\|y\|^p - \|rx+sy\|^p. \end{aligned} \tag{18.36}$$

We consider two cases.

Case 1: $t_0 \le 1 - \varepsilon/2$. By the definition of f at t_0 and by the decreasing property of the function $t \mapsto r + st^p - (r+st)^p$ on $[0,1]$ we have

$$f(t_0) \ge r + st_0^p - (r+st_0)^p \ge r + s\left(1-\frac{\varepsilon}{2}\right)^p - \left(r + s\left(1-\frac{\varepsilon}{2}\right)\right)^p = \varphi_1(r,s,\varepsilon).$$

This and (18.36) give

$$r\|x\|^p + s\|y\|^p - \|rx+sy\|^p \ge \varphi_1(r,s,\varepsilon).$$

Case 2: $t_0 > 1 - \varepsilon/2$. In this case we have

$$\varepsilon_0 = \|x-y_0\| \ge \|x-y\| - \|y-y_0\| = \varepsilon - (1-t_0) > \varepsilon/2.$$

Then, by the definition of f and the nondecreasing property of $\delta_{\|\cdot\|}$ (see Proposition 18.25) we deduce that

$$f(t_0) \ge r + st_0^p - \left(r + st_0 - 2(r\wedge s)t_0\delta_{\|\cdot\|}\left(\frac{\varepsilon}{2}\right)\right)^p =: f_0(t_0).$$

It can be checked that the function $f_0 : [0,1] \to \mathbb{R}$ achieves its minimum at

$$\tau := r\left[s - 2(r\wedge s)\delta_{\|\cdot\|}\left(\frac{\varepsilon}{2}\right)\right]^{1/(p-1)}\left(s^{1/(p-1)} - \left[s - 2(r\wedge s)\delta_{\|\cdot\|}\left(\frac{\varepsilon}{2}\right)\right]^{p/(p-1)}\right)^{-1}$$

satisfying $f_0'(\tau) = 0$, with

$$f_0'(t) = p\left(st^{p-1} - \left[r + st - 2(r\wedge s)t\delta_{\|\cdot\|}\left(\frac{\varepsilon}{2}\right)\right]^{p-1}\left[s - 2(r\wedge s)\delta_{\|\cdot\|}\left(\frac{\varepsilon}{2}\right)\right]\right).$$

It ensues that

$$\left[r + s\tau - 2(r\wedge s)\tau\delta_{\|\cdot\|}(\varepsilon/2)\right]^p = \frac{s\tau^{p-1}\left[r + s\tau - 2(r\wedge s)\tau\delta_{\|\cdot\|}(\frac{\varepsilon}{2})\right]}{s - 2(r\wedge s)\delta_{\|\cdot\}}(\varepsilon/2)},$$

so it follows that

$$\begin{aligned}\inf_{t\in[0,1]} f_0(t) = f_0(\tau) &= r + s\tau^p - \frac{s\tau^{p-1}\left[r + s\tau - 2(r\wedge s)\tau\delta_{\|\cdot\|}(\frac{\varepsilon}{2})\right]}{s - 2(r\wedge s)\delta_{\|\cdot\|}(\varepsilon/2)} \\ &= r - rs\tau^{p-1}\left[s - 2(r\wedge s)\delta_{\|\cdot\|}(\varepsilon/2)\right]^{-1} = \varphi_2(r,s,\varepsilon),\end{aligned}$$

which assures us that

$$r\|x\|^p + s\|y\|^p - \|rx + sy\|^p \geq \varphi_2(r,s,\varepsilon).$$

Setting $\varphi(r,s,\varepsilon) := \min\{\varphi_1(r,s,\varepsilon), \varphi_2(r,s,\varepsilon)\}$ and

$$\mathfrak{d}_p(r,s,\varepsilon) := \min\{\varphi(r,s,\varepsilon), \varphi(s,r,\varepsilon)\}, \tag{18.37}$$

we see that $\mathfrak{d}_p(r,s,\cdot)$ is increasing on $[0,2]$ and that, if either $(x,y) \in \mathbb{S}_X \times \mathbb{B}_X$ or $(x,y) \in \mathbb{B}_X \times \mathbb{S}_X$, we have

$$\|rx + sy\|^p + \mathfrak{d}_p(r,s,\|x-y\|) \leq r\|x\|^p + s\|y\|^p.$$

Now, consider any $x, y \in X$ with $\|x\| \vee \|y\| > 0$. For

$$u := \frac{x}{\|x\| \vee \|y\|} \quad \text{and} \quad v := \frac{y}{\|x\| \vee \|y\|}$$

we have either $(u,v) \in \mathbb{S}_X \times \mathbb{B}_X$ or $(u,v) \in \mathbb{B}_X \times \mathbb{S}_X$, so the latter inequality gives

$$\|ru + sv\|^p + \mathfrak{d}_p(r,s,\|u-v\|) \leq r\|u\|^p + s\|v\|^p.$$

This inequality is equivalent to

$$\|rx + sy\|^p + (\|x\| \vee \|y\|)^p\mathfrak{d}_p\left(r, s, \frac{\|x-y\|}{\|x\| \vee \|y\|}\right) \leq r\|x\|^p + s\|y\|^p,$$

which finishes the proof of the lemma. □

LEMMA 18.47. Let $(X, \|\cdot\|)$ be a normed space which is at least two-dimensional and let $p \in]1, +\infty[$ and $r, s \geq 0$ with $r + s = 1$. Let $\mathfrak{d}_p(r,s,\cdot)$ be defined on $[0,2]$ as in (18.37) with $\varphi(r,s,\cdot)$ and $\varphi_i(r,s,\cdot)$, $i = 1, 2$, as in (18.34) and (18.35). Then there exists a constant $K_p > 0$ such that

$$\limsup_{r\downarrow 0} r^{-1}\mathfrak{d}_p(r,s,\varepsilon) + \limsup_{s\downarrow 0} s^{-1}\mathfrak{d}_p(r,s,\varepsilon) \geq K_p\delta_{\|\cdot\|}(\varepsilon/2)$$

for all $\varepsilon \in [0,2]$.

PROOF. Put $\xi(\varepsilon) := \limsup_{r\downarrow 0} r^{-1}\mathfrak{d}_p(r,s,\varepsilon)$ and note that $\xi(\varepsilon)$ depends only on ε and p since $s \to 1$ as $r \to 0$. By definitions of $\mathfrak{d}_p$, φ_1 and φ_2 we have

$$\xi(\varepsilon) \geq 2\min\left\{\min\left\{\lim_{r\downarrow 0} r^{-1}\varphi_i(r,s,\varepsilon), \lim_{s\downarrow 0} s^{-1}\varphi_i(r,s,\varepsilon)\right\} : i=1,2\right\}. \tag{18.38}$$

Keeping in mind that $r+s=1$, computations through (18.34) and (18.35) give

$$\lim_{r\downarrow 0} r^{-1}\varphi_1(r,s,\varepsilon) = 1-\left(1+\frac{p-1}{2}\varepsilon\right)\left(1-\frac{\varepsilon}{2}\right)^{p-1} =: \ell_1(\varepsilon);,$$

$$\lim_{s\downarrow 0} s^{-1}\varphi_1(r,s,\varepsilon) = \left(1-\frac{\varepsilon}{2}\right)^p - \left(1-\frac{p}{2}\varepsilon\right) =: \ell_2(\varepsilon),$$

$$\lim_{r\downarrow 0} r^{-1}\varphi_2(r,s,\varepsilon) = 1-\left(1+\frac{2p}{p-1}\delta_{\|\cdot\|}\left(\frac{\varepsilon}{2}\right)\right)^{1-p} =: \lambda_1(\delta_{\|\cdot\|}(\varepsilon/2)),$$

$$\lim_{s\downarrow 0} s^{-1}\varphi_2(r,s,\varepsilon) = (p-1)\left(1-\left[1-2\delta_{\|\cdot\|}\left(\frac{\varepsilon}{2}\right)\right]^{p/(p-1)}\right) =: \lambda_2(\delta_{\|\cdot\|}(\varepsilon/2)).$$

By (18.13) and (18.14) we know that

$$\delta_{\|\cdot\|}(\varepsilon/2) \leq \delta_H(\varepsilon/2) = 1-\sqrt{1-\frac{1}{4}\left(\frac{\varepsilon}{2}\right)^2},$$

so putting $\delta := \delta_{\|\cdot\|}(\varepsilon/2)$, we see for any $\varepsilon \in]0,2]$ that $\delta \leq (2-\sqrt{3})/2$ and

$$(\varepsilon/2)^2 \geq 4\delta(2-\delta) \geq (4+2\sqrt{3})\delta.$$

From this it follows for $i=1,2$ that

$$\ell_i(\varepsilon)/\delta_{\|\cdot\|}(\varepsilon/2) \geq (4+2\sqrt{3})\ell_i(\varepsilon)/(\varepsilon/2)^2 \quad \text{for all } \varepsilon \in]0,2],$$

so noticing from the definition of ℓ_i $(i=1,2)$ that ℓ_i is continuous and positive on $]0,2]$ and that $\lim_{\varepsilon\downarrow 0}(\varepsilon)^{-2}\ell_i(\varepsilon)$ exists in $]0,+\infty[$, it results that there exist real constants $C_1, C_2 > 0$ (depending only on p) such that for $i=1,2$

$$\ell_i(\varepsilon)/\delta_{\|\cdot\|}(\varepsilon/2) \geq C_i \quad \text{for all } \varepsilon \in]0,2].$$

On the other hand, since $\delta_{\|\cdot\|}(\varepsilon/2) \leq (2-\sqrt{3})/2$ (as seen above), we have for $i=1,2$

$$\frac{\lambda_i(\delta_{\|\cdot\|}(\varepsilon/2))}{\delta_{\|\cdot\|}(\varepsilon/2)} \geq \inf_{\tau\in]0,(2-\sqrt{3})/2]} \frac{\lambda_i(\tau)}{\tau}.$$

From the definition of λ_i $(i=1,2)$ we can see that λ_i is continuous and positive on $]0,2]$ and that $\lim_{\tau\downarrow 0}\lambda_i(\tau)/\tau$ exists in $]0,+\infty[$, hence there exist real constants $C'_1, C'_2 > 0$ (depending only on p) such that

$$\lambda_i(\varepsilon)/\delta_{\|\cdot\|}(\varepsilon/2) \geq C'_i \quad \text{for all } \varepsilon \in]0,2].$$

According to (18.38) the real constant $K_p := \min\{C_1, C_2, C'_1, C'_2\} > 0$ satisfies the inequality in the lemma. □

PROOF OF THEOREM 18.44. We proceed with the following scheme

$$\begin{array}{ccccc} & & (a) & \Leftarrow & (c') \\ & \nearrow & \Downarrow & & \Uparrow \\ (c) & \Leftarrow & (b) & \Rightarrow & (b') \end{array}$$

First, we observe that the implication (b) ⇒ (b') is evident.

Let us show (a)⇒(b). Assume that (a) holds. Fix $x, y \in X$ not both zero and fix $x^* \in J_{p,\|\cdot\|}(x)$ and $y^* \in J_{p,\|\cdot\|}(y)$. Put

$$u := \frac{x}{\|x\| \vee \|y\|}, v := \frac{y}{\|x\| \vee \|y\|}, u^* := \frac{x^*}{(\|x\| \vee \|y\|)^{p-1}}, v^* := \frac{y^*}{(\|x\| \vee \|y\|)^{p-1}}.$$

Then $\|u\| \vee \|v\| = 1$, $u^* \in J_{p,\|\cdot\|}(u) = \partial(\frac{1}{p}\|\cdot\|^p)(u)$ and $v^* \in J_{p,\|\cdot\|}(v) = \partial(\frac{1}{p}\|\cdot\|^p)(v)$ (see (18.20)). Take any $r, s \geq 0$ with $r + s = 1$. By Lemma 18.46 we have

$$\|ru + sv\|^p + \mathfrak{d}_p(r, s, \|u - v\|) \leq r\|u\|^p + s\|v\|^p,$$

which says that

$$\begin{aligned}
\|u + s(v - u)\|^p - \|u\|^p &\leq s(\|v\|^p - \|u\|^p) - \mathfrak{d}_p(r, s, \|u - v\|) \\
\|v + r(u - v)\|^p - \|v\|^p &\leq r(\|u\|^p - \|v\|^p) - \mathfrak{d}_p(r, s, \|u - v\|).
\end{aligned}$$

This and the inclusions $pu^* \in \partial(\|\cdot\|^p)(u)$ and $pv^* \in \partial(\|\cdot\|^p)(v)$ entail that

$$\begin{aligned}
p\langle u^*, v - u\rangle &\leq \|v\|^p - \|u\|^p - \limsup_{s\downarrow 0} s^{-1}\mathfrak{d}_p(r, s, \|u - v\|) \\
p\langle v^*, u - v\rangle &\leq \|u\|^p - \|v\|^p - \limsup_{r\downarrow 0} r^{-1}\mathfrak{d}_p(r, s, \|u - v\|),
\end{aligned}$$

according the characterization of the subdifferential of a convex function via its directional derivative. Then it ensues that

$$\begin{aligned}
\langle u^* - v^*, u - v\rangle &\geq \frac{1}{p}\left(\limsup_{s\downarrow 0} s^{-1}\mathfrak{d}_p(r, s, \|u - v\|) + \limsup_{r\downarrow 0} r^{-1}\mathfrak{d}_p(r, s, \|u - v\|)\right) \\
&\geq \frac{1}{p}K_p\delta_{\|\cdot\|}(\|u - v\|/2),
\end{aligned}$$

where the latter inequality is due to Lemma 18.47 above. Recalling the definitions of u, v, u^*, v^* we derive that

$$\langle x^* - y^*, x - y\rangle \geq \frac{1}{p}K_p(\|x\| \vee \|y\|)^p\delta_{\|\cdot\|}\left(\frac{\|x - y\|}{2(|x\| \vee \|y\|)}\right).$$

Putting $\phi_p(t) := \frac{1}{p}K_p\delta_{\|\cdot\|}(t/2)$ for all $t \in [0, 2]$, the uniform convexity of $\|\cdot\|$ ensures that the function ϕ_p is increasing on $[0, 2]$ according to (c4) in Proposition 18.25, so $\phi_p \in \mathfrak{S}_{\|\cdot\|}([0, 2], \mathbb{R}_+)$. This and the last inequality above justify the desired implication (a)⇒(b).

Let us prove (b)⇒(c). Assume (b) and let $\phi_p \in \mathfrak{S}_{\|\cdot\|}([0, 2], \mathbb{R}_+)$ be given by (b). Fix $x, y \in X$ not both zero and $x^* \in J_{p,\|\cdot\|}(x)$. Let $j : X \to X^*$ be any selection of $J_{p,\|\cdot\|} = \partial(\frac{1}{p}\|\cdot\|^p)$ such that $j(x) = x^*$. The locally Lipschitz convex function $t \mapsto f(t) := \|x + ty\|^p$ is derivable almost everywhere and $f'(t) = p\langle j(x + ty), y\rangle$ for almost every $t \in \mathbb{R}$. It ensues that

$$\begin{aligned}
\|x + y\|^p - \|x\|^p - p\langle j(x), y\rangle &= p\int_0^1 \langle j(x + ty), y\rangle\, dt - p\langle j(x), y\rangle \\
&= p\int_0^1 \langle j(x + ty) - j(x), y\rangle\, dt.
\end{aligned}$$

Put for all $t \in]0, 1]$

$$F(t) = \frac{(\|x + ty\| \vee \|x\|)^p}{t}\phi_p\left(\frac{t\|y\|}{\|x + ty\| \vee \|x\|}\right).$$

The nonnegative function F is Lebesgue measurable on $]0,1]$ according to the increasing property of ϕ_p. Then, by the assumption property in (b) it follows that

$$\|x+y\|^p - \|x\|^p - p\langle x^*, y\rangle \geq p\int_0^1 F(t)\,dt,$$

which corresponds to the inequality in (c) and confirms the implication (b)⇒(c).

The proof of (b')⇒(c') is similar to that of (b)⇒(c).

Now, let us prove (c')⇒(a). Let j_p be the selection of $J_{p,\|\cdot\|}$ given by (c') and let $j : \mathbb{S}_X \to X^*$ denote the restriction of j_p to $\mathbb{S}_X$. Since $J_{p,\|\cdot\|}$ and the normalized duality multimapping $J_{\|\cdot\|}$ coincide on $\mathbb{S}_X$ (see (18.21)), the mapping j is a selection of $J_{\|\cdot\|}$ on $\mathbb{S}_X$. Further, by (c') we have for $x, y \in \mathbb{S}_X$

$$\begin{aligned} 0 &= \|x+(y-x)\|^p - \|x\|^p \\ &\geq p\langle j(x), y-x\rangle + p\int_0^1 \frac{(\|x+t(y-x)\|\vee\|x\|)^p}{t}\phi_p\left(\frac{t\|y-x\|}{\|x+t(y-x)\|\vee\|x\|}\right)dt \\ &= p\langle j(x), y-x\rangle + p\int_0^1 t^{-1}\phi_p(t\|y-x\|)\,dt \\ &= p\langle j(x), y-x\rangle + p\int_0^{\|y-x\|} \theta^{-1}\phi_p(\theta)\,d\theta. \end{aligned}$$

Then, given $\varepsilon \in]0,2]$ we obtain for any $x, y \in \mathbb{S}_X$ with $\|x-y\| \geq \varepsilon$

$$1 - \langle j(x), y\rangle \geq \int_0^\varepsilon \theta^{-1}\phi_p(\theta)\,d\theta,$$

so it ensues that

$$\gamma_j(\varepsilon) := \inf\{1 - \langle j(x), y\rangle : x, y \in \mathbb{S}_X, \|x-y\| \geq \varepsilon\} \geq \int_0^\varepsilon \theta^{-1}\phi_p(\theta)\,d\theta. \tag{18.39}$$

We deduce that $\gamma_j(\varepsilon) > 0$ for all $\varepsilon \in]0,2]$ since ϕ_p is increasing with $\phi_p(0) = 0$. This and Proposition 18.39 tell us that the norm $\|\cdot\|$ is uniformly convex, hence the implication (c')⇒(a) is justified.

Finally, since (c) obviously entails (c'), the implication (c)⇒(a) also holds true. The proof is complete. □

REMARK 18.48. Notice that the above proof shows that the implication (b)⇒(c) (resp. (b⇒(c')) holds with the same function σ_p in (18.32) of Theorem 18.44, associated with the function ϕ_p furnished by (b) (resp. (b')). □

Let us recall (see Definition 18.37) that the modulus of convexity of a norm $\|\cdot\|$ on a vector space X is of *power type* $q \in [2, +\infty[$ provided that there is a real constant $\kappa > 0$ such that

$$\delta_{\|\cdot\|}(\varepsilon) \geq \kappa\varepsilon^q \quad \text{for all } \varepsilon \in [0,2]. \tag{18.40}$$

COROLLARY 18.49 (Zong-Ben Xu and G.F. Roach). Let $(X, \|\cdot\|)$ be a Banach space which is at least two-dimensional and let $p \in [2, +\infty[$. The following are equivalent:

(a) the norm $\|\cdot\|$ is uniformly convex of power type p;

(b) there exists a real constant $K > 0$ such that for all $x, y \in X$, $x^* \in J_{p,\|\cdot\|}(x)$, $y^* \in J_{p,\|\cdot\|}(y)$

$$\langle x^* - y^*, x - y\rangle \geq K\|x-y\|^p;$$

(b') there exists a real constant $K > 0$ and a selection j_p of $J_{p,\|\cdot\|}$ such that for all $x, y \in X$

$$\langle j_p(x) - j_p(y), x - y\rangle \geq K\|x - y\|^p;$$

(c) there exists a real constant $K > 0$ such that for all $x, y \in X$ and all $x^* \in J_{p,\|\cdot\|}(x)$

$$\|x + y\|^p \geq \|x\|^p + p\langle x^*, y\rangle + K\|y\|^p;$$

(c') there exits a real constant $K > 0$ and a selection j_p of $J_{p,\|\cdot\|}$ such that for all $x, y \in X$

$$\|x + y\|^p \geq \|x\|^p + p\langle j_p(x), y\rangle + K\|y\|^p.$$

PROOF. The implication (a)⇒(b) follows from (a)⇒(b) in Theorem 18.44 and the existence of constants $K', K'' > 0$ such that for ϕ_p in (b) of Theorem 18.44

$$\phi_p(t) \geq K'\delta_{\|\cdot\|}(t/2) \geq K''\|t\|^p,$$

according to the definition of $\mathfrak{S}_{\|\cdot\|}([0, 2], \mathbb{R}_+)$ in (18.31) and to (18.40).

To see that (b)⇒(c), let $K > 0$ be the real constant furnished by (b). The function $t \mapsto \phi_p(t) = Kt^p$ clearly belongs to $\mathfrak{S}_{\|\cdot\|}([0, 2], \mathbb{R}_+)$ and for the corresponding σ_p given by (c) in Theorem 18.44 we have

$$\sigma_p(x, y) = pK \int_0^1 \frac{(\|x + ty\| \vee \|x\|)^p}{t} \frac{t^p\|y\|^p}{(\|x + ty\| \vee \|y\|)^p}\, dt = K\|y\|^p. \tag{18.41}$$

Remark 18.48 assures us that (b)⇒(c), and the proof of (b')⇒(c') is similar.

The implications (b)⇒(b') and (c)⇒(c') being obvious, it remains to justify (c')⇒(a). Let $K > 0$ and j_p be given by (c'). Again $t \mapsto \phi_p(t) = Kt^p$ belongs to $\mathfrak{S}_{\|\cdot\|}([0, 2], \mathbb{R}_+)$ and the associated function σ_p in (c') of Theorem 18.44 satisfies $\sigma_p(x, y) = K\|y\|^p$ as in (18.41). Then we can apply (18.39) in the proof of Theorem 18.44 to obtain for every $\varepsilon \in]0, 2]$

$$\gamma_j(\varepsilon) \geq \int_0^\varepsilon \theta^{-1}\phi_p(\theta)\, d\theta = K\varepsilon^p/p.$$

Fix any $\varepsilon \in]0, 2]$. Noticing that $(\varepsilon/4)^p \leq \varepsilon/4$ and invoking the inequality $\delta_{\|\cdot\|}(\varepsilon) \geq \min\{\varepsilon/4, \gamma_j(\varepsilon/4)\}$ in Proposition 18.39, we deduce that

$$\delta_{\|\cdot\|}(\varepsilon) \geq \min\left\{\frac{\varepsilon^p}{4^p}, \frac{K}{p}\frac{\varepsilon^p}{4^p}\right\} = C\varepsilon^p,$$

with $C := 4^{-p}\min\{1, K/p\}$. The modulus of convexity $\delta_{\|\cdot\|}$ of $\|\cdot\|$ is then of power type p, so the implication (c')⇒(a) holds true. The proof is complete. □

We turn now to characteristic inequalities for uniform smoothness of norms. Such inequalities involve the set $\mathfrak{C}_{\|\cdot\|}([0, 2], \mathbb{R}_+)$ of convex and nondecreasing functions $\phi : [0, 2] \to \mathbb{R}_+$ with $\phi(0) = 0$ and such that there exists a real constant $K > 0$ satisfying

$$\phi(t) \leq K\varrho_{\|\cdot\|}(t) \quad \text{for all } t \in [0, 2]$$

and $\phi(t)/t \to 0$ as $t \downarrow 0$.

THEOREM 18.50 (Zong-Ben Xu and G.F. Roach theorem for uniform smoothness). Let $(X, \|\cdot\|)$ be a Banach space which is at least two-dimensional and let $p \in]1, +\infty[$. The following are equivalent:

(a) the norm $\|\cdot\|$ is uniformly smooth;

(b) $J_{p,\|\cdot\|}$ is single-valued and there exists a function $\phi_p \in \mathfrak{C}_{\|\cdot\|}([0,2],\mathbb{R}_+)$ such that for all $x, y \in X$ with $x \neq y$

$$\|J_{p,\|\cdot\|}(x) - J_{p,\|\cdot\|}(y)\| \leq \frac{(\|x\| \vee \|y\|)^p}{\|x-y\|}\phi_p\left(\frac{\|x-y\|}{\|x\| \vee \|y\|}\right);$$

(c) $J_{p,\|\cdot\|}$ is single-valued and there exists a function $\phi_p \in \mathfrak{C}_{\|\cdot\|}([0,2],\mathbb{R}_+)$ such that for any $x, y \in X$ not both zero

$$\|x+y\|^p \leq \|x\|^p + p\langle J_{p,\|\cdot\|}(x), y\rangle + \mathfrak{s}_p(x,y),$$

where

$$\mathfrak{s}_p(x,y) := p\int_0^1 \frac{(\|x+ty\| \vee \|x\|)^p}{t}\phi_p\left(\frac{t\|y\|}{\|x+ty\| \vee \|x\|}\right);$$

(c') there exist a function $\phi_p \in \mathfrak{C}_{\|\cdot\|}([0,2],\mathbb{R}_+)$ and a selection j_p of $J_{p,\|\cdot\|}$ over X such that for all $x, y \in X$ not both zero

$$\|x+y\|^p \leq \|x\|^p + p\langle j_p(x), y\rangle + \mathfrak{s}_p(x,y),$$

where $\mathfrak{s}_p$ is as defined in (c).

PROOF. Let us first show (a)⇒(b). The uniform smoothness assumption in (a) implies that X is reflexive (see Theorem 18.33) and that $J_{p,\|\cdot\|}$ is single-valued on X and uniformly continuous on bounded subsets (see Proposition 18.41). The uniform smoothness also implies that the dual norm $\|\cdot\|_*$ is uniformly convex (see Theorem 18.30). Denote by $q > 1$ the conjugate exponent of p given by $1/p + 1/q = 1$. The uniform convexity of $\|\cdot\|_*$ assures us by Theorem 18.44 that there is a constant $K_q > 0$ such that for any $u^*, v^* \in X^*$ not both zero and any $u \in J_{q,\|\cdot\|_*}(u^*)$ and $v \in J_{q,\|\cdot\|_*}(v^*)$

$$\langle u - v, u^* - v^*\rangle \geq K_q(\|u^*\|_* \vee \|v^*\|_*)^q \delta_{\|\cdot\|_*}\left(\frac{\|u^* - v^*\|_*}{2(\|u*\|_* \vee \|v^*\|_*)}\right),$$

which entails with $\overline{\delta}_{\|\cdot\|_*}(\varepsilon) := \delta_{\|\cdot\|_*}(\varepsilon)/\varepsilon$ that

$$\|u - v\| \geq \frac{1}{2}K_q(\|u^*\| \vee \|v^*\|_*)^{q-1}\overline{\delta}_{\|\cdot\|_*}\left(\frac{\|u^* - v^*\|_*}{2(\|u^*\|_* \vee \|v^*\|_*)}\right).$$

Let any $x, y \in X$ not both zero. Since $J_{q,\|\cdot\|_*}$ is surjective by the reflexivity of X^* (see (18.19)), there exist $x^*, y^* \in X^*$ such that $x \in J_{q,\|\cdot\|_*}(x^*)$ and $y \in J_{q,\|\cdot\|_*}(y^*)$. Keeping in mind that $J_{p,\|\cdot\|} = Df$ and $J_{q,\|\cdot\|_*} = \partial f^*$ (the subdifferential of f^*, where $f = \frac{1}{p}\|\cdot\|^p$ and f^* is the Legendre-Fenchel conjugate of f), it is readily seen that $x^* = J_{p,\|\cdot\|}(x)$ and $y^* = J_{p,\|\cdot\|}(y)$. Recalling that $\|J_{p,\|\cdot\|}(x)\|_* = \|x\|^{p-1}$ (see (18.19)) and that $(p-1) = 1/(q-1)$, we easily deduce that

$$\|x - y\| \geq \frac{1}{2}K_q(\|x\| \vee \|y\|)\overline{\delta}_{\|\cdot\|_*}\left(\frac{\|J_{p,\|\cdot\|}(x) - J_{p,\|\cdot\|}(y)\|_*}{2(\|x\| \vee \|y\|)^{p-1}}\right),$$

or equivalently

$$\frac{8\|x-y\|}{K_q(\|x\| \vee \|y\|)} \geq 4\overline{\delta}_{\|\cdot\|_*}\left(\frac{\|J_{p,\|\cdot\|}(x) - J_{p,\|\cdot\|}(y)\|_*}{2(\|x\| \vee \|y\|)^{p-1}}\right).$$

The nondecreasing property of $\tau \mapsto \varrho_{\|\cdot\|}(\tau)/\tau =: \overline{\varrho}_{\|\cdot\|}(\tau)$ (see (s3) in Proposition 18.25) gives

$$\overline{\varrho}_{\|\cdot\|}\left(\frac{8\|x-y\|}{K_q(\|x\| \vee \|y\|)}\right) \geq \overline{\varrho}_{\|\cdot\|}\left(4\overline{\delta}_{\|\cdot\|_*}\left(\frac{\|J_{p,\|\cdot\|}(x) - J_{p,\|\cdot\|}(y)\|_*}{2(\|x\| \vee \|y\|)^{p-1}}\right)\right). \tag{18.42}$$

Now, applying Lemma 18.31, we also have for any $\varepsilon \in]0,2]$

$$\varrho_{\|\cdot\|}\left(\frac{4\delta_{\|\cdot\|_*}(\varepsilon)}{\varepsilon}\right) \geq \frac{\varepsilon}{2}\,\frac{4\delta_{\|\cdot\|_*}(\varepsilon)}{\varepsilon} - \delta_{\|\cdot\|_*}(\varepsilon) = \delta_{\|\cdot\|_*}(\varepsilon),$$

which can be rewritten as

$$\varrho_{\|\cdot\|}(4\overline{\delta}_{\|\cdot\|_*}(\varepsilon)) \geq \frac{\varepsilon}{4}\,4\overline{\delta}_{\|\cdot\|_*}(\varepsilon), \quad \text{or equivalently} \quad \overline{\varrho}_{\|\cdot\|}(4\overline{\delta}_{\|\cdot\|_*}(\varepsilon)) \geq \varepsilon/4,$$

hence by (18.42)

$$\begin{aligned}\overline{\varrho}_{\|\cdot\|}\left(\frac{8\|x-y\|}{K_q(\|x\| \vee \|y\|)}\right) &\geq \overline{\varrho}_{\|\cdot\|}\left(4\overline{\delta}_{\|\cdot\|_*}\left(\frac{\|J_{p,\|\cdot\|}(x) - J_{p,\|\cdot\|}(y)\|_*}{2(\|x\| \vee \|y\|)^{p-1}}\right)\right) \\ &\geq \frac{1}{8}\,\frac{\|J_{p,\|\cdot\|}(x) - J_{p,\|\cdot\|}(y)\|_*}{(\|x\| \vee \|y\|)^{p-1}}.\end{aligned}$$

It ensues that

$$\begin{aligned}\|J_{p,\|\cdot\|}(x) - J_{p,\|\cdot\|}(y)\|_* &\leq 8(\|x\| \vee \|y\|)^{p-1}\overline{\varrho}_{\|\cdot\|}\left(\frac{8\|x-y\|}{K_q(\|x\| \vee \|y\|)}\right) \\ &= K_q\frac{(|x\| \vee \|y\|)^p}{\|x-y\|}\varrho_{\|\cdot\|}\left(\frac{8\|x-y\|}{K_q(\|x\| \vee \|y\|)}\right).\end{aligned}$$

We distinguish two cases.

Case 1: $8/K_q \leq 1$. By convexity of $\varrho_{p,\|\cdot\|}$ and by the equality $\varrho_{p,\|\cdot\|}(0) = 0$ we derive

$$\|J_{p,\|\cdot\|}(x) - J_{p,\|\cdot\|}(y)\| \leq 8\frac{(\|x\| \vee \|y\|)^p}{\|x-y\|}\varrho_{\|\cdot\|}\left(\frac{\|x-y\|}{\|x\| \vee \|y\|}\right).$$

Case 2: $8/K_q > 1$. Using the constant $c > 0$ furnished by Proposition 18.28 we can write

$$\begin{aligned}&\|J_{p,\|\cdot\|}(x) - J_{p,\|\cdot\|}(y)\| \\ &\leq \frac{8^2(\|x\| \vee \|y\|)^{p-2}\|x-y\|}{K_q}\varrho_{\|\cdot\|}\left(\frac{8\|x-y\|}{K_q\|x\| \vee \|y\|}\right)\Big/\left(\frac{8\|x-y\|}{K_q\|x\| \vee \|y\|}\right)^2 \\ &\leq c\,\frac{8^2(\|x\| \vee \|y\|)^{p-2}\|x-y\|}{K_q}\varrho_{\|\cdot\|}\left(\frac{\|x-y\|}{\|x\| \vee \|y\|}\right)\Big/\left(\frac{\|x-y\|}{\|x\| \vee \|y\|}\right)^2 \\ &= \frac{8^2c(\|x\| \vee \|y\|)^p}{K_q\|x-y\|}\varrho_{\|\cdot\|}\left(\frac{\|x-y\|}{\|x\| \vee \|y\|}\right).\end{aligned}$$

On the other hand, putting $L_p := \max\{8, 8^2c/K_q\}$ and $\phi_p(t) := L_p\varrho_{\|\cdot\|}(t)$ for all $t \in [0,2]$, the function ϕ_p is null at 0 and it is convex, continuous and nondecreasing on $[0,2]$ (see (s2) in Proposition 18.25). Further, the uniform smoothness assumption of $\|\cdot\|$ ensures by definition (see Definition 18.24) that $\phi_p(t)/t \to 0$ as $t \downarrow 0$. Thus, the function $\phi_p \in \mathfrak{C}_{\|\cdot\|}([0,2],\mathbb{R}_+)$. This combined with Case 1 and Case 2 above yields that the implication (a)⇒(b) holds true.

To show (b)⇒(c), assume that (b) holds and consider the function ϕ_p in $\mathfrak{C}_{\|\cdot\|}([0,2],\mathbb{R}_+)$ furnished by (b). Fix $x, y \in X$ not both zero. The subdifferential $\partial(\frac{1}{p}\|\cdot\|^p) = J_{p,\|\cdot\|}$ being single-valued, the continuous convex function $\|\cdot\|^p$ is Hadamard differentiable on X (see Proposition 2.76). Thus, the locally Lipschitz function $t \mapsto f(t) := \|x + ty\|^p$ is derivable with $f'(t) = p\langle J_{p,\|\cdot\|}(x+ty), y\rangle$. It ensues that $\|x+y\|^p - \|x\|^p = p\int_0^1 \langle J_{p,\|\cdot\|}(x+ty), y\rangle\, dt$, hence

$$\begin{aligned}\|x+y\|^p - \|x\|^p - p\langle J_{p,\|\cdot\|}(x), y\rangle &= p\int_0^1 \langle J_{p,\|\cdot\|}(x+ty) - J_{p,\|\cdot\|}(x), y\rangle\, dt \\ &\le p\|y\| \int_0^1 \left\| J_{p,\|\cdot\|}(x+ty) - J_{p,\|\cdot\|}(x)\right\|_* dt,\end{aligned}$$

which entails by the assumption (b) that for $y \neq 0$

$$\begin{aligned}&\|x+y\|^p - \|x\|^p - p\langle J_{p,\|\cdot\|}(x), y\rangle \\ &\le p\int_0^1 \frac{(\|x+ty\| \vee \|x\|)^p}{t}\phi_p\left(\frac{t\|y\|}{\|x+ty\| \vee \|x\|}\right) dt.\end{aligned}$$

The latter inequality being still true for $y = 0$, the assertion (c) follows.

The implication (c)⇒(c') being evident, it remains to show (c')⇒(a). Let ϕ_p in $\mathfrak{C}_{\|\cdot\|}([0,2], \mathbb{R}_+)$ satisfying the property in (c'). Take any $\tau \in]0,2]$ and take any $x \in \mathbb{S}_X$ and any $y \in X$ with $\|y\| \le \tau$. By the inequality in (c') with $\mathfrak{s}_p(x,y)$ as defined there, we have

$$\|x+y\|^p \le \|x\|^p + p\langle j_p(x), y\rangle + \mathfrak{s}_p(x,y) = 1 + p\langle j_p(x), y\rangle + \mathfrak{s}_p(x,y)$$

as well as

$$\|x-y\|^p \le \|x\|^p - p\langle j_p(x), y\rangle + \mathfrak{s}_p(x,-y) = 1 - p\langle j_p(x), y\rangle + \mathfrak{s}_p(x,-y).$$

Putting $\mu := 1 + \max\{\mathfrak{s}_p(x,y), \mathfrak{s}_p(x,-y)\}$, it ensues that

$$\begin{aligned}\|x+y\| + \|x-y\| &\le (\mu + p\langle j_p(x), y\rangle)^{1/p} + (\mu - p\langle j_p(x), y\rangle)^{1/p} \\ &= \mu^{1/p}\left[\left(1 + \frac{p}{\mu}\langle j_p(x), y\rangle\right) + \left(1 - \frac{p}{\mu}\langle j_p(x), y\rangle\right)\right].\end{aligned}$$

Noticing that $\sup_{\|y\|\le\tau} |(p/\mu)\langle j_p(x), y\rangle| \le p\tau$ (tending to 0 as $\tau \downarrow 0$), there is some $\tau_0 \in]0,1/2]$ such that for any $\tau \in]0,\tau_0]$, any $x \in \mathbb{S}_X$ and any $y \in X$ with $\|y\| \le \tau$,

$$\left(1 + \frac{p}{\mu}\langle j_p(x), y\rangle\right)^{1/p} = 1 + \sum_{n=1}^{\infty} \binom{1/p}{n} \left(\frac{p}{\mu}\langle j_p(x), y\rangle\right)^n$$

$$\left(1 - \frac{p}{\mu}\langle j_p(x), y\rangle\right)^{1/p} = 1 + \sum_{n=1}^{\infty} \binom{1/p}{n} \left(-\frac{p}{\mu}\langle j_p(x), y\rangle\right)^n,$$

where as usual for each $n \in \mathbb{N}$

$$\binom{1/p}{n} := \frac{(1/p)((1/p)-1)\cdots((1/p)-n+1)}{n!}.$$

Since $\left(\frac{p}{\mu}\langle j_p(x), y\rangle\right)^n + \left(-\frac{p}{\mu}\langle j_p(x), y\rangle\right)^n = 0$ for each odd $n \in \mathbb{N}$, it follows with $\tau \in]0,\tau_0]$, $x \in \mathbb{S}_X$ and $y \in \tau\mathbb{B}_X$ that

$$\|x+y\| + \|x-y\| \le 2\mu^{1/p}\left[1 + \sum_{n=1}^{\infty} \binom{1/p}{2n} \left(\frac{p}{\mu}\langle j_p(x), y\rangle\right)^{2n}\right],$$

so, observing that $\binom{1/p}{2n} \le 0$ for all $n \in \mathbb{N}$, we obtain that

$$\|x+y\| + \|x-y\| \le 2\mu^{1/p} \le 2\mu$$

since $\mu^{1/p} \le \mu$ (because $\mu > 1$ and $1/p < 1$). For each $\tau \in]0, \tau_0]$ setting

$$\mathfrak{s}(\tau) := \sup\{\mathfrak{s}_p(x,y) : x \in \mathbb{S}_X, \|y\| \le \tau\},$$

it ensues by definition of μ that for all $x \in \mathbb{S}_X$ and $y \in X$ with $\|y\| \le \tau$

$$\|x+y\| + \|x-y\| \le 2(1+\mathfrak{s}(\tau)),$$

which, by the equality (18.12) for the modulus of smoothness, implies that

$$\varrho_{\|\cdot\|}(\tau) \le \mathfrak{s}(\tau) \quad \text{for all } \tau \in]0, \tau_0]. \tag{18.43}$$

Now, observe that, for $\tau \in]0, \tau_0]$, $x \in \mathbb{S}_X$, $y \in \tau\mathbb{B}_X$ and $t \in [0,1]$ we have

$$1/2 = 1 - 1/2 \le \|x + ty\| \le 1 + 1/2 = 3/2,$$

hence by the nondecreasing property of ϕ_p

$$\mathfrak{s}_p(x,y) \le 2p\left(\frac{3}{2}\right)^p \tau \int_0^1 \frac{1}{2\tau t}\phi_p(2\tau t)\, dt. \tag{18.44}$$

Fix any $\varepsilon > 0$. Since $\phi_p(\tau)/\tau \to 0$ as $\tau \downarrow 0$ by the inclusion $\phi_p \in \mathfrak{C}_{\|\cdot\|}([0,2], \mathbb{R}_+)$, there exists $\tau_1 \in]0, \tau_0/2]$ such that

$$\phi_p(\tau)/\tau \le \frac{1}{2p}\left(\frac{2}{3}\right)^p \varepsilon \quad \text{for all } \tau \in]0, 2\tau_1].$$

This and (18.44) yield that $\mathfrak{s}(\tau)/\tau \le \varepsilon$ for every $\tau \in]0, \tau_1]$, which by (18.43) assures us that $\varrho_{\|\cdot\|}(\tau)/\tau \le \varepsilon$ for all $\tau \in]0, \tau_1]$. This means that $\varrho_{\|\cdot\|}(\tau)/\tau \to 0$ as $\tau \downarrow 0$, so the norm $\|\cdot\|$ is uniformly smooth. □

Given a normed space $(X, \|\cdot\|)$, we recall that the modulus of smoothness $\varrho_{\|\cdot\|}$ is of *power type* $s \in]1,2]$ (see Definition 18.37) if there exists a real constant $c > 0$ such that

$$\varrho_{\|\cdot\|}(\tau) \le c\,\tau^s \quad \text{for all } \tau \ge 0.$$

It is easy to derive from Theorem 18.50 the following inequalities characterizing that the modulus of smoothness is of power type.

COROLLARY 18.51 (Zong-Ben Xu and G.F. Roach). Let $(X, \|\cdot\|)$ be a Banach space which is at least txo-dimensional and let $s \in]1,2]$. The following are equivalent:]

(a) the norm $\|\cdot\|$ is uniformly smooth of power type s;

(b) $J_{s,\|\cdot\|}$ is single-valued and there exists a real constant $L > 0$ such that for all $x, y \in X$,

$$\|J_{s,\|\cdot\|}(x) - J_{s,\|\cdot\|}(y)\|_* \le L\|x-y\|^{s-1};$$

(c) there exists a real constant $L > 0$ such that for all $x, y \in X$ and all $x^* \in J_{s,\|\cdot\|}(x)$

$$\|x+y\|^s \le \|x\|^s + s\langle x^*, y\rangle + L\|y\|^s;$$

(c') there exits a real constant $L > 0$ and a selection j_s of $J_{s,\|\cdot\|}$ such that

$$\|x+y\|^s \le \|x\|^s + s\langle j_s(x), y\rangle + L\|y\|^s.$$

The case $p = 2$ involves the normalized duality mapping. The related result stated below follows directly from the two previous theorems.

THEOREM 18.52 (Zong-Ben Xu and G.F. Roach theorem on moduli of convexity and smoothness). Let $(X, \|\cdot\|)$ be a Banach space which is at least two-dimensional. The following hold:
(a) If the norm $\|\cdot\|$ is uniformly convex, then there exists a real constant $K > 0$ such that, for all $x, y \in X$ not both zero, for all $x^* \in J_{\|\cdot\|}(x)$ and for all $y^* \in J_{\|\cdot\|}(y)$,

$$\langle x^* - y^*, x - y\rangle \geq K(\|x\| \vee \|y\|)^2 \delta_{\|\cdot\|}\left(\frac{\|x - y\|}{2\|x\| \vee \|y\|}\right).$$

(b) If the norm $\|\cdot\|$ is uniformly smooth, then $J_{\|\cdot\|}$ is single-valued and there exists a real constant $L > 0$ such that, for all $x, y \in X$ with $x \neq y$,

$$\left\|J_{\|\cdot\|}(x) - J_{\|\cdot\|}(y)\right\|_* \leq L(\|x\| \vee \|y\|)^2 \frac{1}{\|x - y\|} \varrho_{\|\cdot\|}\left(\frac{\|x - y\|}{\|x\| \vee \|y\|}\right).$$

COROLLARY 18.53 (Zong-Ben Xu and G.F. Roach). Let $(X, \|\cdot\|)$ be a Banach space which is at least two-dimensional.
(a) Assume that the norm $\|\cdot\|$ is uniformly convex. Then for any real $\beta > 0$ there exists a real constant $K_\beta > 0$ such that, for all $x, y \in \beta\mathbb{B}_X$, for all $x^* \in J_{\|\cdot\|}(x)$ and for all $y^* \in J_{\|\cdot\|}(y)$,

$$\langle x^* - y^*, x - y\rangle \geq K_\beta(\|x\| \vee \|y\|)\delta_{\|\cdot\|}\left(\frac{\|x - y\|}{2\beta}\right).$$

If in addition the modulus of convexity of $\|\cdot\|$ is of power type $q \geq 2$, one also has for some real constant $K'_\beta > 0$ that, for all $x, y \in \beta\mathbb{B}_X$, for all $x^* \in J_{\|\cdot\|}(x)$ and for all $y^* \in J_{\|\cdot\|}(y)$,

$$\langle x^* - y^*, x - y\rangle \geq K'_\beta(\|x\| \vee \|y\|)\|x - y\|^q.$$

(b) Assume instead that the norm $\|\cdot\|$ is uniformly smooth. Then $J_{\|\cdot\|}$ is single-valued and for each real $\beta > 0$ there exists a real constant $L_\beta > 0$ such that

$$\left\|J_{\|\cdot\|}(x) - J_{\|\cdot\|}(y)\right\|_* \leq L_\beta \frac{1}{\|x - y\|} \varrho_{\|\cdot\|}\left(\frac{\|x - y\|}{\beta}\right) \quad \text{for all } x \neq y \text{ in } \beta\mathbb{B}_X.$$

If in addition the modulus of smoothness of $\|\cdot\|$ is power type $s \in]1, 2]$, one also has for some real constant $L'_\beta > 0$

$$\left\|J_{\|\cdot\|}(x) - J_{\|\cdot\|}(y)\right\|_* \leq L'_\beta\|x - y\|^{s-1} \quad \text{for all } x, y \in \beta\mathbb{B}_X.$$

PROOF. (a) Assume that $\|\cdot\|$ is uniformly convex. Consider the real constant $K > 0$ in Theorem 18.52(a) and recall that $t \mapsto t^{-1}\delta_{\|\cdot\|}(t)$ is nondecreasing by uniform convexity of $\|\cdot\|$ (see Proposition 18.25(c3)). Then, for any real $\beta > 0$, any $x, y \in \beta\mathbb{B}_X$ with $x \neq y$, any $x^* \in J_{\|\cdot\|}(x)$ and any $y^* \in J_{\|\cdot\|}(y)$, we have

$$\begin{aligned}
\langle x^* - y^*, x - y\rangle &\geq \frac{K}{2}\|x - y\|(\|x\| \vee \|y\|)\left(\frac{2(\|x\| \vee \|y\|)}{\|x - y\|}\right)\delta_{\|\cdot\|}\left(\frac{\|x - y\|}{2(\|x\| \vee \|y\|)}\right) \\
&\geq \frac{K}{2}\|x - y\|(\|x\| \vee \|y\|)\left(\frac{2\beta}{\|x - y\|}\right)\delta_{\|\cdot\|}\left(\frac{\|x - y\|}{2\beta}\right) \\
&= \beta K(\|x\| \vee \|y\|)\delta_{\|\cdot\|}\left(\frac{\|x - y\|}{2\beta}\right),
\end{aligned}$$

which justifies the desired inequality. The second inequality in (a) under the power type property of $\delta_{\|\cdot\|}$ follows from the previous inequality.

(b) Now, assume the norm is uniformly smooth. We know by Theorem 18.52(b) that there is a constant $L > 0$ such that

$$\big\|J_{\|\cdot\|}(x) - J_{\|\cdot\|}(y)\big\|_* \le L(\|x\| \vee \|y\|)^2 \frac{1}{\|x-y\|} \varrho_{\|\cdot\|}\left(\frac{\|x-y\|}{\|x\| \vee \|y\|}\right).$$

Let $c > 0$ be the real constant furnished by Proposition 18.28 such that $\varrho_{\|\cdot\|}(\theta)/\theta^2 \le c\,\varrho_{\|\cdot\|}(\tau)/\tau^2$ for all $0 < \tau \le \theta$. Taking any real $\beta > 0$, we deduce that for any $x, y \in \beta\mathbb{B}_X$ with $x \neq y$

$$\begin{aligned}
\big\|J_{\|\cdot\|}(x) - J_{\|\cdot\|}(y)\big\|_* &\le L\|x-y\| \left(\frac{\|x\| \vee \|y\|}{\|x-y\|}\right)^2 \varrho_{\|\cdot\|}\left(\frac{\|x-y\|}{\|x\| \vee \|y\|}\right) \\
&\le cL\|x-y\| \frac{\beta^2}{\|x-y\|^2} \varrho_{\|\cdot\|}\left(\frac{\|x-y\|}{\beta}\right) \\
&= cL\beta^2 \frac{1}{\|x-y\|} \varrho_{\|\cdot\|}\left(\frac{\|x-y\|}{\beta}\right).
\end{aligned}$$

This translates the first inequality in (b). The second inequality, under the power type condition for $\varrho_{\|\cdot\|}$, directly follows. □

Before closing this subsection and in preparation for the next ones, let us particularize the duality multimapping for the Cartesian product whose one of two spaces is the real line $\mathbb{R}$. Given a normed space $(X, \|\cdot\|)$, the space $X \times \mathbb{R}$ will be endowed with the norm $\|\!|\cdot|\!\|$ defined by

$$\|\!|(x,r)|\!\| = (\|x\|^2 + r^2)^{1/2} \quad \text{for all } (x,r) \in X \times \mathbb{R}.$$

Putting $\varphi(x,r) := (1/2)\|\!|(x,r)|\!\|^2 = (1/2)(\|x\|^2 + r^2)$, we see that

$$J_{\|\!|\cdot|\!\|}(x,r) = \partial\varphi(x,r) = J_{\|\cdot\|}(x) \times \{r\},$$

thus, when the norm $\|\cdot\|$ is Gâteaux differentiable off zero, so is $\|\!|\cdot|\!\|$ and

$$J_{X\times\mathbb{R}}(x,r) = (J(x), r) \quad \text{for all } (x,r) \in X \times \mathbb{R}. \tag{18.45}$$

We also notice by Proposition 18.18 (resp. Proposition 18.34) that the uniform convexity (resp. uniform smoothness) of $\|\cdot\|$ entails the same property for the norm $\|\!|\cdot|\!\|$ on $X \times \mathbb{R}$.

18.2. Proximal normals in normed spaces

Proximal normals of subsets of Hilbert spaces have been defined and largely studied in Section 4.6 of Chapter 4. In the same section, given a subset C of a normed space X, *variational proximal normals* of C at $x \in C$ are defined (see Definition 4.133) as specific elements in the topological dual X^*, that is, specific continuous linear functionals on X. For the theory of prox-regularity of sets in uniformly convex Banach spaces in Sections 18.3, 18.5 and 18.6, we will need to involve proximal normals which are vectors in the initial data space X. Then, in the context of the normed space X, we naturally adopt for definition of proximal normal vector in X the same as Definition 4.131 relative to Hilbert spaces.

DEFINITION 18.54. Let C be a nonempty set in a normed space $(X, \|\cdot\|)$. A vector $v \in X$ is said to be a *proximal normal vector of* (or *perpendicular vector to*) C at $x \in C$ if there is a real $r > 0$ such that

$$x \in \operatorname{Proj}_C(x + rv), \quad \text{or equivalently } x \in \operatorname{Proj}_{\operatorname{cl} C}(x + rv). \tag{18.46}$$

The cone of all such vectors v is called the *proximal normal cone of* C *at* x, and it will be denoted by $N^P(C;x)$ or $N_C^P(x)$. By convention, we put $N^P(C;x) = \emptyset$ for $x \in X \setminus C$.

Lemma 18.8 directly yields the following characterization in terms of unique nearest point.

PROPOSITION 18.55. Assume that $(X, \|\cdot\|)$ is strictly convex. For a subset C of X and $x \in C$, one has $v \in N^P(C;x)$ if and only if there is a real $s > 0$ such that $x = P_{\mathrm{cl}\, C}(x + sv)$.

The concept of proximal normal vector is local in the following sense.

PROPOSITION 18.56. Let $(X, \|\cdot\|)$ be a normed space and C be a nonempty set in X. For any $u \notin \mathrm{cl}\, C$ and any closed ball $V := B[x, \beta]$ centered at $x \in C$ such that $\|u - x\| = d_{C\cap V}(u)$ one has $u - x \in N^P(C;x)$.

PROOF. Let u and V be as in the statement. Put $\rho := d_{C\cap V}(u) > 0$ and $u_t := x + t(u - x)$ for any fixed positive $t < \min\{1, \beta, \frac{\beta}{2\rho}\}$. Observe first that $u_t \in \mathrm{int} V$ according to the inequality $t < \beta/(2\rho)$, and hence $u_t \notin \mathrm{cl}\, C$ because otherwise one would obtain $u_t \in \mathrm{cl}\,(C \cap V)$ and

$$\|u_t - u\| = (1 - t)\|u - x\| < d_{C\cap V}(u),$$

which would be a contradiction. Further $\|u_t - x\| = t\rho$ and, on the one hand, for any $y \in C \cap V$ one can write

$$\|u_t - y\| = \|u + (1 - t)(x - u) - y\| \ge \|u - y\| - (1 - t)\|u - x\| \ge t\rho = \|u_t - x\|.$$

On the other hand, for any $y \in C \setminus V$ one has

$$\|u_t - y\| \ge \|y - x\| - \|u_t - x\| > \beta - t\|u - x\| = \beta - t\rho > t\rho = \|u_t - x\|,$$

the last inequality being due to the choice of t. So $\|u_t - x\| = d_C(u_t)$, and hence by definition of $N^P(C;x)$ we have

$$u - x = t^{-1}(u_t - x) \in N^P(C;x),$$

which finishes the proof. □

Throughout the rest of this chapter, since the norm $\|\cdot\|$ of the vector space X will be fixed, we follow the convention before Proposition 18.39. Otherwise stated, we denote the normalized duality mappings $J_{\|\cdot\|}$ of X and $J_{\|\cdot\|_*}$ of its dual X^* by J and J^* respectively.

Proximal normal functionals are defined via the normalized duality mapping.

DEFINITION 18.57. Let $(X, \|\cdot\|)$ be a normed space whose norm $\|\cdot\|$ is Gâteaux differentiable off zero. A functional $v^* \in X^*$ is said to be a *proximal normal functional* of C at $x \in C$ if $v^* = J(v)$ for some $v \in N^P(C;x)$, or equivalently if there are $u \in X$ and $\rho > 0$ such that $v^* = \rho J(u - x)$ and $\|u - x\| = d_C(u)$. The cone of all proximal normal functionals will be denoted by $N^{*,P}(C;x)$ or $N_C^{*,P}(x)$. As usual, we put $N^{*,P}(C;x) = \emptyset$ for $x \in X \setminus C$.

If $(X, \|\cdot\|)$ is reflexive and both norms $\|\cdot\|$ and $\|\cdot\|_*$ are Gâteaux differentiable off zero, we see that $v^* \in X^*$ is a proximal normal functional of C at $x \in C$ if and only if there exists $r > 0$ such that $x \in \mathrm{Proj}_{\,\mathrm{cl}\, C}(x + rJ^*(v^*))$. Further, under the reflexivity of $(X, \|\cdot\|)$ and the Gâteaux differentiability of $\|\cdot\|$ and

$\|\cdot\|_*$, one easily verifies that if $v \in N^P(C;x)$, then $J(v) \in N^{*,P}(C;x)$, and that, if $v^* \in N^{*,P}(C;x)$, then $J^*(v^*) \in N^P(C;x)$; therefore, $N^P(C;x)$ and $N^{*,P}(C;x)$ completely determines each other.

We now show the inclusion of the cone of proximal normal functionals into the Fréchet normal cone under the Fréchet differentiability of the norm.

PROPOSITION 18.58. Let $(X, \|\cdot\|)$ be a normed space whose norm $\|\cdot\|$ is Fréchet differentiable off zero and let C be a nonempty subset of X. Then, for any $x \in C$ one has

$$N^{*,P}(C;x) \subset N^F(C;x).$$

PROOF. Take $v^* \in N^{*,P}(C;x)$ and choose $r > 0$ and $u \in X$ such that $v^* = rJ(u-x)$ and $x \in \operatorname{Proj}_{\operatorname{cl} C}(u)$. By Fréchet differentiability of $(1/2)\|\cdot\|^2$ at $u-x$ we have for $x' \in C$

$$\|u-x'\|^2 = \|(u-x)+(x-x')\|^2 = \|u-x\|^2+2\langle J(u-x), x-x'\rangle+2\|x'-x\|\,\eta(\|x'-x\|),$$

where $\eta(t) \to 0$ as $t \downarrow 0$. Combining this with the inequality $\|u-x\|^2 \le \|u-x'\|^2$ yields for all $x' \in C$

$$\langle J(u-x), x'-x\rangle \le \|x'-x\|\,\eta(\|x'-x\|),$$

so $J(u-x) \in N^F(C;x)$, which justifies the desired inclusion. $\square$

It is known according to Lau theorem (see Theorem 14.29) that given any nonempty closed subset C in a reflexive Banach space endowed with a Kadec-Klee norm, the set of those points which have a nearest point in C is a dense set; in particular, this property holds in any uniformly convex Banach space $(X, \|\cdot\|)$. Although the reverse inclusion in Proposition 18.58 fails, this density property of elements with nearest points will allow us to show, for large classes of spaces, that any Fréchet normal to a set can be approximated by proximal normal functionals.

PROPOSITION 18.59. Let $(X, \|\cdot\|)$ be a reflexive Banach space such that both the norm $\|\cdot\|$ and its dual norm $\|\cdot\|_*$ are Fréchet differentiable off the origins. Let C be a closed set of X and let $x \in C$ and $x^* \in N^F(C;x)$. Then, for any real $\varepsilon > 0$ there exist $u_\varepsilon \in C$ and $u_\varepsilon^* \in N^{*,P}(C;u_\varepsilon)$ such that

$$\|u_\varepsilon - x\| < \varepsilon \quad \text{and} \quad \|u_\varepsilon^* - x^*\|_* < \varepsilon. \tag{18.47}$$

PROOF. We may suppose that $\|x^*\|_* = 1$. By definition of Fréchet normal there exists a function ρ from $[0,+\infty[$ into $[0,+\infty[$ with $\lim_{t\downarrow 0}\rho(t) = 0$ and such that

$$\langle x^*, y-x\rangle \le \|y-x\|\,\rho(\|y-x\|) \quad \text{for all } y \in C. \tag{18.48}$$

For $h := J^*(x^*)$, since

$$\langle x^*, h\rangle = \|x^*\|_*\,\|h\| = 1, \tag{18.49}$$

we see from (18.48) that $x + th \notin C$ for positive t small enough, say for $0 < t < t_0$. Fix any $t \in]0, t_0[$. By Lau theorem (see Theorem 14.29) there is some $h_t \in X$ with $\|h_t - h\| < t$ and such that the element $x + th_t \notin C$ and admits a nearest point in C, say $u_t \in C$. Writing u_t in the form $u_t = x + tv_t$ we have

$$t\|v_t - h_t\| = d_C(x+th_t) \le t\|h_t\|, \quad \text{and hence} \quad \|v_t\| \le 2\|h_t\| \le 2(1+t). \tag{18.50}$$

Then, taking (18.48) into account we see that

$$\langle x^*, tv_t\rangle \le \|tv_t\|\,\rho(t\|v_t\|) \le 2t(1+t)\,\rho(t\|v_t\|),$$

which gives

$$\langle x^*, v_t \rangle \le 2(1+t)\,\rho(t\|v_t\|). \tag{18.51}$$

Now observe by the first inequality of (18.50) that

$$\|h - v_t\| \le \|h - h_t\| + \|h_t - v_t\| \le \|h - h_t\| + \|h_t\| \le \|h\| + 2\|h - h_t\| < 1 + 2t,$$

and hence for $w_t := (1+2t)^{-1}(h - v_t)$ we have $\|w_t\| \le 1$. Note that

$$(1+2t)^{-1}[1 - 2(1+t)\,\rho(t\|v_t\|)] \le \langle x^*, w_t \rangle \le 1,$$

the first inequality being due to (18.49) and (18.51), and the second being due to the inequality $\|w_t\| \le 1$. The latter inequalities being true for every $t \in]0, t_0[$, it results that

$$\langle x^*, w_t \rangle \to 1 = \langle x^*, h \rangle \quad \text{as } t \downarrow 0.$$

Since the dual norm $\|\cdot\|_*$ is Fréchet differentiable at the point x^* of the unit sphere of X^* and since $\|w_t\| \le 1$ and $\|h\| = 1$, the Šmulian lemma (see Lemma 18.10(b)) entails that $\|w_t - h\| \to 0$ as $t \downarrow 0$, which is equivalent to $\|v_t\| \to 0$ as $t \downarrow 0$.

Fix again any $t \in]0, t_0[$. Since $x + th_t \notin C$, the continuous linear functional

$$u_t^* := \|x + th_t - u_t\|^{-1} J(x + th_t - u_t)$$

is well defined, and by definition it satisfies the inclusion $u_t^* \in N^{*,P}(C; u_t)$ as well as the equality

$$\langle u_t^*, x + th_t - u_t \rangle = \|x + th_t - u_t\|.$$

The latter equality is equivalent to $\langle u_t^*, h_t - v_t \rangle = \|h_t - v_t\|$ according to the equality $u_t = x + tv_t$. Since $\|v_t\| \to 0$ and $h_t \to h$ as $t \downarrow 0$, we derive that $\langle u_t^*, h_t \rangle \to \|h\| = 1$, and hence

$$\langle u_t^*, h \rangle \to 1 = \langle x^*, h \rangle \quad \text{as } t \downarrow 0.$$

Keeping in mind that $\|u_t^*\|_* = 1$ and $\|x^*\|_* = 1$ and that the norm $\|\cdot\|$ is Fréchet differentiable at the point h of the unit sphere of X, the Šmulian lemma again entails that $\|u_t^* - x^*\|_* \to 0$ as $t \downarrow 0$. The proof is then complete because obviously we also have $\|u_t - x\| \to 0$ as $t \downarrow 0$. □

Using the definition of Mordukhovich limiting normal cone $N^L(C; \overline{x})$ along with the inclusion $N^{*,P}(C; x) \subset N^F(C; x)$ in Proposition 18.58, the following corollary immediately follows from Proposition 18.59 above. It extends the corresponding result in Hilbert space seen in Theorem 4.156(d).

COROLLARY 18.60. Let $(X, \|\cdot\|)$ be a reflexive Banach space such that both the norm $\|\cdot\|$ and its dual norm $\|\cdot\|_*$ are Fréchet differentiable off the origins. Let C be a set in X which is closed near $\overline{x} \in C$. Then one has the equality

$$N^L(C; \overline{x}) = {}^{\text{seq}} \operatorname*{Lim\,sup}_{C \ni x \to \overline{x}} N^{*,P}(C; x).$$

Let us now extend the definition of proximal subgradient in Definition 4.131 to general normed spaces.

DEFINITION 18.61. Let $(X, \|\cdot\|)$ be a normed space, $f : X \to \mathbb{R} \cup \{+\infty\}$ be a proper function. We say that $x^* \in X^*$ is a *proximal subgradient* of f at $x \in \operatorname{dom} f$ if $(x^*, -1)$ is a proximal normal functional of the epigraph of f at $(x, f(x))$ with respect to the norm $|\!|\!|\cdot|\!|\!|$ on $X \times \mathbb{R}$ defined by $|\!|\!|(x, r)|\!|\!| := (\|x\|^2 + r^2)^{1/2}$ for all $(x, r) \in X \times \mathbb{R}$. The *proximal subdifferential* of f at x, denoted by $\partial_P f(x)$, consists of all such functionals. Thus, we have $x^* \in \partial_P f(x)$ if and only if $(x^*, -1) \in N^{*,P}_{\mathrm{epi}\, f}(x, f(x))$. By convention we also set $\partial_P f(x) = \emptyset$ whenever $x \notin \mathrm{dom}\, f$.

It is worth noticing from Proposition 18.58 that

$$(x^*, -r^*) \in N^{*,P}_{\mathrm{epi}\, f}(x, f(x)) \implies r^* \geq 0.$$

Recalling that $x^* \in \partial_F f(x)$ if and only if $(x^*, -1) \in N^F(\mathrm{epi}\, f; (x, f(x)))$ (see Proposition 4.4), it is clear that the following inclusion of $\partial_P f(x) \subset \partial_F f(x)$ directly results from Proposition 18.58.

PROPOSITION 18.62. Let $(X, \|\cdot\|)$ be a normed space such that the norm $\|\cdot\|$ is Fréchet differentiable off zero. Then, for any function $f : X \to \mathbb{R} \cup \{+\infty\}$ and any $x \in \mathrm{dom}\, f$, one has $\partial_P f(x) \subset \partial_F f(x)$.

The next proposition establishes the graphical density of $\mathrm{Dom}\, \partial_P f$ in $\mathrm{dom}\, f$ under quite general appropriate assumptions.

PROPOSITION 18.63. Let $(X, \|\cdot\|)$ be a reflexive Banach space such that both the primal norm $\|\cdot\|$ and the dual norm $\|\cdot\|_*$ are Fréchet differentiable off the origins. Let $f : X \to \mathbb{R} \cup \{+\infty\}$ be a function which is both proper and lower semicontinuous function on an open set $U \subset X$. Then, the following hold:
(a) For any $(x, x^*) \in \mathrm{gph}\, \partial_F f$ with $x \in U$ there exists a sequence $(x_n, x_n^*)_n$ in $\mathrm{gph}\, \partial_P f$ with $x_n \in U$ such that $(x_n, f(x_n)) \to (x, f(x))$ and $\|x_n^* - x^*\|_* \to 0$ as $n \to \infty$.
(b) The set $U \cap \mathrm{Dom}\, \partial_P f$ is graphically dense in $U \cap \mathrm{dom}\, f$, that is, for any $x \in U \cap \mathrm{dom}\, f$ there exists a sequence $(x_n)_n$ in $U \cap \mathrm{Dom}\, \partial_P f$ such that $(x_n, f(x_n)) \to (x, f(x))$ as $n \to \infty$.

PROOF. We recall first that $X \times \mathbb{R}$ is endowed with the norm $|\!|\!|\cdot|\!|\!|$ defined by $|\!|\!|(x, r)|\!|\!| := (\|x\|^2 + r^2)^{1/2}$ for all $(x, r) \in X \times \mathbb{R}$. It is clear from the assumptions on $\|\cdot\|$ that both $|\!|\!|\cdot|\!|\!|$ and its dual norm $|\!|\!|\cdot|\!|\!|_*$ are Fréchet differentiable off the origins; it is easily seen that $|\!|\!|(x^*, r)|\!|\!|_* = (\|x^*\|_*^2 + r^2)^{1/2}$ for all $(x^*, r) \in X^* \times \mathbb{R}$. Suppose without loss of generality that $U = X$.
(a) Fix any $(x, x^*) \in \mathrm{gph}\, \partial_F f$. Then $(x^*, -1) \in N^F_{\mathrm{epi}\, f}\big((x, f(x))\big)$, so by Proposition 18.59 there are sequences $(x_n, r_n) \to (x, f(x))$ and $(u_n^*, -r_n^*) \to (x^*, -1)$ with $(x_n, r_n) \in \mathrm{epi}\, f$ and $(u_n^*, -r_n^*) \in N^{*,P}_{\mathrm{epi}\, f}\big((x_n, r_n)\big)$. We may suppose that $r_n^* > 0$ for all n, so putting $x_n^* := (1/r_n^*) u_n^*$ we see that $\|x_n^* - x^*\|_* \to 0$ and $(x_n^*, -1) \in N^{*,P}_{\mathrm{epi}\, f}\big((x_n, r_n)\big)$. Since $N^{*,P}(\cdot;\cdot) \subset N^F(\cdot;\cdot)$ by Proposition 18.58, we have for each $n \in \mathbb{N}$ that $r_n = f(x_n)$ (see Proposition 4.21(b)), and hence $x_n^* \in \partial_P f(x_n)$. Further, the inequality $f(x_n) \leq r_n$ and the lower semicontinuity of f ensures that

$$f(x) \leq \liminf_{n\to\infty} f(x_n) \leq \limsup_{n\to\infty} f(x_n) \leq \lim_{n\to\infty} r_n = f(x),$$

which yields $f(x_n) \to f(x)$.
(b) The assertion (b) readily follows from (a). □

Taking the definition of the above norm $|\!|\!|\cdot|\!|\!|$ on $X \times \mathbb{R}$ into account, one easily verifies the following equality between the cone of proximal normal functionals and the proximal subdifferential of the indicator function Ψ_C of a set C.

PROPOSITION 18.64. Let C be a subset of a normed space $(X, \|\cdot\|)$. For any $x \in C$, one has

$$\partial_P \Psi_C(x) = N^{*,P}(C; x).$$

Like for the variational proximal normal cone (see Proposition 4.153, one can express, in any normed space with Fréchet differentiable norm, the proximal normal functional cone of C in terms of the proximal subdifferential of the distance function d_C. For $\rho \geq 0$, recall that the ρ-enlargement $\mathrm{Enl}_\rho C$ of C is defined by $\mathrm{Enl}_\rho C := \{u \in X : d_C(u) \leq \rho\}$.

PROPOSITION 18.65. Let $(X, \|\cdot\|)$ be a normed space whose norm $\|\cdot\|$ is Fréchet differentiable off zero. For any nonempty subset C of X and any $x \in C$, one has

$$\partial_P d_C(x) = N^{*,P}(C; x) \cap \mathbb{B}_{X^*}.$$

PROOF. The inclusion $x^* \in \partial_P d_C(x)$ means $(x^*, -1) \in N^{*,P}_{\mathrm{epi}\ d_C}(x, 0)$, or equivalently, for any $t > 0$ small enough,

$$\inf_{(y,\lambda)\in \mathrm{epi}\ d_C} \{\|x + tv - y\|^2 + (-t - \lambda)^2\} = t^2\|v\|^2 + t^2, \tag{18.52}$$

where v is taken (according to the definition of $N^{*,P}_{\mathrm{epi}\ d_C}(\cdot)$) such that $(x^*, -1) = J_{|||\cdot|||}(v, -1)$, that is, $x^* = J(v)$ (see (18.45)). Fix $t > 0$ satisfying (18.52). This entails that, for all $y \in C$, since $(y, 0) \in \mathrm{epi}\, d_C$,

$$\|x + tv - y\|^2 + t^2 \geq t^2\|v\|^2 + t^2,$$

so $\varphi(y) := \|x + tv - y\|^2 \geq t^2\|v\|^2$. Noting that $\varphi(x) = t^2\|v\|^2$ along with $x \in C$, we deduce that

$$\inf_{y\in C}\{\|x + tv - y\|^2\} = t^2\|v\|^2, \text{ that is, } x \in \mathrm{Proj}_C(x + tv),$$

hence $v \in N^P_C(x)$. Further, (18.52) again ensures for all $y \in X$ that

$$\|x + tv - y\|^2 + 2td_C(y) + d_C^2(y) \geq t^2\|v\|^2,$$

and hence, since $-2tJ(v) = D_F(-\|\cdot\|^2)(tv)$, for each $\varepsilon > 0$ there exists some positive real $r < \varepsilon$ such that for all $y \in B[x, r]$

$$2t\langle -J(v), x - y\rangle \leq \varepsilon\|x - y\| + 2td_C(y) + d_C^2(y),$$

and taking the inequality $d_C(y) \leq \|x - y\|$ into account we see that

$$2t\langle -J(v), x - y\rangle \leq (\varepsilon + 2t + \|x - y\|)\|x - y\| \leq (2\varepsilon + 2t)\|x - y\|.$$

This easily yields $\|v\| = \|-J(v)\|_* \leq 1$, and hence $x^* \in N^{*,P}_C(x) \cap \mathbb{B}_{X^*}$ (we note that the inequality $\|x^*\| \leq 1$ could also be deduced from the inclusion $\partial_P d_C(x) \subset \partial_F d_C(x)$ in Proposition 18.62).

Take conversely $x^* \in N^{*,P}_C(x) \cap \mathbb{B}_{X^*}$. There exists $v \in N^P_C(x)$ such that $x^* = J(v)$. Choose $t > 0$ small enough that $d_C^2(x + tv) = t^2\|v\|^2$. Noting that $X = \bigcup_{\rho\geq 0} \mathrm{Enl}_\rho C$ and putting $h(\rho) := \inf_{y\in \mathrm{Enl}_\rho C}\{\|x + tv - y\|^2 + (t + \rho)^2\}$, we can write

$$\inf_{y\in X}\{\|x+tv-y\|^2+(t+d_C(y))^2\} = \inf_{\rho\geq 0}\inf_{y\in \mathrm{Enl}_\rho C}\{\|x+tv-y\|^2+(t+d_C(y))^2\} \leq \inf_{\rho\geq 0} h(\rho).$$

On the other hand, for each $y \in X$ we have

$$\|x + tv - y\|^2 + (t + d_C(y))^2 = h(d_C(y)) \geq \inf_{\rho\geq 0} h(\rho),$$

which combined with the previous inequality ensures that

$$\inf_{y\in X}\{\|x + tv - y\|^2 + (t + d_C(y))^2\} = \inf_{\rho\geq 0} h(\rho).$$

According to the obvious equality $h(\rho) = d^2_{\mathrm{Enl}_\rho(C)}(x+tv) + (t+\rho)^2$, we consider two cases:
- If $d_C(x+tv) \le \rho$, then $h(\rho) = (t+\rho)^2 \ge t^2 + d_C^2(x+tv)$.
- If $d_C(x+tv) > \rho$, using Lemma 3.259 we have

$$d_{\mathrm{Enl}_\rho C}(x+tv) = d_C(x+tv) - \rho,$$

and thus

$$h(\rho) = d_C^2(x+tv) + t^2 + 2\rho[\rho + t - d_C(x+tv)] \ge d_C^2(x+tv) + t^2,$$

where the last estimation is due to the fact that $\|v\| \le 1$.

In any case we have $h(\rho) \ge d_C^2(x+tv) + t^2$=h(0), where the latter equality follows from the fact that $d_{\mathrm{Enl}_0 C}(\cdot) = d_C(\cdot)$. It ensues that $\inf_{\rho \ge 0} h(\rho) = d_C^2(x+tv) + t^2$. Making use of the above equality $d_C^2(x+tv) = t^2\|v\|^2$ (given by the choice of t) we obtain

$$\inf_{y \in X} \{\|x + tv - y\|^2 + (-t - d_C(y))^2\} = t^2\|v\|^2 + t^2,$$

or equivalently (with the norm $|\!|\!|(u,\tau)|\!|\!| = \sqrt{\|u\|^2 + \tau^2}$ on $X \times \mathbb{R}$)

$$\inf_{y \in X} |\!|\!|(x+tv, -t) - (y, d_C(y))|\!|\!|^2 = t^2\|v\|^2 + t^2.$$

Noticing that $(x+tv, -t) \notin \operatorname{epi} d_C$ (by the inequality $-t < d_C(x+tv)$ since $t > 0$), we deduce that

$$d^2_{\operatorname{epi} d_C}(x+tv, -t) = t^2\|v\|^2 + t^2, \text{ so } (x,0) \in \operatorname{Proj}_{\operatorname{epi} d_C}\big((x,0) + t(v,-1)\big).$$

This means that $(v,-1)$ is in $N^P_{\operatorname{epi} d_C}(x,0)$ or, in other words, that $x^* \in \partial_P d_C(x)$. $\square$

18.3. Prox-regular sets and J-plr functions

Let C be a nonempty subset of a normed space $(X, \|\cdot\|)$, let $\overline{x} \in C$ and let $r \in]0,+\infty]$ and $\delta \in]0,+\infty]$. Translating Definitions 15.24 and 15.65 we say:

• The set C is *r-prox-regular*, if C is closed in X and if for any $x \in C$ and any $v \in N^P(C;x) \cap \mathbb{B}_X$ one has

$$(\mathcal{D}_{\text{global}}) \qquad x \in \operatorname{Proj}_C(x+tv) \quad \text{for all positive real numbers } t \le r.$$

• The set C is *r-prox-regular at $\overline{x} \in C$ with δ-latitude*, if $C \cap B(\overline{x},\delta)$ is closed relative to $B(\overline{x},\delta)$ and if for any $x \in C \cap B(\overline{x},\delta)$ and any $v \in N^P(C;x) \cap \mathbb{B}_X$ one has

$$(\mathcal{D}_{\text{local}}) \qquad x \in \operatorname{Proj}_C(x+tv) \quad \text{for all positive real numbers } t \le r.$$

When such a $\delta > 0$ exists, C is *r-prox-regular at $\overline{x}$.*

Clearly, a closed set C is r-prox-regular if and only if it is r-prox-regular at any of its points.

Assume that C is a closed convex set in X. Take any $x \in C$ and any $y \in X \setminus C$ with $x \in \operatorname{Proj}_C(y)$ (if any). We already know that $x \in \operatorname{Proj}_C(u)$ for every $u \in [x,y]$. Take any $u \in \{(1-\theta)x + \theta y : \theta > 1\}$. Then $y \in]x,u[$, so there is some $t \in]0,1[$ such that $y = (1-t)x + tu$. Consider any $z \in C$ and put $y_t := (1-t)x + tz$. By convexity of C we have $y_t \in C$. We note that

$$\frac{\|u-x\|}{\|y-x\|} = \frac{1}{t} \quad \text{and} \quad \frac{\|u-z\|}{\|y-y_t\|} = \frac{1}{t},$$

which ensures that

$$\|u - x\| = \frac{\|y - x\|}{\|y - y_t\|}\|u - z\| \le \|u - z\|$$

since $\|y - x\| \le \|y - y_t\|$ according to the inclusions $x \in \mathrm{Proj}\,_C(y)$ and $y_t \in C$. The inequality $\|u - x\| \le \|u - z\|$ being true for any $z \in C$, it ensues that $x \in \mathrm{Proj}\,_C(u)$. Consequently, $x \in \mathrm{Proj}\,_C(u)$ for any $u \in x + \mathbb{R}_+(y - x)$. In other words, for any $v \in N^P(C;x) \cap \mathbb{B}_X$ we have $x \in \mathrm{Proj}\,_C(x + tv)$ for all $t \in [0, +\infty[$. This means:

Any closed convex set C of X is r-prox-regular for $r = +\infty$.

Consider now any $a \in X$ and any real $r > 0$, and put $C := X \setminus B(a,r)$. Take any $x \in C$ with $\|x - a\| = r$. It is clear that $x \in \mathrm{Proj}\,_C(a)$, hence $x \in \mathrm{Proj}\,_C(u)$ for all $u \in [a, x]$. From this we see for any $v \in N^P(C;x) \cap \mathbb{B}_X$ we have $x \in \mathrm{Proj}\,_C(x+tv)$ for any $t \in [0, r]$. It results:

$X \setminus B(a,r)$ is r-prox-regular.

We already showed all the remarkable properties that prox-regular sets enjoy in Hilbert spaces. The development of similar properties in normed spaces, such as Fréchet differentiability of d_C on the open tube $U_r(C) \setminus C$ and others, will require specific conditions on the normed spaces. Let us fix the context which we will work in and which will allow us to find striking characterizations of prox-regularity as in the Hilbert space setting.

In the rest of this chapter, unless otherwise stated, we assume that

(XCS) $\left\{ \right.$ $(X, \|\cdot\|)$ is a uniformly convex Banach space whose norm $\|\cdot\|$ is uniformly smooth, so $\|\cdot\|$ **is both uniformly convex and uniformly smooth**.

The assumption will be written merely in certain main theorems. We recall (see Enflo theorem C.19 in Appendix C) that any uniformly convex Banach space can be renormed with an equivalent norm which is both uniformly convex and uniformly smooth; by Pisier theorem C.20 in Appendix C the new norm can even be chosen such that its moduli of convexity and smoothness are of power type.

Unlike the Hilbert space setting in Chapter 15 where we began with the theory of global prox-regularity of sets, we choose here in general uniformly convex Banach spaces to start with the local aspect. First, let us rephrase the above definition $(\mathcal{D}_{\text{local}})$ of local prox-regularity of sets in the context of the uniformly convex Banach space $(X, \|\cdot\|)$ by using the duality mapping.

DEFINITION 18.66. Let $r \in]0, +\infty]$ and $\delta \in]0, +\infty]$. Let C be a set in X and let $\overline{x} \in C$. The set C is said be *(r, δ)-prox-regular at $\overline{x}$* or *r-prox-regular at $\overline{x} \in C$ with δ-latitude*, if $C \cap B(\overline{x}, \delta)$ is closed relative to $B(\overline{x}, \delta)$ and if for any $x \in C \cap B(\overline{x}, \delta)$ and $v^* \in N^{*,P}(C;x) \cap \mathbb{B}_{X^*}$ one has

$$x \in \mathrm{Proj}\,_C(x + tJ^*(v^*)) \quad \text{for all positive real numbers } t \le r.$$

We say that C is *r-prox-regular* (resp. *prox-regular*) at $\overline{x}$ when it is (r, δ)-prox-regular at $\overline{x}$ for some $\delta > 0$ (resp. for some $r > 0$ and $\delta > 0$).

For certain reasons which will be clear later, we say δ-latitude instead of the expression of α-latitude used in Definition 15.65.

PROPOSITION 18.67. Let C be a nonempty subset of X which is r-prox-regular at $\overline{x} \in C$ with δ-latitude. Then for every positive real $t \le r$, for every $x \in C$ with $\|x - \overline{x}\| < \delta$ and for every $v^* \in N^{*,P}(C;x)$ with $\|v^*\| \le 1$

$$0 \ge \langle J[J^*(v^*) - t^{-1}(x' - x)], x' - x\rangle, \quad \forall x' \in C.$$

PROOF. Suppose that C is prox-regular at $\overline{x}$ with δ-latitude. Fix any positive real $t \le r$, any $x \in C$ with $\|x - \overline{x}\| < \delta$ and any $v^* \in N^{*,P}(C;x)$ with $0 < \|v^*\| \le 1$. By definition, x is a nearest point of $x + tJ^*(v^*)$ in C, that is,

$$\|x + tJ^*(v^*) - x\| \le \|x + tJ^*(v^*) - x'\|, \quad \forall x' \in C.$$

Setting $v := J^*(v^*)$ we rewrite the latter as

$$t\|v\| \le \|x - x' + tv\|, \quad \forall x' \in C. \tag{18.53}$$

Since $J(u)$ is the derivative of $\frac{1}{2}\|\cdot\|^2$ at u, we have for all $\lambda > 0$ and all $x' \in X$ that

$$\begin{aligned}\zeta(x') &:= \langle J[J^*(v^*) - t^{-1}(x' - x)], x' - x\rangle = \langle J(v - t^{-1}(x' - x)), x' - x\rangle \\ &\le (2\lambda)^{-1}\{\|v - t^{-1}(x' - x) + \lambda(x' - x)\|^2 - \|v - t^{-1}(x' - x)\|^2\}.\end{aligned}$$

In particular for $\lambda = t^{-1}$, we obtain

$$\begin{aligned}\zeta(x') &\le \frac{t}{2}\{\|v\|^2 - \|v - t^{-1}(x' - x)\|^2\} = \frac{t}{2}\{\|v\|^2 - t^{-2}\|x - x' + tv\|^2\} \\ &= \frac{1}{2t}\left\{(t\|v\|)^2 - \|x - x' + tv\|^2\right\}.\end{aligned}$$

Taking (18.53) into account we deduce that $\zeta(x') \le 0$ for all $x' \in C$, which is the desired inequality. □

The next definition is concerned with the local prox-regularity with respect to a normal vector.

DEFINITION 18.68. A set $C \subset X$ is called *prox-regular at* $\overline{x} \in C$ *for* $\overline{v}^* \in N^{*,P}(C;\overline{x})$ if there exist reals $\varepsilon > 0$ and $r > 0$ such that $C \cap B(\overline{x}, \varepsilon)$ is closed relative to $B(\overline{x}, \varepsilon)$ and if for every $x \in C \cap B(\overline{x}, \varepsilon)$ and every $v^* \in N^{*,P}(C;x)$ with $\|v^* - \overline{v}^*\| < \varepsilon$ the point x is a nearest point of $x + rJ^*(v^*)$ in $C \cap B(\overline{x}, \varepsilon)$; in such a case we will also say that C is *prox-regular at* $\overline{x}$ *for* $\overline{v}^*$ *with sizes* $\varepsilon > 0$ *and* $r > 0$.

In fact, we can show as follows that the property in the above concept for subsets of X needs to be checked merely for the normal direction $\overline{v}^* = 0$.

PROPOSITION 18.69. Let $C \subset X$ be a nonempty subset of X and let $\overline{x} \in C$. The following hold.
(a) The set C is prox-regular at $\overline{x}$ for every $v^* \in N^{*,P}(C;x)$ if and only if it is prox-regular at $\overline{x}$ for $v_0^* = 0$; the latter is also equivalent to the r-prox-regularity of C at $\overline{x}$ for some $r > 0$ in the sense of Definition 18.66.
(b) If the set C is prox-regular at $\overline{x}$ for $v_0^* = 0$ with sizes ε and r as in Definition 18.68, then for all positive $t \le r$, for all $x \in C$ with $\|x - \overline{x}\| < \varepsilon$ and for all $v^* \in N^{*,P}(C;x)$ with $\|v^*\| \le \varepsilon$

$$0 \ge \langle J[J^*(v^*) - t^{-1}(x' - x)], x' - x\rangle, \quad \forall x' \in C \text{ with } \|x' - \overline{x}\| < \varepsilon.$$

PROOF. (a) Obviously, if C is prox-regular at $\overline{x}$ for all $\overline{v}^* \in N^{*,P}(C;\overline{x})$, then it is so for $v_0^* = 0$. To establish the converse, let us assume that C is prox-regular at $\overline{x}$ for $0 \in X^*$ with sizes $\varepsilon > 0$ and $r > 0$. Take any $\overline{v}^* \in N^{*,P}(C;\overline{x})$ with $\overline{v}^* \neq 0$, and set $\varepsilon' := \min\{\varepsilon/2, \|\overline{v}^*\|/2\}$. Since $C \cap B(\overline{x}, \varepsilon)$ is closed relative to $B(\overline{x}, \varepsilon)$, it is clear that $C \cap B(\overline{x}, \varepsilon')$ is closed relative to $B(\overline{x}, \varepsilon')$. Further, for $x \in C$ and $v^* \in N^{*,P}(C; x)$ with $\|x - \overline{x}\| < \varepsilon'$ and $\|v^* - \overline{v}^*\| < \varepsilon'$ we have

$$\frac{\varepsilon}{2\|\overline{v}^*\|}\|v^*\| \leq \frac{\varepsilon}{2\|\overline{v}^*\|}(\|\overline{v}^* - v^*\| + \|\overline{v}^*\|) \leq \frac{\varepsilon}{2\|\overline{v}^*\|}\varepsilon' + \frac{\varepsilon}{2} \leq \frac{\varepsilon}{4} + \frac{\varepsilon}{2} < \varepsilon.$$

We may rewrite the latter as $\left\| \frac{\varepsilon v^*}{2\|\overline{v}^*\|} - 0 \right\| < \varepsilon$. By prox-regularity of C at $\overline{x}$ for 0 with sizes ε and r, we have that x is a nearest point in $\{x' \in C : \|x' - \overline{x}\| < \varepsilon\}$ to $x + rJ^*\left(\frac{\varepsilon v^*}{2\|\overline{v}^*\|}\right) = x + \frac{r\varepsilon}{2\|\overline{v}^*\|}J^*(v^*)$, hence x is also a nearest point of $x + \frac{r\varepsilon}{2\|\overline{v}^*\|}J^*(v^*)$ in $C \cap B(\overline{x}, \varepsilon')$. This means that C is prox-regular at $\overline{x}$ for $\overline{v}^*$ with sizes ε' and $r' := r\varepsilon/(2\|\overline{v}^*\|)$, which finishes the proof of the first equivalence in (a).

The second equivalence in (a) is easily seen to be due, for $\overline{x} \in C$ and $u \in B(\overline{x}, \eta)$ with $\eta > 0$, to the equalities (see Lemma 2.219)

$$d_C(u) = d_{C \cap B(\overline{x}, 2\eta)}(u) \quad \text{and} \quad \mathrm{Proj}_{\,C}(u) = \mathrm{Proj}_{\,C \cap B(\overline{x}, 2\eta)}(u).$$

(b) The proof of the assertion (b) is similar to that of Proposition 18.67. □

The equivalences in (a) of Proposition 18.69 lead us to particularize the following face of local prox-regularity in the context of uniformly convex Banach spaces.

DEFINITION 18.70. By the second equivalence in Proposition 18.69(a), we see that a nonempty set C of X is prox-regular at $\overline{x} \in C$ in the sense of Definition 18.66 if and only if it is prox-regular at $\overline{x}$ for $\overline{v}^* = 0$ with sizes $\varepsilon > 0$ and $r > 0$ for some reals $\varepsilon, r > 0$, that is, $C \cap B(\overline{x}, \varepsilon)$ is closed relative to $B(\overline{x}, \varepsilon)$ and

$$x \in \mathrm{Proj}_{\,C \cap B(\overline{x}, \varepsilon)}(x + rJ^*(v^*)) \text{ for all } x \in C \cap B(\overline{x}, \varepsilon),\ v^* \in N^{*,P}(C; x) \cap B(0, \varepsilon);$$

in such a case we say for simplicity that the set C is *prox-regular at* $\overline{x}$ *with sizes* $\varepsilon > 0$ *and* $r > 0$.

It is clear that if the set C is prox-regular at $\overline{x}$ with some sizes $\varepsilon > 0$ and $r > 0$, then it is so for any constant sizes $0 < \varepsilon' \leq \varepsilon$ and $0 < r' \leq r$.

The notion corresponding to the inequality in (b) of Proposition 18.69 is introduced for functions in Definition 18.71 below. Let us note that another definition was considered in Definition 11.1, using the "proximal-type" estimation with the square of the norm instead of the inequality in Definition 18.71. In the Hilbert setting, by Proposition 11.3 (with $s = 2$ therein) the definition given below is equivalent to that in Definition 11.1 for the large class of *primal lower regular* (plr) functions (that is, s-lower regular functions with $s = 2$). Recall that Moreau envelopes of plr functions (see Subsection 11.3.2, particularly Theorem 11.40) enjoy diverse remarkable properties.

We will see in the next section that the J-plr concept introduced in Definition 18.71 below for functions also yields, regarding their Moreau envelopes, various important properties which have their own interest. These properties applied to the indicator functions of sets will be among the keys of the development of the study of prox-regularity in the context of uniformly convex spaces in this chapter.

DEFINITION 18.71. A function $f : X \to \mathbb{R} \cup \{+\infty\}$ is called *J-primal lower regular* (*J-plr*) at $\overline{x} \in \operatorname{dom} f$ if there exist a threshold $t_0 \geq 0$ and real constants $\varepsilon > 0$ and $r > 0$ such that f is lower semicontinuous on $B(\overline{x}, \varepsilon)$ and

$$f(y) \geq f(x) + \langle J[J^*(v^*) - t(y-x)], y - x\rangle$$

for any $x, y \in B(\overline{x}, \varepsilon)$, any $v^* \in \partial_P f(x)$, and any real $t \geq t_0$ such that $\|v^*\| \leq \varepsilon r t$. In such a case we say that f is *J-plr at $\overline{x}$ with threshold $t_0 \geq 0$ and real constants $\varepsilon > 0$ and $r > 0$*.

REMARK 18.72. It is easily seen that if f is J-plr at $\overline{x}$ with threshold $t_0 \geq 0$ and some real constants $\varepsilon > 0$ and $r > 0$, then it is so for the threshold t_0 and any constants $\varepsilon' \in]0, \varepsilon]$ and $r' > 0$ such that $\varepsilon' r' \leq \varepsilon r$. □

For indicator functions the threshold t_0 is useless as shown in the following proposition.

PROPOSITION 18.73. Let C be a nonempty subset of X, let $\overline{x} \in C$ and let $\varepsilon, r \in]0, +\infty[$. The following are equivalent:
(a) the indicator function Ψ_C is J-plr at $\overline{x} \in C$ with a threshold $t_0 \geq 0$ and the constants ε, r;
(b) for any real $t_0 \geq 0$, the function Ψ_C is J-plr at $\overline{x}$ with the threshold t_0 and the same constants ε, r;
(c) the function Ψ_C is J-plr at $\overline{x}$ with the threshold $t_0 := 0$ and the same constants ε, r;
(d) the set $C \cap B(\overline{x}, \varepsilon)$ is closed relative to $B(\overline{x}, \varepsilon)$, and for any $x, x' \in C \cap B(\overline{x}, \varepsilon)$ and any $x^* \in N^{*,P}(C; x)$ with $\|x^*\| \leq \varepsilon$ one has

$$0 \geq \langle J[J^*(x^*) - r^{-1}(x' - x)], x' - x\rangle.$$

PROOF. The implications (c)⇒(b)⇒(a) are clearly evident. Let us show that (a)⇒(c). Suppose that (a) holds with some $t_0 > 0$, since the implication is trivial with $t_0 = 0$. Fix any $x, x' \in C \cap B(\overline{x}, \varepsilon)$. For any $x^* \in N^{*,P}(C; x)$ and any real $t \geq t_0$ with $\|x^*\| \leq \varepsilon r t$ one has by Definition 18.71

$$0 \geq \langle J[J^*(x^*) - t(x' - x)], x' - x\rangle.$$

Take any real $t > 0$ and any $x^* \in N^{*,P}(C; x)$ with $\|x^*\| \leq \varepsilon r t$. Since $t^{-1} t_0 x^* \in N^{*,P}(C; x)$ and $\|t^{-1} t_0 x^*\| \leq \varepsilon r t_0$, the above inequality gives

$$0 \geq \langle J[J^*(t^{-1} t_0 x^*) - t_0(x' - x)], x' - x\rangle, \text{ that is, } 0 \geq \langle J[J^*(x^*) - t(x' - x)], x' - x\rangle;$$

and the latter inequality still holds if $t = 0$ and $\|x^*\| \leq \varepsilon r t$. This justifies the desired property with threshold $\tau_0 = 0$, that is, (a)⇒(c) holds.

Suppose that (d) holds. The closedness of $C \cap B(\overline{x}, \varepsilon)$ relative to $B(\overline{x}, \varepsilon)$ ensures that Ψ_C is lower semicontinuous on $B(\overline{x}, \varepsilon)$. On the other hand, taking any real $t \geq r^{-1}$ and any $x^* \in N^{*,P}(C; x)$ with $\|x^*\| \leq \varepsilon r t$, we have $r^{-1} t^{-1} \|x^*\| \leq \varepsilon$, hence under (d) we obtain

$$0 \geq \langle J[J^*(r^{-1} t^{-1} x^*) - r^{-1}(x' - x)], x' - x\rangle, \text{ that is, } 0 \geq \langle J[J^*(x^*) - t(x' - x)], x' - x\rangle.$$

This tells us that (d) implies (a) with $t_0 = r^{-1}$. The proof is then finished, since (c) trivially implies (d). □

The above proposition leads to the following definition.

DEFINITION 18.74. Let C be a nonempty set in X, let $\overline{x} \in C$ and let $\varepsilon, r \in]0, +\infty[$. When anyone of properties (a),$\cdots$,(d) in Proposition 18.73 holds, we will say that the indicator function Ψ_C is J-plr at $\overline{x}$ with real constants $\varepsilon > 0$ and $r > 0$.

If the function f is J-plr at $\overline{x} \in \operatorname{dom} f$ with threshold $t_0 \geq 0$ and real constants $\varepsilon > 0$ and $r > 0$, one readily derives that

$$\langle J[J^*(u^*) - t(y - x)] - J[J^*(v^*) - t(x - y)], y - x\rangle \leq 0 \tag{18.54}$$

for any $x, y \in B(\overline{x}, \varepsilon)$, for any $u^* \in \partial_P f(x)$, $v^* \in \partial_P f(y)$, and any real $t \geq t_0$ such that $\max\{\|u^*\|, \|v^*\|\} \leq \varepsilon r t$. This is the analog of the *hypomonotonicity property* of certain truncations (or localizations) of $\partial_P f$ which characterize plr functions in Hilbert spaces, see Definition 11.47. The hypomonotonicity is no more appropriate in the setting of Banach spaces, and therefore according to (18.54) we introduce the following concept, which is called J-hypomonotonicity.

DEFINITION 18.75. A multimapping $T : X \rightrightarrows X^*$ is said to be *J-hypomonotone of rate $\overline{t} \geq 0$ on a set $U \subset X$* if for any real $t \geq \overline{t}$ and any $(x_i, x_i^*) \in \operatorname{gph} T$ with $x_i \in U$, $i = 1, 2$, one has

$$\langle J[J^*(x_1^*) - t(x_2 - x_1)] - J[J^*(x_2^*) - t(x_1 - x_2)], x_2 - x_1\rangle \leq 0.$$

When $U = X$, we just say that T is J-hypomonotone of rate $\overline{t} \geq 0$.

Throughout, we will also use the following concept of truncation of a multimapping.

DEFINITION 18.76. Let $T : X \rightrightarrows X^*$ be a multimapping and let reals $\varepsilon > 0$ and $t \geq 0$. Its *ε, t-truncation* at a point $\overline{x} \in X$ is defined as the multimapping $T_{\overline{x},\varepsilon,t}$ whose graph is given by

$$\operatorname{gph} T_{\overline{x},\varepsilon,t} := \{(x, x^*) \in \operatorname{gph} T : \|x - \overline{x}\| < \varepsilon,\ \|x^*\| \leq t\}.$$

When there is no risk of ambiguity, $T_{\overline{x},\varepsilon,t}$ will be denoted for simplicity by T_t.

The following proposition shows the J-hypomonotonicity of $(\partial_P f)_{\overline{x},\varepsilon,\varepsilon r \overline{t}}$ with rate $\overline{t}$. This is the main example for our development.

PROPOSITION 18.77. If a function $f : X \to \mathbb{R} \cup \{+\infty\}$ is J-plr at $\overline{x} \in \operatorname{dom} f$ with threshold $t_0 \geq 0$ and real constants $\varepsilon > 0$ and $r > 0$ in Definition 18.71, then for any real $\overline{t} \geq t_0$ the multimapping $(\partial_P f)_{\overline{x},\varepsilon,\varepsilon r \overline{t}}$ is J-hypomonotone of rate $\overline{t}$.

PROOF. Assume that f is J-plr at $\overline{x}$ with threshold $t_0 \geq 0$ and real constants $\varepsilon, r > 0$. Fix any real $\overline{t} \geq t_0$ and take any $(x_i, x_i^*) \in \operatorname{gph} (\partial_P f)_{\overline{x},\varepsilon,\varepsilon r \overline{t}}$. This means that $\|x_i - \overline{x}\| < \varepsilon$ and $x_i^* \in \partial_P f(x_i)$ with $\|x_i^*\| \leq \varepsilon r \overline{t}$, where $i = 1, 2$. Take any real $t \geq \overline{t}$. Then, $t \geq t_0$ and $\|x_i^*\| \leq \varepsilon r t$, hence (18.54) entails that

$$\langle J[J^*(x_1^*) - t(x_2 - x_1)] - J[J^*(x_2^*) - t(x_1 - x_2)], x_2 - x_1\rangle \leq 0.$$

This translates that $(\partial_P f)_{\overline{x},\varepsilon,\varepsilon r \overline{t}}$ is J-hypomonotone of rate $\overline{t}$. □

We are not far from sets since the next proposition shows that the prox-regularity of a set C entails the J-plr property of its indicator function Ψ_C. The equivalence will be obtained later in Theorem 18.96, as well as the equivalence with the J-hypomonotonicity of a certain truncation of the normal cone $N^{*,P}(C; \cdot)$.

It will be convenient, for any $\sigma \geq 0$, to denote below by $N^{*\sigma,P}(C; \cdot)$ (or $N_C^{*\sigma,P}(\cdot)$) the multimapping $N^{*,P}(C; \cdot)$ truncated with $\sigma \mathbb{B}_{X^*}$, that is,

$$N_C^{*\sigma,P}(x) = N^{*\sigma,P}(C; x) := N^{*,P}(C; x) \cap \sigma \mathbb{B}_{X^*} \quad \text{for all } x \in X. \tag{18.55}$$

PROPOSITION 18.78. Let C be a set in X which is prox-regular at $\overline{x} \in C$, with sizes $\varepsilon > 0$ and $r > 0$. The following hold.
(a) The indicator function Ψ_C of C is J-plr at $\overline{x}$ with real constants $\varepsilon > 0$ and $r > 0$; there also exist reals ε' and r' with $0 < \varepsilon' < 1 < \frac{1}{r'}$ such that Ψ_C is J-plr with real contants ε' and r' (in fact it is so for any $0 < \varepsilon' < \min\{1, \varepsilon\}$ and $0 < r' < \min\{1, r\}$).
(b) For any real $t \geq 0$, putting $\sigma := \varepsilon r t$, the restriction of the multimapping $N^{*\sigma,P}(C;\cdot)$ to $B(\overline{x}, \varepsilon)$ is J-hypomonotone of rate t.

PROOF. Since C is prox-regular at $\overline{x}$ for $\overline{v}^* = 0$ with sizes $\varepsilon > 0$ and $r > 0$, the set $C \cap B(\overline{x}, \varepsilon)$ is closed relative to $B(\overline{x}, \varepsilon)$, hence Ψ_C is lower semicontinuous on $B(\overline{x}, \varepsilon)$. Further, Proposition 18.69 tells us that

$$\Psi_C(x') \geq \Psi_C(x) + \langle J[J^*(v^*) - r^{-1}(x' - x)], x' - x\rangle,$$

whenever $x' \in B(\overline{x}, \varepsilon)$, $x \in C \cap B(\overline{x}, \varepsilon)$, and $\|v^*\| \leq \varepsilon$ with $v^* \in N^{*,P}(C; x)$. We know by Proposition 18.64 that $v^* \in N^{*,P}(C; x)$ if and only if $v^* \in \partial_P \Psi_C(x)$. If $x \in C \cap B(\overline{x}, \varepsilon)$ and $v^* \in N^{*,P}(C; x)$ with $\|v^*\| \leq \varepsilon r t$, then $r^{-1}t^{-1}v^* \in N^{*,P}(C; x)$ with $\|r^{-1}t^{-1}v^*\| \leq \varepsilon$, hence for all $x' \in B(\overline{x}, \varepsilon)$

$$\Psi_C(x') \geq \Psi_C(x) + \langle J[J^*(r^{-1}t^{-1}v^*) - r^{-1}(x' - x)], x' - x\rangle,$$

or equivalently

$$\Psi_C(x') \geq \Psi_C(x) + \langle J[J^*(v^*) - t(x' - x)], x' - x\rangle.$$

This means that the function Ψ_C is J-plr at $\overline{x}$ with constants ε and r, according to Definition 18.74 and Proposition 18.73(c).

Regarding the second property in (a), it suffices to apply Remark 18.72.
(b) The assertion (b) is a direct consequence of (a), of Proposition 18.77 and of Proposition 18.73(c). □

We will need another result concerning J-hypomonotone multimappings. It will be one of the crucial steps of the development of the proof of Theorem 18.84. First, recall that a multimapping $T : X \rightrightarrows X^*$ is bounded when its range $T(X) = \bigcup_{x \in X} T(x)$ is a bounded set in X^*.

LEMMA 18.79. Let $T : X \rightrightarrows X^*$ be a bounded multimapping with nonempty domain $\operatorname{Dom} T$ and which is J-hypomonotone of rate $\overline{r}$. Then, for any real $r > 2\overline{r}$ one has that $(I + r^{-1}J^* \circ T)^{-1}$ is single-valued on its domain D, and the mapping that it determines on D is uniformly continuous on any nonempty bounded set of X included in D.

If in addition the modulus of convexity of X is of power type q, then the single-valued mapping restriction of $(I + r^{-1}J^* \circ T)^{-1}$ to D is Hölder continuous with power $1/q$ on any nonempty bounded subset of D.

PROOF. Let us denote by μ any upper bound of $\{\|y\| : y \in T(x), x \in \operatorname{Dom} T\}$. Fix any reals $r > 2\overline{r}$ and $\rho > 0$ with $\rho \mathbb{B}_X \cap \operatorname{Dom}(I + r^{-1}J^* \circ T)^{-1} \neq \emptyset$. Take $x_i \in \operatorname{Dom}(I + r^{-1}J^* \circ T)^{-1}$ with $\|x_i\| \leq \rho$, $i = 1, 2$. Choose any $y_i \in (I + r^{-1}J^* \circ T)^{-1}(x_i)$, that is, $J[r(x_i - y_i)] \in T(y_i)$, $i = 1, 2$, hence $\|x_i - y_i\| \leq \mu/r$. By assumption, for any real $t \geq \overline{r}$ we have

$$0 \geq \langle J\{J^*(J[r(x_1 - y_1)]) - t(y_2 - y_1)\} - J\{J^*(J[r(x_2 - y_2)]) - t(y_1 - y_2)\}, y_2 - y_1\rangle,$$

or equivalently

$$0 \geq \langle J(rx_1 - ty_2 + (t - r)y_1) - J(rx_2 - ty_1 + (t - r)y_2), y_2 - y_1\rangle.$$

At this stage, we see that the result is obvious if X is one-dimensional. So, suppose that X is at least two-dimensional. Then, for any $\lambda \in]0,1[$ such that $\frac{\lambda r}{2} > \overline{r}$, replacing t in the latter inequality by rt_λ, where $t_\lambda := \lambda/2$, we obtain

$$\begin{aligned}0 \geq &\langle J(x_1 - t_\lambda y_2 - (1-t_\lambda)y_1) - J(x_2 - t_\lambda y_1 - (1-t_\lambda)y_2), x_1 - x_2 + (1-2t_\lambda)(y_2 - y_1)\rangle \\ &+ \langle J(x_1 - t_\lambda y_2 - (1-t_\lambda)y_1) - J(x_2 - t_\lambda y_1 - (1-t_\lambda)y_2), x_2 - x_1\rangle := \text{(I)} + \text{(II)}.\end{aligned}$$

Note also that

$$\begin{aligned}\|x_1 - t_\lambda y_2 - (1-t_\lambda)y_1\| &\leq (1-t_\lambda)\|x_1 - y_1\| + t_\lambda\|x_1 - y_2\| \\ &\leq (1-t_\lambda)\|x_1 - y_1\| + t_\lambda(\|x_1 - x_2\| + \|x_2 - y_2\|) \\ &\leq (1-t_\lambda)\frac{\mu}{r} + t_\lambda\left(2\rho + \frac{\mu}{r}\right) \leq \gamma,\end{aligned}$$

where $\gamma := \rho + \frac{\mu}{r}$, and similarly $\|x_2 - t_\lambda y_1 - (1-t_\lambda)y_2\| \leq \gamma$. Considering the constant $K > 0$ furnished by Xu-Roach Theorem 18.52(a) and putting

$$u_1 := x_1 - t_\lambda y_2 - (1-t_\lambda)y_1 \quad \text{and} \quad u_2 := x_2 - t_\lambda y_1 - (1-t_\lambda)y_2,$$

it follows that, for $u_1 \neq u_2$

$$\begin{aligned}\text{(I)} &\geq K(\|u_1\| \vee \|u_2\|)^2 \delta_X\left(\frac{\|u_1 - u_2\|}{2(\|u_1\| \vee \|u_2\|)}\right) \\ &= \frac{1}{2}K\|u_1 - u_2\|(\|u_1\| \vee \|u_2\|)\frac{2(\|u_1\| \vee \|u_2\|)}{\|u_1 - u_2\|}\delta_X\left(\frac{\|u_1 - u_2\|}{2(\|u_1\| \vee \|u_2\|)}\right) \\ &\geq \frac{1}{2}K\|u_1 - u_2\|(\|u_1\| \vee \|u_2\|)\frac{2\gamma}{\|u_1 - u_2\|}\delta_X\left(\frac{\|u_1 - u_2\|}{2\gamma}\right) \\ &= \gamma K(\|u_1\| \vee \|u_2\|)\,\delta_X\left(\frac{\|u_1 - u_2\|}{2\gamma}\right),\end{aligned}$$

where the second inequality is due to the nondecreasing property on $]0,2]$ of the function $\varepsilon \mapsto \delta_X(\varepsilon)/\varepsilon$ (established in Proposition 18.25(c3)). On the other hand, by Theorem 18.52(b) and the constant $L > 0$ therein, we also have

$$\begin{aligned}\text{(II)} &\geq -\|J(u_1) - J(u_2)\|\,\|x_1 - x_2\| \\ &\geq -L\frac{(\|u_1\| \vee \|u_2\|)^2}{\|u_1 - u_2\|}\varrho_X\left(\frac{\|u_1 - u_2\|}{\|u_1\| \vee \|u_2\|}\right)\|x_1 - x_2\|.\end{aligned}$$

Thus, observing that $\frac{\|u_1 - u_2\|}{\|u_1\| \vee \|u_2\|} \leq 2$, the nondecreasing property of the function $\tau \mapsto (1/\tau)\varrho_X(\tau)$ on $]0,+\infty[$ (see Proposition 18.25(s3)) allows us to write

$$\text{(II)} \geq -L(\|u_1\| \vee \|u_2\|)(1/2)\varrho_X(2)\,\|x_1 - x_2\|.$$

This and the above inequality for (I) combined with the inequality $0 \geq \text{(I)} + \text{(II)}$ yield

$$\delta_X\left(\frac{\|(x_1 - x_2) - (1-2t_\lambda)(y_1 - y_2)\|}{2\gamma}\right) \leq \frac{L}{2\gamma K}\varrho_X(2)\,\|x_1 - x_2\|.$$

From this it results that $\left(I + r^{-1}J^* \circ T\right)^{-1}$ is a single-valued mapping on its domain and it is uniformly continuous on the intersection of its domain with the ball $\rho\mathbb{B}$ according to the (strict) increasing property of $\delta_X(\cdot)$ on $[0,2]$ seen in Proposition 18.25(c2) (or according to the fact that δ_X is nondecreasing along with $\delta_X(\varepsilon) = 0 \Leftrightarrow \varepsilon = 0$).

Assume now that $\delta_X(\cdot)$ is of power type q. Then, we have

$$\kappa\,\frac{\|(x_1-x_2)-(1-2t_\lambda)(y_1-y_2)\|^q}{2^q\gamma^q}\le\frac{L}{2\gamma K}\varrho_X(2)\,\|x_1-x_2\|, \tag{18.56}$$

where $\kappa>0$ is the constant involved in the definition of the power type property of the modulus of convexity $\delta_X(\cdot)$ (see 18.40). To proceed further the estimation, we need to consider two cases.

The *first case* is when $(1-2t_\lambda)\|y_1-y_2\|>\|x_1-x_2\|$. In that case, we need to estimate below $\|a-b\|^q$ for $\|a\|>\|b\|$. Since $\|a-b\|\ge\|a\|-\|b\|>0$, we derive

$$\|a-b\|^q\ge(\|a\|-\|b\|)^q=\|a\|^q\left(1-\frac{\|b\|}{\|a\|}\right)^q.$$

This leads us to consider the real-valued function $g(s)=(1-s)^q+qs$ on the interval $s\in[0,1[$. As the derivative $g'(s)=-q(1-s)^{q-1}+q$ is non-negative on this interval, the function g is non-decreasing on $[0,1[$ and then $g(s)\ge g(0)=1$ for all $s\in[0,1[$. Finally, $(1-s)^q\ge 1-qs$ for $s\in[0,1[$. We conclude that

$$\|a-b\|^q\ge\|a\|^q\left(1-\frac{\|b\|}{\|a\|}\right)^q\ge\|a\|^q\left(1-q\frac{\|b\|}{\|a\|}\right)=\|a\|^q-q\|a\|^{q-1}\|b\|.$$

This tells us that

$$\begin{aligned}&\|x_1-x_2+(1-2t_\lambda)(y_2-y_1)\|^q\\ &\ge(1-2t_\lambda)^q\|y_1-y_2\|^q-q(1-2t_\lambda)^{q-1}\|y_1-y_2\|^{q-1}\|x_1-x_2\|\\ &\ge(1-2t_\lambda)^q\|y_1-y_2\|^q-\alpha\,\|x_1-x_2\|,\end{aligned}$$

where α is some non-negative constant depending on μ,ρ,r. Combining this with (18.56) we derive that, for some constant $\beta'>0$ (depending on μ,ρ,r)

$$\|y_1-y_2\|\le\beta'\,\|x_1-x_2\|^{1/q}.$$

The *second case* is when $(1-2t_\lambda)\|y_1-y_2\|\le\|x_1-x_2\|$. Noticing in this case that $\|x_1-x_2\|\le(2\rho)^{1-\frac{1}{q}}\|x_1-x_2\|^{1/q}$, we see that in both cases we have that

$$\|y_1-y_2\|\le\beta\,\|x_1-x_2\|^{1/q},$$

for some constant $\beta>0$ (depending on μ,ρ,r). This, combined with the similar inequality in the first case, justifies that the single-valued mapping $\big(I+r^{-1}J^*\circ T\big)^{-1}$ is Hölder continuous (with power $1/q$) on the intersection of its domain with the ball $\rho\mathbb{B}$. □

18.4. Local Moreau envelopes of J-plr functions

The main tool in this section will be the so-called local Moreau envelope of a function already considered in Section 11.2. Let $f:X\to\mathbb{R}\cup\{+\infty\}$ be a function and $W\subset X$ be a nonempty closed subset over which f is lower semicontinuous, bounded from below and finite at some point. We recall that its *local Moreau envelope of index* $\lambda>0$ (relative to W), is defined as

$$e_{\lambda,W}f(x):=\inf_{y\in W}\left\{f(y)+\frac{1}{2\lambda}\|x-y\|^2\right\}\quad\text{for all } x\in X. \tag{18.57}$$

Note that the infimum in (18.57) may be seen as taken over the whole space X for the function $\tilde f$ given by $\tilde f(x)=f(x)$ if $x\in W$ and $\tilde f(x)=+\infty$ otherwise. By Theorem 11.23 we know that the functions $e_{\lambda,W}f$ are everywhere defined and

Lipschitz on bounded subsets. We also recall that the *proximal mapping of index* λ (relative to W) is defined by

$$\operatorname{Prox}_{\lambda,W} f(x) := \left\{ y \in W : e_{\lambda,W} f(x) = f(y) + \frac{1}{2\lambda}\|x-y\|^2 \right\} \quad \text{for all } x \in X.$$

Whenever there exists some $p_\lambda(x) \in \operatorname{Prox}_{\lambda,W} f(x)$ one has by (11.20)

$$\partial_F(e_{\lambda,W} f)(x) \subset \{\lambda^{-1} J(x - p_\lambda(x))\} \cap \partial_F \tilde{f}(p_\lambda(x)), \tag{18.58}$$

and hence if in addition $\partial_F(e_{\lambda,W} f)(x) \neq \emptyset$ the set $\operatorname{Prox}_{\lambda,W} f(x)$ is a singleton thanks to the one-to-one property of the mapping J. We know by Theorem 11.23(d) that the infimum in (18.57) is attained whenever x is a point of Fréchet subdifferentiability of $e_{\lambda,W} f$. Denoting by G_λ the subset of X where $e_{\lambda,W} f$ is Fréchet subdifferentiable, we obtain for any $x \in G_\lambda$ that $\operatorname{Prox}_{\lambda,W} f(x) = \{p_\lambda(x)\}$ and $\partial_F(e_{\lambda,W} f)(x) = \{\lambda^{-1} J(x - p_\lambda(x))\}$. Note that G_λ is dense in X according to the result in Corollary 4.53 concerning the density of subdifferentiability points in Asplund spaces (keep in mind that any reflexive Banach space is Asplund).

When $\operatorname{Prox}_{\lambda,W} f(x)$ is a singleton, we will generally denote by $P_{\lambda,W} f(x)$ its unique element, that is, $\operatorname{Prox}_{\lambda,W} f(x) = \{P_{\lambda,W} f(x)\}$.

18.4.1. Fréchet differentiability of local Moreau envelope. This subsection is mainly devoted to the study of differentiability of local Moreau envelopes of J-plr functions.

We begin with the following lemma which will be used in the proof of Proposition 18.91.

LEMMA 18.80. Let $f : X \to \mathbb{R} \cup \{+\infty\}$ be a function and $W \subset X$ be a closed set over which f is lower semicontinuous and bounded from below, and with $W \cap \operatorname{dom} f \neq \emptyset$. For $x \in X$ the following assertions are equivalent:
(a) $\partial_F(e_{\lambda,W} f)(x) \neq \emptyset$;
(b) $e_{\lambda,W} f$ is Fréchet differentiable at x.

Further, in the case of (a) or (b), one has

$$\operatorname{Prox}_{\lambda,W} f(x) = \{P_{\lambda,W} f(x)\} \quad \text{and} \quad D_F(e_{\lambda,W} f)(x) = \lambda^{-1} J(x - P_{\lambda,W} f(x)).$$

PROOF. Obviously (b) implies (a). Now suppose that (a) holds. As we have seen above, the set $\operatorname{Prox}_{\lambda,W} f(x)$ is a singleton with a unique element $p_\lambda(x) := P_{\lambda,W} f(x)$ and $\partial_F(e_{\lambda,W} f)(x) = \{\lambda^{-1} J(x - p_\lambda(x))\}$. Then, for any $\varepsilon > 0$ there exists some $\delta > 0$ such that for any $t \in]0, \delta[$ and for any $y \in \mathbb{B}_X$

$$\langle \lambda^{-1} J(x - p_\lambda(x)), ty \rangle \le e_{\lambda,W} f(x + ty) - e_{\lambda,W} f(x) + \varepsilon t,$$

hence

$$t^{-1}\big(e_{\lambda,W} f(x + ty) - e_{\lambda,W} f(x)\big) - \langle \lambda^{-1} J(x - p_\lambda(x)), y \rangle \ge -\varepsilon. \tag{18.59}$$

On the other hand, taking δ smaller if necessary and using the definition of $e_\lambda f$ and the fact that

$$J(x - p_\lambda(x)) = D_F\left(\frac{1}{2}\|\cdot\|^2\right)(x - p_\lambda(x)),$$

we have for any $y \in \mathbb{B}_X$

$$\begin{aligned}
&e_{\lambda,W} f(x+ty) - e_{\lambda,W} f(x) \\
&\le f(p_\lambda(x)) + \frac{1}{2\lambda}\|x+ty-p_\lambda(x)\|^2 - f(p_\lambda(x)) - \frac{1}{2\lambda}\|x-p_\lambda(x)\|^2 \\
&\le \langle \lambda^{-1} J(x-p_\lambda(x)), ty\rangle + \varepsilon t,
\end{aligned}$$

that is,

$$t^{-1}\big(e_{\lambda,W} f(x+ty) - e_{\lambda,W} f(x)\big) - \langle \lambda^{-1} J(x-p_\lambda(x)), y\rangle \le \varepsilon.$$

Combining this and (18.59) we obtain the Fréchet differentiability of $e_{\lambda,W} f$ at x as well as the equality $D_F(e_{\lambda,W} f)(x) = \lambda^{-1} J(x - p_\lambda(x))$. □

The differentiability of the Moreau envelopes is also related to their regularity. This connexion given by the equivalence between assertions (a) and (b) of the next lemma will be needed in Proposition 18.91. The lemma also establishes in preparation for Theorem 18.84 the differentiability of $e_{\lambda,W} f$ under the single valuedness of $\operatorname{Prox}_{\lambda,W} f$ and the continuity of the associated mapping $P_{\lambda,W} f$. Before giving its statement, let us recall that a function f is Fréchet subdifferentially regular at x provided $\partial_F f(x) = \partial_C f(x)$, where we recall that ∂_C stands for the Clarke subdifferential.

LEMMA 18.81. Under the assumptions of Lemma 18.80, for any nonempty open subset U of X, the equivalences (a)⇔(b) and (c)⇔(d) hold for the following properties:
(a) $e_{\lambda,W} f$ is Fréchet subdifferentially regular on U.
(b) $e_{\lambda,W} f$ is Fréchet differentiable on U and its Fréchet derivative $D_F(e_{\lambda,W} f) : U \to X^*$ is norm-to-weak* continuous.
(c) $e_{\lambda,W} f$ is continuously Fréchet differentiable on U (and hence (a) and (b) hold).
(d) The single-valued mapping $P_{\lambda,W} f$ is well defined on U and norm-to-norm continuous on U.

In any one of these cases, $e_{\lambda,W} f$ is Fréchet differentiable on U with

$$D_F(e_{\lambda,W} f)(x) = \lambda^{-1} J(x - P_{\lambda,W} f(x)) \quad \text{for all } x \in U.$$

PROOF. (a)⇒(b): If $\partial_F(e_{\lambda,W} f)(x) = \partial_C(e_{\lambda;W} f)(x)$ for any $x \in U$, then we have that $\partial_F(e_{\lambda,W}) f(x) \neq \emptyset$ for any $x \in U$, and hence $e_{\lambda,W} f$ is Fréchet differentiable on U from Lemma 18.80. Moreover, $\partial_C(e_{\lambda,W} f)(x) = \{D_F(e_{\lambda,W} f)(x)\}$ for any $x \in U$, and by the norm-to-weak* upper semicontinuity of $\partial_C(e_{\lambda,W} f)$ (see Proposition 2.74) we have that $D_F(e_{\lambda,W} f)$ is norm-to-weak* continuous.
(b)⇒(a): Conversely, if $e_{\lambda,W} f$ is Fréchet differentiable and norm-to-weak* continuous on U, we have that $\partial_F(e_{\lambda,W}) f(x) = \{D_F(e_{\lambda,W} f)(x)\} = \partial_L(e_{\lambda,W} f)(x)$ for any $x \in U$, keeping in mind that, for a locally Lipschitz function $g : X \to \mathbb{R}$ the *limiting subdifferential* $\partial_L g$ coincides with the weak* sequential outer limit

$$\text{(18.60)} \qquad {}^{\text{seq}}\operatorname*{Lim\,sup}_{y \to x} \partial_F g(y) := \{w^* - \lim x_n^* : x_n^* \in \partial_F g(x_n), x_n \to x\}.$$

By Theorem 4.120(c) we know that $\partial_C g(x) = \overline{\text{co}}^* \partial_L g(x)$, where we recall that $\overline{\text{co}}^*$ denotes the w^*-closed convex hull in X^*. Thus, we obtain that

$$\partial_F(e_{\lambda,W} f)(x) = \partial_C(e_{\lambda,W} f)(x) \quad \text{for all } x \in U,$$

which is the assertion (a).
(c)⇒(d): The assumption that $e_{\lambda,W} f$ is continuously differentiable on U implies

via Lemma 18.80 that the single valued mapping $P_{\lambda,W} f$ is well defined on U and norm-to-norm continuous therein, that is, the desired implication holds.
(d)$\Rightarrow$(c): Assume now that $P_{\lambda,W} f$ is well defined on U and norm-to-norm continuous therein. This continuity property along with (18.58) and (18.60) gives $\partial_C(e_{\lambda,W} f)(x) = \{\lambda^{-1} J(x - P_{\lambda,W} f(x))\}$, which entails that $e_{\lambda,W} f$ is Gâteaux differentiable on U with $D_G(e_{\lambda,W} f)(x) = \lambda^{-1} J(x - P_{\lambda,W} f(x))$. The norm-to-norm continuity of $P_{\lambda,W} f$ once again yields the existence of $D_F(e_{\lambda,W} f)$ as well as its norm-to-norm continuity on U. The proof of the lemma is then complete. □

18.4.2. Uniform continuity of local proximal mappings of J-plr functions. Throughout this subsection, *we fix a point* $\overline{x} \in \operatorname{dom} f$ *around which* f *is lower semicontinuous and we fix* $\rho > 0$ *such that* f *is bounded from below over* $B[\overline{x}, 4\rho]$ *and lower semicontinuous there. So, we also fix* $W = B[\overline{x}, 4\rho]$. Note that according to the lower semicontinuity of f around $\overline{x}$, one always has some $\rho > 0$ with the desired property.

By the basic localization in Lemma 11.17 with $\beta = c = 0$, there exists some $\lambda_0 > 0$ such that for all $\lambda \in]0, \lambda_0]$

$$\text{Prox}_{\lambda,W} f(x) \subset B(\overline{x}, 3\rho) \quad \text{for all } x \in U := B(\overline{x}, \rho). \tag{18.61}$$

So, (recalling that G_λ denotes the set of points where $e_{\lambda,W} f$ is Fréchet subdifferentiable) for any $x \in U \cap G_\lambda$ the unique element $P_{\lambda,W} f(x)$ of $\text{Prox}_{\lambda,W} f(x)$ belongs to $B(\overline{x}, 3\rho)$. Further, for any $x \in U \cap G_\lambda$, on the one hand by Lemma 18.80

$$D_F(e_{\lambda,W} f)(x) = \lambda^{-1} J(x - P_{\lambda,W} f(x)) \in \partial_F f(P_{\lambda,W} f(x)), \tag{18.62}$$

and on the other hand

$$\|x - P_{\lambda,W} f(x)\| \le \|x - \overline{x}\| + \|\overline{x} - P_{\lambda,W} f(x)\| < \rho + 3\rho = 4\rho. \tag{18.63}$$

In fact, we can make (18.62) more precise by proving in the following lemma that the stronger inclusion $D_F(e_{\lambda,W} f)(x) \in \partial_P f(P_{\lambda,W} f(x))$ holds for $x \in U \cap G_\lambda$.

LEMMA 18.82. Let $\rho > 0$, $W = B[\overline{x}, 4\rho]$ and $\lambda_0 > 0$ be as above such that (18.61) holds for all $\lambda \in]0, \lambda_0]$. Then, for any $\lambda \in]0, \lambda_0]$, $x \in U \cap \operatorname{Dom} \text{Prox}_{\lambda,W} f$, and $p_\lambda(x) \in \text{Prox}_{\lambda,W} f(x)$, we have that $\lambda^{-1} J(x - p_\lambda(x)) \in \partial_P f(p_\lambda(x))$. In other words, for any $x \in U$ and any $\lambda \in]0, \lambda_0]$,

$$\text{Prox}_{\lambda,W} f(x) \subset (I + \lambda J^* \circ \partial_P f)^{-1}(x).$$

PROOF. Fix any $0 < \lambda \le \lambda_0$, $x \in U \cap \operatorname{Dom} \text{Prox}_{\lambda,W} f$, and $p_\lambda(x) \in \text{Prox}_{\lambda,W} f(x)$. Then,

$$f(p_\lambda(x)) + (2\lambda)^{-1} \|x - p_\lambda(x)\|^2 \le f(y) + (2\lambda)^{-1} \|x - y\|^2, \quad \forall y \in W,$$

that is,

$$(2\lambda)^{-1} \|x - p_\lambda(x)\|^2 - (2\lambda)^{-1} \|x - y\|^2 \le f(y) - f(p_\lambda(x)), \quad \forall y \in W.$$

Since $p_\lambda(x) \in B[\overline{x}, 3\rho]$, the latter inequality holds true in particular for all $y \in B[p_\lambda(x), \rho]$. Let us set $v := \lambda^{-1}(x - p_\lambda(x))$. From the same latter inequality we have

$$2^{-1} \lambda \|v\|^2 - (2\lambda)^{-1} \|x - y\|^2 \le f(y) - f(p_\lambda(x)), \quad \forall y \in B[p_\lambda(x), \rho],$$

which entails

$$\frac{\lambda^2}{2} \|v\|^2 - \frac{1}{2} \|\lambda v + p_\lambda(x) - y\|^2 \le \lambda [f(y) - f(p_\lambda(x))], \quad \forall y \in B[p_\lambda(x), \rho],$$

or equivalently

$$\lambda^2\|v\|^2-\|\lambda v+p_\lambda(x)-y\|^2 \le 2\lambda[\beta-f(p_\lambda(x))], \quad \forall(y,\beta)\in \operatorname{epi} f \text{ with } y\in B[p_\lambda(x),\rho].$$

Adding λ^2 to both sides yields, for all $(y,\beta)\in \operatorname{epi} f$ with $y\in B[p_\lambda(x),\rho]$,

$$\lambda^2\|v\|^2-\|\lambda v+p_\lambda(x)-y\|^2+\lambda^2 \le 2\lambda[\beta-f(p_\lambda(x))]+\lambda^2,$$

and using the inequality $2\lambda[\beta-f(p_\lambda(x))]+\lambda^2 \le [\beta-f(p_\lambda(x))+\lambda]^2$ we obtain that, for all $(y,\beta)\in \operatorname{epi} f$ with $y\in B[p_\lambda(x),\rho]$,

$$\lambda^2\|v\|^2+\lambda^2 \le \|\lambda v+p_\lambda(x)-y\|^2+[\beta-f(p_\lambda(x))+\lambda]^2.$$

So, we obtain that for all $(y,\beta)\in \operatorname{epi} f$ with $y\in B[p_\lambda(x),\rho]$

$$|\!|\!|\lambda(v,-1)|\!|\!| \le |\!|\!|(p_\lambda(x),f(p_\lambda(x)))+\lambda(v,-1)-(y,\beta)|\!|\!|.$$

By the local character of the proximal normal cone seen in Proposition 18.56 the latter inequality entails that $(v,-1)\in N^P_{\operatorname{epi} f}(p_\lambda(x),f(p_\lambda(x)))$, which gives by the definition of proximal subdifferential that $J(v)\in\partial_P f(p_\lambda(x))$. This means that

$$\lambda^{-1}J(x-p_\lambda(x))\in\partial_P f(p_\lambda(x)),$$

which entails the inclusion of the lemma. □

REMARK 18.83. (a) In the case when f is the indicator function of a nonempty set $C\subset X$, for any $\lambda>0$ and $W=X$, or equivalently for $\rho=+\infty$, one has $e_\lambda f(x)=\frac{1}{2\lambda}d_C^2(x)$ and $\operatorname{Prox}_\lambda f(x)=\operatorname{Proj}_C(x)$ for all $x\in X$. If C is closed, that is, f is lower semicontinuous, then the above conclusions hold for $W=X$.
(b) Still with $f=\Psi_C$, for any $\overline{x}\in C$, any $W\ni\overline{x}$ and any $\lambda>0$ one has $e_\lambda f(x)=\frac{1}{2\lambda}d^2_{C\cap W}(x)$ and $\operatorname{Prox}_\lambda f(x)=\operatorname{Proj}_{C\cap W}(x)$ for all $x\in X$. Let $W:=B[\overline{x},4\rho]$ for some $\rho>0$. Recalling by Lemma 2.219 that for every $x\in B(\overline{x},\rho)$

$$d(x,C)=d\big(x,C\cap B(\overline{x},2\rho)\big) \quad\text{and}\quad \operatorname{Proj}_C(x)=\operatorname{Proj}_{C\cap B(\overline{x},2\rho)}(x),$$

taking $W:=B[\overline{x},4\rho]$ we derive from the above equalities for $d_C(x)$ and $\operatorname{Proj}_C(x)$ and from the inclusion $B(\overline{x},2\rho)\subset W$ that for every $x\in B(\overline{x},\rho)$

$$d_C(x)=d_{C\cap W}(x)=d(x,C\cap B(\overline{x},2\rho)),$$

$$\operatorname{Proj}_C(x)=\operatorname{Proj}_{C\cap W}(x)=\operatorname{Proj}_{C\cap B(\overline{x},2\rho)}(x)\subset B(\overline{x},2\rho).$$

Therefore, for any real $\lambda>0$ we obtain that for all $x\in U:=B(\overline{x},\rho)$

$$e_{\lambda,W}f(x)=\frac{1}{2\lambda}d_C^2(x) \text{ and } \operatorname{Prox}_{\lambda,W}f(x)=\operatorname{Proj}_C(x)\subset B(\overline{x},2\rho). \tag{18.64}$$

If $C\cap W$ is closed, then the conclusions in Lemma 18.82 hold with $f=\Psi_C$ and $\operatorname{Proj}_C(x)$ in place of $\operatorname{Prox}_{\lambda,W}f(x)$. □

Given a real $\alpha>0$, recall that a function g is of class $\mathcal{C}^{1,\alpha}$ on an open set $U\subset X$ when it is differentiable on U and the derivative Dg is locally Hölder continuous on U with power α.

THEOREM 18.84 (uniform continuity of local proximal mapping). Let $f:X\to\mathbb{R}\cup\{+\infty\}$ be a function which is J-plr at $\overline{x}\in\operatorname{dom} f$ with a threshold $t_0\ge 0$ and real constants ε and r in Definition 18.71, such that $\varepsilon<1<\frac{1}{r}$. Let $\rho\in\,]0,\frac{r\varepsilon}{16}]$ be fixed in such a way that f is bounded from below on $W:=B[\overline{x},4\rho]$. Let $\lambda_0>0$ be such that $4\lambda_0 t_0\le 1$ and for any $0<\lambda\le\lambda_0$

$$\operatorname{Prox}_{\lambda,W}f(x)\subset B(\overline{x},3\rho) \quad \text{for all } x\in B(\overline{x},\rho) \tag{18.65}$$

(such reals λ_0 exist according to (18.61)). Then, for any $\lambda \in]0, \lambda_0]$ the multimapping $x \mapsto \operatorname{Prox}_{\lambda,W} f(x)$ is single-valued over $U := B(\overline{x}, \rho)$, that is, $\operatorname{Prox}_{\lambda,W} f(x) = \{P_{\lambda,W} f(x)\}$ for every $x \in U$, and the mapping $P_{\lambda,W} f$ is uniformly continuous on U. Moreover, the function $e_{\lambda,W} f$ is differentiable on U with $De_{\lambda,W} f$ uniformly continuous on U, and

$$D(e_{\lambda,W} f)(x) = \lambda^{-1} J(x - P_{\lambda,W} f(x)) \quad \text{for all } x \in U.$$

If the modulus of convexity of $\|\cdot\|$ is of power type q, then the mapping $P_{\lambda,W} f$ is Hölder continuous on U with power $1/q$, that is, for some constant $\gamma \geq 0$

$$\|P_{\lambda,W} f(x) - P_{\lambda,W} f(x')\| \leq \gamma \|x - x'\|^{\frac{1}{q}} \quad \text{for all } x, x' \in U;$$

and further, the function $e_{\lambda,W} f$ is of class $\mathcal{C}^{1,\alpha}$ on U with $\alpha := q^{-1}(s-1)$ whenever in addition the modulus of smoothness of $\|\cdot\|$ is of power type s.

PROOF. By Definition 18.71 we note that f is lower semicontinuous on $B(\overline{x}, \varepsilon)$. Let λ_0 with $4\lambda_0 t_0 \leq 1$ be given as in the statement of the theorem for ρ fixed as in the same statement of the theorem. We will work with arbitrary fixed $\lambda \in]0, \lambda_0]$. Put $c := r\varepsilon$ and $T_{ct} := (\partial_P f)_{\overline{x},\varepsilon,ct}$ for any $t \geq 0$. The proof is divided in three steps.

Step 1. Let us prove that $\operatorname{Prox}_{\lambda,W} f$ is single-valued and uniformly continuous on $U \cap \operatorname{Dom} \operatorname{Prox}_{\lambda,W} f$.
We have from Definition 18.75 and from Proposition 18.77 that the multimapping T_{ct} is J-hypomonotone of rate t for any $t \geq t_0$. Hence, for $t_\lambda := c/(4\lambda)$, the multimapping T_{t_λ} is J-hypomonotone of rate $\overline{r} := 1/(4\lambda)$. As $\lambda^{-1} > 2\overline{r}$, Lemma 18.79 entails that $(I + \lambda J^* \circ T_{t_\lambda})^{-1}$ is a single-valued mapping on its domain and that this mapping is uniformly continuous (resp. Hölder continuous with power $1/q$) on any nonempty bounded subset of its domain. From Lemma 18.82, we have that $\operatorname{Prox}_{\lambda,W} f(x) \subset (I + \lambda \partial_P f)^{-1}(x)$ for any $x \in U$. We claim that we even have

$$\operatorname{Prox}_{\lambda,W} f(x) \subset (I + \lambda J^* \circ T_{t_\lambda})^{-1}(x) \quad \text{for any } x \in U. \tag{18.66}$$

Indeed, fixing any $x \in U \cap \operatorname{Dom} \operatorname{Prox}_{\lambda,W} f$ and $p_\lambda(x) \in \operatorname{Prox}_{\lambda,W} f(x)$, we know by the assumption (18.65) that $p_\lambda(x) \in B[\overline{x}, 3\rho]$. So $\|p_\lambda(x) - \overline{x}\| < r\varepsilon < \varepsilon$, and

$$\begin{aligned} &\|\lambda^{-1} J(x - p_\lambda(x))\| \\ &\leq \lambda^{-1} \|x - p_\lambda(x)\| \leq \lambda^{-1}(\|x - \overline{x}\| + \|\overline{x} - p_\lambda(x)\|) \leq 4\rho\lambda^{-1} \leq (c\lambda^{-1})/4 = t_\lambda. \end{aligned}$$

Hence, as also $\lambda^{-1} J(x - p_\lambda(x)) \in \partial_P f(p_\lambda f(x))$ by Lemma 18.81, we have that

$$\lambda^{-1} J(x - p_\lambda(x)) \in T_{t_\lambda}(p_\lambda(x)),$$

which proves the claim. Therefore, we obtain that $\operatorname{Prox}_{\lambda,W} f(x) = \{p_\lambda(x)\}$ is a singleton for all $x \in U \cap \operatorname{Dom} \operatorname{Prox}_{\lambda,W} f$ and that the mapping p_λ is uniformly continuous (resp. Hölder continuous with power $1/q$) over $U \cap \operatorname{Dom} \operatorname{Prox}_\lambda f$, that is, for some modulus of continuity $\omega(\cdot)$ (resp. some real $\gamma > 0$) one has for all $x, x' \in U \cap \operatorname{Dom} \operatorname{Prox}_{\lambda,W} f$

$$\|p_\lambda(x) - p_\lambda(x')\| \leq \omega(\|x - x'\|) \tag{18.67}$$

(resp.

$$\|p_\lambda(x) - p_\lambda(x')\| \leq \gamma \|x - x'\|^{\frac{1}{q}}). \tag{18.68}$$

Step 2. Let us prove that $U \subset \operatorname{Dom} \operatorname{Prox}_{\lambda,W} f$.
Take any $x \in U$ and fix some integer $k \geq 1$ with $B(x, 1/k) \subset U$. According to the

density of the Fréchet subdifferentiability points of $e_{\lambda,W}f$, for any integer $n \ge k$ there exists some point $x_n \in G_\lambda \cap B(x, 1/n)$, where we recall that G_λ denotes the set of points where $e_{\lambda,W}f$ is Fréchet subdifferentiable. By (18.67), for any integers $n, m \ge k$,

$$\|p_\lambda(x_n) - p_\lambda(x_m)\| \le \omega(\|x_n - x_m\|).$$

Hence, $(p_\lambda(x_n))_n$ is a Cauchy sequence in W. Denote by $z_\lambda \in W$ its limit. By the definition of $p_\lambda(x_n)$ we have

$$f(p_\lambda(x_n)) + \frac{1}{2\lambda}\|x_n - p_\lambda(x_n)\|^2 \le f(y) + \frac{1}{2\lambda}\|x_n - y\|^2, \quad \forall y \in W.$$

Since f is lower semicontinuous on W, the latter implies

$$f(z_\lambda) + \frac{1}{2\lambda}\|x - z_\lambda\|^2 \le f(y) + \frac{1}{2\lambda}\|x - y\|^2, \quad \forall y \in W.$$

This means that $z_\lambda \in \operatorname{Prox}_{\lambda,W} f(x)$, which yields $U \cap \operatorname{Dom} \operatorname{Prox}_{\lambda,W} f = U$. Hence, (18.67) (resp. (18.68)) holds for all $x, x' \in U$ and, through Step 1, $\operatorname{Prox}_{\lambda,W} f$ is single-valued on U, and its associated mapping p_λ is uniformly continuous (resp. Hölder continuous with power $1/q$) on U. Then, by Lemma 18.81 the envelope $e_{\lambda,W}f$ is continuously Fréchet differentiable on U with $D_F e_{\lambda,W}f(x) = \lambda^{-1}J(x - p_\lambda(x))$ for any $x \in U$.

Step 3. We will prove that $De_{\lambda,W}f$ is uniformly continuous (resp. Hölder continuous with power $\alpha = q^{-1}(s-1)$ on U).
According to the equality $D_F(e_{\lambda,W}f)(x) = \lambda^{-1}J(x - p_\lambda(x))$, the uniform continuity of $D(e_{\lambda,W}f)$ on U follows from the uniform continuity of J on bounded subsets of X (see Proposition 18.41) and from the uniform continuity of p_λ on U.

Concerning the Hölder property, assume the modulus of smoothness of $\|\cdot\|$ is of power type s. Take any $x, x' \in U$ and, for $\beta := 4\rho$, use Corollary 18.51(b) to estimate

$$\begin{aligned}
&\|D_F(e_{\lambda,W}f)(x) - D_F(e_{\lambda,W}f)(x')\| = \lambda^{-1}\|J(x - p_\lambda(x)) - J(x' - p_\lambda(x'))\| \\
&\le \lambda^{-1}L'_\beta\|x - p_\lambda(x) - x' + p_\lambda(x')\|^{s-1} \le \lambda^{-1}L'_\beta[\|x - x'\| + \|p_\lambda(x) - p_\lambda(x')\|]^{s-1} \\
&\le \lambda^{-1}L'_\beta[\|x - x'\| + \gamma\|x - x'\|^{\frac{1}{q}}]^{s-1} \le \lambda^{-1}KL'_\beta(1+\gamma)^{s-1}\|x - x'\|^{\frac{s-1}{q}},
\end{aligned}$$

where the third inequality follows from (18.68) and the last one from the fact that $\|x - x'\| < 1$. The proof of the theorem is then complete. □

We now state in the next proposition the relation obtained between the proximal mappings and some truncations of the subdifferential of a J-plr function.

PROPOSITION 18.85. Under the assumptions of Theorem 18.84, one has for all $\lambda \in \,]0, \lambda_0]$ and $x \in U = B(\overline{x}, \rho)$

$$P_{\lambda,W}f(x) = (I + \lambda^{-1}J^* \circ T_{t_\lambda})^{-1}(x),$$

where $t_\lambda := \varepsilon r/(4\lambda)$ and $T_{t_\lambda} := (\partial_P f)_{\overline{x},\varepsilon,t_\lambda}$.

PROOF. Theorem 18.84 tells us in particular that the mapping $P_{\lambda,W}f$ is well defined over U. Then, for each $x \in U$, by (18.66) established in the proof of the same theorem we have that $P_{\lambda,W}f(x)$ belongs to the right member of the proposition. Since $(I + \lambda^{-1}J^* \circ T_{t_\lambda})^{-1}$ is at most single-valued as seen in Lemma 18.79, the equality in the proposition holds true with the identification of the right member singleton set with its element. □

In the next corollary and throughout the rest of the chapter, when $\operatorname{Proj}_C(x) = \{p(x)\}$ is a singleton, we will denote its unique element by $P_C(x)$ according to our convention of notation through the book.

COROLLARY 18.86. Let $C \subset X$ be a non-empty set of X, and let $\overline{x} \in C$ and $\varepsilon \in]0,1[$ be such that $C \cap B(\overline{x}, \varepsilon)$ is closed relative to $B(\overline{x}, \varepsilon)$. Assume that the indicator function Ψ_C is J-plr at $\overline{x} \in C$ with real sizes ε and $r \in]0,1[$ (hence $\varepsilon < 1 < \frac{1}{r}$). The following hold.
(a) For $\rho = \frac{\varepsilon r}{16}$ the single-valued mapping $x \mapsto P_C(x)$ is well defined on $U = B(\overline{x}, \rho)$ and uniformly continuous on U. Further, the function d_C^2 is differentiable on U with $D\, d_C^2$ uniformly continuous on U, and

$$D\left(\frac{1}{2}d_C^2\right)(x) = J(x - P_C(x)) \quad \text{for all } x \in U.$$

(b) If the modulus of convexity of $\|\cdot\|$ is of power type q, then there is some constant $\gamma \geq 0$ such that

$$\|P_C(x) - P_C(x')\| \leq \gamma \|x - x'\|^{\frac{1}{q}}, \quad \forall x, x' \in U;$$

and further, the function d_C^2 is of class $\mathcal{C}^{1,\alpha}$ on U with $\alpha := q^{-1}(s-1)$ whenever in addition the modulus of smoothness of $\|\cdot\|$ is of power type s.

PROOF. The corollary follows directly from (18.64) in Remark 18.83(b) and from Theorem 18.84. □

COROLLARY 18.87. Under the assumptions of Corollary 18.86, we have that

$$P_C(x) = (I + J^* \circ N_C^{*\sigma,P})^{-1}(x) \quad \text{for all}\, x \in U = B(\overline{x}, \varepsilon r/16),$$

where $\sigma := \varepsilon r/4$ and where the multimapping $N_C^{*\sigma,P}(\cdot)$ is defined by (18.55).

PROOF. Taking $\rho = \frac{\varepsilon r}{16}$ as in the statement of Corollary 18.86, since $r < 1$ it is easily checked for $T := \partial_P \psi_C$ that for all $x \in U = B(\overline{x}, \rho)$

$$(I + J^* \circ T_{\overline{x},\varepsilon,\frac{\varepsilon r}{4}})^{-1}(x) = (I + J^* \circ N_C^{*\varepsilon r/4,P})^{-1}(x).$$

Note that the properties in (18.64) of Remark 18.83(b) allow us to apply Proposition 18.85 with $\lambda = \lambda_0 = 1$ to $f = \Psi_C$. Doing so and using the above equality for $(I + J^* \circ T_{\overline{x},\varepsilon,\frac{\varepsilon r}{4}})^{-1}(x)$ give the desired equality of the corollary. □

18.5. Characterizations of local prox-regular sets in uniformly convex Banach spaces

In this section we will give diverse characterizations of local prox-regularity for sets in uniformly convex Banach spaces.

18.5.1. Metric projection of local prox-regular sets. Given a set C in X which is prox-regular at $\overline{x} \in C$, we will establish various basic properties of the metric projection to C around the point $\overline{x}$.

First, the following lemma and the next ones and propositions, until Proposition 18.94, consider some general properties of differentiability of distance function in the uniformly convex Banach space X.

LEMMA 18.88. Let C be a nonempty closed subset of X. Then the single valuedness and norm-to-weak continuity of the metric projection mapping P_C over a nonempty open set U imply its norm-to-norm continuity on U.

PROOF. Let $(u_n)_n$ be any sequence in X with $u_n \xrightarrow{\|\cdot\|} u$. By the norm-to-weak continuity assumption of P_C we have $P_C(u_n) \xrightarrow{w} P_C(u)$. Further, from the continuity of the distance function we also have

$$\|u_n - P_C(u_n)\| = d_C(u_n) \longrightarrow d_C(u) = \|u - P_C(u)\|.$$

By the Kadec-Klee property of the norm $\|\cdot\|$ (see Proposition 18.17) we conclude that $P_C(u_n) \xrightarrow{\|\cdot\|} P_C(u)$. □

REMARK 18.89. The proof of Lemma 18.88 makes clear that its statement is still valid if $(X, \|\cdot\|)$ is any normed space whose norm $\|\cdot\|$ is Kadec-Klee. □

The following proposition establishes that the continuity of the metric projection mapping to a set C is equivalent to the continuous differentiability of the distance function d_C, as shown in the Hilbert setting in Chapter 15, where it is proved that those properties characterize the prox-regularity of a set in a Hilbert space. Its proof follows directly from Lemma 18.81 with $f = \Psi_C$.

PROPOSITION 18.90. Let $C \subset X$ be a nonempty closed set and $U \subset X$ be a nonempty open set. Then the following are equivalent:
(a) the single-valued mapping P_C is well defined on U and norm-to-norm continuous on U;
(b) d_C^2 is of class $\mathcal{C}^1$ on U.

In fact, these properties are equivalent to the only Fréchet subdifferentiability of the distance function. This is shown, among diverse features, in the following proposition which extends to uniformly convex Banach spaces the results proved for Hilbert spaces in Proposition 15.19.

PROPOSITION 18.91. Let $(X, \|\cdot\|)$ be a uniformly convex Banach space whose norm $\|\cdot\|$ is also uniformly smooth. For any nonempty closed set $C \subset X$ and any nonempty open set U of X the following are equivalent:
(a) d_C is continuously differentiable on $U \setminus C$;
(b) $\partial_F d_C(x)$ is non-empty for all $x \in U$;
(c) $\partial_F d_C^2(x)$ is non-empty for all x in U;
(d) d_C is Fréchet differentiable on $U \setminus C$;
(e) d_C is Fréchet subdifferentially regular on $U \setminus C$;
(f) d_C is Gâteaux differentiable on $U \setminus C$ with $\|D_G d_C(x)\|_* = 1$ for all $x \in U$.

PROOF. (a) ⇒ (b) is obvious since one always has $0 \in \partial_F d_C(u)$ for any $u \in C$.
(b) ⇒ (c) follows from the fact that for any $x^* \in \partial_F d_C(x)$, one has $2d_C(x)x^* \in \partial_F d_C^2(x)$, which can be seen as in the proof of (a) in Lemma 15.11.
(c) ⇒ (d). By Lemma 18.80 and Remark 18.83(a) we have under (c) that d_C^2 is Fréchet differentiable on U, hence so is d_C on $U \setminus C$.
(d) ⇒ (a). It is clear that (d) entails (b) and hence (c). From Lemma 18.80 and Remark 18.83(a) we get the Fréchet differentiability of d_C^2 on U. The latter implies (see Lemma 18.80) that the single-valued mapping P_C is well defined on U and that

$$D_F d_C^2(x) = 2J(x - P_C(x)) \quad \text{for any } x \in U. \tag{18.69}$$

So, it remains to prove the norm-to-norm continuity of P_C over U and we will obtain that of $D_F d_C^2$. Take any $x_0 \in U$, and $U \ni x_n \xrightarrow{\|\cdot\|} x_0$. Then, we also have that

$$\|x_n - P_C(x_n)\| = d_C(x_n) \longrightarrow d_C(x_0) = \|x_0 - P_C(x_0)\|,$$

which entails that the sequence $(P_C(x_n))_n$ is bounded. The space X being reflexive, let z be the weak limit of any weakly convergent subsequence of $(P_C(x_n))_n$ that we do not relabel. As $\|x_0 - P_C(x_n)\| \longrightarrow \|x_0 - P_C(x_0)\|$, having in mind the Kadec-Klee property of the norm $\|\cdot\|$ (see Proposition 18.17), it suffices to prove that

$$\|x_0 - z\| = \|x_0 - P_C(x_0)\| \tag{18.70}$$

to get that $P_C(x_n) \xrightarrow{\|\cdot\|} z$, and hence that $z \in C$. Then, using (18.70) and the fact that $P_C(x_0)$ is a singleton, we will have that $P_C(x_0) = \{z\}$, which easily gives that all the sequence $(P_C(x_n))_n$ strongly converges to $P_C(x_0)$, so P_C is norm-to-norm continuous at x_0. In order to show (18.70), consider the real sequence $(t_n)_n$ defined by $t_n^2 := \|x_0 - P_C(x_n)\|^2 - d_C^2(x_0)$ for each $n \in \mathbb{N}$, and note that $t_n \geq 0$ for all n along with $t_n \to 0$ as $n \to \infty$. If $t_n = 0$ for n in an infinite subset J of $\mathbb{N}$, then for each $n \in J$ it ensues that $P_C(x_n) = P_C(x_0)$ due to the single-valuedness of Proj_C on U, hence $z = P_C(x_0)$ by the weak convergence of $(P_C(x_n))_n$ to z, so (18.70) holds true in this case. We may then suppose that $t_n > 0$ for every $n \in \mathbb{N}$. Fix any real $\varepsilon > 0$. From the Fréchet differentiability of d_C^2 at x_0 and from the boundedness of the sequence $(P_C(x_n) - x_0)_n$, for n large enough we have

$$\begin{aligned}
\langle D_F d_C^2(x_0), P_C(x_n) - x_0\rangle &\leq \frac{d_C^2(x_0 + t_n(P_C(x_n) - x_0)) - d_C^2(x_0)}{t_n} + \frac{\varepsilon}{4} \\
&\leq \frac{\|x_0 + t_n(P_C(x_n) - x_0) - P_C(x_n)\|^2 - d_C^2(x_0)}{t_n} + \frac{\varepsilon}{4} \\
&= \frac{(1 - t_n)^2\|x_0 - P_C(x_n)\|^2 - d_C^2(x_0)}{t_n} + \frac{\varepsilon}{4},
\end{aligned}$$

which entails for n large enough

$$\begin{aligned}
\langle D_F d_C^2(x_0), P_C(x_n) - x_0\rangle &\leq \frac{\|x_0 - P_C(x_n)\|^2 - d_C^2(x_0)}{t_n} - 2\|x_0 - P_C(x_n)\|^2 + \frac{\varepsilon}{2} \\
&= t_n - 2\|x_0 - P_C(x_n)\|^2 + \frac{\varepsilon}{2} \\
&\leq -2d_C^2(x_0) + \varepsilon.
\end{aligned}$$

Passing to the limit, we obtain $\langle D_F d_C^2(x_0), x_0 - z\rangle \geq 2d_C^2(x_0)$, or equivalently

$$\langle 2J(x_0 - P_C(x_0)), x_0 - z\rangle \geq 2\|x_0 - P_C(x_0)\|^2,$$

which implies that $\|x_0 - z\| \geq \|x_0 - P_C(x_0)\|$. Since $\|x_n - P_C(x_n)\| \leq \|x_n - P_C(x_0)\|$, one also has that

$$\liminf_{n\to\infty} \|x_n - P_C(x_n)\| \leq \lim_{n\to\infty} \|x_n - P_C(x_0)\|,$$

and hence by the weak lower semicontinuity of the norm one gets $\|x_0 - z\| \leq \|x_0 - P_C(x_0)\|$. Consequently, $\|x_0 - z\| = \|x_0 - P_C(x_0)\|$, that is, (18.70) is proved, and hence the implication (d) $\Rightarrow$ (a) is established.
The implications (e) $\Rightarrow$ (d) and (a) $\Rightarrow$ (e) follow from Lemma 18.81, and the equivalence (d) $\Leftrightarrow$ (f) is a direct consequence of Corollary 14.22. The proof is then complete. □

We now proceed to establish two lemmas. The first one is a key result already proved in Lemma 15.20 in the Hilbert context. The principal ideas of the proof are valid in this new setting, so we just sketch the main additional parts below.

LEMMA 18.92. Let C be a nonempty closed subset of X. Assume that d_C is Fréchet differentiable on a neighborhood of a point $\overline{u} \notin C$. Then there exists $\delta > 0$ such that whenever $u \in B(\overline{u}, \delta)$ and $P_C(u) = x$, there exists some $t > 0$ such that the point $u_t := u + t(u - x)$ likewise has $P_C(u_t) = x$.

PROOF. By Propositions 18.91 and 18.90, there exists $\varepsilon > 0$ such that the single-valued mapping P_C is well defined on $B(\overline{u}, 2\varepsilon)$ and norm-to-norm continuous on $B(\overline{u}, 2\varepsilon)$, with d_C continuously Fréchet differentiable on this ball as well. For each $u \in B(\overline{u}, \varepsilon)$ and each $t > 0$ put $u_t := u + t(u - P_C(u))$. Following the proof of Lemma 15.20, we find out some positive numbers $\delta < \varepsilon$ and $s < 1$ such that for all $u \in B(\overline{u}, \delta)$ one has $d_C(u) \geq \delta$, $sd_C(u) < \delta$ and $d_C(u_s) > d_C(u)$. Fix now $u \in B(\overline{u}, \delta)$ and consider the closed set $D := \{w \in X : d_C(w) \geq d_C(u_s)\}$. As $u \notin D$, according to Lau's theorem (see 14.29) there is a sequence $D \not\ni y_n \xrightarrow{\|\cdot\|} u$ with $\operatorname{Proj}_D(y_n) \neq \emptyset$. Choosing $w_n \in \operatorname{Proj}_D(y_n)$ we have $d_C(w_n) = d_C(u_s)$ (because w_n is a boundary point of D). For all n large enough, $w_n \in B(\overline{u}, 2\delta)$ since

(18.71)
$$\|y_n - w_n\| = d_D(y_n) \leq \|y_n - u_s\| \longrightarrow \|u - u_s\| = s\|u - P_C(u)\| = sd_C(u) < \delta.$$

Consequently, d_C is Fréchet differentiable at w_n and by (18.69) we have

$$D_F d_C(w_n) = J(w_n - P_C(w_n))/d_C(w_n) \quad \text{and} \quad \|D_F d_C(w_n)\| = 1.$$

Therefore, by Proposition 2.194 the half-space $E := \{v \in X : \langle -D^F d_C(w_n), v\rangle \leq 0\}$ coincides with the Bouligand-Peano tangent cone of D at w_n, and hence its negative polar cone $-[0, +\infty[\, D_F d_C(w_n)$ contains the Fréchet normal cone of D at w_n. The nonzero functional $J(y_n - w_n)$ being a proximal normal functional to D at w_n, it belongs to the Fréchet normal cone of D at w_n (see Proposition 18.58). Consequently, there exists some $\lambda_n > 0$ such that $J(y_n - w_n) = -\lambda_n D_F d_C(w_n)$, which entails

$$y_n - w_n = -\lambda_n(w_n - P_C(w_n))/d_C(w_n) \quad \text{and} \quad \lambda_n = \|y_n - w_n\|.$$

For n large enough, we have by (18.71) that $\lambda_n < \delta$, and hence

$$\lambda_n < \delta \leq d_C(u) < d_C(u_s) = d_C(w_n).$$

It follows that for $\alpha_n := \lambda_n/d_C(w_n)$ we have $\alpha_n \in]0, 1[$ and $y_n = (1 - \alpha_n)w_n + \alpha_n P_C(w_n)$. Hence, it results that $P_C(y_n) = P_C(w_n)$ and

$$\lambda_n = \|y_n - w_n\| = d_C(w_n) - d_C(y_n) = d_C(u_s) - d_C(y_n).$$

Putting $t_n := \dfrac{\alpha_n}{1 - \alpha_n} = \dfrac{d_C(u_s) - d_C(y_n)}{d_C(y_n)}$, we obtain $w_n = y_n + t_n(y_n - P_C(y_n))$.

As $(t_n)_n$ converges to $t := (d_C(u_s) - d_C(u))/d_C(u) > 0$, we have $w_n \xrightarrow{\|\cdot\|} u_t$ and $u_t \in B(\overline{u}, 2\delta)$ by (18.71) and by the inclusion $u \in B(\overline{u}, \delta)$. So, by continuity of P_C over $B(\overline{u}, 2\delta)$ we get $P_C(w_n) \xrightarrow{\|\cdot\|} P_C(u_t)$. But we also have $P_C(w_n) = P_C(y_n) \xrightarrow{\|\cdot\|} P_C(u)$. Finally, for this number t we have $P_C(u_t) = P_C(u)$, and hence the proof is complete. $\square$

The next lemma follows from the previous one.

LEMMA 18.93. Let C be a closed subset and $\overline{x} \in C$. If the single-valued mapping P_C is well defined over a neighborhood U of $\overline{x}$ and norm-to-norm continuous therein, then there exists some real $\varepsilon > 0$ such that for all $x \in C \cap B(\overline{x}, \varepsilon)$ and all $v \in N^P(C; x)$ with $v \neq 0$ the equality $P_C(x + \varepsilon \frac{v}{\|v\|}) = x$ holds.

PROOF. Let $\delta > 0$ be such that P_C is single-valued on $B(\overline{x}, \delta)$ and norm-to-norm continuous therein. Take $0 < \varepsilon < \delta/2$ and consider any non-zero $v \in N^P_C(x)$ with $\|x - \overline{x}\| < \varepsilon$. By definition of the proximal normal cone and by Proposition 18.55, there exists $\lambda > 0$ such that $P_C(x + \lambda v) = x$. Set

$$\lambda_s := \sup\left\{\lambda \leq \varepsilon : P_C(x + \lambda \frac{v}{\|v\|}) = x\right\}.$$

By the continuity of P_C on $B(\overline{x}, \delta)$ we have that $P_C(x + \lambda_s \frac{v}{\|v\|}) = x$. Suppose that $\lambda_s < \varepsilon$. As $x + \lambda_s \frac{v}{\|v\|}$ belongs to the open set $B(\overline{x}, \delta)$ where d_C is Fréchet differentiable according to Proposition 18.90, by Lemma 18.92 there exists $\eta > 0$ with $\lambda_s + \eta \leq \varepsilon$ such that $P_C\left(x + (\lambda_s + \eta)\frac{v}{\|v\|}\right) = x$. This gives a contradiction with the definition of λ_s. We then conclude that $\lambda_s = \varepsilon$. □

Lemma 18.93 allows us to establish the following proposition. It is concerned with the metric projection to a local prox-regular set and it prepares the theorem on characterizations of local prox-regularity.

PROPOSITION 18.94. Let C be a nonempty set of X which is closed near $\overline{x} \in C$. The following assertions are equivalent:
(a) C is prox-regular at $\overline{x}$;
(b) there exists $\varepsilon>0$ such that the condition $\left.\begin{array}{c} x=P_C(u), x \neq u \\ 0 < \|u-\overline{x}\| < \varepsilon \end{array}\right\}$ implies that $x=P_C(u')$ for $u':=x+\varepsilon\frac{u-x}{\|u-x\|}$;
(c) there exists $\varepsilon > 0$ such that $v \in N^P(C; x)$ with $x \in C \cap B(\overline{x}, \varepsilon)$ and $v \neq 0$ implies that

$$P_C\left(x + \varepsilon \frac{v}{\|v\|}\right) = x.$$

PROOF. Without loss of generality, we may and do suppose that C is closed.
(a) $\Rightarrow$ (b): If C is prox-regular at $\overline{x}$, then by Proposition 18.78 and Corollary 18.86 there is a neighborhood U of $\overline{x}$ over which the single-valued mapping P_C is well defined and uniformly continuous as well. Let a real $\varepsilon > 0$ be given by Lemma 18.93. By continuity of P_C, take $\varepsilon' \in]0, \varepsilon[$ such that $\left.\begin{array}{c} x = P_C(u), x \neq u \\ 0 < \|u - \overline{x}\| < \varepsilon' \end{array}\right\}$ implies that $\|x - \overline{x}\| = \|P_C(u) - P_C(\overline{x})\| < \varepsilon$. Lemma 18.93 applied with $v := u - x$ and Lemma 18.8 ensure that $P_C(x + \varepsilon' \frac{u-x}{\|u-x\|}) = x$.
(b) $\Rightarrow$ (c). Let us suppose that (b) holds with some $\varepsilon > 0$. Let $x \in C \cap B(\overline{x}, \varepsilon/2)$ and $v \in N^P_C(x)$ with $v \neq 0$. By definition of $N^P_C(x)$ and by Lemma 18.8 there exists some $\eta \in]0, \varepsilon/2[$ such that $x = P_C(u)$, where $u := x + \eta\frac{v}{\|v\|}$. Then, we have

$$\|u - \overline{x}\| \leq \|u - x\| + \|x - \overline{x}\| < \varepsilon/2 + \varepsilon/2,$$

which combined with (b) gives $P_C(x+\varepsilon\frac{u-x}{\|u-x\|}) = x$. Hence, we obtain (c) with $\varepsilon/2$.
(c) $\Rightarrow$ (a). We suppose that (c) holds with some $\varepsilon > 0$. Let $0 \neq v^* \in N^{*,P}_C(x)$ with $x \in C \cap B(\overline{x}, \varepsilon)$ and $\|v^*\| \leq \varepsilon$. There exists $u \notin C$ such that

$$\left\langle \frac{v^*}{\|v^*\|}, u - x \right\rangle = \|u - x\|.$$

Thus, $u-x \in N^P_C(x)$ and $\frac{v^*}{\|v^*\|} = J(\frac{u-x}{\|u-x\|})$. We derive by (c) that $P_C(x+\varepsilon\frac{u-x}{\|u-x\|}) = x$, so $P_C(x+\frac{\varepsilon}{\|v^*\|}J^*(v^*)) = x$. Now, for all real numbers $0 < s \leq 1$ (since $1 \leq \frac{\varepsilon}{\|v^*\|}$), we have that $P_C(x + sJ^*(v^*)) = x$. By Definition 18.70 the set C is prox-regular at $\overline{x}$ for 0, thus by Proposition 18.69 it is prox-regular at $\overline{x}$. $\square$

Now, we give a characterization of the local prox-regularity of a set by means of hypomonotonicity of the truncated normal cone (see (18.55)) of proximal normal functionals.

THEOREM 18.95 (J-hypomonotonicity of truncations of normal cones). Let C be a subset of X which is closed near $\overline{x} \in C$. The set C is prox-regular at $\overline{x} \in C$ if and only if, for some reals $\varepsilon, r > 0$, the multimapping $N_C^{*\varepsilon,P} : X \rightrightarrows X^*$ which assigns to each $x \in X$ the truncated cone of proximal normal functionals $N_C^{*\varepsilon,P}(x)$ is J-hypomonotone of rate r on $B(\overline{x}, \varepsilon)$, that is, for all $t \geq r$ and all $x_i \in C \cap B(\overline{x}, \varepsilon)$, $x_i^* \in N_C^{*,P}(x_i) \cap \varepsilon \mathbb{B}_{X^*}$, $i = 1, 2$,

$$\langle J[J^*(x_1^*) - t(x_2 - x_1),] - J[J^*(x_2^*) - t(x_1 - x_2)], x_1 - x_2 \rangle \geq 0.$$

PROOF. By Proposition 18.78, if C is prox-regular at $\overline{x}$, then for some reals $\varepsilon > 0$ and $\rho > 0$, the truncated normal functional cone multimapping $N_C^{*\varepsilon,P}$ is J-hypomonotone of rate ρ on $B(\overline{x}, \varepsilon)$.

Conversely, suppose that $N_C^{*\varepsilon,P}$ is J-hypomonotone of rate ρ. Then the argument of Theorem 18.84 or Corollary 18.86 works as well (since it only makes use of the J-hypomonotonicity of the truncation of $\partial_P f$), to get that P_C is single-valued and continuous on a neighborhood of $\overline{x}$. It just remains to invoke Lemma 18.93 and Proposition 18.94 to conclude. $\square$

18.5.2. Basic characterizations of local prox-regularity in uniformly convex Banach spaces. We are now able to establish the theorem giving several characterizations of the local prox-regularity of a set in the context of uniformly convex Banach spaces. It is convenient here to write in the statement of the theorem, the assumption (XCS) which has been utilized in this section as well as in Section 18.3 and Section 18.4.

THEOREM 18.96 (characterizations of local prox-regularity in uniformly convex space). Let $(X, \|\cdot\|)$ be a uniformly convex Banach space whose norm $\|\cdot\|$ is also uniformly smooth. Let C be a set in X which closed near $\overline{x} \in C$. The following assertions are equivalent:
(a) the set C is prox-regular at $\overline{x}$;
(b) there exists a neighborhood U of $\overline{x}$ such that the mapping P_C is well defined on U and and norm-to-norm uniformly continuous there;
(b') there exists a neighborhood U of $\overline{x}$ such that the mapping P_C is well defined on U and and norm-to-norm continuous there;
(c) there exists a neighborhood U of $\overline{x}$ such that the mapping P_C is well defined on U and norm-to-weak continuous there;
(d) there exists $\varepsilon > 0$ such that $v \in N^P(C; x)$ with $x \in B(\overline{x}, \varepsilon)$ and $v \neq 0$ implies that $P_C(x + \varepsilon \frac{v}{\|v\|}) = x$;
(e) there exists ε>0 such that the condition $\left.\begin{array}{c} x{=}P_C(u), x \neq u \\ 0 < \|u{-}\overline{x}\| < \varepsilon \end{array}\right\}$ implies that one has $x{=}P_C(u')$ for $u'{=}x{+}\varepsilon\frac{u-x}{\|u-x\|}$;
(f) the function d_C^2 is Fréchet differentiable on some neighborhood U of $\overline{x}$ with its derivative uniformly continuous therein;
(f') the function d_C^2 is of class $\mathcal{C}^1$ on some neighborhood of $\overline{x}$;

(g) the function d_C is Fréchet differentiable on $U \setminus C$ for some neighborhood U of the point $\overline{x}$;
(h) there exists some neighborhood U of $\overline{x}$ such that the function d_C is Gâteaux differentiable on $U \setminus C$ with $\|D_G d_C(x)\|_* = 1$ for all $x \in U \setminus C$;
(i) there exists a neighborhood U of $\overline{x}$ such that $\partial_F d_C(u) \neq \emptyset$ for all $u \in U$;
(j) there exists a neighborhood U of $\overline{x}$ such that d_C is Fréchet subdifferentially regular on $U \setminus C$;
(k) there is a neighborhood U of $\overline{x}$ such that $\mathrm{Proj}_C(u) \neq \emptyset$ for all $u \in U$ and $\partial_H d_C(u) \neq \emptyset$ for all $u \in U$;
(l) the indicator function Ψ_C is J-plr at $\overline{x}$;
(m) there exist $\varepsilon, r > 0$ such that the truncated normal functional cone multimapping $N_C^{*\varepsilon,P}$ is J-hypomonotone of rate r on $B(\overline{x}, \varepsilon)$.

If C is weakly closed, one has one more equivalent condition:
(n) the mapping P_C is well defined on some neighborhood U of $\overline{x}$.

PROOF. First we will establish all the equivalences with (b') and (f') in place of (b) and (f) respectively. Without loss of generality, we may suppose that C is closed. The proof follows the scheme:

$$\begin{array}{ccccccccccccccccc}
(m) & \Leftrightarrow & (a) & \Rightarrow & (l) & \Rightarrow & (f') & \Leftrightarrow & (j) & \Leftrightarrow & (h) & \Leftrightarrow & (i) & \Leftrightarrow & (g) & \Leftrightarrow & (k) \\
 & & & & & & \Updownarrow & & & & & & & & & & \\
 & & & & (c) & \Leftrightarrow & (b') & \Rightarrow & (d) & \Leftrightarrow & (e) & \Leftrightarrow & (a). & & & &
\end{array}$$

(m) $\Leftrightarrow$ (a) is Theorem 18.95.
(a) $\Rightarrow$ (l) is established in Proposition 18.78.
(l) $\Rightarrow$ (f') follows from Corollary 18.86.
(f') $\Leftrightarrow$ (j) $\Leftrightarrow$ (h) $\Leftrightarrow$ (i) $\Leftrightarrow$ (g) is Proposition 18.91.
(g) $\Rightarrow$ (k): Assume that (g) holds. This obviously ensures the Hadamard subdifferentiability of d_C on $U \setminus C$ and since one always has $0 \in \partial_H d_C(x)$ for all $x \in C$, we obtain that d_C is Hadamard subdifferentiable on U. The nonvacuity of Proj_C on U follows from the above implication (g) $\Rightarrow$ (f') and from Proposition 18.90.
(k) $\Rightarrow$ (g): Fix any $x \in U \setminus C$ and choose by assumption some $x^* \in \partial_H d_C(x)$. Choose also by assumption some $p(x) \in \mathrm{Proj}_C(x)$. By the definition of the Hadamard subdifferential, for any $\varepsilon > 0$ there is some $\delta > 0$ such that, for any $t \in]0, \delta[$ one has

$$\langle x^*, p(x) - x \rangle \le t^{-1}[d_C(x + t(p(x) - x)) - d_C(x)] + \varepsilon,$$

and hence

$$\begin{aligned}
\langle x^*, x - p(x) \rangle &\ge t^{-1}[d_C(x) - d_C(x + t(p(x) - x))] - \varepsilon \\
&\ge t^{-1}[\|x - p(x)\| - \|x + t(p(x) - x) - p(x)\|] - \varepsilon.
\end{aligned}$$

Then, for any $\varepsilon > 0$, using the equality $D(\|\cdot\|^2)(x - p(x)) = 2J(x - p(x))$ and taking some real $t > 0$ small enough, we obtain

$$\begin{aligned}
\langle x^*, x - p(x) \rangle &\ge \left\langle \frac{J(x - p(x))}{\|x - p(x)\|}, x - p(x) \right\rangle - 2\varepsilon \\
&= \|x - p(x)\| - 2\varepsilon,
\end{aligned}$$

the last equality being due to the fact that $\langle J(y), y\rangle = \|y\|^2$. Therefore, it results that $\|x - p(x)\| \le \langle x^*, x - p(x)\rangle$, thus we have

$$\begin{aligned}\|x - p(x)\| \le \langle x^*, x - p(x)\rangle &\le \liminf_{t\downarrow 0} t^{-1}[d_C(x + t(x - p(x)) - d_C(x)] \\ &\le \limsup_{t\downarrow 0} t^{-1}[d_C(x + t(x - p(x)) - d_C(x)] \le \|x - p(x)\|,\end{aligned}$$

(where the second inequality is due to the inclusion $x^* \in \partial_H d_C(x)$), so

$$\lim_{t\downarrow 0} t^{-1}[d_C(x + t(x - p(x)) - d_C(x)] = \|x - p(x)\|.$$

Further, observe that for each $t \in [-1, 0[$ one has $p(x) \in \operatorname{Proj}_C(x + t(x - p(x))$ (see Lemma 4.123), and hence

$$t^{-1}[d_C(x + t(x - p(x))) - d_C(x)] = \|x - p(x)\|.$$

Therefore, $\lim_{t\to 0} t^{-1}[d_C(x+t(x-p(x))-d_C(x)] = \|x-p(x)\|$, that is, d_C has a bilateral directional derivative in the full direction $x - p(x)$. Consequently, the assertion (g) follows from Corollary 14.23.
(f') $\Leftrightarrow$ (b') is Proposition 18.90.
(b') $\Leftrightarrow$ (c) follows from Lemma 18.88.
(b') $\Rightarrow$ (d) is Lemma 18.93.
(d) $\Leftrightarrow$ (e) $\Leftrightarrow$ (a) is Proposition 18.94.

Under the additional assumption of weak closedness of C, to see that we have (n) $\Leftrightarrow$ (c) we need to prove the implication (n) $\Rightarrow$ (c).
Let us take any $u \in U$ and $u_n \xrightarrow{\|\cdot\|} u$. By weak compactness, taking a subsequence if necessary, we may suppose that $P_C(u_n) \xrightarrow{w} v$. From the weak lower semicontinuity of the norm, we have

$$\|v - u\| \le \liminf_{n\to\infty} \|u_n - P_C(u_n)\| = \lim_{n\to\infty} d_C(u_n),$$

and hence $\|v-u\| \le d_C(u)$. Therefore, as C is weakly closed, $v \in C$ and $P_C(u) = v$. Thus, $P_C(u_n) \xrightarrow{w} P_C(u)$, which gives the norm-to-weak continuity of P_C on U.

Now, we note that the implications (b) $\Rightarrow$ (b') and (f) $\Rightarrow$ (f') are evident. Then, to conclude we apply Corollary 18.86(a) to obtain (a) $\Rightarrow$ (b) and (a) $\Rightarrow$ (f) but now with the uniform character of the continuity that holds on a (possibly smaller) neighborhood of $\overline{x}$. The proof is then complete. □

Using Corollary 18.86(b) the following is easily seen.

PROPOSITION 18.97. If the norm $\|\cdot\|$ of the Banach space $(X, \|\cdot\|)$ has its modulus of convexity of power type q and its modulus of smoothness of power type s, then one can add the two following conditions to the list of equivalences in the above theorem:
(b') there exists a neighborhood U of $\overline{x}$ such that the mapping P_C is well defined on U and Hölder continuous on U with power $1/q$;
(f') there exists a neighborhood U of $\overline{x}$ over which d_C^2 is of class $\mathcal{C}^{1,\alpha}$ with $\alpha := q^{-1}(s-1)$.

The prox-regularity of C can also be characterized in terms of the $J_{p,\|\cdot\|}$ duality mapping (instead of the normalized duality mapping).

PROPOSITION 18.98. Let $(X, \|\cdot\|)$ be a uniformly convex Banach space whose norm is also uniformly smooth and let C be a nonempty subset in X which is closed near $\overline{x} \in C$. Let any real number $p > 1$. The following are equivalent:
(a) the set C is prox-regular at $\overline{x}$;
(l') there exist $\varepsilon > 0$ and $r > 0$ such that for all $x \in C$ and all $x^* \in N^{*,P}(C;x)$ with $\|x - \overline{x}\| < \varepsilon$ and $\|x^*\| \leq 1$, and for all $t \geq r$

$$0 \geq \langle J_p[J_p^{-1}(x^*) - r^{-1}(x' - x)], x' - x\rangle \quad \forall x' \in C \text{ with } \|x' - \overline{x}\| < \varepsilon;$$

(l") there exist $\varepsilon > 0$ and $r > 0$ such that for all $x \in C$, $t \geq 0$, and $x^* \in N^{*,P}(C;x)$ with $\|x - \overline{x}\| < \varepsilon$ and $\|x^*\| \leq r^{p-1}t^{p-1}$

$$0 \geq \langle J_p[J_p^{-1}(x^*) - t(x' - x)], x' - x\rangle \quad \forall x' \in C \text{ with } \|x' - \overline{x}\| < \varepsilon.$$

PROOF. Note first that the plr-property of Ψ_C in property (l) of Theorem 18.96 is equivalent by Proposition 18.73(c) to the existence of $\varepsilon > 0$ and $r > 0$ such that: for any $x, x' \in C \cap B(\overline{x}, \varepsilon)$, any $t \geq 0$, any $x^* \in N^{*,P}(C;x)$ with $\|x^*\| \leq rt$,

$$0 \geq \langle J[J^*(x^*) - t(x' - x)], x' - x\rangle. \tag{18.72}$$

With this formulation, we are able to prove that (a) is equivalent to (l").

Suppose that (l") holds. Observe first that for any non zero $x^* \in X^*$, one has $J_p^{-1}(x^*) = J^*(\|x^*\|^{\frac{2-p}{p-1}} x^*)$ and for any non zero $u \in X$, one has $J_p(u) = J(\|u\|^{p-2}u)$ (see (18.22)). Hence, putting $p' := \frac{2-p}{p-1}$, for any non zero $x^* \in X^*$ and any $t \geq 0$ one has the equivalences

$$\begin{aligned}
\langle J[J^*(x^*) - t(x' - x)], x' - x\rangle \leq 0 \quad &\Leftrightarrow \\
\langle J[\|x^*\|^{-p'} J_p^{-1}(x^*) - t(x' - x)], x' - x\rangle \leq 0 \quad &\Leftrightarrow \\
\langle J[J_p^{-1}(x^*) - t\|x^*\|^{p'}(x' - x)], x' - x\rangle \leq 0 \quad &\Leftrightarrow \\
\langle J_p[J_p^{-1}(x^*) - t\|x^*\|^{p'}(x' - x)], x' - x\rangle \leq 0. &
\end{aligned} \tag{18.73}$$

Fix now any $t > 0$, any $x, x' \in C \cap B(\overline{x}, \varepsilon)$, and any non zero $x^* \in N^{*,P}(C;x)$ such that $rt \geq \|x^*\|$. The latter inequality ensures that $rt\|x^*\|^{p'} \geq \|x^*\|^{\frac{1}{p-1}}$, that is, $\|x^*\| \leq r^{p-1}\theta^{p-1}$ for $\theta := t\|x^*\|^{p'}$, thus (l") with $\theta = t\|x^*\|^{p'}$ in place of t yields (18.73). By the above equivalences we obtain the inequality (18.72).

Conversely, suppose that (18.72) is fulfilled. Fix any $t \geq 0$, $x, x' \in C \cap B(\overline{x}, \varepsilon)$, and any non zero $x^* \in N^{*,P}(C;x)$ such that $\|x^*\| \leq r^{p-1}t^{p-1}$. One has $\|x^*\|^{\frac{1}{p-1}} \leq rt$, and hence $\|x^*\| \leq rt\|x^*\|^{1-\frac{1}{p-1}}$, which by (18.72), with $\theta := t\|x^*\|^{1-\frac{1}{p-1}} = t\|x^*\|^{-p'}$ in place of t (where as above $p' := \frac{2-p}{p-1}$), yields

$$0 \geq \langle J[J^*(x^*) - \theta(x' - x)], x' - x\rangle.$$

By (18.73) this is equivalent to

$$0 \geq \langle J_p[J_p^{-1}(x^*) - \theta\|x^*\|^{p'}(x' - x)], x' - x\rangle,$$

which is exactly

$$0 \geq \langle J_p[J_p^{-1}(x^*) - t(x' - x)], x' - x\rangle.$$

The latter being still true for $x^* = 0$, we obtain (l").

The equivalence between (l') and (l") can be shown with similar arguments as above. □

18.5.3. Tangential regularity. Prox-regular sets in X are normally regular and tangentially regular as well, as the following theorem shows.

THEOREM 18.99 (tangential/normal regularity under prox-regularity). Assume that a set C in X is prox-regular at $\overline{x} \in C$. Then there exists a neighborhood U of $\overline{x}$ such that for any $x \in U \cap C$ one has the following normal regularity

$$N^{*,P}(C;x) = N^F(C;x) = N^L(C;x) = N^C(C;x),$$

and hence

$$\partial_P d_C(x) = \partial_F d_C(x) = \partial_L d_C(x) = \partial_C d_C(x),$$

that is, the distance function itself is subdifferentially regular at all points of $U \cap C$.

So, the set C is in particular tangentially regular at any $x \in U \cap C$.

PROOF. By assumption, there exist positive real numbers ε, r with $\varepsilon < 1/2$ such that $C \cap B(\overline{x}, \varepsilon)$ is closed relative to $B(\overline{x}, \varepsilon)$ and such that for every $x \in C \cap B(\overline{x}, \varepsilon)$ and every $x^* \in N^{*,P}(C;x)$ with $\|x^*\| \le 1$, we have $\|x + tJ^*(x^*) - x'\| \ge \|tJ^*(x^*)\|$ for any $x' \in C \cap B(\overline{x}, \varepsilon)$ and $t \in]0, r]$. Fix any $x, x' \in B(\overline{x}, \varepsilon) \cap C$ and $x^* \in N^{*,P}(C;x)$ with $\|x^*\| \le 1$. It ensues that

$$\|x + tJ^*(x^*) - x'\|^2 \ge \|tJ^*(x^*)\|^2 \quad \text{for any } t \in]0, r]. \tag{18.74}$$

On the other hand, for any $t \in]0, r]$, according to the Fréchet differentiability of $\| \cdot \|^2$, we have

$$\begin{aligned} &\|x + tJ^*(x^*) - x'\|^2 \\ &= \|tJ^*(x^*)\|^2 + 2 \int_0^1 \langle J[tJ^*(x^*) + \theta(x - x')], x - x' \rangle \, d\theta \\ &= \|tJ^*(x^*)\|^2 + 2t\langle x^*, x - x' \rangle + 2 \int_0^1 \langle J[tJ^*(x^*) + \theta(x - x')] - J[tJ^*(x^*)], x - x' \rangle \, d\theta. \end{aligned}$$

Combining this with (18.74) we obtain

$$\langle x^*, x' - x \rangle \le \frac{1}{t} \|x' - x\| \int_0^1 \|J[tJ^*(x^*) + \theta(x - x')] - J[tJ^*(x^*)]\| \, d\theta. \tag{18.75}$$

In Proposition 18.41(c) we saw that the duality mapping J is uniformly continuous over bounded subsets of X. Therefore, denoting by ω_{r+1} the modulus of uniform continuity of J over the bounded set $(r+1)\mathbb{B}_X$, that is,

$$\omega_{r+1}(\tau) := \sup\{\|J(u) - J(u')\| : u, u' \in (r+1)\mathbb{B}_X, \ \|u - u'\| \le \tau\} \quad \text{for } \tau > 0,$$

we have $\omega_{r+1}(\tau) \underset{\tau \downarrow 0}{\longrightarrow} 0$, and (18.75) entails

$$\langle x^*, x' - x \rangle \le \frac{1}{t} \|x' - x\| \, \omega_{r+1}(\|x - x'\|). \tag{18.76}$$

Fix now $x \in C \cap B(\overline{x}, \varepsilon)$ and $x^* \in N^L(C;x)$, and fix also any $\eta > 0$. By Corollary 18.60 let $x_n^* \in N^{*,P}(C;x_n)$ with $x_n \in C \cap B(\overline{x}, \varepsilon)$ such that $x_n \to x$ and $x_n^* \overset{w^*}{\to} x^*$ as $n \to \infty$. Choose a real number $\gamma > 0$ such that $\|x_n^*\| \le \gamma$ for all integers n and choose a positive $\alpha < \varepsilon - \|x - \overline{x}\|$ such that $\omega_{r+1}(\tau) \le \frac{r\eta}{\gamma}$ for all positive $\tau < \alpha$. Take any $x' \in B(x, \alpha) \cap C$. We have $x' \in B(\overline{x}, \varepsilon)$ and, for n large enough $\|x' - x_n\| < \alpha$. By (18.76) for n large enough we then have

$$\langle x_n^*, x' - x_n \rangle \le \frac{1}{r} \|x_n^*\| \, \|x' - x_n\| \, \omega_{r+1}(\|x' - x_n\|) \le \frac{1}{r} \gamma \|x' - x_n\| \frac{r\eta}{\gamma},$$

and hence passing to the limit as $n \to \infty$ we obtain

$$\langle x^*, x' - x \rangle \le \eta \|x' - x\|.$$

The latter inequality being true for all $x' \in B(x, \alpha) \cap C$, it means that $x^* \in N^F(C; x)$. So far, we have proved that for any $x \in B(\overline{x}, \varepsilon) \cap C$, we have $N^L(C; x) \subset N^F(C; x)$. As the reverse inclusion is also true according to the definition of $N^L(C; \cdot)$, we have in fact the equality

$$N^L(C; x) = N^F(C; x) \quad \text{for all } x \in B(\overline{x}, \varepsilon) \cap C. \tag{18.77}$$

Moreover, by Theorem 4.120(a) we know that $N^C(C; \cdot) = \overline{\text{co}}^*(N^L C; \cdot))$, where we recall that $\overline{\text{co}}^*$ denotes the weak* closed convex hull in X^*. Since $N^F(C; x)$ is convex and (strongly) closed (see Proposition 4.8(c)), we deduce from the equality (18.77) that we even have

$$N^F(C; x) = N^L(C; x) = N^C(C; x) \quad \text{for all } x \in B(\overline{x}, \varepsilon) \cap C. \tag{18.78}$$

Let us now prove that the three cones in (18.78) are also equal to the cone of proximal normal functionals to C. In our uniformly convex setting where the norm $\| \cdot \|$ of X is uniformly convex and uniformly smooth, we know according to Proposition 18.59 that for any $x \in B(\overline{x}, \varepsilon) \cap C$, $x^* \in N^F(C; x)$ with $\|x^*\| < 1$, there exists $x_n \to x$ with $x_n \in C$, and $x_n^* \in N^{*,P}(x_n)$ such that $\|x_n^* - x^*\|_* \to 0$ as $n \to \infty$. For n large enough we have $x_n \in B(\overline{x}, \varepsilon)$ and $\|x_n^*\| < 1$. For any such integer n, for any $t \in]0, r]$ and $x' \in B(\overline{x}, \varepsilon) \cap C$, we have by (18.74)

$$\|x_n + tJ^*(x_n^*) - x'\|^2 \ge \|tJ^*(x_n^*)\|^2,$$

which gives, by passing to the limit and by the continuity of J^*,

$$\|x + tJ^*(x^*) - x'\|^2 \ge \|tJ^*(x^*)\|^2,$$

that is, $x^* \in N^{*,P}(C; x)$ thanks to the local character in Proposition 18.56 of (primal) proximal normal vector. So, for any fixed $x \in B(\overline{x}, \varepsilon) \cap C$ we obtain that $N^F(C; x) \subset N^{*,P}(C; x) \subset N^L(C; x)$, which combined with (18.78) gives the equalities

$$N^{*,P}(C; x) = N^F(C; x) = N^L(C; x) = N^C(C; x).$$

These equalities also ensure

$$\partial_C d_C(x) \subset N^C(C; x) \cap \mathbb{B}_{X^*} = N^{*,P}(C; x) \cap \mathbb{B}_{X^*} = \partial_P d_C(x),$$

the last equality being due to Proposition 18.65 (see also Proposition 2.95(c) for the first inclusion). Consequently, we have

$$\partial_P d_C(x) = \partial_F d_C(x) = \partial_L d_C(x) = \partial_C d_C(x).$$

Finally, the equality between $N^F(C; x)$ and $N^C(C; x)$ ensures that C is tangentially regular at x (see Corollary 4.10). □

Taking into account Definition 18.66 and Theorem 18.96, an immediate corollary of Theorem 18.99 is the following:

COROLLARY 18.100. Let $(X, \| \cdot \|)$ be a uniformly convex Banach space whose norm $\| \cdot \|$ is also uniformly smooth. Let C be a set in X which closed near a point $\overline{x} \in C$, and let $\mathcal{N}(C; \cdot)$ be any of normal cones $N^F(C; \cdot)$, $N^L(C; \cdot)$, $N^C(C; \cdot)$. The following are equivalent:

(a) the set C is prox-regular at $\overline{x}$;

(a') there exist $\varepsilon, r > 0$ such that $C \cap B(\overline{x}, \varepsilon)$ is closed relative to $B(\overline{x}, \varepsilon)$ and such that for any $x \in C \cap B(\overline{x}, \varepsilon)$ and any $v^* \in \mathcal{N}(C; x) \cap \mathbb{B}_{X^*}$ one has

$$x \in \operatorname{Proj}_C(x + tJ^*(v^*)) \quad \text{for any positive real } t \le r;$$

(m') there exist reals $\varepsilon, r > 0$ such that the truncated normal cone multimapping $\varepsilon\mathbb{B}_{X^*} \cap \mathcal{N}(C; \cdot)$ is J-hypomonotone on $B(\overline{x}, \varepsilon)$.

18.6. Characterizations and properties of uniformly prox-regular sets in uniformly convex Banach spaces

We turn in this section to uniformly prox-regular sets in uniformly convex Banach spaces. The aim is to provide diverse characterizations as well as many other properties.

18.6.1. Characterizations of uniform prox-regularity of sets in uniformly convex Banach spaces. We start this subsection by transforming the definition of prox-regularity of $C \subset X$ at $\overline{x} \in C$ for $\overline{v} \in N^P(C; \overline{x})$ in Definition 18.70 in a uniform way. This amounts to rephrase via the duality mapping J the definition $(\mathcal{D}_{\text{global}})$ as $(\mathcal{D}_{\text{local}})$ has been rephrased in Definition 18.66 via J. In other words, we follow what has been utilized in Hilbert space in Definition 15.24 with a constant $r \in]0, +\infty]$.

Definition 18.101. Let an extended real $r \in]0, +\infty]$. A nonempty closed set C in X is called *uniformly r-prox-regular* (or simply *r-prox-regular*) if for any $x \in C$, any $v^* \in N^{*,P}(C; x)$ with $\|v^*\| \le 1$ and any positive real $t \le r$, one has

$$x \in \operatorname{Proj}_C(x + tJ^*(v^*)),$$

otherwise stated $C \cap B(x + tJ^*(v^*), t) = \emptyset$ for any $v^* \in N^{*,P}(C; x)$ with $\|v^*\| = 1$.

Theorem 18.102 (characterizations of uniform prox-regularity in uniformly convex spaces). Let $(X, \|\cdot\|)$ be a uniformly convex Banach space whose norm $\|\cdot\|$ is also uniformly smooth. Let C be a nonempty closed set in X and let an extended real number $r \in]0, +\infty]$. The following are equivalent:
(a) the set C is r-prox-regular;
(b) the function d_C is continuously differentiable on $U_r(C) \setminus C$;
(c) the function d_C is Fréchet subdifferentially regular on $U_r(C) \setminus C$;
(d) the function d_C is Fréchet differentiable on $U_r(C) \setminus C$;
(e) the function d_C is Gâteaux differentiable on $U_r(C) \setminus C$ with $\|D_G d_C(x)\| = 1$ for all $x \in U_r(C) \setminus C$;
(f) $\partial_F d_C(u) \neq \emptyset$ for all $u \in U_r(C)$;
(g) for all $u \in U_r(C)$ one has $\operatorname{Proj}_C(u) \neq \emptyset$ and $\partial_H d_C(u) \neq \emptyset$;
(h) the function d_C^2 is differentiable on $U_r(C)$ with its derivative locally uniformly continuous therein;
(h') the function d_C^2 is of class $\mathcal{C}^1$ on $U_r(C)$;
(i) the mapping P_C is well defined on $U_r(C)$ and locally uniformly continuous therein;
(j) the mapping P_C is well defined on $U_r(C)$ and norm-to-norm (or norm-to-weak continuous) therein;
(k) for any $x \in C$ and any non-zero $v \in N^P(C; x)$ one has $x \in \operatorname{Proj}_C(x + t\frac{v}{\|v\|})$ for any nonnegative real $t \le r$;

(l) for any $u \in U_r(C) \cap \operatorname{dom} P_C$, for any nonnegative real $t \le r$, for $x = P_C(u)$ and for $u_t = x + t\frac{u-x}{\|u-x\|}$ one has $x \in \operatorname{Proj}_C(u_t)$.

If C is weakly closed, one has the additional equivalent condition:
(m) the mapping P_C is well defined on $U_r(C)$.

PROOF. It suffices to proceed along the following scheme:

$$\begin{array}{ccccccccccccc} (a) & \Leftrightarrow & (k) & \Leftrightarrow & (l) & \Rightarrow & (i) & \Rightarrow & (h) & \Rightarrow & (d) & \Leftrightarrow & (g) \\ & & & & \Uparrow & & & & & & \Updownarrow & & \\ & & (j) & \Leftrightarrow & (b) & \Leftrightarrow & (c) & \Leftrightarrow & (e) & \Leftrightarrow & (f) & & \end{array}$$

(a) $\Leftrightarrow$ (k) $\Leftrightarrow$ (l): These equivalences are obvious, and anyone of these assertions will be considered below.
(b)$\Leftrightarrow$(c)$\Leftrightarrow$(d)$\Leftrightarrow$(e)$\Leftrightarrow$(f): These equivalences follow from Proposition 18.91.
(d)$\Leftrightarrow$(g) derives from Corollary 14.23 like in the proof of the similar equivalence in Theorem 18.96.
(b)$\Leftrightarrow$(j): This is a consequence of Proposition 18.90 and Lemma 18.88.
(b) $\Rightarrow$ (l): Let $u \in U_C(r) \cap \operatorname{dom} P_C$ and $x = P_C(u)$, and let any positive real $t \le r$. Since (b) holds, according to Lemma 18.92 there exists some real $s_0 > 0$ such that $P_C(u_s) = x$ for all $u_s := u + s(u-x)/\|u-x\|$ with $0 < s < s_0$. Suppose $d_C(u) < t$ and consider the number λ_0 given by the supremum over all $s \in [0, t - d_C(u)]$ such that $x \in \operatorname{Proj}_C(u_s)$. Using the equivalence (note that $x \in C$ and $\|x - u_s\| = d_C(u) + s$)

$$x \in \operatorname{Proj}_C(u_s) \Leftrightarrow \forall x' \in C,\ \|x' - u_s\| \ge d_C(u) + s,$$

it is easily seen that the supremum λ_0 is attained. We now claim that $\lambda_0 = t - d_C(u)$. Assume the contrary, that is, $\lambda_0 < t - d_C(u)$. Then, one would have on the one hand $x = P_C(u_{\lambda_0})$ because of the assumption (b) and Proposition 18.90. Applying Lemma 18.92 again one would obtain a contradiction with the supremum property of λ_0. So, the equality $\lambda_0 = t - d_C(u)$ holds true. As u_s can be written in the form $u_s = x + (d_C(u) + s)(u - x)/\|u - x\|$, taking $s = \lambda_0$ gives $x = P_C(u_t)$ for every real $t \in]d_C(u), r]$. From this it is clear that one also has $x = P_C(u_t)$ for every real $t \in]0, d_C(u)]$. The implication (b)$\Rightarrow$(l) is then established.
(i)$\Rightarrow$(h) This implication follows from Lemma 18.81 and Remark 18.83.
(h) $\Rightarrow$ (d): This implication is obvious.
(l) $\Rightarrow$ (i). For this implication, without loss of generality we may obviously and do suppose $r < +\infty$. We will proceed in five steps.

Step 1. Consider any $x^* \in N_C^{*r,P}(x) = N_C^{*,P}(x) \cap r\mathbb{B}_*$ with $x \in C$ and any $\lambda \in]0, 1]$. From (k) $\Leftrightarrow$ (l) one has $\operatorname{Proj}_C\left(x + r\lambda\frac{J^*(x^*)}{\|x^*\|}\right) \ni x$. This means that, for any $x' \in C$,

$$\left\| x + r\lambda\frac{J^*(x^*)}{\|x^*\|} - x \right\| \le \left\| x + r\lambda\frac{J^*(x^*)}{\|x^*\|} - x' \right\|.$$

On the other hand, as $J = D(\frac{1}{2}\|.\|^2)$, one also has

$$\frac{1}{2}\left\| x + r\lambda\frac{J^*(x^*)}{\|x^*\|} - x' \right\|^2 + \left\langle J(x - x' + r\lambda\frac{J^*(x^*)}{\|x^*\|}), x' - x \right\rangle \le \frac{1}{2}\left\| r\lambda\frac{J^*(x^*)}{\|x^*\|} \right\|^2.$$

So, $\langle J(J^*(x^*) - \frac{\|x^*\|}{r\lambda}(x'-x)), x'-x\rangle \le 0$. It is possible to take any λ in $]0, \frac{\|x^*\|}{r}]$. Hence, for any $x_i \in C, x_i^* \in N_C^{*r,P}(x_i)$, $i = 1, 2$, and any real $t \ge 1$, one has

$$\begin{aligned} &\langle J(J^*(x_1^*) - t(x_2 - x_1)), x_2 - x_1\rangle \le 0 \\ \text{and} \quad &\langle J(J^*(x_2^*) - t(x_1 - x_2)), x_1 - x_2\rangle \le 0. \end{aligned}$$

By adding, one obtains

$$\langle J(J^*(x_1^*) - t(x_2 - x_1)) - J(J^*(x_2^*) - t(x_1 - x_2)), x_1 - x_2\rangle \ge 0,$$

which is the J-hypomonotonicity of $N_C^{*r,P}$ of rate $\bar{t}$ for any real $\bar{t} \ge 1$.

Step 2. For any $\alpha \in]0, 1/2[$, we have by Step 1 and Lemma 18.79 that the multimapping $(I + J^* \circ N_C^{*\alpha r,P})^{-1}$ is uniformly continuous on any nonempty bounded subset of its domain. Further, we also have, for any real $r' > 0$,

$$\text{Proj}_C(x) \subset (I + J^* \circ N_C^{*r',P})^{-1}(x) \quad \text{for any } x \in U_{r'}(C). \tag{18.79}$$

Indeed, for any $x \in U_{r'}(C)$, the inclusion $y \in \text{Proj}_C(x)$ entails that $J(x-y) \in N_C^{*,P}(y)$, and since $\|y - x\| < r'$ it follows that $J(x-y) \in N_C^{*r',P}(y)$. It ensues that for any $\alpha \in]0, 1/2[$, Proj_C is single-valued and uniformly continuous on any nonempty bounded set included in $\text{Dom}\,\text{Proj}_C \cap U_{\alpha r}(C)$. Then, by the arguments of Step 2 in the proof of Theorem 18.84, Proj_C is also nonempty, single-valued on $U_{\alpha r}(C)$. As α can be made as close as one wants to $\frac{1}{2}$, the mapping P_C is well defined on $U_{r/2}(C)$ and locally uniformly continuous on $U_{r/2}(C)$.

Step 3. The step 3 corresponds to Lemma 18.104 below. We recall that the line segment between two points $u, v \in X$ is denoted by $[u, v]$, that is, $[u, v] := \{tu + (1-t)v : t \in [0, 1]\}$. For convenience we put for any real $\rho > 0$,

$$C(\rho) := \text{Enl}_\rho C = \{x \in X : d_C(x) \le \rho\} \text{ and } D_C(\rho) := \{x \in X : d_C(x) = \rho\}.$$

To facilitate the reading of Lemma 18.104 below as well as the development of Step 4, we first put the statements of Lemmas 3.259 and 15.47 together in the following form.

Lemma 18.103. Let C be any nonempty closed subset of any normed vector space $(Y, \|.\|)$. Let a real $\rho \ge 0$ and $u \notin C(\rho)$. Then, the following hold:
(a) $d_C(u) = \rho + d_{C(\rho)}(u) = \rho + d_{D_C(\rho)}(u)$;
(b) If $u_0 \in \text{Proj}_C(u)$ and $y_0 \in [u_0, u] \cap D_C(\rho)$, then $y_0 \in \text{Proj}_{C(\rho)}(u)$;
(c) If $y \in \text{Proj}_{C(\rho)}(u)$ and $z \in \text{Proj}_C(y)$, then $z \in \text{Proj}_C(u)$. Further, if $\text{Proj}_{C(\rho)}(u)$ is a singleton, say $\text{Proj}_{C(\rho)}(u) = \{y\}$, and if $z \in \text{Proj}_C(y)$, then $y \in [z, u]$ and $\text{Proj}_C(u) = \{z\}$.

Lemma 18.104. If C satisfies the assertion (l) of Theorem 18.102 with parameter r and Proj_C is nonempty, single-valued on $U_{\alpha r}(C)$ for some $\alpha \in]0, 1]$, then for any $\alpha' \in]0, \alpha[$, the set $C(\alpha' r) := \{x \in X : d_C(x) \le \alpha' r\}$ satisfies (l) in Theorem 18.102 with parameter $r(1-\alpha')$.

Proof. For convenience we will put, for any $\rho > 0$

$$\mathcal{U}_C(\rho) := U_\rho(C) = \{x \in X : d_C(x) < \rho\}.$$

Take $u \in \mathcal{U}_{C(\alpha' r)}(r(1-\alpha'))$ with $u \notin C(\alpha' r)$ and put $r' := d_C(u)$. Note that $0 < d_{C(\alpha' r)}(u) < r(1-\alpha')$, and hence by (a) of Lemma 18.103 one has $\alpha' r < d_C(u) < r$, which implies in particular $u \in \mathcal{U}_C(r)$. Suppose that $\text{Proj}_{C(\alpha' r)}(u) = \{y\}$. We have to prove that $y \in \text{Proj}_{C(\alpha' r)}(y + r(1-\alpha')\frac{u-y}{\|u-y\|})$. Observing that $y \in \mathcal{U}_C(\alpha r)$, we may put $z := P_C(y)$ according to the assumption on Proj_C over $\mathcal{U}_C(\alpha r)$. By (c) of Lemma 18.103, we have $z \in \operatorname{Proj}_C(u)$. Since $d_C(z) = 0$ and $d_C(u) > \alpha' r$, we may take $y_1 \in [z, u] \cap D_C(\alpha' r) \neq \emptyset$. The assertion (b) of Lemma 18.103 says that $y_1 \in \operatorname{Proj}_{C(\alpha' r)}(u)$, and hence $y_1 = y$. Now, since $y_1 \in \mathcal{U}_C(r)$ and $z = P_C(y_1)$ we have, by our assumption of (l) in Theorem 18.102, that $\operatorname{Proj}_C(u') \ni z$ for

$$u' := z + r\frac{y_1 - z}{\|y_1 - z\|} = z + r\frac{u - z}{\|u - z\|}.$$

This entails by (b) of Lemma 18.103 again that $\operatorname{Proj}_{C(\alpha' r)}(u') \ni y$ since (recall that $y = y_1$)

$$y \in [z, u] \cap D_C(\alpha' r) \subset [z, u'] \cap D_C(\alpha' r).$$

Further, since $\|u' - y\| = \|u' - z\| - \|y - z\| = r - \alpha' r$ (the second equality being due to the definition of u' and to the inclusion $y \in D_C(\alpha' r)$) we see that

$$u' = y + (r - \alpha' r)\frac{u - y}{\|u - y\|}.$$

So, we obtain that $C(\alpha' r)$ satisfies (l) with parameter $r(1 - \alpha')$, and the proof of the lemma is complete. □

Step 4. For $\alpha \in]0, 1]$, let us consider the property

$$\mathcal{P}(\alpha) \begin{cases} C \text{ satisfies (k) with parameter } r \text{ and} \\ \operatorname{Proj}_C \text{ is single-valued on } \mathcal{U}_C(\alpha r), \text{ locally uniformly continuous on } \mathcal{U}_C(\alpha r). \end{cases}$$

We claim that $\mathcal{P}(\alpha) \Rightarrow \mathcal{P}(\frac{\alpha+1}{2})$.
Suppose that the property $\mathcal{P}(\alpha)$ holds. From Step 3 and Steps 1 and 2, we have that for any $\alpha' \in]0, \alpha[$,
(18.80)
$\operatorname{Proj}_{C(\alpha' r)}$ is single-valued, locally uniformly continuous on $\mathcal{U}_{C(\alpha' r)}(r(1 - \alpha')/2)$.

Take any $u \in \mathcal{U}_C\left(\alpha' r + \frac{r(1-\alpha')}{2}\right)$ such that $r' := d_C(u) > \alpha' r$. By (a) of Lemma 18.103 we have $d_{C(\alpha' r)}(u) = r' - \alpha' r$, and so

$$u \in \mathcal{U}_{C(\alpha' r)}(r(1 - \alpha')/2) \tag{18.81}$$

since $r' - \alpha' r < \frac{r(1-\alpha')}{2}$ because of the inclusion $u \in \mathcal{U}_C(\alpha' r + \frac{r(1-\alpha')}{2})$. We may then put $y := P_{C(\alpha' r)}(u)$ according to (18.80) and put $z := P_C(y)$ according to the second assumption in $\mathcal{P}(\alpha)$. Then by (c) of Lemma 18.103 we have $z = P_C(u)$, and so $P_C(u) = z = P_C \circ P_{C(\alpha' r)}(u)$. So for any $\alpha' \in\in]0, \alpha[$, (18.80), (18.81), and $\mathcal{P}(\alpha)$ ensure that Proj_C is single-valued and locally uniformly continuous on $\mathcal{U}_C(\alpha' r + \frac{r(1-\alpha')}{2}) \setminus C(\alpha' r)$. By assumption it is also locally uniformly continuous on $\mathcal{U}_C(\alpha r)$. So Proj_C is single valued, locally uniformly continuous on $\mathcal{U}_C(\alpha' r + \frac{r(1-\alpha')}{2})$ for any $\alpha' \in]0, \alpha[$, and hence also on $\mathcal{U}_C(\alpha r + \frac{r(1-\alpha)}{2}) = \mathcal{U}_C(\frac{\alpha+1}{2} r)$. This establishes the claim and finishes the proof of Step 4.

Step 5. Define $(\alpha_n)_n$ by $\alpha_0 = 1/2$, $\alpha_{n+1} = (\alpha_n + 1)/2$. We have $\alpha_n \to 1$ and by Step 4, $\mathcal{P}(\alpha_n) \Rightarrow \mathcal{P}(\alpha_{n+1})$. As $\mathcal{P}(\alpha_0)$ is true by Step 1, we have that $\mathcal{P}(1)$ is true, which ensures that (i) holds. So, (b)⇒(i) is demonstrated.

Finally, since the additional equivalence (m) can be established like in Theorem 18.96, the proof is now complete. □

If the moduli of convexity and smoothness are of power type, Corollary 18.86(b) allows us, in the proof of Theorem 18.102, to replace the uniform continuity property by appropriate Hölder property. Doing so yields the following proposition.

PROPOSITION 18.105. If the norm $\|\cdot\|$ of the uniformly convex Banach space $(X, \|\cdot\|)$ has its modulus of convexity of power type q and its modulus of smoothness of power type s, then one can add the two following conditions to the list of equivalences in Theorem 18.102 above:
(i') the mapping P_C is well defined on $U_r(C)$ and locally Hölder continuous on $U_r(C)$ with power $1/q$;
(h') the function d_C^2 is of class $\mathcal{C}^{1,\alpha}$ on $U_r(C)$ with $\alpha := q^{-1}(s-1)$.

By Lemma 18.103 the following corollary of Theorem 18.102 directly follows.

COROLLARY 18.106. Let $(X, \|\cdot\|)$ be a uniformly convex Banach space whose norm $\|\cdot\|$ is also uniformly smooth. Let C be a nonempty closed set in X which is r-prox-regular for an extended real $r \in]0, +\infty]$. Then for any real $\rho \in]0, r[$ the set $\mathrm{Enl}_\rho C$ is $(r-\rho)$-prox-regular and $\mathrm{Proj}_{\,\mathrm{Enl}_\rho C}$ is single-valued on $U_r(C)$ and uniformly continuous therein.

Another corollary of Theorem 18.102 is concerned with certain subregularity properties of the distance function of a set characterizing its prox-regularity.

COROLLARY 18.107. Let C be a nonempty closed set of a uniformly convex Banach space $(X, \|\cdot\|)$ whose norm $\|\cdot\|$ is also uniformly smooth. Given an extended real $r \in]0, +\infty]$ the following are equivalent:
(a) The set C is r-prox-regular;
(b) the distance function d_C is differentiable on $U_r(C) \setminus C$ and its derivative is locally uniformly continuous therein;
(c) the distance function d_C is locally semiconvex on $U_r(C) \setminus C$ in the sense of Definition 10.1;
(d) the distance function d_C is subsmooth on $U_r(C) \setminus C$;
(e) the distance function d_C is Fréchet subdifferentially regular on $U_r(C) \setminus C$.

PROOF. First, Proposition 8.16 guarantees that the implication (c)$\Rightarrow$(d) holds true (and this is also clear by definition of semiconvex functions and by the characterization of subsmooth functions in Proposition 8.12).
(b)$\Rightarrow$(c) This implication is a direct consequence of Proposition 10.9(a).
(d)$\Rightarrow$(e) This implication follows from Theorem 8.25(b).
(a)$\Rightarrow$(b) Assume that C is r-prox-regular. From Theorem 18.102 the metric projection P_C is well defined on $U_r(C)$ and from the same theorem and the expression of $De_\lambda f(x)$ in Lemma 18.80 we have the equality

$$D_F d_C(x) = J(x - P_C(x))/d_C(x) \quad \text{for all } x \in U_r(C) \setminus C.$$

Further, (i) in Theorem 18.102 again ensures that the metric projection P_C is locally uniformly continuous on $U_r(C) \setminus C$. This and the previous equality entails that the Fréchet derivative mapping $D_F d_C$ is locally uniformly continuous on $U_r(C)$. The implication (a)$\Rightarrow$(b) is established.
(e)$\Rightarrow$(a) Assume that (e) is satisfied. The Fréchet subdifferential regularity of the Lipschitz function d_C on the open set $U_r(C) \setminus C$ means that for each $x \in U_r(C) \setminus C$ we have $\partial_F d_C(x) = \partial_C d_C(x)$. Then, for each $x \in U_r(C) \setminus C$ by the non-vacuity of the Clarke subdifferential $\partial_C d_C(x)$ (due to the Lipschitz property of d_C) we deduce

that the Fréchet subdifferential $\partial_F d_C(x) \neq \emptyset$. Consequently, $\partial_F d_C(x) \neq \emptyset$ for any $x \in U_r(C)$ (since $0 \in \partial_F d_C(x)$ for every $x \in C$), hence (f) in Theorem 18.102 yields that the set C is r-prox-regular. This finishes the proof of the corollary. □

The next proposition gives the expression of the metric projection mapping in terms of the proximal normal cone.

PROPOSITION 18.108. Let $(X, \|\cdot\|)$ be a uniformly convex Banach space whose norm $\|\cdot\|$ is also uniformly smooth. Let C be a nonempty closed set in X which is r-prox-regular for an extended real $r \in]0, +\infty]$. Then one has

$$D_F d_C(x) = J(x - P_C(x))/d_C(x) \quad \text{for all } x \in U_r(C) \setminus C,$$

and also

$$P_C(x) = (I + J^* \circ N_C^{*r,P})^{-1}(x) \quad \text{for all } x \in U_r(C).$$

PROOF. The equality concerning the derivative has been already justified in the proof of Corollary 18.107 above. Let us prove the equality concerning the projection mapping. We may and do suppose that $r < +\infty$. We know by (18.79) that $\mathrm{Proj}_C(u) \subset (I + J^* \circ N_C^{*r})^{-1}(u)$ for every $u \in U_r(C)$, so it is enough to check that $(I + J^* \circ N_C^{*r,P})^{-1}$ is at most single-valued on $U_r(C)$, or equivalently on $U_r(C) \setminus C$. Suppose that y_1, y_2 are in $(I + J^* \circ N_C^{*r,P})^{-1}(x)$ with $x \in U_C(r) \setminus C$, that is, $J(x - y_i) \in N_C^{*r,P}(y_i)$, $i = 1, 2$. The assertion (k) in Theorem 18.102 tells us that $y_i \in \mathrm{Proj}_C(y_i + r\frac{x-y_i}{\|x-y_i\|})$. Since $\|x - y_i\| < r$, Lemma 18.8 entails that $y_i = P_C(y_i + \|x - y_i\| \frac{x-y_i}{\|x-y_i\|})$, that is, $y_i = P_C(x)$, which finishes the proof. □

The uniform prox-regularity also entails the J-hypomonotonicity of the truncated normal functional cone.

PROPOSITION 18.109. Let $(X, \|\cdot\|)$ be a uniformly convex Banach space whose norm $\|\cdot\|$ is also uniformly smooth. Let C be a nonempty closed set in X which is r-prox-regular for an extended real $r \in]0, +\infty]$. The following property holds.
(l) The truncated normal functional cone mapping $N_C^{*r,P}$ is J-hypomonotone of rate t for any $t \geq 1$.

Conversely, (l) entails the assertions of Theorem 18.102 with parameter $r/2$ instead of r.

PROOF. See Step 1 and Step 2 in the proof of Theorem 18.102. □

The following corollary of Theorem 18.102 provides several characterizations of convex sets in the setting of uniformly convex Banach space.

COROLLARY 18.110. Let $(X, \|\cdot\|)$ be a uniformly convex Banach space whose norm is also uniformly smooth and let C be a nonempty closed set in X. The following are equivalent:
(a) the set C is convex;
(b) the set C is uniformly ∞-prox-regular, or equivalently uniformly r-prox-regular for any real number $r > 0$;
(c) d_C is continuously differentiable on $X \setminus C$;
(d) d_C is Fréchet subdifferentially regular on $X \setminus C$;
(e) d_C is Fréchet differentiable on $X \setminus C$;
(f) d_C is Gâteaux differentiable on $X \setminus C$ with $\|D_G d_C(x)\| = 1$ for all $x \in X \setminus C$;
(g) $\partial_F d_C(x) \neq \emptyset$ for all $x \in X$;

(h) $\mathrm{Proj}_C(x) \neq \emptyset$ and $\partial_H d_C(x) \neq \emptyset$ for all $x \in X$;
(i) d_C^2 is of class $\mathcal{C}^1$ on X and its derivative is locally uniformly continuous on X;
(j) the single-valued mapping P_C is well defined on X and locally uniformly continuous on X;
(k) the single-valued mapping P_C is well defined on X and norm-to-norm continuous on X
(l) the single-valued mapping P_C is well defined on X and norm-to-weak continuous on X;
(m) for any $v \in N_C^P(x)$ with $x \in C$ one has $x \in \mathrm{Proj}_C(x+v)$;
(n) If $u \in X \setminus C$ and $x = P_C(u)$, then $x \in \mathrm{Proj}_C(u')$ for $u' = x + r(u-x)$ and any real $r > 0$.

PROOF. The equivalence between all the assertions from (b) to (n) is easily seen to follow from Theorem 18.102. The implication (a)⇒(g) is obvious according to the convexity of the continuous function d_C under (a). By Proposition 18.109, the condition (g) entails that $N_C^{*r,P}$ is J-hypomonotone of rate 1 on $U_r(C)$ for any $r > 0$. Fix any $x, y \in C$ and any $x^* \in N_C^{*,P}(x)$, $y^* \in N_C^{*,P}(y)$. By definition of J-hypomonotonicity of rate 1, for every real $s > 0$ large enough one has

$$\langle J[J^*(sx^*) - (y-x)] - J[J^*(sy^*) - (x-y)], x - y\rangle \geq 0,$$

or equivalently

$$\left\langle J[J^*(x^*) - \frac{1}{s}(y-x)] - J[J^*(y^*) - \frac{1}{s}(x-y), x - y \right\rangle \geq 0.$$

Using the continuity of J and J^* and passing to the limit as $s \to +\infty$, we obtain

$$\langle x^* - y^*, x - y\rangle \geq 0.$$

This means that $N_C^{*,P}$ is monotone. Since any $(x, x^*) \in \mathrm{gph}\, N_C^F$ is the limit of a sequence in $\mathrm{gph}\, N_C^{*,P}$ (see Proposition 18.59), we derive that N_C^F is monotone, and hence by Corollary 6.69(d) the set C is convex, that is, (a) holds. The proof of the corollary is then complete. □

18.6.2. Connected components of r-prox-regular sets. Concerning union and connected components of r-prox-regular sets, it is clear that all of the results in the context of Hilbert spaces in Subsection 15.2.4 are still valid in the framework of uniformly convex and uniformly smooth Banach spaces. They are stated without proof. It suffices to use Theorem 18.102(j) in place of Theorem 15.28(m).

PROPOSITION 18.111. Let $(X, \|\cdot\|)$ be a uniformly convex Banach space whose norm is also uniformly smooth. Let C be a nonempty r-prox-regular set of X with $\mathrm{diam}\, C < 2r$, where $r \in]0, +\infty]$. Then C is arcwise connected.

Like Proposition 15.63 the union of r-prox-regular sets with all gaps not less than $2r$ is r-prox-regular.

PROPOSITION 18.112. Let $(X, \|\cdot\|)$ be a uniformly convex Banach space whose norm is also uniformly smooth. Let $(C_i)_{i\in I}$ be a family of nonempty closed sets of X such that for a real $r > 0$ the inequality $\mathrm{gap}(C_i, C_j) \geq 2r$ holds for all $i \neq j$ in I. Then the set $C := \bigcup_{i\in I} C_i$ is r-prox-regular if and only if each set C_i is r-prox-regular.

The criteria in Proposition 15.64 for the r-prox-regularity of a closed set of a Hilbert space via its connected components is still valid in the framework of unformly convex and uniformly smooth Banach space.

PROPOSITION 18.113. Let $(X, \|\cdot\|)$ be a uniformly convex Banach space whose norm is also uniformly smooth. Let C be a nonempty closed set of X and let $r \in]0, +\infty[$. The set C is r-prox-regular if and only if all of its connected components are r-prox-regular and $\mathrm{gap}(C', C'') \geq 2r$ for any pair of its distinct connected components C', C''.

18.6.3. Enlargements and exterior points of r-prox-regular sets. In this subsection, we keep $(X, \|\cdot\|)$ as a Banach space whose norm is both uniformly convex and uniformly smooth. In this setting, we discuss, for an r-prox-regular set C, various properties of the ρ-enlargement $\mathrm{Enl}_\rho C$ with $\rho \in]0, r[$ and of the set of ρ-exterior points to C

$$\mathrm{Exte}_\rho C := \{u \in X : d_C(u) \geq \rho\}.$$

The first result concerns the cone of proximal normal functionals to the ρ-enlargement of C.

PROPOSITION 18.114. Assume that $C \subset X$ is r-prox-regular. Then, for any $\rho \in]0, r[$ and any $y \in D_C(\rho) := \{x \in X : d_C(x) = \rho\}$,

$$N^{*,P}(\mathrm{Enl}_\rho \mathrm{C}; y) = N^C(\mathrm{Enl}_\rho \mathrm{C}; y) = \mathbb{R}_+ D_F d_C(y) \subset N^{*,P}(C; P_C(y)).$$

PROOF. By Theorem 18.102(b) and Proposition 18.108 we know that d_C is of class $\mathcal{C}^1$ on $U_r(C)$ and

$$D_F d_C(x) = \frac{J(x - P_C(x))}{\|x - P_C(x)\|} \quad \text{for all } x \in U_r(C).$$

For convenience we will put $C(\rho) := \mathrm{Enl}_\rho C$. Fix $y \in D_C(\rho)$. From Corollary 18.106, the set $C(\rho)$ is $(r-\rho)$-prox-regular and $\mathrm{Proj}_{C(\rho)}$ is single-valued on $U_r(C)$. To see that $D_F d_C(y)$ actually belongs to $N^{*,P}_{C(\rho)}(y)$, put

$$u := y + \varepsilon(y - P_C(y))$$

where ε is small enough that $d_C(u) < r$. As $P_C(u) = P_C(y)$ by Theorem 18.102(k), we have

$$d_C(u) = \|u - P_C(y)\| = \|y + \varepsilon(y - P_C(y)) - P_C(y)\| = (1+\varepsilon)\|y - P_C(y)\| = (1+\varepsilon)\rho > \rho.$$

Then, by Lemma 18.103(b), since $\mathrm{Proj}_{C(\rho)}$ is single-valued on $U_r(C)$, one has $y = P_{C(\rho)}(u)$. So, $u - y \in N^P_{C(\rho)}(y)$, and hence $y - P_C(y) \in N^P_{C(\rho)}(y)$, which entails $D_F d_C(y) \in N^{*,P}_{C(\rho)}(y)$ by Proposition 18.108. Further, by the inclusion $y \in D_C(\rho)$ we have

$$C(\rho) = \{x \in X : d_C(x) \leq d_C(y)\}.$$

Since d_C is $\mathcal{C}^1$ near y along with $\|D_F d_C(y)\| = 1$, we then know that $N^C_{C(\rho)}(y) = \mathbb{R}_+ D_F d_C(y)$ (see Proposition 2.196), and hence

$$\mathbb{R}_+ D_F d_C(y) \subset N^{*,P}_{C(\rho)}(y) \subset N^C_{C(\rho)}(y) = \mathbb{R}_+ D_F d_C(y).$$

So, it remains to see that the inclusion $D_F d_C(y) \in N^{*,P}_C(P_C(y))$ follows from the fact that

$$D_F d_C(y) = \frac{J(y - P_C(y))}{\|y - P_C(y)\|} \in N^{*,P}_C(P_C(y)).$$

□

Before establishing the result related to the set of r-exterior points, let us prove the following lemma which has its own interest. The lemma extends to uniformly convex Banach spaces the result established in Lemma 15.50 for Hilbert spaces.

LEMMA 18.115. Assume that $C \subset X$ is r-prox-regular. Then, for any real $\rho \in]0, r]$ and $y \in U_\rho(C)$ one has

$$d_C(y) + d_{\mathrm{Exte}_\rho C}(y) = \rho.$$

PROOF. Fix $y \in U_\rho(C)$ and put $x := P_C(y)$ according to Theorem 18.102(i). Then for $u := x + \rho \frac{y-x}{\|y-x\|}$ one has $x \in \mathrm{Proj}_C(u)$ by Theorem 18.102(k), and hence $u \in D_C(\rho) \subset \mathrm{Exte}_\rho C$ and

$$d_C(y) + d_{\mathrm{Exte}_\rho C}(y) \leq \|y - x\| + \|y - u\| = \|u - x\| = \rho. \tag{18.82}$$

For any $z \in \mathrm{Exte}_\rho C$ we also have

$$\|z - y\| \geq \|z - P_C(y)\| - \|y - P_C(y)\| \geq d_C(z) - d_C(y) \geq \rho - d_C(y),$$

hence

$$d_{\mathrm{Exte}_\rho C}(y) \geq \rho - d_C(y).$$

It follows from this and (18.82) that $d_C(y) + d_{\mathrm{Exte}_\rho C}(y) = \rho$. □

From this lemma we see, through Theorem 18.102(d), that if C is r-prox-regular, then for any $\rho \in]0, r[$, the set $\mathrm{Exte}_\rho C$ is ρ-prox-regular, because $d_{\mathrm{Exte}_\rho C}(\cdot) = \rho - d_C(\cdot)$ on $U_\rho(C) \setminus C = U_\rho(\mathrm{Exte}_\rho C) \setminus \mathrm{Exte}_\rho C$. On the other hand, the lemma allows us to add one more characterization to the list of Theorem 18.102.

THEOREM 18.116 (metric characterization of r-prox-regularity). Let $(X, \|\cdot\|)$ be a uniformly convex Banach space whose norm is also uniformly smooth. A nonempty closed set $C \subset X$ is r-prox-regular for some real $r > 0$ if and only if

$$d_C(y) + d_{\mathrm{Exte}_r C}(y) = r \quad \text{for all } y \in U_r(C). \tag{18.83}$$

PROOF. The fact that (18.83) is implied by the uniform r-regularity of C follows from Lemma 18.115 with $\rho = r$. Assume now that (18.83) holds and consider any $y \in U_r(C)$ for which $\mathrm{Proj}_C(y)$ is a singleton, say $P_C(y) = x$. Then, for any $y' \in]x, y[$, one has $P_C(y') = x$. This yields for any non zero $t \in]-1, \frac{\|y-y'\|}{\|y'-x\|}[$

$$\begin{aligned} &t^{-1}[d_C(y' + t(y' - x)) - d_C(y')] \\ &= t^{-1}[\|y' + t(y' - x) - x\| - \|y' - x\|] = \|y' - x\|, \end{aligned}$$

which entails that d_C has a (bilateral) Gâteaux directional derivative in the full direction $y' - x$, hence by Corollary 14.23 the function d_C is Fréchet differentiable at y'. Thus, by (18.83) the function $d_{\mathrm{Exte}_\rho C}$ is also Fréchet differentiable at y'. We then deduce by Lemma 18.80(b) that $\mathrm{Proj}_{\mathrm{Exte}_r C}(y')$ is a singleton which will be denoted by u'. Using successively the inclusion $u' \in D_C(r)$ and (18.83) we obtain

$$r = d_C(u') \leq \|x - u'\| \leq \|y' - x\| + \|y' - u'\| = d_C(y') + d_{\mathrm{Exte}_r C}(y') = r,$$

so $y' \in]x, u'[$ (see Lemma 18.2) and $\mathrm{Proj}_C(u') \ni x$. By Theorem 18.102(l), this means that C is uniformly r-prox-regular. □

18.7. Lipschitz continuity of metric projection and radius of prox-regularity

For any r-prox-regular closed set C of a Hilbert space H and any real $0 < \gamma < 1$, we saw in Theorem 15.28 that the metric projection P_C is well defined on $U_r^\gamma(C)$ and Lipschitz on $U_r^\gamma(C)$ with $(1-\gamma)^{-1}$ as a Lipschitz constant therein, that is,

$$\|P_C(x) - P_C(y)\| \le (1-\gamma)^{-1}\|x-y\| \quad \text{for all } x, y \in U_r^\gamma(C).$$

Equivalently, for any real $s \in]0, r[$, taking $\gamma := s/r$ we see that

$$\|P_C(x) - P_C(y)\| \le \frac{r}{r-s}\|x-y\| \quad \text{for all } x, y \in U_s(C),$$

that is, P_C is Lipschitz on $U_s(C)$ with $r/(r-s)$ as a Lipschitz constant.

Conversely, suppose that, for a closed set C of the Hilbert space H and some real number $s > 0$, the metric projection P_C is well defined on $U_s(C)$ and Lipschitz therein with some Lipschitz constant $K > 1$. Considering the real $r > s$ given by the equality $r/(r-s) = K$, say $r = Ks/(K-1)$, the following question arises: Is the set C r-prox-regular? The next theorem provides a positive answer even in the context of a uniformly convex Banach space.

THEOREM 18.117 (M.V. Balashov theorem on prox-regularity radius). Let X be a uniformly Banach space endowed with a norm $\|\cdot\|$ which is both uniformly convex and unformly smooth. Let C be a nonempty closed set of X such that, for some real $s > 0$, the metric projection P_C is well defined on $U_s(C)$ and Lipschitz therein with some Lipschitz constant $K > 1$. Then the set C is $\frac{Ks}{K-1}$-prox-regular.

PROOF. We know by Theorem 18.102(j) that the set C is s-prox-regular. It is enough to show that, for any $\sigma \in]0, s[$ the set C is $(K-1)^{-1}K\sigma$-prox-regular. Suppose that there is some $\sigma \in]0, s[$ such that, for $r := (K-1)^{-1}K\sigma$ the set C is not r-prox-regular. Then there exists some $r_0 \in]0, r[$, $x_0 \in C$ and $v_0 \in N^P(C; x_0)$ with $\|v_0\| = 1$ and such that $C \cap B(x_0 + r_0 v_0, r_0) \neq \emptyset$. According to the s-prox-regularity of C, there is some $u_0 \in X$ with $d_C(u_0) = \sigma$ and such that $x_0 = P_C(u_0)$ and $v_0 = (u_0 - x_0)/d_C(u_0)$, and hence according to what precedes

$$C \cap B\left(x_0 + r_0 \frac{u_0 - x_0}{d_C(u_0)}, r_0\right) \neq \emptyset. \tag{18.84}$$

Put $y_0 := x_0 + \sigma^{-1} r_0 (u_0 - x_0)$ and consider two cases: $y_0 \in C$ and $y_0 \notin C$.

Case 1: $y_0 \in C$. Then, since $x_0 = P_C(u_0)$, we have

$$|1 - \sigma^{-1} r| \, \|u_0 - x_0\| = \|u_0 - y_0\| > \|u_0 - x_0\|,$$

which can be rewritten as $|1 - \sigma^{-1} r_0| > 1$, or equivalently

$$\sigma^{-1} r_0 > 2, \quad \text{that is, } r_0 > 2\sigma. \tag{18.85}$$

Denote by y_1 a nearest point in $[u_0, y_0] \cap C$ of u_0, and set $r_1 := \|x_0 - y_1\|$. Noting that $y_1 \neq u_0$ since $u_0 \notin C$, concerning $]x_0, y_0[$ it appears that $y_1 \in]u_0, y_0]$ and $u_0 \in]x_0, y_0[$ (according to the definition of y_0 and the inequality $\sigma^{-1} r_0 > 1$). It follows that

$$r_1 = \|x_0 - y_1\| \le \|x_0 - y_0\| = r_0 < r \text{ and } \sigma = \|x_0 - u_0\| < \|x_0 - y_1\| = r_1,$$

so $\sigma < r_1 < r$. We can then choose a real $\rho > 0$ such that

$$\rho < \min\left\{\frac{\sigma(r-r_1)}{4(r-\sigma)}, \frac{\sigma}{4}, r_1 - \sigma\right\}.$$

Since the inclusions $y_1 \in [u_0, y_0] \subset [x_0, y_0]$ give

$$\|u_0 - y_1\| = \|y_1 - x_0\| - \|u_0 - x_0\| = r_1 - \sigma > \rho,$$

we can choose $z \in]u_0, y_1[$ such that $\|z - y_1\| = \rho$. From the inclusion $z \in]u_0, y_1[$ we deduce the inequality $\|u_0 - z\| < \|u_0 - y_1\|$ and the inclusion $z \in [u_0, y_0]$, so $z \notin C$ since y_1 is a nearest point in $[u_0, y_0] \cap C$ of u_0. The open set $B(z, \rho) \cap (X \setminus C)$ is then nonempty, and the Lau theorem (see Theorem 14.29) yields some $u_1 \in B(z, \rho) \cap (X \setminus C)$ and some $x_1 \in C$ such that $x_1 = P_C(u_1)$. Since $y_1 \in C$, it follows that

$$\|u_1 - x_1\| \le \|u_1 - y_1\| \le \|u_1 - z\| + \|z - y_1\| < 2\rho < \sigma,$$

which ensures on the one hand $d_C(u_1) < s$, and on the other hand

$$\|z - x_1\| \le \|z - u_1\| + \|u_1 - x_1\| < 3\rho,$$

and hence

$$\begin{aligned}\|x_0 - x_1\| &\ge \|x_0 - z\| - \|z - x_1\| \ge \|x_0 - y_1\| - \|y_1 - z\| - \|z - x_1\| \\ &= r_1 - \rho - \|z - x_1\| > r_1 - 4\rho.\end{aligned}$$

Further, from the equalities

$$\|u_0 - z\| = \|y_1 - x_0\| - \|y_1 - z\| - \|u_0 - x_0\| = r_1 - \sigma - \rho$$

(due to the inclusions $[x_0, u_0] \subset [x_0, z] \subset [x_0, y_1]$), we see that

$$\|u_0 - u_1\| \le \|u_0 - z\| + \|z - u_1\| < r_1 - \sigma.$$

Consequently, it results that $u_i \in U_s(C) \setminus C$ along with $x_i = P_C(u_i)$, $i = 0, 1$, and

$$\frac{\|x_0 - x_1\|}{\|u_0 - u_1\|} > \frac{r_1 - 4\rho}{r_1 - \sigma} > \frac{r}{r - \sigma} = K, \text{ thus } \|x_0 - x_1\| > K\|u_0 - u_1\|,$$

where the inequality $\dfrac{r_1 - 4\rho}{r_1 - \sigma} > \dfrac{r}{r - \sigma}$ comes from the choice $\rho < \dfrac{\sigma(r - r_1)}{4(r - \sigma)}$.

Case 2: $y_0 \notin C$. By (18.84) fix a point $c \in C \cap B(y_0, r_0)$ and put $r_1 := \|c - y_0\| < r_0$. Since $y_0 \notin C$ by the assumption of Case 2, we can choose some real $\sigma_0 > 0$ such that $C \cap B(y_0, \sigma_0) = \emptyset$. For

$$\beta := \frac{(K' - K)(r_0 - r_1)}{K'K + 2(K' - K)}, \text{ where } K' := \frac{r_0}{r_0 - \sigma} > K = \frac{r}{r - \sigma},$$

choose some real $\rho > 0$ such that

$$\rho < \min\left\{\sigma_0, \beta, \frac{1}{2}(r_0 - r_1)\right\}.$$

By Lau theorem choose $y_1 \in B(y_0, \rho)$ and $x_1 \in C$ such that $x_1 = P_C(y_1)$, and note that $x_1 \neq y_0$ and $x_1 \neq y_1$ since $B(y_0, \rho) \cap C = \emptyset$ according to the choice $\rho < \sigma_0$. Writing

$$\|y_1 - x_1\| \le \|y_1 - c\| \le \|y_1 - y_0\| + \|y_0 - c\| < \rho + r_1$$

and taking the choice $\rho < (r_0 - r_1)/2$ into account, it ensues that

$$\|y_0 - x_1\| \le \|y_0 - y_1\| + \|y_1 - x_1\| < r_1 + 2\rho < r_0, \tag{18.86}$$

and hence, in particular, $x_1 \neq x_0$ since $\|y_0 - x_0\| = r_0$ by definition of y_0. Consider the points $u_1 \in]x_1, y_1[$ and $z \in]x_1, y_0[$ such that

$$\frac{\|x_1 - y_1\|}{\|u_1 - y_1\|} = \frac{\|x_1 - y_0\|}{\|z - y_0\|} = K' > 1. \tag{18.87}$$

Since $x_1 = P_C(y_1)$ and $u_1 \in]x_1, y_1[$, we have $x_1 = P_C(u_1)$ (see Lemma 18.8) and

$$\begin{aligned} d_C(u_1) = \|u_1 - x_1\| &= \|x_1 - y_1\| - \|u_1 - y_1\| = \left(1 - \frac{1}{K'}\right)\|x_1 - y_1\| \\ &= \frac{\sigma}{r_0}\|x_1 - y_1\| < \frac{\sigma}{r_0}(\rho + r_1) < \sigma, \end{aligned}$$

the latter inequality being due to the choice $\rho < r_0 - r_1$. Further, since $\|x_0 - y_0\| = r_0$ according to the definition of y_0, and since the inclusion $u_0 \in [x_0, y_0]$ allows us to write $\|u_0 - y_0\| = \|x_0 - y_0\| - \|x_0 - u_0\| = r_0 - \sigma$, we see that $x_0 - y_0 = (r_0 - \sigma)^{-1} r_0 (u_0 - y_0)$, that is, $x_0 - y_0 = K'(u_0 - y_0)$. We deduce that

$$x_1 - x_0 = x_1 - y_0 - (x_0 - y_0) = K'(z - y_0) - K'(u_0 - y_0) = K'(z - u_0),$$

thus in particular

$$\|x_1 - x_0\| = K'\|z - u_0\|. \tag{18.88}$$

Notice also that

$$y_1 - x_1 = K'(y_1 - u_1) = K'(y_1 - x_1) + K'(x_1 - u_1),$$

hence $K'(u_1 - x_1) = (K' - 1)(y_1 - x_1)$, or equivalently $y_1 - x_1 = (K' - 1)^{-1} K'(u_1 - x_1)$. Similarly, one also has $y_0 - x_1 = (K' - 1)^{-1} K'(z - x_1)$. It follows that

$$y_1 - y_0 = y_1 - x_1 - (y_0 - x_1) = \frac{K'}{k' - 1}(u_1 - x_1) - \frac{K'}{K' - 1}(z - x_1) = \frac{K'}{K' - 1}(u_1 - z),$$

which gives by (18.85)

$$\|u_1 - z\| = \left(1 - \frac{1}{K'}\right)\|y_0 - y_1\| = \frac{\sigma}{r_0}\|y_0 - y_1\| < \frac{\sigma}{r_0}\rho < \rho.$$

We deduce that

$$\frac{\|x_1 - x_0\|}{\|u_1 - u_0\|} \geq \frac{\|x_1 - x_0\|}{\|u_1 - z\| + \|z - u_0\|} \geq \frac{\|x_1 - x_0\|}{\rho + \|z - u_0\|} = \frac{K'}{1 + \frac{\rho}{\|z - u_0\|}}.$$

On the other hand, the above inequality $\|y_0 - x_1\| < r_1 + 2\rho$ in (18.86) gives

$$\|x_1 - x_0\| \geq \|y_0 - x_0\| - \|y_0 - x_1\| = r_0 - \|y_0 - x_1\| > r_0 - r_1 - 2\rho,$$

thus by (18.88)

$$\|z - u_0\| = \frac{\|x_1 - x_0\|}{K'} > \frac{r_0 - r_1 - 2\rho}{K'},$$

which implies

$$\frac{K'}{1 + \frac{\rho}{\|z - u_0\|}} > \frac{K'}{1 + \frac{K'\rho}{r_0 - r_1 - 2\rho}} > \frac{K'}{1 + \frac{K'\beta}{r_0 - r_1 - 2\beta}} = K,$$

the latter equality being due to the definition of β. It ensues that $\|x_0 - x_1\| > K\|u_0 - u_1\|$.

In both cases we obtained points $u_i \in U_s(C) \setminus C$ and $x_i = P_C(u_i)$, $i = 0, 1$, such that $\|x_0 - x_1\| > K\|u_0 - u_1\|$, which contradicts the assumption concerning the constant K and finishes the proof. □

COROLLARY 18.118. Let X be a uniformly convex Banach space endowed with a norm $\|\cdot\|$ which is both uniformly convex and unformly smooth. Let C be a nonempty closed set of X such that, for some real $s > 0$, the metric projection P_C is well defined on $U_s(C)$ and Lipschitz therein with some Lipschitz constant $K = 1$. Then the set C is convex.

PROOF. For any real $r > s$, since $K' := \frac{r}{r-s} > 1$, the metric projection P_C is K'-Lipschitz on $U_s(C)$, and hence by Theorem 18.117 the set C is $\frac{K's}{K'-1}$-prox-regular, or equivalently r-prox-regular. Consequently, the closed set C is convex since it is r-prox-regular for all reals $r > 0$ (see Corollary 18.110). □

18.8. Prox-regularity and geometric variational properties of cones

Let K be a cone of a normed space X, that is, $rx \in K$ for any $x \in K$ and any real $r > 0$. It is easily seen that, for $x \in K$ and $r > 0$,

$$T^B(K;rx) = T^B(K;x),\ T^C(K;rx) = T^C(K;x),$$

as well as

(18.89)
$$N^C(K;rx) = N^C(K;x),\ N^F(K;rx) = N^F(K;x),\ N^L(K;rx) = N^L(K;x).$$

In a Hilbert space, the equality $N^P(K;rx) = N^P(K;x)$ also holds true.

Concerning the situation when the reference point is the origin, we have:

PROPOSITION 18.119. Let K be a cone of a normed space X with $0 \in K$.
(a) $T^C(K;0) = \bigcap_{x\in K} T^C(K;x)$ and hence $N^C(K;0) = \bigcup_{x\in K} N^C(K;x)$.
(b) If X is an Asplund space, then $N^L(K;0) = \bigcup_{x\in K} N^L(K;x)$.

PROOF. (a) To prove the first equality in (a) it suffices to show that $T^C(K;0) \subset T^C(K;x)$ for any fixed $x \in K$. Let any $h \in T^C(K;0)$. Consider any sequences $(x_n)_n$ in K converging to x and $(t_n)_n$ tending to 0 with $t_n > 0$. Then $t_nx_n \in K$ for all n and $t_nx_n \to 0$. Since $t_n^2 \downarrow 0$, the sequential characterization of $T^C(K;0)$ furnishes some sequence $(h_n)_n$ converging to h such that $t_nx_n + t_n^2h_n \in K$ for all n, so $x_n + t_nh_n \in K$ for all n, that is, $h \in T^C(K;x)$. This justifies the desired inclusion $T^C(K;0) \subset T^C(K;x)$.

For any fixed $x \in K$, by polarity we deduce $N^C(K;x) \subset N^C(K;0)$, and hence $\bigcup_{x\in K} N^C(K;x) \subset N^C(K;0)$, and the converse inclusion is obviously valid.

(b) Fix any $x \in K$ and take $x^* \in N^L(K;x)$. There are sequences $(x_n)_n$ in K converging to x and $(x_n^*)_n$ converging weakly star to x^* with $x_n^* \in N^F(K;x_n)$. Taking $r_n := 1/n$, by the aforementioned equality we have $x_n^* \in N^F(K;r_nx_n)$. Since $r_nx_n \to 0$, it ensues that $x^* \in N^L(K;0)$, which justifies the inclusion $\bigcup_{x\in K} N^L(K;x) \subset N^L(K;0)$ from which the equality in (b) follows. □

REMARK 18.120. The inclusion $T^B(K;0) \subset T^B(K;x)$ may fail. It is the case for the cone $K := \{(r,s) \in \mathbb{R}^2 : s = |r|\}$ since $(-1,1) \in T^B(K;0)$ but $(-1,1) \notin T^B(K;x)$ for $x := (1,1)$. □

PROPOSITION 18.121. Let K be a closed cone in a uniformly convex Banach space $(X, \|\cdot\|)$ whose norm is also uniformly smooth. Then K is prox-regular at the origin if and only if it is convex.

PROOF. The implication $\Leftarrow$ is clear by the equivalence (a)$\Leftrightarrow$(b) in Corollary 18.110. Conversely, assume that the closed cone K is prox-regular at the origin. Theorem 18.99 furnishes an open neighborhood U of zero such that $N^{*,P}(K;x) = N^C(K;x)$ for all $x \in K \cap U$. Then, by Theorem 18.95 there are reals $r > 0$ and $\varepsilon > 0$ with $B(0,\varepsilon) \subset U$ such that for all $x_i \in K \cap B(0,\varepsilon)$ and $x^* \in N^C(K;x_i)$ with $\|x_i^*\| \le \varepsilon$

$$\langle J[J^*(x_1^*) - r(x_2 - x_1)] - J[J^*(x_2^*) - r(x_1 - x_2)], x_1 - x_2\rangle \ge 0.$$

Take any $x, y \in K$, any $x^* \in N^C(K;x)$ and any $y^* \in N^C(K;y)$. Choose a real $\mu > 0$ such that

$$\mu \max\{\|x\|, \|y\|\} < \varepsilon \quad \text{and} \quad \mu \max\{\|x^*\|, \|y^*\|\} < \varepsilon.$$

Then for any real $\theta \in]0,\mu]$, noticing by (18.89) that $x^* \in N^C(K;\theta x)$ and $y^* \in N^C(K;\theta y)$, we deduce that

$$\langle J[J^*(\mu x^*) - r(\theta y - \theta x)] - J[J^*(\mu y^*) - r(\theta x - \theta y)], \theta x - \theta y\rangle \ge 0,$$

which dividing by $\theta > 0$ gives

$$\langle J[J^*(\mu x^*) - r(\theta y - \theta x)] - J[J^*(\mu y^*) - r(\theta x - \theta y)], x - y\rangle \ge 0.$$

Making $\theta \downarrow 0$ and recalling that J is continuous and that J^* coincides with the inverse of J we obtain

$$\langle \mu x^* - \mu y^*, x - y\rangle \ge 0, \ \text{ or equivalently } \langle x^* - y^*, x - y\rangle \ge 0.$$

The latter inequality being true for any $x, y \in K$, any $x^* \in N^C(K;x)$ and any $y^* \in N^C(K;y)$, it results that the closed set K is convex according to (d) in Corollary 6.69. □

18.9. Comments

The concept of uniformly convex normed space has been introduced in 1936 by J. A. Clarkson [**251**], who also coined this name to such spaces. Proposition 18.18 is due to Clarkson [**251**] and its proof is taken from there. In [**251**] (see pages 400-403 there) one can also find the first statements and proofs that, for any real $p > 1$, the spaces $\ell^p_{\mathbb{R}}(\mathbb{N})$ and $L^p_{\mathbb{R}}(T)$ are uniformly convex when they are endowed with their usual norms. In a footnote in [**251**, p. 404] the name of strictly convex space has been given to normed spaces satisfying the property (d) in Lemma 18.2. M. M. Day offered a series of papers on uniformly convex spaces: in [**317**] he gave the first example of a reflexive Banach space which does not admit an equivalent norm which is uniformly convex. Locally uniformly convex norms were defined by R. A. Lovaglia [**676**, Definition 0.2] in 1955 as a weaker type of convexity for a norm than the uniform convexity. Diverse properties of locally uniformly convex spaces were studied by Lovaglia [**676**]. Proposition 18.23 was established by Lovaglia [**676**, Theorem 2.3].

The results in Lemma 18.10 were first given in diverse forms by V. L. Šmulian [**896**, p. 644] in 1940. Similar characterizations for the Gâteaux differentiability of a norm were also provided by Šmulian in his previous 1939 paper [**895**]. Other results on differentiability of norms are contained in Šmulian's papers [**893, 896**].

The proof of the assertion (c4) of Proposition 18.25 follows that of Lemma 1.e.8 in the 1979 book [**666**] of J. Lindenstrauss and L. Tzafriri.

The result in Proposition 18.28 seems to be first established in the 1979 book [**666**] by J. Lindenstrauss and L. Tzafiri. The proofs of this proposition and of

Lemma 18.27 are taken from Proposition 1.e.5 in this book [**666**]. Notice that Lemma 18.27 was previously almost contained in the lemma in page 251 of the 1963 paper [**663**] by J. Lindenstrauss.

In Remark 18.29 we have used the inequality $\varrho_H(\tau) \leq \varrho_{\|\cdot\|}(\tau)$ for all $\tau \geq 0$. This inequality is, by Lemma 18.31, a direct consequence of the following theorem by G. Norlander:

THEOREM 18.122 (G. Norlander). *Let $(X, \|\cdot\|)$ be a normed space with dimension $\dim X \geq 2$. Then one has*

$$\delta_{\|\cdot\|}(\varepsilon) \leq \delta_H(\varepsilon) \quad \text{for all } \varepsilon \in [0, 2],$$

that is, the modulus of convexity of a Hilbert space is the greatest possible modulus of convexity.

A proof of Theorem 18.122 can be found in Norlander's paper [**773**] in 1960 or in the book of J. Diestel [**334**] (see therein Theorem III.3.1, p. 60).

Theorem 18.30 was established by V. L. Šmulian in his 1940 paper [**896**, p. 648]; it was also independently established in 1944 by M. M. Day in Theorem 6.10 in his paper [**318**]. The proof that we gave is based on Lemma 18.31 shown by J. Lindenstrauss in 1963 in [**663**] (see Theorem 1 in page 242 in [**663**]). Both proofs of this lemma and of Theorem 18.30 are taken from the book of J. Lindenstrauss and L. Tzafiri [**666**] (see Proposition 1.e.2 therein). V. L. Šmulian's method in [**896**] is different. Day's method is based on a specific modulus that he introduced and called modulus of flattening. This deserves to be commented. M. M. Day defined in [**318**, p. 376] the concept of *modulus of flattening for* $(X, \|\cdot\|)$ as the function

$$]0, 2] \ni \tau \mapsto \sup\left\{\frac{1 - \|(x+y)/2\|}{\|x-y\|} : x, y \in \mathbb{S}_X, 0 < \|x-y\| \leq \tau\right\},$$

and he defined $(X, \|\cdot\|)$ to be *uniformly flattened* if this function of τ tends to 0 as $\tau \downarrow 0$. While the modulus of uniform convexity measures the infimum depth below the surface of the unit ball for the middle point of any line segment with endpoints on the sphere, the modulus of flattening measures the supremum ratio of any such a depth to the length of the line segment. After showing diverse estimations between moduli of uniform convexity and flattening of two-dimensional spaces and their duals, Day proved in Theorem 6.7 in [**318**] that a Banach space $(X, \|\cdot\|)$ is uniformly convex (resp. flattened) if and only if the dual Banach space $(X^*, \|\cdot\|_*)$ is uniformly flattened (resp. convex). A slightly modified form of the above modulus implicitly appeared in the book of G. Köthe [**615**, p. 363] as

$$]0, 2] \ni \tau \mapsto \eta_{\|\cdot\|}(\tau) := \sup\left\{\frac{\|x\| + \|y\| - \|x+y\|}{2\|x-y\|} : \|x\| \geq 1, \|y\| \geq 1, \|x-y\| \leq \tau\right\}.$$

Take $\tau \in]0, 1[$ and any $x, y \in \mathbb{S}_X$. Putting $u := (1-\tau)^{-1}(x + \tau y)$ and $v := (1-\tau)^{-1}(x - \tau y)$, we see that $\|u - v\| = 2\tau/(1-\tau)$, $\|u + v\| = 2/(1-\tau)$, $\|u\| \geq 1$, $\|v\| \geq 1$ and

$$\frac{\|x + \tau y\| + \|x - \tau y\| - 2}{2\tau} = \frac{\|u\| + \|v\| - \|u+v\|}{\|u-v\|} \leq \eta_{\|\cdot\|}\left(\frac{2\tau}{1-\tau}\right).$$

From this we obtain for any $\tau \in]0, 1[$

$$\frac{\varrho_{\|\cdot\|}(\tau)}{\tau} \leq \eta_{\|\cdot\|}\left(\frac{2\tau}{1-\tau}\right). \tag{18.90}$$

Now take any $\tau \in]0,2[$ and any distinct $u, v \in X$ with $\|u\| \geq 1$, $\|v\| \geq 1$, $\|u-v\| \leq \tau$. Notice that

$$\|u+v\| \geq \|2v\| - \|u-v\| \geq 2 - \|u-v\| \geq 2-\tau,$$

and define $x := \dfrac{u+v}{\|u+v\|}$ and $y := \dfrac{u-v}{\|u+v\|}$, so $\|y\| \leq \tau/(2-\tau)$. By definition of the modulus of uniform smoothness $\varrho_{\|\cdot\|}$ and by the equality $\|x\| = 1$ we can write

$$\frac{\|u\|+\|v\|-\|u+v\|}{2\|u-v\|} = \frac{\|x+y\|+\|x-y\|-2}{4\|y\|}$$

$$= \frac{1}{2\|y\|}\left[\frac{1}{2}(\|x+y\|+\|x-y\|)-1\right] \leq \frac{1}{2}\frac{\varrho_{\|\cdot\|}(\|y\|)}{\|y\|},$$

so invoking the nondecreasing property of $t \mapsto \varrho_{\|\cdot\|}(t)/t$ (see Proposition 18.25(s3)) we see that

$$\frac{\|u\|+\|v\|-\|u+v\|}{2\|u-v\|} \leq \frac{1}{2}\frac{\varrho_{\|\cdot\|}(\tau/(2-\tau))}{\tau/(2-\tau)}.$$

It ensues by definition of $\eta_{\|\cdot\|}$ that

$$\eta_{\|\cdot\|}(\tau) \leq \frac{1}{2}\frac{\varrho_{\|\cdot\|}(\tau/(2-\tau))}{\tau/(2-\tau)}.$$

This and the previous inequality (18.90) show that the norm $\|\cdot\|$ is uniformly smooth if and only if $\eta_{\|\cdot\|}(\tau) \to 0$ as $\tau \downarrow 0$.

Theorem 18.33 on the reflexivity of uniformly convex Banach spaces is due independently to D. Milman in his 1938 paper [**711**] and to B. J. Pettis in his 1939 paper [**799**]. The proof in the book is that of Proposition 1.e.3 by Lindenstrauss and Tzafiri [**666**].

Concerning Remark 18.26 one can find in V. I. Liokoumovich's paper [**667**]

(18.91) examples of norms whose moduli of convexity are not convex functions.

Proposition 18.39 is due S. Reich; in its proof we followed the arguments of Reich in Lemma 2.4 and the remark in pages 116 and 117 of his 1981 paper [**833**]. Concerning Proposition 18.41 the proofs of (a)⇒(b) and (b)⇒(a) are classical, while for the proof of (b)⇒(c) we used ideas from Remark 3.2 in the 2014 paper [**590**] by A. Jourani, L. Thibault and D. Zagrodny. The proof of Proposition 18.42 utilizes the main ideas of the proof of Lemma 5.1 in the book [**333**] of R. Deville, G. Godefroy and V. Zizler.

All the results in Subsection 18.1.5 are due to Zong-Ben Xu and G. F. Roach and were established in their 1991 paper [**975**]. Theorem 18.44 on characterizations of uniformly convex spaces corresponds to Theorem 1 in [**975**]; its proof as well as the proofs of its preparatory Lemmas 18.45, 18.46 and 18.47 are taken from this paper [**975**] by Zong-Ben Xu and G. F. Roach. Theorem 18.50 on characterizations of uniformly smooth spaces and its proof reproduce the statement and proof of Theorem 2 by Zong-Ben Xu and G. F. Roach in [**975**]. Corollaries 18.49, 18.51 and 18.53 along with Theorem 18.52 are reformulations of other results in [**975**].

Proposition 18.58 has been first observed by J. M. Borwein and H. Strójwas [**144**] and Proposition 18.59 is due to A. D. Ioffe [**518**] (even in a more general setting). Proposition 18.65 has been stated and proved by F. Bernard, L. Thibault and N. Zlateva [**97**].

The concept of J-primal lower regular functions in Definition 18.71 has been introduced by F. Bernard, L. Thibault and N. Zlateva in their 2006 paper [**97**] as an adaptation to uniformly convex spaces of the notion of primal lower nice functions defined in $\mathbb{R}^n$ by R. A. Poliquin [**805**]. Lemma 18.79 appeared in the same paper [**97**] in the context where the norm is both uniform convex with power type and uniformly smooth with power type. The general statement and its proof in the manuscript are slight modifications of [**97**].

All the results in Section 18.4 are due to and taken from the 2006 paper [**97**] by F. Bernard, L. Thibault and N. Zlateva. Lemma 18.92 follows the main lines of the similar lemma in Hilbert space in [**813**]. Theorem 18.96 corresponds to the theorem by Bernard, Thibault and Zlateva [**97**] providing the same characterizations of local prox-regularity of sets when the norm is uniformly convex with power type and uniformly smooth with power type. The proof here without the power type condition follows [**97**]. Proposition 18.29, Theorem 18.99 and corollary 18.100 were shown in the paper [**98**] by F. Bernard, L. Thibault and N. Zlateva.

The characterizations of global prox-regularity in Theorem 18.102 have been also first established by Bernard, Thibault and Zlateva [**97**] under the power type condition for the uniform convexity and uniform smoothness of the norm. The extension of the equivalences between (a), (b) and (j) to the setting where the power type condition is removed has been shown by M. V. Balashov and G. E. Ivanov in Theorem 2.4 of their 2009 paper [**67**]. The approach used in [**67**] to obtain these equivalences is different from that in the book. Since the implications (a)$\Rightarrow$(b)$\Rightarrow$(j) in [**97**] are still valid without the hypothesis of power type, Balashov and Ivanov established geometrical deep arguments showing the implication (j)$\Rightarrow$(a). The proof that we gave in the manuscript driving to the implication (j)$\Rightarrow$(a), and also to all the other equivalent assertions in Theorem 18.102, corresponds to an adaptation of the approach by Bernard, Thibault and Zlateva [**97**] to the general case without the power type condition. Proposition 18.109 and Theorem 18.116 are taken from [**97**]. Another semiconvexity property in the line of (c) in Corollary 18.107 was proved with different arguments by M. V. Balashov [**64**, Theorem 3.5] under the r-prox-regularity of the set. Other results for prox-regular sets in uniformly convex Banach spaces can be found in the paper [**535**] by G. E. Ivanov and in the survey [**444**] by V. V. Goncharov and G. E. Ivanov.

Theorem 18.117 and its proof in the manuscript are due to M. V. Balashov [**65**, Theorem 2.1].

APPENDIX A

Topology

Recall that a preorder $\preceq_J$ on a nonempty set J is a relation on J which is reflexive and transitive, and for such a relation the pair $(J, \preceq_J)$ is called a preordered set. The reflexivity amounts to saying that

$$j \preceq_J j \quad \text{for all } j \in J,$$

while the transivity means that given $j_1, j_2, j_3 \in J$

$$\big(j_1 \preceq_J j_2 \quad \text{and} \quad j_2 \preceq j_3\big) \Longrightarrow j_1 \preceq_J j_3.$$

It is often convenient to write $j_2 \succeq_J j_1$ instead of $j_1 \preceq_J j_2$. When the preorder $\preceq_J$ is such that for any $j_1, j_2 \in J$ there exists $j_3 \in J$ satisfying $j_1 \preceq_J j_3$ and $j_2 \preceq_J j_3$, one generally says that $(J, \preceq_J)$ is a *directed set.*

Given a nonempty set T, any family $(t_j)_{j\in J}$ indexed by a directed set $(J, \preceq_J)$ is called a *net*; one also writes $(t_j)_j$ when there is no ambiguity for the set J. Any sequence in T is clearly a particular net of T. Several other nets (which are not sequences) are already given in Example 1.15. Subsequences are known to be crucial in metric spaces to characterize compactness of sets. The same type of characterization also holds true in topological spaces through the concept of subnet. A *subnet* of the net $(t_j)_{j\in J}$ is a net of the form $(t_{s(i)})_{i\in I}$ where $(I, \preceq_I)$ is a directed set and $s : I \to J$ is a *directed mapping* in the sense that, for each $j_0 \in J$ there is $i_0 \in I$ such that $j_0 \preceq_J s(i)$ for all $i \in I$ satisfying $i_0 \preceq_I i$. Any subsequence of a sequence is obviously a subnet of this sequence, but there are obvious subnets of a sequence which are not subsequences. Particular subnets are achieved via the concept of cofinal set. A nonempty set $J_0 \subset J$ is called *cofinal* in the set $(J, \preceq_J)$ whenever for each $j \in J$ there is some $j' \in J_0$ such $j \preceq_J j'$. Endowing J_0 with the induced preorder we see that $(t_j)_{j\in J_0}$ is a subnet of the net $(t_j)_{j\in J}$.

If T is endowed with a topology τ, one says that the net $(t_j)_{j\in J}$ converges to some t in (T, τ_T) provided that for each neighborhood V of t there is some $j_V \in J$ such that $t_j \in V$ for all $j \in J$ with $j \succeq_J j_V$; in such a case one writes $t_j \underset{j\in J}{\to} t$ or $t_j \to t$. When the topology τ_T is Hausdorff, such an element (if any) is unique; in such a case it is called the limit of $(t_j)_{j\in J}$ and one writes $t = \lim_{j\in J} t_j$.

The following proposition summarizes some fundamental properties of nets and subnets. For its statement as well as for the foregoing concepts we refer, for example, to J. L. Kelley's book [**604**, Chapter 2, p. 65-72].

PROPOSITION A.1. Let (T, τ_T) and (X, τ_X) be topological spaces, S be a subset of T and $f : T \to X$ be a mapping. The following holds.
(a) One has $t \in \operatorname{cl} S$ if and only if there is a net $(t_j)_j$ in S converging to t in (T, τ_T). The set S is thus τ_T closed if and only if S contains any element to which τ_T-converges a net in S.
(b) The mapping f is continuous at a point $\bar{t} \in T$ if and only if $f(t_j) \to f(\bar{t})$ in

(X, τ_X) for any net $(t_j)_{j \in J}$ in T with $t_j \to \bar{t}$.
(c) A net $(t_j)_{j \in J}$ converges to t in (T, τ_T) if and only if everyone of its subnets converges to t.
(d) A net $(t_j)_{j \in J}$ in (T, τ_T) converges to some $t \in T$ if and only if every subnet admits itself a subnet converging to t.
(e) Given another topological space $(T', \tau_{T'})$ and endowing $T \times T'$ with the product topology, a net $(t_j, t'_j)_{j \in J}$ converges to (t, t') in $T \times T'$ if and only if $t_j \to t$ and $t'_j \to t'$.
(f) A subset $S \subset T$ is τ_T-compact if and only if every net in S admits a subnet τ_T-converging to an element in S.

Another remarkable feature of nets is the following property of iterated limits (see, for example, [**604**, Chapter 2, Theorem 4]):

PROPOSITION A.2. Let $(\Lambda, \preceq_\Lambda)$ be a directed set and for each $\lambda \in \Lambda$ let $(I_\lambda, \preceq_\lambda)$ be also a directed set. Let $J := \Pi_{\lambda \in \Lambda} I_\lambda$ and let $\Lambda \times J = \Lambda \times \Pi_{\lambda \in \Lambda} I_\lambda$ be endowed with the product preorder. Assume that for each $\lambda \in \Lambda$ a net $(y_{\lambda,i})_{i \in I_\lambda}$ of a topological space Y converges to some z_λ in Y and that the net $(z_\lambda)_{\lambda \in \Lambda}$ converges to some $z \in Y$. Then putting $\xi_{\lambda,j} := y_{\lambda, j_\lambda}$ for every (λ, j) in $\Lambda \times J$, the net $(\xi_{\lambda,j})_{(\lambda,j) \in \Lambda \times J}$ converges to z.

Let us turn now to paracompact spaces. Let (T, τ) be a Hausdorff topological space. A family $(U_j)_{j \in J}$ is called an *open cover* (or *open covering*) of T if $\bigcup_{j \in J} U_j = T$ and each U_j is an open set in T. The open cover $(U_j)_{j \in J}$ is *locally finite* when each $t \in T$ has a neighborhood intersecting only finitely many U_j. An open cover $(O_i)_{i \in I}$ of T is said to be an *open refinement* of an open cover $(U_j)_{j \in J}$ provided that for each $i \in I$ there is some $j \in J$ such that $O_i \subset U_j$. One says that T is *paracompact* if any open cover admits a locally finite open refinement.

Metric spaces and compact Hausdorff topological spaces are remarkable examples of paracompact spaces.

Recall that a family of real-valued functions $(\varphi_i)_{i \in I}$ on T is called a locally finite continuous partition of unity on T if

(i) each point $t \in T$ has a neighborhood on which all but a finite number of functions φ_i are null;
(ii) all the functions φ_i are non-negative and

$$\sum_{i \in I} \varphi_i(t) = 1 \quad \text{for every } t \in T.$$

If T is a Banach space X and each function φ_i is in addition of class $\mathcal{C}^p$ with $p \in \mathbb{N} \cup \{\infty\}$, one says that $(\varphi_i)_{i \in I}$ is a *locally finite $\mathcal{C}^p$-partition of unity* on X.

A continuous (resp. $\mathcal{C}^p$) partition $(\varphi_i)_{i \in I}$ of unity on T (resp. on X) is said to be *subordinated* to an open cover $(U_j)_{j \in J}$ of T (resp. of X) if each function φ_i is null outside some U_j.

Paracompact spaces enjoy the following fundamental property (see, for example, [**604**, p. 171]).

THEOREM A.3 ($\mathcal{C}^0$-partition of unity under paracompactness). For any open cover $(U_j)_{j \in J}$ of a paracompact Hausdorff topological space T there is a locally finite continuous partition of unity on T which is subordinated to $(U_j)_{j \in J}$.

Theorem 3 in the paper [**932**] of H. Toruńczyk that we recall below established $\mathcal{C}^\infty$-partition of unity in Hilbert spaces.

THEOREM A.4 (H. Toruńczyk). *For any open cover $(U_j)_{j\in J}$ of a Hilbert space H there is a locally finite $\mathcal{C}^\infty$-partition of unity on H which is subordinated to $(U_j)_{j\in J}$.*

APPENDIX B

Topological properties of convex sets

Let us recall first the classical analytic Hahn-Banach theorem. We continue with the setting that *all vector spaces will be considered over the field* $\mathbb{R}$.

THEOREM B.1 (Hahn-Banach theorem: analytic form). Let X be a vector space and $s : X \to \mathbb{R}$ be a positive homogeneous convex function (also called sublinear function). Let $\ell_0 : X_0 \to \mathbb{R}$ be a linear functional on a vector subspace X_0 of X such that $\ell_0(u) \le s(u)$ for all $u \in X_0$. Then ℓ_0 can be extended on a linear functional $\ell : X \to \mathbb{R}$ to the whole space X so that $\ell(x) \le s(x)$ for all $x \in X$.

From this theorem it is classical to derive the geometric Hahn-Banach theorems. The main lines of the proofs will be sketched.

THEOREM B.2 (Hahn-Banach theorem: geometric form I). Let U be a nonempty open convex set in a topological vector space X and let a point $b \in X \setminus U$. Then there exists a continuous linear functional ℓ on X such that $\ell(x) < \ell(b)$ for all $x \in U$.

PROOF. Fix $u \in U$ and observe that $V := -u + U$ is an open convex set containing zero and that $a := b - u \notin V$. Consider the linear functional $\zeta : \mathbb{R}a \to \mathbb{R}$ defined by $\zeta(ta) = tj_V(a)$ for all $t \in \mathbb{R}$, where j_V denotes the Minkowski gauge function of V. It is easy to verify that $\zeta(ta) \le j_V(ta)$ for all $t \in \mathbb{R}$, hence the analytic Hahn-Banach theorem above gives that ζ can be extended to a linear functional $\ell : X \to \mathbb{R}$ such that $\ell(x) \le j_V(x)$ for all $x \in X$. Since j_V is continuous and $V = \{x \in X : j_V(x) < 1\}$, the linear functional ℓ is continuous and $j_V(a) \ge 1$. It results that

$$\ell(v) \le j_V(v) < 1 \le j_V(a) = \zeta(a) = \ell(a) \quad \text{for all } v \in V,$$

which easily yields that $\ell(x) < \ell(b)$ for all $x \in U$. $\square$

THEOREM B.3 (Hahn-Banach theorem: geometric form II). Let X be a topological vector space, C be a nonempty open convex set in X and D be a nonempty convex set in X such that $C \cap D = \emptyset$. Then there exist a nonzero continuous linear functional $\ell : X \to \mathbb{R}$ and a real α such that

$$C \subset \{\ell(\cdot) < \alpha\} \quad \text{and} \quad D \subset \{\ell(\cdot) \ge \alpha\}.$$

PROOF. Since the open convex set $U := C - D$ does not contain zero, the previous theorem furnishes a continuous linear functional $\ell : X \to \mathbb{R}$ such that $\ell(u) < \ell(0) = 0$ for all $u \in U$. It ensues that $\ell(x) < \ell(y)$ for all $x \in C$ and $y \in D$, hence $\sup\limits_{x \in C} \ell(x) \le \inf\limits_{y \in C} \ell(y)$ and both the supremum and infimum are finite. Choosing α between the two latter reals, we obtain $C \subset \{\ell(\cdot) \le \alpha\}$ and $D \subset \{\ell(\cdot) \ge \alpha\}$. Since ℓ is nonzero, it is not difficult to deduce from the latter inclusion that $C \subset \{\ell(\cdot) < \alpha\}$ as desired. $\square$

THEOREM B.4 (Hahn-Banach theorem: geometric form III). Let X be a locally convex vector space, C, D be nonempty disjoint convex subsets such that $C - D$ is closed, which holds in particular if one convex set is compact and the other is closed. Then there exist a nonzero continuous linear functional $\ell : X \to \mathbb{R}$ and reals α, ε with $\varepsilon > 0$ such that

$$C \subset \{\ell(\cdot) \le \alpha - \varepsilon\} \quad \text{and} \quad D \subset \{\ell(\cdot) \ge \alpha + \varepsilon\}.$$

PROOF. Since the convex set $C-D$ is closed and $0 \notin C-D$, there is a symmetric open convex neighborhood V of zero in X such that $(V+V) \cap (C-D) = \emptyset$, or equivalently $(C+V) \cap (D+V) = \emptyset$. Noting that $D+V$ is convex and $C+V$ is convex and open, there exists by Theorem B.3 a continuous linear functional $\ell : X \to \mathbb{R}$ and a real $\alpha > 0$ such that $\ell(x+u) \le \alpha \le \ell(y+v)$ for all $x \in C$, $y \in D$ and $u, v \in V$. From this the conclusion of the theorem easily follows. □

Convex sets admit well-known remarkable closure and interior properties.

PROPOSITION B.5. Let X be a topological vector space and C be a convex set in X. The following hold.
(a) The topological closure $\mathrm{cl}\, C$ and interior $\mathrm{int}\, C$ of C are convex.
(b) If $\mathrm{int}\, C \neq \emptyset$, then

$$[x, y[\, \subset \mathrm{int}\, C \quad \text{for all } x \in \mathrm{int}\, C \text{ and } y \in \mathrm{cl}\, C,$$

along with

$$\mathrm{cl}\,(\mathrm{int}\, C) = \mathrm{cl}\, C \quad \text{and} \quad \mathrm{int}\,(\mathrm{cl}\, C) = \mathrm{int}\, C.$$

PROOF. (b) Assume that $\mathrm{int}\, C \neq \emptyset$. Fix any $x \in \mathrm{int}\, C$. Take any $y \in \mathrm{cl}\, C$ and any $z \in]x, y[$ (if any). There are $s, t \in]0, 1[$ with $s + t = 1$ such that $z = sx + ty$. Choose a neighborhood U of zero such that $x + U + U \subset C$. Note that $y \in C + t^{-1}sU$ (since $y \in \mathrm{cl}\, C$), so there is $c \in C$ such that $y \in c + t^{-1}sU$, hence

$$z + sU = sx + ty + sU \subset s(x + U + U) + tc \subset sC + tC = C.$$

Since sU is a neighborhood of 0, it ensues that $z \in \mathrm{int}\, C$. It then results that $[x, y[\, \subset \mathrm{int}\, C$. The latter inclusion gives $sx + (1-s)y \in \mathrm{int}\, C$ for all $s \in]0, 1[$, hence $y \in \mathrm{cl}\,(\mathrm{int}\, C)$. This justifies the equality $\mathrm{cl}\,(\mathrm{int}\, C) = \mathrm{cl}\, C$. To prove the other equality in (b), take any $u \in \mathrm{int}\,(\mathrm{cl}\, C)$. There is some $\alpha > 1$ such that $v := \alpha u + (1-\alpha)x \in \mathrm{cl}\, C$. By what precedes we have $[x, v[\, \subset \mathrm{int}\, C$, hence $u \in \mathrm{int}\, C$. This justifies the equality $\mathrm{int}\,(\mathrm{cl}\, C) = \mathrm{int}\, C$.
(a) The convexity of $\mathrm{cl}\, C$ is obvious. Concerning $\mathrm{int}\, C$, its convexity is trivial if it is empty, and otherwise its convexity follows from the first property in (b) established above. □

It is known that the closure of intersection of convex sets does not coincide with the intersection of closures even for intervals in $\mathbb{R}$ as confirmed by $C_1 := [0, 1[$ and $C_2 := [1, 2]$. Nevertheless, we have the following proposition (see S. Dolecki [**341**]).

PROPOSITION B.6. Let X be a topological vector space and C, D be two convex subsets in X such that $C \cap \mathrm{int}\, D \neq \emptyset$. Then one has

$$\mathrm{cl}\,(C \cap \mathrm{int}\, D) = \mathrm{cl}\,(C \cap D) = (\mathrm{cl}\, C) \cap (\mathrm{cl}\, D).$$

PROOF. Choose $\overline{x} \in C \cap \mathrm{int}\, D$ and an open neighborhood U_0 of zero in X such that $\overline{x} + U_0 \subset D$. To show that the third member is included in the first one fix any $x \in (\mathrm{cl}\, C) \cap (\mathrm{cl}\, D)$. Consider any neighborhood U of zero and choose a balanced neighborhood V of zero in X satisfying $V + V \subset U \cap U_0$. Choose a real $r \geq 1$ such that $\overline{x} - x \in rV$ and then choose a real $s > 0$ such that $s(1+2r)/(1+2s) < 1$. There exists $y \in (x + sV) \cap D$. Put $t := 2s/(1+2s)$ (so, $0 < t < 1$) and note by Proposition B.5 that

$$(1-t)y + t(\overline{x} + U_0) \subset \operatorname{int} D.$$

Since

$$\begin{aligned}(1-t)(x+sV) + t\overline{x} &\subset (1-t)(y + sV + sV) + t\overline{x}\\ &= (1-t)y + \frac{t}{2}V + \frac{t}{2}V + t\overline{x} \subset (1-t)y + t(\overline{x} + V + V),\end{aligned}$$

it ensues that $(1-t)(x+sV) + t\overline{x} \subset \operatorname{int} D$. Noticing that $(x + sV) \cap C \neq \emptyset$ and using the inclusion $\overline{x} \in C$, we deduce that

$$\big((1-t)(x+sV) + t\overline{x}\big) \cap (C \cap \operatorname{int} D) \neq \emptyset. \tag{B.1}$$

Since $2rs/(1+2s) < 1$ and V is balanced, we observe that

$$t(\overline{x} - x) + s(1-t)V \subset rtV + s(1-t)V \subset \frac{2rs}{1+2s}V + \frac{s}{1+2s}V \subset V + V \subset U,$$

which combined with (B.1) yields $(x + U) \cap (C \cap \operatorname{int} D) \neq \emptyset$. It results that $x \in \operatorname{cl}(C \cap \operatorname{int} D)$, hence the third member of the proposition is included in the first. From this the equalities of the proposition are clear. □

Now let us recall that a subset F of a vector space X is *affine* if by definition it contains any *affine combination* of its elements, that is, for any $x_1, \cdots, x_p$ in F and any reals $t_1, \cdots, t_p$ with $t_1 + \cdots + t_p = 1$ one has

$$t_1x_1 + \cdots + t_px_p \in F.$$

When F is nonempty, taking any $u \in F$ we see that $F - u$ is a vector subspace of X. Consequently, any nonempty affine subset F of X is of the form $u + E$, where $u \in X$ and E is a vector subspace of X; such a vector space E is unique and its dimension is called the dimension of F, which is generally stated as $\dim F := \dim E$. Given any subset S of X, the intersection of all affine subsets of X containing S is an affine set, and it is the smallest affine subset of X containing S. It is called the *affine hull* of S and denoted by $\operatorname{aff} S$. When S is nonempty, it is easily seen that $\operatorname{aff} S$ coincides with the set of all affine combinations of elements in S, that is, $x \in \operatorname{aff} S$ if and only if for some integer p there are $x_1, \cdots, x_p \in S$ and $t_1, \cdots, t_p \in \mathbb{R}$ with $t_1 + \cdots + t_p = 1$ such that $x = t_1x_1 + \cdots + t_px_p$.

Assume now that X is a finite-dimensional normed space and let C be a convex set in X. We recall that the *relative interior* $\operatorname{rint} C$ of C is the interior of C relative to the affine hull $\operatorname{aff} C$ endowed with the induced topology, which can be written in the form

$$\operatorname{rint} C = \operatorname{int}_{\operatorname{aff} C}(C).$$

An affine subset F of X being closed, we see that

$$\operatorname{rint} F = F = \operatorname{cl} F.$$

At the opposite of the classical topological interior, it is worth noticing that, for two convex sets C_1, C_2 in X

$$C_1 \subset C_2 \not\Rightarrow \operatorname{rint} C_1 \subset \operatorname{rint} C_2;$$

indeed, for the subsets $C_1 := \{0\}$ and $C_2 := [0, 1[$ in $\mathbb{R}$, we have $C_1 \subset C_2$ but $\operatorname{rint} C_1 \not\subset \operatorname{rint} C_2$ since $\operatorname{rint} C_1 = \{0\}$ whereas $\operatorname{rint} C_2 =]0, 1[$.

Suppose for a moment that the convex set C in X is nonempty. We note that for any $u \in C$ the vector space $E := \mathrm{Vect}\,(C-u)$ spanned by $C-u$ coincides with $\mathbb{R}(C-u)$, does not depend on $u \in C$ and satisfies the equality

$$-u + \operatorname{rint} C = \operatorname{int}_E(C-u); \tag{B.2}$$

one defines the dimension of C as $\dim C := \dim E$. For $p := \dim E$, it is clear that there are $v_1, \cdots, v_p \in C-u$ linearly independent such that

$$Q := \{\lambda_1 v_1 + \cdots + \lambda_p v_p : \lambda_i \in]0,1[, \lambda_1 + \cdots + \lambda_p < 1\} \subset C - u \subset E.$$

Since the nonempty subset Q of E is open relative to E, it ensues that $\operatorname{int}_E(C-u) \neq \emptyset$. This and (B.2) tell us that $\operatorname{rint} C \neq \emptyset$. Further, taking $u \in \operatorname{rint} C$ the same equality B.2 ensures that the vector space spanned by $-u + \operatorname{rint} C$ coincides with E, hence $\operatorname{int}_E(-u + \operatorname{rint} C) = \operatorname{int}_E(-u + C)$, which means by (B.2) again that $\operatorname{rint}(\operatorname{rint} C) = \operatorname{rint} C$, which also holds true if $C = \emptyset$.

We summarize all those features as follows.

PROPOSITION B.7. Let C, C_1, C_2 be convex sets in a finite-dimensional normed space X. One has $\operatorname{rint}(\operatorname{rint} C) = \operatorname{rint} C$ along with

$$C_1 \subset C_2 \not\Rightarrow \operatorname{rint} C_1 \subset \operatorname{rint} C_2$$

and

$$C \neq \emptyset \Rightarrow \operatorname{rint} C \neq \emptyset.$$

From the second property of the proposition we can derive the following corollary.

COROLLARY B.8. If C is a nonempty convex set of a finite-dimensional normed space X with $\operatorname{int}_X C = \emptyset$, then for any $u \in C$, one has $\mathrm{Vect}(C-u) \neq X$.

PROOF. Fix any $u \in C$ and notice that $\operatorname{int}_X(C-u) = \emptyset$ since $\operatorname{int}_X C = \emptyset$. Denoting $E := \mathrm{Vect}(C-u)$, we have $-u + \operatorname{rint} C = \operatorname{int}_E(C-u)$ by (B.2) above. Since $\operatorname{rint} C \neq \emptyset$ by Proposition B.7, it ensues that $\operatorname{int}_E(C-u) \neq \emptyset$. This and the above equality $\operatorname{int}_X(C-u) = \emptyset$ entail that $E \neq X$. □

By Proposition B.5(a) we know that, for any convex set Q of a locally convex space X, any $x \in \operatorname{int} Q$ and any $y \in \operatorname{cl} Q$, one has $[x, y[\subset \operatorname{int} Q$. This and (B.2) yield:

PROPOSITION B.9. Let C be a nonempty convex set in a finite-dimensional normed space X. The following hold.
(a) For any $x \in \operatorname{rint} C$ and any $y \in \operatorname{cl} C$, one has $[x, y[\subset \operatorname{rint} C$.
(b) One also has the equalities

$$\operatorname{cl}(\operatorname{rint} C) = \operatorname{cl} C \quad \text{and} \quad \operatorname{rint}(\operatorname{cl} C) = \operatorname{rint} C.$$

As a corollary we have the additional equalities in finite dimensions for convex sets.

COROLLARY B.10. If C is a convex set in a finite-dimensional normed space X, then

$$\operatorname{int}_X(\operatorname{cl} C) = \operatorname{int}_X(C) \quad \text{and} \quad \operatorname{bdry}_X(\operatorname{cl} C) = \operatorname{bdry}_X C.$$

PROOF. We may suppose that $C \neq \emptyset$.

Case 1: $\mathrm{int}_X C \neq \emptyset$**.** In this case, the equality $\mathrm{int}_X(\mathrm{cl}\, C) = \mathrm{int}_X(C)$ follows from Proposition B.5(b).

Case 2: $\mathrm{int}_X C = \emptyset$**.** Suppose $\mathrm{int}_X(\mathrm{cl}\, C) \neq \emptyset$. Choose some $c \in \mathrm{int}_X(\mathrm{cl}\, C)$ and some real $r > 0$ with $B_X(c,r) \subset \mathrm{int}(\mathrm{cl}\, C)$. The fact that $\mathrm{int}_X(\mathrm{cl}\, C) \neq \emptyset$, entails that

$$\mathrm{int}_X(\mathrm{cl}\, C) = \mathrm{rint}(\mathrm{cl}\, C) = \mathrm{rint}\, C,$$

where the right equality is due to Proposition B.9(b). It ensues that

$$B_X(c,r) \subset \mathrm{rint}\, C \subset C,$$

hence $c \in \mathrm{int}_X C$, and this contradicts that $\mathrm{int}_X C = \emptyset$. Consequently, $\mathrm{int}_X(\mathrm{cl}\, C) = \emptyset$, thus $\mathrm{int}_X(\mathrm{cl}\, C) = \mathrm{int}_X C$.

In any case we obtain $\mathrm{int}_X(\mathrm{cl}\, C) = \mathrm{int}_X C$.

Finally, the second equality $\mathrm{bdry}\,_X(\mathrm{cl}\, C) = \mathrm{bdry}\,_X C$ is a consequence of the latter. □

In addition to Proposition B.6, we have the following situations where the closure of the intersection of finitely many convex sets coincides with the intersection of closures.

PROPOSITION B.11. Let X be a topological vector space and $(C_i)_{i\in I}$ be a family of convex sets in X. Assume that either $\bigcap\limits_{i\in I} \mathrm{int}\, C_i \neq \emptyset$ or X is a finite-dimensional normed space and $\bigcap\limits_{i\in I} \mathrm{rint}\, C_i \neq \emptyset$. Then one has $\mathrm{cl}\left(\bigcap_{i\in I} C_i\right) = \bigcap\limits_{i\in I} \mathrm{cl}\, C_i$.

PROOF. The left side is obviously included in the right one. Conversely, fix z in $\bigcap\limits_{i\in I} \mathrm{int}\, C_i$ $\left(\text{resp. in } \bigcap\limits_{i\in I} \mathrm{rint}\, C_i\right)$ and take any x in $\bigcap\limits_{i\in I} \mathrm{cl}\, C_i$. For each $t \in]0,1[$ putting $x_t := tz + (1-t)x$, by Proposition B.5(a) (resp. Proposition B.9(a)) we have $x_t \in C_i$ for every $i \in I$, and hence $x_t \in \bigcap\limits_{i\in I} C_i$. Since $x_t \to x$ as $t \downarrow 0$, it follows that $x \in \mathrm{cl}\left(\bigcap\limits_{i\in I} C_i\right)$, which finishes the proof. □

Via the above assertion (a) in Proposition B.9 the next proposition establishes a useful characterization of the relative interior of a convex set. The arguments that we follow here as well as those for Propositions B.14, B.15 and B.18 are taken from R. T. Rockafellar [**852**].

PROPOSITION B.12. Let C be a nonempty convex set in a finite-dimensional normed space X. An element $x \in C$ belongs to $\mathrm{rint}\, C$ if and only if for each $y \in C$ there exists some real $t > 1$ such that $tx + (1-t)y \in C$.

PROOF. By (B.2) the implication $\Rightarrow$ is obvious. Conversely, suppose that $x \in C$ fulfills the property in the statement. Since $\mathrm{rint}\, C \neq \emptyset$ according to Proposition B.7, we can choose $y \in \mathrm{rint}\, C$. There exist some real $t > 1$ such that $z := tx + (1-t)y \in C$. With $\theta := 1/t \in]0,1[$ we see that $x = (1-\theta)y + \theta z$, hence by Proposition B.9(a) we conclude that $x \in \mathrm{rint}\, C$. □

The following corollary is easily derived from the latter proposition. (Notice that it could be also proved via the definition.)

COROLLARY B.13. If $C_1, \cdots, C_p$ are convex sets of finite-dimensional normed spaces $X_1, \cdots, X_p$ respectively, then one has

$$\operatorname{rint}(C_1 \times \cdots \times C_p) = (\operatorname{rint} C_1) \times \cdots \times (\operatorname{rint} C_p).$$

At the opposite of the usual topological interior, the finite intersection operation is not preserved by the relative interior even for convex sets. This is illustrated with the same above intervals $C_1 := [0,1]$ and $C_2 := [1,2]$ in $\mathbb{R}$. However, Proposition B.12 allows us to prove that this preservation holds true under the same condition in Proposition B.11.

PROPOSITION B.14. Let $C_1, \cdots, C_p$ be a finite family of convex sets in a finite dimensional normed space X such that $\bigcap_{i=1}^{p} \operatorname{rint} C_i \neq \emptyset$. Then one has

$$\operatorname{rint}\left(\bigcap_{i=1}^{p} C_i\right) = \bigcap_{i=1}^{p} \operatorname{rint} C_i.$$

PROOF. Denote by K_ℓ and K_r the left and right sides of the equality to prove. We note first by Proposition B.11 and Proposition B.9(b) that both convex sets $\bigcap_{i=1}^{p} C_i$ and $\bigcap_{i=1}^{p} \operatorname{rint} C_i$ have the same closure $\left(\text{equal to } \bigcap_{i=1}^{p} \operatorname{cl} C_i\right)$, hence the same relative interior by Proposition B.9(b). This gives

$$\operatorname{rint}\left(\bigcap_{i=1}^{p} C_i\right) = \operatorname{rint}\left(\bigcap_{i=1}^{p} \operatorname{rint} C_i\right) \subset \bigcap_{i=1}^{p} \operatorname{rint} C_i,$$

which justifies the inclusion $K_\ell \subset K_r$.

To prove the converse inclusion, fix any $x \in K_r$. Take any $y \in \bigcap_{i=1}^{p} C_i$. For each $i = 1, \cdots, p$ by Proposition B.12 there exits a real $t_i > 1$ such that $(1-t_i)x + t_i y \in C_i$. Putting $t := \min\{t_1, \cdots, t_p\}$, we see that $t > 1$ and $(1-t)x + ty \in C_i$ for all $i = 1, \cdots, p$ by convexity of C_i. Then $(1-t)x + ty \in \bigcap_{i=1}^{p} C_i$, so Proposition B.12 again tells us that $x \in \operatorname{rint}\left(\bigcap_{i=1}^{p} C_i\right)$. This translates the inclusion $K_r \subset K_\ell$ and finishes the proof. □

The sequence of intervals $(J_n)_{n\in\mathbb{N}}$ with $J_n := [0, 1 + (1/n)]$ confirms that in general $\operatorname{rint}\left(\bigcap_{i\in I} C_i\right) \neq \bigcap_{i\in I} \operatorname{rint} C_i$ for infinite families of convex sets even when the condition $\bigcap_{i\in I} \operatorname{rint} C_i \neq \emptyset$ is fulfilled.

At the opposite of the topological interior, the relative interior is well-behaved with respect to images of convex sets under linear mappings.

PROPOSITION B.15. Let $A : X \to Y$ be an affine mapping between two finite-dimensional normed spaces. For any convex set C in X one has

$$\operatorname{rint}(A(C)) = A(\operatorname{rint} C).$$

PROOF. We may suppose that $C \neq \emptyset$. We note first by Proposition B.9(b) that

$$A(\operatorname{rint} C) \subset A(C) \subset A(\operatorname{cl} C) = A(\operatorname{cl}(\operatorname{rint} C)) \subset \operatorname{cl}(A(\operatorname{rint} C)),$$

where the latter inclusion is due to the continuity of A. This entails that the convex sets $A(C)$ and $A(\operatorname{rint} C)$ have the same closure, hence the same relative interior by Proposition B.9(b). This ensures that

$$\operatorname{rint}(A(C)) = \operatorname{rint}(A(\operatorname{rint} C)) \subset A(\operatorname{rint} C).$$

To show the converse inclusion, fix any $v \in A(\operatorname{rint} C)$, so $v = A(u)$ for some $u \in \operatorname{rint} C$. For any $y \in A(C)$, choosing $x \in C$ with $y = A(x)$, Proposition B.12 gives some real $t > 1$ such that $tu + (1-t)x \in C$, hence $tv + (1-t)y \in A(C)$. Proposition B.12 again entails that $v \in \operatorname{rint}(A(C))$, and the proof is finished. □

Given convex sets $C_1, \cdots, C_p$ in finite-dimensional normed spaces $X_1, \cdots, X_p$ respectively, we saw in Corollary B.13 that

$$\operatorname{rint}(C_1 \times \cdots \times C_p) = (\operatorname{rint} C_1) \times \cdots \times (\operatorname{rint} C_p).$$

Assuming $X_i = X$, for $i = 1, \cdots, p$, we have $C_1 + \cdots + C_p = A(C_1 \times \cdots \times C_p)$, where $A : X^p \to X$ is the linear mapping defined by $A(x_1, \cdots, x_p) := x_1 + \cdots + x_p$. The following corollary of the above proposition then directly follows.

COROLLARY B.16. Let X be a finite-dimensional normed space.
(a) For any convex set C in X and any real $\alpha \in \mathbb{R}$ one has

$$\operatorname{rint}(\alpha C) = \alpha \operatorname{rint} C.$$

(b) For any convex sets $C_1, \cdots, C_p$ in X one has

$$\operatorname{rint}(C_1 + \cdots + C_p) = \operatorname{rint}(C_1) + \cdots + \operatorname{rint}(C_p).$$

Although the sets $C_1 := [-1, 1]$ and $C_2 := \{0\}$ in $\mathbb{R}$ tell us that the inclusion $0 \in \operatorname{int}(C_1 - C_2)$ does not imply that $(\operatorname{int} C_1) \cap (\operatorname{int} C_2) \neq \emptyset$, the following equivalence for relative interiors is a direct corollary of Corollary B.16.

COROLLARY B.17. Let C_1, C_2 be convex sets in a finite-dimensional normed space X. Then one has

$$0 \in \operatorname{rint}(C_1 - C_2) \iff (\operatorname{rint} C_1) \cap (\operatorname{rint} C_2) \neq \emptyset.$$

Given a linear mapping $A : X \to Y$ between finite-dimensional spaces, we saw in Proposition B.15 that $\operatorname{rint}(A(C)) = A(\operatorname{rint} C)$ for any convex set $C \subset X$. Via a certain condition, the relative interior is also well-behaved under inverse image of convex sets by linear mappings.

PROPOSITION B.18. Let $A : X \to Y$ be an affine mapping between finite-dimensional normed spaces. For any convex set C in Y with $A^{-1}(\operatorname{rint} C) \neq \emptyset$ one has

$$\operatorname{rint}\left(A^{-1}(C)\right) = A^{-1}(\operatorname{rint} C).$$

PROOF. Considering the projector $\pi_X : X \times Y \to X$ (with $\pi_X(x, y) := x$) and the affine set $\operatorname{gph} A$, we see that $A^{-1}(C) = \pi_X Q$, where $Q := (\operatorname{gph} A) \cap (X \times C)$. Choosing by assumption $u \in A^{-1}(\operatorname{rint} C)$, that is, $A(u) \in \operatorname{rint} C$, we obtain $(u, A(u))$ in both $\operatorname{gph} A = \operatorname{rint}(\operatorname{gph} A)$ and $\operatorname{rint}(X \times C)$. We deduce by Proposition B.14 that $\operatorname{rint} Q = (\operatorname{gph} A) \cap (X \times \operatorname{rint} C)$, which yields by Proposition B.15 that

$$\operatorname{rint}\left(A^{-1}(C)\right) = \pi_X(\operatorname{rint} Q) = A^{-1}(\operatorname{rint} C),$$

which completes the proof. □

In addition to the above features in finite dimensions, we present now the class of polyhedral convex sets which in particular enjoy the intrinsic closedness property. We maintain X as a finite-dimensional normed space. A set C in X is called *polyhedral convex* if it is the intersection of a finite collection $(\mathcal{H}_i)_{i\in I}$ of half-spaces in X, that is, there are a finite set I, elements $a_i^* \in X^*$ and $\beta_i \in \mathbb{R}$ such that

$$C = \{x \in X : \langle a_i^*, x\rangle \leq \beta_i,\ \forall i \in I\}.$$

The sets X (with $I = \emptyset$) is the greatest convex polyhedral set in X, and the empty set $\emptyset$ (as the intersection of two disjoint half-spaces) is the smallest. A basic result is the following *vertex/ray representation* for which a complete proof can be found, for example, in R. T. Rockafellar and R. J-B. Wets [**865**, Theorem 3.52 and Corollary 3.53].

THEOREM B.19 (Minkowski-Weyl theorem for convex polyhedra). A nonempty set C in the finite dimensional vector space X is polyhedral convex if and only if there are points $a_1, \cdots, a_p$ in X and directions associated with $a_{p+1}, \cdots, a_q$ in X such that $x \in C$ provided there are $t_1, \cdots, t_p \geq 0$ with $t_1 + \cdots + t_p = 1$ and $t_{p+1}, \cdots, t_q \geq 0$ such that

$$x = t_1 a_1 + \cdots + t_p a_p + t_{p+1} a_{p+1} + \cdots + t_q a_q. \tag{B.3}$$

APPENDIX C

Functional analysis

Let us now recall certain fundamental basic facts concerning continuous linear operators/mappings. We start with a well-known invertibility result.

PROPOSITION C.1. Let $A : X \to X$ be a continuous linear operator from a Banach space X into itself. If $\|\mathrm{id}_X - A\| < 1$, then A is invertible and $A^{-1} = \mathrm{id}_X + \sum_{n=1}^{\infty} A^n$, where $A^n := A \circ \cdots \circ A$.

In the setting of Hilbert space it is also worth recalling the following property of positivity of a linear operator.

PROPOSITION C.2. Let $A : H \to H$ be a continuous linear operator from a Hilbert space H into itself. If $\|\mathrm{id}_H - A\| \le 1$, then the linear operator A is positive, that is $\langle Ax, x\rangle \ge 0$ for all $x \in H$.

PROOF. It suffices to notice that

$$\langle Ax, x\rangle = \langle x, x\rangle + \langle (A - \mathrm{id}_H)x, x\rangle \ge \|x\|^2 - \|\mathrm{id}_H - A\| \, \|x\|^2,$$

and hence $\langle Ax, x\rangle \ge 0$. $\square$

We continue now with the Banach-Schauder open mapping theorem (see, for example, Theorem 2.6 in H. Brezis' book [**178**]).

THEOREM C.3 (Banach-Schauder open mapping theorem). Let $A : X \to Y$ be a surjective continuous linear mapping between Banach spaces X, Y. Then there exits a real constant $c > 0$ such that $c\mathbb{B}_Y \subset A(\mathbb{B}_X)$.

The Krein-Šmulian theorem (also called Banach-Dieudonné theorem in the literature) furnishes a practical way to verify the weak* closedness of convex sets. A proof can be found, for example, in the book of N. Dunford and J. T. Schwartz [**366**, Theorem V.5.7].

THEOREM C.4 (Krein-Šmulian theorem). Let X be a Banach space and X^* its topological dual. A convex set in X^* is $w(X^*, X)$ closed if and only if its intersection with every closed ball in X^* centered at the origin is $w(X^*, X)$ closed.

Given a continuous linear operator/mapping $A : X \to Y$ between two Hausdorff locally convex spaces X and Y, the adjoint $A^* : Y^* \to X^*$ of A is defined by

$$A^*(y^*) := y^* \circ A^* \quad \text{for all } y^* \in X^*.$$

Consider the important situation when X and Y are two normed spaces. It is easily verified that A^* is norm-to-norm continuous and $\|A^*\| = \|A\|$; if A is an isomorphism (resp. isometric isomorphism), then so is A^* and

$$(A^{-1})^* = (A^*)^{-1}. \tag{C.1}$$

Further, it is not difficult to see that

$$\mathrm{Ker}\,(A^*) = (A(X))^\perp \quad \text{and} \quad \mathrm{Ker}\,(A) = (A^*(Y^*))^{\perp_X},$$

where $(A^*(Y^*))^{\perp_X}$ denotes the orthogonal of $(A^*(Y^*))$ in X.

One also has the following theorem (see, for example, Theorem 2.19 in Brezis' book [**178**]) concerning the ranges of A and A^*.

THEOREM C.5 (closedness of the range of a continuous operator). Let $A : X \to Y$ be a continuous linear mapping between two Banach spaces. The following assertions are equivalent:
(a) $A(X)$ is closed in Y;
(b) $A^*(Y^*)$ is norm-closed in X^*;
(c) $A(X)$ coincides with the orthogonal in Y of the kernel of A^*;
(d) $A^*(Y^*)$ coincides with the orthogonal in X^* of the kernel of A.

The surjectivity of a continuous linear mapping between Banach spaces is known (see, for example, Theorem 2.20 in [**178**]) to be characterized with the help of its adjoint as follows.

THEOREM C.6 (surjectivity of continuous operator and its adjoint). Let $A : X \to Y$ be a continuous linear mapping between two Banach spaces X, Y. Then the following are equivalent:
(a) A is surjective;
(b) There exists a real constant $c \geq 0$ such that

$$\|y^*\| \leq c\|A^*y^*\| \quad \text{for all } y^* \in Y^*;$$

(c) $\mathrm{Ker}\,A^* = \{0\}$ and $A^*(Y^*)$ is norm-closed in X^*.

Let us recall now the Goldstine theorem.

THEOREM C.7 (Goldstine theorem). For any normed space $(X, \|\cdot\|)$ the unit ball $\mathbb{B}_X$ of X is $w(X^{**}, X^*)$-dense in the unit ball $\mathbb{B}_{X^{**}}$ of the topological bidual space X^{**}.

The famous James theorem on weak compactness [**552**] is generally stated for subsets of Banach spaces in the following form (see, for example, J. Diestel's book [**334**, Chapter I, Theorem 3, (i) and (iv)] and see also the proof there).

THEOREM C.8 (James characterization theorem of weak compactness in Banach space). A nonempty weakly closed set C in a Banach space X is weakly compact if and only if every continuous linear functional on X attains it supremum on C.

In fact, as stated in Theorem C.10 below, a more general version holds true in certain locally convex spaces. Given a Hausdorff locally convex space (X, τ_X) and its topological dual X^*, we recall that the strong topology $\beta(X^*, X)$ on X^* is the topology generated by the uniform convergence over bounded sets of X. The bidual of X, denoted by X^{**}, is the topological dual of $(X^*, \beta(X^*, X))$. The locally convex space X is called *semi-reflexive* if the canonical embedding (or evaluation mapping) $X \ni x \mapsto \langle \cdot, x \rangle$ from X into X^{**} is surjective. In contrast, the Hausdorff locally convex space (X, τ_X) is called *reflexive* if the canonical embedding is a homeomorphism from (X, τ_X) onto $(X^{**}, \beta(X^{**}, X^*))$, where $\beta(X^{**}, X^*)$ is the topology on X^{**} of uniform convergence over bounded sets in $(X^*, \beta(X^*, X))$ (see

[**881**] for more details). It is worth mentioning that every semi-reflexive normed space is a reflexive Banach space (see [**881**, Corollary 2, IV p. 145]).

THEOREM C.9 (semi-reflexivity). A Hausdorff locally convex space X is semi-reflexive if and only if every bounded subset of X is relatively weakly compact.

We recall next the following James' theorem for which we refer to R. C. James' paper [**551**]; a proof can also be found in J. Diestel's book [**334**, Chapter I, Theorem 5, (i) and (iv)].

THEOREM C.10 (James characterization theorem of weak compactness in locally convex space). Let X be a complete Hausdorff locally convex space. A nonempty weakly closed subset C in X is weakly compact if and only if every continuous linear functional on X attains it supremum on C.

We recall now some renorming features for Banach spaces. Remind that a norm $\|\cdot\|$ on a Banach space X has the (sequential) *Kadec-Klee property* when for any sequence $(x_n)_n$ of X converging weakly to x with $\|x_n\| \to \|x\|$ one has $\|x_n - x\| \to 0$. The following renorming theorem for reflexive Banach spaces can be found for example in R. Deville, G. Godefroy and V. Zizler's book [**333**, Proposition VII-2-1].

THEOREM C.11 (F-differentiable LUC renorm under reflexivity). Any reflexive Banach space can be renormed with an equivalent norm which is both Fréchet differentiable (off zero) and locally uniformly convex, so in particular the renorm is both Kadec-Klee and strictly convex.

Recall that a Banach space $(X, \|\cdot\|)$ is *Asplund* provided that any continuous convex function $f : U \to \mathbb{R}$ on an open convex set U of X is Fréchet differentiable on a dense G_δ set in U. It is known (see, for example, the book of R. Deville, G. Godefroy and V. Zizler [**333**, Theorem I.5.7] and the book of R. R. Phelps [**801**, Theorem 2.34]) that a Banach space is Asplund if and only if the topological dual of any separable subspace is separable. It is also known (see, for example, [**333**, Corollary II.3.3]) that a separable Banach space admits an equivalent Fréchet differentiable (off zero) norm if and only if its topological dual is separable. Consequently, we can reformulate both above results in the following unified way.

THEOREM C.12 (Asplund property via separable subspaces). For a Banach space X, the following are equivalent:
(a) The space X is Asplund;
(b) the topological dual of any separable subspace of X is separable;
(c) any separable subspace of X admits an equivalent norm which is Fréchet differentiable off zero.

By the assertion (b) we have the following corollary with the example of reflexive Banach spaces as Asplund spaces.

COROLLARY C.13. Any reflexive Banach space is an Asplund space.

Note that the corollary can also be justified by Theorem C.11 and the Ekeland-Lebourg Theorem 4.49.

By reflexivity the Lebesgue and Sobolev spaces $L^p(\Omega, \mathbb{R})$ and $W^{m,p}(\Omega, \mathbb{R})$ with $p \in]1, \infty[$ (and Ω open in $\mathbb{R}^n$) are examples of Asplund spaces. However, these spaces with either $p = 1$ or $p = \infty$ as well as the Banach space of continuous functions $\mathcal{C}([0,1],\mathbb{R})$ are not Asplund spaces.

Bounded subsets of topological duals of Asplund spaces enjoy a remarkable weak* sequential compactness property. Given a Banach space X, recall that a subset Q of the topological dual X^* is said to be *weak* sequentially compact* whenever every sequence in Q admits a subsequence weak* converging to some element in Q. A key result related to such a sequential property is the famous Hagler-Johnson theorem (see [**460**]) whose proof can also be found, for example, in J. Diestel's book [**335**, Chapter XIII, Theorem 6].

THEOREM C.14 (J. Hagler and W.B. Johnson). If the topological dual X^* of a Banach space X contains a bounded sequence without a weak* convergent subsequence, then X contains a separable subspace with nonseparable dual.

As a direct consequence of the latter theorem and the characterization (b) in Theorem C.12, one has the following theorem for the situation of Asplund spaces.

THEOREM C.15 (sequential w*-compactness of balls in Asplund spaces). If X is an Asplund space, then any closed ball of the topological dual X^* is weak* sequentially compact, or equivalently, any bounded sequence in X^* contains a weak* convergent subsequence.

It is also worth stating a particular case of a result of J. Hagler and F. Sullivan in [**461**, Theorem 1] related to the weak* compactness of dual balls.

THEOREM C.16 (J. Hagler and F. Sullivan). If a Banach space X admits an equivalent Gâteaux differentiable (off zero) norm, then any bounded sequence in X^* has a weak* convergent subsequence.

Concerning uniformly convex (resp. smooth) norms defined in Definition 18.12 (resp. Definition 18.24) and presented in Subsections 18.1.2 and 18.1.3, some additional details are needed for certain renorming features utilized in the theory developed in Sections 18.5 and 18.6 for prox-regular sets in uniformly convex Banach spaces. The story of renorming uniformly convex Banach spaces with norms possessing improved properties can be briefly presented as follows.

Given a property $(\mathcal{P})$ for Banach spaces, one says that a Banach space $(X, \|\cdot\|_X)$ has super-$(\mathcal{P})$ whenever every Banach space $(Y, \|\cdot\|_Y)$ which is finitely representable in $(X, \|\cdot\|_X)$ possesses the property $(\mathcal{P})$. A Banach space $(Y, \|\cdot\|_Y)$ is *finitely representable* in $(X, \|\cdot\|_X)$ if for any finite dimensional vector subspace Y_0 of Y and any real $\lambda > 1$ there exists an isomorphism T from Y_0 onto a vector subspace of X such that

$$\frac{1}{\lambda}\|y\|_Y \le \|T(y)\|_X \le \lambda\|y\|_Y \quad \text{for every } y \in Y_0;$$

see James' paper [**554**, Definition 1]. Clearly, $(X, \|\cdot\|_X)$ possesses the property $(\mathcal{P})$ whenever it has super-$(\mathcal{P})$. The concept of Banach spaces with super-$(\mathcal{P})$ is mostly known when the property "$\mathcal{P}$"is the "reflexivity property". The Banach space $(X, \|\cdot\|_X)$ is called *super-reflexive* by R. C. James [**554**, Definition 3] provided that any Banach space which is finitely representable in $(X, \|\cdot\|_X)$ is reflexive. Many results of interest were established for super-reflexive spaces by James. In particular, he proved in [**554**, Theorem 2] the following result:

THEOREM C.17 (R.C. James). A Banach space is super-reflexive if and only if its topological dual is.

Evidently, by what precedes Theorem C.17 every super-reflexive Banach space is reflexive, but the converse fails as confirms Enflo's Theorem C.18 below. Via uniformly non-square Banach spaces, James also proved (see [**553**, Lemma C]) that any Banach space admitting an equivalent uniformly convex (resp. uniformly smooth) norm is super-reflexive. The converse implication of the latter result was achieved with the fundamental Enflo theorem [**385**].

THEOREM C.18 (P. Enflo). *A Banach space $(X, \|\cdot\|)$ is super-reflexive if and only if it admits an equivalent norm which is uniformly convex.*

For the development of that result of Enflo we refer also to the book of R. Deville, G. Godefroy and V. Zizler [**333**, p. 139-152]; as stated above the result corresponds to Corollary IV.4.6 in that book [**333**].

Let $(X, \|\cdot\|)$ be a Banach space and let $\|\cdot\|_i$, $i = 1, 2$, be two norms on X equivalent to $\|\cdot\|$. Suppose that the norm $\|\cdot\|_1$ possesses a certain degree of convexity $(\mathcal{C}_1)$ (of type "strict convexity", "uniform convexity", "uniform convexity with power type"q, "local uniform convexity"). Suppose also that the dual norm of $\|\cdot\|_2$ on X^* has a certain degree of convexity $(\mathcal{C}_2)$ (of one of the above types). Recall that E. Asplund averaging procedure (see, for example, J. Diestel's book [**334**, p. 113]) provides a norm $|||\cdot|||$ equivalent to $\|\cdot\|$ and which possesses the degree of convexity $(\mathcal{C}_1)$ and whose dual norm $|||\cdot|||_*$ on X^* possesses the degree of convexity $(\mathcal{C}_2)$. The Asplund's avering procedure is largely and well developed in Disetel's book [**334**, p. 106-113].

According to this Asplund's averaging procedure, to Enflo Theorem C.18 and to James Theorem C.17 recalled above, one can state:

THEOREM C.19 (P. Enflo). *Any uniformly convex Banach space admits an equivalent norm which is both uniformly convex and uniformly smooth.*

G. Pisier [**803**] extended the above Enflo Theorem C.18 in showing that the equivalent renorm can be required to be such that its modulus of convexity be of power type. Pisier's result can also be found, for example, in B. Beauzamy's book [**79**, p. 273] and a proof is developed in pages 273-290 of that book [**79**]. By the Asplund averaging procedure again and by Proposition 18.38 we can state:

THEOREM C.20 (G. Pisier). *Any uniformly convex Banach space admits an equivalent norm which is uniformly convex with modulus of convexity of power type $q \in [2, +\infty[$ and uniformly smooth with modulus of smoothness of power type $s \in]1, 2]$.*

APPENDIX D

Measure theory

Let λ denote the *outer Lebesgue measure* on the Euclidean space $\mathbb{R}^N$ given, for every subset $S \subset \mathbb{R}^N$, by $\lambda(S)$ equal to the infimum of $\sum_{n\in\mathbb{N}} \text{meas}_N(P_n)$ over the sequences $(P_n)_n$ of N-rectangles in $\mathbb{R}^N$ with $S \subset \bigcup_{n\in\mathbb{N}} P_n$. Given a set S in $\mathbb{R}^N$, a point $\overline{x} \in S$ is called a *λ-density point* or an *N-dimensional outer Lebesgue density point* of S whenever

$$\lim_{r\downarrow 0} \frac{\lambda(S \cap B(\overline{x}, r))}{\lambda(B(\overline{x}, r))} = 1.$$

The classical theorem on density points (see, e.g., [**389**, Corollary 3, p. 45], [**695**, Corollary 2.14(1), p. 38]) is generally stated as: λ-almost every point of a Lebesgue measurable set A in $\mathbb{R}^N$ is a Lebesgue density point of A. Consider any nonempty bounded set S in $\mathbb{R}^N$ and take a Lebesgue measurable set $A \supset S$ with $\lambda(S) = \lambda(A)$. For any Lebesgue measurable set B, by outer measure property of λ we have

$$\lambda(A \cap B) + \lambda(A \cap B^c) = \lambda(A) = \lambda(S) \leq \lambda(S \cap B) + \lambda(S \cap B^c),$$

where $B^c := \mathbb{R}^N \setminus B$. This and the inequality $\lambda(A\cap E) \geq \lambda(S\cap E)$ gives $\lambda(S\cap B) = \lambda(A \cap B)$ for any Lebesgue measurable set B in $\mathbb{R}^N$. Using this and the definition of λ-density point, as noticed in [**695**, Remark 2.15(2), p. 39], it is easily seen that the theorem on Lebesgue density points can be reformulated as follows:

THEOREM D.1. Lebesgue almost every point of a set S of the Euclidean space $\mathbb{R}^N$ is a Lebesgue density point of S.

Let (X, d) be a metric space, $s \in [0, +\infty[$ and $0 < \alpha(s) < +\infty$ with $\alpha(0) = 1$. For any set $C \subset X$ and any real $\delta > 0$ let

$$\mathcal{H}_{s,\delta}(C) = \inf \left\{ \sum_{j\in\mathbb{N}} \frac{\alpha(s)}{2^s} (\text{diam}(Q_j))^s : C \subset \bigcup_{j\in\mathbb{N}} Q_j, \text{diam}(Q_j) \leq \delta \right\},$$

where we use the convention $0^0 = 0$ and $\text{diam}(\emptyset) = 0$. One defines $\mathcal{H}_s(C)$ by

$$\mathcal{H}_s(C) = \lim_{\delta\downarrow 0} \mathcal{H}_{s,\delta}(C) = \sup_{\delta>0} \mathcal{H}_{s,\delta}(C).$$

It is known that for any subsets C and C_j, with $j \in \mathbb{N}$, of X satisfying $C \subset \bigcup_{j\in\mathbb{N}} C_j$ one has (see, for example, [**389**, Chapter 2] and [**695**, Chapter 4])

$$\mathcal{H}_s(C) \leq \sum_{j\in\mathbb{N}} \mathcal{H}_s(C_j),$$

so $\mathcal{H}_s$ is an outer measure on the set of all subsets of X. It is called the *s-Hausdorff measure on X relative to $\alpha(\cdot)$*. For $s = 0$ the Hausforff measure $\mathcal{H}_0$ is the *counting measure*, that is,

$$\mathcal{H}_0(C) = \text{card}\, C,$$

and for $0 \leq s < t < +\infty$ (see, for example, [**695**, Chapter 4])

$$\big(\mathcal{H}_s(C) < +\infty \Rightarrow \mathcal{H}_t(C) = 0\big) \quad \text{and} \quad \big(\mathcal{H}_t(C) > 0 \Rightarrow \mathcal{H}_s(C) = +\infty\big).$$

Further, if (Y, d_Y) is another metric space and $f : X \to Y$ is a Lipschitz continuous mapping with Lipschitz constant ℓ, one has

$$\mathcal{H}_s(f(C)) \leq \ell^s \mathcal{H}_s(C). \tag{D.1}$$

When (X, d) is the Euclidean space $X = \mathbb{R}^N$ with its canonical Euclidean distance, one has (see, for example, [**389**, Chapter 2])

(i) $\mathcal{H}_s \equiv 0$ for any real $s > N$;
(ii) $\mathcal{H}_s(a + A(C)) = \mathcal{H}_s(C)$ for any $a \in \mathbb{R}^N$ and any isometric linear mapping $A : \mathbb{R}^N \to \mathbb{R}^N$;
(iii) $\mathcal{H}_s(\rho C) = \rho^s\big(\mathcal{H}_s(C)\big)$ for any real $\rho > 0$.

Furthermore, in the same setting of $\mathbb{R}^N$ and with

$$\alpha(s) = \frac{\pi^{s/2}}{\Gamma(\frac{s}{2} + 1)}, \tag{D.2}$$

where $\Gamma(\cdot)$ is the usual Γ-function, it is known (see, for example, [**389**, Chapter 2, Theorem 2]) that the Hausdorff measure $\mathcal{H}_N$ relative to $\alpha(N)$ with $\alpha(\cdot)$ as in (D.2) coincides with the N-dimensional Lebesgue measure on $\mathbb{R}^N$. Accordingly, the Hausdorff measure $\mathcal{H}_s$ relative to $\alpha(s)$ in (D.2) is called the *normalized s-Hausdorff measure.* When $s = k$ is an integer $0 \leq k \leq N$ and $\alpha(k)$ is taken as in (D.2), one says that $\mathcal{H}_k$ is the *normalized k-dimensional Hausdorff measure* on $\mathbb{R}^N$.

APPENDIX E

Differential calculus and differentiable manifolds

We pass now to three pillars of differential calculus in Banach spaces: inverse mapping theorem, implicit function theorem and local submersion theorem. Let us begin first with the inverse mapping theorem. This requires to recall the concept of diffeomorphism.

Let $f : U \to V$ be a mapping from an open set U of a Banach space X into an open set V of a Banach space Y. One says that f is a *$\mathcal{C}^p$-diffeomorphism* for an integer $p \geq 1$ if f is a bijection from U onto V such that f and its inverse f^{-1} are of class $\mathcal{C}^p$ on U and V respectively. With this we can state the classical inverse mapping theorem which can be found, for example, in Theorem 2.5.2 of the book [**1**] of R. Abraham, J. E. Marsden and T. Ratiu.

THEOREM E.1 (inverse mapping theorem). Let $f : O \to Y$ be a mapping from an open set O of a Banach space X into a Banach space Y and let $\overline{x} \in O$ be a point around which f is of class $\mathcal{C}^p$ (resp. let $\overline{x} \in O$ be a point at which f is strictly Fréchet differentiable) and such that $Df(\overline{x})$ is a bijection from X onto Y. Then there exist an open neighborhood $U \subset O$ of $\overline{x}$ such that $V := f(U)$ is an open neighborhood of $f(\overline{x})$ and the restriction $f_U : U \to V$ of f to U is a $\mathcal{C}^p$-diffeomorphism (resp. and the inverse f_U^{-1} is strictly Fréchet differentiable at $f(\overline{x})$). Further, one has

$$D({f_U}^{-1})\big(f(x)\big) = \big(Df(x)\big)^{-1} \quad \text{for all } x \in U,$$

(resp. $D({f_U}^{-1})\big(f(\overline{x})\big) = \big(Df(\overline{x})\big)^{-1}$).

We state now the implicit function theorem (see, for example, [**1**, Theorem 2.5.5]).

THEOREM E.2 (implicit function theorem). Let X, Y be Banach spaces, $O \subset X \times Y$ be an open set in $X \times Y$, and $f : O \to Z$ be a mapping from O into a Banach space Z. Assume that f is of class $\mathcal{C}^p$ around a point $(\overline{x}, \overline{y}) \in O$ (with an integer $p \geq 1$) and that the derivative with respect to the second variable $D_2 f(\overline{x}, \overline{y})$ is a bijection from Y onto Z. Then there exist open neighborhoods U of $\overline{x}$, V of $\overline{y}$ with $U \times V \subset O$, an open neighborhood W of $f(\overline{x}, \overline{y})$, and a unique $\mathcal{C}^p$ mapping $g : U \times W \to V$ such that

$$f(x, g(x, z)) = z \quad \text{for all } (x, z) \in U \times W.$$

Given a Banach space X, a closed vector subspace X_1 admitting a complement closed vector space X_2, we recall (with $i = 1, 2$) the notation of the mapping $\pi_{X_i} : X \to X_i$ defined for every $x \in X = X_1 \oplus X_2$ by $x = \pi_{X_1}(x) \oplus \pi_{X_2}(x)$ with $\pi_{X_i}(x) \in X_i$. With this we can recall the local submersion theorem for which the reader can consult, for example, Theorem 2.5.13 in [**1**].

THEOREM E.3 (local submersion theorem). Let X, Y be Banach spaces, O be an open set in X and $f : O \to Y$ be a mapping of class $\mathcal{C}^p$ on O with an integer $p \geq 1$. Let $\overline{x} \in O$ be such that $Df(\overline{x})$ is surjective and $X_2 := \operatorname{Ker} Df(\overline{x})$ admits a topological complement vector space X_1 in X with $X = X_1 \oplus X_2$. Then there exist an open neighborhood $U \subset O$ of $\overline{x}$ in X, an open neighborhood V of $(f(\overline{x}), \pi_{X_2}(\overline{x}))$ in $Y \times X_2$, and a $\mathcal{C}^p$-diffeomorphism $\psi : V \to U$ from V onto U such that

$$f \circ \psi(v_1, v_2) = v_1 \quad \text{for all } (v_1, v_2) \in V$$

along with $\psi(v_1, v_2) = \psi_1(v_1, v_2) \oplus v_2$ with a $\mathcal{C}^p$-mapping $\psi_1 : V \to X_1$. Further, for each $(v_1, v_2) \in V$ one has that $D\psi(v_1, v_2)_{|Y \times \{0_{X_2}\}} : Y \times \{0_{X_2}\} \to X_1$ is an isomorphism.

We now recall some basic concepts concerning submanifolds in Banach spaces. To do so, we need first some features on locally uniformly continuous mappings. Given two normed spaces X, Y and a nonempty open set U of X, a mapping $f : U \to Y$ is said to be locally uniformly continuous when for each $\overline{x} \in U$ there exists some neighborhood $U_0 \subset U$ of $\overline{x}$ such that f is uniformly continuous on U_0; otherwise stated, for every real $\varepsilon > 0$ there is $\delta > 0$ such that

$$\|f(x) - f(x')\| \leq \varepsilon \quad \text{for all } x, x' \in U_0 \text{ with } \|x - x'\| \leq \delta.$$

While the class of continuous mappings from U into Y is usually denoted by $\mathcal{C}(U, Y)$ or $\mathcal{C}^0(U, Y)$, the class of locally uniformly continuous mappings from U into Y will be denoted by $\mathcal{C}^{0,0}(U, Y)$, and similarly $\mathcal{C}^{0,1}(U, Y)$ will stand for the class of locally Lipschitz continuous mappings from U into Y. Analogously, recalling (for $p \in \mathbb{N}$) that the class of p-continuously differentiable mappings from U into Y is denoted by $\mathcal{C}^p(U, Y)$, the class of mappings $f \in \mathcal{C}^p(U, Y)$ such that the p-th derivative $D^p f$ is locally uniformly continuous (resp. locally Lipschitz continuous) will be denoted by $\mathcal{C}^{p,0}(U, Y)$ (resp. $\mathcal{C}^{p,1}(U, Y)$).

When the normed space X is finite-dimensional, it is clear according to the compactness of closed balls in X that the classes $\mathcal{C}^{p,0}(U, Y)$ and $\mathcal{C}^p(U, Y)$ coincide for each $p \in \{0\} \cup \mathbb{N}$. Such a coincidence fails in infinite dimensions as shown in the following Izzo example [**550**].

EXAMPLE E.4 (A.J. Izzo's example). The function $f : \ell_2(\mathbb{N}) \to \mathbb{R}$, defined by $f(x) = \sum_{n=1}^{\infty} x_n^2 \cos(n x_n)$ for all $x := (x_n)_{n \in \mathbb{N}} \in \ell_2(\mathbb{N})$, is continuous but fails to be locally uniformly continuous. □

We present, in Lemma E.5 and Proposition E.6, two properties of locally uniformly continuous mappings.

LEMMA E.5. Let X_1, X_2 and X_3 be three normed spaces and let U and V be two nonempty open sets of X_1 and X_2, respectively. Let $f : U \to V$ and $g : V \to X_3$ be such that $f \in \mathcal{C}^{p,0}(U, X_2)$ and $g \in \mathcal{C}^{p,0}(V, X_3)$ (resp. $f \in \mathcal{C}^{p,1}(U, X_2)$ and $g \in \mathcal{C}^{p,1}(V, X_3)$), where p is an integer $p \geq 0$. Then, one has

$$g \circ f \in \mathcal{C}^{p,0}(U, X_3) \quad (\text{resp. } g \circ f \in \mathcal{C}^{p,1}(U, X_3)).$$

PROOF. We consider only the $\mathcal{C}^{p,0}$-case since the $\mathcal{C}^{p,1}$-case is similar. We proceed by induction. The result is clear for $p = 0$. Let us show it for $p = 1$. Suppose that f and g are of class $\mathcal{C}^{1,0}$. By chain rule, we know that for every $u \in U$,

$$D(g \circ f)(u) = Dg(f(u)) \circ Df(u) = \beta(Dg(f(u)), Df(u)),$$

where $\beta : \mathcal{L}(X_2, X_3) \times \mathcal{L}(X_1, X_2) \to \mathcal{L}(X_1, X_3)$ is the continuous bilinear operator given by $\beta(\Lambda_2, \Lambda_1) = \Lambda_2 \circ \Lambda_1$. Define

$$F : U \to X := \mathcal{L}(X_2, X_3) \times \mathcal{L}(X_1, X_2) \quad \text{by} \quad F(u) = (h(u), Df(u))$$

with $h := (Dg) \circ f$, and notice that F is locally uniformly continuous since both mappings h and Df are locally uniformly continuous. The mapping β being locally uniformly continuous (in fact, it is of class $\mathcal{C}^\infty$), the equality $D(g \circ f)(\cdot) = \beta \circ F(\cdot)$ ensures the local uniform continuity of $D(g \circ f)$, which justifies the result for $p = 1$.

Now assume that the result holds for $p \geq 1$ and let us show that it holds for $p + 1$. Suppose that $f \in \mathcal{C}^{p+1,0}(U, X_2)$ and $g \in \mathcal{C}^{p+1,0}(V, X_3)$. Then $Dg \in \mathcal{C}^{p,0}(V, \mathcal{L}(X_2, X_3))$ and by the induction assumption we also has $h := (Dg) \circ f \in \mathcal{C}^{p,0}(U, \mathcal{L}(X_2, X_3))$. Thus, the mapping $D^p F(\cdot) = (D^p h(\cdot), D^p f(\cdot))$ is locally uniformly continuous, which combined with the fact that β is of class $\mathcal{C}^{p,0}$ entails by the induction assumption again that

$$D(g \circ f)(\cdot) = \beta \circ F(\cdot) \in \mathcal{C}^{p,0}(U, \mathcal{L}(X_1, X_3)).$$

This means that $g \circ f \in \mathcal{C}^{p+1,0}(U, X_3)$, and the proof is finished. □

Consider now two nonempty open sets U and V of Banach spaces X and Y respectively, and a $\mathcal{C}^p$-diffeomorphism $F : U \to V$ with $p \in \mathbb{N}$, so by definition $F^{-1} : V \to U$ is also a $\mathcal{C}^p$-diffeomorphism. We have for every $v \in V$

$$DF^{-1}(v) = \big(DF(F^{-1}(v))\big)^{-1} = J \circ DF \circ F^{-1}(v), \tag{E.1}$$

where $J : \mathrm{Iso}(X, Y) \to \mathrm{Iso}(Y, X)$ is the homeomorphism defined by $J(\Lambda) := \Lambda^{-1}$ and $\mathrm{Iso}(X, Y)$ denotes the subset in $\mathcal{L}(X, Y)$ of isomorphisms from X onto Y. It is known (see, for example, [**409**, Theorem 3.1.5]) that J is a $\mathcal{C}^\infty$-mapping, and hence it belongs to $\mathcal{C}^{p,0}(\mathrm{Iso}(X, Y), \mathcal{L}(Y, X))$. Note that F^{-1} is of class $\mathcal{C}^{p-1,0}$ since it is of class $\mathcal{C}^p$. If in fact $F \in \mathcal{C}^{p,0}(U, Y)$, then E.1 and Lemma E.5 give that $DF^{-1} \in \mathcal{C}^{p-1,0}(V, \mathcal{L}(Y, X))$, so $F^{-1} \in \mathcal{C}^{p,0}(V, X)$. Similar arguments are valid when F is a $\mathcal{C}^{p,1}$-diffeomorphism. Then, by changing the roles of F and F^{-1} we obtain:

Proposition E.6. Let U and V be two nonempty open sets of Banach spaces X and Y respectively. For any $\mathcal{C}^p$-diffeomorphism $F : U \to V$ (with $p \geq 1$), one has

$$F \in \mathcal{C}^{p,0}(U, Y) \Leftrightarrow F^{-1} \in \mathcal{C}^{p,0}(V, X) \quad (\text{resp. } F \in \mathcal{C}^{p,1}(U, Y) \Leftrightarrow F^{-1} \in \mathcal{C}^{p,1}(V, X))$$

Remark E.7. Through Proposition E.6 we can see, by the proof of Theorem 2.5.13 in [**1**], that the mapping $\psi : V \to U$ in Theorem E.3 above is a $\mathcal{C}^{p,0}$-diffeomorphism (resp. $\mathcal{C}^{p,1}$-diffeomorphism) whenever the mapping $f : O \to Y$ therein is of class $\mathcal{C}^{p,0}$ (resp. of class $\mathcal{C}^{p,1}$). □

Let us now recall the definition of the three concepts of $\mathcal{C}^p$-submanifolds, $\mathcal{C}^{p,0}$-submanifolds and $\mathcal{C}^{p,1}$-submanifolds.

Definition E.8. Let M be a subset of a Banach space X and let $m_0 \in M$. One says that M is a $\mathcal{C}^p$-*submanifold* with $p \in \mathbb{N}$ at/near m_0 if there exist a closed vector subspace E of X, an open neighborhood U of m_0 in X, an open neighborhood V of zero in X, and a mapping $\varphi : U \to V$ such that

(a) φ is a $\mathcal{C}^p$-diffeomorphism, that is, $\varphi : U \to V$ is bijective and both φ and its inverse are of class $\mathcal{C}^p$;

(b) $\varphi(m_0) = 0$ and $\varphi(M \cap U) = E \cap \varphi(U)$.

In such a case, E is a *model space* and (U, φ) is a *local chart.*

If in addition the condition

(c) $\varphi \in \mathcal{C}^{p,0}(U, X)$ (resp. $\varphi \in \mathcal{C}^{p,1}(U, X)$)

is satisfied, one says that M is a $\mathcal{C}^{p,0}$-submanifold (resp. $\mathcal{C}^{p,1}$-submanifold) at/near $m_0 \in M$.

When M is a $\mathcal{C}^p$-submanifold (resp. $\mathcal{C}^{p,0}$-submanifold, or $\mathcal{C}^{p,1}$-submanifold) at/near each point in M with the same model vector space E, M is called a $\mathcal{C}^p$-submanifold (resp. $\mathcal{C}^{p,0}$-submanifold, or $\mathcal{C}^{p,1}$-submanifold).

If $M \subset X$ is a $\mathcal{C}^1$-submanifold at $m_0 \in M$, its *tangent space* at m_0 is

$$\text{(E.2)} \qquad T_{m_0}M := \left\{ h \in X \; : \; \begin{array}{c} \exists \gamma :]-1,1[\to M \text{ a } \mathcal{C}^1\text{-curve with} \\ \gamma(0) = m_0 \text{ and } \gamma'(0) = h \end{array} \right\}.$$

For any local chart (U, φ) and any model (closed vector) space E representing M as a $\mathcal{C}^1$-submanifold at m_0, one has (see (17.14))

$$\text{(E.3)} \qquad T_{m_0}M = D\varphi(0)^{-1}E = T^C(M; m_0) = T^B(M; m_0),$$

where we recall that $T^C(\cdot;\cdot)$ (resp. $T^B(\cdot;\cdot)$) denotes the Clarke (resp. the Bouligand-Peano) tangent cone.

It is also worth noticing by Proposition E.6 that condition (c) in Definition E.8 guarantees that the inverse mapping φ^{-1} also belongs to $\mathcal{C}^{p,0}(\varphi(U), X)$ (resp. belongs to $\mathcal{C}^{p,1}(\varphi(U), X)$).

Assume that the model vector space E admits a topological vector complement E_c, so $X = E \oplus E_c$. With the local chart (U, φ) associated with E in Definition E.8, define $g : U \to E_c$ by $g(x) = \pi_{E_c} \circ \varphi(x)$. It is clear that g is of class $\mathcal{C}^p$ and that $U \cap M = \{x \in U : g(x) = 0\}$. A certain converse holds as shown in the following proposition.

PROPOSITION E.9. Let U_0 be an open set of a Banach space X with $\overline{x} \in U_0$, let $p \in \mathbb{N}$ and let g be a mapping of class $\mathcal{C}^p$ (resp. $\mathcal{C}^{p,0}$, or $\mathcal{C}^{p,1}$) from U_0 into a Banach space Y such that $Dg(\overline{x})$ is surjective. Assume that $\operatorname{Ker} Dg(\overline{x})$ admits a topological vector complement in X (which is the case, in particular, if X is a Hilbert space). Then, with $\overline{y} := g(\overline{x})$, the level set $M := \{x \in U_0 \; : \; g(x) = \overline{y}\}$ is a $\mathcal{C}^p$-submanifold (resp. $\mathcal{C}^{p,0}$-submanifold, or $\mathcal{C}^{p,1}$-submanifold) of X at $\overline{x}$.

PROOF. Suppose (without loss of generality) that $\overline{x} = 0$ and $\overline{y} = 0$. We know by the local submersion theorem (see foregoing Theorem E.3 and Remark E.7) that, denoting by X_1 a topological vector complement in X of $X_2 := \operatorname{Ker} A$ with $A := Dg(\overline{x})$, there exist an open neighborhood $U \subset U_0$ of $\overline{x} = 0$ in X, an open neighborhood V of $(0,0)$ in $Y \times X_2$, and a $\mathcal{C}^p$-diffeomorphism (resp. $\mathcal{C}^{p;0}$-diffeomorphism, or $\mathcal{C}^{p,1}$-diffeomorphism) $\psi : V \to U$ from V onto U such that for all $(v_1, v_2) \in V$ one has $g \circ \psi(v_1, v_2) = v_1$ along with $\psi(v_1, v_2) = \psi_1(v_1, v_2) \oplus v_2$ with $\psi_1(v_1, v_2) \in X_1$. The continuous linear mapping $A_0 : X_1 \to Y$ with $A_0(x_1) := A(x_1)$ for all $x_1 \in X_1$ is bijective, hence an isomorphism from X_1 onto Y by the closed graph theorem, so the mapping $j : X_1 \oplus X_2 \to Y \times X_2$ defined by $j(x_1 \oplus x_2) = \big(A_0(x_1), x_2\big)$ is also an isomorphism. Let $j_V : j^{-1}(V) \to V$ the bijective restriction from $j^{-1}(V)$ onto V and consider the $\mathcal{C}^p$-diffeomorphism (resp. $\mathcal{C}^{p;0}$-diffeomorphism, or $\mathcal{C}^{p,1}$-diffeomorphism) $\varphi := j_V^{-1} \circ \psi^{-1}$ from U onto $j_V^{-1}(V)$.

Then, with $\pi_i := \pi_{X_i}$ we have

$$\begin{aligned} x \in \varphi(U \cap M) &\Leftrightarrow \psi \circ j_V(x) \in U \text{ and } g \circ \psi(j_V(x)) = 0 \\ &\Leftrightarrow \psi \circ j_V(x) \in U \text{ and } g \circ \psi\big(A_0 \circ \pi_1(x), \pi_2(x)\big) = 0 \\ &\Leftrightarrow \psi \circ j_V(x) \in U \text{ and } A_0 \circ \pi_1(x) = 0. \end{aligned}$$

From this we see that

$$x \in \varphi(U \cap M) \;\Leftrightarrow\; \big(\psi \circ j_V(x) \in U \text{ and } \pi_1 x = 0\big) \;\Leftrightarrow x \in \varphi(U) \cap X_2,$$

which means that $\varphi(U \cap M) = \varphi(U) \cap X_2$. Choose $(\overline{v}_1, \overline{v_2}) \in V$ such that $\psi(\overline{v}_1, \overline{v}_2) = 0$. We have $\overline{v}_1 = g \circ \psi(\overline{v}_1, \overline{v}_2) = 0$. Further, the equalities $0 = \psi(\overline{v}_1, \overline{v}_2) = \psi_1(\overline{v}_1, \overline{v}_2) \oplus \overline{v}_2$ entails that $\overline{v}_2 = 0$. Consequently, $\psi^{-1}(0) = (0, 0)$, so $\varphi(0) = j_V^{-1}(\psi^{-1}(0)) = 0$. Altogether, it results that M is a $\mathcal{C}^p$-submanifold (resp. $\mathcal{C}^{p,0}$-submanifold, or $\mathcal{C}^{p,1}$-submanifold) at $\overline{x} = 0$. $\square$

The next theorem shows that a submanifold as defined in Definition E.8 can be seen as the graph of a suitable mapping.

THEOREM E.10. Let M be a subset of a Hilbert space H and $m_0 \in M$, and let Z be a closed vector subspace in H. Given an integer $p \in \mathbb{N}$, the set M is a $\mathcal{C}^p$-submanifold (resp. $\mathcal{C}^{p,0}$-submanifold, or $\mathcal{C}^{p,1}$-submanifold) at m_0 with Z as model vector space if and only if there exist an open neighborhood U of m_0 in H, an open neighborhood $V_Z \subset Z$ of zero in Z, and a mapping $\theta : V_Z \to Z^{\perp}$ such that

(a) $\theta \in \mathcal{C}^p(V_Z, Z^{\perp})$ (resp. $\theta \in \mathcal{C}^{p,0}(V_Z, Z^{\perp})$, or $\theta \in \mathcal{C}^{p,1}(V_Z, Z^{\perp})$);
(b) $\theta(0) = 0$ and $D\theta(0) = 0$;
(c) $M \cap U = \big(m_0 + L^{-1}(\operatorname{gph} \theta)\big) \cap U$,

where $L : X \to Z \times Z^{\perp}$ is the canonic isomorphism given by the equality $L(x) = (\pi_Z(x), \pi_{Z^{\perp}}(x))$.

Further, under (a), (b) (c) one has $T_{m_0} M = Z$.

PROOF. We will consider only the situation when M is a $\mathcal{C}^{p,0}$-submanifold at m_0. The $\mathcal{C}^p$-submanifold is easier with similar arguments and the locally Lipschitz case is analogous.

To show the implication $\Leftarrow$, assume that (a), (b), (c) hold. Let us first prove that $\operatorname{gph} \theta$ is a $\mathcal{C}^{p,0}$-submanifold at $(0, 0)$. Indeed, consider the open set $U_1 := V_Z \times Z^{\perp}$ in $Z \times Z^{\perp}$ and the mapping $\varphi : U_1 \to Z \times Z^{\perp}$ defined by

$$\varphi(v, z_2) = (v, z_2 - \theta(v)) \quad \text{for all } (v, z_2) \in V_Z \times Z^{\perp}.$$

Obviously, we have $\varphi \in \mathcal{C}^{p,0}(U_1, Z \times Z^{\perp})$ and $D\varphi(0, 0) = \mathrm{id}_{Z \times Z^{\perp}}$. By the local inverse mapping theorem (see Theorem E.1 above) there exists an open neighborhood $U_2 \subset U_1$ of $(0, 0)$ in $Z \times Z^{\perp}$ such that $\phi := \varphi\big|_{U_2} : U_2 \to \varphi(U_2)$ is a $\mathcal{C}^p$-diffeomorphism, and clearly $\phi \in \mathcal{C}^{p,0}(U_2, Z \times Z^{\perp})$. Further, since ϕ is the restriction of φ to U_2, we also have the equivalences

$$\begin{aligned} (v, z_2) \in U_2 \cap \operatorname{gph}(\theta) &\iff (v, z_2) \in U_2 \text{ and } z_2 = \theta(v) \\ &\iff \phi(v, z_2) \in \phi(U_2) \text{ and } \pi_{Z^{\perp}}(\phi(v, z_2)) = 0 \\ &\iff \phi(v, z_2) \in \phi(U_2) \cap (Z \times \{0\}). \end{aligned}$$

Thus, by Definition E.8 we see that $\operatorname{gph} \theta$ is a $\mathcal{C}^{p,0}$-submanifold at $(0, 0)$.

Now consider the affine mapping $\widetilde{L} : X \to Z \times Z^{\perp}$ defined by $\widetilde{L}(x) = L(x - m_0)$. Without loss of generality, we may suppose that $U_3 := \widetilde{L}^{-1}(U_2) = L^{-1}(U_2) + m_0 \subset$

U. Consider the mapping $\widetilde{\phi} : U_3 \to L^{-1}(\phi(U_2))$ defined by $\widetilde{\phi} := L^{-1} \circ \phi \circ \widetilde{L}$. Since the bijective mappings L and $\widetilde{L}$ are of class $\mathcal{C}^\infty$ as well as their inverses, the mapping $\widetilde{\phi} : U_3 \to L^{-1}(\phi(U_2))$, defined by $\widetilde{\phi} := L^{-1} \circ \phi \circ \widetilde{L}$, satisfies the following properties:

(a) $\widetilde{\phi}$ is a $\mathcal{C}^p$-diffeomorphism;

(b) $\widetilde{\phi}(M \cap U_3) = \widetilde{\phi}(\widetilde{L}^{-1}(\operatorname{gph}\theta) \cap \widetilde{L}^{-1}(U_2))) = \widetilde{\phi}(\widetilde{L}^{-1}(\operatorname{gph}\theta \cap U_2)))$
$= L^{-1}(\phi(U_2) \cap Z \times \{0\}) = \widetilde{\phi}(U_3) \cap Z$;

(c) $\widetilde{\phi} \in \mathcal{C}^{p,0}(U_3, X)$ (by Lemma E.5).

This says that M is a $\mathcal{C}^{p,0}$-submanifold at m_0 with model space Z, so the desired implication $\Leftarrow$ is proved. Further, keeping in mind that $D\phi(0,0) = \mathrm{id}_{Z\times Z^\perp}$ the equality (E.3) gives that

$$T_0 M = D\widetilde{\phi}(0)^{-1} Z = (L^{-1} \circ D\phi(0,0)^{-1} \circ L) Z = Z,$$

which justifies the additional tangential equality in the statement of the theorem.

To prove the converse, assume that M is a $\mathcal{C}^{p,0}$-submanifold at m_0 with model space Z. Choose an open neighborhood W of m_0 in H and a $\mathcal{C}^p$-diffeomorphism $\varphi : W \to \varphi(W) \subset H$ such that $\varphi \in \mathcal{C}^{p,0}(W, H)$, $\varphi(m_0) = 0$ and that $\varphi(W \cap M) = \varphi(W) \cap Z$. Replacing φ by $D\varphi^{-1}(0) \circ \varphi$ if necessary and using the equality (E.3), we may and do suppose that $T_{m_0} M = Z$.

Consider the mapping $\phi : \varphi(W) \cap Z \to Z$ defined by $\phi(z) = \pi_Z(\varphi^{-1}(z) - m_0)$. Since $D\phi(0) = D\varphi^{-1}(0)\big|_Z$ is an isomorphism from Z to Z, the local inverse mapping theorem furnishes an open neighborhood O of zero in Z such that $\phi : O \to \phi(O)$ is a $\mathcal{C}^p$-diffeomorphism. Moreover, Lemma E.5 and Proposition E.6 give that $\phi \in \mathcal{C}^{p,0}(O, Z)$.

Choose $\delta > 0$ small enough such that

$$U := \varphi^{-1}(B_X(0,\delta)) \subset W \quad \text{and} \quad Z \cap B_X(0,\delta) \subset O.$$

Put $V := \phi(Z \cap B_X(0,\delta))$ and observe that

$$\varphi(M \cap U) = Z \cap \varphi(U) = Z \cap B_X(0,\delta) = \phi^{-1}(V). \tag{E.4}$$

Now define $\theta : V \to Z^\perp$ by $\theta := \pi_{Z^\perp} \circ (\varphi^{-1} \circ \phi^{-1}(\cdot) - m_0)\big|_V$, and note that it belongs to $\mathcal{C}^{p,0}(V, Z^\perp)$ according to Lemma E.5. Further, we have $\theta(0) = 0$ and

$$D\theta(0) = \pi_{Z^\perp} \circ D\varphi^{-1}(0) \circ D\phi^{-1}(0) = 0,$$

since $D\varphi^{-1}(0) \circ D\phi^{-1}(0) Z = Z$. It remains to show that $U \cap M = (L^{-1}(\operatorname{gph}\theta) + m_0) \cap U$. For any $v = \phi(\varphi(m))$ with $m \in U \cap M$ noticing that

$$v_m = \phi(\varphi(m)) = \pi_Z(\varphi^{-1}(\varphi(m)) - m_0) = \pi_Z(m - m_0),$$

the definition of θ yields

$$L(m - m_0) = (\pi_Z(m - m_0), \pi_{Z^\perp}(\varphi^{-1} \circ \phi^{-1}(v) - m_0)) = (v, \theta(v)). \tag{E.5}$$

This ensures that for any $m \in U \cap M$, we have $L(m - m_0) = (v_m, \theta(v_m))$ with $v_m := \phi(\varphi(m))$, while (E.4) gives $v_m \in V$. We deduce that $U \cap M \subset (L^{-1}(\operatorname{gph}\theta) + m_0) \cap U$. For the converse inclusion, take $v \in V$ such that $L^{-1}(v, \theta(v)) + m_0 \in U$. By (E.4) again there exists $m \in M \cap U$ such that $v = \phi \circ \varphi(m)$, so (E.5) entails the equality $L(m - m_0) = (v, \theta(v))$, hence $L^{-1}(v, \theta(v)) + m_0 \in M \cap U$. This justifies the desired converse inclusion and finishes the proof. $\square$

Bibliography

[1] R. Abraham, J. E. Marsden and T. Ratiu, *Manifolds, Tensor Analysis, and Applications*, Third Edition, Springer-Verlag, New York (2002).

[2] R. A. Adams, *Sobolev Spaces*, Academic Press, New York (1975).

[3] R. A. Adams and J. J. F. Fournier, *Sobolev Spaces*, Volume 140, Academic Press, Cambridge (2003).

[4] S. Adly, *A Variational Approach to Nonsmooth Dynamics, Applications in Unilateral Mechanics and Electronics*, Springer Briefs in Mathematics, Springer, New-York (2017).

[5] S. Adly, F. Nacry and L. Thibault, *Preservation of prox-regularity of sets with applications to constrained optimization*, SIAM J. Optim. 26 (2016), 448-473.

[6] S. Adly, F. Nacry and L. Thibault, *Discontinuous sweeping process with prox-regular sets*, ESAIM: COCV 23 (2017), 1293-1329

[7] S. Adly, F. Nacry and L. Thibault, *Prox-regularity approach to generalized equations and image projection*, ESAIM: COCV 24 (2018), 677-708.

[8] S. Adly, F. Nacry and L. Thibault, *New metric properties for prox-regular sets*, Math. Prog. 189 (2021), 7-36.

[9] N. I. Akhiezer and I. M. Glazman, *Theory of Linear Operators in Hilbert space*, Moscow, Gostekhizdat, (1950) (in Russian); English version Pitman Press (1980).

[10] P. Albano, *Some properties of semiconcave functions with general modulus*, J. Math. Anal. Appl. 271 (2002), 217-231.

[11] P. Albano and P. Cannarsa, *Singularities of semiconcave functions in Banach spaces*, in "Stochastic Analysis, Control, Optimization and Applications", Birkhäuser, Boston (1999), 171-190.

[12] P. Albano and P. Cannarsa, *Structural properties of singularities of semiconcave functions*, Ann. Scuola Norm. Sup. Pisa 28 (1999), 719–740.

[13] G. Alberti, *On the structure of singular points of convex functions*, Calc. Var. Partial Differential Equations, 2 (1994), 17-27.

[14] G. Alberti, L. Ambrosio and P. Cannarsa, *On the singularities of convex functions*, Manuscr. Math. 76 (1992), 421-435.

[15] O. Alvarez, P. Cardaliaguet and R. Monneau, *Existence and uniqueness for dislocation dynamics with positive velocity*, Interfaces Free Bound. 7 (2005), 415-434.

[16] C. Amara and M. Ciligot-Travain, *Lower CS-closed sets and functions*, J. Math. Anal. Appl. 239 (1999), 371-389.

[17] L. Ambrosio, P. Cannarsa and H. M. Soner, *On the propagation of singularities of semiconvex functions*, Ann. Scuola Norm. Sup. Pisa 20 (1993), 597–616.

[18] L. Ambrosio, N. Fusco, D. Pallara, *Functions of Bounded Variation and Free Discontinuity Problems*, Oxford Science Publications, Clarendon, Oxford (2000).

[19] R. D. Anderson and V. L. Klee, *Convex functions and upper semicontinuous collections*, Duke Math. J. 19 (1952), 349-357.

[20] F. J. Aragón Artacho and M. H. Geoffroy, *Characterizations of metric regularity of subdifferentials*, J. Convex Anal. 15 (2008), 365-380.

[21] F. J. Aragón Artacho and M. H. Geoffroy, *Metric regularity of the convex subdifferential in Banach spaces*, J. Nonlinear Convex Anal. 15 (2015), 35-47.

[22] G. Aronsson, *Extension of functions satisfying Lipschitz conditions*, Arkiv Math. 6 (1967), 551-561.

[23] N. Aronszajn, *Differentiability of Lipschitzian mappings between Banach spaces*, Studia Math. 57 (1976), 147-190.

[24] N. Aronszajn and K. T. Smith, *Functional spaces and functional completion*, Ann. Inst. Fourier, 6 (1956), 125-185.

[25] E. Asplund, *Farthest points in reflexive locally uniformly rotund Banach spaces*, Israel J. Math. 4 (1966), 213-216.

[26] E. Asplund, *Sets with unique farthest points*, Israel J. Math. 5 (1967), 201-209.

[27] E. Asplund, *Averaged norms*, Israel J. Math. 5 (1967), 227-233.

[28] E. Asplund, *Fréchet differentiability of convex functions*, Acta. Math. 121 (1968), 31-47.

[29] E. Asplund, *Čebyšev sets in Hilbert spaces*, Trans. Amer. Math. Soc. 144 (1969), 235-240.

[30] E. Asplund and R. T. Rockafellar, *Gradients of convex functions*, Trans. Amer. Math. Soc. 139 (1969), 443-467.

[31] H. Attouch, *Convergence de fonctions, des sous-différentiels et semi-groupes associés*, C. R. Acad. Sci. Paris 284 (1977), 539-542.

[32] H. Attouch, *Variational Convergence for Functions and Operators*, Pitman, Boston (1984).

[33] H. Attouch, J.-B. Baillon, M. Théra, *Variational sum of monotone operators*, J. Convex Anal. 1 (1994), 1-29.

[34] H. Attouch and G. Beer, *On the convergence of subdifferentials of convex functions*, Arch. Mat. 60 (1993), 389-400.

[35] H. Attouch, G. Buttazzo and G. Michaille, *Variational Analysis in Sobolev and BV Spaces: Applications to PDEs and Optimization*, (Second Edition) MPS-SIAM Book Series on Optimization 17, SIAM, Philadelphia (2014).

[36] H. Attouch and R. J-B. Wets, *Isometries for the Legendre-Fenchel transform*, Trans. Amer. Math. Soc. 296 (1986), 33-60.

[37] H. Attouch and R. J-B. Wets, *Quantitative stability of variational systems: I. The epigraphical distance*, Trans. Amer. Math. Soc. 328 (1991), 695-730.

[38] H. Attouch and R. J-B. Wets, *Quantitative stability of variational systems: II. A framework for nonlinear conditioning*, SIAM J. Optim. 3 (1993), 359-381.

[39] H. Attouch and R. J-B. Wets, *Quantitative stability of variational systems: III. ε-approximate solutions*, Math. Prog. 61 (1993), 197-214.

[40] J.-P. Aubin, *Contingent derivatives of set-valued maps and existence of solutions to nonlinear inclusions and differential inclusions*, in Mathematical Analysis and Applications, edited by L. Nachbin, pp. 159-229, Academics Press, New York (1981).

[41] J.-P. Aubin, *Lipschitz behavior of solutions to convex minimization problems*, Math. Oper. Res. 9 (1984), 87-111.

[42] J.-P. Aubin, A.Cellina, *Differential Inclusions. Set-Valued Maps and Viability Theory*, Springer, Berlin (1984).

[43] J.-P. Aubin and I. Ekeland, *Applied Nonlinear Analysis*, Wiley, New-York (1984).

[44] J.-P. Aubin and H. Frankowska, *Set-Valued Analysis*, Birkhäuser, Boston (1990).

[45] A. Auslender and M. Teboulle, *Asymptotic Cones and Functions in Optimization and Variational Inequalities*, Springer Monographs in Mathematics, Springer, New York (2003).

[46] D. Aussel, J.-N. Corvellec and M. Lassonde, *Subdifferential characterization of quasiconvexity and convexity*, J. Convex Anal. 1 (1994), 1-7.

[47] D. Aussel, J.-N. Corvellec and M. Lassonde, *Mean value property and subdifferential criteria for lower semicontinuous functions*, Trans. Amer. Math. Soc. 347 (1995), 4147-4161.

[48] D. Aussel, J.-N. Corvellec and M. Lassonde, *Nonsmooth constrained optimization and multidirectional mean value inequalities*, SIAM J. Optim. 9 (1999), 690-706.

[49] D. Aussel, A. Daniilidis and L. Thibault, *Subsmooth sets: functional characterizations and related concepts*, Trans. Amer. Math. Soc. 357 (2005), 1275-1301.

[50] D. Azagra, *Smooth negligibility and subdifferential calculus in Banach spaces, with applications*, Ph.D. dissertation thesis, Department of Mathematics, Universidad Complutense, Madrid (1997).

[51] D. Azagra and R. Deville, *Subdifferential Rolle's and mean value inequality theorems*, Bull. Austral. Math. Soc. 56 (1997), 317-329.

[52] D. Azagra, J. Ferrera and F. López-Mesas, *Approximate Rolle's theorem for the proximal subgradient and the generalized gradient*, J. Math. Anal. Appl. 283 (2003), 180-191.

[53] D. Azagra and M. Jiménez-Sevilla, *The failure of Rolle's theorem in infinite dimensional Banach spaces*, J. Funct. Anal. 182 (2001), 207-226.

[54] D. Azé, *Duality for the sum of convex functions in normed spaces*, Arch. Math. 62 (1994), 554-561.

[55] D. Azé, *Éléments d'Analyse Convexe et Variationnelle*, Ellipses, Paris (1997).
[56] D. Azé, *A unified theory for metric regularity of multifunctions*, J. Convex Anal. 13 (2006), 225-252.
[57] D. Azé and J.-N. Corvellec, *Characterizations of error bounds for lower semicontinuous functions on metric spaces*, ESAIM Control Optim. Calc. Var. 10 (2004), 409-425.
[58] D. Azé and J.-B. Hiriart-Urruty, *Sur un air de Rolle and Rolle*, Revue des Math. de l'Ens. Supérieur 3-4 (2000), 455-460.
[59] D. Azzam-Laouir, C. Castaing and M. D. P. Monteiro Marques, *Perturbed evolution problems with continuous bounded variation in time and applications*, Set-Valued Var. Anal. 26 (2018), 693-728.
[60] M. Bacák, J. M. Borwein, A. Eberhard and B. S. Mordukhovich *Infimal convolution and Lipschitzian properties of subdifferentials for prox-regular functions in Hilbert spaces*, J. Convex Anal. 17 (2010), 737-763.
[61] R. Baire, *Sur les Fonctions de Variables Réelles*, Thesis (Thèse de Doctorat ès Sciences Mathématiques), Université de Paris (1899).
[62] A. Bakan, F. Deutsch and W. Li, *Strong CHIP, normality, and linear regularity of convex sets*, Trans. Amer. Math. Soc. 357 (2005), 3831-3863.
[63] V. S. Balaganski and L. P. Vlasov, *The problem of convexity of Chebyshev sets*, Russ. Math. Surv. 51:6 (1996), 1127-1190.
[64] M. V. Balashov, *Weak convexity of the distance function*, J. Convex Anal. 20 (2013), 93-106.
[65] M. V. Balashov, *Proximal smoothness of a set with the Lipschitz metric projection*, J. Math. Anal. Appl. 406 (2013), 360-363.
[66] M. V. Balashov and M. O. Golubev, *About the Lipschitz property of the metric projection in the Hilbert space*, J. Math. Anal. Appl. 394 (2012), 545-551.
[67] M. V. Balashov and G. E. Ivanov, *Weakly convex and proximally smooth sets in Banach spaces*, Izvestiya RAN: Ser. Mat. 73:3 (2009), 23-66 (in Russian); English translation in Izvestiya: Mathematics 73:3 (2009), 455-499.
[68] M. V. Balashov and D. Repovš, *Weakly convex sets and modulus of nonconvexity*, J. Math. Anal. Appl. 371 (2010), 113-127.
[69] L. Ban and W. Song, *Duality gap of the conic convex constrained optimization problems in normed spaces*, Math. Program. 119 (2009), 195-214.
[70] S. Banach, *Théorie des Opérations Linéaires*, Monografje Matematyczne, Warsaw (1932).
[71] R. Foygel Barber and W. Ha, *Gradient descent with nonconvex constraints: local concavity determines convergence*, Information and Inference: A Journal of the IMA 4 (2018), 755-806.
[72] V. Barbu and T. Precupanu, *Convexity and Optimization in Banach Spaces*, 4th Edition, Springer Monographs in Mathematics (2012).
[73] M. S. Bazaraa, J. J. Goode and M. Z. Nashed, *On the cone of tangents with applications to mathematical programming*, J. Optim. Theory Appl. 13 (1974), 389-426.
[74] M. Bardi, I. Capuzzo-Dolcetta, *Optimal Control and Viscosity Solutions of Hamilton–Jacobi–Bellman Equations*, Birkhäuser, Boston (1997).
[75] H. H. Bauschke and J. M. Borwein, *On the convergence of von Neumann's alternating projection algorithm for two sets*, Set-Valued Anal. 1 (1993), 185-212.
[76] H. H. Bauschke, J. M. Borwein and W. Li, *Strong conical hull intersection property, bounded linear regularity, Jameson's property (G), and error bounds in convex optimization*, Math. Program., Ser. A, 86 (1999), 135-160.
[77] H. H. Bauschke, J. M. Borwein and P. Tseng, *Bounded linear regularity, strong CHIP, and CHIP are distinct properties*, J. Convex Anal., 7 (2000), 395-412.
[78] H. H. Bauschke and P. L. Combettes, *Convex Analysis and Monotone Operator Theory in Hilbert Spaces*, CMS Books in Mathematics, Springer (second edition), New York (2017).
[79] B. Beauzamy, *Introduction to Banach Spaces and their Geometry*, 2nd Edition, North-Holland, Amsterdam (1985).
[80] E. F. Beckenbach, *Convex functions*, Bull. Amer. Math. Soc. 54 (1948), 439-460.
[81] G. Beer, *Conjugate convex functions and the epi-distance topology*, Proc. Amer. Math. Soc. 108 (1990), 117-126.
[82] G. Beer, *Topologies on Closed and Closed Convex Sets*, Kluwer, Dordrecht (1993).
[83] R. Bellman and W. Karush, *On a new functional transform in analysis: the maximum transform*, J. Soc. Indust. Appl. Math. 67 (1961), 501-503.

[84] R. Bellman and W. Karush, *Mathematical programming and the maximum transform*, Bull. Amer. Math. Soc. 10 (1962), 550-567.

[85] R. Bellman and W. Karush, *On the maximum transform*, J. Math. Anal. Appl. 6 (1963), 67-74.

[86] H. Benabdellah, *Existence of solutions to nonconvex sweeping process*, J. Differential Equations 164 (2000), 286-295.

[87] H. Benabdellah, C. Castaing, A. Salvadori and A. Syam, *Nonconvex sweeping processes*, J. Appl. Anal. 2 (1996), 217-240.

[88] J. Benoist and J.-B. Hiriart-Urruty, *General squeeze theorems in nonsmooth analysis*, Canadian Mathematical Society Conference Proceedings, 24 (2000), 7-17.

[89] Y. Benyamini and J. Lindenstrauss, *Geometric Nonlinear Functional Analysis, Vol. 1*, American Matheamtical Society Colloquium Publications 48, American Mathematical Society, Providence, Rhode Island (2000).

[90] C. Berg, J. P. R. Christensen and P. Ressen, *Harmonic Analysis on Semigroups*, Springer-Verlag, New-York (1984).

[91] C. Berge, *Espaces Topologiques et Fonctions Multivoques*, Dunod, Paris (1959).

[92] E. R. Berkson, *Some metrics on the subspaces of a Banach space*, Pacific J. Math. 13 (1963), 7-22.

[93] F. Bernard and L. Thibault, *Prox-regularity of functions and sets in Banach spaces*, Set-valued Anal. 12 (2004), 25-47.

[94] F. Bernard and L. Thibault, *Prox-regular functions in Hilbert spaces*, J. Math. Anal. Appl. 303 (2005), 1-14.

[95] F. Bernard and L. Thibault, *Uniform prox-regularity of functions in Hilbert spaces*, Nonlinear Analysis 60 (2005), 187-207.

[96] F. Bernard, L. Thibault and D. Zagrodny, *Integration of primal lower nice functions in Hilbert spaces*, J. Optim. Theory Appl. 124 (2005), 561-579.

[97] F. Bernard, L. Thibault and N. Zlateva, *Characterization of prox-regular sets in uniformly convex Banach spaces*, J. Convex Anal. 13 (2006), 525-559.

[98] F. Bernard, L. Thibault and N. Zlateva, *Prox-regular sets in uniformly convex Banach space: Tangential and various other properties*, Trans. Amer. Math. Soc. 363 (2011), 2211-2247 (submitted in 2008).

[99] A. S. Besicovitch, *On tangents to general sets of points*, Fundam. Math. 22 (1934), 49-53.

[100] A. S. Besicovitch, *On singular points of convex surfaces*, Proceedings of Symposia in Pure Mathematics, vol VII, Providence, Rhode Island (1963).

[101] D. N. Bessis and F. H. Clarke, *Partial subdifferentials, derivates and Rademacher's theorem*, Trans. Amer. Math. Soc. 351 (1999), 2899-2926.

[102] Z. Birnbaum and W. Orlicz, *Über die Verallgemeinerung des begriffes der zueinander ko,jugierten Potenzen*, Studia Math. 3 (1931), 1-67.

[103] E. Bishop and R. R. Phelps, *A proof that every Banach space is subreflexive*, Bull. Amer. Math. Soc. 67 (1961), 97-98.

[104] E. Bishop and R. R. Phelps, *The support functional of convex sets*, in "Convexity", edited by V. Klee, Symposia Pure Maths VII, Amer. Math. Soc. Providence (1963), 27-35.

[105] J. Blatter, *Weiteste Punkte und nächste Punkte*, Rev. Roumaine Math. Pures Appl. 14 (1969), 615-621.

[106] V. I. Bogachev, *Measure Theory. Volume I*, Mathematics-Analysis, Springer, (2007).

[107] V. I. Bogachev, *Measure Theory. Volume II*, Mathematics-Analysis, Springer, (2007).

[108] J. Bolte and E. Pauwels, *Conservative set valued fields, automatic differentiation, stochastic gradient method and deep learning*, Math. Program.188 (2021), 19-51.

[109] D. Bongiorno, *Stepanoff's theorem in separable Banach spaces*, Comment. Math. Univ. Carolin. 39 (1998), 323-335.

[110] D. Bongiorno, *Radon-Nikodým property of the range of Lipschitz extensions*, Atti. Sem. Mat. Fis. Univ. Modena 48 (2000), 517-525.

[111] J. F. Bonnans and A. S. Shapiro, *Perturbation Analysis of Optimization Problems*, Springer-Verlag, New-York (2000).

[112] J.-M. Bonnisseau and B. Cornet, *Fixed point theorems and Morse's lemma for Lipschitzian functions*, J. Math. Anal. Appl. 146 (1990), 318-332.

[113] J.-M. Bonnisseau and B. Cornet, *Existence of marginal cost pricing equilibria: The nonsmooth case*, International Economic Review 31 (1990), 685-708.

[114] J.-M. Bonnisseau and B. Cornet, *Existence of equilibria with a tight marginal pricing rule*, J. Mathematical Economics 44 (2008), 613-624.

[115] J.-M. Bonnisseau, B. Cornet and M.-O. Czarnecki, *The marginal pricing rule revisited*, Economic Theory 33 (2007), 579-589.

[116] S. Boralugada and R. A. Poliquin, *Local integration of prox-regular functions in Hilbert spaces*, J. Convex Anal. 13 (2006), 27-36.

[117] Yu. G. Borisovich, B. D. Gelman, A. D. Myshkis and V. V. Obukhovskii, *Multi-valued mappings*, J. Soviet Math. 18 (1982), 719-791.

[118] J. M. Borwein, *A Lagrange multiplier theorem and a sandwich theorem for convex relations*, Math. Scand. 48 (1081), 189-204.

[119] J. M. Borwein, *Convex relations in analysis and optimization in generalized concavity in optimization and economics*, (Academic Press, London 1981), 335-371.

[120] J. M. Borwein, *A note on ε-subgradients and maximal monotonicity*, Pacific. J. Math. 103 (1982), 305-314.

[121] J. M. Borwein, *Continuity and differentiability properties of convex operators*, Proc. London Math. Soc. 44 (1982), 420-444.

[122] J. M. Borwein, *Stability and regular points of inequality systems*, J. Optim. Theory Appl. 48 (1986), 9-52.

[123] J. M. Borwein, *Epi-Lipschitz-like sets in Banach space: theorems and examples*, Nonlinear Anal. 11 (1987), 1207-1217.

[124] J. M. Borwein, *Minimal CUSCOS and subgradients of Lipschitz functions*, Fixed PointTheory and its Applications, Pitman Research Notes 252 (1991), 57-81.

[125] J. M. Borwein, *Proximality and Chebyshev sets*, Optimization letters, (2007), 21-32.

[126] J. M. Borwein, J. V. Burke and A. S. Lewis, *Differentiability of cone-monotone functions on separable Banach space*, Proc. Amer. Math. Soc. 132 (2004), 1067-1076.

[127] J. M. Borwein and M. Fabian, *A note on regularity of sets and of distance functions in Banach space*, J. Math. Anal. Appl. 182 (1994), 566-570.

[128] J. M. Borwein and S. Fitzpatrick, *Existence of nearest points in Banach spaces*, Canad. J. Math. 61 (1989), 702-720.

[129] J. M. Borwein and S. Fitzpatrick, *Weak* sequential compactness and bornological limit derivatives*, J. Convex Anal. 2 (1995), 59-67.

[130] J. M. Borwein and S. Fitzpatrick, *Duality inequalities and sandwiched functions* (Preprint, Simon Fraser University at Burnaby 1998), Nonlinear Anal. Th. Meth. Appl. 46 (2001), 365-380.

[131] J. M. Borwein, S. Fitzpatrick and J. R. Giles, *The differentiability of real functions on normed space using generalized gradients*, J. Math. Anal. Appl. 128 (1987), 512-534.

[132] J. M. Borwein, S. Fitzpatrick and R. Girgensohn, *Subdifferentials whose graphs are not norm×weak* closed*, Canad. Math. Bull. 48 (2003), 538-545.

[133] J. M. Borwein and J. R. Giles, *The proximal normal formula in Banach spaces*, Trans. Amer. Math. Soc. 302 (1987), 371-381.

[134] J. M. Borwein and A. D. Lewis, *Convex Analysis and Nonlinear Optimization: Theory and Examples*, CMS Books in Mathematics, Springer (2005).

[135] J. M. Borwein and W. B. Moors, *Essentially smooth Lipschitz functions*, J. Func. Anal. 149 (1997), 305-351.

[136] J. M. Borwein and W. B. Moors, *A chain rule for essentially smooth Lipschitz functions*, SIAM J. Optim. 8 (1998), 300-308.

[137] J. M. Borwein and W. B. Moors, *Null sets and essentially smooth Lipschitz functions*, SIAM J. Optim. 8 (1998), 309-323.

[138] J. M. Borwein, W. B. Moors and X. Wang, *Lipschitz functions with prescribed derivatives and subderivatives*, Nonlinear Anal. 29 (1997), 53-64.

[139] J. M. Borwein, W. B. Moors and X. Wang, *Generalized subdifferentials: A Baire categorical approach*, Trans. Amer. Math. Soc. 353 (2001), 3875-3893.

[140] J. M. Borwein and D. Preiss, *A smooth variational principle with applications to subdifferentiability and to differentiability of convex functions*, Tran. Amer. Math. Soc. 303 (1987), 517-527.

[141] J. M. Borwein and H. Strójwas, *Directionally Lipschitzian mappings on Baire spaces*, Canad. J. Math. 36 (1984), 95-130.

[142] J. M. Borwein and H. Strójwas, *Tangential approximations*, Nonlinear Anal. 9 (1985), 1347-1366.

[143] J. M. Borwein and H. Strójwas, *Proximal analysis and boundaries of closed sets in Banach space. I: Theory*, Canad. J. Math. 38 (1986), 431-452 (submitted: November 1984).

[144] J. M. Borwein and H. Strójwas, *Proximal analysis and boundaries of closed sets in Banach space. II: Applications*, Canad. J. Math. 39 (1987), 428-472.

[145] J. M. Borwein and H. Strójwas, *The hypertangent cone*, Nonlinear Anal. 13 (1989), 125-144.

[146] J. M. Borwein and J. D. Vanderwerff, *Convex Functions: Constructions, Characterizations and Counterexamples*, Cambridge University Press (2010).

[147] J. M. Borwein and X. Wang, *Distinct differentiable functions may share the same Clarke subdifferential at all points*, Proc. Amer. Math. Soc. 125 (1997), 807–813.

[148] J. M. Borwein and X. Wang, *Lipschitz functions with maximal Clarke subdifferentials are generic*, Proc. Amer. Math. Soc. 128 (2000), 3221-3229.

[149] J. M. Borwein and X. Wang, *Cone-monotone functions, differentiability and continuity*, Canad. J. Math. 57 (2005), 961-982.

[150] J. M. Borwein and Q. J. Zhu, *Viscosity solutions and viscosity subderivatives in smooth Banach spaces with applications to metric regularity*, SIAM J. Control Optim. 34 (1996), 1568–1591.

[151] J. M. Borwein and Q. J. Zhu, *A survey of subdifferential calculus with applications*, Nonlinear Anal. 38 (1999), 687–773.

[152] J. M. Borwein and Q. J. Zhu, *Techniques of Variational Analysis*, CMS Books in Mathematics, Springer-Verlag, New-York (2005).

[153] J. M. Borwein and D. M. Zhuang, *Verifiable necessary and sufficient conditions for openness and regularity for set-valued and single-valued maps*, J. Math. Anal. Appl. 134 (1988), 441-459.

[154] U. Boscain, B. Piccoli, *Optimal Syntheses for Control Systems on 2-D Manifolds*, Springer, Berlin (2004).

[155] R. I. Bot, *Conjugate Duality in Convex Optimization*, Lecture Notes in Economics and Mathematical Systems, Vol. 637, Springer-Verlag, Berlin (2010).

[156] M. Bougeard, *Contributions à la théorie de Morse en dimension finie*, Thesis, Université Paris IX Dauphine, (1978).

[157] G. Bouligand, *Sur quelques points de topologie restreinte du premier ordre*, Bull. Soc. Math. France 56 (1928), 26-35.

[158] G. Bouligand, *Sur l'existence des demi-tangentes à une courbe de Jordan*, Fundamenta Mathematicae 15 (1930), 215-218.

[159] G. Bouligand, *Problèmes connexes de la notion d'enveloppe de M. Geeorges Durand*, C.R. Acad. Sci. Paris 189 (1929), p. 146.

[160] G. Bouligand, *Expression générale de la solidarité entre le problème du minimum d'une intégrale et l'équation correspondante d'Hamilton-Jacobi*, Rendiconti dei Lincei, (1930).

[161] G. Bouligand, *Sur quelques applications de la théorie des ensembles à la géométrie infinitésimale*, Bull. Intern. Acad. Polonaise Sc. L (1930), 407-420.

[162] G. Bouligand *Sur quelques points de méthodologie géométrique*, Revue Générale des Sciences Pures et Appliquées, 41 (1930), 39-43.

[163] G. Bouligand, *Sur une application du contingent à la théorie de la mesure*, Acta Math. 56 (1931), p. 371.

[164] G. Bouligand, *Introduction à la Géométrie Infinitésimale Directe*, Paris (1932).

[165] M. Bounkhel, *Régularité Tangentielle en Analyse Non Lisse*, Ph.D. dissertation thesis, Université de Montpellier (1999).

[166] M. Bounkhel, *On the distance function associated with a set-valued mapping*, J. Nonlinear Convex Anal. 2 (2001), 265-278.

[167] M. Bounkhel, *On arc-wise essentially smooth mappings between Banach spaces*, Optimization 51 (2002), 11-29.

[168] M. Bounkhel, *Scalarization of normal Fréchet regularity for set-valued mappings*, New Zealand J. Math. 33 (2004), 129-146.

[169] M. Bounkhel, *Regularity Concepts in Nonsmooth Analysi. Theory and Applications*, Springer Optimization and Its Applications, Vol. 59, Springer, New York (2011).

[170] M. Bounkhel and D.-L. Azzam, *Existence results on the second-order nonconvex sweeping processes with perturbations*, Set-Valued Anal. 12 (2004), 291-318.

[171] M. Boukhel and L. Thibault, *Scalarization of tangential regularity of set-valued mappings*, Set-Valued Anal. 7 (1999), 33-53.

[172] M. Bounkhel and L. Thibault, *On various notions of regularity of sets in nonsmooth analysis*, Nonlinear Anal. 48 (2002), 223-246.

[173] M. Bounkhel and L. Thibault, *Nonconvex sweeping process and prox-regularity in Hilbert space*, J. Nonlinear Convex Anal. 6 (2005), 359-374.

[174] N. Bourbaki, *Espaces Vectoriels Topologiques*, Fascicule XV, XVII, Hermann, Paris (1966).

[175] D. G. Bourgin, *Approximate isometries*, Bull. Amer. Math. Soc. 52 (1946), 704-714.

[176] H. Brezis, *Propriétés régularisantes de certains semi-groupes non linéaires*, Israel J. Math. 9 (1971), 513-534.

[177] H. Brezis, *Opérateurs Maximaux Monotones et Semi-groupes de Contractions dans les Espaces de Hilbert*, Math. Stud. 5, North-Holland, Amsterdam, 1973.

[178] H. Brezis, *Functional Analysis, Sobolev Spaces and Partial Differential Equations*, Springer, New-York (2011).

[179] A. Brønsted, *Conjugate convex functions in topological vector spaces*, Math. Fys. Medd. Dansk. Vid. Selsk. 34 (1964), 1-27.

[180] A. Brønsted, *On the subdifferential of the supremum of two convex functions*, Math. Scand. 31 (1972), 225-230.

[181] A. Brønsted and R. T. Rockafellar, *On the subdifferentiability of convex functions*, Proc. Amer. Math. Soc. 16 (1965), 605-611.

[182] J. V. Burke, M. C. Ferris and M. Qian, *On the Clarke subdifferential of the distance function of a closed set*, J. Math. Anal. Appl. 166 (1992), 199-213.

[183] J. V. Burke, A. S. Lewis, M. L. Overton, *Approximating subdifferentials by random sampling of gradients*, Math. Oper. Res. 27 (2002), 567-584.

[184] J. V. Burke and R. A. Poliquin, *Optimality conditions for non-finite convex composite functions*, Math. Program. 57 (1992), 103-120.

[185] L. N. H. Bunt, *Bijdrage tot de theorie de convexe puntverzamelingen*, Thesis, Univ. of Groningen, Amsterdam (1934).

[186] F. Cabello Sánchez, *Nearly convex functions, perturbations of norms and K-spaces*, Proc. Amer. Math. Soc. 129 (2001), 753-758.

[187] F. Cabello Sánchez, J.M.F. Castillo and P.L. Papini, *Seven views on approximate convexity and the geometry of K-spaces*, J. Lond. Math. Soc. 72 (2005), 457-477.

[188] A. Cabot, H. Engler and S. Gadat, *On the long time behavior of second order differential equations with asymptotically small dissipation*, Tran. Amer. Math. Soc. 361 (2009), 5983-6017.

[189] A. Cabot and L. Thibault, *Inclusion of subdifferentials, linear well-conditioning and steepest descent equation*, SIAM J. Optim. 23 (2013), 552-575.

[190] A. Cabot and L. Thibault, *Sequential formulae for the normal cone to sublevel sets*, Trans. Amer. Math. Soc. 366 (2014), 6591-6628.

[191] L. Caffarelli, *The regularity of free boundaries in higher dimensions*, Acta Math. 139 (1977), 155-184.

[192] A. Canino, *On p-convex sets and geodesics*, J. Differential Equations 75 (1988), 118-157; submitted 18 February 1987.

[193] A. Canino, *Existence of a closed geodesic on p-convex sets*, Ann. Inst. Henri Poincaré 5 (1988), 501-518.

[194] A. Canino, *Local properties of geodesics on p-convex sets*, Ann. Mat. Pura Appl. 159 (1991), 17-44.

[195] A. Canino, *Periodic solutions of quadratic Lagrangian systems on p-convex sets*, Ann. Fac. Sci. Toulouse Math. 12 (1991), 37-60.

[196] P. Cannarsa, *Regularitiy properties of solutions to Hamilton Jacobi equations in infinite dimensions and nonlinear optimal control*, Diff. Integral Equations 2 (1989), 479-493.

[197] P. Cannarsa and P. Cardialaguet, *Representation of equilibrium solutions to the table problem for growing sandpiles*, J. Eur. Math. Soc. 6 (2004), 435-464.

[198] P. Cannarsa and P. Cardaliaguet, *Perimeter estimates for reachable sets of control systems*, J. Convex Anal. 13 (2006), 253-267.

[199] P. Cannarsa and H. Frankowska, *Interior sphere property of attainable sets and time optimal control problems*, ESAIM: Control Optim. Calc. Var. 12 (2006), 350-370.

[200] P. Cannarsa and C. Sinestrari, *Convexity properties of the minimum time function*, Calc. Var. 3 (1995), 273-298.

[201] P. Cannarsa and C. Sinestrari, *Semiconcave Functions, Hamilton–Jacobi Equations, and Optimal Control*, Birkhäuser, Boston (2004).

[202] P. Cannarsa and H. M. Soner, *On the singularities of the viscosity solutions to Hamilton–Jacobi–Bellman equations*, Indiana Univ. Math. J. 36 (1987), 501–524.

[203] M. J. Cánovas, D. Klatte, M. A. Lopez, and J. Parra, *Metric regularity of convex semi-infinite programming problems under convex perturbations*, SIAM J. Optim. 16 (2007), 717-732.

[204] M. J. Cánovas, M. A. Lopez, J. Parra and F. J. Toledo, *Lipschitz continuity of the optimal value via bunds on the optimal set in linear semi-infinite optimization*, Math. Oper. Res; 31 (2006), 478-489.

[205] I. Capuzzo Dolcetta and I. Ishii, *Approximate solutions of the Bellman equation of deterministic control theory*, Appl. Math. Optim. 11 (1984), 161-181.

[206] A. Carioli and L. Vesely, *Normal cones and continuity of vector-valued convex functions*, J. Convex Anal. 25 (2013), 495-500.

[207] B. Cascales, J. Orihuela and A. Pérez, A., *One-sided James compactness theorem*, J. Math. Anal. Appl. 445 (2017), 1267-1283.

[208] E. Casini and P. L. Papini, *A counterexample to the infinity version of the Hyers-Ulam stability theorem*, Proc. Amer. Math. Soc., 118 (1993), 885-890.

[209] C. Castaing, *Sur les Multi-Applications Mesurables*, Thesis (Thése de Doctorat d'État ès Sciences Mathématiques), Université de Caen (1967).

[210] C. Castaing, *Sur les multi-applications mesurables*, Revue Francaise d'Informatique et de Recherche Opérationnelle, 1 (1967), 91-126.

[211] C. Castaing, A. G. Ibrahim and M. Yarou, *Existence problems in second order evolution inclusions: discretization and variational approach*, Taiwanese J. Math. 12 (2008), 1433-1475.

[212] C. Castaing, A. G. Ibrahim and M. Yarou, *Some contributions to nonconvex sweeping process*, J. Nonlinear Convex Anal. 10 (2009), 1-20.

[213] C. Castaing, A. Salvadori and L. Thibault, *Functional evolution equations governed by nonconvex sweeping process*, J. Nonlinear Convex Anal. 2 (2001), 217-241.

[214] C. Castaing and M. Valadier, *Convex Analysis and Measurable Multifunctions*, Lectures Notes in Mathematics, Vol. 580 springer-Verlag, Berlin (1977).

[215] A. Cellina, A. Marino and C. Olech, Eds, *Methods of Nonconvex Analysis*, Lecture Notes in Mathematics, Springer (Berlin) 1990.

[216] Y. Chabrillac and J.-P. Crouzeix, *Continuity and differentiability of monotone real functions of several variables*, in: Nonlinear Analysis and Optimization (Louvain-la-Neuve, 1983), Math. Prog. Study 30 (1987), 1-16.

[217] L. Chamard, *Sur les Propriétés de la Distance à un Ensemble Ponctuel*, Thesis (Thèse de Doctorat ès Sciences Mathématiques), Université de Poitiers (1933).

[218] T. Champion, *Duality gap in convex programming*, Math. Program. 99 (2004), 487-498.

[219] G. Chavent, *Quasiconvex sets and size $\times$ curvature condition, application to non linear inversion*, Appl. Math. Optim. 24 (1991), 129-169.

[220] G. Chavent, *New size $\times$ curvature conditions for strict quasiconvexity of sets*, SIAM J. Control Optim. 29 (1991), 1348-1372.

[221] G. Chavent, *On p-convex, proximally smooth, quasi-convex, strictly quasi-convex and approximately convex sets*, J. Convex Anal. 22 (2015), 427-446.

[222] N. Chemetov, M. D. P. Monteiro Marques, *Non-convex quasi-variational differential inclusions*, Set-Valued Anal. 15 (2007), 209-221.

[223] P. W. Cholewa, *Remarks on the stability of functional equations*, Aequationes Math. 27 (1984), 76-86.

[224] G. Choquet, *Sur les notions de filtre et de grille*, C. R. Acad. Sciences, 224 (1947), 171-173.

[225] G. Choquet, *Convergences*, Annales de l'université de Grenoble, 23 (1947-1948), 57-112.

[226] G. Choquet, *Ensembles et cônes convexes faiblement complets*, C. R. Acad. Sei. Paris 254 (1962), 1908-1910.

[227] J. P. R. Christensen, *On sets of Haar-measure zero in Abelian groups*, Israel J. Math. 13 (1972), 255-260.

[228] J. P. R. Christensen, *Topology and Borel Structure*, Math. Studies 10, Notas Mathematica, Amsterdam (1974).

[229] R. Cibulka and M. Fabian, *Attainment and (sub)differentiability of the infimal convolution of a function and the square of the norm*, J. Math. Anal. Appl. 368 (2010), 538-550.

[230] M. Ciligot-Travain, *On Lagrange-Kuhn-Tucker multipliers for Pareto optimization problems*, Numer. Funct. Anal. Optim. 15 (1994), 689-693.

[231] M. Ciligot-Travain, *An intersection formula for the normal cone associated with the hypertangent cone*, J. Appl. Anal. 5 (1999), 239-247.

[232] F. H. Clarke, *Necessary Conditions for Nonsmooth Problems in Optimal Control and the Calculus of Variations*, Ph.D. dissertation thesis, University of Washington, Seattle (1973).

[233] F. H. Clarke, *Generalized gradients and applications*, Trans. Amer. Math. Soc. 205 (1975), 247-262.

[234] F. H. Clarke, *The Euler-Lagrange differential inclusion*, J. Differential Equations 19 (1975), 80-90.

[235] F. H. Clarke, *Admissible relaxation in variational and control problems*, J. Math. Anal. Appl. 51 (1975), 557-576.

[236] F. H. Clarke, *The generalized problem of Bolza*, SIAM J. Control Optim. 14 (1976), 682-699.

[237] F. H. Clarke, *The maximum principle under minimal hypotheses*, SIAM J. Control Optim. 14 (1976), 1078-1091.

[238] F. H. Clarke, *Optimal solutions to differential inclusions*, J. Optim. Theory Appl. 19 (1976), 469-478.

[239] F. H. Clarke, *A new approach to Lagrange multipliers*, Math. Oper. Res. 1 (1976), 165-174.

[240] F. H. Clarke, *On the inverse function theorem*, Pac. J. Math. 64 (1976), 97-102.

[241] F. H. Clarke, *Extremal arcs and extended Hamiltonian systems*, Trans. Amer. Math. Soc. 231 (1977), 349-367.

[242] F. H. Clarke, *Inequality constraints in the calculus of variations*, Canad. J. Math. 3 (1977), 528-540.

[243] F. H. Clarke, *Optimization and Nonsmooth Analysis*, Wiley Intersciences, New York (1983). Second Edition: Classics in Applied Mathematics, 5, Society for Industrial and Applied Mathematics, Philadelphia (1990).

[244] F. H. Clarke,*Functional Analysis, Calculus of Variations and Optimal Control*, Graduate Texts in Mathematics, Springer, New York (2013).

[245] F. H. Clarke and Yu. S. Ledyaev *Mean value inequalities*, Proc. Amer. Math. Soc. 122 (1994), 1075-1083.

[246] F. H. Clarke and Yu. S. Ledyaev *Mean value inequalities in Hilbert space*, Trans. Amer. Math. Soc. 344 (1994), 307-324.

[247] F. H. Clarke, Yu. S. Ledyaev, R. J. Stern and P. R. Wolenski, *Nonsmooth Analysis and Control Theory*, Springer, Berlin (1998).

[248] F. H. Clarke and R. M. Redheffer, *The proximal subgradient and constancy*, Canad. Math. Bull. 36 (1993), 30-32.

[249] F. H. Clarke, R. J. Stern and P. R. Wolenski, *Subgradient criteria for monotonicity, the Lipschitz condition, and convexity*, Canad. J. Math. 45 (1993), 1167-1183; submitted 24 August 1992.

[250] F. H. Clarke, R. J. Stern and P. R. Wolenski, *Proximal smoothness and the lower C^2 property*, J. Convex Analysis 2 (1995), 117-144.

[251] J. A. Clarkson, *Uniformly convex spaces*, Trans. Amer. Math. Soc. 40 (1936), 396-414.

[252] S. Cobzas, *Antiproximinal sets in Banach spaces*, Math. Balkanica 4 (1974), 79-82.

[253] S. Cobzas, *Geometric properties of Banach spaces and the existence of nearest and farthest points*, Abst. Appl. Anal. 3 (2005), 259-285.

[254] G. Colombo and V. V. Goncharov, *The sweeping processes without convexity*, Set-Valued Anal. 7 (1999), 357–374.

[255] G. Colombo and V. V. Goncharov, *Variational inequalities and regularity properties of closed sets in Hilbert spaces*, J. Convex Anal. 8 (2001), 197–221.

[256] G. Colombo and V. V. Goncharov, *Continuous selections via geodesics*, Topological methods in Nonlin. Anal. 18 (2001), 171–182.

[257] G. Colombo, V. V. Goncharov and B. S. Mordukhovich, *Well-posedness of minimal time problem with constant dynamics in Banach space*, Set-Valued Var. Anal. 18 (2010), 349-372.

[258] G. Colombo and A. Marigonda, *Differentiability properties for a class of non-convex functions*, Calc. Var. 25 (2006), 1–31.

[259] G. Colombo and A. Marigonda, *Singularities for a class of non-convex sets and functions, and viscosity solutions of some Hamilton–Jacobi equations*, J. Convex Anal. 15 (2008), 105–129.

[260] G. Colombo, A. Marigonda and P. R. Wolenski, *Some new regularity properties for the minimal time function*, SIAM J. Control 44 (2006), 2285–2299.

[261] G. Colombo, A. Marigonda and P. R. Wolenski, *The Clarke generalized gradient for functions whose epigraph has positive reach*, Math. Oper. Res. 38 (2013), 451-468.

[262] G. Colombo and M. D. P. Monteiro Marques, *Sweeping by a continuous φ-convex set*, J. Differential Equations 187 (2003), 46–72.

[263] G. Colombo and T. K. Nguyen, *Quantitative isoperimetric inequalities for a class of nonconvex sets*, Calc. Var. 37 (2010), 141-166.

[264] G. Colombo and T. K. Nguyen, *On the structure of the minimum time function*, SIAM J. Control 48 (2010), 4776-4814.

[265] G. Colombo and L. Thibault, *Prox-regular sets and applications*, Handbook of Nonconvex Analysis and Applications, p. 99-182, D.Y. Gao, D. Motreanu Eds., International Press, Somerville (2010).

[266] G. Colombo and P. R. Wolenski, *The subgradient formula for the minimal time function in the case of constant dynamics in Hilbert space*, J. Global Optim. 28 (2004), 269-282.

[267] G. Colombo and P. R. Wolenski, *Variational analysis for a class of minimal time functions in Hilbert spaces*, J. Convex Anal. 11 (2004), 335-361.

[268] C. Combari, A. Elhilali Alauoi, A. Levy, R. Poliquin and L. Thibault, *Convex composite functions in Banach spaces and the primal lower-nice property*, Proc. Amer. Math. Soc. 126 (1998), 3701-3708.

[269] C. Combari, M. Laghdir and L. Thibault, *Sous-différentiels de fonctions composées*, Ann. Sci. Math. Québec 62 (1994), 119-148.

[270] C. Combari, M. Laghdir and L. Thibault, *A note on subdifferentials of convex composite functionals*, Arch. Math. 67 (1996), 239-252.

[271] C. Combari, M. Laghdir and L. Thibault, *On subdifferential calculus for convex functions defined on locally convex spaces*, Ann. Sci. Math. Québec 23 (1999), 23-36.

[272] C. Combari, S. Marcellin and L. Thibault, *On the graph convergence of ε-subdifferentials of convex functions*, J. Nonlin. Convex Anal. 4 (2003), 309-324.

[273] C. Combari, R. A. Poliquin and L. Thibault, *Convergence of subdifferentials of convexly composite functions*, Canad. J. Math. 51 (1999), 250-265.

[274] C. Combari and L. Thibault, *Epi-convergence of convexly composite functions in Banach spaces*, SIAM J. Optim. 13 (2003), 986–1003.

[275] R. Cominetti, *On Pseudo-differentiability*, Trans. Amer. Math. Soc. 324 (1991), 843-865.

[276] H. O. Cordes and J. P. Labrousse, *The invariance of the index in the metric space of closed operators*, J. Math. Mech. 12 (1963), 693-720.

[277] B. Cornet, *A remark on tangent cones*, CEREMADE Publication (1979), Univ. Paris-Dauphine.

[278] B. Cornet, *Regular properties of tangent and normal cones*, CEREMADE Publication 8130 (1981), Univ. Paris-Dauphine.

[279] B. Cornet, *Existence of equilibria in economies with increasing returns*, in Contributions to Operations Research and Economics: The twentieth anniversary of C.O.R.E., MIT Press, Cambridge, (1982).

[280] B. Cornet, *Existence of slow solutions for a class of differential inclusions*, J. Math. Anal. Appl. 96 (1983), 179-186.

[281] B. Cornet and M.-O. Czarnecki, *Smooth representations of epi-Lipschitzian subsets*, Nonlinear Anal. Th. Meth. Appl. 37 (1999), 139-160.

[282] B. Cornet and M.-O. Czarnecki, *Existence of generalized equilibria*, Nonlinear Anal. Th. Meth. Appl. 44 (2001), 555-574.

[283] R. Correa, P. Gajardo and L. Thibault, *Subdifferential representation formula and subdifferential criteria for the behavior of nonsmooth functions*, Nonlinear Anal. 65 (2006), 864-891.

[284] R. Correa, P. Gajardo and L. Thibault, *Links between directional derivatives through multidirectional mean value inequalities*, Math. Program. 116 Ser. B (2009), 57-77.

[285] R. Correa, P. Gajardo and L. Thibault, *Various Lipschitz-like properties for functions and sets. I. Directional derivative and tangential characterizations*, SIAM J. Optim. 20 (2010), 1766-1785.

[286] R. Correa, P. Gajardo and L. Thibault, *Various Lipschitz-like properties for functions and sets. II. Subdifferential and normal characterizations*, J. Convex Anal. (2020).

[287] R. Correa, A. Hantoute and A. Jourani, *Characterizations of convex approximate subdifferential calculus in Banach spaces*, Trans. Amer. Math. Soc. 368 (2016), 4831-4854.

[288] R. Correa, A. Hantoute, and M. A. López *Valadier-like formulas for the supremum function I* J. Convex Anal., 25 (2018), 1253-1278.

[289] R. Correa, A. Hantoute, M. A. López, *Valadier-like formulas for the supremum function II: the compactly indexed case*, J. Convex Anal. 26 (2019), 299-324.

[290] R. Correa and A. Jofre, *Tangentially continuous directional derivatives in nonsmooth analysis*, J. Optim. Theor. Appl. 61 (1989), 1-21.

[291] R. Correa, A. Jofre and L. Thibault, *Characterization of lower semicontinuous convex functions*, Proc. Amer. Math. Soc. 116 (1992), 67-72.

[292] R. Correa, A. Jofre and L. Thibault, *Subdifferential monotonicity as characterization of convex functions*, Numer. Funct. Anal. Optimiz. 15 (1994), 531-536.

[293] R. Correa, A. Jofre and L. Thibault, *Subdifferential characterization of convexity*, in Recent Advances in Nonsmooth Optimization, edited by D. Du, L. Qi and R. Womersley, World Scientific Publishing, Singapore, 1-23 (1995).

[294] R. Correa, D. Salas and L. Thibault, *Smoothness of the metric projection onto nonconvex bodies in Hilbert spaces*, J. Math. Anal. Appl., 457 (2018), 1307-1322.

[295] R. Correa and L. Thibault, *Subdifferential analysis of bivariate separately regular functions*, J. Math. Anal. Appl., 148 (1990), 157-174.

[296] R. Courant ans D. Hilbert, *Methods of Mathematical Physics*, Volume 2, 1961; German original edition 1938.

[297] M. G. Crandall, H. Ishii and P.-L. Lions, *User's guide to viscosity solutions of second-order partial differential equations*, Bull. Amer. Math. Soc. 27 (1992), 1-67.

[298] M. G. Crandall and P.-L. Lions, *Viscosity solutions of Hamilton-Jacobi equations*, Trans. Amer. Math. Soc. 277 (1983), 1-42; submitted: December 1, 1981.

[299] J.-P. Crouzeix, *Continuity and differentiability of quasi-convex functions*, Handbook of generalized convexity and generalized monotonicity, 121-149, Nonconvex Optim. Appl. 76 Springer, New-York (2005).

[300] M. Csörnyei, *Aronszajn null and Gaussian null sets coincide*, Israel J. Math. 111 (1999), 191-202.

[301] A. Cwiszewski and W. Kryszewski, *Equilibria of set-valued maps: a variational approach*, Nonlinear Anal. Th. Meth. Appl. 48 (2002), 707-746.

[302] A. Cwiszewski and W. Kryszewski, *The constrained degree and fixed point index theory for set-valued maps*, Nonlinear Anal. Th. Meth. Appl. 64 (2006), 2643-2664.

[303] M.-O. Czarnecki, Ph.D thesis, Université Paris 1 Panthéon Sorbonne, (1996).

[304] M.-O. Czarnecki, A. N. Gudovich, *Representation of epi-Lipschitzian sets*, Nonlinear Anal. Th. Meth. Appl. 73 (2010), 2361-2367.

[305] M.-O. Czarnecki and L. Thibault, *Sublevel representations of epi-Lipschitz sets and other properties*, Math. Program. 18 (2018), 555-569.

[306] J. Daneš, *A geometric theorem useful in nonlinear functional analysis*, Boll. Un. Mat. Ital. 6 (1972), 369-375.

[307] J. Daneš, *Equivalence of some geometric and related results of nonlinear functional analysis*, Commentationes Mathematicae Universitatis Carolinae 26 (1985), 443-454.

[308] J. W. Daniel, *The continuity of metric projections as functions of the data*, J. Approximation Theory 12 (1974), 234-239.

[309] A. Daniilidis and P. Georgiev, *Approximate convexity and submonotonicity*, J. Math. Anal. Appl. 291 (2004), 292-301.

[310] A. Daniilidis, W. L. Hare and J. Malick, *Geometrical interpretation of the predictor-corrector type algorithms in structured optimization problems*, Optimization 55 (2006), 481-503.

[311] A. Daniilidis, F. Jules and M. Lassonde, *Subdifferential characterization of approximate convexity: the lower semicontinuous case*, Math. Program. Ser. B 116 (2009), 115-127; submitted: September 18, 2005.

[312] A. Daniilidis, A. S. Lewis, J. Malick and H. Sendov, *Prox-regularity of spectral functions and spectral sets*, J. Convex Anal. 15 (2008), 547-560.

[313] A. Daniilidis and J. Malick, *Filling the gap between lower-C^1 and lower-C^2 functions*, J. Convex Anal. 12 (2005), 315-329.

[314] A. Daniilidis and L. Thibault, *Subsmooth sets and metrically subsmooth sets and functions in Banach space*, Unpublished paper, Univ. Montpellier II, 2008.

[315] J. M. Danskin, *The theory of max-min with applications*, SIAM J. Appl. Math. 14 (1966), 641-644.

[316] G. Dantzig, *Linear programming*, Operations Res. 50 (2002), 42-47.

[317] M. M. Day, *Reflexive Banach spaces not isomorphic to uniformly convex spaces*, Bull. Amer. Math. Soc. 47 (1941), 313-317.

[318] M. M. Day, *Uniform convexity in factor and conjugate spaces*, Ann. Math. 45 (1944), 375-385.

[319] J.-P. Dedieu, *Cône asymptote d'un ensemble non convexe. Application à l'optimisation*, C. R. Acad. Sci., Paris, Sér. A 285 (1977), 501-503.

[320] E. De Giorgi, M. Degiovanni, A. Marino and M. Tosques, *Evolution equations for a class of nonlinear operators*, Atti Accad. Naz. Lincei Rend. Cl. Sci. Fis. Mat. Natur. (8) 15 (1983), 1-8.

[321] M. Degiovanni, A. Marino and M. Tosques, *Evolution equations with lack of convexity*, Nonlinear Anal. 9 (1985), 1401-1443.

[322] E. De Giorgi, *Sulla proprietà isoperimetrica dell'ipersfera, nella classe degli insiemi aventi frontiera orientata di misura finita*, (Italian) Atti Accad. Naz. Lincei. Mem. Cl. Sci. Fis. Mat. Nat., Sez. I 8 (1958), 33-44; also (in English) in De Giorgi, *Selected papers*, L. Ambrosio, G. Dal Maso, M. Forti, M. Miranda, S. Spagnolo (Eds.), Springer, Berlin (2006).

[323] F. Delbaen, *Coherent Risk Measure*, Lectures at the Scuola Normale di Pisa, March 2000.

[324] M. C. Delfour and J.-P. Zolésio, *Shape analysis via oriented distance functions* J. Funct. Anal. 123 (1994), 129-201.

[325] M. C. Delfour and J.-P. Zolésio, *Shapes and Geometries: Metrics, Analysis, Differential Calculus, and Optimization*, SIAM series on Advances in Design and Control, Society for Industrial and Applied Mathematics, Philadelphia, 2nd ed. (2011).

[326] A. Denjoy, *Les quatre cas fondamentaux des nombres dérivés* C.R. Acad. Sci. Paris, 161 (1915), 124-127.

[327] A. Denjoy, *Mémoire sur les nombres dérivés des fonctions continues*, J. Math. Pures Appl. 1 (1915), 105-240.

[328] A. Denjoy, *Sur les fonctions dérivés sommables*, Bull. Soc. Math. France 43 (1915), 161-248.

[329] A. Denjoy, *Mémoire sur la totalisation des nombres dérivés non sommables*, Ann. École Norm. Sup. 33 (1916), 127-222.

[330] A. Denjoy, *Mémoire sur la totalisation des nombres dérivés non sommables*, Ann. École Norm. Sup. 34 (1916), 181-238.

[331] A. Denjoy, *Sur l'intégration des coefficients différentiels d'ordre supérieur*, Fundam. Math. 25 (1935), 273-326.

[332] R. Deville, *A mean value theorem for the non differentiable mappings*, Serdica Math. J. 21 (1995), 59-66.

[333] R. Deville, G. Godefroy and V. Zizler, *Smootness and Renormings in Banach Spaces*, Longman and Wiley, New York (1993).

[334] J. Diestel, *Geometry of Banach Spaces-Selected Topics*, Springer-Verlag, Berlin (1975).

[335] J. Diestel, *Sequences and Series in Banach Spaces*, Springer-Verlag, Berlin (1984).

[336] S. J. Dilworth, R. Howard and J. W. Roberts, *Extremal approximately convex functions and estimating the size of convex hulls*, Advances in Math. 148 (1999), 1-43.

[337] S. J. Dilworth, R. Howard and J. W. Roberts, *On the size of approximately convex sets in normed spaces*, Studia Math. 140 (2000), 213-241.

[338] S. J. Dilworth, R. Howard and J. W. Roberts, *Extremal approximately convex functions and the best constants in a theorem of Hyers and Ulam*, Adv. Math. 172 (2002), 1-14.

[339] U. Dini, *Fondamenti per la Teorica delle Funzioni di Variabili Reali*, Pisa (1878).

[340] A. V. Dmitruk, A. A. Milyutin, and N. P. Osmolovskii, *Lyusternik's theorem and the theory of extrema*, Uspekhi Mat. Nauk 35:6 (1980), 11-46; English transl., Russian Math. Surveys 35:6 (1980), 11-51.

[341] S. Dolecki, *Tangency and differentiation: Some applications of convergence theory*, Ann. Math. Pura Appl. 130 (1982), 281-301.

[342] S. Dolecki and G. H. Greco, *Towards historical roots of necessary conditions of optimality: Regula of Peano*, Control and Cybernetics 36 (2007), 491-518.

[343] S. Dolecki and G. H. Greco, *Tangency vis-à-vis differentiability by Peano, Severi and Guareschi*, J. Convex Anal. 18 (2011), 301-339.

[344] S. Dolecki and S. Kurcyusz, *On Φ-convexity in extremal problems*, SIAM J. Cont. Optim. 16 (1978), 277-300.
[345] E. P. Dolzhenko, *Boundary properties of arbitrary functions* (in Russian), Izv. Akad. Nauk. SSSR ser. Mat. 31 (1967), 3-14.
[346] A. L. Dontchev, *The Graves theorem revisited*, J. Convex Anal. 3 (1996), 45-54.
[347] A. L. Dontchev and H. Frankowska, *Lyusternik-Graves theorem and fixed-points*, Proc. Amer. Math. Soc. 139 (2011), 521-534.
[348] A. L. Dontchev and H. Frankowska, *Lyusternik-Graves theorem and fixed-points 2*, J. Convex Anal. 19 (2012), 955-974.
[349] A. L. Dontchev and W. W. Hager, *An inverse mapping theorem for set-valued maps*, Proc. Amer. Math. Soc. 121 (1994), 481-489.
[350] A. L. Dontchev and W. W. Hager, *Lipschitz, functions, Lipschitz maps and stability in optimization*, Math. Oper. Res. 3 (1994), 753-768.
[351] A. L. Dontchev, A. S. Lewis and R. T. Rockafellar, *The radius of metric regularity*, Trans. Amer. Math. Soc. 355 (2003), 493-517.
[352] A. L. Dontchev, M. Quincampoix and N. Zlateva, *Aubin criterion for metric regularity*, J. Convex Anal. 13 (2006), 281-297.
[353] A. L. Dontchev and R. T. Rockafellar, *Regularity and conditioning of solution mappings in variational analysis*, Set-Valued Anal. 12 (2004), 79-109.
[354] A. L. Dontchev and R. T. Rockafellar, *Implicit Functions and Solution Mappings: A view From Variational Analysis*, Springer Monographs in Mathematics, 2nd edition, New York (2014).
[355] R. Douady, *Petites perturbations d'une suite exacte et d'une suite quasi-exacte*, Seminaire d'Analyse, Nice (1965-1966), 21-34.
[356] A. Douglis, *The continuous dependence of generalized solutions of non–linear partial differential equations upon initial data*, Comm. Pure Appl. Math. 14 (1961), 267-284.
[357] D. Drusvyatskiy and A. D. Ioffe, *Quadratic growth and critical point stability of semi-algebraic functions*, Math. Program. 153 (2015), 635-653.
[358] D. Drusvyatskiy and A. S. Lewis, *Tilt stability, uniform quadratic growth, and strong metric regularity of the subdifferential*, SIAM J. Optim. 23 (2013), 256-267.
[359] D. Drusvyatskiy, B. S. Mordukhovich and T. T. Nghia, *Second-order growth, tilt stability, and metric regularity of the subdifferential*, J. Convex Anal. 21 (2014), 1165-1192.
[360] A. J. Dubovicki and A. A. Milyutin, *Extremal problems with constraints*, U.S.S.R. Comp. Maths. Math. Phys. 5 (1965), 1-80.
[361] J. Duda, *On Gâteaux differentiability of pointwise Lipschitz mappings*, Canad. Math. Bull. 51 (2008), 205-216.
[362] J. Duda, *Cone monotone mapping: Continuity and differentiability*, Nonlinear Anal. 68 (2008), 1963-1972.
[363] J. Duda and L. Zajíček, *Semiconvex functions: representations as suprema of smooth functions and extensions*, J. Convex Anal. 16 (2009), 239-260.
[364] J. Duda and L. Zajíček, *Smallness of singular sets of semiconvex functions in separable Banach spaces*, J. Convex Anal. 20 (2013), 573-598.
[365] R. M. Dudley, *Real Analysis and Probability*, Cambridge University Press, Cambridge (2002).
[366] N. Dunford and J. T. Schwartz, *Linear Operators, part I: General Theory*, Interscience, New York (1958).
[367] G. Durand, *Sur un critère de dénombrabilité*, Acta. Math. 56, 1931, p. 363.
[368] G. Durand, *Sur une généralisation des surfaces convexes*, J. Math. Pures Appl. 10 (1931), 335-414.
[369] S. Dutta, *Generalized subdifferential of the distance function*, Proc. Amer. Math. Soc. 133 (2005), 2949-2955.
[370] M. Edelstein, *Farthest points of sets in uniformly convex Banach spaces*, Israel J. Math. 4 (1966), 171-176.
[371] M. Edelstein, *On nearest points of sets in uniformly convex Banach spaces*, J. Lond. Math. Soc. 43, (1968), 375–377.
[372] M. Edelstein, *A note on nearest points*, Quarterly J. Math. 21 (1970), 403-405.
[373] M. Edelstein, *Weakly proximinal sets*, J. Approx. Theory 18 (1976), 1-8.
[374] J. F. Edmond, *Delay perturbed sweeping processes*, Set-Valued AnaL. 14 (2006), 295-317.

[375] J. F. Edmond and L. Thibault, Inclusions and integration of subdifferentials, J. Nonlin. Convex Anal. 3 (2002), 411–434.

[376] J. F. Edmond and L. Thibault, *Relaxation of an optimal control problem involving a perturbed sweeping process*, Math. Program. ser B 104 (2005), 347-373.

[377] J. F. Edmond and L. Thibault, *BV solution of nonconvex sweeping process with perturbation*, J. Differential Equations 226 (2006), 135-179.

[378] A. V. Efimov, *Linear methods of approximating continuous periodic functions*, Amer. Math. Soc. Transl. 28 (1963), 221-268; translation of Russian original version in Mat.sb. 54 (1961), 51-90.

[379] N. V. Efimov and S. B. Stechkin, *Some supporting properties of sets in Banach spaces and Chebyshev sets*, Dokl. Akad. Nauk SSSR 127:2 (1959), 254–257. (in Russian)

[380] I. Ekeland, *On the variational principle*, J. Math. Anal. Appl. 47 (1974), 324-353.

[381] I. Ekeland, *Nonconvex minimization problems*, Bull. Amer. Math. Soc. 1 (1979), 443-474.

[382] I. Ekeland and J.-M. Lasry, *On the number of periodic trajectories for a Hamiltonian flow*, Ann. Math. 112 (1980), 293-319.

[383] I. Ekeland and G. Lebourg, *Generic Fréchet differentiability and perturbed optimization problems in Banach spaces*, Trans. Amer. Math. Soc. 224 (1976), 193-216.

[384] I. Ekeland and R. Temam, *Analyse Convexe et Problèmes Variationnels*, Dunod, Paris (1974).

[385] P. Enflo, *Banach spaces which can be given an equivalent unformly convex norm*, Israel J. Math. 13 (1972), 281-288.

[386] P. Erdös, *On the Hausdorff dimension of some sets in Euclidean space*, Bull. Amer. Math. Soc. 52 (1946), 107–109.

[387] E. Ernst and M. Volle, *Generalized Courant-Beltrami penalty functions and zero duality gap for conic convex programs*, Positivity 17 (2013), 945-964.

[388] E. Ernst and M. Volle, *Zero duality gap and attainment with possibly non-convex data*, J. Convex Anal. 23 (2016), 615-629.

[389] L. C. Evans and R. F. Gariepy, *Measure Theory and Fine Properties of Functions*, Studies in Adv. Math., CRC Press, Boca Raton (1992).

[390] M. Fabian, *On classes of subdifferentiabiliy spaces of Ioffe*, Nonlinear Anal. Th. Meth. Appl. 12 (1988), 63-74.

[391] M. Fabian, *Subdifferentiability and trustworthiness in the light of a new variational principle of Borwein and Preiss*, Acta Univ. Carolinae 30 (1989), 51-56.

[392] M. Fabian, *Infimal convolution in Efimov–Stečkin Banach spaces*, J. Math. Anal. Appl. 339 (2008), 735-739.

[393] M. Fabian, R. Henrion, A. Y. Kruger and J. V. Outrata, *Error bounds: necessary and sufficient conditions*, Set-Valued Var. Anal. 18 (2010), 121-149.

[394] M. Fabian and A. D. Ioffe, *Separable reduction in the theory of Fréchet subdifferentiability*, Set-Valued Var. Anal. 21 (2013), 661-671.

[395] M. Fabian and A. D. Ioffe, *Separable reduction and rich family in the theory of Fréchet subdifferentials*, J. Convex Anal. 3 (2016), 631-648.

[396] M. Fabian and B. S. Mordukhovich, *Separable reduction and supporting properties of Fréchet-like normals in Banach spaces*, Canad. J. Math. 51 (1999), 51-56.

[397] M. Fabian and B. S. Mordukhovich, *Separable reduction and extremal principles in variational analysis*, Nonlinear Anal. Th. Meth. Appl. 49 (2002), 265-292.

[398] M. Fabian and D. Preiss, *On intermediate differentiability of Lipschitz functions on certain Banach spaces*, Proc. Amer. Math. Soc. 113 (1991), 733-740.

[399] M. Fabian and J. Revalski, *A variational principle in reflexive spaces with Kadec–Klee norm*, J. Convex Anal. 16 (2009) 211-226.

[400] M. Fabian and N. V. Zhivkov, *A characterization of Asplund spaces with the help of local ε-supports of Ekeland and Lebourg*, C. R. Acad. Bulg. Sci. 38 (1985), 671-674.

[401] E. Fabry, *Sur les rayons de convergence d'une série double*, C. R. Acad. Sci. Paris 134 (1902), 1190-1192.

[402] H. Federer, *Curvature measures*, Trans. Amer. Math. Soc. 93 (1959), 418-491.

[403] H. Federer, *Geometric Measure Theory*, Vol Grundlehren der mathematischen Wissenschaften, Springer, Berlin (1969).

[404] W. Fenchel, *On conjugate convex functions*, Canad. J. Math. 1 (1949), 73-77.

[405] W. Fenchel, *Convex Cones, Sets and Functions*, Lecture Notes, Princeton University, Princeton, New Jersey (1951).
[406] J. Ferrer, *Rolle's theorem fails in ℓ_2*, Amer. Math. Monthly (1996), 161-165.
[407] J. Ferrer, *On Rolle's theorem in spaces of infinite dimension*, Indian J. Math., 42 (2000), 21-36.
[408] J. Ferrer, *An approximate Rolle's theorem for polynomials of degree four in a Hilbert space*, Publ. RIMS, Kyoto Univ. 41 (2005), 375-384.
[409] M. Field, *Differential Calculus and Its Applications*, Dover Publications, Mineola (2012).
[410] S. Fitzpatrick, *Metric projections and the differentiability of distance functions*, Bull. Austral. Math. Soc. 22 (1980), 291-312.
[411] S. Fitzpatrick, *Differentiation of real-valued functions and continuity of metric projections*, Proc. Amer. Math. Soc. 91 (1984), 544-548.
[412] S. Fitzpatrick and R. R. Phelps, *Differentiability of the metric projection in Hilbert space*, Tran. Amer. Math. Soc. 170 (1982), 483-501.
[413] S. Fitzpatrick and S. Simons, *The conjugates, compositions and marginals of convex functions*, J. Convex Anal. 8 (2001), 423-446.
[414] S. Flåm, *Upward slopes and inf-convolution*, Math. Oper. Res. 31 (2006), 188-198.
[415] S. Flåm, J.-B. Hiriart-Urruty and A. Jourani, *Feasibility in finite time*, J. Dynamical Control Syst. 15 (2009), 537-555.
[416] A. Fougéres, *Coercivité des intégrandes convexes normales. Application à la minimisation des fonctionnelles intégrales et du clacul des variations*, Sémin. Anal. Convexe Montpellier, exposé 19 (1976).
[417] A. Fougères and A. Truffert, *Régularisation s.c.i. et épiconvergence: Approximations inf-convolutives associées à un référentiel*, Ann. Math. Pura Appl. 152 (1988), 21-51.
[418] H. Frankowska, *Some inverse mapping theorems*, Ann. Inst. H. Poincaré 7 (1990), 183-234.
[419] H. Frankowska and M. Quincampoix, *Hölder metric regularity of set-valued maps*, Math. Program. A 132 (2012), 333-354.
[420] M. Fréchet, *Sur la notion de différentielle*, C. R. Acad. Sci. Paris (1911), 845-847.
[421] M. Fréchet, *Sur la notion de différentielle*, C. R. Acad. Sci. Paris (1911), 1050-1051.
[422] M. Fréchet, Comptes Rendus du Congrès des Sociétés Savantes, Paris, (1912), p. 44.
[423] M. Fréchet, *Sur la notion de différentielle totale*, Nouvelles Annales de Mathématiques (4e série) 12 (1912), 385-433.
[424] M. Fréchet, *Sur la notion de différentielle totale*, Nouvelles Annales de Mathématiques (4e série) 12 (1912), 433-449.
[425] M. Fréchet, *Sur la notion de différentielle d'une fonction de ligne*, Tran. Amer. Math. Soc. 10 (1914), 135-161.
[426] M. Fréchet, *La notion de différentielle dans l'analyse générale*, Ann. Éc. Norm. Sup. 42 (1925), 293-323.
[427] J. H. G. Fu, *Tubular neighborhoods in Euclidean spaces*, Duke Math. J. 52 (1985), 1025-1046.
[428] B. Fuglede, *Stability in the isoperimetric problem for convex or nearly spherical domains in $\mathbb{R}^n$*, Trans. Am. Math. Soc. 314 (1989), 619-638.
[429] L. Gajek and D. Zagrodny, *Geometric mean value theorems for the Dini derivative*, J. Math. Anal. Appl. 191 (1995), 56-76.
[430] D. Gale, H. W. Kuhn and A. W. Tucker, *Linear programming and the theory of games*, in Activity Analysis of Production and Allocation, edited by T. C.Koopmans, Cowles Commission Monographs, Vol. 13 (1951), 317-329. Wiley, New York.
[431] R. Gâteaux, *Sur les fonctionnelles continues et les fonctionnelles analytiques*, C. R. Acad. Sci. Paris 157 (1913), 325-327.
[432] R. Gâteaux, *Fonctions d'une infinité de variables indépendantes*, Bull. Soc. Math. France 47 (1919), 70-96.
[433] R. Gâteaux, *Sur diverses questions de calcul fonctionnel*, Bull. Soc. Math. France 50 (1922), 1-37.
[434] M. Gaydu, M. H. Geoffroy and C. Jean-Alexis, *Metric subregularity of order q and the solving of inclusions*, Cent. Eur. J. Math., 9 (2011), 327-344.
[435] M. H. Geoffroy and M. Lassonde, *On a convergence of lower semicontinuous functions linked with the graph convergence of their subdifferentials*, Canadian Mathematical Society Conference Proceedings, 27 (2000), 93-109.

[436] E. Giner, *On the Clarke subdifferential of an integral functional on L_p, $1 \leq p < \infty$*, Canad. Math. Bull. 41 (1998), 41-48.

[437] E. Giner, *Subdifferential regularity and characterizations of Clarke subgradients of integral functionals*, J. Nonlin. Convex Anal. 9 (2008), 25-36.

[438] E. Giner, *Calmness properties and contingent subgradients of integral functionals on Lebesgue spaces L_p, $1 \leq p < \infty$*, Set-Valued Anal. 17 ((2009), 321-357.

[439] E. Giner, *Lagrange multipliers and lower bounds for integral functionals*, J. Convex Anal. 17 (2010), 301-308.

[440] B. Ginsburg and A. D. Ioffe, *The maximum principle in optimal control of systems governed by semilinear equations*, IMA Vol. Math. Appl. 78 (1996), 81-110.

[441] I. C. Gohberg and M. G. Krein, *The basic propositions on defect numbers, root numbers, and indices of linear operators*, Uspehi Mat. Nauk 12, 2 (74) (1957), 43-118 (in Russian).

[442] I. C. Gohberg and A. S. Markus, *Two theorems on the opening of subspaces of Banach space*, Uspehi Mat. Nauk 14, 5 (89) (1959), 135-140 (in Russian).

[443] G. Godefroy, *Some remarks on subdifferential calculus*, Revista matematica Complutense 11 (1998), 269-279.

[444] V. V. Goncharov and G. E. Ivanov, *Strong and weak convexity of closed sets in a Hilbert space*, in N. Daras and T. Rassias (eds) Operations Research, Engineering, and Cyber Security, Springer Optimization and its Applications, vol 113 (2017), 259-297,Springer.

[445] V. V. Goncharov, F. Pereira, *Neighbourhood retractions of nonconvex sets in a Hilbert space via sublinear functionals*, J. Convex Anal. 18 (2011), 1-36.

[446] L. Górniewicz, *Topological Fixed Point Theory for Multivalued Mappings*, Kluwer Academic Publishers, Dordrecht (1999).

[447] A. S. Granero, M. Jiménez-Sevilla, J. P. Moreno, *Intersections of closed balls and geometry of Banach spaces*, Extracta Mathematicae 19 (2004), 55-92.

[448] L. M. Graves, *Some mapping theorems*, Duke Math. J. 17 (1950), 111-114.

[449] J. W. Green, *Approximately convex functions*, Duke Math. J. 19 (1952), 499-504.

[450] S. Guillaume, *Evolution equations governed by the subdifferential of a convex composite function in finite dimensional equations*, Discrete Contin. Dynam. Systems 2 (1996), 23-52.

[451] V. I.Gurariúı, *Openings and inclinations of subspaces of a Banach space*, Teor. Funktsii, Funktsional. Anal. i Prilozhen. 1 (1965), 194-204 (in Russian).

[452] J. Hadamard, *Étude sur les propriétés des fonctions entiȓes et en particulier d'une fonction considérée par Riemann*, J. Math. Pures Appl. 58 (1893), 171-215.

[453] J. Hadamard, *Sur les fonctions entières*, Bull. Soc. Math. Fr. 24 (1896), 186-187.

[454] J. Hadamard, *Sur certaines propriétés des trajectoires en dynamique*, J. Math. Pures Appl. 62 (1897), 331-388.

[455] J. Hadamard, *La notion de différentielle dans l'enseignement*, Scripta Universitatis atque Bibliothecae Hierosolymitanarum (1923); Reprinted in the Mathematical Gazette 19, no. 236 (1935), 341–342.

[456] T. Haddad, A. Jourani and L. Thibault, *Reduction of sweeping process to unconstrained differential inclusion*, Pac. J. Optim. 4 (2008), 493-512.

[457] T. Haddad, J. Noel and L. Thibault, *Perturbed sweeping processes with a subsmooth set depending on the state*, Linear Nonlin. Anal. 2 (2016), 155-174.

[458] T. Haddad and L. Thibault, *Mixed upper semicontinuous perturbation of nonconvex sweeping process*, Math. Program. 123 (2010), 225–240.

[459] N. Hadjisavvas and D. T. Luc, *Second-order asymptotic directions of unbounded sets with application to optimization*, J. Convex Anal. 18 (2011), 181-202.

[460] J. Hagler and W. B. Johnson, *On Banach spaces whose dual balls are not weak* sequentially compact*, Israel J. Math. 28 (1977), 325–330.

[461] J. Hagler and F. Sullivan, *Smoothness and weak* sequential compactness*, Proc. Amer. Math. Soc. 78 (1980), 497-503.

[462] M. Hamamdjiev and M. Ivanov, *New multidirectional mean value inequality*, J. Convex Anal. 25 (2018), 1319-1334.

[463] A. Hantoute and M. A. López, *A new tour on the subdifferential of suprema. Highlighting the relationship between suprema and finite sums*, to appear.

[464] A. Hantoute, M. A. López, and C. Zalinescu, *Subdifferential calculus rules in convex analysis: A unifying approach via pointwise supremum functions*, SIAM J. Optim. 19 (2008), 863-882.

[465] A. Hantoute and A. Svensson, *A general representation of normal sets to sublevels of convex functions*, Set-Valued Var. Anal. 25 (2017), 651-678.

[466] A. Haraux, *How to differentiate the projection on a convex set in Hilbert space. Some applications to variational inequalities*, J. Math. Soc. Japan 29 (1977), 615-631 (submitted: September 1975).

[467] W. L. Hare and A. S. Lewis, *Identifying active constraints via partial smoothness and prox-regularity*, J. Convex Anal. 11 (2004), 251-266.

[468] W. L. Hare and R. A. Poliquin, *Prox-regularity and stability of the proximal mapping*, J. Convex Anal. 14 (2007), 589-606.

[469] W. L. Hare and C. Sagastizábal, *Computing proximal points of nonconvex functions*, Math. Program. 116 (2009), Ser. B, 221-258.

[470] U. S. Haslam-Jones, *Derivate planes and tangent plane of a measurable function*, Quart. J. Math. Oxford Ser. 3 (1932), 120-132.

[471] U. S. Haslam-Jones, *Tangential properties of a plane set of points*, Quart. J. Math. Oxford Ser. 7 (1936), 116-123.

[472] U. S. Haslam-Jones, *The discontinuities of an arbitrary function of two variables*, Quart. J. Math. Oxford Ser. 7 (1936), 184-190.

[473] F. Hausdorff, *Mengenlehre*, Walter de Gruyter, Berlin (1927).

[474] F. Hausdorff, *Set Theory*, Chelsea, English translation of the third (1937) German edition of *Mengenlehre*.

[475] J. Heinonen, *Lectures on Lipschitz Analysis*, Lectures at the 14th Jyväskylä Summer School in August 2004.

[476] R. Henrion, *The approximation subdifferential and parametric optimization*, Habilitation thesis, Humbold University, Berlin (1997).

[477] R. Henrion and A. Jourani, *Subdifferential conditions for calmness of convex constraints*, SIAM J. Optim. 13 (2002), 520-534.

[478] R. Henrion, A. Jourani and J. V. Outrata, *On the calmness of a class of multifunctions*, SIAM J. Optim. 13 (2002), 603-618.

[479] C. Henry, An existence theorem for a class of differential equations with multivalued right-hand side, J. Math. Anal. Appl. 41 (1973), 179-186.

[480] Ch. Hermite, *Sur deux limites d'une intégrale définie*, Mathesis 3 (1983), p. 82.

[481] J.-B. Hiriart-Urruty, *Contributions à la Programmation Mathématique: Cas Déterministe et Stocastique*, Thesis (Thèse de Doctorat d'État ès Sciences Mathhématiques), Université de Clermont-Ferrand (1977).

[482] J.-B. Hiriart-Urruty, *On optimality conditions in non-differentiable programming*, Math. Program. 14 (1978), 73-86.

[483] J.-B. Hiriart-Urruty, *Gradients généralisés de fonctions marginales*, SIAM J. Control Optim. 16 (1978), 301-316.

[484] J.-B. Hiriart-Urruty, *Tangent cones, generalized gradients and mathematical programming in Banach spaces*, Math. Oper. Res. 4 (1979), 79-97 (submitted: May, 1977).

[485] J.-B. Hiriart-Urruty, *New concepts in non differentiable programming*, Bull. Soc. Math. France Mémoire 60 (1979), 57-85.

[486] J.-B. Hiriart-Urruty, *Refinements of necessary optimality conditions in nondifferentiable programming. I*, Appl. Math. Optim. 5 (1979), 63-82.

[487] J.-B. Hiriart-Urruty, *A note on the mean value theorem for convex functions*, Boll. Un. Mat. Ital. B 17 (1980), 765-775.

[488] J.-B. Hiriart-Urruty, *Mean value theorems in nonsmooth analysis*, Numer. Funct. Anal. Optim. 2 (1980), 1-30.

[489] J.-B. Hiriart-Urruty, *Extension of Lipschitz functions*, J. Math. Anal. Appl. 77 (1980), 539-554.

[490] J.-B. Hiriart-Urruty, *Refinements of necessary optimality conditions in nondifferentiable programming. II*, Math. Program. Study 19 (1982), 120-139.

[491] J.-B. Hiriart-Urruty, *A short proof of the variational principle for approximate solutions of a minimization problem*, Amer. Math. Monthly 90 (1983), 206-207.

[492] J.-B. Hiriart-Urruty, *Images of connected sets by semicontinuous multifunctions*, J. Math. Anal. Appl. 111 (1985), 407-422.

[493] J.-B. Hiriart-Urruty, *Ensembles de Tchebychev vs. ensembles convexes: l'état de la situation vu par l'analyse non lisse*, Ann. Sci. Math. Que. 22 (1998), 47-62.

[494] J.-B. Hiriart-Urruty and C. Lemaréchal, *Convex Analysis and Minimization Algorithms. Part 1: Fundamentals*, Grundlehren der Mathematischen Wissenschaften, Springer-Verlag 305, Berlin (1993).

[495] J.-B. Hiriart-Urruty, M. Moussaoui, A. Seeger and M. Volle, *Subdifferential calculus without qualification conditions, using approximate subdifferentials: a survey*, Nonlinear Anal. th. Meth. Appl. 24 (1995), 1727-1754.

[496] J.-B. Hiriart-Urruty and R. R. Phelps, *Subdifferential calculus using ε-subdifferentials*, J. Funct. Analysis 118 (1993), 154-166.

[497] J.-B. Hiriart-Urruty and Ph. Plazanet, *Moreau's theorem revisited*, Ann. Inst. H. Poincaré 6 (1993), 544-555.

[498] J.-B. Hiriart-Urruty and L. Thibault, *Existence et caractérisation de différentielles généralisées d'applications localement lipschitziennes d'un espace de Banach séparable dans un espace de Banach réflexif séparable*, C. R. Acad. Sci. Paris Sér. A-B 290 (1980), 1091-1094.

[499] A. J. Hoffman, *On approximate solutions of systems of linear inequalities*, J. Res. Nat. Bur. Standards B 49 (1952), 263-265.

[500] M. O. Hölder, *Über einen Mittelwertsatz*, Nachr. Ges. Wiss. Göttingen, 1889, 38-47.

[501] R. B. Holmes, *Smoothness of certain metric projections on Hilbert space*, Trans. Amer. Math. Soc. 184 (1973), 87-100.

[502] L. Hörmander, *Sur la fonction d'appui des ensembles convexes dans un espace localement convexe*, Arkiv Math. 3 (1955), 181-186.

[503] M. M. Hrustalev, *Necessary and sufficient optimality conditions in the form of Bellman's equation*, Soviet Math. Dokl. 19 (1978), 1262-1266.

[504] D. H. Hyers, *On the stability of the linear functional equation*, Proc. Nat. Acad. Sei. U.S.A. vol. 27 (1941), 222-224.

[505] D. H. Hyers and S. M. Ulam, *On approximate isometries*, Bull. Amer. Math. Soc. 51 (1945), 288-292.

[506] D. H. Hyers and S. M. Ulam, *Approximate isometries of the space of continuous functions*, Ann. Math. 48 (1947), 285-289.

[507] D. H. Hyers and S. M. Ulam, *Approximately convex functions*, Proc. Amer. Math. Soc. 3 (1952), 821-828.

[508] A. D. Ioffe, *Regular points of Lipschitz functions*, Trans. Amer. Math. Soc. 251 (1979), 61-69.

[509] A. D. Ioffe, *Approximate subdifferentials of nonconvex functions*, CEREMADE Publication 8120 Université Paris IX Dauphine (1981).

[510] A. D. Ioffe, *Nonsmooth analysis: differential calculus and non-differentiable mappings*, Trans. Amer. Math. Soc. 266 (1981), 1-56.

[511] A. D. Ioffe, *On subdifferentiability spaces*, Ann. N. Y. Acad. Sci. 410 (1983), 107-121.

[512] A. D. Ioffe, *Calculus of Dini subdifferentials of functions and contingent derivatives of set-valued maps*, Nonlin. Anal. Th. Meth. Appl. 8 (1984), 517-539.

[513] A. D. Ioffe, *Necessary conditions in nonsmooth optimization*, Math. Oper. Res. 9 (1984), 159-189.

[514] A. D. Ioffe, *Approximate subdifferentials and applications, I: the finite dimensional theory*, Trans. Amer. Math. Soc. 281 (1984), 389-416; submitted: November 20, 1981.

[515] A. D. Ioffe, *Approximate subdifferentials and applications, II: functions on locally convex spaces*, Mathematika 33 (1986), 111-128.

[516] A. D. Ioffe, *On the local surjection property*, Nonlin. Anal. Th. Meth. Appl. 11 (1986), 565-592.

[517] A. D. Ioffe, *Approximate subdifferential and applications, III: the metric theory*, Mathematika 71 (1989), 1-38.

[518] A. D. Ioffe, *Proximal analysis and approximate subdifferentials*, J. London Math. Soc. 41 (1990), 175-192.

[519] A. D. Ioffe, *Fuzzy principles and characterization of trustworthiness*, Set-Valued Anal. 6 (1998), 265-276.

[520] A. D. Ioffe, *Codirectional compactness, metric regularity and subdifferential calculus*, Amer. Math. Soc., Providence, RI 2000.

[521] A. D. Ioffe, *Metric regularity and subdifferential calculus*, Russian Math. Surveys 55 (2000), 501-558.

[522] A. D. Ioffe, *Three theorems on subdifferentiation of convex integral functionals*, J. Convex Anal., 13 (2006), 759-772.

[523] A. D. Ioffe, *Typical convexity (concavity) of Dini-Hadamard upper (lower) directional derivatives of functions on separable Banach spaces*, J. Convex Anal. 17 (2010), 1019-1032.

[524] A. D. Ioffe, *Separable reduction revisited*, Optimization 60 (2011), 211-221.

[525] A. D. Ioffe, *On the general theory of subdifferentials*, Adv. Nonlinear Anal. 1 (2012), 47-120.

[526] A. D. Ioffe, *Variational Analysis of Regular Mappings: Theory and Applications*, Springer (2017).

[527] A. D. Ioffe, and J. V. Outrata, *On metric and calmness qualification conditions in subdifferential calculus*, Set-Valued Anal. 16 (2008), 199-227.

[528] A. D. Ioffe and Y. Sekiguchi, *Regularity estimates for convex multifunctions*, Math. Program. 117 (2009), 255–270 (submitted: November 2005).

[529] A. D. Ioffe and V. M. Tikhomirov, *Theory of Extremal Problems*, Nauka, Moscow (1974); English transl., Studies in Mathematics and its Applications, no. 6, North-Holland, Amsterdam, New York (1979).

[530] A. Iusem and A. Seeger, *Distances between closed convex cones: old and new results*, 17 (2010), 1033-1055.

[531] G. E. Ivanov, *Weak convexity in the sense of Vial and Efimov-Stechkin*, Izv. Ross. Akad. Nauk Ser. Mat. 69 (2005), 35-60 (in Russian); English tanslation in Izv. Math. (2005), 1113-1135.

[532] G. E. Ivanov, *Weakly convex sets and their properties*, Mat. Zametki 79 (2006), 60–86.

[533] G. E. Ivanov, *Weakly Convex Sets and Functions: Theory and Applications*, Fizmatlit, Moscow (2006) (in Russian).

[534] G. E. Ivanov, *On well posed best approximation problems for a nonsymmetric seminorm*, J. Convex Anal. 20 (2013), 501-529.

[535] G. E. Ivanov, *Weak convexity of sets and functions in Banach spaces*, J. Convex Anal. 22 (2015), 365-398.

[536] G. E. Ivanov, *Continuity and selections of the intersection operator applied to nonconvex sets*, J. Convex Anal. 22 (2015), 939-962.

[537] G. E. Ivanov, *Sharp estimates for the moduli of continuity of metric projections onto weakly convex sets*, Izvestiya Math. 79 (2015), 668-697.

[538] G. E. Ivanov, *Nonlinear images of sets. I: strong and weak convexity*, J. Convex Anal. 27 (2020), 361-380.

[539] G. E. Ivanov and M. V. Balashov, *Lipschitz parametrizations of set-valued maps with weakly convex images*, Izv. Ross. Akad. Nauk Ser. Mat. 71 (2007), 47-68 (in Russian); English translation in Izv. Math. 71 (2007), 1123-1143.

[540] G. E. Ivanov and L. Thibault, *Infimal convolution and optimal time control problem I: Fréchet and proximal subdifferentials*, Set-Valued Var. Anal. 26 (2018), 581-606.

[541] G. E. Ivanov and L. Thibault, *Infimal convolution and optimal time control problem II: Limiting subdifferential*, Set-Valued Var. Anal. 25 (2017), 517-542.

[542] G. E. Ivanov and L. Thibault, *Infimal convolution and optimal time control problem III: Minimal time projection set*, SIAM J. Optim. 28 (2018), 30-44.

[543] G. E. Ivanov and L. Thibault, *Well-posedness and subdifferentials of optimal value and infimal convolution*, Ser-Valued Var. Anal.

[544] M. Ivanov and N. Zlateva, *A Clarke-Ledyaev type inequality for certain nonconvex sets*, Serdica Math. J; 26 (2000), 277-286.

[545] M. Ivanov and N. Zlateva, *On primal lower nice property*, C.R. Acad. Bulgare Sci. 54 (2001), 5-10.

[546] M. Ivanov and N. Zlateva, *On nonconvex version of the inequality of Clarke and Ledyaev*, Nonlinear Anal. 49 (2002), 1023-1036.

[547] M. Ivanov and N. Zlateva, *Subdifferential characterization of primal lower nice functions on smooth spaces*, C.R. Acad. Bulgare Sci. 57 (2004), 13-18.

[548] M. Ivanov and N. Zlateva, *A new proof of the integrability of the subdifferential of a convex function on a Banach space*, Proc. Amer. Math. Soc. 136 (2008), 1787-1793.

[549] M. Ivanov and N. Zlateva, *On characterizations of metric regularity of multi-valued maps*, J. Convex Anal. 27 (2020), 381-388.

[550] A. J. Izzo, *Locally uniformly continuous functions*, Proc. Amer. Math. Soc. 122 (1994), 1095-1100.

[551] R. C. James, *Characterizations of reflexivity*, Studia Math. 23 (1963/1964), 205-216.
[552] R. C. James, *Weakly compact sets*, Trans. Amer. Math. Soc. 113 (1964), 129-140.
[553] R. C. James, *Some self-dual properties of normed linear spaces*, Symposium on Infinite Dimensional Topology, Annals of Mathematics Studies 69 (1972), 159-175.
[554] R. C. James, *Super-reflexive Banach spaces*, Can. J. Math. 24 (1972), 896-904.
[555] R. Janin, *Sur une classe de fonctions sous-linéarisables*, C. R. Acad. Sci. Paris 277 (1973), 265-267.
[556] R. Janin, *Sur la Dualité et la Sensibilité dans les Problèmes de Programmes Mathématiques*, Thesis (Thèse de Doctorat d'État ès Sciences Mathématiques), Université Paris VI, (1974).
[557] R. Janin, *Sur des multi-applications qui sont des gradients généralisés*, C. R. Acad. Sci. Paris 294 (1982), 115-117.
[558] S. Janiszewski, *Sur les continus irreductibles entre deux points*, Journal de l'Ecole Polytechnique 2e série - 16e Cahier (1912), pp. 79-170; (and Thèse de Doctorat ès Scienses Mathématiques, Universitè de Paris (1911)).
[559] J. L. W. V. Jensen *Om konvexe Funktioner og Uligheder mellem Middelvaerdier*, Nyt. Tidsskrift for Mathematik 16 B (1905), 49-69.
[560] J. L. W. V. Jensen *Sur les fonctions convexes et les inégalités entre les valeurs moyennes*, Acta Math. 30 (1906), 175-193.
[561] W. L. Jones, *On conjugate functionals*, Ph.D. dissertation thesis, Columbia University (1960).
[562] M. Jouak and L. Thibault, *Equicontinuity of families of convex and concave-convex operators*, Canad. J. Math. 36 (1984), 883-898.
[563] M. Jouak and L. Thibault, *Directional derivatives and almost everywhere differantiability of biconvex and concave-conve operators*, Math. Scand. 57 (1985), 215-224.
[564] A. Jofre and L. Thibault, *D-representation of subdifferentials of directionally Lipschitz functions*, Proc. Amer. Math. Soc. 110 (1990), 117-123.
[565] E. Jouini, *A remark on Clarke's normal cone and the marginal cost pricing rule*, J. Math. Econ. 18 (1989), 95-101.
[566] E. Jouini, *Functions with constant generalized gradients*, J. Math. Anal. Appl. 148 (1990), 121-130.
[567] E. Jouini, W. Schachermayer, and N. Touzi, *Law invariant risk measures have the Fatou property*, Adv. Math. Econ. 9 (2006), 49-71.
[568] A. Jourani, *On metric regularity of multifunctions*, Bull. Austral. Math. Soc. 44 (1991), 1-9.
[569] A. Jourani, *Regularity and strong sufficient optimality conditions in differentiable optimization problems*, Numer. Funct. Anal. optim. 14 (1993), 69-87.
[570] A. Jourani, *Compactly epi-Lipschitzian sets and A-subdifferentials in WT-spaces*, Optimization, 34 (1995), 1-17.
[571] A. Jourani, *Qualification conditions for multivalued functions in Banach spaces with applications to nonsmooth vector optimization problems*, Math. Program. 66 (1994), 1-23.
[572] A. Jourani, *Open mapping theorem and inversion theorem for γ-para-convex multivalued mappings and applications*, Studia Math. 117 (1996), 123-136.
[573] A. Jourani, *Subdifferentiability and subdifferential monotonicity of γ-para-convex functions*, Control Cybernetics, 25 (1996), 721-737.
[574] A. Jourani, *The role of locally compact cones in nonsmooth analysis*, Communications Appl. Nonlinear Anal. 5 (1998), 1-35.
[575] A. Jourani, *Variational sum of subdifferentials of convex functions*, Proceedings of the IV Catalan Days of Applied Mathematics, Tarragona, (1998), 71-79.
[576] A. Jourani, *Limits superior of subdifferentials of uniformly convergent functions in Banach spaces*, Positivity 3 (1999), 33-47.
[577] A. Jourani, *Weak regularity of functions and sets in Asplund spaces*, Nonlinear Analysis 45 (2006), 660-676.
[578] A. Jourani, *Radiality and semismoothness*, Control Cybernetics, 36 (2007), 669-680.
[579] A. Jourani and M. Sene, *Characterization of the Clarke regularity of subanalytic sets*, Proc. Amer. Math. Soc. 146 (2018), 1639-1649.
[580] A. Jourani and M. Sene, *Geometric characterizations of the strict Hadamard differentiability of sets*, Pure Appl. Funct. Anal., to appear.
[581] A. Jourani and L. Thibault, *Approximate subdifferential and metric regularity: the finite-dimensional case*, Math. Program. 47 (1990), 203-218.

[582] A. Jourani and L. Thibault, *Approximate subdifferentials of composite functions*, Bull. Aust. Math. Soc. 47 (1993), 443-455.

[583] A. Jourani and L. Thibault, *A note on approximate subdifferentials of composite functions*, Bull. Aust. Math. Soc. 49 (1994), 111-116.

[584] A. Jourani and L. Thibault, *Metric regularity and subdifferential calculus in Banach spaces*, Set-Valued Anal. 3 (1995), 87-100.

[585] A. Jourani and L. Thibault, *Metric regularity for strongly compactly Lipschitzian mappings*, Nonlinear Analysis 24 (1995), 229-240.

[586] A. Jourani and L. Thibault, *Verifiable conditions for openness and regularity of multivalued mappings*, Trans. Amer. Math. Soc. 347 (1995), 1255-1268.

[587] A. Jourani and L. Thibault, *Extensions of subdifferentials calculus rules in Banach spaces and applications*, Canadian J. Math. 48 (1996), 834-848.

[588] A. Jourani and L. Thibault, *Coderivatives of multivalued mappings, locally compact cones and metric regularity*, Nonlinear Anal. 35 (1999), 925-945.

[589] A. Jourani, L. Thibault and D. Zagrodny, $C^{1,\omega(\cdot)}$*-regularity and Lipschitz-like properties of subdifferential*, Proc. London Math. Soc. 105 (2012), 189-223.

[590] A. Jourani, L. Thibault and D. Zagrodny, *Differential properties of the Moreau envelope*, J. Funct. Anal. 266 (2014), 1185-1237.

[591] A. Jourani and E. Vilches, *Positively α-far sets and existence results for generalized perturbed sweeping processes*, J. Convex Anal. 23 (2016), 775-821.

[592] A. Jourani and T. Zakaryan, *The validity of the liminf formula and a characterization of Asplund spaces*, J. Math. Anal. Appl. 416 (2014), 824-838.

[593] F. Jules, *Sur la somme de sous-différentiels de fonctions semi-continues inférieurement*, Dissertationes Mathematicae 423 (2003), 62 pp.

[594] F. Jules and M. Lassonde, *Formulas for subdifferentials of sums of convex functions*, J. Convex Anal. 9 (2002), 519-533.

[595] F. Jules and M. Lassonde, *Subdifferential test for optimality*, J. Global Optim. 59 (2014), 101–106.

[596] W. Karush, *A queuing model for an inventory problem*, Operations Res. 5 (1957), 693-703.

[597] W. Karush, *A general algorithm f or the optimal distribution of effort*, Management Science 9 (1962), 50-72.

[598] T. Kato, *Perturbation theory for nullity, deficiency and other quantities of linear operators*, J. Analyse Math. 6 (1958), 261-322.

[599] T. Kato, *Perturbation Theory for Linear Operators*, Vol. 132 Grundlehren der mathematischen Wissenschaften, Springer-Verlag, Berlin, (2006).

[600] G. Katriel, *Are the approximate and the Ckarke subgradients generically equal?*, J. Math. Anal. Appl. 193 (1995), 588-593.

[601] I. Kecis and L. Thibault, *Subdifferential characterization of s-lower regular functions*, Appl. Anal. 94 (2015), 85-98.

[602] I. Kecis and L. Thibault, *Moreau envelopes of s-lower regular functions*, Nonlinear Anal. 127 (2015), 157-181.

[603] I. Kecis and L. Thibault, *Evolution differential inclusions associated to primal lower regular functions*, J. Nonlin. Convex Anal. 20 (2019), 1949-1979.

[604] J.L. Kelley, *General Topology*, D. Van Nostran Company, Princeton, New Jersey (1961).

[605] M.D. Kirszbraun, *Über die zusammenziehenden und Lipschitzchen tranformationen*, Fund. Math. 22 (1934), 77-108.

[606] T.H. Kjeldsen, *A Contextualized historical analysis of the Kuhn-Tucker theorem in nonlinear programming: the Impact of world war II*, Historia Mathematica 27 (2000), 331-361.

[607] D. Klatte and B. Kummer, *Nonsmooth Equations in Optimization: Regularity, Calculus, Methods, and Applications*, Kluwer, Boston, Massachusetts (2002)

[608] D. Klatte and B. Kummer, *Stability of inclusions: characterizations via suitable Lipschitz functions and algorithms*, Optimization 55(5-6) (2006), 627-660.

[609] V. Klee, *Convexity of Chebyshev sets*, Math. Annalen 142 (1961), 292-304.

[610] V. Klee, *On a question of Bishop and Phelps*, Amer. J. Math. 85 (1963), 95-98.

[611] A. Kolmogoroff and J. Verčenko, *Ueber Unstetigkeitspunkte von Funktionen zweier Veränderlichen*, C. R. Acad. Sci. URSS 1 (1934), 105-107.

[612] A. Kolmogoroff and J. Verčenko, *Weitere Untersuchungen über Unstetigkeitspunkte von Funktionen zweier Veränderlichen*, C. R. Acad. Sci. URSS 4 (1934), 361-364.

[613] S. V. Konjagin, *On approximation properties of closed sets in Banach spaces and the characterization of strongly convex spaces*, Soviet Math. Dokl. 21 (1980), 418-422.

[614] M. Konstantinov and N. Zlateva, *Epsilon subdifferential method and integrability*, J. Convex Anal. (2022), to appear.

[615] G. Köthe, *Topological Vector Spaces I*, Grundlehren der mathematischen Wissenschaften, Springer-Verlag, Berlin (1969) (English translation).

[616] S. G. Krantz and H. R. Parks, *Distance to $\mathcal{C}^k$ hypersurfaces*, J. Differential Equations, 40 (1981), 116-120.

[617] M. G. Krein and M. A. Krasnoselskiĭ, *Fundamental theorems concerning the extension of Hermitian operators and some of their applications to the theory of orthogonal polynomials and the moment problem*, Uspekhi Mat. Nauk. 2, 3 (1947), 60-106 (in Russian).

[618] M. G. Krein, M. A. Krasnoselskiĭ, and D. P. Milman, *Concerning the deficiency numbers of linear operators in Banach space and some geometric questions*, Sbornik Trudov Inst. A. N. Ukr. S. S. R., 11 (1948) (in Russian).

[619] A. Y. Kruger, *Epsilon-semidifferentials and epsilon-normal elements*, Deposited in VINITI no 1331-81, Minsk, 1981 (in Russian).

[620] A. Y. Kruger, *Generalized differentials of nonsmooth functions*, Deposited in VINITI no. 1332-81, Minsk, 1981 (in Russian).

[621] A. Y. Kruger, *Generalized differentials of nonsmooth functions and necessary conditions for an extremum*, Siberian Math. J. 26 (1985), 370-379.

[622] A. Y. Kruger, *Properties of generalized differentials*, Siberian Math. J. 26 (1985), 822-832.

[623] A. Y. Kruger, *A covering theorem for set-valued mappings*, Optimization 19 (1988), 763-780.

[624] A. Y. Kruger, *Strict ε-semidifferentials and extremality conditions*, Dokl. Nat. Akad. Nauk Belarus 41 (1997), 21-26 (in Russian).

[625] A. Y. Kruger, *On extremality of sets systems*, Dokl. Nat. Akad. Nauk Belarus 42 (1998), 24-28 (in Russian).

[626] A. Y. Kruger, *Strict (ε, δ)-semidifferentials and extremality of sets and functions*, Dokl. Nat. Akad. Nauk Belarus 44 (2000), 21-24 (in Russian).

[627] A. Y. Kruger, *Strict (ε, δ)-semidifferentials and extremality conditions*, Optimization 51 (2002), 539-554.

[628] A. Y. Kruger, *On Fréchet subdifferentials*, J. Math. Sci. 116 (2003), 3325-3358.

[629] A. Y. Kruger, *Weak stationarity: eliminating the gap between necessary and sufficient conditions*, Optimization 53(2) (2004), 147-164.

[630] A. Y. Kruger, *Stationarity and regularity of set systems*, Pac. J. Optim. 1(1) (2005), 101-126.

[631] A. Y. Kruger, *About regularity of collections of sets*, Set-Valued Anal. 14(2) (2006), 187-206.

[632] A. Y. Kruger, *Stationarity and regularity of real-valued functions*, Appl. Comput. Math. 5(1) (2006), 79-93.

[633] A. Y. Kruger, *About stationarity and regularity in variational analysis*, Taiwanese J. Math. 13 (2009), 1737-1785.

[634] A. Y. Kruger and B. S. Mordukhovich, *Extremal points and the Euler equation in nonsmooth problems of optimization*, Dokl. Akad. Nauk BSSR 24 (1980), 684-687 (in Russian).

[635] A. Y. Kruger and B. S. Mordukhovich, *Generalized normals and derivatives and necessary conditions for an extremum in problems of nondifferentiable programming*, Parts I and II, Deposited in VINITI, I - no. 408-80, II - no. 494-80, Minsk, 1980 (in Russian).

[636] S. N. Kruzhkov, *The Cauchy problem in the large for certain nonlinear first order differential equations*, Soviet. Math. Dokl. 1 (1960), 474-477.

[637] W. Kryszewski and S. Plaskacz, *Periodic solutions to impulsive differential inclusions with constraints*, Nonlin. Anal. Th. Meth. Apl. 65 (2006), 1974-1804.

[638] B. Kummer, *Metric regularity: Characterizations, nonsmooth variations and successive approximation*, Optimization 46 (1999), 247-281.

[639] B. Kummer, *Inverse functions of pseudo regular mappings and regularity conditions*, Math. Program., Ser. B 88, 2 (2000), 313-339.

[640] C. Kuratowski, *Les fonctions semi-continues dans l'espace des ensembles fermés*, Fund. Math. 18 (1932), 148-159; submitted in 1931.

[641] C. Kuratowski, *Topologie I*, Monografje Matematyczne (1933).

[642] C. Kuratowski, *Topologie II*, Monografje Matematyczne, (1933).

[643] A. G. Kusraev and S. S. Kutateladze, *Subdifferentials: Theory and Applications*, Mathematics and its Applications Vol. 323, Kluwer Academic Publishers, Dordrecht (1995).

[644] M. Laczkovich, *The local stability of convexity, affinity and of the Jensen equation*, Aequationes Math. 58/1-2 (1999), 135-142.
[645] A. Largillier, *A note on the gap convergence*, Appl. Math. Lett. 7 (1994), 67–71.
[646] M. Lassonde, *First-order rules for nonsmooth constrained optimization*, Nonlinear Anal. 44 (2001), 1031-1056.
[647] K.-S. Lau, *Farthest points in weakly compact sets*, Israel J. Math. 22 (1975), 168-174.
[648] K.-S. Lau, *Almost Chebyshev subsets in reflexive Banach spaces*, Indiana Univ. Math. J. 27 (1978), 791-795.
[649] P.-J. Laurent, *Approximation et Optimisation*, Hermann (1972).
[650] G. Lebourg, *Valeur moyenne pour gradients généralisés*, C. R. Acad. Sci. Paris Sér. A 281 (1975), 795-798.
[651] G. Lebourg, *Solutions en densité de problèmes d'optimisation paramétrés*, C. R. Acad. Sci. PAris Sér. A 289 (1979), 79-82.
[652] G. Lebourg, *Generic differentiability of Lipschitzian functions*, Trans. Amer. Math. Soc. 256 (1979), 123-144.
[653] A.-M. Legendre, *Mémoire sur l'intégration de quelques équations aux différences partielles*, Histoire de l'Académie royale des sciences (1787), 309-351.
[654] E. S. Levitin and B. T. Polyak, *Constrained minimization methods*, U.S.S.R. Computational Math. and Math. Phys. 6 (1966), 787-823.
[655] A. B. Levy, R. A. Poliquin and R. T. Rockafellar, *Stability of locally optimal solutions*, SIAM J. Optim. 10 (2000), 580-604.
[656] A. B. Levy, R. A. Poliquin and L. Thibault, *A partial extension of Attouch's theorem and its applications to second-order differentiation*, Trans. Amer. Math. Soc. 347 (1995), 1269-1294.
[657] A. B. Levy, R. A. Poliquin, R. T. Rockafellar, *Stability of locally optimal solutions*, SIAM J. Optim. 10 (2000), 580-604.
[658] A. S. Lewis and R. Lucchetti, *Nonsmooth duality, sandwich and squeeze theorems*, (Preprint, Universtà degli studi Milano 1998), SIAM J. Control Optim. 38 (2000), 613-626.
[659] A. S. Lewis, D. R. Luke and J. Malick, *Local linear convergence for alternating and averaged nonconvex projections*, Found. Comput. Math. 9 (2009), 485-513.
[660] A. S. Lewis and A. D. Ralph, *A nonlinear duality result equivalent to the Clarke-Ledyaev mean value inequality*, Nonlinear Anal. Th. Meth. Appl. 26 (1996), 343-350.
[661] A. S. Lewis and S. Zhang, *Partial smoothness, tilt stability, and generalized Hessians*, SIAM J. Optim. 23 (2013), 74-94.
[662] G. Li and B. S. Mordukhovich, *Hölder metric subregularity with applications to proximal point method*, SIAM J. Optim. 22 (2012), 1655-1684.
[663] J. Lindenstrauss, *On the modulus of smoothness and divergent series in Banach spaces*, Michigan Math. J. 10 (1963), 241-252.
[664] J. Lindenstrauss, *On operators which attain their norm*, Israel J. Math. 1 (1963), 139-148.
[665] J. Lindenstaruss, L. Tzafiri, *Classical Banach Spaces I. Sequence Spaces*, Springer, Berlin (1977).
[666] J. Lindenstaruss, L. Tzafiri, *Classical Banach Spaces II. Function Spaces*, Springer, Berlin (1979).
[667] V. I. Liokkoumovich, *The existence of B-spaces with non-convex modulus of convexity*, Izv. Vysš. Učebn. Zadev. Matematika 12 (1973), 13-50 (in Russian).
[668] P. D. Loewen, *The proximal normal formula in Hilbert space*, Nonlin. Anal. Th. Meth. AppL. 11 (1987), 979-995 (submitted: January 1986).
[669] P. D. Loewen, *The proximal subgradient formula in Banach spaces*, Canad. Math. Bull. 31 (1988), 353-361.
[670] P. D. Loewen, *Limit of Fréchet normals in nonsmooth analysis*, Optimization and Nonlinear Analysis (A.D. Ioffe, L. Marcus and S. Reich editors), Pitman Research Notes Math. Ser; 244 (1992), 178-188.
[671] P. D. Loewen, *Optimal Control via Nonsmooth Analysis*, Amer. Math. Soc., Providence, Rhode Island (1993).
[672] P. D. Loewen, *A mean value theorem for Fréchet subgradients*, Nonlin. Anal. Th. Meth. Appl. 23 (1994), 1365-1381.
[673] P. D. Loewen and R. T. Rockafellar, *Optimal control of unbounded differential inclusions*, SIAM J. Control Optim. 32 (1994), 442-470.

[674] P. D. Loewen and X. Wang, *A generalized variational principle*, Canad. J. Math. 53 (2001), 1174-1193.

[675] O. Lopez and L. Thibault, *Sequential formula for subdifferential of integral sum of convex functions*, J. Nonlinear Convex Anal., 9 (2008), 295-308.

[676] R. A. Lovaglia, *Locally uniformly convex Banach spaces*, Trans. Amer. Math. Soc. 78 (1955), 225-238.

[677] R. Lozano, B. Brogliato, O. Egeland and B. M. Maschke, *Dissipative Systems Analysis and Control*, Springer London, CCE Series (2000).

[678] Dinh The Luc, *Theory of Vector Optimization*, Lect. Notes Econ. Math. Sci. 319, Springer, Berlin (1989).

[679] Dinh The Luc, *A strong mean value theorem and applications*, Nonlinear Anal. 26 (1996), 915-923.

[680] Dinh The Luc and S. Swaminathan, *A characterization of convex functions*, Nonlinear Anal. Th. Meth. Appl. 20 (1993), 697-701.

[681] K.R. Lucas, *Submanifolds of dimension $n-1$ in $\mathcal{E}^n$ with normals satisfying a Lipschitz condition*, Studies In Eigenvalue Problems, Technical Report 18, Kansas University, Dept. of Mathematics, 1957. http://hdl.handle.net/2027/mdp.39015017419931

[682] L. A. Lyusternik, *On conditional extrema of functionals*, Mat. Sb. 41 (1934), 390-401.

[683] J. Makó and Z. Páles, *Strengthening of strong and approximate convexity*, Acta Math. Hung. 132 (2011), 78-91.

[684] J. Malý, *A simple proof of the Stepanov theorem on differentiabilty almost everywhere*, Exposition. Math. 17 (1999), 59-61.

[685] J. Malý and L. Zajíček, *On Stepanov type of differentiability theorems*, Acta Mathemaica Hungarica 145 (2015), 174–190.

[686] S. Mandelbrojt, *Sur les fonctions convexes*, C. R. Acad. Sci. Paris, 209 (1939), 977-978.

[687] P. Mankiewicz, *On the differentiabiliy of Lipschitz mappings in Fréchet spaces*, Studia Math. 45 (1973), 15-29.

[688] S. Marcellin and L. Thibault, *Integration of ε-subdifferential and maximal cyclic monotonicity*, J. Global Optim. 32 (2005), 83-91.

[689] S. Marcellin and L. Thibault, *Evolution problems associated with primal lower nice functions*, J. Convex Anal. 13 (2006), 385-421.

[690] J. Marcinkiewicz, *Sur les nombres dérivés*, Fundam. Math. 24 (1935), 305-308.

[691] J. Marcinkiewicz, *Sur les séries de Fourier*, Fundam. Math. 27 (1936), 38-69.

[692] J. Marcinkiewicz and A. Zygmund, *On the differentiability of functions and summability of trigonometrical series*, Fundam. Math. 26 (1936), 1-43.

[693] A. Marigonda, *Differentiability Properties for a Class of Non-Lipschitz Functions and Applications*, Ph.D. dissertation thesis, University of Padova, 2006.

[694] M. Marques-Alves and B. F. Svaiter, *A new proof for maximal monotonicity of subdifferential operators*, J. Convex Anal. 15 (2008), 345-348.

[695] P. Mattila, *Geometry of Sets and Measures in Euclidean Spaces*, Cambridge studies in advanced mathematics, Cambridge University Press (1995).

[696] B. Maury and J. Venel, *A mathematical framework for a crowd motion model*, C. R. Math. Acad. Sci. Paris 346 (2008), no. 23-24, 1245-1250.

[697] M. Mazade and L. Thibault, *Differential variational inequalities with locally prox regular sets*, J. Convex Anal. 19 (2012), 1109–1139.

[698] M. Mazade and L. Thibault, *Regularization of differential variational inequalities with locally prox-regular sets*, Math. Program. Ser. B 139 (2013), 243-269.

[699] M. Mazade and L. Thibault, *Primal lower nice functions and their Moreau envelopes*, In: Computational and Analytical Mathematics, New York (NY): Springer Proceedings in Mathematics Statistics (2013), pp. 521-553.

[700] S. Mazur, *Über konvexe Mengen in linearen normierten Räumen*, Studia Math. 4 (1933), 70-84.

[701] S. Mazur, *Über schwache Konvergentz in den Räumen Lp*, Studia Math. 4 (1933), 128-133.

[702] S. Mazurkiewicz, *Sur les nombres dérivés*, Fund. Math. 23 (1934), 9-10.

[703] E. J. Mc Shane, *Extension of range of functions*, Bull. Amer. Math. Soc. 40 (1934), 837-842 (submitted, June 1934).

[704] R. E. Megginson, *An Introduction to Banach Space Theory*, Graduate Texts in Mathematics, Springer, New York (1998).

[705] E. Michael, *An existence theorem for continuous functions*, Bull. Amer. Math. Soc. 59 (1953), p. 180.

[706] E. Michael, *Continuous selections. I*, Ann. Math. 63 (1956), 361-382 (submitted: December 1954).

[707] E. Michael, *Selected selection theorems*, Amer. Math. Monthly 63 (1956), 233-238.

[708] P. Michel and J.-P. Penot, *A generalized derivative for calm and stable functions*, Diff. Int. Equations 5 (1992), 433-454.

[709] R. Mifflin, *Semismooth and semiconvex functions in constrained optimization*, SIAM J. Control Optim. 15

[710] F. Mignot, *Contrôle dans les inèquations variationelles elliptiques*, J. Funct. Anal. 22 (1976), 130-185.

[711] D. Milman, *On some criteria for the regularity of spaces of the type (B)*, Comptes Rendus (Doklady) de l'Académie des Sciences de l'URSS, new series, vol. 20 (1938), 243-246.

[712] G. J. Minty, *On the monotonicity of the gradient of a convex function*, Pacific J. Math. 14 (1964), 243-247.

[713] G. J. Minty, *On the extension of Lipschitz, Lipschitz Hölder continuous, and monotone functions*, Bull. Amer. Math. Soc. 76 (1970), 334-339.

[714] J. Mirguet, *Nouvelles Recherches sur les Notions Infinitésimales Directes du Premier Ordre*, Thesis (Thèse de Doctorat ès Sciences Mathématiques) Université de Paris (1934).

[715] D. S. Mitrinović and I. B. Lacković, *Hermite and convexity*, Aequationes Mathematicae, 28 (1985), 229-232.

[716] M. D. P. Monteiro Marques, *Differential Inclusions in Nonsmooth Mechanical Problems–Shocks and Dry Friction*, Birkhäuser, Boston (1993).

[717] V. Montesinos, *Drop property equals reflexivity*, Studia Math. 87 (1987), 93-100.

[718] W. B. Moors, *Weak compactness of sublevel sets*, Proc. Amer. Math. Soc. 145 (2017), 3377-3379.

[719] W. B. Moors, *On a one-sided James' theorem*, J. Math. Anal. Appl. 449 (2017), 528-530.

[720] B. S. Mordukhovich, *Maximum principle in the problem of time optimal response with nonsmooth constraints*, J. Appl. Math. Mech. 40 (1976), 960-969.

[721] B. S. Mordukhovich, *Metric approximation and necessary optimality conditions for general classes of nonsmooth extremal problems*, Soviet Math. Dokl. 22 (1980), 526-530.

[722] B. S. Mordukhovich, *Nonsmooth analysis and nonconvex generalized differentials and adjoint mappings*, Dokl. Akad. Nauk BSSR 28 (1984), 976-979 (in Russian).

[723] B. S. Mordukhovich, *Approximation methods in problems of optimization and control*, Nauka, Moscow (1988) (in Russian).

[724] B. S. Mordukhovich, *Complete characterization of openness, metric regularity, and Lipschitzian properties of multifunctions*, Trans. Amer. Math. Soc. 340 (1993), 1-36.

[725] B. S. Mordukhovich, *Variational Analysis and Generalized Differentiation, I: Basic Theory*, Vol. 330 Grundlehren der mathematischen Wissenschaften, Springer-Verlag, Berlin, (2006).

[726] B. S. Mordukhovich, *Variational Analysis and Generalized Differentiation, II: Applications*, Vol. 331 Grundlehren der mathematischen Wissenschaften, Springer-Verlag, Berlin, (2006).

[727] B. S. Mordukhovich and N. M. Nam, *Subgradient of distance functions and applications to Lipschitzian stability*, Math. Program. Ser. B 104 (2005), 635-668.

[728] B. S. Mordukhovich and T. T. Nghia, *Second-order variational analysis and characterizations of tilt-stable optimal solutions in finite and infinite dimensions*, Nonlinear Anal. 86 (2013), 159-180.

[729] B. S. Mordukhovich and Y. Shao, *Differential characterizations of covering, metric regularity, and Lipschitzian properties of multifunctions between Banach spaces*, Nonlinear Anal. 25 (1995), 1401-1424.

[730] B. S. Mordukhovich and Y. Shao, *Nonsmooth sequential analysis in Asplund spaces*, Trans. Amer. Math. soc. 124 (1996), 1235-1280.

[731] B. S. Mordukhovich and Y. Shao, *Nonconvex differential calculus for infinite dimensional multifunctions*, Set-Valued Anal. 4 (1996), 205-256.

[732] B. S. Mordukhovich and Y. Shao, *Stability of set-valued mappings in infinite dimensions: point criteria and apllications*, SIAM J. Control Optim. 35 (1997), 285-314.

[733] J. J. Moreau, *Fonctions convexes en dualité*, (multigraphié), Séminaires de Mathématiques, Faculté des Sciences de Monpellier (1962), 18 pages.

[734] J. J. Moreau, *Fonctions convexes duales et points proximaux dans un espace hilbertien*, C.R. Acad. Sci. Paris 255 (1962), 2897-2899.

[735] J. J. Moreau, *Inf-concolution*, (multigraphié), Séminaires de Mathématiques, Faculté des Sciences de Montpellier (1963), 48 pages.

[736] J. J. Moreau, *Propriétés des applications 'prox'*, C.R. Acad. Sci. Paris, 256 (1963), 1069-1071.

[737] J. J. Moreau, *Inf-convolution des fonctions numériques sur un espace vectoriel*, C.R. Acad. Sci. Paris, 256 (1963), 5047-5049.

[738] J. J. Moreau, *Etude locale d'une fonctionnelle convexe*, (multigrapié) Séminaires de Mathématique, Faculté des Sciences de Montpellier (1963), 25 pages.

[739] J. J. Moreau, *Fonctionnelles sous-différentiables*, C. R. Acad. Sci. Paris 257 (1963), 4117-4119.

[740] J. J. Moreau, *Sur la fonction polaire d'une fonction semicontinue supérieurement*, C. R. Acad. Sci. Paris 258 (1964), 1128-1131.

[741] J. J. Moreau, *Proximité et dualité dans un espace hilbertien*, Bull. Soc. Math. France 93 (1965), 273-299.

[742] J. J. Moreau, *Semi-continuité du sous-gradient d'une fonctionnelle*, C. R. Acad. Sci. Paris 260 (1965), 1067-1070.

[743] J. J. Moreau, *Quadratic programming in mechanics: dynamics of one-sided constraints*, SIAM J. Control 4 (1966), 153-158.

[744] J. J. Moreau, *Fonctionnelles Convexes*, Lecture Notes, Collège de France (1966-1967); Second edition by Facoltà di ingegneria, Università di Roma "Tor Vergata" (2003).

[745] J. J. Moreau, *Inf-convolution, sous-additivité, convexité des fonctions numériques*, J. Math. Pures Appl. 49 (1970), 109-154.

[746] J. J. Moreau, *Rafle par un convexe variable I.* Sém. Anal. Convexe Montpellier, Exposé 15, (1971).

[747] J. J. Moreau, *Rafle par un convexe variable II.* Sém. Anal. Convexe Montpellier, Exposé 3, (1972).

[748] J. J. Moreau, *On unilateral constraints, friction and plasticity*, Corso tenuto a Bressanone dal 17 al 26 giugno 1973, Centro Internazionale Matematico Estivo (1973) p. 172-322.

[749] J. J. Moreau, *Multi-applications à rétraction finie*, Ann. Scuola Norm. Sup. Pisa 1 (1974), 169-203.

[750] J. J. Moreau, *On unilateral constraints, friction and plasticity*, in New Variational Techniques in Mathematical Physics, Proceedings from CIME (Capriz and Stampacchia eds.), Cremonese, Rome (1974), 173-322.

[751] J. J. Moreau, *Evolution problem associated with a moving convex set in a Hilbert space*, J. Differential Equations 26 (1977), 347-374.

[752] U. Mosco, *Convergence of convex sets and of solutions of variational inequalities*, Advances Math. 3 (1969), 510-585.

[753] T. S. Motzkin, *Sur quelques propriétés caractéristiques des ensembles convexes*, Att. R. Acad. Lincei Rend. 21 (1935), 562-567.

[754] F. Nacry, *Truncated nonconvex state-dependent sweeping process: implicit and semi-implicit adapted Moreau's catching-up algorithms*, J. Fixed Point Theory Appl. 20 (2018), Paper No 121.

[755] F. Nacry and L. Thibault, *Regularization of sweeping process: old and new*, Pure Appl. Funct. Anal. 4 (2019), 59-117.

[756] F. Nacry and L. Thibault, *BV prox-regular sweeping process with bounded truncated variation*, Optimization 69 (2020), 1391-1437.

[757] F. Nacry and L. Thibault, *Distance function associated to a prox-regular set*, Set-Valued Var. Anal., to appear.

[758] F. Nacry and L. Thibault, *Farthest points and strong convexity*, in progress.

[759] N. M. Nam, *Subdifferential formulas for a class of nonconvex infimal convolutions*, Optimization 64 (2015), 2213-2222.

[760] N. M. Nam and D. V. Cuong, *Generalized differentiation and characterizations for differentiability of infimal convolutions*, Set-Valued Var. Anal. 23 (2015), 333-353.

[761] J. Nash, *Real algebraic manifolds*, Ann. Math. 56 (1952), 405-421.

[762] J. Nečas, *Les Méthodes Directes en Théorie des Équations Elliptiques*, Masson, Paris (1967).

[763] A. Nekvinda and L. Zajíček, A simple proof of the Rademacher theorem, Časopis Pěst. Mat., 113 (1988), 337-341.
[764] A. Nekvinda and L. Zajíček, *Gâteaux differentuability of Lipschitz functions via directional derivatives*, Real Analysis Exchange 28 (2002/2003), 287-320.
[765] J. D. Newburg, *A topology for closed operators*, Ann. Math. 53 (1951), 250-255.
[766] K. F. Ng and W. H. Yang, *Regularities and their relations to error bounds*, Math. Program., Ser. A, 99 (2004), 521-538.
[767] H. V. Ngai, D. T. Luc and M. Théra, *Approximate convex functions*, J. Nonlinear Convex Anal. 1 (2000), 155-176.
[768] H. V. Ngai and J.-P. Penot, *Approximately convex sets*, J. Nonlinear Convex Anal. 8 (2007), 337-355.
[769] H. V. Ngai and J.-P. Penot, *Approximately convex functions and approximately monotone operators*, Nonlinear Anal. 66 (2007), 547-564; submitted 6 January 2005.
[770] H. V. Ngai and J.-P. Penot, *Subdifferentiation of regularized functions*, Set-Valued Var. Anal., 24 (2016), 167-189.
[771] H. V. Ngai and M. Théra, *φ-regular functions in Asplund spaces*, Control Cyber. 36 (2007), 755-774.
[772] T. K. Nguyen, *Hypographs satisfying an external sphere condition and the regularity of the minimum time function*, J. Math. Anal. Appl. 372 (2010), 611-628.
[773] G. Norlander, *The modulus of convexity in normed linear spaces*, Ark. Mat. 4 (1960), 15-17.
[774] C. Nour, R. J. Stern and J. Takche, *The union of uniform closed balls conjecture*, Control and Cybernetics 38 (2009), 1525-1534.
[775] C. Nour, R. J. Stern and J. Takche, *Proximal smoothness and the exterior sphere condition*, J. Convex Anal. 16 (2009), 501-514.
[776] C. Nour, R. J. Stern and J. Takche, *Validity of the union of uniform closed balls conjecture*, J. Convex Anal. 18 (2011), 589-600.
[777] J. Orihuela, *Conic James' compactness theorem*, J. Convex Anal. 25 (2018), 1335-1344.
[778] M. I. Ostrovskiĭ, *Topologies on the set of all subspaces of a Banach space and related questions of Banach space geometry*, Quaest. Math. 17 (1994), 259-319.
[779] P. Painlevé, *Sur les Lignes Singulières des Fonctions Analytiques*, Thesis (Thèse de Doctorat ès Sciences Mathématiques) Université de Paris (1887).
[780] P. Painlevé, *Sur les lignes singulières des fonctions analytiques*, Annales de la Faculté des Sciences de Toulouse 1re série, tome 2 (1888), p. B1-B130
[781] Z. Páles, *On approximately convex functions*, Proc. Amer. Math. Soc. 131 (2002), 243-252.
[782] G. Peano *Applicazioni Geometriche del Calcolo Infinitesimale*, Fratelli Bocca, Torino (1887).
[783] G. Peano *Lezioni di Analisi Infinitesimale*, Candeletti, Torino (1893).
[784] G. Peano *Formulario Mathematico*, Fratelli Bocca, Torino (1908).
[785] J.-P. Penot, *Calcul sous-différentiel et optimisation*, J. Funct. Anal. 27 (1978), 248-276.
[786] J.-P. Penot, *A characterization of tangential regularity*, Nonlinear Anal. Th. Meth. Appl. 5 (1981), 625-643.
[787] J.-P. Penot, *The drop theorem, the petal theorem and Ekeland's variational principle*, Nonlinear Anal. Th. Meth. Appl. 10 (1986), 459-468.
[788] J.-P. Penot, *Metric regularity, openness and Lipschitzian behavior of multifunctions*, Nonlin. Anal. Th. Meth. Appl. 13 (1989), 629-643.
[789] J.-P. Penot, *The cosmic Hausdorff topology, the bounded Hausdorff topology, and continuity of polarity*, Proc. Amer. Math. Soc. 113 (1991), 275-286.
[790] J.-P. Penot, *On the interchange of subdifferentiation and epi-convergence*, J. Math. Anal. Appl. 196 (1995), 676-698.
[791] J.-P. Penot, *Favorable classes of mappings and multimappings in nonlinear analysis and optimization*, J. Convex Anal. 3 (1996), 97-116.
[792] J.-P. Penot, *Subdifferential calculus without qualification assumptions*, J. Convex Anal. 3 (1996), 207-219.
[793] J.-P. Penot, *Compactness properties, openness criteria and coderivatives*, Set-Valued Anal. 6 (1998), 363-380.
[794] J.-P. Penot, *A short proof of the separable reduction theorem*, Demonstratio Math. 4 (2010), 653-663.
[795] J.-P. Penot, *Calculus Without Derivatives*, Graduate Texts in Mathematics, Springer, Berlin (2013).

[796] J.-P. Penot and M. Bougeard, *Approximation and decomposition properties of some classes of locally d.c. functions*, Math. Program. 41 (1988), 195-227.

[797] P. Pérez-Aros and L. Thibault, *Weak compactness of sublevel sets in complete locally convex spaces*, J. Convex Anal. 26 (2019), 739-751.

[798] P. Pérez-Aros and E. Vilches, *Moreau envelope of supremum functions with applications to infinite and stochastic programming*, to appear.

[799] B. J. Pettis, *A proof that every uniformly convex space is reflexive*, Duke Math. J. 5 (1939), 249-253.

[800] R. R. Phelps, *Gaussian null sets and differentiability of Lipschitz maps on Banach spaces*, Pacific J. Math. 18 (1978), 523-531.

[801] R. R. Phelps, *Convex Functions, Monotone Operators and Differentiability*, 2nd ed. Lectrure Notes in Mathematics, vol. 1364, Springer-Verlag, New York (1993).

[802] J. Pierpont, *Theory of Functions of Real Variables*, t. I, Boston, (1905).

[803] G. Pisier, *Martingales with values in uniformly convex spaces*, Israel J. Math. 20 (1975), 326-350.

[804] R. A. Poliquin, *Subgradient monotonicity and convex functions*, Nonlinear Anal. 14 (1990), 305-317.

[805] R. A. Poliquin, *Integration of subdifferentials of nonconvex functions*, Nonlinear Anal. 17 (1991), 385-398.

[806] R. A. Poliquin, *An extension of Attouch's theorem and its application to second-order epi-differentiation of convexly composite functions*, Trans. Amer. Math. Soc. 332 (1992), 861-874.

[807] R. A. Poliquin and R. T. Rockafellar, *Amenable functions in optimization*, in Nonsmooth Optimization Methods and Applications, edited by F. Gianessi, (1992) pp. 338-353, Gordon and Breach, Philadelphia.

[808] R. A. Poliquin and R. T. Rockafellar, *A calculus of epi-derivatives applicable to optimization*, Canad. J. Math. 45 (1993), 879-896.

[809] R. A. Poliquin and R. T. Rockafellar, *Prox-regular functions in variational analysis*, Trans. Amer. math. Soc. 348 (1996), 1805-1838.

[810] R. A. Poliquin and R. T. Rockafellar, *Generalized Hessian properties of regularized non-smooth functions*, SIAM J. Optimization 6 (1996), 1121-1137.

[811] R. A. Poliquin, R. T. Rockafellar, *Tilt stability of a local minimum*, SIAM J. Optim. 8 (1998), 287-299.

[812] R. A. Poliquin and R. T. Rockafellar, *A calculus of prox-regularity*, J. Convex Anal. 17 (2010), 203-210.

[813] R. A. Poliquin, R. T. Rockafellar and L. Thibault, *Local differentiability of distance functions*, Trans. Amer. Math. Soc. 352 (2000), 5231-5249; submitted 17 June 1997.

[814] R. A. Poliquin, J. Vanderwerff and V. Zizler, *Renormings and convex composite representation of functions*, Bull. Polish Acad. Sci. Math. 42 (1994), 9-19.

[815] E. S. Polovinkin and M. V. Balashov, *Elements of Convex and Strongly Convex Analysis*, Fizmatlit, Moscow (2004). (in Russian)

[816] J.-B. Poly and G. Raby, *Fonction distance et sigularités*, Bull. Sc. Math. (2me Série) 108(2) (1984), 187-195.

[817] B. T. Polyak, *Existence theorems and convergence of minimizing sequences in extremum problems with restrictions*, Dokl. Akad. Nauk SSSR, 166 (1966), 287 - 290 (in Russian); English translation in Soviet Math. Dokl. 7 (1966), 72-75.

[818] D. Pompeiu, *Sur la continuité des fonctions complexes*, Annales de la Faculté des Sciences de Toulouse, tome 7, no 3 (1905), 265-315.

[819] D. Preiss, *Differentiability of Lipschitz functions on Banach spaces*, J. Func. Anal. 91 (1990), 312-345.

[820] D. Preiss and J. Tišer, *Two unexpected examples concerning differentiability of Lipschitz functions on Banach spaces*, in Collection: *Geometric Aspects of Functional Analysis, Operator Theory: Advances and Applictions* 77, Birkhäuser, Boston (1995), 219-238.

[821] D. Preiss and L. Zajíček, *Sigma-porous sets in products of metric spaces and sigma-directionally porous sets in Banach spaces*, Real Anal. Exchange 24 (1998/9), 295-314.

[822] D. Preiss and L. Zajíček, *Directional derivatives of Lipschitz functions*, Israel J. Math; 125 (2001), 1-27.

[823] V. Pták, *A quantitative refinement of the closed graph theorem*, Czechoslovak Math. J. 24 (1974), 503-506.

[824] V. Pták, *A nonlinear subtraction theorem*, Proc. Roy. Irish Acad. Sect. A 82 (1982), 47-53.

[825] M. Quincampoix, *Differential inclusions and target problems*, SIAM J. Control Optim. 30 (1992), 324-335.

[826] G. Rabaté, *Sur les Notions Originelles de la Géométrie Infinitésimale Directe*, Thesis (Thèse de Doctorat ès Sciences Mathématiques) Université de Toulouse (1931)

[827] P. J. Rabier, *Points of continuity of quasiconvex functions on topological vector spaces*, J. Convex Anal. 20 (2013), 701-721.

[828] P. J. Rabier, *Differentiability of quasiconvex functions on separable Banach spaces*, Israel J. Math. 207 (2015), 11-51.

[829] H. Rademacher, *Über partielle und totale Dijfferenzierbarkeit*, Math. Ann. 79 (1918), 340-359.

[830] H. Rådström, *An embedding theorem for spaces of convex sets*, Proc. Amer. Math. Soc. 3 (1952), 165-169.

[831] J. Rataj and L. Zajíček, *On the structure of sets with positive reach*, Math. Nachr. 290 (2017), 1806-1829.

[832] R. Redefer and W. Walter, *The subgradient in* $\mathbb{R}^n$, Nonlinear Anal. Th. Meth. Appl. 20 (1993), 1345-1348; submitted 1 January 1992.

[833] S. Reich, *On the asymptotic behavior of nonlinear semigroups and the range of accretive operators*, J. Math. Anal. Appl. 79 (1981), 113-126.

[834] P. Redont, Personal communication.

[835] Yu. G. Reshetnyak, *On a generalization of convex surfaces*, Mat. Sb. N.S. 40 (1956), 381-398 (in Russian).

[836] F. Riesz, *Sur l'existence de la dérivée des fonctions d'une variable réelle et des fonctions d'intervalle*, Verhandl. Internat. Math. Kongress Zurich, I (1932), 258-259.

[837] S. M. Robinson, *Normed convex processes*, Trans. Amer. Math. Soc. 174 (1972), 127-140.

[838] S. M. Robinson, *Stability theory for systems of inequalities in nonlinear programming I: linear systems*, SIAM J. Numer. Anal. 12 (1975), 754-769.

[839] S. M. Robinson, *Regularity and stability for convex multivalued functions*, Math. Oper. Res. 1 (1976), 130-143.

[840] S. M. Robinson, *Stability theory for systems of inequalities in nonlinear programming II: Differentiable nonlinear systems*, SIAM J. Numer. Anal. 13 (1976), 497-513.

[841] S. M. Robinson, *Strongly regular generalized equations*, Math. Oper. Res. 5 (1980), 43-62.

[842] S. M. Robinson, *Localized normal maps and the stability of variational conditions*, Set-Valued Anal. 12 (2004), 259-274.

[843] S. M. Robinson, *Aspects of the projector on prox-regular sets*, Variational analysis and applications, 963-973, Nonconvex Optim. Appl., 79, Springer, New York, (2005).

[844] R. T. Rockafellar, *Convex Functions and Dual Extremum Problems*, Ph.D. dissertation, Harvard University (1963).

[845] R. T. Rockafellar, *Duality theorems for convex functions*, Bull. Amer. Math. Soc. 70 (1964), 189-192.

[846] R. T. Rockafellar, *Characterization of the subdifferentials of convex functions*, Pac. J. Math. 17 (1966), 497-510.

[847] R. T. Rockafellar, *Level sets and continuity of conjugate convex functions*, Trans. Amer. math. Soc. 123 (1966), 46-63.

[848] R. T. Rockafellar, *Extension of Fenchel's duality theorem for convex functions*, Duke Math. J., 33 (1966), 81-90.

[849] R. T. Rockafellar, *Conjugates and Legendre transforms of convex functions*, Canadian J. Math., 19 (1967), 200-205.

[850] R. T. Rockafellar, *Duality and stability in extremum problems involving convex functions*, Pacific J. Math. 21 (1967), 167-187.

[851] R. T. Rockafellar, *On the maximal monotonicity of subdifferential mapping*, Pac. J. Math. 33 (1970), 209-216.

[852] R. T. Rockafellar, *Convex Analysis*, Princeton University Press, Princeton, New Jersey (1970).

[853] R. T. Rockafellar, *Conjugate Duality and Optimization*, Conference Board of Mathematical Sciences Series, 16, SIAM Publications, Philadelphia (1972).

[854] R. T. Rockafellar, *Clarke's tangent cone and the boundaries of closed sets in* $\mathbb{R}^n$, Nonlin. Anal. Th. Meth. Appl. 3 (1979), 145-154.

[855] R. T. Rockafellar, *Directional Lipschitzian functions and subdifferential calculus*, Proc. London Math. Soc. 39 (1980), 331-355.

[856] R. T. Rockafellar, *Generalized directional derivatives and subgradients of nonconvex functions*, Canad. J. Math. 32 (1980), 157-180 (submitted: March, 1978).

[857] R. T. Rockafellar, *The Theory of Subgradients and its Applications to Problems of Optimization: Convex and Nonconvex Functions*, Heldermann Verlag, Berlin (1981).

[858] R. T. Rockafellar, *Proximal subgradients, marginal values and augmented Lagrangians in nonconvex optimization*, Math. Oper. Res. 6 (1981), 424-436 (sumitted: july 1980).

[859] R. T. Rockafellar, *Favorable classes of Lipschitz continuous functions in subgradient optimization*, in Progress in Nondifferentiable Optimization, edited by E.A. Nurminskii, pp. 125-143, IIASA, Laxenburg, Austria (1982).

[860] R. T. Rockafellar, *Extensions of subgradient calculus with applications to optimization*, Nonlin. Anal. Th. Meth. Appl. 9 (1985), 665-698.

[861] R. T. Rockafellar, *Lipschitzian properties of multifunctions*, Nonlin. Anal. Th. Meth. Appl. 9 (1985), 867-895.

[862] R. T. Rockafellar, *First and second order epi-differentiability in nonlinear programming*, Trans. Amer. Math. Soc. 307 (1988), 75-108.

[863] R. T. Rockafellar, *Proto-differentiability of set-valued mappings and its applications in optimization*, Ann. Inst. H. Poincaré (1989), 449-482.

[864] R. T. Rockafellar, *Generalized second derivatives of convex functions and saddle functions*, Trans; Amer. Math. Soc. 322 (1990), 51-77.

[865] R. T. Rockafellar and R. J-B. Wets, *Variational Analysis*, Vol 317 Grundlehren der mathematischen Wissenschaften, Springer, Berlin (1998).

[866] B. Rodriguez and S. Simons, *Conjugate functions and subdifferentials in non-normed situations for operators with complete graphs*, Nonlinear Anal. 12 (1988), 1062-1078.

[867] F. Roger, *Les propriétés tangentielles des ensembles euclidiens de points*, 69 (1937), 99-133.

[868] F. Roger, *Sur l'extension à l'ordre n des théorèmes de M. Denjoy sur le nombres dérivés du premier ordre*, C. R. Acad. Sci. Paris 209 (1939), 11-14.

[869] S. Rolewicz, *On paraconvex multifunctions*, Oper. Res. Verfahren 31 (1979), 540-546.

[870] S. Rolewicz, *On γ-paraconvex multifunctions*, Math. Japon. 24 (1979), 293-300.

[871] S. Rolewicz, *On drop property*, Studia Math. 85 (1987), 27-35.

[872] S. Rolewicz, *On $\alpha(\cdot)$-paraconvex and strongly $\alpha(\cdot)$-paraconvex functions*, Control and Cybernetics 29 (2000), 367-377.

[873] S. Rolewicz, *Paraconvex Analysis*, Control and Cybernetics 34 (2005), 951-965.

[874] J. Saint Raymond, *Weak compactness and variational characterization of the convexity*, Mediterr. J. Math. 10 (2013), 927-940.

[875] S. Saks, *Sur quelques propriétés métriques d'ensembles*, Fundam. Math. 26 (1936), 234-240.

[876] S. Saks, *Theory of the integral*, Monographie Matematyczne, Tom VIII, Second revised edition, Hafner Publ. Comp., New-York (1937).

[877] S. Saks and A. Zygmund, *Sur les faisceaux de tangentes à une courbe*, Fundam. Math. 6 (1924), 117-121.

[878] D. Salas and L. Thibault, *On characterizations of submanifolds via smoothness of the distance function in Hilbert spaces*, J. Optim. Theory Appl. 182 (2019), 189-210.

[879] D. Salas and L. Thibault, *Quantified characterizations of bodies with smooth boundaries*, unpublished paper.

[880] D. Salas and L. Thibault, *Quantitative characterizations of nonconvex bodies with smooth boundaries via the metric projection in Hilbert spaces*, J. Math. Anal. Appl. 494 (2021), 1-21.

[881] H. H. Schaefer and M. P. Wolff, *Topological Vector Spaces*, Second edition. Graduate Texts in Mathematics, 3. Springer-Verlag, New York, (1999).

[882] J. Schauder, *Uber die Umkerhrung linearer, stetiger Funktionaloperationen*, Studia Math. 2 (1930), 1-6.

[883] M. Sebbah and L. Thibault, *Metric projection and compatibly parameterized families of prox-regular sets in Hilbert space*, Nonlinear Anal. 75 (2012), 1547-1562.

[884] O. Serea, *On reflecting boundary problem for optimal control*, SIAM J. Control Optim. 42 (2003), 559-575.

[885] O. Serea and L. Thibault, *Primal lower nice property of value functions in optimization and control problems*, Set-Valued Var. Anal. 18 (2010), 569-600.

[886] F. Severi, *Su alcune questioni di topologia infinitesimale*, Ann. Polon. Math. 9 (1930), 97-108.

[887] A. Shapiro, *Existence and differentiability of metric projections in Hilbert spaces*, SIAM J. Optimization 4 (1994), 130-141.

[888] S. H. Shkarin, *On Rolle's theorem in infinite dimensional Banach spaces*, Math. Zam. 51 (1992), 128-136.

[889] V. Sierpinski and A. N. Singh, *On derivatives of discontinuous functions*, Fund. Math. 36 (1949), 283-287.

[890] S. Simons, *Minimax and Monotonicity*, Lecture Notes in Mathematics 1693, Springer, Berlin (1998).

[891] S. Simons, *A new proof of the maximal monotonicity of subdifferentials*, J. Convex Anal. 16 (2009), 165-168.

[892] A. N. Singh, *On infinite derivates*, Fund. Math. 33 (1945), 106-107.

[893] V. L. Šmulian, Rec. Math., 6(48):1, (1939).

[894] V. L. Šmulian, *On some geometrical properties of a sphere in the space of the type (B)*, C. R. Acad. Sci. U.R.S.S. XXIV (1939), 648-652

[895] V. L. Šmulian, *On some geometrical properties of the unit sphere in the space of the type (B)*, Mat. Sb. 6 (1939), 77-94 (in Russian).

[896] V. L. Šmulian, *Sur la dérivabilité de la norme dans l'espace de Banach*, C.R. Acad. Sci. U.R.S.S., XXVII, (1940).

[897] V. L. Šmulian, *Sur la structure de la sphère unitaire dans l'espace de Banach*, Rec. Math. [Mat. Sbornik] N.S. Vol. 9(51), No 3 (1941), 545-561.

[898] J. E. Spingarn, *Submonotone subdifferentials of Lipschitz functions*, Trans. Amer. Math. Soc. 264 (1981), 77-89 (submitted, November 1979).

[899] S. B. Stechkin, *Approximative properties of sets in normed linear spaces*, Revue Math. Pures Appl. 8 (1963), 5-18.

[900] C. Stegall, *Optimization of functions on certain subsets of Banach spaces*, Math. Ann. 236 (1978), 171–176.

[901] J. Steiner, *Über parallele Flächen*, Monatsber. Preuss. Akad. Wiss. (1840), 114–118, [Ges. Werke, Vol II (Reimer, Berlin, 1882), 245–308].

[902] E. Steinitz, *Bedingt konvergente Reihen und konvexe Systeme*, J. Reine Angew. Math. 143 (1913), 128-175.

[903] O. Stolz, *Grundzüge der Differential und Integral-Rechnung*, t. I, Leipzig, (1893).

[904] J. J. Stoker, *Unbounded convex point sets*, Amer. J. Math. 62 (1940), 165-179.

[905] Ch. Thäle, *50 Years Sets with Positive Reach - A Survey*, Surveys in Mathematics and its Applications, Vol. 3 (2008), pp. 123-165.

[906] L. Thibault, *Problème de Bolza dans un espace de Baanach séparable*, C. R. Acad. Sci. Paris Sér. I Math. 282 (1976), 1303-1306.

[907] L. Thibault, *Subdifferentials of compactly Lipschitz functions*, Sémin. Anal. Convexe Montpellier, Exposé 5 (1978).

[908] L. Thibault, *Mathematical programming and optimal control problems defined by compactly Lipschitzian mappings*, Sémin. Anal. Convexe Montpellier, Exposé 10 (1978).

[909] L. Thibault, *Subdifferentials of compactly Lipschitzian vector valued functions*, Ann. Mat. Pura Appl. 125 (1980), 157-192.

[910] L. Thibault, *On generalized differentials and subdifferentials of Lipschitz vector-valued functions*, Nonlinear Anal. 6 (1982), 1037-1053.

[911] L. Thibault, *Tangent cones and quasi-interiorly tangent cones to multifunctions*, Trans. Amer. Math. Soc. 277 (1983), 601-621.

[912] L. Thibault, *V-subdifferentials of convex operators*, J. Math. Anal. Appl. 115 (1986), 442-460.

[913] L. Thibault, *On subdifferentials of optimal value functions*, SIAM J. Control Optim. 29 (1991), 1019-1036.

[914] L. Thibault, *A generalized sequential formula for subdifferentials of sum of convex functions defined on Banach spaces*, in Recent Developments in Optimization, 7th French-German Conference on Optimization, Dijon, France, june 27-july 2, 1994, R. Durier, C. Michelot eds.; Lecture Notes Econ. Math. Syst. 429 (1995), 1434-1444, Springer-Verlag, Berlin.

[915] L. Thibault, *A note on the Zagrodny mean value theorem*, Optimization 35 (1995), 127-130.

[916] L. Thibault, *Sequential convex subdifferential calculus and sequential Lagrange multipliers*, (preprint 1994) SIAM J. Control Optim. 39 (1997), 331-355.

[917] L. Thibault, *A direct proof of a sequential formula for the subdifferential of the sum of two convex functions*, Unpublished paper, Univ. de Montpellier, Montpellier (1994).

[918] L. Thibault, *On compactly Lipschitzian mappings*, in Recent Advances in Optimization, Lecture Notes Econ. Math. Syst. 456 (1997) Springer-Verlag, Berlin.

[919] L. Thibault, *Equivalence between metric regularity and graphical metric regularity for set-valued mappings*, Unpublished note, Univ. de Montpellier, Montpellier (1999).

[920] L. Thibault, *Various forms of metric regularity*, Unpublished note, Univ. de Montpellier, Montpellier (1999).

[921] L. Thibault, *Limiting convex subdifferential calculus with applications to integration and maximal monotonicity of subdifferential*, Canadian Mathematical Society Conference Proceedings, 27 (2000), 279-289.

[922] L. Thibault, *Sweeping process with regular and nonregular sets*, J. Differential Equations 193 (2003), 1-26.

[923] L. Thibault, *Regularization of nonconvex sweeping process in Hilbert space*, Set-Valued Anal. 16 (2008), 319-333.

[924] L. Thibault, *Jean Jacques Moreau: a selected review of his mathematical works*, J. Convex AnaL. 23 (2016), 5-22.

[925] L. Thibault, *Subsmooth functions and sets*, Linear Nonlinear Anal. 4 (2018), 157-269.

[926] L. Thibault and D. Zagrodny, *Integration of subdifferentials of lower semicontinuous functions on Banach spaces*, J. Math. Anal. Appl. 189 (1995), 33-58; submitted 19 February 1993.

[927] L. Thibault and D. Zagrodny, *Enlarged inclusion of subdifferentials*, Canad. Math. Bull. 48 (2005), 283–301.

[928] L. Thibault and D. Zagrodny, *Subdifferential determination of essentially directionally smooth functions in Banach space*, SIAM J. Optim. 20 (2010), 2300-2326.

[929] L. Thibault and D. Zagrodny, *Determining functions by slopes*, to appear.

[930] L. Thibault and N. Zlateva, *Integrability of subdifferentials of directionally Lipschitz functions*, Proc. Amer. Math. Soc. 133 (2005), 2939–2948.

[931] A. A. Tolstonogov, *Differential Inclusions in a Banach Space*, Kluwer, Dordrecht (2000).

[932] H. Toruńczyk, *Smooth partitions of unity on some non-separable Banach spaces*, Studia Math. 46 (1973), 43-51.

[933] J. S. Treiman, Ph.D. dissertation, University of Washington, Seattle (1983).

[934] J. S. Treiman, *Characterization of Clarke's tangent and normal cones in finite and infinite dimensions*, Nonlinear Anal. Th. Meth. Appl. 7 (1983), 771-783.

[935] J. S. Treiman, *Generalized gradients, Lipschitz behavior and directional derivatives*, Canad. J. Math. 37 (1985), 1074-1084.

[936] J. S. Treiman, *Generalized gradients and paths of descent*, Optimization 17 (1986), 181-186.

[937] J. S. Treiman, *Clarke's gradients and epsilon-subgradients in Banach spaces*, Trans. Amer. Math. Soc. 294 (1986), 65-78.

[938] J. S. Treiman, *Shrinking generalized gradients*, Nonlinear Anal. Th. Meth. Appl. 12 (1988), 1429-1450.

[939] J. S. Treiman, *Finite dimensional optimality conditions: B-gradients*, J. Optim. Theory Appl. 62 (1989), 771-783.

[940] J. S. Treiman, *The linear nonconvex generalized gradients and Lagrange multipliers*, SIAM J. Optim. 5 (1995), 670-680.

[941] J. S. Treiman, *Lagrange multipliers for nonconvex generalized gradients with equality, inequality, and set constraints*, SIAM J. Control Optim. 37 (1999), 1313-1329.

[942] J. S. Treiman, *The linear generalized gradient in infinite dimensions*, Nonlinear Anal. Th. Meth. Appl. 48 (2002), 427-443.

[943] C. Ursescu, *Multifunctions with closed convex graphs*, Czech. Math. J. 25 (1975), 438-441.

[944] M. Valadier, *Sous-différentiel d'une borne supérieure et d'une somme continue de fonctions convexes*, C. R. Acad. Sci. Paris Sér. A-B, 268 (1969), A39-A42.

[945] M. Valadier, *Contribution à l'Analyse Convexe*, Thesis (Thèse de Doctorat d'État ès Sciences Mathématiques), Université de Paris (1970).

[946] M. Valadier, *Sous-différentiabilité de fonctions convexes dans un espace vectoriel ordonné*, Math. Scand. 30 (1972), 65-74.

[947] M. Valadier, *Quelques résultats de base concernant le processus de rafle*, Sémin. Anal. Convexe Montpellier (1988), exp. 3 (30 pages).

[948] M. Valadier, *Quelques problèmes d'entrainement unilatéral en dimension finie*, Sémin. Anal. Convexe Montpellier (1988), exp. 8 (21 pages).

[949] M. Valadier, *Lignes de descente de fonctions lipschitziennes non pathologiques*, Sémin. Anal. Convexe Montpellier (1988), exp. 9 (10 pages).

[950] M. Valadier, *Entrainement unilateral, lignes de descente, fonctions lipschitziennes non pathologiques*, C. R. Acad. Sc. Paris 308 (1989), 241-244.

[951] F. A. Valentine, *On the extension of a vector function so as to preserve a Lipschitz condition*, Bull. Amer. Math. Soc. 49 (1943), 100-108.

[952] F. A. Valentine, *A Lipschitz condition preserving extension for a vector function*, Amer. J. Math. 67 (1945), 83-93.

[953] F. Vasilesco, *Essai sur les Fonctions Multiformes de Variables Réelles*, Thesis (Thèse de Doctorat ès Sciences Mathématiques), Université de Paris (1925).

[954] L. Veselý, *On the multiplicity points of monotone operators ob separable Banach spaces*, Comment. Math. Univ. Carolinae 27 (1986), 551-570.

[955] L. Veselý, *On the multiplicity points of monotone operators ob separable Banach spaces II*, Comment. Math. Univ. Carolinae 28 (1987), 295-299.

[956] L. Veselý and L. Zajíček, *Delta-convex mappings between Banach spaces and applications*, Dissertationes Mathematicae 289, Warszawa (1989), 48 pp.

[957] J.-P. Vial, *Strong convexity of sets and functions* J. Math. Econom. 9 (1982), 187-205; submitted in 1978.

[958] J.-P. Vial, *Strong and weak convexity of sets and functions*, Math. Oper. Res. 8 (1983), 231-259; submitted in 1981.

[959] R. Vinter, *Optimal Control*, Birkhäuser, Boston, Massachusetts (2000).

[960] L.P. Vlasov, *On Chebyshev sets*, Soviet Math. Dokl. 8 (1967), 401-404.

[961] M. Volle, *On the subdifferential of an upper envelope of convex functions*, Acta Math. Vietnam. 19 (1994), 137-148.

[962] M. Volle, J.-B. Hiriart-Urruty and C. Zalinescu, *When some variational properties force convexity*, ESAIM: COCV 19 (2013), 701-709.

[963] J. von Neumann, *Zur Theorie der Gesellschaftsspiele*, Math. Ann. 100 (1928), 295-320.

[964] J. von Neumann, *Discussion of a maximum problem*, in John von Neumann Collected Works, edited by A. H. Taub, Vol. 6, pp. 89-95, Pergamon, Oxford.

[965] G. Wachsmuth, *A Guided Tour of Polyhedric Sets: Basic Properties, New Results on Intersections and Applications*, J. Convex Anal. 26 (2020, 153-188.

[966] G. Wachsmuth, *No-gap second-order conditions under n-polyhedric constraints and finitely many nonlinear constraints*, J. Convex Anal. 27 (2020), 733-751.

[967] A. J. Ward, *On the differential structure of real functions*, Proc. Lond. Math. Soc. 39 (1935), 339-362.

[968] D. W. Walkup and R. J-B. Wets, *Continuity of some convex-cone valued mappings*, Proc. Amer. Math. Soc. 18 (1967), 229-235.

[969] X. Wang, *On Chebyshev functions and Klee functions*, J. Math. Anal. Appl. 368 (2010), 293-310.

[970] V. Weckesser, *The subdifferential in Banach spaces*, Nonlin. Anal. Th. Meth. Appl. 20 (1993), 1349-1354; submitted 1 January 1992.

[971] H. Weyl, *On the volume of tubes*, Amer. J. Math. 61 (1939), 461-472.

[972] H. Whitney, *Analytic extension of differentiable functions*, Trans. Amer. Math. Soc. 36 (1934), 63-89 (submitted, December 1932).

[973] M. D. Wills, *Hausdorff distance and convex sets*, J. Convex Anal. 14 (2007), 109-117.

[974] P. R. Wolenski and Y. Zhuang, *Proximal analysis and the minimal time function*, SIAM J. Control Optim. 36 (1998), 1048-1072.

[975] Zong-Ben Xu and G. F. Roach, *Characteristic inequalities of uniformly convex and uniformly smooth Banach spaces*, J. Math. Anal. Appl. 157 (1991), 189-210.

[976] J. C. Yao, X. Y. Zheng and J. Zhu, *Stable minimizers of φ-regular functions*, SIAM J. Optim. 27 (2017), 1150-1170.

[977] G. C. Young, *On the derivates of a function*, Proc. London Math. Soc. 15 (1916), 360-384.

[978] W. H. Young, *The Fundamental Theorems of Differential Calculus*, Cambridge (1910).

[979] W. H. Young, *On classes of summable functions and their Fourier series*, Proc. Royal Society, A (1912), 225-229.

[980] W. H. Young, *La symétrie de structure des fonctions de variables réelles*, Bull. Sci. Math., 52 (1928), 265-280.

[981] D. Zagrodny, *Approximate mean value theorem for upper subderivatives*, Nonlinear Anal. 12 (1988), 1413-1428.

[982] D. Zagrodny, *A note on the equivalence between the mean value theorem for the Dini derivative and the Clarke-Rockafellar derivative*, Optimization 21 (1990), 179-183.

[983] D. Zagrodny, *Some recent mean value theorems in nonsmooth analysis*, Nonsmooth Optimization: Methods and Applications (Erice, 1991), Gordon and Breach, Montreux (1992), 421-428.

[984] L. Zajíček, *On the points of multiplicity of monotone operators*, Comment. Math. Univ. Carolinae 19 (1978), 179-189.

[985] L. Zajíček, *On the points of multivaluedness of metric projections in separable Banach spaces*, Comment. Math. Univ. Carolinae 19 (1978), 513-523.

[986] L. Zajíček, *On the differentiation of convex functions in finite and infinite dimensional spaces*, Czechoslovak Math. J. 29 (1979), 340-348.

[987] L. Zajíček, *Differentiability of the distance function and points of multivaluedness of the metric projection in Banach space*, Czechoslovak Math. J. 33 (1983), 292-308.

[988] L. Zajíček, *Strict differentiability via differentiability*, Acta Univ. Carolinae 28 (1987), 157-159.

[989] L. Zajíček, *Porosity and σ-porosity*, Real Analysis Exchange 13 (1987/88), 314-350.

[990] L. Zajíček, *Fréchet, strict differentiability and subdifferentiability*, Czechoslovak Math. J. 41 (1991), 471-489.

[991] L. Zajíček, *On differentiability properties of Lipschitz functions on a Banach space with a Lipschitz uniformly Gâteaux differentiable bump function*, Comment. Math. Univ. Carolinae 38 (1997), 329-336.

[992] L. Zajíček, *A note on intermediate differentiability of Lipschitz functions*, Comment. Math. Univ. Carolinae 40 (1999), 1501–1505.

[993] L. Zajíček, *On σ-porous sets in abstract spaces*, Abstr. Appl. Anal. (2005), 509-534.

[994] L. Zajíček, *On Lipschitz and D.C. surfaces of finite codimension in a Banach space*, Czechoslovak Math. J. 58 (2008), 849-864.

[995] L. Zajíček, *Differentiability of approximately convex, semiconcave and strongly paraconvex functions*, J. Convex Anal. 15 (2008), 1-15.

[996] L. Zajíček, *Singular points of order k of Clarke regular and arbitrary functions*, Comment. Math. Univ. Carolinae 53 (2012), 51-63.

[997] L. Zajíček, *Remarks on Fréchet differentiability of pointwise Lipschitz, cone-monotone and quasi-convex functions*, Comment. Math. Univ. Carolinae 55 (2014), 203-213.

[998] L. Zajíček, *Hadamard differentiability via Gâteaux differentiabilty*, Proc. Amer. Math. Soc. 143(2015), 279-288.

[999] L. Zajíček, *Properties of Hadamard directional derivatives: Denjoy-Young-Saks theorem for functions on Banach spaces*, J. Convex Anal. 22 (2015), 161-176.

[1000] C. Zalinescu, *Convex Analysis in General Vector Spaces*, World Scientific (2002).

[1001] E. H. Zarantonello, *Projections on convex sets in Hilbert space and spectral theory I and II*, in E. H. Zarantonello, ed., Contributions to Nonlinear Functional Analysis, Academic Press, New York (1971), pp. 237-424.

[1002] R. Zhang and J. S. Treiman, *Upper-Lipschitz multifunctions and inverse subdifferentials*, Nonlinear Anal. Theory Meth. Appl. 24 (1995), 273-286.

[1003] X. Y. Zheng and Q. H. He, *Characterization for metric regularity for σ-subsmooth multifunctions* Nonlinear Anal. 100 (2014), 111-121.

[1004] X. Y. Zheng and K. F. Ng, *Linear regularity for a collection of subsmooth sets in Banach spaces*, SIAM J. Optim., 19 (2008), 62-76.

[1005] X. Y. Zheng and K. F. Ng, *Calmness for L-subsmooth multifunctions in Banach spaces*, SIAM J. Optim. 19 (2009), 1648-1673.

[1006] X. Y. Zheng and K. F. Ng, *Metric subregularity for proximal generalized equations in Hilbert spaces*, Nonlinear Anal. 75 (2012) 1686-1699.

[1007] X. Y. Zheng and K. F. Ng, *Hölder stable minimizers, tilt stability and Hölder metric regularity of subdifferential*, SIAM J. Optim., 25 (2015), 416-438.

[1008] X. Y. Zheng and K. F. Ng, *Hölder weak sharp minimizers and Hölder tilt-stability*, Nonlinear Anal., 120 (2015), 186-201.

[1009] X. Y. Zheng and J. Zhu, *Stable well-posedness and tilt stability with respect to admissible functions*, ESAIM: COCV 23 (2017), 1397-1418.

[1010] N. V. Zhivkov, *Metric projections and antiprojections in strictly convex normed spaces*, C.R. Acad. Bulgare Sci. 31 (1978), 369-372.

[1011] N. V. Zhivkov, *Generic Gâteaux differentiability of locally Lipschitzian functions*, Collection: Constructive function theory 81 (Varna, 1981), 590-594.

[1012] N. V. Zhivkov, *Generic Gâteaux differentiability of directionally differentiable mappings*, Rev. Roumaine Math. Pures Appl., 32 (1987), 179-188.

[1013] Q. J. Zhu, *Clarke-Ledyaev mean value inequalities in smooth Banach spaces*, Nonlinear Anal., 32 (1998), 315-324.

[1014] Q. J. Zhu, *The equivalence of several basic theorems for subdifferentials*, Set-Valued Anal. 6 (1998), 171-185.

[1015] N. Zlateva, *Integrability through infimal regularization*, Compt. Rend. Acad. Bulg. Sci. 68 (2015), 551-560.

[1016] L. Zoretti, *Sur les fonctions analytiques uniformes qui possèdent un ensemble parfait discontinu de points singuliers*, J. Math. Pure. Appl., 6e série - tome I - Fasc. I (1905), p. 1-51.

Index